Volume 1

Australian Beetles

Morphology, Classification and Keys

John F. Lawrence and Adam Ślipiński

CSIRO PUBLISHING

National Library of Australia Cataloguing-in-Publication entry

Lawrence, John F. (John Francis), 1934– author.

Australian beetles. Volume 1 : morphology, classification
and keys / by John F. Lawrence and Adam Ślipiński.

9780643097285 (hardback)
9780643097292 (epdf)
9780643106673 (epub)

Includes bibliographical references and index.

Beetles – Australia.
Beetles – Australia – Identification.

Ślipiński, Adam author.

595.760994

Published by
CSIRO Publishing
36 Gardiner Road, Clayton VIC 3168
Private Bag 10, Clayton South VIC 3169
Australia

Telephone: [+613] 9545 8555
Email: csiropublishing@csiro.au
Website: www.publishing.csiro.au

Cover images by Cate Lemann and Adam Ślipiński

Cover design by Andrew Weatherill
Typeset by Thomson Digital
Printed by Ingram Lightning Source

Jan26_RP_ILS

CONTENTS

PREFACE

This is the first in a series of three volumes on Australian Coleoptera. It is based in part on Lawrence and Britton (1994), which, in turn, was an expanded version of the chapter on Coleoptera in the *Insects of Australia* (Lawrence and Britton 1991). It differs from the 1994 book in the following respects: (1) the family sections are greatly expanded, with the abbreviated family diagnoses replaced by full descriptions of adults and larvae, comments added on the world distribution of each family, and a complete list included of Australian genera with their authors; (2) the section on fossil history is expanded to include all significant fossil literature on Coleoptera from various regions, but with an emphasis on the Mesozoic Era; and (3) ~1175 figures are included of adults, larvae and anatomical structures. The volume contains keys to all families and many subfamilies, tribes and distinctive genera, based on adults and larvae (292 and 281 couplets, respectively), but there are no subfamily or tribal keys included in the family sections. The following abbreviations throughout the text and keys for Australian states and territories: ACT (Australian Capital Territory), NSW (New South Wales), NT (Northern Territory), QLD (Queensland), SA (South Australia), VIC (Victoria) and WA (Western Australia).

Volumes 2 and 3 of this series will be multiauthored with invited specialists from various parts of the world providing sections similar to those in Arnett and Thomas (2001) and Arnett *et al.* (2002), with adult keys to all genera occurring in Australia and some subfamily or tribal keys based on larvae when possible. It is planned that Volume 2 will include Archostemata, Myxophaga, Adephaga, Hydrophiloidea, Staphylinoidea and Scarabaeoidea, and Volume 3 will include Scirtoidea, Dascilloidea, Buprestoidea Byrrhoidea, Elateroidea, Derodontoidea, Bostrichoidea, Lymexyloidea, Cleroidea, Cucujoidea, Tenebrionoidea and Chrysomeloidea. Curculionoidea will not be covered in these volumes, since some curculionoid families have been treated in the published volumes of *Australian Weevils* (Zimmerman 1991, 1992, 1993, 1994a, 1994b) and others will be covered in further volumes authored by R. G. Oberprieler.

ACKNOWLEDGMENTS

We gratefully acknowledge CSIRO Ecosystem Sciences for support of research and for allowing the use of copyrighted images. We are also in debt to the many institutions and individuals who have supported our various research projects over the years by supplying specimens, literature and information. Those mentioned below have been particularly helpful during the preparation of the present work, as well as its forerunners (Lawrence and Britton 1991 and 1994). A portion of this project was supported by National Science Foundation Grant (Division of Environmental Biology) #0531768. ATOL: Collaborative Research: Assembling the Beetle Tree of Life (1 Oct. 2005–30 Sept. 2010). Principal Investigator: Brian Farrell. Some of the character state images were taken with support from a National Science Foundation Research Grant 7502606. Anne Hastings is acknowledged for her artwork and graphics skills, Cate Lemann for photography and shipment of specimens, and Hermes Escalona for assistance of various kinds, but particularly scanning elecron microscopy.

The following individuals, institutions and institutional curators have provided loans and gifts of specimens, literature and advice: Elizabeth Arias, Essig Museum, University of California, Berkeley, CA, USA (Chilean fiueld support); Martin Baehr, Zoologische Staatssammlung, Munich, Germany (Carabidae review); Robert Beattie, Department of Education and Training, NSW (information on Australian fossils); Rolf Beutel, Institut für Spezielle Zoologie und Evolutionsbiologie. Friedrich-Schiller-Universität, Jena Germany (morphological information and advice); Ladislav Bocak, Palaký University, Department of Zoology, Olomouc, Czech Republic (Lycidae review); Bohart Museum, University of California, Davis, CA, USA (Lynn Kimsey); California Department of Food and Agriculture, Sacramento, CA, USA (Charles Bellamy, Andrew Cline); Charles A. Triplehorn Insect Collection, Ohio State University, Colymbus OH, USA (Norman F. Johnson, Luciana Musetti) (*Lecontia* larva); Essig Museum of Entomology, Department of Environmental Science, Policy and Management, University of California, Berkeley, CA, USA (Cherly Barr); Field Museum of Natural History, Division of Insects, Chicago, IL, USA (Margaret K. Thayer, Alfred F. Newton); Martin Fikáček, Department of Entomology, National Museum, Prague, Czech Republic (Hydrophilidae review); Nadine Guthrie, Department of Wildlife and Conservation, Wanneroo, WA (*Acanthocnemus* larva); Terry Houston, Western Australian Museum, Perth, WA (image of female *Riekella*); Insect Research Collection, Department of Entomology, University of Wisconsin-Madison, Madison, WS, USA (Daniel K. Young); Peter Jel, Queensland Department of Mines (information on Australian fossil sites); Sarah Martin, Geological Survery of Western Australia, Perth, WA (information on Australian fossils); Museum of Comparative Zoology, Harvard University, Cambridge, MA, USA (Philip D. Perkins, Duane McKenna, Brian Farrell); New Zealand Arthropod Collection, Landcare Research, Auckland, New Zealand (Richard A. B. Leschen); Rolf Oberprieler, CSIRO Ecosystem Sciences (Curculionoidea review; provision of specimens); Queensland Museum, Brisbane, QLD, Australia (Christine Lambkin, Geoff Monteith, Geoff Thompson); United States Department of Agriculture Systematic Entomology Laboratory and Smithsonian Institution, Washington, DC, USA (Terry Erwin, David Furth, Gary Hevel, Alexander Konstantinov, Steven Lingafelter, Lisa Roberts, Warren Steiner, Natalia Vandenberg); Tom Weir, CSIRO Ecosystem Sciences (Hydradephaga and Scarabaeoidea reviews; identification of illustrated specimens); Kip Will, Essig Museum, University of California, Berkeley, CA, USA (Chilean field support; Carabidae review).

We thank the following photographers for providing photos of living beetles: Arthur Anker, Farhan Bokari, Peter Chew, Katarina Christenson, Anthony Daley, Kristi Ellingsen, C. K. and L. Fisher, Owen Kelly, Bromwyn King, Jiří and Marie Lochman (Lochman Transparencies, Innaloo, WA), Geoff Monteith, Rolf Oberprieler, David Pike, Graham Prichard, David Rentz, and Paul Zborowski (Close-Up Photolibrary, Kuranda, QLD). The following artists were responsible for borrowed illustrations: Natalie Banks, A. Dabrowska, Anthony Davies, Tess Feltes, Jaroslaw Kania, Tatiana Kiselyova, E. P. Roberts, Jolanta Swiętojańska, Geoff Thompson, Wioletta Tomaszewska, Federica Turco and Marlene Werner.

Special thanks are given to Margaret Thayer, Al Newton and Rich Leschen for carefully reviewing various portions of the manuscript.

1. INTRODUCTION

When a budding Liverpool rock band was looking for a name, it is not surprising that they chose such a well known entity as the beetle (with just a twist of Lennon). There are ~420 000 named species of Coleoptera in the world (Ślipiński *et al.* 2011), more species than there are vascular plants or fungi and 75 times as many as there are mammals. Beetles occupy virtually every non-marine habitat in the world and are enormously diverse in structure and in size; the largest of them, *Titanus giganteus* (Linnaeus) from South America and *Xixuthrus heros* Heer from Fiji (both Cerambycidae: Prioninae) may attain a length of 200 mm, more than 600 times as long as the smallest ones, the nanoselline Ptiliidae, which may be 0.3 mm in length and fall well within the size range of protozoans like *Paramecium.* J. B. S. Haldane's famous comment on the Coleoptera has been quoted or misquoted many times (Hutchinson 1959; Crowson 1981; Fisher 1988; Gould 1993; Farrell 1998). The only printed statement we could find is in a report on a lecture on the biological aspects of space flight, given in 1951 before the British Interplanetary Society. Apparently, Haldane remarked that 'the Creator, if He exists, has a special preference for beetles, and so we might be more likely to meet them than any other type of animal on a planet that would support life' (Slater 1951; Clarke 1968).

The English common name 'beetle' comes from the Middle English *bityl* or *betyll* and the Old English *bitula,* meaning 'little biter' (from the word *bitan,* to bite). Two equally common English words, which have come to be used for specific beetle groups are 'weevil', (now applied to the superfamily Curculionoidea) and 'chafer' (now used for the scarabaeid subfamily Melolonthinae). The former is derived from the Middle English word for beetle, *wivil* or *wevel*, and the Old English *wifel*, and it was often used to refer to stored products pests. The latter comes from the Middle English *cheafer*, the Old English *ceafor* and the Old High German *kevar.* This last word also gave rise to the modern German word for beetle, 'Käfer'. Like the English common name, the German one also relates to biting, since it derives from the Middle High German word *kivel* or *kiver*, meaning 'jaw'.

The technical name for beetle is Coleoptera, which dates from the writings of Aristotle in the fourth century B.C.; he described those insects with wing cases (or elytra) as Κολεοπτεροσ (coleopteros), literally meaning 'sheath winged'. In some modern languages, this classical word is used as the common name for beetle (e.g. coléoptère in French). The Greeks and Romans had many terms for particular kinds of beetles and several these were adopted by Linnaeus (1758) as generic names (used today primarily for Northern Hemisphere species that often are types of the major beetle families); examples are *buprestis, carabus* (*karabos*), *curculio, kantharis, kerambyx, scarabaeus, silphe* and *staphylinos.*

Although Aristotle's definition is probably sufficient for the average person to identify many common beetles, an 'air tight' definition of the order that covers all exceptions, is much more difficult to provide. Beetles are **holometabolous** or **endopterygote** insects; that is they have **complete metamorphosis** with distinct larval, pupal and adult stages. The larvae usually undergo several moults, but the different instars tend to resemble one another and are quite different from the adult. In contrast, the **hemimetabolous** or **exopterygote** insect orders (such as the Blattodea, Dermaptera or Hemiptera) undergo an incomplete or partial metamorphosis, where larvae or nymphs usually resemble the adult, with wing pads and genitalia that increase in size with each moult. Beetle larvae never have wing pads, genitalia, compound eyes or more than a single tarsal segment (combined with the tibia in two of the four suborders) and they rarely have more than four antennal segments. Some female adults, however, may retain larval characteristics. The evolution of insect metamorphosis has been reviewed by Sehnal *et al.* (1996).

The following are usually given as diagnostic for adult Coleoptera:

1. The head is usually inserted into the anterior part of the prothorax, concealing the basal membrane.
2. The antennae almost always have 11 or fewer segments, exceptions being Rhipiceridae and a few Cerambycidae, where males have modified multi-segmented antennae, and a few other groups that have the last antennal segment subdivided.
3. The mouthparts are almost always **mandibulate** or of the chewing type with mandibles moving in a horizontal plane (rarely suctorial or reduced), and with maxillae bearing maxillary palps. This feature distinguishes adult beetles from bugs (Hemiptera).
4. The body is usually more or less depressed, so that coxae and pleural regions lie ventrally; the procoxae and mesocoxae are usually deeply recessed into coxal housings called **coxal cavities** and the metacoxae are usually slightly recessed and transversely oriented, sometimes with posterior excavations for the reception of the hind femora. The coxae are only weakly recessed in Archostemata and coxal cavities are secondarily lost in some groups of Polyphaga, but in most other orders the coxae tend to be completely exposed and more laterally placed.
5. The prothorax is well developed, almost always free, and forms with the head a distinct fore body, contrasting with the hind body (the reduced mesothorax, the enlarged metathorax and the abdomen). In most endopterygote orders, the prothorax is reduced and more closely associated with the mesothorax, which is at least as well developed as the metathorax.
6. At least the mesosternum and metasternum are invaginated at the midline, so that they are represented externally only by a groove, the **discrimen** (secondary lost in some groups), and internally by apodemes or furcae called **endosternites**, and compound ventral sclerites, referred to as **mesoventrite** and **metaventrite** and composed of pleural and precoxal elements, are expanded to occupy all of the ventral surface in front of and between the meso- and metacoxae. The invagination of the meso- and metasterna

occurs throughout the Endopterygota, but the formation of the meso- and metaventrites occurs only in Coleoptera.

7. The mesothoracic wings or fore wings are modified into more or less hardened, non-foldable, rigid **elytra** or wing covers, which lack discrete venation, partly or completely cover the hind-wings and abdomen and usually meet edge to edge in a straight line when at rest. In some other insect groups that have the fore wings partly or completely sclerotised, the two tegmina (Blattodea and Orthoptera) or hemelytra (Hemiptera) usually overlap one another at rest, rather than meeting in a straight line. The Dermaptera have similar elytra, but wing-folding, metamorphosis and other features are different.
8. The mesoscutellum is relatively small and usually visible from above as a scutellary shield, which lies between the elytral bases. Some groups of Hemiptera (Scutelleridae) and wingless Diptera are occasionally confused with beetles because the large scutellum looks like a pair of elytra, but that structure is never divided at the midline.
9. The metathoracic or hind wings, when developed, are membranous, folded longitudinally and almost always transversely, and are alone used for propulsion in flight (**posteromotorism**). Other Endopterygota, with the exception of the Strepsiptera, use both wings or the fore wings only for flight. The mechanism for folding and unfolding of the hind wing is unique to the order and will be discussed in more detail below (**see** pp. 46–49).
10. The abdominal sternites are almost always more heavily sclerotised than the tergites (exceptions occurring in taxa with secondarily shortened elytra); sternite I is lost, and sternite II is usually reduced and concealed by the metacoxae.
11. The lateral portions of the elytra are closely coadapted with those of the pterothorax and abdomen, with various types of interlocking mechanisms.
12. The terminalia (genital and pregenital structures) are usually telescoped into the apex of the abdomen: This is one of the features separating modern beetles from their Palaeozoic predecessors.

Coleoptera larvae are less easily distinguished from those of other endopterygotes, but the following features are usually diagnostic for the order:

1. Well developed and usually sclerotised head capsule;
2. Absence of adfrontal ridges (paired endocarinae forming a V between the ecdysial lines);
3. Antennae with four or fewer segments;
4. Stemmata on each side six or fewer;
5. Mandibles almost always of the chewing type, opposable in a transverse plane;
6. Absence of a median labial silk gland or spinneret;
7. Abdomen almost always without ventral prolegs bearing crochets;
8. Respiratory system usually peripneustic and never with a functional metathoracic spiracle;
9. Spiracles, if cribriform, never with the ecdysial scar completely enclosed by the sieve plate.

The first instars of Strepsiptera resemble triungulins of Meloidae and Ripiphoridae, but they differ in lacking mandibles, antennae, labial palps and trochanters. The complete and sclerotised head capsule with antennae and mandibulate mouthparts (rarely suctorial, entognathous or reduced) separates beetle larvae from those of most Diptera and some Hymenoptera. Larvae of Trichoptera, Megaloptera, Raphidioptera and Neuroptera differ from most beetle larvae in having 6-segmented legs, and those beetles that do have an additional leg segment differ from members of the first three orders in having a sclerotised ligula (Archostemata) or fused labrum (Adephaga), and from the last in the possession of mandibulate mouthparts and labial palps. Larvae with 6-segmented legs also occur in Nannomecoptera, but they have more than six pairs of stemmata and panorpoid spiracles (with the sieve plate surrounding the ecdysial scar). Lepidopterous larvae are most often confused with those of beetles, but they almost always have a median labial gland developed into a spinneret, paired adfrontal ridges, and crochet-bearing prolegs on sterna III to V or VI, and on X. It is very rare to find a lepidopteran without at least one of these three features. The primitive moth family Micropterigidae, in which adfrontal ridges and spinneret are both lacking, do have prolegs, but they are atypical, being simple and similar to those in Mecoptera; these larvae may be recognised by their unusual form (hexagonal in cross-section), retracted head, 3-segmented legs and characteristic, thickened setae. Larvae of Siphonaptera, nematoceran Diptera and higher Hymenoptera are often confused with legless larvae of Curculionidae and certain other families. Those of the first two groups differ from most apodous beetle larvae in being elongate and narrow, with a prognathous head and a different type of spiracular system (holopneustic, amphipneustic or metapneustic and with a panorpoid type of spiracle or a cribriform type with three openings). Apodous Hymenoptera, on the other hand, often differ in having a labial gland, two pairs of thoracic spiracles, and reduced maxillary and labial palps; the characteristic trident sclerite on the prementum of most weevil larvae also distinguishes them from larvae of Hymenoptera or other orders. Larvae of Symphyta (Hymenoptera) usually have well developed legs and resemble many beetle larvae; most of them are distinguished by the presence of paired prolegs on segments II to VII or VIII (and usually X). Some Pergidae lack prolegs but have a single pair of large stemmata, dome-like, unsegmented antennae and an obvious metathoracic spiracle. Xiphydriid and siricid larvae have vestigial legs and a spine-like process on the abdominal apex, as in the beetle family Mordellidae; they differ from mordellids in having elongate–elliptical spiracles.

Beetle pupae are almost always **adecticous** (without functional mandibles) and **exarate** (with legs and wings free from the body). **Obtect** pupae (with appendages tightly appressed to body) are known only in staphylinine Staphylinidae.

More information on the structure of beetles is given in the sections on adult and larval morphology, eggs and pupae.

It is generally conceded that the single most important factor contributing to the success of Coleoptera is the development of sclerotised elytra, resulting in the protection of the folded hind wings when not in use and permitting the occupation of enclosed spaces and cryptic habitats by adults. This was accompanied by a series of other morphological changes (noted above) and resulted in the general reduction of exposed membrane, all of which contributed to increased survival in cryptic habitats by protecting the body from both predation and infection by microorganisms. Another important result of elytral development was the formation of a series of interlocking devices joining the elytra, pterothorax and abdomen, so that an enclosed space, the **subelytral cavity**, was formed beneath the elytra, enclosing the metathoracic and abdominal spiracles and reducing water loss through transpiration (Cloudsley-Thompson 1965). Although beetles must have originated in the warm, humid environments existing at the boundary of the Carboniferous and Permian, the first significant radiation took place in more xeric and temperate environments of the Upper Permian and Triassic, where their unique combination of structural features gave the group a decided selective advantage (**see** Fossil History below).

For the past twenty years or more beetles have been considered to be most closely related to the Neuropterida (Megaloptera, Raphidioptera and Neuroptera) but not to any single order within that group. The morphological synapomorphies (shared, derived features) of Neuropterida + Coleoptera were discussed by Kristensen (1991) and Kukalová-Peck (1991) and preliminary phylogenetic analyses of the Holometabola using molecular data (Wheeler 1989; Wheeler *et al.* 2001) supported a Coleoptera-Neuroptera sister-group relationship. In two later papers, Kristensen (1995, 1999) claimed that there was at least weak support for two major holometabolan clades: Neuropterida + Coleoptera and Mecopterida + Hymenoptera. In most recent works based on both morphological and molecular data (Kukalová-Peck and Lawrence 2004; Wiegmann *et al.* 2009; Beutel *et al.* 2010e; Friedrich and Beutel 2010; Longhorn *et al.* 2010; McKenna and Farrell 2010) Hymenoptera forms the sister group of the remaining Holometabola and Coleoptera forms a clade with Neuropterida (Wiegmann *et al.* 2009), Neuropterida + Mecopterida (Kukalová-Peck and Lawrence 2004) or Antliophora (Mecoptera + Diptera + Siphonaptera) (Beutel *et al.* 2010, Friedrich and Beutel 2010). Coleoptera was sister to Mecopterida in a cladogram produced by Longhorn *et al.* (2010), but no Neuropterida were included. The position of Strepsiptera in a clade with Coleoptera to form the superorder Coleopterida is supported in almost all of the more recent morphological and molecular works (but **see** below). The Halteria (Strepsiptera + Diptera), proposed by Whiting *et al.* (1997); Whiting (1998) and Wheeler *et al.* (2001) based on 18S rDNA has not been supported in subsequent analyses and was probably based on long-branch attraction (Wiegmann *et al.* 2009). A Coleoptera–Strepsiptera clade is based partly on hind wing dominance, free prothorax, heavily sclerotised abdominal sternites and furcate metendosternite. Similarities based on the first instar larvae are certainly homoplasious (convergent). Kukalová-Peck (1991) and Kukalová-Peck and Lawrence (1993, 2004) have shown that strepsipteran wing venation could be derived from a common ancestor of Coleoptera that had not yet evolved the unique wing-folding mechanism that characterises all beetles. In an unusual result based on nine nuclear genes (McKenna and Farrell 2010), Strepsiptera formed a clade well within the beetle suborder Polyphaga, a position formerly proposed by Crowson (1981), but results of a recent study based on both genomic and morphological evidence (Niehuis *et al.* 2012) overwhelmingly support a sister-group relationship between Strepsiptera and Coleoptera.

The basal branching pattern in Coleoptera has been the subject of much controversy, but the subordinal relationships proposed by Crowson (1955; 1960) – Archostemata (Adephaga (Myxophaga + Polyphaga)) – has been the best supported, especially with the additional morphological evidence provided by Beutel and Haas (2000). An alternate scheme – Polyphaga (Archostemata (Myxophaga + Adephaga)) was proposed by Kukalová-Peck and Lawrence (1993, 2004) based on wing venation, and several other four-taxon statements have been published based on morphological and/or molecular data (Lawrence *et al.* 2011). A phylogeny based on the Crowson-Beutel scheme but with major fossil families included was published by Beutel (1997, 2005a).

The Australian beetle fauna includes more than 23 000 described species in ~3300 genera and 117 families, but several thousand more are known in collections and the true number may exceed 50 000 (an estimate of 80 000 to 100 000 was given by Yeates *et al.* 2003). Curculionidae is the largest family, with more than 6000 Australian species. Scarabaeidae and Chrysomelidae each have about half that number, Carabidae ~2500, Staphylinidae with almost 2000, and Tenebrionidae and Cerambycidae each between 1200 and 1500 species.

The most important early taxonomists working exclusively or largely on Australian Coleoptera were as follows (dates in parentheses refer to the dates of their first and last publications): W. J. Macleay (1863–1888), F. P. Pascoe (1859–88), T. Blackburn (1888–1912), T. G. Sloane (1881–1923), A. M. Lea (1894–1932) and H. J. Carter (1905–1939). A history of work carried out by the staff of the Australian National Insect Collection from 1926 to 1991 may be found in Upton (1997). In addition to the authors of the present volume, CSIRO scientists who have contributed to the knowedge of the Australian beetle fauna over the years include E. B. Britton (melolonthine Scarabaeidae), A. A. Calder (Elateridae, Cerambycidae, Curculionoidea), P. B. Carne (ruteline and dynastine Scarabaeidae), E. G. Matthews (Scarabaeidae), B. P. Moore (Carabidae, Lucanidae), R. G. Oberprieler (Curculionoidea), J. A. L. Watson (aleocharine Staphylinidae), T. A. Weir (Scarabaeidae, Buprestidae) and E. C. Zimmerman (Curculionoidea). Other coleopterists, both in Australia and overseas, have contributed greatly to our understanding of the Australian fauna. Notable among these are P. G. Allsopp (Scarabaeidae), M. Baehr (Carabidae), P. Bouchard (Tenebrionidae), M. S. Caterino (Histeridae), D. S. Chandler (pselaphine Staphylinidae), P J. Darlington (Carabidae), J. T. Doyen (Tenebrionidae), D. G. H. Halstead (Silvanidae), M. Hansen (Hydrophilidae), L. Hendrich (Hygrobiidae, Dytiscidae), H. F. Howden (Geotrupidae, Hybosoridae), W. F. Humphreys (Dytiscidae), Z. Kaszab (Archeocrypticidae, Tenebrionidae), Z. Kejval (Anthibidae), A. G. Kirejtschuk (Nitidulidae), D. H. Kistner (Staphylinidae), J. Kolibáč (Cleridae), G. Kuschel (Nemonychidae), R. A. B. Leschen (Erotylidae, Cryptophagidae), E. G. Matthews (Tenebrionidae), B. M. May (larvae of Curculionoidea), G. S. Medvedev (Tenebrionidae), O. Merkl (Tenebrionidae), G. B. Monteith (Scarabaeidae), J Muona (Eucnemidae), A. Neboiss (Elateridae), A. F. Newton (Leiodidae, Staphylinidae), H. Pang (Coccinelllidae), P. D. Perkins (Hydraenidae), D. A. Pollock (Pyrochroidae, Pythidae, Boridae), J. Poorani (Coccinellidae), R. D. Pope (Coccinellidae), C. A. M. Reid (Chrysomelidae), W. Schawaller (Prostomidae), C. H. Scholtz (Trogidae), T. Sen Gupta (Boganiidae, Erotylidae, Cerylonidae), J. A. Smith (Psephenidae), Z. T. Stebnicka (aphodiine Scarabaeidae), R. I. Storey (Scarabaeidae), D. Telnov (Anthicidae). M. K. Thayer (Staphylinidae), M. C. Thomas (Silvanidae), R. T. Thompson

(Curculionoidea, Phalacridae), W. Tomaszewska (Endomychidae, Corylophidae), V. G. Volkovtsh (larvae of Buprestidae), J. M. Vrydagh (Bostrichidae), M. Wanat (apionine Brentidae), Q. Wang (Cerambycidae), J. C. Watt (Tenebrionidae), C. H. S. Watts (Dytiscidae, Hydrophilidae, Scirtidae), Q. D. Wheeler (Lymexylidae), W. Wittmer (Cantharidae, Melyridae) and P. Zwick (Leiodidae, Scirtidae).

Modern textbooks dealing exclusively with Coleoptera are those of Crowson (1981), which covers in detail most aspects of beetle biology, and the family by family coverage in recent volumes of the *Handbook of Zoology* (Beutel and Leschen 2005a; Leschen *et al.* 2010a; Leschen and Beutel unpublished data). Basic references on beetle classification include those of Crowson (1955, 1960, 1967) and Lawrence and Newton (1982, 1995). Lawrence (1982) gives brief descriptions of families and higher taxa based on both adults and larvae, the interactive keys and databases produced by Lawrence *et al.* (1999a, 1999b) provide detailed information on all beetle families at a world level, and the recent cladistic analysis by Lawrence *et al.* (2011) is based on a taxon–character matrix of 516 adult and larval characters coded for 359 exemplar genera representing 314 families or subfamilies.

The only complete set of global family catalogues is the *Coleopterorum Catalogus* published by W. Junk between 1910 and 1939, with supplements published for some families between 1950 and 1981. The Nearctic and Neotropical faunas were originally catalogued by Leng (1920) and Blackwelder (1944–1957), and an updated catalogue for North and Central America was edited by Arnett (1983). The most recent catalogue covering the entire Palaearctic fauna is that edited by Löbl and Smetana (2003–2011). The Australian fauna has been treated by Britton (1970, 1974a) and Lawrence and Britton (1991, 1994), regional manuals on Australian beetles have been produced by Matthews (1980, 1982, 1984, 1985, 1987, 1992a, 1997) and Matthews and Reid (2002) for SA and Moore (1980–1992) for south-eastern Australia, and the related New Zealand beetle fauna has been treated by Klimaszewski and Watt (1997). Australian species catalogues have been published for Archostemata, Myxophaga and Adephaga (Lawrence *et al.* 1987), Scarabaeoidea (Houston 1992), Buprestoidea (Bellamy 2002) and Elateroidea (Calder 1998). Popular accounts of the Australian beetle fauna include those of Hawkeswood (1987a) and Hangay and Zborowski (2010). Other popular beetle books with a global coverage are those of Klausnitzer (1983), Paulian (1988) and Evans and Bellamy (1996). Meanings of scientific terms used in describing beetles or other insects may be found in general texts or in the glossaries of Nichols (1989) or Gordh and Headrick (2011). Scientific and common names of well known or economically important Australian insects may be found in Carne (1987).

FOSSIL HISTORY

The oldest beetle-like fossils were found in the Lower Permian, from the Artinskian (~270 million years ago or mya) of Moravia (Kukalová 1969) and North America (Lubkin and Engel 2005; Beckemeyer and Engel 2008) and the Kungurian (260 mya) of the Ural Mountains in Russia (Ponomarenko 1969b). These earliest species in the family Tshekardocoleidae were placed by Ponomarenko (1969b) in an expanded beetle suborder Archostemata, including all Permian and most early Mesozoic beetle families, as well as extant Ommatidae and Cupedidae. Kukalová-Peck (1991) considered the tshekardocoleids to be part of a coleopterid stem group, representing a paraphyletic assemblage of related orders probably coexisting with true Coleoptera in the Upper Permian. Crowson (1975) used the name Protocoleoptera for these Lower Permian taxa and Archecoleoptera for Upper Permian forms apparently intermediate between these and true beetles; however, no clear distinctions were made between the two groups and we prefer to use the term Protocoleoptera for the entire assemblage. Although most of these early coleopterids were replaced by the beginning of the Mesozoic, the family Labradorocoleidae from the Lower Cretaceous (120 mya) of Canada (Ponomarenko 1969a) appears to have the characteristics of this early group. The Cretaceous family Umenocoleidae (Chen and T'an 1973) was also considered to be a very primitive type of beetle, but the group apparently belongs to the order Blattaria (Vršanský 2003).

Adiphlebia lacoana Scudder from the Carboniferous (Kasimovian) of Mazon Creek, Illinois, USA, placed in the protorthopteran family Stephocladidae by Carpenter (1992), was claimed to be the earliest 'stem-coleopteran' by Béthoux (2009). Although these insects may be among the stem groups of Endoneoptera (Holometabola), perhaps ancestral to the Neuropterida + Mecopterida, they do not appear to be part of the Coleopterida (Kukalová-Peck and Beutel 2012). Ponomarenko (1972) described a complete hind wing from the Upper Permian beds near Chepanikha in the Ural Mountains, which he considered to be a primitive beetle wing. Kukalová-Peck (1991, p. 176, fig. 6.28) reinterpreted the wing venation, which lacked the coleopteran synapomorphies listed by her, and considered the wing to be that of an independent derivative of the coleopterid stem group.

In order to grasp the enormous amount of data that has accumulated concerning fossil Coleoptera, it is necessary to examine those characters traditionally used to distinguish fossil taxa. All important Permian and Mesozoic taxa are based on impression fossils, usually consisting of elytra only, and those specimens that are more complete have often become distorted over time. Ponomarenko (1969b) keyed out the archostematan families mainly based on elytral characters. His major assumption appears to be that the large window-punctures on the elytra are remnants of the original wing membrane and the longitudinal and transverse divisions separating these cells are derived from wing veins and cross-veins. When some of the longitudinal divisions are raised and stand out from the others, these are considered to be **principal veins**, while the others are **interstitial veins**. The elytra are of two types: (1) those with **clathrate** venation, with cells (window punctures) wider than the veins and cross-veins separating them, and without a ventral interlocking process; and (2) those that are either smooth or with punctate striae and with or without a ventral interlocking process. Within the clathrate group, the principal and interstitial veins are indistinguishable and the cells are subquadrate in ten fairly regular rows in members of the family Taldycupedidae, but the two vein types are clearly distinguishable in the remaining families, which may be separated as follows: (1) elytron with three long principal veins and no less than three cells between any two of them – Tricoleidae; (2) elytron with three long principal veins but with five rows of cells in the middle of elytron – Asiocoleidae; (3) elytron with four or five veins (rarely with three, but then most areas with two rows of cells) – Cupedidae (*sensu lato*, including Ommatidae); and (4) elytron with no less than six veins – Tshekardocoleidae, Moravocoleidae and Permocupedidae. The last three families were distinguished mainly by the number of cells at the base of the anterior vertical area of the elytron (four in the Lower Permian Tshekardocoleidae and Moravocoleidae, three in the Upper Permian Permocupedidae). Homologisation of elytral veins in the Protocoleoptera was given in the recent paper by Kukalová-Peck and Beutel (2012). Those elytra of the non-clathrate group with a ventral interlocking process (**schiza**) include those forms with rows of punctures or punctate-striae – the Upper Permian Rhombocoleidae, and those which are smooth or irregularly punctate – the Mesozoic Catiniidae and Schizophoridae and the Upper Permian parataxonomic form family Schizocoleidae. Those with non-clathrate, usually striate elytra lacking an interlocking process include members of the Upper Permian parataxonomic or form family Permosynidae and the Mesozoic Ademosynidae. Fortunately, some of the earliest of these fossils are more complete, so that additional characters are avilable. Tshekardocoleidae and Permocupedidae, for instance, differ from true beetles in having more than 11 antennomeres, broadly closed procoxal cavities located well away from the posterior edge of the prothorax, a short abdomen not fitting tightly against the elytra and an external ovipositor. A similar prothoracic structure also occurs in *Rhombocoleites adumbratus* Ponomarenko (Rhombocoleidae?) from the Upper Permian of southern Russia (Ponomarenko 1969b). Changes between the stem-group assemblage and more typical Coleoptera include a decrease in the number of antennal segments to 11, development of a more regular subparallel ribbing on the elytra, close coadaptation of the elytra and abdomen, invagination of the external ovipositor, and modifications of the hind wing to accommodate transverse wing folding.

The climate in the very early Permian in Moravia is thought to have been warm and humid but most of the Late Permian fossils are associated with more xeric conditions. Permian floras have an abundance of basal gymnosperm groups, such as the cordaite *Noeggerathiopsis* and various seed ferns (pteridosperms), as well as ferns and horsetails (White 1986). Among the land vertebrates, most of the dominant groups of Palaeozoic Amphibia were present, as well as the early reptilian lineages such as captorhinids, stem diapsids and pelycosaurs, but the dominant archosaurs of the Mesozoic had not yet evolved (Carroll 1988). The majority of the Permian coleopterids are from sites in Russia and represent inhabitants of the ancient land mass known as Angara, which extended from eastern Europe to Central Asia and Siberia (Ponomarenko 1969b, 2000, 2003b, 2004). Gondwanan fossils of a similar age have been found in the Belmont Beds of Australia (Tillyard 1924; Beattie 2007). Permian fossils are also known from the Irati Formation in Brazil (Pinto 1987) and Whitehill and Volksrust Formations in South Africa (Geertsema *et al.* 2002; Ponomarenko and Mostovski 2005). [Rohdendorf 1961; Rohdendorf and Ponomarenko 1962].

In the Mesozoic Era, Coleoptera were much more common and diverse than in the late Palaeozoic. According to Ponomarenko (1995), there are two main stages of beetle evolution in the Mesozoic: one in the Triassic and first half of the Jurassic and another in the second half of the Jurassic and early Cretaceous. Beetles of the late Cretaceous are considered by him to be more closely related to the Cenozoic and Recent faunas. Triassic plants included the first of the forked-leaved pteridosperms (*Dicroidium, Lepidopteris*), the unusual lycopod *Cyclomeia,* conifers like *Voltziopsis* and *Walkomiella,* cycads and ginkgos, while Triassic land vertebrates included mammal-like reptiles (Therapsida) and early dinosaurs (Archosauria) but no true mammals or birds. The largest Triassic assemblage is from the Madygen Formation in Fergana Valley of Central Asia (Ladinian or Carnian), which is richer and more complete than any from the Southern Hemisphere, and probably includes representatives of all four modern suborders. The most common beetles in these beds were placed in Cupedidae (*sensu lato*), as defined above, and were divided into three subfamilies, which were elevated to family rank by Lawrence (1999) and distinguished by the separation of the procoxal cavities as follows: procoxae contiguous (Ommatidae) or procoxae separated by a prosternal process that is narrower than (Cupedidae) or broader than (Triadocupedidae) a coxal diameter. Seven genera of triadocupedids occurred in these Central Asian beds, but Ponomarenko also placed in this group the Mid-Triassic *Moltenocupes* Zeuner from South Africa. The Ommatidae was divided into three tribes by Ponomarenko, which were placed in four subfamilies by Lawrence (1999). Lithocupedinae (*Lithocupes* Ponomarenko) were restricted to these Triassic beds, while Notocupedinae extended into the Cretaceous; the extant subfamilies Ommatinae and Tetraphalerinae originated in the Jurassic. Cupedidae also contained three subfamilies, only one of which (*Mesocupoides* Ponomarenko) occurs in the Triassic. Another important group of Madygen fossils consists of the diverse family Schizophoridae with ten Triassic genera and the related Catiniidae with three; these groups are often considered to be the ancestors of Adephaga and/or Myxophaga. The adephagan family Triaplidae, with enlarged metacoxal plates like those of Haliplidae, and the first Trachypachidae (*Sodgodromus* Ponomarenko) occur in the Madygen beds, as well as several Ademosynidae, which may be near the base of the Polyphaga. *Peltosyne triassica* Ponomarenko appears to be a true polyphagan with a prothoracic cryptopleuron and a metanepisternum that does not extend to the edge of the mesocoxal cavity, but other Triassic Polyphaga are known but not yet described. The family Obrieniidae (Zherikhin and Gratshev 1993), also described from Central Asian Triassic fossils, was placed tentatively by its authors in the superfamily Curculionoidea, but later more definitely moved to Archostemata (Zherikhin 2002; Kuschel 2003). The adephagan family Colymbothetidae was described by Ponomarenko (1993) based on larval specimens from the Upper Triassic Tologoy Formation in Kazakhstan. The first staphyliniform beetle, *Leeheremania prorova* Chatzimanolis *et al.* (2012) was found in Late Triassic beds in eastern North America. Major Triassic sites in the Southern Hemisphere include the Molteno Formation in South Africa and the Ipswich Coal Measures (Denmark Hill and Mount Crosby) in southern Queensland (Dunstan 1923), which are of about the same age as the Madygen Formation. Undescribed Mid-Triassic (Anisian) beetles are known from the Cloyna Fauna, near Murgon and the Esk Beds on the shores of Wivenhoe Dam, southern Queensland. Almost 60 species in 20 genera were described from the Ipswich site, but most are represented by single elytra and appear to be misplaced in several polyphagan families. The genus *Simmondsia* Dunstan was transferred to Taldycupedidae by Ponomarenko (1969b), who also commented on a possible relationship between *Mesothoris* Tillyard and the Triadocupedidae. Several species from Denmark Hill were included in the genus *Ademosyne* Handlirsch and a related genus

Ademosynoides Dunstan, but these genera are problematic. Although *Ademosyne major* Handlirsch was based on a more or less complete specimen, most *Ademosyne* species are based on elytra only and should be placed in the form genus *Permosyne* Tillyard (Ponomarenko 2003b; Martin 2010). The type species of *Ademosynoides* is also based on a single elytron, although another species, *A. magnifica* Dunstan, is based on a complete specimen, the pterothorax of which appears to be of the polyphagan type with the metanepisternum not or barely reaching the mesocoxal cavity. As suggested by Martin (2010), this genus should be re-examined. The Triassic and Jurassic ademosynid genera *Dolichosyne*, *Gnathosyne*, *Rhanis* and *Petrosyne* of Ponomarenko (1969b) belong to Archostemata in the broad sense, with the mesanepisterna broadly meeting the mesocoxal cavities and the propleuron extending to the anterior edge of the prothorax, although these features are not evident in his three *Ademosyne* species or in his genera *Cephalosyne* and *Sphaerosyne*. Other Triassic fossils from the Southern Hemisphere have been found in the Molteno beds of Natal, South Africa (Zeuner 1961; Riek 1974) and from Mount Flora, Grahamland, Antarctica (Zeuner 1958). The elytron of *Ademosyne speciosa* Riek from the Molteno Formation of South Africa beds was illustrated by Kukalová-Peck (1991), who considered the venation to be that of a complete and primitive coleopterid type.

The Jurassic period (210 to 145 mya) was a time of great diversification of the order, and thousands of specimens have been studied from more than 150 localities, mainly in Europe and northern Asia. The largest early Jurassic fossil beds occur in Central Asia (mainly Kirghizstan). As in the Triassic, Archostemata are abundant, but Triadocupedidae are absent, notocupedine Ommatidae are dominant, and the extant genera *Omma* and *Tetraphalerus* make their first appearance. Other common Archostemata included Ademosynidae and Schizophoridae. Adephaga are relatively rare in these early Jurassic beds, although *Necronectus avus* Ponomarenko from the Kenderlyk site in eastern Kazakhstan is an unusual aquatic adephagan of uncertain relationships. Well preserved Polyphaga are rare in the Lower Jurassic, but include two genera of the family Praelateridae and two genera in the elaterid subfamily Protagrypninae (Dolin 1973). Lower Jurassic fossils from Australia include those recently described by Martin (2010) from the Mintaja site in Western Australia. Middle Jurassic fossil sites are primarily in Siberia, where Archostemata are virtually absent and Adephaga are represented by the aquatic families Parahygrobiidae, Liadytidae, Coptoclavidae and Gyrinidae and by dytiscoid larvae in the genera *Parahygrobia* Ponomarenko, *Angaragabus* Ponomarenko and *Stygeonectes* Ponomarenko. Geadephaga are absent from the Siberian sites, and the Polyphaga are represented by a large number of specimens of *Mesecanus communis* (Ponomarenko), an agyrtid-like beetle with paired ocelli, plus some undescribed hydrophilids and staphylinids. A notable Mid-Jurassic fossil from the Transbaikal region of Russia is *Jurodes ignoramus* Ponomarenko (1985a), which was placed by its author in an adephagan family Jurodidae but later found to be closely related to an extant beetle, *Sikhotealinia zhiltovae* Lafer (Sikhotealiniidae), from Far Eastern Russia, a taxon with a mixture of subordinal features currently placed as *incertae sedis* within Archostemata (Lafer 1996; Kirejtshuk 1999).

A relatively recently worked Middle Jurassic site is that from Daohugou (Jiulongshan Formation), Inner Mongolia, China (Bathonian, 163–167 mya) (Ren *et al.* 2010; Kirejtshuk *et al.* 2010c). The Daohugou biota included a relatively small number of Archostemata, including two schizophorid genera, two more species of *Jurodes* (Yan *et al.* 2013) and a few Tricoleidae, and several eodromeine Trachypachidae. Interesting polyphagan groups include the earliest well documented scarabaeoid, placed by Bai *et al.* (2012b) in a new family Alloioscarabaeidae, the oldest Ochodaeidae (Nikolajev and Ren 2010), *Lasiosyne* Tan *et al.* (2007b), described as an ademosynid but later made the type of the elateriform family Lasiosynidae by Kirejtshuk *et al.* (2010a), *Paraelateriformius* Yan and Wang (2010), an elateriform of uncertain affinities, protagrypnine Elateridae (Chang *et al.* 2009), the genus *Cervicatinius* Tan and Ren (2007), placed in Catiniidae but later moved to Cleroidea by Kirejtshuk *et al.* (2010c), the earliest member of the family Trogossitidae (Yu *et al.* 2012) and the earliest tenebrionoid genus possibly related to Melandryidae (Wang and Zhang 2011). Another interesting fossil from this site is *Tarsomegamerus mesozoicus* Zhang (2005), considered to be a chrysomeloid by its author, but transferred to Lasiosynidae by Kirejtshuk *et al.* (2010a).

By far the richest assemblage of Late Jurassic Coleoptera is from Karatau (Karabastau Formation) in southern Kazakhstan (Kimmeridgian–Oxfordian, 152–158 mya). Archostemata comprise ~10% of all taxa, but three-fourths of these are Ommatidae or Cupedidae. Schizophoroid Archostemata are not abundant but include unique, probably aquatic forms, including *Tersus* Martynov, *Xyphosternum* Ponomarenko and *Catinius* Ponomarenko. Adephaga also comprise ~10% of the total diversity, with Gyrinidae, Dytiscidae, Coptoclavidae (several genera), Trachypachidae and Carabidae. The family Coptoclavidae was a large and diverse group of aquatic beetles, first occurring in the Late Triassic of England and Early Jurassic of Europe and North America but most common in the Mid to Late Jurassic and Early Cretaceous of Central Asia, eastern Asia and southern Europe (Ponomarenko 1977, 2003a; Soriano *et al.* 2007; Wang *et al.* 2010). Some species were almost 50 mm in length, the adults with transversely divided eyes, long and slender fore legs and paddle-like mid and hind legs, as in modern Gyrinidae, and the larvae with mid and hind legs modified for swimming, the eighth spiracles dorsally placed and the ninth segment terminal and consisting of two movable urogomphi, as in modern Amphizoidae, Noteridae and Dytiscidae. Polyphaga are distinctly dominant at this site and include species of Hydraenidae and Hydrophilidae (Ponomarenko 1977); Staphylinidae (Tikhomirova 1968, 1973); two scarabaeoids in the genera *Juraclopus* Nikolaev and *Antemnacrassa* Gomez Pallerola (Nikolaev 1992, 2004, 2005a; Krell 2006, 2007); Mesocinetidae (Kirejtshuk and Ponomarenko 2010); Buprestidae (Alexeev 1993, 1995); Lasiosynidae (Yan 2009; Kirejtshuk *et al.* 2010a); Cerophytidae (Dolin 1980; see below); Elateridae (Dolin 1975a, 1976); and a variety of Cucujiformia. The cucujiform family Parandrexidae (Kirejtshuk 1994), later found in the Lower Cretaceous of Spain (Soriano *et al.* 2006b) was transferred from Cucujoidea to Cleroidea (Kirejtshuk *et al.* 2010c). The chrysomelid subfamily Protoscelidinae (Medvedev 1968), which should probably be elevated to family rank in line with current treatments of extant groups such as Megalopodidae and Orsodacnidae, was also found in these beds, along with several Curculionoidea placed in Eobelidae by Arnoldi (1977) but belonging to a stem group of the family Nemonychidae (Gratshev and Zherikhin 1996, 2003;

R. Oberprieler *in litt.*). Other important Late Jurassic sites are Solnhofen in Bavaria (Tithonian, 151–153 mya) (Ponomarenko 1971b, 1980, 1985b) and Shar-Teg, Gobi-Altai region, Mongolia (Tithonian) (Kirejtshuk and Ponomarenko 2010). According to Ponomarenko (1995), the Solnhofen assemblage was dominated by several genera of Coptoclavidae, with the next most diverse group being the Cupedidae. The Shar-Teg fauna was similar to that of the Karatau site, but Cupedidae, Coptoclavidae, Elateridae and Curculionoidea were rare, Mesocinetidae were abundant, and Lasiosynidae (Kirejtshuk *et al.* 2010a) were dominant among the Polyphaga. Several undescribed Australian Coleoptera are known from the Upper Jurassic (late Kimmeridgian) Talbragar fish beds in New South Wales (Beattie and Avery 2012).

The Cretaceous fauna (145 to 65 mya) includes many more modern forms and lacks the abundance of Archostemata that characterises the early Mesozoic. As noted above, the Lower Cretaceous fauna resembled that of the Jurassic rather than that of the Upper Cretaceous. Some important sites include the Gurvan-Eren Formation in Khvod, western Mongolia; the Laiyang Formation, Shandong Province, China; the Zaza Formation at Baissa, Transbaikalia, Russia; the Yixian Formation in Liaoning and Hebei Provinces, China; the La Pedrera de Rubies and Calizas de la Huerguina Formations in Spain; the Khurilt Formation at Bon-Tsagan, central Mongolia; the Koonwarra fossil bed in Victoria, Australia; the Crato Formation in Santana; Brazil, and the Emanra Formation at the Khetana River, north-eastern Siberia. Although all of these sites appear to be Lower Cretaceous, opinions vary as to the actual ages of some of them, including the Laiyang, Yixian and Zaza Formations (Zhang and Rasnitsyn 2006). The Baissa locality is noted for the large number of Coptoclavidae (mainly *Coptoclava longipoda* Ping) (Ponomarenko 1961). An interesting larval form is *Cretotaenia pallida,* which was placed in Adephaga *incertae sedis;* this larva resembles some Carabidae in head structure and in the presence of segmented urogomphi and a terminal 10th segment, but the 8th spiracles are enlarged and placed dorsally as in many Hydradephaga. Cupedidae and Carabidae represent 4% of the Baissa fauna, and polyphagan families include Hydrophilidae, Staphylinidae, Scarabaeidae, and Elateridae (Ponomarenko 1995). [Ponomarenko 1985a, 1986, 1990; Nikolajev 1993a, 1993b, 1995, 1996, 1998, 2000a, 2000b, 2002].

The Yixian Formation in Liaoning Province, China has produced a rich selection of beetle fossils many of which are yet to be described. The age of this site is either Late Jurassic or Early Cretaceous (Tithonian–Berriasian), and the beetle fauna was summarised briefly by Kirejtshuk *et al.* (2010c). Archostemata are much more numerous at this site than at the Mid Jurassic site at Daohugou, Inner Mongolia. Several papers dealing primarily with Ommatidae and Cupedidae have been published by Ren and Tan (2006), Ren *et al.* (2006), Tan and Ren (2006a, 2006b, 2006c, 2009), Tan *et al.* (2005, 2006b, 2006c, 2007a, 2007b, 2007c, 2012), the last of these including the first cladistic analysis of the family Ommatidae, based on both extant and extinct taxa. *Sinorhombocoleus* Tan and Ren (2009) was placed by its authors in the Permian family Rhombocoleidae, but transferred to Schizophoridae by Kirejtshuk *et al.* (2010). The first fossil myxophagan (Hydroscaphidae) was reported from these beds by Cai *et al.* (2012). Adephaga include several eodromeine Trachypachidae, a few Coptoclavidae and several genera of Dytiscidae. Several Hydrophilidae have been described by Prokin *et al.* (2010), but most Polyphaga are members of the Staphylinidae and Elateriformia, with smaller numbers of Cleroidea, including the genus *Mathesius* Kolibáč & Huang (2011) of uncertain family, Tenebrionoidea, including the famlily Liaoximordellidae Wang (1993), and Curculionoidea (Liu *et al.* 2006). According to Kirejtshuk *et al.* (2010) a unique feature of this fauna is the large number of scarabaeoid fossils. Those described so far include members of the Geotrupidae (Nikolajev 2008), Glaresidae (Bai *et al.* 2010), Lucanidae (Bai *et al.* 2012a), Glaphyridae (Nikolajev *et al.* 2011), Scarabaeidae: Aclopinae (Bai *et al.* 2011) and a new family Septemventeridae (Bai *et al.* 2012c). The family Magnocoleidae was described by Hong (1998) from early Cretaceous beds in Hebei Province. The latter was based on a single elytron, which was compared to the Permo-Triassic families Asiocoleidae and Tricoleidae; its true relationships, however, are far from clear. The Spanish fossil beds at Las Hoyas and El Montsec have produced several interesting fossils of early Cretaceous (Barremian) age (Zherikhin and Gratshev 1997; Ponomarenko and Martínez-Delclòs 2000; Soriano and Delclòs 2006; Soriano *et al.* 2006a, 2006b). The Coptoclavidae are more diverse than in any other locality, with representatives of four of the five subfamilies, including the endemic Hispanoclavinae; two of the genera have the forelegs modified for filter-feeding (Soriano *et al.* 2007). The Koonwarra Cretaceous (Aptian) fossil beds in Victoria, Australia (Jell 2004) contain relatively few beetle fossils (mainly Dytiscidae, Hydraenidae, Hydrophilidae, Staphylinidae and Scirtidae).

Cretaceous amber sites in Lebanon and Myanmar have produced the earliest records of Micromalthidae and liparochrine Hybosoridae from the former (Kirejtshuk *et al.* 2009, 2011) and Lepiceridae (as Haplochelidae) (Kirejtshuk and Poinar 2006, Ge *et al.* 2010), Sphaeriusidae (Kirejtshuk 2009), Prostomidae (Engel and Grimaldi 2008) and the staphylinid subfamilies Scydmaeninae (Chatzimanolis *et al.* 2010) and Solieriinae (Thayer *et al.* 2012) from the latter. The justifications for and placements of the scirtoid family Elodophthalmidae and several new cucujiform subfamilies described from Lebanese amber by Kirejtshuk and Azar (2008) and Kirejtshuk *et al.* (2009) are not convincing. Upper Cretaceous fossil resins (retinite) from the Taimyr Peninsula in northern Siberia produced two species of *Aphytocerus* Zherikhin (1977), then considered to be the oldest record for the family Cerophytidae; however, recent work by Chang *et al.* (2011) has shown that the family was common in the Jurassic and Cretaceous (**see** above) and included genera formerly misplaced in Silphidae (*Mercata festiva* Lin 1986), Elateridae (*Idiomerus* Dolin 1980) and Oedemeridae (*Necromera* Martynov 1926).

The largest and most important collection of Cenozoic fossil insects are those embedded in Baltic amber of mid Eocene age. A general summary of this fauna may be found in Larsson (1978), a list of described Coleoptera was published by Spahr (1981a, 1981b), and numerous new taxa have been described in later works (e.g. Winkler 1987; Muona 1993a; Hőrnschemeyer 1998; Jałoszyński 2012b). Early Eocene amber from the Oise region in northern France has produced a new genus of trinodine Dermestidae (Kirejtshuk *et al.* 2010b) and several new Cucujiformia (Kirejtshuk and Nel 2008, 2009). Several other sites are known throughout the world (Penney 2010), but a detailed discussion of Cenozoic beetle faunas is beyond the scope of this work. In Australia, a few Tertiary beetles have turned up at three sites in QLD: the limestones at Riversleigh (Duncan *et al.* 1998), ferruginous mudstone at Redbank Plains (Jel 2004), and amber on the eastern Cape York Peninsula (Hand *et al.* 2011). Porch and Elias (2000), Porch *et al.* (2009) and Porch (2010) have discussed the use of Australian Quaternary subfossil beetles in the reconstruction of past climates.

AUSTRALIAN BEETLES AND THEIR HABITATS

The following section contains a general survey of the types of habitats and microhabitats most frequented by beetles in Australia and the ecological roles played by various beetle groups within these habitats, as shown primarily by their feeding behaviour. The main breakdown will be by specific habitat type (e.g. on leaf surfaces, under bark, in flowers) as they occur in the eucalypt-dominated forests of south-eastern Australia, but major variations within other biomes (e.g. mallee or tropical rainforest) will be mentioned where appropriate. In general, the beetle fauna will be discussed by superfamily or family (as they appear in the taxonomic sections) and mainly at the generic level.

Vegetative Surfaces – Forest Canopy

Carabidae do not usually inhabit leaf surfaces, but in more tropical areas adults and larvae of *Chlaenius* and some members of the Lebiini are common predators in this habitat.

Scarabaeidae commonly feed on leaves as adults but most do so at night and are less likely to be encountered by collectors. The most obvious foliage-feeding scarabs are the Christmas beetles (*Anoplognathus* species in the Rutelinae), which are capable of seriously defoliating eucalypts, but the greatest number of nocturnal feeders belong to genera such as *Sericesthis, Heteronyx, Liparetrus* and *Scitala* in the subfamily Melolonthinae. Howden and Howden (1991) have shown that several dung-feeding Scarabaeinae perch on leaves at night.

Among the Staphyliniformia, Elateriformia and Cleroidea, several species may occur in small numbers on vegetative surfaces, but most are either general predators or incidentals normally occurring elsewhere. Scirtidae may be abundant in this habitat, especially in riparian situations, but they are more likely to be encountered in TAS and the cooler parts of mainland Australia.

Several small Cucujoidea are common on leaf surfaces and many probably feed on phylloplane fungi, which are the first organisms to invade the tissue of dying leaves; examples may be found among the Nitidulidae, Phalacridae, Silvanidae, Corylophidae and Latridiidae. Beating dead leaves on eucalypts sometimes produces large numbers of *Phalacrinus* (Phalacridae) and *Cryptamorpha* (Silvanidae). Coccinellidae may also occur on leaf surfaces, where they feed on a variety of phytophagous Hemiptera or mites. There are several tenebrionoid taxa whose adults may be found in the canopy while their larvae occur in dead wood or leaf litter; examples include Lagriini, Cnodalonini and Stenochiini among the Tenebrionidae.

Chrysomeline Chrysomelidae feed on leaves and other vegetative parts of a variety of plant species as both larvae and adults. The paropsine genera *Paropsis, Chrysophtharta, Trachymela* and *Paropsisterna* (Fig. 10B) are particularly common on eucalypts, whereas *Calomela* (Fig. 10G) and *Dicranosterna* feed on *Acacia.* Cryptocephalinae and Eumolpinae are also common leaf-feeders but only as adults; larvae occur in leaf litter and soil respectively (**see** below). Common eucalypt-feeding cryptocephalines belong to genera such as *Cadmus, Aporocera* and *Ditropidus,* while *Edusella* species are common eumolpines in this habitat. Galerucinae are also-major leaf feeders, sometimes as both larvae and adults, but often only in the adult stage.

Adult curculionoids (weevils and their relatives) are particularly common on vegetative surfaces both during the day and at night. Examples may be found among the Belidae, Attelabidae, Brentidae and in many groups of Curculionidae. Species of the attelabid genus *Euops* (Fig. 12C) are leaf-rollers, laying their eggs in a nest constructed by cutting and rolling a portion of the leaf. Among the Curculionidae, adults of *Myllocerus, Perperus, Polyphrades, Leptopius, Neolaemosaccus, Haplonyx* and many other genera can be beaten from foliage. *Chrysolopus spectabilis* (Fabricius) (Fig. 11E) are often found on *Acacia* foliage; the larvae live in stems and roots of *Acacia.* Only one group, the Gonipterinae (*Gonipterus,* Fig. 12A, and *Oxyops*), are external leaf feeders in both larval and adult stages; most other weevil larvae either occur in the soil or feed within plant tissue (**see** below).

Flowers and Male Cones – Pollen-Feeding

Adephaga and Staphyliniformia are not usually common in flowers. Carabid beetles, such as Lebiini, certainly prey on flower-inhabiting insects, and *Plagiotelum* occurs in flowers in TAS. The hydrophilid genus *Pseudohydrobius* is commonly found on *Leptospermum* flowers. The staphylinid genus *Paraphloeostiba* has been associated with flowers of a variety of unusual plants, including the cunjevoi, *Alocasia macrorrhizai* (Araceae), *Myristica insipida* (Myristicaceae), Xanthorrhoea (Xanthorrhoeaceae) and *Balanophora* (Balanophoraceae) but these same staphylinids may feed in rotting fruits and fermenting *Araucaria* cones. Species of the aleocharine genus *Leucocraspedum* are known to be floricolous and have been taken in *Eucryphia* flowers in TAS, and the omaliine genera *Austrolophrum, Leaskia* and *Anthobiomimus* are obligate pollen-feeders mainly on Proteaceae. Among the Scarabaeidae, the major flower-feeding species are diurnal members of the Melolonthinae and Cetoniinae. In the former group, species of *Phyllotocus, Diphucephala* (Fig. 68K) and *Automolius* often occur in large numbers in flowers, while common cetoniines include *Eupoecila australasiae* (Donovan) (Fig. 2H) and species of *Diaphonia;* melolonthine larvae occur in the soil, while many cetoniine larvae feed in rotten wood or leaf mould (**see** below). Scirtidae are commonly found in flowers, as well as on foliage, and several species of *Pseudomicrocara,* as well as *Macrohelodes crassus,* have been found in *Eucryphia* flowers in TAS.

Buprestidae or jewel beetles are some of Australia's better-known flower beetles. A great variety of species, usually ranging in size from a few mm to 5 or 6 cm in length, are particularly common in flowers, especially Myrtaceae (*Eucalyptus, Angophora* and *Leptospermum*). The larger and more common genera include *Stigmodera* (Fig. 4E), *Temognatha, Castiarina* (Fig. 4F), *Melobasis* and *Curis.* Larvae of these pollen-feeding beetles occur in living and dead woody plant tissue (Fig. 5B).

Lycid beetles are commonly found in flowers as adults, while larvae occur in leaf litter and rotting wood. Several Australian lycids, including species of *Porrostoma* and *Metriorrhynchus*, are black with bright orange elytra (Fig. 5E). They are known to produce a variety of defensive chemicals, including warning odours, bitter principles and antifeedants, and they form a mimicry complex with various Buprestidae, Oedemeridae (Fig. 7D), Meloidae (Fig. 75E), Pyrochroidae (Fig. 95I), Cerambycidae (Fig. 9A) and Belidae (Fig. 79A) (Moore and Brown 1989). Among the Cantharidae, species of *Chauliognathus* (Fig. 6C) and *Heteromastix* may be abundant on flowers. Species of the latter genus are often black with a pink or red prothorax and form part of another mimicry complex involving smaller species of Scirtidae, Melyridae, Mordellidae, Oedemeridae, Pyrochroidae, Scraptiidae, Aderidae and Chrysomelidae.

Dermestid beetles, including species of *Trogoderma* and *Anthrenocerus,* may frequent flowers, especially in drier areas. Their larvae occur in litter, animal nests, spider webs and other accumulations of organic debris. Cleridae are usually predators as larvae and adults. Among the flower-inhabiting species, larvae of the smaller hydnocerines in the genus *Lemidia* occur in leaf litter, while those of the larger clerines, such as *Phlogistus* and *Eleale,* usually prey on subcortical or wood-boring insects. Melyridae currently included in the genus *Dasytes* may also be common on flowers (Hawkeswood 1987b).

There are numerous pollen-feeding species in the families Kateretidae and Nitidulidae. Adults of *Pria*, *Epuraea*, *Aethina* and *Notobrachypterus* are commonly beaten from flowers, and larvae of the last two are also pollen feeders. Some members of *Aethina* (*Circopes*) occur in male cycad cones, while species of *Aethina* (*Idaethina*) are specialists on Malvaceae (*Hibiscus* and *Alogyne*). *Platychoropsis* species are known to breed in *Pandanus* flowers in Micronesia (Gillogly 1962) and it is likely that the species in northern Australia do the same. Both larvae and adults of a *Brachypeplus* species have been collected in the flowers of *Alocasia* (Araceae), but it is possible that, like the staphylinids mentioned above, the species is also attracted to fermenting plant material.

Members of the family Boganiidae may have an ancient association with pollen-producing organs, since both larvae and adults of the Paracucujinae feed on cycad pollen in Australia and South Africa, while those of Boganiinae feed on angiosperm pollen in both areas (Crowson 1991). In Australia, *Paracucujus rostratus* Sen Gupta and Crowson (Fig. 85J) is associated with *Macrozamia* in WA, *Boganium* species (Fig. 85F) are known from eucalypt flowers, and *Athertonium* is known from the flowers of rainforest plants (Myrtaceae and Elaeocarpaceae). Within the family Erotylidae, *Xenocryptus tenebroides* Arrow in WA and some species of *Hapalips* in northern QLD feed as both larvae and adults on cycad pollen. Latridiidae and Corylophidae may be as abundant in flowers as they are on vegetative surfaces and are often included in flower samples.

Mordellidae are common in a variety of flower types, while their larvae are internal feeders usually in wood or herbaceous stems. Ripiphorine Ripiphoridae, such as *Macrosiagon* (Fig. 76G), are also flower visitors; the adults feed on pollen and deposit eggs in the flowers where the triungulins will have access to host species of Hymenoptera. All Australian Meloidae are parasites of Hymenoptera, with adults commonly occurring in flowers; species of *Tmesidera* and *Palaestra* (Fig. 75E) exhibit black-and-orange warning coloration and belong to the *Metriorrhynchus* mimicry complex. The tenebrionid genus *Ulomoides* is also known to be associated with cycad pollen on which both larvae and adults feed. The Oedemeridae are well known as pollen feeders and the Pyrochroidae occur on flowers or vegetation; *Pseudolycus* (Fig. 7D) in the first group and *Morpholycus* (Fig. 95I) in the second also belong to the lycid mimicry complex.

In the Cerambycidae there are several lycid-mimicking flower inhabitants, including species of *Tritocosmia, Tropocalymma, Stenoderus* (Fig. 9A) and *Eroschema*. In addition, some flower-inhabiting cerambycids, notably species of *Hesthesis* (Fig. 9F), have very short elytra and mimic wasps. Another group of floricolous species are the Clytini (e.g. *Chlorophorus* and *Aridaeus*). Members of the megalopodid subfamily Palophaginae (Fig. 96A) feed exclusively on the pollen of Araucariaceae. Several species of Chrysomelidae are flower specialists. Well known examples are species of the chrysomeline genus *Pyrgoides,* which feed on *Acacia* flowers as adults and larvae. *Stethopachys formosa* Baly (Fig. 76K) is a specialist on orchids; larvae feed on the flowers and are known to cause damage to nursery stock.

The primitive curculionoid family Nemonychidae appears to be associated exclusively with the pollen of native conifers (*Araucaria* and *Agathis*) on which both larvae and adults feed. Belidae are relatively common flower inhabitants, especially species of *Rhinotia.* Some of these, including *R. haemoptera* (Fig. 79A), are mimics of *Metriorrhynchus.* Other flower-inhabiting curculionoids belong to the attelabid genus *Auletobius* and many curculionid genera, such as *Haplonyx* (Fig. 80J), *Elleschodes, Emplesis, Storeus* and *Cyttalia.* The genus *Tranes* (Fig. 12F) is frequently found feeding on cycad pollen but the larvae feed within the tissue of the male cone as it decays (Forster *et al.* 1994; Oberprieler 1995).

Cones, Seeds and Seed Pods

Several beetles feed on developing female cones, seeds and seed pods, and others attack these items after they have fallen to the ground. As mentioned above, some staphylinids have been collected in fermenting conifer cones but are not restricted to this microhabitat. Members of the anobiid genus *Dryophilodes* are specialists on the seed pods of eucalypts (gum nuts) and some other Myrtaceae, but at least one species is known to breed in the female cones of *Araucaria.*

In the Nitidulidae, larvae of *Neopocadius froggatti* (Kirejtshuk and Lawrence) feed in the seed pods of *Brachychiton,* while the related species *N. ursula* (Kirejtshuk and Lawrence) is associated with proteaceous seeds in northern QLD. Cryptophagine Cryptophagidae have been found breeding in gum nuts on the ground but were probably feeding at least in part on fungi that had invaded the pods. Among the Curculionoidea, several Anthribidae are seed feeders, as are many Curculionidae.

Bark and Wood Surfaces

Baehr (1990a) discussed the community of carabid beetles living under the bark of Australian eucalypts; he included among those commonly found in this habitat species of *Arthropterus* (Paussinae), *Adelotopus* and *Sphallomorpha* (Pseudomorphinae), *Amblytelus* (Moriomorphinae), *Sarothrocrepis* (Cyclosomini) and several Lebiini, including *Philophloeus, Agonocheila, Demetrida* and *Trigonothops.* Members of this assemblage, which occur primarily in dry sclerophyll forests, are surface-active predators; they are not subcortical in the sense used below in that they do not feed or lay their eggs under bark. They are really part of the surface fauna and more easily seen at night, when the observer is using a head lamp. In rainforests, bark and wood surfaces are often covered with algae, lichens and mosses, which form the habitat for yet another suite of beetles. Among the Carabidae are the peculiar tiger beetles (Cicindelinae) in the genera *Tricondyla*, *Distipsidera* and *Rhysopleura.* In addition to Carabidae, various other groups may be found on these surfaces. In TAS and the more mountainous parts of the mainland, species of Byrrhidae may be found on mossy surfaces, and predaceous Elateridae and Cantharidae may occur in the same habitats. Clerid predators, such as *Trogodendron fasciculatum* (Schreibers) (Fig. 8F), may be seen on eucalypt surfaces.

On tree trunks in rainforests species of·*Propalticus* (Propalticidae) (Fig. 85B) may be seen, apparently feeding on lichens or algae; when disturbed these minute beetles will jump long distances. Adults and larvae of small predaceous Coccinellidae are encountered in similar situations, as are those of unusual, coccinellid-like Tenebrionidae (Leiochrinini) (Fig. 75J). Other surface-active tenebrionids include adult Heleini and Cnodalonini and larvae and adults of Titaenini and some Amarygmini. In TAS, moss-covered log surfaces may be inhabited by some adult Melandryidae and the unusual larvae and adults of the promecheilid genus *Sirrhas* (Fig. 93C–D). The peculiar lamiine cerambycid *Gnoma australis* Schwarzer is a tree trunk dweller in wet tropical forests. Some Chrysomelidae and many Curculionidae also inhabit moss-covered woody surfaces in cool temperate rainforests.

One peculiar microhabitat–located on living tree trunks, often near the base–is the slime flux. This consists of a cavity, usually the site of an old wound, where a thick, dark liquid indicates the presence of sap fermentation produced by yeasts, such as *Dipodascus* and *Endomyces,* as well as bacteria. Common inhabitants of these slime fluxes are adults and larvae of the genus *Nosodendron* (Nosodendridae) (Figs 35J, 81G–H), which probably feed on the fungal by-products, but may also take in fly larvae (Lawrence 1989).

Subcortical Habitats – Under Bark of Living and Dead Trees and Logs

A large number of beetle larvae and adults occur under the bark of trees and logs in a variety of environments whose characteristics depend on moisture, as well as the degree and nature of decay taking place in the phloem and cambium. Very loose, dry bark will house beetles of a very different kind than that which forms a tight, moist environment beneath it. Living or recently killed trees will also have a different subcortical fauna from old, well rotted logs. This microhabitat tends to grade into the next one, and many subcortical species move into rotten wood. The succession of decay organisms in woody substrates is discussed by Hudson (1968), Käärik (1975) and Shigo (1967).

Carabidae are commonly encountered hiding under loose bark, as mentioned above, but a few groups, such as *Tachys, Tachyta* and their relatives, are subcortical predators. Many Staphylinidae are subcortical, the most common of which are members of the osoriine tribes Eleusini (*Zeoleusis*) and Lispinini (*Nacaeus, Lispinus*). Although most staphylinids are predaceous, these are thought to be saprophagous, feeding on decaying cambium or the products of bacterial or fungal action. Predaceous Staphylinidae also occur in this habitat and, with Histeridae such as *Hololepta* (Fig. 77C) and *Platysoma,* commonly feed on fly larvae and other soft-bodied insects.

Relatively few scarabaeoids are subcortical but they may be encountered in old logs where they are part of the rotten wood fauna (**see** below). Some Aphodiinae, such as *Saprus* and *Saprosites,* may breed in rotten cambium. Eucinetidae and Clambidae are occasionally present in fungal growths under bark, and larvae of some Scirtidae occur in wet and slimy subcortical regions. Larvae of Buprestidae are common under bark in a variety of situations but are usually found in living or recently killed trees. The larvae of Lycidae occur in moist, fermenting situations under bark and may be associated with slime mould plasmodia.

Among the Trogossitidae, Trogossitinae, like *Leperina* (Fig. 8E) and *Egolia* (Fig. 82H), are subcortical predators as larvae and adults but the smaller Lophocaterini may feed on fungus or rotting cambium. Clerid larvae may be common predators in this habitat. A large number of Cucujoidea are subcortical and the majority of them feed on fungi or fungal by-products, as well as dead plant matter. Species of *Brachypeplus* and *Carpophilus terminalis* (Murray) (Nitidulidae) are often abundant as larvae and adults under fermenting bark. Other cucujoids in this habitat include species of *Mimemodes* (Monotomidae) (Fig. 83F), *Hymaea* (Phloeostichidae) (Fig. 83C) and *Silvanus* (Silvanidae) (Fig. 85D), and various Laemophloeidae, Cryptophagidae, Biphyllidae and Cerylonidae. Species of Brontini (Silvanidae) (Fig. 71C) are most often encountered in somewhat drier situations, along with their active larvae (Fig. 34J). The strongly flattened larvae (Fig. 84D) and adults (Fig. 71E) of the cucujid genus *Platisus* are subcortical predators, often found under very tightly fitting eucalypt bark. Phalacrid larvae have been found within thick stringy bark that has been penetrated by fungi, while the endomychid *Stenotarsus* (Figs 88G, 89E) is associated with subcortical fungi. Parasitic larvae of some Bothrideridae (*Deretaphrus, Pseudobothrideres*) may be found attached to some of the larger cambial borers, such as Cerambycidae.

Several Zopheridae feed on rotten cambium or subcortical fungi, as do larvae (Fig. 91J) and adults (Fig. 92A) of the ulodid *Meryx rugosa* Latreille. Among the Tenebrionidae, *Adelium* adults (Fig. 74K) may occur under loose bark but *Platycotylus* and some species of *Palorus* are subcortical specialists. The flattened larvae of eurypine Mycteridae are found under tightly fitting bark, often of living trees; their adults may occur in flowers. Larvae of *Morpholycus* (Fig. 91L), *Temnopalpus* and other pilipalpine Pyrochroidae are regular inhabitants of the subcortical habitat, where they feed on dead plant material; the adults are also surface-active, occurring on vegetation or in flowers. Among the Salpingidae, larvae in the subfamilies Othniinae, Inopeplinae and Prostominiinae are all subcortical and adults of the last two groups may also be found in this habitat.

Cerambycidae, like Buprestidae, includes many of the larger subcortical borers, the adults of which may be collected on foliage or flowers. Common eucalypt-feeding cerambycids are *Epithora dorsalis* (Macleay) and various species of *Phoracantha* (Cerambycinae). The commonest subcortical Curculionidae are in the subfamilies Cossoninae and Scolytinae, which tunnel under bark in both adult and larval stages.

Living and Dead Wood – Wood Borers

The subcortical habitat just discussed grades into the dead wood just beneath it. Wood-borers consist mainly of larvae, although certain highly specialised adults, such as those of many Bostrichidae, Anobiidae and Curculionidae are capable of tunnelling through wood as adults. A distinction may be made between those species that occur in the more or less solid wood of living or recently killed trees and those which inhabit wood that has been broken down by the action of fungi. The two faunas do overlap and, within the second, there may be another breakdown based on the extent or type of wood decay involved.

Larvae of Cupedidae live mainly in relatively dry but well rotted wood, and have been taken in building materials with dry rot. Rhysodine Carabidae occur in rotten logs and may be associated with slime moulds. Osoriine Staphylinidae (*Holotrochus, Osorius, Plastus*, Fig. 67C) are saprophagous, while teretriine Histeridae (*Teretrius* and *Teretriosoma*) are predators of small wood-boring beetles, such as Bostrichidae.

Lucanidae and Passalidae are larger and more obvious inhabitants of old rotten logs, and the latter may be found in sub-social colonies. Rotten wood feeders among the Scarabaeidae include larvae of most Dynastinae and Cetoniinae. In Australia, larvae of some Scirtidae may be found in old, red-rotten logs well away from water. Callirhipid larvae tunnel in well rotted wood, usually that which is infected with white rot. Among the Elateridae, species of *Paracalais* (Fig. 70C) may live in the tunnels of wood-boring beetles on which they prey and various other groups, such as Lissominae (Figs 64G, 65D), occur in this habitat. Many eucnemid larvae occur in dead wood, including that which is still relatively sound. They are able to move through this apparently solid substrate without chewing by wedging their extremely flattened bodies between the wood fibres.

Adult Bostrichidae and some Anobiidae are able to bore into relatively solid wood in order to lay their eggs in the sapwood beneath the surface; the highly modified larvae then tunnel through the sapwood. Several highly specialised clerid larvae may be found as predators within these bostrichid galleries; included among these are *Paratillus carus* (Newman) and species of *Cylidrus* (Fig. 71J). Larvae of *Trogodendron fasciculatum* (Schreibers) and *Stigmatium* species may be found attacking wood-boring Cerambycidae. Lymexylidae are one of three groups of ambrosia beetles (**see** Curculionidae: Xyleborini and Platypodinae below), which bore into solid wood as larvae and feed on yeasts that line the tunnel walls.

Cucujoidea are rarely capable of excavating galleries in dead wood, even that in a late stage of decay, but some cucujoids live in the tunnels of other insects, where they may be predacious or feed on fungi cultivated by ambrosia beetles. *Shoguna* species (Monotomidae) (Fig. 72A) are often found inhabiting tunnels in rotten logs and they may be predators. Members of the Passandridae are known to be ectoparasites and have a triungulin-like first instar that seeks out prey larvae. Most Bothrideridae are also associated with the tunnels of wood-borers; some like *Deretaphrus, Sosylus* and the Bothriderini are definitely predaceous, but Teredinae, which occur in ambrosia tunnels, may actually be mycophagous.

Wood-boring is a dominant way of life among larvae of many Tenebrionoidea. Pycnomerine and zopherine Zopheridae feed in white-rotten wood. Colydiine Zopheridae have varied habits but some (*Pseudendestes*, Fig. 92I), *Munaria* and *Nematidium*, Fig. 92C) are known to be predators in the tunnels of platypodine weevils. Within the Tenebrionidae, many genera are known to feed in rotten wood as larvae and examples may be found in most of the subfamilies: *Aspasis* and *Beplegenes* (Lagriinae); *Dioedus* and *Scolytocaulus* (Phrenapatinae); *Meneristes, Cyphaleus, Uloma, Achthosus* and *Amarygmus* (Tenebrioninae); *Tanychilus* (Alleculinae); *Menimus* (Diaperinae); and *Promethis* (Fig. 91K), *Zophophilus* and *Strongylium* (Stenochiinae). *Lyphia* and *Corticeus* adults and larvae live in the tunnels of some ambrosia beetles (platypodine and scolytine Curculionidae). Oedemeridae is another family whose larvae are rotten wood feeders. Larvae (Fig. 35K) and adults of *Prostomis* (Fig. 93E) and *Dryocora* (Fig. 76D) (Prostomidae) occur within localised patches in old red-rotten logs.

Some species of Cerambycidae tunnel through wood, rather than being restricted to the cambial area. Larvae of Parandrinae and those of several Prioninae favour older, decaying wood, whereas others, like the European house borer, *Hylotrupes bajulus* (Linnaeus), may occur in seasoned wood. Larvae of Belidae also occur in dead or dying wood that may be relatively solid. *Rhinotia* larvae have been taken in dead and dying *Acacia* branches and those of *Cyrotyphus* were found in a dead eucalypt log. Anthribidae are common as larvae in rotten wood, as are many Brentidae (Brentinae). Within the family Curculionidae there are many wood-boring species, including two groups of ambrosia beetles: the Platypodinae; and species of *Xyleborus* and related genera among the Scolytinae. Larvae of *Strongylorhinus* are known to breed in large galls, which they form in the stems of some *Eucalyptus*. Cossoninae and Cryptorhynchini contain many wood-boring species. Grove *et al.* (2008) studied the succession of 311 saproxylic beetle species in *Eucalyptus* logs in TAS.

Herbaceous Plant Tissue – Stem Borers, Gall Makers and Leaf Miners

The habit of feeding as larvae in the living tissue of herbaceous plants, whether it be stems, galls or leaves, occurs in only a few specialised phytophagous families or subfamilies. In the family Buprestidae, it is mainly members of the subfamily Agrilinae (especially Trachydini) that have this habit. Among the Cucujoidea, species of languriine Erotylidae (*Anadastus* and *Caenolanguria* in northern Australia) are known to be stem borers as larvae, but adults are usually collected from vegetative surfaces. Larvae of the mordellid genus *Mordellistena* also occur in stems. Some of the smaller Cerambycidae, such as the lucerne crown-borers, *Zygrita diva* Thomson and *Corrhenes stigmatica* (Pascoe), attack herbaceous plants. Leaf-mining chrysomeloids are known in the megalopodid subfamily Zeugophorinae (*Zeugophora*) and chrysomelid subfamilies Hispinae (*Notosacantha, Octotoma*) and Galerucinae. Members of the weevil genus *Rhynchaenus* are leaf-miners, and many other curculionid larvae feed in a variety of herbaceous tissues (stems, galls, shoot tips, flower buds, roots). Some Anthribidae, like *Araecerus palmaris* (Pascoe), feed as larvae in galls formed by the rust *Uromycladium* (Uredinales) on *Acacia.*

Palm Fronds

A unique fauna has developed at the bases of living fronds and within the tissue of dying or dead fronds of ferns and of palms, *Pandanus,* and some other monocotyledonous angiosperms. All of these beetles are very elongate and usually strongly flattened. Nitidulid beetles in the subfamily Cillaeinae, some species of *Hapalips* (Erotylidae), the genera *Lorelus* (Tenebrionidae: Lagriinae) and *Hemipeplus* (Mycteridae: Hemipeplinae), undescribed Zopheridae and Salpingidae, some hispine Chrysomelidae (*Brontispa, Ceratispa*) and several cossonine weevils occur in this habitat, some preferring the leaf bases and others the rotting fronds.

Leaf Litter and Ground Debris

There is a large and diverse fauna associated with leaf litter and other plant debris accumulated on the forest floor, in cavities in old logs and stumps and as flood debris bordering rivers and small streams. The litter in more mesic habitats is obviously richer and many groups drop out in drier areas; those taxa peculiar to dry litter are mentioned where appropriate.

The family Carabidae includes most of the larger predators and many of the smaller ones within this habitat. Species of *Pamborus* (Fig. 66B) (Pamborini), *Carenum* (Fig. 66A) (Scaritini) and *Notonomus* (Fig. 54G) (Pterostichini) are near the upper end of the range (up to 20 mm or more in length), whereas species of *Amblystomus* (Harpalini), *Mecyclothorax* (Psydrini) and various Trechini (Fig. 54H) are small, ground-dwelling predators (usually less than 5 mm in length). Sphaeridiine Hydrophilidae, such as *Notocercyon,* occur in moist accumulations of ground debris, where they are scavengers as adults and predators as larvae. Small Histeridae in genera such as *Bacanius* and *Acritus* are also common litter-inhabitants, predaceous as both larvae and adults. The minute species of Ptiliidae are more common in mesic litter but at least some species are more widely distributed; all are spore feeders. All of the subfamilies of Leiodidae are represented in the litter fauna and some may be very abundant locally; many are scavengers, but others feed on specific types of fungi (**see** below). Pitfall trapping in forest litter often produces large numbers of leiodids, including *Colenisia* and *Zeadolopus* (Leiodinae) and various members of the Cholevinae. Staphylinidae is certainly the most common and diverse family within the forest floor environment. Saprophages occur among the Oxytelinae (*Anotylus*, *Oxytelus*) and Proteininae (*Anepius*), and Scaphidiinae in the genus *Baeocera* may feed on slime moulds, but the majority of litter-inhabiting species of Tachyporinae, Euaesthetinae, Paederinae and Staphylininae are probably predators, and larvae and adults of the dominant subfamilies Pselaphinae and Scydmaeninae prey on small arthropods, including mites. Most Staphylinidae belong to the large and poorly known Aleocharinae, which includes at least some spore-feeders, as well as known predators.

Scarabaeoid adults in the leaf litter belong mainly to those species with soil-dwelling larvae that are emerging from the site of pupation and preparing to fly to adult feeding or mating sites in the forest canopy; these would include many Melolonthinae and Rutelinae and some Cetoniinae. In addition various dung-feeding species, in the genera *Aphodius* and *Onthophagus,* will turn up in litter collections. Eucinetidae and Clambidae may be taken by sampling litter, where they are feeding on slime moulds. Byrrhidae (Fig. 69I–J) occur mainly in wetter litter and are more abundant in TAS. They feed primarily on mosses and but some may be associated with other plant material in the litter or soil. Members of the limnichid genus *Byrrhinus* (Fig. 69H) are common litter-inhabitants in the warmer, wetter parts of Australia, but most of the other limnichid genera are restricted to riparian habitats. Larvae of Elateridae (Fig. 35F), Lycidae (Fig. 42C) and Cantharidae (Figs 6E, 35A) are all relatively common in litter samples. Cantharid larvae are known to be predators and lycid larvae feed on decaying plant material or possibly slime mould plasmodia. Elateridae have varied habits and both predatory and phytophagous larvae may be encountered in forest litter.

Megatomine Dermestidae (e.g. *Anthrenocerus, Trogoderma*) are particularly common in drier litter and ground debris in mallee or desert environments; larvae (Fig. 63J–K) are common saprophages in the litter, while adults (Figs 6D, 63G) are more likely to be collected at flowers. Some small lophocaterine Trogossitidae are also inhabitants of dry litter, but the feeding habits of these are unknown. Larvae of *Lemidia* (Cleridae: Hydnocerinae) (Fig. 82C) are one of the commonest types of small predators in forest leaf litter. Melyrid larvae (Fig. 34I) are also common predators in litter and they are far more abundant than *Lemidia* in drier habitats.

Several Nitidulidae are taken in litter samples and many of them (e.g. *Lasiodactylus, Stelidota,* some *Epuraea*) are usually concentrated near accumulations of dead fruit or other decaying plant material. Some groups, like *Thalycrodes,* are locally abundant as adults but may breed in subterranean fungi (**see** below). The genus *Monotoma* (Fig. 87Q) (Monotomidae), which was introduced from the Northern Hemisphere, is common in grass piles and compost heaps in urban situations but may also occur in nearby forest litter. The priasilphid genus *Priastichus* (Figs 84O, 85A) is an unusual inhabitant of litter in TAS. *Cryptamorpha* species (Fig. 85H) (Silvanidae: Brontinae) and some Silvaninae occur in litter, as do various Cryptophagidae. Introduced cryptophagids, like *Atomaria lewisi* (Reitter), are usually restricted to compost heaps. Cerylonids in the genera *Australiorylon, Mychocerus* and *Hypodacnella* (Fig. 72D) are commonly taken in litter samples, as are Endomychidae belonging to the genus *Erotendomychus* (in wetter forests in the north). In mountainous regions from TAS to QLD, Corylophidae of the subfamily Periptyctinae (Figs 71F, 88H) are reasonably common litter inhabitants.

Many of the smaller Archeocrypticidae, like *Australenneboeus,* occur in litter, especially in the northern parts of Australia. Species of *Enhypnon, Caanthus* and *Mnionychus* (Zopheridae: Colydiinae) occur in wet habitats along the east coast and in TAS. Among the Tenebrionidae, larvae and adults of several groups, including Adeliini, Lagriini, Opatrini, Crypticini and Ectychini, are scavengers inhabiting the forest litter or ground debris; adults of several other groups, such as Heleini and Alleculinae, may be collected in litter samples but their larvae are usually found in the soil. Larval and adult Anthicidae and Aderidae and larval Scraptiidae are also litter inhabitants.

Few Cerambycidae regularly inhabit the forest floor; exceptions are flightless Lamiinae, such as the species of *Athemistus, Mesolita, Somatidia* and *Microtragus*. Among the Chrysomelidae, larvae of the Cryptocephalini are very common in litter samples and are easily recognised by the faecal cases in which they live; unlike most chrysomelids these larvae feed on dead leaves. Some Eumolpinae, Chrysomelinae and Galerucinae also occur in leaf litter, but either their biology is unknown or they have root-feeding larvae. Curculionidae may be extremely abundant as adults in wet forest litter samples, where Cryptorhynchini is the dominant group. Species of *Mandalotus* (Entiminae) are commonly taken in litter samples from open eucalypt forests.

Animal Nests

Mammal nests and burrows of various kinds provide shelter for several insects, but some species are found only in these habitats, where they feed on dung or other accumulations of organic material. Aphodiine Scarabaeidae are well known dung-feeding inhabitants of underground mammal nests. At the entrances to wombat burrows in SA, three aphodiine species, *Proctophanes* sp., *Ataenius* sp. and the introduced *Aphodius granarius* (Linnaeus) were found and larvae of one species were collected in partly buried wombat dung. Also feeding on the wombat dung were larvae and adults of a species of *Helea* (Tenebrionidae). Bird nests also provide habitats for a variety of beetles, as indicated by the numerous records compiled by Hicks (1959, 1962, 1971). Particularly common are Histeridae, Staphylinidae, Leiodidae, Trogidae, Dermestidae, Ptinidae (Ptininae), Cryptophagidae, Latridiidae, Mycetophagidae and Tenebrionidae. In Australia, New Zealand and southern South America, specialised nest inhabitants have been described in the family Cavognathidae. These beetles occur in bird nests as both larvae and adults and have occasionally been found on nestlings (**see** p. 316). Various Australian dermestids have been found in bird nests, as have the tenebrionid species *Ulomoides heroni* (Carter) and *U. tetraspilotus* (Hope) Many beetles inhabit the nests of ants and termites; these are discussed in a later section (**see** Symbiotic Relationships, p. 90).

Caves

Caves and other subterranean enclosures in limestone and basaltic rock form a unique type of habitat for a variety of organisms, including Coleoptera. Although cave inhabiting beetles are much less common in Australia than they are in the Alpine and Appalachian regions of Europe and eastern North America, several species have been recorded, particularly from caves in TAS, south-eastern SA, south-western VIC, coastal WA and northern QLD. Cave inhabitants are usually placed in one of three groups based on the degree of association with the subterranean environment: (1) **troglobites** or **obligatory cavernicoles**, which are unable to exist outside of the cave environment; (2) **troglophiles** or **facultative cavernicoles**, which usually occur in caves and frequently complete the life cycle within the cave environment but which may be found outside caves as well; and (3) **trogloxenes** or **adventive cavernicoles**, which spend only a part of the life cycle underground (Moore 1964). Because of the limited amount of data available on distribution and biology, true troglobites are often recognised by their obvious cave adaptations, such as long legs and antennae, lack of pigmentation, reduced eyes, slender bodies and long, tactile hairs. Some of these adaptations, however, also occur in litter-inhabiting species and so are not always indicative of troglobitic habits. Some cave-dwelling arthropods feed on bat guano or the dung of other small mammals, while others are predators or general scavengers, utilising carrion or organic nutrients washed into the cave system from the surface. The highly adapted forms living in deep crevices may subsist on autotrophic bacteria and fungi.

Fewer than 50 species of beetles belonging to 15 families have been recorded from caves in Australia. The majority of these are in the family Carabidae, which is represented by 8 tribes, 16 genera and ~25 species; three species of *Idacarabus* (Zolini), one *Tasmanotrechus* and two *Goedetrechus* (Fig. 54B) (Trechini) and one *Teraphis* (Moriomorphinae) are troglobitic. Among the troglophiles are harpalines of the genus *Notospeophonus* and lebiines of the genera *Speotarus* and *Anomotarus.* The family Staphylinidae includes an eyeless, cave dwelling paederine, *Tripectenopus occultus* Britton, but a related QLD species also lacks eyes and occurs outside caves. Another species found in caves is *Myotyphlus jansoni* (Matthews), a staphylinine associated with rats. Some species of the jacobsoniid genus *Derolathrus* have been found in bat guano, and members of the tenebrionid genus *Brises* are often found in caves, where they feed on dung and carrion. **See** p. 16 for Coleoptera occurring in subterranean aquifers. [Moore 1967b, 1972a, 1972b, 1977, 1978b; Goede 1967; Richards 1971; Britton 1974b; Howarth 1983, 1988; Matthews 1986; Moore and Humphreys 2001].

Grasses and Grasslands

The grassland insect fauna has been imperfectly studied, and there is some overlap with the forest litter fauna discussed above. At least two groups are known to specialise on smuts that grow on grass heads, namely *Leucohimatium* (Erotylidae: Pharaxonothinae) and *Phalacrus* (Phalacridae). Among the Chrysomelidae, some hispines (*Eurispa*, Fig. 96I, and *Hispellinus*, Fig. 96L) are specialised grass feeders. Some amycterine weevil larvae feed on grasses as adults and their larvae are root feeders (Howden 1986), as are those of several other groups discussed in the next section.

Soil

The soil insect fauna in Australia is also poorly known because of the specialised techniques necessary for adequate sampling, but several soil organisms are collected in pitfall traps or Berlese samples. Very small Staphylinidae, such as Euaesthetinae, Leptotyphlinae (as yet undescribed) and *Typhlobledius* (Osoriinae), are known as adults from soil core samples. Several medium to large larvae are dominant members of the soil fauna, most of them feeding on roots and rhizomes or taking in humus and soil and extracting nutrients from it in much the same way as earthworms. Among the Scarabaeoidea, larvae of Trogidae and Geotrupidae may be found in burrows, where they feed on material provided by the adults: bits of vertebrate carcasses in the case of trogids; dead leaves, fungi or dung in the case of geotrupids. Whitegrubs, cockchafers and various other types of scarabaeid larvae make up the bulk of soil-dwelling beetles. These are mainly members of the subfamilies Melolonthinae (*Phyllotocus*, *Antitrogus*, *Lepidiota*, *Rhopaea*, *Sericesthis*, *Heteronyx*, *Scitala*), Rutelinae (*Anoplognathus*) and Dynastinae (*Cheiroplatys*, *Heteronychus*, *Adoryphorus*), but some aphodiines, such as the blackheaded pasture cockchafer, *Podotenus tasmaniae* (Hope), are also common in soil.

Larvae of *Notodascillus* (Dascillidae) have habits similar to soil-dwelling scarabs, and those of the related *Rhipicera*

(Rhipiceridae) are probably parasites of cicada nymphs. Some Buprestidae may also be encountered in the soil around the bases of the trees on which they feed. Elaterid larvae, known as wireworms, are common soil inhabitants. These are mainly phytophagous members of the subfamilies Agrypninae (*Agrypnus, Conoderus*), Dendrometrinae (*Hapatesus*) and Elaterinae, but some of them may also be predacious. Within the family Tenebrionidae, the false wireworms (e.g. *Pterohelaeus, Helea, Saragus, Gonocephalum*) are also common phytophagous soil-dwellers. Some Cerambycidae are root feeders, as are various Chrysomelidae, particularly members of the Eumolpinae and Galerucinae. Finally, the family Curculionidae contains a very large number of species with soil-dwelling larvae, such as the amycterines mentioned above and the large subfamily Entiminae.

Dung

The beetles inhabiting dung of native and introduced mammals belong to relatively few groups and include those which are specialised dung feeders and those which prey on fly larvae. In the former group are a few Leiodidae, adult Hydrophilidae, oxyteline Staphylinidae (especially *Anotylus* species), a few Geotrupidae, scarabaeine and aphodiine Scarabaeidae, and probably ptinine Ptinidae. Predacious beetles in dung include hydrophilid larvae (Sphaeridiinae, such as *Cercyon*), histerid adults and larvae (e.g. *Saprinus* and *Carcinops*) and Staphylinidae (mainly Staphylininae, such as *Philonthus* and *Creophilus*). Cattle dung is utilised mainly by introduced Aphodiinae (*Aphodius*) and Scarabaeinae (*Sisyphus, Onitis, Euoniticellus, Onthophagus*). The dung of native mammals is fed upon by several *Onthophagus* species. The ecology of native and introduced dung beetles in Australia is discussed by Doube *et al.* (1991). Some Australian species of *Ptinus* may also feed on small mammal dung, as this is a common habit in other parts of the world.

Carrion

The carcasses of vertebrates also provide food for a relatively small number of beetle groups. As in dung, fly larvae are abundant and serve as prey for species of Histeridae (*Saprinus*, Fig. 2D) and Staphylinidae, especially the devil's coach-horse, *Creophilus erythrocephalus* (Fabricius) (Fig. 1G). Carrion feeders occur in the families Leiodidae, Silphidae (*Ptomaphila*, Fig. 2A, and *Diamesus*, Fig. 2B), Trogidae (Fig. 3H), Hybosoridae, Scarabaeidae (a few scarabaeines), Dermestidae (*Dermestes*), Cleridae (*Necrobia*) and beach-inhabiting Phycosecidae (Fig. 8H). Trogids feed on very old carcasses and are able to break down keratin from hair and feathers. Phycosecids occur on beaches and often feed on dead fish.

Fungal Fruiting Bodies

A large number of beetles, particularly in the superfamilies Staphylinoidea, Cucujoidea and Tenebrionoidea, are associated with the fruiting bodies of a wide variety of fungi, feeding on the spores and/or hyphal masses or in some cases preying on fungivorous insects. A general discussion of mycophagy in Coleoptera is given by Lawrence (1989) and another on mycophagy in Staphylinoidea by Newton (1984). Summaries of insect–fungus associations were given by Hammond and Lawrence (1989) and Lawrence and Milner (1996).

Myxomycetes

Several groups of beetles are specialists on the spores of slime moulds, especially species of *Fuligo, Stemonitis* or *Arcyria*, whose fruiting bodies are either large or abundant. Most of those species feeding on myxomycete spores are specialists, occurring in no other habitat. The main Australian taxa are Leiodidae: Camiarinae (*Neopelatops*), Staphylinidae: Scaphidiinae (*Baeocera* and *Scaphobaeocera*), Eucinetidae (Fig. 60C), Clambidae, Sphindidae (*Notosphindus*, Fig. 83K, and *Aspidiphorus*, Fig. 83J) and Latridiidae (*Enicmus*).

Ascomycetes (particularly Xylariaceae)

The so-called pyrenomycete or carbon fungi have a large number of associated beetles that may feed on the dark, brittle tissue comprising the ascocarps or, in some cases, graze on the external spore fields (Lawrence 1977a). Among the Cucujoidea, species of Monotomidae, Laemophloeidae, Phalacridae, Cryptophagidae, Biphyllidae, Bothrideridae (*Xylariophilus*) and Corylophidae feed on these Ascomycetes. Adults of the erotylid beetle *Episcaphula bifasciata* (Macleay) have been seen feeding externally at night on *Xylaria* fruiting bodies in northern QLD. Species of the tenebrionoid families Mycetophagidae, Colydiidae and Salpingidae have also been associated with these fungi, as have several Anthribidae. A unique beetle fauna occurs in association with sooty moulds (Capnodiaceae and related families) growing on tree trunks in New Zealand and a few beetle groups occur in similar situations within temperate rain forests of south-eastern Australia and TAS. The derodontid *Nothoderodontus darlingtoni* (Fig. 81C) and some species of *Orchesia* (Melandryidae) have been associated with Australian sooty moulds. According to Leschen (2000), it is likely that this habit gave rise to predation on Adelgidae in the related Northern Hemisphere derodontid genus *Laricobius* Rosenhauer.

Wood-Rotting Mushrooms and Bracket Fungi

This habitat, formed by the fruiting bodies of various wood-rotting Agaricales (e.g. Tricholomataceae) and Aphyllophorales (e.g. Polyporaceae, Hymenochaetaceae), is particularly rich in species of Coleoptera, including those grazing the fresh pore surfaces, those boring into the relatively solid hyphal mass comprising the fresh fruiting body and those feeding on decaying hyphal tissues.

Minute members of the ptiliid subfamily Nanosellinae are capable of entering the pore tubess of bracket fungi in order to feed on the spores. Camiarine Leiodidae (*Agyrtodes, Paragyrtodes*) are often found on old rotting polypores, as are some Staphylinidae (*Thoracophorus*). Members of the staphylinid subfamily Scaphidiinae are usually associated with bracket fungi, mushrooms and other Hymenomycetes; species of *Scaphisoma* and *Scaphidium* (Fig. 67F) graze on fresh pore surfaces. Among the Staphylinidae, some *Sepedophilus* species graze on pore surfaces, as do *Gyrophaena* and related groups of Aleocharinae.

Dorcatomine Ptinidae commonly bore into the harder fruiting bodies of some bracket fungi like *Ganoderma* and *Phellinus*. Thaneroclerine Cleridae (Fig. 71M) are small-sized predators in macrofungi, feeding on the larvae of Anobiidae, Ciidae and other groups in the right size range.

Nitidulidae tend to feed only on fresh and relatively soft fruiting bodies of some polypores and wood-rotting mushrooms. *Gaulodes* is known to breed in fresh *Piptoporus australiensis,* while species of *Pallodes* and *Cychramus* usually feed in mushrooms. Hobartiidae are fungivorous: *Hydnobioides* breeds in soft polypores and mushrooms, and *Hobartius* in old, well rotted fruiting bodies. The endemic family Lamingtoniidae is also associated with soft polypores. Erotylidae is a common and conspicuous group of fungus feeders. Adults often feed externally and are aposematically coloured with black and red and sometimes yellow; larvae may be lightly pigmented internal feeders, like those of *Thallis, Cnecosa* (Fig. 73F) and *Microsternus,* or conspicuously coloured and often spinose external grazers, like those of *Episcaphula.* External grazing on polypores is known in several exotic members of the endomychid subfamily Eumorphinae but larval habits are unknown in the Australian species. Species of *Stenotarsus* (Fig. 72C) are associated with fungi, often under bark. Some Corylophidae and Latridiidae may be common on large bracket fungi (conks), but they are usually feeding on moulds that tend to cover the surface of old brackets.

Tenebrionoids are also common on macrofungi, especially as internal feeders. Some of the softer polypores, such as *Piptoporus* species, are attacked by archeocrypticids (*Enneboeus* and *Enneboeopsis*). Most species of Ciidae are restricted to the fruiting bodies of polypores, and both adults and larvae are capable of boring into hard and woody hyphal masses. Similar habits are found in some Melandryidae (*Orchesia*). The zopherid beetle *Zopherosis georgei* White (Fig. 92L) feeds on polypore conks as an adult but the larvae probably feed in rotten wood, as do those of exotic relatives. Ulodidae are often associated with soft fruiting bodies of *Piptoporus* species, in which the larvae feed. Several Tenebrionidae in the tribes Toxicini (*Toxicum*, Fig. 74C, *Orcopagia*), Bolitophagini (*Byrsax, Rhipidandrus*) and Diaperini (*Platydema, Pentaphyllus*) are specialists on wood-rotting fungi as larvae and adults. [Hawkeswood 1989, Webb and Simpson 1991].

Ground Mushrooms, Puffballs and Stinkhorns

Ground mushrooms have a much more restricted beetle fauna and some of these, such as the larger Staphylinidae, are predators of fly larvae, which are a more conspicuous part of the fauna. Some Aleocharinae, especially *Gyrophaena* and their allies, feed on mushroom spores. Some scarabaeine Scarabaeidae, such as *Onthophagus dunningi* feed in mushrooms (Bornemissza 1971b), and it is possible that various other scarabaeines and possibly Geotrupidae use mushrooms for larval provisioning.

Puffballs (Lycoperdaceae and Sclerodermataceae) form the food for several beetle groups, including Leiodidae (*Nargomorphus filitarsis* Scymczakowski), Erotylidae (*Neothallis*), Cryptophagidae (*Cryptogasterus*), Nitidulidae (*Pocadius*), and Latridiidae. The nitidulid genus *Gymnocychramus* (Fig. 73H) is the only Australian beetle group recorded from stinkhorns (Phallaceae).

Hypogean Fruiting Bodies

There are probably several Australian Coleoptera restricted to underground or hypogean fruiting bodies belonging to various groups of Ascomycetes (Tuberales) and Gasteromycetes (Hymenogastrales etc.), but because of the cryptic nature of this habitat direct evidence of these associations is scarce. Likely candidates among the Leiodidae are the highly modified burrowers in the genus *Dietta* (Fig. 67I), undescribed genera of Sogdini, and species of the genus *Colon.* Some bolboceratine Geotrupidae apparently use hypogean fungi to provision their larvae. The nitidulid genus *Thalycrodes* is thought to be a specialist on subterranean fungi, although no association of larvae with fungal fruiting bodies has yet been made. According to Kirejtshuk and Lawrence (1992b) adults have been taken in association with mushrooms (*Coprinus*), as well as Lycoperdaceae and Clathraceae, but the unusual structure of the antennal club (with enclosed cavities for the sensilla) suggests an underground feeding site. Larvae of Tenebrionidae (*Lepispilus*) and Curculionidae (*Phaspharus*) have been found in association with the hypogean fungus *Mesophellia* (Lycoperdales: Mesophelliaceae). It is likely that further collecting will increase the known number of beetles with hypogean fungi as their larval food source.

Marine and Freshwater Littoral Zones

Littoral zones present a gradient of microhabitats from strictly terrestrial ones to those which are aquatic or subaquatic and require special respiratory adaptations. Sphaeriusidae are minute beetles that live in the spaces among grains of sand and gravel on the banks of streams. These alga feeders are constantly inundated with water and the larvae have tracheal gills. Carabid beetles are particularly common in sandy and muddy areas at the edges of ponds and streams; some of the representative genera in this habitat are *Bembidion, Tachys* and *Clivina.* Among the Hydrophilidae, species of *Georissus* (Fig. 57K) are restricted to sand banks but various other groups may be found there as well. Hydraenidae are primarily littoral, as are many Staphylinidae. On ocean beaches, the staphylinid genera *Sartallus, Carpelimus* and *Cafius* may be found among littoral debris. Adults and larvae of *Bledius* occur in tunnels in wet sand banks near fresh or salt water, where they feed on algae.

Scirtid larvae (Fig. 60H) are common in boggy areas and along river banks, where they use a complex filter apparatus to sieve organic particles from a muddy substrate. Many ptilodactylids live along stream margins and the larvae have a variety of mechanisms for obtaining oxygen during flooding. Heteroceridae and Limnichidae both construct tunnels in sand banks along the margins of freshwater and sometimes salt-water bodies; larvae and adults appear to feed on algae, diatoms and other small organisms, which are taken in with sand and mud. The limnichid genus *Hyphalus* is restricted to intertidal rocks.

Sandy beaches are home to several Melyridae, whose larvae may be scavengers or predators. The related family Phycosecidae (Fig. 8H) is restricted to sandy ocean beaches, where they feed on dead fish and other organic debris. The questionable anthicid *Lagrioida australis* Champion (Fig. 95C) is also a beach inhabitant, living in the sand beneath dune plants. Tenebrionidae are common scavengers in most littoral situations; *Sphargeris physodes* Pascoe (Fig. 74D) is usually found in masses of drying seaweed, while species of *Caediomorpha, Scymena* and *Trachyscelis* may be found in sand beneath dune plants. The beach weevils of the genus *Aphela* (Notiomimetini) also occur in this habitat.

Standing Fresh Water (Lentic Habitats)

Lakes, ponds and billabongs, as well as temporary ditches, form habitats for several aquatic beetles. Dytiscidae is the largest group of swimming predators, while the Noteridae appear to be detritus feeders. Adults and larvae of *Haliplus* (Haliplidae) feed on stoneworts (Charophyceae) and other green algae. *Hygrobia* species (Hygrobiidae) are slow-moving bottom-feeding predators, known to take tubificid worms. Gyrinidae occur mainly on the surface as adults, while the larvae are bottom-feeding predators.

Hydrophilidae are scavengers as adults and predators as larvae; many species are aquatic, and these occur primarily in still water habitats. Adults of *Stenus* (Staphylinidae: Steninae) are capable of moving on the water's surface but they may also be found at the water's edge, along with their larvae. Scirtid larvae (Fig. 60H) are also to be found in still water, often near the edges of ponds.

Several Chrysomelidae are aquatic, in that larvae feed on emergent aquatic plants. *Donacia australasiae* Blackburn feeds on water lilies, and the larva has spine-like posterior spiracles that can pierce the plant stem in order to obtain oxygen. Species of *Altica* may be associated with emergent vegetation, such as *Ludwigia*, and *Agasicles hygrophila* Selman and Vogt is an aquatic species introduced to control alligatorweed (*Alternanthera*). The brentid genus *Austronanodes* (Fig. 80D) is associated with aquatic plants, as are various Curculionidae, including native species of *Echinochaetus* and *Bagous,* as well as species introduced for the control of noxious aquatic weeds (*Cyrtobagous* for *Salvinia, Neohydronomus* for *Pistia* and *Neochetina* for *Eichhornia*).

Subterranean Aquifers

This is a special type of aquatic habitat consisting of underground waters in caves or wells. Several stygobitic or stygobiontic species in the family Dytiscidae occur in Australia, and have been described in papers by Watts and Humphreys (1999, 2000, 2001, 2003, 2004), Balke and Ribera (2004) and Balke *et al.* (2004).

Moving Fresh Water – Lotic and Hygropetric Habitats

Australian aquatic beetles associated with moving water occur in the families Hydraenidae, Psephenidae, Ptilodactylidae and Elmidae. Species of the hydraenid genus *Tympanogaster* (Fig. 77D) are the only beetles so far recorded from hygropetric habitats (those formed when a film of water flows over a rock surface); the alga-feeding larvae of these hydraenids have spiracular siphons on the thorax, which project above the water film. The families Torridincolidae and Hydroscaphidae, which may occur in this habitat in other parts of the world, are absent from Australia.

Australia has a single genus of water-pennies (Psephenidae), *Sclerocyphon* (Figs 62J, 65I), species of which occur from TAS to northern QLD; members of the other three subfamilies of Psephenidae do not occur in Australia. Aquatic ptilodactylid larvae (*Byrrocryptus*, Fig. 65C) usually occur in slower streams, where they feed on decaying vegetation. The family Elmidae is well represented in Australia and includes both Larainae (*Hydora, Ovolara, Stetholus*, Fig. 62G, *Potamophilinus*) and Elminae (Austrolimnius, Figs 63E, 65H, *Coxelmis, Kingolus, Notriolus, Simsonia* and *Stenelmis*). Habits vary somewhat and larvae of some groups are less likely to occur in riffles than they are in water-logged wood.

COLLECTING BEETLES

Although the traditional methods of hand collecting or netting insect specimens are still commonly used, a large number of specialised collecting devices and techniques have been developed over the years to increase yield and efficiency. Some of these are discussed below, especially those which are particularly useful for collecting beetles. Hand collecting is still the best way to capture insects within their microhabitats and often to obtain living adults, pupae and larvae for rearing; it is usually carried out with forceps and collecting bottle but except in very wet habitats, a far more efficient device for collecting small and active specimens is the **aspirator**, sometimes referred to as a 'pooter'. Designs vary greatly, but the simplest one (Upton and Mantle 2010: figs 14–15) consists of a glass tube blocked at one end by fine gauze and attached to a flexible (rubber or plastic) tube; small insects are sucked into the glass tube and then blown back out into a collecting jar or vial. In another type (Upton and Mantle 2010: fig. 16), the intake tube and suction tube are both fit into a rubber stopper on a collecting bottle so that the insects are sucked into the bottle, which can be exchanged with another when filled.

Beating and Brushing

Branches and shrubs may be beaten with a stick so as to dislodge surface-active insects, which fall on to the flat surface of a beating tray. The tray usually consists of a piece of heavy cloth spread out on a wooden or metal frame (Upton and Mantle 2010: figs 12–13), but a plastic or light metal tray may also be used. The insects may be hand collected with forceps or picked up with an aspirator. A variation of this method, which may be used on tree trunks, involves a heavy bristle or wire brush to dislodge the insects and a tray adapted to fit against the curved bark or wood surface. A softer painters brush may be used to dislodge surface active microcoleopterans. Instead of a tray, a sifter may be used to collect the specimens (**see** below).

Sweeping

Like beating, this method is designed to collect surface active insects from vegetation but it is most efficient for collecting in grasslands and among herbaceous plants. A heavy gauge insect net is literally swept through the vegetation and the insects collect at the bottom, along with bits of vegetation. The collected debris may be spread out on a tray or placed in a container and sorted later in the laboratory. This method has been used in ecological surveys, where comparable samples are taken based on the number of sweeps. A recent modification of the sweeping technique has been used to sample insects in flight; this device, discussed and illustrated by Peck and Cook (1992), consists of nets mounted on the front of a vehicle. On a series of moderately warm evenings in Canada, the authors collected an average of 1200 beetles an hour.

Sifting Litter and Other Debris

One of the most important aids in collecting beetles is the **sifter** or litter concentrator, which consists of a heavy cloth cone or cylinder. An open metal frame with a handle attached is sewn into the top end and another similar frame, to which a metal 1 cm grid is soldered, attached ~25 cm below the first one (Upton and Mantle 2010: fig. 39). The bottom end of the cone or cylinder is tied shut, so that a

Fig. 1. Beetles in their natural habitats. A, *Omma rutherfordi* Lawrence (Ommatidae), © Farhan Bokhari. B, *Philoscaphus* sp. (Carabidae: Scaritinae), © Jiří Lochman. C, *Demetrida brachynodera* (Chaudoir) (Carabidae: Harpalinae: Lebiini), © Katarina Christenson. D, *Eretes australis* (Erichson) (Dytiscidae: Dytiscinae), © Jiří Lochman. E, *Cicindela* (*Myriochile*) *semicincta* Brullé (Carabidae: Cicindelinae), © Jiří Lochman. F, *Scaraphites* sp. (Carabidae: Scaritinae), © Jiří Lochman. G, *Creophilus erythrocephalus* (Fabricius) (Staphylinidae: Staphylininae), © Marie Lochman. H, *Actinus macleayi* Olliff (Staphylinidae: Staphylininae), © Jiří Lochman.

Fig. 2. Beetles in their natural habitats. A, *Ptomaphila lacrymosa* (Schreibers) (Silphidae), © Jiří Lochman. B, *Diamesus osculans* (Vigors) (Silphidae), © Paul Zborowski. C, *Hydrophilus* sp. (Hydrophilidae: Hydrophilinae), © Paul Zborowski. D, *Saprinus* sp. (Histeridae: Saprininae) © Paul Zborowski. E, *Ceratognathus niger* Westwood (Lucanidae: Ceratognathinae), © Kristi Ellingsen. F, *Callodes grayianus* (White) (Scarabaeidae, Rutelinae), © Owen Kelly. G, *Lamprima* sp. (Lucanidae: Lampriminae), © Graham Prichard. H, *Eupoecila australasiae* Donovan (Scarabaeidae: Cetoniinae), © Owen Kelly.

Fig. 3. Beetles in their natural habitats. A, *Monoplistes* sp. (Scarabaeidae: Scarabaeinae), with dung ball, © Jiří Lochman. B, *Mentophilus hollandiae* Laporte (Scarabaeidae: Scarabaeinae), © Arthur Anker. C, *Xylonichus* sp. (Scarabaeidae: Melolonthinae), © Kristi Ellingsen. D, *Anoplognathus smaragdinus* Ohaus (Scarabaeidae: Rutelinae), © Marie Lochman. E, *Geotrupes spiniger* (Marsham) (Geotrupidae: Geotrupinae), © Kristi Ellingsen. F, *Xylotrupes ulysses australicus* (Thomson) (Scarabaeidae: Dynastinae), © Jiří Lochman. G, *Aulacocyclus* sp. (Passalidae: Aulacocyclinae), © Peter Chew. H, *Omorgus* sp. (Trogidae), © Jiří Lochman.

Fig. 4. Beetles in their natural habitats. A-H: Buprestidae: A, *Cyrioides imperialis* (Fabricius) (Buprestinae), © Graham Prichard; B, *Nascioides parryi* (Hope) (Buprestinae), © Kristi Ellingsen; C, *Selagis* sp. (Buprestinae), © Owen Kelly; D, *Temognatha duponti* (Boisduval) (Buprestinae), © Jiří Lochman; E, *Stigmodera macularia* Donovan (Buprestinae), © Owen Kelly; F, *Castiarina sanguinolenta* (Gory & Laporte) (Buprestinae), © Jiří Lochman; G, *Julodimorpha saundersi* Thomson (Buprestinae), males on stubby beer bottle, © Jiří Lochman; H, *Stigmodera roei* Saunders (Buprestinae) © Jiří Lochman.

Fig. 5. Beetles in their natural habitats. A, *Temognatha heros* (Gehin) (Buprestidae: Buprestinae) being consumed by *Phellus tiliferus* Dakin & Fordham (Diptera: Asilidae), © Jiří Lochman. B, Genus ? sp. (Buprestidae) larva within prepupal chamber, © Paul Zborowski. C, *Rhipicera* sp. (Rhipiceridae), © Jiří Lochman. D, *Lingana* sp. (Elateridae: Elaterinae), © Jiří Lochman. E, *Porrostoma* sp. (Lycidae: Lycinae), in copulo, © Jiří Lochman. F, *Atyphella scintillans* Olliff (Lampyridae: Luciolinae), © Jiří Lochman. G, *Brachypsectra* sp. (Brachypsectridae) late and early instar larvae, © Geoff Monteith. H, *Atyphella* sp. (Lampyridae Luciolinae), brachypterous female, © Jiří Lochman.

Fig. 6. Beetles in their natural habitats. A, *Lasioderma serricorne* (Fabricius) (Ptinidae: Xyletininae), © Paul Zborowski. B, *Mesoxylion cylindricus* (Macleay) (Bostrichidae: Bostrichinae), © Katarina Christenson. C, *Chauliognathus lugubris* (Fabricius) (Cantharidae: Chauliognathinae), © Katarina Christenson. D, *Anthrenus* sp. (Dermestidae: Megatominae), © Katarina Christenson. E, *Chauliognathus* sp. (Cantharidae: Chauliognathinae), larva, © Kristi Ellingsen. F, *Encymon immaculatus* (Montrouzier) (Endomychidae: Lycoperdininae), feeding on mushrooms, © Paul Zborowski. G, *Rodolia cardinalis* (Mulsant) (Coccinellidae: Coccinellinae), © Bronwyn King. H, *Brittonema pygidiatum* Kirejtshuk (Nitidulidae: Cillaeinae), © Dave Rentz.

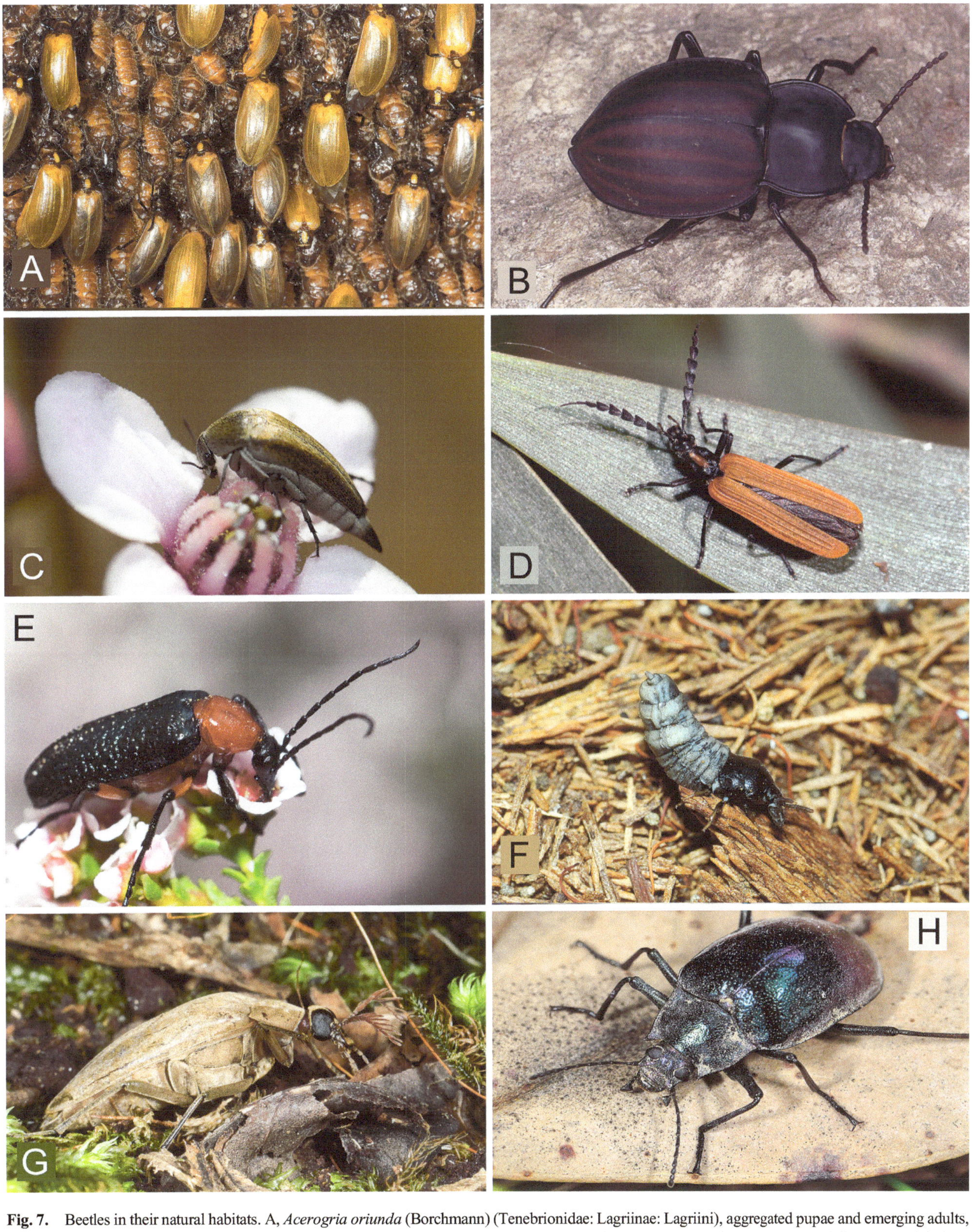

Fig. 7. Beetles in their natural habitats. A, *Acerogria oriunda* (Borchmann) (Tenebrionidae: Lagriinae: Lagriini), aggregated pupae and emerging adults, © C K & L Fisher 2012. B, *Nyctozoilus vittatus* (Blackburn) (Tenebrionidae: Tenebrioninae: Heleini), © Owen Kelly. C, *Austromordella* sp. (Mordellidae), © Jiří Lochman. D, *Pseudolycus* sp. (Oedemeridae: Oedemerinae), © Peter Chew. E, Genus ? sp. (Meloidae: Nemognathidae), © Jiří Lochman. F, *Riekella nitidioides* Selander (Ripiphoridae: Ripidiinae, female, © David Pike. G, *Trigonodera* sp. (Ripiphoridae: Ptilophorinae), male, © Jiří Lochman. H, *Cyphaleus* sp. (Tenebrionidae: Tenebrioninae: Heleini), © Peter Chew.

Fig. 8. Beetles in their natural habitats. *Dicranolaius bellulus* (Guerin-Meneville) (Melyridae: Malachiinae), © Jiří Lochman. B, *Raractocetus kreuslerae* (Pascoe) (Lymexylidae: Atractocerinae), © Owen Kelly. C, *Dicranolaius* ? sp. (Melyridae: Malachiinae), © Farhan Bokhari. D, *Carphurus* sp. (Melyridae: Malachiinae), © Bronwyn King. E, *Leperina decorata* (Erichson) (Trogossitidae: Trogossitinae), © Owen Kelly. F, *Trogodendron fasciculatum* (Schreibers) (Cleridae: Clerinae), © Donald Hobern. G, *Eleale cuprea* Mjöberg (Cleridae: Clerinae), © Jiří Lochman. H, *Phycosecis litoralis* Pascoe (Phycosecidae), © Kristi Ellingsen.

Fig. 9. Beetles in their natural habitats. A, *Stenoderus sp.* (Cerambycidae: Cerambycinae), in copulo, © Marie Lochman. B, *Cyocyphax praonethoides* Thomson (Cerambycidae: Lamiinae), © Owen Kelly. C, *Penthea pardalis* (Newman) (Cerambycidae: Lamiinae), © Jiří Lochman. D, *Tragocerus formosus* Pascoe (Cerambycidae: Cerambycinae), © Owen Kelly. E, *Eurynassa* sp. (Cerambycidae: Prioninae), © Jiří Lochman. F, *Hesthesis vesparia* Pascoe. (Cerambycidae: Cerambycinae), © Peter Chew. G, *Agrianome spinicollis* (Macleay) (Cerambycidae: Prioninae), © Jiří Lochman. H, *Tricheops ephippigera* Newman (Cerambycidae: Cerambycinae), © Owen Kelly.

Fig. 10. Beetles in their natural habitats. A. *Aporocera consors* (Boisduval), female coating egg with faeces (Chrysomelidae: Cryptocephalinae), © Bronwyn King. B, *Paropsisterna* sp. (Chrysomelidae: Chrysomelinae), © Jiří Lochman. C, *Hoplostines laporteae* (Weise) (Chrysomelidae: Galerucinae), © Owen Kelly. D, *Aproida balyi* Pascoe (Cassidinae), © Owen Kelly. E, *Diaphanops westermanni* Boheman (Chrysomelidae: Sagrinae), © Jiří Lochman. F, *Phyllocharis cyanicornis* (Fabricius) (Chrysomelidae: Chrysomelinae), © Owen Kelly. G, *Calomela* sp. (Chrysomelidae: Chrysomelinae), in copulo, © Katarina Christenson. H, *Cydippa* sp. (Chrysomelidae: Galerucinae), © Jiří Lochman.

Fig. 11. Beetles in their natural habitats. A, *Mesetia amoena* Blackburn (Brentidae: Brentinae), © Owen Kelly. B, *Rhinotia* sp. (Belidae), © Jiří Lochman. C, *Orthorhinus cylindrirostris* (Fabricius) (Curculionidae: Molytinae), © Katarina Christenson. D, *Iphthimorhinus australasiae* Roelofs (Curculionidae: Dryophthorinae), © Paul Zborowski. E, *Chrysolopus spectabilis* (Fabricius) (Curculionidae: Cyclominae: Aterpini), © Katarina Christenson F, *Myllorhinus bicaudatus* Boisduval (Curculionidae: Curculioninae), © Anthony Daley. G, *Platypus* sp. (Curculionidae: Platypodinae), © Jiří Lochman. H, *Meriphus* sp. (Curculionidae: Curculioninae), © Kristi Ellingsen.

Fig. 12. Beetles in their natural habitats. A. *Gonipterus* sp. (Curculionidae: Subfamily:?: Gonipterini), © Bronwyn King. B, Genus ? sp. (Curculionidae: Molytinae: Aedemonini ?), © Jiří Lochman. C, *Euops* sp. (Attelabidae: Attelabinae), © Bronwyn King. D, *Eurhynchus laevior* (Kirby) (Brentidae: Eurhynchinae), © Rolf Oberprieler, CSIRO. E, *Leptopius* sp. (Curculionidae: Entiminae), with *Ningaui ridei* Archer (Marsupialia: Dasyuridae), © Jiří Lochman. F, *Tranes lyterioides* (Pascoe) (Curculionidae: Molytinae), © Rolf Oberprieler, CSIRO; G, *Enteles vigorsi* Gyllenhal (Curculionidae: Molytinae: Cryptorhynchini), © Owen Kelly. H, *Gagatophorus* sp. (Curculionidae: Cyclominae: Amycterini), © Jiří Lochman.

bag is formed. Debris placed in the top of the bag rests on the grid, and when the sifter is vigorously shaken, fine debris, including small arthropods, falls through the grid and collects at the bottom of the bag. This material is then spread out on a tray, so that the beetles can be aspirated, or placed in one of the extraction devices discussed below.

Litter Extraction

There are two major types of extraction methods for sifted litter samples: the **Tullgren** (modified **Berlese**) **Funnel** and the **Winkler (Moczarski) Apparatus**. Designs vary but both are devices in which a mass of sifted litter is suspended over a funnel with a collecting jar (usually containing ethanol) at the end. In a Tullgren funnel (Upton and Mantle 2010: fig. 40), the litter is placed on a fine grid and either subjected to slow air drying or heated more rapidly by means of a light bulb placed directly over it. Arthropods and other organisms move away from the heat source, dropping through the grid and into the collecting jar. A **Winkler Apparatus** consists of a series of large-mesh cloth bags filled with litter, suspended over the collecting jar and contained within a cloth-covered wire frame (Upton and Mantle 2010: fig. 41). No device is used to drive the organisms from the samples; it is their movement through the substrate alone, that causes them to drop into the jar. The advantages of this over the Berlese funnel are discussed by Besuchet *et al.* (1987); disadvantages include slowr and more labour-intensive processing.

Other kinds of extraction techniques exist. One involves the placing of a litter or soil sample in a container of water and collecting the beetles that float to the top. A more efficient method, which has been particularly useful in extracting organisms from sand and gravel along water courses, is the **elutriator** (Kingsbury and Beveridge 1967), which involves pumping both air and water through a sample and collecting the small arthropods on a fine grid.

Pitfall Traps

Pitfall trapping has been particularly useful in recent years, especially for ecological surveys requiring uniform collecting methods. Pitfalls usually consist of plastic, glass or metal containers placed into the ground so that their open ends are flush with or slightly below ground level (Upton and Mantle 2010: fig. 25); small organisms moving on the ground will fall into the pit. The catch is usually improved if the trap is placed next to a metal or gauze fence, which tends to direct movement towards the pitfall. The traps may be empty or filled with some wetted tissue, if living insects are required, or filled with a variety of collecting fluids, such as ethanol, detergent and water or propylene glycol. In order to keep rain water from filling the pits, they are usually covered with some kind of a roof. Another version of the pitfall is the **trough trap**, which consists of a long gutter placed in the ground; these act in the same way as a normal pitfall and are discussed below under flight interception. These traps are meant to sample all beetles and other organisms as they move along the ground, but special purpose traps may also be designed that attract certain types of beetles by suspending an attractant like malt, honey, carrion or dung from a grid over the pitfall (Upton and Mantle 2010: fig. 26).

Other kinds of traps for attracting insects include yellow pans containing alcohol; some insects are attracted to yellow colour and others are drawn to the ethanol.

Light Sheets and Light Traps

Collecting at bright lights for Coleoptera is not as productive as it is for Lepidoptera, but on warmer nights large numbers of beetles may be obtained in this way. In general, mercury vapour lamps or ultraviolet fluorescent tubes are most attractive to insects, and these may be placed near a white sheet (Upton and Mantle 2010: figs 31, 33) to increase the attraction or incorporated into a light trap (Upton and Mantle 2010: figs 32, 34) Adults of many Carabidae and Staphylinidae, water beetles, such as Dytiscidae, Hydrophilidae and Heteroceridae, and various nocturnal Cerambycidae and Curculionidae may dominate the catch, but this depends somewhat on the type of environment.

Flight Interception

In recent years the use of various types of devices for collecting flying insects by trapping them during flight have greatly increased collections of insects formerly thought to be rare. These **flight intercept** or **window traps** use a barrier made of clear plastic or fine gauze which, when placed in a flight path, causes poor fliers, like most beetles, to collide with it and fall into collecting fluid placed below. The traps may be placed in the canopy or, more often, on the ground, in which case, the trough, placed at ground level, also serves as a pitfall. Such traps are illustrated by Upton and Mantle (2010: figs 22–23) and Peck and Davies (1980). A plastic roof is usually placed over the window, not only to keep rain and falling debris out of the trough, but also to capture some of the more efficient fliers that would otherwise be able to fly over the pane. A special type of flight interception device is known as a **Malaise trap** and includes a series of baffles, as well as a barrier, so that those insects that would normally avoid the barrier are funnelled into an attached collecting jar (Upton and Mantle 2010: figs 18–19).

Canopy Fogging

Fogging or misting with insecticides is the most recent and perhaps the most rewarding of the mass collecting techniques in current use. The detailed design and methodology, which has been discussed by Erwin (1983) and Arias *et al.* (2008), involves the spraying of pyrethrin-based insecticides on to canopy foliage and collecting the insects as they drop into a series of large funnels suspended from tree trunks or sheets on the ground. Although special spraying devices hoisted into the canopy are used for large-scale canopy sampling, the same technique can be used on tree trunks, logs or lower canopy strata by spreading sheets on the ground, spraying with commercial aerosols and hand collecting the specimens.

Other specialised collecting techniques, more detail on the above methods, and techniques for the preservation and study of insect specimens may be found in the following general works: Peterson (1959), Cogan and Smith (1974), Walsh and Dibb (1975), Martin (1977), Smithers 1981; Booth *et al.* (1990), Kuschel (1990), and Upton and Mantle (2010).

ADULT MORPHOLOGY

Body Form and Size

Beetles vary in length from 0.3 mm to 200 mm with a likely average size of less than 5 mm, although some groups, such as Scarabaeoidea, Buprestidae and Cerambycidae are commonly much larger than the mean (Crowson 1981). According to Kaiser *et al.* (2007), the upper size limit in beetles and other insects, appears to be related to the amount of space available for the tracheal system. The lower size limit, on the other hand, may be determined by several factors, including the egg size and brain size. The effects of miniaturisation in Coleoptera have been discussed by Grebennikov and Beutel (2002), Grebennikov (2008a), Polilov (2005, 2008) and Polilov and Beutel (2009, 2010).

Cuticle and Surface Features

The cuticle of beetles is similar in structure and composition to that of other insects, except for differences in hardness and in the thickness of some layers, consisting of a fine basement membrane lying above an ectodermal hypodermis, a moderately thick but flexible **endocuticle**, an **exocuticle**, which is often very thick, rigid and composed of chitin and tanned proteins, and a thin **epicuticle** consisting of waxy molecules overlain by a very thin cement layer. The term **sclerotisation** refers to the stabilisation or hardening of the exocuticle by the addition of phenolic and quinonoid compounds, which cross-link proteins in a polymerisation-like process called **tanning**. The former synonym **chitinisation** was based on the erroneous assumption that **chitin**, a component of all cuticle, was responsible for the hardening process. In most parts of the adult cuticle microfibrils are oriented regularly in one direction, but with successive layers rotated 90° in the manner of plywood; while intersegmental membranes exhibit a helicoidal microstructure similar to that of most larval cuticles. [Crowson 1981; Hackman 1984; Vincent and Wegst 2004].

The cuticle of most adult beetles forms a relatively thick exoskeleton, providing in itself an effective defensive mechanism against a variety of predators and parasites, especially in those forms that have evolved one or more types of internal strengthening mechanisms (apodemes and endocarinae) and/or interlocking devices between sclerites that conceal all membrane when engaged. These features, in turn, make longer adult life spans possible. A secondary reduction in cuticular thickness has occurred several times in the order, often correlated with reduction in adult life span and/or development of chemical defensive mechanisms. Since the cuticle represents the interface with the environment, it may be provided with a wide variety of modified setae, tubercles, ridges or spines and glandular openings, as well as exudate containment and distribution systems. Most of these features occur in the adult, but some surface-inhabiting larvae (e.g. some Coccinellidae and Chrysomelidae) have similar features. Setae may be slightly thickened or apically expanded forming **bristles** to greatly expanded and flattened forming **scales**. The latter are particularly common in many groups of Tenebrionidae and Curculionidae, but also occur in scattered families, such as Phycosecidae. Scales of many weevils and tenebrionids are various shades of brown and black, producing concealing colouration or **crypsis** against a background of bark, while the grey to white scales of phycosecids match the sandy background of their marine littoral habitat. Some weevil species are clothed with iridescent green scales produced by three-dimensional photonic crytals (see below). Most trinodine Dermestidae possess detachable spicisetae, which are similar to those of many dermestid larvae and form an effective defence against ant predators. Other complex types of setae often occur on the ventral surfaces of aquatic beetles forming a plastron (**see** p. 89). The layers of the cuticle itself may also produce a variety of optical effects produced by multilayer reflectors and diffraction gratings. Brilliant iridescent colours are produced by several beetle groups, especially Scarabaeidae and Buprestidae. A detailed study of iridescence and structural colour mechanisms in beetles was published recently by Seago *et al.* (2009). The functions of these various optical effects are still uncertain, but the iridescent greens would certainly produce crypsis in those beetles living on vegetation in a tropical forest. mimicking not only the leaf colours but also the effects produced by water droplets.

Integumentary glands, ducts and subsurface reservoirs occur in various locations on many beetle species, as noted for Cryptophagidae and Erotylidae by Leschen (1996, 2003, 2010c) and Leschen *et al.* (2010b) or in Chrysomelidae by Deroe and Pasteels (1982) and Pasteels and Rowell-Rahier (1989). The structure of integumentary glands was discussed by Noirot and Quennedey (1974) and Barbier *et al.* (1992). Most of these glands are probably defensive in nature, serving as protection from both macropredators and microorganisms, but some are involved in sexual attraction or host appeasement (**see** below). The cuticle may also contain one or more kinds of exoskeletal cavities whose functions have been variously interpreted. Several of these cavities are known to function as mycangia (e.g. those in scolytine Curculionidae), but others have been assumed to have a mycangial function without sufficient evidence. A review of this subject has been published recently by Grebennikov and Leschen (2010). Deep, glabrous cavities may be the external evidence of an internal apodeme used for strengthening or muscle attachment, but the function of those cavities characteristic of pselaphine Staphylinidae remains unclear. Some cavities containing numerous setae and glandular pores function as sexual attractants (**see sex patches**, p. 91), while others are associated with myrmecophily and serve to appease the host species (**see trichomes**, p. 90). Cuticular surfaces in some beetles contain a series of features serving to distribute and contain secretions. Lawrence and Hlavac (1979) discussed the presence of elevated ridges, channels and bridges formed by paired evaginations on the heads of some Derodontidae, but also on the pronotum, elytra, or ventral surfaces in various beetle families. The fluids distributed by and contained within these systems may be defensive in funcion, but in the case of myrmecophilous beetles, they are more likely to be host appeasing.

Head Capsule

The head capsule or **cranium** (Fig. 13A) is a rigid structure, which is often divided into areas for descriptive purposes but of little significance in terms of homology. The head is usually described as being **prognathous** when the mouthparts face forward and **hypognathous** when they are ventrally oriented. These terms, however, may be confusing, since ventrally-facing mouthparts may result from: (1) the entire head being ventrally oriented due to the angle of attachment between the head base and the anterior edge of the prothorax as in Scirtidae (Fig. 15J);

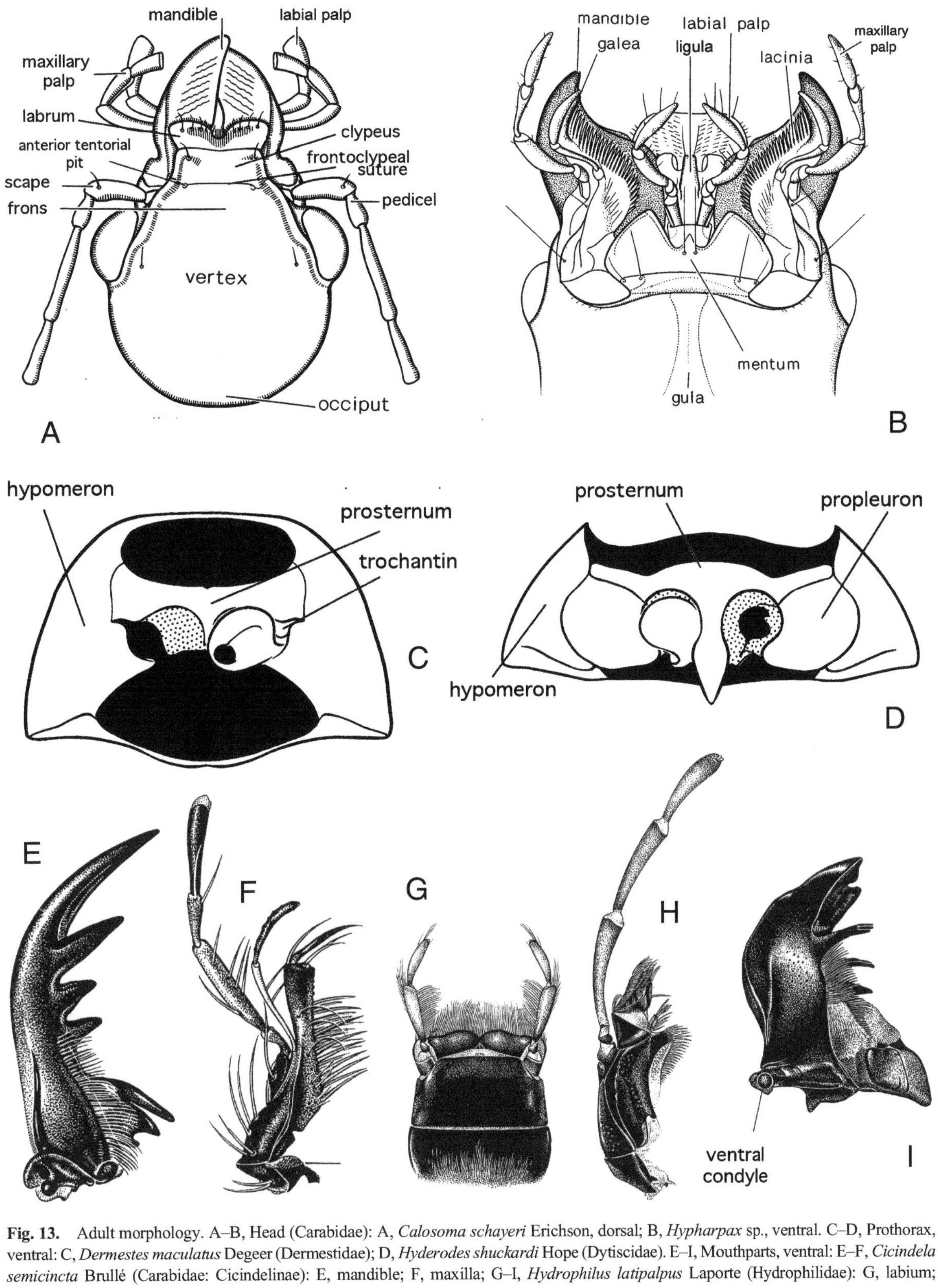

Fig. 13. Adult morphology. A–B, Head (Carabidae): A, *Calosoma schayeri* Erichson, dorsal; B, *Hypharpax* sp., ventral. C–D, Prothorax, ventral: C, *Dermestes maculatus* Degeer (Dermestidae); D, *Hyderodes shuckardi* Hope (Dytiscidae). E–I, Mouthparts, ventral: E–F, *Cicindela semicincta* Brullé (Carabidae: Cicindelinae): E, mandible; F, maxilla; G–I, *Hydrophilus latipalpus* Laporte (Hydrophilidae): G, labium; H, maxilla; I, mandible.

(2) the entire head being ventrally oriented due to a distortion of the prothorax so that its anterior edge is ventral as in some Bostrichidae (Fig. 81E, J); (3) the vertex being abruptly declined between the eyes as in some Cerambycidae and Chrysomelidae (Fig. 15A, D–F); or (4) the frontal or frontoclypeal region being abruptly declined between the antennal insertions as in some Elateridae (Fig. 14K).

The posteriormost portion of the head capsule (Fig. 13A) adjacent to the prothorax is referred to as the **occiput** and lies immediately above the posterior opening of the head capsule, the **occipital foramen**. The occiput is more or less continuous with the **vertex**, which extends between the eyes and continues anteriorly as the **frons**, which usually lies between the antennal insertions or anterior portions of the eyes; these three regions are not separated by sutures and have no clear-cut boundaries. The area immediately in front of the frons is the **clypeus**, which may be separated from the frons by a **frontoclypeal suture** or **epistomal suture**, representing an internal invagination, the **epistomal ridge**, to which the anterior tentorial arms are attached forming the anterior tentorial pits (**see** below). The frontoclypeal suture is usually straight or slightly curved (Fig. 13A) and may be weakly indicated or absent; in some groups, however, the suture may be strongly curved (Figs 44E, 83J) or distinctly angulate (Fig. 57A) and is occasionally attached to a longitudinal **midcranial suture** (Fig. 57B). The clypeus may be divided into a posterior, sclerotised **postclypeus** and an anterior, membranous **anteclypeus**. The **labrum** is usually a transverse flap, clearly visible from above and attached by membrane at or beneath the anterior edge of the clypeus; in some beetles, however, it may be membranous and concealed beneath the clypeus (e.g. in Scarabaeinae and Aphodiinae), abutting the clypeus and immovable with no connecting membrane or solidly fused to the clypeus forming a **clypeolabrum**. The area lying beneath the labrum and forming part of the upper wall of the preoral cavity is called the **epipharynx** and the area immediately behind this is called the **pharynx** or **cibarium**. A pair of sclerites, the **tormae**, often lie near the basolateral angles of the labrum-epipharynx and may extend posteriorly onto the pharynx.

The posterior portion of the head capsule behind the eyes may be abruptly constricted forming distinct **temples** or **tempora** (sing. **tempus**), and a **neck** (Figs 15K, 94A, 95J) that may be broad or narrow and is sometimes inflated to form a ball and socket joint with the anterior portion of the prothorax. The occipital region sometimes bears a low **transverse occipital carina** (Fig. 44G) or an elevated **transverse occipital ridge**, which in some cases abuts the anterior edge of the prothorax, thus concealing the neck region (Fig. 28C, T). A longitudinal line or groove indicating the presence of a **median occipital endocarina** (Fig. 15G) may also occur in this region, beginning at the edge of the occipital foramen and extending anteriorly. Other longitudinal or X-shaped grooves (Fig. 15A, C) may occur on the frons or vertex, but these do not extend to the posterior edge of the head. Also occurring on the vertex in a very few beetle groups are simple eyes or **ocelli** (sing. **ocellus**). Paired ocelli (Figs 15B, 67G) occur in a few Agyrtidae and Leiodidae, several Hydraenidae, in some Staphylinidae (e.g. Glypholomatinae, Microsilphinae, Neophoninae, nearly all Omaliinae and the scydmaenine genus *Neseuthia* Scott) and in all Derodontidae (Fig. 81C), while a single median ocellus occurs in the staphylinid genus *Metopsia* Wollaston (Proteininae) and in most Dermestidae (Fig. 28F) (but not Dermestinae). The presence of three ocelli has been observed in teratological specimens of a few omaliine Staphylinidae, but also occurs in the very rare Eastern Palaearctic beetle *Sikhotealinia zhiltovae* Lafer (Jurodidae). [Naomi 1987a; Newton and Thayer 1995; Lafer 1996; Kirejtshuk 1999; Leschen and Beutel 2004].

The compound eyes are highly varied in size and shape and may be absent or so large that they meet or almost meet above and/or below the head (Figs 44D, 64L, 82A). In several families the eyes are anteriorly emarginate or partly divided by a lobe (**canthus**), and in Gyrinidae, some Scarabaeoidea and a few Cerambycidae the division into an upper and lower portion may be complete (Figs 14A, 55H, 55J). The facets of the eye tend to be small and flat in diurnal species and coarse and convex in those forms active in reduced light. The structure of the **ommatidium**, which has been surveyed by Caveney (1986) for most families, has proven useful in characterising some major groups. A **eucone** eye, possessing a crystalline lens, is considered to be primitive in Coleoptera, and those eyes that have lost this lens are termed **acone**; determining the difference between a eucone and acone eyes, however, requires special preparation, since the crystalline lens is dissolved when specimens are cleared with KOH. In the **exocone** eye (Fig. 29H), there is no crystalline lens, but the cornea is expanded inwardly to form a different type of lens, which is not destroyed by clearing. Exocone eyes occur in Pleocomidae, Trogidae (some *Omorgus* species), Passalidae, Geotrupidae (some Bolboceratinae), Dascillidae, Rhipiceridae, Byrrhoidea (excluding Dryopidae and Lutrochidae), Elateroidea (excluding Eucnemidae), Nosodendridae, Dermestidae (*Dermestes*), Endecatomidae, Bostrichidae and Ptinidae. The retinular cells comprising the **rhabdom** of each ommatidium may be well separated from one another (**open rhabdom**) or fused together (**fused rhabdom**). Acone eyes with an open rhabdom are found throughout the series Cucujiformia (Wachmann 1977).

Antennal insertions usually occur on the frons, often in a lateral position in front of the eyes, where they are concealed from above by **frontal ridges**, which may extend mesally and unite at the midline (Fig. 14K) or extend posteriorly as **supraorbital ridges**. In several families, however, the antennal insertions are exposed from above, and may be located between the eyes and near the midline (Fig. 44B–D) or sometimes within emarginations of the eyes (Figs 14D, 15D, F). In a few groups (e.g. lissomine Elateridae) the antennal sockets are countersunk forming saucer-like **antennal fossae** (Fig. 14K). The area immediately below the eye may be concave and shiny forming a **subantennal groove** for housing the base of the antenna (Fig. 14L), and in some groups these grooves extend well behind the eyes. The frontoclypeal region in many Curculionoidea and a few members of other superfamilies may be produced forward forming a short **muzzle** (Figs 98F, 98I, K) or a longer, slender **rostrum** (Figs 95F, 98A, D–E, J, L–M); in the latter case, the subantennal groove, if present lies posterior to the antennal socket and is usually referred to as a **scrobe** (Fig. 97L) (not to be confused with the mandibular scrobe, **see** below).

Those portions of the head capsule below the eyes on each side are called the **genae**, and these may extend mesally to meet the **gular sutures** (Fig. 13B), which extend from the posterior tentorial pits to the posterior edge of the head capsule. In some beetles, there is a pair of longitudinal sharp ridges extending posteriorly from the maxillary articulations and separating the genae from remainder of the ventral head surface; these are called **subgenal ridges**,

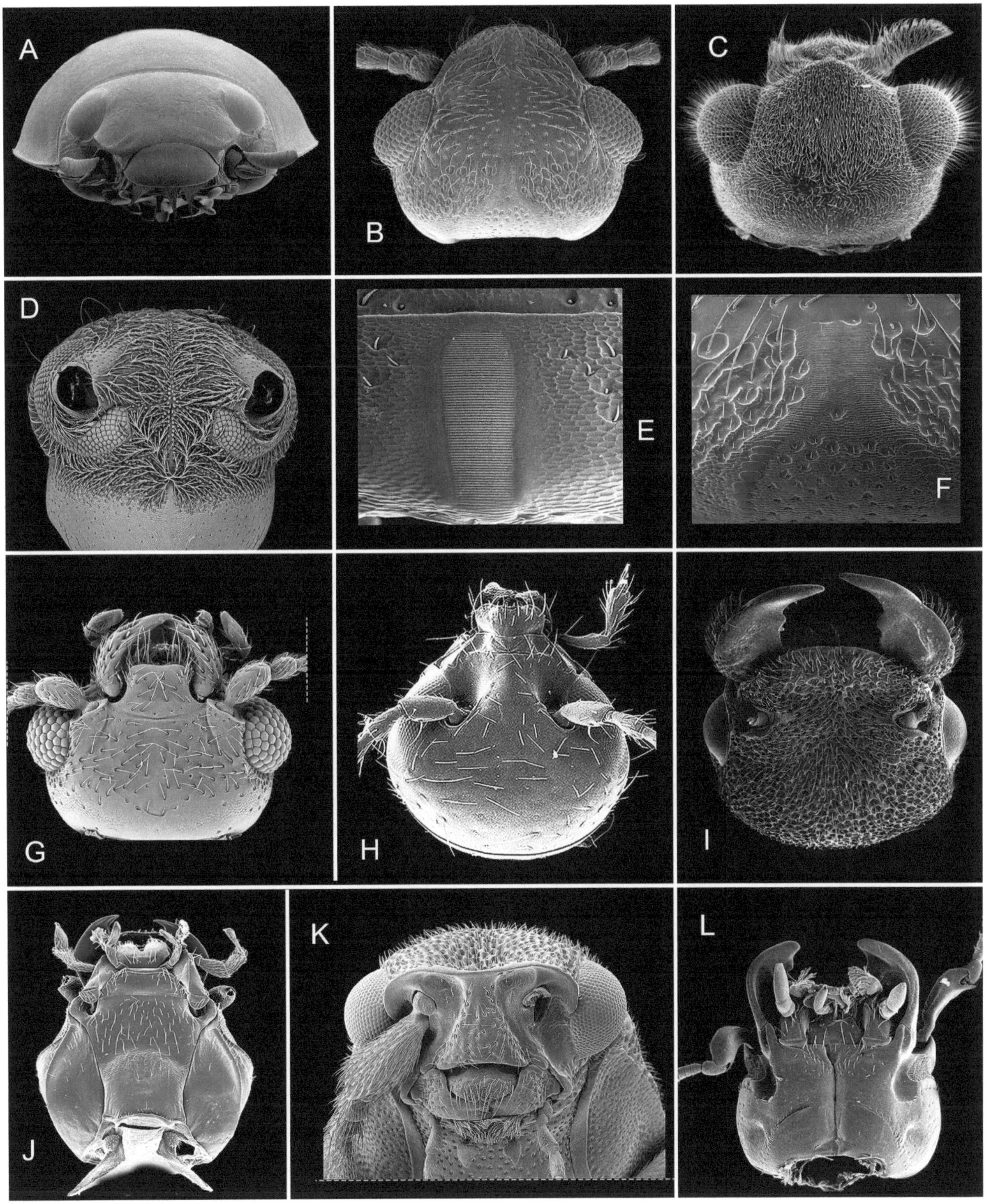

Fig. 14. Adult morphology. A–L, Head: A, *Macrogyrus* sp. (Gyrinidae), anterior; B, *Cnecosa insueta* (Crotch) (Erotylidae), dorsal; C, *Pelonomus* sp. (Dryopidae), dorsal; D, *Sybra* sp. (Cerambycidae: Lamiinae), dorsal; E, *Leucohimatium arundinaceum* (Forskål) (Erotylidae: Pharaxonothinae), occipital region, dorsal, with single stridulatory file; F, *Cnecosa insueta* (Crotch) (Erotylidae: Erotylinae), occipital region, dorsal, with paired stridulatory files; G, *Boganium* sp. (Boganiidae), dorsal; H, *Aleochara* sp. (Staphylinidae), dorsal; I, *Hemiops* sp. (Eucnemidae: Macraulacinae), dorsal; J, *Pseudomicrocara* sp. (Scirtidae), ventral; K, *Rousia dumbrelium* Calder (Elateridae: Dendrometrinae), anterior; L, *Platysoma* sp. (Histeridae: Histerinae), ventral.

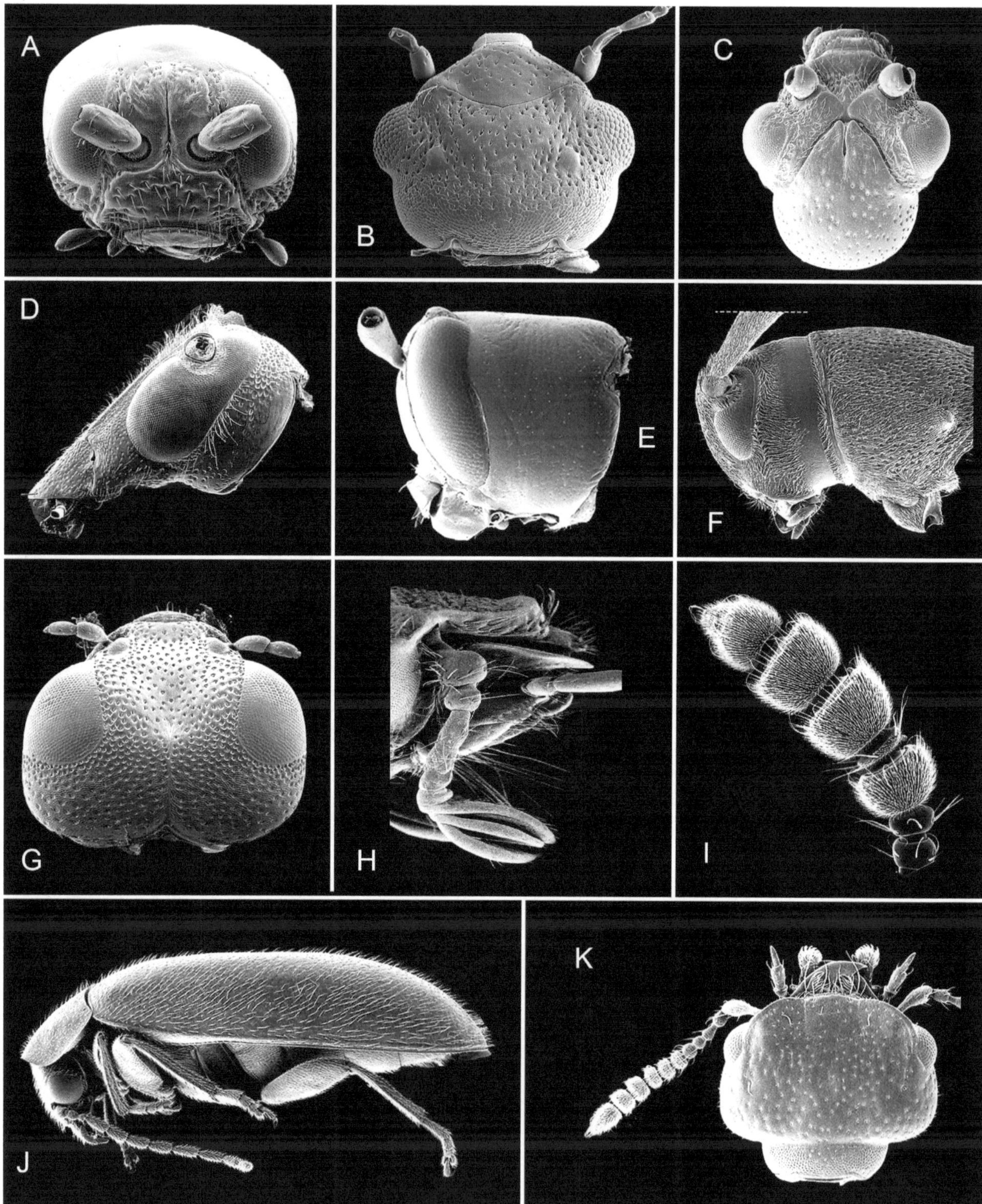

Fig. 15. Adult morphology. A–E, Head: A, *Chelymorpha* sp. (Chrysomelidae: Cassidinae), anterodorsal; B, *Glypholoma pecki* Thayer & Newton (Staphylinidae), dorsal; C, *Crioceris asparagi* (Linnaeus) (Chrysomelidae), dorsal; D, *Hesthesis* sp. (Cerambycidae: Cerambycinae), lateral; E, *Aspidimorpha* sp. (Chrysomelidae: Cassidinae), lateral; F, *Dectes* sp. (Cerambycidae: Lamiinae), adult head and prothorax, lateral; G, *Castiarina* sp. (Buprestidae: Buprestinae), head, dorsal; H, *Phyllopertha horticola* (Linnaeus) (Scarabaeidae: Rutelinae), anterior portion of head and antenna, lateral; I, *Macrohydnobius crestonensis* (Hatch) (Leiodidae: Leiodinae), antennal club; J, *Pseudomicrocara variabilis* Armstrong (Scirtidae), habitus, lateral; K, *Anotylus* sp. (Staphylinidae: Oxytelinae), head, dorsal.

(Figs 14J, 44F), and in groups with hypognathous heads (e.g. Scirtidae) they may abut the procoxae. The area in between the gular sutures is referred to as the **gula**, which is obliterated when the two sutures fuse together; however, there is no anterior line of demarcation between the gula and the region lying between it and the base of the labium (**see** below), and the gular sutures may be vaguely defined or absent. The anterior ventral region of the head capsule in front of the gular sutures is usually called the **submentum** and it often terminates in a **submental peduncle** to which the labium is attached. The term **gulamentum** is sometimes used for the entire ventral portion of the head between the labium and posterior edge of the head capsule. In some cases, each gular suture appears to continue anteriorly beyond the posterior tentorial pit; this anterior suture, termed the **subgenal sulcus** by Lyal (1995), may extend to the base of the submental peduncle or continue dorsally to the anterior tentorial pit (Fig. 14J).

The tentorium consists of a pair of broad **posterior arms**, originating at the posterior tentorial pits, a narrower pair of **anterior arms** which extend forward and terminate at the anterior tentorial pits and sometimes a third pair of **dorsal arms**, which terminate on the roof of the head capsule. The posterior arms are usually joined by a transverse bridge, the **corpotentorium**, which may bear an anterior process and the anterior tentorial arms may be expanded mesally to form **laminatentoria**, which in some cases fuse at the midline to form a secondary **anterior bridge**. The details of tentorial structure vary considerably (Stickney 1923) and may provide useful taxonomic characters.

In many members of the suborder Polyphaga a pair of **cervical sclerites** occurs in the membrane between the head and prothorax. Each cervical sclerite usually consists of two articulated sclerites, the anterior one attached to the posterior edge of the head capsule and the posterior one to that portion of the prothoracic venter extending in front of the procoxal cavities (**see** below). [Matsuda 1965].

Antennae

The antennae are normally 11-segmented, but reduction to ten, nine or eight antennomeres is relatively common. A few groups (e.g. some pselaphine Staphylinidae, Hydrophilidae, Scarabaeidae and Coccinellidae) may have seven or fewer antennomeres, and males of *Rhipicera* (Fig. 62D) may have more than 30. The length of the antenna varies considerably and may be much longer than the body in some Cerambycidae and Anthribidae. The first antennal segment or **antennomere** is called the **scape** and the second one, which contains a chordotonal organ called **Johnston's organ**, is referred to as the **pedicel**. The remaining elements in the antenna are not considered to be true segments and are sometimes referred to individually as **flagellomeres** or collectively as the **flagellum**, but these terms are not very commonly used in Coleoptera. When one or more terminal antennomeres are enlarged to form a club (**see** below), those lying between the scape and the club are referred collectively as the **funicle** and individually as **desmomeres**; but these terms are used mainly in the Curculionoidea. The scape is often longer and/or wider than the pedicel and may be much longer, especially in Curculionidae. The pedicel may be partly or entirely enclosed within the scape in a few groups (e.g. Gyrinidae, Paussini, some malachiine Melyridae). When the attachment of the pedicel to the scape is subterminal, so that the funicle forms an angle with the scape, the antennae are said to be **geniculate** or elbowed (Figs 97L, 98K–M); however, there are some instances where a sharp bend in the scape results in a superficially similar condition (referred to as **subgeniculate** in Curculionoidea). The scape may be greatly enlarged and variously modified in males (e.g. some Laemophloeidae, Pterogeniidae).

The form of the flagellum is highly varied in beetles and is often used as a diagnostic feature. Five major flagellar types were delimited by Lawrence *et al.* (2011): (1) those in which the flagellomeres are more or less equal in width and either slender and elongate (**filiform**, Figs 54E–I) or short and bead-like (**moniliform**, Figs 54J, 85I); (2) those in which the flagellomeres are gradually enlarged apically (**incrassate** or **clavate**, Figs 59M–N, 60C); (3) those in which most flagellomeres (but usually not the apical one) are asymmetrically expanded to form rami, which may be short and saw-like (**serrate**, Figs 62A, F, 64E), comb-like (**pectinate**, Figs 62B, 64G) or fan-like (**flabellate**, extreme development of pectinate type, Fig. 62C–D); (4) those in which one or more apical antennomeres are distinctly wider and/or longer than the preceding ones forming a club, which may be more or less symmetrical or only slightly asymmetrical (**capitate**, Figs 82F, H–I, 83K, 86A–L, 87C–K) or highly asymmetrical with more or less flattened segments (**lamellate**, Figs 15H, 60K–L, 61I); and (5) those in which flagellomeres are more or less equal in width but distinctly wider than long (**perfoliate**, Figs 29J, 54K). Antennae are frequently longer (e.g. Cerambycidae) or more elaborate with longer rami (e.g. Callirhipidae, Fig. 62B–C) in the male than in the female. The antennae are primarily chemosensory organs and the more distal antennomeres may bear a variety of sensilla. Compound, subdivided or multipronged sensilla, usually referred to as **sensoria** occur in almost all Ciidae (Lawrence 1971) and in several groups of Tenebrionidae (Medvedev 1977). In some groups a pubescent antennal club may be preceded by a short, glabrous, cup-like segment called a **cupule**; in aquatic Hydrophilidae and Hydraenidae, these modified antennae (Figs 28R, 57D–H) are utilised in respiration by breaking the surface tension of the water and allowing air to enter the bubble trapped beneath the body by a plastron (**see** p. 90). Modified antennae in the males of some beetles may be used in courtship (e.g. in Melyridae, Meloidae) or possibly in male combat (e.g. in pselaphine Staphylinidae, Laemophloeidae, Pterogeniidae).

Mandibles

The mandible (Figs 13E, I) is usually in the form of a tetrahedron, with dorsal, ventral and external faces, a triangular base, and articulations with the head capsule consisting of a dorsolateral socket and ventrolateral condyle. Mandibles move in a transverse plane, through the action of large adductor and smaller abductor muscles, and the apices are opposable; in some Corylophidae, however, the stylet-like mandibular apex is separated by membrane from the base, and the same muscles produce a forward-backward movement. The outer edge of the mandible is often flattened and setose, but may be concave forming a **mandibular scrobe** in Carabidae. The dorsal surface of the mandibular base may bear a carina or tubercle, which interlocks in various ways with the clypeus or labrum, and a cavity, which may be glabrous or setose and is more or less concealed when the mandibles are in closed position. A setose cavity (Figs 16E, 44E) occurs in *Omma*, *Rhinorhipus*, Boganiidae, Sphindidae and palophagine Megalopodidae, and a glabrous cavity in Phloeostichidae and some Silvanidae. A mandibular tubercle fitting into a clypeal depression occurs in Sphindidae, Hobartiidae and Phloeostichidae, but different types of interlocking mechanisms between mandible and clypeus or labrum occur in some other groups. The basal portion of the mandible's internal (mesal) edge, when thickened, enlarged and usually armed with ridges or tubercles, is called the **mola**. The two molae act together to grind or otherwise process food and/or move food particles into the alimentary canal, and they are absent in most predacious forms, as well as in those species feeding on fresh leaves or wood, in which the mandibular apices scrape or cut solid food material. The mandibular apex is usually armed with one or more fixed teeth or lobes, and in the region lying between the mandibular apex and the molar region (**incisor edge**), there may be a tooth-like **retinaculum**, one or more small teeth, and/or a membranous **prostheca**. An articulated, subapical process is occasionally present on both mandibles (e.g. in Hydraenidae, Fig. 16H and Clambidae) or on one mandible only in Myxophaga. Mandibles attain their maximum development in males of some Lucanidae, where they are used in combat with other males; by contrast some Scarabaeidae have the mandibular apex reduced and more or less membranous, but the basal mola large and well developed. Highly reduced, flattened, simple mandibles (Fig. 28S) occur in Psephenidae and Chelonariidae, while slender, falcate mandibles (Fig. 13E) are present in some predaceous groups, such as Staphylinidae, Phengodidae, Lycidae, Lampyridae, Cantharidae and various Adephaga.

Ventral Mouthparts

The ventral mouthparts include the paired maxillae and the labium. The adult **maxilla** (Figs 13A, F, H, 16A–D, G) usually consists of a small, basal, articulating sclerite, the **cardo**, the major sclerite or **stipes**, which is usually divided by an oblique suture into a **basistipes** and **mediostipes**, a **palpifer** that is attached to the basistipes and usually articulated, a **maxillary palp**, almost always consisting of three or four **palpomeres** (appearing 5-segmented in some Aleocharinae; apparently absent or combined to form a single median projection in *Rhipidioides*), and two **apical lobes** attached to the mediostipes: an external (lateral) **galea**, which is usually articulated and consists of a proximal **basigalea** and distal **distigalea** and an internal (mesal) **lacinia**, which is undivided and only rarely articulated. The apical palpomere varies considerably throughout the order and may be **cylindrical** (Fig. 16B), **fusiform** (wider at the middle and narrower at each end) (Fig. 16C–D, G), **conical** or slightly to strongly **expanded apically** forming a triangle. When the palpomere is unevenly expanded, it is referred to as **securiform** (Fig. 16F) and an extreme form of this (more or less knife-like) is referred to as **cultriform**. In some cases, the apical palpomere is both shorter than and narrower than the preapical one, so that the palp is awl-like or **subulate** (Figs 14H, 16A). Palpal sensilla may be of various types, but two are notable. In Archostemata the terminal palpomere bears a small group of projecting sensilla, which may be located on the surface in Cupedidae and Crowsoniellidae, but within a deep cavity in Ommatidae (Fig. 16F) and Micromalthidae (Pace 1976). In Adephaga and Polyphaga, there may be one or more flattened, **digitiform sensilla** (Honomichl 1980), each of which is located in a groove; these are often difficult to see with light microscopy. Maxillary lobes also vary considerably. The galea and lacinia may be more or less equal in length and setose at the apex (Fig. 16C) in many polyphagan groups, but in most Adephaga the galea is palpiform and much narrower than the lacinia (Fig. 13B, F). In a few taxa (e.g. Nitidulidae) the galea is entirely absent (Fig. 16B) and in others (e.g. *Ripiphorus*) the lacinia is highly reduced or absent. In *Chelonarium*, the galea is divided into two acute, membranous lobes, and in some Meloidae it is a long slender lobe. The lacinia in some groups bears at or near the apex a sclerotised hook referred to as an **uncus** (Fig. 16C), which may be subdivided into two or three lobes.

The **labium** is attached at the anterior end of the submentum or gulamentum and usually consists of a basal **mentum**, an apical **prementum**, and paired **palpigers**, to which the 1-to 3-segmented **labial palps** are attached. The apex of the prementum, known as the **ligula**, is formed from the **glossae** (Fig. 13B) or the glossae and **paraglossae** of primitive insects. It is often emarginate and sometimes subdivided into two or more lobes. In some forms, including *Hydrophilus* (Fig. 13G), a distinct sclerite, the **submentum**, lies between the gula and mentum, but the term has also been applied to the undifferentiated anterior portion of the gulamentum. In a very few taxa (*Myrmecholeva*, Fig. 67J, and *Cautomus*, Fig. 87F, in Australia), the mandibles and maxillae are stylet-like and enclosed within a **tubular sheath** formed by the labrum and labium. [Williams 1938].

Prothorax

The prothorax in Coleoptera is always well developed, forming with the head an obvious fore body separated by intersegmental membrane from the hind body, which consists of the mesothorax, metathorax and abdomen. The prothorax is composed of a large dorsal sclerite, the pronotum, which extends ventrally on each side, a pair of pleural sclerites (internalised in Polyphaga), and a ventral sclerite, the prosternum, which lies between and anterior to the procoxae, forming the anterior and mesal walls of the coxal cavities. These are discussed in more detail below.

The main, dorsal portion of the pronotum is called the **pronotal disc**, and this is usually separated on each side by a sharp edge or carina, the **lateral pronotal carina**, from a deflexed portion, the **hypomeron** (sometimes referred to as **pronotal epipleuron**), which lies adjacent to the pleuron or sternum (**see** below). The lateral carinae are complete to the anterior edge of the prothorax (Figs 19D, 20A) in the archostematan genera *Tetraphalerus* and *Crowsoniella*, all Myxophaga, most Adephaga and many polyphagan groups, but incomplete anteriorly (Fig. 19A) or entirely absent (Figs 19C, 20E–F, H) in most Archostemata, rhysodine, paussine and cicindeline Carabidae and several Polyphaga, especially within the superfamilies Tenebrionoidea, Chrysomeloidea and Curculionoidea. The carinae may be simple (Fig. 20A) or bear margins or beads, (Fig. 19D) and the edges may be crenulate, denticulate or dentate (Fig. 18G). In some cases the pronotal disc may be laterally expanded or **explanate** and either horizontal, as in some Trogossitidae (Fig. 71A–B), Priasilphidae (Fig. 85A) or Nitidulidae (Fig. 73I) or obliquely vertical, as in Phalacridae (Fig. 85E), a few Nitidulidae (Fig. 87A–B) and some Coccinellidae (Fig. 87M–N). The surface of the pronotal disc is often simple, but

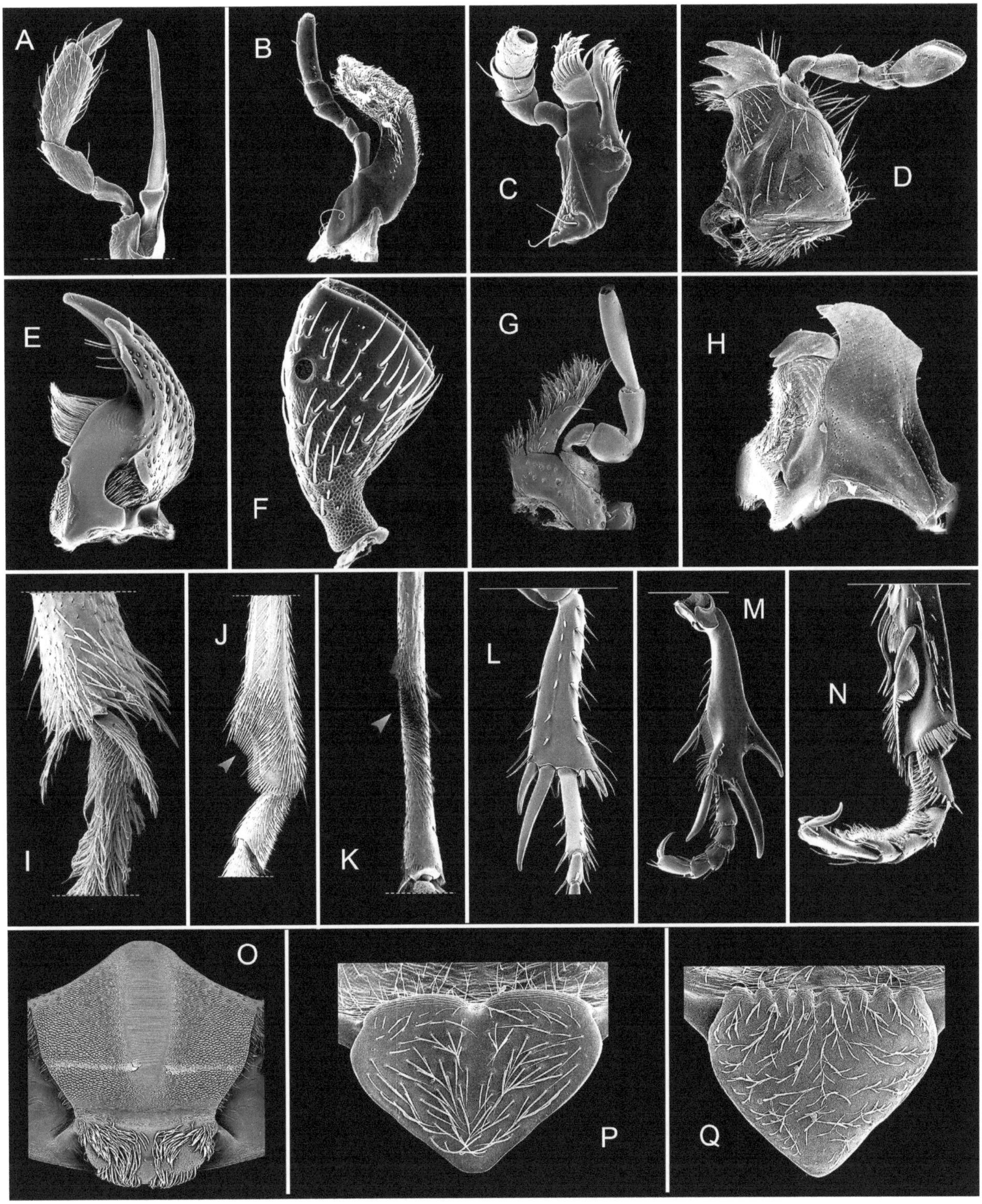

Fig. 16. Adult morphology. A–D, Maxilla: A, *Glyptolopus* sp. (Cerylonidae: Ceryloninae); B, *Glischrochilus* sp. (Nitidulidae: Cryptarchinae); C, *Chiliotis* sp. (Cryptophagidae: Cryptophaginae); D, *Phyllopertha horticola* (Linnaeus) (Scarabaeidae: Rutelinae). E, *Paracucujus rostratus* Sen Gupta & Crowson (Boganiidae), mandible, dorsal. F, *Omma rutherfordi* Lawrence (Ommatidae), apical maxillary palpomere. G, *Ceratognathus niger* Westwood (Lucanidae), maxilla. H, *Meropathus* sp. (Hydraenidae: Ochthebiinae), mandible, ventral. I–N, Tibial apex: I, *Scraptia* sp. (Scraptiidae), showing pubescent spurs; J, *Rhyparida* sp. (Chrysomelidae: Eumolpinae), mesotibia, showing antenna cleaner (arrow); K, *Typodryas* sp. (Disteniidae), mesotibia, showing antenna cleaner (arrow); L, *Ataenius picinus* Harold. (Scarabaeidae: Aphodiinae); M, *Carenum* sp. (Carabidae: Scaritinae), protibia; N, *Amblytelus* sp. (Carabidae: Moriomorphinae), protibia. O, *Temnosternus* sp. (Cerambycidae: Lamiinae), mesonotum with scutellar shield and stridulatory file. P–Q, Scutellar shield: P, *Ptilodactyla* sp. (Ptilodactylidae); Q, *Sclerocyphon* sp. (Psephenidae).

may have various modifications, including a longitudinal groove, sublateral carinae (Figs 85K, 86L) or paired sub-basal pits (Fig. 90K). The anterior edge is usually simple, but may have a marginal bead or in some cases a broad, transverse collar. The meeting of the lateral carinae on each side with the anterior and posterior edges of the disc forms the **anterior angles** and **posterior angles**, respectively. The anterior angles may be produced forward and acute to broadly rounded (depending on the extent of the marginal bead). In a few groups (e.g. Chelonariidae, Fig. 28E), the lateral carinae form a continuous curve with the anterior edge. The posterior angles may also be rounded or angulate to acute, and in some cases produced posteriorly beyond the elytral humeri, as in many Elateroidea (Figs 44M, 64A–H). The posterior edge of the disc may be truncate, rounded, or bisinuate, with or without a marginal bead, and sometimes complexly lobed and often crenulate (Figs 19H, 20C) as part of a **pro-mesothoracic interlocking mechanism**, which may also involve ventrally located posterior projections on the hypomera.

The pronotal hypomeron in Polyphaga provides useful taxonomic characters in some groups. It may be broadly and shallowly impressed anteriorly to receive the antenna, as in some cerylonine Cerylonidae (Figs 19G, 26B), deeply excavate anteriorly to receive the antennal club, with the cavity sometimes extending onto the upper surface of the pronotum, as in *Murmidius* (Fig. 19F) or some Histeridae (Fig. 18H), or broadly, shallowly impressed posteriorly to receive the leg, with the impression sometimes extending anteriorly as a narrow groove housing the protarsi (Fig. 20B). Narrow and deep antennal grooves may also occur either along the lateral edges of the hypomera, as in some Eucnemidae (Fig. 19H). In some Elateridae the anterior portion of the notosternal suture is more or less expanded to form an anterior groove receiving the antenna, and in the subfamilies Lissominae and Thylacosterninae these are greatly expanded to form cavities beneath the surface of the hypomeron (Fig. 18C, I).

The morphology of the prothoracic **pleuron** and **sternum** is complex and has been discussed by Matsuda (1960, 1970); Larsén (1966), Evans (1971, 1974), Hlavac (1973, 1975), Baehr (1979), and more recently by Beutel and Haas (2000), Beutel and Komarek (2004), Beutel and Lawrence (2005), Ge *et al.* (2007), Friedrich *et al.* (2008) and Lawrence *et al.* (2010b, 2011). The **prosternum**, as defined in most recent works, including Lawrence *et al.* (2010b, 2011) and used in the present text for keys and family descriptions, is probably composed of both pleural and subcoxal elements, as in the mesothorax and metathorax, although there is no indication of a median invagination (the paired endosternites are always well separated) or a transverse katepisternal suture. Although we continue to use prosternum for this region, the alternate term, **proventrite**, was used by Friedrich *et al.* (2008) for Archostemata and may be preferable since it is likely that this prothoracic sclerite is serially homologous to the mesoventrite and metaventrite, as discussed below. In almost all Archostemata, Adephaga and Myxophaga, the pleuron on each side is visible externally, extends both in front of and well behind the procoxae, and may form a major part of the lateral thoracic wall. It extends to the anterior edge of the prothorax, completely separating notum from sternum, in the priacmine Cupedidae (Fig. 17E) and in the myxophagan families Lepiceridae (Fig. 17A), Sphaeriusidae (Fig. 17J) and Hydroscaphidae, and is separated from the notum by the **notopleural** suture and from the prosternum (or proventrite) by the **pleurosternal suture** (or anapleural cleft). Unlike the meso- and metapleura, the propleuron is not clearly subdivided. In the ommatid genus *Tetraphalerus* Waterhouse, cupedine Cupedidae and almost all Adephaga, the propleuron ends at least slightly before the anterior edge of the prothorax, so that notum and sternum meet anteriorly forming a short **notosternal suture**, and in most Carabidae (Fig. 17D, G) the posterior portion of the pleuron extends mesally to meet the prosternal process, closing the procoxal cavities from behind. In the genus *Omma* and a few cicindeline Carabidae (e.g. *Tricondyla*), the prosternum and propleuron are fused, so that the pleurosternal suture is absent. In the myxophagan family Torridincolidae (Fig. 17I) the anterior pleural extension is minimal, so that notum and sternum are broadly joined anteriorly, and in the archostematan family Micromalthidae all the prothoracic sclerites are fused together. The exposure and extent of the propleuron in the archostematan *Crowsoniella relicta* Pace (Crowsoniellidae) or *Sikhotealinia zhiltovae* Lafer (Jurodidae) is not clearly shown in the figures of Pace (1976) and Lafer (1996), and we have not had the opportunity to examine specimens of either taxon. The propleuron never extends behind the procoxae in the suborder Polyphaga and extends slightly in front of the coxal cavities only in a few Derodontidae (Fig. 17B), Decliniidae and Scirtidae.

The prothoracic **trochantin** in Archostemata and Adephaga is a small sclerite articulating by a single condyle with the coxa and separated by membrane from the pleuron; it is externally visible (Fig. 17E) in almost all Archostemata (absent in Micromalthidae), but concealed (Fig. 17G) in Adephaga. In both Myxophaga (Fig. 17I–J) and Polyphaga, the trochantin has apparently fused completely with the pleuron forming a **pleurotrochantin**, and in the latter group, it has become more or less internalised, forming an **endopleuron** or **cryptopleuron** (dashed lines in Fig. 17B-C, F, H). The visible portion of the pleurotrochantin in Polyphaga is usually referred to as the **trochantin** although there is no obvious demarcation between it and the endopleuron. The notopleural suture in Polyphaga is absent, the coxa articulates with the pleurotrochantin at two points, and the deflexed portion of the notum (or hypomeron), is attached directly to the sternum and separated from it by the notosternal suture. In some Polyphaga, the trochantin may be reduced, concealed, and/or fused to the notal wall (Fig. 17H), and in others the suture may be absent (Fig. 19E, I).

The prosternum usually consists of a **prosternal process** lying between the procoxae and an anterior portion, which extends in front of the coxae forming the body of the sternite; the lateral extensions of the prosternum are sometimes referred to as **alae**. The nature and extent of the prosternum in front of the coxae varies considerably among Polyphaga, being highly reduced forming a slender transverse band in some Staphylinoidea (Fig. 20G) and most Scirtoidea, well developed and often longer than the coxal cavities in many groups, and produced forward forming a **chin piece** concealing the mouthparts in some Histeridae (Fig. 18H), Elateridae (Fig. 18B–C) and a few other families. The anterior portion of the prosternum may also have a median carina or paired carinae (Fig. 19L), usually extending onto the prosternal process, and in some beetles the anterior edge of the prosternum may be excavate to receive a hypognathous head (Fig. 63B). An **incomplete** prosternal process terminates before the posterior ends of the coxae; in some cases it may be very short and apically rounded or angulate (Figs 18F, 20G, 44Q). A **complete** prosternal process

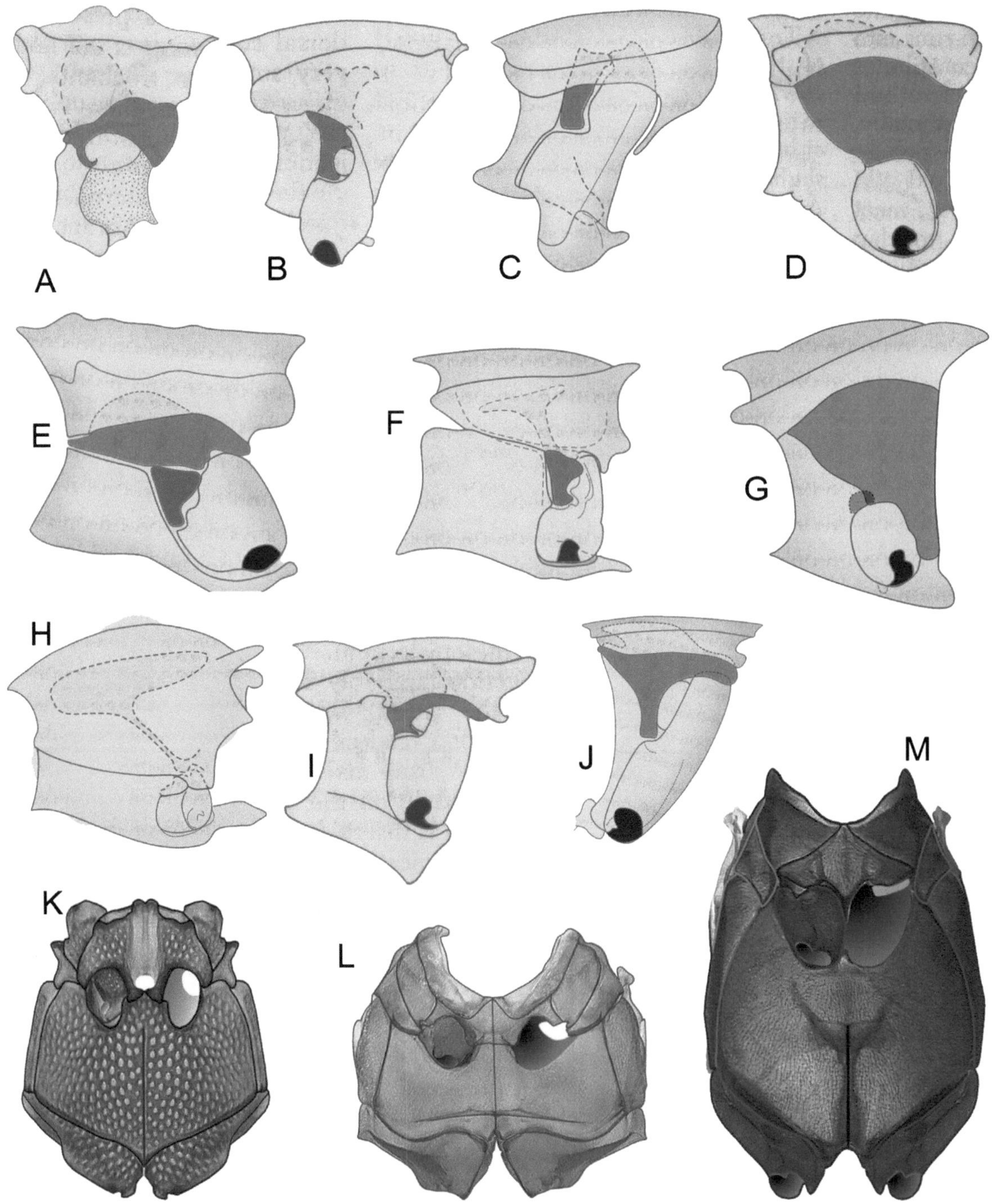

Fig. 17. Adult morphology. A–J, Prothorax, lateral, showing external and internal extent of propleuron, trochantin and/or propleuro-trochantin (darker shading): A, *Lepicerus inaequalis* Motschulsky (Lepiceridae); B, *Laricobius rubidus* Leconte (Derodontidae); C, *Pleocoma* sp. (Pleocomidae); D, *Ozaena elevata* Bänninger (Carabidae: Paussinae); E, *Priacma serrata* LeConte (Cupedidae), with exposed trochantin; F, *Pelonomus obscurus* LeConte (Dryopidae); G, *Metrius contractus* Eschscholtz (Carabidae: Paussinae), with concealed trochantin; H, *Perothops muscida* (Say) (Eucnemidae), with concealed pleurotrochantin fused to notum; I, *Iapir britskii* (Reichardt & Costa) (Torridincolidae); J, *Sphaerius politus* Horn (Sphaeriusidae). K–M, Pterothorax, ventral: K, *Agrypnus* sp. (Elateridae);. L, *Schizopus* sp. (Buprestidae: Schizopodinae); M, *Pyrochroa* sp. (Pyrochroidae: Pyrochroinae).

extends at least to the ends of the coxae (Figs 18E, G-H, 19G–L), and in some cases it may extend well beyond this (Figs 19H, 44M), in which case it usually fits into the mesoventral cavity (**see** below). In Haliplidae (Fig. 20I) the prosternal process extends beyond the apex of the metaventral process, completely concealing the mesoventrite. Sometimes the prosternal process appears to be

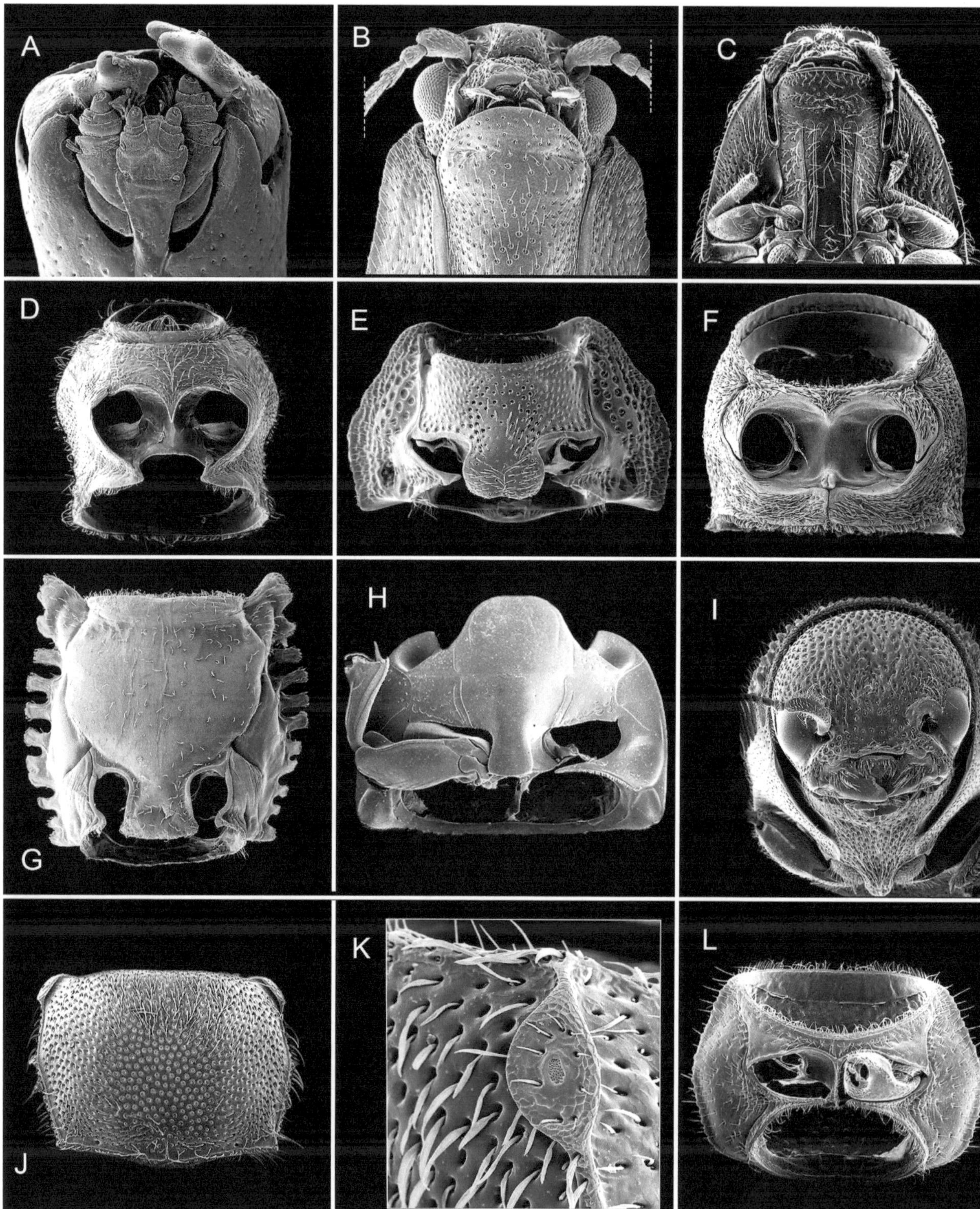

Fig. 18. Adult morphology. A, *Car* sp. (Caridae), apex of rostrum, ventral. B, *Parasaphes* sp. (Elateridae: Pityobiinae), head and anterior portion of prothorax, ventral. C, *Drapetes vicinus* Fleutiaux (Elateridae: Lissominae), head and prothorax, ventral. D–H, Prothorax, ventral: D, *Pergetus campanulatus* (LeConte) (Anthicidae: Eurygeniinae); E, *Priasilpha obscura* Broun (Priasilphidae); F, *Ithycerus noveboracensis* (Forster) (Brentidae: Ithycerinae); G, *Brontopriscus* sp. (Silvanidae: Brontinae); H, *Platysoma* sp. (Histeridae: Histerinae). I, *Cussolenis mutabilis* (Bonvouloir) (Elateridae: Thylacosterninae), head and prothorax, anterior. J, *Cryptophagus cellaris* (Scopoli) (Cryptophagidae: Cryptophaginae), prothorax, dorsal; K, same, anterior pronotal angle with callosity and glandular openings. L, *Necrobia rufipes* (Degeer) (Cleridae: Korynetinae), prothorax, ventral.

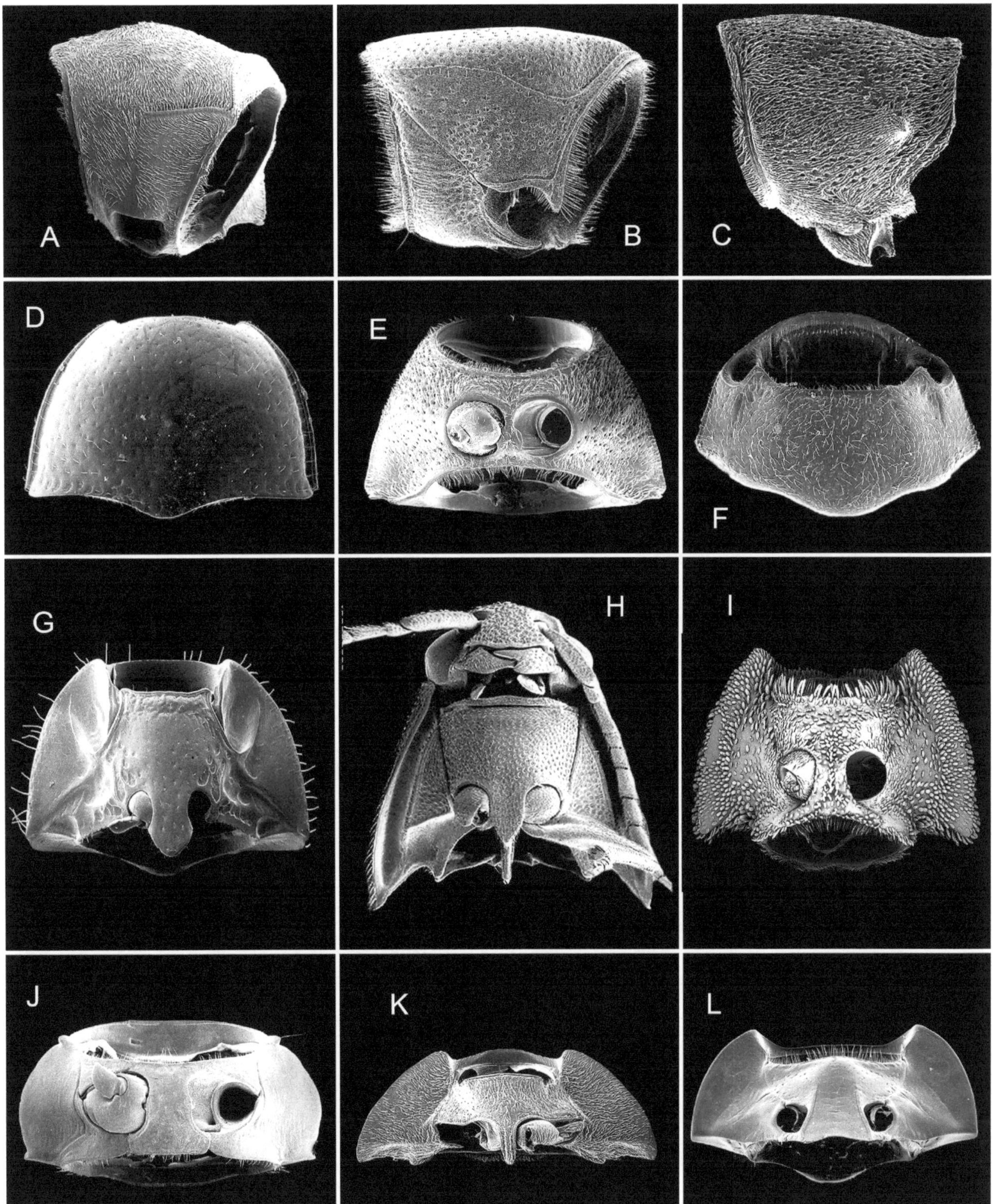

Fig. 19. Adult morphology. A–G, Prothorax: A, *Euparius* sp. (Anthribidae), ventrolateral; B, *Synercticus heteromerus* Newman (Boridae: Synercticinae), lateral; C, *Dectes* sp. (Cerambycidae: Lamiinae), lateral; D, *Mychocerus discretus* (Casey) (Cerylonidae: Ceryloninae), dorsal; E, *Araecerus coffeae* (Fabricius) (Anthribidae), ventral; F, *Murmidius ovalis* (Beck) (Cerylonidae: Murmidiinae), dorsal; G, *Mychocerus discretus* (Casey) (Cerylonidae), ventral. H, *Fornax* sp. (Eucnemidae), head and prothorax, ventral. I–L, Prothorax, ventral: I, *Ulodes* sp. (Ulodidae); J, *Rhyparida* sp. (Chrysomelidae); K, *Sclerocyphon* sp. (Psephenidae); L, *Aulacochilus* sp. (Erotylidae).

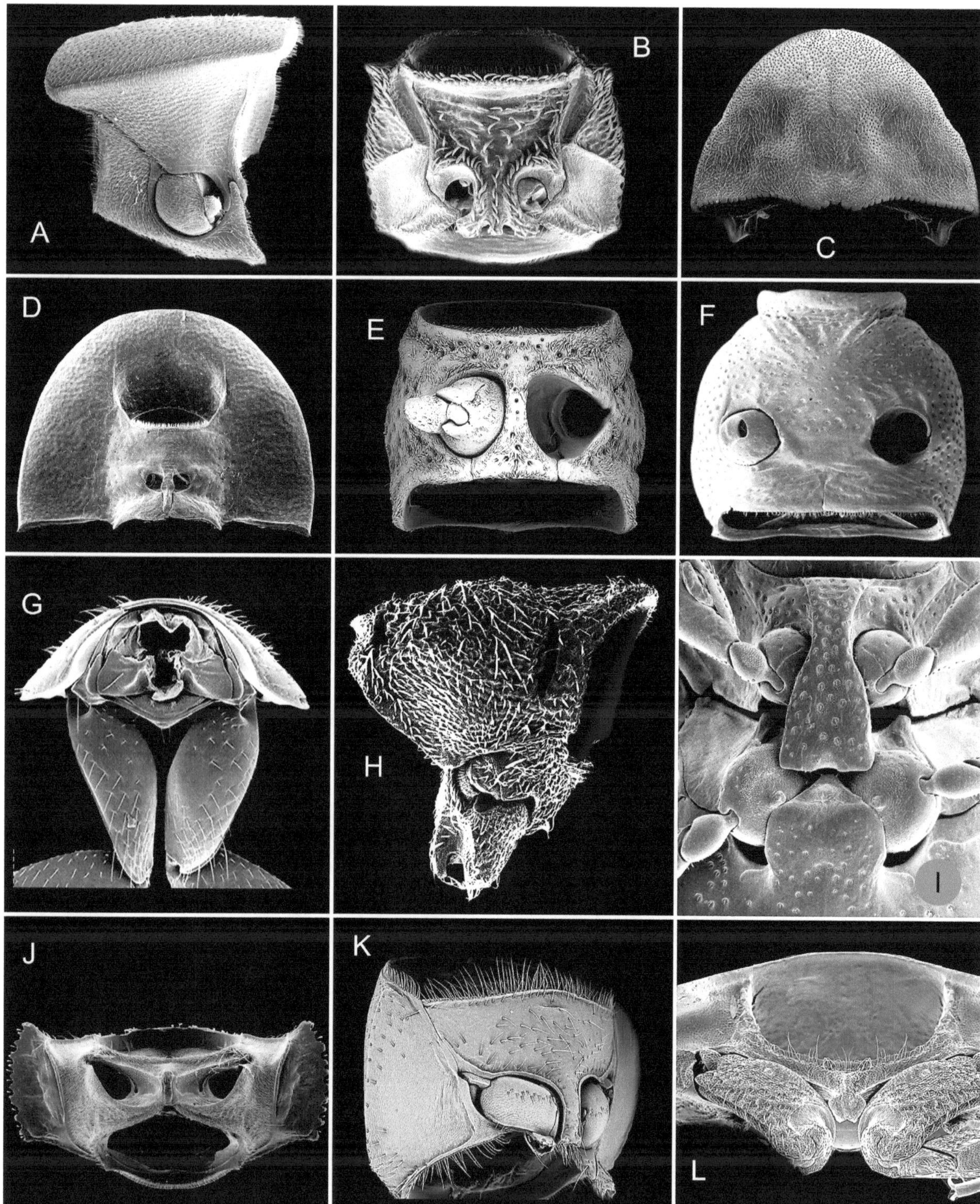

Fig. 20. Adult morphology. A–H, Prothorax: A, *Archeocrypticus topali* Kaszab (Archeocrypticidae), lateral; B, *Latometus* sp. (Zopheridae: Zopherinae), ventral; C, *Callirhipis reticulata* Lea (Callirhipidae), dorsal; D, *Cossyphus* sp. (Tenebrionidae: Lagriinae: Cossyphini), ventral; E, *Apomecyna* sp. (Cerambycidae), ventral; F, *Cossonus* sp. (Curculionidae: Cossoninae), ventral; G, *Aleochara* sp. (Staphylinidae: Aleocharinae), anterior; H, *Rhinorhipus tamborinensis* Lawrence (Rhinorhipidae), lateral. I, *Haliplus* sp. (Haliplidae), prothorax and part of pterothorax, ventral, showing secondary mesocoxal articulations and concealed mesoventrite. J–L, Prothorax: J, *Spercheus platycephalus* Macleay (Hydrophilidae: Spercheinae), ventral; K, *Paracucujus rostratus* Sen Gupta & Crowson (Boganiidae), ventrolateral; L, *Orphilodes australis* Lawrence & Slipinski (Dermestidae: Orphilinae), ventral, showing secondary procoxal articulations.

incomplete because its middle section has been obliterated to accommodate contiguous procoxae, but its apex remains just beyond the ends of the coxae; this **interrupted** type (Fig. 18F) occurs in Curculionoidea and the apical piece has been called the 'sternellum', although it is highly unlikely to be homologous to that structure in other insect orders. The shape, curvature and apex of the prosternal process are also useful diagnostic features in some groups. The process may be more or less parallel-sided (Figs 18L, 19K), apically narrowed (Figs 13C, 18D, 44I, R, W), gradually widened apically (Figs 18G–H, 44K) or abruptly widened at the apex (Figs 19I, J, 20A–B, E), and it may be more or less level (Figs 20E), evenly curved and elevated posteriorly (Fig. 25C), or abruptly elevated at the apex (Fig. 20A). The apex is usually simple and acute to truncate, but may be more complex (Fig. 20B).

The **procoxae** may be strongly transverse (Figs 18L, 19K, 44H–I, W) to globular (Figs 18G, 19E, I–J, L, 20D–F) and are sometimes conical and projecting (Fig. 20G–H); they are normally capable of limited circular movement, but may be highly motile in soft-bodied, surface-inhabiting beetles, such as Cantharidae, Lycidae, Melyridae and some Staphylinidae. In almost all beetles, the procoxae are contained within countersunk cuticular housings called **procoxal cavities**. In Holometabola, including Strepsiptera, the coxae are attached by membrane to the surface and there are no coxal cavities, but the cavities occur in basal members of all beetle suborders and are secondarily reduced only in taxa considered derived within Archostemata (e.g. Micromalthidae) and Polyphaga (e.g. atractocerine Lymexylidae, many soft-bodied Elateroidea, some malachiine Melyridae and Ripiphoridae). The nature of the procoxal cavities, including their shape, separation, external and internal closure, has been used extensively to define and distinguish beetle taxa at various levels. The shape varies as indicated above for the procoxae, and the cavities may be confluent (Figs 18D, F) or narrowly (Figs 18L, 19E, K) to widely (Figs 18E, G, 19J, L) separated. In some cases, there may be a narrow lateral extension of the cavity where the prosternum and hypomeron are slightly separated at the posterior end of the notosternal suture (Figs 13C, 20K). The cavities may be **externally closed** by the meeting of the postcoxal processes of the hypomeron with the prosternal process (Figs 19I–J, 20E, 44J, L, N) or sometimes by the meeting of the two postcoxal proceses at the midline (Figs 18F, 20F). They may also be **internally closed** by the meeting of the internal lining of the cavity with the inner wall of the hypomeron (Figs 13D, 18D, F, 44N, R). Special modifications affecting the motility of the procoxae occur in various groups. A secondary sterno-coxal articulation consisting of a coxal knob fitting into a fossa in the prosternal process occurs in most Adephaga and in scattered groups of Polyphaga. A similar articulation consisting of a lateral projection on the prosternal process and a coxal fossa occurs in Cerophytidae and the dermestid subfamily Orphilinae (Fig. 20L). Modifications of the protrochanter allowing greater motility of the foreleg are known in cryptocephaline Chrysomelidae (Chamorro-Lacayo and Konstantinov 2004) and a unique type of screw-and-nut system producing a rotary leg movement has been described in the cryptorhynchine weevil genus *Trigonopterus* Fauvel (Van de Kamp *et al.* 2011).

Pterothorax and Elytra

The meso-and metathorax are united and more or less immovable ventrally to form the **pterothorax**, containing the housings for the mid and hind legs and dorsolateral articulations with the elytra and hind wings (Fig. 21D–F). The mesothorax is reduced in comparison with the metathorax, except in highly-derived flightless forms. The dorsal surface of each of the two segments is divided into an anterior **prescutum**, a median **scutellum** and lateral **scuta** (the divided **scutum**). Behind these sclerites in the metathorax is the transverse **postnotum**. The **mesoscutellum** is usually subtriangular and visible from above between the elytral bases, forming a **scutellar shield**. The shield may be gradually (Fig. 16O) or abruptly (Fig. 16P, Q) elevated, and its anterior edge is sometimes notched (Fig. 16P) or crenulate (Fig. 16Q) forming part of the pro-mesothoracic interlocking mechanism. The prescutum in some Chrysomeloidea has a median stridulatory file (Fig. 16O). The **mesothoracic spiracles** are normally concealed in the membrane between the prothorax and mesothorax.

The **elytra** are modified mesothoracic wings, which are characteristically rigid, fitting over the abdomen when at rest with their inner (posterior) edges in contact. In most Polyphaga studied (Schneider 1975, 1978) and in Archostemata (Atkins 1958), the elytra are held out at an angle during flight and beat synchronously with the wings, thus contributing to the aerodynamic lift. In some Scarabaeidae (Scarabaeinae, Cetoniinae), however, the elytra are barely lifted during flight and exhibit very little movement; in Adephaga they are held rigidly out to the side; and in Silphidae they are held vertically in a roof-like position (Schneider 1975). An **elytron** consists of a **disc**, which is the general dorsal surface, a mesal **sutural edge**, which meets the corresponding edge of the other elytron, and usually a defined incurved lateral portion known as the **epipleuron**, which may be separated from the disc by a sharp ridge or carina; the anterolateral portion of the disc is usually somewhat elevated and angulate, forming the **humerus**, which embraces the mesopleuron. The edge-to-edge fitting of the elytra at rest is often assisted by a flange on the edge of one elytron and a groove on the opposed edge; in some flightless species, however, the interlocking device may be more complex. The·elytra also fit jointly into a groove on the metanotum and may interlock with the metapleura and abdominal segments by a variety of tongue-in-groove devices or **friction patches** consisting of fine spicules. There is a small, 2-headed condyle at the base of each elytron, which articulates with the lateral part of the notum by means of axillary sclerites; this articular process carries blood, nerves and tracheae and acts as a point of attachment for muscles that move the elytron. The elytral disc is usually marked by a longitudinal series of punctures, which may be contained in impressed grooves called **striae**; these **strial punctures** correspond to sclerotised pillars connecting the upper and lower surfaces of the elytron. Erwin (1974) suggested the use of **interneur** as a generic term applying to all longitudinal elytral elements, whether they be striae or puncture rows, but the term is not commonly used. The usual number of striae or puncture rows is nine, ten or sometimes eleven, but larger numbers occur in some groups, such *Calosoma* (Carabidae) and *Exangeltus* (Tenebrionidae), and in others the punctation is irregular or rarely absent. The spaces between striae are known as **intervals**, **interstices** or **interstriae**, and these are numbered from the suture outwards; the sutural interval is number 1, the first few intervals correspond to veins of the cubito-anal group, the lateral edge is the costal margin, and the epipleural area appears to be precostal in origin. A short stria or puncture row, known as the **scutellary striole**, sometimes occurs near the scutellar shield and

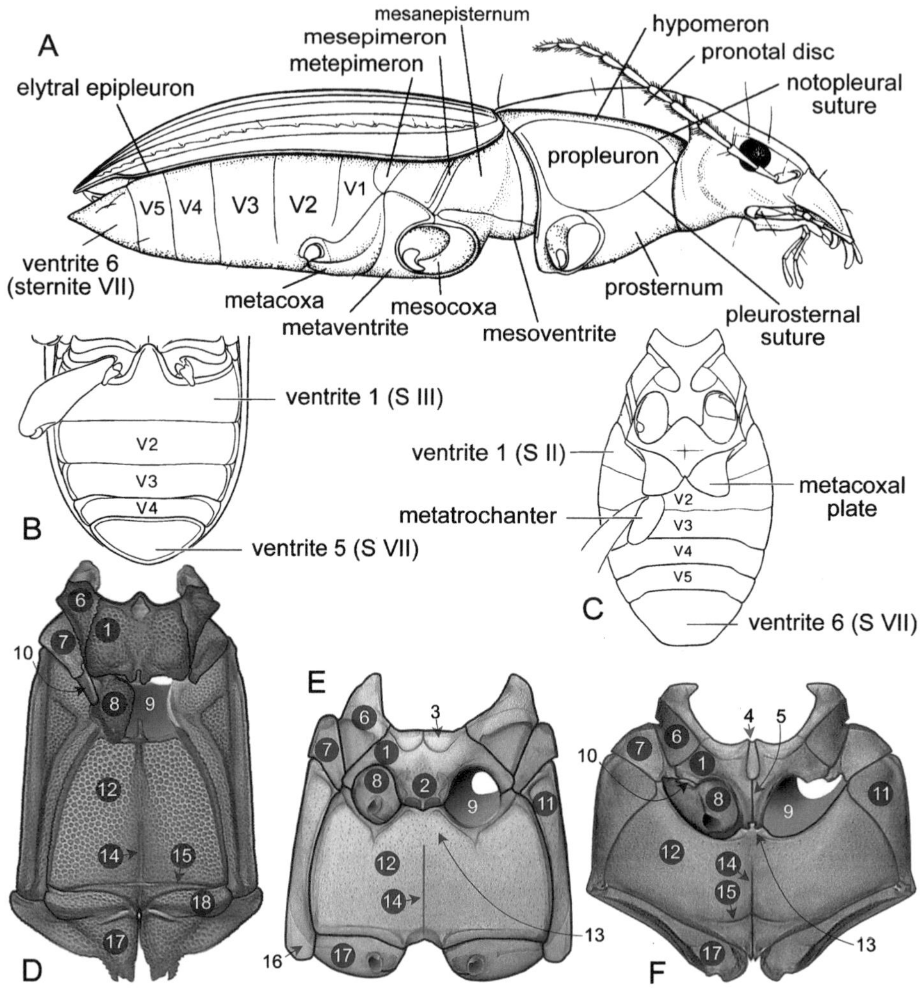

1. Mesoventrite
2. Mesoventral Process
3. Procoxal Rest
4. Mesoventral Cavity
5. Mesothoracic Discrimen
6. Mesanepisternum
7. Mesepimeron
8. Mesocoxa
9. Mesocoxal Cavity
10. Mesotrochantin
11. Metanepisternum
12. Metaventrite
13. Metaventral Process
14. Metathoracic Discrimen
15. Metakatepisternal Suture
16. Metepimeron
17. Metacoxa
18. Metatrochantin

Fig. 21. Adult morphology. A, *Notonomus violaceus* (Laporte) (Carabidae), lateral. B, *Meneristes australis* (Boisduval) (Tenebrionidae), abdomen, ventral. C, *Pamborus guerinii* Gory (Carabidae), pterothorax and abdomen, ventral. D–F, pterothorax, ventral, with major features labelled below: D, *Distocupes varians* (Lea) (Cupedidae); E, *Episcaphula australis* (Boisduval) (Erotylidae); F, *Sclerocyphon* sp. (Psephenidae).

adjacent to the sutural edge. In a few groups, especially among the Cucujoidea (e.g. Cryptophagidae, Priasilphidae, Phloeostichidae, Cavognathidae, Passandridae), the sutural flange on one side is exposed when the elytra are fully closed. The underside of each elytral apex bears a friction patch in those taxa with a medial fleck or binding patch on the hind wing (**see** below), and an apical interlocking tongue occurs in a few groups (e.g. Artematopodidae and some Byrrhidae).

The ventral surface of the pterothorax is composed of the two major sclerites, the **mesoventrite** and **metaventrite**, each flanked by two pleural elements, the **anepisternum** and **epimeron**. Until recently these ventral sclerites were considered to represent the mesosternum and metasternum, although some morphologists had demonstrated more than 50 years ago (Ferris 1940; Campau 1940; Matsuda 1960) that the true sternum in Coleoptera has been invaginated along the midline forming a longitudinal sulcus called the **discrimen** and that portion of the venter on each side of the invagination consists of two parts joined by a transverse suture (**katepisternal suture**): anteriorly a subcoxal element, the **preepisternum**, and posteriorly a pleural element, the **katepisternum**. This interpretation was adopted by Lawrence (1999), Lawrence *et al.* (1999b) and Beutel and Haas (2000), who proposed the neutral terms mesoventrite and metaventrite to replace mesosternum and metasternum (**see also**

Lawrence *et al.* 2010b, 2011). The discrimen and katepisternal suture are most easily seen in the metathorax, where they occur together (Figs 17L, 21D, F, 28V) in most Archostemata, some Myxophaga and at least basal members of most adephagan and polyphagan lineages; the metathoracic discrimen occurs alone (Fig. 21E) in various groups, but has been reduced or lost in many lineages. In the mesothorax, the transverse katepisternal suture occurs only as an internal ridge in a few Archostemata and Adephaga, but the discrimen (Figs 17L, 21F), occurs in most Archostemata, Gyrinidae, most Scirtoidea, Dascillidae, most Buprestidae, several families of Byrrhoidea and the elateroid genus *Artematopus*.

In most beetles, the anterior edge of the mesoventrite is more or less truncate, and is separated from the mesanepisternum on each side by a suture, the **anapleural cleft**, but in some families (e.g. Hydrophilidae, Lymexylidae, Mordellidae, Meloidae, Pyrochroidae, Salpingidae, Anthicidae, Scraptiidae) the mesanepisterna meet at the midline, so that the anterior edge of the ventrite is more or less acute (Figs 17M, 44S–T). The anterior edge of the mesoventrite may have a pair of concave areas, the **procoxal rests** (Figs 17K, 21E, 26K, 28V) into which the procoxae fit, and the mesanepisterna may be similarly modified to house portions of the procoxae, as well as posterior hypomeral projections; these structures are usually present when the procoxal cavities are externally open behind. When the procoxal cavities are externally closed, the coxal rests may be absent or replaced by a mesothoracic **peduncle** (Fig. 44V) on which the entire prothorax is capable of rotating; this condition is particularly common in beetles with digging fore legs, such as scaritine Carabidae (Figs 54E, 66A) or members of the Scarabaeoidea (Figs 60D, J). The mesoventrite may also bear a shallow to very deep **mesoventral cavity** that receives the apex of the prosternal proces. A weak or shallow cavity (Figs 21F, 26B) occurs in several groups, including Cupedinae, most Hydradephaga, most Scirtoidea, Dascillidae, Byrrhidae, Elmidae, Ptilodactylidae, Psephenidae, Callirhipidae, Rhinorhipidae, Brachypsectridae, Nosodendridae, some Dermestidae and Ptinidae, and a variety of cucujoid families, while a larger, deeper cavity (Figs 17K, 28V, 44Z) occurs in Haliplidae, most Buprestidae, Dryopidae, Lutrochidae, Limnichidae, Chelonariidae, Eulichadidae, Cerophytidae, Eucnemidae, Throscidae and most Elateridae. In the families Cerophytidae, Throscidae, Eucnemidae and Elateridae, the mesoventral cavity is part of a complex of pro-mesothoracic structures responsible for the **clicking manoeuvre**, which has been discussed by Evans (1972, 1973). The anterior wall of the mesoventral cavity (Fig. 17K) forms an oblique ramp and the anterior edge of the opening bears a slight elevation or boss, which blocks the movement of the long prosternal process. The prothorax in these groups is enlarged with small, globular procoxae, more or less fused to the pronotal wall, and contains large dorsal longitudinal muscles (M4 of Larsén 1966) that are broadly attached to the dorsal wall of the pronotum and extend to the inflexed edge of the mesonotum and serve as depressors of the notum; the contraction of these muscles causes a build up of pressure on the apex of the prosternal process, which is eventually released and plunges into the mesoventral cavity with some force, causing the body of the beetle to flip into the air and produce the familiar clicking noise.

Mesocoxae are usually more or less globular but may be strongly transverse (Fig. 25A) in some groups, and strongly oblique (Fig. 25I) in others. The coxae are almost always located in distinct cavities, formed anteriorly and mesally by the mesoventrite, posteriorly and mesally by the metaventrite, and laterally by pleural sclerites (usually the mesepimeron alone or in combination with the mesanepisternum) or by the meeting of lateral portions of the mesoventrite and metaventrite. Less commonly (e.g. in Archostemata, Myxophaga, some Scirtidae and Derodontidae) they may be partly or even primarily closed by the metanepisterna. The coxal cavity is said to be **laterally open** when it is partly closed by pleural elements (Figs 17L–M, 21F, 26K, 44S–T, Z) and **laterally closed** when closed by the meso- and metaventrites alone (Figs 17K, 21E, 25L, 44V). Laterally closed mesocoxal cavities occur in many Carabidae, some Hydraenidae, Ptiliidae and Staphylinidae, Passalidae and Trogidae, cardiophorine Elateridae, Jacobsoniidae, Bostrichidae, Ptinidae, Phycosecidae, Erotylidae, Cryptophagidae, Passandridae, Phalacridae, Bothrideridae, Cerylonidae and a few other cucujoid families, Zopheridae, several Tenebrionidae, Prostomidae, Mycteridae, Salpingidae, some Cerambycidae and most Curculionoidea. Mesal separation of the coxal cavities varies greatly from contiguous (Figs 17M, 21D, 44S, V) to very widely separated (Figs 17K, 25B, I, 26E, 44Z). The mesotrochantin is often visible (Fig. 25A–B, D), but may be concealed, especially when the mesocoxal cavities are laterally closed (Fig. 25L). When a mesocoxa is removed, the separation of the mesothorax and metathorax may be visible within the cavity as a suture or sometimes as a strip of membrane. The separation of the meso- and metathorax by a membranous strip is characteristic of all Archostemata and Adephaga, but also occurs in at least some members of the polyphagan families Agyrtidae, Leiodidae, Staphylinidae, Decliniidae, Clambidae, Scirtidae, Dascillidae, Byrrhidae, Artematopodidae, Rhinorhipidae, Derodontidae, Nosodendridae, Acanthocnemidae, Boganiidae, Erotylidae, Hobartiidae, Agapythidae, Phloeostichidae, Cucujidae, Cavognathidae and Laemophloeidae. The **mesendosternites** are at least narrowly separated in almost all Coleoptera, but in Ommatidae they are contiguous at the base.

The mesometaventral junction between the mesocoxal cavities varies considerably, in line with the development of the mesoventral and metaventral processes and separation of the coxal cavities. The two ventrites may be well separated (Figs 44S, V), meet at a point (Fig. 44T) or abut one another (Figs 25I, 26B, 44Z), but in some cases one lies above the other, forming an oblique junction in lateral view (Fig. 25B), or a complex **monocondylic** or **dicondylic** joint may be formed (Fig. 25A, D). A complete fusion of the meso- and metaventrites with loss of the suture occurs in scattered groups, including some Lucanidae, Scarabaeidae, Dryopidae and a few Elateridae.

As indicated above, the metathoracic discrimen is often well developed, but it may be reduced or absent in some groups (e.g. in most Staphylinoidea). The transverse **metakatepisternal suture** occurs widely throughout the order, but sometimes it is short, crossing the discrimen but not extending very far laterally, or located very close to the metacoxae, so that the katepisternum is highly reduced or apparently absent; the suture is absent or rarely present in most Hydradephaga, Hydrophiloidea, Staphylinoidea, Scarabaeoidea, Elateroidea, Tenebrionoidea and Curculionoidea. Lines or ridges called **femoral lines** may occur just behind the mesocoxae (Figs 26B, 28V, 87N) and are often associated with the positioning of the mid femora when at rest. The space between the femoral line, the posterior edge of the mesocoxal cavity and the lateral edge of the metaventrite has been called an axillary space (especially in Nitidulidae). The metanepisterna are usually narrow and elongate, flanking the metaventrite on each side, but the metepimera are largely concealed beneath the elytra, with a small section (Figs 21E, 87N) often exposed posteriorly. The large **metathoracic spiracle** on each side is located above the metepimeron (Fig. 32C) and is thus concealed by the elytron. Metacoxae are usually transverse and less motile than the anterior two pairs, being articulated to the metasternum both laterally and near the midline; their trochantins are well developed and externally visible in most Archostemata (not Micromalthidae), while in all other beetles they are represented only by an internal muscle disc. In Adephaga the metacoxae are more or less immovable, and in the Hydradephaga they are fused to the metaventrite and fused together at the midline. The posterior edges of these coxae may be excavated to receive the femora and distinct plates may be formed that conceal the femoral bases; this condition is common in Adephaga and Elateriformia, and occurs in some other groups as well.

The **metendosternite**, also called the **metafurca**, is a complex fork-like structure to which several sets of muscles are attached; Crowson (1938, 1944, 1955) made extensive use of this feature as a basis for determining phylogenetic relationships among the families of Coleoptera. This structure (Fig. 17M; see also Lawrence *et al.* 2011: fig. 47) usually consists of a moderately long and narrow **stalk**, a pair of **lateral arms** bearing flattened **laminae** and an **anterior process** bearing a pair of **anterior tendons**. In some, the lateral arms may be very short and the anterior process long, the anterior tendons separated by a deep emargination, or the laminae absent and the anterior tendons located at the ends of the arms. A relatively uncommon feature is the presence of a pair of ventrolateral processes, as in some Scirtoidea, some Ptilodactylidae, Chelonariidae, Rhinorhipidae and Brachypsectridae.

Hind Wings

The metathoracic wings of Coleoptera, where functional, are nearly always longer than the elytra and at rest are folded longitudinally and transversely so that they can be concealed beneath the elytra. The wing is rotated forwards on its base into the flight position by the action of direct wing muscles; the same action spreads the wing, opening the longitudinal folds, which leads to the automatic opening of the transverse folds. Structural modifications connected with the folding and unfolding of the wing have made the homologisation of veinal systems more difficult. An attempt to homologise all veins with their counterparts in other Neoptera was made by Kukalová-Peck and Lawrence (1993), utilising the general venational scheme proposed by Kukalová-Peck (1978, 1983, 1991), modified by Kukalová-Peck and Lawrence (2004), and based in part on the work of Haas and Kukalová-Peck (2001). This system of veinal nomenclature, which is used here and illustrated in Figs 22–24, differs in several respects from that proposed by Forbes (1922, 1926) and followed with some minor changes by Crowson (1955), Ponomarenko (1972), Hörnschemeyer (2005) and Fedorenko (2009). The major differences between the two systems concern the identity of RA, RP, MP and the veins immediately following MP. RA_{3+4} in Archostemata and Adephaga has been called the 'radial sector' or 'sector' (Hamilton 1971, 1972; Ward 1979); but this term in Polyphaga has been applied to a cross-vein (**see** below). The major vein, MP_{1+2} has been called the cubitus by most workers, but there has been little agreement on the names of those veins immediately following it. What is here referred to as MP_{3+4} has been called Cu_2 by Ponomarenko (1972) and 'plical' by Hamilton (1972). The true Cu, which immediately follows, was called 'empusal' by Hamilton. Forbes (1922) and Crowson (1955, 1981) refer to all veins behind our MP_{1+2} as anal veins, and Crowson called those veins lying between MP_{1+2} and the anal fold as 'anal veins in the main group'.

The main veins normally visible in beetle wings are: **Costa (C)** and **Subcosta Posterior (ScP)**, which are separated at the base but adjacent or fused for most of their lengths, thus strengthening the anterior margin; **Radius Anterior (RA)**, which is closely associated with ScP and C and divided into two branches beyond the middle of the wing; **Radius Posterior (RP)**, which is deeply concave and tends to become desclerotised proximally, so that its basal connection is often lost; **Media Posterior (MP)**, a long and strong vein which is primitively forked twice; **Cubitus (Cu)**, which is also forked to form the wedge cell; and the **Anal Anterior (AA_{3+4})** and **Anal Posterior (AP_{3+4})**, which are separated by the **anal fold**. The **Precosta (PC)** is rarely visible as a soft, flat strip anterior to the costa near the base. The **Subcosta Anterior (ScA)** is strongly sclerotised, superimposed on ScP and fused with the Costa near the base. The **Media Anterior (MA)** in beetles, as in all endopterygotes, is fused with the radius and has become part of RA. The separation of RA and RP near the base, coupled in a special way with the entry of MA into the base of RA, is an apomorphy of beetles. The base of RP is often reduced, and its connection with RA may be partly concealed by the base of MA, which is transformed into a brace called the **Medial Bridge** (Figs 23F, 24A–C).

As shown by Haas and Kukalová-Peck (2001), the anterior branch of AA is absent in Coleoptera. Also, the **Claval Fold** is very short or absent, the base of the anal system (**Anal Basivenale**) is almost completely obscured by a fold, and the **Jugal** veins are reduced to one or frequently missing.

In order to better understand coleopteran wing venation, one must recognise some basic landmarks and have at least a superficial knowledge of the process of wing folding and unfolding in beetles. There are two reinforced struts in all beetle wings: the **Anterior Strut** or **Radial Bar** consisting of the basal portions of ScP and RA combined and the **Posterior Strut** or **Medial Bar**, the basal portion of MP plus MP_{1+2}. These are joined near the base by the medial bridge mentioned above. For descriptive purposes, the wing membrane may be divided into five regions or **fields** (Fig. 23E): **radial**, between the radial bar and RP but basal to the radial cross-veins; **central**, an area of complex folding involving the radial and radio-medial cross-veins and associated cells; **apical**, the area apical to the cross-veins and cells; **medial**, between the medial bar and the anal fold; and **anal**, posterior to the anal fold. Another landmark in some, but not all, beetle wings is the **pterostigma**, a blood sinus, which appears as a darkened area, located between the two branches of RA.

When discussing wing folding, we will emphasise that taking place in basal members of the four suborders, and discuss folding in the more derived groups of Polyphaga in a following section.

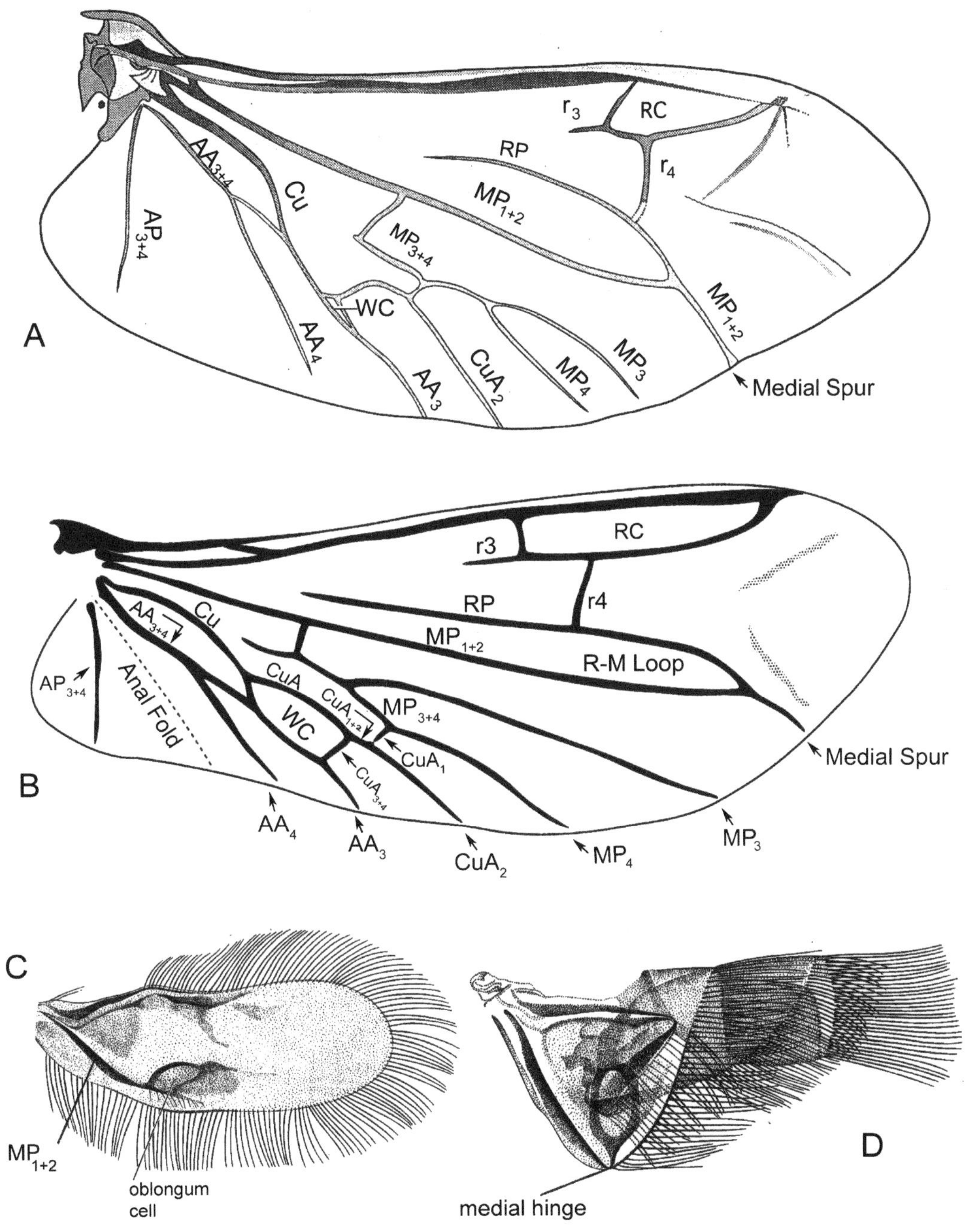

Fig. 22. Adult morphology. A–D, Hind wing: A, *Austrolichas monteithi* Lawrence & Stribling (Ptilodactylidae: Cladotominae); B, *Stichotomus* sp. (Elateridae: Dendrometrinae); C, *Sphaerius ovensensis* (Oke) (Sphaeriusidae); D, same, folded.

Critical to the folding-unfolding process are: (1) the sub-basal joining of the two wing struts by the medial bridge; (2) the nature of the bending zones; and (3) the apical connection of the two wing struts by a series of cross-veins and cells.

As indicated above, the medial bridge connects the two wing struts near their bases; this bridge serves as a spring clamp, which holds the two struts together until acted upon by indirect flight muscles operating through the axillary sclerites and the radial and medial basivenalia, which causes the struts to diverge. The median bridge is spring loaded and automatically returns to the closed position if the indirect flight muscles are relaxed. The opening and closing of the medial bridge also effects the rotation of the two struts, which, in turn changes the position of the bending zones.

The concept of **bending zone** (Biegungzone) was first proposed by Schneider (1978) and utilised by Beutel and Haas (2000) in their study on the relationship of the four beetle suborders. Kukalová-Peck and Lawrence (1993) referred to these bending zones as 'springs', implying that resilin plays a major role in the bending,

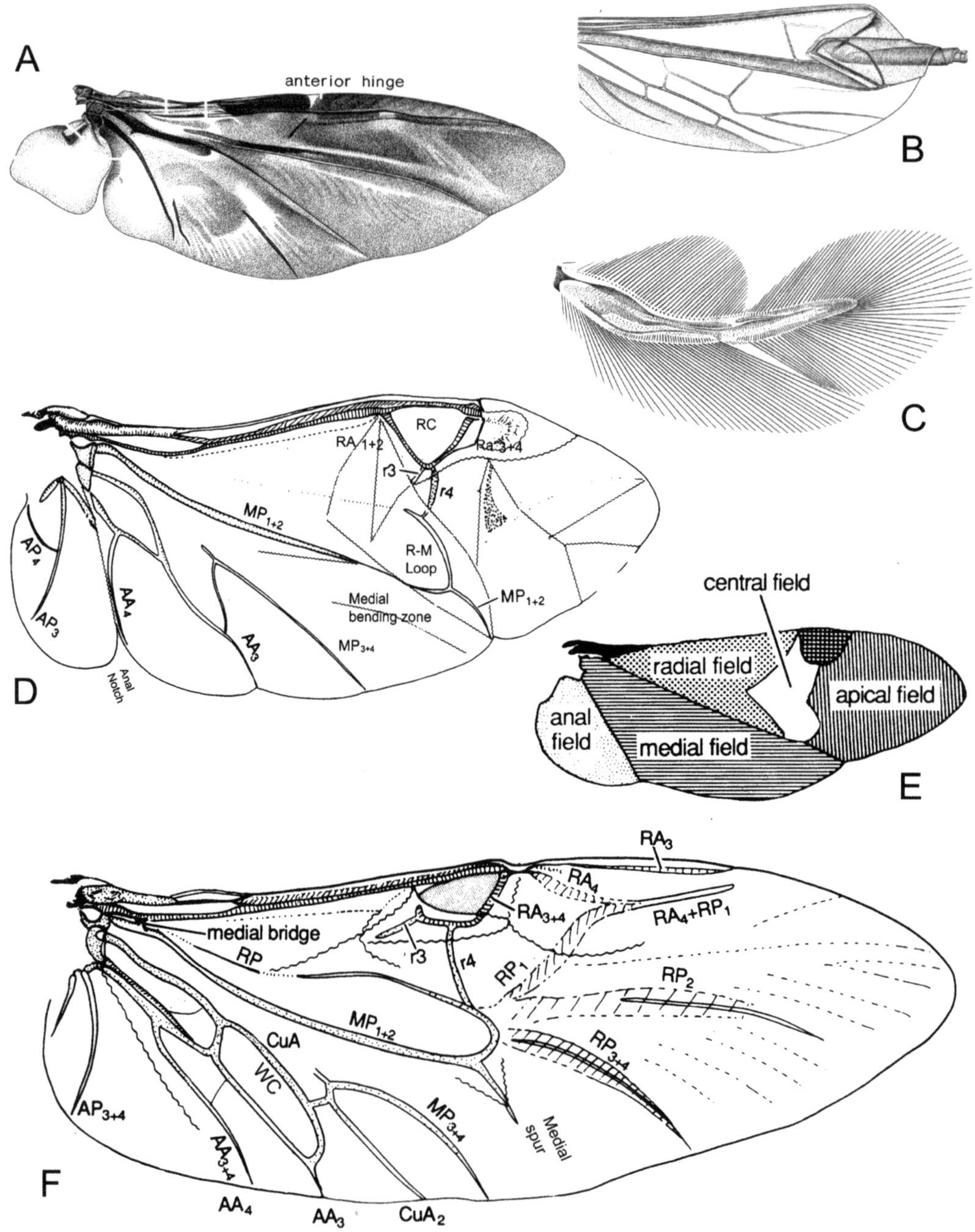

Fig. 23. Adult morphology. A–F, Hind wing: A, *Creophilus erythrocephalus* (Fabricius) (Staphylinidae: Staphylininae); B, *Adinolepis mathesonae* (Neboiss) (Cupedidae), folded; C, Genus ? sp. (Ptiliidae); D, *Macrohelodes* sp. (Scirtidae); E, hypothetical polyphagan wing, showing radial cell and five fields of wing membrane; F, *Pseudohydrobius flavus* Lea (Hydrophilidae: Sphaeridiinae).

which was not confirmed by Haas *et al.* (2000), although resilin was found to be important in movable wing joints and especially in the medial bridge. The function of the bending zones in wing folding was clarified by Kukalová-Peck and Lawrence (2004). These zones, when present, are asymmetrically weakened, either by flattening or by an asymmetrical arrangement of transverse crimping; thus the radial and medial bars are capable of bending towards one another only when each bar is rotated so that the weakened zones face one another; if the two bars are rotated 90° from this position, this bending is impossible and the two bars become straightened. The **radial bending zone** is relatively long and the bend relatively gradual in most Adephaga (Fig. 24A) and in some polyphagan groups, such as basal Staphyliniformia, Scarabaeoidea, Nosodendridae (Fig. 29F) and some Bostrichidae; it is, in contrast, very short and almost hinge-like in Archostemata, Myxophaga and the polyphagan family Clambidae.

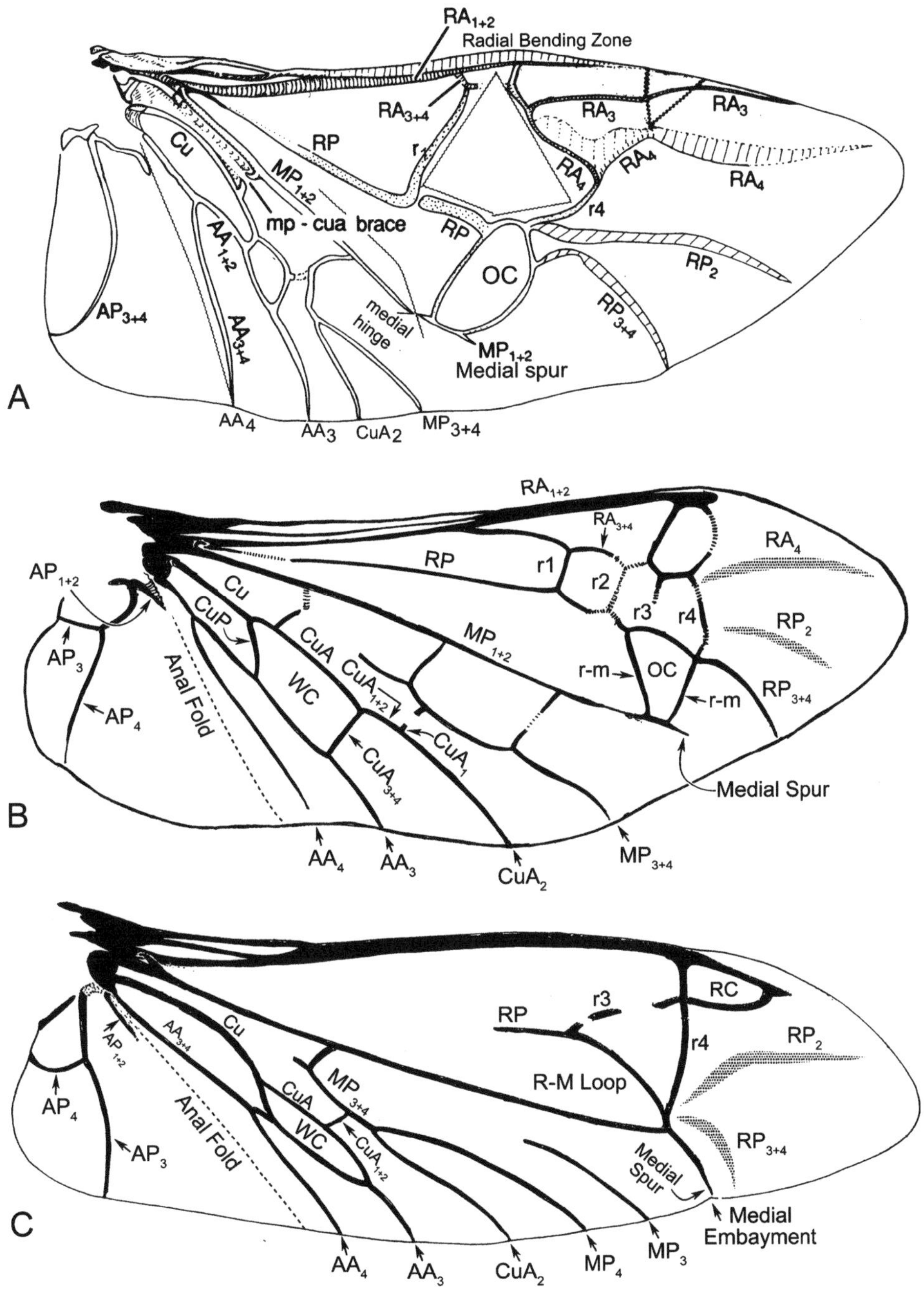

Fig. 24. Adult morphology. A–C, Hind wing: A, *Hygrobia nigra* (Clark) (Hygrobiidae); B, *Omma stanleyi* Newman (Ommatidae); C, *Notodascillus sublineatus* Carter (Dascillidae).

In Hydrophilidae, Sphaeritidae and Synteliidae (but not Histeridae) and in all Scarabaeoidea, the radial bending zone may be accompanied by a distinct hinge occurring beyond the end of the radial bar. We use the term **hinge** here to refer to an narrow zone of weakness or a joint in a reinforced vein or strut (such as the radial and medial bars). When a fold extends to the wing margin, without sclerotizations or reinforcements on either side of it, that is called a **folding point**. In most groups of Polyphaga, including the more derived groups of Staphylinoidea, several smaller bostrichoids and most of Elateriformia and Cucujiformia, the radial bending zone is absent, and all transverse wing folding occurs in the apical field. In Silphidae and Staphylinidae, there is no bending zone, but there is a distinct hinge proximal to the radial cell or the end of the radial bar.

The **medial bending zone**, is relatively long and gradual only in a few basal Polyphaga (Fig. 29F), whereas it is abrupt and hinge-like in Archostemata (Fig. 29E), Myxophaga (Fig. 22D) and Adephaga (Fig. 24A) and in the polyphagan family Clambidae.

This bending zone is entirely absent in the remaining Polyphaga, where all transverse folds, if present, occur beyond the end of MP_{1+2}.

The apical end of the radial bar is usually connected by a series of cross-veins with the apex of the medial bar, and in some cases one or two cells may also occur in this region. This connection is responsible for the fact that when the median bridge is opened and the two struts, now in the non-bending position, diverge from one another, the apical portion of the wing automatically unfolds. The above three processes are entirely responsible for the folding and unfolding processes in the major portion of the hind wing, but the folding of the anal lobe along the anal fold is effected by a separate, simpler mechanism.

In Archostemata (Fig. 24B), the radial vein forks just beyond mid-wing, forming RA_{1+2} and RA_{3+4}, which extend well into the apical field, and a pair of cross-veins may join the two forming what has been called a **brachial cell** (Atkins 1963). RP is more or less complete, although somewhat weakened basally, and usually extends into the apical field as RP_2 and RP_{3+4}.There are as many as four cross-veins between RA_{3+4} and RP, although one or two of them may be lost, and there are two cross-veins between RP and MP_{1+2}, which form the **oblongum cell** (OC). The radial venation in Adephaga and Myxophaga is very similar to the archostematan type in most respects; but a brachial cell is only rarely present and RA_3 and RA_4 extend into the apical field as strong independent veins (Fig. 24A). The basal portion of RA_{3+4} may be partly or entirely obliterated in Myxophaga and there is a distinct pterostigma, not obvious in most Adephaga. In almost all Polyphaga, the radial venation is quite different. RA forks to form RA_{1+2} and RA_{3+4}, which reunite to form an eyelet-like cell, called the **radial cell** (RC), and beyond this RA_4 (and sometimes a short RA_3) extends into the apical field (Fig. 23D, 23F). The pterostigma, when present, is usually located within the radial cell. Cross-veins r1 and r2 are never present, and r3 is almost always incomplete and sometimes longitudinally oriented (and thus considered by some authors to be a true vein called the 'radial sector' or 'radial recurrent'). The basal portion of RP is never complete to the base but may be very long to quite short, and is absent in some groups. The oblongum cell is never present, and a single cross-vein joins RP and MP_{1+2} forming the **radio-medial loop** ('medial recurrent' of Forbes 1922 and 'medio-cubital loop' of Crowson 1955). In some polyphagan groups, such as Scarabaeoidea, the loop may be narrow and acute or absent.

The **apical field**, beyond the radio-medial loop, usually contains as many as four radial extensions. In most Archostemata, Adephaga and Myxophaga, RA_3 and RA_4 extend towards the anterior portion of the wing apex, while RP_2 and RP_{3+4} extend from the oblongum cell into the posterior portion (Figs 24A–B). In Polyphaga, the apical field occasionally receives a short RA3, usually a long RA4, which often meets or fuses with RP_1 (Fig. 23F), and RP_2, which extends posteroapically, and sometimes RP_{3+4}, which may be strongly curved posteriorly (e.g. in Dascillidae, Fig. 24C) or closely associated with an an apical extension of MP_{1+2} (medial spur) (e.g. in some Staphyliniformia and Scarabaeoidea). At least one and usually two radial extensions are present in Clambidae, Eucinetidae, Dascilloidea, Derodontidae, most Bostrichoidea, some groups of Cleroidea, Cucujoidea and Tenebrionoidea, and most Chrysomeloidea and Curculionoidea, but these are absent or replaced by vague sclerotisations or oblique linear sclerites (occasionally forming an epsilon figure) in Scirtidae, most Elateriformia and some bostrichoid and many cucujiform groups. The **post-radial sclerite**, located just beyond the radial cell, occurs in several Cucujiformia. [Saalas 1936; Jolivet 1957, 1959; Zherikhin and Gratshev 1995; Lawrence *et al.* 2011].

As mentioned above, the medial bar or MP_{1+2} in Archostemata, Adephaga and Myxophaga always has a distinct hinge, which is just basal to the beginning of the oblongum cell (Figs 22D, 24A). This hinge is absent in all Polyphaga, except the family Clambidae. A short, straight extension of MP_{1+2} may occur in Archostemata and Myxophaga; this is the **medial spur**, which is better developed and usually posteriorly curved in Polyphaga (Fig. 24C). The wing margin opposite the end of the medial spur may have a distinct **medial embayment** in some groups (Fig. 24C).

The **medial field** in most Archostemata and Adephaga (Figs 24A–B) contains four free veins, which usually extend to the wing margin: MP_{3+4}, CuA_2, AA_3 and AA_4. The origin of MP_{3+4} is usually obscure or absent, but there may be one or two cross-veins between this branch and MP_{1+2} and another between it and CuA. The cubital vein forks to form CuA and CuP and the latter is immediately joined by AA_{3+4}. The Cubitus forks again to form CuA_{1+2} and CuA_{3+4}, the first of which divides again to form CuA_1, which joins MP_{3+4}, and CuA_2, which extends to the wing margin, and the second of which joins AA_3 to complete a cell called the **wedge cell** (WC in Figs 23F, 24C). The wedge cell is often given the name 'second cubito-anal cell', implying that the area basal to CuP and between the cubital and anal systems forms another cell (first cubito-anal cell). AA_{3+4} divides to form AA_3, which meets the wedge cell and AA_4 which usually extends to the wing margin. The anastomosis of the medial, cubital and anal systems has caused some nomenclatural problems. We have chosen to simplify the name of a wing vein terminus based on one of its components, although it represents the fusion of two or more different veins (e.g. MP_{3+4}, rather than $CuA_1 + MP_{3+4}$; AA_3 rather than $CuP + AA_3 + CuA_{3+4}$). The medial field in Adephaga differs from that in Archostemata mainly in being somewhat simpler, with the wedge cell reduced or absent and CuA_1 and CuA_2 often being captured by MP, and MP_{3+4} forked to form MP_3 and MP_4. In the suborder Myxophaga, the medial field is highly reduced, usually containing three or fewer veins and lacking a wedge cell. In basal Polyphaga, the medial field may contain five free veins (MP_3, MP_4, CuA_2, AA_3 and AA_4, Fig. 24C) and rarely contains six (a few Buprestidae and Dermestidae, Fig. 29K); however, reduction in the number of veins is common and in some wings the entire medial field is highly reduced. Because of the anastomosis of medial, cubital and anal veins in this region, often combined with subsequent reductions, veinal homologies may be difficult. The wedge cell may be well developed, reduced or absent, and its apex varies from truncate (Figs 22B, 24B) to acute (Figs 22A, 24C). In some Polyphaga, a binding patch is located between MP_{1+2} and MP_3, while in some adephagan families, there is a similar structure (Fig. 29L) located between MP_3 and MP_4. This type of structure was called a 'subcubital fleck' by Crowson (1955) and 'sub-cubital binding patch' by Hammond (1979), who stated that it may serve to hold that portion of the wing in place during folding by attaching to similar patch beneath the elytra. Considering the proximity to MP as defined here, the term **medial fleck** or **medial binding patch** are more appropriate names. The adephagan binding patch, occurring in Aspidytidae, Noteridae, Amphizoidae, Dytiscidae and Trachypachidae, is not homologous to that in Polyphaga, which is widely distributed within Cucujoidea and Tenebrionoidea and also occurs in some Chrysomeloidea. It is interesting that the medial fleck remains intact in some brachypterous wings, in which almost all venation has disappeared.

The **anal field**, lying behind the anal fold, is relatively simple, usually containing an undivided AP3+4, but rarely with AP_3 and AP_4 (e.g. in most Archostemata and in the polyphagan family Dascillidae) and sometimes with a short stub representing AP_{1+2}. An **anal embayment** may occur at the end of the anal fold, marking the beginning of the **anal lobe** or **ano-jugal lobe**, and this may be deep and notch-like especially in smaller wings, where it has been called an **anal notch** (Fig. 23D).

Although the subject of wing-folding and unfolding has been discussed above, specific folding patterns and their distribution throughout the families of Coleoptera is a topic too complex to be included in this general chapter. Forbes (1926) made the first detailed study of wing folding and defined a series of folds and folding areas based on a large number of beetle groups, and Fedorenko (2009) has recently published a detailed discussion of this topic with numerous illustrations. One of the more important of the folding areas is the **triangular folding area**, as defined by Kukalová-Peck and Lawrence (2004) and Lawrence *et al.* (2011), equivalent to Areas C plus D of Forbes (1926) and Field B of Fedorenko (2009). This area is relatively large and bisects RA3+4 in Archostemata, Myxophaga and Adephaga, and Ponomarenko (1969b, 1973) distinguished the hing wing of fossil Schizophoridae and Catiniidae from those of the remaining Archostemata by the skewing of this folding area so that its anterior angle lies basal to the apical and posterior angles. The wing folding in the apical field is unusual in Archostemata and Myxophaga in that the wing membrane is rolled (Figs 22D, 23B, 29E), rather than folded. In Hydrophiloidea and Scarabaeoidea, the apical wing membrane may contain numerous vein-like sclerotisations ('ghost branches' of Kukalová-Peck and Lawrence 1993) that apprently strengthen the extensive wing apex.

The nature of the axillary region, homologies of the axiillary sclerites and their attachments to the basivenalia have been discussed in some detail by Kukalová-Peck and Lawrence (2004) and will not be repeated here. It must be said, however, that our knowledge of variation in this complex region within the order Coleoptera is still based on relatively few taxa, and there are disagreements concerning homologies of the various sclerites involved.

Wings in Coleoptera are frequently atrophied to a greater or lesser extent, and this is a common phenomenon in island or mountain-top species (Darlington 1943). The wings may be shortened and not folded (**brachypterous**), reduced to mere vestiges (**micropterous**), or completely absent (**apterous**). Wing polymorphism, which has been reviewed by Hammond (1985), Wagner and Liebherr (1992) and Thayer (1992), is common in various families, and is most frequently under genetic control. Some beetles with fully developed wings may be unable to fly because of the degeneration of flight muscles during the life of the insect associated with the development of the gonads (Jackson 1956; Langor 1987). In very small beetles, such as Sphaeriusidae (Figs 22C–D) and Ptiliidae (Fig. 58H) the wings are commonly fringed with long setae; the wing surface in these species is highly reduced and the setae have taken over its function.

Legs

The legs of beetles are normally adapted for walking or running, and they normally increase in size from front to rear; but in some families one or more pairs may be modified for burrowing in soil (Scarabaeidae, many Tenebrionidae), tunneling in wood (Bostrichidae, scolytine and platypodine Curculionidae), swimming (Gyrinidae, Dytiscidae, some Hydrophilidae) or jumping (Eucinetidae, alticine Chrysomelidae). As in most other insects, the beetle leg consists of a coxa, trochanter, femur, tibia, subdivided tarsus and pretarsus; however, the coxae are more conveniently treated above in the sections dealing with the individual thoracic segments.

The trochanter is usually largest in the hind leg, and its junction with the femur is usually slightly oblique in the first two pairs of legs and strongly oblique in the hind pair. The trochanterofemoral junction of the mid leg in some groups (e.g. most Tenebrionoidea, Biphyllidae, Bothrideridae) is strongly oblique, so that there is direct contact between the femur and coxa; this has been called a **heteromeroid trochanter** by Crowson (1955, 1981). Very long and narrow trochanters occur in a few groups (e.g. some pselaphine Staphylinidae, lissomine Elateridae, Latridiidae). The femora are usually similar in shape on all three pairs of legs, but the hind femora are inflated in saltatorial species of Scirtidae and galerucine Chrysomelidae and in other beetles that are not known to jump, such as bruchine and sagrine Chrysomelidae. The tibiae are usually more or less expanded towards the apex, where they often bear **ctenidia** (combs of spines) and usually a pair of enlarged, articulated spines known as **tibial spurs**. The outer edges of the protibiae are often strongly expanded and toothed externally in digging or burrowing forms, such as Scarabaeidae (Figs 30H, 61A–B, E–F), Geotrupidae (Fig. 60K–L) or scaritine Carabidae (Figs 1B, 54E). Tibial **antenna cleaners** may occur on the fore legs of most Carabidae (Fig. 16N), the fore and mid legs of some lamiine Cerambycidae and all Disteniidae (Fig. 16K), and the mid and hind legs of some eumolpine Chrysomelidae (Fig. 16J). The nature of the tibial apex varies considerably in some groups and has played an important in the their classification. In Ciidae, the outer apical angle, especially that of the protibia may bear an outwardly directed tooth or spinose lobe, and curved tibial processes, sometimes referred to as **unci**, occur in families such as Passandridae, Bothrideridae and Zopheridae. The tibial apex is particularly important in the family Curculionidae and its complex variation was discussed in detail by Thompson (1992). In this group an **uncus** often arises at the outer apical angle and projects inwardly, a separate process called a **mucro** may arise from the inner apical angle, and a smooth, shiny area called a **corbel** may surround the tarsal articulation. Tarsi are normally 5-segmented, each of which is referred to as a **tarsomere**, but the number is reduced to four in some groups and three or rarely two in others; in most Tenebrionoidea the hind pair of legs has four tarsomeres while the anterior two pairs have five, and the same condition occurs in males of certain cucujoid families (e.g. Sphindidae, Monotomidae, Hobartiidae, cryptophagine Cryptophagidae, Agapythidae, Priasilphidae, Phloeostichidae, Cucujidae, Myraboliidae, Laemophloeidae, Cyclaxyridae, some Phalacridae). In most Chrysomeloidea and Curculionoidea, and a few members of other groups, the 4th tarsomere is reduced and concealed at the base of the 3rd, which is lobed beneath (Fig. 96H, J); these tarsi are known as the **pseudotetramerous** type. In cassidine Chrysomelidae and some lamiine Cerambycidae, however, this reduced penultimate tarsomere becomes completely fused to the terminal one, so that the tarsi are truly 4-segmented. In a few families, like Coccinellidae and Endomychidae, the tarsi are **pseudotrimerous**, in that they are 4-segmented with a similarly reduced third tarsomere. A few other patterns of reduction in number of tarsomeres occur in various Staphylinoidea (some Leiodidae, many aleocharine and some other Staphylinidae). The basal two to four tarsomeres are often clothed beneath with **tarsal pads** composed of densely packed hairs or specialised adhesive setae. Adhesive setae occur on the fore tarsi of many male Adephaga, where they assist in copulation, and similar male setae occur on the fore and/or mid tarsi in some Staphylinoidea. In other groups (e.g. Dascillidae, Cantharidae, Cleridae, Coccinellidae, Chrysomeloidea and Curculionoidea), the adhesive setae occur in both sexes and are used in climbing (Stork 1980). The pretarsus normally consists of two claws, which may be equal or unequal, connate or reduced to one, and simple, dentate (with basal tooth), bifid or pectinate; sometimes each claw bears a membranous lobe or bristle beneath it. An **empodium** may arise between the tarsal claws (Fig. 27E–F), but this is reduced or absent in many groups. Details of arthropod attachment systems have been discussed by Gorb (2004) and Beutel and Gorb (2006).

Abdomen

The abdomen is usually composed of ten segments, but segment IX is modified forming the **genital segment** and X may be highly reduced or fused to IX. If one excludes the genital segment, eight terga can be counted on the dorsal surface of the abdomen in most adult beetles. Although the terga are less easily traceable in some flightless species, their number is readily determined by counting the spiracles starting with the enlarged one at the base of the abdomen (Figs 30E, 32C). Each tergum has a major plate called a **tergite**, which may be flanked by an additional pair of sclerites called **laterotergites**. The spiracles are usually located in the pleural membrane, and there may be an additional sclerite (**parasternite**) between the spiracle and the inflexed portion of the sternum (**laterosternite**) (Fig. 30E). The first five or six tergites in most beetles are lightly sclerotised, but the last few may be hardened when they are exposed beyond the elytral apices; in these cases, the apparent terminal segment (VII or VIII) is called the **pygidium** and the one preceding it the **propygidium**. More tergites are heavily sclerotised in beetles with very short elytra (e.g. Staphylinidae, Fig. 59F, L–O) but also in some having complete elytra (e.g. some Buprestidae). Patches of spicules occur on at least some abdominal tergites in several taxa across all suborders, but are most common in Polyphaga and more likely to occur in smaller beetles. These patches apparently allow movements of the abdomen to assist in the wing folding process (Hammond 1979).

In almost all beetles, sternites III to VII are sclerotised and visible externally in the intact specimen (Figs 21B–C, 30E); sternite I is absent or rarely represented by a small sclerite concealed by the hind coxae, II is often visible only laterally (often hidden beneath the edges of the elytra, Fig. 32C), and VIII is often retracted along with the pregenital segments (IX and X) (Fig. 32A). Those sternites which are externally visible are called **ventrites**, and their number is often used as a key character. Most commonly five (III to VII) are visible; six occur when II is visible or VIII exposed at the abdominal apex, seven when both conditions occur together, and eight in males of some groups (e.g. *Pheropsophus* in Carabidae, Hydraenidae, Lycidae, Cantharidae) in which sternite IX is also exposed or in larviform females.

Ventrites may be joined by membrane (exposed or concealed by the overlapping of the sternites) and capable of free movement, but in many families the basal two or three ventrites are **connate** (immovably joined together, Figs 25J–K, 30E) and in a few groups, such as Elateridae (Fig. 26C) or Anthribidae, four or even five may be connate. The degree of connation varies considerably and there is a graded series from free and overlapping ventrites to those which are abutting and solidly fused or with a weak or incomplete suture between them. In most members of the family Aderidae, the suture between the first two ventrites is virtually absent, so that there appear to be only four ventrites. The lateral edges of the ventrites in many groups are modified to form interlocking devices with the elytra; these and also the nature of ventrite junctions have been discussed in detail by Kasap and Crowson (1975). Ventrite 1 often bears straight (Figs G–H) or curved (Fig. 25E) **femoral lines** like those on the metaventrite, and in males of some species **sex patches** (groups of secretory setae) occur on the basal one or more ventrites or occasionally elsewhere (Faustini and Halstead 1982; Holloway 1985). The last ventrite is also modified in many species. The free edge may be crenulate to serve as an interlocking device with the elytral apex (as in Cerylonidae, Fig. 44AA) or variously modified in males to accommodate the extrusion of the aedeagus during copulation. In females of cryptocephaline Chrysomelidae, a cavity on the last ventrite (sternite VII) assists in the formation of the faecal egg case. Segment VIII in males is usually unmodified, except for a short basal strut called the **spiculum relictum** (Fig. 28L), which occurs in scattered taxa. In females, however, there is usually a long basal strut, the **spiculum ventrale**, which may be fixed or articulated at the base (Fig. 29D).

The basic number of functional spiracles on the abdomen is eight (Fig. 30E), but in several groups those on segment VIII become reduced or absent, and those on the preceding one or two segments may also be atrophied; in Hydrophilidae and some Staphylinoidea, however, apical spiracles may be functional while one or more intermediate ones have become atrophied. There are never more than seven pairs of functional spiracles in Cucujiformia and Hydrophiloidea, many Staphylinoidea and Scarabaeoidea, most Byrrhoidea, Scirtidae and a few other groups. Although spiracles are usually located in pleural membrane, they may be on the tergites in some Staphylinoidea and Scarabaeoidea, or on the sternites in some members of the latter group. [Matsuda 1976].

Male Genitalia

The male ano-genital complex in its most complete form (Figs 30E, 31A, 32A–B), consists of tergum X, the **proctiger** (or **epiproct**), lying just above the anus, and a tube-like or ring-like segment IX to which the aedeagus is attached. Segment VIII is usually unmodified, but in some males, sternite VIII may be emarginate, excavate or complexly lobed. The combined genital segments are sometimes

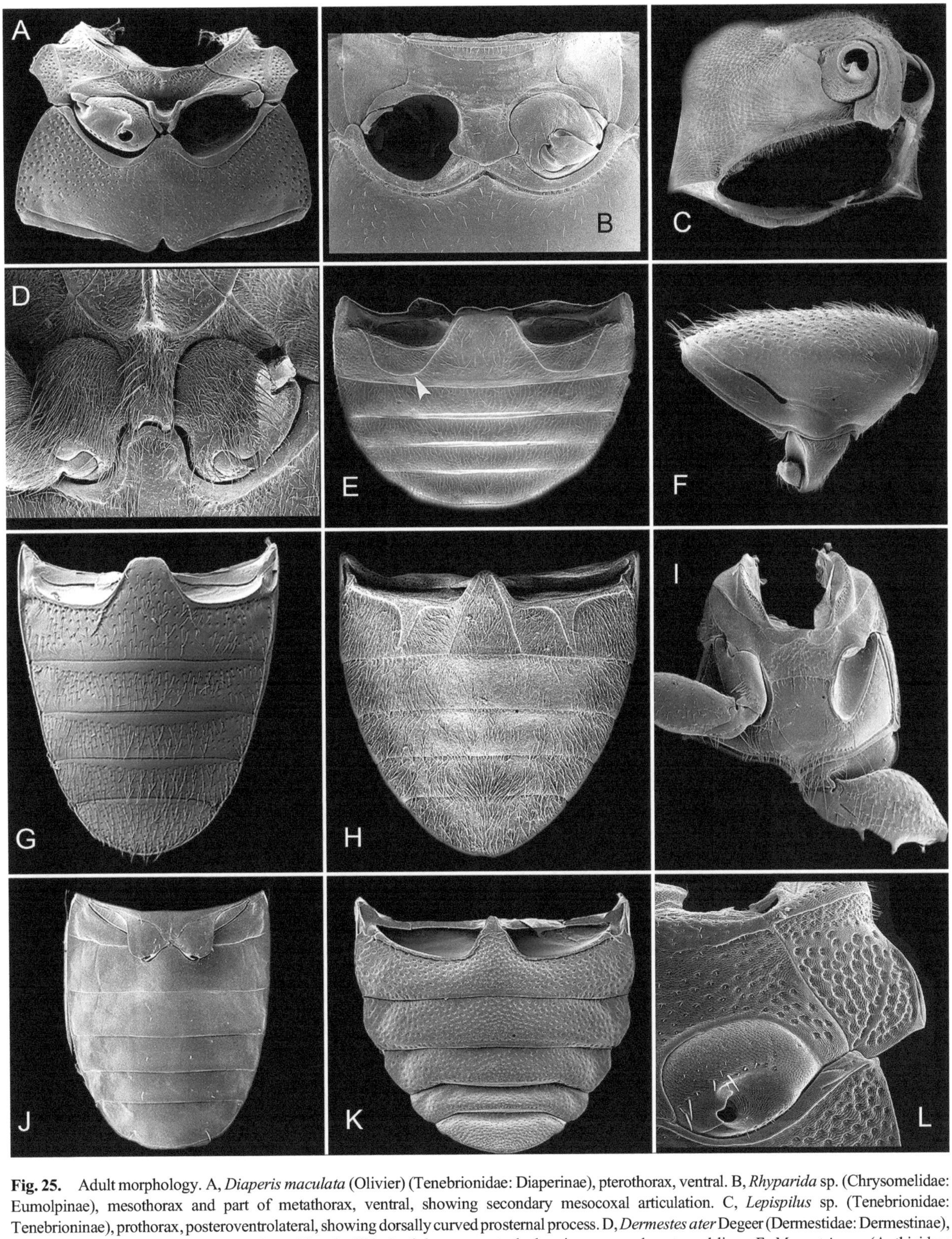

Fig. 25. Adult morphology. A, *Diaperis maculata* (Olivier) (Tenebrionidae: Diaperinae), pterothorax, ventral. B, *Rhyparida* sp. (Chrysomelidae: Eumolpinae), mesothorax and part of metathorax, ventral, showing secondary mesocoxal articulation. C, *Lepispilus* sp. (Tenebrionidae: Tenebrioninae), prothorax, posteroventrolateral, showing dorsally curved prosternal process. D, *Dermestes ater* Degeer (Dermestidae: Dermestinae), part of pterothorax, ventral. E, Genus? sp. (Coccinellidae), abdomen, ventral, showing recurved postcoxal lines. F, *Macratria* sp. (Anthicidae: Macratriinae), prothorax, lateral. G–H, Abdomen, ventral: G, *Cryptophilus* sp. (Erotylidae: Cryptophilinae); H, *Diplocoelus fasciatus* (Macleay) (Biphyllidae). I, *Dietta* sp. (Leiodidae: Leiodinae), pterothorax, ventral, with left mid leg and right hind leg removed. J–K, Abdomen, ventral: J, *Amblytelus* sp. (Carabidae: Moriomorphinae); K, *Eleates depressus* (Randall) (Tenebrionidae: Tenebrioninae). L, *Corticeus* sp. (Tenebrionidae: Diaperinae), portion of pterothorax, ventral, showing lateral closure of mesocoxal cavity by meso- and metaventrites.

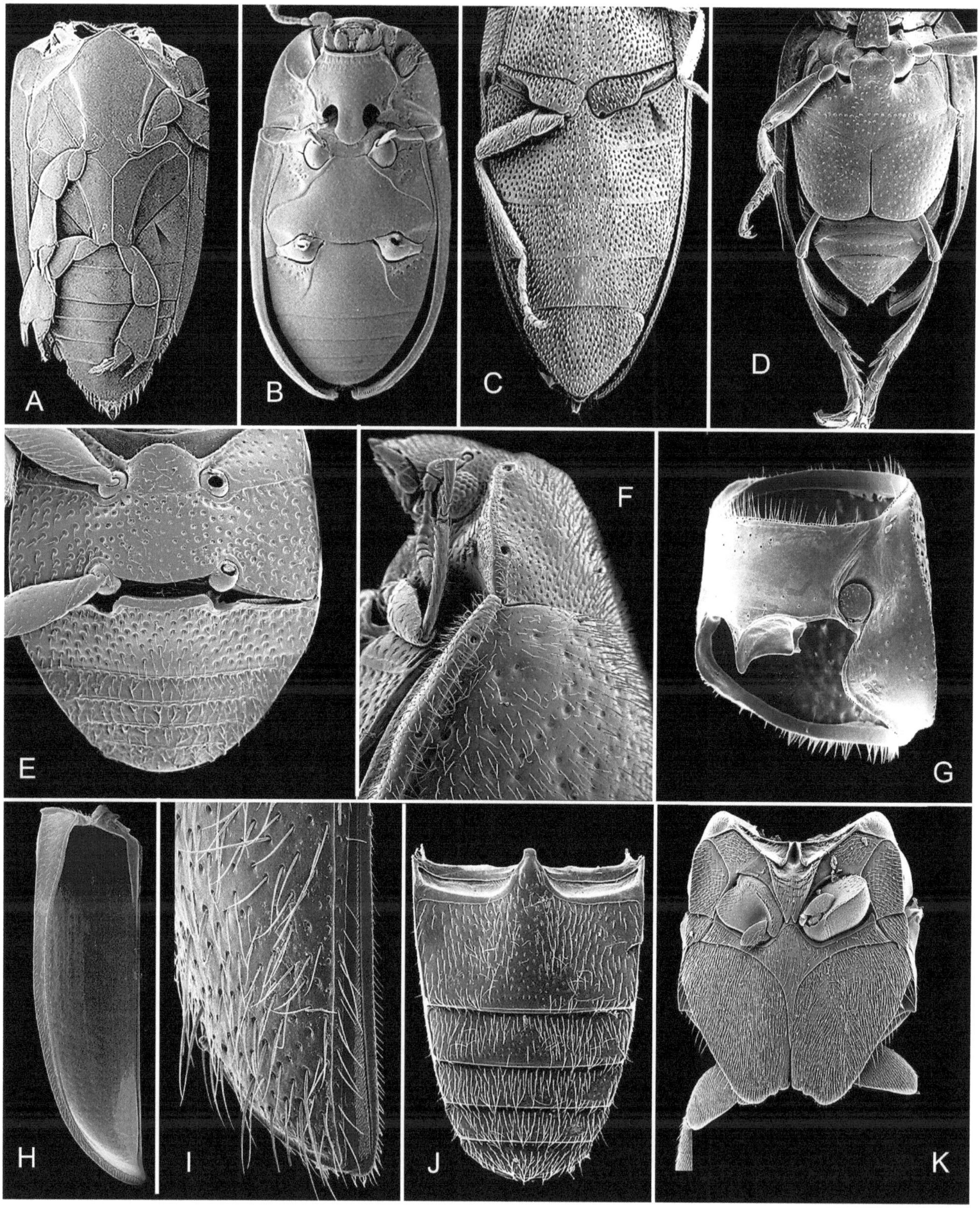

Fig. 26. Adult morphology. A, *Macrogyrus* sp. (Gyrinidae), pterothorax and abdomen, ventral. B, *Mychocerus* sp. (Cerylonidae: Ceryloninae), habitus, ventral. C, *Agrypnus* sp. (Elateridae: Agrypninae), abdomen and part of metathorax, ventral. D, *Haliplus* sp. (Haliplidae), pterothorax and abdomen, ventral. E, *Aphanocephalus* sp. (Discolomatidae), pterothorax and abdomen, ventral. F, Same, anterior portion of body, lateral. G, *Acanthocnemus nigricans* (Hope) (Acanthocnemidae), prothorax, lateroventral, showing prothoracic heat receptor organ. H, *Pharaxonotha kirschi* Reitter (Erotylidae: Pharaxonothinae), right elytron, ventral. I, *Cryptophagus cellaris* (Scopoli) (Cryptophagidae: Cryptophaginae), apex of left elytron, dorsal. J, *Myrabolia* sp. (Myraboliidae), abdomen, ventral. K, *Eucinetus* sp. (Eucinetidae), pterothorax, ventral.

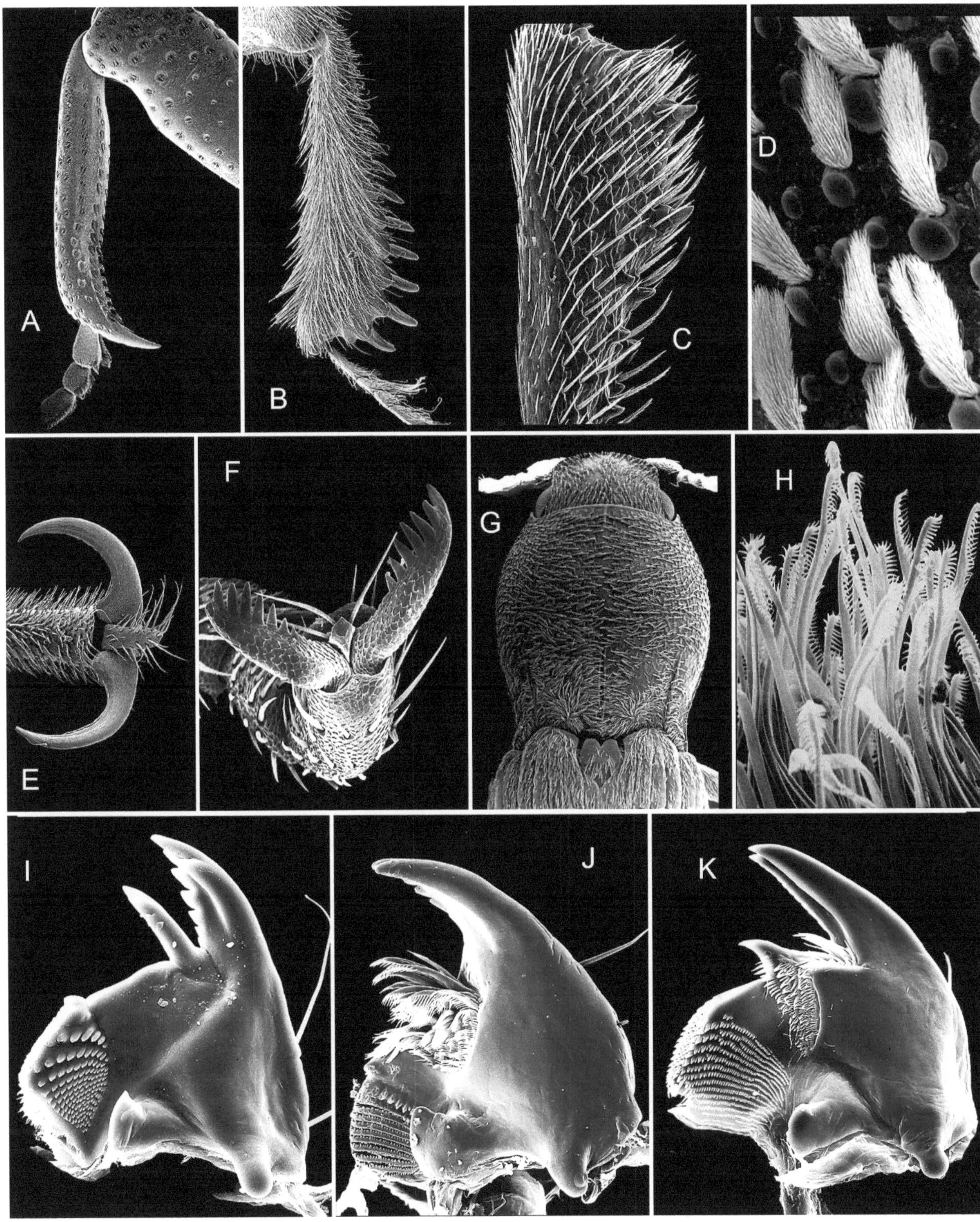

Fig. 27. Adult and larval morphology. A–B, Adult protibia: A, *Cosmopolites sordidus* (Germar) (Curculionidae: Dryophthorinae); B, *Heterocerus* sp. (Heteroceridae). C, *Hylurgops* sp. (Curculionidae: Scolytinae), adult mesotibia. D, *Austrolichas monteithi* Lawrence & Stribling (Ptilodactylidae: Cladotominae), larva, portion of abdominal cuticle. E–F, Adult pretarsus: E, *Callirhipis reticulata* Lea (Callirhipidae); F, *Rhinorhipus tamborinensis* Lawrence (Rhinorhipidae). G, *Cardiophorus* sp. (Elateridae: Cardiophorinae), anterior portion of adult body, dorsal, showing notched scutellar shield and sublateral pronotal carinae. H, *Prionocyphon* sp. (Scirtidae), maxillary apex, showing comb–hairs. I–K, Larval mandibles, ventral: I, *Monotoma americana* Aubé (Monotomidae); J, *Anchorius lineatus* Casey (Biphyllidae); K, *Derodontus* sp. (Derodontidae).

Fig. 28. Adult morphology. A, *Passandra* sp. (Passandridae), head, ventral. B, *Laemophloeus* sp. (Laemophloeidae), head and prothorax, dorsal. C, *Prionochaeta opaca* (Say) (Leiodidae: Cholevinae), head and prothorax, lateral. D, *Cybister* sp. (Dytiscidae), head and thorax, ventral. E, *Chelonarium* sp. (Chelonariidae), head and prothorax, anterior. F, *Orphilodes australis* Lawrence and Ślipiński (Dermestidae: Orphilinae), head, dorsal, showing median ocellus. G–K, Abdomen, ventral:, G, *Sphaerius* sp. (Sphaeriusidae);. H, *Heteromastix* sp. sp. (Cantharidae: Dysmorphocerinae); I, *Clivina* sp. (Carabidae: Scaritinae); J, *Tranes lyterioides* (*Pascoe*) (Curculionidae: Molytinae); K, *Glypholoma pecki* Thayer & Newton (Staphylinidae: Glypholomatinae). L, *Eurhynchus laevior* (Kirby) (Brentidae: Eurhynchinae), male sternite VIII, ventral. M, *Adinolepis* sp. (Cupedidae: Cupedinae), section of elytron, dorsal. N, *Echinocnemus* sp. (Curculionidae: Brachycerinae), mandible, dorsal. O, *Chelonarium* sp. (Chelonariidae), tibia and tarsus. P, *Nothoderodontus* sp. (Derodontidae: Laricobiinae), elytron, dorsal. Q, *Ochthebius* sp. (Hydraenidae), antenna, dorsal. R, *Heteromastix* sp. (Cantharidae), abdominal segments V and VI, lateral. S, *Chelonarium* sp. (Chelonariidae), mandible. T, *Prionochaeta opaca* (Say) (Leiodidae: Cholevinae), head, dorsal. U, *Colon hirtale* (Broun) (Leiodidae: Coloninae), antenna. V, *Byrrhinus* sp. (Limnichidae: Limnichinae), pterothorax, ventral.

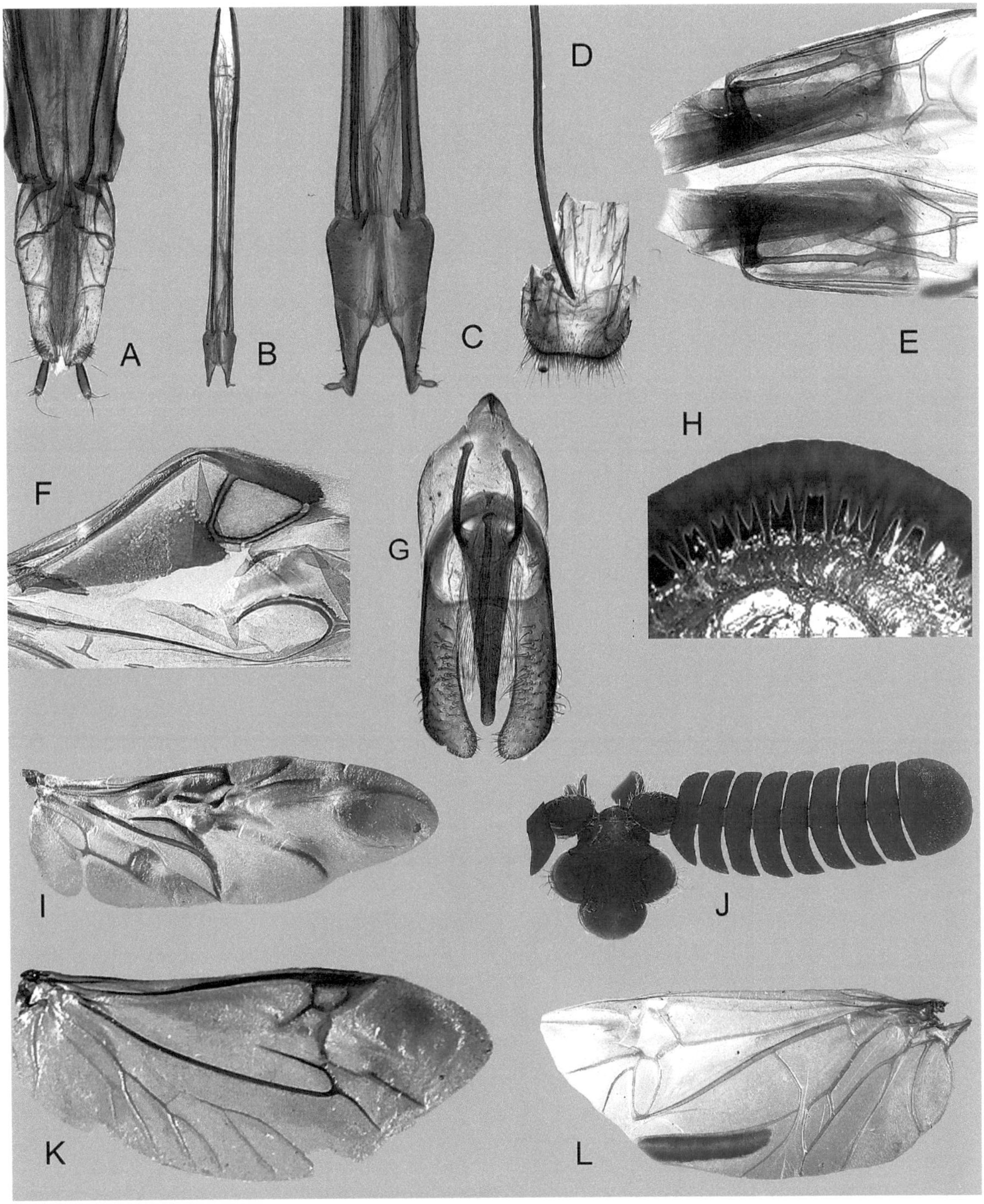

Fig. 29. Adult morphology. A–C, Ovipositor, ventral: A, *Pristoderus antarcticus* White (Zopheridae: Colydiinae); B, *Ampedus sanguineus* (Linnaeus) (Elateridae: Elaterinae); C, same, posterior portion only. D, *Pristoderus antarcticus* White (Zopheridae: Colydiinae), female segment VIII, ventral. E, *Priacma serrata* (LeConte) (Cupedidae), folded wing apices. F, *Nosodendron californicum* Horn (Nosodendridae), partly folded wing, showing two bending zones. G, *Anthrenus scrophulariae* (Linnaeus) (Dermestidae: Megatominae), aedeagus, ventral. H, *Rhinorhipus tamborinensis* Lawrence (Rhinorhipidae), section of compound eye. I, *Harmonia* sp. (Coccinellidae), hind wing. J, *Arthropterus* sp. (Carabidae), head and right antenna, dorsal. K–L, Hind wings: K, *Attagenus pellio* (Linnaeus) (Dermestidae: Attageninae); L, *Hyderodes* sp. (Dytiscidae: Dytiscinae).

referred to as the **genital capsule** when they form a tube surrounding the aedeagus, but that term has also been applied to a capsule formed by tergite VIII in some Cucujoidea (e.g. Nitidulidae). In most males, tergite IX is either divided at the midline (Fig. 31A) or separated into two apparently disconnected pieces called **paraprocts**. Further modifications in the male include the fusion of

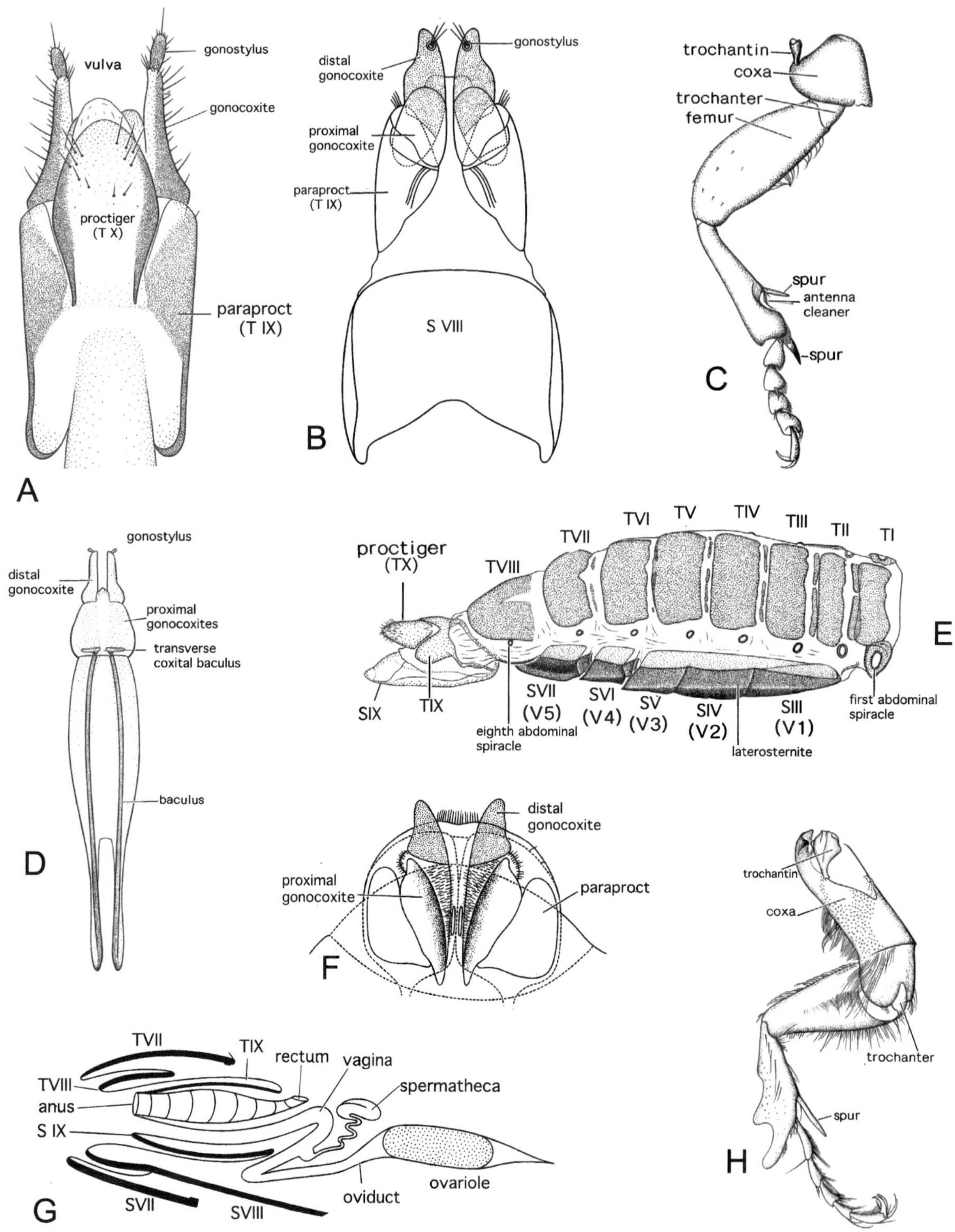

Fig. 30. Adult morphology. A–B, Ovipositor: A, *Rhinorhipus tamborinensis* Lawrence (Rhinorhipidae), dorsal; B, *Ptomaphila lacrymosa* (Schreibers) (Silphidae), ventral. C, *Hypharpax* sp. (Carabidae: Harpalinae), fore leg. D, *Byrrocryptus variegatus* (Carter) (Ptilodactylidae: Anchytarsinae), ovipositor, ventral. E, *Rhinorhipus tamborinensis* Lawrence (Rhinorhipidae), male abdomen, dorsolateral. F, *Calosoma schayeri* Erichson (Carabidae: Carabinae), ovipositor, ventral. G, *Atomaria* sp. (Cryptophagidae: Atomariinae), diagrammatic sagittal section of female abdominal apex. H, *Colpochila* sp. (Scarabaeidae: Melolonthinae), fore leg.

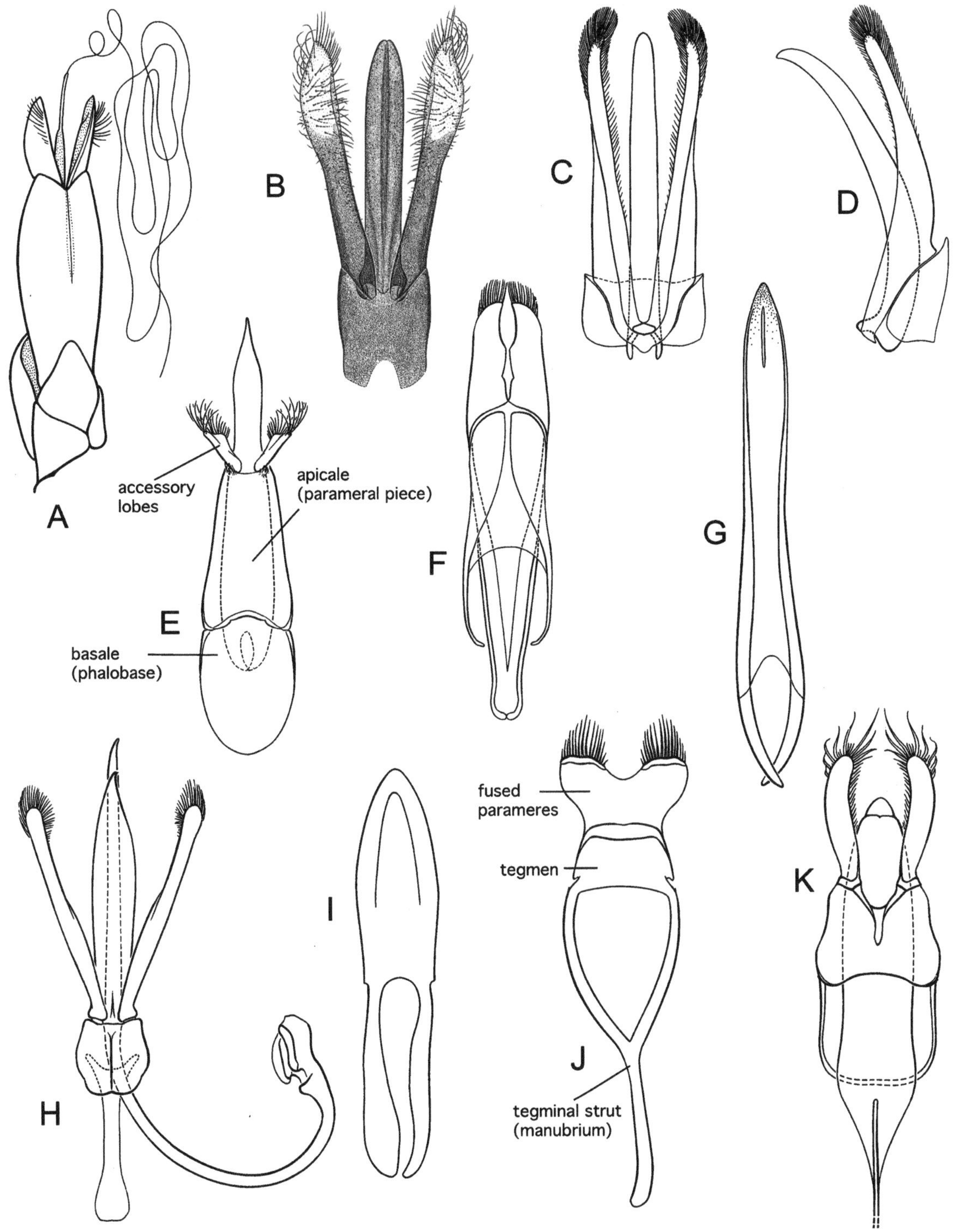

Fig. 31. Adult morphology. A, *Lamprima aurata* Latreille (Lucanidae), male abdominal apex, dorsolateral, with aedeagus extended and endophallus everted. B–E: aedeagus: B, *Rhinorhipus tamborinensis* Lawrence (Rhinorhipidae), ventral; C, *Dermestes maculatus* Degeer (Dermestidae), dorsal; D, same, lateral; E, *Morpholycus costipennis* Lea (Pyrochroidae: Pilipalpinae), dorsal; F, *Leperina decorata* (Erichson) (Trogossitidae: Trogossitinae), tegmen, dorsal. G, Same, penis, dorsal. H, *Epilachna guttatopustulata* (Fabricius) (Coccinellidae: Coccinellinae), aedeagus, dorsal. I, *Eurhynchus laevior* (Kirby) (Brentidae: Eurhynchinae), penis, dorsal. J, same, tegmen, dorsal. K, *Passandra* sp. (Passandridae), aedeagus, dorsal, with apex of penile strut removed.

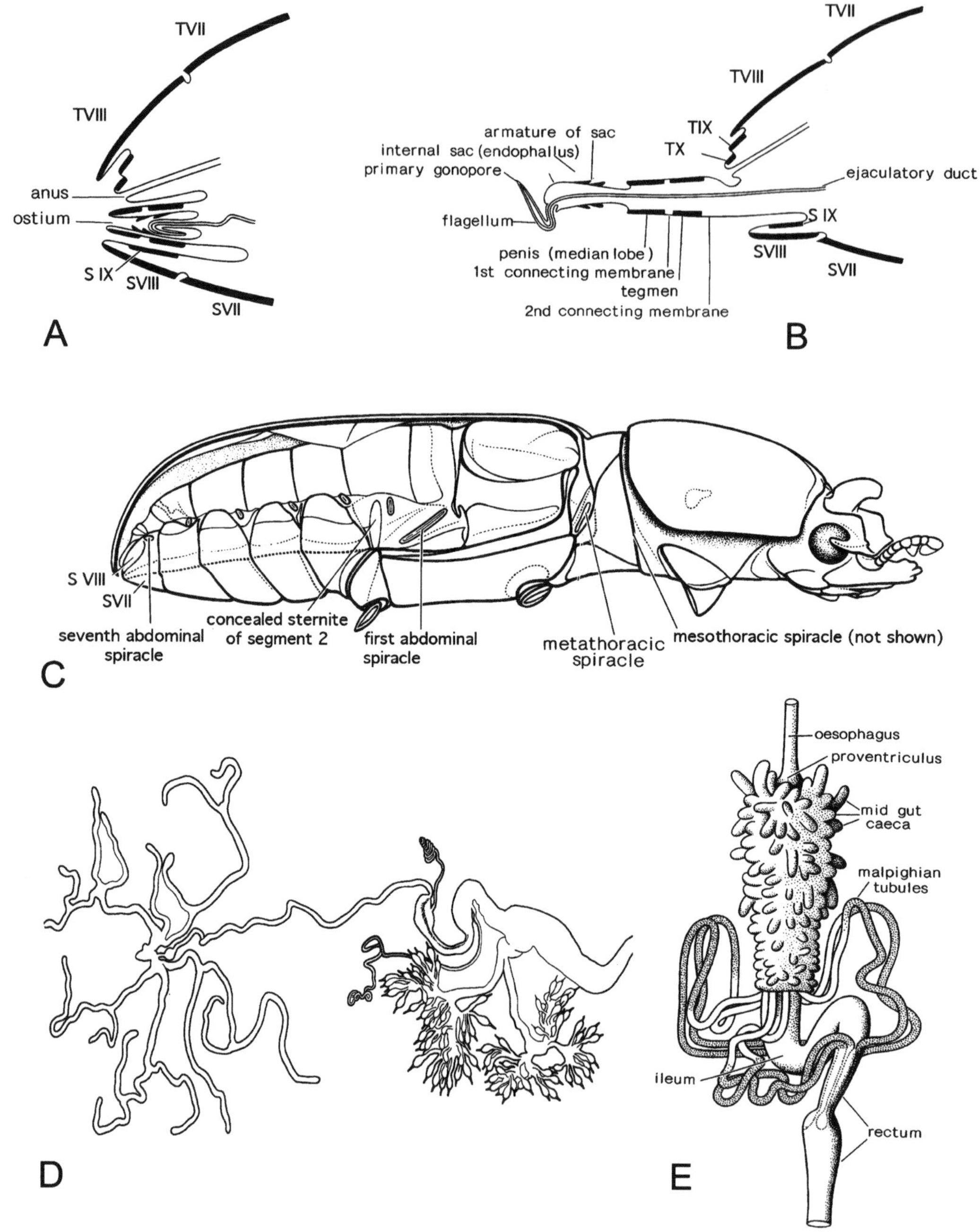

Fig. 32. Adult morphology. A–B, Diagrammatic sagittal section of male abdominal apex: A, with aedeagus retracted; B, with aedeagus exposed. C, *Aulacocyclus collaris* Blackburn (Passalidae), body, lateral with legs and one elytron removed. D, *Agrypnus caliginosus* (Guérin-Méneville) (Elateridae: Agrypninae), female reproductive system. E, *Atomaria* sp. (Cryptophagidae: Atomariinae), alimentary canal.

tergites IX and X forming a simple **genital ring**, and the anterior extension of segment IX forming a ventral strut, the **spiculum gastrale**; in some groups (e.g. Chrysomeloidea and Curculionoidea) the tergal portion of the genital ring becomes desclerotised so that only the ventral strut remains. Crowson (1981) disagreed with the interpretation given here; he considered the proctiger to represent tergite IX and the lateral sclerites to be laterotergites of segment IX which are sometimes fused in front of the proctiger (see Hieke 1966, 1971).

The male copulatory organ or **aedeagus** is a development of the posterior end of the ejaculatory duct, which opens behind sternite IX (Fig. 32A–B) and consists of four parts: the **phallobase (basal piece)**, the paired **parameres (lateral lobes)**, the **penis (median lobe)**, and the **endophallus (internal sac)**, which is primarily membranous but may be armed with a variety of spicules and sometimes a long, sclerotised **flagellum**, at the end of which is the **primary gonopore**. The penis is attached to the phallobase by means of the **first connecting membrane**, and the phallobase is attached to the ninth segment by means of the **second connecting membrane**. During copulation, the internal sac is extruded through an opening, the **ostium**, at or near the end of the penis, and enters the female vagina with the flagellum extending anteriorly to reach the spermatheca (**see** below). Although the term aedeagus is used here in the broad sense, following Crowson (1981), Snodgrass (1963) restricted the term to what is here called the penis and used the term **phallus** for the entire copulatory organ. Some authors (Michener 1944; Wood 1952) have homologised the phallobase with the fused gonocoxites and the parameres with the gonostyli of the female ovipositor.

The **trilobate** aedeagus, generally considered to be the most primitive type in beetles, consists of a ventrally sclerotised phallobase to which a pair of freely articulated parameres are attached, and a simple penis lying above it (Fig. 31A–B). From this basic type, it is possible to derive the variety of aedeagal types present within the various coleopteran lineages. A common modification has been the reduction or loss of the phallobase, so that the parameres appear to articulate directly with the penis; this condition occurs in most Archostemata, Myxophaga and Adephaga, and in many Staphylinoidea. Another common modification is the fusion of the two parameres to form a single tubular piece enveloping the penis, as in most Histeridae and Buprestidae, or a single, flattened piece appressed to the penis, as in some Staphylinidae. In most members of the Bostrichoidea, the phallobase is retained but more or less overlaps the bases of the parameres, which are more or less joined together basally or sub-basally and also fused to the base of the penis (Figs 29G, 31C–D). The greatest diversity of genitalic types occurs within the Cucujiformia, and Crowson (1955) proposed a series of possible evolutionary changes by means of which such an array of structures could have been derived. The phallobase forms a tubular structure enveloping the penis, and the parameres are usually reduced and often fused to the phallobase; this is the **cucujiform** or **sheath type** of aedeagus, and the phallobase in this type and in those apparently derived from it is usually referred to as the **tegmen**. The attached parameres, when present, are considered by some to be secondary developments not homologous to the parameres of the trilobate aedeagus, and the term tegmen is often applied to the combined phallobase and parameres, especially when the latter are non-articulated. The anterior edge of the tegmen or phallobase often bears an anterior strut or apodeme, which may be called a **tegminal strut** or **manubrium** (Fig. 31J). There may also be a pair of anterior tegminal struts resulting in the **double tegmen** (Fig. 31F) of Crowson (1964b). This condition is characteristic of Cleroidea but also occurs in the cucujoid families Biphyllidae and Byturidae.

A more common type, occurring throughout the Cucujoidea, is the **ring type** or **cucujoid** aedeagus, in which the phallobase forms a ring around the penis, often with the parameres articulating posteriorly with the dorsal portion (Fig. 31K) and/or a median **tegminal strut** extending anteriorly from the ventral portion (Fig. 31J). In Coccinellidae (Fig. 31H), there is also a median tube-like process extending posteriorly between the parameres and serving as a **penis guide**; coccinellid workers refer to this guide as a **basal lobe**, the entire tegmen as the phallobase, the median tegminal strut as the **trabes**, and the penis as the **sipho** (Gordon 1985). Further modifications of the cucujoid aedeagus include the reduction of the ring segment, fusion or disappearance of the parameres, as in many Chrysomeloidea and Curculionoidea, and finally in these same groups the desclerotisation of the dorsal region, so that only a posteriorly-forked strut remains beneath the penis. The penis in this type of aedeagus may be divided into a ventral **pedon** making up the body and the **tectum**, which forms a dorsal lid over the ostium (Fig. 31I) (Holloway 1982). In addition, the anterior edge of the penis may bear one or two **penile struts** (Fig. 31G, I, K).

In the **tenebrionoid** or **heteromeroid** type of aedeagus (Fig. 31E) the ventral portion of the tegmen is desclerotised, so that the sclerotised portion lies entirely above the penis; this results in something like an inverted trilobate type, although in most cases the parameres are fused into a single piece, sometimes called the **parameral piece** or **apicale**, in which case the phallobase is sometimes called the **basale**. Further inversion by rotation about the longitudinal axis may also occur, so that the tegmen comes to lie below the penis in some tenebrionoid groups, such as Mordellidae, Ripiphoridae, Ciidae, Zopheridae, some Tenebrionidae, Prostomidae, Oedemeridae, Mycteridae and Aderidae. In Pythidae and Salpingidae, the apicale bears a pair of **accessory lobes** (Fig. 31E) attached to and usually articulated with the tegmen but not considered to be homologous to parameres. [Crowson 1984a; Iablokoff-Khnzorian 1980]

Female Genitalia

The female ovipositor consists of a pair of appendages, the **gonocoxites** or **coxites**, which are attached ventrally or distally to the **paraprocts** or **laterotergites IX**, which, in turn flank a median **proctiger** or **epiproct** (tergite X) lying immediately above the **anal opening** (Fig. 30A–B, F). Each gonocoxite may be divided into two parts, usually referred to as **proximal** and **distal gonocoxite** and the latter may have an articulated **gonostylus** or **stylus** at its apex. In a few groups (e.g. some Tenebrionidae), additional coxital divisions may be present. Various other terms, such as **valvifer** and **hemisternite** have been proposed for the basal element of the gonocoxite, but these have not been used consistently by different authors.

In very long ovipositors, which usually occur in beetles that lay their eggs deep into a substrate, the entire ninth segment is membranous and retractable, and both proctiger and paraprocts may be strengthened by a long pair of sclerotised rods or **baculi** (singular **baculum**) (Figs 29A–C, 30D). In many cases the proximal gonocoxite or the basal portion of the single gonocoxite may bear a pair of transverse or oblique baculi, which articulate with the distal ends of the longitudinal paraproctal baculi. Based on this articulation, it is possible that the proximal gonocoxite is homologous to the **gonangulum** of Scudder (1961). Short ovipositors with heavily sclerotised coxites and reduced styli (Fig. 30B, F) are often used to dig in soil, while longer, sclerotised ovipositors may serve to cut egg cavities into plant tissue. Highly reduced ovipositors are usually found in those species which lay their eggs on an exposed surface. The classical comparative work on the beetle ovipositor is that of Tanner (1927), while several recent works (Bils 1976; Burmeister 1976, 1980) have concentrated on the suborder Adephaga. In addition, Mickoleit (1973) compared the structure of the beetle ovipositor with that of various neuropteroid orders. The internal female tract, although ectodermal in origin, is discussed below under Internal Anatomy. [Lindroth 1957; Iablokoff-Khnzorian 1974].

Musculature

Beetle musculature has been used far less than cuticular structure in descriptions or phylogenetic studies mainly because of the difficulties involved in properly preserving muscle tissue and determining muscle origins and insertions, especially for slender muscle strands often partly concealed by larger muscle masses. Nevertheless, several comparative myological works have been published, primarily in Germany, first using traditional techniques and later employing improved sectioning technology, scanning electron microscopy and, most recently micro-computer tomography (Friedrich and Beutel 2008a). Larsén (1966) proposed a system of homologizing thoracic muscles, as did Baehr (1975) and Matsuda (1976); these have been revised and updated for all Coleoptera by Beutel and Haas (2000), with modificaations by Ge *et al.* (2007) and for a wider selection of insect orders by Friedrich and Beutel (2008b). A partial list of works dealing with beetle musculature is given below.

Larval Head: Das 1937, Dorsey 1943, Beutel 1986b (Haliplidae), 1993 (Adephaga), 1995 (Elateriformia), Beutel and Molenda 1997 (Staphylinoidea), Beutel and Haas 1998 (Hydroscaphidae), Beutel and Pollock 2000 (Phycosecidae), Beutel and Slipinski 2001 (Sphindidae and Protocucujidae), Beutel and Hörnschemeyer 2002 (Micromalthidae).

Larval Thorax and Abdomen: Craighead 1916 (Cerambycidae).

Adult Head: Beutel 1986a (Hygrobiidae), Beutel 1989a (Spanglerogyrinae), Belkaceme 1991 (Noterinae), Beutel 1994 (Hydrophilidae), Beutel *et al.* 2001 (Spercheinae), Hörnschemeyer *et al.* 2002 (Priacminae), Anton and Beutel 2004 (Helophorinae), Anton and Beutel 2006 (Lepiceridae), Hörnschemeyer *et al.* 2006 (Cupedinae), *Beutel et al.* 2007 (Tetraphalerinae), Dressler and Beutel 2010 (Adephaga), Anton and Beutel 2012 (Dascilloidea and Scarabaeoidea).

Adult Thorax: Larsén 1966 (Gyrinidae and Coleoptera in general), Baehr 1975 (Priacminae), Baehr 1979 (Coleoptera in general), Belkaceme 1986 (Haliplidae), Beutel 1986a (Hygrobiidae), Beutel 1989b (Spanglerogyrinae), Belkaceme 1991 (Noterinae), Beutel 1999 (Gyrinidae), Beutel and Haas 2000 (Coleoptera in general), Beutel and Komarek 2004 (Hydrophiloidea), Friedrich and Beutel 2006 (Scirtoidea), Ge *et al.* 2007 (Derodontidae and Nosodendridae), Friedrich *et al.* 2009 (Archostemata).

Adult Abdomen: Kasap and Crowson 1975 (Elateriformia); Blum 1979 (mainly Silphidae and Staphylinidae).

Male Abdominal Apex: Hieke 1966 (Carabidae, Geotrupidae, Curculionidae), Hünefeld *et al.* 2011 (Tetraphalerinae).

Female Abdominal Apex: Burmeister 1976 (Hydradephaga), Bils 1976 (Geadephaga), Burmeister 1980 (Adephaga), Burmeister 1990a (Spanglerogyrinae), Burmeister 1990b (Amphizoidae).

Internal Anatomy

The space enclosed dorsally by the clypeus and labrum–epipharynx, ventrally by the labium–hypopharynx, and laterally by the maxillae and mandibles is called the **preoral cavity**, since it lies outside the true opening of the alimentary canal. The proximal portion of this cavity, the **cibarium**, and the anterior end of the fore gut, the **pharynx**, are capable of being distended by a series of dilator muscles attached to the walls of the clypeus and frons, respectively, and one or both may form a pump in liquid-feeding species. Salivary glands of various types have been reported opening near the bases of the mandibles and maxillae (Srivastava 1959), and paired silk-producing glands occur at the base of the labium in larvae of the South African tenebrionid *Parastizopus armaticeps* (Schulze 1975). The pharynx is joined to the **oesophagus**, which may extend to the mid gut without modification or form a distended storage organ, the **crop**. The end of the fore gut is usually blocked by a valve, the **proventriculus**, which controls movement of food into the mid gut, but may also contribute to food processing. Forsythe (1982) has shown that there is a correlation between proventricular structure and the nature of the food mass (e.g. animal or plant, solid or liquid) in Carabidae.

The mid gut varies considerably in length and complexity. In many beetles the entire mid gut may be covered with small papillae known as **regenerative crypts** or with larger diverticula called **gastric caeca** (Fig. 32E). A second valve, the **pyloric valve**, indicates the beginning of the hind gut, which consists of an **ileum**, **rectum** and **anus**. The Malpighian tubules, which open into the hind gut near the pyloric valve, are usually four or six in number. The ends of the tubules may be free, but in Cucujiformia and most Bostrichoidea they are attached to the wall of the rectum, an arrangement known as **cryptonephric** or **cryptonephridial** (Saini 1964). In most cryptonephridial systems the Malpighian tubules are attached independently to the rectum, but in Bostrichoidea they are attached in a single bundle. **Pygidial glands**, which are usually paired, occur in most Adephaga, some Staphylinidae, many Tenebrionidae and members of various other groups; these glands are usually defensive in nature, producing a variety of organic acids, aldehydes, ketones, quinones and other substances. Their reservoirs usually open just behind segments VII, VIII or IX. In the carabid subfamilies Paussinae and Brachininae, a pair of complex, 2-chambered glands eject a pungent, corrosive fluid, which is expelled explosively (p. 167). In Gyrinidae (p. 154) and some stenine Staphylinidae (p. 190), the pygidial glands secrete surfactants that are used to propel the beetle on the surface of water.

The basic number of ganglia comprising the central nervous system is three thoracic and eight abdominal with double connectives between them, but in most beetles the number is reduced due to the fusion of the apical abdominal ganglia and the first abdominal with the metathoracic; the number of abdominal ganglia may be as few as one in Histeridae, Scarabaeidae and Curculionidae.

In the male reproductive system, each vas deferens may be enlarged to form a **seminal vesicle** and may be associated with a **spermatophoral gland** and **accessory gland**. The presence of these glands and their form varies considerably within the order. The testes consist of a single coiled tube in Adephaga, but in Polyphaga and Archostemata there are several separate follicles, which may be sessile or pedicellate (joined to the vas deferens by individual vasa efferentia).

The female reproductive tract begins at the **vulva** and continues as the **vagina** (Fig. 30G), which may be enlarged and modified to form a **bursa copulatrix**; the common oviduct, one or more **spermathecae** and usually an **accessory gland** or **spermathecal gland** open into the bursa or vagina. In Elateridae and some other groups (Fig. 32D), the bursa is often armed with a series of spines or sclerites and the accessory gland is very large and complex. There may be considerable variation in the shape and sclerotisation of the spermatheca and in the positions of duct attachments. Spermathecal structure is illustrated for various families by De Marzo (2009), while numerous family specialists, such Holloway (2007, Lucanidae), Calder (1996, Elateridae) and Tschinkel and Doyen (1980, Tenebrionidae), have utilised characters of the spermatheca and other structures of the female tract. An additional pair of glands, the **colleterial glands**, may be located near the opening of the genital tract, and are associated with chorion formation. The ovarioles in Adephaga are **polytrophic**, with each oocyte being provided with its own nutritive cells; those in all Polyphaga studied (Büning 1979) and in the archostematan genus *Micromalthus* (Scott 1941) are **telotrophic**, with a single mass of nutritive cells located at the apex of the ovariole.

Comparative studies of internal anatomy have added significantly to the number of characters available for phylogenetic analysis. Some recent works include: Calder (1989, 1990a) (Australian Curculionoidea); Ekis and Gupta (1971) (alimentary canal of Cleridae, see Crowson 1972a); Kasap and Crowson (1975) (Elateriformia); Kasap and Crowson (1977) (Curculionoidea); and Mann and Crowson (1983) (Chrysomeloidea).

Karyotype

Chromosome numbers and sex-determining mechanisms had been recorded for more than 2100 species of Coleoptera in 50 families, before the general survey by Smith and Virkki (1978). Since that time many more species have been studied cytologically, especially in some of the larger families. Smith and Virkki included ~100 species of Carabidae; but recent work on this family has increased the number to more than 750 (Galian and Moore 1994). In Chrysomelidae, Petitpierre *et al.* (1988) included karyotypic data for almost 750 species, more than doubling the number in Smith and Virkki, and a comparable increase has occurred for Tenebrionidae (Juan and Petitpierre 1991). A recent update of cytogenetic data for Chrysomelinae has been published by Petitpierre (2011).

Diploid numbers in Coleoptera range from 4 (1 + XY) in an elaterid (*Chalcolepidius*) to 69 (34 +X) in a carabid (*Ditomus*) (Petitpierre 1988). The number of species examined and the ranges of and mean diploid numbers for major beetle groups were given by Petitpierre (1988): Carabidae – 564/8–69(30.5); Scarabaeidae – 240/12–30(19.8); Elateridae – 76/4–23(18.6); Coccinellidae – 172/12–25(18.7), Tenebrionidae – 136/14–37(21.2); Cerambycidae – 157/10–36(21.7); Chrysomelidae – 706/8–64 (25.6); Curculionidae – 538/11–54(26.6).

Karyotypes are known for two Archostemata. In *Distocupes varians* the diploid number is 19 with an XO sex-determining system (Galian and Lawrence 1993), and in *Micromalthus debilis* (Scott 1936) it is 20, the male being haploid (**see** below). In the myxophagan *Ytu zeus* the diploid nunber is 20 with an XY system (Mesa and Fontanetti 1985). Adephaga have an XO or XY system, and the most frequent and possibly ancestral diploid number is 37 (n = 18 + X) (Galian and Moore 1994; Serrano 1981). In Archostemata and some Adephaga an 'achiasmatic' meiosis pattern is present (Galian and Lawrence 1993). In Polyphaga the most frequent karyotype – and one which was considered by Smith (1950) to be ancestral for the order – is $n = 9 + Xyp$ ($2n = 20$), that is, 9 pairs of autosomes and a very small y chromosome, which unites with the much larger X at meiosis to form a parachute-like figure. The number varies considerably within the group and several types of sex-determining mechanisms exist, with the y chromosome being lost in several groups, including Elateridae, Cantharidae, Lampyridae and Lycidae.

Thelytoky or female parthenogenesis has been recorded in ~50 species of weevils, and some 7 more families, including Ptiliidae, Ciidae and Chrysomelidae (Smith and Virkki 1978). Arrhenotoky or haplo-diploidy occurs in the tribe Xyleborini of Curculionidae: Scolytinae and in the Micromalthidae. Pseudo gamy (gynogenesis), in which females reproduce by parthenogenesis but only after insemination without subsequent syngamy, occurs in the ptinid species *Ptinus clavipes.* Normal individuals of *P clavipes* have a diploid chromosome number of 18 but the females (form *mobilis*) formed by this type of parthenogenesis have a triploid chromosome number ($3n = 27$) (Moore 1957; Woodroffe 1958).

LARVAL MORPHOLOGY

The following section is based in large part on Lawrence (1991a), with some modifications based on more recent publications, such as Lawrence *et al.* (1999a), Beutel and Lawrence (2005), Lawrence *et al.* (2010b) and Lawrence *et al.* (2011). Other general works on larval morphology include Bertrand (1972, 1977), Crowson (1981) and Costa *et al.* (1988).

Body Form

The beetle larva has a well developed head, three thoracic segments and usually ten abdominal segments (sometimes reduced to nine or eight). The thoracic segments are rather similar, although the first is often larger and more heavily sclerotised. Abdominal segment X is usually relatively small, surrounding the anus, sometimes bears a pair of ventral pygopods and occasionally is elongate and subconical. Segment IX is often armed with paired processes, called urogomphi, which have evolved several times in Coleoptera and are not homologous to the cerci of some other insect orders.

Several purely descriptive terms are commonly used for general body types associated with particular modes of life. **Campodeiform** larvae (Figs 34G, 56B) (less commonly referred to as **lepismatoid** or **thysanuriform**) are active and usually predatory, with a prognathous head, long thoracic legs and well developed, more or less posteriorly oriented, 1-segmented to multi-segmented urogomphi; most Carabidae and Staphylinidae have this type. **Eruciform** larvae (Figs 33G, 34E), including those of many Chrysomelidae, are less active, subcylindrical or slightly flattened forms with short legs and without or with short urogomphi, and **cyphosomatic** larvae, also found in Chrysomelidae, have a convex dorsum and flat venter. Narrowly cylindrical larvae, such as **wireworms** in the families Elateridae (Fig. 35B) and Tenebrionidae (Fig. 91F), may be referred to as **orthosomatic** or occasionally **vermiform**. The **scarabaeiform** type (Figs 33F, 34F, M) is a C-shaped, rather long-legged grub characteristic of Scarabaeoidea and often found in soil or rotten wood. The **apodous** type (Fig. 33E) lacks thoracic legs and urogomphi and has reduced antennae and palps; examples are found throughout the Curculionoidea. These general categories do not readily accommodate several peculiar larval types, such as the flattened and disc-like (**platyform**) larvae of Psephenidae (Fig. 65I), some Corylophidae (Figs 88C–D) or murmidiine Cerylonidae (Fig. 88A); the **onisciform** type, which is

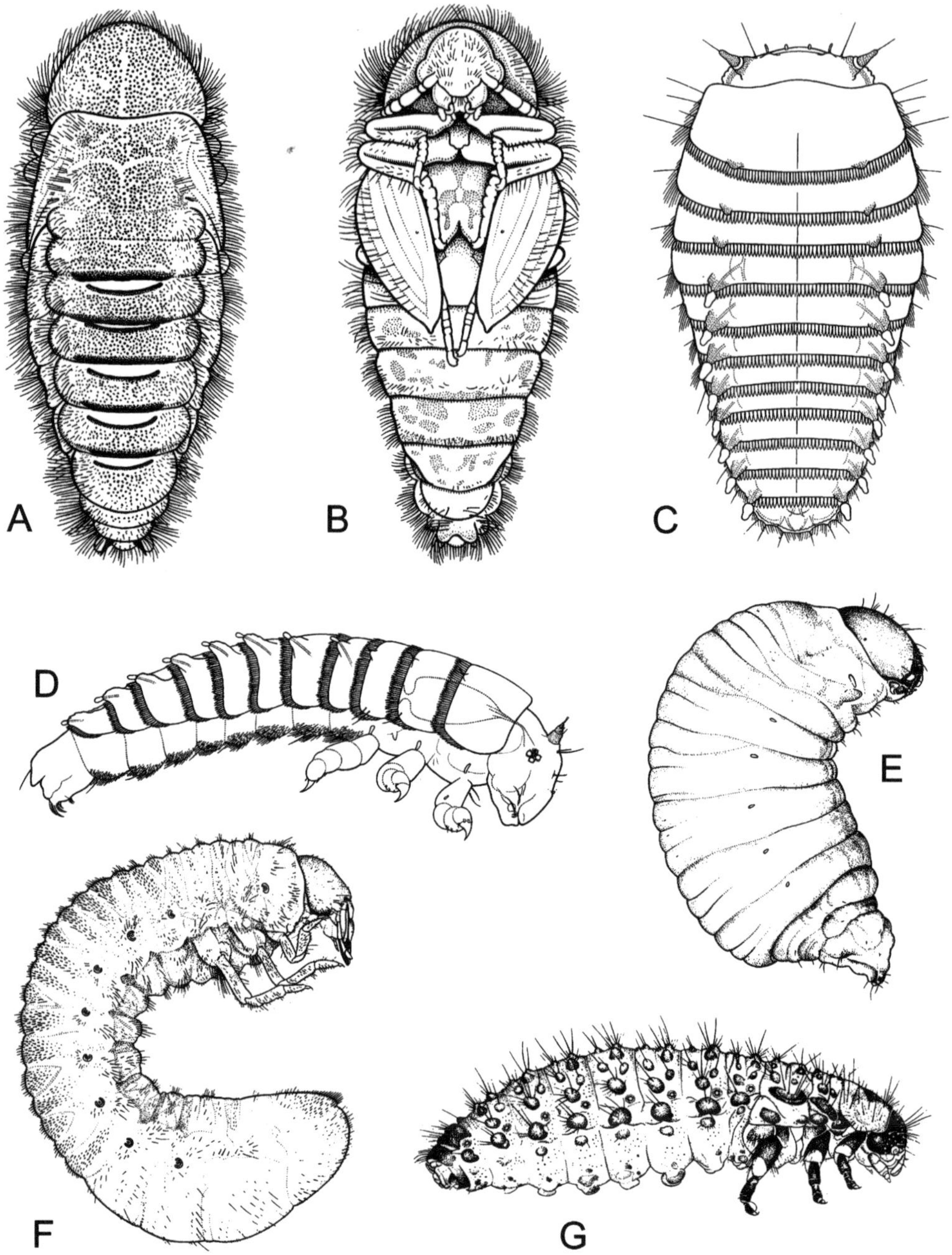

Fig. 33. A, *Dermestes maculatus* Degeer (Dermestidae: Dermestinae), pupa, dorsal. B, Same, ventral. C–G, Larva: C, *Sphaerius ovensensis* (Oke) (Sphaeriusidae), dorsal; D, same, lateral; E, *Trigonotarsus rugosus* (Boisduval) (Curculionidae: Dryophthorinae), lateral; F, *Anoplognathus pindarus* Carne (Scarabaeidae: Rutelinae), lateral; G, *Paropsisterna beata* (Newman) (Chrysomelidae: Chrysomelinae), lateral.

ovate and flattened with lateral processes, as in Brachypsectridae (Fig. 65K) or Priasilphidae (Fig. 84O); or the elongate, strongly flattened, subcortical larvae found in the families Cucujidae (Fig. 84D) or Pyrochroidae (Fig. 91L).

Cuticle and Surface Features

The larval integument or cuticle is usually softer, more and flexible and more lightly pigmented than that of the adult, but there is considerable variation and the degree of pigmentation is not always correlated with sclerotisation. Although tanning, as discussed above in connection with adult morphology, is usually accompanied by a darkening of the cuticle, due to the presence of quinonoid compounds, several larval cuticles (those of some tenebrionid and elaterid wireworms and head capsules in various groups) may be hard and rigid due to a hardening process not involving pigmented compounds (Hackman 1984 and personal communication). Following common usage, the term **sclerotised**

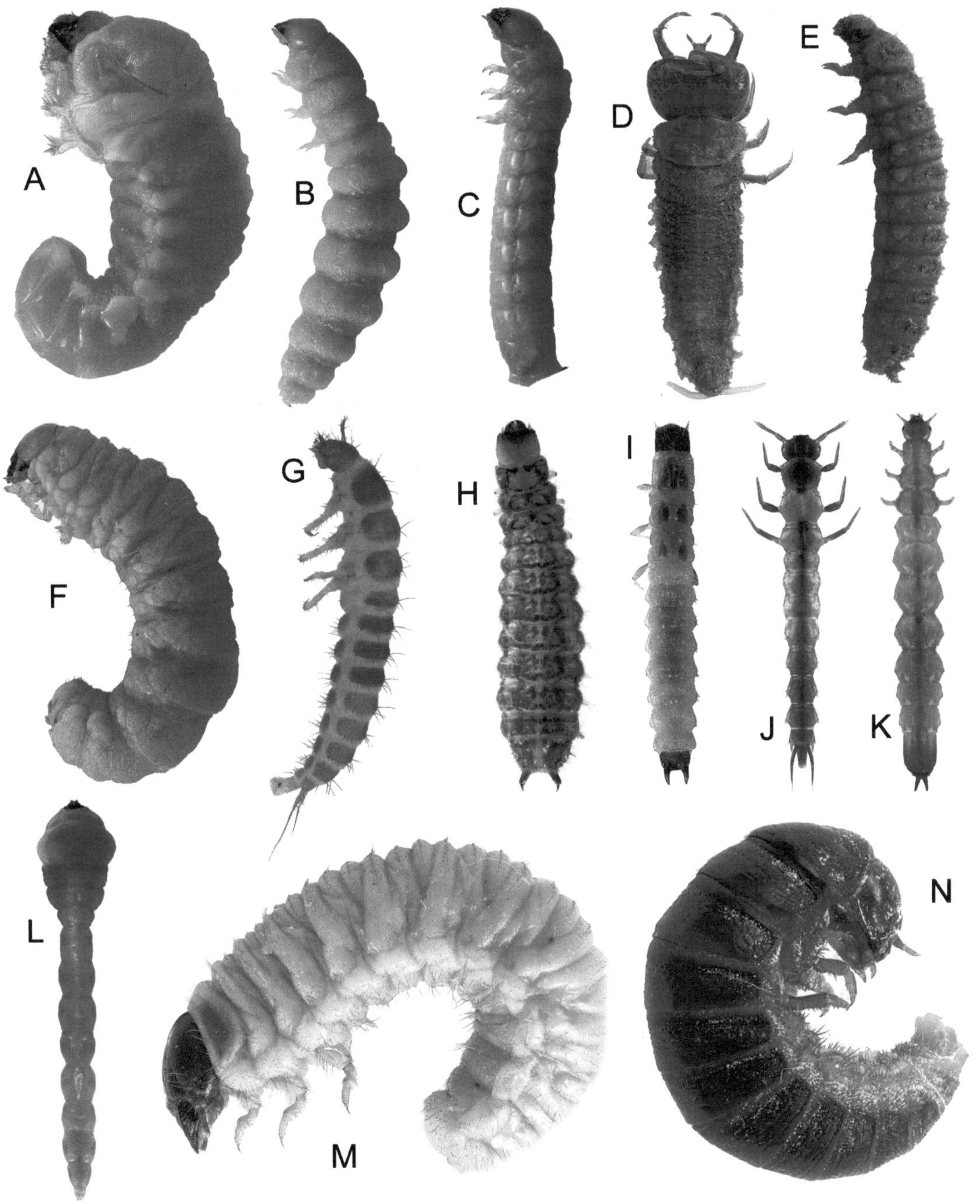

Fig. 34. A–N, Larva: A, Genus ? sp. (Bostrichidae: Bostrichinae), lateral; B, Genus ? sp. (Melandryidae), lateral; C, *Orchesia* sp. (Melandryidae), lateral; D, *Hydrophilus* sp. (Hydrophilidae), dorsal; E, *Episcaphula australis* (Boisduval) (Erotylidae: Erotylinae), lateral; F, *Ceratognathus* sp. (Lucanidae: Ceratognathinae), lateral; G, *Paragyrtodes candens* Szymczakowski (Leiodidae: Camiarinae), H, Genus ? sp. (Cleridae: Clerinae), dorsal; I, Genus ? sp. (Melyridae: Malachiinae), dorsal; J, *Macrohyliota australis* (Erichson) (Silvanidae: Brontinae), dorsal; K, *Placonotus* sp. (Laemophloeidae), dorsal; L, *Melobasis* sp. (Buprestidae), dorsal; M, *Omorgus* sp. (Trogidae), lateral; N, '*Pedilophorus*' sp. (Byrrhidae: Byrrhinae).

used here will imply the presence of yellowish to black pigment, while the term **hyaline** has been used to describe rigid structures, such as mandibular prosthecae (**see** below) that lack pigment and are more of less translucent; this is to distinguish them from truly soft, flexible, membranous structures.

The cuticle is relatively smooth in most beetle larvae, with a vestiture consisting of scattered, simple setae, which may be of fixed position within a particular taxon. In some groups, however, the dorsal surface in particular is more complex and the vestiture may include modified setae of various types. The

Fig. 35. A–L, Larva: A, *Chauliognathus* sp. (Cantharidae), dorsal; B, *Melanotus* sp. (Elateridae: Elaterinae), dorsal; C, *Carenum* sp. (Carabidae: Scaritinae), dorsal; D, *Euryplatus dimidiatus* (Waterhouse) (Salpingidae: Inopeplinae), dorsal; E, *Elacatis delusa* (Pascoe) (Salpingidae: Othniinae), dorsal; F, *Conoderus* sp. (Elateridae: Agrypninae), dorsal; G, *Dysthaeta* sp. (Cerambycidae: Lamiinae), dorsal; H, *Hister* sp. (Histeridae), dorsal; I, *Dromaeolus* sp. (Eucnemidae: Macraulacinae), dorsal; J, *Nosodendron* sp. (Nosodendridae), dorsal; K, *Prostomis* sp. (Prostomidae), dorsal; L, *Neaspis* sp. (Trogossitidae: Lophocaterinae), dorsal.

entire dorsal surface may be covered with fine granules or these may be more sparsely distributed, forming definite patterns (as in the psephenid genus *Sclerocyphon*, Fig. 65I, Priasilphidae, Fig. 84O, or some genera of Nitidulidae). More prominent setiferous tubercles occur in several families, including Eucinetidae, Derodontidae, Hobartiidae and some Erotylidae, while complex dorsal sculpturing is known in Erotylidae, Endomychidae, Cerylonidae, Coccinellidae and Chrysomelidae. A specific terminology was developed by Gage (1920) for cuticular prominences in Coccinellidae and used in several more recent works (Rees *et al.*

1994; Ślipiński 2007). Terms included (from simple to complex) are: **seta** (pleural **setae**) for an articulated hair arising from a flat surface; **chalaza (ae)** for a minute, pimple-like projection bearing a seta (Fig. 46E); **verruca (ae)** for a mound-like projection bearing several setae (Fig. 51C); **struma (ae)** for a mound-like projection bearing a few chalazae; **sentus (i)** for an elongate, unbranched projection bearing a few short setae along its trunk; and **scolus (i)** and **parascolus (i)** for longer and shorter, branched projections, respectively. The term **tubercle** is usually used for any simple protuberance, with or without a seta.

A seta is always articulated and consists of a single cell called a **trichogen** closely associated with another type of cell, the **tormogen**, which forms the setal socket, and interally a nerve cell. Other articulated structures derived from simple setae include **bristles** and **scales** and various modified setae described below. The word **spine** refers to a multicellular fixed process, and terms such as **spinule**, **asperity** or **microtrichia** are used for smaller tooth-like or hair-like fixed processes. The term **spur** refers to a multicellular, articulated process. Although most larvae have a relatively sparse covering of setae, a few, such as tetralobine Elateridae, most Dermestidae, and some Ptinidae and lagriine Tenebrionidae, are densely clothed with long hairs. Cantharid larvae have a velvety appearance, but this is caused not by hairs, but by minute cuticular protubences. There are several modified types of setae occurring in beetle larvae. **Frayed setae** (Fig. 46D), expanded and frayed at the apex, occur in several Staphylinoidea and Cucujoidea. **Pubescent setae** (Fig. 46F), covered with numerous hair-like microtrichia, may be found in larvae of the ptilodactylid genus *Austrolichas* (Fig. 27D). Some **glandular setae** may be hollow with an opening at the apex, while others are simple with a glandular pore at the base. Setae vary considerably in thickness, and terms such as bristle or scale-like seta are sometimes used. Greatly expanded setae, which are sometimes apically flattened with a slender stalk, are referred to as **scales**; these occur in *Brachypsectra* (Fig. 65K), the coccinellid genus *Scymnodes*, and a few Cantharidae. **Spicisetae** (Fig. 46G), lined with barb-like projections, occur in larvae of most Dermestidae and in the cerylonid genus *Murmidius*. A more complex type of seta, found in megatomine Dermestidae, is the **hastiseta** (Fig. 46H) which is not only lined with barbs, but has a pair of 3-pronged processes at the apex. Both hastisetae and spicisetae are weakened at the base, so that they detach easily and entangle potential predators. An unusual type of setal base is found on the dorsum of some elaterid larvae (*Austrelater*, Protelaterini and Lissomini); groups of specialised setae, which are short, moderately thick and attached in such as way that they are capable of moving anteriorly or posteriorly, occur on tergum II and the following three to six terga (Calder *et al.* 1993).

Other important features of the larval cuticle are the **defensive glands**. These glands are probably widespread among beetle larvae, but they have been reported in only a few groups, partly because the openings may be very small or obscured by debris. Paired dorsolateral gland openings on most thoracic and abdominal terga (Fig. 51L, Q) are known in the families Staphylinidae, Cerophytidae, Phengodidae, Cantharidae, Trogossitidae, Melyridae, Coccinellidae, Tenebrionidae and Chrysomelidae, and there are large paired openings on abdominal segments I and VIII (Fig. 45L) or I–VII in some Corylophidae. A single median gland opening occurs on tergum VIII in most aleocharine Staphylinidae and on tergum IX in some Tenebrionidae, while head glands have been reported in Staphylinidae and Tenebrionidae (Doyen and Lawrence 1979; Frank and Thomas 1984; Kasule 1966). In some Tenebrionidae (e.g. Hypophlaeini, Scaphidemini, Leiochrinini) median head glands and paired thoracic and abdominal glands are represented by flat plates containing numerous micropores.

Head Capsule

The head capsule or **cranium** in beetle larvae is always distinct and usually moderately to strongly sclerotised, although variably pigmented. As in the adult, the head has two large openings: the posterior **occipital foramen**, leading to the cervical region and prothorax, and the **mouth cavity** or **preoral cavity**, which may be anterior or ventral depending upon the type of head attachment.

The most common type of larval head is **prognathous**, with the mouthparts directed anteriorly, and **protracted**, with the head not or only slightly retracted into the prothorax (Figs 45B, 56A–F). This type is widely distributed throughout the order, with the exception of a few Byrrhoidea and all Scarabaeoidea, Bostrichoidea, Chrysomeloidea and Curculionoidea. In a few groups, such as Buprestidae, Bostrichidae, a few Ptinidae, most Cerambycidae, some Chrysomelidae (e.g. Sagrinae, Bruchinae, leaf-mining Cassidinae and Galerucinae), a few Anthribidae, Belidae, Attelabidae and some leaf-mining Curculionidae, the head is prognathous but strongly **retracted** into the thorax (Figs 34A, 35G, 45A). In several other groups, including Scarabaeoidea, Byrrhidae, Dermestidae, Endecatomidae, Ptinidae, Lymexylidae, some Endomychidae, Erotylidae and Coccinellidae, cerylonine Cerylonidae, Mordellidae, later instars of Meloidae and Ripiphoridae, a few Tenebrionidae and most Chrysomelidae and Curculionoidea, the head is distinctly declined or **hypognathous**, with the mouth-parts directed ventrally (Figs 33D–G, 34F, M–N, 45C). In some strongly hypognathous heads, like those of most cerylonine Cerylonidae (Figs 46M–N, 88J), the mouthparts are directed posteriorly; these are sometimes referred to as **opisthognathous**, but that term is more suitably applied to the condition in Hemiptera.

The cranium, as viewed from above, is usually divided by an **ecdysial line** (often incorrectly called the **epicranial suture**), which may be V-shaped, U-shaped (Fig. 36C), Y-shaped (Figs 45D, 46C) or **lyriform** (shaped like a lyre with sinuous, doubly curved arms, with or without a stem) (Figs 36D, 45E–F). The stem, when present, is called the **epicranial stem** (or **coronal suture**), while the lateral arms are referred to as **frontal arms** (or **frontal sutures**). The epicranial suture divides the cranium into three major areas: a **frons (front)** or **frontoclypeal region** in front of and between the two frontal arms, and two **epicranial plates** (**epicranial halves**) or **parietal plates**, which are behind and lateral to the two frontal arms and on either side of the epicranial stem, when this is present. When there is no stem, the frontal arms may be contiguous or well separated at the base, so that the frons extends posteriorly to the occipital foramen, as in some Carabidae and Nitidulidae. The dorsal part of the combined epicranial plates is sometimes called the **vertex**, as in adult beetles. When complete, the frontal arms may extend to the mandibular articulations or the antennal sockets, but they are often incomplete anteriorly and sometimes the ecdysial lines are indistinct or completely absent

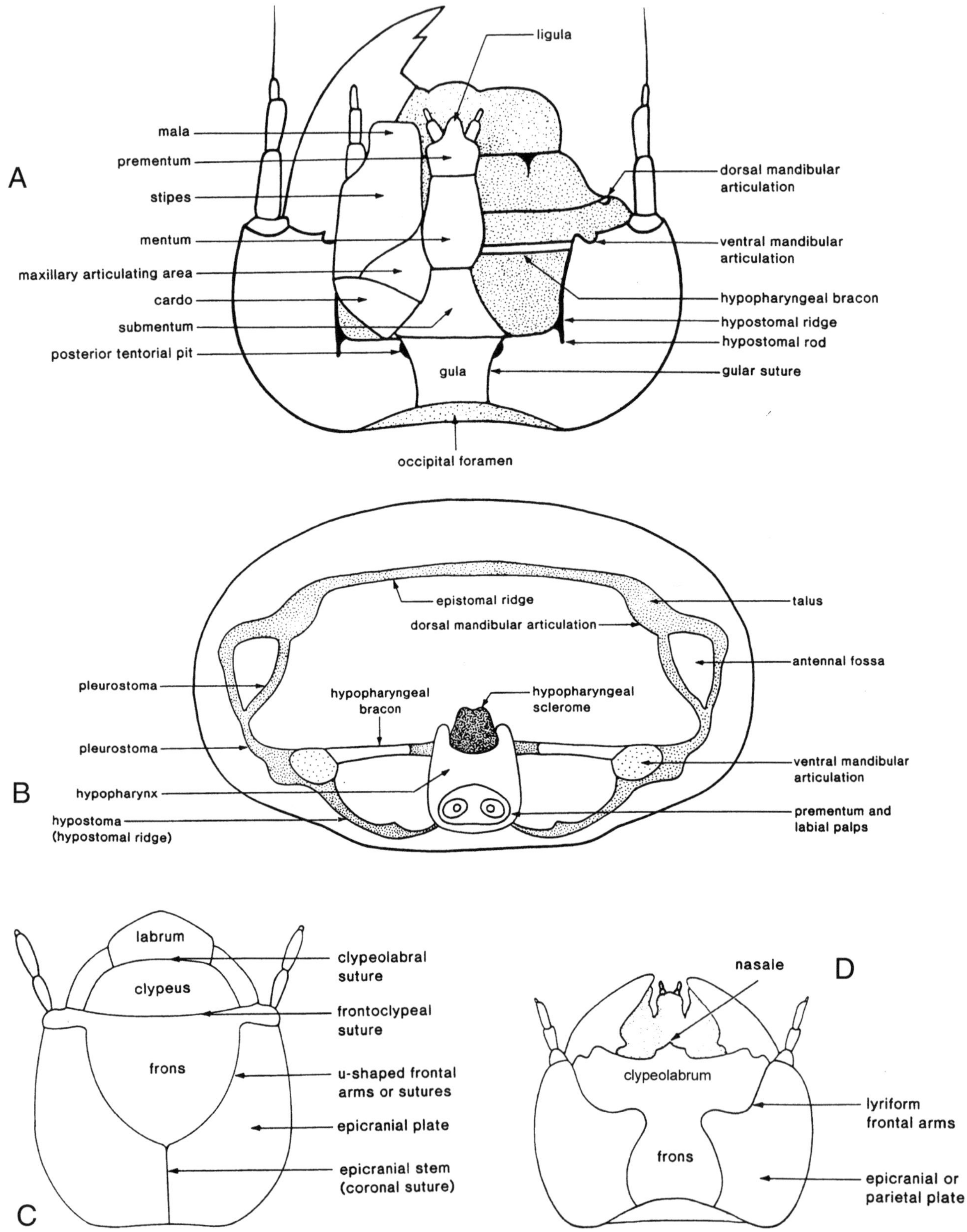

Fig. 36. Larval morphology. A, Head capsule, ventral, with left mandible and left maxilla removed. B, Head capsule, anterior, with mandibles and maxillae removed. C–D, Head capsule, dorsal: C, tenebrionid type, with long epicranial stem, U–shaped frontal arms, frontoclypeal suture and free labrum; D, elaterid type, with no epicranial stem, separated, lyriform frontal arms, no frontoclypeal suture, fused labrum and anteriorly projecting nasale.

(e.g. second instar Histeridae). In a few tenebrionoid families (e.g. Pterogeniidae, pisenine and hallomenine Tetratomidae, Chalcodryidae, toxicine Tenebrionidae, anaspidine Scraptiidae), there is an additional transverse ecdysial line joining the two frontal arms near their anterior ends (Fig. 45K). In Pterogeniidae the epicranial stem is strongly bent to the left (Fig. 45K), which is correlated with the highly asymmetrical mandibles (**see** below).

In addition to the epicranial 'suture', the dorsal part of the head capsule may have one or occasionally two internal ridges or cuticular thickenings called **endocarinae**, which strengthen the head and may provide additional muscle attachment surface. These are usually visible as dark lines, but when they are coincident with portions of the epicranial stem and/or frontal arms they may be difficult to observe. Since the epicranial suture is a line of weakness related to moulting, any section of it which appears darker or thicker is almost certainly associated with an endocarina. The most obvious type of endocarina is a median one lying beneath the epicranial stem, if present, and often extending anteriorly between the frontal arms (Fig. 45D) or present in the absence of an epicranial suture (Fig. 45I); this type occurs in several families, including Cupedidae, most Scarabaeoidea, Buprestidae, Byrrhidae, Dryopidae, Chelonariidae, Callirhipidae, some Dermestidae, Bostrichidae, Ptinidae, Lymexylidae, Chaetosomatidae, most Trogossitidae and Cleridae, pharaxonothine and languriine Erotylidae, some Phalacridae, Ciidae, Melandryidae, Mordellidae, a few Tenebrionidae, anthicine Anthicidae, and various Cerambycidae, Chrysomelidae and Curculionoidea. An endocarina may be Y-shaped and coincident with both epicranial stem and at least the bases of the frontal arms (Fig. 46C); this is known in larvae of leptochirine Staphylinidae, Passalidae, some Melandryidae and Zopheridae, Oedemeridae, Stenotrachelidae (except Stenotrachelinae) and beline Belidae. Paired endocarinae may also occur beneath the bases of the frontal arms when an epicranial stem is absent (Fig. 46A); this type is known in trachyine Buprestidae, some Eucnemidae, the nitidulid genus *Pocadius*, Laemophloeidae, some Melandryidae and Zopheridae, Salpingidae and a few other groups.

The frontoclypeal region, or **frontoclypeus**, may be divided by a transverse groove called the **frontoclypeal suture** (Fig. 36C), frontoclypeal sulcus or epistomal suture, represented internally by an **epistomal ridge**. This suture separates the posterior or dorsal **frons** from the anterior or ventral **clypeus**, and when the suture is absent these areas may be arbitrarily delimited by an imaginary line connecting the dorsal mandibular articulations (**see** below). The clypeus may be further subdivided into a posterior, more heavily sclerotised **postclypeus** and an anterior, lightly sclerotised **anteclypeus**; the term **epistoma** has been applied by different workers to one or the other subdivision of the clypeus or to the clypeus as a whole. The **epistomal ridge** forms the dorsal part of the **mouth frame** or **peristoma** (Fig. 36B). At either end of this ridge is a condyle, which is the **dorsal mandibular articulation**, and it may be surrounded by a heavily sclerotised area sometimes called the **talus**. The attachments of the anterior tentorial arms, if present, may also be located at the lateral ends of the epistomal ridge. The lateral portion of the peristoma on each side, between the dorsal mandibular articulation and the ventral one is called the **pleurostoma**. The **antennal fossa** usually lies very close to the pleurostoma, separated from the mandibular articulations by a slender cuticular bar, but in some larvae (Sphaeriusidae, some Endomychidae, nilionine and leiochrinine Tenebrionidae), the antennae may be separated by a wide strip of cuticle and in others there may only be membrane between the antennal fossa and mandibular articulation.

The **ventral mandibular articulation** is a fossa or acetabulum into which the ventral mandibular condyle is inserted. The lower part of the mouth frame, between the ventral articulation on each side and the median attachment of the labium, is called the **hypostoma** or **hypostomal ridge**, and it is to this that the maxilla is attached. The combined hypostomata may be almost transverse in larvae with **protracted ventral mouthparts**, so that maxillary and mandibular articulations are more or less at the same level (Figs 40L, 43C, 48N, 49A). This condition is characteristic of most Adephaga, most Hydrophiloidea, a few Staphylinidae, Cleridae and Chaetosomatidae, Passandridae, Laemophloeidae, Phalacridae, Smicripidae, some Corylophidae and Latridiidae, Prostomidae, most Cerambyhcidae and some Chrysomelidae. In larvae **with retracted ventral mouthparts**, the maxillary articulations are located well behind those of the mandibles (Figs 37A–B, 38D, G, 48D, H, L), and the hypostomata form a ventral cavity, the **hypostomal cavity**, more or less surrounding the maxillae and labium. This is the normal condition in most beetle larvae. Intermediate situations occur in some groups, like Cantharidae. (Fig. 48P) and ptilodactyline Ptilodactylidae (Fig. 38C), where the maxillae and labium are only slightly retracted. The points of attachment of the maxillae do not necessarily reflect the degree to which the ventral mouthparts extend forward, since this can be influenced by the length of the stipites and length and orientation of the cardines. In almost all Nitidulidae, for instance, the ventral mouthparts are retracted but the maxillae project forward due to the longitudinally oriented cardines (Fig. 49F).

The epicranial plates extend ventrally on each side as the **genae**, and where they continue behind the hypostomal ridge, they may be referred to as **postgenae**. In some groups, a pair of distinct ridges originates just lateral to the ventral mandibular articulations and extends posteriorly more or less parallel to the hypostomata and forming with them a housing for the ventral mouthparts; these are called the **ventral epicranial ridges** (Figs 37B, 38C–D, G, 49N). They are well developed in most Byrrhoidea, those Elateroidea with normal biting mandibles, Lymexylidae, trogossitine Trogossitidae, Derodontidae, Nosodendridae, most Erotylidae (not Cryptophilinae), Monotomidae, Ciidae, toxicine Tenebrionidae, Pythidae, othniine Salpingidae, cryptocephaline Chrysomelidae and some Anthribidae. A pair of cuticular thickenings may reinforce the hypostomata, and where they extend posteriorly beyond the maxillary articulations and onto the postgenae, they are called the **hypostomal rods**; they are most conspicuous in larvae with protracted ventral mouthparts. The most common type of hypostomal rods are relatively long and posteriorly diverging (Figs 48H, 49A), as in many Cucujoidea and Tenebrionoidea, but they are subparallel (Fig. 37A) in several cleroid families, as well as in Laemophloeidae, Smicripidae, murmidiine Cerylonidae, salpingine and inopepline Salpingidae and some Cerambycidae. A few groups like anaspidine Scraptiidae and monommatine Zopheridae have converging hypostomal rods (Fig. 48L) and eustrophine Tetratomidae may have two pairs, one parallel and the other diverging (Fig. 48D).

The area behind the ventral mouthparts is sometimes called the **ventral head closure** and the terms **hypostomal region** and **gular region** have also been used. The term ventral head closure

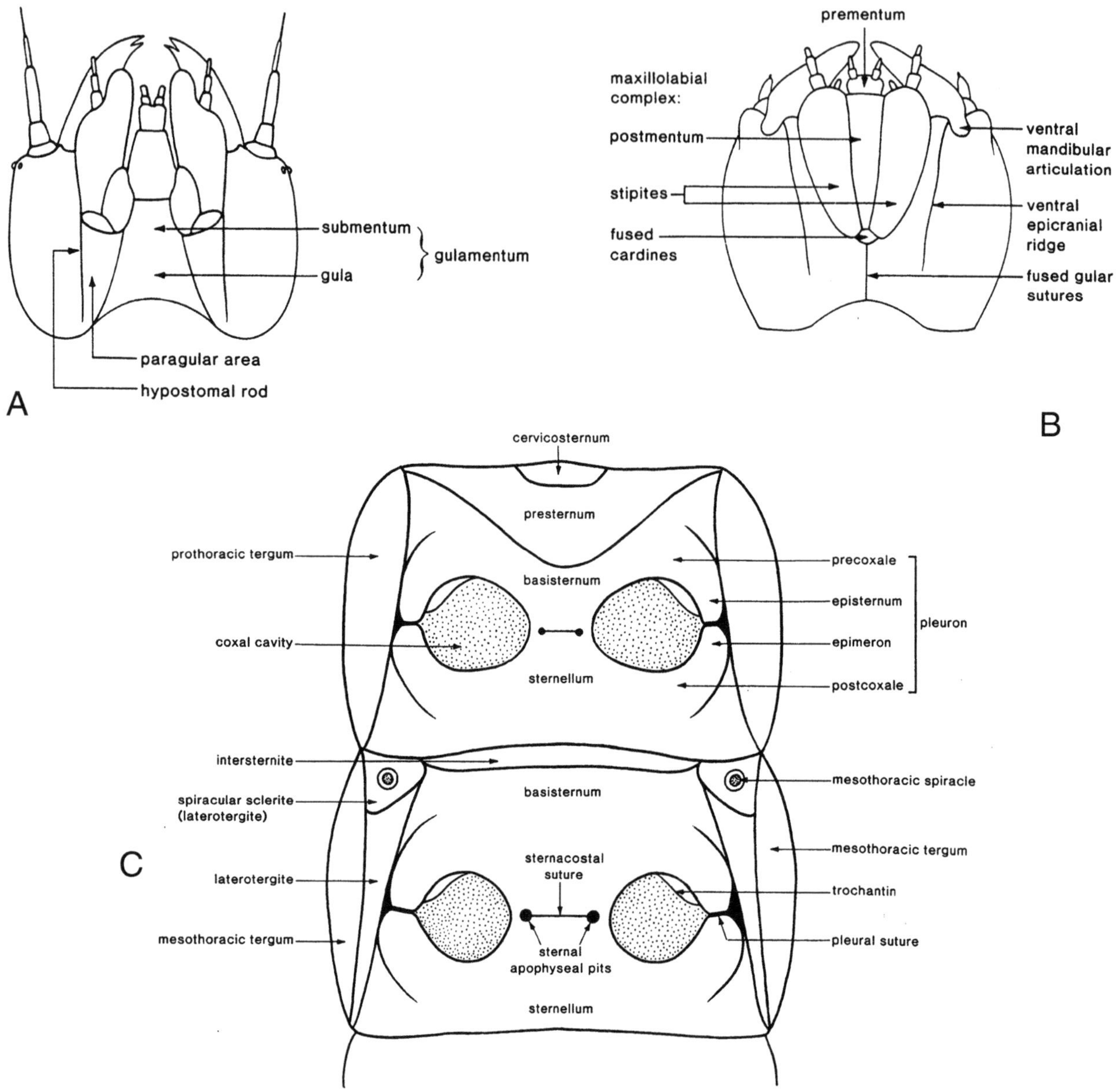

Fig. 37. Larval morphology. A–B, Head capsule, ventral: A, with retracted, freely articulated ventral mouthparts and parallel hypostomal rods; B, with retracted maxillolabial complex and ventral epicranial ridges. C, Prothorax and mesothorax, ventral, with legs removed.

implies that the area is at least moderately sclerotised, so that the type of head occurring in Melyridae, in which this area is very lightly sclerotised, is often said to lack a ventral head closure. The **posterior tentorial pits** (Fig. 36A), representing invaginations of the tentorium, are usually located just behind the maxillary articulations, but they may be located well behind these in some prognathous heads (Dytiscidae, Carabidae, Cleridae). Paired gular sutures may extend from these posteriorly to the occipital foramen, and the area between them is called the **gula**. In some larvae, however, there are paired ecdysial lines extending anteriorly from the tentorial pits or not associated with them at all (Hinton 1963). It has been customary among larval taxonomists to refer to all such lines as gular sutures and the area behind the labium as the gula or gular region, even when the lines are absent or ill defined. Gular sutures may be fused, so that the epicranial plates meet at the midline (some Carabidae, Fig. 48N), or they may be fused posteriorly, so that a triangular area called the **pregula** is formed (some Hydrophilidae). In some larvae an area is defined on either side of the gula by a pair of long, parallel hypostomal rods; these areas have been called **paragular areas** or **paragular plates**. The most common condition in beetle larvae is a reduced gular or hypostomal region with indistinct gular sutures, and in some groups (e.g. Ptinidae, Mordellidae, Chrysomelidae, Curculionoidea) this region is completely absent and the labium is attached directly to the thorax or to cervical membrane (Fig. 49M). In most cerambycid larvae (Fig. 49J) there is an area behind the labium but well in front of the tentorial pits, which is called the **hypostomal bridge**.

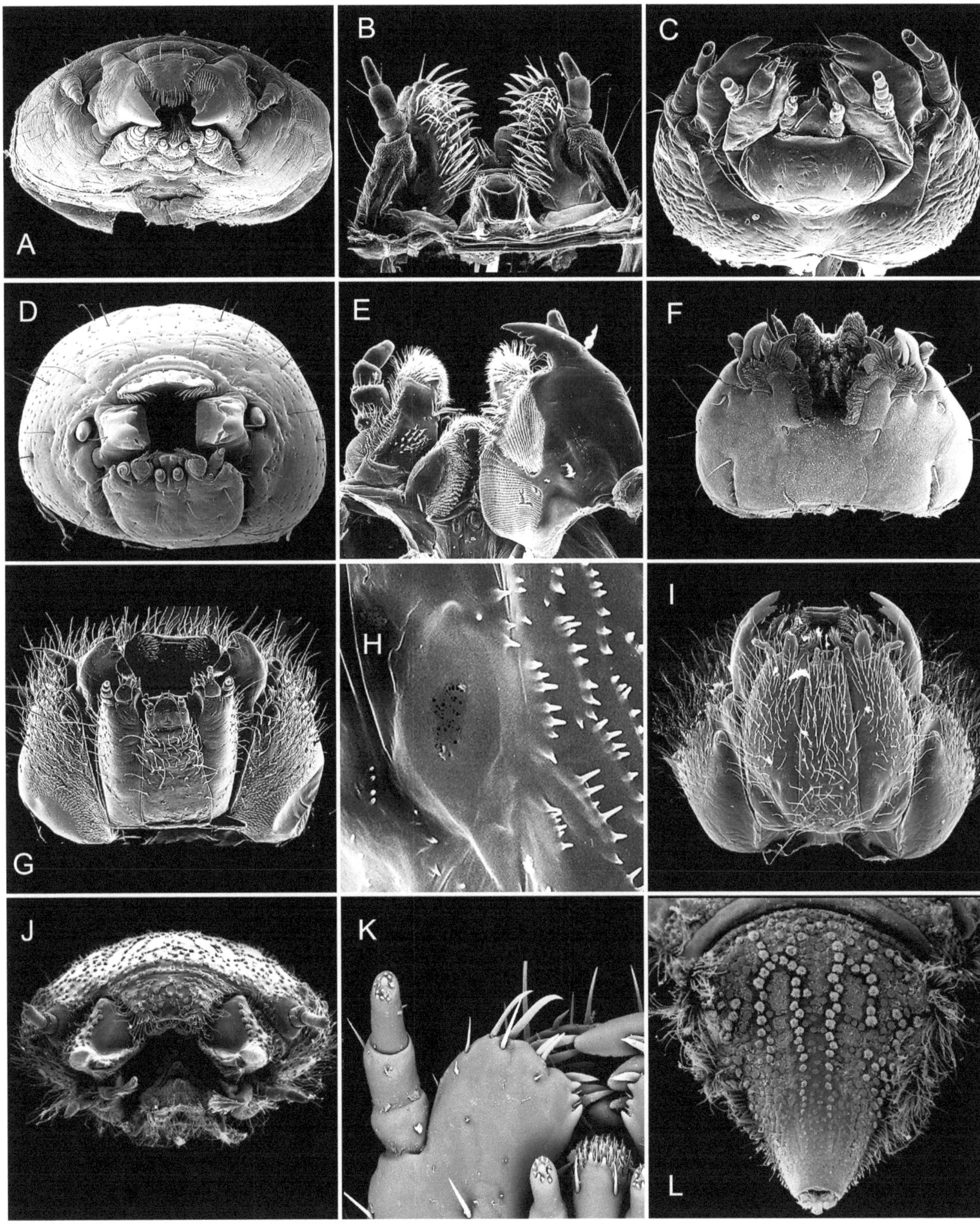

Fig. 38. Larval morphology. A, *Neandra brunnea* (Fabricius) (Cerambycidae: Parandrinae), head, anteroventral. B, *Platydema ellipticum* (Fabricius) (Tenebrionidae: Diaperinae), maxillae and hypopharynx, dorsal. C, *Ptilodactyla serricollis* Say (Ptilodactylidae: Ptilodactylinae), head, anteroventral. D, *Callirhipis* sp. (Callirhipidae), head, anterior. E, *Glischrochilus* sp. (Nitidulidae: Cryptarchinae), mandible, hypopharynx, maxilla, dorsal. F, *Lissomus* sp. (Elateridae: Lissominae), head, dorsal. G, *Byrrhinus* sp. (Limnichidae: Limnichinae), head, ventral. H, *Heterocerus gemmatus* Horn (Heteroceridae), abdominal spiracle, lateral. I, Same, head, ventral. J, *Nosodendron* sp. (Nosodendridae), head, anterior. K, Genus ? sp. (Salpingidae), maxilla and labium, ventral. L, *Nosodendron* sp. (Nosodendridae), abdominal apex, posterodorsal.

The larval **tentorium** or **cephalic endoskeleton** is usually simpler than that of the adult, but a **tentorial bridge** or **corpotentorium** is often present, as well as a pair of **anterior tentorial arms**. Another endoskeletal structure is the **hypopharyngeal bracon** (Fig. 36A), a slender bridge extending between the ventral mandibular articulations and attached to the hypopharynx. This structure is absent in Adephaga but occurs in Archostemata and Myxophaga. Within the Polyphaga, it is widely distributed, but absent in Hydrophiloidea, Dascillidae, Buprestidae, most Byrrhoidea, some Elateroidea and Bostrichoidea and all Chrysomeloidea. The presence of this bracon is one of the characters distinguishing curculionoid larvae from those of Chrysomelidae.

Stemmata

The lateral eyes of larval endopterygotes are sometimes referred to as ocelli (sing. ocellus), but since that term has also been used for the one to three simple, dorsal eyes in adult insects, the word **stemmata** (sing. **stemma**) is now used. Stemmata in beetles vary from one to six on a side and many larval types lack them altogether. Each stemma may have a well developed lens and retina, or it may be reduced, so that it is little more than a pigment spot. In cicindeline Carabidae, one or two of the stemmata may be much larger than the others and contain numerous sensory cells (Paulus 1979). It is advisable to observe the number and positions of stemmata before clearing a larval head, since the pigment often disappears after treatment with potassium hydroxide. They may be well separated in many Adephaga, and in Dryopidae and Byrrhidae), close together in a single group in Elmidae or Ptilodactylidae, or sometimes forming one or two, straight or curved rows. A full complement of six stemmata occurs in most families of Adephaga, the myxophagan family Lepiceridae, and within Polyphaga, the family Agyrtidae, some Silphidae and Staphylinidae, many Hydrophilidae, the eucinetid genus *Nycteus*, Byrrhidae, Dryopidae and some related families, Derodontidae, some Dermestidae, a few Cleroidea, many Cucujoidea, the stenotrachelid genus *Cephaloon*, palophagine Megalopodidae, some cassidine, criocerine, chrysomeline and cryptocephaline Chrysomelidae and some rhynchitine Attelabidae and beline Belidae. The most common number in Polyphaga is five, but it varies considerably in some families and reduction has occurred in many lineages. Some families, like Nitidulidae and Endomychidae, never have more than four on each side, while Coccinellidae have three or fewer. Except for those of *Cephaloon*, tenebrionoid larvae do not have more than five, and most Curculionoidea have no more than two. Within the polyphagan superfamily Elateroidea, larvae either lack stemmata or have a single large stemma on each side. In the family Eulichadidae, there may be a single lens with two pigment spots beneath it.

Labrum

A separate sclerite, the **labrum**, is usually attached to the anterior edge of the clypeus and the suture between them is the **clypeolabral suture** (Figs 36C, 45D–G). The area on the underside (ental surface, or that facing the inside of the mouth cavity) of the labrum is called the **epipharynx**, and the term **labrum–epipharynx** is sometimes used for the entire structure. The base of the labrum bears a pair of lateral sclerites called **tormae**, to which the lateral labral muscles are attached. Tormae vary considerably in length and shape, may be symmetrical or asymmetrical, separate or joined medially (Fig. 46O) and simple or with accessory processes. Various terms applied to the tormae and associated structures in Scarabaeoidea are discussed and illustrated by Hayes (1928), Böving (1936) and Ritcher (1966). In some groups, including Dermestidae (Kiselyova and McHugh 2006) and Curculionoidea (Emden 1938, 1952; May 1993, 1994), there may be another pair of sclerites in this region: the **labral rods** or **epipharyngeal rods** (Fig. 46J) these structures are paramedian in position and extend from near the apex or the middle of the labrum-epipharynx posteriorly to the base or beyond to the pharyngeal region.

The shape of the labral apex (Fig. 46K–L) has been used as a taxonomic character in Scarabaaeoidea and various other groups, while the type and distribution of labral setae is important in Tenebrionidae. The surface of the epipharynx may be very complex; the anterior edge usually bears a row of setae or fine hairs, and in addition there may be patches of setae or spines, groups of sensilla of different types, and sclerotised plates or small rods. A set of terms describing the epipharynx in Scarabaeoidea has been developed, including six regions: the median **corypha**, **haptomerum**, **pedium** and **haptolachus**, from front to back, and a pair of lateral **paria**. Other terms referring to specific structures are included in the three references given above for the tormae. The area immediately behind the epipharynx is the **pharynx** or **cibarium**, which is often provided with a complex armature, including a series of obliquely transverse **cibarial plates**, which are fringed with microtrichia and are located above the microtrichial armature at the base of the mandibles (**see** below).

In various families of beetles, the labrum has become partly or completely fused to the clypeus forming a **clypeolabrum**. Complete fusion (Figs 36D, 38F, 40L) occurs in Hydroscaphidae, all Adephaga, Hydrophiloidea, Elateroidea (except Artematopodidae and Brachypsectridae), nicrophorine Silphidae, many Staphylinidae (including the subfamilies Pselaphinae, Scydmaeninae, Euaesthetinae, Steninae, Paederinae and Staphylininae), Rhipiceridae, Nosodendridae, Boganiidae, some Byturidae (*Xerasia*), Lamingtoniidae, Passandridae, Cyclaxyridae, some Phalacridae (*Litochrus*), meligethine Nitidulidae, some Bothrideridae (*Sosylus*), Cerylonidae (*Cerylon*) and Corylophidae (Periptyctinae, *Holopsis*), Ripiphoridae, Meloidae (some triungulins), the anthicid genus *Ischalia*, Oxypeltidae and some cryptocephaline Chrysomelidae, while partial fusion (Fig. 45M) is known in silphine Silphidae, the eucinetid genus *Nycteus*, Dascillidae, some Sphindidae, Helotidae, Biphyllidae, Phloeostichidae, some Silvanidae and Phalacridae and cybocephaline Nitidulidae. In some groups with a completely fused labrum, the anterior part of the head forms a median projection (Fig. 36D) called the **nasale**, sometimes with an additional pair of lateral lobes called **adnasalia** (sing. **adnasale**) by some workers (Emden 1942b) and **paranasal lobes** by others (Glen 1950).

Antennae

The antennae arise from a pair of **antennal fossae** located on the epicranial plates. The number of antennomeres in beetle larvae is usually three: a basal **scape**, which is often short and broad; a **pedicel**, which may be more elongate and usually bears at or near its apex one (or occasionally more than one) hyaline vesicle called a

sensorium (sensory or sensorial appendage or appendix, supplementary joint); and a **flagellum**, which almost always consists of a single segment with a group of sensilla and one or more setae at the apex (Fig. 46U). Four-segmented antennae (Figs 40A, H, L, 45I, 56A–B) occur in most Adephaga (except for a few derived Carabidae), in the archostematan families Ommatidae, Cupedidae, and Micromalthidae (although in the last group the basal antennomere is highly reduced), and the polyphagan families Leiodidae (*Prionochaeta*), Staphylinidae (some Scydmaeninae, Steninae, Paederinae, Staphylininae), Ochodaeidae, Glaphyridae and many Scarabaeidae. The occurrence of more than four antennomeres is known in some specimens of the cupedid genus *Distocupes*, a few Carabidae, some Dytiscidae and aphodiine Scarabaeidae. The additional segments are almost always added at the base, since the sensorium remains on the penultimate antennome (Fig. 46U). Larvae of the family Scirtidae are unique in having the flagellum divided into numerous segments (occasionally as few as six) (Fig. 60H), while in some Dytiscidae all antennomeres, except the last one are subdivided. The majority of beetle larvae have 3-segmented antennae, but reductions to two or one are not uncommon. Two-segmented antennae occur in first instars of *Priacma* and some other Cupedidae, Sphaeriusidae (Fig. 33D), Hydroscaphidae, Torridincolidae, scydmaenine and pselaphine Staphylinidae, Passalidae, Buprestidae, Throscidae, some Eucnemidae, Lycidae, Ptinidae, silvanine Silvanidae, Passandridae, bothriderine Bothrideridae, Discolomatidae, and some Coccinellidae, Tenebrionidae, Cerambycidae, Chrysomelidae, and Belidae, while antennae consisting of a single segment are known in Rhipiceridae, Callirhipidae, Ptinidae, Coccinellidae, Mordellidae, Ripiphoridae, Meloidae, Chrysomelidae and most Curculionoidea. When the terminal segment is lost, the sensorium comes to lie at the antennal apex, usually accompanied by one or more setae or sensilla representing the true apical antennomere. The lack of a seta at the apex of the sensorium distinguishes it from a true antennomere (Fig. 46U). Fusion of the apical and subapical antennomeres may result in the location of the sensorium near the middle of the apical one, as in some Myxophaga and almost all Ciidae (Fig. 46Q). An unusual situation occurs in some Phalacridae (*Phalacrus* and *Phalacropsis*), where the basal segment has been lost and the apical one subdivided, resulting in a 3-segmented antenna with the sensorium arising from the apex of the first antennomere (Fig. 46P).

The size and shape of the antennal sensorium may be useful in distinguishing some family groups, but this feature is extremely variable in some families (e.g. Elateridae). The sensorium is usually elongate and conical, palpiform or in rare cases almost setiform, and it may be very short to longer than the terminal antennomere (e. g. in Hydroscaphidae, Heteroceridae, some Cerylonidae, Endomychidae, Ciidae). Short, dome-like sensoria (Fig. 46R) occur in several families, including Carabidae (Rhysodinae), Staphylinidae (Scydmaeninae), Scarabaeidae, Dascillidae, Lycidae, Lampyridae, Cucujidae, Erotylidae, Tenebrionidae, Pyrochroidae and Scraptiidae. Bifurcate sensoria are known in some pselaphine Staphylinidae, and tenebrionid larvae exhibit a variety of complex, dome-like sensoria, which may be C-shaped (Fig. 91B), sinuate or broken up into several parts. In at least a few larvae of Hydrophilidae, Scarabaeidae and Elateridae, and in most Histeridae, there are multiple sensoria. In some Scarabaeidae (Fig. 46T) there may be additional antennal vesicles on the terminal antennomere which have been called **dorsal sensory spots** by Ritcher (1966). As mentioned above, the sensorium is usually placed at the end of the penultimate antennomere; an exception occurs in most Staphylinoidea, where the apex of this antennomere is strongly oblique and the large sensorium is attached well before the apex (Fig. 46S). Some of the large, dome-like sensoria may also be subapical (Fig. 46R).

Mandibles

With few exceptions, beetle larvae have a pair of well developed, **opposable** mandibles, which move in a **horizontal plane** and articulate with the **dorsal** and **ventral mandibular articulations** of the head capsule by means of a **dorsal acetabulum** (Fig. 39A, G–H) and **ventral condyle** (Fig. 39D, F). Each mandible is moved by a large **adductor muscle** inserted at the inner angle (just lateral to the mola, if present) and a smaller **abductor muscle** inserted near the outer edge; both muscles originate on the walls of the cranium, but the former occupies most of the cranial cavity, and may be attached as well to the dorsal endocarina. In exceptional cases, such as the mandibles of some Eucnemidae, which are not opposable, the abductor may be larger than the adductor (Ford and Spilman 1979).

Features of the larval mandibles have been used extensively in larval taxonomy at all levels, and their structure strongly reflects the feeding habits of immature Coleoptera. In the following paragraphs mandibular structure and function will be dealt with in some detail; however, most functional statements are based not on behavioural observations, but on a combination of detailed morphological study, knowledge of feeding habits, examination of gut contents, and manipulation of cleared and bleached specimens. For descriptive purposes, the mandible may be divided into four areas (Fig. 47A): the **apex**, the subapical **incisor area** (scissorial area), the **base** of the mesal surface, including the mola, prostheca and associated structures, and the **outer edge**. The mesal portion of the mandibular base is often called the molar region, but this imples the presence of a definite basal structure, the **mola**, which acts in conjunction with the mola of the opposite mandible (and sometimes with a sclerome on the hypopharynx, see below) to process and ingest food.

Mandibular structure varies considerably, but the basic type (Figs 39C, 48A), which may represent that of an ancestral beetle larva, has a relatively wide base with a distinct mola, and a narrow, relatively weak apex. There are usually two apical teeth, which are oriented obliquely subparallel to the plane of movement of the mandible, sometimes with an additional one or two teeth in the incisor area and a thin, hyaline or translucent structure called a prostheca located just distal to the mola. The outer edge usually bears a pair of setae of fixed position. The mesal surface of the mola is covered with small tubercles or asperities, which often continue onto the ventral surface (Fig. 39C, D), while the dorsal surface of the mandibular base has a patch of microtrichia arranged in oblique rows (Fig. 39A, G). This type of mandible occurs in larvae that feed on small particles (spores, pollen, algal cells) or very soft tissue (loose fungal hyphae, highly decayed animal or plant matter). This type of feeding is referred to as **microphagy**. The molae are usually symmetrical, and their surfaces meet so that the asperities or tubercles interdigitate to move small particles into the mouth cavity, while simultaneously shearing or crushing larger pieces

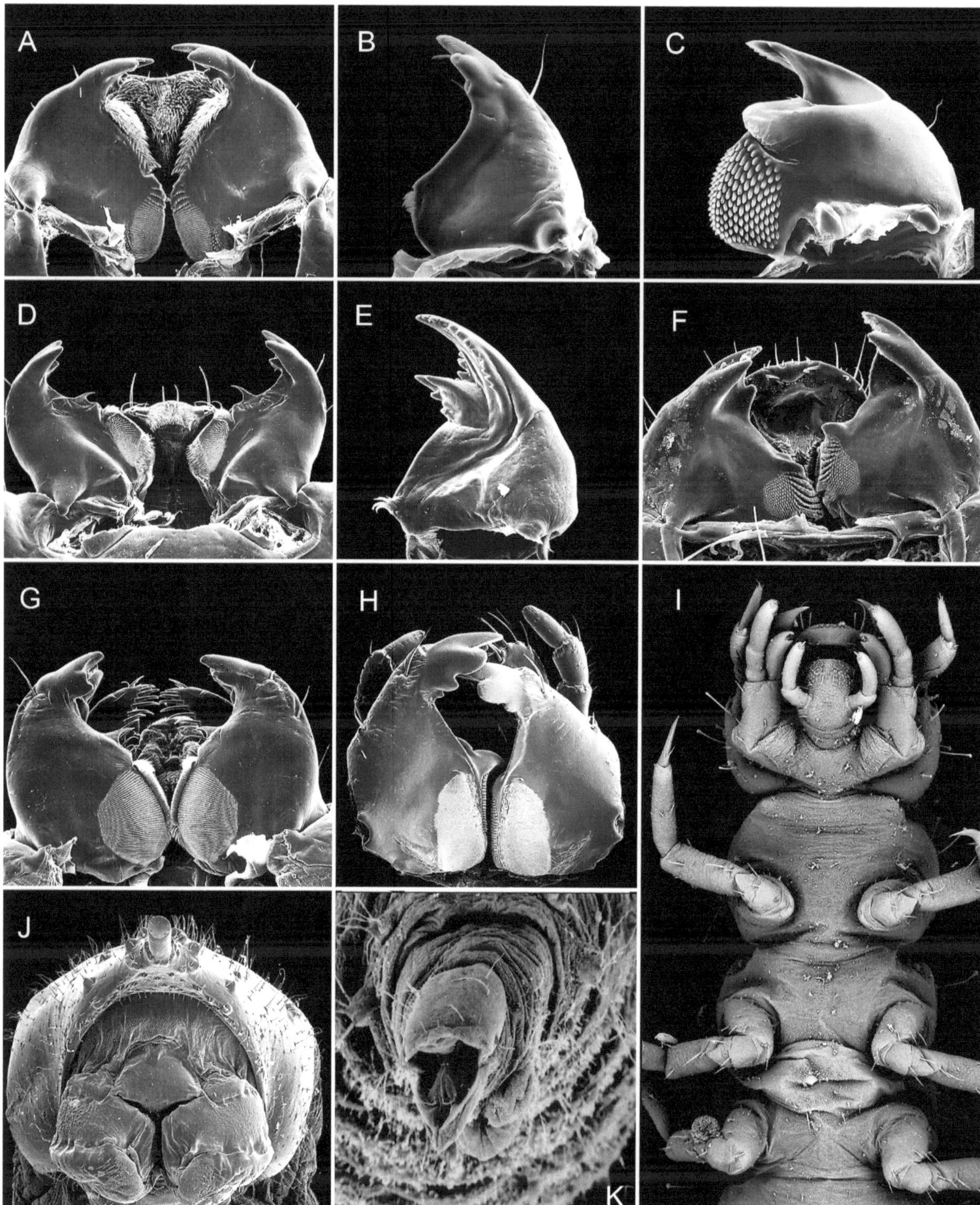

Fig. 39. Larval morphology. A, *Glischrochilus* sp. (Nitidulidae: Cryptarchinae), right and left mandibles, ventral, with ventral mouthparts removed. B, *Salpingus* sp. (Salpingidae: Salpinginae), mandible, dorsal. C, *Eucinetus morio* LeConte (Eucinetidae), mandible, ventral. D, *Cucujus clavipes* Fabricius (Cucujidae), right and left mandibles, open, ventral, with ventral mouthparts removed. E, *Epilachna argus* (Geoffroy) (Coccinellidae Coccinellinae), mandible, dorsal. F, *Mycetophagus punctatus* Say (Mycetophagidae: Mycetophaginae), right and left mandibles, closed, ventral, with ventral mouthparts removed. G, *Cucujus clavipes* Fabricius (Cucujidae), right and left mandibles, dorsal, with labrum–epipharynx removed. H, *Dendroides* sp. (Pyrochroidae: Pyrochroinae), right and left mandibles, closed, dorsal, with labrum–epipharynx removed. I, *Paracucujus rostratus* Sen Gupta & Crowson (Boganiidae), head and prothorax, ventral. J, *Tenomerga cinerea* (Say) (Cupedidae), abdominal apex, posterior. K, *Spercheus* sp. (Hydrophilidae: Spercheinae), abdominal apex, posterior.

into smaller ones, a process called comminution. The ventral asperities may also contribute to the process, since they pass over the hypopharyngeal sclerome beneath them, while the dorsal microtrichial rows act together with the cibarial or pharyngeal plates to move particles caught above the basal edges of the mandibles towards the molae and away from the articulations. The ventral part of the mandibular base has, in addition to the ventral condyle, a second shelf-like projection called the **accessory ventral process** (accessory condyle, accessory ventral condyle, second condyle) (Figs 41B, D, L, 47A, 48A, E) located near the insertion of the adductor muscle, just lateral to the mola and projecting mesally. It is not a condyle, but its true function is unknown. This type of mandible occurs in Myxophaga, basal Staphylinoidea, Eucinetidae, Clambidae, Derodontidae and many Cucujoidea, while the same basic design with modifications may be found in Scirtidae, Dascillidae, Scarabaeoidea, Lymexylidae, various Cucujoidea, many Tenebrionoidea and a few basal Curculionoidea.

Of the many specialised mandibles that can be derived from this basal type, most fall into two categories: (1) the cutting, biting, gouging or scraping mandible of a phytophagous or xylophagous larva; and (2) the grasping, piercing mandible of a larva which is always liquid feeding (with extraoral digestion) and usually predaceous. In both types the mola has been lost and food processing occurs at the mandibular apex, rather than at the base. Mandibles of the first type usually have a very broad base with a greater cross-sectional area and a gradually narrowed apex, which is sometimes not much narrower than the base; the gap between the mandibular apices is small and much greater force can be exerted at the apex and along the incisor edge. The second group usually includes mandibles which are narrow and **falcate** (falciform) with a narrow base and wide gap; these mandibles can grab and often pierce prey objects but cannot generate the force necessary to cut through heavily reinforced plant or fungal tissue. Details of these and other specialised mandible types are covered below.

The mandibular apex may be simple, consisting of a single tooth (**unidentate**), but more commonly there are two (**bidentate**), three (**tridentate**) or more (**multidentate**) apical teeth. Each tooth may be acute or rounded, but this may change with wear in a single individual. In most larvae the apical teeth are oriented obliquely to or almost parallel with the plane of movement of the mandibles (Fig. 40E, G–I, K–L), while in others the teeth are perpendicular to the plane of movement and with the incisor area may form a broad scoop (Figs 38D, 40F). Tridentate or multidentate mandibles with a flat or concave inner surface are sometimes described as **palmate**. Multidentate mandibles of some Byrrhidae and chrysomelidae have the teeth arranged in an oblique row (Figs 40D, 41I), while in a few Eucinetidae and Coccinellidae (Epilachnini and fungivorous Coccinellini) there are numerous, irregularly placed teeth or spines (Figs 39E, 47E). Simple and rounded or flattened and chisel-like apices occur in many substrate-inhabiting larvae, like those of Cerambycidae, bruchine or sagrine Chrysomelidae or Curculionoidea (Figs 38A, 40C, 41P), while acute apices are usually found in predators (Fig. 40K–L). In some larvae with well developed molae, the apex may be highly reduced, as in some Endomychidae and Latridiidae.

The incisor area of the mandible often bears one or more subapical teeth, which may be located on a carina extending basally from the dorsalmost apical tooth and sometimes this **dorsal carina** may be serrate (Figs 41E, 47K). A second **ventral carina** occurs in some mandibles lacking a mola, and a concave, scoop-like area may be formed between the two carinae (Figs 40B, 47B). In certain predaceous larvae an open or partly closed groove may be formed along the inner surface (Fig. 47O); such a condition was probably a precursor to the evolution of perforate mandibles discussed below. Another feature of the incisor area is the presence of a heavily sclerotised tooth called a **retinaculum** (Fig. 47R) (a term sometimes misapplied to the prostheca). A single retinaculum (or occasionally more than one) is usually found on the falcate mandibles of predatory larvae (Carabidae; Hydrophilidae, Histeridae, Elateridae), but similar structures, called **scissorial teeth**, may also occur in non-predatory larvae of Dascillidae (Fig. 41D) or Scarabaeoidea. The subapical region of the mandible may also bear a mola-like area, consisting of tubercles, teeth or fine ridges (Fig. 38A); this structure is called a **pseudomola**, but this term has been applied to a variety of basal or subbasal mola-like structures thought to have evolved secondarily from larvae which have lost the true mola.

As mentioned above, the mola is the processing area at the base of many mandibles which acts in conjunction with the mola of the opposing mandible to feed small particles into the mouth, to cut and shear large particles into smaller ones, or sometimes to compact small particles into pellets to be ingested. The molae considered to be plesiomorphic in the order are symmetrical, relatively narrow, and extend to the base of the mandible; examples may be found in many Cucujoidea (Fig. 39A, D, G), as well as in Derodontidae, Eucinetidae and basal Staphylinoidea. From this type may have evolved the slightly asymmetrical molae (Fig. 39F) of Mycetophagidae or Archeocrypticidae, and the strongly asymmetrical types (Fig. 39H, 40E) found in many Tenebrionoidea (e.g. Pterogeniidae, bolitophagine Tenebrionidae, Prostomidae, Oedemeridae, Synchroidae, Pythidae and Pyrochroidae), where each mola consists of numerous transverse or oblique ridges, which meet at an angle forming multiple shearing surfaces for handling rotten wood and reinforced fungal hyphae. In most cases the left mola is more vertical, with a flat or concave surface and often a tooth or lobe at the distal end, while the right mola is inclined away from the midline and has a convex surface. In another derived type of mandible, both molae are concave, usually with a smooth surface, as in Heteroceridae (Fig. 41O) and some Scarabaeoidea (Fig. 41B) or Tenebrionidae; these serve to compact loose fungal material, and may also shear larger pieces with the sharp outer edges; the tenebrionid type may act in conjunction with a concave hypopharyngeal sclerome (Fig. 38B). The concave molae in Heteroceridae compact masses of wet sand and mud, which are then swallowed whole for the extraction of diatoms, algae and other organic matter; the molae are not basal and are considered to be pseudomolae, since heterocerids have almost certainly evolved from ancestors lacking a larval mola. Another molar type found in the endomychid genus *Lycoperdina* is broad and finely tuberculate for crushing fungal spores. A tubercular or asperate molar surface may be reinforced by the formation of simple, continuous ridges, which are usually transverse or oblique (Figs 40E, 41A, J) but sometimes longitudinal (Dascillidae, Fig. 41D) or complex ridges separated by deep channels (Figs 41J); the ridges may become interconnected with the formation of a poroid superstructure (Fig. 41H), which appears to act as a fluid press, since the channels communicate with the dorsal and ventral surfaces. This fluid press is often associated with

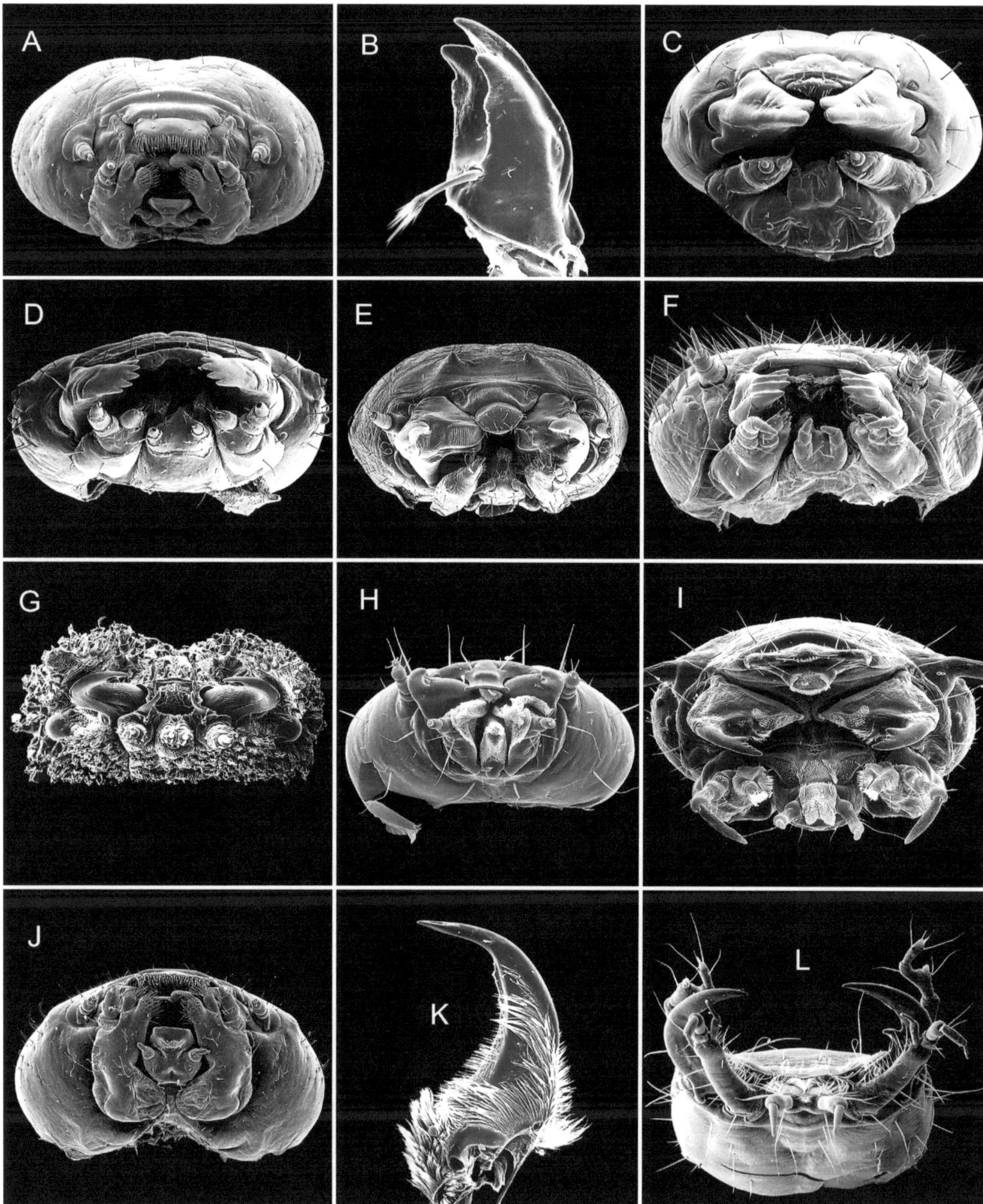

Fig. 40. Larval morphology. A, *Tenomerga cinerea* (Say) (Cupedidae), head, anterior. B, *Ptilodactyla serricollis* (Say) (Ptilodactylidae), mandible, mesodorsal. C–I, Head, anterior: C, *Curculio* sp. (Curculionidae); D, *Leptinotarsa lineolata* Stål (Chrysomelidae: Chrysomelinae); E, *Bolitotherus cornutus* (Panzer) (Tenebrionidae: Tenebrioninae); F, *Lioon* sp. (Byrrhidae: Byrrhinae); G, *Brachypsectra fulva* LeConte (Brachypsectridae); H, *Omoglymmius hamatus* (LeConte) (Carabidae: Rhysodinae); I, *Necrophilus pettitii* Horn (Agyrtidae). J, *Tenomerga cinerea* (Say) (Cupedidae), head, anteroventral. K, Genus ? sp. (Lampyridae), mandible, dorsal. L, *Agonum* sp. (Carabidae: Harpalinae), head, anterior.

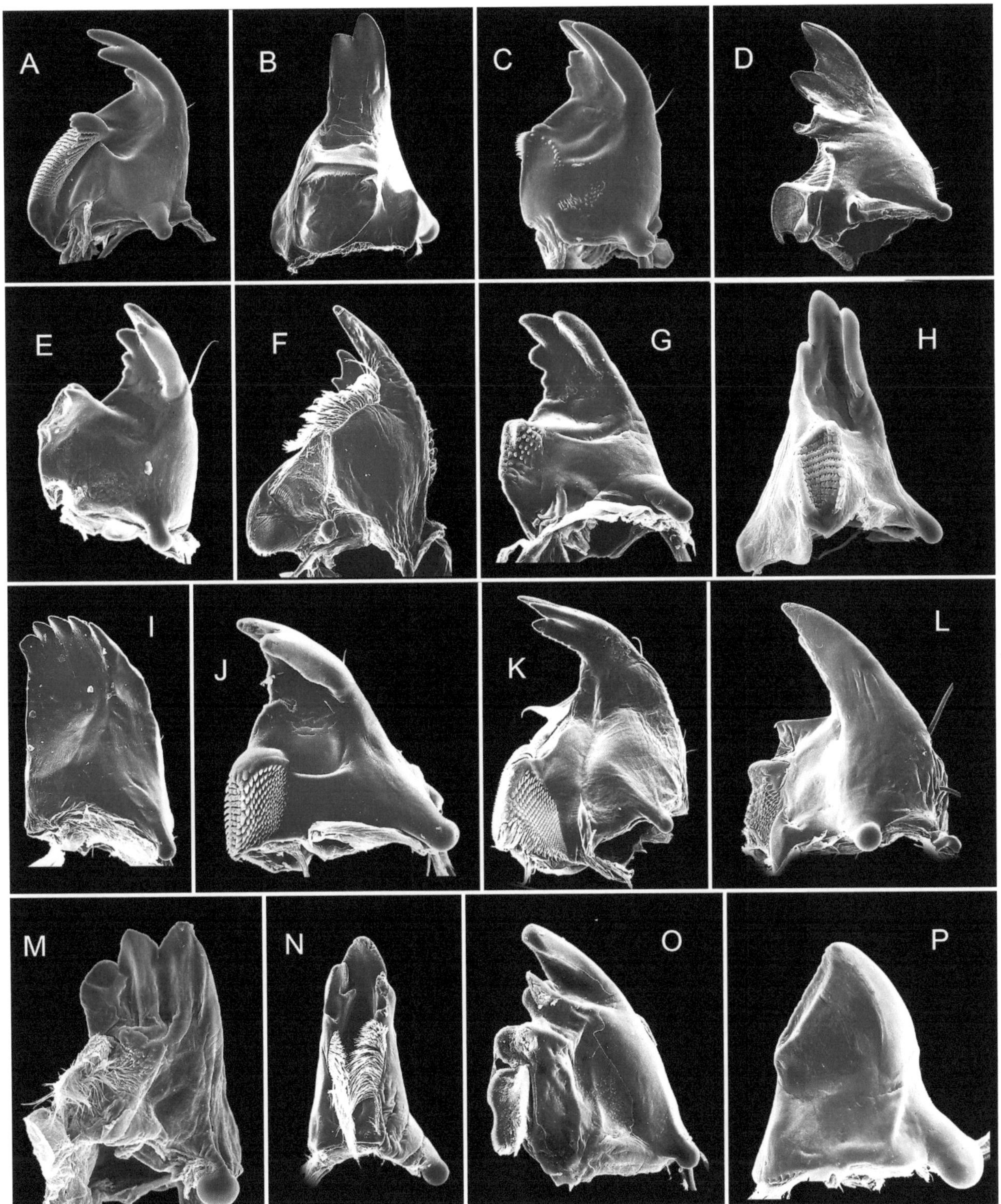

Fig. 41. Larval morphology. A–P, Mandible. A, *Elacatis* sp. (Salpingidae: Othniinae), mesoventral; B, *Sinodendron rugosum* Mannerheim (Lucanidae: Syndesinae), mesal; C, *Hyporhagus gilensis* Horn (Zopheridae: Zopherinae: Monommatini), ventral; D, *Dascillus davidsoni* LeConte (Dascillidae), mesoventral; E, *Lacconotus pinicola* Horn (Mycteridae: Eurypinae), ventral; F, *Nosodendron unicolor* Say (Nosodendridae), ventral; G, *Namunaria pacifica* (Horn) (Zopheridae: Colydiinae), mesoventral; H, *Enneboeus barrocolorado* Merkl (Archeocrypticidae), mesal; I, *Leptinotarsa lineolata* Stål (Chrysomelidae: Chrysomelinae), mesoventral; J, *Aulonium longum* LeConte (Zopheridae: Colydiinae), mesoventral; K, *Austrohyliota chilensis* (Blanchard) (Silvanidae: Brontinae), ventral; L, *Hapalips* sp. (Erotylidae: Languriinae), ventral; M, *Megalodacne* sp. (Erotylidae: Erotylinae), mesoventral; N, *Anchytarsus* sp. (Ptilodactylidae: Anchytarsinae), mesal; O, *Heterocerus gemmatus* Horn (Heteroceridae), mesoventral; P, *Mordella* sp. (Mordellidae), mesoventral.

Fig. 42. Larval morphology. A, *Lara avara* LeConte (Elmidae: Larainae), abdominal apex, lateral. B, *Hygrobia* sp. (Hygrobiidae), habitus, lateral. C, Genus ? sp. (Lycidae), habitus dorsal. D, Genus ? sp. (Mordellidae), head and thorax, lateral. E, *Stanus* sp. (Corylophidae: Corylophinae), abdomen, lateral. F, *Distocupes varians* (Lea) (Cupedidae), habitus, lateral. G, Genus ? sp. (Lycidae), head and prothorax, lateral. H, *Pseudomenes bakewelli* (Blackburn) (Eucnemidae: Pseudomeninae), head, dorsal. I, *Prodirhagus* sp. (Eucnemidae: Melasinae), head, dorsal. J, *Passalus* sp. (Passalidae), head and thorax, ventral. K, *Geotrupes spiniger* (Marsham) (Geotrupidae: Geotrupinae), head and thorax, ventral. L, *Acrossidius tasmaniae* (Hope) (Scarabaeidae: Aphodiinae), habitus, lateral.

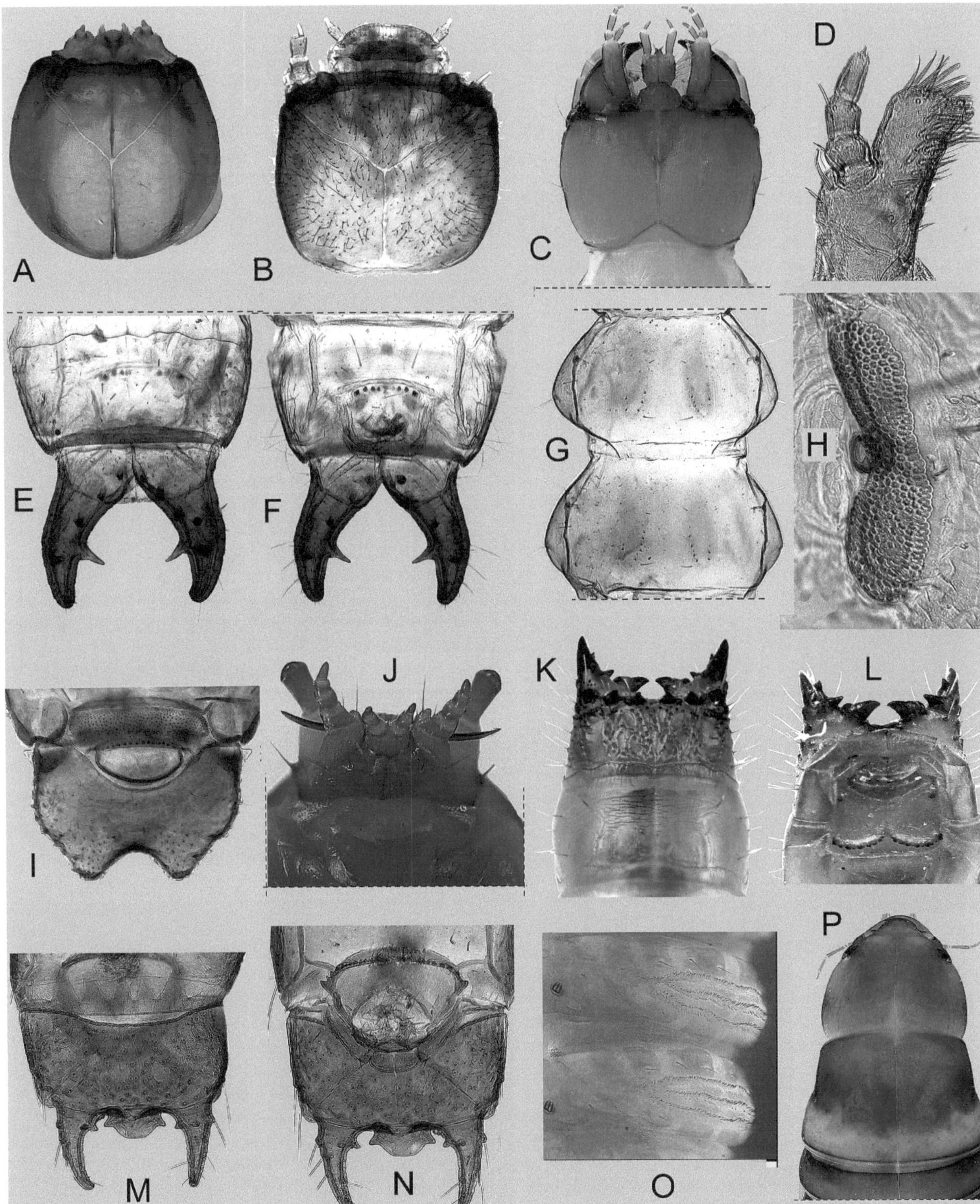

Fig. 43. Larval morphology. A–C, Head, dorsal: A, *Tranes lyterioides* (Pascoe) (Curculionidae: Molytinae), head, dorsal; B, *Collops* sp. (Melyridae: Malachiinae), head, dorsal; C, Genus? sp. (Carabidae: Harpalinae), head, ventral. D, *Elacatis* sp. (Salpingidae: Othniinae), maxilla, ventral. E, *Euryplatus dimidiatus* (Waterhouse) (Salpingidae: Inopeplinae), abdominal apex, dorsal. F, Same, ventral. G, *Lacconotus pinicola* Horn (Mycteridae: Eurypinae), two abdominal terga, dorsal. H, *Grynoma* sp. (Trogossitidae), abdominal spiracle. I, *Prostomis* sp. (Prostomidae), abdominal apex, ventral. J, Genus ? sp. (Lycidae), head, ventral. K–N, Abdominal apex: K, *Anaplopus tuberculatus* Blackburn (Pythidae), dorsal; L, Same, ventral; M, *Temnopalpus* sp. (Pyrochroidae: Pilipalpinae), dorsal. N, Same, ventral. O, *Hister nomas* Erichson (Histeridae), two abdominal terga, dorsolateral. P, *Hygrobia* sp. (Hygrobiidae), head and prothorax, dorsal.

a type of filter feeding occurring in larvae living in fluid substrates, such as slime fluxes (Nosodendridae) or artiodactyl dung (scarabaeine Scarabaeidae).

There are three types of accessory structures associated with the basal mola: a premolar tooth, a submolar lobe and a prostheca. The **premolar tooth** (Figs 39F, H, 41A) is a heavily sclerotised process which is actually part of the mola and often occurs on one mola only. The **submolar lobe** is often a hyaline area, usually setose, at the base of the mola in Biphyllidae, Anthicidae, Aderidae and some Scraptiidae, but in Byturidae (Fig. 48C), it may be a flexible, projecting setose lobe. A brush of hairs or **penicillus** may also occur at the molar base, as in some Bostrichidae (Fig. 48G), *Orphilus* (Dermestidae), some cucujoids and scarabaeoids. The **prostheca** is an important taxonomic character in beetle larvae, but there has been some confusion in the use of the term. We have restricted the term to any of the structures located just distal to the basal mola, but not part of it, which are almost always hyaline, at least basally, or composed of membrane or groups of hairs of different types. Structures similar to the prostheca may occur in mandibles lacking a mola and they are discussed below.

The most common type of prostheca is a fixed hyaline process, which is relatively rigid and more or less acute at the apex; it is sometimes said to be articulated, but this is rarely the case. A common type has a narrow base and simple, acute apex, as in Agyrtidae, Leiodidae, Cucujidae or Silvanidae (Figs 39G, 41K) but it may be serrate (Monotomidae, Fig. 48A, F) or bifid (Cryptophagidae, Fig. 48B), or may have a broader base and more or less angulate apex (Figs 41L, 47A). Other types of prosthecae may consist of a series of fringed membranes (Nitidulidae, Figs 38E, 39A), a brush of simple hairs associated with an acute process (Derodontidae), a group of comb-hairs (Biphyllidae, Fig. 48E) or a membranous lobe (Fig. 48G, M), as in some Bostrichidae and Endomychidae. Other structures which may be referred to as prosthecae include the comb-like process in Scirtidae (Fig. 48I) or the hyaline area and small, articulated process in Dascillidae (Fig. 41D).

In mandibles lacking a mola, there may be a variety of structures at the base of the mesal edge which are useful in identification. Commonly, there is a **brush** of hairs (Lampyridae, Fig. 40K, Byrrhidae, Fig. 47J; Histeridae, Fig. 47R); this is commonly called a **penicillus**, but that term has also been restricted to a slender, 'pencil' of hairs. There may be an acute, hyaline process similar to the simple prostheca (and sometimes called a 'retinaculum'), as in some paussine Carabidae, proteinine and omaliine Staphylinidae, Passandridae, some Bothrideridae (e.g. *Deretaphrus*), some Coccinellidae (Fig. 47Q) and some Ciidae (Fig. 47N), or a more complex bifid, serrate process, usually referred to as a **lacinia mobilis** or **lacinia mandibulae**, as in various Cleroidea (Fig. 47I). A group of two to five hyaline processes occurs in some Trogossitidae, Laemophloeidae, Phalacridae (Fig. 47L), Tetratomidae, Promecheilidae and Salpingidae (Fig. 39B), while a row of hyaline teeth occurs in monommatine Zopheridae (Fig. 41C). These could represent a highly reduced mola, a prostheca retained after the loss of a mola, or independently derived structures. Articulated processes also occur, and they may be covered with fine hairs, as in *Araeopidius* (Ptiliodactylidae), or have a brush of fine hairs at the apex, as in *Ptilodactyla* (Fig. 40B). Some mandibles have two types of basal structures, such as a brush and articulated process in some Elmidae (Fig. 47B) or a fixed, hyaline lobe and brush of hairs in *Dermestes* (Dermestidae) (Fig. 47M). A large, fringed, membranous lobe occurs without a mola in some Erotylidae (Fig. 41M), eustrophine Tetratomidae and Endomychidae (*Endomychus*).

The outer edge of the mandible in most larvae bears a pair of setae of fixed position, but in some groups (e.g. pimeliine Tenebrionidae), there are numerous setae that may be stout and bristle-like. In some ground-dwelling larvae (e.g. cebrionine Elateridae, pimeliine Tenebrionidae), the outer edges of the mandibles are carinate and serve as wedges used in burrowing. As mentioned above, the dorsal portion of the mandibular base may bear a patch of microtrichia arranged in rows; these occur in several families of Cucujoidea and Tenebrionoidea (Fig. 39A, G–H) and are part of the microphagous feeding system. In some Scarabaeidae (Rutelinae, Dynastinae, Cetoniinae and some Melolonthinae), the ventral surface of the mandible bears an oval patch of transverse striae, which forms a **stridulitrum**, lying opposite a **plectrum** consisting of a group of teeth on the dorsal surface of the maxillary stipes (**see** below).

As mentioned above, mandibles lacking a mola usually fall into two groups: the broader, cutting, biting, gouging or scraping type with a simple (Figs 41P, 47G) or multidentate (Fig. 41I, M) apex and the narrower, grasping or piercing type, which are usually falcate (Figs 40L, 47O, R) and may be perforate (Figs 40K, 47P). These, in turn, are associated with two different feeding strategies, namely **macrophagy**, or feeding on large particles (which are often cut or otherwise removed from a larger food mass and then swallowed whole) and **liquid-feeding** or **extraoral digestion** (where digestive enzymes are injected into a large food mass of the surrounding medium and sucked back into the gut through an **oral filter**, which excludes larger particles (**see** below)). While narrower falcate mandibles and extraoral digestion usually occur in predators, there are liquid-feeding phytophagous larvae in the family Elateridae and some predaceous larvae have relatively broad mandibles. Among the macrophagous types of larvae, wood-feeding or **xylophagous** forms are most common in the families Buprestidae, Callirhipidae, Melandryidae, Mordellidae, Cerambycidae and Curculionidae, while **phytophagous** forms or those specialising on leaves and herbaceous tissue (both fresh and decomposed) occur in Byrrhidae, some Erotylidae, epilachnine Coccinellidae, Chrysomelidae and some Curculionidae. Also falling into this group are those larvae which scrape algae from rock surfaces (e.g, Psephenidae and Elmidae). Larvae specialising on the tougher fruiting bodies of higher Badisiomycetes are in the families Trogossitidae (Peltini, Calitini), Erotylidae, Ciidae, and some Mordellidae, while seed feeders occur in bruchine chrysomelidae, Anthribidae and Curculionidae.

Narrow and falcate mandibles occur in most Adephaga, most Hydrophiloidea, many Staphylinoidea, many Elateroidea, and in a few cucujoid and tenebrionoid triungulins. One development in mandibles of this group is the formation of an open groove or closed perforation, which serves to channel digestive fluid into the food mass; the internal channel is sometimes called a **blood channel**. An open or partly closed mandibular groove occurs in larvae of some Noteridae and Dytiscidae, spercheine Hydrophilidae, Omalisidae, Drilidae, Cantharidae and a few Cleridae, while **perforate mandibles** (Figs 40G, K, 47P) are characteristic of most Dytiscidae, Gyrinidae, Brachypsectridae, Phengodidae and Lampyridae. A different type of perforate mandible with broad base and relatively large internal perforation occurs in Haliplidae (Fig. 47H) and apparently serves to suck algal cells into the mouth (Hickman 1931). Another unusual type of mandible apparently associated with liquid feeding is that in the larvae of Lycidae (Figs 43J, 47C); the mandibles are non-opposable and each one is split longitudinally to form two blades, which fit together to form a channel; it is now thought that one of the two blades is actually of clypeal origin. Styliform mandibles are also non-opposable, but they are extremely narrow and may be capable of anteroposterior movement; they are usually enclosed together with styliform maxillary lobes in a median tube formed from the labrum and labium, as in some camiarine Leiodidae and most cerylonine Cerylonidae (Fig. 46M); in the genus *Cerylon*, however, they are endognathous or enclosed within the mouth cavity (Fig. 46N). Non-opposable mandibles also occur in the families Cerophytidae, Throscidae, Eucnemidae and cardiophorine Elateridae, but these are not associated with a median tube. In some eucnemids the mandibles are movable but exodont (with laterally projecting teeth), in some other Eucnemidae, Throscidae and Cardiophorinae, they are fused to the head capsule, and in Cerophytidae they are styliform and enclosed with the maxillary lobes in two separate channels.

Ventral Mouthparts

The **ventral mouthparts** (or **maxillolabial complex**) consist of the paired maxillae and unpaired labium, which may be more or less independent or closely coadapted to move as a single unit (**see** below). The maxilla is composed of the same basic parts as in the adult, namely the **cardo** (pl. cardines), **stipes** (pl. stipites), **palpifer**, **palp**, **galea** or outer lobe and **lacinia** or inner lobe; but in many cases the galea and lacinia are fused to form a single **mala**. In many groups, the maxilla is movable and is joined basally to the labium by a **maxillary articulating area** (basimaxillary membrane of Das 1937) (Figs 36A, 37A); this area is usually absent in protracted ventral mouthparts (Figs 48N, 49A) and in those retracted types in which the maxillae and labium are connate or solidly fused (Figs 37B, 38D, G, I, 48P).

The cardo may be divided into two distinct sclerites, but most often it is either undivided or bisected by an internal ridge; areas mesad and distal to this ridge have been called **proxicardo** and **disticardo**, repsectively. In most larvae the cardo is more or less transversely oriented or slightly oblique, but in almost all Nitidulidae (Fig. 49F), it is longitudinal. In some larvae with protracted mouthparts (Phalacridae, 49A, Smicripidae, Laemophloeidae, Histeridae), the cardo may be absent or represented by a membranous area at the maxillary base. In some other groups, like Coccinellidae, the cardo is fused to the stipes. The two cardines are usually widely separated, lying on either side of the labium, but in some larvae they may be approximate, displacing the labium anteriorly (Callirhipidae, Fig. 49I; Heteroceridae), contiguous, meeting behind the labium (Limnichidae, Fig. 49E), some Elateridae) or fused into a single sclerite (some Elateridae, Fig. 37B; Chelonariidae; Brachypsectridae). In lamiine Cerambycidae, the entire area at the base of the ventral mouthparts is lightly sclerotised, without distinct separations of the labium, cardines and stipites.

The stipes is the main body of the maxilla, housing the muscles that operate the palp and apical lobes. It is usually longer than wide, but in some groups (e.g. Cleridae, Phalacridae, Fig. 49A), it may be subquadrate or wider than long. The stipes is occasionally divided into basal and apical sections (**proxistipes** and **dististipes**, respectively), and Ghilarov (1964) used the term **costistipes** for the structure in Hydrophiloidea here referred to as the palpifer. Another term, juxtistipes, was used by Böving and Craighead (1931) for an area extending mesally from the base of the stipes. The dorsal surface of the stipes near the base in various Scarabaeidae may have a patch or oblique row of stridulatory teeth, but only in Dynastinae, Cetoniinae, Rutelinae and some Melolonthinae is there an opposing stridulitrum on the ventral surface of the mandible (**see** above).

The **maxillary articulating area** is well developed and broadly triangular in most basal groups, loosely joining the base of the maxilla with that of the labium, so that the two maxillae are capable of independent movement (Figs 36A, 37A). In some larvae, however, this area may become narrowly oval, as in Ciidae or Ptinidae, so that movement is more restricted. The area is usually membranous, but in some groups it may be partly sclerotised and occasionally it is divided into two parts. In many Ptilodactylidae, and to some degree in Scirtidae and Psephenidae, the area may be well developed but partly or entirely concealed behind a laterally expanded labium (Figs 38C, 48O). In other groups, there is no articulating area, since the maxillae and labium have become closely coadapted and more or less connate, with further restriction of independent movement. Further consolidation of the maxillolabial complex through fusion is discussed below.

The **palpifer** (maxillary palpiger) is usually a shelf-like area located laterally near the apex of the stipes and bearing the maxillary palp. There has been some confusion about the number of palp segments or **palpomeres**, because an enlarged palpifer may be segment-like and even articulated, and so considered to be an additional palpomere by Lawrence and Newton (1982), Lawrence (1991a) and Lawrence *et al.* (1999a). It now seems more likely to us that a maxillary palpifer is usually present and may be articulated and segment-like. No distinction was made between 3-segmented and 4-segmented palps in the recent matrix published by Lawrence *et al.* (2011), and the development and basal articulation of the palpifer were coded independently (**see** below).

A 3-segmented maxillary palp is probably a plesiomorphic feature in beetle larvae, but an extra palpomere may be present in some Adephaga, Staphylinidae and Scirtidae. Reduction to two palpomeres occurs in several groups, including Hydroscaphidae, Sphaeriusidae, the clambid genera *Calyptomerus* and *Acalyptomerus*, late instar Rhipiceridae, Buprestidae, eubrianacine Psephenidae, some Eucnemidae, a few Phengodidae, Bostrichidae and Ptinidae, rentoniine Trogossitidae, some Passandridae, Smicripidae, cybocephaline Nitidulidae, the bothriderid genus *Sosylus*, Discolomatidae, some Corylophidae and Coccinellidae, Mordellidae, some Ripiphoridae, some Chrysomelidae and most Curculionoidea. Single-segmented palps are known in first instar Rhipiceridae, intermediate and late instar Ripiphoridae, a few bruchine and cassidine Chrysomelidae and some Curculionidae, and maxillary palps are absent in coarctate Meloidae and some Eucnemidae. The apical maxillary palpomere may have a group of sensilla at the apex (the 'bouquet sensoriel' of Corbiere-Tichané 1974). At the base of this palpomere there are often one or more **digitiform sensilla**, which are flattened and located in a shallow cavity, where they are appressed to the surface and difficult to see; these stuctures, which occur in both maxillary and labial palps in larvae and adults, apparently function as tactile mechanoreceptors (Zacharuk 1962; Zacharuk *et al.* 1977; Honomichl 1980).

As in the adult, there may be two apical lobes on the larval maxilla: an articulated **galea** or **outer lobe**, which may be subdivided and a fixed **lacinia** or **inner lobe**, often appearing to be a mesal extension of the stipes. A 2-segmented galea occurs in Carabidae, Gyrinidae, Trogidae, some Byrrhidae, some Ptilodactylidae, Eulichadidae, Artematopodidae, Brachypsectridae, various Elateridae and a few soft-bodied Elateroidea, but in most of these groups both apical lobes are relatively small and the galea sometimes palpiform (Figs 38C–D, G, 48J, 49C, E), while in Trogidae they are both falcate with apical teeth. The major evolutionary trends exhibited by the maxillary apex include: (1) loss of the basal segment of the galea and its basal articulation resulting in two fixed lobes (Fig. 49S–T); (2) loss of one lobe, usually the lacinia, resulting in a single process, often referred to as a **mala** (Fig. 49L); (3) basal fusion and then complete fusion of galea and lacinia, resulting in a combined structure, also called a **mala** (Figs 48D, H, L, 49H, U); and (4) loss of both apical lobes, so that only the palp arises from the palpifer or apex of the stipes. A basally articulated galea and fixed lacinia occurs in Archostemata, Dascillidae, the more basal groups of Scarabaeoidea, most Byrrhoidea and Elateroidea, Endecatomidae and Belidae, while an articulated galea, without a lacinia, characterises some Carabidae, Dytiscidae, Noteridae, Amphizoidae, Hydrophiloidea, paederine Staphylinidae, Buprestidae, Callirhipidae, some Elateridae and most soft-bodied elateroid groups. In Hydrophiloidea, the small palpiform galea is attached, not to the apex of the stipes, but to the palpifer (Fig. 49P), and this structure is referred to in many works (including Lawrence 1991a, on which this section is largely based) as an independently derived appendage of the first palpomere. The complete loss of apical lobes occurs in some Dytiscidae, Hygrobiidae, Eucnemidae, Ripiphoridae and Meloidae. A maxilla with both galea and lacinia fixed occurs in Haliplidae, several more basal groups of Staphylinoidea, some Scarabaeoidea, most Scirtoidea, Rhipiceridae, Psephenidae, Derodontoidea, Bostrichoidea, and a few Cucujoidea, while a single fixed mala occurs in rhysodine Carabidae, Haliplidae, Dytiscidae, most Staphylinoidea, Clambidae, more derived subfamilies of Scarabaeidae, some Buprestidae, Throscidae, some Cantharidae, Lampyridae, most Cucujoidea and Tenebrionoidea, Chrysomeloidea and Curculionoidea. Intermediate conditions may occur where the mala is apically cleft (Fig. 49H).

The galea is usually broader, more rounded or truncate and more setose than the lacinia, except in those groups where both lobes have become reduced and palpiform, but the galeal structure varies considerably and it may be falcate like the lacinia in Dascillidae (Fig. 49O) and some Scarabaeoidea. In Hydraenidae and Ptiliidae, the galea is said to be **fimbriate**, with one or more dense rows of modified seteae along the outer edge of the apex (Fig. 49T), while in Scirtidae it is very broad and its apex is covered with comb-hairs (Fig. 27H). The lacinia is often narrower and more falcate than the galea and may have one or more apical teeth or spines at the apex; these apical teeth are usually called **unci**, but the term **uncus** is used for a variety of lobes or teeth at the inner edge of the maxillary apex (Figs 48L, 49D, H). A rounded **laciniar lobe** occurs in some Ptinidae and a similar structure located dorsally and subapically in Ciidae (Fig. 49G) may represent the lacinia or possibly a secondary structure that has developed on the mala. In most Bostrichoidea, the lacinia has become reduced to a simple, spine-like process. (Fig. 49K).

The form of the fixed mala also varies considerably, but it is usually rounded or truncate. A falciform mala (Fig. 49U) occurs in some Staphylinidae, Throscidae, languriine Erotylidae, Biphyllidae, Monotomidae, Phloeostichidae, Cucujidae, Silvanidae, some Bothrideridae, a few Endomychidae and Corylophidae, zeugophorine Megalopodidae and donaciine Chrysomelidae. In *Donacia* there appears to be two maxillary lobes, but one of them may be a modified seta. A styliform mala (Fig. 49Q) occurs in camiarine Leiodidae, proteinine and a few omaliine Staphylinidae, Cerophytidae, the eucnemid genus *Phyllocerus*, and the cerylonine Cerylonidae. In some groups with a truncate mala, there is a cleft at the apex (Fig. 49H), and within this cleft there may be a malar sclerome; it is not certain whether this cleft represents an incomplete fusion of galea and lacinia or a secondary subdivision of the mala. A cleft mala is most commonly found among the Tenebrionoidea, but it also occurs in Erotylidae, Helotidae, Lymexylidae and some staphylinoids. The malar apex sometimes has a distinct lobe at its inner angle; this structure, which occurs in Nitidulidae, Trictenotomidae, anaspidine Scraptiidae and a few other groups, is sometimes called an **uncus**, but as pointed out above, that term may apply to a variety of structures at the inner apical angle. Other structures occurring at the malar apex include **spatulate setae**, as in some spore-feeding species, and the **pedunculate seta** (an elongate tubercle with one or more setae at the apex) found in all Cleroidea, except Phloiophilidae and basal Trogossitidae, and in Coccinellidae. The inner edge of the mala often bears one or more rows of spines, which may continue along the mesal edge of the stipes, and its dorsal surface may be armed with spines, setae or a dense brush of hairs forming part of an **oral filter** (Figs 38B, E, 40L).

The **labium** is composed of an apical **prementum** and basal **postmentum** (Fig. 37B), but the latter is often subdivided into a basal **submentum** and a **mentum**, which lies between it and the prementum (Fig. 36A). The submentum is often fused with the gula forming a combined area called the **gulamentum** (Fig. 37A) is formed, but the limits of the submentum and gula may be arbitrarily defined by an imaginary line extending between the bases of the cardines. These terms do not necessarily represent true homologies, and alternative terminology may be found in Anderson (1936), Das (1937), Dorsey (1943), and Crowson (1981). The presence of only two labial sclerites occurs throughout the order, but is most commonly found in Adephaga, Byrrhoidea, Elateroidea, Cleroidea, Chrysomeloidea and Curculionoidea. A single labial plate occurs in some Adephaga, Hydrophiloidea, Ripiphoridae and some Curculionoidea.

The prementum represents the fusion of paired labial sclerites that are serially homologous to the maxillary stipites, and it usually bears a pair of **palpigers**, to which the labial palps are attached, rarely a pair of **glossae**, which are homologous to the maxillary laciniae, and most often a medial structure, the **ligula**, which represents the fused glossae. **Paraglossae**, found in many other insect orders, may not occur in beetle larvae, although there are paired lobes on either side of the ligula in Staphylinoidea that have been called paraglossae by Böving and Craighead (1931). In some larvae (Agyrtidae and other basal staphylinoids, some Eucinetidae,

Dascillidae), the ligula is broad and distinctly bilobed, but more often it is undivided and in various groups it is reduced or absent. In Archostemata (Fig. 40A, J) and Callirhipidae (Figs 38D, 49I), the ligula has become heavily sclerotised and wedge-like forming a **ligular sclerome**.

The usual number of labial palpomeres is two, but reduction often occurs and in a few cases (some Carabidae and Dytiscidae, Gyrinidae, some staphylinine Staphylinidae, some Histeridae), there is an apparent increase to three or four (some Dytiscidae). Single-segmented labial palps occur in first instar *Priacma* (Cupedidae), Micromalthidae, rhysodine Carabidae, spercheine Hydrophilidae, most Clambidae, Rhipiceridae, ceratocanthine Hybosoridae, most Buprestidae, many Eucnemidae, dinoderine and lyctine Bostrichidae, rentoniine Trogossitidae, monotomine Monotomidae, Hobartiidae, cryptophagine Cryptophagidae, a few Endomychidae and Coccinellidae, the corylophid genus *Holopsis*, many Latridiidae, the mycetophagid genus *Thrimolus*, a few colydiine Zopheridae, triungulins of the meloid genus *Tetraonyx*, eurypine and hemipepline Mycteridae, leiochrinine Tenebrionidae, the pyrochroid genus *Cononotus*, ripidiine Ripiphorinae, various chrysomelidae, a few Brentidae and some Curculionidae. In Buprestidae, the labial palps are represented by minute papillae or setose areas, which may not be true palps at all, and labial palps are absent in larvae of *Cerophytum*, some Eucmenidae, coarctate instars of Meloidae and ripiphorine Ripiphoridae, bruchine Chrysomelidae, the anthribid subfamily Urodontinae and the anthribid genus *Holostilpna* Jordan.

The mentum or postmentum is very broad in some groups, like Scirtidae, Ptilodactylidae (Fig. 38C) and Psephenidae, where it may be expanded laterally to conceal the maxillary articulating area. In anchytarsine Ptilodactylidae, it is divided into three parts by longitudinal sutures (Fig. 48O). In some other groups, like trogossitine Trogossitidae and Elateridae, it may be much narrower than the stipes, and in the latter group it may be narrowed posteriorly and separated from the gular region by contiguous or fused cardines (Fig. 37B).

The ental surface of the labium, posterior to the ligula, is called the **hypopharynx** (Fig. 36B). In some larvae it is a simple membranous lobe, usually clothed with short hairs, but in others it may have a more complex structure. The hypopharynx is often supported by a **hypopharyngeal bracon** (**see** above) and sometimes by a pair of **hypopharyngeal rods** extending ventrally on each side to the base of the mentum or a **hypopharyngeal suspensorium** consisting of one or more pairs of rods extending dorsally on either side of the cibarium and inserting in its dorsal wall. The detailed structure of this suspensorium and its variation within the order have never been investigated. The hypopharynx is simple in larvae with the mandibular mola reduced or absent, but in those with a well developed mola, there is usually a sclerotised bar or cup-like or molar-tooth-like structure called the **hypopharyngeal sclerome** (Figs 36B, 38B, 50B, D), which acts in conjunction with the mandibles in processing food. In some families, like Elateridae, there may be a bar-like sclerotisation which is not associated with mandibular molae, but its function is not clear. In Dascillidae and Scarabaeoidea, there may be two or more scleromes called **oncyli** (sing. **oncylus**), while in Dascillidae and Scirtidae, the hypopharynx may also have two or more hypopharyngeal combs. According to Beier (1952), these combs in Scirtidae are associated with a filter-feeding apparatus: fine particles are gathered by the maxillary comb-hairs, which are then brushed over the combs, where food particles are collected, moved posteromesally towards the back of the hypopharynx and compacted into a solid pellet by the hypopharyngeal sclerome and the mandibular molae (Fig. 48I). The compacting function of the hypopharyngeal sclerome is also seen in the mycophagous tenebrionid *Platydema ellipticum* (Fabricius) (Fig. 38B), where the larva shears masses of reinforced fungal hyphae from a fungal fruiting body and compacts them between the concave molae and the concave sclerome, probably with the aid of the setose maxillae. In addition to the sclerome, the hypopharynx may have well defined brushes of hairs or lateral lobes, which have been called **maxillulae** by Böving and Craighead (1931). In larval Oedemeridae, there may be a distinct columnar structure, the **prehypopharynx**, between the sclerome and the ligula (Rozen 1960).

In several groups of beetles, there has been a consolidation of the maxillae and labium by reduction and loss of the articulating area and connation of the cardines and stipites with the postmentum or mentum. The resulting maxillolabial complex can be moved only as a single unit. This process is usually accompanied by the loss of the mandibular mola and reduction of the apical maxillary lobes. Connation of the maxillae and labium (Figs 37B, 49E) occurs mainly in some Byrrhoidea, Elateroidea and Cleroidea, but it is also known in a few cucujoids, some Chrysomelidae (especially Cassidinae and Cryptocephalinae) and Curculionoidea. Partial fusion of these elements (partial loss of the sutures) is known in Callirhipidae (Fig. 38D), Dryopidae, Brachypsectridae (Fig. 40G) and Lycidae, while complete fusion (loss of sutures) occurs in Chelonariidae and various Eucnemidae. This consolidation is often associated with liquid-feeding, and the ental surface of the maxillolabial complex may be densely clothed with short hairs forming – with a similar brush on the epipharynx – an **oral filter**, which serves to keep solid particles out of the gut. There are no labial silk glands in beetles, as there are in Lepidoptera, but maxillary glands, located between the bases of the maxillae and labium, have been reported by Srivastava (1959), and paired tubular glands originating in the thorax and opening in the hypopharyngeal region were considered by Beutel and Pollock (2000) and Beutel and Ślipiński (2001) to be a synapomorphy for the series Cucujiformjia.

Thorax

As in the adult, the thorax consists of three segments: **prothorax**, **mesothorax** and **metathorax**, but unlike that of the adult the larval prothorax is usually larger rather than smaller than either mesothorax or metathorax or in a few taxa (e.g. most Buprestidae and some Cerambycidae) larger than the two combined. Each thoracic segment usually bears a pair of walking legs, but these may be highly reduced or absent in Buprestidae, most Eucnemidae, some Cerambycidae and Chrysomelidae and most Curculionoidea. The **cervical region** or neck is usually membranous, but sometimes there are dorsal or ventral sclerotisations, the latter of which are sometimes confused with anterior plates of the prothoracic venter. In cebrionine Elateridae, there is a large **cervical membrane** lined with ridges and asperities, which is capable of being everted, displacing the head dorsally and posteriorly. A pair of **lateral cervical sclerites**, resembling those of adult Polyphaga, may be present in the larvae of some Scirtidae.

The three thoracic segments and abdominal segments I to VII or VIII may be similar in general structure, except for thoracic modifications associated with the head and leg attachments. The **dorsum** of each segment usually consists of a large plate called the **tergum** (also called a tergite or mediotergite) and one or more pairs of smaller plates or **laterotergites**. Each tergum is usually divided at the midline by a narrow ecdysial line, but in some larvae, there is a broader division, with the formation of paired smaller tergites; paired tergites occur on all thoracic and most abdominal segments in some Sphindidae and in early instar Boganiidae, but on mesothorax, metathorax and usually segment IX in some Cleridae and Trogossitidae. In some lightly pigmented larvae (Gyrinidae, Histeridae, Trogidae, some Elateridae, many Cleroidea, various Chrysomeloidea and Curculionoidea), only the prothorax bears a tergal plate. The **pleuron** is usually reduced and membranous, except on the thorax, where is consists of sclerites partly surrounding the leg articulations, and the **venter** usually consists of a single plate, except on the thorax, where poorly understood subdivisions usually occur.

The protergum is usually simple, consisting of no more than a sclerotised plate, but in some Lymexylidae, Buprestidae, Oedemeridae and Cerambycidae there are patches of asperities, in some Belidae there may be a prominent transverse protergal carina, in platypodine Curculionidae there is a transverse row of ring-like sclerites, and in some Eucnemidae and Bostrichidae there are paired rod-like sclerotisations. Larvae with patches of asperities on the metatergum or meso- and metaterga occur in the families Scarabaeidae, Lucanidae, Ptinidae, Zopheridae, Synchroidae, while rows of asperities or transverse carinae on these segments are known in rhysodine Carabidae, Trogidae, Dascillidae, Elateridae, Zopheridae, and Stenotrachelidae. In some soft-bodied larvae, there may be a transverse folding of the tergal region forming two to four **plicae**. When the tergum is divided transversely into two parts, the anterior division is sometimes called the **prescutum** and the posterior one the **postscutum**, and when 3-folds are present, they may be referred to as prescutum, **scutum** and **scutellum**; however, these, terms based on the morphology of the adult tergum, are probably not very meaningful.

The thoracic pleuron (Fig. 37C) is divided by a short **pleural suture**, extending dorsally or laterally from the coxal articulation and representing the internal **pleural apophysis**. This suture divides the pleuron into an anterior **episternum** (=prehypopleurum) and posterior **epimeron** (=posthypopleurum); sometimes there are an additional two sclerites: the **precoxale** (pl. **precoxalia**) in front of the episternum and the **postcoxale** (pl. **postcoxalia**) behind the epimeron, but these are often fused with the episternum and epimeron or with adjacent sternal plates. In some Tenebrionidae, the postcoxale on each side is enlarged and extends mesally to fuse with the **sternellum**, forming a crescentic sclerite behind the coxae, sometimes called the **postcoxal bridge**, but that term has also been used as a synonym of postcoxale.

The thoracic venter (Fig. 37C) is more lightly sclerotised and its subdivisions are neither clearly defined nor clearly separated from the pleural sclerites. In many larvae there is a pair of **sternal pits** lying between the coxae and often joined by a transverse **sternacostal suture**, which separates the anterior **basisternum** and posterior **sternellum**. There is often a large sclerotised plate on the prothoracic venter, which is set off from the basisternum and extends to the anterior edge of the thorax; this has been called the **presternum** (or prosternum), and it may be variously subdivided or joined laterally to the precoxalia. In addition, there may be a smaller sclerite, often beneath the head, which is thought to be part of the cervical region; this was called the **cervicosternum** by Watt (1970), but was considered by St. George (1939) to be the presternum. Specialised ventral thoracic structures, usually the prothorax, include single teeth or spines, patches of asperities (Fig. 56H), or supporting rods of various kinds (Fig. 45J). Other terms used for regions of the thorax (both dorsum and venter) may be found in Craighead (1916, 1923), Wade and St. George (1923), St. George (1924, 1926, 1939), Böving and Craighead (1931), Hyslop and Böving (1935), Glen (1950), Watt (1970, 1974b) and Švácha and Danilevsky (1987).

The basic number of larval leg segments in Archostemata and Adephaga (Fig. 50G–H) is six (**coxa**, **trochanter**, **femur**, **tibia**, **tarsus** and **pretarsus** (consisting of 1 or 2 **claws**), whereas in Myxophaga and Polyphaga (Fig. 50F, I, K–L), the tibia and tarsus have fused to form a **tibiotarsus**, so that the legs never have more than five segments. Some authors, including Crowson (1964b) and Lawrence and Britton (1991), considered the last claw-like segment in Polyphaga to be a **tarsungulus** (fused tarsus and pretarsus) but this has been refuted based on developmental evidence (Tikhomirova 1983; Švácha 1992).

Six-segmented legs occur in all Archostemata and Adephaga, with the exception of the legless forms of Micromalthidae, some myrmecophilous paussine Carabidae (three segments), and ectoparasitic lebiine and brachinine Carabidae (four segments). Five-segmented legs are characteristic of all Myxophaga and many Polyphaga, but reduction in the number of segments has occurred throughout the latter group. Within Polyphaga, four-segmented legs occur in micropepline Staphylinidae, most sphaeridiine Hydrophilidae, some Geotrupidae, a few Callirhipidae and Eucnemidae, euderiine Bostrichidae, Cerambycidae, some megalopodine Megalopodidae, and various groups of Chrysomelidae. Reduction to two or three segments has occurred in georissine Hydrophilidae, Rhipiceridae, Geotrupidae, scarabaeine Scarabaeidae, Passandridae, Mordellidae, Meloidae, Cerambycidae, palophagine and some megalopodine Megalopodidae, some bruchine Chrysomelidae, Anthribidae, Caridae and Brentidae. Larvae having single-segmented legs are known in rare Buprestidae and Eucnemidae, and in a few Cerambycidae, leaf-mining cassidine Chrysomelidae, Nemonychidae and Anthribidae. Legless larvae occur in several Polyphaga, including a few sphaeridiine Hydrophilidae, most Buprestidae and Eucnemidae, a few Ptinidae, most bothriderine Bothrideridae, some Ripiphoridae and Meloidae, some Cerambycidae, zeugophorine Megalopodidae, a few Chrysomelidae and most Curculionoidea. In some of these, however, there may be paired thoracic protuberances, called **pedal lobes**, which are thought to be leg remnants.

In many substrate-inhabiting larvae, the prothoracic legs are enlarged and modified for digging; examples may be found among the Tenebrionidae (Alleculinae, Pimeliinae, Helaeini, Blaptini) and in the Cerophytidae and cebrionine Elateridae. Prehensile prothoracic legs are formed by the opposition of the femur (Noteridae) or the tibia (Haliplidae, matine Dytiscidae) to the more apical leg segments, In several groups of Scarabaeoidea, **stridulatory organs** are formed by the meso- and metathoracic legs (Fig. 50I, K) (Pleocomidae, Passalidae, Geotrupidae, Lucanidae and ceratocanthine Hybosoridae) or pro- and mesothoracic legs (hybosorine Hybosoridae). Sound is produced by a **plectrum** consisting of a carina or row of tubercles, located on the trochanter, femur or apex of the posterior leg, rasping against a **stridulitrum (pars stridens)** consisting of a variable group of tubercles or ridges located on the coxa of the anterior leg. In Passalidae (Figs 42J, 50N) and some Geotrupidae (42K), the metathoracic leg is highly reduced and functions only as a plectrum.

Although the pretarsus is usually claw-like, it may be rounded, reduced or absent in larvae with reduced legs (Figs 42D, 50C). Long pretarsi occur in some triungulins, and in those of some Meloidae, the claw is **spatulate**, with two large setae at its base, so that there appear to be three claws (Fig. 51A). The prothoracic pretarsus may be divided into a lightly sclerotised basal portion and heavily sclerotised apical digging claw in some Tenebrionidae, while in the exotic family Cerophytidae, the pretarsus is enlarged, heavily sclerotised and bifurcate. The polyphagan pretarsus usually bears a pair of basal setae placed side by side, but in certain families the setae are placed in a line with one distal to the other. Plurisetose pretarsi occur in some staphylinine Staphylinidae, spercheine Hydrophilidae, Dascillidae, Scarabaeidae, Eulichadidae, Cantharidae, Ptinidae and Lymexylidae. Reduction to a single pretarsal seta has occurred several times, occurring in all Myxophaga, the byrrhoid families Dryopidae, Lutrochidae, Limnichidae, Heteroceridae, Elmidae, Psephenidae, Cneoglossidae and Ptilodactylidae, all Cleroidea, and the cucujoid families Byturidae, Biphyllidae, Bothrideridae, Cerylonidae, Alexiidae, Discolomatidae, Endomychidae, Coccinellidae, Corylophidae and Latridiidae. Special structures at the apices of the larval legs for adhering to surfaces include **adhesive setae** at the tibiotarsal apex in Coccinellidae or on the pretarsus of meligethine and cybocephaline Nitidulidae, and the **pulvillus** or **paronychial appendix** at the base of the pretarsus in some Chrysomelidae. A basal pretarsal tooth occurs in some Coccinellidae and Chrysomelidae, and a pectinate pretarsal claw occurs in the exotic family Meruidae.

Abdomen

The abdomen of a beetle larva usually consists of ten segments, with the first seven more or less similar in form, segment IX and sometimes VIII variously modified and X often reduced. In the elaterid subfamily Cardiophorinae the abdomen is very long with each segment transversely subdivided, so that it appears to have more than ten segments. A typical abdominal segment consists of a dorsal **tergum** (or dorsum), a reduced, membranous **pleuron** and a ventral **sternum** (or venter). The tergum may be subdivided into a median plate (**tergite** or **mediotergite**) with one or more **laterotergites** on each side, often extending onto the ventral surface, and the spiracles often located in one of the laterotergites. The sternum is usually not subdivided, although in some carabid and staphylinid larvae two to several sternal sclerites may occur.

Abdominal terga of many beetle larvae bear special kinds of armature used to gain leverage when burrowing into soil or wood or beneath tree bark. Patches of asperities or spinules, similar to those on the thoracic terga (Fig. 51J–K, M), occur in various groups, including Histeridae, many Scarabaeoidea, some Elateridae, Ptinidae and Lymexylidae, Synchroidae, most Oedemeridae, some Cerambycidae and rhynchitine Attelabidae. There may be transverse rows of asperities, sometimes forming carinae (Fig. 51U), in rhysodine Carabidae, many Elateridae, monommatine Zopheridae and some Stenotrachelidae, or curved rows forming complete or incomplete circles (Figs 43G, 51N, R–T) on the terga and sometimes sterna in Laemophloeidae, some Phalacridae (e.g. *Litochrus*), Phloeostichidae and Mycteridae. Another type of structure performing a similar function are the **ampullae** (also called **ambulacral warts**), hump-like projections sometimes armed at the apex with asperities; these occur in Archostemata (Fig. 56D), Histeridae (Fig. 43O), some trogossitine Trogossitidae, Cleridae, Melandryidae, Mordellidae, Oedemeridae and some Curculionoidea. In larvae of cicindeline Carabidae (Figs 51B, 56G) a large protuberance on tergum V bears three pairs of hooks, and in the carabid genus *Sphallomorpha* there are rows of stout spines on terga V to VII (Figs 51E, 56F); these structures serve to anchor the larva in its burrow. Some larval Eucnemidae have matte patches consisting of numerous minute asperities, which apparently assist their locomotion in woody substrates. Many soft-bodied larvae have the abdominal terga subdivided into two to 4-folds or **plicae** (Figs 34M, 51D), but the names applied to these vary with the taxon studied (Ritcher 1966 for Scarabaeoidea, Parkin 1933 or Böving 1954 for Ptinidae, Anderson 1952 or Thomas 1957 for Curculionoidea).

Ambulatory modifications on the abdominal sterna are more or less similar to those on the terga and include ampullae, plicae, patches or rows of asperities and matte patches. In some larvae, however, there are distinct ventral **prolegs** similar to those found in Lepidoptera, but usually without the apical hooks or **crochets**. Prolegs with simple or setose apices (Fig. 51D) occur on segments I to VII or VIII in larvae of various Chrysomelidae (Criocerinae, Eumolpinae, Galerucinae) and Curculionidae (*Cionus*, Hyperini), while most Oedemeridae have asperity-bearing prolegs on segments II to III, IV or V (Fig. 51I). Larvae of some Hydrophilidae (e.g. *Enochrus*) have prolegs with crochets at the apex on several abdominal sterna. In some byrrhoid larvae the sternal regions are very narrow and may be absent posteriorly, so that one or more posterior abdominal segments form a continuous cylinder without sutures (Fig. 51H).

Abdominal Apex

Variations in the abdominal apex include the relative size and position of segment X and various modifications of both IX and X. A type of abdominal apex that may have been ancestral in Coleoptera is characterised by having an unmodified, subterminal ninth segment and a terminal tenth segment, with distinct tergum and sternum and a transverse anal opening; such a condition may be found in the larvae of Dascillidae, Scarabaeidae (Fig. 34M) and Byrrhidae (Fig. 34N), and with slight modifications in other families. Major changes that have taken place in the abdominal apex in various lineages include: (1) transformation of segment X into a cylindrical or subconical pygopod, sometimes with hooks, asperities or holdfast organs at the apex; (2) posterior expansion of tergum X forming a median process and resulting in the ventral position of the anal region (some Haliplidae); (3) formation of lobes, pads, setose areas or sclerites around the anus in soft-bodied, internal-feeding larvae; (4) reduction of segment IX and concealment of its sternum, with retention of a well developed, terminal segment X (some Cucujidae and Silvanidae); (5) ventral movement of the anal region, so that tergum IX becomes terminal, extending partly onto the ventral surface, and segment X becomes posteroventrally or ventrally oriented; (6) sclerotisation of tergum IX and development of various types of armature, including articulated or fixed urogomphi; (7) transformation of tergum IX into a hinged sclerotised plate articulating with tergum VIII and often accompanied by the partial enclosure of sternum IX within an emargination of sternum VIII or the concealment of sternum IX; (8) fusion of tergum and sternum IX forming a complete sclerotised cylinder; (9) development on sternum X of paired pygopods, sometimes armed with hooks or asperities; (10) formation of a dorsally-hinged operculum (tergum IX) articulating with tergum VIII and concealing sternum IX and segment X (Callirhipidae); (11) formation of a ventrally-hinged operculum (sternum X) articulating with the ventral portion of segment IX and concealing the anal region in some aquatic and riparian larvae; (12) complete concealment of segment X between the edges of tergum and sternum IX; and (13) reduction of segments IX and X and movement of tergum VIII to a terminal position (**see** below under Spiracles and Respiratory Modifications).

Segment VIII in most larvae is not differentiated from anterior abdominal segments, although it may have sclerotised areas or carinae associated with modifications in segment IX and in some cases tergum VIII may be partly or completely fused to IX (e.g. in some Chrysomelidae). A median defensive gland occurs on tergum VIII in most members of the staphylinid subfamily Aleocharinae; these are particularly prominent in the subtribe Bolitocharina (Seevers 1978) but they occur in other groups as well (Frank and Thomas 1984). Tergum VIII is produced posteriorly to form a spiracular siphon in Nosodendridae and several groups of aquatic Adephaga (**see** below under Spiracles and Respiratory Modifications), but in larvae of Hygrobiidae (Fig. 42B) a narrow projection is formed that lacks spiracles and resembles one of the urogomphi on tergum IX.

The simplest type of modification of tergum IX involves its sclerotisation, which is often greater than that of preceding abdominal terga. In some Trogossitidae, Cleridae, Corylophidae and Chrysomelidae there may be a distinct tergal plate similar to that on the prothorax. The most obvious and consistent feature of tergum IX is the presence of paired **urogomphi** (sing. **urogomphus**), which may be articulated or fixed at the base and may consist of one to several segments. The term urogomphi has been used for almost any paired prominences on tergum IX, with the exception of setae or articulated spines. The structures so designated are not necessarily homologous. The term cerci (sing. cercus) has also been applied to them, but this term should be restricted to appendages on segment X in some other orders, which are unknown in Coleoptera. Other terms that have been applied to urogomphi are pseudocerci and cornuculi. In most campodeiform larvae, which inhabit surfaces or interstitial spaces (e.g. leaf litter, loose soil, under stones, beneath loose bark) the urogomphi are relatively long and narrow, project posteriorly or posterodorsally, have relatively little curvature and bear several setae. These are basally articulated and/or segmented (Figs 34G, 35H, 52D, 52T) in some Carabidae and most Hydradephaga, Staphylinoidea and Hydrophiloidea, and they appear to be tactile in function. In those larvae that inhabit narrower spaces beneath bark or bore into substrates like rotten wood or fungi, the urogomphi are more solidly built, never articulated or segmented, and often recurved (Figs 35D–F, L, 50M, 52K–M). They appear to fix the larva in its burrow and assist in locomotion.

Although most urogomphi are simple or at most tuberculate, some may be bifurcate or have several accessory branches or spines. Bifurcate urogomphi (Figs 52M, O, 53E) occur in several families, including Elateridae (Dendrometrinae), Melyridae, Nitidulidae, Tetratomidae, Pyrochroidae and Salpingidae, while more complex forms with **accessory processes** (Figs 52I, 53B, D) occur in some Trogossitidae (e.g. *Calitys* and *Thymalus*), Nitidulidae (e.g. *Epuraea*, *Cryptarcha*), Erotylidae (Xenoscelinae, Cryptophilinae), Cerylonidae (Euxestinae), Pythidae and Salpingidae (Othniinae, Inopeplinae). Sometimes urogomphi may be accompanied by tubercles or teeth extending onto the surface of the tergum or smaller, paired processes called **pregomphi** (Fig. 52M). Urogomphi may be approximate to widely separated, and sometimes there may be a **median tergal process** (Fig. 52K), a **median interurogomphal pit** or cavity (Figs 52F, I) or two **interurogomphal pits** (Fig. 52O) between them.

In some larval types, tergum IX bears a single median process, which may be spine-like and posteriorly projecting (Figs 42F, 52W, 56D) as in most Archostemata, some elaterine Elateridae, erotyline Erotylidae (*Microsternus*), colydiine Zopheridae (*Cicones*) and Mordellidae, rounded and posteriorly projecting, as in some galerucine Chrysomelidae, or hook-like, as in some Dryopidae. A forked median process (Fig. 84D) occurs in some Cucujidae (*Pediacus*, *Platisus*). In scraptiine Scraptiidae, tergum IX bears a large, rounded, lightly sclerotised, setose **caudal process** (Fig. 90D), which is deciduous or detachable, so that these larvae are often recognisable, not by the presence of the process, but by the truncate tergal apex, to which the process was formerly attached (Švácha 1995).

Another distinctive type of ninth tergum is that bearing a **concave disc** or **pygidium**, often surrounded by teeth or serrations (Fig. 52X). This feature is found in larvae of some Byrrhidae (e.g. *Microchaetes*, Fig. 65B), Lymexylidae, Tenebrionidae (Amarygmini), Ciidae, and Chrysomelidae (Spilopyrinae), and appears to be an adaptation for blocking larval burrows to prevent the entrance of predators or parasites (similar to the elytral declivity in some adult beetles). A more complex pygidium, lined with various kinds of teeth or spines, occurs in stenochiine Tenebrionidae and in the pythid genus *Priognathus*. In paussine Carabidae, tergum IX forms a complex, multilobed disc, the function of which is connected with the capturing of prey. In various flattened larvae living under bark (Cucujidae, Laemophloeidae, some Phalacridae (*Litochrus*), Prostomidae, Mycteridae and Pyrochroidae), the entire ninth tergum forms a hinged plate, sometimes called a **urogomphal plate** (Figs 34K, 35D, 43E–F, 52A–B, L), articulated to segment VIII and apparently used to wedge the larva between layers of bark or wood. In the family Brachypsectridae, tergum IX forms an **articulated spine** (Fig. 65K), which is capable of being extended over the back and assists in prey capture (Crowson 1973b; Costa *et al.* 2006, 2010a).

Sternum IX is usually relatively simple, but in various Tenebrionoidea, there are teeth or asperities variously distributed along its anterior edge or occasionally at the apex. In hallomenine Tetratomidae, Synchroidae, salpingine and aglenine Salpingidae, and pediline Pyrochroidae, there is a single tooth at each basal (anterolateral) angle (Fig. 53E), while in the oedmerid genus *Calopus* and various members of the families Pythidae, Trictenotomidae,

Pyrochroidae and Salpingidae, there may be from two to several asperities on each side of the anterior edge (Figs 43F, 53B, D), sometimes forming a complete basal row (Figs 43N, 53A). Two posteromesal asperities occur in the only known larva of the anthicid subfamily Eurygeniinae (Fig. 53C), while in *Sphindocis* Fall (Ciidae) and Prostomidae (Fig. 43I), there is an apical row of asperities. The ninth sternum is sometimes partly enclosed within an emargination in sternite VIII in those larvae with a hinged urogomphal plate, but an extreme condition is known in the families Boridae and Mycteridae, where sternum IX is U-shaped, almost completely enclosing segment X and bearing two or more teeth at the posterior edge (Fig. 52L); a similar condition occurs in some elaterid larvae. In some larval types, sternum IX is reduced, and it may be concealed by sternum VIII, as in some Phalacridae (Fig. 52B) or Silvanidae (Fig. 52U).

Modifications of segment X are particularly striking in those larvae using this segment as an ambulatory device (pygopod or paired pygopods) and in those soft-bodied larvae in which the abdominal apex assists in locomotion within substrates. In some active larvae with a terminal **pygopod**, such as those of various Carabidae, Staphylinoidea and Lampyridae, there is an anal holdfast organ composed of several eversible, asperated tubes (Fig. 53J) (Brass 1914; Kemner 1918). **Paired pygopods** on either side of the anal opening occur in several groups; they may be simple (Archostemata, first instar brachinine Carabidae, cardiophorine Elateridae, Boganiidae, some Tenebrionidae) (Figs 50M, 84A) or armed with one (Byrrhidae) or several (Ptilodactylidae, Lymexylidae, Tenebrionidae) asperities, teeth or hooks (Fig. 52S). Paired hooks may also occur in the absence of pygopods (e.g. in Hydroscaphidae, Fig. 52C; Ptiliidae; Hydraenidae, Fig. 52T; some Clambidae; Limnichidae; some Elateridae), Gyrinidae have two pairs of long hooks (Fig. 52J), and *Sphaerius* has three hooks on each side (Fig. 33D), and some elaterid and eucnemid larvae may have the anus surrounded by a ring of teeth.

The anus may be surrounded by two or more membranous lobes, which do not project as pygopods, and in various soft-bodied larvae there are a variety of supporting structures in the vicinity of the anus. In Bostrichidae and Ptinidae there is a longitudinal groove lying below the anal opening and separating a pair of oval pads or cushions (Fig. 52V); this structure was called the **nates** by Böving (1954), who used the term **bow** for a curved sclerite below this. The anal opening varies considerably throughout the order, but the condition is often not recorded. In most Scarabaeoidea, the opening may be transverse, Y-shaped or longitudinal and is surrounded by a **dorsal anal lobe** and one or two **ventral anal lobes** or a pair of **lateral anal lobes**. Terms used for these lobes vary even among scarab workers (Ritcher 1966). In most Lucanidae, the dorsal lobe is usually absent or highly reduced, and the lateral lobes each may bear a glabrous area surrounded by a fine sclerotised line; this area is called an **anal pad** (Fig. 53F). In some Geotrupidae and scarabaeinae Scarabaeidae, the area below the anus may be subdivided forming several ventral lobes. The sternal area immediately anterior to the anal region in Scarabaeoidea often bears a group of hairs, bristles or hooked setae, which is called the **raster** (Fig. 33F) and is used in identification of scarab grubs. The raster may consist of a pair of **palidia** (sing. **palidium**), each of which consists of a row of heavy spines or **pali**, surrounded by a bare area called the **septula** and sometimes bordered laterally on each side by a **tegillum**, which is a continuous patch of shorter setae; in some cases, there is a single palidium, which is curved with the pali facing towards the anus, or the pallidium may be replaced by a **teges**, which is a continuous patch of short, straight or hooked setae. Other associated structures include the **campus**, a bare area between the raster and the anterior edge of sternum X, and a **barbula**, which is a tuft of hairs or short setae at the side of the abdomen near the anus. In the genus *Cephalodesmius* (Scarabaeinae), sternum X bears a coarsely striate, longitudinal ridge, which acts as a **plectrum** by rubbing against a finely striate **stridulitrum** on the gular region (Monteith and Storey 1981).

Other modifications of the anal region include the paired spine-like processes (Fig. 52P) on either side of the vertical anal opening in agriline Buprestidae and the complex anal lobes and sclerotised plates found in larvae of some eumolpine Chrysomelidae and entimine Curculionidae (Ghilarov and Medvedev 1964; Arnoldi and Byzova 1964). Concealment of the anal region may be accomplished in several ways. In some Elateridae, segment X is very small, circular, and located near the base of the undivided ninth segment, where it may be partly concealed by the apex of sternum VIII. In some Tenebrionidae, the entire tenth segment may be concealed beneath tergum IX and in Callirhipidae it is concealed by the **dorsally-hinged operculum** (Fig. 52Q) formed by tergum IX. In various larvae belonging to the families Dryopidae, Limnichidae (Hyphalinae), Lutrochidae, Elmidae, Ptilodactylidae (Araeopidiinae, Cladotominae), Chelonariidae, Psephenidae (Eubriinae, Psephenoidinae), a **ventrally-hinged operculum** (Figs 52Z, 65H, J) is formed from sternum X, which articulated with the ventral part of the ninth segment. The operculum is operated by one or two sets of muscles and it may be equipped with a pair of hooks (Lutrochidae, Elmidae, Fig. 42A). In Elmidae, hyphaline Limnichidae, and eubriine Psephenidae, the operculum conceals the retractable anal gills (Fig. 51H). Both segment X and sternum IX may be reduced and at least partly concealed in larvae of Brachypsectridae, Nosodendridae, some Silvanidae, Laemophloeidae, Phalacridae and in various Hydradephaga.

Respiratory Modifications

Although the larval tracheal system in beetles is usually **peripneustic**, with functional spiracles located on the mesothorax and abdominal segments I to VIII, some larvae may have a non-functional spiracular remnant on the metathorax. The mesothoracic spiracle is often located between the meso-and metathorax and it has migrated onto the prothorax in some Ptinidae and Curculionidae. Hinton (1947) has shown that there is considerable variation in the number and position of functional spiracles and final instars of some aquatic larvae differ in this respect from earlier instars. A **hemipneustic** system, with spiracles on the mesothorax and abdominal segments I–VII, is present in the final instar of Haliplidae, Aspidytidae, Hygrobiidae, and some aquatic Lampyridae, as well as in the coarctate phase of parasitic nemognathine Meloidae. An **amphipneustic** system, with mesothoracic and eighth abdominal spiracles only, occurs in all instars of the exotic, gill-bearing Psephenidae (Psepheninae and Eubrianacinae). A **metapneustic** system, with a single functional pair of spiracles on segment VIII, is known in the family Scirtidae, the exotic ptilodactylid genus *Araeopidius*, eubriine Psephenidae (including the Australian *Sclerocyphon*), and early and intermediate instars of

Noteridae, Amphizoidae, Dytiscidae and donaciine Chrysomelidae. An **apneustic** system, with no functional spiracles, is found in early instars of Gyrinidae, Haliplidae, Hygrobiidae, berosine Hydrophilidae, Elmidae, Lutrochidae, some Ptilodactylidae and some eubriine Psephenidae and in all instars of the exotic psephenid subfamily Psephenoidinae. In a few groups, the type of larval respiratory system differs from any of those mentioned; examples are the final instar of Noteridae and all instars of Sphaeriusidae and Torridincolidae, with eight abdominal spiracles and none on the thorax; Hydroscaphidae, with spiracles on the mesothorax and abdominal segments I and VIII; Meruidae, with spiracles on abdominal segments I and VIII only; final instar Gyrinidae with spiracles on abdominal segments I–III; and some exotic meloid triungulins with an enlarged pair of spiracles on abdominal segment I and minute pairs on II–V. (Hinton 1947, 1967a; MacSwain 1956; Parker and Böving 1914).

There is considerable variation in spiracular structure. The simplest is the **annular** (or uniforous) spiracle (Figs 53I, L), which consists of a circular or oval opening, called the **peritreme**, a spiracular chamber or **atrium** and a **spiracular closing apparatus**, lying between the atrium and the end of the tracheal trunk and consisting of a sclerotised ring and apodeme to which a muscle attaches. The peritreme may be simple or crenulate (Fig. 53L). The inside of the atrium may be variously modified with internally projecting hairs or plates sometimes forming a filter, but detailed comparative studies of atrial structure are lacking. During **ecdysis**, the old spiracle is pulled out through the opening of the newly formed one. The term bilabiate was used by Schiödte (1862–1883) and Rymer-Roberts (1930) for a type of annular spiracle with an elongate-oval or elliptical peritreme, as in many Cerambycidae, but that term is likely to be confused with biforous or bicameral (**see** below). The annular spiracle with a closing apparatus occurs in Archostemata, Myxophaga and Adephaga, and among the Polyphaga, it occurs in most Staphylinoidea, all Scirtoidea, Brachypsectridae, most Bostrichoidea, some Cleridae, all Melyridae, and many Cucujoidea, Tenebrionoidea, Chrysomeloidea and Curculionoidea. The closing apparatus is widespread throughout the order, but absent in Dascillidae, most Scarabaeoidea (except Lucanidae, Passalidae, Trogidae and Glaphyridae), and Elateriformia (except Buprestidae, Eucnemidae, Throscidae, Cerophytidae and Brachypsectridae).

Another type of spiracle is one with a main opening plus one or more **accessory openings**. The terms **annular–uniforous** (Fig. 53H), **annular–biforous** (also known as **bicameral**) and **annular–multiforous** (Fig. 53K) refer to those with one, two or multiple accessory openings (Steinke 1919; Roberts 1930; Crowson 1981). Annular–uniforous spiracles are known in a few groups of Polyphaga, including some Bostrichidae and Ptinidae, a few Melandryidae and some Curculionoidea. Annular-multiforous spiracles, with several small accessory chambers distributed around the peritreme, occur in Lymexylidae, a few Ptinidae, several groups of Tenebrionidae, some Oedemeridae and several Cerambycidae.

The **annular–biforous** spiracle (Fig. 53M, U) with two accessory openings is widely distributed within the Polyphaga, occurring in some basal groups such as Agyrtidae, Leiodidae, Derodontidae and Nosodendridae and common in the Cucujiformia. In some cases the accessory openings are very small, so that the spiracle appears to be annular, and in others (e.g. Hydrophiloidea, Nitidulidae), the main opening is small and the accessory openings very long, so that it resembles the biforous type (but lacks a spiracular scar – **see** below). In some Cerambycidae and Curculionoiodea, first instars have annular-biforous spiracles, while later instars have the annular or annular–multiforous type.

The **biforous** spiracle (Fig. 53P, R) has its opening blocked by a median septum with a narrow slit-like opening on either side of it; these two openings are accompanied by a **spiracular scar** (ecdysial or stigmatic scar), which represents the opening through which the old spiracle is pulled during ecdysis. This specialised type of moulting, called by Hinton (1947) the **elateroid ecdysial process** and involving the withdrawal of the old tracheal cuticle through an ecdysial tube, which then leaves a scar adjacent to the functional spiracular opening, is also associated with cribriform spiracles (**see** below). Biforous spiracles occur in late instar Rhipiceridae, Trogidae, the geotrupid genus *Eucanthus*, schizopodine Buprestidae, all Byrrhoidea with functional spiracles and all Elateroidea. The biforous spiracle in Chelonariidae is sometimes called **undulate**, based on the shape of the paired openings (Fig. 53Q), and those of Brachypsectridae and Cantharidae may be reduced or modified.

The **cribriform** spiracle has the main opening blocked, not by a single division, but by a **sieve plate** bearing numerous fine perforations (Fig. 53N, T). The cuticular area immediately adjoining the sieve plate and usually partly surrounded by it is called the **bulla**, and it is here that the spiracular scar is located. Cribriform spiracles with a typical kidney-shaped or reniform sieve plate (Fig. 53N) occur in Dascillidae and almost all Scarabaeoidea and Buprestidae, while in Trogidae and Heteroceridae (Fig. 53S) the shape is different, as are those occurring on segment VIII in some eubriine Psephenidae (Hinton 1955).

Another spiracular type was referred to as **annular–cribriform** by Lawrence *et al.* (2011). This type of spiracle involves a sieve plate, which is adjacent to but does not block the spiracular opening (Figs 43H, 53G); these occur in scattered unrelated taxa, including *Grynoma* (Trogossitidae), Promecheilidae, *Hemipeplus* (Mycteridae) and some cryptocephaline Chrysomelidae (Crowson 1964b; Kasap and Crowson 1976; LeSage 1982; Lawrence 1991b).

The normal location of thoracic spiracles is in membrane between the prothorax and mesothorax or on a mesosternal laterotergite, but in some soft-bodied larvae (lyctine Bostrichidae, Ptinidae) it may be on the prothorax. Abdominal spiracles are normally situated on laterotergites or on lateral extensions of the tergal plate. In many aquatic larvae, those living in wet terrestrial environments subjected to flooding, or in wet, sappy areas in wood or under bark, various mechanisms have evolved for allowing at least some spiracles access to free air. These include: (1) placement of some or all spiracles at the ends of spiracular tubes raising them above the surface of the integument; (2) dorsal placement of some spiracles (usually thoracic and first and eighth abdominal); (3) dorsal or posterior location of spiracles on segment VIII; (4) placement of the eighth spiracles at the end of a posteriorly projecting, median process or siphon; and (5) formation of a terminal respiratory chamber. Spiracular tubes of varying length occur in the hydraenid genus *Tympanogaster*, Chelonariidae, Derodontidae, various Cucujoidea and some cassidine Chrysomelidae. In Hydroscaphidae, some Nitidulidae and Biphyllidae the eighth spiracles are each placed at the end of a tube located at the posterior end of the segment and facing posteriorly (Fig. 52M), while in the chrysomelid subfamily Donaciinae, the eighth spiracles are located on spine-like structures (Fig. 52R) used to puncture the stems of aquatic plants and obtain the air within them. The displacement of the eighth spiracles dorsally and posteriorly, so that they come to lie relatively close to one another and face posteriorly (Fig. 52N), is known in Meruidae, Amphizoidae, some Ptilodactylidae, Chelonariidae and Dryopidae, while their placement at the end of a **spiracular siphon** occurs in Dytiscidae (Fig. 52Y), Noteridae, cladotomine Ptilodactylidae (e.g. *Austrolichas*, Fig. 65J) and Nosodendridae (Fig. 38L).

The **respiratory chamber** (also known as **stigmatic atrium** or **breathing pocket**) (Figs 34D, 39K, 58J, L), which occurs in most members of the more derived subfamilies of Hydrophilidae (Hydrochinae, Spercheinae, Hydrophilinae, Sphaeridiinae), is composed of terga VIII and IX and encloses a large pair of posteriorly oriented spiracles; this chamber may be complexly lobed and may have as many as three pairs of urogomphus-like processes (**mesocerci**, **paracerci** and **acrocerci**) (Böving and Henriksen 1938). A pair of **procerci** may also be present at the lateral edges of tergum VIII, above the spiracles. Segment X is ventrally oriented and located below the floor of the respiratory chamber. A terminal respiratory chamber, which is simpler in structure, also occurs in the family Scirtidae.

In aquatic, semiaquatic and riparian larvae there may be, in addition to spiracles, various structures which allow oxygen to pass into the tracheal system or directly into the haemocoel when the cuticle is entirely submerged. The general term **gill** is used for these structures, and the term **plastron** refers to a specific situation in which a gas layer is held in position by hydrofuge structures which prevent the entry of water under pressure. Most gills are thin-walled processes which project from the venter, the lateral portions of the thorax and abdomen, or the anal region. All gills in Coleoptera larvae appear to be tracheal gills in that they contain tracheae or tracheoles. Gills associated with spiracular openings are called **spiracular gills**. Typical gills consisting of several fine branches radiating from a common stem include the **abdominal gills** (Fig. 51G) on segments I–VII in Eulichadidae and either I–IV or II–VI in some exotic Psephenidae and **anal gills** (Fig. 51H, 65H, 65L–M) in *Hyphalus* (Limnichidae), Lutrochidae, Elmidae, *Drupeus* Lewis (Ptilodactylidae) and eubriine Psephenidae. The gills in the larva of *Hygrobia* are clusters of simple or bifurcate filaments arising from the thorax and first three abdominal segments (not clearly visible in Fig. 42B). Paired **lateral or dorsolateral gills**, which are usually thin-walled and unbranched, occur on abdominal or sometimes thoracic and abdominal segments in some Dytiscidae (*Coptotomus*), some Haliplidae (*Peltodytes*), Gyrinidae (Fig. 56E), some Hydrophilidae (e.g. Berosini, Fig. 58B) and aquatic Lampyridae, and a cluster of unbranched gills occurs in the anal region of Scirtidae. Another type of gill-like structures in the anal region are the **anal papillae** (Fig. 52S) which apparently function as osmoregulatory organs; these occur in some Ptilodactylidae and eubriine Psephenidae.

Spiracular gills are of three types: **segmented** (Fig. 51F), as in Torridincolidae, **vesicular** (Fig. 51P), as in Sphaeriusidae and most Hydroscaphidae, and **tufted** (Fig. 51O), as in the hydroscaphid genus *Scaphydra* (Reichardt 1974). The first type consists of a 2- or 3-segmented process, with the spiracular opening at the end of the first or second segment and the apical portion of the gill covered with a plastron mesh. The vesicular gill consists of a thin-walled extension of the peritreme to form a balloon-like structure with the opening at the apex, and the last type resembles a minute gill tuft.

The North American ptilodactylid genus *Araeopidius* has a peculiar type of respiratory structure; thoracic spiracles and the first seven abdominal spiracles are atrophied, and each is partly surrounded by a plate-like structure (Fig. 50J) covered with minute pores (0.05–0.13 microns in diameter); this **plastron plate** apparently functions as a respiratory organ during periods of submergence in the same way as the plastron of a housefly egg (Hinton 1967b; Lawrence 1991b). Some aquatic larvae are capable of breathing directly through portions of the cuticle with tracheoles near the surface. In some Haliplidae, tracheoles actually penetrate minute cuticular processes called **microtracheal gills** (Seeger 1971).

EGGS

In most cases, the eggs are simple, ovoid and without surface ornamentation, but a sculptured chorion is known in Cupedidae, and in some groups, like some chrysomeline Chrysomelidae (De Little 1979) that lay their eggs on surfaces. Eggs of cryptocephaline Chrysomelidae are enclosed by a layer of faeces called a **scatoshell**. Hatching is often initiated by **egg-bursters** or **hatching spines**, which may be present on the head (e.g. in Hydraenidae, Hydrophilidae, some Staphylinidae) or on thoracic or abdominal terga (e. g. in some Staphylinidae, Histeridae, Scarabaeoidea and many Cucujiformia) of the 1st instar larva (Emden 1946). [Hinton 1981].

PUPAE

As mentioned above, beetle pupae (Fig. 33A, B) are adecticous and nearly always exarate (obtect in Ptiliidae, staphylinine Staphylinidae, Clambidae, Coccinellidae, hispine Chrysomelidae and a few other groups). The abdomen usually includes nine terga and eight sterna, and the number of spiracles is usually reduced in comparison with the larva and adult. The head and body often bear various prominences and setae, which serve to hold the pupa away from the walls of the pupal cell or other surfaces. The abdominal apex may have paired appendages similar to larval urogomphi. In several groups, including Dermestidae (Fig. 33A), there are local sclerotisations on opposable edges of some abdominal segments, which are possibly used to pinch mites and some other predators; these were called **gin traps** by Hinton (1946). In surface-feeding beetles, the pupa often remains partly enclosed by the larval skin, but more commonly pupation occurs within the food plant or in earthen cells below the ground. A cocoon is sometimes constructed. Passalidae, some Scarabaeidae and Tenebrionidae (*Platydema*) use faecal material, and cryptocephaline Chrysomelidae form a closed capsule from the faecal case of the larva. Silken cocoons are formed by some Staphylinidae (Ashe 1982; Frank and Thomas 1984; Thayer 2005), and have also been reported in several other families (e.g. Brachypsectridae, Bothrideridae, Tenebrionidae, Curculionidae). The calcified remains of certain weevil cocoons (e.g. *Leptopius*) are sometimes found in the soil (Lea 1925). Further information on immature Coleoptera may be found in Bertrand (1972, 1977; aquatic immatures), Böving and Craighead (1931), Costa *et al.* (1988), Crowson (1981), Lawrence (1991a), Steiner (1995; tenebrionid pupae), Bouchard and Steiner (2004, tenebrionid pupae), and various chapters in Beutel and Leschen (2005a) and Beutel *et al.* (2010a).

BIOLOGY

General

Beetles occupy almost every available terrestrial and freshwater habitat and a few marginal marine habitats as well. They may be found on the foliage of plants of all kinds; in flowers, fruits and seeds; within living plant tissue in galls, stems, branches, trunks and roots; in soil, humus, leaf litter and flood debris; beneath the bark of living and dead trees; in decomposing wood; under stones and logs; in or on the fruiting bodies of various fungi and slime moulds; in dung and carrion; in the nests of vertebrates and social insects; in stored foodstuffs; in fresh water, from temporary pools or billabongs to cool mountain streams or thermal pools; in the brackish water of salt lakes or estuarine mud flats; in sand, gravel or mud at the edges of streams or ponds; in high-water debris on the seashore; among rocks or coral in the intertidal zone; and in caves. These habitats are discussed in more detail in a previous section (**see** Australian Beetles and Their Habitats).

Members of several beetle families have become adapted to living in fresh water, either in the larval stage only or throughout the life cycle (except during pupation and adult dispersal). Many adult water beetles (e.g. Hydrophilidae, Elmidae) have on the ventral surface a covering of fine hydrofuge hairs, which serves to trap an air bubble; this is called a **plastron** and has been discussed above. Aquatic beetle larvae have evolved several respiratory adaptations, including spiracular gills, lateral feathery gills, ventral and anal gill tufts, enlarged terminal spiracles and siphon-like spiracular tubes (**see** discussions of individual families). Within Australia, all Sphaeriusidae, Haliplidae, Hygrobiidae and Noteridae, almost all Dytiscidae, most Hydrophilidae and Elmidae, and a few Hydraenidae (*Tympanogaster*), Chrysomelidae (*Donacia*) and Curculionoidea are adapted to freshwater existence as both larvae and adults, whereas *Sclerocyphon* (Psephenidae), *Byrrocryptus* (Ptilodactylidae) and most Scirtidae are aquatic only in the larval stage. Some other groups (e.g. Heteroceridae, the staphylinid genus *Bledius*) live in wet mud or gravel near the water's edge; and *Hyphalus* species (Linmichidae) occur in marine intertidal habitats.

Most species of beetles are phytophagous in the narrow sense, feeding on living plant tissue, but a large number have saprophagous habits and consume only vegetable matter which has been substantially altered by the action of bacteria and fungi. The majority of phytophagous beetles belong to the superfamilies Scarabaeoidea, Chrysomeloidea and Curculionoidea, the families Buprestidae, Elateridae, Bostrichidae and Anobiidae, and the coccinellid tribe Epilachnini. Mycophagous or fungivorous beetles may feed on a wide variety of fungal tissue, from spores to heavily reinforced fungal hyphae comprising woody sporocarps, and certain beetle groups specialise on particular fungal groups (Myxomycetes, various types of Ascomycetes and Basidiomycetes). The subject of mycophagy in Coleoptera has been reviewed by Lawrence (1989), Hammond and Lawrence (1989) and Lawrence and Milner (1996), and is discussed in more detail in above sections dealing with fungal habitats.

Carnivorous habits occur in several groups, including almost all Adephaga, and several polyphagan families, such as Hydrophilidae (larvae), Histeridae, Staphylinidae, Elateridae, Lampyridae, Cantharidae, Cleridae and Coccinellidae. Most of these are general predators on insects and other small invertebrates, but some are limited in their choice of prey. The Coccinellidae, for example, feed largely on aphidoid and coccoid Hemiptera, and Lampyridae usually specialise on snails. Parasitoid habits have evolved in a few representatives of the families Carabidae, Staphylinidae, Rhipiceridae, Passandridae, Bothrideridae, Ripiphoridae and Meloidae.

Symbiotic Relationships

Coleoptera may be involved in several symbiotic relationships with other organisms, including **endosymbiosis** with microorganisms and **ectosymbiosis** with certain fungi, with mites and nematodes (**phoresy**), or with social insects, such as ants, bees and termites (**inquilinism**). Many beetles harbour aggregations of endosymbionts, which may be extracellular in the lumen of the gut or in various diverticula, or intracellular in the cytoplasm of cells known as **mycetocytes** and sometimes forming special organs called **mycetomes** (Koch 1967). The symbionts are primarily yeasts in the family Saccharomycetaceae (Ascomycetes) and have been found most often in species that feed on wood, fungal fruiting bodies, cereals and other plant material in families such as Lucanidae, Passalidae, Scarabaeidae, Bostrichidae, Ptinidae, Erotylidae, Silvanidae, Endomychidae, Ciidae, Tenebrionidae, Cerambcyidae, Chrysomelidae and Curculionidae (Suh *et al.* 2003, 2004, 2005; Blackwell 2006; Suh and Blackwell 2006). Mechanisms by which symbionts are transferred to the next generation include: (1) entry into newly hatched larvae which consume the contaminated chorion; (2) entry into the egg before secretion of the chorion; and (3) invasion of the testes, passage into the female with the sperm and infection of the egg through the micropyle.

A highly specialised form of ectosymbiosis involves the so-called ambrosia beetles or pinhole borers (family Lymexylidae and weevil subfamilies Scolytinae and Platypodinae) and yeasts of the genera *Ascoidea* and *Endomycopsis.* The beetles bore into wood and, in doing so, introduce these ambrosia fungi which grow on the tunnel walls and serve as food for both larvae and adults. The fungi are always associated with the beetles and are transported from tree to tree by the adult, which often has specialised organs or **mycangia** for the purpose (Francke-Grosmann 1967).

Several mites and nematodes utilise beetles for transportation from one suitable habitat to another. Examples include saprophagous nematodes of the genus *Golfartia,* which are phoretic on both larvae and adults of Heteroceridae in riparian habitats (Steffan 1967), and several mesostigmatid mites (Hunter and Rosario 1988). Members of the mite genus *Macrocheles* are phoretic on dung-feeding Scarabaeidae and predacious on Diptera larvae and other small invertebrates encountered in the dung pads (Krantz 1983).

Beetles in many families have evolved a more or less obligatory relationship with termites or social Hymenoptera, such as ants or bees. These **inquilines** live within the nest for at least part of their life cycle and are usually either tolerated by the host or in some cases protected or even fed by them. Many of the more specialised inquilines possess glands of various kinds, which secrete substances imbibed by the host. Some of these glands are associated with tufts of hairs called **trichomes**. Ant-associated inquilines or **myrmecophiles** in the Australian fauna include members of the following families: Carabidae (*Adelotopus*), Ptiliidae (*Rodwayia*,

Fig. 59A–B), Staphylinidae (various Aleocharinae, some Pselaphinae), Histeridae (Chlamydopsinae, Fig. 77B), Dermestidae (*Myrmeanthrenus*), Ptinidae (ptinine genera *Ectrephes, Enasiba, Diplocotes* (Fig. 70I) and *Polyplocotes*), Nitidulidae (Lawrencerosini), Silvanidae (*Nepharis* and *Nepharinus*), Tenebrionidae (*Euclarkia* and *Kershawia*), Salpingidae (*Tretothorax* Fig. 95F) and Brentidae (*Cordus* and *Amorphocephalus*). Some species live in close proximity with ants and may prey on them, but do not occur in the nest. Larvae of *Sphallomorpha* (Carabidae: Pseudomorphinae), for instance, live in burrows around the periphery of *Iridomyrmex* nests and feed on worker ants. *Arthropterus* species (Carabidae: Paussinae) (Fig. 54K) have only rarely been found in ant nests, but their highly modified antennae and legs suggest that they have evolved protective mechanisms against ant attack. Although adults lack glandular trichomes present in other paussines known to be myrmecophilous, larvae have a similar body form with a terminal phragmotic disc. Di Giulio and Vigna Taglianti (2001) and Moore and Di Giulio (2006) have described prey capture behaviour in larvae of New World genera in the related tribe Ozaenini, and it is likely that *Arthropterus* larvae exhibit similar behaviour.

Australian **termitophiles** occur primarily in the staphylinid subfamily Aleocharinae, where symbiotic associations with termites have evolved several times; however, they also occur in Hybosoridae (*Cyphopisthes*), Scarabaeidae (*Microvalgus, Rhyparus, Cryptodus,* and *Maechidius* and related melolonthines), Tenebrionidae (*Eutermicola, Pseudeba*) and Aderidae (*Megaxenus* with *Microcerotermes,* Lawrence *et al.* 1990). The aleocharine *Austrospirachtha mimetes* Watson, which lives in *Nasutitermes* colonies, is capable of expanding the abdomen so that it takes on the appearance of a termite worker (Fig. 59J, K) (Watson 1973). The highly specialised larvae of *Drepanoxenus* (Fig. 58F) species, which live in the nests of harvester termites (*Drepanotermes*), have reduced suctorial mouthparts, long geniculate antennae, and lateral intersegmental glands on the thorax and abdomen. These larvae are treated as colony members and are groomed by the termite workers, which appear to gain some benefit from the glandular secretions (Kistner and Watson 1972; Watson and Kistner 1972). Other Australian termitophilous Aleocharinae include members of the genera *Australoptochus, Lauella, Austrointhus, Microceroxenus, Nasutiphilus, Coptotermoecia, Philobrunneus, Hetairotermes, Termitoecia* and *Mastopsenius* (Watson and Howick 1975; Watson and Kistner 1985).

Defensive Adaptations

In Coleoptera, defence against attack by bacteria and fungi, as well as predators, is commonly achieved by means of chemical compounds secreted by special glands on various parts of the body, or sometimes located in the haemolymph and released by **reflex bleeding** or crushing of the cuticle. The subject of defensive chemistry in beetles has been reviewed by Dettner (1987) and discussed in general texts (Crowson 1981). Chemical defence mechanisms have evolved at least 30 times in the order, and have been demonstrated in various groups, including all Adephaga and various members of the families Histeridae, Silphidae, Staphylinidae, Scarabaeidae, Buprestidae, Elateridae, Cantharidae, Lampyridae, Lycidae, Trogossitidae, Melyridae, Endomychidae, Coccinellidae, Meloidae, Oedemeridae, Pyrochroidae, Pythidae, Tenebrionidae, Cerambycidae and Chrysomelidae. Glands are often located at or near the abdominal apex, with their reservoirs opening in the membrane between apical or subapical tergites or sternites; the **pygidial glands** of Adephaga are of this type, as are those of Tenebrionidae, Scarabaeidae and some Staphylinidae. In other groups the glands may be generally distributed on the dorsal surfaces of the pronotum and elytra (chrysomeline, criocerine and galerucine Chrysomelidae) (Deroe and Pasteels 1982), near the anterior edge of the prothorax (some Dytiscidae and Tenebrionidae), or on most abdominal segments (larvae and adults of Cantharidae). Larvae of *Paropsis* and related Chrysomelidae secrete a lethal mixture containing hydrogen cyanide stored in eversible vesicles on tergum VIII (Moore 1967a). Hinton (1951) showed that glandular reservoirs present in some chrysomeline larvae were utilised by the pupae as well, since the latter remain partly enclosed within the larval skin. Lawrence and Hlavac (1979) suggested that the complex cuticular topography (canals, basins, pores and bridges) exhibited by adults of some Derodontidae, Hydraenidae, Hydrophilidae, micropepline and pselaphine Staphylinidae, colydiine Zopheridae and Tenebrionidae may function as a drainage and containment system for defensive chemicals. Many distasteful or poisonous species are **aposematically** coloured and belong to mimetic assemblages discussed by Moore and Brown (1989). Among the non-chemical types of defence are the specialised hairs (**spicisetae** and **hastisetae**, Fig. 46G–H) found in the larvae of most Dermestidae. These hairs are easily detached and adhere to the appendages of ants and other predators, thereby neutralising their attack (Nutting and Spangler 1969). Faeces and other debris that adhere to the cast skins impaled on the long fork arising from tergum IX in tortoise beetle larvae (Chrysomelidae: Cassidinae) have been observed to discourage predators (Eisner *et al.* 1967). Larvae of some Coleoptera will aggregate in large numbers before pupation (Fig. 7A), often with heads or abdominal apices facing the periphery of the mass; this type of behaviour, called cycloalexy, was recently reviewed by Jolivet and Verma (2011) and is usually associated with larval and/or pupal defensive glands and various types of defensive behaviours. Beetles lacking chemical or other types of defensive weapons may be capable of concealment by means of colour and/or cuticular topography. Cryptic adults mimicking bark or dung are commonly found among species of Zopheridae (Fig. 92B, D), Tenebrionidae (Fig. 74I) and Curculionidae (Fig. 12A–B).

Reproduction and Life History

The antennae of male Coleoptera are often larger and more elaborate than those of females; this is presumably correlated with the emission of pheromones by the female, which stimulate the male to fly upwind in search of a mate. Glandular setose areas (**sex patches**) occur on various parts of the body (e.g. frons, mentum, prosternum, abdomen, legs) in males only or, rarely, in females; at least some of these are known to produce aggregation pheromones, which attract both sexes (Faustini and Halstead 1982; Holloway 1985; Houston 1986).

In some groups (e.g. Scarabaeidae, Ciidae, Tenebrionidae), males have horns on the head and/or pronotum, which exhibit allometric variation and are used in combat with males of the same species (Eberhard 1979, 1980, 1982; Cook 1987; Emlen *et al.*

2005, 2007). Complex courtship behaviour preceding copulation has been reported for several families, including Melyridae, Meloidae and Cerambycidae (Crowson 1981). In copulation, the male usually assumes a position on the back of the female, sometimes using modified fore or mid tibiae or tarsi to stay in position. The necessity to curve the abdomen down and forwards to reach the apex of the female's abdomen is often obviated by a curvature in the penis; the latter is inserted into the vagina, with the internal sac being everted subsequently into the female tract.

The number of eggs deposited varies greatly throughout the order. Minute beetles (e.g. Sphaeriusidae and Ptiliidae) produce only one or two eggs at a time, while very large numbers of small eggs may be found in the abdomens of some Lymexylidae and Meloidae. **Oothecae** or cases enclosing several eggs occur in various Chrysomelidae (e.g. Sagrinae, Cassidinae, Spilopyrinae, Cryptocephalinae, Eumolpinae) and some other groups. Subsocial behaviour, i.e. parental care of eggs or larvae, is known in at least ten beetle families (Wilson 1971; Eickwort 1981). In *Spercheus* (Hydrophilidae), eggs are carried beneath the abdomen of the female until hatching, and some Carabidae and Staphylinidae place the eggs in a special chamber, which is defended. Females of *Bledius* (Staphylinidae) (Fig. 67L) maintain and defend the larval tunnels and provide algal food, while some harpaline Carabidae store seeds for their larvae. Males and females of Geotrupidae and scarabaeine Scarabaeidae often co-operate in digging nests and providing the larvae with plant material or dung, and progressive provisioning during the life of the larvae occurs in the latter group and also in silphid beetles of the exotic genus *Nicrophorus.* Some degree of subsociality also occurs in gregarious wood-feeding Passalidae (Reyes-Castillo and Halffter 1983), in the tenebrionid genus *Phrenapates,* and in some ambrosia beetles (scolytine and platypodine Curculionidae). **Eusociality** has been reported in *Austroplatypus incompertus* Schedl by Kent and Simpson (1992).

Unusual life cycles have evolved in various families. **Ovoviviparity**, in which the eggs are retained in the female reproductive tract until hatching, is known in some aleocharine Staphylinidae, chrysomeline Chrysomelidae, Tenebrionidae and pseudomorphine Carabidae (Liebherr and Kavanaugh 1985). Larval **heteromorphosis** or **hypermetamorphosis**, usually involving an active first instar or **triungulin** (Fig. 93J) and grub-like, inactive, sometimes legless, later instars, occurs in several groups of parasitoids (e.g. lebiine Carabidae, some aleocharine Staphylinidae, Rhipiceridae, Passandridae, bothriderine Bothrideridae, Ripiphoridae and Meloidae). Various Coleoptera may have distinctive first instars or pre-pupal resting stages (**prepupae**). In at least some exotic Cantharidae (*Cantharis* Linnaeus and *Rhagonycha* Eschscholtz), the first two instars are inactive and non-feeding with vestigial legs; this phenomenon has been referred to as **foetometamorphosis** (Verhoeff 1917; Janssen 1963). **Paedogenesis** or the retention of larval features in adult females occurs in Micromalthidae (Scott 1938; Pollock and Normark 2002), Phengodidae (Costa and Zaragoza-Caballero 2010) and the Asian lycid tribe Platerodrilini (Bocákova *et al.* 2007, Bocák and Bocákova 2010b); the pupal stage is eliminated and the females are indistinguishable from larvae, except for the reproductive organs. In other groups, such as Dascillidae (Karumiinae), Elateridae (Cebrioninae), Drilidae, Lampyridae, Dermestidae (*Thylodrias*, Fig. 63M) and Ripiphoridae (Ripidiinae) (Fig. 7F), the females lack hind wings and often elytra but have adult legs, antennae and mouthparts.

Parthenogenesis is relatively uncommon in beetles, but it occurs in several weevils (e.g. *Otiorhynchus*), and in some other families, such as Chrysomelidae, Ptiliidae, Ciidae and Bothrideridae (*Anommatus*). Pseudogamous parthenogenesis occurs in the triploid form of *Ptinus clavipes;* these females reproduce only after mating with males of *P. clavipes*, *P. pusillus* or *P. fur*, and the offspring are all females of the form *mobilis* (Woodroffe 1958). The exotic species *Micromalthus debilis* LeConte (Micromalthidae) has a more complex life cycle than any other insect species, involving paedogenesis, thelytoky, arrhenotoky, haplodiploidy and male cannibalism.

Natural Enemies

Coleoptera are subject to attack by the usual array of predators, parasitoids and diseases. Eggs may be attacked by various Hymenoptera (e.g. Mymaridae, Pteromalidae) and larvae may be preyed upon by spiders, wasps (e.g. *Cerceris*), other beetles and vertebrates or parasitised by Chalcidoidea, Braconidae and Ichneumonidae (Hymenoptera) and Tachinidae (Diptera). Larvae living in the soil are attacked by mites and nematodes. Tiphiid wasps of the subfamily Thynninae and the large rutiliine Tachinidae are common parasitoids of soil-dwelling scarabaeid larvae in Australia, and members of the Pyrgotidae (Diptera) and some Braconidae parasitise adult beetles. Aquatic beetles form an appreciable part of the food of fish, and most amphibians and many reptiles and birds feed to a large extent on beetles and other insects. Day-flying and surface-inhabiting adults are preyed upon by many species of birds, lizards, spiders and wasps, as well as asilid flies (Fig. 5A), while nocturnal species are more likely to be fed upon by frogs, cane toads and some smaller marsupials (Fig. 12E). Coleoptera are also subject to various diseases caused by fungi, bacteria and viruses. Common obligate ectoparasites of Coleoptera, particularly Adephaga and Staphyliniformia, are the Laboulbeniales (Ascomycetes) (Benjamin 1973; Tavares 1979; Lawrence and Milner 1996). Another parasitic group, the gregarines (Protista: Apicomplexa), are restricted to the internal organs and coelom (Perkins *et al.* 2000).

Economic Significance

The economic importance of Coleoptera results mostly from the damage some species inflict on materials of value to humans, such as agricultural crops, pasture plants, timber resources, and stored products, and the activities of other species in controlling insect or plant pests. They are of little or no importance in transmitting diseases to humans or other mammals and only a few transmit plant diseases or cause dermatitis (**see** below). Pasture plants, cereal crops and sugarcane are most seriously affected by soil-dwelling, root-feeding larvae of Scarabaeidae (whitegrubs, canegrubs, cockchafers) such as *Adoryphorus couloni* (Burmeister), *Lepidiota* spp., *Rhopaea* spp., *Sericesthis* spp., *Heteronychus arator* (Fabricius), *Acrossidius tasmaniae* (Hope), Elateridae (wireworms) such as species of *Agrypnus* and *Conoderus*) and Tenebrionidae (false wireworms) such as species of *Gonocephalum* and *Pterohelaeus*). Cucurbits may be damaged by species of *Aulacophora* (Chrysomelidae) and *Epilachna* (Coccinellidae). *Cosmopolites sordidus* (Germar) (Curculionidae: Dryophthorinae) is a serious pest of bananas, and a variety of fruit trees and garden crops are attacked by introduced weevils, such as *Asynonychus cervinus* (Boheman), *Graphognathus leucoloma* (Boheman), *Listroderes difficilis* Germain, *Otiorhynchus* spp. and *Phlyctinus callosus* Boheman. Some Cerambycidae, such as *Uracanthus cryptophagus* Olliff, attack the wood of fruit trees, while others (e.g. *Zygrita* and *Corrhenes*) damage soybean and lucerne. The wood of native timber or shade trees is subject to attack by various Cerambycidae (e.g. *Phoracantha*), and their foliage is sometimes severely reduced by Chrysomelidae (*Paropsis* and related genera), Scarabaeidae (adult *Anoplognathus*) or Curculionidae (*Gonipterus*). Ambrosia beetles (e.g. *Platypus* and *Xyleborus*) also cause blue staining of timber, and *Platypus subgranosus* Schedl carries a fungal pathogen fatal to Tasmanian rainforest trees (Elliot and De Little 1985). A potentially serious future threat is posed by the introduced bark beetle *Ips grandicollis,* which attacks *Pinus radiata* plantations (Morgan 1967).

Stored products attacked by beetles include a variety of foodstuffs, leather goods, wood, tobacco and crude drugs. Pests of stored products tend to be adapted to dry environments, and some of them, such as the ptinids *Stegobium paniceum* (Linnaeus) (Fig. 81L) and *Ptinus tectus* Boieldieu, are capable of breeding in a wide variety of materials, while others may be more restricted in their food requirements (e.g. *Dermestes* species, which attack dry animal material of high protein content, and *Sitophilus* species, which infest whole grains of wheat, rice, barley or maize). Several Tenebrionidae occur in stored grains and cereal products, whereas wooden articles are often attacked by Ptinidae or Bostrichidae. About 400 beetle species have been reported from stored products, but only a small proportion of these feed directly on the stored material; some are mycophagous, some predacious, and others of accidental occurrence. Information on stored product beetles may be found in Lepesme (1944), Hinton (1945a, 1945b), Aitken (1975) and Freeman (1980).

Coleoptera have played an important part in the biological control of insect pests and noxious plant species. Various Coccinellidae have been used to control plant pests belonging to the Aphidoidea and Coccoidea (Hemiptera). Some Australian species, such as *Rodolia cardinalis* (Mulsant) (Fig. 6G), *Cryptolaemus montrouzieri* Mulsant and *Rhyzobius* spp.) have been imported into the United States, India, China and the Mediterranean region for the control of certain scale insects and mealybugs (Hemiptera). Exotic beetles have also been imported into Australia to control introduced insect pests, recent examples being *Temnoscheila virescens* (Fabricius) (Trogossitidae) and *Thanasimus formicarius* (Linnaeus) (Cleridae), imported from the Northern Hemisphere for the control of *Ips grandicollis* (Eichhoff). Other importations have involved the control of noxious plants. The classic example is the introduction of the European chrysomelids *Chrysolina hyperici* (Forster) and *C. quadrigemina* (Suffrian) for the control of St John's wort (*Hypericum perforatum*). Other notable examples of biological control agents and their hosts are as follows (Chrysomelidae, except where noted): *Octotoma* and *Uroplata* for *Lantana camara; Moneilema* (Cerambycidae) for *Opuntia* (prickly pear); *Cyrtobagous salviniae* Calder and Sands (Curculionidae) for *Salvinia molesta; Longitarsus jacobaeae* (Waterhouse) for *Senecio jacobaea* (ragwort); *Agasicles hygrophila* Selman and Vogt for *Alternanthera philoxeroides* (alligator weed); and *Acanthoscelides* and *Chlamisus* for *Mimosa pigra* (giant sensitive plant) (Harley *et al.* 1985; Julien 1987). Coprophagous Coleoptera (*Geotrupes* and scarabaeine Scarabaeidae) have been imported from South Africa and the Mediterranean Region for the control of cattle dung and dung-breeding flies (Waterhouse 1974; Bornemissza 1976).

Species of *Paederus* (Staphylinidae) (Fig. 67B) cause severe skin lesions, often referred to as whip-lash dermatitis, when the adult beetles land on humans and are crushed when swatted or brushed, causing an irritating compound, pederin, to be released on to the skin (Frank and Kanamitsu 1987; Whelan and Weir 1987). Similar reactions may be caused by another chemical, cantharidin, produced by beetles of the families Meloidae and Oedemeridae (Moore and Brown 1989).

SPECIAL FEATURES OF THE AUSTRALIAN FAUNA

Of the more than 500 generally recognised families and subfamilies of Coleoptera, about two-thirds are represented in Australia, and these include all of the larger and better known taxa. Among the large, dominant groups that have radiated extensively during the latter part of the Tertiary in association with *Eucalyptus* and *Acacia* communities are the carenine and pseudomorphine Carabidae, chlamydopsine Histeridae, bolboceratine Geotrupidae, heteronychine, liparetrine and scitaline and anoplognathine Scarabaeidae, stigmoderine Buprestidae, agrypnine Elateridae, metriorhynchine Lycidae, malachiine Melyridae, megatomine Dermestidae, adeliine and heleine Tenebrionidae, phoracanthine Cerambycidae, cryptocephaline, eumolpine and chrysomeline Chrysomelidae, beline Belidae and cyclomine Curculionidae.

The degree of endemism is very high, and endemic taxa include the families Rhinorhipidae, Acanthocnemidae (introduced elsewhere), Myraboliidae, Lamingtoniidae and Tasmosalpingidae, the subfamilies Ommatinae (Ommatidae), Elythomerinae (Heteroceridae), Periptyctinae (Corylophidae) Synercticinae (Boridae) Eurhynchinae (Brentidae), several tribes, including Cychramptodini and Lawrencerosini (Nitidulidae: Nitidulinae), Ectychini (Tenebrionidae: Diaperinae), Carpophagini (Chrysomelinae: Sagrinae) and Amycterini (Curculionidae: Cyclominae), and numerous genera.

A few taxa, which are relatively common in most parts of the world, are either absent in Australia or represented by a small number of native species; these include brachinine Carabidae (1), geotrupine Geotrupidae (0), Dryopidae (0), Lampyridae (25, all in the subfamily Luciolinae), meligethine Nitidulidae (1), Tetratomidae (0), pimeliine Tenebrionidae (3), Meloidae (60, all in the subfamily Nemognathinae), bruchine Chrysomelidae (20) and lepturine Cerambycidae (0). A few families, such as Dryopidae and Pterogeniidae, occur in New Guinea and may be found eventually in northern Australia.

The Australian flora is sometimes grouped into three or four components based on age and origin (Crisp *et al.* 1999), and the continent has been divided into as few as three (Torresian, Bassian and Eyrean) or as many as 17 faunal regions (Allsopp 1995). Matthews (2000) discussed four faunal elements: (1) Indo-Malayan, composed of a relatively recent influx of northern species across an archipelago of Gondwanan fragments; (2) Austral Gondwanan or Southern, with sister taxa in southern South America, New Zealand and New Caledonia; (3) Tethyan or Pangean, an older assemblage with connections to India, the Mediterranean region, Africa and the New World; and (4) Amphitropical or Amphipolar, consisting of groups occurring in temperate regions of both Northern and Southern Hemispheres.

Indo-Malayan (or Torresian) elements in the fauna are most common in the rainforests of northern QLD, and include such genera as *Tricondyla* (Carabidae: Cicindelinae), *Clidicus* (Staphylinidae: Scydmaeninae), *Priochirus* (Staphylinidae: Osoriinae), *Niponius* (Histeridae: Niponiinae), *Trypeticus* (Histeridae: Trypanaeinae), *Onthophagus* (Scarabaeaidae: Scarabaeinae), *Xylotrupes* and *Haploscapanes* (Scarabaeidae: Dynastinae), *Cyphogastra* and *Paracupta* (Buprestidae), *Pseudeucinetus* (Limnichidae), *Galbites* (Eucnemidae), *Melanotus* and *Cussolenis* (Elateridae), *Pteroptyx* (Lampyridae), *Sarothrias* (Jacobsoniidae), *Encaustes* and *Aulacochilus* (Erotylidae), *Horia* (Meloidae), *Batocera* (Cerambycidae), *Sagra* and *Donacia* (Chrysomelidae), *Ithystenus* (Brentidae), *Baris, Mecopus* and *Sipalinus* (Curculionidae) and major groups like Melolonthini (Scarabaeidae), Cnodalonini (Tenebrionidae) and Galerucinae (Chrysomelidae).

There are three main groups of Australian taxa within the southern Gondwanan element: (a) those shared with southern South America only; (b) those shared with New Zealand and/or New Caledonia, but not South America; and (c) those found in both areas. The first group includes genera like *Plagiotelum* (Carabidae), *Cochliarion* (Ptiliidae), *Neopelatops* (Leiodidae), *Glypholoma* (Staphylinidae), *Syndesus* and *Sphaenognathus* (Lucanidae), *Necrobiopsis* (Trogossitidae), *Lemidia* (Cleridae), *Ericmodes* (Protocucujidae), *Rhopalobrachium* (Phloeostichidae), *Hobartius* (Hobartiidae) and *Archeophthora* (Tenebrionidae), generic pairs such as *Ptomaphila–Oxelytrum* Gistel (Silphidae), *Austrorhysus-Megarthroides* Steel (Staphylinidae), *Cacostomus–Casignetus* Macleay (Lucanidae), *Sclerocyphon–Tychepsephus* Waterhouse (Psephenidae), *Egolia–Acalanthis* Erichson (Trogossitidae), and some family group taxa like Ommatidae, Aclopinae (Scarabaeidae), Anoplognathini (Scarabaeidae: Rutelinae), Stigmoderini (Buprestidae), Promecheilidae, Palophaginae (Megalopodidae) and Caridae. Taxa shared by Australia and New Zealand and/or New Caledonia include *Hyphalus* (Limnichidae), *Phycosecis* (Phycosecidae) and *Dryocora* (Prostomidae), the generic pairs *Pamborus–Maoripamborus* Brookes (Carabidae), *Anepius–Eupsorus* Broun (Staphylinidae), *Trichelodes–Hexanodes* Blair (Dermestidae) and *Tanylypa–Zolodinus* Blanchard (Tenebrionidae), and the tribes Choerodini, Heleini and Titaenini (Tenebrionidae). The intertidal genus *Hyphalus* has a broader distribution, including Albabra and the Ryukyu Archipelago (Hernando and Ribera 2000), and the littoral genus *Phycosecis* has also been seen from Vanuatu. Taxa occurring in Australia, New Zealand/New Caledonia and southern South America include the genera *Sphaerothorax* (Clambidae), *Rhipicera* (Rhipiceridae), *Hydora* (Elmidae), *Nothoderodontus* (Derodontidae), *Archeocrypticus* (Archeocrypticidae) and *Lagrioida* (Anthicidae), and family-group taxa such as Migadopini and Zolini (Carabidae), Camiarinae (Leiodidae), Microsilphini and Corneolabiini (Staphylinidae), Lampriminae (Lucanidae), Rentoniini (Trogossitidae), Priasilphidae, Cavognathidae, Ulodidae, Adeliini (Tenebrionidae), Spilopyrinae (Chrysomelidae), Belinae (Belidae) and Eugnomini (Curculionidae). Amphipolar groups (Crowson 1980; amphitropical groups of Darlington 1965) occurring in Australia include the Broscini (Carabidae), Syndesinae (Lucanidae), Scirtidae, Byrrhidae, Phloeostichidae, Boridae, Pythidae and Nemonychidae.

Matthews (2000) characterised the members of his Tethyan element as inhabiting deserts and coastal dunes, lacking forest-inhabiting relatives, having no areas of endemism east of the Great Dividing Range, and lacking relatives in southern South America. He included the tenebrionid tribes Belopini, Cnemeplatiini, Vacronini, Hyociini, Ectychini and Trachyscelini, but other possible examples from other families might include the western Australian *Oxypius* (Staphylinidae: Oxytelinae), whose sister taxon *Euphanias* Fairmaire & Laboulbène occurs in the western Palaearctic and Brazil (Newton 1985), *Sphaericus*, a Mediterranean and Macaronesian genus of Ptinidae (Ptininae) with two species from North-west Cape and Barrow Island, WA (Bellés 1998), and the dermestid subfamily Dermestinae, with *Dermestes* occurring primarily in the Palaearctic and Nearctic regions and extending into South America and Africa, *Mariouta* Pic and *Rhopalosilpha* Arrow occurring in Central Asia and Asia Minor, and *Derbyana* from the north-western part of Australia (Lawrence and Ślipiński 2005).

Taxa that do not fit well into the above groups are usually members of old Gondwanan or Pangean lineages with relatives in Africa or tropical America, but with at least some centers of diversity in rainforests of eastern Australia. An obvious example involves the dung beetles in the tribes Canthonini and Dichotomiini, whose closest relatives are Afrotropical, with *Lepanus*, *Demarziella* and *Amphistomus* possible sister taxa to the Malagassy *Arachnodes* Westwood and the African *Pedaria* Laporte and *Anachalcos* Hope, respectively (Davis *et al.* 2002; Philips *et al.* 2004). Basal Scarabaeinae are widely distributed in Australia, but several lineages have radiated in eastern montane rainforests and in New Guinea. Insights into the evolution and biogeography of these and other rainforest Coleoptera have been made possible by intensive studies conducted in the Australian Wet Tropics, an area of high regional endemism between Townsville and Cooktown (Yeates *et al.* 2002; Bouchard *et al.* 2005a). Cladistic and biogeographical studies of the scarabaeine genera *Aptenocanthon*, *Temnoplectron* and *Coptodactyla* (Reid 2000a; Reid and Storey 2000; Storey and Monteith 2000; Bell *et al.* 2004) and cnodalonine Tenebrionidae (Bouchard and Yeates 2001) have done much to clarify areas of endemism in this region. Much of the above discussion on the Australian beetle fauna represents an updated version of that included in Austin *et al.* (2004).

2. CLASSIFICATION AND KEYS

CLASSIFICATION

A history of early beetle classifications may be found in Lawrence *et al.* (1995b). Of the several major classifications of Coleoptera which claim to reflect the phylogeny of the order, the most widely accepted one during the latter half of the 20th Century is that of R. A. Crowson, first published in its entirety in 1955 and variously modified in later papers (Crowson 1960, 1967, 1971a, 1971b, 1973b, 1981). A comparison of Crowson's classification with those of Ganglbauer (1903), Lameere (1903), Kolbe (1908) and Peyerimhoff (1933) (see Crowson 1971b) shows that he incorporated many of the features of the previous systems (especially that of Peyerimhoff, who, like Crowson, was strongly influenced by the larval work of Böving and Craighead 1931), but made several significant innovations. These include the following: (1) recognition of a fourth suborder, Myxophaga, for a few small families of aquatic beetles; (2) rejection of the concept of Haplogastra/Symphiogastra, in which Polyphaga are divided into two groups based on the loss or concealment of abdominal sternite II (see Jeannel and Paulian 1944); (3) placement of the Staphylinoidea, Histeroidea and Hydrophiloidea in a series Staphyliniformia, and treatment of the last two as independent superfamilies; (4) formation of a series Scirtiformia (as Eucinetiformia) for the basal polyphagan families Eucinetidae, Clambidae and Scirtidae; (5) incorporation of the Dascillidae and Rhipiceridae into a superfamily Dascilloidea, considered to be sister group of the Scarabaeoidea (series Scarabaeiformia); (6) combining the Byrrhoidea, Dryopoidea, Buprestoidea, Elateroidea and Cantharoidea into a series Elateriformia, together with elements formerly included in Dascilloidea; (7) inclusion of Dermestidae, Bostrichidae and a few isolated groups (Derodontidae, Nosodendridae and Jacobsoniidae) into a series Bostrichiformia; and (8) inclusion of the order Strepsiptera (as Stylopoidea) in the most derived series Cucujiformia, along with the Lymexylidae, Cleroidea, Clavicornia, Heteromera, Phytophaga and Rhynchophora of older systems.

Some alternate family alignments were proposed by Lawrence and Newton (1982) and Lawrence (1988b), so that the classification used in the second edition of *Insects of Australia* (Lawrence and Britton 1991) and the book *Australian Beetles* by Lawrence and Britton (1994) differed from that of Crowson mainly in: (1) the inclusion of Histeridae and related families in the Hydrophiloidea; (2) incorporation of Hydraenidae into Staphylinoidea); (3) inclusion of the dryopoid families and Byrrhidae in Byrrhoidea; (4) union of Artematopodidae, Elateroidea and Cantharoidea into one superfamily, Elateroidea; (5) placement of Jacobsoniidae and Derodontidae in a superfamily Derodontoidea; (6) placement of Dermestidae and Nosodendridae in Bostrichoidea; (7) the separation of Tenebrionoidea (Heteromera) from Cucujoidea (Clavicornia); and (8) the removal of Strepsiptera from the order. The following additional changes were madein the family list of Lawrence and Newton (1995): (1) inclusion of Pselaphidae within the family Staphylinidae (following Newton and Thayer 1995); (2) the expansion of Elateriformia to include the superfamily Scirtoidea (following Lawrence *et al.* 1995a); (3) the addition of the new scirtoid family Decliniidae (Nikitsky *et al.* 1994); (4) the placement of Podabrocephalidae (Pic 1930) and Rhinorhipidae Lawrence (1988b) as Elateriformia *incertae sedis*; (5) recognition of Ulodidae at the family level following Lawrence (1994a); and (6) recognition of Megalopodidae and Orsodacnidae at the family level. A few minor modifications were made in the interactive keys to beetles (Lawrence *et al.* 1999a, 1999b), including the addition of the family Jurodidae (= Sikhotealiniidae) (Lafer 1996; Kirejtshuk 1999), the elevation of Geotrupidae: Bolboceratinae to family rank (Scholtz and Browne 1996) and the inclusion of the Monommatidae and Colydiidae in the family Zopheridae (Ślipiński and Lawrence 1999), and these were incoporated into the classification used in the first beetle volume of the *Handbook of Zoology* (Beutel and Leschen 2005a). The elateroid subfamily Anischiidae, recognised by Lawrence *et al.* (1999b) and placed in Elateridae by Beutel and Leschen (2005a), was considered to be a subfamily of Eucnemidae by Lawrence *et al.* (2007).

Further modifications introduced in the second Coleoptera volume of the *Handbook of Zoology* (Leschen *et al.* 2010a) include: (1) the recognition of two new adephagan families Meruidae (Spangler and Steiner 2005) and Aspidytidae (Ribera *et al.* 2002a; Ribera and Beutel 2003); (2) replacement of the family name Microsporidae with Sphaeriusidae (International Commission on Zoological Nomenclature 2000); (3) placement of Nosodendridae and Jacobsonidae in Derodontiformia: Derodontoidea; (4) replacement of the family name Anobiidae with Ptinidae based on priority; (5) reduction of the subfamilies of Trogossitidae to two (Peltinae and Trogossitinae) following Kolibáč (2006) and Kolibáč and Leschen (2010a); (6) recognition of the cleroid families Metaxinidae (Kolibáč 2004b), Thanerocleridae (Kolibáč 1992, 1998b) and Mauroniscidae (Majer 1995); (7) use of a new arrangement of cucujoid families following Leschen *et al.* (2005); (8) recognition of the family Cyclaxyridae (Gimmel *et al.* 2009); (9) use of a new subfamly classification of Endomychidae based on Tomaszewska (2000, 2010); (10) recognition of two subfamilies of Coccinellidae, Microweiseinae and Coccinellinae, based on Ślipiński (2007); recognition of two subfamilies of Corylophidae, Periptyctinae and Corylophinae, based on Ślipiński *et al.* (2001, 2009); (11) transfer of the subfamilies Eustrophinae and Hallomeninae from Melandryidae to Tetratomidae following Nikitsky (1998); (12) placement of the Chilean genus *Promecheilus* Solier in the Perimylopidae and chage of the name of that family to Promecheilidae based on priority; (13) use of a new subfamily classification of Tenebrionidae based on Bouchard *et al.* (2005) and Matthews and Bouchard (2008); (14) recognition of an oedemerid subfamly Polypriinae based on Lawrence (2005g); and (15) recognition of the borid subfamily Synercticinae based on Lawrence and Pollock (1994).

The following additional changes are made in the list below: (1) treatment of Rhysodidae as subfamily of Carabidae as suggested by several workers, including Liebherr and Will (1988) (see discussion in Beutel 2005c); (2) re-expansion of the family Geotrupidae to include Bolboceratinae following Smith (2006); (3) combination of the families Ceratocanthidae and Hybosoridae and recognition of several hybosorid subfamilies following Nikolajev (1999a), Ocampo (2006b) and Ocampo and Hawks (2006); (4) elevation of the trogossitid tribe Lophocaterini to subfamily rank following Kolibáč and Zaitsev (2010), but excluding the Decamerini; (5) recognition of the family Akalyptoischiidae following Lord *et al.* (2010); and (6) recognition of the chrysomeloid families Oxypeltidae, Vesperidae and Disteniidae following Švácha *et al.* (1997).

It would be comforting to think that the significant number of systematic works on beetles over the past 50 years has increased the stability of the resulting family-level classification and made it less likely that further rearrangements will be made, but several recent phylogenetic analyses based on molecular and/or morphological data have indicated that several more changes at superfamily, family or subfamily level may be necessary in the future (Maddison *et al.* 2000, 2009; Shull *et al.* 2001; Marvaldi *et al.* 2002, 2009; Caterino *et al.* 2002, 2005; Hunt *et al.* 2007; Robertson *et al.* 2008; Lekanicová 2009; Longhorn *et al.* 2010; McKenna and Farrell 2010; Pons *et al.* 2010; Song *et al.* 2010; Lawrence *et al.* 2011).

The following list includes 174 beetle families for the world, 117 of which occur in Australia. It differs from that published by Ślipiński *et al.* (2011) in the omission of fossil families and the inclusion of Rhysodidae within Carabidae.

LIST OF BEETLE FAMILIES

Figures in parentheses are approximate numbers of Australian species based on described taxa plus estimates of undescribed forms in collections.

Order COLEOPTERA
Suborder ARCHOSTEMATA
001. Ommatidae (4)
002. Cupedidae (6)
Micromalthidae (0)
Crowsoniellidae (0)
Jurodidae (0)

Suborder MYXOPHAGA
Lepiceridae (0)
Torridincolidae (0)
Hydroscaphidae (0)
003. Sphaeriusidae (2)

Suborder ADEPHAGA
004. Gyrinidae (25)
005. Haliplidae (15)
Meruidae (0)
Aspidytidae (0)
006. Noteridae (7)
Amphizoidae (0)
007. Hygrobiidae (4)
008. Dytiscidae (265)
Trachypachidae (0)
009. Carabidae (3000)

Suborder POLYPHAGA
STAPHYLINIFORMIA

HYDROPHILOIDEA
010. Hydrophilidae (230)
Sphaeritidae (0)
Synteliidae (0)
011. Histeridae (280)

STAPHYLINOIDEA
012. Hydraenidae (230)
013. Ptiliidae (80)
Agyrtidae (0)
014. Leiodidae (150)
015. Silphidae (3)
016. Staphylinidae (3000)
SCARABAEIFORMIA

SCARABAEOIDEA
Pleocomidae (0)
017. Geotrupidae (175)
Belohinidae (0)
018. Passalidae (50)
019. Trogidae (55)
Glaresidae (0)
Diphyllostomatidae (0)
020. Lucanidae (110)
Ochodaeidae (0)
021. Hybosoridae (50)
Glaphyridae (0)
022. Scarabaeidae (3500)
SCIRTIFORMIA

SCIRTOIDEA
023. Clambidae (25)
024. Eucinetidae (10)
Decliniidae (0)
025. Scirtidae (200)
ELATERIFORMIA

DASCILLOIDEA
026. Dascillidae (3)
027. Rhipiceridae (15)

BUPRESTOIDEA
028. Buprestidae (1500)

BYRRHOIDEA
029. Byrrhidae (50)
Dryopidae (0)
Lutrochidae (0)
030. Limnichidae (50)
031. Heteroceridae (20)
032. Elmidae (200)
033. Psephenidae (20)
Cneoglossidae (0)
034. Ptilodactylidae (25)
Podabrocephalidae
035. Chelonariidae (1)
Eulichadidae (0)
036. Callirhipidae (5)

ELATEROIDEA
037. Rhinorhipidae (1)
Artematopodidae (0)
038. Brachypsectridae (1)
Cerophytidae (0)
039. Eucnemidae (200)
040. Throscidae (25)
041. Elateridae (1000)
Plastoceridae (0)

Drilidae (0)
Omalisidae (0)
042. Lycidae (300)
Telegeusidae (0)
Phengodidae (0)
Rhagophthalmidae (0)
043. Lampyridae (25)
Omethidae (0)
044. Cantharidae (200)

DERODONTIFORMIA

DERODONTOIDEA

045. Derodontidae (1)
046. Nosodendridae (3)
047. Jacobsoniidae (5)

BOSTRICHIFORMIA

BOSTRICHOIDEA

048. Dermestidae (200)
Endecatomidae (0)
049. Bostrichidae (120)
050. Ptinidae (300)

CUCUJIFORMIA

LYMEXYLOIDEA

051. Lymexylidae (15)

CLEROIDEA

Phloiophilidae (0)
052. Trogossitidae (70)
Chaetosomatidae (0)
Metaxinidae (0)
053. Thanerocleridae (10)
054. Cleridae (400)
055. Acanthocnemidae (1)
056. Phycosecidae (3)
Prionoceridae (0)
Mauroniscidae (0)
057. Melyridae (350)

CUCUJOIDEA

058. Boganiidae (4)
Byturidae (0)
Helotidae (0)
059. Protocucujidae (4)
060. Sphindidae (10)
061. Biphyllidae (50)
062. Erotylidae (120)
063. Monotomidae (15)
064. Hobartiidae (5)
065. Cryptophagidae (100)
Agapythidae (0)
066. Priasilphidae (3)
067. Phloeostichidae (6)
068. Silvanidae (75)
069. Cucujidae (8)
070. Myraboliidae (20)
071. Cavognathidae (2)
072. Lamingtoniidae (5)
073. Passandridae (8)
074. Phalacridae (100)
075. Propalticidae (4)
076. Laemophloeidae (75)
077. Tasmosalpingidae (2)
Cyclaxyridae (0)
078. Kateretidae (15)
079 Nitidulidae (400)
Smicripidae (0)
080. Bothrideridae (70)
081. Cerylonidae (50)
Alexiidae (0)
082. Discolomatidae (20)
083. Endomychidae (75)
084. Coccinellidae (500)
085. Corylophidae 75)
Akalyptoischiidae (0)
086. Latridiidae (75)

TENEBRIONOIDEA

087. Mycetophagidae (10)
088. Archeocrypticidae (25)
Pterogeniidae (0)
089. Ciidae (150)
Tetratomidae (0)
090. Melandryidae (75)
091. Mordellidae (150)
092. Ripiphoridae (80)
093. Zopheridae (150)
094. Ulodidae (20)
095. Promecheilidae (3)
Chalcodryidae (0)
Trachelostenidae (0)
096. Tenebrionidae (1700)
097. Prostomidae (8)
Synchroidae (0)
Stenotrachelidae (0)
98. Oedemeridae (100)
99. Meloidae (100)
100. Mycteridae (30)
101. Boridae (1)
Trictenotomidae (0)
102. Pythidae (3)
103. Pyrochroidae (25)
104. Salpingidae (50)
105. Anthicidae (250)
106. Aderidae (120)
107. Scraptiidae (30)

CHRYSOMELOIDEA

Oxypeltidae (0)
Vesperidae (0)
Disteniidae (0)
108. Cerambycidae (1500)
109. Megalopodidae (8)
Orsodacnidae (0)
110. Chrysomelidae (3500)

CURCULIONOIDEA

111. Nemonychidae (25)
112. Anthribidae (200)
113. Belidae (200)
114. Attelabidae (100)
115. Caridae (4)
116. Brentidae (250)
117. Curculionidae (8000)

ADULT AND LARVAL KEYS

This following keys are long and may be difficult to use for the beginner. An attempt has been made to begin with structures that are relatively easy to examine without dissection, but this has not always been possible. The division of the keys into sections based on higher taxa, such as that in Crowson (1955), was found to be unworkable, since the limits of some superfamilies are still in doubt and the character states defining them are often difficult or impossible to examine without dissection or special preparation. An interesting compromise may be found in the North American key by Ivie (2002) where suborders are keyed out first and all Polyphaga included in a single key, with notes following couplets where highly modified Archostemata (Micromalthidae) or Myxophaga (Hydroscaphidae and Sphaeriusidae) might key out if a mistake were made in the suborder key. Also, given the number of recent changes in the constitution of families and subfamilies, the following keys are to taxonomic units of varying rank (families to distinctive genera). Since both adult and larval keys are particularly long (293 and 281 couplets, respectively), character–state combinations are given in bold type at various points in the key. The keys are primarily dichotomous, but the following multiple choices should be noted: in the adult key, there are three choices in 'couplets' 42, 74, 185, 225 and 230, and five choices in 37; in the larval key there are three choices in 3, 38, 42, 46, 88, 92, 99, 102, 115, 120, 143, 146, 187, 195, 198 and 202, four in 178 and 230, and seven in 222.

ADULTS

1 Abdomen with three ventrites (Fig. 28G); tarsi 3-segmented; body globular, glabrous, at most 1 mm in length; antennae with compact, 3-segmented club (Fig. 55A); metacoxae with large plates concealing hind femora .... MYXOPHAGA Sphaeriusidae (*Sphaerius*)

– Abdomen with four or more ventrites; without other features combined **2**

2(1) Metatrochantins visible; mesocoxae conical and projecting, their cavities mesally contiguous and laterally in broad contact with metanepisterna (Fig. 21D); protrochantin free and exposed and propleuron extending well in front of coxal cavities, forming part of thoracic wall (Fig. 17E); elytral surface distinctly tuberculate with large window punctures, each bordered by several tubercles (Figs 28H, 54D); abdomen with five ventrites (sternites III–VII), the first of which is entire .... **ARCHOSTEMATA 3**

– Metatrochantins never visible (reduced to concealed muscle discs); IF mesocoxae conical and projecting, THEN not in broad contact with mesanepisterna (Fig. 17M); protrochantin concealed or fused to pleuron forming pleurotrochantin (Fig. 17F, H); elytra never with window punctures bordered by tubercles; if propleuron forms part of thoracic wall (Fig. 17G), then abdomen with six ventrites, the first of which is divided into two or three parts by metacoxae (Figs 21C, 28I) **4**

3(2) Antennal insertions lateral, well separated; antennae shorter than head and prothorax combined (Figs 54C, 68B); labrum not distinct; procoxal cavities contiguous; tarsomere 4 not lobed; ventral surfaces without grooves for reception of legs; pleurosternal suture absent; apical maxillary palpomere apically expanded,with cluster of sensilla located in deep cavity (Fig. 16F) .... **Ommatidae**

– Antennal insertions dorsal and more or less approximate; antennae longer than head and prothorax combined (Fig. 54D); labrum free and visible; procoxal cavities separated by prosternal process; tarsomere 4 lobed beneath; ventral surfaces of prothorax and pterothorax with distinct grooves for reception of legs; pleurosternal suture present; apical maxillary palpomere with cluster on sensilla on surface .... **Cupedidae**

4(2) Propleuron extending well in front of procoxal cavitiy forming major part of thoracic wall; notopleural and pleurosternal sutures distinct and separate for most of their lengths (meeting near anterior edge of prothorax) (Figs 13D, 17D); protrochantin free but concealed (Fig. 17G); abdomen with six ventrites, the first three solidly fused together and the first divided into two or three parts by metacoxae, which are solidly fused to metaventrite or at least capable of very limited free movement (Figs 21A, C, 28I) .... **ADEPHAGA 5**

– Propleuron mainly internal and solidly fused to trochantin forming pleurotrochantin, which may be exposed but never extends in front of procoxal cavity (Figs 13C, 17B–C, F, H); first ventrite never divided by metacoxae, which are not solidly fused to metaventrite (Fig. 21B) **10**

5(4) Metacoxae not extending laterally to meet elytra, so that junction of metathorax and first abdominal ventrite usually visible (Fig. 21C); antennae distinctly pubescent and usually filiform (Figs 21A, 54F, L) (moniliform in Rhysodinae, Fig. 54J, flattened and perfoliate in Paussini, Figs 29J, 54K, short and clavate in some Pseudomorphini); dorsal surfaces usually with well developed, erect, tactile setae at fixed positions (Fig. 54E, G); protibia with antenna cleaner (Fig. 16N); metaventrite almost always with transverse (katepisternal) suture (Fig. 21C) (absent in Rhysodinae) .... **Carabidae**

– Metacoxae extending laterally to meet elytra, so that junction of metathorax and first abdominal ventrite is not visible (Fig. 55C, F); antennae entirely or almost entirely glabrous (Fig. 55B–J); dorsal surfaces without fixed tactile setae; protibia without antenna cleaner **6**

Fig. 44. Adult morphology. A–G, Head, dorsal or anterodorsal: A, *Meneristes intermedius* Pascoe (Tenebrionidae:Tenebrioninae); B, *Chauliognathus apterus* (Olliff) (Cantharidae); C, *Porrostoma rufipennis* (Fabricius) (Lycidae); D, *Atyphella olivieri* Lea (Lampyridae); E, *Aspidiphorus* sp. (Sphindidae); F, *Pseudomicrocara* sp. (Scirtidae); G, *Microchaetes* sp. (Byrrhidae: Syncalyptinae). H–N, Prothorax, ventral: H, *Peltonyxa* sp. (Trogossitidae: Lophocaterinae); I, *Acanthocnemus nigricans* Hope (Acanthocnemidae); J, *Austronausibius* sp. (Silvanidae: Silvaninae); K, *Priastichus tasmanicus* Crowson (Priasilphidae); L, *Syndesus cornutus* (Fabricius) (Lucanidae: Syndesinae); M, *Conoderus* sp. (Elateridae: Agryopninae); N, *Scymena variabilis* Pascoe (Tenebrionidae: Tenebrioninae). O, *Eurhynchus laevior* (Kirby) (Brentidae: Eurhynchinae), prothorax, ventrolateral. P–R, Prothorax, ventral: P, *Meryx rugosa* Latreille (Ulodidae); Q, *Melittomma* sp. (Lymexylidae); R, *Loboglossa* sp. (Mycteridae: Eurypinae). S–T, Pterothorax, ventral: S, *Melittomma* sp. (Lymexylidae); T, *Morpholycus costipennis* Lea (Pyrochroidae: Pilipalpinae). U, *Ptilodactyla* sp. (Ptilodactylidae: Ptilodactylinae), prothorax, lateral. V, *Omorgus candidus* (Harold) (Trogidae), pterothorax, ventral. W, *Byrrocryptus variegatus* (Carter) (Ptilodactylidae: Anchytarsinae), prothorax, ventral. X–Y, Tarsi: X, *Notodascillus* sp. (Dascillidae); Y, *Ennometes* sp. (Callirhipidae). Z, *Nascioides parryi* (Hope) (Buprestidae: Buprestinae), pterothorax, ventral. AA, *Australiorylon longipilis* (Carter & Zeck) (Cerylonidae: Ceryloninae), abdominal ventrite 5 (sternite VII). BB, *Ecnolagria grandis* (Gyllenhal) (Tenebrionidae: Lagriinae), tarsus.

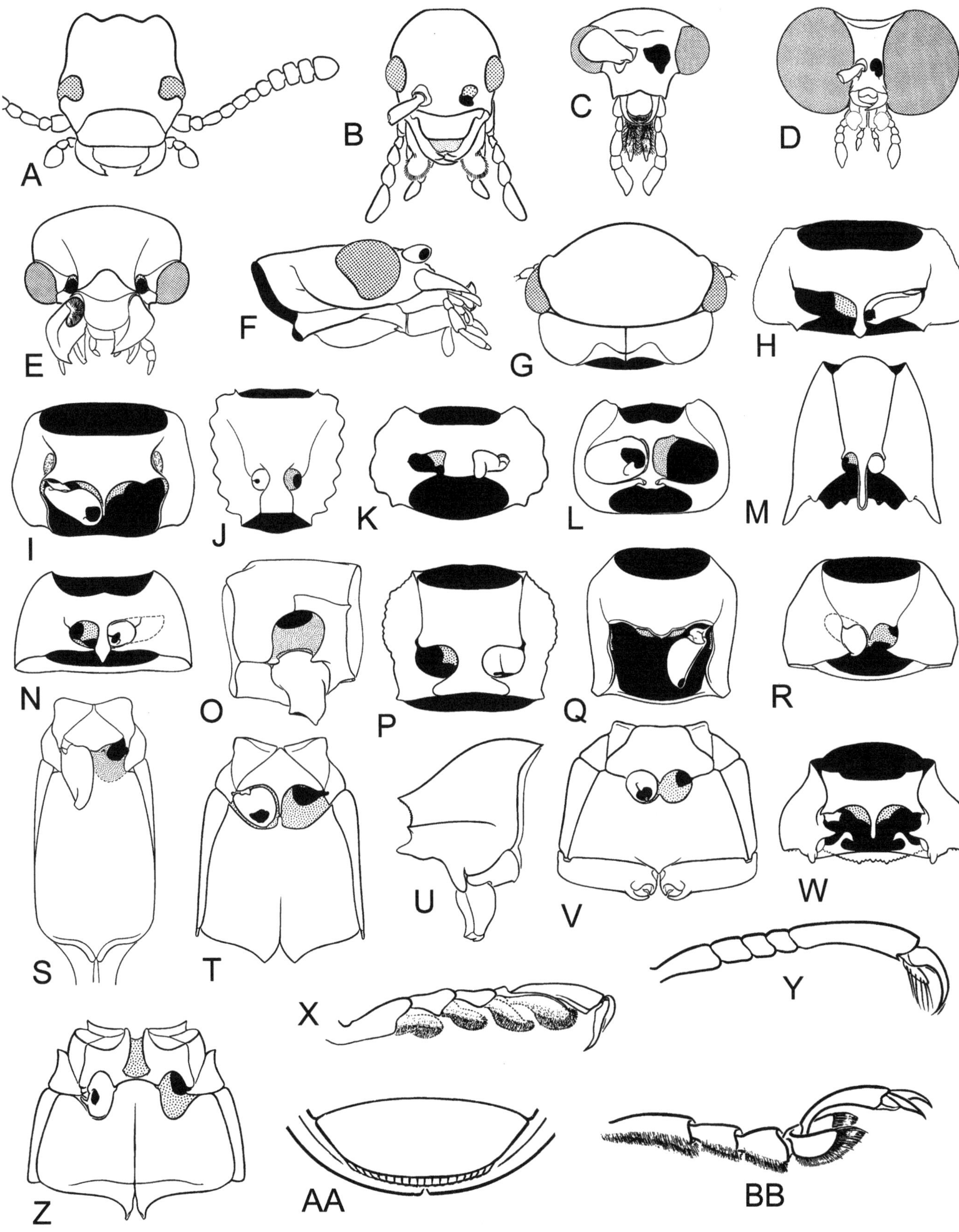
A
B
C
D
E
F
G
H
I
J
K
L
M
N
O
P
Q
R
S
T
U
V
W
X
Y
Z
AA
BB

6(5) Metacoxae with very large plates concealing basal ventrite and most of metafemora; metaventrite with distinct transverse (katepisternal) suture (Fig. 55E); elytra with large punctures forming regular rows (Figs 26D, 55D) ..... **Haliplidae**

– Metacoxal plates absent or consisting of narrow mesal portion only, not concealing basal ventrites or femora (Fig. 55C, F); if metakatepisternal suture present (Fig. 55C), then elytra without regular puncture rows (Fig. 55B) **7**

7(6) Metaventrite with transverse (katepisternal) suture (Fig. 55C); head more exserted, eyes protuberant and well separated from anterior edge of pronotum (Fig. 55B); metacoxae relatively small, their lateral portions shorter than metaventrite ..... **Hygrobiidae**

– Metaventrite without transverse (katepisternal) suture (Figs 26A, 55F); head less exserted, eyes not protuberant and usually in contact with anterior edge of pronotum (Fig. 55F–J); metacoxae much larger, their lateral portions longer than metaventrite (Figs 26A, 55F) **8**

8(7) Eyes completely divided into dorsal and ventral portions (Fig. 14A); antennae very short and thick, with very large pedicel; elytra truncate, exposing most of one abdominal tergite; fore legs long and raptorial, mid and hind legs short and paddle-like (Figs 14A, 55H, J, 66L) ..... **Gyrinidae**

– Eyes not completely divided; antennae longer, without enlarged pedicel; elytra concealing abdominal tergites; fore legs not long and raptorial, mid and hind legs longer and narrower (Fig. 55F, I) **9**

9(8) Dorsal surface strongly convex and scutellum not visible (Fig. 55G); ventral surface flattened; metacoxae with characteristic longitudinal plates covering bases of trochanters; junction between metaventrite and metacoxae angulate in middle; metanepisterna clearly not reaching mesocoxal cavities (Fig. 55F) ..... **Noteridae**

– Dorsal surface hardly more convex than ventral surface and scutellum sometimes visible; metacoxae without plates; junction of metaventrite and metacoxae arcuate; metanepsterna reaching or almost reaching mesocoxal cavities ..... **Dytiscidae**

10(4) Antennae with a 3- to 8-segmented, strongly asymmetrical club, which is usually lamellate (Figs 15H, 61D–F, I), but may be pectinate (Fig. 60D, J) or cupuliform (Fig. 60K–L); head visible from above, not covered by prothorax; procoxae very large, strongly transverse or conical and projecting below prosternum, with concealed trochantins and externally closed procoxal cavities (Fig. 44L); protibiae usually flattened with one or more teeth on outer edge; tarsi with five distinct tarsomeres, none of which is lobed or densely pubescent (Fig. 61A–B, E–F, I) ..... **SCARABAEOIDEA 11**

– Antennae rarely with strongly asymmetrical club; if so, then club with at least 20 antennomeres, procoxae smaller with exposed trochantins, procoxal cavities open behind, head concealed from above by prothorax or protibiae and tarsi different **16**

11(10) Antennae 11-segmented; antennal club circular or oval, cupuliform (Fig. 60K–L); body highly convex and hemispherical, often with extensive armature on head and prothorax of male (Fig. 60L) ..... **Geotrupidae**

– Antennae with fewer than 11 antennomeres **12**

12(11) Mesocoxal cavities closed laterally by broad meeting of mesoventrite and metaventrite (Fig. 44V) **13**

– Mesocoxal cavities partly closed laterally by mesepimera (Fig. 44T, Z) **14**

13(12) Antennomeres comprising antennal club capable of close apposition (Fig. 60L); abdomen with 5 ventrites; mesocoxal cavities contiguous (Fig. 44V); elytra dull and more or less tuberculate, without striae (Fig. 60E) ..... **Trogidae**

– Antennomeres comprising antennal club not capable of close apposition (Fig. 60J); abdomen with 6 ventrites; mesocoxal cavities distinctly separated; elytra shiny and distinctly striate (Fig. 60J) ..... **Passalidae**

14(12) Antennomeres comprising antennal club not capable of close apposition (Fig. 60D); abdomen with 5 ventrites; males usually with enarged mandibles (Figs 60D, 68C–E) but only rarely with tubercles or horns on head or pronotum ..... **Lucanidae**

– Antennomeres comprising antennal club capable of close apposition (Fig. 68J); abdomen with 6 ventrites; mandibles in male not larger than those in female; male armature, if present, involving head and/or pronotum (Fig. 61B–D) **15**

15(14) Labrum and mandibles clearly visible from above, the latter more or less opposable (Fig. 60N); antennae 10-segmented; antennal club cupuliform; metatibiae without obliquely transverse ridges or groups of spines ..... **Hybosoridae**

– If labrum visible from above, then antennal club not cupuliform and metatibiae with obliquely transverse ridges or groups of spines (Fig. 61I); IF mandibles visible with opposable apices, THEN antennae 9-segmented ..... **Scarabaeidae**

16(10) Elytra absent; larviform females (Fig. 63M) **17**

– Elytra present (rarely reduced to small lobes) **18**

17(16) Head with median ocellus (Fig. 63M); tarsi 5-5-5; antennae very slender, 9-segmented ..... **Dermestidae: Trinodinae (*Thylodrias*)**

– Head without median ocellus; tarsi 5-5-4 or 4-4-4; antennae stouter, with 11 or fewer than 5 antennomeres (Figs 7F, 93K) ..... **Ripiphoridae: Ripidiinae**

18(16) Minute, somewhat flattened beetles, rarely more than 1 mm in length; antennae filamentous with antennomeres elongate, narrow and clothed with whorls of long hairs; hind wings almost always lined with fringes of long, fine hairs, often visible at edges of elytra (Fig. 58H) (wings absent in a few highly compact and eyeless forms, Fig. 59A–B); elytra usually more or less truncate exposing one or more tergites; metacoxae moderately to widely separated or with well developed plates; abdomen with 6 or 7 ventrites ..... **Ptiliidae**

– Length usually more than 1 mm; antennae rarely filamentous; if wings fringed with long hairs, body more convex and abdomen with fewer than 6 ventrites **19**

19(18) Elytra almost always exposing 3 or more abdominal tergites (Figs 59F, L–O, 67A–E, H, K–L); if only 1 or 2 exposed, then head with paired ocelli (Fig. 15B) or first exposed tergite longer than elytra **20**

– Elytra almost always completely concealing abdominal tergites or exposing apex of one only; if 1 or 2 complete tergites exposed, then head without paired ocelli and first exposed tergite much shorter than elytra **37**

Non-scarabaeoid Polyphaga with elytra (but excluding Ptiliidae). Exposed abdominal tergites 3 or more

20(19) Tarsi 5-5-4 ..... **21**

– Tarsi not 5-5-4 (usually 5-5-5, occasionally 4-4-4 or 3-3-3) **23**

21(20) Head abruptly constricted posteriorly to form broad or narrow neck; metacoxae contiguous or very narrowly separated; body slightly flattened to moderately convex; antennae serrate to pectinate or flabellate (Figs 76G–H, 93L) ..... **Ripiphoridae (part)**

– Head not abruptly constricted posteriorly, (sometimes gradually narrowed); metacoxae distinctly separated; body strongly flattened; antennae filiform (Fig. 95E) **22**

22(21) Head and pronotum longer than wide; body more than 5 times as long as wide; antennal insertions exposed ..... **Salpingidae (*Diagrypnodes*)**

– Head and pronotum shorter than wide (Fig. 95E); body less than 4 times as long as wide; antennal insertions concealed .... **Salpingidae: Inopeplinae (*Euryplatus*)**

23(20) Antenna with distinct club (Fig. 67G) **24**

– Antenna without distinct club (sometimes gradually clavate) **30**

24(23) Head with paired ocelli (Fig. 15B); abdomen with fewer than 3 exposed tergites **25**

– Head without paired ocelli; abdomen with at least 3 exposed tergites (Fig. 68B) **27**

25(24) Tarsi 4-4-4; body elongate and more or less parallel-sided; elytra without puncture rows; outer edges of tibiae not lined with spines .... **Staphylinidae: Omaliinae (*Metacorneolabium*)**

– Tarsi 5-5-5; body shorter and broader; outer edges of tibiae lined with spines **26**

26(25) Elytral with distinct puncture rows (Fig. 67G) .... **Staphylinidae: Glypholomatinae (*Glypholoma*)**

– Elytra without puncture rows .... **Staphylinidae: Microsilphinae (*Microsilpha*)**

27(24) Antenna with 3-segmented, loose club, which is finely, densely and uniformly pubescent and preceded by two strongly transverse, glabrous, cup-like antennomeres (Fig. 59E); scutellum very large; elytra carinate (Fig. 68B); length greater than 25 mm .... **Silphidae (*Diamesus*)**

– Without above combination of features **28**

28(27) Procoxal cavities externally closed (Fig. 44L); metathoracic discrimen present; antennal club 3-segmented and compact; first 3 tarsomeres more or less lobed and ventrally setose; body parallel-sided, flattened, with prothorax and elytra of equal width (Fig. 83D) .... **Nitidulidae: Cillaeinae**

– Procoxal cavities externally open (Fig. 44Q–R); metathoracic discrimen absent; without other features combined **29**

29(28) Abdomen without dorso-ventral flexibility (Fig. 59I, L); tarsi 3-3-3; pretarsus usually with single claw or with 2 unequal claws; apical maxillary palpomere enlarged and usually variously modified; body often bearing deep foveae at various points .... **Staphylinidae: Pselaphinae**

– Abdomen more or less flexible in dorso-ventral plane; without other features in combination .... **Staphylinidae: Steninae, Euaesthetinae, Leptotyphlinae**

30(23) Mesanepisterna broadly joined at midline, so than mesoventrite is separated from anterior edge of mesothorax (Fig. 44S); hind wing without transverse folds and almost completely exposed by minute elytra (Fig. 82A) .... **Lymexylidae: Atractocerinae**

– Mesanepisterna distinctly separated at midline, so that mesoventrite extends to anterior edge of mesothorax (Fig. 21E–F); hind wing with normal transverse folds and usually folded beneath elytra **31**

31(30) Tarsi pseudotetramerous, (5-segmented with tarsomere 4 highly reduced and partly concealed at base of ventral lobe on tarsomere 3) (Fig. 96J); antennal insertions somewhat elevated (Fig. 77K) and often located within eye emargination (Fig. 14D); hind wings not folded beneath elytra, lying flat 'over abdomen (Fig. 10F) .... **Cerambycidae: Cerambycinae (part)**

– Tarsi not pseudotetramerous; antennal insertions not elevated; hind wings completely concealed beneath elytra **32**

32(31) Tarsomeres simple, not lobed beneath (Fig. 67A–D); metathoracic discrimen absent (Fig. 44S); pretarsal claws simple .... **Staphylinidae (part)**

– All tarsi with at least one tarsomere lobed beneath (Fig. 44BB); metathoracic discrimen present (Fig. 44Z), if indistinct, then pretarsal claws toothed or pectinate **33**

33(32) Antennal insertions concealed (Fig. 15K); labrum completely concealed; prosternal process complete (although laminate) (Fig. 18L); pretarsal claws pectinate (with 2 or more teeth) (Fig. 44A); antennae gradually expanded and apically flattened (Fig. 71J) .... **Cleridae: Tillinae (*Cylidrus*)**

– Antennal insertions exposed (Fig. 44B–E); labrum at least partly exposed; prosternal process incomplete (Fig. 44R) or absent (Fig. 44Q); pretarsal claws simple or with one tooth only; antennae filiform, serrate or pectinate **34**

34(33) Antennal insertions approximate, distance between them less than or only slightly greater than diameter of one of them (Fig. 44C–D) **35**

– Antennal insertions moderately to widely separated, distance between them at least 1.5 times and usually at least 2 times diameter of one of them (Fig. 44A–B) **36**

35(34) Apical maxillary palpomere fusiform (Fig. 44D); abdomen with luminous organ (indicated by yellowish-white colour of ventrite 5) (Figs 5H, 64L) .... **Lampyridae (some *Atyphella* females)**

– Apical maxillary palpomere slightly to strongly expanded at apex (Fig. 44A); abdomen without luminous organ .... **Lycidae (some *Porrostoma* females)**

36(34) Labrum membranous; pretarsal claws simple or weakly toothed at base; vestiture consisting of short, fine hairs; basal protarsomere simple in both sexes .... **Cantharidae (some *Chauliognathus*)**

– Labrum sclerotised; pretarsal claws strongly lobed at base; vestiture of short, fine hairs mixed with stiff, erect, dark setae (Fig. 83I); basal protarsomere in male enlarged and bearing a fine, dark comb .... **Melyridae: Malachiinae (Carphurini)**

Exposed abdominal tergites none (rarely 1 or 2)

37(19) Mesotarsus with 5 distinct tarsomeres **38**

– Mesotarsus pseudotetramerous (with 4 distinct tarsomeres plus a highly reduced penultimate one more or less concealed at base ventral lobe on tarsomere 3) (Fig. 96J) **184**

– Mesotarsus with 4 distinct tarsomeres **223**

– Mesotarsus pseudotrimerous (with 3 distinct tarsomeres plus a highly reduced penultimate one more or less concealed at base of ventral lobe on tarsomere 2) (Figs 87L–M, 89E) **274**

– Mesotarsus with 3 distinct tarsomeres (or rarely fewer) **277**

38(37) Mesocoxal cavities open laterally (not closed by meeting of mesoventrite and metaventrite) (Figs 21F, 25A–B, 44S–T) **39**

– Mesocoxal cavities closed laterally by meeting of mesoventrite and metaventrite (Figs 17K, 25L, 44V) **149**

39(38) Protrochantins at least partly exposed (Figs 18L, 19K, 20K, 44W) **40**

– Protrochantins completely concealed or apparently absent (Figs 19I, 44J, M–N) **103**

40(39) Antenna without distinct club (sometimes gradually clavate); antennomere 8 not smaller than both 7 and 9; transverse occipital carina never combined with 6 ventrites **41**

– Antenna with distinct club consisting of from 1 to 7 antennomeres; if club weakly developed, then antennomere 8 smaller (either shorter or narrower or both) than antennomeres 7 and 9 (Figs 15I, 29C) (in marginal cases the head has a transverse occipital carina and there are 6 ventrites) **74**

Mesotarsus 5-segmented. Mesocoxal cavities laterally open. Protrochantins exposed. Antenna without club

41(40) Tarsi 5-5-5 .... **42**

– Tarsi 5-5-4 .... **70**

42(41) Antenna 9-segmented, very slender, with last 3 antennomeres greatly elongated (Fig. 63L); head with median ocellus; body yelllowish-brown .... **Dermestidae: Trinodinae (*Thylodrias* males)**

– Antennae 10-segmented; basal antennomeres in male often enlarged and modified; soft-bodied and brightly coloured (red or yellow on blue or black) (Fig. 83A) .... **Melyridae: Malachiinae (part)**

– Antennae 11-segmented (or rarely with more than 11 antennomeres) **43**

43(42) Head with paired genal ridges beneath (Figs 14J, 44F) and strongly hypognathous, so that ridges rest against edges of procoxae (Fig. 15J); pronotum very short and broad (Fig. 60A–C) **44**

– Head without paired genal ridges beneath **45**

44(43) Transverse occipital ridge present (Fig. 44G); tarsomeres simple, not lobed beneath; frontoclypeal suture distinct; body more or less fusiform, strongly tapered posteriorly (Fig. 60C); metacoxal plates large and oblique; mid and hind legs with fringes of dark spines at apices of tibiae and on basal 4 tarsomeres (Fig. 26K) .... **Eucinetidae**

– Transverse occipital ridge absent; penultimate tarsomere lobed beneath (Fig. 15J); frontoclypeal suture absent or indistinct; without other features combined (Fig. 60A) .... **Scirtidae**

45(43) Mesanepisterna meeting or very narrowly separated at midline (Fig. 44S–T) **46**

– Mesanepisterna distinctly separated at midline (Figs 17K, 21E) **47**

46(45) Intercoxal process on abdominal ventrite 1 absent (Fig. 28H); head abruptly constricted posteriorly to form neck (Fig. 82G); antennae serrate to pectinate; eyes shallowly emarginate; length greater than 7.5 mm .... **Lymexylidae (part)**

– Intercoxal process on abdomen present and apically acute (Fig. 26J); head not abruptly constricted posteriorly; antennae slightly incrassate (Fig. 44A); eyes entire; length less than 4 mm .... **Melyridae: Dasytinae (part)**

47(45) Labrum not visible, concealed beneath clypeus or solidly fused to clypeus (Fig. 19H) **8**

– Labrum free and mostly visible (Figs 14K, 44A–D) **51**

48(47) Body less than 3 times as long as wide, more or less ovoid (Fig. 63A); posterior edge of pronotum crenulate; head hypognathous, fitting into cavity on anterior edge of prosternum (Fig. 63B); antennal insertions concealed; metacoxal plates well developed and complete; tibiae flattened, widest at middle (Fig. 28O); metaventrite and abdominal ventrite 1 with paired curved lines delimiting crural impressions .... **Chelonariidae (*Chelonarium*)**

– Body more than 3 times as long as wide, more or less parallel-sided; posterior edge of pronotum not crenulate; head, if hypognathous, not fitting into cavity on anterior edge of prosternum; antennal insertions not concealed (Fig. 44B–D); metacoxal plates weakly developed or absent; metaventrite and abdominal ventrite 1 without curved lines **49**

49(48) Body moderately convex, especially ventrally; prothorax distinctly narrower at base than combined elytral bases, without lateral carinae (Fig. 62K); head hypognathous with clypeus elongate and labrum membranous and setose; procoxae conical and projecting (Fig. 20H); mandible moderately elongate with dorsal setose cavity at base .... **Rhinorhipidae (*Rhinorhipus*)**

– Body more or less parallel-sided and slightly to strongly flattened; prothorax about as wide at base as combined elytral bases, with distinct lateral carinae (Figs 71E, 77J); head prognathous; clypeus not elongate; mandibles simple at base **50**

50(49) Head about as long as wide, not abruptly constricted posteriorly (Fig. 77J); procoxal and mesocoxal cavities relatively narrowly separated; procoxal cavities very narrowly open, the notal projections long, acute and almost reaching prosternal process (Figs 18E, 44K); all tibiae with outer apical angle produced to form tooth; tarsomere 4 distinctly shorter than 3 .... **Cerambycidae: Parandrinae**

– Head distinctly shorter than wide, abruptly constricted posteriorly to form short neck and temples (Fig. 71E); procoxal and mesocoxal cavities more widely separated; procoxal cavities broadly open, the notal projections short and angulate (Fig. 18G); outer apical angles of tibiae not produced and acute; tarsomere 4 not distinctly shorter than 3 .... **Cucujidae (*Platisus*)**

51(47) Body very short and broad, less than 1.5 times as long as wide (Fig. 62J) **52**

– Body more than 1.5 times as long as wide (Fig. 62A–B) **54**

52(51) Posterior edge of pronotum distinctly crenulate (Fig. 20C); antennal insertions exposed (Fig. 44B); mesocoxal cavities separated by less than 0.4 times width of one cavity (Figs); apical maxillary palpomere slightly to strongly expanded at apex (Fig. 44A); body more or less flattened and clothed with decumbent hairs forming a pattern (Fig. 62J) .... **Psephenidae: Eubriinae (*Sclerocyphon*)**

– Posterior edge of pronotum not crenulate; antennal insertions concealed (Fig. 44A); mesocoxal cavities separated by more than 0.4 times width of one cavity (Figs 21E, 26B, E); apical maxillary palpomere cylindrical to fusiform (Figs 16B, 44D); body moderately to strongly convex **53**

53(52) Metaventrite without transverse (katepisternal) suture (Fig. 62I); frontoclypeal suture absent; base of labrum on lower plane than apex, so that a deep groove separates labrum from clypeus (Fig. 62I); tarsomeres ventrally pubescent and often with membranous lobe on penultimate tarsomere (Fig. 44BB); prosternum in front of coxae shorter than intercoxal process; body highly convex, often shiny andmetallic in colour and glabrous (Fig. 69J) or clothed with one or more types of setae; antennae more or less clavate .... **Byrrhidae: Byrrhinae**

– Metaventrite with transverse (katepisternal) suture (Figs 17L, 21F, 28V); frontoclypeal suture present (Figs 15B, 44A); without deep groove at base of labrum; tarsomeres simple and slender, usually with scattered setae only (Fig. 63D); prosternum in front of coxae longer than intercoxal pocess (Fig. 63C); body usually black, shiny, and clothed with suberect hairs (Fig. 69H) or with decumbent hairs forming a pattern (Fig. 63D) .... **Limnichidae: Limnichinae (part)**

54(51) Mesocoxal cavities separated by less than shortest diameter of one cavity (Figs 21F, 25D, 44T) **55**

– Mesocoxal cavities separated by more than shortest diameter of one cavity (Figs 26B, 28V, 44Z) **66**

55(54) Intercoxal process on abdominal ventrite 1 present (Fig. 25E, K) **56**

– Intercoxal process on abdominal ventrite 1 absent (Fig. 28H) **62**

56(55) Pronotum with produced and acute posterior angles (Fig. 70H); procoxae and mesocoxae narrowly separated and metacoxae contiguous; moderately broad, flattened, yellowish-brown beetles with clavate or subserrate antennae (Fig. 70H); occurring in the more arid parts of central and northern Australia .... **Brachypsectridae (*Brachypsectra*)**

– Pronotum without produced and acute posterior angles; (if posterior angles slightly produced, then all coxae widely separated) **57**

57(56) All ventrites free or first 2 ventrites connate (Figs 26J, 28K); pronotum without paired basal pits or impressions; head prognathous or only slightly declined (Figs 62A, 83B) **58**

– Ventrites 1 to 3 connate; pronotum with paired basal pits or impressions (Figs 20C, 62C, E); head moderately to strongly hypognathous (Fig. 62C, E) **59**

58(57) Metacoxal plates well developed and complete (Fig. 28V); procoxae transverse, not projecting below prosternum (Fig. 19K); prosternal process complete, extending behind coxae; procoxal cavities narrowly separated but not contiguous (Fig. 44W); basal tarsomeres with membranous appendages (Fig. 44X) .... **Dascillidae (*Notodascillus*)**

– Metacoxal plates weakly developed or absent; procoxae globular to conical and projecting well below prosternum (Fig. 20G–H); prosternal process incomplete, terminating between contiguous procoxae (Fig. 18); tarsi simple, usually clothed with scattered setae only .... **Melyridae: Dasytinae (part)**

59(57) Body more than 3 times as long as wide; empodium large and multisetose (Figs 27E, 44Y); large beetles (length more than 10 mm) with antennae flabellate in male (Fig. 62C) and pectinate in female (Fig. 62B) .... **Callirhipidae**

– Body less than 3 times as long as wide (Fig. 62E–G); empodium bisetose or absent **60**

60(59) Posterior edge of pronotum simple, not crenulate; antennae shorter than head and prothorax combined (Fig. 62G); metaventrite with transverse (katepisternal) suture; mesocoxal cavities moderately to widely separated (by more than 0.4 times shortest diameter of coxal cavity) (Fig. 21E) .... **Elmidae: Larainae (part)**

– Posterior edge of prontum distinctly crenulate; antennae much longer than head and thorax combined (Fig. 62E–F); metaventrite without transverse (ketepisternal) suture; mesocoxal cavities narrowly separated (by less than 0.4 times shortest diameter of coxal cavity) (Fig. 21F) **61**

61(60) Pronotum strongly inflated anteriorly, with complete lateral carinae (Fig. 62E); frontoclypeal suture absent; frontoclypeal region with median longitudinal ridge; apical maxillary palpomere parallel-sided, not apically expanded (Fig. 16B, G) .... **Ptilodactylidae: Cladotominae (*Austrolichas*)**

– Pronotum not strongly inflated anteriorly; lateral pronotal carinae incomplete anteriorly (Fig. 44U); frontoclypeal suture present (Fig. 44A); frontoclypeal region without longitudinal ridge; apical maxillary palpomere apically expanded, more or less securiform (Fig. 16F) .... **Ptilodactylidae (part)**

62(55) Posterior edge of pronotum distinctly crenulate; empodium enlarged and multisetose (Fig. 27E); procoxal cavities separated by prosternal process (Fig. 44W); antennae flabellate, often with more than 30 antennomeres (Fig. 62D) .... **Rhipiceridae (*Rhipicera*)**

– Posterior edge of pronotum simple, not crenulate; empodium bisetose or absent; procoxal cavities contiguous (Fig. 44Q); antennae, if flabellate, with 11 or fewer antennomeres **63**

63(62) Antennal insertions moderately to widely separated (distance between them at least 1.5 times and usually 2 times the diameter of one of them) (Fig. 44B) **64**

– Antennal insertions approximate (distance between them less than or only slightly greater than the diameter of one of them) (Fig. 44C–D) **65**

64(63) Frontoclypeal suture present (Fig. 44B); labrum membranous and usually partly concealed beneath clypeus; pretarsal claws simple or toothed, but without membranous appendages; tarsomere 4 strongly bilobed; pubescence more or less uniform, without scattered stiff, erect setae; without eversible membranous vesicles (Figs 64J, 70G) .... **Cantharidae (part)**

– Frontoclypeal suture absent; labrum fully exposed and sclerotised; pretarsal claws usually with membranous apprendages; IF tarsomere 4 lobed, THEN tarsomere 3 lobed as well; vestiture with stiff, erect, dark setae scattered among the finer hairs (Fig. 83A); prothorax and base of abdomen often with eversible, membranous vesicles .... **Melyridae: Malachiinae (part)**

65(63) Abdomen with luminous organ (visible as yellowish white colour of last two ventrites) (Fig. 64L); eyes very large and approximate (Fig. 44D); metatrochanter with strongly oblique femoral joint .... **Lampyridae (part)**

– Abdomen without luminous organ; eyes moderately large but well separated (Fig. 44C); metatrochanter with transverse or slightly oblique femoral joint .... **Lycidae (part)**

66(54) One or more basal tarsomeres distinctly lobed beneath; ventrites 1–2 or 1–5 connate **67**

– All tarsomeres simple, without ventral lobes; ventrites free or 1–3 connate **68**

67(66) Ventrites 1–2 connate, the sutures between them usually very weak or incomplete (Fig. 28J); tarsal lobes membranous (Fig. 44X); eye entire (Fig. 15G); procoxae globular (Fig. 19E, H–I); body often large and usually brightly coloured or with metallic sheen (Fig. 69B–G) .... **Buprestidae**

– Ventrites 1–5 connate with all sutures distinct; tarsal lobes not membranous; eyes partly divided by canthus; procoxae strongly transverse; length not more than 3 mm; colour reddish-brown .... **Limnichidae: Limnichinae (*Paralimnichus*)**

68(66) Body narrowly elongate, more than 3 times as long as wide, strongly flattened and glabrous (Fig. 71E); head abruptly constricted posteriorly to form broad neck; mesoventral cavity absent; metacoxal plates absent .... **Cucujidae (*Platisus* females)**

– Body less than 3 times as long as wide, neither strongly flattened nor glabrous; head not abruptly constricted posteriorly; mesoventrite with distinct cavity into which prosternal process is inserted (Figs 21F, 28V); metacoxal plates well developed and complete **69**

69(68) Antennomeres 5 to 10 transverse; base of prothorax distinctly narrower than combined elytral bases (Fig. 62G); pronotum constricted anteriorly or with paired longitudinal carinae at base; metacoxae moderately widely separated; abdominal ventrite 1 without postcoxal lines .... **Elmidae: Larainae (part)**

– Antennomeres 5 to 10 longer than wide; base of prothorax not narrower than combined elytral bases; pronotum without anterior constriction or basal longitudinal carinae;metacoxae contiguous or very narrowly separated; ventrite 1 with pair of curved postcoxal lines .... **Limnichidae (part)**

Mesotarsus 5-segmented. Mesocoxal cavities laterally open. Protrochantins exposed. Antenna without club. Tarsi 5-5-4

70(41) Mesanepisterna very narrowly separated, contiguous or broadly joined at midline, so that mesoventrite is either anteriorly acute (Fig. 44T) or separated from anterior edge of mesothorax (Fig. 44S) **71**

– Mesanepisternal distinctly separated at midline (distance between them more than 0.25 times width of mesocoxal cavity), so that anterior edge of mesoventrite truncate (Figs 21E–F, 44V) **73**

71(70) Head not abruptly constricted posteriorly to form narrow neck (gradually narrowed only) (Fig. 95I, K)................... **Pyrochroidae: Pilipalpinae (part)**
– Head abruptly constricted posteriorly to form neck (Figs 75E–F, 94A, 95H, J).. **72**
72(71) All tarsomeres simple; pretarsal claws pectinate, each with bristle-like appendage beneath it; abdomen with 6 ventrites; eyes not or only shallowly emarginate; tibial spurs glabrous.. **Meloidae**
– Penultimate tarsomere lobed beneath (Fig. 44BB); pretarsal claws simple or toothed, not pectinate, without appendages; abdomen with 5 ventrites; eyes deeply emarginate (Fig. 94C); tibial spurs pubescent(Fig. 16I).. **Scraptiidae (part)**
73(70) Mesocoxae separated by less than width of a coxal cavity (Fig. 44T); lateral pronotal carinae absent (Fig. 19C); head not abruptly constricted posteriorly (sometimes gradually narrowed); procoxae conical and projecting well below prosternum (Fig. 20G–H); body convex or slightly flattened, pubescent (Fig. 75A–D).. **Oedemeridae**
– Mesocoxal separated by more than width of a coxal cavity (Fig. 26B, E); lateral pronotal carinae complete and sharply defined; head abruptly constricted posteriorly to form broad neck; procoxae globular and not projecting; body strongly flattened and glabrous (Fig. 71E) .... **Cucujidae (*Platisus* males)**

Mesotarsus 5-segmented. Mesocoxal cavities laterally open. Protrochantins exposed. Antenna with club

74(40) Antennae 7- to 9-segmented.. **75**
– Antennae 10-segmented.. **77**
– Antennae 11-segmented.. **80**
75(74) Scape always short and broad, irregularly shaped; antennomeres each with several to many conspicuous setae; antennal club not preceded by cupule; maxillary palp rarely more than half as long as antenna; body oval to slightly elongate, strongly flattened (Fig. 71B) .... **Trogossitidae: Lophocaterinae (part)**
– Scape always much longer than wide; antennomeres preceding club glabrous, that immediately preceding club almost always transverse and cup-like (Fig. 57D–H); maxillary palp much more than half as long as antenna (Fig. 57A–C, I); body usually moderately to strongly convex (Fig. 57A–C, K–L) **76**
76(75) Antennal club 3-segmented (Fig. 57D–G); abdominal ventrites almost always 5 (if 6, then body oval and elytra completely concealing abdominal apex .... **Hydrophilidae (part)**
– Antennal club 5- or 6-segmented (Figs 28Q, 57H); abdominal ventrites 6 or 7; if body oval, then elytra truncate exposing abdominal apex .... **Hydraenidae (part)**
77(74) Abdominal ventrites 6; antennal insertions concealed from above; antennal club 4-segmented................................ **Leiodidae: Leiodinae (*Zeadolopus*)**
– Abdominal ventrites 5; antennal insertions exposed from above; antennal club with 3 or fewer segments.. **78**
78(77) Mandible with dorsal tubercle and setose cavity; frontoclypeal suture distinct and strongly curved (Fig. 44E); metathoracic discrimen moderately to very long; procoxal cavities narrowly open; elytra glabrous with scutellary striole and 10 distinct puncture rows (Fig. 83K); VIC and TAS; in myxomycete fruiting bodies.. **Sphindidae (*Notosphindus*)**
– Mandible without dorsal tubercle or setose cavity; frontoclypeal suture absent or indistinct; metathoracic discrimen absent.. **79**
79(78) Head prognathous or slightly declined; transverse occipital ridge absent; prosternal process complete, extending behind coxae; outer edge of mesotibia bearing spines or teeth along most of its length; body ovate, strongly flattened .... **Trogossitidae: Lophocaterinae (part)**
– Head hypognathous; transverse occipital ridge present (Fig. 44G); prosternal process incomplete, terminating between coxae; outer edge of mesotibia simple; minute, globular beetles clothed with decumbent hairs which form whorls (Fig. 82D)............ **Trogossitidae: Peltinae: Rentoniini (part)**
80(74) Abdomen with 6 or 7 ventrites.. **81**
– Abdomen with fewer than 6 ventrites.. **84**
81(80) Abdomen with 7 ventrites; antenna with 3-segmented, finely, evenly pubescent club preceded by glabrous, cup-like antennomere; body length at least 14 mm; flattened, yellowish-brown betles with laterally explanate pronotum and elytra and raised tubercles on elytra (Figs 2A, 59E) .... **Silphidae (*Ptomaphila*)**
– Abdomen with 6 ventrites; antennal club not preceded by cup-like antennomere; body less than 12 mm long; without other features combined....... **82**
82(81) Antennal club 5-segmented with 2nd club segment smaller (shorter or narrower or both) than 1st one (Figs 15I, 59C)........................ **Leiodidae (part)**
– If antennal club 5-segmented, then 2nd club segment subequal to or larger than 1st.. **83**
83(82) Procoxal cavities externally closed; procoxae strongly transverse, not projecting and well separated by prosternal process (Fig. 18L); abdominal intersegmental membranes without brick wall pattern of minute sclerites.. **Nitidulidae (part)**
– Procoxal cavities externally open; procoxae globular to conical, projecting well below prosternum and not separated by prosternal process, which is incomplete (Fig. 20G); abdominal intersegmental membranes with brick wall pattern of minute sclerites .... **Leiodidae: Coloninae (*Colon*)**
84(80) Head moderately to strongly declined (hypognathous).. **85**
– Head prognathous or only slightly declined.. **87**
85(84) Antennal insertions located laterally under slight ridge, not visible from above (Fig. 44A); body less than 1.5 times as long as wide, strongly convex (Fig. 69I–J); mesoventrite with distinct cavity into which apex of prosternal process is inserted (Fig. 28V)................................. **Byrrhidae (part)**
– Antennal insertions located dorsally between eyes and easily visible from above (Fig. 44B–D); if body less than 1.5 times as long as wide, then mesoventral cavity absent **86**
86(85) Body minute (less than 1.5 mm long) and globose (Fig. 82D); metacoxae flat, not excavate, without coxal plates; abdominal ventrite 1 longer than following 2 ventrites combined; head without median ocellus.. **Trogossitidae: Peltinae: Rentoniini (part)**
– Body larger (usually more than 1.5 mm long); more elongate and/or flattened; metacoxae excavate, with coxal plates which conceal at least bases of femora (Fig. 63F); abdominal ventrite 1 not as long as following 2 combined; head usually with median ocellus (Fig. 28F)............... **Dermestidae (part)**
87(84) Antennal club 5-segmented with 2nd club segment smaller (shorter or narrower or both) than 1st one (Fig. 59C–D).. **88**
– Antennal club not 5-segmented or with 2nd club segment subequal to or larger than 1st one **89**
88(87) Mandible with dorsal tubercle which fits into lateral cavity at base of clypeus (Figs 14G, 44E); procoxae transverse and not projecting below prosternum; procoxal process complete (Fig. 20K); mesocoxal cavities separated by more than 0.4 times shortest diameter of coxal cavity.. **Hobartiidae (*Hydnobioides*)**

– Mandible without dorsal tubercle; procoxae globular to conical, projecting well below prosternum; prosternal process incomplete, terminating between coxae (Fig. 44R); mesocoxal cavities contiguous or subcontiguous .... **Leiodidae (part)**

89(87) Intercoxal process on abdominal ventrite 1 broadly rounded to broadly angulate or truncate (Fig. 25E, G) **90**

– Intercoxal process on abdominal ventrite 1 acute to narrowly rounded (Figs 25K, 26J) **94**

90(89) Metacoxal plates well developed and complete (Fig. 17K); antennal insertions concealed from above by frontal projections; metaventrite with more or less complete transverse (metakatepisternal) suture (Fig. 17L) .... **Nosodendridae (*Nosodendron*)**

– Metacoxal plates absent (Fig. 86F); antennal insertions exposed (Fig. 44B); metaventrite without transverse (katepisternal) suture **91**

91(90) Mesocoxal cavities separated by more than shortest diameter of one cavity (Fig. 18E); sides of pronotum more or less explanate and lateral carinae broadly undulate; body broad and somewhat flattened; elytra with tubercles covered with stiff, curled setae, which are often concealed beneath debris (Fig. 85A); TAS .... **Priasilphidae (*Priastichus*)**

– Mesocoxal cavities separated by less than shortest diameter of one cavity (Figs 20K, 44N); lateral pronotal carinae simple or absent; without other features combined **92**

92(91) Lateral pronotal carinae absent (Fig. 83C); procoxal cavities internally open; abdominal ventrites 1–3 connate; mandible with dorsal tubercle which fits into lateral cavity at base of clypeus (Fig. 44E); flightless; TAS .... **Phloeostichidae (*Hymaea succinifera* Pascoe)**

– Lateral pronotal carinae complete; procoxal cavities internally closed; abdominal ventrites free; mandible without dorsal tubercle **93**

93(92) Pronotum with large setose cavities at posterior angles (Fig. 85G); elytral sutural flange distinctly deflected near apex (Fig. 26I); procoxal cavities broadly open .... **Tasmosalpingidae (*Tasmosalpingus*)**

– Pronotum without large setose cavities; elytral sutural flange not deflected near apex; procoxal cavities very narrowly open or closed (Figs 18L, 44N) .... **Nitidulidae (part)**

94(89) Lateral pronotal carinae absent; surfaces of pronotum and elytra tuberculate and shiny, with scattered, erect hairs (Fig. 83C) .... **Phloeostichidae**

– Lateral pronotal carinae complete and sharply defined (Fig. 83D–H) **95**

95(94) Procoxal cavities externally closed **96**

– Procoxal cavities externally open **98**

96(95) Fronotoclypeal suture distinctly arcuate (Fig. 44E); tarsi 5-5-4 in male, 5-5-5 in female; elytral punctation seriate, occasionally with several short, longitudinal carinae (Fig. 83G); all tibiae simple; antennal club relatively weak, more or less symmetrical .... **Protocucujidae (*Ericmodes*)**

– Frontoclypeal suture, if present, not arcuate; without other features combined **97**

97(96) Antennal club ovate, as long as or slightly longer than wide, symmetrical; body usually ovate or oblong; prothorax not constricted posteriorly, its base as wide as or slightly narrower than combined elytral bases (Fig. 73C–D); lacinia absent (Fig. 16B); vestiture variable but never including flattened scales .... **Nitidulidae (part)**

– Antennal club at least 2 times as long as wide, asymmetrical; body almost always elongate and parallel-sided; prothorax constricted posteriorly, so that its base is distinctly narrower than combined elytral bases; body glabrous or clothed with flattened scales (Fig. 71A) .... **Trogossitidae: Trogossitinae (part)**

98(95) Antennal club 4- or 5-segmented (Fig. 28U); protarsi often strongly expanded and setose; body oblong, convex and densely clothed with decumbent hairs frontoclypeal suture absent .... **Leiodidae: Coloninae (*Colon*)**

– Antennal club 3-segmented (Fig. 82F); protarsi not strongly expanded and setose; if frontoclypeal suture absent, then body more elongate and flattened or glabrous **99**

99(98) Procoxal cavities contiguous or subcontiguous; pronotal hypomera each with a large, flat-bottomed pit, concealing a heat sensing organ (Fig. 44I); vestiture including scattered, erect, stiff setae (Fig. 82F) .... **Acanthocnemidae**

– Procoxal cavities distinctly separated by prosternal process; without hypomeral heat-sensing organ or scattered, stiff setae **100**

100(99) Body more or less flattened, with pronotum laterally explanate (Fig. 71B); procoxae strongly transverse with widely exposed trochantins (Fig. 44H); metacoxae extending laterally to meet elytral epipleura, so that metapleural sclerites are separated from anterolateral angles of abdominal ventrite 1 .... **Trogossitidae: Lophocaterinae (part)**

– If body flattened, pronotum not laterally explanate; procoxae less strongly transverse and trochantins only partly exposed (sometimes only barely visible within slit-like extension of procoxal cavity); metacoxae separated from epipleura by extensions of metapleural sclerites, which meet anterolateral angles of abdominal ventrite 1 .... **101**

101(100) Frontoclypeal suture distinct (Fig. 44AA); mandible with dorsal tubercle, fitting into lateral cavity at base of clypeus (Fig. 44E); body oblong, moderately convex, pubescent (Fig. 86J) .... **Hobartiidae (*Hobartius*)**

– Frontoclypeal suture absent; mandible without dorsal tubercle; body flattened and parallel-sided or glabrous **102**

102(101) Sides of prothorax more or less parallel, not distinctly converging anteriorly; elytral punctures fine, confused and bearing decumbent hairs; eyes prominent with deep, transverse impression behind them; metathoracic discrimen moderately long; North QLD .... **Cucujidae (*Pediacus*)**

– Sides of prothorax distinctly converging anteriorly (Fig. 86A–B); elytral punctures coarser, forming rows and bearing minute hairs which are barely visible at lower magnifications; eyes not prominent, without transverse impression behind them; metathoracic discrimen absent .... **Lamingtoniidae (*Lamingtonium*)**

Mesotarsus 5-segmented. Mesocoxal cavities laterally open. Protrochantins concealed

103(39) Antennae 11-segmented (or rarely with more than 11 antennomeres) **104**

– Antennae with fewer than 11 antennomeres **142**

104(103) Tarsi 5-5-5 .... **105**

– Tarsi 5-5-4 .... **119**

105(104) Pronotum with strongly produced and acute posterior angles (Fig. 64A–H) **106**

– Pronotum without produced and acute posterior angles **109**

106(105) Labrum entirely or almost entirely concealed beneath clypeus (Fig. 14I) .... **Eucnemidae (part)**

– Labrum free and mostly visible (Fig. 14K) **107**

107(106) Metacoxae widely separated, without coxal plates; metaventrite with two pairs of postcoxal lines; pronotum with paired basal grooves .... **Eucnemidae: Anischiinae (*Anischia*)**

– Metacoxae contiguous or narrowly separated, with distinct coxal plates (Fig. 17K); metaventrite usually without postcoxal lines (rarely with one pair) **108**

108(107) Ventrites 1 to 4 connate (Fig. 26C); antennae never with club; metaventrite and abdomen without tarsal grooves; body usually elongate with sides subparallel or with elytra narrowing posteriorly (Fig. 64A–H) .... **Elateridae (part)**

– Ventrites 1 to 5 connate; antennae almost always with 3-segmented club; body almost always somewhat flattened and tapering anteriorly and posteriorly from junction of pronotum and elytra (Fig. 64D); if antennae not capitate and body parallel-sided, then metaventrite and abdomen with tarsal grooves .... **Throscidae**

109(105) Abdomen with 6 entrites; at least 3 tarsomeres with ventral lobes (Fig. 44X); base of pronotum usually distinctly narrower than combined elytral bases; body more or less elongate, often brightly coloured and either setose or with strongly protruding eyes (Figs 71G–M, 82J) .... **Cleridae (part)**

– Abdomen with 5 ventrites; without other features combined **110**

110(109) Prosternal process strongly expanded at apex (Figs 19I–J, 20A–B, E) **111**

– Prosternal process not strongly expanded at apex **112**

111(110) Mesocoxal cavities separated by more than shortest diameter of one cavity; intercoxal process on abdominal ventrite 1 broadly rounded to truncate (Fig. 25E, G); pronotal disc with lateral, submarginal lines or ridges (Figs 28B, 85K, 87R); lateral pronotal carinae not denticulate; head without ocelli; abdominal ventrite 1 without postcoxal lines .... **Laemophloeidae (females)**

– Mesocoxal cavities separated by less than shortest diameter of one cavity; intercoxal process on abdominal ventrite 1 acute (Fig. 26J); pronotal disc without lateral submarginal lines; lateral pronotal carinae denticulate; head with paired ocelli (Fig. 81C); abdominal ventrite 1 with pair of short, straight postcoxal lines .... **Derodontidae (*Nothoderodontus*)**

112(110) Procoxal cavities externally closed (Fig. 44J) **113**

– Procoxal cavities externally open (Fig. 18G) **115**

113(112) Pregular area with pair of laterally opening, setose cavities; metaventrite with postcoxal lines (Fig. 25H); prosternum in front of coxa at least slightly shorter than procoxal cavity, which is slightly transverse .... **Biphyllidae (part)**

– Pregular area without paired cavities; metaventrite without postcoxal lines; prosternum in front of coxa distinctly longer than procoxal cavity, which is more or less circular (Fig. 44J) .... **Silvanidae: Silvaninae (part)**

114(112) Antennae without distinct club (may be incrassate or clavate) **115**

– Antennae with distinct club **117**

115(114) Body narrowly elongate, more than 3 times as long as wide; head abruptly constricted posteriorly to form broad neck; antennal scape more than 3 times as long as pedicel (Fig. 71C) .... **Silvanidae: Brontinae: Brontini**

– Body less than 3 times as long as wide; head not abruptly constricted posteriorly (sometimes gradually narrowed); antennal scape less than 3 times as long as pedicel **116**

116(115) Mesocoxal cavities separated by less than shortest diameter of one cavity (Fig. 25D); head with median ocellus (Fig. 28F); metaventrite without transverse (katepisternal) suture; metacoxal plates well developed and complete (Fig. 63F) antennae pubescent .... **Dermestidae: Megatominae (part)**

– Mesocoxal cavities separated by more than shortest diameter of one cavity (Fig. 26B); head without median ocellus; metaventrite with transverse (katepisternal) suture; metacoxal plates weakly developed or absent; antennae glabrous (Fig. 63E) .... **Elmidae: Elminae**

117(114) Mesocoxal cavities separated by more than shortest diameter of one cavity (Fig. 26B); antenna geniculate; scape more than 3 times as long as pedicel (Fig. 14L); elytra truncate, exposing at least one abdominal tergite; body glabrous and shiny (Figs 57L, 77A–C) .... **Histeridae (part)**

– Mesocoxal cavities separated by less than shortest diameter of one cavity (Fig. 25D); antenna not geniculate; scape less than 3 times as long as pedicel; elytra concealing abdominal tergites or exposing apex of one only; body pubescent **118**

118(117) Metacoxal plates well developed and complete; head hypognathous, with median ocellus (Fig. 28F); frons without cavities or groove .... **Dermestidae (part)**

– Metacoxal plates absent; head prognathous, without median ocellus; frons with pair of small cavities usually joined by strongly arched groove (Fig. 86I) .... **Cavognathidae (*Taphropiestes*)**

119(104) Mesocoxal cavities separated by more than shortest diameter of one cavity **120**

– Mesocoxal cavities separated by less than shortest diameter of one cavity **124**

120(119) Prosternal process strongly and usually abruptly expanded at apex (Fig. 20A) **121**

– Prosternal process not or only slightly, gradually expanded at apex (Fig. 18G–H) **122**

121(120) Metacoxae contiguous or narrowly separated and intercoxal process on abdominal ventrite 1 acute or narrowly rounded (Figs 25K, 26J); antennal insertions concealed from above (Fig. 44A); prosternum in front of coxae shorter than prosternal process (Fig. 18L); procoxal cavities internally closed (Fig. 44N); body ovoid to oblong and moderately convex (Fig. 89H–I) .... **Archeocrypticidae (part)**

– Metacoxae moderately to very widely separated and intercoxal process on ventrite 1 broadly rounded or truncate (Fig. 25G); antennal insertions exposed (Fig. 44B–D); prosternum in front of coxae longer than prosternal process (Fig. 18E, G); procoxal cavities internally open (Fig. 44P); body more elongate, parallel-sided and flattened (Figs 585A, 87R) .... **Laemophloeidae (males)**

122(120) Procoxal cavities externally closed (Fig. 25C); abdominal ventrites 1 to 3 connate (Fig. 25K); elytra concealing all abdominal tergites or exposing apex of one only; body elongate, more than 2 times as long as wide, and strongly flattened .... **Tenebrionidae (*Platycotylus*)**

– Procoxal cavities externally open (Figs 18H); all abdominal ventrites free or 1 and 2 connate; elytra exposing most of one abdominal tergite (Figs 57L, 93B); body either less than 2 times as long as wide or laterally compressed, posteriorly narrowed and wedge-like **123**

123(122) Antenna without distinct club (sometimes gradually clavate); abdominal apex produced beyond elytral apices to form spine; metacoxae large, flattened, contiguous and fused to metaventrite; head abruptly constricted posteriorly to form narrow neck, but closely adpressed to prothorax, so that neck is concealed; body laterally compressed, narrowed posteriorly and wedge-like, pubescent (Figs 7C, 93B) .... **Mordellidae**

– Antenna with distinct club consisting of 1 or 3 antennomeres; abdominal apex without spine; metacoxae smaller, freely movable and widely separated; head not abruptly constricted posteriorly; body glabrous, neither laterally compressed nor wedge-like (Fig. 77A–C) .... **Histeridae (part)**

124(119) Procoxal cavities externally open (Figs 13C, 19B, 20K) **125**

– Procoxal cavities externally closed (Figs 19I, 20B, D–E, 25C) **137**

125(124) Antennae without distinct club (sometimes gradually clavate) **126**
– Antennae with distinct club consisting of 3 antennomeres **136**
126(125) Mesotarsi without lobed tarsomeres (Fig. 44Y) **127**
– Mesotarsi with at least one lobed tarsomere (Fig. 44BB) **131**
127(126) Lateral pronotal carinae complete but very fine, curved ventrally and then mesally at posterior third, not visible for their entire lengths from above (Fig. 19B); antennomeres 9 and10 each with circular apical channel containing sensilla; body glabrous (Fig. 94D) .... **Boridae (*Synercticus*)**
– Lateral pronotal carinae either complete and sharply defined, incomplete anteriorly or absent; antennomeres 9 and 10 without sensory channels **128**
128(127) Lateral pronotal carinae complete and sharply defined (Fig. 19D) **129**
– Lateral pronotal carina incomplete (Fig. 19A) or absent (Fig. 19C) **130**
129(128) Body narrowly elongate, more than 3 times as long as wide, and glabrous (Fig. 74G); antennal scape less than 3 times as long as pedicel; procoxae externally globular; abdomen with 3 connate ventrites; TAS .... **Tenebrionidae: Zolodininae (*Tanylypa*)**
– Body less than 3 times as long as wide and pubescent; antennal scape more than 3 times as long as pedice; procoxae externally transverse; abdomen with 2 connate ventrites; not known from Australia but occurring in New Guinea .... **Pterogeniidae (*Histanocerus*)**
130(128) Head prognathous or only slightly declined, not abruptly constricted posteriorly; lateral pronotal carinae absent ; pretarsal claws simple; body not laterally compressed, usually with irregular tubercles and carinae on pronotum and elytra (Fig. 94B) .... **Pythidae (*Anaplopus*)**
– Head moderately to strongly declined (hypognathous) and abruptly constricted posteriorly to form narrow neck; lateral pronotal carinae incomplete or absent; pretarsal claws pectinate or deeply bifid; body laterally compressed and wedge-shaped (Figs 7G, 93I) .... **Ripiphoridae (part)**
131(126) Mesotarsomere 3 lobed beneath and tarsomere 4 simple (Fig. 96J); abdomen usually with only 4 ventrites or if 5 are present, then the suture between the first 2 is weak or incomplete; length rarely exceeding 3.5 mm (Fig. 90I) .... **Aderidae (part)**
– Mesotarsomere 4 lobed beneath (Fig. 44BB); abdomen with 5 distinct ventrites **132**
132(131) Head abruptly constricted posteriorly to form narrow neck (sometimes concealed) (Fig. 95H, J) **133**
– Head not abruptly constricted posteriorly (sometimes gradually narrowed) **134**
133(132) Eyes deeply emarginate (Fig. 94C); tibial spurs pubescent (Fig. 16I); intercoxal process on abdominal ventrite 1 absent .... **Scraptiidae**
– Eyes not or only shallowly emarginate (Fig. 95H); tibial spurs not pubescent; intercoxal process on abdominal ventrite 1 acute to broadly rounded .... **Anthicidae (part)**
134(132) Head moderately to strongly declined (hypognathous); eyes shallowly emarginate; base of prothorax not or only slightly narrower than combined elytral bases (Fig. 93A); tibial spurs serrate, those on the hind leg sometimes large and unequal (Fig. 76A–C) .... **Melandryidae**
– Head prognathous or only slightly declined; eyes not emarginate; base of prothorax distinctly narrower than combined celytral bases; tibial spurs simple and equal **135**
135(134) Antennal insertions concealed from above (Fig. 44A); pronotum distinctly longer than wide (Fig. 95C); ventrites 1 and 2 connate .... **Anthicidae: Lagrioidinae (*Lagrioida*)**
– Antennal insertions exposed from above (Fig. 44B); pronotum not as long as wide (Fig. 95I, K); all ventrites free .... **Pyrochroidae: Pilipalpinae (part)**
136(125) Lateral pronotal carinae very fine, simple, curved ventrally at posterior third, not visible for their entire lengths from above (Fig. 19B); mesanepisterna distinctly separated at midline; body glabrous (Fig. 94D) .... **Boridae (*Synercticus*)**
– Lateral pronotal carinae absent; mesanepisterna meeting or very narrowly separated at midline (Fig. 44T); body pubescent (Fig. 95C) .... **Anthicidae: Lagrioidinae (*Lagrioida*)**
137(124) Prosternal process not or only slightly, gradually expanded at apex (Figs 18D, 25C) **138**
– Prosternal process strongly and usually abruptly expanded at apex (Fig. 19I) **139**
138(137) Antennal insertions exposed from above; prosternal process incomplete, terminating between coxae (Fig. 18D); all ventrites free .... **Anthicidae: Eurygeniinae (part)**
– Antennal insertions concealed from above (Fig. 44A); prosternal process complete, extending behind coxae (Fig. 25C); ventrites 1 to 3 connate (Fig. 25K) .... **Tenebrionidae (part)**
139(137) Body more than 3 times as long as wide; mesanepisterna meeting or very narrowly separated at midline (Fig. 44T); apical maxillary palpomere slightly to strongly expanded at apex; body subglabrous; elytra with distinct pattern of spots or blotches (Fig. 93C); TAS .... **Promecheilidae (*Sirrhas*)**
– Body less than 3 times as long as wide; mesanepisterna distinctly separated at midline (Fig. 44V); apical maxillary palpomere cylindrical to fusiform. **140**
140(139) Lateral pronotal carinae absent; metaventrite as long as or shorter than first abdominal ventrite; body glabrous; TAS .... **Promecheilidae (*Melytra*)**
– Lateral pronotal carinae complete and sharply defined; metaventrite distinctly longer than first abdominal ventrite; body densely clothed with hairs and/or scales **141**
141(140) Antennal insertions exposed from above; frontoclypeal suture absent or indistinct; abdominal ventrites 1 to 3 connate (Fig. 25K); procoxal cavities internally open; body more elongate and vestiture consisting of coarser setae or scales (Fig. 76L) .... **Ulodidae (part)**
– Antennal insertions concealed from above; frontoclypeal suture distinct; abdominal ventrites 1 and 2 connate; procoxal cavities internally closed; body ovoid and vestiture consisting of fine, recumbent hairs (Fig. 89H–I) .... **Archeocrypticidae (part)**
142(103) Procoxal cavities externally closed (Fig. 44J) **143**
– Procoxal cavities externally open (Fig. 18H) **145**
143(142) Antennae 8- or 9-segmented; head with rostrum which is longer than wide; pronotum with lateral, submarginal lines or carinae .... **Silvanidae: Silvaninae (*Nepharis* & *Nepharinus*)**
– Antennae 10-segmented; head without elongate rostrum; pronotum without submarginal lines or carinae **144**
144(143) Abdominal ventrites 5; procoxae externally globular and not projecting below prosternum; antennal club 1- or 2-segmented and globular (Fig. 83F) .... **Monotomidae (part)**
– Abdominal ventrites 6; procoxae externally conical and projecting below prosternum; antennal club 1-segmented and much longer than wide .... **Cleridae: Tillinae (*Monophylla*)**

145(142) Mesocoxal cavities separated by less than shortest diameter of one cavity; metacoxae contiguous or narrowly separated; antenna 9-segmented with 3-segmented club preceded by strongly transverse, glabrous, cup-like antennomere (cupule) (Fig. 57D–G); tibiae usually spiny (Fig. 57A); maxillary palp usually longer than antennae (Fig. 57C)........ **Hydrophilidae (part)**

– Mesocoxal cavities separated by more than shortest diameter of one cavity; metacoxae moderately to very widely separated........ **146**

146(145) Elytra truncate, exposing at least 1 abdominal tergite; antennae geniculate with scape more than 3 times as long as pedicel (Fig. 57L); upper surfaces usually smooth, shiny and glabrous (Fig. 77A–C)........ **Histeridae (part)**

– Elytra concealing abdominal tergites or exposing apex of one only; antennae not geniculate; scape less than 3 times as long as pedicel........ **147**

147(146) Antennal club 6-segmented and pectinate; prosternum in front of coxae very long, extending beneath head as chin piece; tarsomere 5 longer than 1 to 4 combined; not known from Australia, occurring in New Guinea, New Zealand, New Caledonia........ **Dryopidae**

– Antennal club 1- to 3-segmented; prosternum relatively short, not forming chin piece; tarsomere 5 much shorter than 1 to 4 combined........ **148**

148(147) Head moderately to strongly declined (hypognathous); antenna 7- or 9-segmented with compact club, which may be undivided or indistinctly divided into 3 parts; upper surfaces tuberculate; elytra costate (Fig. 57L); tergite VII without median groove........ **Hydrophilidae: Georissinae (*Georissus*)**

– Head prognathous or only slightly declined; antenna 10-segmented with clearly 3-segmented, elongate, densely pubescent club; upper surfaces smooth, pubescent; elytra not costate (Fig. 83J); tergite VII with median groove........ **Sphindidae (*Aspidiphorus*)**

Mesotarsus 5-segmented. Mesocoxal cavities laterally closed

149(38) Tarsi 5-5-5........ **150**

– Tarsi 5-5-4........ **172**

150(149) Procoxal cavities externally open (Figs 19H, 20K, 44Q–R)........ **151**

– Procoxal cavities externally closed (Figs 18F, L, 19E, L, 44P)........ **164**

151(150) Abdominal ventrites 6........ **152**

– Abdominal ventrites 5........ **153**

152(151) Mesocoxal cavities separated by more than 0.4 times shortest diameter of one cavity (Fig. 25B); elytra truncate, exposing at least one complete tergite; metathoracic discrimen present; pronotal disc without basal impressions; body glabrous, without constriction between prothorax and elytra (Fig. 67F)........ **Staphylinidae: Scaphidiinae**

– Mesocoxal cavities separated by less than 0.4 times shortest diameter of one cavity (Fig. 25D); elytra usually concealing abdomen or exposing only part of one tergite; metathoracic discrimen absent; body pubescent, usually constricted at junction of prothorax and elytra, producing ant-like appearance (Fig. 59H, P) (if not, then prothorax with paired basal impressions)........ **Staphylinidae: Scydmaeninae**

153(151) Antennae without distinct club (sometimes gradually clavate) (85I)........ **154**

– Antennae with distinct club consisting of from 1 to 7 antennomeres (Fig. 81E)........ **156**

154(153) Pronotum with strongly produced and acute posterior angles (Fig. 70A–F); antennal insertions concealed from above (Fig. 14B, K)........ **Elateridae (part)**

– Pronotum without produced and acute posterior angles; antennal insertions exposed (Fig. 44B–D)........ **155**

155(154) Body elongate, parallel-sided and glabrous (Figs 72J, 85I); head prognathous; prosternum in front of coxae much longer than shortest diameter of coxal cavity (Fig. 44P); length almost always greater than 10 mm........ **Passandridae (part)**

– Body usually shorter and broader, or if elongate, then not glabrous; head strongly declined (hypognathous) (Fig. 81L); prosternum in front of coxae shorter than shortest diameter of coxal cavity (Fig. 44L); length less than 8 mm........ **Ptinidae (part)**

156(153) Head moderately to strongly declined (hypognathous) (Fig. 81E–F, I–J, L)........ **157**

– Head prognathous or only slightly declined (Figs 85B, 86I)........ **158**

157(156) Outer edge of mesotibia bearing spines or teeth along most of its length (Fig. 81E); frontoclypeal suture distinct (Fig. 44A); metatrochanter with strongly oblique femoral attachment........ **Bostrichidae (part)**

– Outer edge of mesotibia simple; frontoclypeal suture absent; metatrochanter with transverse or slightly oblique femoral attachment (Fig. 81F)........ **Ptinidae (part)**

158(156) Body less than 1.5 times as long as wide, broadly ovate and strongly flattened (Fig. 85B); pronotum with dorsal median endocarina (visible as distinct line)........ **Propalticidae**

– Body more than 1.5 times as long as wide; pronotum without median endocarina........ **159**

159(158) Abdominal ventrite 1 more than 2 times as long as 2 (Fig. 26J); prosternal process distinctly expanded apically; procoxal cavities with narrow lateral 'extensions, so that trochantins are barely exposed (Fig. 20K); elytral punctation seriate with 10 rows of primary punctures separated by secondary rows of smaller punctures (Fig. 83E)........ **Myraboliidae (*Myrabolia* females)**

– Abdominal ventrite 1 not more than 2 times as long as 2; procoxal cavities without narrow lateral extensions; elytral punctaton different........ **160**

160(159) Elytral epipleura incomplete or absent; occipital region without stridulatory file; abdominal ventrite 1 usually distinctly longer than 2 (Fig. 86F); elytral apices usually with distinct gap when in place (caused by deflection of sutural flange) (Fig. 26I); anterior edge of lateral pronotal carina or anterolateral portion of pronotal disc often with one or more glandular openings usually placed on a distinct callosity (Fig. 18J–K); procoxal cavities usually broadly open, without or with very short postcoxal projections (Fig. 86F); postcoxal lines on abdominal ventrite 1 usually absent (or IF present, THEN recurved)........ **Cryptophagidae (part)**

– Elytral epipleura complete to apex or almost to apex; ventrite 1 not or only slightly longer than 2; occipital region usually with one or two stidulatory files (Fig. 14B, E–F) (if not then ventrite 1 with pair of straight, diverging postcoxal lines, (Fig. 25G); elytral apices without deflected sutural flange; prothorax rarely with glandular callosites; procoxal cavities narrowly open, with moderately long, acute postcoxal projections (Fig. 18E, H)........ **161**

161(160) Abdominal ventrite 1 with pair of straight, diverging postcoxal lines (Fig. 25G); occiput without stridulatory file; body more or less elliptical, with elytra distinctly curved laterally and narrowed posteriorly; length less than 2 mm........ **Erotylidae: Cryptophilinae: Toramini (*Toramus*)**

– Abdominal ventrite 1 without postcoxal lines; occiput with one or two stridulatory files (Figs 1E–F, 14B); body more elongate, not elliptical; length more than 2 mm........ **162**

162(161) Lateral pronotal carinae microdenticulate with glandular callosity at anterior end; body slender, parallel-sided, with base of pronotum not much narrower than combined elytral bases .... **Erotylidae: Pharaxonothinae (*Leucohimatium*)**
– Lateral pronotal carinae simple or slightly undulate, without glandular callosity; if body slender and parallel-sided, then base of pronotum distinctly narrower than combined elytral bases **163**
163(162) Pronotum distinctly narrower than combined elytral bases, with sides subparallel (Fig. 86L); pregular area with large fovea; tibiae slender, not apically expanded; tarsi simple; occipital region with single stridulatory file (Fig. 14E) .... **Erotylidae: Xenoscelinae (*Protoloberus*)**
– Pronotum not narrower than combined elytral bases, with sides slightly curved (Fig. 72H); pregular area without fovea; tibiae distinctly expanded at apex; tarsomeres 1–3 with ventral, setose lobes (Fig. 44X); occipital region with paired stridulatory files (Fig. 44B, F); WA .... **Erotylidae: Xenoscelinae (*Xenocryptus*)**
164(150) Protrochantins at least partly exposed (Figs 18L, 20K); procoxal cavities strongly transverse (Fig. 18L); prosternal process distinctly expanded apically (Fig. 44P); antennae with 10 or fewer antennomeres and a 1-segmented club (Fig. 82H); metacoxae extending laterally to meet elytral epipleura; empodium well developed and bisetose (Fig. 27F) .... **Trogossitidae: Trogossitinae: Egoliini**
– Protrochantins completely concealed; without other features combined **165**
165(164) Labrum fused to clypeus and not apparent; lateral pronotal carinae incomplete or absent (Fig. 98G–H); antennal scape more that 3 times as long as pedicel; antennal club compact or 1-segmented; outer edges of tibiae dentate (sometimes with socketed spines) (Fig. 27C) or tarsi filiform and distinctly longer than tibiae (Fig. 98H) **166**
– Labrum free and mostly visible (Fig. 44A); lateral pronotal carinae complete; antennal scape less than 3 times as long as pedicel; antennal club at least 2-segmented and usually not compact; outer edges of tibiae not dentate or spinose and tarsi always shorter than tibiae **167**
166(165) Eyes more or less circular (Fig. 98H); tarsi with basal tarsomere longer than next 2 combined; length 5–10 mm .... **Curculionidae: Platypodinae**
– Eyes dorsoventrally elongate (Fig. 98G); tarsi with basal tarsomere shorter than next 2 combined; length less than 5 mm .... **Curculionidae: Scolytinae (part)**
167(166) Antenna 7-segmented with 3-segmented pubescent club preceded by glabrous, cup-like antennomere (cupule) (Fig. 57F); antennae shorter than or only slightly longer than maxillary palps; frontoclypeal suture distinct, angulate and attached at midline to mid-cranial suture; dorsal surfaces glabrous, ventral surfaces densely clothed with very short fine hairs; head and pronotum granulate or confluently punctate (Fig. 57B) .... **Hydrophilidae: Hydrochinae (*Hydrochus*)**
– Antenna much longer than maxillary palps, with at least 9 antennomeres and no cupule; frontoclypeal suture absent or, if present, then slightly curved and without attached mid-cranial suture; dorsal surfaces glabrous or pubescent; ventral surfaces without dense covering of very short hairs **168**
168(167) Antenna 9-segmented with compact 2-segmented club; metacoxae extending laterally to meet elytral epipleura; body strongly convex and clothed with long, fine, erect setae (Fig. 82I); tarsomere 1 much shorter than 2 .... **Trogossitidae: Trogossitinae: Larinotini (*Larinotus*)**
– Antenna 11-segmented with 2- or 3-segmented club; metacoxae not extending laterally to meet elytral epipleura; if tarsomere 1 distinctly shorter than 2 then body elongate and more or less parallel-sided **169**
169(168) Frontoclypeal suture present (Fig. 44A); elytral epipleuron incomplete or absent; procoxal cavities broadly closed internally (Fig. 18D); tibiae not apically expanded; tarsomere 1 distinctly shorter than 2 .... **Bostrichidae: Lyctinae**
– Frontoclypeal suture absent (Fig. 86G); elytral epipleuron complete to apex (Fig. 26H); procoxal cavities internally open (Fig. 44K) or closed by slender bar (Fig. 44N); tibiae more or less expanded apically; tarsomere 1 not shorter than 2 .... **170**
170(169) Antennomere 11 shorter and narrower than 10; body distinctly flattened; pronotum widest anteriorly (Fig. 86G); head without stridulatory file .... **Erotylidae: Cryptophilinae (*Cathartocryptus*)**
– Antennomere 11 not both short than and narrower than 10; body not distinctly flattened; pronotum widest at middle or posteriorly (Figs 73E–G, 86K); occiput often with one or two stridulatory files (Fig. 14B, E–F) **171**
171(170) Occiput with single stridulatory file (Fig. 14E); edge of lateral pronotal carina finely denticulate (Fig. 86K); length less than 3 mm .... **Erotylidae: Cryptophilinae (*Cryptophilus*)**
– Occipital stridulatory files either paired (Fig. 14F) or absent; edge of lateral pronotal carina simple, undulate or crenulate (Figs 73E–G, 86C); length more than 3 mm .... **Erotylidae: Erotylinae: Dacnini (part)**
172(149) Procoxal cavities externally open (Fig. 44R) **173**
– Procoxal cavities externally closed (Figs 20B, 25C) **181**
173(172) Antennae without distinct club (sometimes gradually clavate) **174**
– Antenna with distinct 2- or 3-segmented club **176**
174(173) Body more than 4 times as long as wide and strongly flattened (Fig. 93G); mesanepisterna distinctly separated at midline and mesoventrite broadly rounded anteriorly .... **Mycteridae: Hemipeplinae (*Hemipeplus*)**
– Body less than 4 times as long as wide, not strongly flattened; IF mesanepisterna separated at midline, THEN mesoventrite truncate anteriorly (Fig. 44V) **175**
175(174) Mesanepisterna contiguous or very narrowly separated at midline, so that mesoventrite is anteriorly acute (Fig. 44T); abdominal ventrites all free .... **Salpingidae: Salpinginae (part)**
– Mesanepisterna well separated at midline; mesoventrite not anteriorly acute; abdominal ventrites 1 to 3 or 4 connate, ventrite 2 with setiferous patch in male .... **Mycteridae: Eurypinae**
176(173) Prosternal prosternal process as wide as shortest diameter of coxal cavity, strongly but gradually expanded and truncate at apex (Fig. 19L); abdominal ventrite 1 more than 2 times as long as 2 (Fig. 26J), usually with paired, depressed pore plates; punctation seriate with 10 rows of primary punctures separated by secondary rows of smaller punctures (Fig. 83E) .... **Myraboliidae (*Myrabolia* males)**
– Prosternal process either not expanded apically (Fig. 86F) or abruptly expanded and emarginate at apex (Fig. 20B); ventrite 1 less than twice as long as 2, without pore plates; IF elytral punctation seriate, THEN without distinct secondary puncture rows **177**
177(176) Antennal insertions exposed from above (Fig. 44B–D) **178**
– Antennal insertions concealed from above (Fig. 44A) **180**
178(177) Metatrochanter more than 2 times as long as wide, with transverse of slightly oblique femoral attachment (Fig. 86F); transverse occipital ridge or carina present (Fig. 44G); upper surfaces pubescent (Fig. 86E) .... **Cryptophagidae: Cryptophaginae (males)**

– Metatrochanter less than 2 times as long as wide, with strongly oblique femoral attachment; transverse occipital ridge or carina absent; upper surfaces glabrous or subglabrous, clothed with few minute setae only **179**

179(178) Mesocoxal cavities separated by less than shortest diameter of a cavity (Fig. 25A, D); prosternal process not apically expanded (Fig. 44K) .... **Salpingidae: Salpinginae (part)**

– Mesocoxal cavities separated by more than shortest diameter of a cavity (Fig. 26B); prosternal process strongly and abruptly expanded at apex (Figs 19I, 44P) .... **Salpingidae: Prostominiinae (*Prostominia*)**

180(177) Body ovate; base of prothorax as wide as combined elytral bases; eyes large, almost meeting dorsally; surfaces smooth, glabrous; occurrence in Australa somewhat doubtful .... **Zopheridae: Zopherinae: Monommatini (*Monomma*)**

– Body more elongate; base of prothorax distinctly narrower than combined elytral bases; eyes small and well separated; pronotal hypomera without antennal cavities; upper surfaces usually tuberculate or costate and clothed with scale-like setae (Fig. 92G) .... **Zopheridae: Zopherinae: Latometini (*Latometus*)**

181(172) Antennae 10-segmented; head with rostrum which is longer than wide; pronotum with 3 longitudinal grooves (Fig. 95F) .... **Salpingidae: Dacoderinae (*Tretothorax*)**

– Antennae 11-segmented; head without elongate rostrum; pronotum without longitudinal grooves **182**

182(181) Prosternal process not or only slightly, gradually expanded at apex (Figs 25C, 44N) .... **Tenebrionidae (part)**

– Prosternal process strongly and abruptly expanded at apex (Fig. 19I) **183**

183(182) Antennae fitting into grooves on pronotal hypomera; all coxae moderately to widely separated; dorsal surfaces tuberculate (Fig. 92L); length more than 20 mm .... **Zopheridae: Zopherinae: Zopherini (*Zopherosis*)**

– Antennae not fitting into hypomeral grooves; procoxae and metacoxae contiguous; dorsal surfaces not tuberculate (Fig. 95B); length less that 5 mm .... **Salpingidae: Othniinae (*Elacatis*)**

Mesotarsus pseudotetramerous

184(37) Metatarsus with 4 tarsomeres **185**

– Metatarsus with 5 tarsomeres **186**

185(184) Body form ovate to subcircular (with lateral edes of pronotum and elytra forming continuous arc); lateral edges of pronotum obliquely explanate, with sharp, simple lateral carinae; procoxae not projecting, oval or subcircular, their cavities externally open .... **Phalacridae: Phalacrinae (part)**

– Body form elongate with sides subparallel and not forming continuous arc; lateral edges of pronotum not explanate, with weak lateral carinae, each bearing 3 denticles; procoxae not projecting, strongly transverse, their cavities externally closed; abdomen with 5 distinct ventrites .... **Phloeostichidae (*Bunyastichus* males)**

– Body form elongate with sides subparallel and not forming continuous arc; lateral edges of pronotum not explanate, without lateral carinae; procoxae projecting, their cavities externally open; abdomen usually with 4 ventrites (IF 5 present THEN suture between first 2 weak or incomplete) .... **Aderidae (part)**

186(184) Head without elongate rostrum **187**

– Head with rostrum which is longer than wide (Fig. 98D–E, J–M) **211**

187(186) Antenna without distinct club (sometimes gradually clavate) **188**

– Antenna with distinct club consisting of from 1 to 7 antennomeres **199**

188(187) Mesanepisterna meeting or very narrowly separated at midline (Fig. 44T) .... **Cleridae: Korynetinae (part)**

– Mesanepisterna well separated at midline (Fig. 44V) **189**

189(188) Metacoxae excavate forming coxal plate (Fig. 21F); lateral pronotal carinae well developed posteriorly but obsolete near anterior edge (Fig. 44U); posterior edge of pronotum finely crenulate and with 4 excavations; scutellum deeply notched anteriorly (Fig. 16P); tibial spurs pubescent (16I); antennomeres 4–10 in male with articulated appendages .... **Ptilodactylidae: Ptilodactylinae**

– Metacoxae not excavate, without plates; without other features combined **190**

190(189) Mandible with setose cavity, partly covered by dorsal tubercle, which fits into lateral impression at base of clypeus (Fig. 44G); head with long median endocarina attached anteriorly to epistomal ridge; upper surfaces glabrous (Fig. 85J); WA .... **Boganiidae (*Paracucujus*)**

– Mandible without dorsal tubercle and setose cavity, but sometimes with mandibular cavity; head without long endocarina **191**

191(190) Prosternal process incomplete, terminating bewteen coxae (Fig. 44R); head strongly declined (hypognathous); prothorax narrower anteriorly, with incomplete lateral carinae (Fig. 96E); metafemur strongly inflated with 1 or more teeth on inner edge; pretarsal claws toothed .... **Chrysomelidae: Bruchinae (part)**

– Prosternal process complete, extending behind coxae; without other features combined **192**

192(191) Antennae borne on raised tubercles; usually more than two-thirds body length, with scape more than 3 times as long as pedicel (Figs 77E–H, K, 97A–B) .... **Cerambycidae (part)**

– If antennal scape more than 3 times as long as pedicel, antennae less than two-thirds body length and not borne on raised tubercles **193**

193(192) Mesocoxal cavities partly closed laterally by mesepimeron (Fig. 44T) **194**

– Mesocoxal cavities closed laterally by meeting of mesoventrite and metaventrite (Fig. 44V) **198**

194(193) Antennal insertions exposed from above (Fig. 44B–D) **195**

– Antennal insertions concealed from above (Fig. 44A) **197**

195(194) Mesonotum with stridulatory file (Fig. 16O); upper surfaces distinctly pubescent; pronotum distinctly narrower than combined elytral bases; head without grooves; elytra irregularly punctate, without striae or puncture rows; all tibiae with paired spurs; penis with long basal struts; tegmen with articulated parameres or head distinctly wider than prothorax; female with articulated spiculum ventrale and ovipositor reduced and enclosed in thickened membranous sheath **196**

– Mesonotum without stridulatory file; IF upper surfaces pubescent, pronotum narrower than combined elytral bases and tibial spurs paired, THEN head with distinct grooves; tegmen without articulated parameres; head never wider than prothorax; penis without long basal struts; spiculum ventrale not articulated and ovipositor not reduced .... **Chrysomelidae (part)**

196(195) Head wider than prothorax; lateral pronotal carinae absent; pronotum with sub-basal constriction and tubercle (Fig. 78A); antennae slightly clavate; ligula truncate; pretarsal claws toothed; tegmen without articulated parameres .... **Megalopodidae: Zeugophorinae (*Zeugophora*)**

– Head not wider than thorax; lateral pronotal carinae present; pronotum with weak prominence at about middle (Fig. 96A); antennae filiform; ligula bilobed; pretarsal claws simple; tegmen with articulated parameres .... **Megalopodidae: Palophaginae**

197(194) Labrum visible; sides of pronotum usually dentate or denticulate; procoxal cavities distinctly separated; head with transverse occipital carina and paired longitudinal frontal grooves (Fig. 85H) .... **Silvanidae: Brontinae: Telephanini**

– Labrum not visible; sides of pronotum simple, without lateral carina (Fig. 98D); procoxal cavities contiguous (Fig. 18F); head without occipital carina or frontal groves .... **Belidae: Belinae (part)**

198(193) Abdominal ventrites 1 to 4 connate; head almost always with at least short rostrum (Fig. 98B–C); lateral pronotal carinae distinct posteriorly but absent anteriorly (Fig. 19A); maxillae usually reduced and partly concealed by genal projections (Fig. 18A) .... **Anthribidae (part)**

– Basal abdominal ventrites free or first 2 connate; 2 or 3 sub-basal ventrites connate in some cryptocephaline Chrysomelidae; head without rostrum; maxillae well developed and not concealed by genal projections; lateral pronotal carinae complete or absent .... **Chrysomelidae (part)**

199(187) Mesocoxal cavities partly closed laterally by mesepimeron (Fig. 44T) **200**

– Mesocoxal cavities closed laterally by meeting of mesoventrite and metaventrite (Fig. 44V) **204**

200(199) Procoxae projecting below prosternum (Fig. 18L) .... **Cleridae: Korynetinae (part)**

– Procoxae not projecting below prosternum (Fig. 20K) **201**

201(200) Procoxal cavities broadly open externally and open internally; mandible with setose cavity, partly covered by dorsal tubercle, which fits into lateral impression at base of clypeus (Fig. 14G); anterior edge of lateral pronotal carina usually with glandular callosity (Figs 18K, 85F) .... **Boganiidae (part)**

– Procoxal cavities usually closed externally (Fig. 44J), OR IF narrowly open, THEN internally closed; mandible without setose cavity and tubercle; glandular callosity never present **202**

202(201) Procoxal cavities more or less circular, broadly closed externally and open internally, with trochantin concealed (Fig. 44J); maxilla with galea and lacinia (Fig. 16C) .... **Silvanidae: Silvaninae (part)**

– Procoxal cavities distinctly transverse, narrowly open to broadly closed externally, with trochantin exposed (Fig. 18L) **203**

203(202) Lateral pronotal carinae each with 3 denticles; maxilla with galea and lacinia (Fig. 16C) .... **Phloeostichidae (*Bunyastichus*)**

– Lateral pronotal carinae not dentate; maxilla with single lobe (Fig. 16B) .... **Nitidulidae (part)**

204(199) Procoxal cavities externally open (Fig. 19G) **205**

– Procoxal cavities externally closed (Fig. 19L) **208**

205(204) Body very short and broad, less than 1.5 times as long as wide, with sides uniformly curved (Fig. 85C); head distinctly transverse; sides of pronotum obliquely explanate; pretarsal claws toothed .... **Phalacridae: Phalacrinae**

– Body elongate, more than 1.5 times as long as wide, and more or less parallel-sided (Figs 73L, 86D); head about as long as wide; sides of pronotum not explanate; pretarsal claws simple **206**

206(205) Frontoclypeal suture distinct (Fig. 15A); antennal club somewhat flattened; body subglabrous, with blue or black elytra and reddish head and pronotum (Fig. 86D) .... **Languriidae: Languriinae: Languriini**

– Frontoclypeal suture absent; antennal club not flattened; body usually pubescent and coloration different (Fig. 73I) **207**

207(206) Elytra more than 2.5 times as long as wide (Fig. 73I); upper surfaces clothed with moderately long, erect and suberect hair; pronotal disc simple, without paired pits or transverse groove .... **Erotylidae: Languriinae: Hapalipini (*Hapalips*)**

– Elytra less than 2.5 times as long as wide; upper surfaces glabrous or clothed with very short, fine hairs; pronotal disc with pair of sub-basal pits joined by transverse groove .... **Erotylidae: Loberinae (*Loberus*)**

208(204) Lateral pronotal carinae complete and sharply defined (Fig. 19D); apical maxillary palpomere well developed and slightly to strongly expanded apically (Fig. 16F) .... **Erotylidae: Erotylinae: Tritomini**

– Lateral pronotal carinae incomplete anteriorly (Fig. 19A) or absent (Fig. 19C); apical maxillary palpomere reduced and cylindrical or fusiform (Fig. 18A) **209**

209(208) Labrum free and mostly visible; antennae not geniculate and with scape less than 3 times as long as pedicel (Fig. 98B–C); abdominal ventrites 1 to 4 connate .... **Anthribidae (part)**

– Labrum solidly fused to clypeus; antennae geniculate andwith scape more than 3 times as long as pedicel (Fig. 98J–M); ventrites 1 and 2 connate (Fig. 28J) **210**

210(209) Outer edge of mesotibia with spines or teeth along most of its length (Fig. 27C) .... **Curculionidae: Scolytinae**

– Outer edge of mesotibia simple .... **Curculionidae (part)**

Mesotarsus pseudotetramerous. Rostrum longer than wide

211(186) Antennae 11-segmented **212**

– Antennae with fewer than 11 antennomeres **221**

212(211) Labrum free and mostly visible **213**

– Labrum fused to clypeus **214**

213(212) Mesocoxal cavities separated by less than 0.4 times shortest diameter of a cavity (Fig. 21F); all ventrites free; pronotum without lateral carinae (Fig. 19C) .... **Nemonychidae**

– Mesocoxal cavities separated by more than 0.4 times shortest diameter of a cavity (Fig. 21E); ventrites 1 to 4 connate; pronotum almost always with lateral carinae (Fig. 19A) (sometimes very short or indistinct) .... **Anthribidae (part)**

214(212) Antennae without distinct club, sometimes gradually clavate; elytra more than twice as long as combined width (Fig. 98D, J) **215**

– Antennae with distinct club consisting of from 1 to 3 or 4 antennomeres; IF antennal club weak, THEN elytra less than twice as long as combined width **216**

215(214) Mesocoxal cavities partly closed laterally by mesepimeron (Fig. 44T); abdominal ventrites 1 and 2 free and not or only slightly longer than ventrites 3, 4 or 5; head not abruptly constricted posteriorly (Fig. 98D) .... **Belidae: Belinae (part)**

– Mesocoxal cavities closed laterally by meeting of mesoventrite and metaventrite (Fig. 44V); abdominal ventrites 1 and 2 connate or completely fused (with suture between them incomplete or absent) and much longer than ventrites 3, 4 or 5 (Fig. 28J); head abruptly constricted posteriorly to form neck (Fig. 98J) .... **Brentidae: Brentinae (part)**
216(214) Body more than 3 times as long as wide (Figs 79C, E, 98K–L) **217**
– Body less than 3 times as long as wide **218**
217(216) Antennae not geniculate and with scape less than 3 times as long as pedicel (Fig. 79C, E) .... **Brentidae: Eurhynchinae**
– Antennal geniculate and with scape more than 3 times as long as pedicel (Fig. 98K–L) .... **Curculionidae (part)**
218(216) Metatrochanter more than 2 times as long as wide, with transverse or slightly oblique femoral attachment; body usually characteristic shaped, widest across middle of elytra and gradually narrowing anteriorly (Fig. 98E) .... **Brentidae: Apioninae**
– Metatrochanter less than 2 times as long as wide, with strongly oblique femoral attachment **219**
219(218) Antennae geniculate and with scape more than 3 times as long as pedicel (Fig. 98K–M; ventrites 1 and 2 connate (Fig. 28J) .... **Curculionidae (part)**
– Antennae not geniculate and with scape less than 3 times as long as pedicel (Figs 79D, 98I); ventrites 1 to 3 or 4 connate or rarely all ventrites appear to be free **220**
220(219) Antennal insertions exposed from above (Fig. 98I); ventrites 1 to 3 or 4 connate .... **Attelabidae**
– Antennal insertions concealed from above (Fig. 79D); all ventrites apparently free .... **Caridae**
221(211) Antennae not geniculate and with scape less than 3 times as long as pedicel; antennae 10-segmented with 1-segmented antennal club which is more than 5 times as long as wide (Fig. 80L) .... **Brentidae: Brentinae (*Cylas*)**
– Antennae geniculate and with scape more than 3 times as long as pedicel; antennal club less than 5 times as long as wide **222**
222(221) Metatrochanter more than twice as long as wide, with transverse or slightly oblique femoral attachment; antennae 9-segmented with loose, 3-segmented club (Fig. 80D) .... **Brentidae: Nanophyinae**
– Metatrochanter less than twice as long as wide, with strongly oblique femoral attachment; antennal club compact (Figs 98K–M, 80A–L) .... **Curculionidae (part)**

Mesotarsus with 4 distinct tarsomeres

223(37) First 3 tarsomeres expanded and ventrally pilose, the 3rd deeply bilobed (Fig. 96L); vertex strongly and abruptly declined so that mouthparts ventrally or posteroventrally directed, although base of head is not declined (Fig. 15A, E); body more than twice as long as wide (Figs 78B, D, 96I) OR body subcircular and head much narrower than base of pronotum (Fig. 78I); length greater than 3 mm **224**
– If first 3 tarsomeres expanded and ventrally pilose, then 3rd not deeply bilobed, head almost as wide as base of pronotum, and length less than 2 mm; vertex not strongly declined; IF mouthparts ventrally directed, THEN head hypognathous (Fig. 60G) **225**
224(223) Antennal insertions well separated, distinctly elevated and often located within emarginaion of eye (Fig. 15F); antennae extending well beyond middle of elytra (Figs 77F–H, 97B); protibia with subapical antenna cleaner (setose groove) .... **Cerambycidae: Lamiinae (part)**
– Antennal insertions approximate and not elevated (Figs 15A, 96I); antennae never extending to middle of elytra (Figs 78I, 96I); protibia without antenna cleaner .... **Chrysomelidae: Cassidinae**
225(223) Protrochantins at least partly exposed **226**
– Protrochantins completely concealed **232**
226(225) Antennae 7-segmented with pubescent cupule and 3-segmented club (Fig. 57G); pronotum short, broad and anteriorly widened; lateral pronotal carinae denticulate (Fig. 57J) .... **Hydrophilidae: Spercheinae (*Spercheus*)**
– Antennae 10-segmented with 2-segmented club; metacoxae with large coxal plates concealing femora (Fig. 60F); body capable of conglobation); head strongly declined and fitting against mesocoxae (Fig. 60G); tarsi simple and filamentous .... **Clambidae (part)**
– Antennae 11-segmented; if body capable of conglobation, then metacoxae without plates and tarsi with ventral setose lobes **227**
227(226) Antennae much shorter than head width, incrassate or with last 7 antennomeres expanded forming weak, asymmetrical club; body more than twice as long as wide, somewhat flattened and pubescent (Fig. 62H); outer edges of tibiae with long spines (Fig. 27B) .... **Heteroceridae**
– Antennae with distinct club consisting of 1 to 5 antennomeres (Figs 15I, 28U); IF antennae slightly shorter than head width, THEN body less than 1.5 times as long as wide, strongly convex and glabrous **228**
228(227) Antennal club 5-segmented with 2nd club segment (antennomere 8) smaller than 1st or 3rd club segment (Fig. 15I); procoxae externally globular to conical and projecting well below prosternum (Fig. 28C) .... **Leiodidae (part)**
– Antennal club not 5-segmented or with 2nd club segment subequal to or larger than 1st one; procoxae transverse and not projecting below prosternum **229**
229(228) Mesocoxal cavities contiguous or subcontiguous (Fig. 44V); mesoventrite without cavity **230**
– Mesocoxal cavities moderately to widely separated (Fig. 21E–F); mesoventrite with distinct cavity into which apex of prosternal process is inserted (Fig. 17L) **231**
230(229) Body very short and broad, less than 1.5 times as long as wide, strongly convex, glabrous, and capable of conglobation, with strongly declined head fitting against mesocoxae (Fig. 60G); metacoxae moderately to widely separated (Fig. 21E); basal tarsomeres lobed beneath (Fig. 87B) .... **Nitidulidae: Cybocephalinae (*Cybocephalus*)**
– Body more than 1.5 times as long as wide; slightly to strongly flattened and densely setose (Fig. 90G, K); metacoxae contiguous or narrowly separated (Fig. 26C); tarsomeres not lobed beneath; protarsomere of males 3-segmented .... **Mycetophagidae: Mycetophaginae**
231(229) Metacoxae transverse and narrowly separated, with complete, transverse coxal plates (Fig. 62I); all tibiae spinose; head with transverse occipital ridge or carina (Fig. 44G); eyes lateral, well separated; antennal club 2-segmented; vestiture of scale-like setae (Fig. 69I) .... **Bytrrhidae: Syncalyptinae (*Microchaetes*)**
– Metacoxae subglobular and widely separated, without coxal plates; tibiae not spinose; transverse occipital ridge absent; eyes lateral and well separated; antennal club 3-segmented; vestiture of fine, decumbent hairs .... **Limnichidae: Hyphalinae (*Hyphalus*)**

– Metacoxae large and oblique, approximate, with well developed, longitudinal coxal plates; metatibiae spinose; transverse occipital ridge absent; eyes dorsal, approximate; antennal club 4-segmented; vestiture of fine, decumbent hairs .... **Limnichidae: Thaumastodinae**

Mesotarsus with 4 distinct tarsomeres. Basal tarsomeres not expanded, 3 not bilobed. Protrochantins concealed

232(225) Head hypognathous and concealed from above by pronotum; antennae short, 9-segmented, with 3-segmented club; procoxae very large, concealing prosternum, which is short and concave, forming head rest; upper surfaces tuberculate; elytra costate (Fig. 57K) .... **Hydrophilidae: Georissinae (*Georissus*)**

– Head not hypognathous or concealed from above; antennae with more than 9 antennomeres or, if fewer, then club 1-segmented; without other features combined **233**

233(232) Procoxal cavities externally open (Figs 13C, 19A) **234**

– Procoxal cavities externally closed (Figs 19L, 20B) **250**

234(233) Antennal insertions exposed from above (Fig. 44B–E) **235**

– Antennal insertions concealed from above (Fig. 44A) **243**

235(234) Abdomen with 6 ventrltes **236**

– Abdomen with 5 ventrites **237**

236(235) Antennae 4-segmented with 1-segmented club; body more than 1.5 times as long as wide (Fig. 72G); mesocoxae separated by less than shortest diameter of a cavity; sides of pronotum and elytra not forming continuous curve; abdominal ventrite 1 without postcoxal lines .... **Endomychidae: Pleganophorinae (*Trochoideus*)**

– Antennae 11-segmented with 3-segmented club; body less than 1.5 times as long as wide (Fig. 89A); mesocoxae separated by more than shortest diameter of a cavity (Fig. 26E); sides of pronotum and elytra forming continuous curve; adominal ventrite 1 with pair of recurved postcoxal lines .... **Corylophidae: Corylophinae: Peltinodini (*Holopsis*)**

237(235) Intercoxal process on abdominal ventrite 1 narrowly rounded or acute (Fig. 26J) **238**

– Intercoxal process on abdominal ventrite 1 broadly truncate (Fig. 26B) **241**

238(237) Body less than 1.5 times as long as wide; sides of pronotum and elytra forming continuous curve and strongly, obliquely explanate (Fig. 85E); frontoclypeal suture absent; tarsomeres 1 to 3 with setose ventral lobes; pretarsal claws dentate .... **Phalacridae: Phaenocephalinae**

– Body more than 1.5 times as long as wide; sides of pronotum and elytra not forming continuous curve and not explanate; frontoclypeal suture present (Fig. 44A); tarsomeres without ventral setose lobes (Fig. 44Y); pretarsal claws simple **239**

239(238) Antennal with loose, 3-segmented club; subantennal groove well developed body at least 4 times as long as wide .... **Bothrideridae: Teredinae (*Sysolus*)**

– Antenna with compact, 2-segmented club ; subantennal groove absent; body less than 4 times as long as wide **240**

240(239) Mesocoxal cavities distinctly closed laterally by meeting of mesoventrite and metaventrite (Figs 25L, 44V); body more than 3 times as long as wide, subcylindrical (Fig. 87J) .... **Bothrideridae: Teredinae (*Teredolaemus*)**

– Mesocoxal cavities very narrowly open laterally, with slender portion of mesepimeron extending between the two ventrites; body less than 3 times as long as wide and slightly flattened .... **Bothrideridae: Xylariophilinae (*Xylariophilus*)**

241(237) Posterior edge of abdominal ventrite 5 crenulate (Fig. 44AA); frontoclypeal suture absent; apical maxillary palpomere aciculate, much narrower than and usually shorter than preapical one (Fig. 16A) .... **Cerylonidae: Ceryloninae (part)**

– Prosterior edge of abdominal ventrite 5 not crenulate; frontoclypeal suture present (Fig. 44AQ); apical maxillary palpomere not aciculate, not or only slightly narrower than and usually longer than preapical one **242**

242(241) Antennal club 1-segmented (Fig. 87K); abdominal ventrite 1with pair of curved postcoxal lines (Fig. 26B); transverse occipital ridge or carina present (Fig. 44G); metathoracic discrimen more than 0.25 times as long as body of ventrite; prothorax with pair of cavities at anterior angles for receiving antennal clubs (Fig. 26B) .... **Cerylonidae: Murmidiinae (*Murmidius*)**

– Antennal club 3-segmented (Fig. 89E); ventrite 1 without postcoxal lines; transverse occipital ridge or carina absent; prothorax without cavities at anterior angles .... **Endomychidae (part)**

243(234) Mesocoxal cavities partly closed laterally by mesepimeron (Fig. 44T) **244**

– Mesocoxal cavities closed laterally by meeting of mesoventrite and metaventrite; IF ventrites slightly separated, THEN mesepimeron does not extend between them to meet cavity (Fig. 44V) **246**

244(243) Abdominal ventrite with broadly truncate intercoxal process and a pair of curved postcoxal lines (Fig. 26B); mesocoxal cavities separated by more than shortest diameter of cavity; metaventrite with curved postcoxal lines; antennae 10-segmented with 2-segmented club .... **Endomychidae: Eupsilobiinae (*Eidoreus*)**

– Abdominal ventrite with acute intercoxal process and no postcoxal lines (Fig. 26J); mesocoxal cavities separated by less than half the shortest diameter of a cavity (Figs 20K, 25D); metaventrite without postcoxal lines **245**

245(244) Antennae 11-segmented with 3-segmented club (Fig. 90G, K); procoxal cavities broadly open externally and closed internally (Fig. 44R); tarsi 3-4-4 in male .... **Mycetophagidae: Mycetophaginae (part)**

– Antennae 10-segmented with 2- or 3-segmented club (Fig. 89J); procoxal cavities narrowly open externally and open internally (Fig. 44K); tarsi 4-4-4 in both sexes .... **Ciidae: Ciinae (part)**

246(243) Procoxal cavities internally closed by slender bar (Fig. 44N); all abdominal ventrites free; ventrite 5 without transverse groove **247**

– Procoxal cavities internally open (Fig. 44P); basal 3 or more ventrites connate (sometimes apparently free when separated by distinct grooves); ventrite 5 often with curved, subapical, transverse groove **249**

247(246) Body not parallel-sided; metathoracic discrimen present; upper surfaces densely setose (Fig. 90L) .... **Mycetophagidae: Mycetophaginae (*Typhaea*)**

– Body parallel-sided (Fig. 95D); metathoracic discrimen absent; upper surfaces glabrous **248**

248(247) Eyes absent; lateral pronotal carinae simple; prosternal process parallel-sided, not expanded at apex (Fig. 20K) .... **Salpingidae: Agleninae (*Aglenus*)**

– Eyes present; lateral pronotal carinae slightly undulate posteriorly or with few weak teeth (Fig. 95D); prosternal process strongly expanded at apex (Fig. 44P) .... **Salpingidae: Prostominiinae (*Ocholissa*)**

249(246) Antennomeres 2–11 all transverse and clothed with expanded setae covered with microdenticles (Fig. 92G); pronotal hypomeron, metaventrite and first abdominal ventrite with crural impressions; submentum with small setose fovea in male .... **Zopheridae: Zopherinae: Latometini (*Orthocerodes*)**

– Antennomeres not covered with expanded microdenticulate setae and at least some of them longer than wide; crural impressions, if present, not on pronotum, metaventriteand abdomen; male without submental fovea .... **Zopheridae: Colydiinae (part)**

Mesotarsus with 4 distinct tarsomeres. Basal tarsomeres not expanded, 3 not bilobed. Protrochantins concealed. Head not hypognathous. Procoxal cavities externally closed

250(233) Mesocoxal cavities partly closed laterally by mesepimeron (Fig. 44T) **251**

– Mesocoxal cavities closed laterally by meeting of mesoventrite and metaventrite (Fig. 44V) **256**

251(250) Antennal insertions exposed (Fig. 44B–E); antenna 11-segmented with very weak 3-segmented club; lateral pronotal carinae weakly defined and denticulate; base of pronotum distinctly narrower than combined elytral bases (Fig. 92A); prosternal process straight in lateral view and strongly expanded at apex (Fig. 44P); abdomen with 3 connate ventrites .... **Ulodidae (*Meryx*)**

– Antennal insertions concealed from above (Fig. 44A); IF antennae 11-segmented, THEN antennal club 2- or 5-segmented; lateral pronotal carinae well defined; base of pronotum not or only slightly narrower than combined elytral bases; prosternal process either curved in lateral view (Fig. 25C) or not apically expanded **252**

252(251) All abdominal ventrites free **253**

– Abdominal ventrites 1 and 2 or 1 to 3 connate **254**

253(252) Antennae 8-segmented with 1-segmented club; metacoxae contiguous but slightly transverse and well separated from elytral epipleura; frontoclypeal suture absent; empodium well developed; abdominal ventrites strongly flattened and all on same plane .... **Trogossitidae: Peltinae: Colydiopeltini (*Parapeltis*)**

– Antennae 10-segmented, with 2- or 3-segmented club (Fig. 89J); metacoxae strongly transverse and extending laterally almost to elytral epipleura; frontoclypeal suture present (Fig. 44A); empodium reduced or apparently absent; abdominal ventrites curved and on slightly different planes .... **Ciidae (part)**

254(252) Antenna 8- or 3-segmented, with 1-segmented club; sides of pronotum subparallel and not explanate; lateral pronotal carinae simple; prosternal process not expanded apically; pronotum and elytra with longitudinal costae .... **Tenebrionidae: Lagriinae: Belopini (*Kershawia* and *Euclarkia*)**

– Antenna 11-segmented with 2- or 5-segmented club; sides of pronotum more or less rounded and either explanate or with dentate edges; prosternal process apically expanded (Fig. 20A); pronotum and elytra without costae **255**

255(254) Abdomen with 2 connate ventrites; antennal club 2-segmented; lateral pronotal carinae explanate with simple edges (Fig. 89I); prosternal process straight but oblique in lateral view (Fig. 20A) .... **Archeocrypticidae (*Nothenneboeus* part)**

– Abdomen with 3 connate ventrites (Fig. 25K); antennal club 5-segmented, asymmetrical; lateral pronotal carinae dentate; prosternal process distinctly curved in lateral view (Fig. 25C) .... **Tenebrionidae: Tenebrioninae (*Byrsax egenus* Pascoe)**

Mesotarsus with 4 distinct tarsomeres. Basal tarsomeres not expanded, 3 not bilobed. Protrochantins concealed. Head not hypognathous. Procoxal cavities externally closed. Mesocoxal cavities laterally closed

256(250) Antennal insertions exposed from above (Fig. 44B–E) **257**

– Antennal insertions concealed from above (Fig. 44A) **265**

257(256) Lateral pronotal carinae incomplete anteriorly (Fig. 19A) or absent (Fig. 19C) **258**

– Lateral pronotal carinae complete and sharply defined (Fig. 19D) **259**

258(257) Antennal club 2-segmented; mesocoxal cavities separated by less than shortest diameter of a cavity; prosternal process not or only slightly; gradually expanded apically; body more than 3.5 times as long as wide, subcylindrical (Fig. 92C) .... **Zopheridae: Colydiinae (*Nematidium*)**

– Antennal club 3-segmented; mesocoxal cavities separated by more than shortest diameter of a cavity; prosternal process strongly and abruptly expanded at apex; body less than 3.5 times as long as wide and distinctly flattened; head either wider than prothorax (Fig. 76D) or with strongly projecting mandibles and paired, anteriorly projecting genal processes (Fig. 93E) .... **Prostomidae**

259(257) Posterior edge of ventrite 5 crenulate (Fig. 44AA); apical maxillary palpomere aciculate, much narrower than and usually shorter than preapical one (Fig. 16A); frontoclypeal suture absent .... **Cerylonidae: Ceryloninae (part)**

– Posterior edge of ventrite 5 simple; apical maxillary palpomere not or only slightly narrower than and usually longer than preapical one; frontoclypeal suture present (Fig. 44A) **260**

260(259) Trochanter highly reduced and concealed within notch at base of femur and not visible in ventral view; protibial spurs usually unequal in length, one being enlarged and strongly curved or hooked **261**

– Trochanter not reduced, always visible in ventral view; protibial spurs more or less equal in length, neither greatly enlarged not hooked **264**

261(260) Antennal club 3-segmented; lateral pronotal carinae well developed and visible for their entire lengths from above, but ending well before rounded anterior angles; procoxal cavities separated by less than half the shortest diameter of a cavity; pronotal disc with slender longitudinal groove; base of pronotum distinctly narrower than combined elytral bases; body more than 3.3 times as long as wide .... **Bothrideridae: Bothriderinae (*Deretaphrus*)**

– Antennal club 1- or 2-segmented; lateral pronotal carinae complete to anterior angles, but sometimes very fine and not entirely visible from above; without other features combined **262**

262(261) Body less than 2.5 times as long as wide; upper surfaces clothed with expanded, scale-like setae .... **Bothrideridae: Bothriderinae (*Dastarcus*)**

– Body more than 2.5 times as long as wide); upper surfaces glabrous or clothed with fine hairs (Fig. 87G) **263**

263(262) Body cylindrical; lateral pronotal carinae very fine and not visible for their entire lengths from above; pronotal disc simple; procoxal cavities subcontiguous or very narrowly separated (Fig. 44R) .... **Bothrideridae: Bothriderinae: Sosylini**

– Body more or less flattened above; lateral pronotal carinae usually distinct and visible from above; pronotal disc usually with grooves or impressions (Fig. 87G); procoxal cavities moderately to widely separated (by at least half the width of a cavity) prosternal process often with transverse groove .... **Bothrideridae: Bothriderinae (part)**

264(260) Body more than twice as long as wide and more or less parallel-sided (Fig. 87C); postcoxal lines on abdominal ventrite 1 straight (Fig. 87D); lateral expansions present on abdominal ventrites 3 and 4, which fit into epipleural depressions .... **Cerylonidae: Euxestinae (*Metacerylon*)**

– Body less than twice as long as wide with sides distinctly curved (Fig. 72D); postcoxal lines on ventrite 1 absent or distinctly curved; ventrites without lateral expansions or with one on ventrite 4 only .... **Cerylonidae: Euxestinae (part)**

265(256) Abdomen with all ventrites free (overlapping and similarly articulated, sometimes with exposed connecting membrane) (Fig. 26J) **266**

– Abdomen with at least first 3 ventrites connate (Fig. 25K) **267**

266(265) Body shorter and broader, less than 3 times as long as wide; pronotum produced anteriorly concealing head from above; antennae 10-segmented with 1-segmented club; metaventrite as long as or shorter than first ventrite; body clothed with flattened scales (Fig. 82E) .... **Phycosecidae (*Phycosecis*)**

– Body elongate, more than 3 times as long as wide; pronotum not produced anteriorly over head; antennae 11-segmented with 3-segmented club; metaventrite distinctly longer than first abdominal ventrite; body clothed with fine setae (Fig. 71M) .... **Thanerocleridae**

267(265) Prosternal process flat, not curved dorsally (Fig. 19I); ventrites 3 and 4 on same plane (Fig. 25G) **268**

– Prosternal process distinctly curved dorsally (Fig. 25C); ventrite 3 on different plane than 4 (Fig. 25K) **269**

268(267) Pronotal process abruptly expanded laterally at apex (Fig. 19I); procoxal cavities separated by more than a coxal diameter (Fig. 19G); outer apical angle of protibia produced and acute; metacoxae widely separated and intercoxal process on ventrite 1 truncate (Fig. 26B); body elongate, parallel-sided, subglabrous .... **Zopheridae: Zopherinae: Pycnomerini (*Pycnomerus*)**

– If pronotal process abruptly expanded, then procoxal cavities separated by less than coxal diameter (Fig. 20K; without other features combined .... **Zopheridae: Colydiinae (part)**

269(267) Antenna 6-segmented with 1-segmented antennal club bearing several compound sensoria .... **Tenebrionidae: Diaperinae: Gnathidiini (*Tyrtaeus*)**

– Antenna with at least 10 antennomeres and club, if present, more than 1-segmented **270**

270(269) Lateral pronotal carinae absent (Fig. 19C); antennal club 4- or 5-segmented, relatively weak .... **Tenebrionidae: Diaperinae: Myrmechixenini (*Myrmechixenus*)**

– Lateral pronotal carinae complete (Fig. 20A); antennal club, if present, strongly developed **271**

271(270) Antenna not capitate, more or less expanded at middle and narrowed apically; pronotal disc with longitudinal carinae; epipleura incomplete or absent; tibiae somewhat expanded and flattened .... **Tenebrionidae: Tenebrioninae: Palorini (*Eutermicola*)**

– Antennae capitate; pronotal disc without longitudinal carinae; epipleura complete to apex (Fig. 26H); tibiae not expanded and flattened **272**

272(271) Antennae 10-segmented with 2-segmented club; subantennal grooves extending well behind eyes; anterior angle of pronotum with deep cavity for antennal club .... **Tenebrionidae: Phrenapatinae (*Archaeoglenes*)**

– Antennae 11-segmented with club of 3 or more antennomeres; subantennal grooves, if present, not extending behind eyes; anterior angle of pronotum without antennal cavity **273**

273(272) Antennal club 6-segmented, strongly asymmetrical, pectinate or flabellate; body strongly convex .... **Tenebrionidae: Tenebrioninae: Bolitophagini (*Rhipidandrus*)**

– Antennal club 3-segmented, symmetrical; body somewhat flattened .... **Tenebrionidae: Diaperinae: Hyociini (part)**

274(37) Mesocoxal cavities closed laterally by meeting of mesoventrite and metaventrite (Fig. 44V); frontoclypeal suture absent; antennae more than twice as long as head width, with large, loosely jointed, 3-segmented club bearing sensory vesicles; head usually concealed beneath anterior edge of pronotum (Fig. 89B) OR, if not, THEN prosternum in front of coxae more than twice as long as shortest diameter of procoxal cavity; apical maxillary palpomere cylindrical or apically narrowed (Fig. 16B–D) **275**

– Mesocoxal cavities partly closed laterally by mesepimeron (Fig. 44T); IF frontoclypeal suture absent, THEN antennae less than twice as long as head width, with weak, relatively compact club; head rarely concealed beneath anterior edge of pronotum OR, if so, THEN apical maxillary palpomere apically expanded and securiform (Fig. 16F) **276**

275(274) Body more than 1.5 mm long; prosternum in front of coxa more than twice as long as shortest diameter of procoxal cavity and produced to form chin piece (Fig. 18B); head exposed from above (Figs 71F, 89F) .... **Corylophidae: Periptyctinae**

– Body less than 1.5 mm long; prosternum much less than twice as long as shortest diameter of coxal cavity (Fig. 89B); IF chin piece present, THEN head concealed from above by pronotum .... **Corylophidae: Corylophinae**

276(274) Frontoclypeal suture present; antennae more than twice as long as head width, usually with large and loosely jointed club (Fig. 89E); first abdominal ventrite without postcoxal lines; apical maxillary palpomere cylindrical or fusiform, not apically expanded; pronotum usually with paired basal impressions, basal carinae or submarginal carinae (Figs 72E, 89E) .... **Endomychidae (part)**

– Frontoclypeal suture absent; antennae less than twice as long as head width, with club short, weak and compact club (Fig. 87L–N); first abdominal ventrite withcurved postcoxal lines (Fig. 25E); apical maxillary palpomere usually expanded and truncate at apex, but apically narrowed in Microweiseinae, Sticholotidini and Scymnillini; pronotum without paired basal impressions or carinae .... **Coccinellidae**

277(37) Protrochantins at least partly exposed (Fig. 20K); mesocoxal cavities partly closed laterally by mesepimeron (Fig. 21F); metacoxae contiguous or narrowly separated; antennal club 5-segmented with 2nd club segment smaller (shorter and/or narrower) than either 1st or 3rd one (Fig. 15I) .... **Leiodidae: Leiodinae: Scotocryptini**

– Protrochantins completely concealed or apparently absent; mesocoxal cavities closed laterally by meeting of mesoventrite and metaventrite (Fig. 25L); metacoxae moderately to very widely separated (Figs 21E, 26B, E); 2nd antennal club segment always larger than 1st one **278**

278(277) Procoxal cavities externally open (Fig. 44K, R) **279**

– Procoxal cavities externally closed (Fig. 44J, N) **289**

279(278) Body minute (less than 1 mm long) and narrowly elongate (~4 times as long as wide); metaventrite distinctly longer than abdomen; antenna 10-segmented with 1-segmented club; apical maxillary palpomere aciculate (Fig. 16A) .... **Jacobsoniidae (*Derolathrus*)**

– Without above features combined **280**

280(279) Mesocoxal cavities partly closed laterally by mesepimeron (Fig. 44T) **281**

– Mesocoxal cavities closed laterally by meeting of mesoventrite and metaventrite (Fig. 21E, 44V) **284**

281(280) Metacoxal plates well developed and complete to lateral edge of coxa; body very short and broad (less than 1.5 times as long as wide) with sides of pronotum and elytra strongly explanate; procoxal cavities broadly open externally and broadly closed internally; antennae slender, 11-segmented with 2-segmented club .... **Clambidae: Acalyptomerinae (*Acalyptomerus*)**

– Metacoxal plates absent; without other features combined **282**

282(281) Frontoclypeal suture distinctly impressed (Fig. 44A); antennae 11-segmented, distinctly longer than head width, with loose, 3-segmented club; head abruptly constricted behind eyes to form broad neck and temples; body more than twice as long as wide, parallel sided........ **Mycetophagidae: Berginae (*Nesolathrus*)**

– Frontoclypeal suture absent; antennae 6- to 9-segmented, much shorter than head width, with 1- or 4-segmented club; head not abruptly constricted posteriorly; body broadly ovate with sides distinctly curved **283**

283(282) Antennal insertions concealed (Fig. 44A); antenna less than one-third as long as head width, 6- or 7-segmented, with weak 4-segmented club; upper surfaces pubescent .... **Coccinellidae: Coccinellinae (*Telsimia*)**

– Antennal insertions exposed (Fig. 44B–E); antenna more than one-third as long as head width, 8- or 9-segmented, with elongate, 1-segmented club; upper surfaces glabrous .... **Coccinellidae: Microweiseinae (*Serangium* part)**

284(280) Posterior edge of ventrite 5 crenulate (Fig. 44AA); labrum elongate, anteriorly narrowed and acute, forming a sucking tube containing stylet-like mandibles and maxillae (Fig. 87F); antennae 9-segmented with 2-segmented club; apical maxillary palpomere aciculate (Fig. 16A) .... **Cerylonidae: Ceryloninae (*Cautomus*)**

– Posterior edge of ventrite 5 not crenulate; mouthparts not forming sucking tube; antennae with more than 9 antennomeres OR apical maxillary palpomere not aciculate **285**

285(284) Body more than twice as long as wide; sides of pronotum subparallel or sinuate, with pronotum distinctly narrowed posteriorly; anteior pronotal angles not or barely produced forward; surfaces subglabrous **286**

– Body less than twice as long as wide; sides of pronotum distinctly curved; anterior pronotal angles distinctly produced forward; surfaces pubescent **287**

286(285) Sides of pronotum straight and subparallel (Fig. 87I); lateral pronotal carinae crenulate; antennae 10-segmented with 1-segmented club; metaventrite between mesocoxae and metacoxae shorter than abdominal ventrite 1; hind wings absent; outer apical angle of tibia dentate .... **Bothrideridae: Anommatinae (*Anommatus*)**

– Sides of pronotum sinuate, posteriorly converging (Fig. 89D); lateral prontal carinae simple; antennae 11-segmented with 2-segmented club; metaventrite between mesocoxae and metacoxae longer than abdominal ventrite 1; hind wings present; outer apical angle of tibia not dentate .... **Endomychidae: Merophysiinae (*Holoparamecus*)**

287(285) Antennal insertions concealed from above (Fig. 44A); antennae 8-segmented with 1-segmented club; eyes reduced (with fewer than 10 facets); abdominal ventrite 1 with curved postcoxal lines (Fig. 26B); upper surfaces glabrous .... **Endomychidae: Merophysiinae (*Cholovocera*)**

– Antennal insertions exposed from above (Fig. 44B–E); antennae with more than 8 antennomeres; eyes well developed; abdominal ventrite 1 without postcoxal lines; upper surfaces pubescent **288**

288(287) Antennae 10-segmented with 1-segmented club; procoxal cavities internally open; Christmas Island .... **Cerylonidae: Ostomopsinae (*Ostomopsis*)**

– Antennae 9- or 10-segmented with large, loose, 3-segmented club; procoxal cavities internally closed .... **Endomychidae: Anamorphinae (part)**

289(278) Lateral pronotal carinae absent; metaventrite strongly convex; abdominal ventrite 1 more or less fused to metaventrite; antennae 11-segmented with weak 3-segmented club; antennomeres 4–10 transverse and most antennomeres clothed with pubescent setae; body more than 2.5 times as long as wide; surfaces mostly glabrous but with patches of silvery pruinosity (Fig. 81D) .... **Jacobsoniidae (*Sarothrias*)**

– Lateral prontal carinae complete; metaventrite flat or slightly convex; without other features combined **290**

290(289) Head with rostrum which is much longer than wide (Fig. 80I); labrum fused to head capsule; antennae geniculate; abdomen with first 2 ventrites solidly fused with incomplete suture between them; pretarsal claws absent and apical tarsomere highly reduced, so that preapical tarsomere appears apical and tarsi appear 3-segmented .... **Curculionidae: Brachycerinae (*Aonychus*)**

– Head without rostrum; labrum free; antennae not geniculate; all ventrites free; pretarsal claws present **291**

291(290) Posterior edge of ventrite 5 crenulate (Fig. 44AA); frontoclypeal suture absent; pronotum and elytra more or less parallel sided (Fig. 87E); upper surfaces glabrous and shiny; elytra with 4 to 6 impressed striae .... **Cerylonidae: Ceryloninae (*Cerylonopsis*)**

– Posterior edge of ventrite 5 not crenulate; frontoclypeal suture present (Fig. 44A); pronotum and elytra either independently curved or conjointly curved; elytra not striate **292**

292(291) Body less than 1.5 times as long as wide with sides strongly, evenly curved (Fig. 87K); exposed portions of metacoxae globular and widely separated (major portion of coxa concealed) (Fig. 26E); antennae 9-segmented with 1-segmented club .... **Discolomatidae (*Aphanocephalus*)**

– Body more than 1.5 times as long as wide; sides not strongly, evenly curved; metacoxae not partly concealed, distinctly transverse; antennae 10- or 11-segmented with 2- or 3-segmented club **293**

293(292) Procoxal cavities very broadly closed, with hypomeral projections often wider than shortest diameter of coxal cavity and sometimes at midline or fused together and to prosternal process, which is narrow but continuous and sometimes carinate; clypeus on lower plane than frons; frontoclypeal suture relatively deeply impressed sides of prothorax often constricted at basal third; lateral pronotal carinae usually well defined and pronotal disc often with longitudinal carinae; upper surfaces glabrous, sometimes partly covered with waxy exudate (Fig. 87P) .... **Latridiidae: Latridiinae**

– Procoxal cavities narrowly closed, with hypomeral projections narrower than shortest diameter of coxal cavity and well separated at midline by prosternal process, which is often concave or interrupted, so that coxae are contiguous; clypeus on same plane as frons and frontoclypeal suture not deeply impressed; ides of prothorax not constricted, evenly curved; lateral pronotal carinae usually indistinct, often denticulate sometimes absent; pronotal disc simple or with one or two shallow impressions; upper surfaces pubescent (Fig. 87H) .... **Latridiidae: Corticariinae**

LARVAE

1(0) Thoracic legs absent or highly reduced (less than 0.2 times as long as width of mesothorax) and with fewer than 3 segments, except in palophagine Megalopodidae (see couplet 10); body almost always lightly sclerotised, or IF NOT, THEN either strongly flattened or with antennae and mouthparts reduced and unsegmented **2**

– Thoracic legs moderately well developed, almost always at least 0.2 times as long as width of mesothorax, OR if shorter, THEN body well sclerotised and cylindrical........ **27**

2(1) Mouthparts and antennae absent; body very lightly sclerotised and strongly curved ventrally (C-shaped); endoparasites of cockroaches .... **Ripiphoridae: Ripidiinae (late instar)**

– Mouthparts and antennae present........ **3**

3(2) Head prognathous and protracted (Fig. 45B) **4**

– Head prognathous and retracted (Fig. 45A) **11**

– Head hypognathous (Fig. 45C)........ **17**

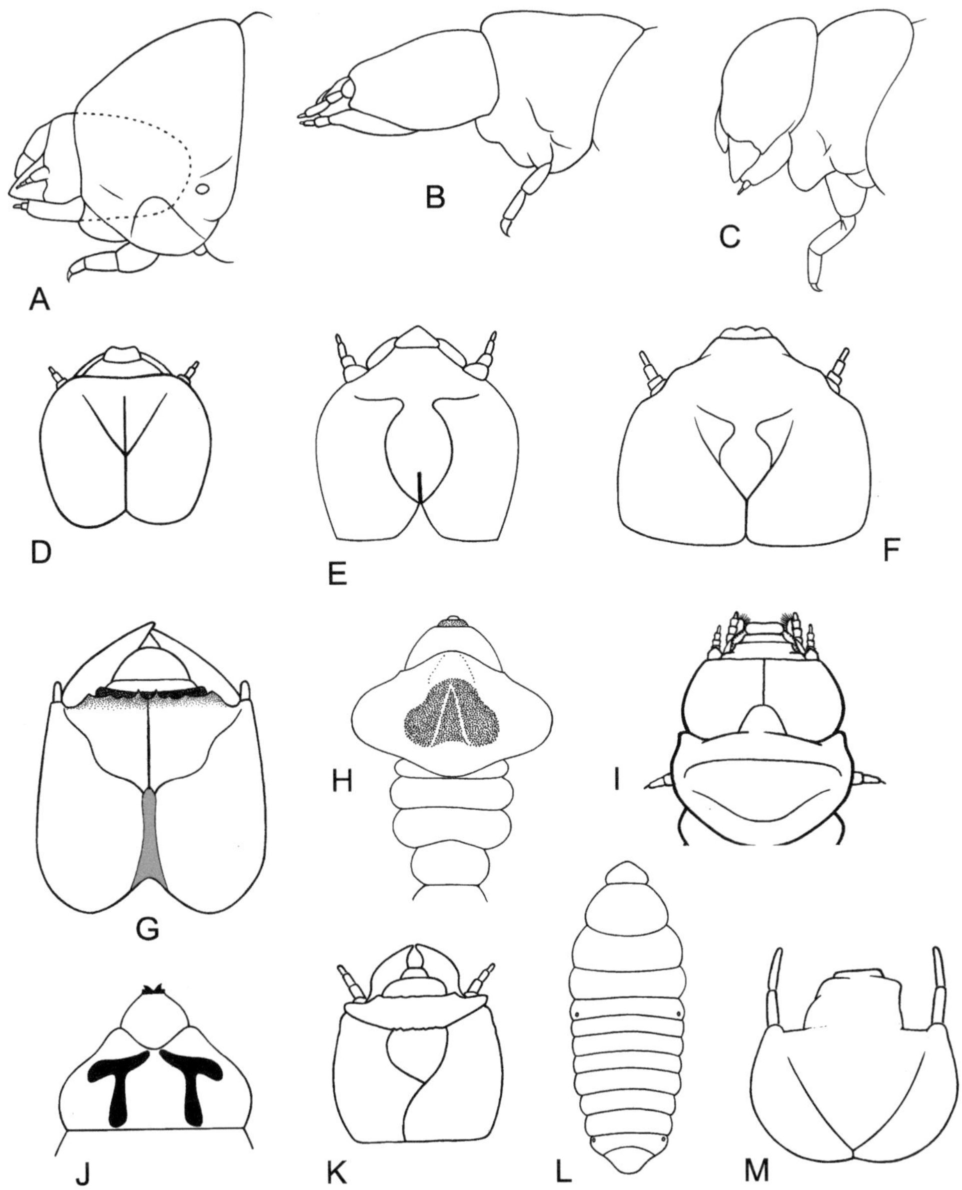

Fig. 45. Larval morphology. A–C, Head and prothorax, lateral: A, prognathous, retracted head; B, prognathous, protracted head); C, hypognathous head. D–F, Head, dorsal, showing variation in frontoclypeal suture, epicranial stem, frontal arms, and median endocarina. G, Prionine Cerambycidae, head, dorsal. H, Buprestidae, head, thorax and base of abdomen, dorsal. I, *Distocupes varians* (Lea), head, dorsal. J, Melasine Eucnemidae, head and prothorax, ventral. K, Pterogeniidae, head, dorsal. L, Corylophidae, body, dorsal. M, Dascillidae, head, dorsal.

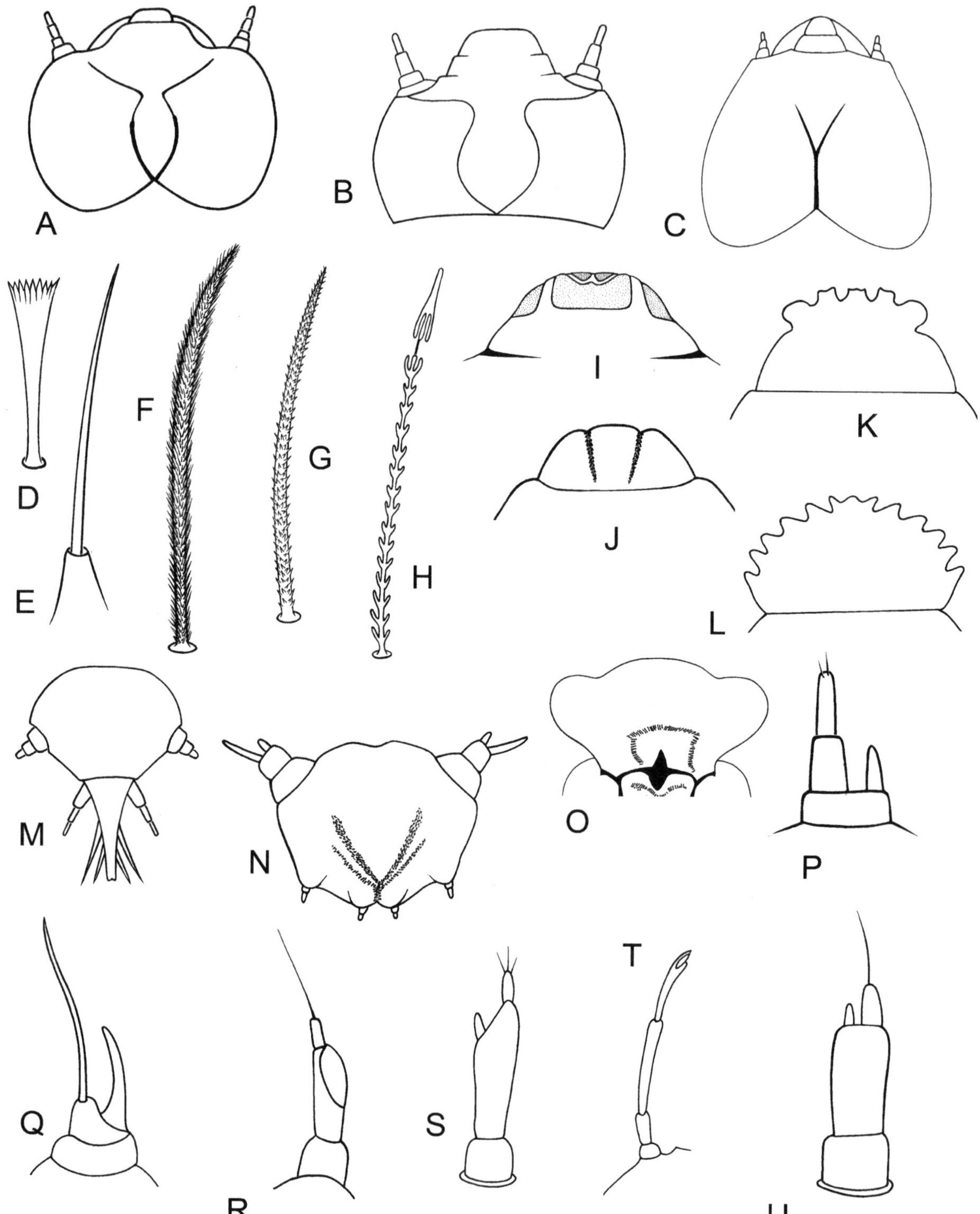

Fig. 46. Larval morphology. A–C, Head, dorsal: A, with paired endocarinae, no epicranial stem and no frontoclypeal suture; B, with no epicranial stem and partly fused labrum; C, with Y–shaped endocarina. D–H, Setal types: D, frayed; E, simple; F, pubescent; G, spiciseta; H, hastiseta. I–O, Labrum–epipharynx: I, subdivided (Silphidae); J, with labral rods; K, with 3 truncate lobes (Hybosoridae); L, serrate (Hybosoridae: Ceratocanthinae). M, Cerylonine head with tubular sucking mouthparts. N, *Cerylon* head with endognathous mouthparts. O, Lucanid labrum–epipharynx with mesally united tormae. P–U, Antennae: P, 3–segmented with sensorium on antennomere 1; Q, 2–segmented with long sensorium at base of antennomere 2; R, with large, dome–like sensorium; S, Ciidae, with oblique antennomere 2 and sensorium located well basad of antennomere 3; T, 3–segmented with sensorium on apical antennomere; U, with truncate antennomere 2 and sensorium and antennomere 3 arising together.

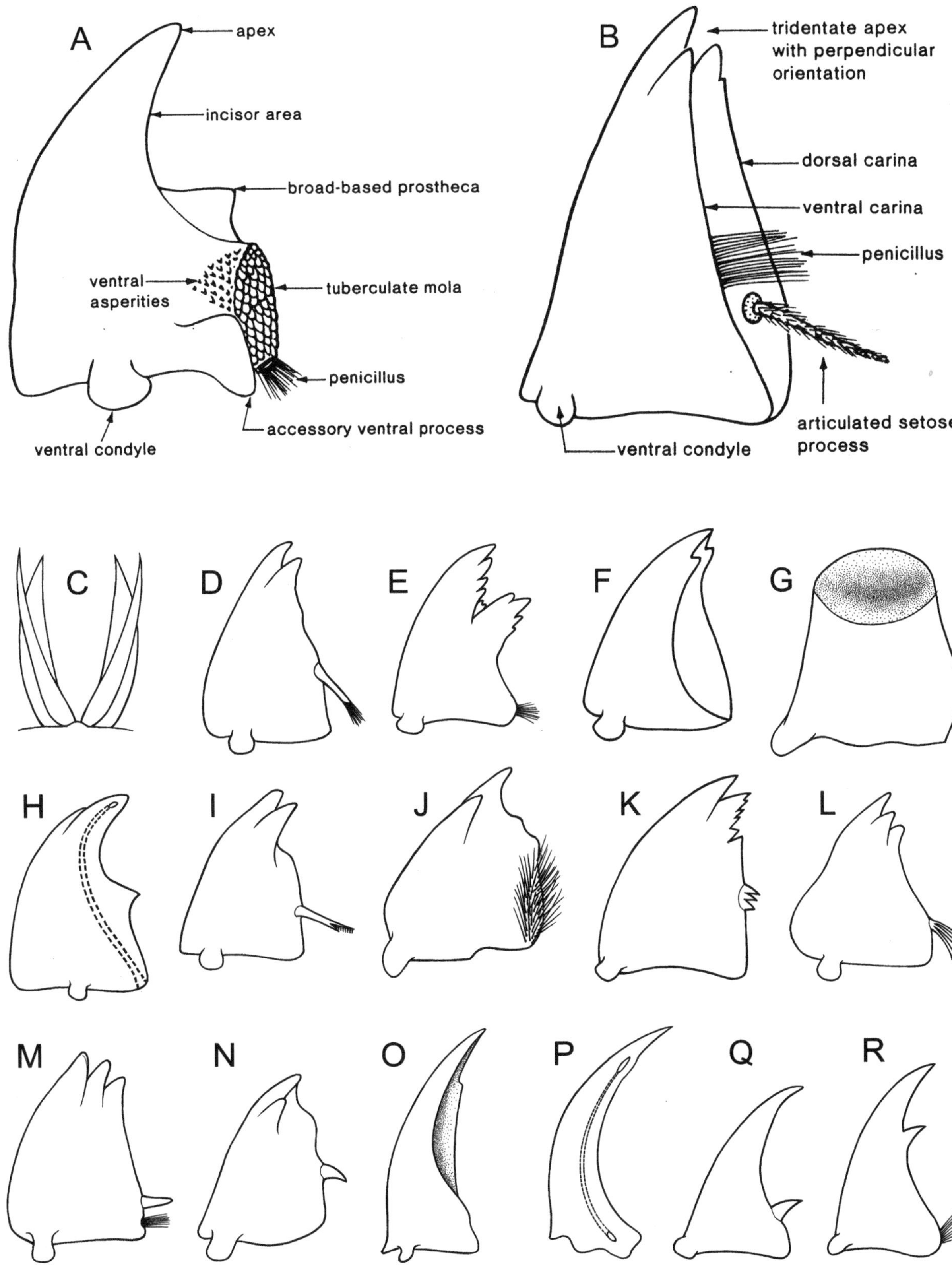

Fig. 47. Larval morphology. A–B, Mandible, ventral: A, with accessory ventral process, basal mola and broad prostheca; B, with no mola and basal penicillus and articulated, pubescent process. C, Lycidae, subdivided, non–opposable mandibles. D–R, Mandibles without basal mola, ventral.

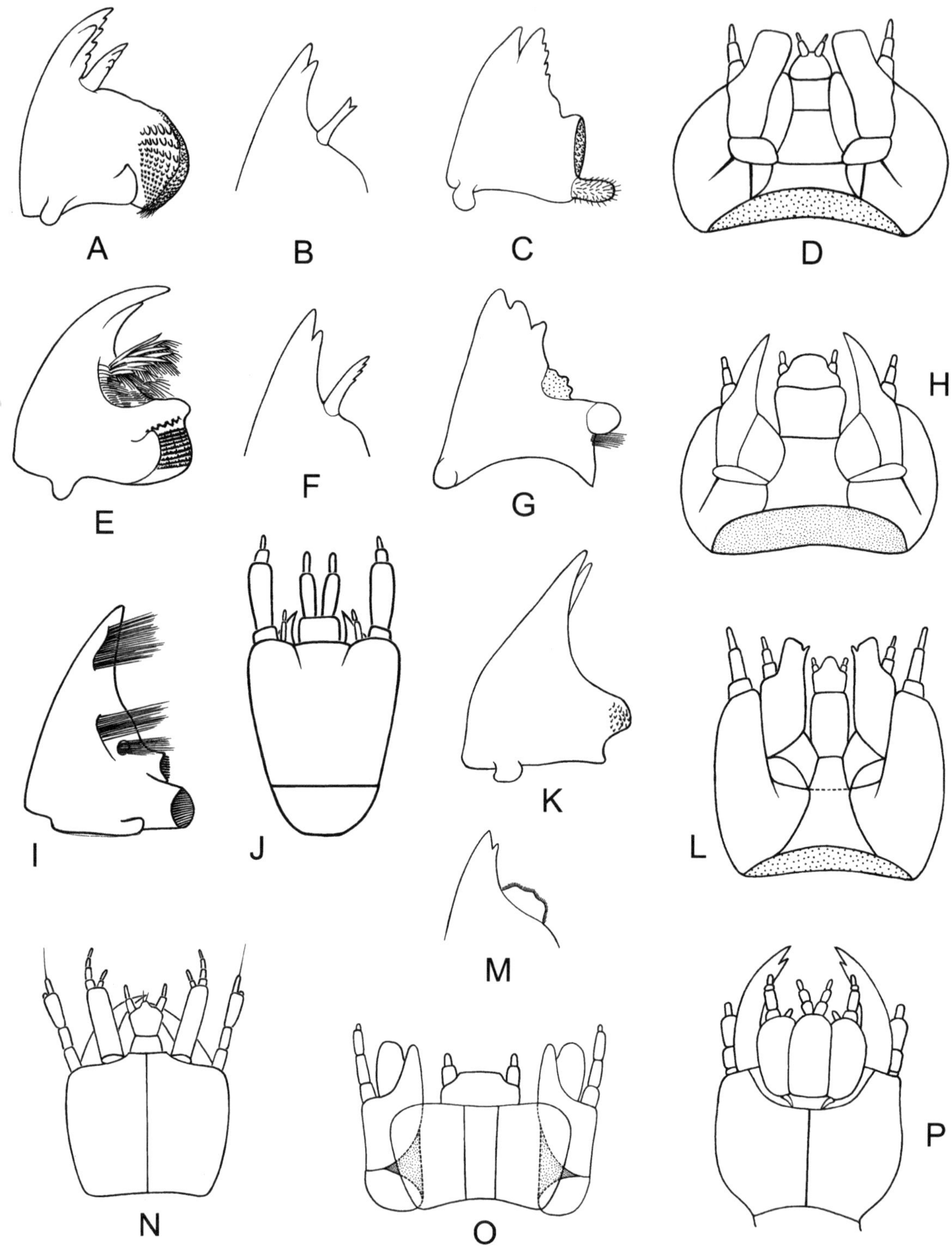

Fig. 48. Larval morphology. A, Mandible, ventral, with serrate incisor lobe, serrate prostheca, well developed basal mola and accessory ventral process; B, Portion of mandible with apically bifurcate prostheca. C, Mandible, ventral, with hyaline, pubescent lobe at base of mola. D, Head, ventral, with truncate malar apex and two pairs of hypostomal rods. E, Mandible, ventral, with prosthecal comb–hairs. F, Portion of mandible with serrate prostheca. G, Mandible, ventral, with sub–basal mola and membranous prostheca. H, Head, ventral, with falcate malar apex and diverging hypostomal rods. I, Scirtidae, mandible, ventral. J, *Brachypsectra* sp. (Brachypsectridae), ventral mouthparts. K, Coccinellidae, mandible, ventral, with sub–basal mola. L, Head, ventral, with malar uncus, retracted, freely articulated ventral mouthparts and short, convergent hypostomal rods. M, Portion of mandible with membranous, ciliate prostheca. N–O, Ventral mouthparts: N, Carabidae; O, anchytarsine Ptilodactylidae. P, Cantharidae, head, ventral, with slightly retracted ventral mouthparts.

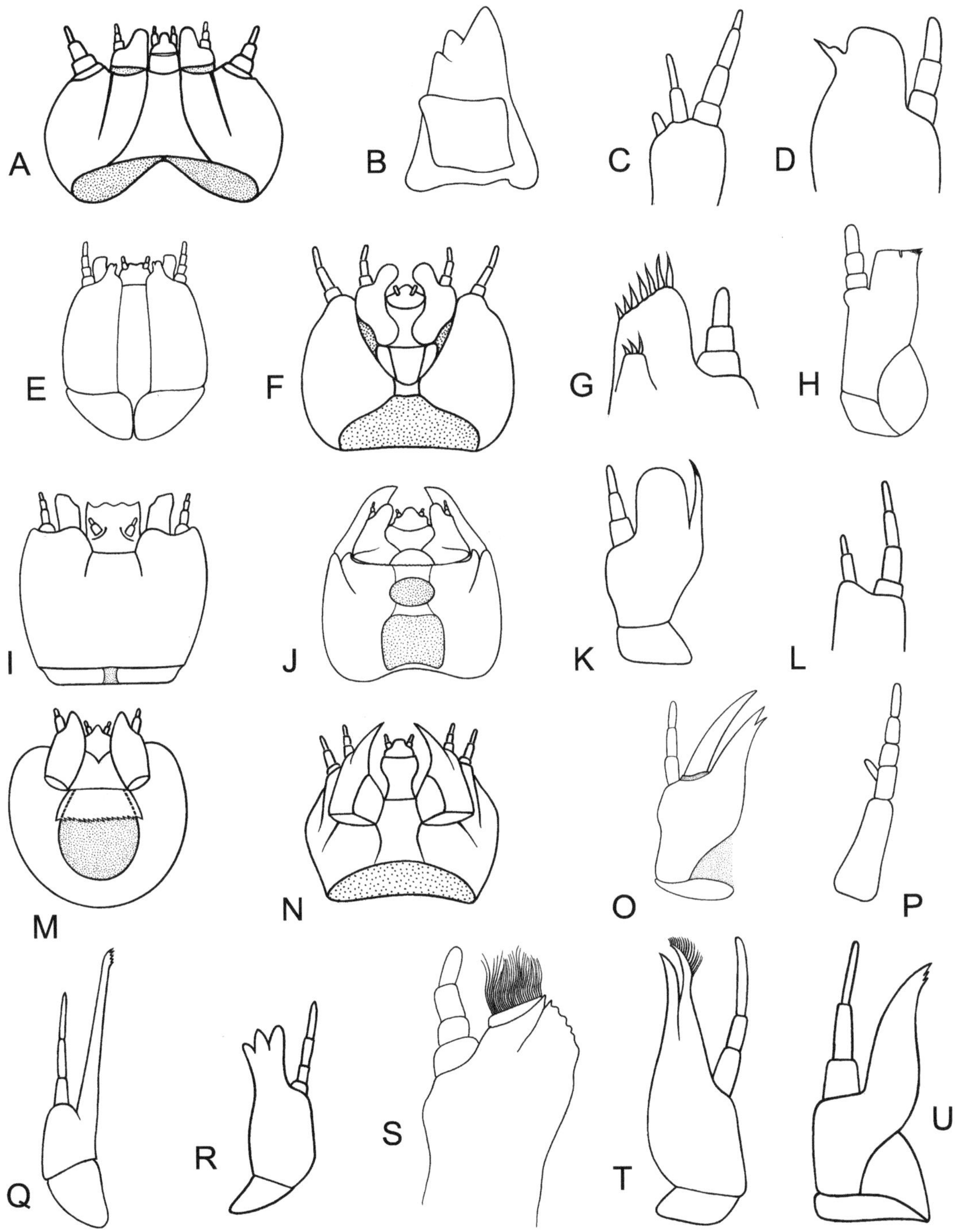

Fig. 49. Larval morphology. A, *Litochrus* sp. (Phalacridae), head, ventral, with protracted ventral mouthparts and long, diverging hypostomal rods. B, Mandible, mesal, with simple mola. C–D, Apical portion of maxilla: C, with articulated, 2–segmented galea and articulated lacinia; D, with fixed mala and distinct uncus. E, Consolidated ventral mouthparts (maxillolabial complex), with articulated galea, fixed labinia and contiguous cardines. F, Epuraeine Nitidulidae, head, ventral, with retracted ventral mouthparts and longitudinally oriented cardines. G, Ciine Ciidae, portion of maxilla, dorsal, with laciniar lobe. H, Maxilla with truncate, cleft mala and toothed inner apical angle. I, Callirhipidae, ventral mouthparts, with cardines approximate, stipites almost fused to postmentum and prementum with ligular sclerome; J, Prionine Cerambycidae, head, ventral, with occipital foramen appearing to be divided (due to position of tentorial bridge); K, Maxilla with laciniar spine. L, Portion of maxilla with 2–segmented, articulated galea. M–N, Head, ventral: M, with gular region absent and labium connected to thoracic or cervical membrane; N, with both diverging hypostomal rods and ventral epicranial ridges. O–U, Maxilla or portion of maxilla: O, Articulated galea and fixed lacinia; P, Hydrophiloid type, with galea near apex of palpifer; Q, with styliform mala; R, with trilobed mala; S, with galea and lacinia partly fused together; T, with fixed, fimbriate galea and fixed, falcate lacinia; U, with falcate mala bearing small teeth near apex.

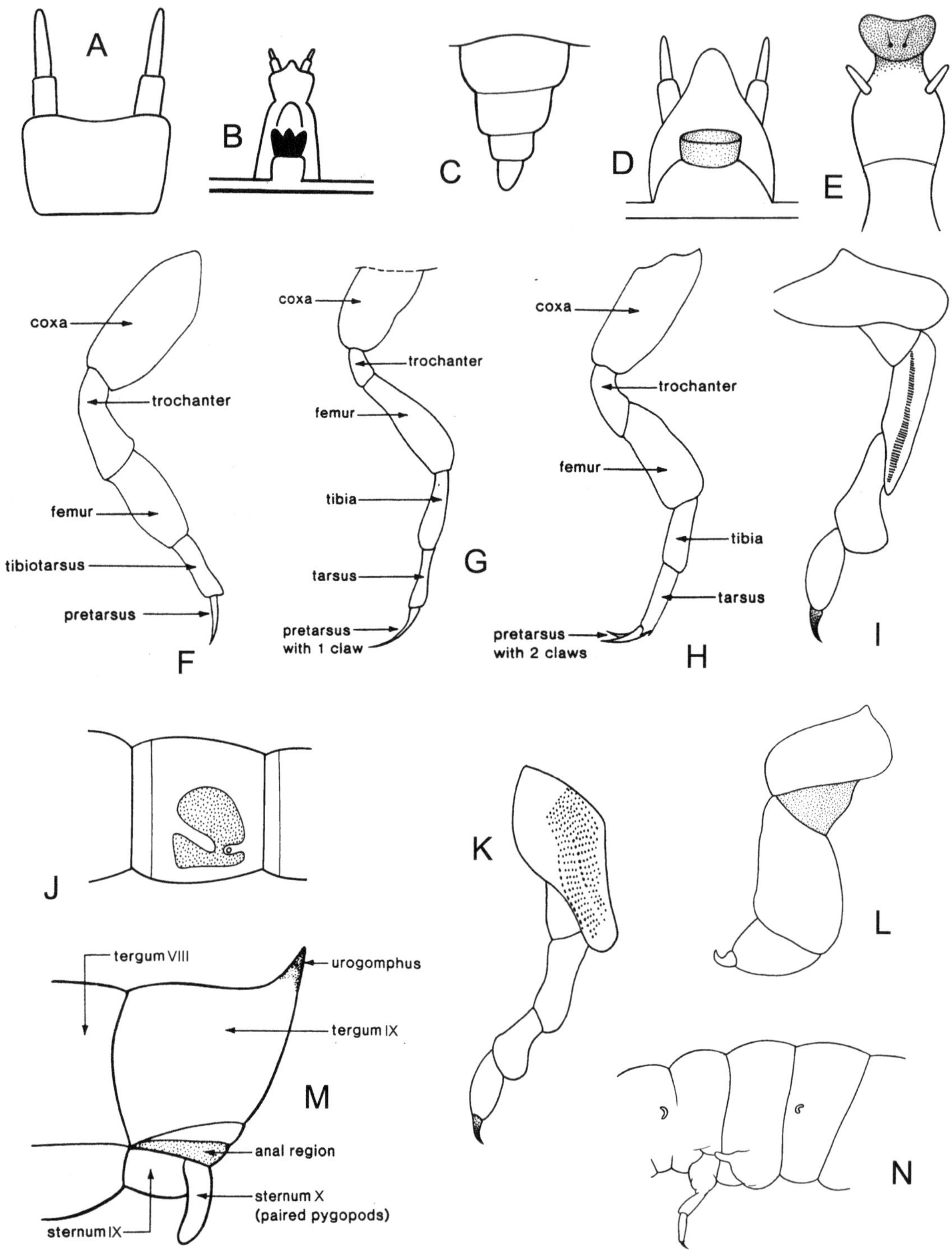

Fig. 50. Larval morphology. A, Prementum with widely separated labial palps and no ligula. B, Labium–hypopharynx, dorsal, with trilobed hypopharyngeal sclerome. C, Reduced leg without claw–like pretarsus. D, Labium–hypopharynx, dorsal, with more or less cup–like sclerome. E, Micromalthidae, labium, ventral, with ligular sclerome and 1–segmented labial palps. F–H, Leg: F, typical polyphagan type with tibiotarsus and undivided, claw–like pretarsus; G, adephagan type with tibia, tarsus and undivided pretarsus; H, adephagan type with divided pretarsus. I, Lucanidae, hind leg, with trochanteral plectrum. J, *Araeopidius monachus* (LeConte) (Ptilodactylidae: Araeopidiinae), abdominal segment, lateral, with plastron plate. K, Lucanidae, mid leg, with coxal stridulitrum. L, Reduced chrysomeloid leg. M, Tenebrionidae, abdominal apex, lateral, with partly concealed segment X and with paired pygopods. N, Passalidae, portion of thorax and abdomen, lateral, with stridulatory organ formed from reduced hind leg and mesocoxa.

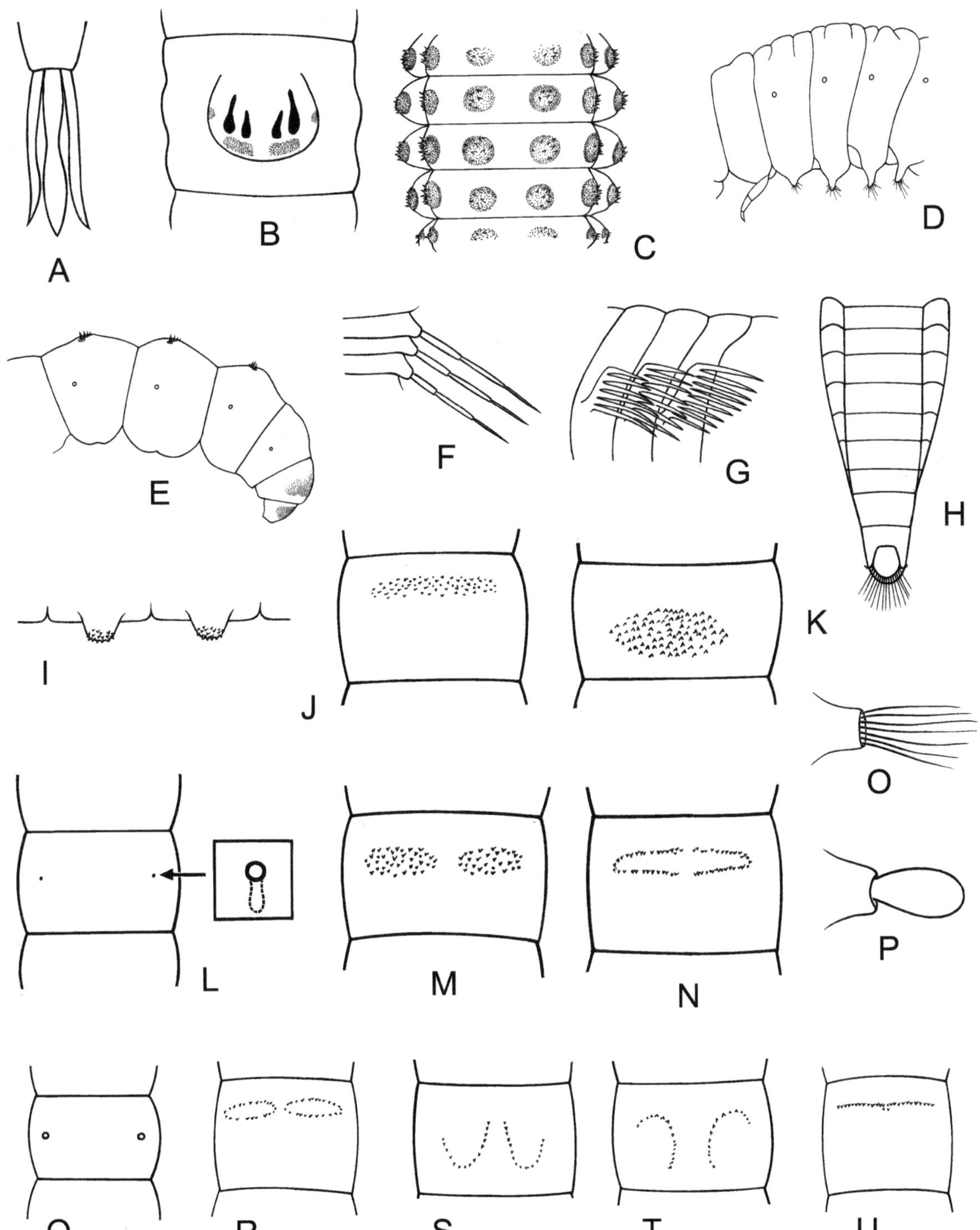

Fig. 51. Larval morphology. A, Meloidae, first instar, with long pretarsal setae giving appearance of 3–segmented claw. B, Cicindeline Carabidae, segment V, dorsal, with set of hooks. C, Coccinellidae, portion of abdomen, dorsal, with rows of verrucae. D, Eumolpine Chrysomelidae, metathorax and portion of abdomen, lateral, with three dorsal plicae and ventral, setose prolegs. E, *Sphallomorpha* sp. (Carabidae), abdominal apex, lateral, with group of spines on terga V–VII. F–G, Portion of abdomen: F, Torridincolidae, with segmented tracheal gills; G, psephenine Psephenidae, with abdominal gill tufts. H, Elmidae, abdomen, ventral. I, Ventral portion of two abdominal segments, with setose prolegs. J–N, One abdominal tergum: J–K, with single transverse patch of asperities; L, with paired gland openings; M, with paired patches of asperities; N, with transversely oriented curved rows of asperities. O–P, Tracheal gills: O, tufted type; P, vesicular type. Q–U, One abdominal tergum: Q, with paired gland openings; R–T, with paired, curved patches of asperities; U, with anterior row of asperities, usually broken in middle.

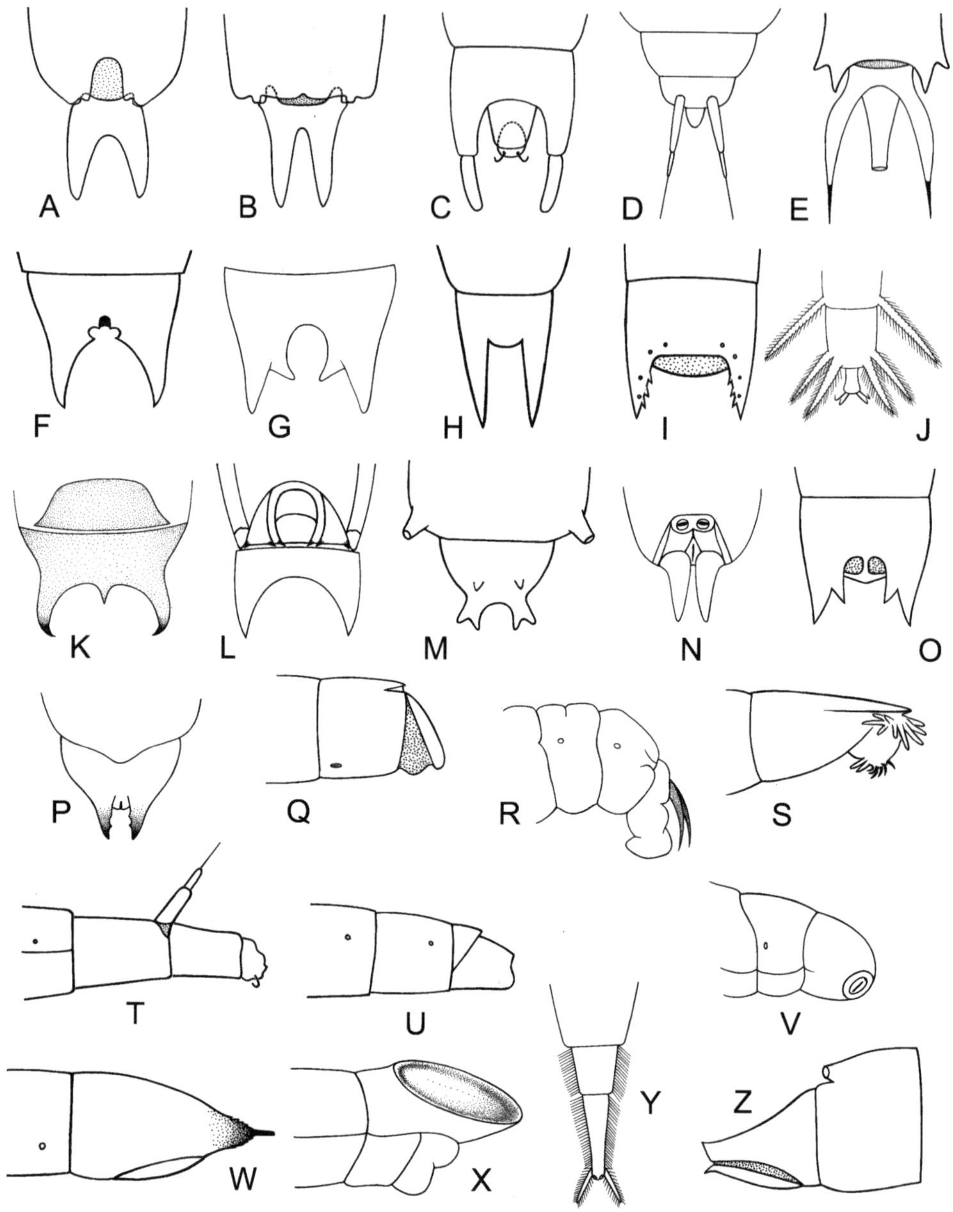

Fig. 52. Larval morphology. A–B, Abdominal apex, ventral: A, Genus ? sp. (Laemophloeidae); B, *Litochrus* sp. (Phalacridae). C–K, Abdominal apex, dorsal: C, *Hydroscapha natans* LeConte (Hydroscaphidae); D, Staphylinoidea, with 2–segmented, articulated urogomphi; E, brontine Silvanidae, with reduced tergum IX bearing long urogomphi and with long segment X; F, with small pit between urogomphi; G, with bifurcate urogomphi; H, Archeocrypticidae, with straight, posteriorly directed, acute urogomphi; I, *Pytho* sp. (Pythidae), with single interurogomphal pit ; J, Gyrinidae, with two pairs of lateral gills on segment IX and two pairs of hooks on segment X; K, lophocaterine Trogossitidae, with divided ninth tergal plate and median process between widely separated urogomphi. L–Z, Abdominal apex: L, Mycteridae, ventral, with tergum IX articulated and sternum IX U–shaped, with several posterior teeth; M, cryptarchine Nitidulidae, dorsal, with bifurcate urogomphi plus pregomphi and with eighth spiracles projecting posteriorly; N, *Amphizoa* sp. (Amphizoidae), posterodorsal, with terminal spracles and articulated urogomphi. ; O, pediline Pyrochroidae, dorsal, with bifurcate urogomphi and two interurogomphal pits; P, *Agrilus* sp. (Buprestidae), dorsal, with paired acute processes on segment X; Q, Callirhipidae, lateral, with dorsal operculum; R, *Donacia* sp. (Chrysomelidae: Donaciinae), lateral with spine–like eighth spiacles; S, anchytarsine Ptilodactyldae, lateral, with osmoregulatoy papillae and with hooks on paired pygopods; T, Hydraenidae, lateral, with articulated, 2–segmented urogomphi and with paired hooks on segment X; U, silvanine Silvanidae, with reduced segment IX and terminal segment X; V, Ptinidae, lateral, with paired, elongate–oval lobes bordering anal region; W, with acute process at apex of tergum IX; X, with concave disc on tergum IX; Y, Dytiscidae, with eighth spiracles at end of terminal siphon and reduced segment IX with articulated urogomphi; Z, Dryopidae, with posteriodorsal spiracles on segment VIII and with ventral operculum lacking hooks.

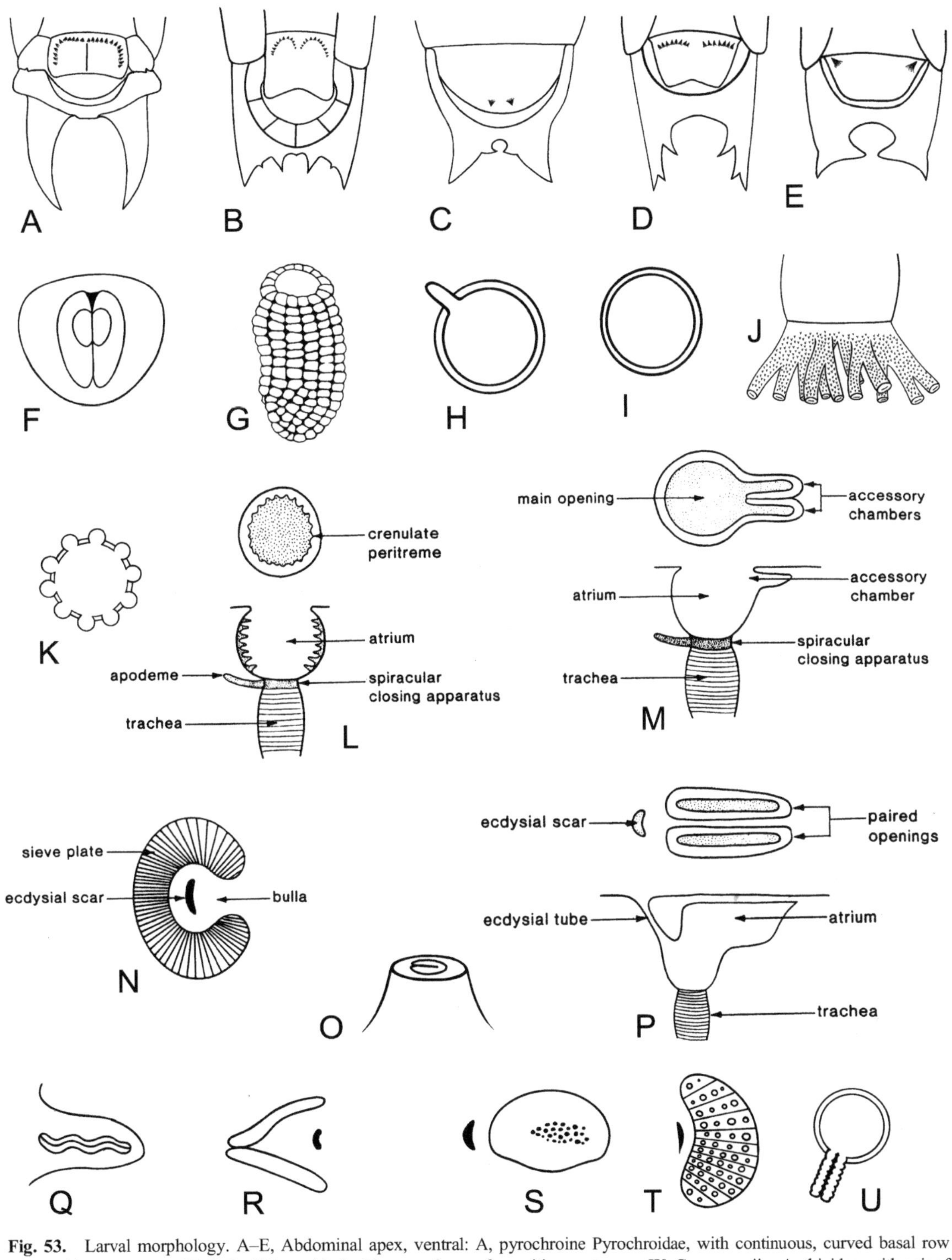

Fig. 53. Larval morphology. A–E, Abdominal apex, ventral: A, pyrochroine Pyrochroidae, with continuous, curved basal row of asperities on sternum IX; B, Pythidae, with doubly curved row of asperities on sternum IX; C, eurygeniine Anthicidae, with pair of apical asperities on tergum IX; D, *Elacatis* sp. (Salpingidae: Othniinae), with several teeth on each urogomphus and with two separate rows of asperities on sternum IX; E, *Salpingus* sp. (Salpingidae), with one pair of basal asperities on sternum IX. F, Lucanidae, abdominal apex, posterior, with paired oval pads flanking anal region. G–I, Spiracles: G, *Hemipeplus* sp. (Mycteridae: Hemipeplinae), with annular–cribriform type; H, annular–uniforous type; I, annular type. J, Lampyridae, abdominal apex, with multitubular, microasperate, eversible holdfast organ. K–U, Spiracles: K, annular multiforous type; L, annular type, with crenulate peritreme (section showing atrial details and spiracular closing apparatus); M, annular–biforous type with section; N, cribriform type, showing bulla and ecdysial scar; O, spiracle located at end of spiracular tube; P, biforous type with section; Q, undulate type of Chelonariidae; R, biforous type with converging paired openings and ecdysial scar; S, cribriform type of Heteroceridae; T, cribriform type of Buprestidae; U, annular–biforous type with long accessory chambers.

4(3) Head highly modified, forming an anteriorly flattened, wedge-shaped plate, which is heavily sclerotised or with dorsal and/or ventral supporting plates or rods (Figs 35I, 42I); antennae small to minute, 2-segmented, sometimes enclosed in deep cavities; body either elongate, parallel-sided and strongly flattened (Fig. 35I) or with prothorax enlarged and bearing paired longitudinal, supporting rods (Fig. 45J)............5

– Head capsule not forming wedge-like plate; antennae variable but not enclosed within deep cavities; body lightly sclerotised, without supporting rods on head or prothorax............8

5(4) Head with large posterior and ventral supporting rods; labrum and mandibles fused to head capsule; antennae not enclosed in cavities; body lightly sclerotised, usually dorsally curved; prothorax ventrally with paired converging and diverging supporting rods; urogomphi minute or absent; segment X posterioventrally oriented, subconical; spiracles annular–biforous; in dead wood or in soil on ectotrophic mycorrhizae............**Throscidae**

– Head entirely sclerotised or with large dorsal and ventral plates (Fig. 42H–I); antennae minute and enclosed in deep cavities; prothoracic supporting rods, if present, T-shaped (Fig. 45J), subparallel and usually both dorsal and ventral; body slightly to strongly flattened, more or less parallel-sided (Fig. 35I); segment X ventrally oriented and ovoid; spiracles biforous (Fig. 53P)............6

6(5) Mandibles either fused to head capsule or non-opposable with divergent apices; labrum always fused to head capsule; anterior edge of head capsule usually serrate (Fig. 35I); in dead wood............**Eucnemidae (part)**

– Mandibles free and opposable; labrum free or fused; anterior edge of head capsule either simple or with several anteriorly projecting lobes (Fig. 42H)............7

7(6) Labrum free and anterior edge of head capsule with several anteriorly projecting lobes (Fig. 42H); mandibles abruptly curved mesally; spicule patches on abdominal terga ovoid, those on sterna hour-glass shaped; tergum IX with pair of small, posteriorly projecting, laterally compressed, bifurcate urogomphi; in rotten wood............**Eucnemidae: Pseudomeninae (*Pseudomenes*)**

– Labrum fused to head capsule, which is anteriorly simple; mandibles gradually curved mesally; abdominal terga evenly sclerotised, sterna with hour-glass shaped spicule patches; tergum IX without urogomph; in rotten wood or fungi............**Eucnemidae: Anischiinae (*Anischia*)**

8(4) Antenna 2-segmented; stemmata absent; abdominal terga simple; sometimes with deep indentation between meso- and metathorax; ectoparasites of various wood-boring insects............**Bothrideridae: Bothriderinae (late instars)**

– Antenna 3-segmented; stemmata on each side 6; abdominal terga with transverse plicae or ambulatory ampullae (Fig. 58I); without deep between meso- and metathorax............9

9(8) Head not strongly transverse and somewhat elevated (hyperprognathous) (Figs 56G, 58B); median endocarina absent; ventral mouthparts protracted (Fig. 58C); maxilla with segment-like palpifer and displaced galea, so that palp appears to be 4-segmented with an articulated lobe on first palpomere (Figs 49P, 58C); anterior spiracles non-functional, terminal spiracles enlarged and located in spiracular chamber at abdominal apex (Figs 39K, 58L); in rotting vegetation, leaf litter, dung............**Hydrophilidae: Sphaeridiinae (*Cercyon*)**

– Head strongly transverse, not elevated, with median endocarina (Fig. 45D–E); ventral mouthparts retracted (Figs 49M); maxilla with mala, `normal palpifer and 3-segmented palp; spiracles all functional, annular (Fig. 53L) or annular–biforous (Fig. 53M, U); abdominal apex without spiracular chamber............10

10(9) Body strongly flattened; stemmata on each side 3; labrum free (Fig. 45D–E); labial palps 1-segmented; protergum without sclerotised plate; legs absent; spiracles annular (Fig. 53L) leaf miners............**Megalopodidae: Zeugophorinae (*Zeugophora*)**

– Body slightly flattened (Fig. 97E–F); stemmata on each side 6; labrum fused to head capsule (Fig. 45F); labial palps 2-segmented; protergum with sclerotised plate; legs 4-segmented; spiracles annular–biforous (Fig. 53M, U); in male cones of *Araucaria*............**Megalopodidae: Palophaginae**

11(3) Spiracles cribriform (Fig. 53T); body flattened; prothorax usually much wider than abdomen, which is narrow and parallel-sided (Fig. 65A) or only slightly wider than abdomen, which is gradually narrowed posteriorly (leaf mining forms); protergum sometimes with plate bearing one or two impressed rods (Fig. 45H); labial palps highly reduced or absent; antennae 2-segmented; under bark or mining leaves............**Buprestidae**

– Spiracles never cribriform; if labial palps absent, antennae 1-segmented or 3-segmented............12

12(11) Head deeply emarginate posteriorly; body flattened; leaf miners............13

– Head not deeply emarginate posteriorly............14

13(12) Antennae 1-segmented; labral rods present (Fig. 46J)............**Curculionidae: Rhynchaenini**

– Antennae 2- or 3-segmented; labral rods absent............**Chrysomelidae: Cassidinae (part)**

14(12) Body relatively short and stout, strongly curved ventrally (C-shaped) (Fig. 33E); maxillary palps 1-segmented; labial palps absent; in seed pods (usually of Fabaceae)............**Chrysomelidae: Bruchinae**

– Body more elongate; maxillary palps with 2 or more segments; labial palps present............15

15(14) Body relatively straight, not curved ventrally (Figs 35G, 97I); maxillary articulating area well developed (Fig. 36A); hypopharyngeal bracon absent; gular region present (separating labium from thorax) (Fig. 49J); in living or dead wood............**Cerambycidae**

– Body usually curved ventrally (C-shaped) (Fig. 33E); maxillary articulating area absent (Fig. 37B); hypopharyngeal bracon present (Fig. 36A); gular region absent (labium contiguous with thoracic membrane) (Fig. 49M)............16

16(15) Antennae 2-segmented with sensorium on basal antennomere; labrum without long, slender epipharyngeal rods; maxillary palps 3-segmented; body not or only slightly curved ventrally; protergum sometimes enlarged and hump-like with transversely keeled plate; boring in dead wood, twigs or branches............**Belidae**

– Antennae 2-segmented with sensorium on apical antennomere; labrum with long, slender epipharyngeal rods (Fig. 46J); maxillary palps 2-segmented; body distinctly curved ventrally, without enlarged, hump-like protergum; in leaf rolls, shoots or flower buds............**Attelabidae**

17(3) Maxilla with fixed galea and lacinia (Fig. 49S–T); antenna 1-segmented; abdominal tergum IX with pair of small upturned urogomphi (Fig. 63K) OR segment X with paired elongate-oval lobes separated by a longitudinal groove (Fig. 52V)............18

– Maxilla with mala (fused galea and lacinia) (Fig. 36A); tergum IX simple and segment X without paired lobes and longitudinal groove............19

18(17) Labrum free (Fig. 36C); mandible with mola (Fig. 41G–H); lacinia with sclerotised inner edge (Fig. 49S); tergum IX simple and segment X with paired elongate-oval lobes separated by longitudinal groove (Fig. 52V); spiracles annular (Fig. 53L); in fungal fruiting bodies............**Ptinidae: Dorcatominae (*Caenocara*)**

– Labrum fused to head capsule (Fig. 36D); mandible without mola (Fig. 47F); lacinia without sclerotised inner edge; tergum IX with small pair of upturned urogomphi (Fig. 63K); segment X without paired oval lobes separated by groove; spiracles annular–biforous (Fig. 53M); in soil parasitising nymphs of cicadas (Hemiptera: Cicadidae)............**Rhipiceridae (late instars)**

19(17) Cuticle heavily sclerotised; antennae and mouthparts highly reduced, unsegmented and fused to body; spiracles annular, those on segment VIII absent .... **Meloidae (coarctate)**

– Cuticle lightly sclerotised, except for head capsule and protergum in some groups **20**

20(19) Body straight or slightly curved ventrally; abdominal terga undivided, without transverse plicae; antennae and mouthparts reduced and unsegmented; tergum IX with pair of short blunt prominences; ectoparasites of wood-boring Coleoptera .... **Ripiphoridae: Pelecotominae**

– Body distinctly curved ventrally; abdominal terga divided to form two or more transverse plicae (Fig. 33E); antennae 1-segmented with terminal sensorium; mandibles well developed and heavily sclerotised; maxillary and labial palps well developed, the former with 2 or 3 segments, the latter with 1 or 2 segments **21**

21(20) Mandible with reduced mola but with accessory ventral process (Fig. 47A); hypopharyngeal sclerome molar-like (Fig. 50D); frontoclypeal suture absent or vaguely indicated; inner apical angle of mala with one or two teeth (Fig. 49H); head with distinct median endocarina (Fig. 45D); maxillary palp 3-segmented ; in male cones of Araucariaceae .... **Nemonychidae: Rhinorhynchinae**

– Mandible without accessory ventral process; if mandibular mola present, then maxilla with laciniar spine; maxillary palp usually with 2 or fewer segments **22**

22(21) Maxilla with distinct laciniar spine; mandibular mola usually present; median endocarina usually absent; in seeds and dried fruits .... **Anthribidae**

– Maxilla without laciniar spine; mandibular mola always absent; median endocarina usually present (Figs 43A, 45D) **23**

23(22) Frontal arms not extending to mandibular membrane, contiguous with antennal insertions; wide variety of habitats, mainly within plant tissue .... **Curculionidae**

– Frontal arms extending to mandibular membrane and not contiguous with antennal insertions **24**

24(23) Abdominal segments II–VII with 3 or 4 dorsal folds (Fig. 33E); thoracic spiracle on prothorax, not in contact with intersegmental membrane **25**

– Abdominal segments II–VII with 2 dorsal folds; thoracic spiracle intersegmental **26**

25(24) Head with median endocarina extending anteriorly between frontal arms (Fig. 45D); stemmata on each side 2; legs not protruding, consisting of 2 flat, circular plates; labial palps 2-segmented; in subterranean tubers of *Ipomoea* (Convolvulaceae) .... **Brentidae: Brentinae (*Cylas*)**

– Endocarina, if present, not extending anteriorly between frontal arms; stemmata absent; legs more or less protruding and distinctly 2-segmented or labial palps 1-segmented; in dead wood .... **Brentidae: Brentinae (part)**

26(24) Labial palps wider than long; spiracles annular–uniforous (Fig. 53H), those on segment VIII vestigial; anal opening transverse; in seeds .... **Brentidae: Apioninae: Apionini**

– Labial palps longer than wide; spiracles annular–biforous (Fig. 53M), those on segment VIII not vestigial; anal opening X-shaped; in stems of *Ludwigia* .... **Brentidae: Apioninae: Nanophyini**

Larva: Well developed thoracic legs

27(1) Mesothoracic leg with 6 segments, including single or paired pretarsal claws (Fig. 50G–H) **28**

– Mesothoracic leg with 5 or fewer segments, including pretarsus (Fig. 50F, K–L) **40**

28(27) Labrum free (Fig. 36C); mandible with well developed mola (Fig. 49B); ligula with wedge-like sclerome (Figs 40A, J, 56H) dorsal surface of head with well developed linear (Fig. 45I) or forked endocarina; body lightly sclerotised, with ambulatory ampullae (Fig. 42F) **29**

– Labrum completely fused to head capsule (Fig. 36D); mandible without mola (Figs 40H, L, 47O, P); ligula without wedge-like sclerome; head without endocarina; body usually more heavily sclerotised dorsally and without ambulatory ampullae (Fig. 56A–D, I) **30**

29(28) Stemmata on each side 4; endocarina on head forked; tergum IX without medial process .... **Ommatidae (*Omma*)**

– Stemmata absent; endocarina on head not forked (Fig. 45I); tergum IX with median, posteriorly projecting, narrow, apically truncate process (Figs 42F, 56D) .... **Cupedidae**

30(28) Adominal terga VIII and IX forming large, flat or concave disc, bearing expanded setae or branched processes; under bark, in ant nests .... **Carabidae: Paussinae**

– Abdominal terga VIII and IX not forming large disc **31**

31(30) Segment VIII terminal, usually produced posteriorly forming short and subconical to long and subacute median terminal process (Figs 52Y, 56I); articulated urogomphi on segment IX usually present, but their attachments more or less concealed beneath segment VIII **32**

– Segment VIII not terminal, without median process **34**

32(31) Spiracles in segment VIII absent; median process long and apically acute (Fig. 42B); coxal bases and abdominal sterna I and II with gill tufts; in ponds or lakes .... **Hygrobiidae (*Hygrobia*)**

– Spiracles on segment VIII well developed, terminal, usually at apex of median process (Figs 52Y, 56I) **33**

33(32) Head strongly transverse, without epicranial stem; mandible relatively short and broad at base, without distinct groove or internal perforation; legs relatively short, without swimming hairs; urogomphi very short or apparently absent (Fig. 58E); in lakes and ponds .... **Noteridae**

– Head usually as long as or longer than wide, with well developed epicranial stem; mandible narrow and falcate, with open groove (Fig. 47O) or internal perforation (Fig. 47P); legs relatively long, often with swimming hairs (Fig. 56I); urogomphi usually well developed and visible from above (Figs 52Y, 56I); in a wide variety of aquatic habitats .... **Dytiscidae**

34(31) Mandible with internal perforation (Fig. 47P); dorsal surfaces granulate or microspinose OR abdominal segments with lateral gills (Fig. 56E); aquatic **35**

– Mandible without internal perforation; dorsal surfaces smooth or occasionally with rows or patches of asperities; abdominal segments without lateral gills; terrestrial **36**

35(34) Mandible narrow and falcate (Fig. 47O); antennae more than half as long as head width; maxillary palps 4-segmented; labial palps 3-segmented; dorsal surfaces smooth; abdominal segment X with two pairs of hooks (Fig. 52J); segments I–VIII each with one pair of lateral gills, segment IX with two pairs (Fig. 56E); in ponds or pools .... **Gyrinidae**

– Mandible broad at base and narrow at apex (Fig. 47HJ); antennae less than half as long as head width; maxillary palps 3-segmented; labial palps 2-segmented dorsal surfaces granulate or microspinose; abdominal segment X without hooks; abdominal segments I–IX without lateral gills; in ponds .... **Haliplidae**

36(34) Tergum IX with paired urogomphi (Fig. 56A–C) .... **Carabidae (major part)**
– Tergum IX without paired urogomphi (Fig. 56F–G) 37
37(36) Ventral mouthparts retracted (Fig. 40H); apex of maxilla without distinct lobes, its surface and than of labium sometimes covered with fine fringed membranes labial palps 1-segmented; abdominal terga I–VII or VIII each with two rows of asperities, which may form incomplete rings (Fig. 51N); in rotten wood .... **Carabidae: Rhysodinae**
– Ventral mouthparts protracted (Figs 43C, 48N); apex of maxilla with one or two distinct lobes (Fig. 49C); labial palps 2-segmented; adbominal terga without rows of asperities 38
38(37) Abdominal terga V to VII each with transverse row of four spines (Figs 51E, 56F); maxilla with galea and lacinia; in burrows around ant nests .... **Carabidae: Pseudomorphinae (*Sphallomorpha*)**
– Abdominal tergum V with distinct protuberance bearing two or three pairs of hooks (Figs 51B, 52G); maxilla with galea only; in burrows in soil .... **Carabidae: Cicindelinae**
– Abdominal terga without hooks or spines 39
39(38) Frontal arms distinctly separated at base (Fig. 36D; stemmata absent; abdomen pale and white, somewhat physogastric; in ant nests .... **Carabidae: Pseudomorphinae (*Adelotopus*)**
– Frontal arms contiguous (Fig. 36C); stemmata present; thoracic and abdominal terga heavily pigmented, sometimes bicolored or metallic (Fig. 35C); labial palps greatly enlarged; in soil or leaf litter .... **Carabidae: Scaritinae: Carenini**
40(27) Labrum partly or entirely fused to head capsule (clypeolabral suture incomplete or absent) (Figs 36D, 45F, M) 41
– Labrum entirely free (clypeolabral suture complete) (Figs 36C, 45D–F) 90
41(40) Mandibular mola absent (Fig. 47B–R) 42
– Mandibular mola present (Fig. 47A–C, E–G, I) 83

Larva: Legs well developed. Legs with 5 or fewer segments. Labrum partly or entirely fused to head capsule. Mola absent

42(41) Maxilla with galea and lacinia (the latter sometimes concealed by brush of hairs) (Fig. 49E, T) 43
– Maxilla with single lobe (lacinia or mala) at apex of stipes (Figs 36A, 37A, 48D, H, L) 44
– Maxilla without lobes at apex of stipes or apparently without apical lobes (when the articulated galea is located on palpifer rather than at apex of stipes) (Fig. 49P); body usually lightly sclerotised (sometimes with pigment pattern or with rows of asperities) 77
43(42) Head without ecdysial lines; body lightly sclerotised (sometimes with pigment pattern) (Fig. 35A); ventral mouthparts slightly retracted and ventral epicranial ridges absent (Fig. 48P; thoracic terga and abdominal terga I–VIII or IX each with pair of glandular openings (Fig. 51Q); spiracles highly reduced .... **Cantharidae**
– Head with frontal arms (Fig. 36D) and sometimes with short epicranial stem; ventral mouthparts strongly retracted and ventral epicranial ridges present (Fig. 37B); body almost always well sclerotised at least dorsally (Figs 35B, 65D) or at apex of abdomen (Fig. 35F); thoracic and abdominal terga without paired gland openings spiracles well developed, biforous (Fig. 53P) .... **Elateridae**
44(42) Antennae 2-segmented 45
– Antennae 3-segmented or 4-segmented 58
45(44) Tergum IX without urogomphi 46
– Tergum IX with paired urogomphi (Fig. 50M) 54
46(45) Maxillary palps 1-segmented; maxillary and labial palps large and tusk-like; body minute, less than 1 mm in length; in bark crevices, soil .... **Rhipiceridae (first instar)**
– Maxillary palps 2-segmented 47
– Maxillary palps 3-segmented or 4-segmented 49
47(46) Mesal surface of mandibular base with fixed, rigid, hyaline process (Fig. 47Q); antennae less than 0.2 times head width; legs 3- or 4-segmented; hypostomal rods present (Fig. 49A); in tunnels of wood-boring beetles or in stored products .... **Passandridae (part)**
– Mesal surface of mandibular base without fixed hyaline process; antennae more than 0.5 times head width; legs 5-segmented; hypostomal rods absent 48
48(47) Body broadly ovate, strongly flattened and disc-like (Fig. 88D); head concealed from above by prothorax; dorsal surfaces more heavily pigmented than ventral surfaces and clothed with modified setae or scales; in leaf litter .... **Corylophidae: Corylophinae (*Gloeosoma*)**
– Body neither broadly ovate nor strongly flattened; head not concealed from above; dorsal and ventral surfaces similarly pigmented and armed with posteriorly-projecting spines (Fig. 93J); in tunnels of ambrosia beetles .... **Bothrideridae (Sosylus triungulins)**
49(46) Mandibles approximate at base and divided longitudinally, so that 2 pairs of diverging blades or stylets are formed (Figs 42G, 43J, 47C); in leaf litter, rotten wood .... **Lycidae**
– Mandibles distant at base, converging apically and not divided longitudinally into 2 parts 50
50(49) Head strongly declined (hypognathous) (Fig. 45C); body strongly curved ventrally (C-shaped) (Fig. 97D); living in case constructed of faeces and debris (Fig. 97C); in leaf litter or on plant surfaces .... **Chrysomelidae: Cryptocephalinae (part)**
– Head prognathous (Fig. 45B); body not strongly curved or C-shaped 51
51(50) Epicranial suture absent; stemmata on each side 4; legs 4-segmented; ectoparasitic on beetle pupae .... **Carabidae: Harpalinae: Lebiini (late instar)**
– If epicranial suture absent, then stemmata absent; legs 5-segmented 52
52(51) Epicranial suture and stemmata absent; antennae geniculate, without sensorium at apex; mandibles highly reduced; abdominal segment X forming slender tube (Fig. 58F); in nests of harvester termites (*Drepanotermes*) .... **Staphylinnidae: Aleocharinae (*Drepanoxenus*)**
– Epicranial suture present (Fig. 36C); stemmata present or absent; antennae straight, with sensorium at apex; mandibles well developed 53
53(52) Stipes short and broad; thoracic and most abdominal segments with lateral tergal processes; antennae more than half as long as head width; in leaf litter, moss .... **Staphylinidae: Scydmaeninae (part)**
– Stipes elongate; thoracic and abdominal segments without lateral tergal processes; antennae usually less than half as long as head width; in leaf litter, moss .... **Staphylinidae: Pselaphinae (part)**

54(45) Epicranial suture absent; maxillae and labium completely or almost completely connate with maxillae (Fig. 48P); hypostomal rods present (Fig. 37A, 48H) OR spiracles annular–biforous (Fig. 53M) or biforous (Fig. 53P) **55**

– Epicranial suture present (Fig. 36C); labium completely free from or basally connate with maxillae (Fig. 48H); hypostomal rods absent; spiracles annular (Fig. 53L) **57**

55(54) Ventral mouthparts strongly protracted (Fig. 49A); mandible tridentate (Fig. 47F, M); mesal surface of mandibular base with fixed, rigid, acute hyaline process (Fig. 47N, Q); hypostomal rods present (Fig. 37A); spiracles annular (Fig. 53L); in tunnels of wood-boring beetles .... **Passandridae (part)**

– Ventral mouthparts retracted (Figs 36A, 37A); mandible unidentate (Fig. 47O); mesal surface of mandibular base simple; spiracles annular–biforous (Fig. 53M) or biforous (Fig. 53P) **56**

56(55) Mandibles long and falcate, approximate basally, each longitudinally divided into 2 parts, so that 4 diverging blades or styli are formed (Figs 42G, 43J, 47C); body usually heavily sclerotised with lateral tergal processes on most segments (Fig. 42C); prothorax without ventral supporting rods; under bark, in rotten wood, leaf litter .... **Lycidae (part)**

– Mandibles flattened and rounded or subtriangular, not divided; body very lightly sclerotised, with ventral supporting rods present on prothorax; in rotten wood, soil associated with roots (ectomycorhizal fungi) .... **Throscidae (part)**

57(54) Stemmata absent; antennal sensorium dome-like (Fig. 46R); labial palps separated by more than 2 basal diameters (Fig. 50A); in leaf litte .... **Staphylinidae: Scydmaeninae (part)**

– Stemmata on each side 2 or 3; antennal sensorium palpiform (Fig. 46S), setiform or sometimes bifid; labial palps separated by less than 2 basal diameters; in leaf litter .... **Staphylinidae: Pselaphinae (part)**

58(44) Antennae 4-segmented **59**

– Antennae 3-segmented **60**

59(58) Urogomphi absent; mesothoracic leg 4-segmented; epicranial stem very short; stemmata absent; in soil .... **Carabidae: Brachininae**

– Urogomphi present and articulated (Fig. 52T); mesothoracic leg 5-segmented; epicranial stem moderately long (Fig. 36C); stemmata usually present; under bark, in leaf litter, carrion .... **Staphylinidae (part)**

60(58) Tergum IX without urogomphi (Figs 58L, 88J) **61**

– Tergum IX with urogomphi (Figs 35F, F, 52T, 58G, 65D) **73**

61(60) Labial palps absent; body minute (length less than 1 mm), fusiform, clothed with posteriorly projecting stout setae; antennae more than half as long as head width, located near posterior edge of head, with long seta at apex; head prognathous (Fig. 45B); ventral mouthparts strongly retracted (Fig. 37A); in flowers, on vegetation, or attached to native bees or wasps .... **Ripiphoridae: Ripiphorinae (first instar)**

– Labial palps present; without other characters in combination **62**

62(61) Head strongly declined (hypognathous) (Fig. 45C) **63**

– Head prognathous (Fig. 45B) **64**

63(62) Mouthparts forming tubular beak containing stylet-like mandibles and maxillae (Fig. 46M); stemmata absent; body somewhat flattened, not strongly curved ventrally (Fig. 88J); under bark of rotting logs .... **Cerylonidae: Ceryloninae**

– Mouthparts not forming tubular beak; mandibles not stylet-like; stemmata on each side at least 5; body strongly curved ventrally (C-shaped) (Fig. 97D); living in case constructed of faeces and debris (Fig. 97C); in leaf litter or on plant surfaces .... **Chrysomelidae: Cryptocephalinae (part)**

64 (62) Maxillary palpifer segment-like with articulated galea at its apex (Fig. 49P); lacinia subtriangular and lined with stout setae; body tapering posteriorly (Fig. 58L), thoracic and anterior abdominal spiracles reduced and non-functional, those on segment VIII enlarged and located in small respiratory chamber formed by segments VIII and IX (Fig. 39K); stemmata on each side 6; in ponds and lakes .... **Hydrophilidae: Spercheinae (*Spercheus*)**

– Maxillary palpifer, if segment-like, not bearing articulated galea; stemmata on each side 4 or fewer; thoracic and abdominal spiracles all functional and and respiratory chamber absent **65**

65(64) Body minute, less than 1.5 mm long, and fusiform with thorax widest and abdomen tapering posteriorly (Fig. 93J); head either longer than wide and widest posteriorly or with Y-shaped epicranial suture; stemmata on each side 2; incisor edge of mandible more or less denticulate; abdominal segment X terminal; tergum IX usually with long, paired setae; antenna with very long-terminal seta and no distinct sensorium; spiracles on segment VIII sometimes placed at ends of conical tubes (Fig. 53O); in flowers or attached to native bees .... **Meloidae: Nemognathinae (first instar)**

– Body usually larger; if less than 1.5 mm long, then without other features combined **66**

66(65) Abdominal segments I–VIII or VII and VII with paired lateral, posteriorly directed, spiniform processes; body elongate, widest at middle and slightly flattened; head strongly transverse; stemmata on each side 3 or 4; antenna less than half as long as head width, with antennomere 3 at least as long as 1 and 2 combined and sensorium at least half as long as antennomere 3; cardines elongate; labial palps 1-segmented; on plant surfaces (feeding on coccids) .... **Nitidulidae: Cybocephalinae (*Cybocephalus*)**

– Abdomen without posteriorly directed spiniform processes; without other features combined **67**

67(66) Body more or less flattened; head concealed from above by prothorax (Fig. 88C, H); epicranial stem absent; frontal arms well separated at base (Fig. 36D); total length less than 7 mm **68**

– If body more or less flattened and head concealed from above by prothorax, then frontal arms contiguous at base (Figs 36C) and total length usually greater than 7 mm **69**

68(67) Maxillary palp 2-segmented; body broadly oval, strongly flattened and disc-like (Fig. 88C); labial palp 1-segmented; glandular openings present on abdominal segments I and VII; in leaf litter, on fungus fruiting bodies .... **Corylophidae: Corylophinae (*Holopsis*)**

– Maxillary palp 3-segmented; body widest anteriorly, narrowing posteriorly, with most terga laterally expanded (Fig. 88H); labial palp 2-segmented; glandular openings absent; in leaf litter .... **Corylophidae: Periptyctinae**

69(67) Body oblong to elongate, usually widest at middle; length less than 4 mm; epicranial stem moderately long (Fig. 36C); antennae more than half as long as head width; mandibles slender, falcate (Fig. 47O) and widely separated at base; stipites not closely adpressed to labium (Figs 36A, 37A); cardo well developed and more or less transverse; maxillary lobe (mala) well developed and fixed; maxillary palp 3-segmented with apical palpomere slender and usually much longer than preapical one; spiracles annular (Fig. 53L) **70**

– Body narrowly elongate and/or length greater than 4 mm; epicranial stem short or absent; antennae less than half as long as head width; stipites closely adpressed to labium (Figs 37B); cardo reduced and usually 'more or less longitudinally oriented; maxillary lobe a reduced, fixed lacinia or an articulated galea; maxillary palp apparently 4-segmented (3-segmented with segment-like palpifer) with apical palpomere not or only slightly longer than preapical one; spiracles biforous (Fig. 53P, R) (highly reduced in Cantharidae)........ **71**

70(69) Stipes short and broad; thoracic and most abdominal segments with lateral tergal processes; antennae more than half as long as head width; in leaf litter, moss .... **Staphylinidae: Scydmaeninae (part)**

– Stipes elongate; thoracic and abdominal segments without lateral tergal processes; antennae usually less than half as long as head width; in leaf litter, moss .... **Staphylinidae: Pselaphinae (part)**

71(69) Thoracic and abdominal terga laterally expanded, so that body is somewhat flattened (Fig. 65E); head usually withdrawn into prothorax and not or only slightly visible from above; mandible falcate with internal perforation and brush of hairs at base (Fig. 40K); segment X with series of asperate, eversible tubes forming a holdfast organ (Fig. 53J); in leaf litter .... **Lampyridae**

– Thoracic and abdominal terga not laterally expanded; head not partly concealed within prothorax; mandible without internal perforation (sometimes with open groove), without basal brush of hairs; segment X without eversible tubes........ **72**

72(71) Body lightly sclerotised and flexible (but usually with dark pigment pattern), its surface densely clothed with very short fine hairs producing a velvety appearance (Fig. 35A); epicranial suture absent; ventral mouthparts slightly retracted (Fig. 48P); maxillary lobe (lacinia) weakly developed, short, rounded and densely clothed with hairs; paired dorsal glands present on thoracic terga and abdominal terga I–IX (Fig. 51Q); in leaf litter........ **Cantharidae (*Chauliognathus*)**

– Body without velvety appearance, usually well sclerotised at least dorsally or at abdominal apex (Fig. 35B, F); epicranial suture present (Fig. 65P); ventral mouthparts strongly retracted (Fig. 37B); maxillary lobe (galea) well developed, articulated and palpiform; paired dorsal glands not present on thorax and abdomen; in leaf litter, under bark, in dead wood........ **Elateridae (part)**

73(60) Ventral mouthparts strongly protracted; stipes wider than long; hypostomal rods long and diverging (Fig. 49A); mesal surface of mandibular base with cluster of acute, hyaline processes (Fig. 47L); spiracles on segment VIII located at ends of long, posteriorly-projecting tubes (Fig. 84C); in fungal fruiting bodies .... **Lamingtoniidae (*Lamingtonium*)**

– Ventral mouthparts retracted (Figs 37B); stipes longer than wide; hypostomal rods absent; mesal surface of mandibular base simple or with brush of hairs (Fig. 47R); spiracles on segment VIII laterally placed and not at ends of tubes **74**

74(73) Maxillae and labium connate for most of their lengths forming maxillolabial complex (Fig. 37B); ventral epicranial ridges present; epicranial stem very short; maxillary palpifer segment-like, so that palps appear 4-segmented (Figs 48P, 49E); spiracles biforous (Fig. 53P)........ **Elateridae (part)**

– Maxillae and labium free or only basally connate (Fig. 37A); ventral epicranial ridges absent; maxillary palps 3-segmented with normal palpifer; spiracles annular (Fig. 53L)........ **75**

75(74) Urogomphi articulated at base (Fig. 52T); stemmata on each side 6; ligula present (Fig. 48H, L); under bark, in leaf litter, dung, carrion........ **Staphylinidae (part)**

– Urogomphi fixed at base (Figs 50M, 52F); stemmata on each 3 or fewer; ligula absent (Fig. 50A)........ **76**

76(75) Antennae more than half as long as head width; stipes short and broad; in leaf litter........ **Staphylinidae: Scydmaeninae (part)**

– Antennae less than half as long as head width; stipes elongate; in leaf litter, moss .... **Staphylinidae: Pselaphinae (part)**

77(42) Maxillary palpifer not segment-like, maxillary palps with 3 or fewer segments (Fig. 49U); antennae 1- or 2-segmented; head hypognathous (Fig. 45C) **78**

– Maxillary palpifer segment-like, so that maxillary palp appears 4-segmented (Figs 48P, 49E); antennae 3-segmented; head prognathous (Fig. 45B)........ **79**

78(77) Antennae more than half as long as head width, 3-segmented (Fig. 93J); stemmata on each side 4; maxillary palps 2-segmented; labial palps 1-segmented; ectoparasites of cockroaches .... **Ripiphoridae: Ripidiinae (first instar)**

– Antennae less than a fourth as long as head width, 1-segmented; stemmata absent; maxillary palps one segmented; labial palps absent; body covered with large, conical protuberances; in nests of native bee .... **Ripiphoridae: Ripiphorinae (late instar)**

79(77) Abdominal apex with respiratory chamber (pocket formed by terga VIII and IX and enclosing enlarged spiracles) (Figs 34D, 39K, 58J); stemmata on 'each side 6; in ponds, streams, leaf litter and dung .... **Hydrophilidae (part)**

– Abdominal apex without respiratory chamber........ **80**

80(79) Tergum IX without urogomphi (Fig. 58A–B)........ **81**

– Tergum IX with urogomphi (Figs 35H, 58I)........ **82**

81(80) Mesal surface of mandibular base simple or slightly expanded; labial palps 2-segmented; maxillary palps 4-segmented; stemmata on each side 6; cardo present; abdominal segments I–VII each with pair of long, narrow, lateral processes (Fig. 58B); in ponds........ **Hydrophilidae: Hydrophilinae (*Berosus*)**

– Mesal surface of mandibular base with brush of hairs (Fig. 47R); labial palps 3-segmented; maxillary palps 5-segmented; stemmata on each side 1 or 0; cardo apparently absent; abdominal segments without long processes (Fig. 35H); under bark, in leaf litter, dung, decaying vegetation, ant nests........ **Histeridae (part)**

82(80) Mesal surface of mandibular base simple or slightly expanded; legs with 3 or 4 segments; urogomphi 1-segmented (Fig. 58I); in sand and mud along streams .... **Hydrophilidae: Georissinae (*Georissus*)**

– Mesal surface of mandibular base with brush of hairs (Fig. 47R); legs 5-segmented including pretarsus; urogomphi usually 2-segmented (Fig. 35H); under bark, in leaf litter, dung, decaying vegetation, ant nests .... **Histeridae (part)**

Larva: Legs well developed. Legs with 5 or fewer segments. Labrum partly or entirely fused to head capsule. Mola present

83(41) Abdominal tergum IX without or with highly reduced urogomphi (Figs 33C, 84G)........ **84**

– Abdominal tergum IX with well developed urogomphi (Fig. 90B–C)........ **88**

84(83) Abdominal terga I–VIII with posterolateral balloon-like tracheal gills borne on tubular projections (Fig. 33C–D); body more or less fusiform, slightly flattened and minute (less than 1.7 mm in length); head large and hypognathous, with endognathous mandibles and very short antennae located well away from mandibular articulations; in sand, gravel or mud at edges of streams .... **Sphaeriusidae (*Sphaerius*)**

– Abdomen without balloon-like tracheal gills; without other characters combined **85**

85(84) Tergum VIII forming posteriorly-projecting, median process with spiracles at apex (Fig. 38L); bases of frontal arms distinctly separated (Fig. 36D); 'mandibular mola with complex armature (filter-feeding apparatus); prostheca consisting of a dense fringe of apically subdivided hairs (Fig. 41F); ventral mouthparts retracted; galea and lacinia fixed; dorsal surfaces distinctly 'tuberculate (Fig. 35J); in slime fluxes or fermenting sap in tree wounds .... **Nosodendridae (*Nosodendron*)**

– Tergum VII without spiracle-bearing process; without other characters combined **86**

86(85) Labial palps 2-segmented; maxillary articulating area present (Fig. 36A); body circular in cross-section (Fig. 84G); ventral epicranial ridges present (Fig. 37B); prostheca consisting of a brush of comb-hairs (Fig. 48E); spiracles annular–biforous (Fig. 53M), those on segment VIII located dorsally near posterior edge of tergum; under bark, in rotten wood or associated with ascomycete fungi .... **Biphyllidae (part)**

– Labial palps 1-segmented; maxillary articulating area absent (Fig. 37B); body slightly to strongly flattened; ventral epicranial ridges absent; prostheca simple or absent; spiracles on segment VIII laterally placed **87**

87(86) Body broadly ovate, strongly flattened and disc-like (Fig. 88C); ventral mouthparts protracted and cardines transverse (Fig. 49A); stemmata on each side 2; antennae more than half as long as head width; spiracles annular (Fig. 53L); in leaf litter .... **Corylophidae: Corylophinae (part)**

– Body elongate and slightly flattened; ventral mouthparts strongly retracted and cardines longitudinal (Fig. 49F); stemmata on each side 4; antennae less than half as long as head width; spiracles annular–biforous (Fig. 53M); in flower .... **Nitidulidae: Meligethinae (*Pria*)**

88(83) Abdominal spiracles annular–biforous (Fig. 53M); body elongate, parallel-sided and strongly flattened, with dorsal and ventral rows of abdominal asperities forming incomplete circles (Fig. 51S–T); urogomphi well developed (Fig. 52H); under bark .... **Phloeostichidae (*Hymaea* and *Rhopalobrachium*)**

– Abdominal spiracles cribriform (Fig. 53T); body elongate, convex and somewhat curved ventrally; urogomphi minute; antennae 3-segmented with 'very small terminal segment; pretarsus with 4 or more setae; epipharynx and hypopharynx with complex series of sclerites; in soil, under rocks .... **Dascillidae (*Notodascillus*)**

– Abdominal spiracles annular (Fig. 53L) **89**

89(88) Antenna less than 0.15 times head width, 3-segmented with sensorium on antennomere 1 (Fig. 46P); ventral mouthparts protracted and stipes wider than long (Fig. 49A); epicranial stem absent; abdominal segment X completely concealed (Fig. 52B); on plant surfaces affected by rusts or smuts .... **Phalacridae (*Phalacrus*)**

– Antenna more than 0.5 times head width, 3-segmented with sensorium on antennomere 2 (Fig. 46U); ventral mouthparts retracted (Figs 36A, 37A); stipes longer than wide; epicranial stem present (Fig. 36C); abdominal segment X distinct and visible from above (Fig. 52E) .... **Silvanidae: Brontinae**

Larva: Legs well developed. Legs with 5 or fewer segments. Labrum entirely free

90(40) Mandibular mola absent (Figs 47B–R) **91**

– Mandibular mola present (Figs 39G–H, 47A, 48A–C, E–G) **177**

91(90) Tergum IX without paired urogomphi (Figs 33F–G, 34M–N) **92**

– Tergum IX with paired urogomphi (Figs 50M, 52D-I, 52K-M) **142**

92(91) Antennae 1-segmented **93**

– Antennae 2-segmented **99**

– Antennae 3-segmented **108**

93(92) Legs 2- or 3-segmented (Fig. 42D); head hypognathous, body curved ventrally (C-shaped) (Fig. 33E); most abdominal terga transversely grooved to form 2 plicae; without paired longitudinally oval pads below anus; spiracles annular–biforous (Fig. 53M) **94**

– Legs 5-segmented (Figs 50F); IF head hypognathous, body C-shaped and terga transversely grooved, THEN segment X below anus with paired longitudinally oval pads separated by groove (Fig. 52V); spiracles not annular–biforous **95**

94(93) Legs 2-segmented, without distinct pretarsal claw; hypopharyngeal sclerome present (Fig. 50D); maxilla with laciniar spine (Fig. 49K); labium without trident sclerite; in dead wood, stems, vines, fungal fruiting bodies .... **Anthribidae (part)**

– Legs 3-segmented with distinct pretarsal claw; hypopharyngeal sclerome absent; maxilla without laciniar spine; labium with trident sclerite; in female cones of *Callitris* (Cupressaceae) .... **Caridae**

95(93) Body elongate, cylindrical and heavily sclerotised; ventral epicranial ridges present and ligula forming sclerotised, wedge-like plate (Fig. 38D); abdominal apex with dorsally-hinged operculum (Fig. 52Q); spiracles biforous (Fig. 53P); in rotten wood .... **Callirhipidae**

– Body form different; ventral epicranial ridges absent; ligula without wedge-like plate; abdominal apex without operculum; spiracles annular (Fig. 53L) or annular–uniforous (Fig. 53H) **96**

96(95) Maxilla with single mala (Figs 49H, 49U); abdominal segment X without longitudinally oval pads beneath anus; frontal arms present (Fig. 45D); body not C-shaped **97**

– Maxilla with galea and lacinia (Fig. 49K); abdominal segment X with paired longitudinally oval pads separated by groove below anus (Fig. 52V); frontal arms absent; body strongly curved ventrally (C-shaped) (Fig. A42J) **98**

97(96) Body oblong to ovate (Fig. 45L); median endocarina absent; apex of mandible bidentate (Fig. 48K); stemmata on each side 3; on plant surfaces .... **Coccinellidae (part)**

– Body elongate and parallel-sided (Fig. 33G); median endocarina extending anterior to epicranial stem and between frontal arms (Fig. 45D); apex of mandible with 3 or more teeth or lobes (Fig. 41I); stemmata on each side 2 or fewer; in leaf litter and soil or on plant surfaces .... **Chrysomelidae: Galerucinae (part)**

98(96) Thoracic spiracle located near anterior end of prothorax; abdominal terga without bands of asperities; in dung, animal nests, stored products .... **Ptinidae: Ptininae**

– Thoracic spiracle located near posterior end of prothorax or between prothorax and mesothorax; abdominal terga usually with bands of asperities (Fig. 51J); in dead wood, twigs, fungus fruiting bodies, stored products .... **Ptinidae (part)**

99(92) Labial palps absent or represented by minute papillae; head strongly retracted (Fig. 45A); body strongly curved ventrally (C-shaped) (Fig. 33E); spiracles annular–biforous (Fig. 53M); in seeds or pops of various plants (usually Fabaceae) .... **Chrysomelidae: Bruchinae**

– Labial palps 1-segmented 100
– Labial palps 2-segmented 101
100(99) Stemmata on each side 1; body more or less elongate and parallel-sided, without narrow lateral processes or elongate, terminal process; in soil, leaf litter, on leaf surfaces .... **Chrysomelidae: Galerucinae (part)**
– Stemmata on each side 5 or 6; body either ovate with several long, narrow lateral processes (Fig. 97H) or more elongate with single, long, narrow process at abdominal apex; on leaf surfaces or in leaf axils .... **Chrysomelidae: Cassidinae (part)**
101(99) Abdominal apex with ventrally-hinged operculum (without hooks) and with paired gill tufts (Fig. 51H); mesal surface of mandibular base with articulated, pubescent process (Fig. 47D); maxilla with articulated galea and fixed lacinia (Fig. 49E); beneath rocks in intertidal zone .... **Limnichidae: Hyphalinae (*Hyphalus*)**
– Abdominal apex without operculum or gill tufts; mesal surface of mandibular base without articulated pubescent process; maxilla usually with mala (Figs 37A, 49H, 49U); galea, if present, not articulated 102
102(101) Head with median endocarina coincident with epicranial stem, not extending between frontal arms or present in the absence of frontal arms (Fig. K19); maxilla with dorsal, subapical, laciniar lobe (Fig. 49G); head globular and hypognathous (Fig. 45C); body subcylindrical, without transverse plicae (Fig. 90B); ventral epicranial ridges and hypopharyngeal bracon present; in fungal fruiting bodies .... **Ciidae: Ciinae (part)**
– Head with median endocarina extending between frontal arms (Fig. 45G); maxilla without laciniar lobe; IF head hypognathous THEN, body with transverse plicae (Fig. 51D); ventral epicranial ridges and hypopharyngeal bracon absent 103
– Head without median endocarina; IF maxilla with a laciniar lobe, THEN body strongly curved ventrally, with transverse plicae (Fig. 81M); ventral epicranial ridges absent 105
103(102) Head strongly retracted (Fig. 45A); gular region present, separating labium from thorax (Fig. 49J); under bark, in living or dead wood .... **Cerambycidae (part)**
– Head protracted or slightly retracted (Fig. 45B–C); gular region absent (labium contiguous with thoracic membrane) (Fig. 49M); on leaf surfaces, in leaf litter, soil 104
104(103) Maxillary palps 2-segmented; abdomen less than twice as long as thorax (Fig. 97A); abdominal apex with forked process (tail shovel); on leaf surfaces .... **Chrysomelidae: Cassidinae (part)**
– Maxillary palps 3-segmented; abdomen more than twice as long as thorax (Fig. 33G); abdominal apex without forked process .... **Chrysomelidae: Galerucinae (part)**
105(102) Body strongly curved ventrally (C-shaped) (Fig. 81M); abdominal terga with 2 or 3 transverse plicae (Fig. 51D); abdominal segment X below anus with paired oval pads separated by longitudinal groove (Fig. 52V); in dead wood, fungus fruiting bodies, stored products .... **Ptinidae (part)**
– Body relatively straight or slightly curved ventrally; terga without transverse plicae; segment X without oval pads separated by longitudinal groove 106
106(105) Body elongate with enlarged abdomen (physogastric); gula longer than wide; head prognathous (Fig. 45B); in tunnels of wood-boring beetles .... **Bothrideridae: Bothriderinae (part)**
– Body not physogastric, elongate and cylindrical or oblong to ovate and somewhat flattened; gula wider than long or absent 107
107(106) Body oblong to ovate and somewhat flattened (Fig. 45L); abdominal terga usually with transverse rows of 6 protuberances, which may bear setose or branched processes (Fig. 51C); head prognathous and gula present; stemmata on each side 3; on plant surfaces .... **Coccinellidae (part)**
– Body elongate and more or less cylindrical, lightly sclerotised (Figs 42D, 91D); head hypognathous and gula absent; stemmata absent; tergum IX usually with median, posteriorly-projecting process (Figs 52W, 91D); in rotten wood, stems, fungal fruiting bodies .... **Mordellidae (part)**

Larva: Legs well developed. Legs with 5 or fewer segments. Labrum free. Mola absent. Urogomphi absent. Antennae 3-segmented

108(92) Abdominal apex with ventrally-hinged operculum (Figs 42A, 51H, 52Z) 109
– Abdominal apex without ventrally-hinged operculum 113
109(108) Body broadly ovate, strongly flattened and disc-like (Fig. 65I); head completely concealed from above by prothorax; antenna more than half as long as head width; frontal arms absent; anal gill tufts present (Figs 42A, 51H) .... **Psephenidae: Eubriinae (*Sclerocyphon*)**
– Body not broadly ovate and strongly flattened; head not concealed from above by prothorax; antenna less than half as long as head width; frontal arms present (Fig. 36D) 110
110(109) Operculum with paired hooks and anal gill tufts present (Fig. 42A); mesal surface of mandibular base usually with articulated, pubescent process as well as brush of hairs (Fig. 47B); on rocks, submerged wood, debris in streams, sometimes in lakes or ponds .... **Elmidae**
– Operculum without paired hooks (Fig. 52Z, 65J); anal gill tufts absent; mesal surface of mandibular base simple; in leaf litter, flood debris 111
111(110) Epicranial stem absent and frontal arms well separated at base (Fig. 36D); stipes free from labium but with narrow articulating area concealed by laterally expanded postmentum (Fig. 48O); spiracles on segment VIII placed at end of single dorsal, posteriorly-projecting process (Fig. 65J); in leaf litter, flood debris .... **Ptilodactylidae: Cladotominae (*Austrolichas*)**
– Epicranial stem present (sometimes very short) and frontal arms contiguous at base (Fig. K20); stipes completely fused to postmentum, the sutures between them incomplete or absent (Fig. 38D); spiracles on segment VIII not located on median, dorsal process 112
112(111) Cardines completely fused together forming single plate (Fig. 38D); spiracles on segment VIII laterally placed; sterna I–VIII well developed and transverse; body somewhat flattened above, with lateral tergal processes, and clothed with modified hairs (Fig. 27H); in leaf litter, flood debris .... **Chelonariidae**
– Cardines well separated by membranous area; spiracles on segment VIII dorsally placed (Fig. 52Z); sterna I–V narrowed, VI–VIII absent (fused with terga forming complete rings); body cylindrical and subglabrous; in leaf litter, flood debris; not occurring in Australia but known from Papua New Guinea, New Zealand and New Caledonia .... **Dryopidae**
113(108) Labial palps absent; body strongly curved ventrally (C-shaped) and lightly sclerotised (Fig. 34A); head strongly retracted (Fig. 45A); in seeds or pods (mainly Fabaceae) .... **Chrysomelidae: Bruchinae**
– Labial palps present 114

114(113) Median endocarina absent, coincident with epicranial stem (rarely Y-shaped at apex) or present in the absence of frontal arms **115**
– Median endocarina extending anteriorly between frontal arms (Fig. 45D–E) **136**
115(114) Maxilla without apical lobes; body very lightly sclerotised; stemmata on each side 1; in cells of native bees .... **Meloidae: Nemognathinae (part)**
– Maxilla with single mala (Figs 36A, 37A, 48D, L, 49M, U) **116**
– Maxilla with galea and lacinia (Fig. 49E, O, T) **129**
116(115) Labial palps 1-segmented **117**
– Labial palps 2-segmented **118**
117(116) Epicranial stem very short or absent; stipes longer than wide; gular region present (separating labium from thorax) (Fig. 48L); mesal surface of mandibular base with fixed, rigid hyaline or pigmented process; apex of mandible with single tooth (Fig. 47Q); stemmata on each side 3; on plant surfaces .... **Coccinellidae (part)**
– Epicranial stem moderately long (Fig. 45D); stipes wider than long; gular region absent (labium contiguous with thoracic membrane) (Fig. 49M); mesal surface of mandibular base simple; apex of mandible with 4 or more teeth in row (Fig. 41I); stemmata on each side 6; body often covered in slime; on plant surfaces .... **Chrysomelidae: Criocerinae**
118(116) Body strongly curved ventrally (C-shaped) (Fig. 33E); head strongly retracted (Fig. 45A); gular region absent (labium contiguous with thoracic membrane) (Fig. 49M); in soil among roots, in stems, in egg cases on vegetation .... **Chrysomelidae: Sagrinae**
– Body relatively straight or slightly curved ventrally; head protracted or slightly retracted (Fig. 45B) **119**
119(118) Head concealed from above by prothorax (Fig. 88J); mouthparts forming sucking tube enclosing stylet-like mandibles and maxillae (Fig. 46M); in leaf litter, rotting wood, fungi .... **Cerylonidae: Ceryloninae**
– Head not concealed from above; mouthparts not forming sucking tube **120**
120(119) Tergum IX with short to long, simple median process, body subcylindrical, head hypognathous and legs short, with fewer than 5 segments (Figs 52W, 91D) in rotten wood, stems, fungal fruiting bodies .... **Mordellidae (part)**
– Tergum IX with long, forked process; body strongly flattened; head prognathous and wider rhan thorax (Fig. 84D); legs well developed, 5-segmented; under bark .... **Cucujidae (*Platisus*)**
– Tergum IX without median process **121**
121(120) Head prognathous (Fig. 45B) **122**
– Head hypognathous (Fig. 45C) **128**
122(121) Stipes longer than wide **123**
– Stipes wider than long **127**
123(122) Epicranial stem absent (Figs 36D, 45E) **124**
– Epicranial stem present (Figs 36C, 45D, F) **125**
124(123) Hypostomal rods absent; frontal arms absent; pretarsus without setae; stemmata each side 1; abdominal terga simple; in cells of native bees .... **Meloidae (part)**
– Hypostomal rods present (Fig. 48H); frontal arms present (Fig. 45E); pretarsus with 1 seta; stemma on each side 3; abdominal terga with transverse rows of 6 protuberances, which may bear branched or setose processes (Fig. 51C); on plant surfaces .... **Coccinellidae (part)**
125(123) Frontoclypeal suture absent; hypostomal rods absent; abdominal segment X visible from above (Fig. 52D, T); apex of mandible not multidentate; apex of antennomere 2 oblique, so that sensorium arises proximal to antennomere 3 (Fig. 46S); usually under bark, in leaf litter or in rotten wood .... **Staphylinidae (part)**
– Frontoclypeal suture distinct (Fig. 36C); hypostomal rods present (Fig. 37A) OR mandibular apex multidentate (Fig. 39E; apex of antennomere 2 truncate, so that sensorium and antennomere 3 arise together (Fig. 36U) **126**
126(125) Hypostomal rods present, usually long and diverging posteriorly; body elongate, more or less parallel-sided, lightly sclerotised, sometimes with paired dorsal and ventral ambulatory ampullae; mandibular apex bidentate; stemmata on each side 5 or 0; in rotten wood, fungal fruiting bodies .... **Melandryidae (part)**
– Hypostomal rods absent; body oblong or ovate, without urogomphi; abdominal terga each with transverse row of 6 protuberances (Fig. 51C), which bear branched processes; apex of mandible multidentate (Fig. 39E); stemmata on each side 3; on plant surfaces .... **Coccinellidae: Coccinellinae (*Epilachna*)**
127(122) Ventral mouthparts strongly protracted (Fig. K92); gula much longer than wide (Fig. 49A); apex of antennomere 2 truncate, so that sensorium and antennomere 3 arise together (Fig. 46U); spiracles annular–biforous (Fig. 53M); tergum IX with sclerotised plate; in fungal fruiting bodies and stored products (predaceous) .... **Thanerocleridae**
– Ventral mouthparts retracted (Figs 36A, 37A–B); gula not or only slightly longer than wide; apex of antennomere 2 oblique, so that sensorium arises proximal to antennomere 3 (Fig. 46S); spiracles annular (Fig. 53L); tergum IX without sclerotised plate; under bark, in leaf litter, rotten wood, carrion, dung .... **Staphylinidae (part)**
128(121) Mandible broad and stout, more or less wedge-like, with single lobe or tooth at apex (Fig. 41P); body elongate, cylindrical and lightly sclerotised (Fig. 91D); gula absent (Fig. 49M); in rotten wood, stems, fungal fruiting bodies .... **Mordellidae (part)**
– Mandible more flattened, with 2 or more apical teeth; body oblong or ovate and somewhat flattened, often with transverse rows of dorsal protuberances (Fig. 51C); under loose bark, on surfaces of vegetation .... **Coccinellidae (part)**
129(115) Maxillary articulating area well developed but concealed behind lateral edge of expanded postmentum, which is divided longitudinally into 3 parts (Figs 48O, 65F); anal hooks 3 or more on each side (Fig. 65M); gular sutures separate; mesal surface of mandibular base with brush of hairs (Fig. 41N); body elongate, subcylindrical, with tergum IX rounded at apex (Fig. A39E); in plant debris in or bordering streams .... **Ptilodactylidae (*Byrrocryptus*)**
– Postmentum not divided longitudinally into 3 parts; if maxillary articulating area concealed behind expanded postmentum, gular sutures fused (Fig. 38C) and mesal surface of mandibular base with articulated process (Fig. 40B) **130**
130(129) Body broadly ovate, strongly flattened and disc-like, lined with lateral branched processes; tergum IX forming articulated spine (Fig. 65K); mandibles narrow and falcate, with internal perforation (Fig. 40G); under bark or stones in dry areas .... **Brachypsectridae (*Brachypsectra*)**
– Body neither broadly ovate nor strongly flattened; tergum IX not forming spine; mandibles never perforate **131**

131(130) Body lightly sclerotised and strongly curved ventrally (C-shaped) (Fig. 33E–F); head protracted and prognathous (Fig. 45B); anterior abdominal spiracles absent or non-functional, those on segment VIII forming pair of projecting spines (Fig. 52R); on roots or stems of aquatic plants .... **Chrysomelidae: Donaciinae (*Donacia*)**
– If body C-shaped, head strongly retracted (Fig. 45A) or hypognathous (Fig. 45C); all abdominal spiracles functional, those on segment VIII not forming projecting spines **132**
132(131) Spiracles annular (Fig. 53L) or annular–uniforous (Fig. 53H) **133**
– Spiracles biforous (Fig. 53P) **134**
133(132) Body strongly curved ventrally (C-shaped) and lightly sclerotised (Fig. 34A); head strongly retracted (Fig. 45A); most abdominal terga with 2 or more transverse plicae (Fig. 51D); abdominal segment X with pair of oval pads separated by longitudinal groove (Fig. 52V); vestiture consisting of fine hairs or setae; in dead wood .... **Bostrichidae (part)**
– Body relatively straight or slightly curved ventrally, usually more heavily sclerotised dorsally than ventrally (Fig. 63J–K); head protracted or slightly retracted (Fig. 45B); abdominal terga without transverse plicae; segment X without oval pads; vestiture including spicisetae(Fig. 46G) and/or hastisetae (Fig. 46H); in a variety of animal and plant products .... **Dermestidae (part)**
134(132) Epicranial stem very short or absent (Fig. 45M); stemmata on each side 5 or 6 forming a tight cluster; abdominal apex with with 3 or more anal hooks on each side (Fig. 65L); gular sutures fused (Fig. 38C); mesal surface of mandibular base with articulated process (Fig. 40B); in leaf litter .... **Ptilodactylidae: Ptilodactylinae (*Ptilodactyla*)**
– Epicranial stem moderately long (Fig. 36C); stemmata on each side 5 or 6 well separated and not forming tight cluster; abdominal apex without anal hooks or with 1 on each side; gular sutures separate (Fig. 36A); mesal surface of mandibular base simple or with brus of hair **135**
135(134) Maxillary articulating area present (Fig. 36A); cardines separated from each other by labium; pretarsal setae 2; mesal surface of mandibular base with brush of hairs (Fig. 47J); body either oblong, convex and heavily pigmented dorsally with tergum IX simple (Fig. 34N) or elongate, cylindrical and lightly pigmented with tergum IX forming concave plate (Fig. 65B); mosses, liverworts, soil, among roots .... **Byrrhidae**
– Maxillary articulating area absent (Fig. 37B); cardines contiguous, not separated by labium (Fig. 49E); pretarsal setae 1; mesal surface of mandibular base simple; body form not as above; in leaf litter, soil, flood debris .... **Limnichidae**
134(112) Head strongly retracted (Fig. 45A) **135**
– Head protracted or slightly retracted (Fig. 45B) **136**
135(134) Body relatively straight (Fig. 35G); occipital foramen apparently divided into 2 parts by tentorial bridge (Fig. 49J); epicranium with broad furrow for attachment of retractor muscles (Fig. 45G); gular region present (separating labium from thorax); under bark, in living or dead wood .... **Cerambycidae: Prioninae**
– Body strongly curved ventrally (C-shaped) (Fig. 33E–F); occipital foramen not apparently divided into two parts; epicranium without broad furrow; gular region absent (labium contiguous with thoracic membrane) (Fig. 49M); in stems, in soil among roots, in egg cases on vegetation .... **Chrysomelidae: Sagrinae**
136(134) Head prognathous (Fig. 45B); posterior edge of head capsule distinctly emarginate **137**
– Head strongly declined (hypognathous) (Fig. 45C); posterior edge of head capsule not emarginate **138**
137(136) Ventral mouthparts protracted (Fig. 49A); gular region elongate; hypostomal rods parallel (Fig. 37A); maxillary palps 3-segmented; labial palps 2-segmented .... **Cleridae (part)**
– Ventral mouthparts retracted (Figs 36A, 37A); gular region short and broad (Fig. 48D); hypostomal rods diverging (Fig. 48H); maxillary palps 2-segmented; labial palps 1-segmented .... **Chrysomelidae: Cassidinae (part)**
138(136) Maxillary mala cleft at apex (Fig. 49H); mesal surface of mandibular base with membranous, setose lobe (Fig. 41M); apex of tergum IX with median, spine-like process (Fig. 52W); in fungal fruiting bodies .... **Erotylidae: Erotylinae (*Microsternus*)**
– Maxillary mala not cleft (Fig. 48D, H); mesal surface of mandibular base simple or with brush of hairs (Fig. 47J); tergum IX without spine-like process **139**
139(138) Labial palps 1-segmented; dorsal and lateral surfaces very lightly sclerotised, without pigmented plates; abdominal sterna usually with projecting, setose lobes (Fig. 51D); in soil, associated with plant roots .... **Chrysomelidae: Eumolpinae**
– Labial palps 2-segmented; dorsal and lateral surfaces often with pigmented plates (Fig. 33G); abdominal sterna without projecting, setose lobes **140**
140(139) Terga VIII and IX more or less fused together and sclerotised forming terminal plate bearing dorsally placed and posteriorly facing spiracles; stemmata on each side 6; on plant surfaces .... **Chrysomelidae: Spilopyrinae**
– Terga VIII and IX not forming spiracle-bearing sclerotised plate; spiracles on segment VIII facing laterally **141**
141(140) Stemmata on each side 1 or 0; in leaf litter, soil, on plant surfaces .... **Chrysomelidae: Galerucinae (part)**
– Stemmata on each side 5 or 6; on leaf surfaces .... **Chrysomelidae: Chrysomelinae**

Larva: Legs well developed. Legs with 5 or fewer segments. Labrum free. Mola absent. Urogomphi present

142(91) Median endocarina absent, coincident with epicranial stem or present when epicranial suture is absent (Figs 36C–D, 45I, M, 46A–C) **143**
– Median endocarina extending between frontal arms (Figs 43A, 45D–E, G) **164**
143(142) Mesal surface of mandibular base simple or slightly expanded **144**
– Mesal surface of mandibular base with 1 acute hyaline processes (sometimes on one mandible only) (Fig. 47N, Q) **155**
– Mesal surface of mandibular base with 2 or more hyaline processes, which may be narrowly acute (Fig. 47L) or short and tooth-like (Fig. 47K) .. **160**
– Mesal surface of mandibular base with membranous lobe (Fig. 41M) **163**
144(143) Apex of mala or galea falcate (Figs 48H, 49O, T–U) **145**
– Apex of mala or galea rounded or truncate (Figs 36A, 37A, 48D, 49H) **146**
145(144) Epicranial stem very short or absent (Fig. 45M); maxilla with separate galea and lacinia (Fig. 49T); usually in leaf litter and fungi .... **Leiodidae (part)**

– Epicranial stem moderately long (Fig. 36C); maxilla with single mala (Figs 49H, 49U); in a variety of terrestrial habitats .... **Staphylinidae (part)**

146(144) Antennae 2-segmented; body elongate and subcylindrical, lightly sclerotised (Figs 90B, 91D); head strongly declined (hypognathous) (Fig. 45C) **147**

– Antennae 3-segmented **148**

– Antennae 4-segmented; urogomphi articulated at base (Fig. 52D, T); under bark, in leaf litter, dung, carrion, rotten wood .... **Staphylinidae (part)**

147(146) Legs with 3 or 4 segments, which are indistinctly separated, without pretarsal claw (Fig. 50C); ventral epicranial ridges absent; maxilla without laciniar lobe; antennal sensorium on preapical antennomere and not or only lightly longer than apical antennomere (Fig. 46U); in rotten wood, stems and some fungal fruiting bodies .... **Mordellidae (part)**

– Legs 5-segmented, including distinct pretarsal claw; ventral epicranial ridges present; maxilla with dorsal, subapical, laciniar lobe (Fig. 49G); antennal sensorium apical and much longer than apical antennomere (Fig. 46Q); in fungal fruiting bodies, under rotten bark or in rotten vines .... **Ciidae (part)**

148(146) Maxilla with separate galea and lacinia (Fig. 49O, S–T) **149**

– Maxilla with single mala (Fig. 49H, U) **150**

149(148) Epicranial stem very short or absent; mandibles broad at base, not falcate (Fig. 45M); labrum not subdivided; abdominal segments without lateral plates (Fig. 34G); galea without large, dense brush of hairs but often fimbriate (with fringe of setae) (Fig. 49T); in leaf litter, carrion, fungi .... **Leiodidae (part)**

– Epicranial stem moderately long (Fig. 36C); mandibles narrow and falcate (Fig. 47O); labrum subdivided into 3 or more sclerites (Fig. 46I); abdominal segments produced laterally forming tergal plates (Fig. 58K); galea with large, dense brush of hairs (Fig. 49S); in carrion .... **Silphidae**

150(148) Apex of antennomere 2 oblique, so that sensorium arises proximal to antennomere 3 (Fig. 46S); abdominal segment X distinct and usually visible from above and urogomphi usually articulated at base (Fig. 52T); abdominal terga bearing distinct plates, which are set off from those on sterna; under bark, in leaf litter, rotten wood, carrion, dung .... **Staphylinidae (part)**

– Apex of antennomere 2 truncate, so that sensorium and antennomere 3 arise together (Fig. 46U); abdominal segment X reduced and not visible from above (Figs 50M, 52W); urogomphi fixed at base; abdominal terga and sterna not forming distinct plates **151**

151(150) Head hypognathous (Fig. 45C), with long median endocarina and no frontal arms (Fig. 45I); gular region absent (labium contiguous with thoracic membrane) (Fig. 49M) **152**

– Head prognathous or slightly declined (Fig. 45B); gular region gular region present (separating labium from thorax) (Fig. 49N) **153**

152(151) Body somewhat flattened and widest anteriorly; dorsal surfaces with several tubercles bearing long, apically frayed setae (Fig. 46D); stemmata on each side 6; mandible elongate, apically acute, with serrate incisor edge; ventral mouthparts protracted (Figs 49A); stipes wider than long (Fig. K92); urogomphi long, slender, diverging and apically rounded and pubescent; on leaf surfaces .... **Phalacridae: Phaenocephalinae (*Phalacrinus*)**

– Body elongate and cylindrical; dorsal surfaces smooth and simple, with fine setae only; stemmata on each side 3 or fewer; mandible short and stout with simple incisor edge (Fig. 41P); ventral mouthparts strongly retracted and stipes longer than wide (Fig. 37A); urogomphi minute and hook-like; in stems .... **Mordellidae (part)**

153(151) Epicranial stem absent; paired endocarinae lying beneath frontal arms (Fig. 46A); frontoclypeal suture absent; hypostomal rods long and parallel (Fig. 37A); under bark, in rotten stems .... **Salpingidae: Salpinginae (part)**

– Epicranial stem moderately to very long (Fig. 36C); paired endocarinae absent; hypostomal rods short and divergent (Fig. 48H) or absent **154**

154(153) Frontal arms lyriform (Fig. 45F); endocarinae absent; antennae more than a third as long as head width; apex of mala cleft (Fig. 49H); upper surfaces with distinct pattern formed by darker and lighter pigment; urogomphi apically forked and with 3 pregomphi on each side (Fig. 93D); spiracles annular–cribriform; on plant surfaces, in leaf litter .... **Promecheilidae (part)**

– Frontal arms V-shaped (Fig. K20); endocarina Y-shaped, beneath epicarnial stem and frontal arms (Fig. 46C); antennae less than a third as long as head width apex of mala not cleft (Fig. 48D); body very lightly pigmented, often with paired dorsal and ventrolateral ampullae (Fig. 34B); urogomphi apically simple, small and hook-like or long and diverging, without pregomphi (Figs 34B–C, 91E); spiracles annular (Fig. 53L) .... **Melandryidae (part)**

155(143) Antennae 2-segmented **156**

– Antennae 3-segmented **158**

156(155) Head hypognathous (Fig. 45C); epicranial stem moderately long; labial palps separated by less than width of first palpomere; ventral epicranial ridges present (Fig. 37B); body subcylindrical, without enlarged abdomen (Fig. 90B); in fungal fruiting bodies .... **Ciidae (part)**

– Head prognathous or slightly declined (Fig. 45B); epicranial stem absent; labial palps separated by more than width of first palpomere; ventral epicranial ridges absent; body with enlarged abdomen (physogastric) **157**

157(156) Ventral mouthparts strongly protracted (Fig. 49A, J); hypostomal rods distinct, diverging; in tunnels of wood-boring insects (ectoparasitic) .... **Passandridae (late instar)**

– Ventral mouthparts retracted (Figs 36A, 37A); hypostomal rods absent; in tunnels of wood-boring insects (ectoparasitic) .... **Bothrideridae: Bothriderinae (*Deretaphrus*)**

158(155) Urogomphi articulated at base (Fig. 52T); mala stylet-like (Fig. 49Q); vestiture including frayed setae (Fig. 46D); head neither flattened nor parallel-sided and not much more heavily pigmented above than below; in leaf litter .... **Staphylinidae: Proteininae**

– Urogomphi fixed at base (Fig. 34I); mala rounded or truncate (Fig. 37A); vestiture consisting of simple setae; head flattened and subquadrate, much more heavily pigmented above than below so that head capsule appears to be open posteriorly behind ventral mouthparts (Fig. 43B) **159**

159(158) Hypostomal rods well developed and extending almost to posterior edge of head (Fig. 37A); stemmata on each side 6; paired dorsal abdominal glands absent; body relatively lightly sclerotised above and not parallel-sided; In sand dunes among roots .... **Phycosecidae**

– Hypostomal rods absent; stemmata on each side 5 or fewer; abdominal segments with paired dorsal glands (Fig. 51L); body usually parallel-sided, with paired pigment patches on thorax (Fig. I); in leaf litter, soil, under bark, in stems .... **Melyridae**

160(143) Ventral mouthparts strongly protracted and stipes wider than long (Fig. 49A); frontal arms distant at base, with paired subparallel endocarinae; tergum IX forming articulated plate (Fig. 52A); body elongate, narrow and flattened (Fig. 34K); under bark, on surfaces of fungi, in stored grains .... **Laemophloeidae (part)**

– Ventral mouthparts retracted (Figs 36A, 37A); stipes longer than wide; IF frontal arms distant at base, THEN endocarinae absent; tergum IX not forming articulated plate (Fig. 84E) **160**

161(160) Paired endocarinae absent; frontal arms distant at base (Fig. 36D); mesal surface of mandibular base with 2 hyaline processes, one acute and the other rounded and pubescent; thoracic terga and abdominal terga I–IX each with distinct pigmented plate (Fig. 84E); in nests of birds .... **Cavognathidae (*Taphropiestes*)**

– Paired endocarinae present and frontal arms contiguous at base (Fig. 46A); basal mandibular processes different; without pigmented plates on thoracic and abdominal terga **162**

162(161) Paired endocarinae coincident with frontal arms (Fig. 46A); abdominal terga without rows of asperities; sternum IX with pair of small anterolateral teeth (Fig. 53E); spiracles annular (Fig. 53L); under bark, in rotten stems, leaf litter .... **Salpingidae: Salpinginae (part)**

– Paired endocarinae lying between frontal arms; abdominal terga with paired rows of asperities forming open or closed rings (Fig. 51S–T); urogomphi with pit between them (Fig. 52F); spiracles annular–biforous (Fig. 53M); in rotting stems or fruits (possibly not Australian) .... **Zopheridae: Zopherinae: Monommatini**

163(143) Urogomphi articulated at base and abdominal segment X distinct and visible from above (Fig. 52T); mandibles more or less flattened, usually broad at base and narrow at apex (Figs 40I, 41K); antennomere 2 oblique at apex so that sensorium arises proximal to antennomere 3 (Fig. 46S); in leaf litter, rotten wood, fungi .... **Leiodidae (part)**

– Urogomphi fixed at base (Fig. K171); abdominal segment X concealed from above (Figs 34E, 50M); mandibles broad and wedge-like, not flattened (Fig. 41H, L); antennomere 2 truncate at apex so that sensorium and antennomere 3 arise together (Fig. 46U); in on on surfaces of fungal fruiting bodies .... **Erotylidae: Erotylinae (part)**

164(142) Head strongly retracted (Fig. 45A); in dead wood, under bark .... **Cerambycidae (part)**

– Head protracted or slightly retracted (Fig. 45B-C) **165**

165(164) Epicranial stem moderately long (Fig. 36C) **166**

– Epicranial stem absent or very short (Figs 36D, 45M) **170**

166(165) Maxilla with separate galea and lacinia (Fig. 49T); abdominal segment X distinct and visible from above (Fig. 63K); vestiture including barbed hairs (spicisetae) (Fig. 46G); mesal surface of mandibular base with narrow hyaline process and brush of hairs (Fig. 47M); spiracles annular (Fig. 53L); on animal products, including carrion .... **Dermestidae: Dermestinae (*Dermestes*)**

– Maxilla with single mala (Fig. 49H, U); abdominal segment X reduced and not visible from above; vestiture consisting of simple setae only **168**

168(167) Head prognathous (Fig. 45B); mesal surface of mandibular base with cluster of hyaline processes (Fig. 47K–L); dorsal surfaces lightly sclerotised and smooth; urogomphi minute; in fungus fruiting bodies .... **Trogossitidae: Trogossitinae (*Larinotus*)**

– Head hypognathous (Fig. 45C); mesal surface of mandibular base simple or with membranous lobe (Fig. 41M); dorsal surfaces granulate, tuberculate (Fig. 34E) or transversely plicate (Fig. 97H); urogomphi more well developed **169**

169(168) Mesal surface of mandibular base simple; antennae 2-segmented; abdominal terga transversely plicate and urogomphi long, slender and contiguous at base (Fig. 97H); spiracles annular (Fig. 53L); on plant surfaces .... **Chrysomelidae: Cassidinae (part)**

– Mesal surface of mandibular base with membranous lobe (Fig. 41M); antennae 3-segmented; dorsal surfaces granulate–tuberculate (Figs 34E, 84I); urogomphi short and stout; in or on fungus fruiting bodies .... **Erotylidae: Erotylinae (part)**

170(165) Stipes longer than wide; ventral mouthparts retracted (Fig. 37A) **171**

– Stipes wider than long; ventral mouthparts strongly protracted (Fig. 49A) **176**

171(170) Tergum IX with transversely divided plate and with median process between urogomphi (Figs 35L, 52K); mesal surface of mandibular base with hyaline process finely subdivided apically (Fig. 47L); ligula absent and labial palps very narrowly separated (Fig. 48D); stemmata on each side 2; under bark, fungal fruiting bodies, leaf litter, stored products .... **Trogossitidae: Lophocaterinae**

– Plate on tergum IX undivided, without median process between urogomphi; without other features combined **172**

172(171) Ventral epicranial ridges present (Fig. 37B); meso- and metatergum each with paired sclerites (Fig. 34I) **173**

– Ventral epicranial ridges absent; meso- and metaterga without paired sclerites **174**

173(172) Plate on tergum IX bordered by raised rim; stemmata on each side 3; under bark .... **Trogossitidae: Trogossitinae: Egoliini**

– Plate on tergum IX not bordered by raised rim; stemmata on each side 5; under bark, in rotten wood, fungal fruiting bodies, stored products .... **Trogossitidae: Trogossitinae: Trogossitini**

174(172) Urogomphi contiguous at base, arising from common base; labial palps 1-segmented; mesal surface of mandibular base with short, acute hyaline process (Fig. 47Q); in leaf litter .... **Trogossitidae: Peltinae: Rentoniini**

– Urogomphi not contiguous; labial palps 2-segmented; mesal surface of mandibular base with several hyaline processes (Fig. 47L) or apically subdivided process (Fig. 47I) **175**

175(174) Urogomphi more or less parallel (Fig. 52H); mesal surface of mandibular base with several acute hyaline processes (Fig. 47K); number of stemmata on each side 5; labial palps separated by more than width of first palpomere; spiracles annular–biforous (Fig. 53M) with short accessory chambers; in litter or on bark in burned areas .... **Acanthocnemidae (*Acanthocnemus*)**

– Urogomphi apically converging; mesal surface of mandibular base with apically subdivided hyaline process (Fig. 47I); number of stemmata on each side 2; labial palps separated by less than width of first palpomere; spiracles annular–biforous (Fig. 53M), with very long, curved accessory chambers; in leaf litter .... **Trogossitidae: Peltinae (*Parapeltis*)**

176(170) Hypostomal rods long and diverging (Fig. 49A); sternum IX completely concealed by sternum VIII (Fig. 52B); apex of mandible tridentate and mesal surface of mandibular base with several hyaline processes (Fig. 47L); tergum IX forming articulated plate; in rotting wood, fungi .... **Phalacridae (*Litochrus*)**

– Hypostomal rods, if present, short or subparallel (Fig. 48L); sternum IX partly or entirely exposed; apex of mandible with fewer than 3 teeth or lobes; tergum IX not forming articulated plate (Fig. 34H) .... **Cleridae (part)**

Larva: Legs well developed. Legs with 5 or fewer segments. Labrum free. Mola present

177(90) Tergum IX without urogomphi (Figs 33F–G, 34A, F, M, 35A–C, G) 178
– Tergum IX with urogomphi (Fig. 35D–F, L) 222
178(177) Antennae 1-segmented 179
– Antennae 2-segmented 181
– Antennae 3-segmented 187
– Antennae with 4 or more segments 219
179(178) Abdominal terga with patches of asperities (Fig. 51J–K); frontal arms absent; abdominal segment X beneath anus with pair of oval pads separated 'by longitudinal `groove (Fig. 52V); in fungus fruiting bodies, rotting wood, stems .... **Ptinidae (part)**
– Abdominal terga without patches of asperities; frontal arms present; abdominal segment X without oval pads separated by longitudinal groove 180
180(179) Epicranial stem very short (Fig. 45M); frontal arms lyriform (Fig. 45E–F); molar surface with numerous fine transverse ridges (Fig. 40E); ligula present (Fig. 48H, L); body broadly ovate, heavily pigmented dorsally and not strongly curved ventrally; on surfaces of rocks and bark covered with algae, moss and lichens .... **Tenebrionidae: Diaperinae: Leiochrinini**
– Epicranial stem moderately long and frontal arms V- or U-shaped (Figs 36C, 45F); molar surface simple (Fig. 41B); ligula absent (Fig. 48D); body subcylindrical, strongly curved ventrally (C-shaped) and lightly pigmented (Fig. 33E); in dead wood, stems, vines, fungal fruiting bodies .... **Anthribidae (part)**
181(178) Baloon-like spiracular gills present on raised tubes on abdominal segments I–VIII (Fig. 33C–D); length less than 2 mm; in sand, gravel or mud at edges of streams .... **Sphaeriusidae (*Sphaerius*)**
– Spiracular gills absent; length usually more than 2 mm; terrestrial habits 182
182(181) Body broadly ovate, strongly flattened and disc-like (Fig. 88B); head concealed from above by prothorax; maxillary palps 2-segmented; under bark, in fungi, leaf litter .... **Discolomatidae**
– Body not broadly ovate, flattened or not; head not concealed from above 183
183(182) Metathoracic leg reduced and 1-segmented (Fig. 42J); maxilla with galea and lacinia (Fig. 49O); spiracles cribriform (Fig. 53N); in rotten wood .... **Passalidae**
– Metathoracic leg not reduced; maxilla with single mala (Fig. 49D, H, U); spiracles not cribriform 184
184(183) Frontoclypeal suture present (Fig. 36C); hypostomal rods absent 185
– Frontoclyperal suture absent; hypostomal rods long and diverging (Fig. 48H) 186
185(184) Antennomere 2 distinctly setose (Fig. 91B); mandibular mola simple, with one or two ridges only; paired ventral gland openings present on abdominal segments I–III; without median gland opening near abdominal apex; in leaf litter .... **Tenebrionidae: Lagriinae (*Pseudolyprops*)**
– Antennomere 2 with few setae only; mandibular mola with numerous transverse ridges (Fig. 40E); abdominal segments I–III without paired gland openings; with dorsal, median gland opening between abdominal terga VIII and IX; in rotten wood .... **Tenebrionidae: Diaperinae (Gnathidiini)**
186(184) Frontal arms separated at base (Fig. 36D); mala rounded or truncate; abdomen with large, paired gland openings on abdominal segments I and VIII (Fig. 45L); body more or less oblong, fusiform; vestiture including expanded or clavate setae; in leaf litter, mouldy surfaces .... **Corylophidae: Corylophinae (*Sericoderus*)**
– Frontal arms contiguous at base (Fig. 45M); mala falcate (Fig. 49U); abdomen without gland openings; body elongate and parallel-sided; abdominal segment IX reduced, its sternum usually concealed and its tergum much shorter than those of segment VIII or X (Fig. 52U); vestiture consisting of simple setae only; under bark, in leaf litter, stored products .... **Silvanidae: Silvaninae**
187(178) Tergum IX with concave terminal disc, sometimes combined with complex armature (Fig. 52X) 188
– Tergum IX with posteriorly-projecting, median process (Fig. 52W) 189
– Tergum IX simple (Fig. 52V) 192
188(187) Head strongly declined (hypognathous) (Fig. 45C); antennae minute, less than 0.15 times as long as head width (Fig. 82B); mandibular molae more or less symmetrical; maxillary mala cleft (Fig. 49H); protergum with patches of asperities (Fig. 51K); boring in wood .... **Lymexylidae (part)**
– Head prognathous or slightly declined (Fig. 45B); antennae more than 0.15 times as long as head width; mandibular molae asymmetrical; maxillary mala not cleft (Figs 37A, 48D); protergum without patches of asperities .... **Tenebrionidae (part)**
189(187) Tergum IX forming articulated plate bearing median, forked process (Fig. 84D); sternum IX almost completely concealed by sternum VIII; body narrowly elongate, parallel-sided and strongly flattened; ventral mouthparts strongly protracted; mala falcate (Fig. 48H); under bark .... **Cucujidae (*Pediacus*)**
– Tergum IX not forming articulated plate; ventral mouthparts retracted and mala not falcate 190
190(189) Paired endocarinae present beneath frontal arms (Fig. 46A); frontoclypeal suture absent; hypostomal rods present (Fig. 37A); spiracles annular–biforous (Fig. 53M); median process on tergum IX sclerotised and more or less acute (Fig. 52W); under bark in rotten wood .... **Zopheridae: Colydiinae (part)**
– Paired endocarinae absent; frontoclypeal suture distinct (Fig. 36C); hypostoma rods absent; spiracles annular (Fig. 53L) 191
191(190) Process on tergum IX large, lightly pigmented, apically rounded, pubescent and deciduous (Fig. 90D); frontal arms lyriform (Figs 36D); maxillary mala cleft (Fig. 36D, 36H); in leaf litter .... **Scraptiidae: Scraptiinae**
– Process on tergum IX not as above; frontal arms V- or U-shaped (Fig. 36C); maxillary mala not cleft (Fig. 48D); under bark, in rotten wood, leaf litter .... **Tenebrionidae (part)**
192(187) Maxilla with distinct galea and lacinia (Fig. 49K, S–T) or with mala distinctly cleft (Fig. 49H) 193
– Maxilla with simple, undivided mala (Figs 36A, 37A, 48D) 200
193(192) Head retracted (Fig. 45A); body strongly curved ventrally (C-shaped) and lightly pigmented (Fig. 34A); most abdominal terga with transverse plicae (Fig. 51D); abdominal segment X with pair of oval pads separated by longitudinal groove (Fig. 52V); laciniar spine present (Fig. 49K); in dead wood, cane furniture 194
– Head not retracted (Fig. 45B–C) without other characters in combination 195
194(193) Spiracles on segment VIII much larger than those on I–VII .... **Bostrichidae: Lyctinae**

– Spiracles on segment VIII not larger than those on I–VII .... **Bostrichidae: Dinoderinae**

195(193) Maxilla with articulated galea and fixed lacinia (Fig. 49E, O); spiracles cribriform (Fig. 53N, S) **196**

– Maxilla with fixed galea and lacinia (Figs 49S–T) or mala distinctly cleft; spiracles never cribriform **199**

– Maxilla with fixed entire mala (Figs 36A, 37A, 49J, U) **201**

196(195) Head prognathous (Fig. 45B); epicranial stem and endocarina absent; mandible without accessory ventral process (Fig. 41O); cardines contiguous, not separated by labium, which is almost completely connate with maxillae (Fig. 49E); body not strongly curved ventrally, well pigmented and clothed with stiff, dark hairs; in sand and mud along streams .... **Heteroceridae**

– Head hypognathous (Fig. 45C); epicranial stem moderately long and median endocarina extending between frontal arms (Fig. 45D); stemmata absent or with single highly reduced pair; mandible with accessory ventral process (Fig. 41B); labium and maxillae not connate and cardines not contiguous; body strongly curved ventrally (C-shaped) and lightly pigmented, clothed with scatered fine setae (Fig. 34F) **197**

197(196) Anterior edge of labrum with 3 truncate lobes (Fig. 46K); antennomere 3 almost as long as 2, bearing sensorium and apical process representing fused antennomere 4 (Fig. 46T); stridulatory device formed from pro- and mesothoracic legs; in soil, dung .... **Hybosoridae**

– Anterior edge of labrum without 3 truncate lobes; antennomere 3 much shorter than 2, which bears 1 or more sensoria; stridulatory device, if present, formed from meso- and metathoracic legs (Fig. 42K) **198**

198(196) Tergum III with 3 transverse plicae, each bearing one or more rows of short, stiff setae; stridulatory device absent; head much more darkly pigmented than body (exept for protergal shield) (Fig. 34M); in animal carcasses .... **Trogidae**

– Tergum III with 2 transverse plicae, which do not bear rows of stiff setae; head and body very lightly pigmented (Fig. 34F); metathoracic legs sometimes greatly reduced and forming part of stridulatory device (Fig. 42K); in deep burrows in ground .... **Geotrupidae**

– Tergum III without distinct transverse plicae (Fig. 34F); meso- and metathoracic legs always forming stridulatory device (asperities on mesocoxa and metatrochanter) (Fig. 50I, K); head not much darker than body; in rotten wood or soil beneath dead wood .... **Lucanidae**

199(195) Head hypognathous (Fig. 45C); frontal arms V-shaped; median endocarina coincident with epicranial stem; body distinctly curved ventrally, with heavily sclerotised dorsal surfaces; prothorax with deep lateral fovea on each side; spiracles on segment VIII large, transverse and ventrally placed; in leaf litter (possibly living with ants) .... **Dermestidae: Orphilinae (*Orphilodes*)**

– Head prognathous or slightly declined (Fig. 45B); frontal arms lyriform (Fig. 46B); median endocarina absent; body widest anteriorly and narrowing posteriorly, with lightly sclerotised dorsal surfaces; prothorax without lateral foveae; spiracles on segment VIII laterally placed and not enlarged. **200**

200(199) Antenna less than 0.2 times as long as head width; sensorium much longer than antennomere 3; maxillary palps 2-segmented; labial palps 1-segmented; stemmata on each side 1; in leaf litter, hay stacks .... **Clambidae: Acalyptomerinae (*Acalyptomerus*)**

– Antenna more than 0.2 times head width; sensorium not much longer than antennomere 3; maxillary palps 3-segmented; labial palps 2-segmented; stemmata on each side 5; in myxomycete fruiting bodies, leaf litter .... **Eucinetidae**

201(195) Frontal arms distinctly separated at base (Fig. 36D); ventral mouthparts protracted (Fig. 49A); maxillary articulating area absent (Fig. 37B); prostheca consisting of 2 hyaline processes; body more or less parallel-sided and slightly flattened; dorsal surfaces lightly sclerotised (sometimes with vague pigment pattern) except for sclerotised plate on tergum IX; abdomen with paired gland openings on terga I–VII (Fig. 45L); in leaf litter on mouldy surfaces .... **Corylophidae: Corylophinae (part)**

– Frontal arms contiguous at base (Fig. 45F, M) or absent; ventral mouthparts almost always retracted (Figs 36A, 37A); IF protracted, THEN tergum IX without sclerotised plate); prostheca, if present, of a different type; paired gland openings not present on abdominal terga I–VII **202**

202(201) Prostheca a brush of simple or complex hairs (Fig. 48E) or a membranous lobe fringed with simple or complex hairs (Fig. 48M) **203**

– Prostheca a fixed, rigid, hyaline process (Figs 41K–L, 47A, 48A) **204**

– Prostheca absent (Fig. 39F, H) **213**

203(202) Stemmata on each side 6; body elongate, parallel-sided and circular in cross-section (Fig. 84G); ventral epicranial ridges present (Fig. 37B); spiracles on segment VII located dorsally near posterior edge of tergum; under bark, in rotten wood, fungi .... **Biphyllidae**

– Stemmata on each side 4 or fewer; body ovate to oblong and at least slightly flattened (Fig. 88G); ventral epicranial ridges absent; under bark, in fungal fruiting bodies, on bark surfaces .... **Endomychidae (part)**

204(202) Mala articulated (Fig. 39I); abdominal segment X with pair of long tubular pygopods (Fig. 84A); in flowers of Myrtaceae .... **Boganiidae (*Boganium*)**

– Mala fixed (Figs 36A, 37A); abdominal segment X without long, pygopods **205**

205(204) Abdominal segment X with pair of hooks (Fig. 52T); body elongate and cylindrical; length less than 1.2 mm; head hypognathous (Fig. 45C); stemmata absent; in pore tubes of bracket fungi .... **Ptiliidae: Nanosellinae**

– Abdominal segment X without paired hooks; without other features combined **206**

206(205) Body broadly ovate, strongly flattened and disc-like, its edges lined with barbed setae (Fig. 88A); head completely concealed from above by prothorax, with median endocarina (Fig. 45I); in leaf litter, decaying vegetation, stored products .... **Cerylonidae: Murmidiinae (*Murmidius*)**

– Body not broadly ovate and strongly flattened; head not conealed from above **207**

207(206) Prostheca broad and obtusely angulate (Fig. 47A) or somewhat rounded (Fig. 48M); apex of mandible often reduced and hyaline; outer edge of mandible with pair of long setae; ventral mouthparts often more or less protracted with hypostomal rods laterally situated; in leaf litter, stored products, associated with moulds and other fungi .... **Latridiidae (part)**

– Prostheca narrow with apex acute (Fig. 41K) or bifid (Fig. 48B), sometimes with serrate basal edge (Fig. 48A, F); mandibles different and hypostomal rods, if present, ventrally situated **208**

208(207) Mala more or less truncate (Fig. 48D) **209**

– Mala falcate (Fig. 48H) **211**

209(208) Head broad at base and strongly narrowed anteriorly (Fig. 88G); stemmata on each side 4; head moderately to strongly hypognathous (Fig. 45C); abdominal segment X ventrally oriented and not visible from above; under bark, on wood surfaces associated with fungi .... **Endomychidae (part)**

– Head not strongly narrowed anteriorly; stemmata on each side 6, occasionally weak or absent; head prognathous or slightly declined (Fig. 45B); abdominal segment X posteriorly or terminally oriented and more or less visible from above **210**

210(209) Antennomere 3 much longer than 2; pretarsus with single seta; in leaf litter, hay stacks, fruiting bodies of Myxomycetes .... **Clambidae: Clambinae**

– Antennomere 2 much longer than 3; pretarsus with 2 setae; in fruiting bodies of Myxomycetes .... **Sphindidae**

211(208) Body more or less ovate; thoracic and abdominal terga produced laterally; body densely clothed with long, fine hairs; mandible without accessory ventral process; in leaf litter, wood and fungi .... **Endomychidae: Anamorphinae (part)**

– Body elongate and more or less parallel-sided or widest anteriorly and narrowed posteriorly; thoracic and abdominal terga not produced laterally; mandible with accessory ventral process (Fig. 47A); body not densely clothed with long hairs **212**

212(211) Antennomere 3 shorter than sensorium; stemmata on each side 5; TAS, in leaf litter, hay stacks .... **Clambidae: Calyptomerinae (*Calyptomerus*)**

– Antennomere 3 much longer than sensorium; stemmata on each side 1; in leaf litter, grass piles .... **Cryptophagidae: Atomariinae (part)**

213(202) Head hypognathous, globular; prothorax enlarged and swollen dorsally (hump-like); body very elongate, with abdomen more than 5 times as long as thorax (Fig. 82B); patches of asperities on lateral portions of all thoracic segments and on apex of tergum IX, which is blunt and rounded; living and dead wood .... **Lymexylidae: Atractocerinae**

– Head prognathous or slightly declined (Fig. 45B); body not as elongate; without other characters combined **214**

214(213) Hypostomal rods absent (Fig. 36A); frontoclypeal suture present (Fig. 36C) OR head asymmetrical with long epicranial stem and Y-shaped endocarina (Fig. 46C) and mandibular molae highly asymmetrical with fine transverse ridges (Fig. 39H); length usually more than 10 mm **215**

– Hypostomal rods present (Fig. 37A); frontoclypeal suture absent; head symmetrical, without or with very short epicranial stem and lyriform frontal arms (Fig. 45F); mandibular mola occasionally asymmetrical, but without transverse ridges; length usually less than 10 mm **216**

215(214) Head asymmetrical; epicranial suture with Y-shaped endocarina lying beneath stem and bases of frontal arms (Fig. 46C); frontoclypeal suture absent or vaguely indicated; mandibular molae with fine transverse ridges (Fig. 39H); paired ventral prolegs (asperity-bearing ampullae) present on sterna II and III or II–IV (Fig. 51I); in rotten wood .... **Oedemeridae**

– Head symmetrical; Y-shaped endocarina absent; frontoclypeal suture present (Fig. 36C); mandibular molae usually without transverse ridges (Fig. 41B); abdominal prolegs absent; under bark, in rotten wood, soil, leaf litter, fungi .... **Tenebrionidae (part)**

216(214) Labial palps 1-segmented; spiracles annular–biforous (Fig. 53M); in flowers .... **Kateretidae**

– Labial palps 2-segmented; spiracles annular (Fig. 53L) **217**

217(216) Stemmata on each side 1, with well developed lens; antennomere 2 more than 2.5 times as long as 1; sensorium dome-like and partly surrounding antennomere 3; mola sub-basal, reduced, usually with hyaline lobe or brush of hairs at its base; pretarsal setae 2; dorsal surfaces smooth, clothed with scattered, long, simple setae; usually in leaf litter .... **Anthicidae: Tomoderinae**

– Stemmata on each side 3 or 0; antennomere 2 less than 2.5 times as long as 1; sensorium conical, often longer than segment 3; mola well developed and extending to base OR dorsal body surfaces more complex, usually with transverse rows of protuberances bearing groups of setae or simple or branched processes (Fig. 51C); pretarsal setae 1 .... **218**

218(217) Cardo and stipes fused; mola reduced, sub-basal and simple or asperate (Fig. 48K); dorsal surfaces of abdominal segments usually with transverse row of 6 protuberances, which may bear setose or branched processes (Fig. 51C); tibiotarsus enlarged with group of spatulate setae at apex; on plant surfaces .... **Coccinellidae**

– Cardo and stipes separate; mola well developed, tuberculate, and extending to base of mandible; dorsal body surfaces smooth, without protuberances of any kind, clothed with scattered, simple setae (Fig. 88K); tibiotarsus not enlarged, without spatulate setae; in soft fungi, under bark, in leaf litter .... **Endomychidae: Merophysiinae (*Holoparamecus*)**

219(178) Antennae with 6 or more segments (usually many) (Fig. 60H); body not strongly curved ventrally, somewhat flattened; prostheca complex, with sclerotised, comb-like process and brush of hairs (Fig. 48I); segment VIII more or less terminal, usually concealing IX from above and bearing enlarged spiracles; anterior spiracles absent or non-functional; in ponds, lakes, tree holes, wet rotten wood .... **Scirtidae**

– Antennae with 4 or 5 segments; body strongly curved ventrally (C-shaped) (Fig. 33F); prostheca absent (Fig. 41B); abdominal segment X terminal; spiracles all functional, those on segment VIII not enlarged and laterally placed **220**

220(219) Labrum with multidentate or strongly serrate apex (Fig. 46L); labial palps 1-segmented; length usually less than 10 mm; under bark, in rotten wood, termite nests .... **Hybosoridae (Ceratocanthinae)**

– Labral apex rounded, truncate or weakly trilobed (Fig. 46O); labial palps 2-segmented **221**

221(220) Stridulatory device consisting of patches or rows of asperities on mesocoxa (Fig. 50K) and metatrochanter (Fig. 50I); tergum III without distinct transverse plicae (sometimes weakly divided into 2 parts); anterior abdominal terga with patches of asperities or short, stout setae (Fig. 34F); galea and lacinia distinct; tormae united mesally (Fig. 46O); anal opening more or less vertical, bordered by 2 or 3 anal pads (Fig. 53F); in rotten wood .... **Lucanidae (part)**

– Stridulatory device absent on meso- and metathoracic legs; tergum III with 3 transverse plicae (Fig. 33F); IF anterior terga with patches of asperities, THEN galea and lacinia fused into single mala and tormae not united mesally; anal opening transverse or Y-shaped; in soil, roots, dung, rotten wood, animal nests .... **Scarabaeidae**

Larva: Legs well developed. Legs with 5 or fewer segments. Labrum free. Mola present. Urogomphi present

222(177) Sternum IX with row of apical asperities (Fig. 43I); body elongate, strongly flattened and lightly pigmented and head wider than thorax (Fig. 35K); stemmata absent; mandibles asymmetrical with well developed, finely, transversely ridged molae; tergum IX forming short, lightly pigmented, articulated plate with short widely separated urogomphi; in brown rotten wood .... **Prostomidae**

– Sternum IX with 2 apicomesal asperities (Fig. 53C); body slightly flattened and lightly pigmented; stemmata absent; mandibular mola with only 2 or 3 ridges; tergum IX not forming articulated plate, urogomphi each with accessory mesal process near base; in soil, leaf litter (based on exotic genus) .... **Anthicidae: Eurygeniinae**

– Sternum IX U-shaped with 1–3 apicolateral teeth at each end of the U (Fig. 52L); body strongly flattened; tergum IX forming articulated plate **223**

– Sternum IX with 1 basal asperity on each side (Fig. 53E) **224**

– Sternum IX with 2 to 6 basal asperities on each side (Fig. 53D) **225**

– Sternum IX with more than 6 basal asperities on each side, forming row (Fig. 53A) **226**
– Sternum IX simple **230**
223(222) Antenna less than half as long as head width; abdominal terga and sterna II–VI usually with paired longitudinal rows of asperities, usually curved and sometimes forming incomplete rings (Fig. 51T); mandibular molae somewhat reduced and sub-basal (Fig. 41E); under bark .... **Mycteridae: Eurypinae**
– Antenna more than half as long as head width; abdominal terga without rows of asperities; mandibular molae well developed, extending to base of mandible (Fig. 41J); in leaf axils of monocots .... **Mycteridae: Hemipeplinae (*Hemipeplus*)**
224(222) Urogomphi strongly, evenly curved dorsally and relatively lightly, evenly pigmented, each with 1 small tooth mesally near base; stemmata absent; mala not cleft; length less than 3.5 mm; in rotting vegetation .... **Salpingidae: Agleninae (*Aglenus*)**
– Urogomphi lightly pigmented and posteriorly projecting basally, heavily pigmented and dorsally curved at apex, each with 2 small teeth mesally near base (Fig. 52F); stemmata on each side 5; mala cleft (Fig. 49H); length greater than 3.5 mm; in sand or litter beneath vegetation on ocean beaches .... **Anthicidae: Lagrioidinae (*Lagrioida*)**
225(222) Paired endocarina contiguous with frontal arms (Fig. 46A); ligula absent and labial palps subcontiguous; body of tergum IX with urogomphi contiguous at base and distinctly diverging, with at least one accessory process (Figs 35D, 43E–F); under bark .... **Salpingidae: Inopeplinae (*Euryplatus*)**
– Paired endocarinae absent; ligula present and labial palps distinctly separated; tergum IX with bifurcate urogomphi (Fig. 91G) or multiple lobes; under bark, in rotten fern fronds, leaf litter .... **Salpingidae: Prostominiinae**
226(222) Row of basal asperities on sternum IX strongly, doubly curved (Figs 43L, 53B); tergum IX with complexly lobed urogomphi, a row of pregomphal tubercles and an interurogomphal pit (Fig. 43K); sternum IX with a pair of apicolateral asperities, in addition to basal ones; in rotten wood .... **Pythidae (*Anaplopus*)**
– Row of basal asperities on sternum IX straight or slightly sinuate (Fig. 43N) **227**
227(226) Tergum IX heavily sclerotised, forming articulated plate (Fig. 43M); urogomphi variable but always with 2 interurogomphal pits between them (Fig. 43O); antennae more than half as long as head width, with dome-like sensorium; body strongly flattened (Fig. 91L); hypostomal rods absent or short and diverging; mala cleft (Fig. 49H); spiracles annular (Fig. 53L); under bark .... **Pyrochroidae: Pilipalpinae**
– Tergum IX not heavily sclerotised, not forming articulated plate; urogomphi without or with single interurogomphal pit; antennae less than half as long as head width, with conical or palpiform sensorium (46U); body slightly flattened; hypostomal rods moderately long; IF spiracles annular (Fig. 53L), THEN hypostomal rods subparallel (Fig. 37A) **228**
228(227) Tergum IX with widely separated, bifurcate urogomphi, two rounded, posteriorly projecting lobes between them and a curved row of pregomphal tubercles; labial palps contiguous; spiracles annular (Fig. 53); under bark, in dead branches .... **Salpingidae (*Diagrypnodes*)**
– Interurogomphal lobes, if present, minute; tergum IX without curved row of pregomphal tubercles; labial palps not contiguous; spiracles annular-biforous (Fig. K197) **229**
229(228) Urogomphi narrowly separated at base, apically bifurcate, with one large mesal and two small dorsal accessory processes, and a pit between them (Fig. 35E); stemmata on each side 5; hypostomal rods diverging (Fig. 48H); under bark, in rotten stems .... **Salpingidae: Othniinae (*Elacatis*)**
– Urogomphi widely separated at base, apically bifurcate, with 2 min interurogomphal tubercles between them; accessory processes absent; interurogomphal pit absent; stemmata absent; hypostomal rods subparallel (Fig. 37A); in dead frons of fan palms .... **Salpingidae (part)**
230(222) Prostheca consisting of a brush of simple or complex hairs, a series of fringed membranes (Fig. 39A), or a fringed membranous lobe; cardines longitudinally oriented, labial palps 1-segmented and hypostomal rods absent (Fig. 49F); under bark, in leaf litter, carrion, rotting fruit, fungi .... **Nitidulidae (part)**
– Prostheca consisting of a simple membranous lobe (Fig. 48G) **231**
– Prostheca consisting of a fixed, rigid, hyaline process, sometimes partly sclerotised or appearing articulated (Figs 41K, 47A, 48A–B) **232**
– Prostheca absent (Figs 39F, H, 41B, H) **256**
231(230) Head strongly retracted (Fig. 45A); body strongly curved ventrally (C-shaped) (Fig. 81K); spiracles on segment VIII much larger than others; in dead wood .... **Bostrichidae: Lyctinae (first instar)**
– Head protracted or slightly retracted (Fig. 45B–C); body not strongly curved ventrally spiracles on segment VIII not enlarged; in fungus fruiting bodies .... **Erotylidae: Erotylinae (part)**
232(230) Urogomphi articulated at base (Fig. 52T); galea often with fringe of setae at outer apex (fimbriate galea) (Fig. 49T) **233**
– Urogomphi fixed at base (Figs 49M, 52F–H); maxilla usually with single, fixed mala (Fig. 49U) **236**
233(232) Abdominal segment X with pair of hooks (Fig. 52T) **234**
– Abdominal segment X without pair of hooks **235**
234(233) Urogomphi 2-segmented (Fig. 52T); stemmata on each side 3 to 5; epicranial suture present; in sand, on rocks, or in vegetation in or near streams or ponds .... **Hydraenidae**
– Urogomphi 1-segmented; stemmata on each side 0 or occasionally 1; epicranial suture absent; in leaf litter, decaying vegetation, rotten wood, fungi .... **Ptiliidae (part)**
235(233) Stemmata present; epicranial suture present; vestiture often including expanded setae (Fig. 46D); in leaf litter, fungi, carrion .... **Leiodidae (part)**
– Stemmata absent; epicranial suture absent; vestiture consisting of simple setae (Fig. 46F); body length 1.2 mm or less; in leaf litter, associated with ants .... **Ptiliidae: Cephaloplectinae**
236(232) Maxilla with fixed galea and lacinia (Fig. 49S–T) **237**
– Maxilla with single fixed mala (Figs 36A, 37A, 49H, U) **239**
237(236) Dorsal surfaces coarsely granulate or tuberculate, usually with distinct setiferous processes; stemmata on each side 6; mandible with accessory ventral process (Fig. 39C); abdominal spiracles placed at ends of spiracular tubes (Fig. 46E); on bark in association with sooty moulds, in leaf litter .... **Derodontidae: Laricobiinae (*Nothoderodontus*)**
– Dorsal surfaces smooth or very finely granulate, without setiferous processes; stemmata on each side 5 or fewer; mandible without accessory ventral process; abdominal spiracles not placed at ends of spiracular tubes **238**

238(237) Dorsal surfaces more heavily pigmented than ventral ones (Fig. 84G); antenna less than 0.2 times as long as head width; apex of antennomere 2 truncate, so that sensorium and antennomere 3 arise together (Fig. 46U); minute (usually less than 1.3 mm in length); under bark, in leaf litter, rotten wood .... **Jacobsoniidae**

– Dorsal and ventral surfaces similarly pigmented (Fig. 34G); antenna more than 0.2 `times as long as head width; apex of antennomere 2 oblique, so that sensorium arises proximal to antennomere 3 (Fig. 46S); in leaf litter, fungi, carrion .... **Leiodidae (part)**

239(236) Prostheca broad, with apex obtusely angulate or rounded (Fig. 47A) **240**

– Prostheca narrow, with apex acute or bifid (Fig. 48A–B) **246**

240(239) Mala truncate or rounded (Figs 48D, 49U), often cleft at apex (Fig. 49H); head moderately to strongly declined (hypognathous) (Fig. 45C); hypostomal rods very short (Fig. 48L); body form variable but usually with upper surfaces distinctly tuberculate (Fig. 34E) and urogomphi with accessory processes; in or on surfaces of fungal fruiting bodies .... **Erotylidae: Erotylinae (part)**

– Mala falcate (Fig. 49U) **241**

241(240) Median endocarina lying between frontal arms (Fig. 45E); body slightly flattened and upper surfaces relatively smooth, with few tubercles on tergum IX only; in smut-infested inforescences .... **Erotylidae: Pharaxonothinae (*Leucohimatium*)**

– Median endocarina absent **242**

242(241) Dorsal surfaces smooth; body more or less cylindrical and lightly sclerotised; epicranial stem moderately long (Fig. 36C); mandible without accessory ventral process; in stems .... **Erotylidae: Languriinae: Languriini**

– Dorsal surfaces granulate or tuberculate (Fig. 84K); body slightly flattened; epicranial stem very short or absent; mandible with accessory ventral process (Fig. 47A) **243**

243(242) Ventral epicranial ridges absent; frontal arms contiguous at base (Fig. 46B); pretarsus with 2 setae of unequal length lying one distal to the other, or occasionally with one seta only; thorax and abdomen usually with flattened lateral processes and urogomphi acute and approximate (Fig. 84K); in leaf litter, mouldy surfaces, decaying vegetation .... **Erotylidae: Cryptophilinae**

– Ventral epicranial ridges present (Fig. 37B); frontal arms separated at base (Fig. 36D); pretarsus with 2 subequal setae lying side by side **244**

244(243) Tergum IX with 2 pairs of long, slender processes; urogomphi moderately long, basally contiguous and diverging, each bearing several short, tooth-like accessory processes; in decaying flowers or seed pods .... **Erotylidae: Loberinae (*Loberus*)**

– Tergum IX mesally near base with pair of short, acute, hook-like processes; urogomphi shorter, separated at base; accessory processes longer or located at base only **245**

245(244) Urogomphi complex, with at least 5 accessory processes; hook-like processes near base of tergum IX joined by transverse row of small rounded tubercles; head with rows of granules forming pattern; in dead fronds of tree ferns, palms or cycads, in male cones of *Cycas* .... **Erotylidae: Languriinae (*Hapalips*)**

– Urogomphi relatively simple, with 2 small accessory processes near base; hook-like processes near base of tergum IX not joined by transverse row of tubercles; head without rows of tubercles; WA, in male cones of cycads (*Macrozamia*) .... **Erotylidae: Xenoscelinae (*Xenocryptus*)**

246(239) Mala truncate (Fig. 49H) **247**

– Mala falcate (Fig. 49U) **248**

247(246) Antenna less than 0.2 times as long as head width; labial palps 2-segmented; prostheca apically bifid (Fig. 48B); stemmata on each side 2; dorsal surfaces not tuberculate, clothed with expanded or frayed setae (Fig. 88E); in mouldy stored products .... **Endomychidae: Mycetaeinae (*Mycetaea*)**

– Antenna more than 0.2 times as long as head width; labial palps 1-segmented; prostheca apically acute (Fig. 41K); stemmata on each side 5; dorsal surfaces with numerous setiferous tubercles; vestiture of simple setae only; in fungal fruiting bodies .... **Hobartiidae (*Hydnobioides*)**

248(246) Labial palps 1-segmented **249**

– Labial palps 2-segmented **251**

249(248) Hypostomal rods absent; ventral epicranial ridges present (Fig. 37B); dorsal surfaces granulate or tuberculate and urogomphi not strongly upturned (Fig. 84L); in leaf litter, decaying vegetation, under bark, on spore-covered fungal surfaces .... **Monotomidae**

– Hypostomal rods long and diverging posteriorly (Fig. 48H); ventral epicranial ridges absent; dorsal surfaces smooth (Fig. 84B) or with elongate, setiferous tubercles (Fig. 84M); urogomphi strongly upturned (Fig. 50M) **250**

250(249) Dorsal surfaces smooth, without setiferous tubercles (Fig. 84B); spiracles not placed at ends of spiracular tubes; in leaf litter, rotten wood, fungi, stored products .... **Cryptophagidae: Cryptophaginae**

– Dorsal surfaces covered with elongate, setiferous tubercles (Fig. 84M); spiracles placed at ends of short spiracular tubes (Fig. 53O); in rotting fungal fruiting bodies .... **Hobartiidae (*Hobartius*)**

251(248) Thoracic and abdominal segments with lateral tergal processes; body more or less ovate (Fig. 84O) **252**

– Thoracic and abdominal segments without lateral tergal processes; body elongate and more or less parallel-sided (Fig. 34J) **253**

252(251) Abdominal spiracles located at ends of tergal processes; dorsal surfaces more or less granulate (often covered with dirt) (Fig. 84O); stemmata on each side 5; mandible with accessory ventral process; TAS, in leaf litter .... **Priasilphidae (*Priastichus*)**

– Abdominal spiracles not located at ends of tergal processes; dorsal surfaces not granulate; stemmata on each side 1 or 0; mandible without accessory ventral process; in fungi, rotten wood, lead litter .... **Endomychidae: Anamorphinae (part)**

253(251) Tergum IX (excluding urogomphi) much shorter than tergum VIII (Fig. 52E); segment X posteriorly oriented; body distinctly flattene; urogomphi approximate and more or less parallel (Fig. 34J); in leaf litter, stored products .... **Silvanidae: Brontinae (part)**

– Tergum IX (excluding urogomphi) about as long as tergum VIII; segment X posteroventrally oriented; body not or only slightly flattened **254**

254(253) Tergum IX with series of elongate, tubercles distributed around periphery, in addition to complex urogomphi (Fig. 84N); anterior abdominal terga bearing several shorter, flattened tubercles; spiracles placed at ends of spiracular tubes (Fig. 53O); in tree fern crowns, on vegetation .... **Protocucujudae (*Ericmodes*)**

– Tergum IX simple, except for pair of short, simple urogomphi; dorsal surfaces smooth (Fig. 84A); spiracles not placed at ends of spiracular tubes **255**

255(254) Stemmata on each side 6; mala articulated at base (Fig. 39I); segment X with pair of long, tubular pygopods (Fig. 84A); WA, in male cones of cycads (*Macrozamia*) .... **Boganiidae (*Paracucujus*)**

– Stemmata on each side 1 or 0; mala fixed at base (Fig. 48H); segment without long, tubular pygopods; in leaf litter, grass piles, stored products .... **Cryptophagidae: Atomariinae (part)**

256(230) Antennae 2-segmented **257**

– Antennae 3-segmented **258**

257(256) Head prognathous or slightly declined (Fig. 45B); antenna more than 0.2 times as long as head width, setose, with sensorium short and dome-like (Fig. 91B); frontoclypeal suture distinct (Fig. 36C); median endocarina absent; hypopharyngeal sclerome present (Fig. 50B, D); under bark, in rotten wood, soil, leaf litter .... **Tenebrionidae: Lagriinae (part)**

– Head moderately to strongly declined (hypognathous) (Fig. 45C); antenna less than 0.2 times head width, with palpiform sensorium longer than antennome 2 (Fig. 46Q); frontoclypeal suture absent or vaguely indicated; median endocarina coincident with epicranial stem; hypopharyngeal sclerome absent; in fungus fruiting bodies .... **Ciidae (part)**

258(256) Ventral mouthparts strongly protracted (Fig. 49A); stipes wider than long; cardo absent or membranous; sternum IX concealed (52B) or retracted within apex of sternum VIII (Fig. 52A) **259**

– Ventral mouthparts retracted (Figs 36A, 37A); stipes longer than wide (Fig. 48L); cardo distinct; sternum IX visible and not retracted within apex of sternum VIII (Fig. 53A–E) **260**

259(258) Head with paired endocarinae basally contiguous with frontal arms but extending anteriorly beyond them; body narrowly elongate and flattened (Fig. 34K); tergum IX forming articulated plate, heavily sclerotised; abdominal terga and sterna usually with paired, curved rows of asperities (Fig. 51S–T); sternum IX almost completely enclosed within an apical emargination of sternum VIII (Fig. 52A); hypostomal rods parallel; under bark, on surfaces of fungi, in stored grains .... **Laemophloeidae (part)**

– Head without paired endocarinae; median endocarina usually present between frontal arms (Fig. 45E) (apparently absent in *Phalacrus uniformis*); body usually more broadly elongate and wider at middle and only slightly flattened; tergum IX not forming articulated plate, relatively lightly sclerotised; terga and sterna without curved rows of asperities; sternum IX concealed by sternum VIII (Fig. 52B); hypostomal rods diverging (Fig. 49A); on surfaces associated with smuts, rusts and other fungi .... **Phalacridae (part)**

260(258) Mandibles symmetrical, left and right molae more or less similar in shape (Fig. 39A, G) **261**

– Mandibles asymmetrical, left mola differing considerably from right one (Fig. 39F, H) **275**

261(260) Labial palps 1-segmented; hypostomal rods very short or absent; body more or less cylindrical .... **Zopheridae: Colydiinae (*Nematidium* & *Pseudendestes*)**

– Labial palps distinctly 2-segmented **262**

262(261) Abdominal segment X distinct and visible from above (Figs 34G, 52T); urogomphi narrow, straight and almost always basally articulated (Fig. 52D); dorsal surfaces smooth; mola with pubescent lobe at base; antennae usually more than half as long as head widt; apex of antennomere 2 oblique, so that palpiform sensorium arises proximal to antennomere 3 (Fig. 46S); in leaf litter, carrion, fungi .... **Leiodidae (part)**

– If segment X visible from above, then dorsal surfaces granulate and mola without pubescent lobe at base; urogomphi usually curved and always fixed at base; antennae less than half as long as head width; apex of antennomere 2 truncate, so that sensorium and antennomere 3 arise together (Fig. 46U) OR sensorium dome-like (Fig. 46R) **263**

263(262) Mandibular mola sub-basal (Figs 41E, 48K), usually with hyaline lobe or brush of hairs at its base (Fig. 48C); stemmata on each side 1 or occasionally 0; pretarsus with 2 setae, which are either strongly asymmetrical or placed in line with one distal to the other **264**

– Mandibular mola basal, without hyaline lobe or brush of hairs; usually with more than one stemma on each side; pretarsus with 1 seta or with 2 setae which are more or less equidistant from pretarsal base **266**

264(263) Sensorium dome-like (Fig. 46R); hypostomal rods moderately long and diverging posteriorly; under bark or associated with fungi .... **Anthicidae: Lemodinae (*Lemodes*)**

– Sensorium conical or cylindrical (Fig. 46U); hypostomal rods absent, barely indicated or converging posteriorly (Fig. 48L) **265**

265(264) Median endocarina absent or located beneath frontal arms; hypostomal rods absent or barely indicated; urogomphi well separated at base, simple; in leaf litter, under bark .... **Aderidae (part)**

– Median endocarina extending between frontal arms (Fig. 45E); hypostomal rods long and converging; urogomphi more or less approximate at base, usually bearing tubercles or with a small interurogomphal pit (Fig. 52F); under bark, stones, in leaf litter .... **Anthicidae (part)**

266(263) Mandible with accessory ventral process (Fig. 47A); pretarsus with 1 seta; spiracles usually placed on long or short spiracular tubes (Fig. 53O) **267**

– Mandible without accessory ventral process; pretarsus with 2 setae; spiracles not placed on tubes **270**

267(266) Lateral tergal processes present on abdominal segments VI–IX; urogomphi short, widely separated and strongly upturned, with 2 pairs of truncate lobes between them; stemmata absent; vestiture of simple setae; in rotten wood, leaf litter .... **Bothrideridae: Anommatinae (*Anommatus*)**

– Lateral tergal processes, if present, on segments I–IX; urogomphi without 2 pairs of truncate lobes between them; vestiture including some expanded or frayed setae (Fig. 46D) **268**

268(267) Hypostomal rods absent; stemmata on each side 2; urogomphi long 'and narrow, straight or slightly curved, usually approximate at base and diverging apically, with 2 or more accessory processes (Fig. 88F); in rotten wood, leaf litter, fungi, ant nests, guano .... **Cerylonidae: Euxestinae**

– Hypostomal rods well developed, diverging posteriorly (Fig. 48H); stemmata on each side 5 or 6; urogomphi widely separated at base and strongly curved at apex **269**

269(268) Tergum IX with single round protuberance between urogomphi; spiracles placed on long tubes visible from above; dorsal surfaces without large tubercles; in fruiting bodies of Asocmycetes .... **Bothrideridae: Xylariophilinae (*Xylariophilus*)**

– Tergum IX without rounded protuberance between urogomphi; spiracular tubes much shorter, not visible from above; dorsal surfaces with pairs of large tubercles; in tunnels of ambrosia beetles(Curculionidae: Platypodinae) .... **Bothrideridae: Teredinae (*Teredolaemus*)**

270(266) Maxillary mala falcate (Fig. 49U); body elongate, cylindrical and lightly sclerotised; in stems .... **Erotylidae: Languriinae: Languriini (part)**

– Maxillaey mala rounded or truncate (Figs 36A, 37A) **271**

271(270) Frontoclypeal suture present (Fig. 36C); hypostomal rods absent; frontal arms always V- or U-shaped; spiracles annular (Fig. 53L), annular–multiforous (Fig. 53K) or occasionally annular–biforous (Fig. 53M), with very short accessory chambers **272**

– Frontoclypeal suture absent; hypostomal rods present (Fig. 37A); frontal arms lyriform (Fig. 36D); spiracles annular–biforous with well developed accessory chambers (Fig. 53M, U) **273**

272(271) Maxillary mala cleft (Fig. 49H); cardo distinctly divided; median endocarina Y-shaped, coincident with epicranial stem and frontal arms (Fig. 46C); spiracles annular–biforous (Fig. 47M), with very short accessory chambers; prosternum with patch of asperities (Fig. 56H); thoracic and abdominal terga with transverse patches of asperities (Fig. 51K); dead wood; based on exotic genus *Zopherus* .... **Zopheridae: Zopherinae: Zopherini**

– Maxillary mala simple (Fig. 48D, H); cardo not divided; median endocarina absent (Fig. 36C); spiracles annular (Fig. 53L) or annular–multiforous (Fig. 53K); prosternum without patch of asperities; thoracic and abdominal terga without transverse patches of asperities; under bark, in dead wood, leaf litter, soil, fungal fruiting bodies .... **Tenebrionidae (part)**

273(271) Median endocarina Y-shaped, lying beneath short epicranial stem and frontal arms, which meet at an angle less than 60° (Fig. 91H); hypostomal rods short and subparallel or slightly converging (Fig. 48L); labial palps separated by less than width of basal palpomere; thoracic and abdominal terga very lightly sclerotised, except for apex of tergum IX; metatergum and first few abdominal terga each with transverse patch of asperities near anterior edge (Fig. 91H); in dead wood .... **Zopheridae: Zopherinae: Pycnomerini**

– Median endocarina absent; frontal arms meeting at an angle greater than 60° (Fig. 36C); labial palps well separated; metatergum and abdomen without transverse patches of asperities **274**

274(273) Epicranial stem short but distinct; ligula as long as or longer than labial palp; all thoracic and abdominal terga evenly pigmented; tergum IX covered with small tubercles and urogomphi posteriorly oriented, slightly upturned and converging at apex (Fig. 91J), without pit between them; under bark, in rotten logs .... **Ulodidae (*Meryx*)**

– Epicranial stem absent; ligula shorter than labial palp; thoracic and abdominal terga usually unevenly pigmented, often with pigmentation restricted to tergum IX; urogomphi variable, often with pit between them (Fig. 52F); under bark, in rotten wood, leaf litter .... **Zopheridae: Colydiinae (part)**

275(260) Frontoclypeal suture distinct (Fig. 36C) **276**

– Frontoclypeal suture absent or vaguely indicated **278**

276(275) Hypostomal rods absent; abdominal spiracles annular (Fig. 53I, L) or annular–multiforous (Fig. 53K); under bark, in rotten wood, soil, leaf litter .... **Tenebrionidae (part)**

– Hypostomal rods present (Fig. 48H, L); abdominal spiracles annular-biforous (Fig. 53M) **277**

277(276) Epicranial stem very short (Fig. 45M); urogomphi straight, narrow, acute, lightly pigmented and posteriorly oriented, without accessory processes (Fig. 52H); hypostomal rods subparallel (Fig. 37A); ligula absent; molar surface tuberculate or asperate (Fig. 41H); dorsal surfaces smooth; anal region posteriorly or terminally oriented; in fungus fruiting bodies, leaf litter .... **Archeocrypticidae**

– Epicranial stem moderately long (Fig. 36C); urogomphi with accessory processes (Fig. 53D); hypostomal rods diverging posteriorly; ligula present; molar surface simple dorsal surfaces generally granulate or tuberculate; anal region ventrally or posteroventrally oriented; in fungus fruiting bodies, under bark .... **Ulodidae (part)**

278(275) Hypstomal rods absent **279**

– Hypostomal rods present (Figs 37A, 48H) **281**

279(278) Urogomphi with accesory processes and tergum IX with several additional spines or tubercles; anal region posteroventrally oriented; in rotten wood .... **Ulodidae (part)**

– Urogomphi simple and tergum IX without additional tubercles or processes; anal region posteriorly or terminally oriented **280**

280(279) Labial palps contiguous or separated by less than width of first palpomere; abdominal segment X concealed from above, distinct or more or less fused to segment IX (Fig. 50M); mola with hyaline lobe at base; urogomphi fixed at base, often heavily pigmented and apically acute (Fig. 90E); in leaf litter, under bark .... **Aderidae (part)**

– Labial palps separated by more than width of first palpomere; abdominal segment X distinct and visible from above (Fig. 34G); mola without hyaline lobe at base; urogomphi usually articulated at base; in leaf litter, fungi, carrion .... **Leiodidae (part)**

281(278) Epicranial stem long and bent to the left (Fig. 45K); frontal arms joined anteriorly by transverse ecdysial line; spiracles on segment VIII larger than those on VII; in fungus fruiting bodies; not occurring in Australia but known from New Guinea .... **Pterogeniidae**

– Epicranial stem shorter and straight; frontal arms not joined anteriorly by transverse line; spiracles on segment VIII not larger than those on VII; terga usually evenly pigmented and urogomphi dorsally curved and apically acute (Fig. 90F); in fungus fruiting bodies, leaf litter, stored products .... **Mycetophagidae**

3. FAMILY TREATMENTS

ARCHOSTEMATA

Archostemata shares with members of the Myxophaga and Adephaga the presence of an external propleuron (somewhat reduced in Torridincolidae, fused to the notum and sternum in Micromalthidae and not clearly indicated in available images of Jurodidae), the absence of cervical sclerites, and a type of hind wing venation (Fig. 24B) in which RA_{1+2} and RA_{3+4} do not meet apically to form a radial cell, cross-veins between RP and MP_{1+2} form an obongum cell, the radial bar bears a short, relatively abrupt bending zone (Fig. 29E), and a major transverse fold crosses the medial bar (MP_{1+2}) forming a sharp hinge. Most Archostemata differ from all other beetles in having a well developed, visible metatrochantin (Fig. 21D), but this absent in Micromalthidae and not clearly indicated in the available images of Jurodidae and Crowsoniellidae. As in Myxophaga, the wing apex is spirally rolled (Figs 23B, 29E). In contrast to Adephaga, adults have movable metacoxae and five (rather than six) ventrites and larvae are wood-boring, with large mandibular molae and a ligular sclerome. In Cupedidae, the testes are not tubular and there are four Malpighian tubules. Myxophaga differ in having the larval tibia and tarsus fused to form a tibiotarsus and in a suite of larval and adult adaptations to an aquatic or riparian existence. The suborder includes five families, two of which occur in Australia. Micromalthidae contains a single species, *Micromalthus debilis* LeConte, which is native to North America but introduced to various parts of the world in decayed structural timber (Philips and Young 2000; Beutel and Hörnschemeyer 2002b; Hörnschemeyer 2005, 2009). Crowsoniellidae is also monotypic with *Crowsoniella relicta* Pace known from a single locality in central Italy (Pace 1976; Crowson 1976; Hörnschemeyer 2005, 2009). Jurodidae is based on a few Jurassic fossils from the eastern Palaearctic and the extant *Sikhotealinia zhiltovae* Lafer from Far Eastern Russia. Although tentatively included in this suborder, the wing venation (based on photos and drawings) is of the polyphagan type and access to the single known specimen is limited (Lafer 1996; Kirejtshuk 1999; Hörnschemeyer 2005). [Ponomarenko 1969; Crowson 1975; Lawrence *et al.* 1987; Neboiss 1989; Galian and Lawrence 1993; Grebennikov 2004b].

1. OMMATIDAE

Adult (Figs 1A, 54A, C, 68A)

Length 6–26 mm. Body ~2.4–3 times as long as wide, usually widest at posterior third; sides of pronotum and elytra slightly, independently rounded, slightly flattened dorsally, distinctly so ventrally; colour uniformly dark brown to black; upper surfaces usually tuberculate, the tubercles spine-like in *Omma rutherfordi* Lawrence, and clothed with short, slightly thickened, decumbent setae in *O. stanleyi* Newman or with flattened, ribbed scales which differ in colour and form a distinct pattern in other species of *Omma*, and interspersed with stout, erect setae in *O. rutherfordi*.

Head prognathous, always constricted posteriorly to form a neck, but distinct temples absent in *O. rutherfordi*. Occipital region with moderately long endocarina. Eyes usually relatively small and not protuberant, larger and strongly protuberant in *O. stanleyi*, entire, very finely facetted, without interfacetal setae. Antennal insertions lateral, barely concealed by frontal ridges, which are strongly produced in *O. rutherfordi*; subantennal groove absent. Frontoclypeal suture absent. Labrum fused to clypeus. Antennae usually short (moderately long in *O. rutherfordi*), 11-segmented and filiform. Mandible short and broad, broadly tridentate, without mola or prostheca; dorsal surface near base with setose cavity. Maxilla highly reduced and partly concealed by mentum, galea and lacinia slender, hyaline and setose, the latter shorter, without uncus; palps moderately well developed; apical palpomere expanded and securiform with small cavity bearing sensilla near outer edge of upper surface (Fig. 16F). Mentum about as long as wide, concealing small, rounded ligula and palpal insertions; apical palpomere expanded and secufirom, with similar sensorial cavity. Gular sutures well separated; gula longer than wide. Corpotentorium weakly developed or absent. Cervical sclerites absent.

Pronotum ~0.75–1.0 times as long as wide, widest anteriorly with sides more or less straight posteriorly and weakly rounded anteriorly. Lateral pronotal carinae absent, anterior angles not produced forward; posterior angles right or slightly rounded; posterior edge straight; disc slightly convex with shallow transverse impression anteriorly and sometimes an additional one posteriorly. Prosternum in front of coxae distinctly longer than mid length of procoxal cavity, moderately convex, with transverse impression in *O. stanleyi*, partly or completely fused to pleuron (pleurosternal suture incomplete or absent), without paired crural impressions. Prosternal process incomplete, narrowed apically and acute at apex. Notopleural suture complete to anterior edge, never joined by pleurosternal suture. Procoxae subglobular, not strongly projecting, without concealed lateral extensions, without or with reduced coxal plates; trochantin large, well sclerotised, broadly exposed and abutting sternum and pleuron. Procoxal cavities slightly transverse, contiguous, broadly open externally and internally.

Scutellar shield somewhat abruptly elevated, anteriorly simple, posteriorly expanded and truncate. Elytra ~1.9–2.4 times as long as combined width and 3.6–4.8 times as long as pronotum, complete with independently subacute to acute apices (almost contiguous in *O. stanleyi*); disc with ten rows of window punctures and no scutellary striole; epipleura moderately broad and complete. Mesoventrite separated by complete sutures from mesanepisterna, which are well separated from one another; anterior edge on same plane as metaventrite; at middle with acute anterior projection separating paired procoxal rests extending onto mesanepisterna, mesoventral cavity absent; discrimen moderately long and transverse (mesokatepisternal) suture incomplete (not always visible externally). Mesoventral process divided into two short, acute processes.

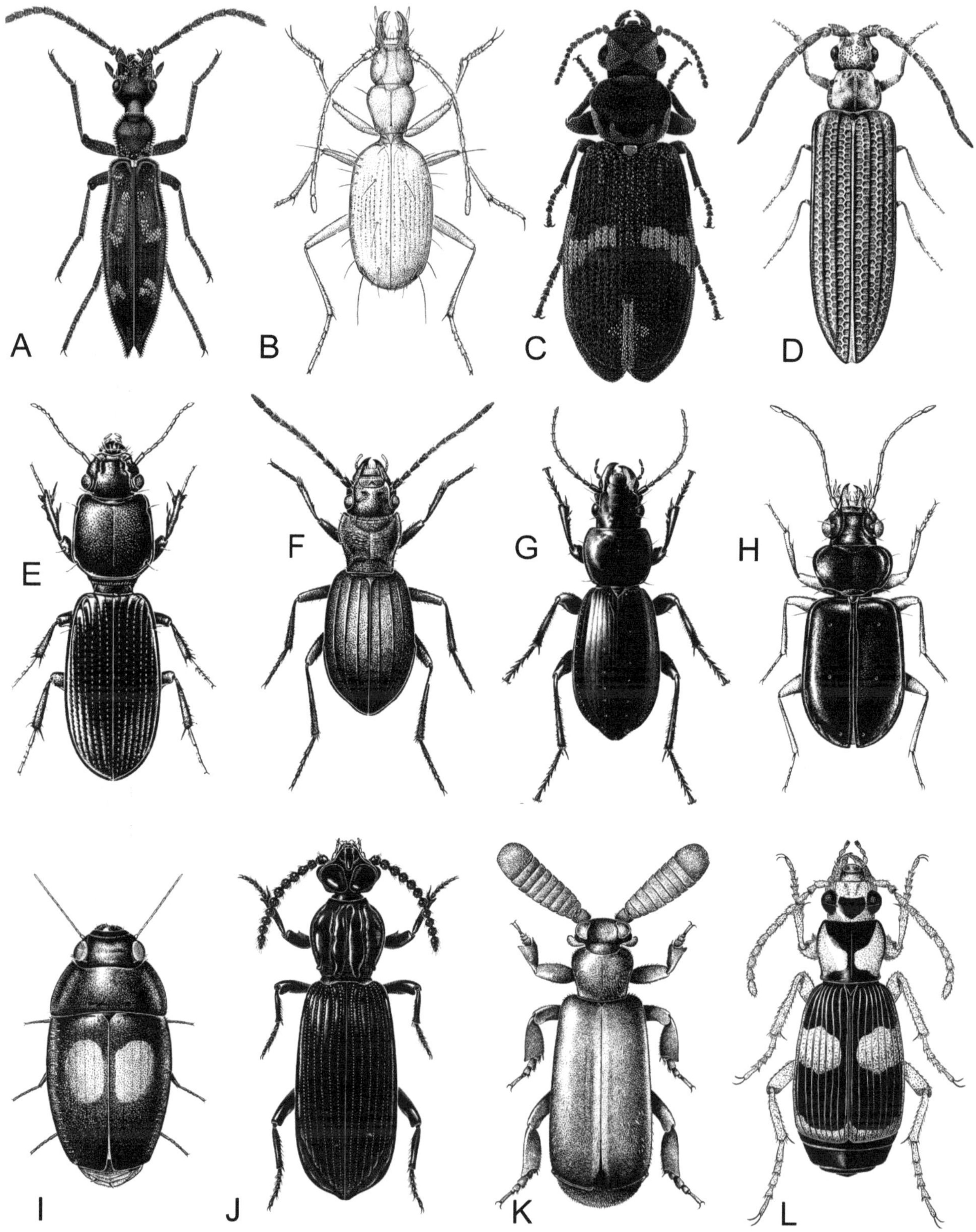

Fig. 54. Adult, dorsal: A, *Omma rutherfordi* Lawrence (Ommatidae); B, *Goedetrechus mendumae* Moore (Carabidae: Trechinae: Trechini); C, *Omma mastersi* Macleay (Ommatidae); D, *Distocupes varians* (Lea) (Cupedidae); E–L, Carabidae: E, *Clivina basalis* Chaudoir (Scaritinae: Clivinini); F, *Mystropomus subcostatus* Chaudoir (Paussinae: Ozaenini); G, *Notonomus* sp. (Harpalinae: Pterosichini); H, *Trechodes bipartitus* (Macleay) (Trechinae: Trechini); I, *Sphallomorpha biplagiata* (Laporte) (Pseudomorphinae); J, *Leoglymmius lignarius* (Olliff) (Rhysodinae); K, *Arthropterus* sp. (Paussinae: Paussini); L, *Pheropsophus verticalis* (Dejean) (Brachininae).

Mesocoxae more or less globular, not or slightly projecting, without or with short plates and broadly exposed trochantins. Mesocoxal cavities slightly transverse, contiguous, partly closed laterally by mesepimera and metanepisterna; meso- and metathoracic joint within coxal cavities membranous. Mesometaventral junction absent. Metaventrite slightly transverse, slightly convex, widest at posterior end and strongly narrowed anteriorly; discrimen about half as long as ventrite and transverse (metakatepisternal) suture well developed; metanepisternum completely exposed and about twice as long as widest point (near anterior end); exposed portion of metepimeron narrowly elongate and widest at posterior end; metatrochantins completely free and exposed. Metacoxae strongly transverse with subconical mesal projections, contiguous, extending laterally to meet elytra, with weak coxal plates. Metendosternite with a moderately long, slender stalk, short, broad ventrolateral processes, moderately long and distally slender lateral arms, a moderately long, apically emarginate anterior process with tendons moderately widely separated on either side of emargination. Hind wing (Fig. 24B) almost always well developed and moderately broad (reduced in males of *O. rutherfordi*), with very short bending zone near apex of radial bar (anterior wing strut), which does not extend to the wing apex; apical field very short, with three radial extensions (one RA and two RP); RA_{1+2} and RA_{3+4} separating just beyond mid-wing, the former extending to the end of the radial bar and the latter almost to the wing apex; pterostigma absent; two cross-veins join RA_{1+2} and RA_{3+4} to form brachial cell (Atkins 1963); RP complete almost to wing base, joined to RA_{3+4} by four cross-veins: r1 (basal and complete), r2 and r3 (incomplete) and r4 (apical and complete), portions of radial veins and cross-veins weakly sclerotised where crossed by triangular folding area; RP and MP joined by two cross-veins forming the oblongum cell, from the apical edge of which RP_2 and RP_{3+4} emerge; MP_{1+2} with very short bending zone or hinge and a very short, straight medial spur; medial field with four free veins and no medial fleck; base of MP_{3+4} complete, with two cross-veins between it and MP_{1+2}; one cross-vein between base of MP_{3+4} and CuA broken in middle by longitudinal fold; wedge cell well developed and apically truncate, CuA_1 joining MP_{3+4} but broken by fold, CuA_2, AA_3 reaching wing margin and AA_4 almost so; anal lobe well developed, without embayment; AP_{3+4} forked. Trochanterofemoral joint only slightly oblique. Legs short and slender; tibia subequal in length to femur; protibia with small setose excavation at basal third in males; tibial spurs well developed, usually unequal and paired on all legs. Tarsi 5-5-5; tarsomeres without ventral lobes; pretarsal claws simple; empodium bisetose.

Abdomen with five ventrites, which are flattened, more or less connate, separated by very narrow grooves and all on the same plane, with edges aparently abutting; ventrite 1 subequal in length to 2, with very weakly developed intercoxal process (sternite II with distinct but slender intercoxal process concealed from below by the mesal projections of the metacoxae). Abdominal tergites lightly sclerotised or membranous. Functional spiracles on abdominal segments I–VIII. Anterior edge of segment IX in male subtruncate; apex broadly rounded; tergite IX apically rounded; segment X free, forming a cylindrical protrusion bearing the anus. Aedeagus of modified trilobate type, with parameres completely fused to phallobase and partially fused together, each with a mesal notch near apex; penis with short basal struts. Sternite VIII in female with long, articulated spiculum ventrale. Ovipositor narrowly elongate, lightly pigmented except for baculi; paraprocts much longer than coxites, each of which is narrowed apically but not clearly subdivided, with well developed apical styli. [Neboiss 1960, 1989; Lawrence 1999; Lawrence *et al.* 1999b; Hörnschemeyer 2005].

Larva

Body elongate, widest near anterior end and gradually narrowed posteriorly, not ventrally curved, very lightly pigmented except for mandibles; surfaces smooth, clothed with scattered long setae.

Head prognathous and slightly retracted, narrower than prothorax, slightly transverse, weakly rounded laterally. Posterior edge of head capsule deeply emarginate; epicranial stem absent and frontal arms not apparent; endocarina thick and V-shaped with each arm of V forked at apex and with a slender arm at middle extending posterolaterally almost to the posterior edge of the head capsule. Stemmata four on each side, well separated. Frontoclypeal suture indistinct or absent. Labrum free, strongly transverse. Antennae moderately long, 4-segmented, with distinct antennifer; sensorium on antennomere 3 about half as long as antennomere 4. Mandibles short and broad, tridentate, with large, slightly asymmetrical and irregular molae, which lack asperities or transverse ridges; prostheca absent. Ventral mouthparts retracted forming maxillolabial complex, with well developed maxillary articulating area; cardo transverse and divided; stipes short and broad; galea articulated and palpiform, lacinia fixed and apically rounded with several stout spines and setae; maxillary palp 3-segmented with distinct palpifer. Labium consisting of prementum and postmentum, the former with a well developed ligular sclerome, slightly narrowed and emarginate at apex and continuous dorsally with hypopharyngeal sclerome; labial palps 2-segmented and widely separated. Hypostomal rods long, converging for anterior third, then diverging and extending almost to posterior edge of head capsule; ventral epicranial ridges absent. Gular region reduced.

Prothorax slightly longer than meso- and metathorax combined; meso-and metatergum with strongly transverse ambulatory ampulla; thoracic venter without distinct armature. Legs reduced, 6-segmented, with two pretarsal claws; coxae widely separated. Abdomen more than twice as long as thorax; transverse tergal ampulae present on first three segments; ampullae paired or reduced on posterior segments. Tergum IX subequal in length to VIII, apically rounded, without armature; sternite IX exposed, simple, not distinctly separated from segment X, which is posteriorly oriented; anal opening Y-shaped with dorsal and paired ventrolateral lobes. Spiracles annular, with closing apparatus. [Lawrence 1999; Lawrence *et al.* 1999a; Grebennikov 2004].

Classification and World Distribution

Recent members of the family Ommatidae occur only in Australia (Ommatinae) and in parts of southern South America (Tetraphalerinae). The South American *Tetraphalerus* Waterhouse is definitely known from the relatively arid parts of northern Argentina and Bolivia but may also occur in Brazil. [Monrós and Monrós 1952; Beutel *et al.* 2008; Hünefeld *et al.* 2011].

Australian Fauna

There are four species of *Omma* Newman in Australia. *Omma stanleyi* Newman occurs along the eastern coast but does not extend into the wet tropics, *O. mastersi* Macleay occurs mainly in the central and southern part of the continent, and *O. rutherfordi* Lawrence and *O. sagitta* Neboiss are southern and south-western in distribution. Ommatids are rarely collected. A male *Omma rutherfordi* was taken in a trough trap, while a female was found in a swimming pool filter basket. Adults of *O. stanleyi* have been found under loose bark, while a specimen *O. mastersi* was found running on a dead tree trunk. The scale colour patterns of *O. mastersi* and *O. sagitta* are similar to those of certain mutilid wasps and clerid beetles occurring in the same areas, suggesting a mimicry ring. The only known *Omma* larva, which could be that of *O. sagitta* or *O. rutherfordi*, was taken in soil at Red Hill, near Perth, and may have been feeding in underground roots or buried wood (Lawrence 1999).

2. CUPEDIDAE

Adult (Fig. 54D)

Length 6–17 mm. Body ~3.2–3.65 times as long as wide, parallel-sided and distinctly flattened; colour yellowish-brown to dark brown; upper surfaces densely clothed with decumbent and erect, expanded, scale-like setae which differ in colour and form a distinct pattern.

Head prognathous, always constricted posteriorly to form a narrow neck and distinct temples, which are somewhat shorter than eye length as seen from above. Vertex with moderately long endocarina beginning at anterior edge of neck. Eyes relatively large and more or less protuberant, entire, very finely facetted, without interfacetal setae. Antennal insertions dorsal and relatively narrowly separated, not concealed by frontal ridges, which are produced dorsally; subantennal groove absent. Frontoclypeal suture absent. Labrum free, very small, transverse to subquadrate and apically truncate. Antennae moderately long, 11-segmented and filiform. Mandible short and broad, subtriangular and ventrally concave, unidentate with one or two subapical teeth; without mola or prostheca. Maxilla highly reduced and partly concealed by mentum, galea and lacinia hyaline and setose, the former very slender, the latter broad, without uncus; palps moderately well developed; apical palpomere expanded and securiform with patch of sensilla lying on surface and not in a cavity. Mentum distinctly transverse, with curved transverse carina delimiting a subapical concavity; palpal insertions and prementum concealed by mentum, the latter consisting of two trilobed, membranous processes; apical palpomere expanded and secufirom. Gular sutures well separated; gula longer than wide. Corpotentorium weakly developed or absent. Cervical sclerites absent.

Pronotum ~0.7–0.75 times as long as wide, widest at about middle, and distinctly narrower at base than combined elytral bases; sides straight or sinuate and anterior edge produced forward and rounded. Lateral pronotal carinae complete, anterior angles usually slightly to moderately produced laterally (strongly produced forward in *Ascioplaga* Neboiss); posterior angles right or slightly rounded; posterior edge straight; disc with median longitudinal ridge separating a pair of broad lateral impressions. Prosternum in front of coxae more than twice as long as mid length of procoxal cavity, moderately convex, with paired oblique crural impressions extending mesally almost to midline, where separated by a short carina, and laterally across the pleuron and onto the pronotal hypomeron. Notopleural and pleurosternal sutures separated posteriorly on either side of elongate, slender, anteriorly acute pleuron, the sutures uniting just before the anterior edge and continuing as a short notosternal suture. Prosternal process complete, narrowed posteriorly with narrowly rounded apex. Procoxae subglobular, not strongly projecting, without concealed lateral extensions, with distinct coxal plates; trochantin large, well sclerotised, broadly exposed and abutting sternum and pleuron. Procoxal cavities slightly transverse, narrowly separated, broadly open externally and internally.

Scutellar shield abruptly elevated, anteriorly simple, posteriorly expanded and rounded to truncate. Elytra ~2.8–3.15 times as long as combined width and 6–6.85 times as long as pronotum, complete with independently or contiguously subacute; disc with nine rows of window punctures (Fig. 28M), with or without scutellary striole, usually with three longitudinal carinae (excluding lateral and sutural edges); epipleura narrow and complete. Mesoventrite separated by complete sutures from mesanepisterna, which are well separated from one another; anterior edge biemarginate, on same plane as metaventrite; at middle with acute anterior process behind which is a small mesoventral cavity discrimen short. Mesoventral process divided into two short, acute processes. Mesocoxae conical and projecting, with short plates and broadly exposed trochantins. Posterior portion of mesanepisterna, and entire mesepimeron concave, forming part of oblique crural impression for mid leg. Mesocoxal cavities about as long as wide, contiguous, partly closed laterally by mesepimera and metanepisterna; meso- and metathoracic joint within coxal cavities membranous. Mesometaventral junction absent. Metaventrite slightly as long as or longer than wide, slightly convex, widest posterior end and strongly narrowed anteriorly; discrimen more than half as long as ventrite and transverse (metakatepisternal) suture well developed; metanepisternum completely exposed and about three times as long as widest point (near anterior end), anterior concavity and longitudinal groove forming part of crural impression for mid leg; exposed portion of metepimeron narrowly elongate and slightly wider at posterior end; metatrochantins completely free and exposed. Metacoxae strongly transverse with subconical mesal projections, contiguous, extending laterally to meet elytra, with well developed coxal plates. Metendosternite with a long, slender stalk, short, narrow ventrolateral processes, short, broad lateral arms, and a very long and slender anterior process bearing subapproximate anterior tendons. Hind wing well developed and moderately broad, with very short bending zone near apex of radial bar (anterior wing strut), which does not extend to the wing apex; apical field very short, with three radial extensions (one RA and two RP); RA_{1+2} and RA_{3+4} separating just beyond mid-wing, the former extending to the end of the radial bar, the latter almost to the wing apex; pterostigma absent; two cross-veins join RA_{1+2} and RA_{3+4} to form brachial cell (Atkins 1963); RP complete to wing base, usually joined to RA_{3+4} by two cross-veins: r1 (very short and almost obliterated in *Adinolepis* Neboiss),and r4, portions of radial veins and cross-veins weakly sclerotised where crossed by triangular folding area; RP and MP joined by two cross-veins forming the oblongum cell, from the apical edge of which RP_2 and RP_{3+4} emerge; MP_{1+2} with very short bending zone or hinge and a short, straight medial spur extending to wing margin where there is a weak medial embayment; medial field with four or five free veins and no medial fleck; base of MP_{3+4} complete, with one cross-veins between it and MP_{1+2} and another one between base of MP_{3+4} and CuA broken in middle by longitudinal fold; wedge cell well developed and apically truncate, CuA_1 joining MP_{3+4} but broken by fold, CuA_2, AA_3 reaching wing margin and AA_4 almost so; anal lobe well developed, without embayment; AP_{3+4} forked. Trochanterofemoral joint only slightly oblique. Legs short and slender; tibia distinctly longer than femur; tibial spurs small and more or less concealed by apical setae, paired on all legs. Tarsi 5-5-5; tarsomere 4 lobed beneath and tarsomeres 1–4 densely setose beneath; pretarsal claws simple; empodium not apparent.

Abdomen with five ventrites, which are flattened, more or less connate, but slightly overlapping, with ventrites 2 to 5 each on a slightly higher plane than the preceding one; ventrite 1 distinctly shorter than 2, with well developed, narrowly acute intercoxal process (intercoxal process on sternite II less well developed). Abdominal tergites lightly sclerotised or membranous. Functional spiracles on abdominal segments I–VIII. Anterior edge of segment IX in male broadly rounded; posterior edge narrowly rounded or acute; tergite IX at least partly fused to X. Aedeagus of modified trilobate type, with parameres completely fused to phallobase and partially fused together, each with a small hook near apex; penis with short basal struts. Sternite VIII in female with long, fixed spiculum ventrale. Ovipositor narrowly elongate, lightly pigmented except for baculi; paraprocts much longer than coxites, each of which is narrowed apically but not clearly subdivided, with well developed apical styli. [Neboiss 1984, 1987; Lawrence 1999; Lawrence *et al.* 1999b; Young 2000; Hörnschemeyer 2005, 2009; Hörnschemeyer *et al.* 2006].

Larva (Figs 42F, 56D)

Body elongate, more or less parallel-sided, slightly wider near abdominal apex, not ventrally curved, very lightly pigmented except for mandibles and apex of tergum IX; surfaces smooth, clothed with scattered long setae.

Head (Figs 40A, J, 45I) prognathous and protracted, narrower than prothorax, slightly transverse, strongly rounded laterally. Posterior edge of head capsule deeply emarginate; epicranial suture not apparent; median endocarina simple, not forked, extending as far as clypeus. Stemmata usually absent (one on each side in first instar of *Distocupes* Neboiss). Frontoclypeal suture indistinct. Labrum free, strongly transverse. Antennae in late instars moderately long and usually 4-segmented, with distinct antennifer and sensorium on antennomere 3 about half as long as antennomere 4; antennae of first instar *Distocupes* 2-segmented with sensorium almost as long as apical antennomere. Mandibles short and broad, tridentate or broadly bidentate with an additional subapical tooth, with large, slightly asymmetrical and irregular molae, which lack asperities or transverse ridges; prostheca absent. Ventral mouthparts retracted forming maxillolabial complex, with well developed maxillary articulating area; cardo transverse; stipes slightly longer than wide; galea articulated and palpiform, lacinia fixed and apically rounded with several stout spines and setae; maxillary palp 3-segmented with distinct palpifer. Labium consisting of prementum and postmentum, the former with a well developed ligular sclerome (Fig. 40A, J), slightly narrowed and emarginate at apex and continuous dorsally with hypopharyngeal sclerome; labial palps 2-segmented and widely separated. Hypostomal rods long and diverging; ventral epicranial ridges absent. Gular sutures separated; gula transverse.

Prothorax slightly longer than meso- and metathorax combined; meso- and metatergum with strongly transverse ambulatory ampulla; thoracic venter with large field of asperities (Fig. 56H). Legs reduced, 6-segmented, with two unequal pretarsal claws or with claws fused together; coxae widely separated. Abdomen more than twice as long as thorax; transverse tergal ampulae present on most abdominal segments. Tergum IX slightly longer than VIII, apically narrowed to form narrow truncate process (Fig. 39J), with several tubercles at or near its base, extending onto ventral surface; sternite IX exposed, simple; segment X posteroventrally oriented, with pair of large, globular projecting lobes; anal opening Y-shaped, between these lobes and a weak transverse dorsal lobe (Fig. 39J). Spiracles annular, with closing apparatus. [Böving and Craighead 1931; Neboiss 1968; Lawrence 1991b; 1999; Lawrence *et al.* 1999a; Young 2000; Beutel and Hörnschemeyer 2002a; Grebennikov 2004; Hörnschemeyer 2005, 2009].

Classification and World Distribution

The family Cupedidae includes 36 extant species, placed in nine genera. The subfamily Priacminae includes *Priacma serrata* (LeConte) from north-western North America and *Paracupes brasiliensis* (Kolbe) from Brazil and eastern Ecuador. Cupedine genera include *Cupes* Fabricius from eastern North America, *Prolixocupes* Neboiss, with one species from the from Chile and Argentina and another from south-western North America, *Tenomerga* Neboiss, with species distributed throughout the Holarctic, Afrotropical and Oriental regions, the Afrotropical genus *Rhipsideigma* Neboiss, *Ascioplaga* Neboiss from Australia and New Caledonia and the Australian genera *Distocupes* Neboiss and *Adinolepis* Neboiss. The genera *Tenomerga* and *Distocupes* were synonymised with *Cupes* by Kirejtshuk (2005). [Vulcano and Pereira 1975; Neboiss 1984, 1987, 1989; Young 2002; Hörnschemeyer 2005, 2009].

Australian Fauna

Distocupes varians (Lea) occurs in eastern Australia, from south--eastern QLD to TAS, *Ascioplaga scalena* (Neboiss) is known only from North QLD, and *Adinolepis* species occur in both eastern and south-western Australia. Cupedid larvae bore into dead wood that has been attacked by wood-rotting fungi, but they may occur in relatively dry habitats. Larvae of *Distocupes varians* (Lea) have been collected in structural timber and also in old rotten pine logs on the ground. [Neboiss 1960, 1968, 1984, 1987; Lawrence 1999].

MYXOPHAGA

This small group of minute beetles includes four families, only one of which occurs in Australia. Hydroscaphidae occur in the Holarctic, Neotropical and Oriental regions and in North Africa and Madagascar, Lepiceridae are restricted to the Neotropical region (from Mexico to northern South America), and Torridincolidae are known from eastern and northern South America, southern and central Africa, Madagascar, Japan and China. Crowson (1960) considered this taxon to be the sister group of Polyphaga based mainly on the fusion of the larval tibia and tarsus; Ponomarenko

(1971a) and Lawrence and Newton (1982) suggested that Recent Myxophaga may be descendants of the schizophoroid Archostemata (Ponomarenko 1969), especially the Triassic and Jurassic Catiniidae. Myxophaga occur in various riparian or aquatic situations, such as in mud and sand in and at the edges of streams and ponds, in hot springs (some *Hydroscapha*) and on the surfaces of rocks which are covered with a moving film of water or constantly sprayed by waterfalls (hygropetric habitats). Adults of Torridincolidae may have a ventral plastron (Hinton 1969) and larvae of Torridincolidae, Sphaeriusidae and Hydroscaphidae possess spiracular gills (tracheal gills) (Hinton 1967a). [Reichardt 1973; Spangler 1980b; Satô 1982; Lawrence *et al.* 1987; Beutel and Haas 1998; Beutel 1999b, 2005b; Beutel *et al.* 1999; Hall 2001b].

3. SPHAERIUSIDAE

Adult (Fig. 55A)

Length 0.7–1.0 mm. Body ~1.34–1.37 times as long as wide, widest at middle; sides of pronotum and elytra uniformly rounded, strongly convex dorsally, somewhat flattened ventrally; colour uniformly black; upper surfaces subglabrous (sparsely clothed with very short, fine hairs) and shiny.

Head prognathous, short, broad and slightly flattened, not constricted posteriorly; posterior edge straight, without emarginatons. Occipital region without median endocarina or transverse ridge. Eyes well developed, longitudinally oval, entire, not or barely protruding and coarsely facetted, without interfacetal setae. Antennal insertions exposed and moderately widely separated; subantennal grooves extending below eyes. Frontoclypeal suture distinct. Labrum well developed, free, apically rounded. Antennae moderately long, 11-segmented with compact, 3-segmented club. Mandible short and broad, more or less flattened and partly concealed in lateral view by labrum and mentum, weakly, obliquely bidentate; incisor edge with a few serrations; mola well developed, asperate; prostheca well developed, membranous with several acute hyaline processes at apex process and left mandible with an additional forked sclerotised process. Maxilla reduced and partly concealed by labrum and mentum, with single, falciform apical lobe (mala or lacinia); palps moderately well developed, subulate, with apical palpomere much shorter and narrower than preapical one. Mentum about as long as wide, concealing bilobed, papillate igula and palpal insertions; apical palpomere shorter and narrower than preapical one. Gular sutures well separated; gula slightly transverse. Corpotentorium well developed, straight, located at posterior end of head capsule. Cervical sclerites absent.

Pronotum ~0.37–0.4 times as long as wide, as wide at base as combined elytral bases; sides strongly converging anteriorly and obliquely explanate, so that hypomera and propleura are concave. Lateral pronotal carinae complete, anterior angles produced forward and acute; posterior angles right to slightly acute; posterior edge straight; disc strongly convex and simple. Prosternum in front of coxae much shorter than mid length of procoxal cavity, concave laterally on each side of mesal shelf continuing posteriorly as prosternal process, which is more or less complete and narrowly rounded at apex. Notopleural and pleurosternal sutures not meeting anteriorly. Procoxae subcylindrical, not strongly projecting, without concealed lateral extensions, without coxal plates; trochantin fused to pleuron forming trochantinopleuron, which is external, abutting both sternum and notum and extending to anterior edge (Fig. 17J). Procoxal cavities strongly transverse, narrowly separated, broadly open externally, closed internally by slender bar.

Scutellar shield small, not abruptly elevated, anteriorly simple, posteriorly acute. Elytra ~1.03–1.07 times as long as combined width and 3.25–3.5 times as long as pronotum, complete with conjointly rounded apices; with lateral internal interlocking tongue at about middle; disc with irregular punctation and no scutellary striole; epipleura narrow and incomplete. Mesoventrite separated by complete sutures from mesanepisterna, which are well separated from one another; anterior edge on much higher plane than metaventrite, with large, vertical paired procoxal rests extending onto mesanepisterna; mesoventral cavity, discrimen and transverse (mesokatepisternal) suture absent. Mesoventral process short and truncate. Mesocoxae not projecting, without plates, with exposed trochantins. Mesocoxal cavities strongly transverse, very widely separated, partly closed laterally by mesanepisterna, mesepimera and metanepisterna, which are concave forming crural impressions for the mid legs; meso- and metathoracic joint within coxal cavities solidly fused, without suture. Mesometaventral junction a straight line (mesoventrite and metaventrite fused). Metaventrite slightly transverse, flattened and more or less parallel-sided; discrimen and transverse (metakatepisternal) suture absent; exposed postion of metanepisternum long and very slender; metepimeron concealed; metatrochantins absent. Metacoxae slightly transverse and contiguous, extending laterally to meet elytra, with large coxal plates, which are capable of concealing hind legs. Metendosternite with a short, broad stalk and long, slender lateral arms; anterior tendons not apparent. Hind wing (Fig. 22C–D) somewhat reduced, with very short radial bar (extending to about mid wing) with a short hinge-like bending zone near its apex; pterostigma absent; apical field moderately long, with three or four radial extensions; RA_{1+2} and RA_{3+4} separating near wing base, the former very short and extending only slightly beyond the hinge (no cross-veins join the two branches), the latter more or less obliterated basally by the triangular folding area but clearly indicated beyond this area and forking to form RA_3 (extending to anterior wing margin) and RA_4 (extending almost to wing apex); basal portion of RP completely absent, but apical extensions present, one extending almost to wing apex; RA and RP joined by single cross-vein (r4) and oblongum cell formed between RP and MP, which is more or less longitudinally oriented and gives rise to a long RP_2 and a very short RP_{3+4}; M_{1+2} with very short bending zone or hinge and no medial spur; medial field highly reduced, with two free veins (probably Cu and AA_3) and anal field absent; entire wing margin lines with long, specialised fringe hairs. Trochanterofemoral joint only slightly oblique. Legs short; profemora and mesofemora expanded at middle; tibia shorter than femre; protibia slightly expanded apically; tibial spurs single on all legs. Tarsi 3-3-3; tarsomeres very slender, without ventral lobes; pretarsal claws simple; empodium not apparent.

Abdomen (Fig. 28G) with three free, flattened ventrites, ventrites 1 and 3 much longer than 2, the former undivided but probably representing a fusion of sternites II–V, with large coxal cavities combined with crural impressions and with acute intercoxal process. Abdominal tergites lightly sclerotised or membranous. Functional spiracles on abdominal segments I–VIII, those on VIII

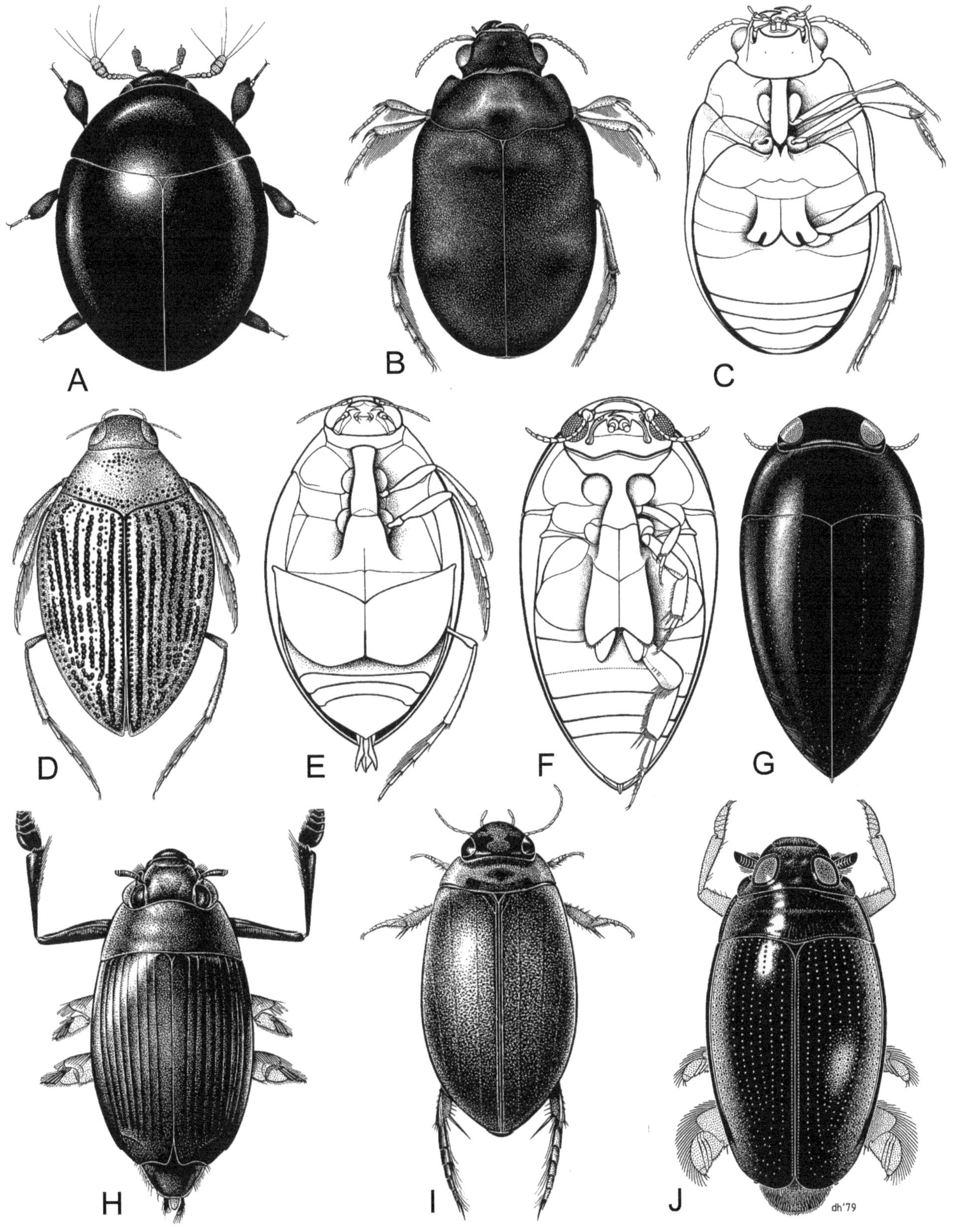

Fig. 55. Adult. A, *Sphaerius ovensensis* (Oke) (Sphaeriusidae), dorsal; B, *Hygrobia nigra* (Clark)(Hygrobiidae), dorsal; C, same, ventral; D, *Haliplus testudo* Clark (Haliplidae), dorsal; E, same, ventral; F, *Hydrocanthus australasiae* Wehnke (Noteridae), ventral; G, same, dorsal; H, *Macrogyrus oblongus* (Boisduval) (Gyrinidae), dorsal; I, *Rhantus suturalis* (Macleay) (Dytiscidae: Colymbetinae), dorsal; J, *Gyrinus convexiusculus* Macleay (Gyrinidae), from Watt *et al*. 1997; © Landcare Research, New Zealand.

enlarged and posteriorly oriented. Anterior edge of segment IX in male with paired struts; tergite IX completely divided into two separate laterotergites; segment X free. Aedeagus of modified trilobate type, with parameres completely fused to phallobase or apparently absent; penis without anterior struts. Sternite VIII in female without spiculum ventrale. Ovipositor short and broad, lightly pigmented except for baculi; paraprocts slightly longer than and situated lateral to coxites, which are undivided with longitudinal baculi and highly reduced, palpiform, apical styli. [Löbl 1995; Hall 2000a; Beutel and Arce-Pérez 2005].

Larva (Fig. 33C–D)

Body oblong, widest near anterior end and gradually narrowed posteriorly, not ventrally curved, dorsal surfaces lightly pigmented but somewhat darker than ventral ones; surfaces smooth except for scattered, short, fine transverse serrate ridges, with numerous long, lateral contact hairs.

Head large, prognathous and protracted at base, but with vertex distinctly declined, slightly transverse, widest behind middle, weakly rounded laterally. Posterior edge of head capsule not emarginate; epicranial stem absent; frontal arms lyriform and contiguous at base; endocarina absent. Stemmata four on each side, located in single tight cluster on lateral projections. Frontoclypeal suture indistinct or absent. Labrum free, large, transverse, broadly rounded. Antennae short, with well developed antennifer, posteriorly directed, 2-segmented, well separated from mandibular articulation, conical sensorium located at about middle of apical antennomere and surpassing its apex. Mandibles somewhat endognathous, short and broad, unidentate, with serrate incisor edge and large, symmetrical molae; prostheca absent. Ventral mouthparts moderately retracted forming maxillolabial complex, with narrow maxillary articulating area; cardo transverse and undivided; stipes slightly elongate; apex of mala falcate and cleft; maxillary palp 2-segmented without distinct palpifer. Labium consisting of prementum and postmentum; ligula bilobed with numerous microvilli, longer than labial palps, which are 1-segmented and widely separated. Hypostomal rods and ventral epicranial ridges absent. Gular region reduced.

Prothorax longer than meso- and metathorax combined; posterior edges of all thoracic terga and abdominal cterga I–VIII with dense fringe of posteriorly directed flattened setae; thoracic venter without distinct armature. Legs short, 5-segmented, with one pretarsal claw bearing a single seta; coxae widely separated. Abdomen less than twice as long as thorax. Tergum IX subequal in length to VIII, apically truncate, extending onto ventral surface, without armature; sternite IX exposed, simple; segment X posteroventrally oriented, with three membranous lobes, each bearing a pair of hooks. Thoracic spiracles non-functional or absent; abdominal spiracles on segments I–VIII forming balloon-shaped spiracular gills (Figs 33C–D, 51P), all dorsal and posterolaterally directed. [Britton 1966; Hinton 1967a; Beutel *et al.* 1999; Hall 2000a; Beutel and Arce-Pérez 2005].

Classification and World Distribution

The family Sphaeriusidae includes the single genus *Sphaerius* Waltl with ~25 species occurring on all continents and on Madagascar (Endrödy-Younga 1997; Beutel and Raffaini 2003; Beutel and Arce-Pérez 2005). A sister-group relationship between Sphaeriusidae and Hydroscaphidae was supported by cladistic analyses based on adults and larvae (Beutel 1999b; Beutel *et al.* 1999). The replacement of the family Microsporidae with Sphaeriusidae is based on a decision of the International Commission on Zoological Nomenclature (2000).

Australian Fauna

There are two described Australian species: *Sphaerius ovensensis* (Oke) from VIC and NSW and *S. coenensis* (Oke) from northern QLD. Both adults and larvae usually occur in wet mud, sand or gravel at the edges of streams, where they probably feed on algae. The adult has no plastron but may store air beneath the elytra. Löbl (1995) described four Himalayan species collected well away from water on the floor of a broad-leaf forest or from moss above tree line.

ADEPHAGA

Adephaga is a relatively large, monophyletic group of highly specialised beetles that differ from Archostemata and Myxophaga in having a secondary mesal articulation of the procoxae, relatively immobile metacoxae, six abdominal ventrites, the first three of which are connate and the first of which is divided by the metacoxae, and pygidial defence glands in the adult, and liquid-feeding mouthparts with a fused labrum and no mandibular molae in the larva. There are also four Malpighian tubules, tubular, coiled testes and polytrophic ovarioles. Most of species are predacious, with exceptions in the mycophagous rhysodine Carabidae, the algophagous Haliplidae and Noteridae and some phytophagous species among the harpaline Carabidae. Adephaga have been the subjects of several major comparative anatomical studies, such as Dressler and Beutel (2010) on the adult head, Forsythe (1982) and Evans and Forsythe (1985) on the adult feeding system, Hlavac (1971) and Regenfuss (1975) on the antenna cleaner, Baehr (1979), Bell (1966, 1967), Beutel (1986), Beutel and Belkaceme (1986), Evans (1977, 1982, 1985), Hlavac (1973, 1975) and Larsén (1966) on the thorax and locomotory adaptations, Hammond (1979) and Ward (1979) on the hind wing, Forsyth (1968, 1970, 1972) on the pygidial glands, Bils (1976) and Burmeister (1976, 1980) on the ovipositor, Ruhnau (1986), Beutel (1993), and Arndt (1993, 1998) on larval morphology. In addition, the chemistry of defence glands has been studied comparatively by Moore (1979) and Dettner (1985). These and other studies have been accompanied by cladograms or have stimulated by new phylogenetic hypotheses, and in recent years morphological cladograms have been combined with those based on molecular data (e.g. Maddison *et al.* 2000, 2008; Shull *et al.* 2001; Ober 2002; Ribera *et al.* 2002b; Hunt *et al.* 2007). Traditionally, the suborder has been divided inito Geadephaga and Hydradephaga, comprising the terrestrial and aquatic forms, respectively; however, the Trachypachidae (including the Holarctic *Trachypachus* Motschulsky and the Chilean *Systolosoma* Solier) appears to have at least some adult features of Hydradephaga combined with a terrestrial carabid-like larva. Furthermore, trachypachids are among the earliest Adephaga in the fossil record, occurring in the Triassic along with haliplid and gyrinid-like forms (Ponomarenko 1977). Evans (1982, 1985) suggested that aquatic Adephaga may have arisen three times (Haliplidae, Gyrinidae, remaining Hydradephaga) from a terrestrial ancestor and the Triassic eodromeine Trachypachidae may have given rise to both aquatic dytiscoids and all modern terrestrial Adephaga. Ponomarenko presented a similar scheme but considered the ancestral adephagans to have been aquatic. In a scenario proposed by Beutel and Roughley (1988), Gyrinidae was the first group to invade the aquatic habitat, followed by Haliplidae and then the dytiscoid complex (including Trachypachidae), the modern members of which are considered to be secondarily terrestrial. Other workers have proposed phylogenies in which Trachypachidae is the sister group of all Hydradephaga (Bell 1966, 1967; Ruhnau 1986) or Hydradephaga excluding Haliplidae (Beutel and Belkaceme 1986; Roughley 1981); while Nichols (1985) considered Trachypachidae and all aquatic groups to form a clade sister to *Omophron* Latreille, which in turn was placed within a derivative group of Carabidae. In most recent molecular phylogenies of Adephaga (Shull *et al.* 2001; Maddison *et al.* 2008) Geadephaga is monophyletic with Trachypachidae at its base. The following adephagan families do not occur in Australia: Meruidae (northern South America) (Spangler and Steiner 2005; Beutel *et al.* 2006, 2010d; Balke *et al.* 2008), Aspidytidae (China and southern Africa) (Ribera *et al.* 2002a; Ribera and Beutel 2003; Beutel *et al.* 2010c), Amphizoidae (western North America, China and North Korea) (Kavanaugh 1986; Burmeister 1990a; Xie 2000; Philips and Xie 2000; Dettner 2005b) and Trachypachidae (Holarctic and Chile) (Arndt and Beutel 1995; Ball 2000; Beutel and Arndt 2005). [Lawrence *et al.* 1987; Beutel 1995a, 1997, 1998; Arndt *et al.* 2005; Beutel and Ribera 2005; Beutel *et al.* 2010b].

4. GYRINIDAE

Adult (Figs 55H, J, 66L)

Length 4–18 mm. Body ~1.7–1.95 times as long as wide, more or less elliptical, streamlined, widest at about middle; sides evenly curved; slightly flattened to moderately convex dorsally, flattened to slightly convex ventrally; colour usually uniformly black or black with yellow sides, sometimes coppery-brown or with blue or green tinge; upper surfaces smooth and shiny, glabrous.

Head prognathous and protracted, strongly transverse, not abruptly constricted posteriorly, with short occipital region; short, weak median endocarina in *Aulonogyrus* Regimbart and *Gyrinus* Müller. Eyes (Fig. 14A) large, completely divided into upper and lower sections by broad strip of cuticle, finely facetted, without interfacetal setae. Antennal insertions concealed from above. Frontoclypeal suture distinct. Labrum free, transverse, broadly rounded. Antennae 8-, 10- or 11-segmented, short and highly modified, with cup-like scape, slightly longer, flattened pedicel narrowly rounded and densely lined with long, fine hairs, and compact flagellum consisting of six to nine segments. Mandible short and broad, broadly bidentate with two teeth widely separated and joined by a transverse ridge, incisor edge sometimes with an additional tooth; mola and prostheca absent. Maxilla with slender galea in Gyrinini, absent in Enhydrini; lacinia with falciform apex and numerous setae along inner edge; palps short, with apical palpomere fusiform or somewhat expanded apically. Mentum transverse and deeply emarginate apically, enclosing prementum; apical labial palpomere fusiform or slightly expanded apically. Gular sutures well separated, connected by oblique sutures to posterolateral corners of mentum; gula slightly longer than wide; median endocarina very short in the form of an apodeme and located anteriorly, close to the base of the mentum. Corpotentorium very slender. Cervical sclerites absent.

Pronotum ~0.3–0.36 times as long as wide, widest at posterior end with sides weakly rounded and converging anteriorly. Lateral pronotal carinae well developed, with distinct bead; anterior edge usually bisinuate (not in *Gyrinus*); anterior angles distinctly produced forward and acute; posterior angles slightly acute; disc slightly to strongly convex and simple. Prosternum in front of coxae distinctly shorter than mid length of procoxal cavity, produced forward and angulate at midline and strongly tumid or carinate. Prosternal process relatively long but incomplete, parallel-sided and narrowly rounded at apex. Procoxae not or slightly projecting, with or without short, concealed lateral extensions, without coxal plates; trochantin free, small, exposed or concealed. Procoxal cavities slightly transverse, narrowly separated, broadly open externally and closed internally by slender bar.

Scutellar shield not abruptly elevated, anteriorly simple, posteriorly broadly angulate to acute. Elytra ~1.15–1.45 times as long as combined width and 4.1–4.95 times as long as pronotum, incomplete with apices usually subtruncate (independently broadly rounded in *Aulonogyrus*), exposing most or all of tergite VIII and sometimes part of tergite VII; disc usually impunctate, irregularly punctate or with numerous transverse impressions (with eleven distinct puncture rows and no scutellary striole in *Gyrinus*), often with about eight weak longitudinal impressions; epipleura well developed, usually moderately broad, but complete only to posterior truncation, variously excavate as part of crural impression for fore legs. Mesoventrite large and about as long as wide, somewhat diamond-shaped, separated by complete sutures from mesanepisterna, which are well separated from one another; anterior edge on different plane than metaventrite, with paired, vertical procoxal rests extending onto mesanepisterna; main portion of disc moderately convex, with anterior angulate process on each side of which is an oblique carina extending posterolaterally from its base and delimiting a portion of the crural impression for the fore leg; mesoventral cavity absent; discrimen moderately long; transverse (mesokatepisternal) suture absent. Mesoventral process divided into two short, subacute processes. Mesocoxae subcylindrical, not projecting, without plates and with concealed trochantins. Mesocoxal cavities strongly transverse and oblique, narrowly separated, partly closed laterally by mesepimera; meso- and metathoracic joint within coxal cavities membranous. Mesometaventral junction monocondylic or dicondylic. Metaventrite reduced and strongly transverse, subdivided into three sections joined by narrow strips of cuticle: the mesal one diamond-shaped and the lateral ones triangular; discrimen short to long; transverse (mesokatepisternal) suture absent; exposed portion of metanepisternum relatively short and broad, much wider anteriorly; metepimeron usually concealed; metatrochantins concealed. Metacoxae (Fig. 26A) greatly enlarged, slightly transverse, their bases transverse and apices oblique, mesally fused together, extending laterally to meet elytra, with weak, oblique coxal plates. Metendosternite with a long, slender stalk, no lateral arms, a moderately long, slender, anterior process with tendons moderately widely separated on either side of a deep emargination. Hind wing with long bending zone occupying most of radial bar (anterior wing strut); apical field moderately long, with as many as six radial extensions (three RA and three RP); RA_{1+2} and RA_{3+4} separating just beyond mid-wing, the former extending along anterior edge of wing, the latter abruptly curving posteriorly and giving rise to RA_3 and RA_4, both of which extend to wing apex; one cross-vein joins RA_{1+2} and RA_{3+4}; pterostigma well developed, occupying most of the area beyond this cross-vein and between RA_{1+2} and RA_{3+4} and RA_3; RP complete almost to wing base, joined near mid-wing to RA_{3+4} by cross-vein r1; portions of RA_{3+4} and a middle section of RP weakly sclerotised or obliterated where crossed by triangular folding area; cross-vein r2 absent and r3 and r4 incomplete stubs; RP and MP joined by two cross-veins forming oblongum cell, from the apical edge of which RP1, RP_2 and RP_{3+4} emerge; MP_{1+2} with very short bending zone or hinge and a very short, straight medial spur; medial field with four free veins and no medial fleck; MP1+2 near base of wing with cross-vein to Cu, broken in middle by longitudinal fold; MP_{3+4} with long basal cross-vein but not complete to origin; wedge cell usually absent (present in *Dineutus* Macleay), anal lobe well developed, without embayment; AP1+2 present; AP_{3+4} not forked. Fore legs very long and slender, fitting into crural impressions involving mesothorax, metathorax and elytral epipleura; with femur slightly longer than tibia, usually with row of long spines on inner edge; tibial spurs absent; tarsomeres short, slightly longer than wide to transverse, simple in females, with ventral adhesive disks in males. Mid and hind legs short, broad and flattened, with trochanterofemoral joint strongly oblique; femur and tibia subequal in length, apically expanded and subtriangular, the latter with swimming hairs at the outer angle and paired spurs at the inner angle; tarsi also flattened and paddle-like with fringe of swimming lamellae; tarsi 5-5-5, tarsomere 1 subtriangular, 2 and 3 strongly transverse, 4 and 5 elongate, but the latter much shorter; pretarsal claws simple.

Abdomen with seven ventrites (sternites II–VIII), the first two of which are solidly fused, usually with the suture more or less obliterated, the second two connate but separated by a complete suture, and the remaining ones freely articulated and overlapping; lateral portion of ventrite 1 shorter than 2, which extends to midline but lacks a distinct intercoxal process. Abdominal tergites VII–VIII well sclerotised. Functional spiracles on abdominal segments I–VIII, the last two located on tergites. Anterior edge of segment IX in male broadly rounded; tergite IX and segment X apparently fused and membranous. Aedeagus of modified trilobate type, with parameres completely fused to phallobase and partially fused together, apically rounded; phallobase with short manubrium; penis without or with very short basal struts. Sternite VIII in female simple, without spiculum ventrale. Ovipositor short and broad, lightly pigmented except for except for paraproctal baculi; coxites well sclerotised and setose, without styli. [Hatch 1925; Ochs 1949; Larsén 1966; Forsyth 1968; Beutel and Roughley 1988, 2005; Beutel 1989a, 1989b, 1990b; Burmeister 1990b; Roughley 2000a].

Larva (Fig. 56E)

Body narrowly elongate, more or less parallel-sided and somewhat flattened, lightly pigmented except for head capsule and protergal plate; surfaces smooth, clothed with scattered long setae.

Head prognathous and protracted, distinctly longer than wide, about as long as prothorax; slightly flattened with sides weakly rounded; occipital constriction present or absent. Posterior edge of head capsule distinctly emarginate; epicranial stem moderately long and frontal arms V-shaped; median endocarina absent. Stemmata six on each side, arranged in a tight cluster consisting of two vertical rows of three. Frontoclypeal suture absent. Labrum completely fused to head capsule; paranasalia weakly developed; nasale with two to four teeth. Antennae more than half as long as head width, slender, 4-segmented, without sensorium on antennomere 3. Mandibles long, slender and curved, with simple, acute apex and internal channel; mola and prostheca absent. Ventral mouthparts protracted; maxillae and labium well separated at base, without maxillary articulating area; cardo usually longer than wide (about as long as wide in *Gyrinus*); stipes short and broad; galea articulated, palpiform, 2-segmented; lacinia slender and falcate; maxillary palp 3-segmented with segment-like palpifer. Labium consisting of prementum and postmentum; labial palps 2-segmented with elongate, segment-like palpigers. Hypostomal rods and ventral epicranial ridges absent. Gular region very long; gular sutures fused together.

Thoracic segments about equal in length, prothorax slightly narrower than meso- or metathorax, with pigmented plate; protergal plate and its venter with distinct, undivided presternum. Legs very long, 6-segmented, with elongate coxa and two pretarsal claws; coxae widely separated. Abdomen more than three times as long as thorax; segments I–VIII each with pair of long, slender, posterolaterally projecting, articulated gills, segment IX with two pairs of similar gills. Segment IX slightly shorter and narrower than VIII, its tergum not extending onto ventral surface; sternite IX exposed, simple. Segment X shorter and narrower than IX, more or less tubular and posteriorly oriented, bearing 4 long hooks at apex (Fig. 52J). Spiracles annular, with closing apparatus, present on segments I–III in final instar. [Böving and Craighead 1931; Bertrand 1972; Spangler 1991; Beutel and Roughley 2005].

Classification and World Distribution

Gyrinidae contains three subfamilies: Spanglerogyrinae, with a single species *Spanglerogyrus albiventris* Folkerts from southern Alabama, USA, Heterogyrinae with a single species from Madagascar, and Gyrininae, occurring in most regions of the world and containing eleven genera in three tribes: Gyrinini, Enhydrini and Orectochilini. [Steiner and Anderson 1981; Beutel 1989a, 1989b, Beutel 1990b; Beutel and Roughley 1988, 1994; Burmeister 1990b; Roughley 2000a; Miller and Bergsten 2012].

Australian Fauna

Gyrinidae are widely distributed in Australia and are represented by *Gyrinus convexiusculus* Macleay, *Aulonogyrus strigosus* (Fabricius), two species of *Dineutus*, and 15 species of *Macrogyrus*. The last genus, including our largest species, was divided into seven subgenera by Ochs (1949) and five species groups by Watts and Hamon (2010). Adults and larvae usually inhabit still water of ponds and lakes. Adults are gregarious, usually active in daylight, and known for the way in which they gyrate, or swim in tight circles on the water surface, with the aid of surfactants secreted by pygidial glands (Vulinec 1987). They feed mainly on insects which have fallen into the water, and they locate prey using the Johnston's organ (located in the antennal pedicel), which is sensitive to surface waves. Larvae are bottom predators, which remain submerged and

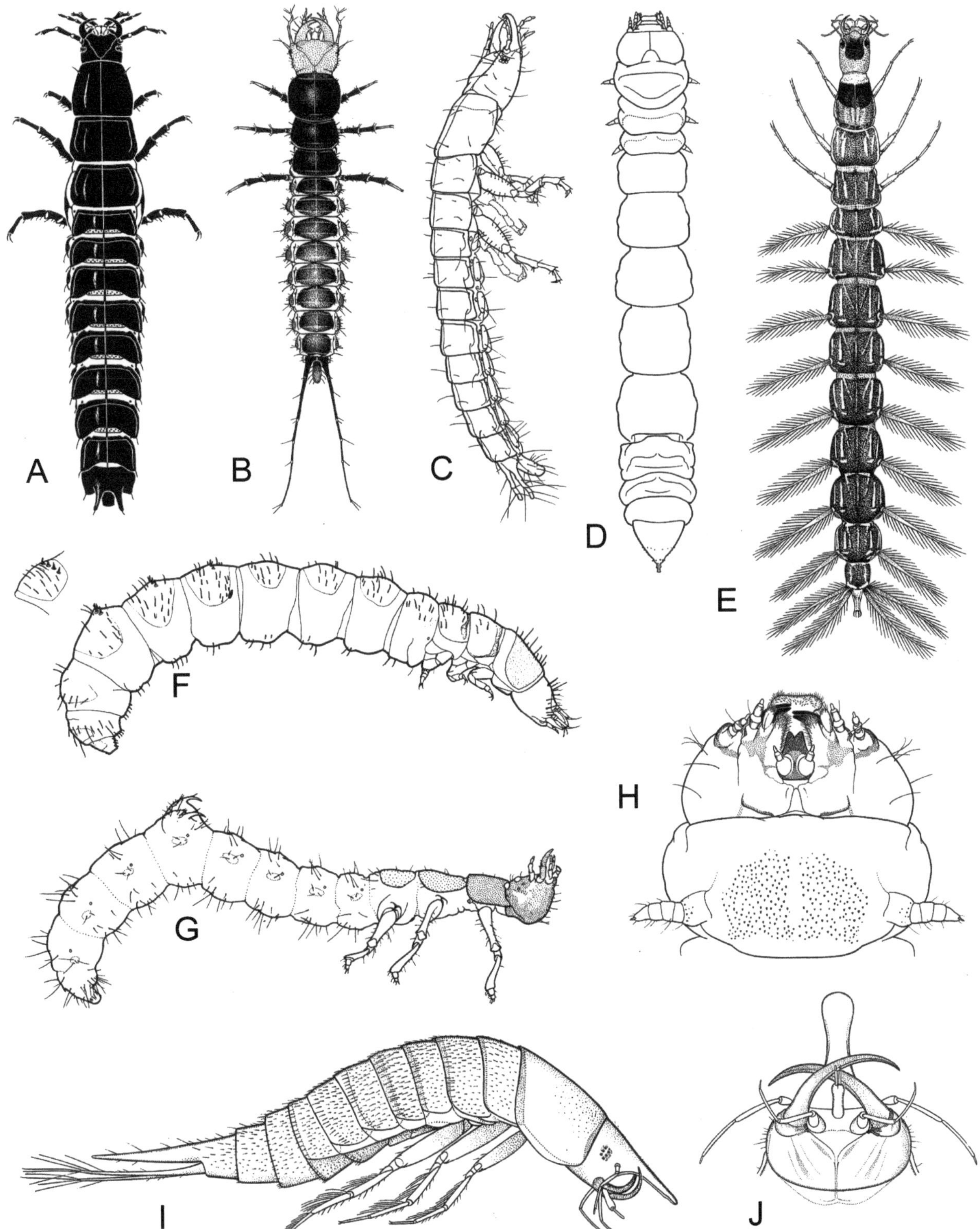

Fig. 56. Larva. A, *Pamborus macleayi* Laporte (Carabidae: Carabinae), dorsal; B, *Eudalia macleayi* Bates (Carabidae: Harpalinae: Odacanthini), dorsal; C, *Eurylynchus blagravei* Laporte (Carabidae: Broscinae), lateral; D, *Distocupes varians* (Cupedidae), dorsal; E, *Macrogyrus oblongus* (Boisduval) (Gyrinidae), dorsal; F, *Sphallomorpha colymbetoides* (Westwood) (Carabidae), lateral; G, *Cicindela* sp. (Carabidae: Cicindelinae), lateral; H, *Distocupes varians*, head and prothorax, ventral; I, Genus ? sp. (Dytiscidae: Hydroporinae), lateral; J, same, head, ventral.

obtain oxygen through their abdominal gills. The final instar leaves the water to pupate within a cocoon. [Ochs 1949; Larson 1993].

Australian Subfamilies, Tribes and Genera

Gyrininae: *Aulonogyrus* Motschulsky, *Gyrinus* Müller.
Enhydrinae: *Macrogyrus* Régimbart, *Dineutus* Macleay.

5. HALIPLIDAE

Adult (Fig. 55D–E)

Length 2.4–4.2 mm. Body ~1.65–1.75 times as long as wide, broadly elliptical, widest at about middle; sides not evenly curved, with sides of pronotum straight or with curvature differing from that of elytra; slightly convex dorsally and strongly convex ventrally; colour yellow to reddish-brown, sometimes with dark markings; upper surfaces smooth and shiny, subglabrous.

Head prognathous, about as long as wide, globular not abruptly constricted posteriorly, without median endocarina. Eyes large, moderately protuberant, entire and finely facetted, without interfacetal setae. Antennal insertions exposed located between eyes and separated by less than 2 diameters; subantennal grooves absent. Frontoclypeal suture absent. Labrum free, transverse, broadly rounded. Antennae 11-segmented, filiform, glabrous; pedicel short and broad, not or slightly longer than pedicel. Mandible short and broad, unidentate; incisor edge with at least one weak tooth and setal fringe; mola absent. Maxilla with 2-segmented, palpiform galea and falcate lacinia with several teeth along inner edge; palp 4-segmented with distinct palpifer and apical palpomere cylindrical, distinctly shorter and slightly narrower than preapical one. Mentum strongly transverse and apically biemarginate; ligula large, well sclerotised and truncate; labial palp 3-segmented and similar to that of maxilla. Gular sutures widely separated, gula almost as long as wide, without median endocarina. Corpotentorium slender, arched, with long anterior projection. Cervical sclerites absent.

Pronotum ~0.45–0.5 times as long as wide, widest posteriorly, with sides straight to weakly rounded and converging anteriorly. Lateral pronotal carinae well developed, without distinct bead; anterior angles not produced forward; posterior angles right to slightly acute; posterior edge produced to form mesal lobe concealing scutellum; disc slightly convex, simple or with pair of short, basal, longitudinal grooves. Lateral walls of prothorax (pleuron and sternum) steep, almost vertical; propleuron ending well before anterior edge of prothorax. Prosternum in front of coxae slightly shorter than shortest diameter of procoxal cavity, its anterior edge abruptly elevated at middle. Prosternal process complete, parallel-sided, its sharp lateral edges extending anteriorly to form median shelf on different plane than remainder of prosternum and its posterior edge truncate and abutting metaventrite (Fig. 20I). Procoxae globular, not projecting, without concealed lateral extensions or coxal plates, with mesal knob fitting into cavity on prosternal process; trochantin free, small, concealed. Procoxal cavities circular, widely separated, broadly open externally and closed internally by slender bar.

Scutellar shield absent. Elytra ~1.25–1.35 times as long as combined width and 3.2–3.8 times as long as pronotum, complete with apices conjointly narrowly rounded or subacute; disc with ten distinct puncture rows and no scutellary striole; epipleura moderately broad basally, but incomplete, with distinct notch at about middle for reception of metacoxal knob; ventrally with longitudinal apicolateral tongue fitting into groove on abdominal segment V. Mesoventrite about as long as wide, separated by complete sutures from mesanepisterna, which are well separated from one another; anterior edge on different plane than metaventrite, with large paired, oblique procoxal rests, not extending onto mesanepisterna; mesoventral cavity large and deep; discrimen absent. Mesoventral process undivided, rounded. Mesocoxae globular, not projecting, with mesal knob fitting into cavity near apex of metaventral process (Fig. 20I); trochantins concealed. Mesocoxal cavities moderately widely separated, partly closed laterally by mesepimera; meso- and metathoracic joint within coxal cavities membranous. Mesometaventral junction with mesoventrite overlapping metaventrite. Metaventrite transverse; discrimen complete; transverse (mesokatepisternal) suture well developed; exposed portion of metanepisternum relatively short and broad, much wider anteriorly, subacute posteriorly; metepimeron concealed; metatrochantins concealed. Metacoxae greatly enlarged, slightly transverse, firmly attached to metaventrite but not fused together at midline, their bases slightly oblique; extending laterally to meet elytra, with very large coxal plates concealing bases of femora and basal three ventrites (Fig. 26D). Metendosternite with a long, slender stalk, no lateral arms, and a moderately long anterior process with tendons moderately widely separated on either side of a deep emargination. Hind wing with long bending zone occupying most of radial bar (anterior wing strut); apical field moderately long, with four radial extensions; RA_{1+2} and RA_{3+4} separating at basal third, the former extending along anterior edge of wing, the latter abruptly curving posteriorly and giving rise to RA_3 and RA_4, both of which extend almost to wing apex; two cross-veins join RA_{1+2} and RA_{3+4}; pterostigma well developed, in between cross-veins; RP complete almost to wing base, joined near mid-wing to RA_{3+4} by cross-vein r1; portions of RA_{3+4} and a middle section of RP weakly sclerotised or obliterated where crossed by triangular folding area; cross-vein r2 and r3 absent, RP and MP joined by two cross-veins forming short, broad oblongum cell, the posterior edge of which is partly obliterated; RP_2 and RP_{3+4} emerge from apical edge of cell; MP_{1+2} with very short bending zone or hinge and a very short, straight medial spur; medial field with four free veins and no medial fleck; MP_{1+2} without cross-vein to Cu; MP_{3+4} with long basal cross-vein and no spur; wedge cell absent; anal lobe well developed, with deep, notch-like embayment; AP_{1+2} present; AP_{3+4} not forked. Legs relatively long and slender, fore legs shortest, with protibia slightly expanded at apex, with short spines along outer edge and 2 longer ones at outer apical angle; hind legs longest, very slender, with femora distinctly widened at base and tibia slightly widened at apex; all tibiae with swimming hairs; tibial spurs more or less serrate; tarsi 5-5-5; tarsomeres long and slender, simple, with some swimming hairs; pretarsal claws simple; trochanters short and broad with transverse femoral joint.

Abdomen with six ventrites (sternites II–VII), the first three of which are solidly fused, usually with the suture incomplete or obliterated, the remaining ones freely articulated and overlapping; most of first three ventrites occupied by large metacoxal cavities, the first two completely divided and ventrite 3 (sternite IV) extending across abdomen and bearing a short, angulate intercoxal process. Abdominal tergites lightly sclerotised, several bearing patches of wing-folding microtrichia. Functional spiracles on abdominal segments I–VIII, the last pair located on tergite. Tergite VIII in male divided by median longitudinal suture; sternite VIII divided into two parts. Segments IX and X in male forming genital capsule, which is broadly rounded anteriorly; tergite IX divided into paired laterotergites and segment X moderately well sclerotised and free. Aedeagus of modified trilobate type, in which phallobase is not apparent and asymmetrical parameres articulate with base of penis, which lacks basal struts. Tergite VIII in female similar to that of male, divided by median suture; sternite VIII divided into two lateral sclerites (gonocoxosternites). Ovipositor short and broad, lightly pigmented, except for paraproctal baculi; and short, oblique basal coxital baculum; coxites lightly sclerotised, undivided, apically narrowed, and setose, without styli; a ventral, slender, apically cleft sclerite may be a remnant of sternite IX. [Zimmermann 1924; Belkaceme 1986; Watts 1988a; Beutel and Ruhnau 1990; Vondel 1995, 2005; Roughley 2000b].

Larva

Body narrowly elongate, more or less parallel-sided and slightly flattened, yellowish-brown in colour; surfaces finely tuberculate, clothed with setiferous tubercles representing microtracheal gills and with scattered long setae.

Head prognathous and protracted, about as long as wide, slightly flattened with sides weakly rounded. Posterior edge of head capsule not emarginate; epicranial suture not apparent; median endocarina absent. Stemmata six on each side, arranged in a tight cluster. Frontoclypeal suture absent. Labrum completely fused to head capsule; anterior edge of clypeolabrum slightly emarginate, without nasale or paranasalia. Antennae about half as long as head width, slender, 4-segmented, with sensorium on antennomere 3 almost as long as antennomere 4. Mandibles broad at base, with narrow, strongly curved, unidentate apex and complete internal channel (Fig. 47H); mola and prostheca absent. Ventral mouthparts protracted; maxillae and labium well separated at base, without maxillary articulating area; cardo transverse; stipes elongate; galea short, broad. apically rounded and fixed; lacinia reduced or absent; maxillary palp 3-segmented without palpifer. Labium consisting of prementum and postmentum; labial palps 2-segmented without distinct palpigers. Hypostomal rods and ventral epicranial ridges absent. Gular region moderately long; gular sutures absent.

Thoracic segments about equal in length, each tergum with pair of lateral and paramedian, subacute projections, directed posterolaterally and posteriorly, respectively; prothoracic venter with undivided presternum. Legs long, 6-segmented, with elongate coxa and one pretarsal claw; fore legs not modified as grasping organs; coxae moderately separated. Abdomen about five times as long as thorax; segments I–IX each with lateral and paramedial projections similar to those on thorax. Tergum IX slightly shorter than VIII, not extending onto ventral surface; sternite IX exposed, simple. Segment X more than 5 times as long as IX, broad at base, narrowing apically, with very slender, narrowly forked apex. Spiracles annular–biforous, with closing apparatus, present on thorax and abdominal segments I–VII in final instar. [Böving and Craighead 1931; Seeger 1971; Bertrand 1972; Beutel 1986b; Spangler 1991; Strand and Spangler 1994; Vondel 2004, 2005].

Classification and World Distribution

The family occurs in all major zoogeographic regions and includes five genera: the worldwide genus *Haliplus*, *Peltodytes* Régimbart, which is widely distributed but absent from the Neotropical and Australian regions, the Holarctic genus *Brychius* Thomson, *Algophilus* Zimmermann from South Africa and *Apteraliplus* Chandler from north-western North America.

Australian Fauna

The Australian haliplid fauna consists of 15 species of *Haliplus* Latreille belonging to three subgenera but mainly *Haliplus* (*Liaphlus*) Guignot. The group is widely distributed and occur among aquatic vegetation at the edges of ponds, lakes or slow-moving streams. Adults usually crawl along the bottom, but they are able to swim using alternate leg movements. Adult and larval haliplids commonly feed on stoneworts (Charophyceae) and some other kinds of green algae. Larvae are capable of sucking algal cells through a perforation in each mandible. Spiracles are lacking except in the final instar, and respiration takes place through microtracheal gills. [Hickman 1931; Seeger 1971; Watts 1988a; Vondel 2004, 2005].

6. NOTERIDAE

Adult (Fig. 55F–G)

Length 1.4–6 mm. Body ~1.28–1.55 times as long as wide, broadly to narrowly elliptical, widest at about middle; sides evenly curved with pronotum and elytra forming continuous outline; colour yellow to reddish-brown or black, with elytra slightly darker than head and pronotum in Notomicrini, with yellow maculae on dark elytra in *Canthydrus*; upper surfaces smooth, shiny, glabrous.

Head prognathous, strongly transverse, not abruptly constricted posteriorly, without median endocarina. Eyes moderately large but not protuberant, entire or very slightly emarginate anteriorly, finely facetted, without interfacetal setae. Antennal insertions concealed from above; subantennal grooves absent. Frontoclypeal suture absent. Labrum free, transverse, subtruncate. Antennae 11-segmented, filiform or slightly serrate, glabrous, with antennomeres 5 or 6 to 9 or 10 asymmetrically widened and terminal one or two narrower; scape constricted, so that it appears to be 2-segmented, pedicel elongate. Mandible short, broad and asymmetrically bidentate, with cutting edge between two apical teeth; with weak or strong retinaculum; mola absent. Maxilla with 2-segmented, palpiform galea and falcate lacinia with several stout setae along inner edge; palp 4-segmented with distinct palpifer and apical palpomere fusiform, distinctly longer and slightly wider than preapical one. Mentum strongly transverse and apically deeply emarginate; ligula large, well sclerotised and truncate; labial palp 3-segmented with apical palpomere fusiform or slightly expanded. Gular sutures widely separated, gula almost as long as wide, with median gular apodeme (except in *Notomicrus*). Corpotentorium very slender, not arched. Cervical sclerites absent.

Pronotum ~0.43–0.5 times as long as wide, widest posteriorly, with sides usually moderately curved, converging anteriorly (almost straight in *Neohydrocoptus*) and obliquely explanate. Lateral pronotal carinae well developed, with distinct bead; anterior angles produced and acute; posterior angles right to slightly acute; posterior edge produced to form short, angulate mesal lobe concealing scutellum in Hydrocanthini, not in Notomicrini; disc simple. Prosternum in front of coxae slightly shorter than shortest diameter of procoxal cavity, anterior edge slightly produced and rounded, surface somewhat tumid. Prosternal process complete, extending well behind procoxae and abutting metaventrite, gradually expanded and apically truncate in Hydrocanthini, expanded and then narrowed with rounded apex in Notomicrini. Procoxae globular, not projecting, without concealed lateral extensions or coxal plates, with mesal knob fitting into cavity on prosternal process; trochantin free, small, concealed. Procoxal cavities circular, widely separated, broadly open externally and closed internally by slender bar.

Scutellar shield usually absent (partly exposed in *Notomicrus*). Elytra ~1.23–1.55 times as long as combined width and 3–3.75 times as long as pronotum, complete with apices conjointly narrowly rounded or subacute; humeri sharply angulate; disc without puncture rows or with a single sutural row only; epipleuron moderately broad and obliquely explanate basally, but ending at about basal third; each elytron ventrally with long, slender endocarina beginning near end of epipleuron and extending almost to apex. Mesoventrite separated by complete sutures from mesanepisterna, which are well separated from one another; anterior edge on different plane than metaventrite, with large paired, oblique procoxal rests, not extending onto mesanepisterna; mesoventral cavity large and deep; discrimen absent. Mesoventral process undivided, truncate or emarginate. Mesocoxae globular, not projecting, with mesal cavity receiving metaventral knob; trochantins concealed. Mesocoxal cavities moderately widely separated externally (more or less contiguous internally), partly closed laterally by mesepimera; meso- and metathoracic joint within coxal cavities membranous. Mesometaventral junction with mesoventrite overlapping metaventrite, but concealed by meeting of prosternal and metaventral processes. Metaventrite transverse; discrimen complete; transverse (mesokatepisternal) suture absent; exposed portion of metanepisternum relatively short and broad, much wider anteriorly, subacute posteriorly; metepimeron concealed; metatrochantin concealed. Metacoxae greatly enlarged, slightly transverse, fused together and to metaventrite but separated by sutures in Hydrocanthini, but incompletely separated from metaventrite (sutures incomplete) in Notomicrini; coxal bases horizontal but angulate; extending laterally to meet elytra, with elongate, posteriorly projecting coxal plates occupying less than half of coxal width and concealing bases of femora only (Fig. 55F). Metendosternite with a long, slender stalk, short to long lateral arms, and a moderately long anterior process with tendons moderately widely separated on either side of a deep emargination. Hind wing (*Hydrocanthus*) with long bending zone occupying most of radial bar (anterior wing strut); apical field moderately long, with four radial extensions; RA_{1+2} and RA_{3+4} separating at mid wing, the former extending along anterior edge of wing, the latter abruptly curving posteriorly and giving rise to RA_3 and RA_4, both of which extend almost to wing apex; two cross-veins join RA_{1+2} and RA_{3+4}; pterostigma well developed, in between cross-veins; RP complete to wing base, joined near mid-wing to RA_{3+4} by cross-vein r1; portions of RA_{3+4} obliterated where crossed by triangular folding area; cross-vein r2 and r3 absent, RP and MP joined by two cross-veins forming short, broad oblongum cell; RP_2 and RP_{3+4} emerge from apical edge of cell; MP_{1+2} with very short bending zone or hinge and a very short, straight medial spur; medial field with four free veins and with elongate medial fleck posterior to and attached to MP_{3+4}; MP_{1+2} with cross-vein to Cu; MP_{3+4} with long basal cross-vein and no spur; wedge cell small and apically acute; anal lobe well developed, with very slight embayment; AP_{1+2} present; AP_{3+4} not forked. Hind wing in *Notomicrus* with similar venation but with no wedge cell, a reduced anal lobe and deep, notch-like anal embayment. Legs usually relatively short and stout (somewhat longer and narrower in *Notomicrus*), fore legs shortest, with protibia slightly expanded near apex and with two spurs in Notomicrini, produced apically beyond tarsal attachment and with longitudinal fringe of spines and a single, curved apical spur in Hydrocanthini; mid and hind legs longer, with swimming hairs, stout spines and paired spurs, tibiae in Hydrocanthini somewhat flattened; one metatibial spur in *Hydrocanthus* and least some spurs in *Notomicrus* more or less serrate; tarsi 5-5-5; tarsomeres simple, with swimming hairs in Hydrocanthini; basal tarsomere much longer than others; pretarsal claws simple; pro- and mesotrochanters short and broad with slightly oblique femoral joint; metatrochanter enlarged with strongly oblique joint.

Abdomen with six ventrites (sternites II–VII), the first three of which are connate; suture between ventrites 1 and 2 complete, that between 2 and 3 vaguely indicated, incomplete or absent; ventrite 1 completely divided by coxal cavities; ventrite 2 without or with very weak, rounded intercoxal process; ventrite 6 narrowly rounded to subacute. Most abdominal tergites lightly sclerotised, tergites VII and VII more heavily so, IV–VII with patches of wing-folding microtrichia. Functional spiracles on abdominal segments I–VIII, the last pair reduced and located on tergite. Tergite VIII in both sexes undivided and sternite VIII divided into two parts (gonocoxosternites). Segments IX and X in male forming genital capsule, which is slightly asymmetrical, broadly rounded anteriorly and produced posteriorly to form broadly rounded lobe in *Notomicrus* or subacute apically cleft lobe in *Hydrocanthus*; tergite IX undivided and apically emarginate; segment X membranous. Aedeagus of modified trilobate type, in which phallobase is not apparent and asymmetrical parameres articulate with base of penis, which is laterally compressed and strongly curved, without basal struts. Ovipositor in *Notomicrus* slightly longer than wide, lightly pigmented, with short, subparallel paraproctal baculi and coxites undivided, partly fused together and finely setose, without styli; ovipositor in Hydrocanthini more elongate, with long diverging baculi and coxites contiguous but not fused, narrowed anteriorly and well sclerotised, without styli. [Belkaceme 1991; Roughley 2000c; Dettner 2005a].

Larva (Fig. 58E)

Body narrowly elliptical, widest at about middle, slightly flattened, yellowish-brown in colour; surfaces smooth and glabrous.

Head prognathous and protracted, about as long as wide, widest posteriorly, slightly flattened, with sides very weakly rounded. Posterior edge of head capsule not emarginate; epicranial stem absent, frontal arms V-shaped; median endocarina absent.

Stemmata six on each side, arranged in a partial circle. Frontoclypeal suture absent. Labrum completely fused to head capsule; anterior edge of clypeolabrum without nasale or paranasalia. Antennae about a third as long as head width, slender, 4-segmented, without sensorium on antennomere 3. Mandibles broad at base, with narrow, strongly curved, unidentate apex and short open groove; mola and prostheca absent. Ventral mouthparts protracted; maxillae and labium well separated at base, without maxillary articulating area; cardo transverse; stipes slightly elongate; galea 2-segmented, articulated, palpiform; maxillary palp 3-segmented without distinct palpifer. Labium consisting of prementum and postmentum; labial palps 2-segmented without distinct palpigers. Hypostomal rods and ventral epicranial ridges absent. Gular region moderately long; gular sutures short, widely separated.

Protergum about as long as meso- and metatergum combined; sides of all thoracic terga obliquely explanate; prothoracic venter without distinct sclerites. Legs moderately long, 6-segmented, armed with a few stout setae, with elongate coxa, tibia and tarsus equal in length, with two pretarsal claws; coxae relatively narrowly separated. Abdomen less than twice as long as thorax; segment I with incomplete lateral edge, I–VIII without lateral edges. Tergum VIII ~1.67 times as long as VII, narrowed apically and more or less conical, with pair of terminal spiracles present in all instars and with segment IX more or less enclosed within apex of VIII and bearing a pair of very short, articulated urogomphi; segment X membranous and not apparent. Anterior abdominal spiracles annular, with closing apparatus, present on segments I–VII in final instar. [Bertrand 1972; Spangler 1991; Roughley 2000c; Dettner 2005a].

Classification and World Distribution

Noteridae are widely distributed, with the highest diversity in tropical and subtropical regions. The subfamily Phreatodytinae includes the single genus *Phreatodytes* Uéno with a few species inhabiting subterranean waters in Japan. Notomicrinae includes *Notomicrus* Sharp with species in the Nearctic, Neotropical, Oriental and Australian regions and *Speonoterus* Spangler from Sulawesi. The subfamily Noterinae includes four tribes: Tonerini with a single species of *Tonerus* Miller from Venezuela, Neohydrocoptini with the genus *Neohydrocoptus* Satô from China and Australia, Pronoterini with the New world genus *Pronoterus* Sharp, and Noterini with the remaining genera, including *Canthydrus* Sharp, Hydrocanthus Say and the Palaearctic *Noterus* Clairville. [Ueno 1957, 1996; Beutel and Roughley 1987; Nilsson *et al.* 1989; Belkaceme 1991; Miller 2009].

Australian Fauna

The Australian fauna includes the minute *Notomicrus tenellus* (Clark) (Notomicrinae) and six species of Noterinae: *Neohydrocoptus subfasciatus* (Sharp), *Canthydrus bovillae* Blackburn, *C. ephemeralis* Watts, *Hydrocanthus australasiae* Wehncke, *H. pederzanii* Toledo & Hendrich and *H. waterhousei* Blackburn. All are known from the northern part of the continent. Noterids usually inhabit the margins of shallow ponds, where adults and larvae are commonly found among the roots of floating or emergent aquatic plants. Larvae of some species obtain air by tapping plants with the aid of the spiracle-bearing siphon at the end of the abdomen. Little is known of the feeding habits, but it is generally thought that noterids are detritus feeders. Known pupae are contained in air-filled cocoons attached to the roots of aquatic plants. [Balfour-Browne and Balfour-Browne 1940; Larson 1993; Dettner 2005a].

Australian Subfamilies, Tribes and Genera

Notomicrinae: *Notomicrus* Sharp.
Noterinae: Neohydrocoptini: *Neohydrocoptus* Satô; **Noterini:** *Canthydrus* Sharp, *Hydrocanthus* Say.

7. HYGROBIIDAE

Adult (Fig. 55B–C)

Length 8–12 mm. Body ~1.5–1.6 times as long as wide, oblong, widest behind middle; sides weakly but not continuously curved; colour yellowish-brown to reddish-brown or black, with elytra and/or pronotum sometimes bicolored; upper surfaces smooth, mostly clothed with very short, fine setae, but with much longer setae laterally.

Head prognathous, about as long as wide, not abruptly constricted posteriorly, without median endocarina. Eyes large and strongly protuberant, entire, finely facetted, without interfacetal setae. Antennal insertions barely concealed from above; subantennal grooves absent. Frontoclypeal suture well developed, more or less angulate. Labrum free, strongly transverse, subtruncate. Antennae 11-segmented, filiform or slightly serrate, glabrous; scape about twice as long as pedicel, which is slightly elongate. Mandible short, broad, unidentate, with retinaculum almost as large as apical tooth; base of mandible somewhat expanded and mola-like, but slender and without armature. Maxilla with 2-segmented, palpiform galea with bifid apical segment; lacinia slender and falcate; palp 4-segmented with distinct palpifer and apical palpomere fusiform. Mentum strongly transverse and apically deeply emarginate; ligula well sclerotised and broadly rounded; labial palp 3-segmented with elongate palpiger and fusiform apical palpomere. Anterior edge of gulamentum emarginate, with produced anterolateral angles forming with genal processes a pocket partly concealing maxillary bases. Gular sutures widely separated, gula longer than wide, without median gular apodeme. Corpotentorium very slender, with anterior median process. Cervical sclerites absent.

Pronotum ~0.4–0.5 times as long as wide, widest at base, which is slightly narrower than combined elytral bases, with sides very slightly curved, converging anteriorly. Lateral pronotal carinae well developed, with distinct bead; anterior angles produced and acute; anterior edge with fringe of setae; posterior angles right to slightly obtuse; posterior edge more or less trisinuate. Prosternum in front of coxae somewhat shorter than shortest diameter of procoxal cavity, anterior edge not produced forward, its surface almost vertical. Prosternal process complete, extending well behind procoxae and beyond metaventrite, strongly curved in lateral view, slightly expanded to middle, then parallel-sided and apically rounded. Procoxae subcylindrical, not projecting, with short concealed lateral extension and no coxal plate, with mesal knob fitting into cavity on prosternal process; trochantin free, small, concealed.

Procoxal cavities obliquely transverse, widely separated, broadly open externally and closed internally. Scutellar shield small, not elevated anteriorly, acute posteriorly. Elytra ~1.2–1.3 times as long as combined width and 3.6–4.2 times as long as pronotum, complete with apices conjointly narrowly rounded; humeri well developed; disc without puncture rows, with distinct anteromesal elevation and weak posterolateral one; epipleuron moderately broad to anterior third, then narrowing and incomplete; each elytron ventrally with distinct interlocking tongue at about middle. Mesoventrite separated by complete sutures from mesanepisterna, which are well separated from one another; anterior edge on different plane than metaventrite, with large, contiguous, strongly oblique procoxal rests, not extending onto mesanepisterna; mesoventral cavity large and deep; discrimen absent. Mesoventral process undivided, acute. Mesocoxae slightly transverse, not projecting, without mesal articulation; trochantins concealed. Mesocoxal cavities moderately widely separated externally (more or less contiguous internally), partly closed laterally by mesepimera; meso- and metathoracic joint within coxal cavities membranous. Mesometaventral junction with mesoventrite and metaventrite not or barely meeting, but concealed by overlapping of prosternal and metaventral processes. Metaventrite strongly transverse; discrimen complete to base of metaventral process, which has a shallow cavity at its apex to receive tip of prosternal process; transverse (mesokatepisternal) suture short and doubly curved (Fig. 55C); exposed portion of metanepisternum only slightly longer than wide, much wider anteriorly, subacute posteriorly; metepimeron slightly exposed; metatrochantin concealed. Metacoxae moderately enlarged, about as long as metaventrite, slightly transverse, fused together and fused to metaventrite but separated by sutures, with coxal bases horizontal; coxae extending laterally to meet elytra, with reduced, elongate coxal plates occupying about a fifth of coxal width and concealing bases of femora only. Metendosternite with a long, slender stalk, short lateral arms, and a moderately long anterior process with tendons moderately widely separated on either side of a deep emargination. Hind wing (Fig. 24A) with long bending zone occupying most of radial bar (anterior wing strut); apical field moderately long, with four radial extensions; RA_{1+2} and RA_{3+4} separating at mid wing, the former extending along anterior edge of wing, the latter abruptly curving posteriorly and giving rise to RA_3 and RA_4, both of which extend almost to wing apex; two cross-veins join RA_{1+2} and RA_{3+4}; pterostigma well developed, extending to end of RA_3; RP complete to wing base, joined near mid-wing to RA_{3+4} by cross-vein r1; portions of RA_{3+4} obliterated where crossed by triangular folding area; cross-vein r2 and r3 absent, RP and MP joined by two cross-veins forming slightly, obliquely elongate oblongum cell; RP_2 and RP_{3+4} emerge from apical edge of cell; MP_{1+2} with very short bending zone or hinge and a very short, straight medial spur; medial field with four free veins and no medial fleck; MP_{1+2} with cross-vein to Cu; MP_{3+4} with long basal cross-vein and no spur; wedge cell small and apically acute; anal lobe well developed, without embayment; AP_{1+2} present; AP_{3+4} not forked. Legs moderately long and slender, with swimming hairs on all tibiae and tarsi; fore legs shortest; protibia and mesotbia slightly expanded near apex, hind legs less so; all tibiae with two apical spurs; tarsi 5-5-5; tarsomeres simple, with basal tarsomere much longer than others; pretarsal claws simple; pro- and mesotrochanters short and broad with slightly oblique femoral joint; metatrochanter enlarged with strongly oblique joint.

Abdomen with six ventrites (sternites II–VII), the first three of which are connate; suture between ventrites 1 and 2 complete, that between 2 and 3 incomplete; ventrite 1 completely divided by coxal cavities; ventrite 2 with undulate or subacute intercoxal process; ventrite 6 broadly rounded. Most abdominal tergites lightly sclerotised, tergites VII and VII more heavily so; patches of wing-folding microtrichia absent. Functional spiracles on abdominal segments I–VIII, the last two pairs located on tergites. Tergite VIII in both sexes divided by median suture and sternite VIII divided into two parts (gonocoxosternites). Segments IX and X in male forming genital capsule, which is slightly asymmetrical, broadly rounded anteriorly; tergite IX divided apically into two parts; segment X membranous. Aedeagus of trilobate type; phallobase asymmetrical, with curved anterior strut (manubrium); parameres symmetrical, separated by membrane from phallobase and articulated with base of penis with base of penis, which is laterally compressed and strongly curved, without basal struts. Ovipositor shorter than wide, lightly pigmented, with laterally located paraprocts shorter than coxites, which are undivided and setose, without styli. [Forsyth 1970; Burmeister 1976; Britton 1981; Beutel 1986a; Dettner 2005c].

Larva (Fig. 42B)

Body broadly elliptical, widest at middle of head and narrowed posteriorly, slightly flattened, yellowish-brown in colour; surfaces smooth, clothed with scattered, long, fine hairs.

Head (Fig. 43P) prognathous and protracted, about as long as wide, widest at middle, slightly flattened, with sides weakly rounded. Posterior edge of head capsule not emarginate; epicranial stem long, frontal arms V-shaped; median endocarina absent. Stemmata six on each side, arranged in two vertical rows of 3. Frontoclypeal suture absent. Labrum completely fused to head capsule; anterior edge of clypeolabrum broadly rounded, without nasale or paranasalia. Antennae about more than half as long as head width, slender, 4-segmented, with sensorium on antennomere 3more than half as long as antennomere 4. Mandibles slender, strongly curved, with unidentate apex and moderately long, open mesal groove; retinaculum, mola and prostheca absent. Ventral mouthparts protracted; maxillae and labium well separated at base, without maxillary articulating area; cardo fused to stipes, which is ~3 times as long as wide; galea and lacinia absent; maxillary palp 3-segmented without distinct palpifer. Labium consisting of prementum and postmentum; labial palps 2-segmented without distinct palpigers. Hypostomal rods and ventral epicranial ridges absent. Gular region moderately long; gular sutures widely separated; gula longer than wide.

Protergum about as long as meso- and metatergum combined; all thoracic terga and abdominal terga I–VII moderately well sclerotised with sides slightly, more or less obliquely produced; prothoracic venter without distinct sclerites; ventral portions of abdominal segments IV–VII well sclerotised. Legs moderately long, 6-segmented, with narrowly elongate coxa; metatibia and metatarsus with swimming hairs; pretarsus with single claw; coxae relatively narrowly separated. Abdomen without appendages shorter than thorax; segments I–VII of equal length but with gradually decreasing widths. Tergum VIII produced posteriorly forming long, narrow process bearing long setae but without spiracles at the apex; sternum VIII highly reduced. Body of segment IX not visible from above, more or less enclosed within apex of VIII, but bearing a pair of slender, setose, articulated urogomphi which are as long as median process on of tergum IX; segment X membranous and not apparent. Reduced annular spiracles present on mesothorax and abdominal segments I–VII in final instar. Branched gills present on coxal bases and on abdominal sterna I–III. [Balfour-Browne 1922; Alarie *et al.* 2004; Dettner 2005c].

Classification and World Distribution

The family contains the single genus *Hygrobia* Latreille with one species in Europe and North Africa, one in south-western China and four in Australia. [Jäch 1995].

Australian Fauna

Two of the Australian species, *H. australasiae* (Clark) and *H. nigra* (Clark), occur mainly in the south-eastern part of the continent, the former occurring in TAS, as far west as SA and as far north as southern QLD, and the latter found primarily in VIC and NSW but recently recorded from west of Innisfail in North QLD. *H. maculata* Britton is known from North QLD and the NT, while *H. wattsi* Hendrich (2001) was recently described from the south-western part of WA. Hygrobiids are relatively slow-moving, bottom-feeding predators, occurring in stagnant water. Adults swim using alternate leg movements and periodically come to the surface for air, which is stored beneath the elytra. Adults stridulate using the abdominal apex and a file on the undersurface of each elytron. Both adults and larvae are predaceous, and the European *Hygrobia hermanni* is known to feed on tubificid worms (Balfour-Browne 1922). [Britton 1981].

8. DYTISCIDAE

Adult (Figs 1D, 55I)

Length 1–38 (usually 3–32) mm. Body usually ~1.35–2.05 times as long as wide (as much as 2.35, especially in stygobitic forms), usually broadly to narrowly elliptical and widest at about middle, with sides evenly curved and pronotum and elytra forming continuous outline (pronotum and elytra independently curved or sides of pronotum straight and those of elytra curved in several Hydroporinae and in the genus *Eretes*); usually slightly to moderately convex dorsally and ventrally, rarely strongly convex; colour yellow to reddish-brown or black, sometimes with yellow maculae on dark elytra, pale yellow in subterranean forms; upper surfaces usually smooth, shiny and glabrous (clothed with short, fine, decumbent hairs in some Hydroporinae and with scattered long setae in subterranean forms).

Head prognathous, strongly transverse, not abruptly constricted posteriorly, without median endocarina. Eyes usually moderately large but not protuberant, following the lateral outline of the head (large and slightly protruding in *Eretes*), entire or very slightly emarginate anteriorly, finely facetted, without interfacetal setae; eyes absent in subterranean forms. Antennal insertions almost always concealed from above (barely exposed in some Dytiscinae); subantennal grooves absent. Frontoclypeal suture usually incomplete or absent. Labrum free, strongly transverse, subtruncate to emarginate, sometimes almost vertical and not visible from above. Antennae moderately long, 11-segmented, filiform, glabrous; scape elongate and more or less parallel-sided, usually about as long as pedicel (much longer than pedicel in Dytiscinae). Mandible short, broad and unidentate or asymmetrically bidentate, often with cutting edge between two apical teeth; with or without retinaculum; mola absent. Maxilla with 2-segmented, palpiform galea and falcate lacinia with several stout setae along inner edge; palp 4-segmented with distinct palpifer and apical palpomere fusiform. Mentum strongly transverse and apically deeply emarginate; ligula well sclerotised and truncate; labial palp 3-segmented with apical palpomere fusiform. Gular sutures widely separated, gula transverse or about as long as wide, without median gular apodeme. Corpotentorium slender, not arched. Cervical sclerites absent.

Pronotum usually ~0.22–0.50 times as long as wide (rarely as much as 0.65), usually widest posteriorly with sides moderately curved, converging anteriorly and obliquely explanate (widest at middle in some Bidessini and widest anteriorly in *Carabhydrus* and some stygobitic taxa). Lateral pronotal carinae well developed, with or without distinct bead; anterior angles produced and acute; posterior angles right to distinctly acute; posterior edge broadly rounded to bisinuate or produced to form mesal lobe; disc usually simple (with paired longitudinal grooves at base in Bidessini and Carabhydrini). Prosternum in front of coxae (Fig. 13D) slightly to much shorter than shortest diameter of procoxal cavity, anterior edge not or slightly produced forward, sometimes forming abruptly vertical wall, surface flat to strongly tumid. Prosternal process complete, extending well behind procoxae and abutting anterior edge of metaventrite (Fig. 28D), narrow at base, then expanded and narrowed again to narrowly rounded or acute apex, flat to strongly curved in lateral view, longitudinally grooved in Matinae. Procoxae globular or slightly transverse, not projecting below prosternum, but often obliquely vertical, without concealed lateral extensions or coxal plates, without distinct coxal knob or secondary sternocoxal joint; trochantin free, small, concealed. Procoxal cavities circular, narrowly separated, broadly open externally and closed internally.

Scutellar shield, when present, not abruptly elevated anteriorly and broadly angulate to narrowly rounded or acute posteriorly, absent in Laccophilinae and Hydroporinae. Elytra usually ~1.15–1.67 times as long as combined width (as much as 1.85 especially in stygobitic forms) and 2.8–6.40 times as long as pronotum, complete with apices conjointly broadly rounded to subacute; humeri well developed; disc almost always irregularly punctate, with 7 distinct striae in some *Copelatus*, with short paired longitudinal grooves at base in Bidessini and with two pairs of longer grooves in *Carabhydrus*; epipleuron moderately broad basally, but narrowing apically and incomplete; each elytron ventrally with short posterolateral interlocking tongue in Hyphydrini. Mesoventrite separated by complete sutures from mesanepisterna, which are well separated from one another; anterior edge on different plane than metaventrite, with large paired, slightly to strongly oblique procoxal rests, not extending onto mesanepisterna; mesoventral cavity small to moderately large, extending to posterior edge of ventrite; discrimen absent. Mesoventral process undivided, truncate or roundedd. Mesocoxae globular, not projecting, without sternocoxal articulation; trochantins concealed. Mesocoxal cavities narrowly to moderately widely separated, usually partly closed laterally by mesepimera and metanepisternum (closed by mesepimeron alone in Laccophilinae); meso- and metathoracic joint within coxal cavities membranous. Mesometaventral junction with mesoventrite overlapping metaventrite, but concealed by meeting of prosternal and metaventral processes. Metaventrite about as long as wide or somewhat transverse, often with diamond-shaped mesal section and paired lateral wings; discrimen complete to base of metaventral process, which is anteriorly notched or concave to receive prosternal process; transverse (mesokatepisternal) suture absent; exposed portion of metanepisternum short and broad, subtriangular, subacute posteriorly; metepimeron usually concealed (exposed in *Eretes*); metatrochantin concealed. Metacoxae greatly enlarged, slightly transverse, fused together and to metaventrite but usually separated by sutures, their anterior edges oblique and strongly curved, posterior edges straight to slightly oblique; extending laterally to meet elytra, with coxal plates highly reduced or absent. Metendosternite with a long, slender stalk, short lateral arms, and a moderately long anterior process with tendons moderately widely separated on either side of a deep emargination. Hind wing with long bending zone occupying most of radial bar (anterior wing strut); apical field moderately long, with four radial extensions; RA_{1+2} and RA_{3+4} separating beyond mid wing, the former extending along anterior edge of wing, the latter abruptly curving posteriorly and giving rise to RA_3 and RA_4, both of which extend almost to wing apex; one cross-vein joins RA_{1+2} and RA_{3+4}; pterostigma well developed; RP complete to wing base, joined near mid-wing to RA_{3+4} by cross-vein r1; portions of RA_{3+4} obliterated where crossed by triangular folding area; cross-vein r2 and r3 absent, RP and MP joined by two cross-veins forming short, broadly or narrowly obliquely transverse oblongum cell; RP_2 and RP_{3+4} emerge from apical edge of cell; MP_{1+2} with very short bending zone or hinge, without or with very short, straight medial spur; medial field with four free veins and with narrowly elongate medial fleck (Fig. 29L) posterior to and atached to MP_{3+4}; MP_{1+2} with cross-vein to Cu; MP_{3+4} with long basal cross-vein and no spur; wedge cell usually well developed and apically acute; anal lobe well developed, with very slight embayment; AP_{1+2} present; AP_{3+4} not forked; venation reduced in some forms and wings sometimes vestigial. Fore legs and mid legs relatively short and usually slender, but tibiae sometimes flattened, subapically widened, and spinose; swimming hairs often on mid legs; hind legs usually greatly enlarged and modified for swimming, with tarsi longer than tibiae, somewhat flattened, broadest at base and gradually narrowing apically and often with single pretarsal claw; swimming hairs usually lining tibiae and tarsi; trochanterofemoral joints more or less oblique, those on hind legs strongly so; tibial spurs well developed, usually paired, sometimes asymmetrical; tarsi usually clearly 5-5-5; pro- and mesotarsi in Hydroporinae pseudotetramerous, with tarsomere 4 reduced and embedded at base of lobe on tarsomere 3); protarsomeres 1–3 in males (Fig. 28D) and in females of Hydroporinae expanded with ventral adhesive setae; mesotarsomeres 1–3 expanded with adhesive setae in some males.

Abdomen with six ventrites (sternites II–VII), the first three of which are connate but separated by complete sutures; ventrite 1 completely divided by coxal cavities; ventrite 2 without or with very weak, rounded intercoxal process; ventrite 6 more or less rounded. Most abdominal tergites lightly to moderately heavily sclerotised, with or without patches of wing-folding microtrichia on some tergites. Functional spiracles on abdominal segments I–VIII, the last pair located on tergite. Tergite VIII in both sexes divided by median suture and sternite VIII divided into two parts (gonocoxosternites). Segments IX and X in male forming genital capsule, which is usually asymmetrical and broadly rounded anteriorly; tergite IX undivided and apically emarginate or divided into two parts; segment X moderately sclerotised and free or membranous. Aedeagus of modified trilobate type, in which phallobase is not apparent and parameres articulate with base of penis. Ovipositor variable but usually with paraprocts lying lateral to undivided coxites, which are usually narrowed apically, without styli. [Forsyth 1968; Watts 1978; Pederzani 1995; Miller 2001; Roughley and Larson 2001; Balke and Ribera 2004; Hendrich and Watts 2004, 2009; Balke 2005; Hendrich *et al.* 2009].

Larva (Fig. 56I–J)

Body elongate and elliptical, widest at about middle, slightly flattened; head, thoracic terga and abdominal terga I–VIII usually well sclerotised (terga I–VI weakly sclerotised in *Cybister*); ventral surfaces usually lightly sclerotised, except for segment VIII and usually VII; cuticle usually smooth, sometimes with anterior transverse carinae on most terga; vestiture of scattered fine setae.

Head prognathous and protracted, usually as long as or longer than wide (occasionally slightly transverse), usually widest at middle, slightly flattened, with sides weakly to strongly rounded; head occasionally constricted posteriorly to form broad neck. Posterior edge of head capsule not emarginate; epicranial stem long, frontal arms U-shaped or V-shaped; median endocarina absent. Stemmata almost always six on each side (absent in stygobitic forms), arranged in a circle or in two vertical rows of 3. Frontoclypeal suture absent. Labrum completely fused to head capsule; anterior edge of clypeolabrum often relatively simple and rounded, trilobed in *Cybister*, produced to form nasale of varying shapes in Hydropoprinae; adnasalia sometimes present. Antennae usually more than half as long as head width, slender, 4-segmented, with antennomeres 1–3 subdivided in Lancetinae and Dytiscinae; sensorium on antennomere 3 short and inconspicuous to elongate (sometimes as long as antennomere 4). Mandibles slender and falcate, strongly upturned in Hydroporinae; almost always with closed groove or sucking channel (absent in *Copelatus* in which mesal edge is serrate), mola and prostheca absent. Ventral mouthparts protracted; maxillae and labium well separated at base, without maxillary articulating area; cardo small and transverse (sometimes fused to stipes); stipes slightly elongate to more than 5 times as long as wide; galea usually articulated and palpiform (falcate in Copelatus), absent in Hydroporinae and and *Cybister*; maxillary palp 3-segmented (palpomeres subdivided in *Cybister*), with segment-like palpifer. Labium consisting of prementum and postmentum; ligula usually absent (small and horn-like in *Eretes* and *Cybister*); labial palps 2-segmented, usually widely separated, palpomeres subdivided in *Cybister*, without distinct palpigers. Hypostomal rods and ventral epicranial ridges absent. Gular region moderately long; gular sutures widely separated, or absent.

Protergum longer than meso- or metatergum, often longer than the two combined; sides of thoracic terga not explanate; meso- and metaterga occasionally subdivided; prothoracic venter sometimes with presternal sclerite (subdivided in *Cybister*). Legs usually long and slender, 6-segmented, often lined with swimming hairs, with elongate coxa, tibia and tarsus equal in length, and two, simple pretarsal claws; coxae relatively narrowly separated. Abdomen usually more than twice as long as thorax, gradually narrowing posteriorly; terga with or without sharp lateral edges. Tergum VIII usually much longer than VII, narrowed apically and more or less conical, with pair of terminal spiracles present in all instars (Fig. 52Y); sides of segments VII and VIII with fringes of swimming hairs in Dytiscinae; segment IX more or less enclosed within apex of VIII and bearing a pair articulated 1- or 2-segmented urogomphi, which may be short to very long; segment X membranous and not apparent. Annular spiracles present on mesothorax and segments I–VII in final instar. [Böving and Craighead 1931; Watts 1963, 1964; Bertrand 1972; De Marzo 1979; Alarie 1991; Spangler 1991; Alarie *et al.* 2000, 2001, 2002, 2004; Balke 2005].

Classification and World Distribution

Dytiscidae is a large, worldwide family with ~4000 species in 175 genera. The most recent classification (Miller 2001) divides the group into ten subfamilies: Matinae, Laccophilinae (with Laccophilini and Agebetini), Coptotominae, Copelatinae, Hydrodytinae, Hydroporinae (with nine tribes), Agabinae (with Agabini and Hydronebriini), Lancetinae, Colymbetinae (with three tribes) and Dytiscinae (with seven tribes). [Nilsson *et al.* 1989; Nilsson 2001].

Australian Fauna

The Australian dytiscid fauna includes eight of the ten subfamilies (Coptotominae and Hydrodytinae are each based on a single New World genus). The family is widely distributed in Australia with the greatest number of species in the south-eastern part of the continent. Dytiscids inhabit a variety of lentic and some lotic freshwater habitats, but they are most abundant in the littoral zone at the edges of lakes and ponds. They may also occur in water-filled ditches, billabongs, pools in intermittent streams, and even inland saline lakes. Both adults and larvae are predators on other aquatic animals, including insects, crustaceans, worms, leeches, molluscs, tadpoles or even small fish. Adults and some larvae are capable of taking in solid and liquid food through the mouth opening, but most larvae have a closed mouth opening and use the mandibular channels both for injecting digestive enzymes into the prey and for sucking in the resulting fluids by means of a cibarial-pharyngeal pumping apparatus. Eggs are often deposited in slits made by the ovipositor in the stems of aquatic plants, but this habit is not universal. Pupation takes place in a cell formed by the larva in damp soil near, but not in, water. Both adults and larvae come to the surface for air, which they obtain through the terminal pair of spiracles, and adults may also store air beneath the elytra. Adults are normally capable of sustained, crepuscular or nocturnal flight, which is necessary for dispersal to isolated habitats; they are often attracted to lights and may mistake artificial shining surfaces, such as glass, for water.

Eretes australis (Erichson) is a common and widespread inland species and *Cybister godeffroyi* Wehncke from Cape York Peninsula is the largest Australian species. Although almost all dytiscids are aquatic, two species of *Terradessus* are blind and terrestrial, occurring in high altitude rainforests in northern QLD (Watts 1982; Brancucci and Monteith 1997). Hydroporinae is the largest subfamily, with more than 20 genera, and included within this group are some unusual forms, such as the species of *Carabhydrus*, which lack the usual streamlined body form and look more like Carabidae. Most *Carabhydrus* have been collected from small streams under in closed canopy vine forests (Larson and Storey 1994), but *C. stephanieae* Watts *et al.* (2007) is stygobitic, an obligate inhabitant of groundwater. Other dytiscids from the central and western parts of Australia, including many species of *Limbodessus* and *Paroster*, as well as *Exocelina abditus* (Balke *et al.*), also occur in underground waters (Watts and Humphreys 1999, 2000, 2001, 2003, 2004, 2008; Balke and Ribera 2004; Balke *et al.* 2004). [Larson 1993; Balke 2005].

Australian Subfamilies, Tribes and Genera

Matinae: *Batrachomatus* Clark.
Laccophilinae: *Australphilus* Watts, *Laccophilus* Leach.
Copelatinae: *Copelatus* Erichson, *Exocelina* Broun.
Hydroporinae: Bidessini: *Allodessus* Guignot, *Clypeodytes* Regimbart, *Gibbidessus* Watts, *Hydroglyphus* Motschulsky, *Kakadudessus* Hendrich & Balke, *Leiodytes* Guignot, *Limbodessus* Guignot, *Neobidessodes* Hendrich & Balke, *Uvarus* Guignot. **Carabhydrini:** *Carabhydrus* Watts. **Hydroporini:** *Antiporus* Sharp, *Barretthydrus* Lea, *Chostonectes* Sharp, *Megaporus* Brinck, *Necterosoma* Sharp, *Paroster* Sharp, *Petrodessus* Miller, *Sakaliporus* Watts, *Sternopriscus* Sharp, *Tiporus* Watts. **Hydrovatini:** *Hydrovatus* Motschulsky. **Hyphydrini:** *Hyphydrus* Illiger. **Tribe ?:** *Terradessus* Watts.
Agabinae: *Platynectes* Sharp.
Lancetinae: *Lancetes* Sharp.
Colymbetinae: *Rhantus* Dejean.
Dytiscinae: Aciliini: *Rhanaticus* Sharp, *Sandracottus* Sharp. **Cybistrini:** *Austrodytes* Watts, *Cybister* Curtis, *Onychohydrus* Schaum & White, *Spencerhydrus* Sharp, *Sternhydrus* Brinck. **Hyderodini:** *Hyderodes* Hope. **Eretini:** *Eretes* Laporte. **Hydaticini:** *Hydaticus* Leach.

9. CARABIDAE

Adult (Figs 1B–C, E–F, 54B, L, 66A–K)

Length 1.0–60 (usually 2.5–35) mm. Body ~1.5–4.2 times as long as wide, broadly elliptical to narrowly elongate; sides rarely evenly curved (e.g. *Stichonotus*, *Sphallomorpha*), usually narrowest between prothorax and elytra; strongly flattened to moderately convex; colour usually black, sometimes metallic, occasionally yellow to red or bicolored; upper surfaces usually smooth, shiny and subglabrous, almost always with long setae at fixed positions on head, prothorax and elytra (Fig. 21A) (reduced in Rhysodinae, absent in Paussini); surfaces occasionally with dense clothing of decumbent hairs (e.g. *Chlaenius*) or suberect hairs (e.g. *Arthropterus*, *Craspedophorus*, *Helluonidius*).

Head (Fig. 13A) almost always prognathous (hypognathous in some Pseudomorphinae), usually more or less elongate and not abruptly constricted posteriorly to form neck, but with a moderately abrupt constriction and neck present in Paussini, Rhysodinae and various Harpalinae (e.g. *Ophionea*, *Zuphium*). Eyes usually relatively large, protuberant, entire and finely facetted, with or without interfacetal setae; variously reduced in some forms, occasionally absent (e.g. *Illaphanus*). Antennal insertions usually exposed or barely concealed from above (distinctly concealed in Pseudomorphinae and Scaritini) subantennal grooves usually absent (short in Scaritini, long and extending well behind eyes in *Sphallomoprha*). Paired frontal grooves sometimes present. Frontoclypeal suture present or absent. Labrum free, usually transverse, apically truncate, rounded, emarginate, bilobed or multidentate. Antennae 11-segmented (apparently 10-segmented in Paussini due to enclosure of pedicel within apex of scape) and almost always filiform (perfoliate in *Arthropterus*, Fig. 29J; serrate in *Megalopaussus*, Fig. 66H; moniliform in Rhysodinae, Fig. 54J; incrassate in some Pseudomorphinae), with antennomeres 3–11 or 4–11 densely pubescent and basal antennomeres glabrous or with few scattered setae only (except in Rhysodinae, where pubescence is more uniform and not very dense); scape usually elongate and often much longer than pedicel (short and broad in Paussini and Rhysodinae). Mandible short and broad to moderately elongate, almost always unidentate (bidentate in Rhysodinae); outer edge usually with concavity (scrobe) which may contain a seta; inner edge usually with one or more teeth (terebral and retinacular teeth and ridges of Acorn and Ball 1991) and fringe or brush of setae at base; mola reduced or absent (Fig. 13E). Maxilla almost always with 2-segmented, palpiform galea and falcate lacinia with several teeth along inner edge; palp 4-segmented with distinct palpifer and apical palpomere usually cylindrical to fusiform (Fig. 13F), sometimes slightly expanded apically (*Calosoma*), subtriangular or securiform (*Pamborus*, Peleciini, Panagaeini, some Lebiini), or aciculate (some Trechinae). Mentum usually strongly transverse and apically biemarginate (Fig. 13B) (enlarged, fused with submentum and partly fused laterally to head capsule in Rhysodinae); ligula large, well sclerotised and truncate; labial palp 3-segmented and usually similar to that of maxilla. Gular sutures widely separated, gula moderately elongate, with or without median apodeme. Corpotentorium usually well developed; laminatentorium often complete. Cervical sclerites absent.

Pronotum ~0.35–2.25 times as long as wide, usually widest anteriorly or at middle and slightly to much narrower at base than combined elytral bases, widest basally in a few groups (e.g. Pseudomorphinae, Oodini, *Stichonotus*), with sides weakly rounded or sinuate. Lateral pronotal carinae usually well developed and complete, with or without distinct bead, and anterior angles usually at least weakly produced forward (carinae indistinct, incomplete or absent and anterior angles rounded and not produced in Cicindelinae, Paussini, *Setodyschirius*, *Apotomus*, and some derived Harpalinae (e.g. *Ophionea*); posterior angles right to obtuse or rounded, rarely slightly acute (e.g. *Stichonotus*); posterior edge almost always straight or slightly rounded; disc almost always flat to slightly convex and simple, often with slender, median longitudinal impression (with transverse impressions in Cicindelinae, longitudinal grooves in Rhysodinae). Lateral walls of prothorax (pleuron and sternum) usually oblique, sometimes almost vertical (transverse and somewhat concave in Pseudomorphinae); propleuron ending well before anterior edge of prothorax (Fig. 17D, G). Prosternum in front of coxae almost always at least slightly longer and often much longer than shortest diameter of procoxal cavity and slightly to moderately convex. Prosternal process always complete, slightly to strongly, sometimes abruptly, curved in lateral view and slightly to strongly expanded at apex, which is rounded to truncate (parallel-sided and not curved behind coxae in Carabini, interrupted at middle and concave in lateral view in Paussini). Procoxae usually globular or slightly transverse, projecting below prosternum in Cicindelinae and Paussini, without concealed lateral extensions or coxal plates, almost always with mesal knob (sternocoxal articulation) fitting into cavity on prosternal process (absent in Paussini); trochantin free, small, concealed (Fig. 17G). Procoxal cavities circular to slightly transverse, narrowly to widely separated (more or less contiguous in Paussini), almost always closed externally (open in Carabinae), internally open or closed by slender bar.

Scutellar shield usually exposed, sometimes absent (e.g. Rhysodinae, *Illaphanus*) or concealed by pronotum (*Mystropomus*, *Pheropsophus*, some Cicindelinae), not abruptly elevated at base, broadly rounded to acute apically. Elytra ~1.1–3.0 times as long as combined width and 1.6–5 times as long as pronotum, usually complete with apices conjointly rounded or angulate, occasionally obliquely truncate, rarely spined, exposing a partial tergite (squarely truncate, exposing at least one entire tergite in Paussini and Brachininae; with elevated groove for directing defensive spray at posterolateral angle of each elytron); elytral bases with sharp humeri or more rounded, sometimes abruptly constricted to form a peduncle, the anterior edge of which fits within the prothorax; scutellum restricted to peduncle and not extending onto main portion of disc in Scaritinae and Broscinae; disc sometimes impunctate, irregularly punctate or with longitudinal grooves or costae, but usually with eight distinct puncture rows or striae (nine in Migadopini, 16 in *Calosoma*); base commonly with short scutellar striole; epipleura almost always incomplete, ending apicolaterally and not at sutural angle (extending to or almost to sutural angle in *Mystropomus*, some Rhysodinae and Carenini); elytron sometimes with ventral (internal) longitudinal ridge near lateral edge. Mesoventrite about as long as wide, usually at least partly separated from mesanepisterna, which are well separated from one another; anterior edge usually on at least a slightly different plane than metaventrite, with large paired, strongly inclined procoxal rests and shallow mesoventral cavity in Carabinae, but usually with smooth collar-like region fitting within posterior edge of prothorax; discrimen absent. Mesoventral process undivided, emarginate. Mesocoxae globular, not projecting, with mesal knob fitting into cavity near apex of metaventral process; trochantins concealed. Mesocoxal cavities usually moderately to widely separated (narrowly separated in *Mystropomus*, contiguous in Paussini), usually closed laterally by meeting of mesoventrite and metaventrite (partly closed by mesepimera in Carabinae, Rhysodinae and Migadopinae); metanepisternum never reaches mesocoxal cavity; meso- and metathoracic joint within coxal cavities membranous. Mesometaventral junction usually monocondylic. Metaventrite usually transverse or subquadrate (elongate in Rhysodinae) with long discrimen usually extending to base of metaventral process and complete transverse (katepisternal) suture (discrimen short and katepisternal suture absent in Rhysodinae); exposed portion of metanepisternum short, broad, and subquadrate, to elongate and narrowed posteriorly (very long and narrow or concealed in Rhysodinae); metepimera exposed or concealed; metatrochantins concealed. Metacoxae slightly to strongly transverse, closely adpressed to metaventrite but capable of limited movement, their bases straight to moderately oblique; contiguous or well separated, not mesally connate, extending laterally to meet elytral epipleura only in Rhysodinae, with reduced coxal plates on mesal portion only. Metendosternite usually with a long, slender stalk, short, broad lateral arms (sometimes absent) and a long anterior process with tendons moderately widely separated on either side of a deep emargination. Hind wing highly variable and sometimes absent, but basically of the adephagan type with long bending zone occupying most of radial bar (anterior wing strut), moderately long apical field, with four radial extensions, RA_{1+2} and RA_{3+4} separating at about mid-wing but partly obliterated by triangular folding area, one or two cross-veins joining RA_{1+2} and RA_{3+4}, pterostigma well developed, RP complete almost to wing base, joined to RA_{3+4} by cross-vein r1, cross-vein r2 and r3 absent, oblongum cell usually present and transversely oriented (absent in Cicindelinae and Rhysodinae), MP_{1+2} with very short bending zone or hinge, medial field with four free veins and no medial fleck, wedge cell absent in Paussini, Cicindelinae and Rhysodinae and anal lobe well developed, without embayment. Legs almost always long and slender, often armed with long setae or spines (shorter in Rhysodinae, with expanded femora in some Pseudomorphinae or expanded and flattened tibiae in *Arthropterus*); protibia usually with subapical antenna cleaner (Fig. 16N, 30C) (apical in Cicindelinae, Carabinae and Rhysodinae, absent in Paussini); outer edge and outer apical angle of protibia in Scaritini with distinct teeth (Fig. 16M), and tibial apices in Rhysodinae with distinct teeth which may be curved and hook-like on protibia; tibial spurs usually paired, sometimes asymmetrical (obsolete or absent in Rhysodinae and Paussini), one spur almost always located well before apex (both spurs terminal in *Mystropomus* and Cicindelinae); tarsi 5-5-5; tarsomeres usually relatively long and slender, without ventral lobes and armed with sparse and often stout setae; basal one to four protarsomeres in some males expanded and clothed beneath with adhesive setae (in some males expanded basal tarsomeres also occur on mid legs, and in at least some *Arthropterus* they occur on all legs); penultimate tarsomere bilobed in some Moriomorphinae, Lebiini and Odacanthini; pretarsal claws usually simple (serrate or pectinate in some Lebiini); pro- and mesotrochanters relatively short with slightly oblique femoral joint, but metatrochanters enlarged and attached with a strongly oblique joint, so that the metafemur is attached directly to the coxa.

Abdomen usually with six ventrites (sternites II–VII) (seven or eight in Brachininae), the first three of which are fused; suture between ventrites 2 and 3 often weak or absent; ventrite 1 (sternite II) usually divided into two or occasionally three parts by metacoxal cavities (Figs 21C, 28I). Basal abdominal tergites usually lightly sclerotised with the last one or two more heavily so, sometimes with most tergites well sclerotised. Functional spiracles on abdominal segments I–VIII, the last one or two pairs located on tergite. Defensive gland openings located at posterior edge of tergite VIII. Tergite VIII in male usually undivided; sternite VIII sometimes divided into two parts by median suture. Segments IX and X in male forming genital capsule, which is usually asymmetrical and broadly rounded to acute anteriorly; tergite IX more or less fused to segment X. Aedeagus of modified trilobate type, in which phallobase is not apparent and parameres articulate with base of penis, which lacks basal struts; parameres usually unequal; endophallus commonly with sclerites of varied shapes. Tergite VIII in female sometimes divided by median suture; sternite VIII divided into two lateral sclerites (gonocoxosternites). Ovipositor short and broad (Fig. 30F) to slightly elongate, lightly to moderately heavily pigmented, often without baculi, undivided or divided; coxites usually more heavily sclerotised apically, without styli. [Darlington 1943, 1961, 1971; Moore 1967b, 1972a, 1977, 1978b; Erwin 1970, 1974, 1980; Grandi 1971; Forsyth 1972; Evans 1977; Bell and Bell 1978, 1979, 1982, 1991; Forsythe 1982; Baehr 1984a, 1984b, 1985a, 1985b, 1986a, 1986b, 1986c, 1986d, 1987a, 1987b, 1988a, 1988b, 1989, 1990a, 1990b, 1990c, 1990d, 1992, 1994, 1996, 1997, 1999, 2003, 2004a, 2004b, 2005, 2006a, 2006b, 2006c, 2007, 2008, 2009a, 2009b, 2011, 2012a, 2012b; Straneo and Ball 1989; Liebherr 1990; Acorn and Ball 1991; Moore and Lawrence 1994; Liebherr and Will 1998; Roig-Juñent 1998; Ball and Bousquet 2000; Arnett and Ivie 2000; Arndt *et al.* 2005].

Larva (Fig. 56A–C, F–G)

Body usually narrowly elongate, more or less parallel-sided and slightly flattened; lightly to heavily sclerotised dorsally and usually lightly sclerotised ventrally, sometimes with well sclerotised ventral plates in some groups (e.g. Carabini, Scaritinin); yellowish-brown to black in colour, rarely bicolored; surfaces usually smooth and clothed with scattered long setae (expanded or clubbed setae present in Paussini and some Pseudomorphinae).

Head prognathous (sometimes hyperprognathous) and protracted, usually about as long as wide, slightly flattened with sides subparallel or weakly rounded (occasionally transverse or strongly rounded laterally). Head often slightly constricted posteriorly forming a pair of lateral grooves (cervical grooves). Posterior edge of head capsule not to deeply emarginate; epicranial stem short to moderately long, sometimes absent; frontal arms V-shaped, often undulate, usually contiguous at base (well separated in *Arthropterus*); median endocarina absent. Stemmata usually six on each side, arranged in two vertical rows (occasionally reduced in number or absent); sometimes with ocular groove behind stemmata. Frontoclypeal suture absent. Labrum completely fused to head capsule; anterior edge of clypeolabrum usually at least slightly produced forward to form nasale, which may be flanked by paranasalia; nasale sometimes with 2 to 4 or more teeth. Antennae almost always moderately long and 4-segmented, with lateral, palpiform or dome-like sensorium at apex of antennomere 3 (reduced in Cicindelinae). Mandibles usually broad at base, with narrow, strongly curved, unidentate apex, with or without small, basal penicillus and usually with large retinaculum. Ventral mouthparts almost always protracted (Fig. 40L) (somewhat retracted in Rhysodinae), with maxillae and labium well separated at base or contiguous, without maxillary articulating area; cardo short and transverse; stipes usually elongate; lacinia usually short, fixed, sometimes absent; galea 2-segmented and palpiform (galea and lacinia highly reduced, fixed and covered with fringed membranes in Rhysodinae, Fig. 40H); maxillary palp 3-segmented with well developed, segment-like palpifer. Labium consisting of prementum and postmentum; ligula present or absent; labial palps usually

2-segmented without distinct palpigers (1-segmented in Rhysodinae). Hypostomal rods and ventral epicranial ridges absent. Gular region usually long (shorter in Rhysodinae); gular sutures very narrowly separated or fused together.

Prothorax usually somewhat larger than meso- or metathorax (greatly enlarged in Cicindelinae); thoracic and abdominal terga often well sclerotised, with anterior transverse carina on meso- and metaterga and most abdominal terga; terga somewhat expanded laterally in Carabinae; prothoracic venter with undivided presternum. Legs usually long (relatively short in Rhysodinae), 6-segmented, with elongate coxa and usually two pretarsal claws (a single claw in Rhysodinae, Paussini, most Trechinae, some Brachininae and a few Scaritinae). Abdomen much longer than thorax; segments I–IX usually with tergal plates and often with a series of lateral and ventral sclerites; abdominal terga I–VII and usually meso- and metaterga lightly sclerotised but with paired transverse rows of asperities in Rhysodinae; tergum V with several large hooks in Cicindelinae (Fig. 56G); terga V–VII transverse rows of small hooks in some Pseudomorphinae (Fig. 56F). Tergum IX usually slightly shorter than VIII, extending onto ventral surface, and bearing a pair of urogomphi, which may be fixed or basally articulated and undivided or subdivided into two or more segments (urogomphi absent in Paussinae, Cicindelinae, Rhysodinae, Carenina and Pseudomorphinae); sternite IX exposed, simple. Segment X usually well developed, visible from above, cylindrical or subconical and pygopod-like (shorter and lightly sclerotised in Cicindelinae, Rhysodinae and some Pseudomorphinae, reduced and located ventrally beneath terminal plate formed by segments VIII and IX in Paussinae). Spiracles annular, with closing apparatus, present on thorax and abdominal segments I–VIII in all instars, those on segment VIII neither enlarged nor displaced. [Gardner 1933a; Emden 1942a; Grandi 1971; Burakowski 1975a; Thompson 1979; Bousquet and Goulet 1984; Bousquet 1986, 1987, 1991; Bell 1991; Arndt 1993, 1998; Kasahara and Miyano 1997; Ball and Bousquet 2000; Arndt *et al.* 2005; Beutel 2005c].

Classification and World Distribution

The Carabidae is a large, worldwide family of terrestrial, primarily carnivorous beetles, including more than 40 000 species in more than 1500 genera. The classification used here differs from that used by Lawrence and Britton (1994), which was based on that used in Moore *et al.* (1987), in that the family Rhysodidae is considered to be a carabid subfamily, several tribes often placed in the subfamily Carabinae, including Scaritini, Clivinini, Apotomini, Psydrini, Migadopini, Broscini, Zolini, Trechini, Bembidiini and Pogonini, have been reorganised to form several subfamilies, and the majority of carabine tribes are now placed in the subfamily Harpalinae. For the most part, this follows Lawrence and Newton (1995), Arndt *et al.* (2005), and is more or less consistent with cladograms produced from morphological (Liebherr and Will 1998) and molecular (Maddison *et al.* 2000) analyses. Discussions of carabid classification may be found in Ball (1979), Ball *et al.* (1998) and Arndt *et al.* (2005). Australian subfamilies recognised here are listed below with their included tribes and genera. Additional non-Australian subfamilies include Nototylinae (Brazil), Cicindisinae (Argentina, Asia Minor), Gehringiinae (north-western North America), Notokasiinae (Neotropical), Elaphrinae (Holarctic), Nebriinae (Holarctic), Omophroninae (North to Central America, Palaearctic, Afrotropical, Oriental), Loricerinae (Palaearctic), Hiletinae (South America, Africa, Madagascar, South-east Asia), Melaeninae (South America, Asia Minor, Central Asia, South-east Asia, Africa), Promecognathinae (North America, southern Europe, South Africa) and Siagoninae (northern South America, southern Palaearctic, Africa, India, Malay Archipelago, Philippines).

Australian Fauna

Carabidae, or predaceous ground beetles, are widely distributed in Australia, occurring in all environments and usually easily recognisable by their prominent mandibles and palps, shortened metaventrite with a distinct transverse (metakatepisternal) suture, striate elytra, long, slender legs, and a characteristic series of punctures of fixed position bearing tactile setae. Most carabids have an antennal cleaning organ consisting of an excavation lined with a comb-like setal fringe and located near the apex of the protibia. Carabid antennae are usually more or less uniformly pubescent beginning with the third, fourth or fifth antennomere. Many adult ground beetles are black and shiny, but some are brightly coloured and a few pubescent. Larvae (Fig. 56A–C) are usually campodeiform, with well developed legs, prominent antennae and mandibles and almost always fixed urogomphi, which may be segmented. As a dominant group of terrestrial predators, carabids occur in a wide variety of habitats. Darlington (1961, 1971) placed them in three main ecological groups: **geophiles** living on the ground but not associated with water, **hydrophiles** living at the edges of streams or ponds or in swamps, and **arboricoles** living above the ground on tree trunks, logs or leaves. The proportion of geophiles: hydrophiles: arboricoles in Australia was found to be 2 : 1 : 1, while that in New Guinea was 1.2 : 1 : 1.

The subfamily **Paussinae** includes two tribes occurring in Australia, Ozenini and Paussini, which are quite different in appearance but share a unique crepitating system and lack several features which characterise the more derived carabid groups. Ozaenini are represented by a few species of *Mystropomus* (Fig. 54F) occurring mainly in coastal rainforests of NSW and QLD (Monteith and Turco 2010); these have filiform antennae, slender legs and fixed tactile setae, as in most Carabidae, and a plesiomorphic type of antenna cleaner, which does not displace one of the tibial spurs. The Australian Paussini include the northern Queensland *Megalopaussus amplipennis* Lea (Fig. 66H), which has been found in a chamber within the nest of a species of the ant genus *Notostigma*, and the genus *Arthropterus* (Figs 29J, 54K), which is distributed throughout Australia, especially in the more arid regions. Several species of *Arthropterus* have been associated with ants (Lea 1910), and two species in western VIC were abundant on the ground in the evening accompanied by a species of *Camponotus*. Adult Paussini are unique in the family for having flattened legs, flattened and serrate or perfoliate, apparently 10-segmented antennae (with the pedicel concealed within the scape) and no protibial antenna cleaner or fixed, tactile setae. Larvae of both tribes (as well as those of the exotic Metriini) have a highly modified, disc-like abdominal apex (Bousquet 1986; Moore 2004, 2009; Moore *et al.* 2011). The use of the terminal disc in prey capture has been described by Di Giulio and Vigna Taglianti (2001) and Moore and Di Giulio (2006) for the exotic paussines *Pachyteles* species and *Goniotropis kuntzeni* Bänninger, respectively.

Cicindelinae (tiger beetles, Figs 1E, 66F) are usually brightly coloured, with a large labrum, large, protruding eyes, long, slender legs and non-striate elytra. Adults of most species are capable of running and flying rapidly in open areas, such as river banks, beaches, forest clearings and salt pans, but others occur on tree trunks and a few are wingless. Larvae (Fig. 56G) are highly modified ambush predators living in vertical tunnels in the soil or sometimes in wood; they lack urogomphi, have a flattened head which serves as a lid or plug, and are armed with two or three pairs of dorsal abdominal hooks, which serve to anchor them in the tunnel. Species of *Cicindela* inhabit sandy heaths, beaches, sand banks or salt pans and are diurnal, while *Megacephala* species are nocturnal and are metallic blue, violet or green in colour. *Dystipsidera* and *Rhysopleura* are active on tree trunks, as is the wingless, black *Tricondyla aptera* (Olivier), which occurs from Cape York to the Moluccas and Vanuatu. Australian *Cicindela* were revised by Freitag (1979), subgenera of *Cicindela* were treated by Sumlin (1982), and a revision of Australian species excluding *Cicindela* was published by McCairns *et al.* (1997). [Pearson 1988].

The subfamily **Carabinae** is primarily a Northern Hemisphere group, and the Australian fauna includes only the worldwide genus *Calosoma* (Carabini) and the autochthonous genus *Pamborus* (Pamborini) (Fig. 66B). *Calosoma* species are widely distributed predators of caterpillars, while *Pamborus* species feed on snails and earthworms in eastern coastal forests (Takami and Sota 2006). Although sometimes placed in a separate family, the **Rhysodinae** (Fig. 54J) are now generally considered to be Carabidae, based on the presence of tactile setae, a carabid-like protibial antenna cleaner, and the structure of the pygidial defence system (Forsyth 1972). The position of the group within the family is still uncertain, but Beutel (1990a), Bell (1998) and Liebherr and Will (1998) have all suggested that the group is related to Scaritini or Clivinini. The subfamily is represented in Australia by the endemic *Leoglymmius* and *Sloanoglymmius* and three widely distributed genera *Kaveinga*, *Rhyzodiastes* and *Omoglymmius* (Bell and Bell 1978, 1979, 1982, 1991). Both adults and larvae of Rhysodinae live in old, decayed logs and are thought to be associated with Myxomycetes. Larvae differ considerably from those of other carabids in having retracted ventral mouthparts, reduced gula, reduced maxillary lobes, sometimes covered with fringed membranes (Fig. 40H), transverse rows of tergal asperities, no urogomphi and a ventrally directed segment X. The subfamily **Scaritinae** (Figs 1B, F, 54E, 66A) include several small to very large fossorial carabids with a pedunculate body (large prothorax capable of rotating on a pterothoracic pedicel) and modified digging protibiae (Fig. 16M). Several genera of smaller species are usually included in a tribe Clivinini and the larger ones in Scaritini; however, Moore and Lawrence (1994) suggested that many of the larger Australian taxa, such as *Carenum*, *Euryscaphus* and related genera (but not *Scaraphites*, *Geoscaptus* or *Steganomma*) belong to a separate lineage (Carenini) based on an unusual type of larva (Fig. 35C) plus several adult features. [Baehr 2006a, 2008].

Migadopinae is a small group of Notogean carabids, which are relatively rare in Australia and have relatives in New Zealand and Chile (Baehr 2009b, 2012e). **Broscinae**, as here delimited, includes Apotomini, based on Roig-Juñent (1998, 2000), who considers this tribe and the exotic Melaenini to be sister to the remaining broscines. Broscini is an amphipolar group (Crowson 1980) restricted to the temperate parts of Northern and Southern Hemispheres; they are somewhat pedunculate like the Scaritinae, but differ in other respects and have a distinctive type of larva (Fig. 56C). Broscines are most widely distributed in arid and semi-arid regions (Erwin 1980, Baehr 1990d). The subfamily **Trechinae** includes many small to minute carabids (Fig. 54H) most abundant in the mountains of south-eastern Australia and TAS and often frequenting damp situations; *Perileptus constricticeps* (Sloane) is found in gravel at the edges of streams, and *Geodetrechus mendumae* Moore (Fig. 54B) is an eyeless, depigmented troglobite (Moore 1972a). The tribe Bembidiini include some of the smallest carabids (*Illaphanus* spp.), while Zolini include the cave-inhabiting *Idacarabus troglodytes* Lea. [Baehr 1984a, 1986a, 1987a, 1988a, 1990c; Giachino 2005; Grebennikov 2008b]. **Moriomorphinae** includes most of the genera formerly placed in the tribe Psydrini, while *Laccocenus* remains in **Psydrinae**, along with the Northern Hemisphere genera *Psydrus* LeConte and *Nomius* Laporte. [Baehr 1999, 2003, 2004a, 2007, 2011; Liebherr 2011].

Of the more than 500 known species of **Brachininae** or bombardier beetles, only seven species of the widely distributed genus *Pheropsophus* (Figs 54L, 66J) occur in Australia (Erwin 1970; Baehr 2012c). Adults are black with yellow spots, and when disturbed they emit a small cloud of visible vapour from the abdominal apex, accompanied by an audible 'pop'. The crepitating mechanism responsible for this defensive discharge and the chemistry of the spray are similar to that found in Paussinae (Aneshansley *et al.* 1983). Although some larval Brachininae are known to be ectoparasitoids of beetle pupae (Erwin 1967), the Neotropical *Pheropsophus aequinoctalis* (Linnaeus) and the Japanese *Stenaptinus jessoensis* (Morawitz) are known to prey on the egg masses of mole crickets (Orthoptera: Gryllotalpidae) (Habu and Saganaga 1965; Frank *et al.* 2009).

Pseudomorphinae is a predominantly Australian group of oblong to ovoid and sometimes strongly flattened beetles, whose antennae and legs may be completely concealed beneath the body when at rest. Adults of *Sphallomorpha* (Fig. 54I) have filiform antennae as in most carabids, but in other Australian genera, such as *Adelotopus*, they are short and club-like. Larvae of *Sphallomorpha* (Fig. 56F) live in vertical tunnels at the periphery of ant nests, and, like cicindelines, they are ambush predators and have abdominal hooks to secure them in the tunnel (Moore 1974). Larvae of *Adelotopus* and at least some exotic *Pseudomorpha* are physogastric inquilines in ant nests (Erwin 1981). An unusual feature in this subfamily is ovovivipárity, which has been reported for *Pseudomorpha* (Liebherr and Kavanaugh 1985) and most Australian genera except *Sphallomorpha* (Baehr 1997). The Australian pseudomorphines have been revised by Baehr (1992, 1994, 1997 and supplements, **see** Baehr 2009a).

The subfamily **Harpalinae**, as here delimited includes most of the other tribes of Carabidae, as indicated below, and these will be covered only briefly in the present account. The tribe Pterostichini forms the largest group of typical, forest-dwelling geophiles in Australia. Species of *Notonomus* (Fig. 54G) are common in coastal forests of eastern Australia, and like some other pterostichines, may exhibit brood care (Brandmayr and Zetto-Brandmayr 1979; Moore 1980). The related tribe Morionini includes several gigantic species, such as species of *Catadromus* (Fig. 66I) and *Hyperion* *schroetteri* (Schreibers), which feeds on scarab larvae in rotten wood. Harpalini are not as well represented in Australia as in other parts of the world. Some harpaline carabids tend to be phytophagous and both adults and larvae may have more heavily built mandibles than their predaceous relatives (Acorn and Ball 1991). Lebiini (Figs 1C, 66G) includes many small, flattened, yellow and black carabids common beneath loose bark or running on bark and wood surfaces at night. Larvae of some exotic lebiines are predators or parasitoids on the pupae of other beetles (Erwin 1979). Although the habits of Australian Helluonini (Fig. 66C) are unknown, exotic species are known to be associated with ants (Kistner 1982) or termites (Gardner 1933a). Other harpaline tribes illustrated here include Dryptini (Fig. 66D), Chlaeniini (Fig. 66E) and Panagaeini (Fig. 66K). Taxonomic works on various groups of Australian Harpalinae include: Straneo and Ball 1989; Baehr 2012f) (Peleciini), Baehr 1984b, 1985a, 1985b, 1986c, 1986d (Zuphiini); Baehr 1988b (Leleupidini); Baehr 1986b, 1996, 2004b; Liebherr 1990 (Odacanthini); Baehr 2010, 2012 (Pentagonicini); Baehr 2012d (Hexagoniini); Baehr 1987b, 1990b, 2006b, 2006c, 2012b (Lebiini); Baehr 1989 (Physocrotaphini); Baehr 2005 (Helluonini). [Moore *et al.* 1987].

Australian Subfamilies, Tribes and Genera

Paussinae: Ozaenini: *Mystropomus* Chaudoir. **Paussini:** *Arthropterus* Macleay, *Megalopaussus* Lea

Cicindelinae: Collyridini: *Tricondyla* Latreille. **Megacephalini:** *Megacephala* Latreille. **Cicindelini:** *Cicindela* Linnaeus, *Distipsidera* Westwood, *Nickerlea* Horn, *Rhysopleura* Sloane.

Carabinae: Carabini: *Calosoma* Weber. **Pamborini:** *Pamborus* Latreille.

Rhysodinae: *Kaveinga* Bell and Bell, *Leoglymmius* Bell and Bell, *Omoglymmius* Ganglbauer, *Ryzodiastes* Fairmaire, *Sloanoglymmius* Bell and Bell.

Scaritinae: Clivinini: *Clivina* Latreille, *Clivinarchus* Sloane, *Platysphyrus* Sloane, *Pseudoclivina* Kult, *Rhysocara* Sloane, *Setodyschirius* Bulirsch, *Syleter* Andrewes. **Scaritini:** *Geoscaptus* Chaudoir, *Parathlibops* Basilewsky, *Scaraphites* Westwood, *Steganomma* Macleay. **Carenini:** *Carenum* Bonelli, *Conopterum* Chaudoir, *Epilectus* Blackburn, *Euryscaphus* Macleay, *Laccopterum* Macleay, *Monocentrum* Chaudoir, *Neocarenum* Laporte, *Neoscaphus* Sloane, *Philoscaphus* Macleay, *Trichocarenum* Blackburn.

Psydrinae: *Laccocenus* Sloane.

Migadopinae: *Calyptogonia* Sloane, *Decogmus* Sloane, *Migadopiella* Baehr, *Nebriosoma* Laporte, *Stichonotus* Sloane.

Broscinae: Apotomini: *Apotomus* Illiger. **Broscini**: *Adotela* Laporte, *Brithysternum* Macelay, *Cerotalis* Laporte, *Chylnus* Sloane, *Eurylychnus* Bates, *Gnathoxys* Westwood, *Percolestus* Sloane, *Percosoma* Schaum, *Promecoderus* Dejean.

Trechinae: Zolini: *Idacarabus* Moore, *Percodermus* Sloane, *Pterocyrtus* Sloane, *Sloaneana* Csiki. **Trechini:** *Austrotrechus* Moore, *Bothynotrechus* Moore, *Cyphotrechodes* Jeannel, *Eutrechopsis* Moore, *Eutrechus* Moore, *Goedetrechus* Moore, *Mimanillus* Moore, *Mimotrechus* Moore, *Nototrechus* Moore, *Paratrechodes* Jeannel, *Perileptus* Schaum, *Pogonoschema* Jeannel, *Sloanella* Jeannel, *Tasmanitachoides* Erwin, *Tasmanorites* Jeannel, *Tasmanotrechus* Moore, *Trechiella* Jeannel, *Trechimorphus* Jeannel, *Trechistus* Moore, *Trechobembix* Jeannel, *Trechodes* Blackburn, *Tropidotrechus* Jeannel. **Bembidiini:** *Austranillus* Giachino, *Bembidion* Latreille, *Elaphropus* Motschulsky, *Illaphanus* Macleay, *Lymnastis* Motschulsky, *Pericompsus* LeConte, *Philipis* Erwin, *Polyderis* Motschulsky, *Pseudillaphanus* Giachino, *Tachys* Stephens, *Tachyta* Kirby, *Tasmanillus* Giachino. **Pogonini:** *Pogonus* Nicolai, *Syrdenoidius* Baehr & Hudson.

Moriomorphinae: *Amblytelus* Erichson, *Celanida* Laporte, *Dystrichothorax* Blackburn, *Epelyx* Blackburn, *Mecyclothorax* Sharp, *Melisodera* Westwood, *Meonis* Laporte, *Moriodema* Laporte, *Moriomorpha* Laporte, *Neonomius* Moore, *Paratrichothorax* Baehr, *Pseudamblytelus* Baehr, *Pterogmus* Sloane, *Raphetis* Moore, *Rhaebolestes* Sloane, *Sitaphe* Moore, *Teraphis* Laporte, *Theprisa* Moore, *Trephisa* Moore, *Trichamblytelus* Baehr.

Brachininae: *Pheropsophus* Solier.

Pseudomorphinae: *Adelotopus* Hope, *Cainogenion* Notman, *Cryptocephalomorpha* Ritsema, *Paussotropus* Waterhouse, *Sphallomorpha* Westwood.

Harpalinae: Morionini: *Catadromus* Macleay, *Hyperion* Laporte, *Morion* Latreille. **Pterostichini:** *Abacetus* Dejean, *Acanthoferonia* Moore, *Brachidius* Chaudoir, *Caelostomus* macleay, *Castelnaudia* Tschitscherine, *Conchitella* Moore, *Cosmodiscus* Sloane, *Cratoferonia* Tschitscherine, *Cratogaster* Blanchard, *Cuneipectus* Sloane, *Cyrtoderus* Hope, *Darodilia* Laporte, *Delinius* Westwood, *Eurystomis* Chaudoir, *Feronista* Moore, *Fouquetius* Maindron, *Gastrogmus* Sloane, *Geobaenus* Dejean, *Leiradira* Laporte, *Lesticus* Dejean, *Liopasa* Tschitscherine, *Litarthrum* Sloane, *Loxandrus* LeConte, *Loxodactylus* Chaudoir, *Loxogenius* Sloane, *Mecynognathus* Macleay, *Megadromus* Motschulsky, *Nelidus* Chaudoir, *Notabax* Moore, *Notolestus* Sloane, *Notonomus* Chaudoir, *Nurus* Motschulsky, *Paranurus* Tschitscherine, *Pediomorphus* Chaudoir, *Phaenaulax* Tschitscherine, *Platycoelus* Blanchard, *Prosopogmus* Chaudoir, *Pseudoceneus* Tschitscherine, *Rhabdotus* Chaudoir, *Rhytisternus* Chaudoir, *Sarticus* Motschulsky, *Secatophus* Laporte, *Setalimorphus* Sloane, *Setalis* Laporte, *Simodontus* Chaudoir, *Terophа* Laporte, *Trichosternus* Chaudoir, *Zeodera* Laporte. **Platynini:** *Colpodes* Macleay, *Dicranoncus* Chaudoir, *Euplynes* Schmidt-Gobel, *Laemostenus* Bonelli, *Lorostema* Motschulsky, *Notagonum* Darlington, *Notoplatynus* Moore, *Violagonum* Darlington. **Perigonini:** *Perigona* Laporte. **Harpalini:** *Acupalpus* Latreille, *Amblystomus* Erichson, *Batoscelis* Dejean, *Cenogmus* Sloane, *Chydaeus* Baehr, *Coleolissus* Bates, *Egadroma* Motschulsky, *Euthenarus* Bates, *Gnathaphanus* Macleay, *Haplaner* Chaudoir, *Harpalus* Latreille, *Hypharpax* Macleay, *Lecanomerus* Chaudoir, *Loxoncus* Schmidt-Goebel, *Nornalupia* Kataev, *Notiobia* Perty, *Notospeophonus* Moore, *Parophonus* Ganglbauer, *Phorticosomus* Schaum, *Stenolophus* Dejean, *Trichotichnus* Morawitz, *Veradia* Laporte. **Peleciini:** *Agonica* Sloane, *Pseudagonica* Moore. **Panagaeini:** *Craspedophorus* Hope, *Dischissus* Bates, *Microcosmodes* Strand, *Tinognathus* Chaudoir, *Trichisia* Motschulsky. **Chlaeniini:** *Chlaenius* Bonnelli, *Hololeius* Laferte-Senectere. **Oodini:** *Anatrichis* LeConte, *Coptocarpus* Chaudoir, *Nanodiodes* Bousquet, *Oodes* Bonnelli. **Licinini:** *Badister* Clairville, *Dircochile* Guérin-Méneville, *Dilonchus* Andrewes, *Horma-*

crus Sloane, *Lacordairia* Laporte, *Lestignathus* Erichson, *Microferonia* Blackburn, *Omestes* Andrewes, *Physolesthus* Chaudoir, *Platylytron* Macleay, *Siagonyx* Macleay. **Odacanthini:** *Anasis* Laporte, *Andrewesia* Csiki, *Archicolliuris* Liebke, *Aulacolius* Sloane, *Basistichus* Sloane, *Clarencia* Sloane, *Deipyrodes* Bousquet, *Dicraspeda* Chaudoir, *Eudalia* Laporte, *Gestroania* Liebke, *Giachinoana* Baehr, *Homethes* Newman, *Lachnothorax* Motschulsky, *Myrmecodemus* Sloane, *Neoeudalia* Baehr, *Ophionea* Klug, *Porocara* Sloane, *Renneria* Baehr, *Tricharnhemia* Baehr. **Pentagonicini:** *Parascopodes* Darlington, *Pentagonica* Schmidt-Goebel, *Scopodes* Erichson. **Ctenodactylini:** *Plagiotelum* Solier. **Hexagoniini:** *Hexagonia* Kirby. **Cyclosomini:** *Aephnidius* Macleay, *Australomasoreus* Baehr, *Caphora* Schmidt-Goebel, *Sarothrocrepis* Chaudoir, *Tetragonoderus* Dejean. **Lebiini:** *Abaditicus* Ball & Hilchie, *Agonocheila* Chaudoir, *Anomotariella* Do, *Anomotarus* Chaudoir, *Aristolebia* Bates, *Austrovelinda* Baehr, *Barrymooreana* Baehr, *Brigalowia* Baehr, *Catascopus* Kirby, *Celaenephes* Schmidt-Goebel, *Cheilagona* Baehr, *Coptodera* Dejean, *Coptoglossus* Chaudoir, *Crassagena* Baehr, *Demetrida* White, *Diabaticus* Bates, *Dolichoctis* Schmidt-Goebel, *Endynomena* Chaudoir, *Geoffreyella* Baehr, *Habutarus* Ball & Hilchie, *Holcoderus* Chaudoir, *Lachnoderma* Macleay, *Lebia* Latreille, *Microlestodes* Baehr, *Minuthodes* Andrewes, *Miscelus* Klug, *Mochtheroides* Andrewes, *Mochtherus* Schmidt-Goebel, *Nototarus* Baehr, *Oxyodontus* Chaudoir, *Parena* Motschulsky, *Philophloeus* Chaudoir, *Phloeocarabus* Macleay, *Physodera* Eschscholtz, *Rugitarus* Baehr, *Somotrichus* Seidlitz, *Speotarus* Moore, *Stricklandia* Macleay, *Syntomus* Hope, *Trigonothops* Macleay. **Dryptini:** *Dendrocellus* Schmidt-Goebel, *Drypta* Latreille. **Zuphiini:** *Acrogenys* Macelay, *Parazuphium* Jeannel, *Planetes* Macelay, *Pseudaptinus* Laporte, *Speothalpius* Moore, *Speozuphium* Moore, *Zuphium* Latreille. **Leleupidiini:** *Colasidia* Basilewsky. **Physocrotaphini:** *Pogonoglossus* Chaudoir. **Helluonini:** *Aenigma* Newman, *Ametroglossus* Sloane, *Creagris* Nietner, *Dicranoglossus* Chaudoir, *Epimicodema* Sloane, *Gigadema* Thomson, *Helluapterus* Sloane, *Helluarchus* Sloane, *Helluo* Bonelli, *Helluodema* Laporte, *Helluonidius* Chaudoir, *Helluosoma* Laporte, *Neohelluo* Sloane, *Platyhelluo* Baehr.

POLYPHAGA

The suborder Polyphaga contains ~90% of beetle families and at least 90% of the species. Adult Polyphaga may be distinguished from almost all members of the other three suborders by: (1) fusion of the protrochantin to the propleuron and the reduction and at least partial internalisation of the resulting trochantinopleuron, so that it barely extends in front of the coxal cavity only in a few Derodontidae and Scirtoidea and never extends behind the coxal cavity (Fig. 17B–C, F, H); (2) presence of lateral cervical sclerites (Fig. 14J) (secondarily reduced in several lineages); (3) formation of a radial cell in the hind wing due to the separation and subsequent fusion of RA_{1+2} and RA_{3+4} (Figs 22A–B, 23D, 24C, 29F) (secondary reduction or loss of the cell in various lineages); (4) absence of an abrupt hinge in the medial bar (MP_{1+2}) (Fig. 29F); and (5) mesocoxal cavities rarely partly closed laterally by a small segment of the metanepisternum but never broadly closed by this sclerite. Larval polyphagans are distinguished by the fusion of tibia and tarsus to form a tibiotarsus and the fusion of the two pretarsal claws to form a single claw (Fig. 50F); both of these features, however, occur also in Myxophaga. [Beutel and Leschen 2005b].

STAPHYLINIFORMIA

Staphyliniformia is used here in the sense of Crowson (1960) in that it includes the superfamilies Hydrophiloidea, Histeroidea and Staphylinoidea, but differs from Crowson's concept in the inclusion of Hydrophiloidea and Histeroidea in a single superfamily and the placement of Hydraenidae in Staphylinoidea (rather than Hydrophiloidea), as suggested by Böving and Craighead (1931), followed by Lawrence and Newton (1982), and accepted by most recent workers. The hind wings of staphyliniforms have the radial cell incomplete at base or highly reduced and followed by an apical hinge-like structure and the apical field extensive with three or four radial extensions (Fig. 23F), and in all but the more basal hydrophiloids, the radio-medial loop and basal portion of RP are reduced or absent and the number of veins in the medial field is reduced to four or usually fewer. In silphid and staphylinid wings there is a distinct secondary hinge well before the apex of the radial bar (Fig. 23A), a synapomorphy of these two families. Larval staphyliniforms usually have articulated urogomphi (reduced in various derived groups) and a spiracular closing apparatus. Hansen (1997c) also proposed that Scarabaeoidea be included within Staphyliniformia, as suggested by Kukalová-Peck and Lawrence (1993) (hydrophiloid lineage) based on wing characters, thus making the series more or less equivalent to Haplogastra of Kolbe (1908), as modified by Jeannel and Paulian (1944). This hypothesis was also supported by a cladistic analysis of adult and larval characters by Beutel and Leschen (2005g). Most of the wing features mentioned above also apply to scarabaeoids, although the latter group tend to have a well developed radial bending zone, in addition to the apical hinge. Other adult features characterising the Staphyliniformia in this expanded sense include: (1) the concealed but distinct lateral sclerite representing segment II, which is not visible ventrally (secondary lost in some and sternite II more extensively developed in others); (2) tarsi usually 5-5-5 and simple; (3) antennae filiform or capitate but rarely serrate or pectinate; (4) protibia often strongly spinose or toothed outwardly; (5) aedeagus never of cucujoid type; (6) metendosternite never of hylecoetoid type; and (7) Malpighian tubules free, usually 4 in number. Scarabaeoid larvae, on the other hand never have articulated urogomphi, and the spiracular closing apparatus, if present, appears to be of a different type than that found in staphyliniforms in the restricted sense. Anton and Beutel (2011) found several possible synapomorphies supporting the clade Staphyliniformia including Scarabaeoidea in the morphology of the adult head and the molecular analyses of Caterino *et al.* (2005) and Song *et al.* (2010) also supported this clade. In the molecular cladogram of Hunt *et al.* (2007), however, Scarabaeoidea, Histeroidea, Ptilidae + Hydraenidae and Agyrtidae + Leiodidae form four separate basal poyphagan clades, while most Staphylinoidea form with Hydrophiloidea a clade sister to most of the remaining Polyphaga. [Hansen 1997a; Newton and Thayer 1992; Beutel and Molenda 1997; Korte *et al.* 2004; Beutel and Leschen 2005d].

HYDROPHILOIDEA

This superfamily may not be monophyletic, especially if one accepts the premise that Staphylinoidea and possibly Scarabaeoidea could have evolved from an ancestor with adult features of this group. The inclusion of Histeridae and related families within this group was supported by Archangelsky *et al.* (2005) and Bernhard *et al.* (2009), but several hydrophilid specialists prefer to restrict the group to those taxa here included within the family Hydrophilidae (**see** below). Adult synapomorphies include: (1) eighth abdominal spiracles atrophied; (2) antennomere 8 cupuliform; (3) antenna with 3-segmented club; and (4) tubular mandibular glands (lost in some Hydrophilidae); while those of the larva include: (1) head prognathous; (2) frontoclypeal suture absent; (3) ecdysial lines missing in final instar; (4) labrum fused to head capsule; (5) antennal sensorium lateral; (6) mandibular mola reduced or absent; (7) mandible with penicillus (lost in some Hydrophilidae); (8) lacinia usually absent; (9) galea located on palpifer; (10) abdomen largely membranous; and (11) predatory habits. The families Sphaeritidae, Synteliidae and Histeridae form a monophyletic group and are often placed in a distinct superfamily. Wing venation of hydrophilids, in particular, is relatively complete for Staphyliniformia, with a well developed radial cell and radiomedial loop. The hydrophiloid aedeagus is basically of the primitive trilobate type, with a phallobase and articulated parameres. In histerids and their relatives, however, the parameres form a complete tube enclosing the penis. [Scholtz *et al.* 1994; Beutel 1999a; Beutel and Komarek 2004; Archangelsky *et al.* 2005].

Two hydrophiloid families do not occur in Australia: Sphaeritidae (north-western North America, northern Europe, Eastern Russia, Mongolia, Japan and central China) (Nikitsky 1976b; Newton 2000, 2005a) and Synteliidae (eastern Russia, Japan and China to India and Myanmar, Mexico, Guatemala) (Mamaev 1974; Newton 2005b).

10. HYDROPHILIDAE

Adult (Fig. 57A–C, J–K)

Length ~0.9–42 (usually 2–35) mm. Body ~1.2–2.1 times as long as wide (as much as 2.6 times in *Hydrochus*), with sides more or less evenly rounded in Hydrophilinae (except Berosini) and in Sphaeridiinae, well sclerotised and usually moderately to strongly convex (more or less parallel-sided and slightly flattened in *Hydrochus*; colour yellow to brown or black, occasionally bicolored, often with metallic sheen in *Hydrochus*; dorsal surfaces usually smooth and glabrous, sometimes with fine pubescence in Berosini and Megasternini, with stout, erect setae in *Spercheus*; tuberculate and often costate in *Georissus* and *Spercheus*; ventral surfaces with hydrofuge pubescence (plastron), which is partly reduced in Sphaeridiinae, some Hydrophilinae and Spercheinae and completely absent in Georissinae.

Head more or less prognathous, deeply inserted into prothorax. Eyes usually well developed, protuberant or not, circular and entire to slightly emarginate anteriorly (deeply emarginate in *Sphaeridium*), divided into dorsal and ventral parts in *Amphiops*, finely facetted, without interfacetal setae. Antennal insertions almost always concealed from above (but somewhat exposed due to constriction of clypeus in Megasternini and Omicrini); subantennal groove usually present. Frontoclypeal suture often distinct, angulate, with attached mid-cranial suture; clypeus large, usually truncate or rounded anteriorly (emarginate in *Spercheus*). Labrum usually strongly transverse, well sclerotised and apically truncate or rounded (concealed from above in *Spercheus* and concealed and largely membranous in many Sphaeridiinae). Antennae short, usually 8- or 9-segmented (7-segmented in *Hydrochus*, *Spercheus*, *Berosus*, some *Paracymus* and some *Georissus*), with basal antennomeres glabrous and club 3-segmented (1-segmented in some *Georissus*), pubescent and almost always preceded by cup-like, often asymmetrical antennomere or cupule (Fig. 57D–F) (cupule pubescent in *Spercheus*). Mandibles usually large (highly reduced in Omicrini) and more or less concealed beneath clypeus and labrum; apex unidentate or bidentate; mola well developed and often asymmetrical; prostheca well developed and variable, sometimes including articulated, sclerotised process (Fig. 13I). Maxilla usually with setose galea and lacinia, the former often flattened with apical curved setae arranged in regular rows (slender and somewhat falcate in *Spercheus*), the later with or without uncus, sometimes reduced; maxillary palps slender and usually longer than antennae; apical palpomere fusiform (Fig. 13H). Mentum usually transverse and apically truncate (triangular in *Georissus*); ligula usually bilobed; apical labial palpomere usually fusiform (Fig. 13G) (expanded in *Georissus* and *Hydrochus*). Corpotentorium narrow; laminatentoria well developed, sometimes meeting at midline. Cervical sclerites well developed.

Pronotum ~0.25–0.65 times as long as wide (0.75 times as long in *Georissus* and as much as 1.1 times as long in *Hydrochus*), usually widest posteriorly or at middle (widest anteriorly in some *Hydrochus*), about as wide as combined elytral bases (narrower in *Georissus* and *Hydrochus*); sides slightly to strongly curved and often obliquely explanate (straight or sinuate in *Hydrochus*). Anterior edge strongly produced forward and rounded in *Georissus*. Lateral pronotal carinae complete and usually simple (denticulate in *Spercheus*); anterior angles usually not or slightly produced forward and rounded (strongly produced and acute in *Spercheus*); posterior angles right to rounded and sometimes embracing elytral bases; posterior edge straight to slightly sinuate, with underlying ventral ridge, which interlocks with elytra and scutellar shield. Prosternum in front of coxae about as long as to much shorter than shortest diameter of procoxal cavity, usually tumid to carinate with anterior edge sometimes produced and angulate or acute (flat in *Hydrochus*, biconcave in *Spercheus* and strongly concave in *Georissus*). Prosternal process almost always incomplete or complete but very narrow and apically acute (complete and apically expanded and rounded or truncate in *Hydrochus* and *Spercheus*, absent in *Georissus*). Notosternal sutures complete, but sometimes vaguely indicated. Hypomeron sometimes with glabrous and shiny lateral portion separated by ridge from pubescent mesal portion; with or without anterior groove for reception of antennal club. Procoxae more or less conical and projecting, with short to moderately long concealed lateral extensions; trochantins concealed or exposed; endopleuron not fused to notum. Procoxal cavities slightly to strongly transverse (subcircular in *Hydrochus*), narrowly separated or contiguous, usually broadly open externally (closed in *Hydrochus* and *Spercheus*), often with distinct narrow anterolateral extension, internally closed.

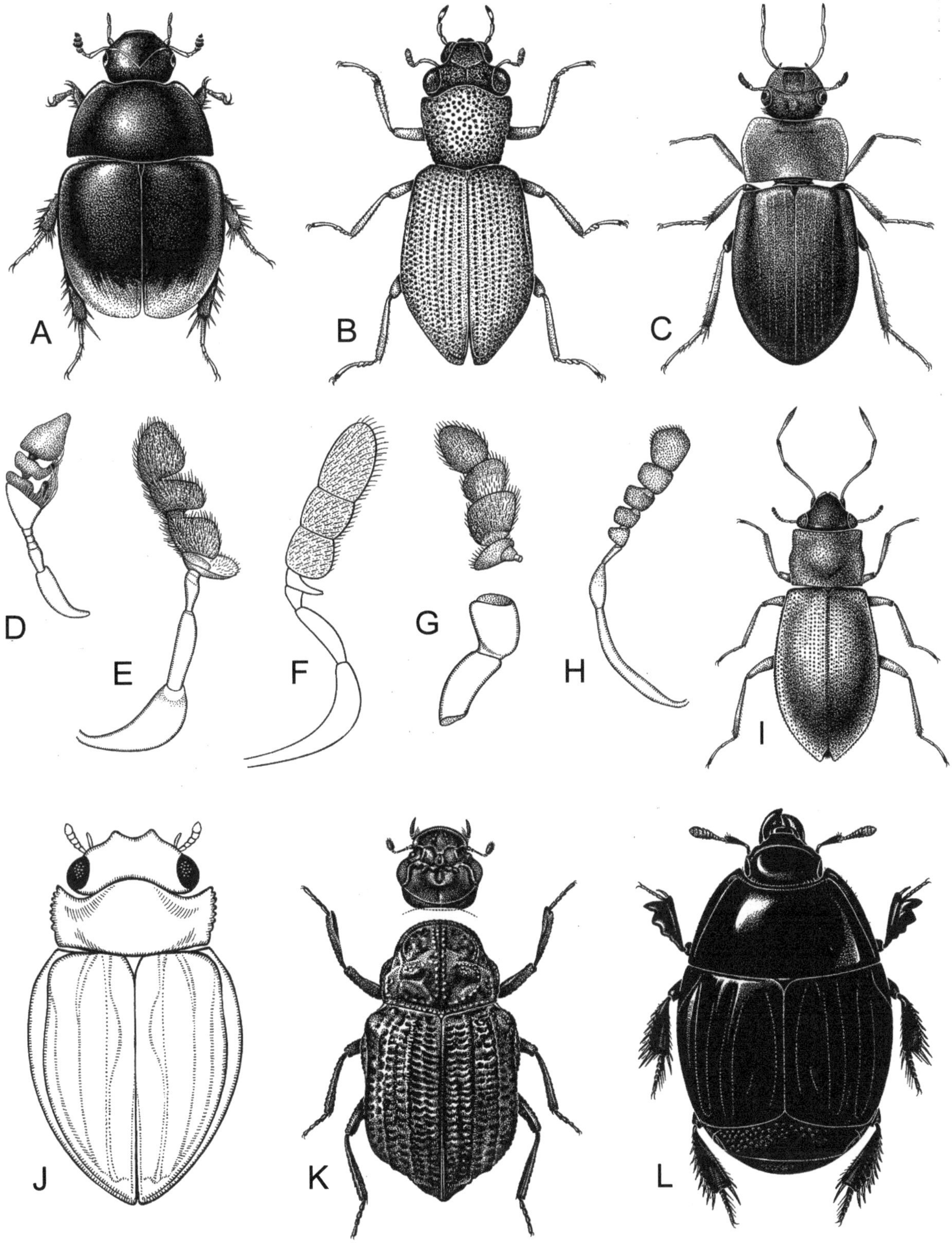

Fig. 57. A–C, Adult, dorsal: A, *Sphaeridium discolor* Orchymont (Hydrophilidae: Sphaeridiinae), dorsal; B, *Hydrochus serricollis* Lea (Hydrophilidae: Hydrochinae), dorsal; C, *Helochares australis* Blackburn (Hydrophilidae: Hydrophilinae), dorsal. D–H, Adult antennae: D, *Hydrophilus* sp. (Hydrophilidae: Hydrophilinae); E, *Berosus australiae* (Hydrophilidae: Hydrophilinae); F, *Hydrochus* sp. (Hydrophilidae: Hydrochinae); G, *Spercheus platycephalus* Macleay (Hydrophilidae: Spercheinae); H, *Tympanogaster* ? sp. (Hydraenidae). I–L, Adult, dorsal: I, *Hydraena luridipennis* Macleay (Hydraenidae); J, *Spercheus platycephalus* Macleay (Hydrophilidae: Spercheinae); K, *Georissus australis* (King) (Hydrophilidae: Georissinae), with hypognathous head removed; L, *Hister walkeri* Lewis (Histeridae: Histerinae).

Scutellar shield abruptly elevated, posteriorly narrowly rounded to acute (minute in *Georissus*). Elytra ~1.0–1.65 times as long as combined width (as much as 1.95 times in *Hydrochus*) and 1.0–5.0 times as long as pronotum; sides usually moderately curved and apices conjointly rounded (subparallel with apices subtruncate in *Hydrochus*); humeri well developed, often somewhat impressed where overlapped by pronotum; disc usually with ten (less frequently with nine) distinct puncture rows or striae, with or without scutellary striole, punctation sometimes irregular or with puncture rows indistinct or fewer in number; epipleura usually incomplete (complete in some Sphaeridiinae), sometimes obliquely or vertically explanate basally. Mesoventrite usually separated by complete sutures from mesanepisterna, which are usually narrowly separated or contiguous at midline (mesanepisternal widely separated at midline in *Georissus* and *Hydrochus*, fused with mesoventrite in some Hydrophilinae and Sphaeridiinae), anterior edge often on different plane than metaventrite, without procoxal rests; mesal portion often with longitudinal carina or ventral projection; mesoventral cavity absent; discrimen absent. Mesocoxae usually slightly to strongly transverse with exposed trochantins (subglobular with concealed trochantins in *Hydrochus*). Mesocoxal cavities usually narrowly to moderately widely separated (contiguous in *Spercheus*, widely separated in *Georissus*), partly closed laterally by mesepimeron or mesepimeron and mesanepisternum. Mesometaventral junction usually consisting of abutting or overlapping ventrites (sometimes monocondylic). Metaventrite usually distinctly transverse (more elongate in *Hydrochus*), flat to slightly convex, but often with posteromesal portion somewhat raised and often delimited by curved lines; discrimen and transverse (metakatepisternal) suture present or absent; exposed portion of metanepisternum moderately broad to very narrow, sometimes concealed beneath elytra. Metacoxae usually strongly transverse and subcontiguous, extending laterally to meet elytra (slightly transverse in *Georissus* and *Hydrochus*, widely separated in *Georissus*); plates absent. Metendosternite usually with long, slender stalk (shorter in *Georissus*), and long, lateral arms; anterior process either absent or very short and broadly emarginate, so that anterior tendons are on widely separated lobes. Hind wing usually with long radial bending zone at the end of which is a distinct hinge; apical field usually with three radial extensions, sometimes with several secondary linear sclerites, and with a small sclerite just beyond radial cell, which is short and broad, partly enclosing pterostigma and with basal edge incomplete or absent; cross-vein r3 usually long and slightly oblique, sometimes short or absent; cross-vein r4 complete; basal portion of RP moderately long to short or absent; medial spur straight and ending well before wing margin, sometimes very short; medial field usually with four free veins; MP_{3+4} sometimes with basal cross-vein and rarely with trace of spur; wedge cell often well developed, obliquely truncate at apex; anal lobe well developed to reduced, sometimes with notch-like embayment; AP absent or undivided. Femoral attachment of mid trochanter oblique with base of femur in contact with coxa; protrochanter fused to coxa in *Georissus*. Tibiae cylindrical to broadly flattened, often lined with teeth or spines, swimming hairs often present on mesotibiae or meso- and metatibiae; tibial spurs usually well developed (reduced in *Georissus* and *Spercheus*). Tarsi almost always 5-5-5 (4-4-4 in *Spercheus* and *Georissus*), tarsomeres simple, tarsomere 1 sometimes reduced; pretarsal claws simple; empodium usually well developed and bisetose (multisetose in *Spercheus*).

Abdomen with almost always with five free ventrites (6 ventrites in *Laccobius* and some *Berosus*, 4 ventrites in *Regimbartia* and *Allocotocerus*; basal 2 ventrites connate in *Georissus*); ventrite 1 sometimes with median carina; intercoxal process almost always acute (truncate in *Georissus*). Functional spiracles on abdominal segments I–VI; abdominal tergites relatively lightly sclerotised, usually with patches of wing-folding asperities on one or more segments. Anterior edge of sternite VIII in male usually without median strut (present in some Megasternini). Anterior edge of segment IX in male broadly rounded or truncate, without spiculum gastrale; tergite IX emarginate; tergite X free. Aedeagus of the trilobate type, usually symmetrical, with short and broad to elongate phallobase and simple parameres; penis usually with basal struts. Ovipositor moderately elongate, lightly sclerotised; paraprocts much shorter than coxites, which are subdivided with broad proximal lobe, slender distal lobe and small terminal styli. [Watts 1987, 1988b, 1989, 1990, 1999; Newton 1989; Hansen 1990, 1991a, 1995; Beutel 1994; Hebauer 1997; Archangelsky 1998, 2004; Beutel *et al.* 2001; Van Tassell 2000; Archangelsky *et al.* 2005].

Larva (Fig. 58A–C, I–J, L)

Body elongate, parallel-sided, slightly wider at middle, or widest at anterior end and tapering posteriorly, slightly flattened; head and protergum (and sometimes meso- and metatergum) well sclerotised with remaining surfaces more or less membranous, except for sclerotised plate on tergum VIII and/or small tergal sclerites on most abdominal segments; surfaces usually pubescent and sometimes granulate or asperate, or with lateral abdominal projections.

Head prognathous (hyperprognathous in Sphaeridiinae and most Hydrophilinae), usually somewhat elongate (shorter and broader in *Hydrochus* and *Spercheus*), more or less parallel-sided or with sides only slightly curved; posterior edge usually straight (distinctly emarginate in *Georissus*); epicranial stem short or absent; frontal arms V-shaped with bases contiguous to well separated; median endocarina absent. Stemmata on each side usually 6 (5 in *Spercheus*), usually well separated or in two vertical rows, sometimes fused to form two tight groups. Frontoclypeal suture absent and labium fused to head capsule forming clypeolabrum, the anterior edge of which may be produced to form a medial nasale and paired adnasalia. Antennae well developed, almost always 3-segmented (4-segmented in second and third instars of *Hydrophilus*); antennomere 1 usually longest, antennomere 2 with laterally placed sensorium (two sensoria in *Georissus*). Mandibles symmetrical or asymmetrical, moderately broad at base with narrow, strongly curved and unidentate apex (apparently bidentate in *Spercheus*), usually with 1 to 3 retinacula (absent in some Sphaeridiinae); mola or prostheca absent, but *Hydrochus* with a basal mola-like structure; basal penicillus present in *Georissus*. Ventral mouthparts usually protracted with maxillary articulating area absent (distinctly retracted with articulating area in *Hydrochus* and *Spercheus*); cardines distinct, usually divided (undivided in *Spercheus*); stipes usually elongate and cylindrical (about as long as wide in *Spercheus*); galea small, palpiform, articulated and located at apex of palpifer (membranous and reduced to various extents in some groups); lacinia usually

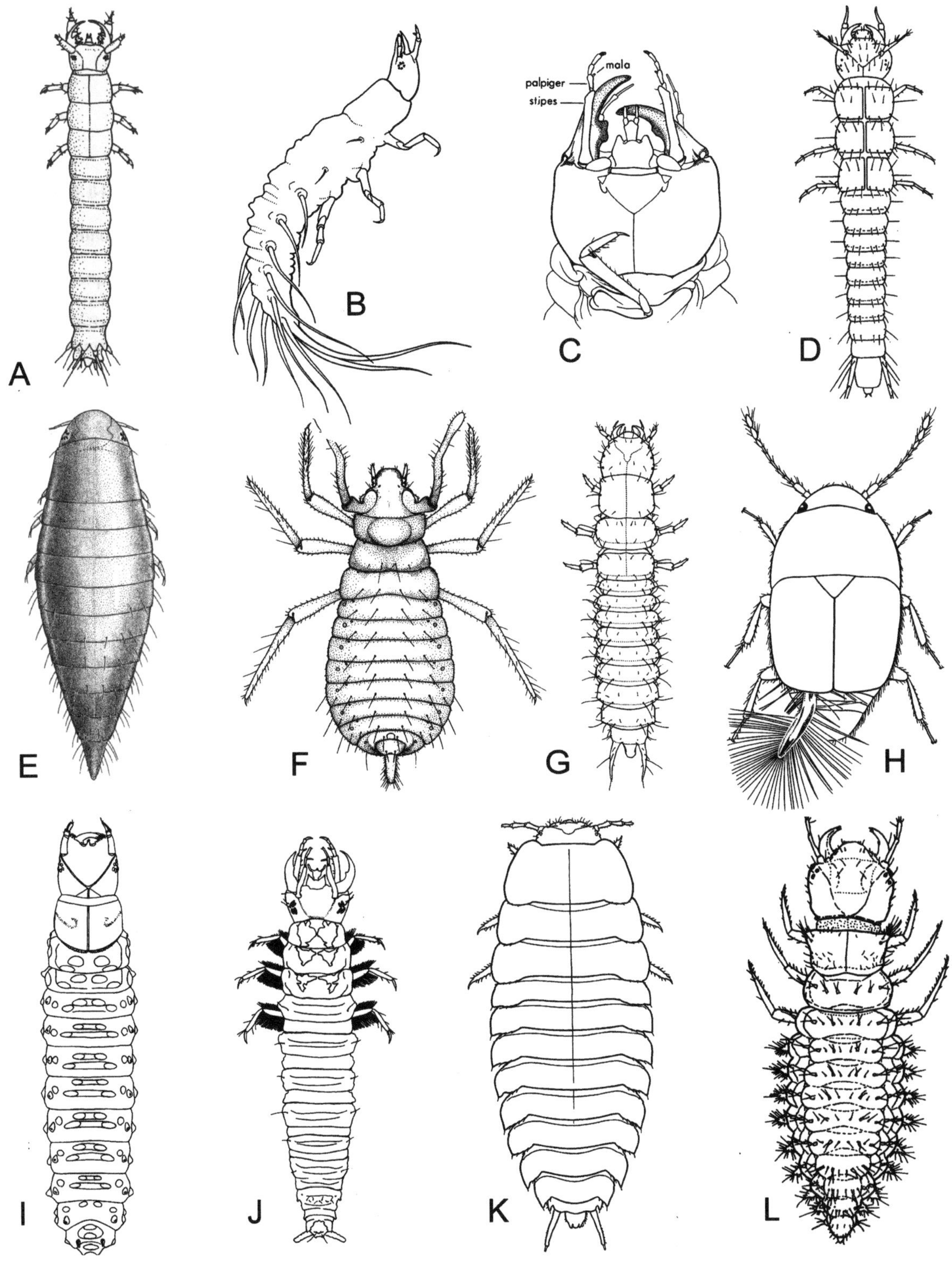

Fig. 58. A–G, Larva: A, *Hydrochus* squamifer LeConte (Hydrophilidae: Hydrochinae), dorsal; B, *Berosus* sp., lateral; C, *Hydrophilus* sp. (Hydrophilidae), head, ventral. D, *Hydraena pennsylvanica* Kiesenwetter (Hydraenidae), dorsal. E, *Hydrocanthus* sp. (Noteridae), dorsal. F, *Drepanoxenus ardea* Watson (Staphylinidae: Aleocharinae), dorsal. G, *Omalium* sp. (Staphylinidae: Omaliinae), dorsal. H, *Acrotrichis* sp. (Ptiliidae), adult, dorsal. I–L, larva: I, *Georissus crenulatus* Rossi (Hydrophilidae: Georissinae), dorsal; J, *Hydrophilus* sp. (Hydrophilidae), dorsal. K, *Ptomaphila lacrymosa* (Schreibers) (Silphidae), dorsal. L, *Spercheus emarginatus* (Shaller) (Hydrophilidae: Spercheinae), dorsal. A, D from Richmond 1920; E, I, J from Spangler 1991; L from Böving and Henriksen 1938.

absent (broad at base, apically acute and lined with setae in *Spercheus*, short and rounded in some *Hydrochus*); palp 3-segmented, with segment-like palpifer. Labium consisting of prementum and postmentum; palps almost always 2-segmented (apparently 1-segmented in *Spercheus*); ligula present or absent. Hypopharyngeal sclerome and hypostomal rods absent; ventral epicranial ridges absent (except in *Spercheus*). Gular region usually long, with gular sutures fused (gular sutures well separated and gula transverse in *Hydrochus* and *Spercheus*).

Prothorax usually distinctly longer than meso- or metathorax, with single tergal plate divided by median ecdysial line; meso- and metaterga often with paired plates. Prothoracic venter usually with divided or undivided presternal plate. Legs usually well developed and 5-segmented with pretarsus forming distinct claw (3-segmented in *Georissus* and *Chaetarthria*, without distinct pretarsus; with fewer than 3 segments or absent in some Sphaeridiinae), with long swimming hairs on femur in some Hydrophilini; coxae moderately widely separated. Abdominal segments I–VIII usually lightly sclerotised, each with transverse row of small, round sclerites in *Georissus*; terga I–VII sometimes with transverse plicae; segments I–VII with long, lateral gills in *Berosus*; lateral membranous projections sometimes present; abdominal sterna III–VII in *Enochrus* and *Agraphydrus* with paired, asperity-bearing prolegs. Abdominal apex usually with stigmatic atrium formed by modifications of segments VIII–X (tergum VIII forming sclerotised dorsal valve at apex of abdomen, segment IX usually trilobed and forming ventral valve, which bears a pair of small 1-segmented urogomphi; large posteriorly oriented spiracles located between the two, and segment X reduced and ventral). In *Berosus* atrium absent but tergum VIII forms terminal plate, and in *Georissus*, segment VIII not terminal, tergum IX completely visible with 1-segmented urogomphi, and segment X terminal. Functional spiracles absent except for an enlarged annular pair in spiracular atrium of most Hydrophilidae; functional biforous spiracles present on mesothorax and abdominal segments I–VIII in *Georissus*; all spiracles non-functional in *Berosus*. [Böving and Henriksen 1938; Emden 1956; Anderson 1976; Spangler 1991; Hansen 1991a, 2000; Beutel 1994; Archangelsky 1994, 1996, 1998, 1997, 1999, 2004; Hansen and Richardson 1998; Van Tassell 2000; Watts 2002; Shepard 2003; Archangelsky *et al.* 2005; Fikáček *et al.* 2011; Minoshima and Hayashi 2011].

Classification and World Distribution

The family Hydrophilidae is a worldwide group of ~160 genera and more than 3350 described species, usually divided into nine subfamilies, five of which occur in Australia (**see** below). Exotic subfamilies include Helophorinae from the Holarctic and Afrotropical regions; Epimetopinae from the Nearctic, Neotropical, Oriental and Afrotropical regions, Horelophinae from New Zealand and Horelophopsinae from New Guinea and the Ryukyu Islands (Hansen 1997b; Satô and Yoshitomi 2004). Some authors (*e.g.*, Hansen 1997c, 1999; Short and Hebauer 2006; Short and Fikáček 2011) treat the subfamilies Helophorinae, Epimetopinae, Georissinae, Hydrochinae and Spercheinae as separate families. In a combined morphological and molecular analyses by Bernhard *et al.* (2009), Helophorinae, Georissinae and Hydrochinae usually formed a clade with Epimetopinae, but the position of Spercheinae varied and Sphaeridiinae was always embedded within the Hydrophilinae. [Fikáček *et al.* 2011].

Australian Fauna

The Australian **Georissinae** includes several species of *Georissus* (Figs 57K), which are minute, black and tuberculate beetles, with a deflexed head concealed from above, a 1-segmented or compact 3-segmented antennal club, large procoxae fused to their trochanters, and two connate ventrites. Adults and larvae may be found in wet soil at the edges of streams. **Hydrochinae** includes several species of *Hydrochus* (Fig. 57B), which resemble hydraenids in being elongate, narrow and somewhat metallic and which may be found attached to plants in ponds or slow-moving streams. The Australian **Spercheinae** includes two species: *Spercheus platycephalus* Macleay (Fig. 57J) and *S. wattsi* Hebauer, both of which are brown beetles with coarsely punctate and costate elytra and a distinctive type of antenna (Fig. 57G). Spercheines inhabit stagnant ponds and adults and larvae are reported to walk on the underside of the surface film. The remaining Hydrophilidae belong to the primarily terrestrial **Sphaeridiinae** and the aquatic Hydrophilinae. The sphaeridiines are small to minute and may be abundant in decaying vegetable matter, dung, carrion and damp soil. The introduced *Cercyon haemorrhoidalis* (Fabricius) and *Sphaeridium discolor* Orchymont (Fig. 57A) are common dung inhabitants; species of the *Notocercyon* and other megasternine genera may be abundant in forest litter; and *Pseudohydrobius* species have been found in *Leptospermum* flowers. Most **Hydrophilinae** are oval to globose, smooth and glabrous, often resembling Dytiscidae, from which they differ in having short, clubbed antennae and usually long maxillary palps (somewhat shorter in *Chaetarthria*, *Anacaena* and *Paracymus*). Hydrophilines may be small to quite large, with the largest Australian species belonging to the genus *Hydrophilus*. They also have a ventral plastron, which communicates with the subelytral air reservoir; when the beetle rises to the surface, it breaks the surface tension using the specialised antennal club, allowing the flow of air between the atmosphere and the plastron. Adult and larval hydrophilids, although occupying similar habitats, have very different habits; all adults are phytophagous or saprophagous, while larvae are predaceous (except for a few exotic species of *Helophorus* Fabricius). Hydrophilid larvae have prognathous or hypoprognathous heads with six pairs of stemmata (five in *Spercheus*) and most have a metapneustic respiratory system with only the eighth spiracles functional and located in a respiratory chamber or atrium formed by segments VIII and IX (Fig. 39K). Larvae of Georissinae (as well as those of the exotic Helophorinae and Epimetopinae) differ in having eight pairs of functional abdominal spiracles. Larvae of the hydrophiline tribe Berosini (Fig. 58B) are apneustic, lacking functional spiracles but having seven pairs of lateral gills.

Australian Subfamilies, Tribes and Genera

Georissinae: *Georissus* Latreille.
Hydrochinae: *Hydrochus* Leach.
Spercheinae: *Spercheus* Kugelann.
Hydrophilinae: Acidocerini: *Agraphydrus* Régimbart, *Chasmogenus* Sharp, *Enochrus* C.G. Thomson, *Helochares* Mulsant.
Anacaenini: *Anacaena* Thomson, *Crenitis* Bedel, *Notohydrus*

Balfour-Browne, *Paracymus* Thomson, *Phelea* Hansen. **Berosini:** *Berosus* Leach, *Regimbartia* Zaitzev, *Allocotocerus* Kraatz. **Chaetarthriini:** *Amphiops* Erichson, *Chaetarthria* Stephens. **Hydrobiusini**: *Hybogralius* Orchymont, *Limnoxenus* Motschulsky. **Hydrophilini:** *Hydrobiomorpha* Blackburn, *Hydrophilus* Geoffroy, *Sternolophus* Solier. **Laccobiini:** *Laccobius* Erichson.

Sphaeridiinae: Rygmodini: *Eurygmus* Hansen, *Pseudohydrobius* Blackburn, *Rygmostralia* Orchymont; **Andotypini:** *Coelostomopsis* Hansen, undescribed genus (genus A of Hansen 1991). **Borborophorini:** *Borborophorus* Hansen, *Petasopsis* Hansen. **Coelostomatini:** *Coelostoma* Brullé, *Dactylosternum* Wollaston. **Omicrini:** *Paromicrus* Scott, *Psalitrus* Orchymont, *Mircogioton* Orchymont, *Noteropagus* Orchymont. **Megasternini:** *Cercyodes* Broun, *Ercycodes* Hansen, *Notocercyon* Blackburn, *Ceronocyton* Hansen, *Cenebriophilus* Hansen, *Chledocyon* Hansen, *Pseudoosternum* Hansen, *Pilocnema* Hansen, *Cetiocyon* Hansen, *Cercyon* Leach, *Australocyon* Hansen, *Pachysternum* Motschulsky, *Cryptopleurum* Mulsant. **Sphaeridiini:** *Sphaeridium* Fabricius.

11. HISTERIDAE

Adult (Figs 2D, 57L, 77A–C)

Length ~1.2–12 mm. Body ~1.1–3 times as long as wide (as much as 3.5 times in *Niponius* and *Trypeticus*), more or less compact and well sclerotised, with sides usually either evenly rounded or subparallel (sides of pronotum and elytra usually differing in Chlamydopsinae and Onthophilinae); usually moderately to strongly convex (distinctly flattened in Hololeptini, most Paromalini and some Platysomatini); subcylindrical in *Trypeticus*, *Niponius* and Tetetriini; colour usually black, but sometimes red, metallic blue or green or black with red markings; dorsal surfaces usually smooth, shiny, impunctate or finely punctate and subglabrous (highly variable and often coarsely punctate, setose and/or with elytral trichomes in Chlamydopsinae).

Head (Fig. 14L) usually hypognathous and retracted into prothorax (prognathous in *Hololepta* and *Trypeticus*) and deeply inserted into prothorax (non-retractile in *Hololepta* and *Niponius*), not abruptly constricted posteriorly; posterior edge sometimes deeply biemarginate; transverse occipital ridge occasionally present; median occipital endocarina present in *Niponius*. Eyes usually well developed, not or barely protuberant, and often vertically elongate, posteriorly flattened or slightly concave, finely facetted and without interfacetal setae. Antennal insertions usually at least slightly concealed from above (exposed in Teretriini, Saprininae and Chlamydopsinae); subantennal groove present. Frontal region with 2 anteriorly projecting horns in *Niponius*. Frontoclypeal suture absent. Labrum usually transverse, anteriorly rounded, truncate or emarginate. Antennae relatively short and geniculate, usually 11-segmented with 3-segmented club (9-segmented with 7-segmented funicle when club segments completely fused together, 10-segmented with 6-segmented funicle and 3-segmented club in *Niponius*, 8-segmented with 6-segmented funicle and 1-segmented club in *Trypeticus*); scape usually elongate and slightly to strongly curved (short and expanded in Chlamydopsinae). Mandibles slightly to moderately elongate apex, strongly projecting in *Hololepta*, usually unidentate (bidentate in *Trypeticus*), with or without subapical teeth or retinaculum; mola usually absent (present in *Trypeticus* and Teretriini); prostheca usually consisting of a fringe of setae. Maxilla with setose galea and lacinia; galea usually elongate (very long and slender in *Hololepta*), sometimes bearing spatulate or complex setae; lacinia much shorter than galea, sometimes with uncus; palp 4-segmented with apical palpomere fusiform. Mentum variably shaped, concealing ventral mouthparts in *Hololepta*; ligula usually bilobed; apical labial palpomere usually fusiform. Corpotentorium very narrow and arched or absent. Cervical sclerites present.

Pronotum ~0.45–1.33 times as long as wide, usually as wide at base as combined elytral bases (narrower in some Chlamydopsinae); usually widest posteriorly with sides slightly to strongly curved and anteriorly converging (more or less parallel sided and narrowed only near anterior end in *Hololepta*, *Platysoma* and Teretriini, not at all narrowed in *Niponius*). Lateral pronotal carinae complete and simple; anterior angles often produced forward and rounded to acute; posterior angles right to obtuse or rounded; posterior edge usually straight or somewhat angulate or sinuate, usually with underlying ventral ridge, which interlocks with elytra and scutellar shield. Prosternum (Fig. 18H) in front of coxae shorter than to distinctly longer than shortest diameter of procoxal cavity, central portion of prosternum flat to strongly tumid or forming elevated platform often with marginal striae; anterior edge often produced forming chin piece, which is produced laterally, concealing antennal cavities from below in Chlamydopsinae, Onthophilinae, Tribalinae and Histerinae. Prosternal process almost always complete, not dorsally curved, moderately broad and apically rounded, truncate or concave (incomplete and apically notched in *Trypeticus* and Teretriini). Notosternal sutures usually complete. Procoxae strongly transverse and not projecting, without or with short concealed lateral extensions; trochantins concealed; endopleuron fused to notum. Procoxal cavities strongly transverse, narrowly to moderately widely separated, broadly to narrowly open externally, without narrow anterolateral extension, internally open.

Scutellar shield reduced and often absent; when present abruptly elevated anteriorly, posteriorly rounded to narrowly acute. Elytra ~0.6–1.5 times as long as combined width and 0.7–2.33 times as long as pronotum; sides subparallel or slightly curved and apices almost always squarely truncate, exposing one or two tergites, conjointly rounded in a few groups (e.g. *Bacanius*); humeri well developed; mesepimeron sometimes visible from above between pronotum and elytral humeri (e.g. in *Hololepta* and *Hister*); disc usually irregularly punctate, sometimes with as many as 8 striae, without scutellary striole, but usually with fewer than 8 and these often incomplete; anterolateral portions of elytra usually with trichomes in Chlamydopsinae; epipleura incomplete, sometimes absent; elytra ventrally (internally) usually with anterior and posterior ridges along mesal edge (interlocking with metanotum and abdominal tergite IV) and one anterolaterally (interlocking with metapleuron). Mesoventrite separated by complete sutures from mesanepisterna, which are widely separated at midline; anterior edge usually on different plane than metaventrite, with strongly declined or vertical procoxal rests and sometimes with weakly concave, transverse prosternal rest; mesal portion flat; mesoventral cavity absent; discrimen absent. Mesocoxae usually subglobular (slightly transverse and slightly oblique in *Niponius*), with concealed trochantins, sometimes (e.g. in *Niponius* and *Trypeticus*) with mesal knobs fitting into cavities in mesoventral process, which is more or less truncate at apex. Mesocoxal cavities widely separated, partly closed laterally by mesepimeron, which is often partly or completely fused to mesanepisternum. Mesometaventral junction consisting of abutting or overlapping ventrites, which may be fused together. Metaventrite usually distinctly transverse (slightly transverse in *Niponius*, elongate in *Trypeticus*), flat to moderately convex; usually with paired curved or straight postcoxal lines; discrimen usually complete to anterior edge of ventrite; transverse (metakatepisternal) suture absent; exposed portion of metanepisternum usually narrow, sometimes concealed beneath elytra; metepimera exposed or concealed. Metacoxae usually globular are very widely separated (slightly transverse and moderately widely separated in *Niponius*), not extending laterally to meet elytra; plates absent. Metendosternite without or with very short and broad stalk and long, lateral arms; anterior process either absent or located on arms. Hind wing without radial bending zone, but with distinct hinge; apical field very long (more than two-thirds total wing length), with as many as 3 but usually only 1 or 2 radial extensions; radial cell, cross-vein r3 and usually r4 absent; R-M loop, absent or forming an acute angle, but usually with a long RP extension; medial spur straight and usually extending to wing margin; medial field with as many as 3 free veins, but usually with 2 or fewer; wedge cell absent; anal lobe well developed to reduced, usually with notch-like embayment; AP usually absent. Femoral attachment of mid trochanter usually strongly oblique with base of femur often in contact with coxa (trochanter fused to femur in *Chlamydopsis*). Tibiae usually at least slightly expanded at apex and commonly strongly expanded subapically and flattened, with outer edge lined with broad lobes, teeth, fixed spines, or socketed spines; tibial spurs usually well developed, often unequal in length. Tarsi almost always 5-5-5 (5-5-4 in *Acritus*); tarsomeres simple; pretarsal claws simple; empodium reduced or absent.

Abdomen with five free ventrites; ventrite 1 about as long as 2, with intercoxal process usually truncate (more or less rounded in *Niponius* and *Trypeticus*). Functional spiracles on abdominal segments I–V; abdominal tergites I–III lightly sclerotised, IV–VII well sclerotised, IV with median interlocking groove, VI and VII forming propygidium and pygidium, the latter of which is enlarged and usually strongly declined, vertical or more or less ventrally oriented (only slightly declined in *Hololepta*); patches of wing-folding asperities absent. Segments VIII to X lightly sclerotised and completely concealed within abdominal apex. Anterior edge of sternite VIII in male usually with paired basal struts. Anterior edge of segment IX in male broadly rounded to acute or with spiculum gastrale; tergite IX slightly emarginate; tergite X free. Aedeagus of a modified trilobate type, in which parameres are fused for most of their lengths and sometimes fused to phallobase, which is short to moderately long and sometimes asymmetrical; penis without basal struts. Sternite VIII in female with spiculum ventrale or with paired struts. Ovipositor moderately elongate; proctiger lightly sclerotised, without baculi; paraprocts lightly sclerotised with distinct baculi, much longer than coxites, which are undivided, heavily sclerotised, often setose, sometimes with apical or subapical teeth and usually with subapical styli. [Gardner 1935a; Hinton 1945b; Ôhara 1994, 1999; Kovarik and Caterino 2000, 2005; Kanaar 2003].

Larva (Fig. 35H)

Body elongate, parallel-sided, slightly flattened; head and protergum well sclerotised and remaining surfaces more or less membranous, but often with a variable number of smaller sclerites; surfaces usually clothed with scattered long and short, fine setae.

Head prognathous (sometimes hyperprognathus) and usually more or less elongate and parallel-sided, sometimes transverse with slightly rounded sides, posterior edge straight or emarginate; ecdysial lines absent in second (final) instar; epicranial stem short to moderately long; frontal arms V-shaped or U-shaped, with bases contiguous; median endocarina absent. Stemmata either absent or with one on each side. Frontoclypeal suture absent and labium fused to head capsule forming clypeolabrum, the anterior edge of which is produced to form a nasale, which is usually irregular or asymmetrically toothed. Antennae moderately long, 3-segmented with 1 or 2 laterally placed sensoria on antennomere 2. Mandibles symmetrical, unidentate and narrowly falcate, usually with 1 or 2 retinacula; mola or prostheca absent; basal penicillus present. Ventral mouthparts protracted with maxillary articulating area absent; cardines fused to stipes, which is elongate; galea small, palpiform, articulated and located at apex of palpifer; lacinia absent; palp 3-segmented or rarely 4-segmented, with segment-like palpifer. Labium consisting of prementum and postmentum; ligula absent; palps 2-segmented or rarely 3-segmented, narrowly separated, with segment-like palpiger. Hypopharyngeal sclerome, hypostomal rods and ventral epicranial ridges absent. Gular region long, with or without single tentorial pit; gular sutures absent; median ecdysial line sometimes present.

Prothorax distinctly longer than meso- or metathorax, with single tergal plate divided by median ecdysial line; meso- and metaterga often with single or paired plates, and sometimes with additional small sclerites. Prothoracic venter usually with presternal plate subdivided longitudinally into 3 parts. Legs small, slender and 5-segmented with pretarsus forming distinct claw, which is bisetose or glabrous; coxae very widely separated. Abdominal segments I–VIII lightly sclerotised and often bearing several small sclerites; terga and sterna I–VIII sometimes with transverse plicae, transverse ambulatory ampullae and/or transverse rows of asperities (Fig. 43O); lateral membranous projections absent. Tergum IX about as long as VIII, usually extending onto ventral surface and almost always with a pair of slender, diverging, 1- or 2-segmented, articulated urogomphi (absent in *Trypeticus* and some *Teretrius*); segment X usually not visible from above. Functional biforous spiracles with closing apparatus present on mesothorax and abdominal segments I–VIII. [Gardner 1930; Nikitsky 1976a; Newton 1991; Kovarik and Passoa 1993; De Marzo 2000; Kovarik and Caterino 2000, 2005].

Classification and World Distribution

Histeridae is a large and worldwide family containing more than 390 genera and 4250 species (Mazur 2011) currently placed in ten subfamilies, of which all but the New World and Palaearctic

Haeteriinae occur in Australia. Subfamily classification has varied among authors (Ôhara 1994; Ślipiński and Mazur 1999; Caterino and Vogler 2002).

Australian Fauna

Adult Histeridae are often easy to recognise by their compact form, black and shiny body, deeply inserted head, large, transverse procoxae, dentate protibiae, antennae which are almost always geniculate with a compact, pilose club, and elytra with six or fewer striae and truncate apices exposing one or two abdominal tergites. The North QLD species *Niponius interstitialis* Lewis (**Niponiinae**) is atypical in being narrowly elongate with a strongly produced, elongate head bearing a pair of anterior projections, while myrmecophilous species of *Chlamydopsis* (**Chlamydopsinae**) (Fig. 77B) are often red in colour with setal tufts or trichomes and long, modified legs, and the genus *Epiechinus* (**Onthophilinae**) is clothed with scale-like setae. Other elongate and subcylindrical adults occur in the subfamilies **Dendrophilinae** (*Teretrius*) and **Trypeticinae** (*Trypeticus*). The typical histerid body form is characteristic of Saprininae (Fig. 77A) and Histerinae (Fig. 57L), but some members of the latter group (e.g. *Hololepta*, Fig. 77C, *Platysoma*) are strongly flattened. Larval histerids resemble those of Hydrophilidae in general form and head structure, but they have a full set of functional spiracles, a penicillus of hairs at the base of each mandible, 2-segmented urogomphi (sometimes lacking) and no apical respiratory chamber. Both adults and larvae are carnivorous, feeding mainly on the larvae of other insects. They are commonly found in carrion, dung and decaying vegetable matter, or under the bark of dying or dead trees, where they usually prey on fly larvae. Niponiinae, Trypeticinae and some Teretriini inhabit the tunnels of wood-boring insects, such as Bostrichidae and scolytine weevils. Some Saprininae and Histerinae may be found in carrion and dung; *Hypocaccus*, *Halacritus* and some *Saprinus* are common in beach drift; and *Hololepta* and *Platysoma* occur under bark. The myrmecophilous chlamydopsines have setose secretory structures (trichomes) which apparently produce an appeasement substance (Wilson 1971) necessary for their acceptance within an ant colony. [Froggatt 1927; Gardner 1930; Hinton 1945b; Nikitsky 1976a; Mazur 2011; Caterino 2003, 2006; Caterino and Dégallier 2007].

Australian Subfamilies, Tribes and Genera

Niponiinae: *Niponius* Lewis.
Chlamydopsinae: *Chlamydopsis* Westwood, *Ectatommiphila* Lea, *Eucurtia* Mjöberg, *Orectoscelis* Lewis, *Pheidoliphila* Lea, *Teretriopsis* Caterino & Dégallier.
Onthophilinae: *Epiechinus* Lewis, *Onthophilus* Leach.
Dendrophilinae: Bacaniini: *Australanius* Gomy, *Bacaniomorphus* Thérond, *Bacanius* Leconte. **Dendrophilini:** Dendrophilus Leach; **Paromalini:** *Carcinops* Marseul, *Eulomalus* Cooman, *Platylomalus* Cooman.
Abraeinae: Acritini: *Acritus* Le Conte, *Aeletes* Horn, *Halacritus* Schmidt. **Acritomorphini:** *Acritomorphus* Wenzel. **Teretriini:** *Teretrius* Erichson.
Trypeticinae: *Trypeticus* Lewis.
Saprininae: *Chalcionellus* Reichardt, *Gnathoncus* Jacquelin du Val, *Hypocacculus* Bickhardt, *Hypocaccus* C.G. Thomson, *Notosaprinus* Kryzhanovskij, *Saprinodes* Lewis, *Saprinus* Erichson, *Tomogenius* Marseul.
Tribalinae: *Parepierus* Bickhardt, *Stictostix* Marseul, *Tribalus* Erichson.
Histerinae: Exosternini: *Arbolister* Mazur, Asterix Mazur, *Diabletes* Reichardt, *Pseudister* Bickhardt. **Histerini:** *Hister* Linnaeus, *Nasaltus* Mazur & Węgrzynowicz, *Pachylister* Lewis. **Hololeptini:** *Hololepta* Paykull. **Platysomatini:** Eurylister Bickhardt, Kanaarister Mazur, Platylister Lewis, *Platysoma* Leach, Sunilis Mazur & Ôhara.

STAPHYLINOIDEA

The superfamily Staphylinoidea includes the families Ptiliidae, Hydraenidae, Agyrtidae, Leiodidae, Silphidae and Staphylinidae. Several formerly recognised families, such as Micropeplidae, Dasyceridae, Pselaphidae, Scaphidiidae and Scydmaenidae, are now considered subfamilies of the megadiverse family Staphylinidae. This large and variable group is characterised by a further reduction in the wing venation from that found in basal staphyliniforms and a high grade of wing folding often assisted by abdominal tergal asperities. Another abdominal feature found mainly in Staphylinidae is the pattern of microsclerites in the membrane separating abdominal segments – this often resembles a brick wall. The aedeagus is derived from the trilobate type, with the phallobase becoming membranous or reduced, so that the parameres appear to be attached directly to the penis. Hansen (1997c) listed nine staphylinoid autapomorphies: (1) pronotum without accessory ridge below posterior edge; (2) ventral face of elytra without mediolateral binding patch; (3) hindwing without apical hinge distal to radial cell; (4) hindwing without radio-medial loop; (5) medial field of hind wing with only one vein; (6) phallobase membranous or apparently absent; (7) larval abdominal segment X with numerous fine teeth on eversible anal lobes; (8) larval head without frontoclypeal suture; and (9) urogomphi present, articulated, 2-segmented. Lawrence & Newton (1982) also noted 4 Malpighian tubules as possible character. The hind wing of Silphidae and Staphylinidae also has a secondary hinge well proximal to the radial cell (or end of the radial bar). Although Staphylinoidea is usually considered to be monophyletic, the Hydraenidae + Ptiliidae and Agyrtidae + Leiodidae clades are well separated from the bulk of staphylinoids in some recent analyses (Hunt *et al.* 2007; Lawrence *et al.* 2011). [Naomi 1985; Newton and Thayer 1988; Beutel and Molenda 1997; Betz *et al.* 2003; Grebennikov 2005; Caterino *et al.* 2005]. The only staphylinoid family not occurring in Australia is Agyrtidae (Holarctic and New Zealand) (Newton 1997, 2005c; Peck 2000a).

12. HYDRAENIDAE

Adult (Figs 57I, 77D, I)

Length ~0.8–2.6 mm. Body ~1.6–2.5 times as long as wide, sides almost always subparallel or with sides of pronotum and elytra differing in lateral curvature (with distinct pronoto-elytral angle) (sides uniformly rounded, without pronoto-elytral angle in *Limnebius*), slightly convex to strongly flattened; colour yellow to brown or black, usually with legs and palps paler, sometimes with metallic sheen in *Ochthebius*; dorsal surfaces often with distinct impressions, grooves or costae in Octhebiinae; dorsal surfaces usually sparsely setose (glabrous in *Limnebius*, with stout, bristle-like setae in *Tympallopatrum*); ventral surfaces usually with hydrofuge pubescence (plastron).

Head more or less prognathous, usually abruptly constricted immediately behind eyes forming a neck region (Fig. 28Q), which is delimited by posteriorly curved ridge on vertex (not in *Limnebius*). Paired ocelli sometimes present (absent in *Limnebius* and *Hydraena*). Eyes usually well developed, usually protuberant (not in *Limnebius*), circular and entire. Antennal insertions concealed from above; subantennal groove well developed, usually extending behind eye. Frontoclypeal suture distinct, angulate or curved, without attached mid-cranial suture; clypeus moderately large, usually truncate anteriorly. Labrum short and strongly transverse (*Limnebius*) or longer and deeply emarginate, with basolateral notches in *Hydraena*. Antennae short, usually 9-segmented with 5-segmented, pubescent club preceded by glabrous cupule (Fig. 28Q) (8-segmented in *Tympanogaster*, Fig. 57H, *and Tympallopatrum*, with preceding antennomere not cupule-like); scape slender, curved, usually much longer than pedicel (scape and pedicel subequal in *Limnebius*); pedicel at least slightly elongate in *Limnebius*, *Hydraena* and most Ochthebiinae, forming asymmetrical cupule in *Tympanogaster* and *Tympallopatrum*; antennomere 3 usually minute (well developed and elongate in *Limnebius*). Mandibles fairly small, flattened, more or less asymmetrical and largely concealed between labrum, maxillae and mentum; apex unidentate or narrowly bidentate; mola well developed and asymmetrical; prostheca well developed and including articulated, sclerotised process (Fig. 16H); lateral edge with prominent labro-mandibular interlocking device in *Hydraena*. Maxilla sometimes with fimbriate galea, lacinia with bifid uncus in some Ochthebiinae; maxillary palps 4- segmented, long and slender in Hydraeninae, with apical palpomere longer than subapical one, shorter and subulate in Ochthebiinae, with apical palpomere shorter and narrower than preapical one. Mentum often large, concealing prementum and labial palps, which are 3-segmented, minute and widely separated; ligula not distinctly bilobed. Corpotentorium narrow; laminatentoria well developed and broadly joined at midline. Cervical sclerites present.

Pronotum ~0.43–0.85 times as long as wide, usually widest at middle or anteriorly and distinctly narrower at base than combined elytral bases (widest at base, which is not narrower than combined elytral bases in *Limnebius*); sides slightly to strongly curved or sinuate, horizontally explanate and occasionally with lateral or posterolateral hyaline area in some Ochthebiinae, obliquely explanate in *Limnebius*. Lateral pronotal carinae complete and usually simple (sometimes finely crenulate or denticulate); anterior angles not produced forward; posterior angles right to rounded; posterior edge straight, without underlying ventral ridge; disc with various ridges and impressions in Ochthebiinae. Prosternum in front of coxae about as long as or slightly shorter than shortest diameter of procoxal cavity, slightly convex to carinate, usually with micro-sculpture and hydrofuge pubescence; anterior edge with acute median projection in *Limnebius* and *Hydraena*. Prosternal process complete, usually narrow, parallel-sided and apically acute or narrowly rounded (strongly abruptly expanded and truncate at apex in *Hydraena*). Notosternal sutures complete. Hypomeron with anterior antennal pocket, sometimes including antenna-cleaning brush, and usually with longitudinal hypomeral ridge. Procoxae usually more or less conical and projecting (transversely cylindrical in *Limnebius*), with short to moderately long concealed lateral extensions; trochantins concealed or exposed; endopleuron not fused to notum. Procoxal cavities slightly to strongly transverse, narrowly separated, usually narrowly open externally (broadly open in *Limnebius*, distinctly closed in *Hydraena*), with distinct narrow anterolateral extensions, internally open or closed.

Scutellar shield usually present (scutellum concealed in *Tympallopatrum*), not abruptly elevated, posteriorly rounded to acute. Elytra ~1.18–1.95 times as long as combined width and 2.1–3.5 times as long as pronotum; sides usually slightly curved and apices conjointly rounded, concealing all abdominal tergites (apices subtruncate in *Limnebius*); humeri well developed; disc usually with ten distinct puncture rows and no scutellary striole (elytra with few irregular punctures in *Limnebius*, with more than 10 rows in *Hydraena*); epipleura usually narrow but complete almost to apex (broad basally but incomplete in Hydraeninae). Mesoventrite usually at least partly fused to mesanepisterna (separated by complete sutures in *Limnebius*); mesanepisterna widely separated at midline; anterior edge of mesoventrite on same plane as metaventrite, without procoxal rests, but usually with broad prothoracic rest or collar; main portion of ventrite usually with paired longitudinal carinae and sometimes with additional median longitudinal carina; mesoventral cavity absent; discrimen absent. Mesocoxae usually globular or slightly transverse with exposed or concealed trochantins. Mesocoxal cavities usually contiguous to narrowly separated (more widely separated in *Limnebius*), partly closed laterally by mesanepisternum in Hydraeninae, closed by meeting of mesoventrite and metaventrite in Ochthebiinae. Mesometaventral junction, if present, consisting of abutting or overlapping ventrites. Metaventrite slightly to strongly transverse, flat to slightly convex, with posteromesal portion raised and impunctate (lacking hydrofuge pubescence) in *Tympanogaster*; discrimen usually absent (present in *Tympallopatrum*); transverse (metakatepisternal) suture absent; exposed portion of metanepisternum elongate and very narrow. Metacoxae strongly transverse and subcontiguous or narrowly separated, sometimes extending laterally to meet elytra; plates absent. Metendosternite with moderately long, slender stalk; long, lateral arms; and anterior process either short and broadly emarginate, so that anterior tendons are widely separated, or absent (*Limnebius*). Hind wing without radial bending zone but with distinct hinge at end of radial bar; apical field about two-thirds as long as total wing length, usually with two radial extensions; radial cell absent or with lumen entirely obliterated by pterostigma; cross-vein r3 absent and r4 incomplete; R-M loop and basal portion of RP absent, but apical extension at least vaguely indicated; medial spur undulate and extending almost to wing margin; medial field with one free vein; wedge cell absent; anal lobe highly reduced or virtually absent; posterior wing margin with moderately long fringe hairs. Legs moderately long and slender. Tibiae more or less cylindrical; tibial spurs usually well developed. Tarsi 5-5-5, with tarsomeres 1–3 reduced (especially tarsomere 1), tarsomere 5 usually very long; tarsomeres simple; pretarsal claws simple; empodium absent or not apparent.

Abdomen usually with 7 free ventrites, but ventrite 1 (sternite II) visible only between metacoxae as part of intercoxal process (lateral portions of sternite II variously developed but concealed by coxae). Functional spiracles on abdominal segments I–VIII present in pleural membrane. Abdominal tergites relatively lightly sclerotised, with patches of wing-folding asperities on one or more segments. Anterior edge of sternite VIII in male without median strut. Anterior edge of segment IX with articulated spiculum gastrale; tergite IX with broad basal struts, its posterior edge slightly emarginate; tergite X free. Aedeagus of modified trilobate type, usually asymmetrical, with phallobase and parameres fused to penis, which lacks basal struts. Tergite IX in female not divided into two parts; proctiger sclerotised. Ovipositor short and broad; coxites lightly sclerotised, undivided, without baculi, sometimes fused together to form single plate; styli absent or represented by terminal tufts of setae. [Zwick 1977; Perkins 1980, 1981, 1997, 2000, 2004a, 2004b, 2004c, 2005, 2006, 2007a, 2007b; Hansen 1991b; Beutel and Jäch 1995; Jäch 2001; Beutel *et al.* 2003; Jäch *et al.* 2005].

Larva (Fig. 58D)

Body elongate, parallel-sided, narrowing posteriorly, slightly flattened; head and all terga well sclerotised and variously pigmented, ventral surfaces usually with more lightly sclerotised plates; vestiture of scattered long and short, fine setae.

Head prognathous, or slightly declined, more or less transverse, distinctly rounded laterally and slightly flattened; posterior edge straight; epicranial stem short to moderately long; frontal arms V-shaped or U-shaped, with bases contiguous; median endocarina absent. Stemmata usually five on each side, in two oblique rows. Frontoclypeal suture present; labium free, about as long as wide, broadly rounded at apex. Antennae moderately long, 3-segmented, with setiform sensorium on antennomere 2 more than half as long as antennomere 3. Mandibles symmetrical, unidentate or narrowly bidentate with 2 or more subapical teeth or serrations; mola well developed; prostheca hyaline and apically acute. Ventral mouthparts moderately retracted, with maxillary articulating area; cardines free and undivided; stipes elongate; galea articulated and palpiform in Ochthebiinae, fimbriate in Hydraeninae; lacinia falcate or narrowly truncate, spinose or setose; palp 3-segmented. Labium consisting of prementum, mentum and submentum; ligula with numerous papillae; palps 2-segmented and moderately widely separated. Hypopharyngeal bracon present; sclerome absent. Hypostomal rods and ventral epicranial ridges absent. Gular region short; gular sutures well separated.

Prothorax not much longer than meso- or metathorax; thoracic terga and all abdominal terga with single tergal plate (unless divided by median ecdysial line); thoracic terga and abdominal terga I–VIII in *Tympanogaster* with flattened, multilobed projections, posterolaterally directed on abdomen. Prothoracic venter with undivided presternal plate. Legs well developed, slender and 5-segmented, with pretarsus forming distinct bisetose claw; coxae narrowly to moderately widely separated. Anterior abdominal terga and sterna usually separated by lateral membranous areas, with separated laterotergites in early instars; segments IX and X, and often preceding two or more segments, forming continuous sclerotised ring; surfaces of terga simple or with transverse carina anteriorly, sometimes with paired gland openings. Tergum IX about as long as or slightly shorter than VIII, not extending onto ventral surface, with a pair of slender, diverging, parallel-sided, 2-segmented, articulated urogomphi; segment X visible from above, more or less cylindrical, armed with pair of apical hooks (Fig. 52T). Functional annular spiracles with closing apparatus usually present laterally on mesothorax and abdominal segments I–VIII; spiracles usually placed at ends of lateral spiracular tubes (those in *Tympanogaster* at posterior end of prothorax and at ends of lateral tergal projections); non-functional spiracles often present on metathorax. [Spangler 1991; Delgado and Archangelsky 2005; Jäch *et al.* 2005].

Classification and World Distribution

The family Hydraenidae contains ~2000 species in 44 genera and is widely distributed throughout the world, including the Subantarctic Islands. The group is divided into four subfamilies, two of which, the Afrotropical Prosthetopinae and New Zealand Orchymontiinae, do not extend into Australia. [Hansen 1998; Jäch *et al.* 2005; Trizzino *et al.* 2013].

Australian Fauna

The Australian hydraenid fauna includes *Limnebius acupunctus* Perkins and many *Hydraena* species in the subfamily **Hydraeninae** and six genera of **Ochthebiinae**: the widely distributed *Ochthebius*, *Gymnochthebius*, with New World and Australian species, and the autochthonous genera *Tympallopatrum*, *Tympanogaster*, *Gymnanthelius* and *Hughleechia*. Hydraenids are small beetles with relatively inconspicuous antennae, often concealed beneath the head and prothorax (Beutel and Jäch 1995). The maxillary palps may be as long as or longer than the antennae, as in Hydrophilidae, but unlike hydrophilids Hydraenidae have at least a 5-segmented antennal club. Larvae are usually elongate and campodeiform, with well developed legs, articulated urogomphi and a pair of hooks on segment X; those of *Tympanogaster*, however, are more or less flattened and limuloid with a pair of spiracular tubes projecting dorsally between the prothorax and mesothorax. Hydraenids feed on algae in a variety of aquatic, riparian and littoral habitats, including streams, waterfalls, wet rock faces, ponds and ditches, marine rock pools, seabird nests, and inland salt lakes. Most adults are capable of breathing in water by means of a ventral plastron formed by hydrofuge hairs, but most larvae have no aquatic modifications. Exceptions occur in the genus *Tympanogaster*, where both larvae and adults live in the splash zone beside waterfalls or among boulders. *Hughleechia giulianii* Perkins is a marine intertidal species living in rock crevices and pools within the high tide splash zone. [Perkins 1981; Newton 1985; Jäch 2001].

Australian Subfamilies, Tribes and Genera

Hydraeninae: Hydraenini: *Hydraena* Kugelann; **Limnebiini**: *Limnebius* Leach.

Ochthebiinae: **Ochthebiini**: *Gymnanthelius* Perkins, *Gymnochthebius* Orchymont, *Hughleechia* Perkins, *Ochthebius* Leach, *Tympallopatrum* Perkins, *Tympanogaster* Janssens.

13. PTILIIDAE

Adult (Figs 58H, 59A–B)

Length ~0.3–1.2 mm. Body ~1.5–5.6 times as long as wide, sides subparallel, uniformly rounded or with sides of pronotum

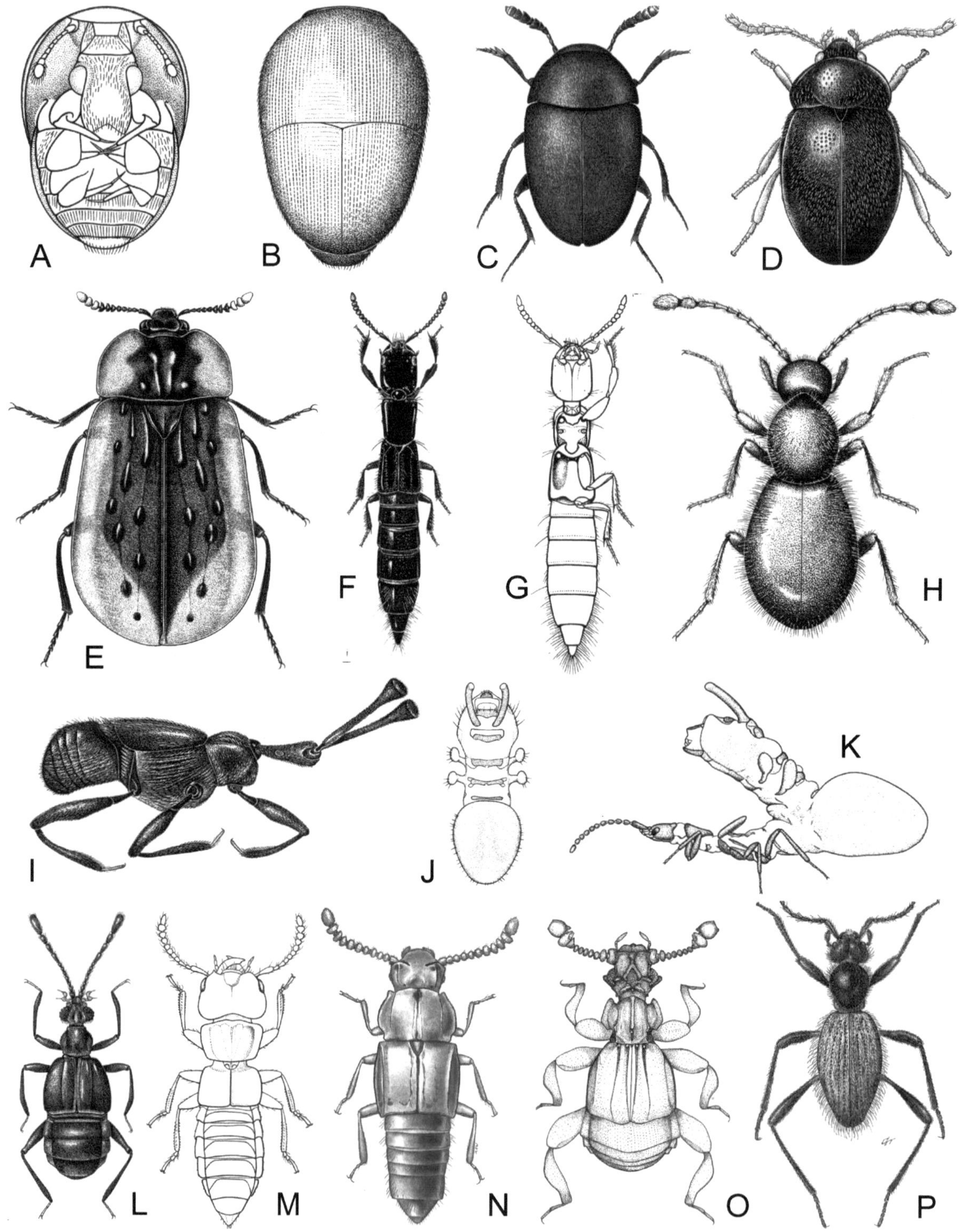

Fig. 59. Adult. A, *Rodwayia orientalis* Lea (Ptiliidae), ventral; B, same, dorsal; C, *Pseudonemadus australis* (Erichson)(Leiodidae: Cholevinae), dorsal. D, *Eublackburniella* sp. (Leiodidae: Camiarinae), dorsal. E, *Ptomaphila lacrymosa* (Schreibers) (Silphidae), dorsal. F–I, Staphylinidae: F, *Thyreocephalus chalcopterus* (Erichson) (Staphylininae), dorsal; G, same, ventral; H, *Scydmaenus myrmecophilus* Lea (Scydmaeninae), dorsal; I, *Tiracerus foveicollis* Raffray (Pselaphinae), lateral. J, *Austrospirachtha mimetes* Watson (Staphylinidae: Aleocharinae), abdomen, dorsal; K, same, habitus, lateral. L–P, Staphylinidae, dorsal: L, *Ctenicellus major* Raffray (Pselaphinae); M, *Anotylus* sp. (Oxytelinae); N, *Myrmelibia kingi* (Lea) (Osoriinae), from Newton 1990; O, *Malleecola myrmecophila* Oke (Pselaphinae), from Chandler 2001; P, *Clidicus abbotensis* O'Keefe (Scydmaeninae), from O'Keefe and Monteith 2000, G. Thompson, © Queensland Museum.

and elytra differing in lateral curvature (with distinct pronoto-elytral angle); strongly convex to moderately flattened; colour yellow to reddish-brown or black; dorsal surfaces simple and usually clothed with fine, recumbent hairs, occasionally glabrous or with suberect hairs; ventral surfaces never with hydrofuge pubescence.

Head more or less prognathous or somewhat hypognathous, usually transverse or about as long as wide, not constricted behind eyes to form neck. Paired ocelli absent. Eyes prominent to reduced or absent, with or without interfacetal setae. Antennal insertions concealed from above; subantennal groove well developed. Frontoclypeal suture absent; anterior edge of clypeus truncate. Labrum transverse and broadly rounded. Antennae short to moderately long, usually 11-segmented (10-segmented in *Rodwayia*) with 2- or 3-segmented club; scape and pedicel enlarged, funicular segments slender and elongate, each with whorl of long setae at its apex, and antennal club relatively slender and loosely jointed, with similar whorls of setae. Mandibles very short and broad, largely concealed between labrum, maxillae and mentum; with well developed mola and reduced unidentate apex; outer edge dorsally with ridge which may represent a labro-mandibular interlocking device similar to that found in some Hydraenidae. Maxilla apically setose (fimbriate in *Acrotrichis*), lacinia with or without uncus; maxillary palps 4-segmented, subulate with apical palpomere shorter and narrower than preapical one. Mentum usually subquadrate; labial palps sometimes reduced, with fewer than 3 palpomeres; ligula not bilobed. Corpotentorium slender, sometimes arched; laminatentoria absent. Cervical sclerites absent.

Pronotum ~0.5–1.6 times as long as wide, usually widest at base or at middle, with sides slightly to strongly curved (rarely widest apically or with subparallel sides), always as wide at base as combined elytral bases; sides sometimes obliquely explanate (e.g. in *Acrotrichis*, *Cochlearion*, *Rodwayia*). Lateral pronotal carinae complete and simple, usually without distinct margin or bead; anterior angles sometimes produced forward; posterior angles often right to obtuse (strongly produced and acute in *Acrotrichis*, *Cochlearion* and *Rodwayia*); posterior edge straight or curved, without underlying ventral ridge; disc almost always simple. Prosternum in front of coxae shorter to slightly longer than shortest diameter of procoxal cavity, flat to slightly convex; anterior edge not or only slightly projecting forward. Prosternal process, when complete, usually narrow, parallel-sided and apically acute or narrowly rounded; sometimes incomplete and acute at apex or absent (in *Rodwayia*, Fig. 59A, laterally expanded at middle and extending posterad beyond anterior edge of metaventrite, partly concealing both procoxae and mesocoxae from below). Notosternal sutures absent. Hypomeron with anterior antennal pocket and posterior crural impression in *Rodwayia*. Procoxae slightly transverse to globular, sometimes projecting, with short to moderately long concealed lateral extensions; trochantins concealed or exposed; endopleuron fused to notum. Procoxal cavities circular to slightly transverse, narrowly separated to contiguous (appearing broadly separated in *Rodwayia* due to lateral expansions of prosternal process, narrowly to broadly open externally, with or without narrow anterolateral extensions, internally narrowly open to broadly closed.

Scutellar shield usually exposed, sometimes very large (largely concealed in *Rodwayia*), not abruptly elevated anteriorly, posteriorly broadly angulate to acute. Elytra ~0.55–3.5 times as long as combined width and 0.85–3.1 times as long as pronotum; sides usually slightly curved and apices conjointly rounded concealing all or nearly all abdominal tergites (apices truncate exposing most of pygidium in *Rodwayia* or 3–5 tergites in some Acrotrichinae and Ptinellini); disc with irregular punctation and no scutellary striole, with transverse sculpturing in *Rodwayia*; epipleura weakly developed and incomplete or absent. Mesoventrite partly or completely fused to mesanepisterna, its anterior edge on same plane as metaventrite, without procoxal rests, but usually with broad prothoracic rest or collar; main portion of ventrite usually tumid or with longitudinal carina; mesoventral cavity absent; discrimen absent. Mesocoxae usually globular or slightly transverse with concealed trochantins. Mesocoxal cavities narrowly to moderately widely separated, probably closed laterally by meeting of mesoventrite and metaventrite (not evident when mesoventrite completely fused to mesanepisternum; mesepimera slender and not normally visible. Mesometaventral junction usually consisting of overlapping ventrites (ventrites possibly fused in *Rodwayia*). Metaventrite slightly to strongly transverse, flat to slightly convex, sometimes with post-coxal lines; discrimen absent; transverse (metakatepisternal) suture absent; exposed portion of metanepisternum elongate and very narrow. Metacoxae strongly transverse to slightly transverse or subconical and subcontiguous to widely separated, sometimes extending laterally to meet elytra; plates well developed. Metendosternite without or with very short stalk and long lateral arms, which may be widely separated at base; anterior tendons, if present, on lateral arms. Hind wing highly modified, with slender stalk and long narrow membrane completely lined with very long hairs (Fig. 23C); radial bending zone, radial cell, R-M loop and anal lobe absent; usually with 2 or 3 vein remnants; wings absent in *Rodwayia* and some Ptiliinae. Legs short and usually slender; femora expanded and grooved to receive flattened tibiae in *Rodwayia* and *Cochliarion*; tibial spurs paired but often not distinguishable from apical spines. Tarsi 3-3-3 or apparently 2-2-2; tarsomeres very slender, with apical much longer than first two combined and often with tarsomere 2 vestigial; pretarsal claws simple; empodium with 1 or 2 setae.

Abdomen usually with 6 or 7 free ventrites (sternites III–IX). Functional spiracles present on abdominal segments I–VI, I–VII or I–VIII. Abdominal tergites moderately well sclerotised, with patches of wing-folding asperities on one or more segments and tergite VII usually with palisade fringe along posterior edge. Anterior edge of sternite VIII in male without median strut. Anterior edge of segment IX with spiculum gastrale; tergite IX with broad basal struts, its posterior edge slightly emarginate; tergite X free. Aedeagus of modified trilobate type, symmetrical or asymmetrical, with parameres usually reduced or fused to phallobase and penis without basal struts. Sternite VIII in female without spiculum ventrale; tergite IX not divided into two parts; proctiger sclerotised. Ovipositor short and broad; coxites lightly sclerotised, undivided, without baculi, fused together to form single plate; styli absent. Spermatheca well developed and sclerotised. [Deane 1930, 1931; Seevers and Dybas 1943; Johnson 1982, 1985; Dybas 1990; Sörensson 1997; Hall 1999, 2000c, 2005; Polilov 2005, 2008; Grebennikov 2008a; Polilov and Beutel 2009].

Larva

Body elongate, more or less parallel-sided or slightly narrowing posteriorly, slightly flattened; head and all terga lightly sclerotised but sometimes with lightly pigmented tergal plates on thorax and abdominal segments I–IX; vestiture of scattered long and short, fine setae.

Head prognathous, or slightly declined, more or less transverse, distinctly rounded laterally and slightly flattened; posterior edge straight; epicranial suture absent; median endocarina absent. Stemmata usually absent. Frontoclypeal suture absent or weakly defined; labium free, slightly transverse, apically truncate. Antennae moderately long, 3-segmented, with palpiform sensorium arising well before apex of antennomere 2 and more than half as long as antennomere 3, which is apically acute with several long subapical setae. Mandibles symmetrical, unidentate with several subapical teeth or serrations; mola well developed, asperate; prostheca a slender, apically acute, hyaline process. Ventral mouthparts moderately retracted, with large maxillary articulating area; cardines free and undivided; stipes usually at least slightly elongate; galea and lacinia fixed, often fused together, the former laterally fimbriate, the latter setose and with one or more spines along inner edge; palp 3-segmented, without distinct palpifer. Labium consisting of prementum and postmentum; ligula well developed, undivided; palps 2-segmented and moderately widely separated. Hypopharyngeal bracon present; sclerome absent or consisting of transverse bar. Hypostomal rods and ventral epicranial ridges absent. Gular region short; gular sutures well separated or absent.

Prothorax not much longer than meso- or metathorax; thoracic terga and all abdominal terga with single tergal plate but often lightly pigmented and not apparent. Legs well developed, slender and 5-segmented with pretarsus forming distinct bisetose claw; coxae narrowly to moderately widely separated. Anterior abdominal terga and sterna usually separated by lateral membranous areas. Tergum IX about as long as VIII, not extending onto ventral surface, almost always with a pair of short, articulated but unsegmented, dorsolaterally-projecting urogomphi (absent in Nanosellini); segment X visible from above, more or less cylindrical, usually armed with pair of apical hooks (absent in Cephaloplectinae). Functional annular spiracles with closing apparatus usually present laterally on mesothorax and abdominal segments I–VIII. [Dybas 1976, 1990; Hall 2000c, 2005; Grebennikov and Beutel 2002; Grebennikov 2008a; Polilov and Beutel 2009].

Classification and World Distribution

Ptiliidae is a cosmopolitan family with nearly 700 described species in more than 90 genera. There are three recognised subfamilies, all of which occur in Australia.

Australian Fauna

Ptiliidae are usually recognised by their extremely small size, filamentous antennae clothed with whorls of long hairs and narrow hind wings fringed with very long hairs. The body is usually somewhat flattened and pubescent, the elytra often truncate, exposing one to three abdominal tergites, and tarsi are almost always 2-segmented and may appear to be 1-segmented. Species of *Cochliarion* and *Rodwayia* (Fig. 59A–B) differ from other Australian ptiliids in having the body more convex and compact, the hind wings and eyes absent, the head more deflexed and the antennae shorter. Larvae are elongate and lightly sclerotised, with short articulated urogomphi (rarely absent). Ptiliids are relatively abundant in decaying organic matter, including leaf litter, compost heaps, rotten logs, tree holes and dung, where their major food source appears to be fungal spores and hyphae. Some species of Nanosellini live within the pore tubes of bracket fungi (Polyporaceae), where they feed on developing spores. *Actinopteryx fucicola* Allibone occurs in decaying seaweed on sea coasts throughout the world. Species of *Rodwayia* (Cephaloplectinae) occur in the nests of ants; they are largely ignored by their hosts and appear to feed on larval exudations. [Deane 1930, 1931; Dybas 1976; Seevers and Dybas 1943].

Australian Subfamilies, Tribes and Genera

Acrotrichinae: Acrotrichini: *Acrotrichis* Motschulsky, *Actinopteryx* Matthews, *Neotrichopteryx* Deane. **Nephanini:** *Etronia* Deane, *Ptiliodes* Matthews.

Cephaloplectinae: *Rodwayia* Lea.

Ptiliinae: Discheramocephalini: *Cissidium* Motschulsky (=*Epibaptus* Deane). **Nanosellini:** *Garicaphila* Hall, *Isolumpia* Deane, *Mikado* Matthews, *Nellosana* Johnson, *Nellosanoides* Hall, *Tasmangarica* Hall. **Ptenidiini:** *Cochliarion* Deane, *Ptenidium* Erichson (=*Epoptia* Deane). **Ptiliini:** *Actidium* Matthews (=*Ptenidula* Deane), *Ptilium* Gyllenhal. **Ptinellini:** *Leaduadicus* Deane, *Ptinella* Motschulsky (=*Achosia* Deane, *Cnemadoxia* Deane).

14. LEIODIDAE

Adult (Figs 59C–D, 67I–J)

Length ~1.2–5 mm. Body ~1.22–2.35 times as long as wide, sides usually more or less rounded, with pronotum and elytra differing in lateral curvature (sides uniformly rounded in some Leiodinae), slightly to strongly convex; colour yellow to reddish-brown or black, rarely bicolored; dorsal surfaces densely setose to subglabrous.

Head prognathous or hypognathous; in Cholevinae only, abruptly constricted immediately behind eyes forming a broad neck and very short temples, in combination with transverse occipital carina more or less abutting anterior edge of pronotum (Fig. 28C, T). Paired ocelli present only in *Neopelatops*. Eyes usually well developed, protuberant, circular and entire, finely or coarsely facetted, with or without interfacetal setae (eyes reduced or not projecting beyond lateral margins of head in some Leiodinae and Cholevinae). Antennal insertions exposed or concealed from above; subantennal groove present in some Leiodinae. Frontoclypeal suture curved or angulate, with or without attached mid-cranial suture, or absent; clypeus relatively short and anteriorly truncate. Labrum usually transverse and truncate to emarginate (elongate and apically acute in *Myrmecholeva*). Antennae moderately long, almost always 11-segmented, with weak to strong, 5-segmented club with antennomere 8 shorter and narrower than 7 and 9 (Fig. 15I) (antennae 10-segmented in *Zeadolopus* and club 4-segmented in *Colon*, Fig. 28U, and *Zeadolopus*); antennomeres 7,

9 and 10 (7 to 9 in *Zeadolopus*, 8–10 in *Colon*) each with periarticular gutter containing sensilla and usually opening through narrow slit to surface; scape usually elongate and not modified. Mandibles short and broad to moderately elongate, symmetrical or asymmetrical; apex usually unidentate without or with one to several subapical teeth; mola usually small or absent; prostheca weakly developed or absent. Maxilla with 2-segmented galea; lacinia setose or spinose, sometimes with uncus; maxillary palps 4-segmented, with apical palpomere usually fusiform and often somewhat narrower than preapical one (enlarged in Camiarini). Mentum usually trapezoidal and transverse; labial palps 3-segmented, sometimes with apical palpomere shorter or narrower than preapical one; ligula usually bilobed. Mouthparts modified to form sucking tube in *Myrmecholeva* (Fig. 67J). Corpotentorium narrow, arched or not; laminatentoria well developed and often broadly joined at midline. Cervical sclerites usually well developed (absent in *Colon*).

Pronotum ~0.38–0.90 times as long as wide, usually widest posteriorly or at middle, never distinctly narrower at base than combined elytral bases; sides slightly to strongly curved, obliquely or vertically explanate in some Leiodinae. Lateral pronotal carinae complete, simple, usually with raised margin or bead; anterior angles produced forward and broadly rounded to acute in some Leiodinae; posterior angles usually right to obtuse, sometimes slightly acute; posterior edge almost always straight or evenly rounded, (produced forming mesal lobe in *Myrmecholeva* and *Eublackburniella*); without underlying ventral ridge; disc usually simple or with transverse striolations (with paired basal impressions in *Myrmecholeva* and *Eublackburniella*). Prosternum in front of coxae slightly to much shorter than shortest diameter of procoxal cavity (sometimes virtually absent), usually flat to slightly convex (apparently carinate in some Leiodinae, but carina probably represents internal septum between coxal cavities). Prosternal process almost always incomplete and apically acute or angulate (complete and narrowly rounded at apex in *Neopelatops* and Sogdini). Notosternal sutures complete, incomplete or absent. Hypomeron without antennal pockets. Procoxae transverse to more or less conical and projecting below prosternum, without or with short concealed lateral extensions; trochantins exposed; endopleuron not fused to notum. Procoxal cavities slightly to strongly transverse, usually contiguous (narrowly separated in *Neopelatops* and Sogdini), either moderately broadly open externally, with relatively long and acute postcoxal projections, and internally open (*Neopelatops*, Sogdini, *Dietta*, *Eublackburniella* and *Myrmecholeva*) or very broadly open externally, with short, angulate postcoxal projections and internally closed (*Agyrtodes*, *Colon*, most Leiodinae and Cholevinae).

Scutellar shield well developed, not abruptly elevated, posteriorly narrowly rounded to acute. Elytra ~0.80–1.75 times as long as combined width and 1.78–4.20 times as long as pronotum; sides slightly to strongly curved and apices usually conjointly rounded concealing all abdominal tergites (apices subtruncate exposing pygidium in some Cholevinae); humeri well developed; disc often with 9 distinct puncture rows and no scutellary striole; punctation sometimes irregular; transverse striolations sometimes present; epipleura sometimes broad at base, but usually narrow and always incomplete. Mesoventrite often at least partly fused to mesanepisterna, which are widely separated at midline; anterior edge of mesoventrite usually on same plane as metaventrite (on different plane in some Leiodinae), with paired procoxal rests or a broad prothoracic rest or collar; main portion of ventrite sometimes with median longitudinal carina; mesoventral cavity absent; discrimen absent. Mesocoxae slightly to strongly transverse and not to moderately oblique (strongly oblique in *Dietta,* Fig. 25I), usually with exposed trochantins. Mesocoxal cavities usually narrowly to moderately widely separated (very widely separated in *Dietta*) and partly closed laterally by mesepimeron. Meso-metathoracic joint within coxal cavities usually absent (membranous in *Neopelatops*, solid in Sogdini and some Agyrtodini). Mesometaventral junction, if present, consisting of overlapping ventrites or a monocondylic joint. Metaventrite slightly to strongly transverse, flat to slightly convex, sometimes with transverse carina tangential to hind edges of mesocoxal cavities; transverse (metakatepisternal) suture sometimes weakly developed but very close to posterior edge of ventrite; discrimen absent; exposed portion of metanepisternum elongate and very narrow, more or less concealed in some Leiodinae). Metacoxae strongly transverse and subcontiguous or narrowly separated, not extending laterally to meet elytra; plates weakly developed or absent. Metendosternite usually with long, slender stalk; long, transverse or oblique, lateral arms; and anterior process usually very short with widely separated anterior tendons, but sometimes longer with narrowly separated tendons. Hind wing of staphylinoid type without radial bending zone but with distinct hinge at end of radial bar; apical field about half to two-thirds as long as total wing length, usually with 2 or 3 radial extensions; radial cell absent or incomplete basally, with lumen more or less obliterated by pterostigma; cross-vein r3 absent and r4 incomplete or absent; R-M loop and basal portion of RP absent but apical extension usually well developed; medial spur short to very long; medial field usually with one free vein ($Cu\text{-}AA_3$), often with one or two unattached cross-veins; wedge cell absent; anal lobe highly reduced or absent. Legs moderately long and slender to short and stout. Femora and tibiae sometimes modified in males; tibiae often spinose, sometimes strongly widened; tibial spurs usually well developed. Tarsi usually 5-5-5 (5-5-4 in *Zeadolopus*, 5-4-4 in Pseudoliodini, 3-3-3 in Scotocryptini); basal tarsomeres sometimes expanded in males (in males and females of *Colon*); pretarsal claws simple; empodium bisetose.

Abdomen usually with 5 or 6 free ventrites (III–VII or VIII) (4 ventrites in female *Colon* due to invagination of lightly sclerotised sternite VII); lateral portions of sternite II variously developed but concealed by coxae; ventrite 1 with recurved postcoxal lines in Scotocryptini; intersegmental membranes with brick-wall pattern of microsclerites in *Colon*. Functional spiracles on abdominal segments I–VIII in Neopelatopini and Leiodinae, I–VII in Agyrtodini and Cholevinae and I–VI in *Colon*. Abdominal tergites usually relatively lightly sclerotised, sometimes with pygidium well sclerotised; with patches of wing-folding asperities on one or more tergites. Anterior edge of sternite VIII in male usually without median strut (present in Sogdini). Anterior edge of segment IX variable, sometimes with anteriorly acute subgenital plate, short spiculum gastrale or separated basal struts; tergite IX entire, divided into two sclerites by tergum X or fused with tergum X. Aedeagus of modified trilobate type, usually symmetrical, with or without distinct phallobase; parameres free, fused to penis or absent. Sternite VIII in female with or without spiculum ventrale. Tergite IX in female divided into two parts; proctiger sclerotised. Ovipositor short and broad; coxites usually divided, without baculi, partly fused together in *Colon*; styli usually present. [Corbiere-Tichané 1974; Peck 1977, 1990, 2000b, 2003; Zwick 1979; Newton 1998, 2005d; Peck and Cook 2003, 2009; Seago 2009].

Larva (Fig. 34G)

Body elongate, parallel-sided or slightly narrowing posteriorly, slightly curved dorsally in Sogdini and Leiodini, slightly flattened; head and all terga weakly to strongly sclerotised; vestiture of scattered long and short, fine setae, sometimes including expanded or frayed setae.

Head prognathous and protracted, more or less transverse, distinctly rounded laterally and slightly flattened; posterior edge straight; epicranial stem short to moderately long; frontal arms V-shaped or U-shaped to lyriform, with bases contiguous or separate; median endocarina absent. Stemmata 5 on each side in Camiarinae, usually 3 or fewer. Frontoclypeal suture absent; labium free, usually transverse and broadly rounded at apex (elongate and apically acute in *Myrmecholeva*). Antennae usually moderately long (longer than head width in *Myrmecholeva*, sometimes less than half as long as head width), 3-segmented with palpiform sensorium on antennomere 2 usually subapically anterior to antennomere 3, which is much shorter than basal two antennomeres. Mandibles symmetrical or asymmetrical, unidentate or bidentate; incisor edge simple, with 1 or 2 subapical teeth or serrate; mola usually well developed and transversely ridged or asperate (mola absent in *Myrmecholeva*); prostheca consisting of a hyaline lobe, which may be serrate, or absent. Ventral mouthparts strongly retracted, with maxillary articulating area; cardines free and divided; stipes about as long as wide or slightly elongate; mala usually deeply cleft, forming distinct but fixed galea and lacinia (with single styliform mala in *Myrmecholeva*); galea with outer fringe (fimbriate); lacinia falcate with several spines along inner edge; palp 3-segmented. Labium consisting of prementum, mentum and submentum; ligula usually elongate, bilobed or truncate; palps 2-segmented and usually moderately widely separated (narrowly so in *Myrmecholeva*). Hypopharyngeal bracon present, with pair of mesally directed apodemes; sclerome present or absent. Hypostomal rods and ventral epicranial ridges absent. Gular region short; gular sutures well separated or absent.

Prothorax not much longer than meso- or metathorax; thoracic terga and all abdominal terga with single tergal plate; thoracic terga laterally expanded in *Myrmecholeva*. Prothoracic venter without presternal plate. Legs well developed, slender and 5-segmented with pretarsus forming distinct bisetose claw; coxae narrowly to moderately widely separated. Tergum IX about as long as or slightly shorter than VIII, not extending onto ventral surface, with a pair of urogomphi, which are usually 2-segmented and articulated, with long, slender, sometimes multiannulate terminal segment (rarely short or fixed); segment X visible from above, more or less cylindrical, usually with pair of eversible spiculate lobes. Functional spiracles annular or annular–biforous, with closing apparatus, placed laterally on mesothorax and abdominal segments I–VIII. [Newton 1991, 2005d; Peck 2000b; Kilian 2007, 2012].

Classification and World Distribution

The family Leiodidae contains ~3900 species in 360 genera and is usually divided into six subfamilies, four of which occur in Australia. Catopocerinae and Platypsyllinae are both Holarctic groups. Within the subfamily Leiodinae, only the tribe Agathidiini is absent from Australia; and within the large subfamily Cholevinae, only the subtribe Eunemadina of the tribe Anemadini occurs in Australia. [Newton 1985, 1998, 2005d; Perreau and Růžička 2007; Peck and Cook 2011].

Australian Fauna

Leiodidae are relatively small beetles with more or less spiny legs and simple tarsi (protarsi and occasionally mesotarsi modified in some males and in *Colon*), differing from most other staphylinoids in having complete elytra and almost always recognisable by the type of antennal club, which is 5-segmented with the second club segment (antennomere 8) smaller (shorter or shorter and narrower) than either the first (antennomere 7) or the third (antennomere 9). However, the club may be very weak (*Eublackburniella*, Fig. 59D, or *Nargiotes*) with antennomere 8 only slightly reduced or very strong (*Dietta*, Fig. 67I) with antennomere 8 greatly reduced. The genera *Zeadolopus* and *Colon* are exceptional in having only four segments in the antennal club. Leiodids also possess a unique type of internal sensory vesicle usually opening on the distal surfaces of antennomeres 7, 9 and 10 (Corbiere-Tichané 1974; Peck 1977). Larvae, like those of most staphylinoids, are of the active campodeiform type with urogomphi which are almost always articulated at base; they usually differ from staphylinid larvae in having a mandibular mola and two maxillary lobes. Leiodidae are abundant in decaying organic matter and also occur in carrion, fungal fruiting bodies, some types of dung and nests of vertebrates; they are abundant in caves in most parts of the world, but have not yet been recorded from Australian caves. Many are general scavengers, but certain groups are associated with particular fungi (*Neopelatops* with Myxomycetes; some *Nargomorphus* with puffballs; some leiodines and probably *Colon* with hypogean fungi) (Newton 1984). The subfamily **Camiarinae** is a Notogean group, occurring in New Zealand, Australia, South Africa and southern South America. Australian species are mainly in the tribes Neopelatopini and Agyrtodini, but an undescribed species of Camiarini is known. Species of the neopelatopine genus *Myrmecholeva* (Figs 46M, 67J) are very unusual in that both larvae and adults have suctorial mouthparts consisting of a labro-labial tube and styliform mandibles and maxillae. Members of the subfamily **Leiodinae** are usually glabrous and strongly convex and may have burrowing adaptations, such as enlarged legs (*Dietta*, *Zeadolopus*). The leiodid tribe Sogdini has not yet been recorded from Australia, but at least three undescribed sogdine genera have been seen. Also in this group are the minute species of Pseudoliodini and Scotocryptini. Members of the subfamily **Cholevinae** are relatively common and belong to several genera, and the widely distributed genus *Colon* (**Coloninae**) is represented by several species.

Australian Subfamilies, Tribes and Genera

Camiarinae: Camiarini: New genus sp. **Neopelatopini:** *Eublackburniella* Jeannel, *Myrmicholeva* Lea, *Neopelatops* Jeannel. **Agyrtodini:** *Agyrtodes* Portevin, *Agyrtolasia* Szymczakowski, *Cholevomorpha* Blackburn, *Paragyrtodes* Szymczakowski, *Zeagyrtes* Broun ?.

Leiodinae: Sogdini: New genera spp. **Estadiini:** *Dietta* Sharp. **Leiodini:** *Zeadolopus* Broun. **Pseudoliodini:** *Colenisia* Fauvel, *Dermatohomoeus* Hlisnikovsky. **Scotocryptini:** *Aglyptinus* Cockerell, New genus sp.

Coloninae: *Colon* Herbst.

Cholevinae: Anemadini: *Austrocholeva* Zwick, *Austronargus* Zwick, *Austronemadus* Zwick, *Catoposchema* Jeannel, *Nargiotes* Jeannel, *Nargomorphus* Jeannel, *Paranemadus* Zwick, *Pseudonemadus* Portevin, *Rangiola* Jeannel.

15. SILPHIDAE

Adult (Figs 2A–B, 59E, 68B)

Length ~14–28 mm (as much as 45 mm in *Diamesus*). Body ~1.45–2.2 times as long as wide (as much as 2.7 times in *Diamesus*), sides slightly rounded to subparallel; slightly to strongly flattened; colour yellowish-brown with dark markings (*Ptomaphila*) or black with reddish markings (*Diamesus*); surfaces of pronotum and elytra clothed with extremely short, fine hairs.

Head prognathous, abruptly constricted immediately behind eyes forming very short temples and a short, broad neck delimited dorsally by a raised transverse occipital ridge. Eyes well developed, protuberant, elongate-oval, shallowly emarginate posteriorly, and finely facetted, without interfacetal setae. Antennal insertions exposed from above; subantennal grooves absent. Frontoclypeal suture absent; clypeus anteriorly truncate. Labrum strongly transverse and slightly emarginate. Antennae moderately long, 11-segmented with strong, 3-segmented, finely, evenly pubescent club with scattered longer setae, preceded by two glabrous and more or less cup-like antennomeres; scape more than twice as long as wide, slightly curved. Mandibles ~1.5 times as long as wide, symmetrical; apex usually unidentate without or with one subapical tooth; mola and prostheca well developed. Maxilla with 2-segmented, densely setose galea; lacinia with strong, unidentate and densely setose inner edge; palp 4-segmented with apical palpomere cylindrical and not longer than preapical one. Mentum strongly transverse; ligula broad and bilobed; labial palps 3-segmented with cylindrical apical palpomere. Corpotentorium absent; laminatentoria well developed and broadly joined at midline. Cervical sclerites usually well developed.

Pronotum ~0.55–0.90 times as long as wide, widest posteriorly, not much narrower at base than combined elytral bases; sides slightly to strongly curved, horizontally explanate in *Ptomaphila*. Lateral pronotal carinae complete, simple, with raised margin or bead; anterior angles slightly produced forward and broadly rounded in *Ptomaphila*; posterior angles broadly rounded; posterior edge evenly rounded in *Diamesus*, biemarginate in *Ptomaphila*, without underlying ventral ridge; disc with paired basal impressions in *Diamesus*. Prosternum in front of coxae slightly to distinctly shorter than shortest diameter of procoxal cavity, flat to slightly convex. Prosternal process incomplete and apically acute or angulate. Notosternal sutures complete. Hypomeron without antennal pockets. Procoxae conical and projecting below prosternum, without concealed lateral extensions; trochantins exposed; endopleuron not fused to notum. Procoxal cavities slightly transverse, contiguous, broadly open externally, with postcoxal projections angulate in *Ptomaphila*, absent in *Diamesus*, internally open.

Scutellar shield well developed, not abruptly elevated, posteriorly narrowly rounded to acute, very large with weak median carina in *Diamesus*. Elytra ~0.95–1.50 times as long as combined width and 1.9–3.0 times as long as pronotum; sides moderately curved and apices conjointly rounded concealing all abdominal tergites in *Ptomaphila*, sides subparallel and apices truncate exposing more than 3 abdominal tergites in *Diamesus*; disc with irregular fine punctation and each with 3 longitudinal carinae, those in *Ptomaphila* broken by placement of irregular carinal tubercles; epipleura incomplete, narrow in *Diamesus*, broadly explanate in *Ptomaphila*. Mesoventrite separated by complete sutures from mesanepisterna, which are widely separated at midline; anterior edge of mesoventrite on same plane as metaventrite, without paired procoxal rests or broad prothoracic collar; mesoventral cavity absent; discrimen absent. Mesocoxae slightly to strongly transverse and slightly oblique, with exposed trochantins. Mesocoxal cavities moderately widely separated and partly closed laterally by mesanepisternum and mesepimeron. Meso-metathoracic joint within coxal cavities absent. Mesometaventral junction consisting of abutting ventrites. Metaventrite slightly to moderately convex, without postcoxal lines; transverse (metakatepisternal) suture weakly developed but very close to posterior edge of ventrite; discrimen absent; exposed portion of metanepisternum moderately elongate. Metacoxae strongly transverse and subcontiguous or narrowly separated, not extending laterally to meet elytra; plates weakly developed or absent. Metendosternite usually with long, slender stalk, long, transverse lateral arms, long, slender anterior process and narrowly separated tendons. Hind wing with moderately well developed radial bending zone combined with distinct hinge at end of radial bar; apical field about half as long as total wing length, with 2 or 3 radial extensions; radial cell absent or with lumen more or less obliterated by pterostigma; cross-vein r3 absent and r4 incomplete or absent; R-M loop and basal portion of RP absent but apical extension well developed; medial spur not curved posteriorly, extending to wing margin, where there is a medial embayment; medial field with 4 free veins; MP_{3+4} very short, not extending to base and without cross-vein or spur, broadly forked to form long MP_3 and MP_4, the latter connected by cross-vein to apex of Cu; wedge cell absent; AA_{3+4} not forked; anal lobe well developed, without embayment; AP_{3+4} not forked. Legs moderately long and slender. Metafemora enlarged and toothed in males of *Diamesus*; tibiae lined with longitudinal rows of setae or socketed spines; tibial spurs well developed. Tarsi 5-5-5; basal protarsomeres (*Ptomaphila*) or pro- and mesotarsomeres (*Diamesus*) expanded in males; pretarsal claws simple or slightly expanded basally; empodium well developed and bisetose.

Abdomen with 7 free ventrites (sternites II–VIII) in female or 8 (sternites II–IX) in male; ventrite 1 (sternite II) divided into 2 or 3 parts by coxae, without postcoxal lines; intersegmental membranes with brick-wall pattern of microsclerites. Functional spiracles on abdominal segments I–VIII. Abdominal tergites I–III relatively lightly sclerotised, posterior tergites more heavily so (especially in *Diamesus*); with patches of wing-folding asperities on tergites III and IV or III–V. Anterior edge of sternite VIII in male without median strut. Anterior edge of segment IX with spiculum gastrale;

tergite IX almost completely divided into two laterotergites and tergum X free in *Diamesus*; tergites IX and X fused in *Ptomaphila*. Aedeagus of modified trilobate type, symmetrical; phallobase reduced, mostly membranous, with lateral struts (*Diamesus*) or a basal curved sclerite (*Ptomaphila*); parameres free, fused to penis or absent. Sternite VIII in female without spiculum ventrale. Tergite IX in female divided into two parts; proctiger sclerotised. Ovipositor (Fig. 30B) short and broad; paraprocts slightly shorter than coxites, with baculi; coxites divided, proximal lobe with basal oblique baculi, distal lobe more heavily sclerotised; styli reduced to subapical tuft of setae. [Peck 2000c, 2000d; Sikes 2005].

Larva (Fig. 58K)

Body oblong, widest anteriorly and narrowing towards abdominal apex, more or less flattened; head and all terga strongly sclerotised and darkly pigmented; vestiture of scattered setae.

Head prognathous and protracted, more or less transverse, slightly rounded laterally and slightly flattened; posterior edge straight; epicranial stem moderately long; frontal arms V-shaped or U-shaped, with bases contiguous; median endocarina absent. Stemmata 6 on each side. Frontoclypeal suture incomplete at middle; labium separated by suture but immovable, without tormae, transverse, subdivided and broadly rounded at apex (Fig. 46I). Antennae more than half as long as head width, 3-segmented, with antennomeres subequal in length; antennomere 2 with more or less oblique apex bearing short, subonical sensorium. Mandibles symmetrical, unidentate, incisor edge with one subapical tooth; mola and prostheca absent. Ventral mouthparts strongly retracted, with maxillary articulating area; cardines free and internally divided; stipes slightly elongate; mala deeply cleft forming distinct but fixed galea and lacinia; galea densely setose; lacinia falcate with several spines along inner edge; palp 3-segmented with segment-like palpifer. Labium consisting of prementum, mentum and submentum; ligula well developed, bilobed; palps 2-segmented without distinct palpiger, widely separated. Hypopharyngeal bracon present; sclerome absent. Hypostomal rods very short and converging; ventral epicranial ridges present but more or less coincident with hypostomal rods. Gular region short; gular sutures well separated, converging anteriorly.

Prothorax slightly longer than meso- or metathorax; thoracic terga and abdominal terga with single tergal plate; thoracic terga and abdominal terga I–VIII expanded to form distinct, posterolaterally directed plates. Prothoracic venter with presternal plate divided longitudinally into 3 sclerites. Legs moderately long and slender, spinose, 5-segmented with pretarsus forming distinct bisetose claw; coxae moderately widely separated. Abdominal sterna II–VIII with lightly to moderately sclerotised plates and slight posterolateral projections; sternum II with reduced plate. Tergum IX slightly shorter than VIII, not extending onto ventral surface, with a pair of 2-segmented, articulated urogomphi; segment X visible from above, slightly flattened and slightly narrowing apically, with pair of eversible spiculate lobes. Functional spiracles annular, with closing apparatus, placed laterally on mesothorax and abdominal segments I–VIII; non-functional spiracle on metathorax. [Newton 1991; Peck 2000c; Sikes 2005].

Classification and World Distribution

Silphidae is a relatively small, widely distributed family containing 15 genera and almost 200 species. There are two subfamilies: Nicrophorinae is primarily a Northern Hemisphere group, with some species extending into the Malay Archipelago, New Guinea and montane habitats in the Neotropical region (Sikes *et al.* 2002). Silphinae are more widely distributed but with most genera in the Holarctic region.

Australian Fauna

Two genera of **Silphinae** occur in Australia, *Ptomaphila* Hope (Figs 2A, 59E) and *Diamesus* Hope (Figs 2B, 68B). They are relatively large, somewhat flattened beetles with a distinctive antenna (large 3-segmented, slightly asymmetrical and finely pilose antennal club preceded by at least one glabrous, strongly transverse, concave segment (Fig. 59E) reminiscent of the hydrophilid cupule. The body is either glabrous or clothed with very short, fine hairs, the eyes are moderately large and protuberant, the procoxae are large and projecting, and all tarsomeres are clothed beneath with yellow hairs. In *Ptomaphila*, the sides of both prothorax and elytra are explanate and the elytra conceal the abdomen and are provided with short longitudinal ridges or tubercles, while in *Diamesus* the elytra are carinate and truncate, exposing four or five abdominal tergites. *Ptomaphila* larvae (Fig. 58K) are broad, depressed and heavily sclerotised, with lateral plate-like expansions on the thorax and abdomen. Australian Silphidae, like most members of the famiy, are associated with vertebrate carcases, but adults of *Ptomaphila lacrymosa* (Schreibers) have been observed feeding on maggots. In other parts of the world, there are a few phytophagous species of silphines, as well as some that feed on snails. *Diamesus osculans* (Vigors) is widely distributed from India and Indonesia, through New Guinea to northern and eastern Australia. *Ptomaphila lacrymosa* occurs in the south-eastern and south-western parts of the continent, while *P. perlata* Kraatz is confined to the eatern coast, from East Gippsland to Cape York. The genus *Oxelytrum* Gistel, the apparent sister group of *Ptomaphila*, occurs in Central and South America. [Young 1983; Peck and Anderson 1985; Peck and Kaulbars 1987; Peck 2000d].

16. STAPHYLINIDAE

Adult (Figs 1G–H, 59F–P, 67A–H, 67K–L)

Length ~0.8–20 (usually 2–8) mm. Body ~1.5–10 times as long as wide, but usually more than twice as long as wide, sides usually subparallel, but unevenly controured laterally (sides more or less uniformly curved in *Microsilpha*, *Glypholoma* and Scaphidiinae); usually slightly flattened to moderately convex; colour usually reddish-brown to dark brown or black, sometimes with yellow or red markings, rarely iridescent; dorsal surfaces subglabrous or clothed with hairs or bristles (rarely scales in some Pselaphinae).

Head prognathous to hypognathous, with or without neck constriction. Eyes usually well developed, not to strongly protuberant, usually more or less circular, sometimes shallowly emarginate anteriorly (strongly so in some Scaphidiinae), posteriorly emarginate in some Pselaphinae, finely to coarsely facetted, with or without interfacetal setae; eyes absent in Leptotyphlinae and some members of other subfamilies. Paired ocelli present in *Glypholoma*,

Fig. 15B, *Microsilpha* and Omaliinae. Antennal insertions usually exposed from above, located between eyes in Aleocharinae (Fig. 14H), Steninae and Scaphidiinae, concealed from above in some groups (Fig. 15K); subantennal grooves usually absent. Frontoclypeal suture present or absent, rarely with attached mid-cranial stem; clypeus usually rounded or truncate anteriorly (tridentate in *Megalopinus*). Labrum usually free, visible and strongly transverse, less commonly as long as wide, and apically convex to emarginate (longer than wide and deeply emarginate in *Megalopinus*). Antennae usually moderately long and 11-segmented (10-segmented in some Aleocharinae, 9- or 10-segmented in some Euaesthetinae, from 3- to 10-segmented in some Pselaphinae), filiform, moniliform or incrassate-clavate, or weakly capitate (club strongly developed in *Glypholoma*, *Microsilpha*, *Megalopinus*, and some Pselaphinae, Scaphidiinae, Scydmaeninae, Euaesthetinae and Leptotyphlinae); scape usually elongate; antenna rarely geniculate. Mandibles short and broad to very narrow and elongate, symmetrical or asymmetrical; apex usually unidentate, often bidentate (multidentate in some Osoriinae); without or with one or two subapical teeth; mola and prostheca well developed, reduced or absent. Maxilla almost always with distinct galea and lacinia (with one lobe in some Aleocharinae and Pselaphinae); galea usually setose and lacinia with or without apical or subapical hook(s) or spines; palp usually 4-segmented (apical palpomere subdivided in Microsilphinae and some Aleocharinae; with 3 or fewer palpomeres in some Pselaphinae); apical palpomere usually cylindrical or fusiform, but may be expanded and truncate, securiform or aciculate. Ligula undivided to bilobed or deeply emarginate; labial palps usually 3-segmented (1- or 2-segmented in some Pselaphinae) with apical palpomere cylindrical to fusiform, sometimes much smaller than preapical one, expanded apically in some Staphylininae and in Trichopseniini. Corpotentorium narrow to incomplete or absent, rarely with median process. Cervical sclerites usually well developed.

Pronotum ~0.5–1.5 times as long as wide, widest posteriorly, at middle or anteriorly, not to distinctly narrower at base than combined elytral bases; sides subparallel to strongly curved, sinuate or variously lobed or dentate; obliquely or vertically explanate in *Microsilpha*, *Trichopsenius* and some Tachyporinae. Lateral pronotal carinae usually complete, or incomplete (absent in Leptotyphlinae, many Pselaphinae and Scydmaeninae, and some Aleocharinae, *Stenus* and Euaesthetinae), usually simple, finely crenulate or denticulate, with distinct teeth in *Megalopinus*, with or without raised margin or bead; anterior angles sometimes produced forward and acute to broadly rounded; posterior angles usually broadly rounded to right or slightly acute (distinctly acute in a few Aleocharinae, Scaphidiinae and Scydmaeninae); posterior edge usually straight or evenly rounded, strongly produced to form mesal lobe in some Osoriinae and Scaphidiinae, without underlying ventral ridge; disc with one or more sub-basal foveae or a transverse furrow in *Edaphus*, some Scydmaeninae and many Pselaphinae and Scydmaeninae, with median furrow in *Plastus*, *Prognathoides* and various other genera. Prosternum in front of coxae usually much shorter than to slightly longer than shortest diameter of procoxal cavity (distinctly longer in some groups (e.g. *Clidicus*, *Megalopinus*, *Stenus*, Euaesthetinae, Leptotyphlinae, some Osoriinae), usually flat to moderately convex. Prosternal process almost always either absent or incomplete (Figs 20G, 59G) and apically acute or angulate (complete and parallel-sided or only slightly expanded apically in Cephenniini, some Osoriinae and Scaphidiinae). Notosternal sutures present or absent. Hypomeron without antennal pockets. Procoxae usually more or less conical and projecting well below prosternum, without or with short concealed lateral extensions (strongly transverse in *Glypholoma* and *Microsilpha*; with long concealed extensions in *Plestus*, *Megalopinus*, *Stenus*, Euaesthetinae, Leptotyphlinae and some Pselaphinae); trochantins usually at least partly exposed (completely concealed in Pselaphinae, Scaphidiinae, *Plastus*, Scydmaeninae, *Megalopinus*, *Stenus*, Euaesthetinae and Leptotyphlinae); endopleuron usually free. Procoxal cavities usually slightly transverse to circular, narrowly separated or contiguous, almost always broadly open externally (closed in *Plastus* and some Paederini), with postcoxal projections short and angulate or absent (long and acute in *Glypholoma*).

Scutellar shield not abruptly elevated, posteriorly broadly rounded to acute, sometimes absent. Elytra ~0.95–1.50 times as long as combined width and 1.9–3.0 times as long as pronotum; sides usually subparallel or weakly rounded (more strongly rounded in many Scydmaeninae, Scaphidiinae and *Glypholoma*, converging anteriorly in some Pselaphinae), almost always truncate, exposing at least 3 abdominal tergites (exposing fewer tergites in *Glypholoma*, Fig. 67G), *Microsilpha* and Scaphidiinae, Fig. 67F), with apices conjointly rounded concealing all abdominal tergites in most Scydmaeninae; disc usually impunctate, irregularly punctate or with fewer than 5 puncture rows (with 11 puncture rows in *Glypholoma*); epipleura incomplete or absent. Mesoventrite usually separated by complete sutures from mesanepisterna, which are widely separated at midline, or partly of completely fused to them; anterior edge of mesoventrite usually on same plane as metaventrite (on different plane in *Glypholoma* and some *Microsilpha*, Pselaphinae and Scaphidiinae), with or without paired procoxal rests or broad prothoracic collar; mesoventral cavity absent; discrimen absent. Mesocoxae globular to strongly transverse and slightly to strongly oblique, projecting or not, with exposed or concealed trochantins. Mesocoxal cavities contiguous to widely separated and open laterally (partly closed by mesepimeron and sometimes mesanepisternum) or closed laterally by meeting of meso- and metaventrites; cavities sometimes weakly impressed or almost absent. Meso-metathoracic joint within coxal cavities usually absent or represented by a solid suture (membranous in *Glypholoma* and *Microsilpha* and some Omaliinae). Mesometaventral junction, if present, variable (ventrites sometimes fused together). Metaventrite flat to strongly convex, with or without postcoxal lines; with postcoxal pits in *Megalopinus* and most Pselaphinae; transverse (metakatepisternal) suture usually absent (present in *Glypholoma* and *Microsilpha*); discrimen usually absent or very short (long in some Scaphidiinae); exposed portion of metanepisternum usually narrowly elongate (shorter and broader in some Scaphidiinae, often concealed in Scydmaeninae). Metacoxae slightly strongly transverse to subconical and subcontiguous or narrowly separated (more widely separated in some Pselaphinae and Scaphidiinae), often extending laterally to meet elytra or sides of body; plates usually weakly developed or absent (well developed and complete in *Glypholoma*, well developed mesally in *Microsilpha*). Metendosternite with stalk long to short, lateral arms usually long and horizontal, anterior process usually short and distinctly emarginate or absent, and anterior tendons widely separated, sometimes on lateral arms.

Hind wing of staphylinoid type (Fig. 23A), without or with short radial bending zone and with distinct hinge at end of radial bar; apical field often much more than half total wing length, usually with 2 or 3 radial extensions (apical region usually absent and wing truncate in *Glypholoma*); radial cell absent or with lumen more or less obliterated by pterostigma; cross-vein r3 absent and r4 usually incomplete or absent; R-M loop and basal portion of RP absent but apical extension well developed; medial spur absent or not curved posteriorly and often extending to wing margin; medial field with 4 or fewer free veins; wedge cell absent; anal lobe reduced or absent, with or without deep embayment; wings reduced or absent in various taxa. Legs usually moderately long and slender; tibiae sometimes lined with longitudinal rows of setae or socketed spines; outer edge of protibia dentate in some Osoriinae and Oxytelinae; tibial spurs usually well developed and paired (absent in *Stenus* and Euaesthetinae). Tarsi usually 5-5-5 (rarely 5-5-4 in Euaesthetinae, 5-4-4 in *Atanygnathus*, 4-5-5 or 4-4-5 in many Aleocharinae, 4-4-4 in *Metacorneolabium*, *Bledius*, the aleocharine tribes Corotocini and Oligotini, and most Euaesthetinae, 3-3-3 in some Oxytelinae and Leptotyphlinae and most Pselaphinae, 2-2-2 in some Leptotyphlinae and apparently 2-2-2 in some Pselaphinae); tarsomeres almost always simple (rarely with penultimate tarsomere lobed); pretarsal claws usually simple, equal and paired (unequal or single in some Pselaphinae); empodium usually bisetose.

Abdomen usually with 6 free ventrites (sternites III–VII), with 7 (sternites II–VIII) in most Oxytelinae, 7 (sternites III–IX) in some male Pselaphinae, sometimes with only 5 (III–VII) in some Pselaphinae females and in *Microsilpha* and Scaphidiinae) or 3 (fused III–VI plus VII and VIII) in *Cyathiger*; usually with same number of tergites (IV–VI fused into plate in *Cyathiger*, *Tiracerus* and *Clavigeropsis*); sternite II usually apparently absent or present as concealed or partly exposed lateral sclerites; sternite III divided into 2 or 3 parts by coxae in some Pselaphinae; sternite III much longer than IV in Scaphidiinae and Pselaphinae, rarely with postcoxal lines; intersegmental membranes often with brick-wall pattern of microsclerites (Fig. 28K) (absent in some subfamilies, of other shapes in *Pseudophloeocharis* and Staphylininae); usually at least tergites IV–VIII well sclerotised (fewer in taxa with complete elytra, e.g. most Scydmaeninae) and with at least some segments bearing one or two pairs of paratergites (absent in all Osoriinae and some Pselaphinae, Steninae, Euaesthetinae and Paederinae); sometimes with tergosternal sutures absent forming ring-like abdominal segments (e.g. most Osoriinae); commonly with wing-folding spicule patches on some tergites (usually some or all of III–V). Functional spiracles usually on abdominal segments I–VIII (absent on IV–VI in Pselaphinae, Proteininae and Tachyporinae, on IV–VII or VIII in Scaphidiinae, and VIII in *Microsilpha*). Defensive glands located on various parts of abdomen (see Thayer 2005). Anterior edge of sternite VIII in male without median strut. Anterior edge of segment IX with spiculum gastrale in *Microsilpha* and *Glypholoma*; tergite IX apically truncate to almost completely divided into two laterotergites; tergum X usually free, well sclerotised to membranous, sometimes partly or entirely fused to tergite IX. Aedeagus of modified trilobate type, usually symmetrical (asymmetrical in most Pselaphinae and Paederinae and in some other groups); phallobase reduced, mostly membranous, parameres free, fused to penis or absent. Sternite VIII in female without spiculum ventrale, sometimes with paired struts. Tergite IX in female usually divided into two parts, but sometimes undivided and apically emarginate; proctiger usually evenly sclerotised (sometimes longitudinally divided, posteriorly emarginate, reduced or membranous. Ovipositor usually short and broad; paraprocts usually subequal in length to and lying lateral to coxites, without baculi (except in *Glypholoma* and some Omaliinae); coxites usually divided with apical styli (sometimes undivided, partly or completely fused together or with styli reduced or absent). [Franz 1975; Thayer and Newton 1978; Thayer 1985, 1997, 2005; Naomi 1987a, 1987b, 1988a, 1988b, 1988c, 1988d, 1989a, 1989b, 1989c, 1989d, 1990; Chandler 2000; Newton *et al.* 2000; O'Keefe and Monteith 2000; O'Keefe 2000, 2005; Grebennikov and Newton 2009].

Larva (Fig. 58G)

Body elongate and more or less parallel-sided (more or less oblong-ovate and often strongly curved ventrally in some Scydmaeninae), slightly to strongly flattened; head and all terga very lightly to moderately heavily pigmented; ventral surfaces usually more lightly pigmented; vestiture usually of scattered simple setae (frayed or otherwise modified setae present in Proteininae, and some Pselaphinae, Omaliinae, Tachyporinae, Alerocharinae, Oxytelinae and Staphylininae); sometimes with cuticular glands in various positions (see Thayer 2005).

Head prognathous or slightly declined and protracted, usually without distinct neck (broad or narrow neck present in Euaesthetinae, Leptotyphlinae, *Megalopinus*, Paederinae and Staphylininae), slightly transverse to elongate, distinctly rounded to straight laterally and slightly flattened; posterior edge straight; epicranial stem usually moderately long; frontal arms V-shaped or U-shaped, with bases contiguous; median endocarina usually absent (coincident with epicranial stem in some Osoriinae). Stemmata on each side a maximum of 6, with fewer in most groups. Frontoclypeal suture absent; labium usually free (partly fused to clypeus in *Megalopinus*, completely fused in Pselaphinae, Scydmaeninae, *Stenus*, Euaesthetinae, Leptotyphlinae, Paederinae and Staphylininae). Antennae usually moderately long and 3-segmented (2-segmented in some Pselaphinae and Scydmaeninae, 4-segmented in *Stenus*, Paederinae and Staphylininae); basal articular membrane with eversible organ in most Pselaphinae; sensorium on antennomere 2 usually located anterior or mesal to antennomere 3 (located posterior or lateral to antennomere 2 in *Stenus*, Leptotyphlinae, some Pselaphinae and a few Paederinae, dorsal in some Paederinae, ventral in in *Microsilpha*, xantholinine Staphylininae and a few Omaliinae, posteroventral in Euaesthetinae, some Pselaphinae, anteroventral in faronine Pselaphinae); sensorium located on apical antennomere in those pselaphines with 2-segmented antennae; apex of antennomere 2 usually more or less oblique, so that sensorium arises proximal to antennomere 3 (Fig. 46S) (truncate so that sensorium and apical antennomere arise together in *Megalopinus* and some Pselaphinae); sensorium of various lengths and usually conical or palpiform (bifid in a few Pselaphinae, dome-like in some Scydmaeninae). Mandibles usually symmetrical to slightly asymmetrical, (strongly asymmetrical in Piestinae and Osoriinae), usually unidentate (bidentate or tridentate in Proteininae, Scaphidiinae, some Aleocharinae, and most Oxytelinae and Osoriinae); incisor edge simple, serrate or with one or more subapical teeth or retinacula, with subapical pseudomola in some Scaphidiinae; mola absent; mesal surface of mandibular base with hyaline process in *Glypholoma* and some Omaliinae. Ventral mouthparts usually strongly retracted, with maxillary articulating area (protracted without articulating area in some Euaesthetinae, Paederinae and Staphylininae); cardines free, not or only internally divided; stipes longer than wide to wider than long; with single mala, usually fixed (articulated in Leptotyphlinae, Paederinae and Staphylininae); apex cleft in *Glypholoma*; apex of mala rounded, truncate, falciform, trilobed or stylet-like, usually setose or spinose (less often glabrous, with simple seta or spine, with comb-hairs or spatulate hairs, or with 2 or 3 spurs or bifid spur); palp usually 3-segmented (4-segmented in *Stenus*, Paederinae and some Staphyliniinae). Labium usually consisting of prementum, mentum and submentum (with prementum and postmentum in all Osoriinae, Paederinae and Staphylininae and most Pselaphinae; occasionally with single plate); ligula usually shorter than palp and usually simple (bilobed in *Stenus*, absent in Pselaphinae, Scaphidiinae and Scydmaeninae); palps usually 2-segmented (3-segmented in some Staphylininae), with or without distinct palpiger, usually widely separated. Hypopharyngeal bracon present; sclerome usually absent (a sclerotised bar in Osoriini). Hypostomal rods and ventral epicranial ridges absent. Gular region short to long; gular sutures well separate or partly to completely fused; with ventral ecdysial lines in *Stenus*, Paederinae and Staphylininae.

Prothorax slightly to distinctly longer than meso- or metathorax; thoracic terga and abdominal terga usually with single tergal plate (sometimes divided by median ecdysial line, without plates in *Plastus* and Osoriini); meso- and metaterga usually with transverse anterior carina. Prothoracic venter with or without presternal plate. Legs moderately long and slender, spinose, 5-segmented with pretarsus forming distinct claw, which is usually bisetose but may bear one seta, 3 or more or none; coxae narrowly to widely separated. Tergum VIII in most Aleocharinae with posteromesal projection associated with internal glands. Tergum IX not or only slightly longer than VIII, not extending onto ventral surface, usually with a pair of 1- or 2-segmented, articulated urogomphi (fixed in Pselaphinae, a few Aleocharinae and some Scydmaeninae and Leptotyphlinae; absent in some Pselaphinae, Aleocharinae, Scaphidiinae and Scydmaeninae); segment X visible from above (concealed in some Pselaphinae), transverse to elongate, cylindrical, usually with pair of eversible spiculate lobes. Functional spiracles annular, with closing apparatus, placed laterally on mesothorax and abdominal segments I–VIII; opening at ends of spiracular tubes in *Stenus*; spiracles on segments I–IV reduced in some Scydmaeninae. [Böving and Craighead 1931; Besuchet 1956b; Kasule 1966; Schuster 1966; De Marzo 1986, 1987; Newton 1991; Thayer 2000, 2005; O'Keefe and Monteith 2000; O'Keefe 2000, 2005; Newton *et al.* 2000; Leschen and Newton 2003; Carlton and Leschen 2008; Grebennikov and Newton 2008; Jałoszyński and Kilian 2012].

Classification and World Distribution

The Staphylinidae is a very large, cosmopolitan family containing over 58 000 species in ~3600 genera. The group is currently divided into 32 subfamilies, divided into four major groups: Omaliine (10 subfamilies), Tachyporinae (6), Oxyteline (6) and Staphylinine (10) (Thayer 2005; Grebennikov and Newton 2009). Twenty of the subfamilies occur in Australia, the exceptions being: Empelinae (Nearctic), Micropeplinae (most regions), Neophoninae (southern South America), Dasycerinae (Holarctic and Oriental), Protopselaphinae (Oriental), Olisthaerinae (Holarctic), Habrocerinae (Holarctic, Neotropical, Oriental), Trigonurinae (Holarctic, Oriental), Apateticinae (Palaearctic, Oriental), Pseudopsinae (Holarctic, Neotropical, Oriental, New Zealand), Oxyporinae (Holarctic, Neotropical, Oriental) and Solieriinae (southern South America).

Australian Fauna

Most Australian staphylinids are easily recognised by their narrowly elongate body with shortened, truncate elytra, usually exposing more than half of the abdomen; all but the first one or two abdominal tergites are heavily sclerotised and often flanked by paratergites and the abdominal instersegmental membranes frequently have a pattern of minute sclerites. In three of the largest subfamilies (Aleocharinae, Paederinae, Staphylininae), the body is rather loosely organised and flexible, with projecting procoxae, reduced notal projections and much exposed membrane (Fig. 59G). Staphylinid larvae are usually of the active, campodeiform type, with well developed legs, characteristic antennae with the apex of antennomere 2 oblique so that the sensorium arises before the apex, no mandibular mola, and a pair of articulated urogomphi on tergum IX. A few forms, however, are onisciform, and some may have fixed or no urogomphi.

Two unusual subfamilies in the Omaliine Group are **Glypholomatinae** (*Glypholoma*) (Fig. 67G) and **Microsilphinae** (*Microsilpha*), both of which have nearly complete elytra. The subfamilies **Omaliinae** and **Proteininae** include several small and usually flattened species occurring in the southern part of the continent and, like the members of the first two subfamilies, they have affinities with the New Zealand and South American faunas. (Newton 1985; Thayer and Newton 1978; Thayer 1985; Newton and Thayer 1995; Thayer 1997, 2000). Another member of the omaliine group is the **Pselaphinae** (Fig. 59I, L, O), which have been traditionally considered to be a distinct family. Unlike typical staphylinds, pselaphines have a shorter, broader, non-flexible abdomen, enlarged apical maxillary palpomeres, foveae on the head, prothorax and other parts of the body, 2- or 3-segmented tarsi, and pretarsal claws that are sometimes unequal or reduced to one. Pselaphines are usually reddish- or yellowish-brown in colour and the antennae are almost always incrassate or variously clubbed. Larvae are elongate to fusiform and clothed with erect setae, the antennal sensorium is usually enlarged or complex, paired eversible glands arise from the antennal sockets (except in Faronini), and the urogomphi are reduced and fixed or absent. Pselaphines are abundant in leaf litter, rotten wood, moss, tree holes, caves and animal nests; except for some myrmecophilous forms, adults and larvae are predators on minute organisms. The larva of the European *Batrisodes oculatus* Aubé was observed feeding on a collembolan, which it captured by sticking the prey to its abdomen, which was covered with a viscous substance secreted by dorsal glands (De Marzo 1986); the same species formed a silken cocoon for pupation. The subfamily is well represented in Australia and following a recent generic revision by Chandler (2000), the Australian fauna includes 164 genera in six supertribes and 21 tribes. The most basal supertribe Faronitae includes two genera, Batrisitae eleven, Euplectitae 65, Goniaceritae 35, Pselaphitae 48 and Clavigeritae three. The last of these includes highly specialised myrmecophiles which are tolerated in ant nests and even fed by workers; they bear setose secretory structures (trichomes), which in an American species *Adranes taylori* Wickham, were observed to be fed upon by larvae of the host ant *Lasius sitkaensis* (Akre and Hill 1973). [Besuchet 1956b, 2008; Newton and Chandler 1989].

Within the Tachyporine Group, **Phloeocharinae** includes a single genus *Pseudophloeocharis* and **Trichophyinae** with one adventive species of *Trichophya*, while **Tachyporinae** are a moderately large subfamily with a characteristic fusiform shape, inserted head and tapered abdomen with the segments capable of being telescoped. The most diverse and common genus is *Sepedophilus*, some species of which graze on the undersides of bracket fungi. **Aleocharinae** are the largest group of staphylinids with more than 300 described Australian species and many more awaiting study. Most of aleocharines are free-living predators, but *Aleochara* species (Fig. 67A) are ectoparasitoids of Diptera pupae (Drea 1966) and *Gyrophaena* and their relatives feed on Basidiomycetyes (Ashe 1984). Inquilinism has evolved several times in aleocharines. In Australia, *Polylobus* (Oxylopodini) and many Myrmedoniini occur with ants, while seven tribes (including Corotocini, Drepanoxenini and Trichopseniini) and three subtribes of Athetini are composed of termitophiles. Two very interesting members of the latter group are *Austrospirachtha mimetes* Watson (Corotocini) (Fig. 59J–K) and *Drepanoxenus* species (Drepanoxenini) (Fig. 58F), both of which are discussed on p. 91. [Watson and Howick 1975; Kistner 1979, 1982; Frank and Thomas 1984; Watson and Kistner 1985 Ashe 2003].

There are four Australian subfamilies in the Oxyteline Group. **Piestinae** is represented by *Prognathoides mjobergi* (Bernhauer), a distinctive flattened staphylinid with paired horns on the head. **Scaphidiinae** (Fig. 67F) were often considered to be a separate family because of the distinctive, wedge-like form and relatively long elytra, concealing all but the last one or two abdominal tergites. Most scaphidiines feed on the spores and hyphae of various Basidiomycetes, but species of *Baeocera* and *Scaphobaeocera* are myxomycete spore feeders. A common genus associated with polypores is *Scaphisoma*. [Löbl 1997; Löbl and Leschen 2003, 2010; Leschen and Löbl 2005]. **Osoriinae** are primarily tropical, saprophagous, and found under bark and in rotten wood. Included are the large, flattened *Plastus miles* Bernhauer (Fig. 67C) from northern QLD, cylindrical species of *Osorius*, and the very small, flattened Eleusinini. *Thoracophorus* are thought to be mycetophagous. Newton (1990) provided a key to all Australian osoriine genera. **Oxytelinae** are primarily saprophagous and some are abundant in leaf litter and dung; our commonest genus is *Anotylus* (Fig. 59M). The flightless, litter-inhabiting *Oxypius peckorum* Newton from south-western Australia is a relict form most closely related to species of *Euphanias* Fairmaire and Laboulbène from the Mediterranean Region and South America (Newton 1982). *Sartallus signatus* Sharp (Fig. 67K) occurs under seaweed on beaches, and species of *Bledius* (Fig. 67L) construct burrows in moist sand adjacent to rivers, lakes and oceans and feed on algae and diatoms (Herman 1986).

The Staphylinine Group includes the seven remaining Australian subfamilies. **Megalopsidiinae** includes the genus *Megalopinus* with a few small, compact species with protruding eyes. **Steninae** includes several species of *Stenus*, which also have large, protruding eyes but a slender, less compact body, live in marshes and at the edges of streams, where they are able to move on the surface of water by secreting from the pygidial glands a surfactant, which reduces the surface tension behind them (Schildknecht 1970; Schildknecht *et al.* 1975). *Stenus* also possess a unique adult prey capture mechanism, consisting of a protrusible, adhesive labium (Weinreich 1968; Koerner *et al.* 2012). **Euaesthetinae** include several minute, often depigmented, soil-inhabiting species. The subfamily **Scydmaeninae** were considered to be a separate family until relatively recently (Grebennikov and Newton 2009) when it was shown that they belong to this derived group of staphylinid subfamilies. Scydmaenines are small to minute beetles usually distinguished by their 'waisted' body form, with a distinct constriction at the junction of the prothorax and elytra and usually another at the base of the head, so that a neck is formed. Most species are reddish-brown and pubescent, with coarsely facetted eyes, large maxillary palps with a reduced apical palpomere, approximate procoxae, distant metacoxae, no lateral pronotal carinae and clavate femora. Species of the tribe Cephenniini are exceptional in lacking the neck and waist, but in most respects they are typical scydmaenids. Larvae may be elongate to ovate and setose, with large, club-like antennae and usually without or with very small urogomphi. Scydmaenines occur in a variety of habitats, such as leaf litter, rotten wood, moss, tree holes, sawdust piles and ant nests. They are predaceous beetles, feeding on both soft-bodied and hard-bodied minute arthropods (Jałoszyński 2012a, 2012c; Jałoszyński and Kilian 2012). Schmid (1988) described structures on adult legs and mouthparts of some scydmaenines which are used for grasping and penetrating the cuticle of heavily armored oribatid mites. Schuster (1966) and Newton (1991) have described adhesive discs in scydmaenine larvae which are also used for capturing mites, and Jałoszyński and Beutel (2012) described these in more detail. The subfamily is represented in Australia by *Clidicus abbotensis* O'Keefe (Clidicini) (Fig. 59P) from tropical northern QLD, the cephenniine genera *Cephennodes* and *Cephennomicrus*, the scydmaenine genera *Scydmaenus* and *Palaeoscydmaenus*, and several genera of Cyrtoscydmini, including *Euconnus* (Fig. 59H) and *Horaeomorphus*. [Franz 1975; Newton and Franz 1998; O'Keefe and Monteith 2000; Monteith 2011]. The subfamily **Leptotyphlinae** includes several minute, eyeles, soil-inhabiting staphylinids in various parts of the world (Pace 1996); at least two undescribed genera are known from Australia. Paederinae and Staphylininae are the two largest groups of highly motile, specialised predators. Among the **Paederinae**, *Paederus* (Fig. 67B) is a well known genus of mostly brightly coloured species that produce a fluid that causes severe skin blistering (Frank and Kanamitsu 1987; Whelan and Weir 1987). Other common genera are *Astenus*, *Lathrobium*, *Scopaeus*, *Hyperomma*, *Ochthephilum* and *Pinophilus*. Among the **Staphylininae**, *Creophilus erythrocephalus* (Fabricius), the devil's coach-horse (Figs 1G, 67H), which is black with a red head bearing a central black spot, is often associated with cadavers in open or disturbed areas, where it preys on maggots; the similar species *C. lanio* (Erichson) occurs in more forested areas. Large and metallic species of *Actinus* (Fig. 1H) occur in northern QLD, while *Cafius* are predators in seaweed on beaches. Other major groups are the genera *Philonthus*, *Quedius* and *Heterothops* and the tribe Xantholinini. *Myotyphlus jansoni* (Matthews) sometimes lives on the fur of native *Rattus*, but does not seem (as once thought) to be closely related to South American species with similar habits (Hamilton-Smith and Adams 1966). [Kasule 1966;

Moore and Legner 1979; Newton 1984, 1985; Naomi 1985; Klimaszewski *et al.* 1996; Herman 2001; Solodovnikov 2006; Solodovnikov and Newton 2009].

Australian Subfamilies, Tribes and Genera

Glypholomatinae: *Glypholoma* Jeannel.

Microsilphinae: *Microsilpha* Broun.

Omaliinae: Omaliini: *Anthobiomimus* Steel, *Austrolophrum* Steel, *Ischnoderus* Fauvel, *Leaskia* Steel, *Omaliopsis* Jeannel, *Omalium* Gravenhorst, *Paraphloeostiba* Steel, *Phloeonomus* Heer, *Phloeostiba* Thomson, *Stenomalium* Blackwelder, *Xylodromus* Heer. **Corneolabiini:** *Metacorneolabium* Steel.

Proteininae: Silphotelini: *Alloproteinus* Steel. **Austrorhysini:** *Austrorhysus* Steel. **Anepiini:** *Anepius* Blackburn.

Pselaphinae: Faronitae: Faronini: *Logasa* Chandler, *Sagola* Sharp. **Euplectitae: Euplectini:** *Austroeuplectus* Chandler, *Dorrigo* Chandler, *Euplectus* Leach, *Okella* Chandler, *Tyxs* Chandler. **Dimerini:** *Tuberoplectus* Park. **Bythinoplectini:** *Dichocoryna* Comellini, *Orlandia* Comellini, *Zethopsus* Reitter. **Tyrogastrini:** *Austrorhexius* Chandler, *Mesoplatus* Raffray, *Platomesus* Chandler, *Tomeplasus* Chandler. **Trychonychini:** *Aphilia* Reitter, *Apoterus* Raffray, Ararat Chandler, *Armidale* Chandler, *Aulaxus* Raffray, *Barrengarry* Chandler, *Bartle* Chandler, *Bithongabel* Chandler, *Bruxner* Chandler, *Calarus* Raffray, *Chichester* Chandler, *Cleland* Chandler, *Clyde* Chandler, *Dandenong* Chandler, *Deroplectus* Raffray, *Dungog* Chandler, *Eungella* Chandler, *Eupinion* Oke, *Euplectops* Reitter, *Gayundah* Chandler, *Gordon* Chandler, *Gubarra* Chandler, *Jindabyne* Chandler, *Joalah* Chandler, *Kapalga* Chandler, *Kerplectus* Chandler, *Kioloa* chandler, *Kyogle* Chandler, *Limoniates* Raffray, *Macroplectus* Raffray, *Meithton* Chandler, *Miallo* Chandler, *Millaa* Chandler, *Namssom* Chandler, *Noota* Chandler, *Paraplectus* Raffray, *Perditadens* Chandler, *Plectostenus* Lea, *Plectusodes* Raffray, *Praeruptifrons* Chandler, *Protoplectus* Raffray, *Quotidiaplectus* Chandler, *Ramussia* Kurbatov, *Scabritia* Chandler, *Sealy* Chandler, *Stenoplectus* Raffray, *Tooloom* Chandler, *Torvicia* Chandler, *Unumgar* Chandler, *Washpool* Chandler, *Whyanbeel* Chandler, *Xyts* Chandler. **Batrisitae: Batrisini:** *Amblybatrisus* Chandler, *Ancistromus* Jeannel, *Batrisocenus* Raffray, *Coryphomodes* Jeannel, *Daintree* Chandler, *Gadgarra* Chandler, *Hypochraeus* Raffray, *Mossman* Chandler, *Neurum* Chandler, *Spurgeon* Chandler, *Tinaroo* Chandler. **Goniaceritae: Brachyglutini:** *Anabaxis* Raffray, *Anarmoxys* Raffray, *Antipodebaxis* Chandler, *Araneabaxis* Chandler, *Batraxis* Reitter, *Booloumba* Chandler, *Briara* Reitter, *Bundjulung* Chandler, *Eupinella* Raffray, *Eupines* King, *Eupinolus* Oke, *Eupinopsis* Raffray, *Iluka* Chandler, *Mallanganee* Chandler, *Malleecola* Oke, *Mundaring* Chandler, *Paluma* Chandler, *Pedisinops* Newton & Chandler, *Physoplectus* Reitter, *Reichenbachia* Leach, *Rybaxis* Saulcy, *Startes* Broun, *Storeyella* Chandler, *Vasse* Chandler, *Wataranka* Chandler, *Wiangaree* Chandler, *Wollomombi* Chandler, *Woodenbong* Chandler. **Iniocyphini:** *Apoplectus* Raffray, *Simplicorfa* Chandler, *Sunorfa* Raffray. **Proterini:** *Euparops* Kurbatov & Cuccodoro, *Gnesion* Raffray, *Pseudopterus* Raffray. **Cyathigerini:** *Plagiophorus* Motschulsky (= *Cyathiger* King). **Pselaphitae: Tyrini:** *Abascantodes* Strand, *Agatyrus* Broun, *Chalcoplectus* Oke, *Chasoke* Chandler, *Collacerothorax* Lea, *Daveyia* Lea, *Durbos* Sharp, *Gerallus* Sharp, *Googarna* Chandler, *Hamotopsis* Raffray, *Hamotulus* Schaufuss, *Kiera* Chandler, *Narrabeen* Chandler, *Palimbolus* Raffray, *Rytus* King, *Somatipion* Schaufuss, *Spilorhombus* Raffray, *Swan* Chandler, *Tasmanityrus* Chandler, *Tyrogetus* Broun, *Tyromacrus* Chandler, *Tyromorphus* Raffray. **Schistodactylini:** *Leanymus* Raffray, *Schistodactylus* Raffray. **Odontalgini:** *Warrumbungle* Chandler. **Tmesiphorini:** *Eudranes* Sharp, *Jardine* Chandler, *Raphitreus* Sharp, *Stethotaphrus* Newton & Chandler, *Syntectodes* Reitter, *Tmesiphorus* LeConte, *Tyrocarius* Chandler. **Ctenistini:** *Ctenicellus* Raffray, *Ctenisophus* Raffray, *Edocranes* Reitter. **Hybocephalini:** *Stipesa* Sharp. **Arhytodini:** *Tolga* Chandler. **Pselaphini:** *Bellenden* Chandler, *Curculionellus* Westwood, *Kakadu* Chandler, *Mareeba* Chandler, *Margaris* Schaufuss, *Maydena* Chandler, *Peckiella* Chandler, *Pselaphaulax* Reitter, *Pselaphogenius* Reitter, *Pselaphophus* Raffray, *Tyraphus* Sharp. **Clavigeritae: Clavigerini:** *Clavigeropsis* Raffray, *Tapas* Besuchet. **Tiracerini:** *Tiracerus* Besuchet.

Phloeocharinae: *Pseudophloeocharis* Steel.

Tachyporinae: Tachyporini: *Austrotachinus* Steel, *Coproporus* Kraatz, *Leucotachinus* Coiffait & Sàiz, *Mimocyptus* Cameron, *Sepedophilus* Gistel, *Tachinomorphus* Kraatz, *Tachyporus* Gravenhorst. **Mycetoporini:** *Bryoporus* Kraatz, *Ischnosoma* Stephens, *Lordithon* Thomson.

Trichophyinae: *Trichophya* Mannerheim.

Aleocharinae: Aenictoteratini: *Dentaphila* Kistner, *Weiria* Ashe. **Aleocharini:** *Aleochara* Gravenhorst (= *Correa* Fauvel). **Athetini:** *Acrotona* Thomson, *Aloconota* Thomson, *Amischa* Thomson, *Apphiana* Olliff, *Atheta* C.G. Thomson, *Australoconota* Pace, *Australoplatyola* Scheerpeltz, *Australopelioptera* Pace, *Brundinia* Totenham, *Coptolimulus* Kistner & Pasteels, *Coptophilus* Kistner & Pasteels, *Coptotermoecia* Oke, *Dalotia* Casey, *Gastropaga* Bernhauer, *Geostiba* Thomson, *Giachinusa* Pace, *Iotarphia* Cameron, Ischyrodyodoys Pace, *Leptostiba* Pace, *Microceroxenus* Kistner, *Mimacrotona* Cameron, *Nasutiphilus* Kistner, *Pelioptera* Kraatz, *Philobrunneus* Kistner & Pasteels, *Psammopora* Pace, *Thamiaraea* C. G. Thomson. **Corotocini:** *Australoptochus* Kistner, *Austrospirachtha* Watson, *Eutermitoptochus* Silvestri, *Neoguinella* Pasteels & Jacobson, *Termitoptochus* Silvestri, *Termitoptocinus* Silvestri, *Tumulipcinus* Jacobson & Kistner. **Deinopsini:** *Adinopsis* Cameron, *Deinopsis* Matthews. **Diestotini:** *Diestota* Mulsant & Rey. **Drepanoxenini:** *Drepanoxenus* Kistner & Watson. **Falagriini:** *Anaulacaspis* Ganglbauer, *Cordalia* Jacobs, *Leptagria* Casey, *Myrmecocephalus* MacLeay, *Myrmecopora* Saulcy. **Homalotini:** Adelarthra Cameron, *Anomognathus* Gemminger & Harold, *Bolitochara* Mannerheim, *Brachida* Mulsant & Rey, *Brachyglyptaglossa* Pace, *Coenonica* Kraatz, *Drepanomastax* Ashe, *Eumecognathus* Ashe, *Gyrophaena* Mannerheim, *Heterota* Mulsant & Rey, *Paralinoglossa* Pace, *Phanerota* Casey, *Pseudoligota* Cameron, *Siagotanyx* Ashe, *Stenomastax* Cameron, Sternotropa Cameron. **Hoplandriini:** *Microlia* Casey (= *Tinotoma* Cameron), *Platandria* Casey. **Hypocyphtini:** *Microtachyporus* Oke, *Oligota* Mannerheim. **Leucocraspedini:** *Leucocraspedum* Kraatz. **Lomechusini:** *Amblyoponiphilus* Oke, *Dabra* Olliff, *Dabro-*

soma Lea, *Daccordiusa* Pace, Pella Stephens, *Tetrabothrus* Bernhauer, *Trisporusa* Pace, *Zyras* Stephens. **Myllaenini:** Myllaena Erichson. **Oxypodini:** *Antistydatusa* Pace, *Apimela* Mulsant & Rey, *Australiusa* Pace, *Aylikusa* Pace, *Blepharhymenus* Solier, *Calodera* Mannerheim, *Dymerinx* Pace, *Halobrecta* Thomson, *Leichusa* Pace, *Melvilloxenus* Kistner, *Ocalea* Erichson, *Oxypoda* Mannerheim, *Phloeopora* Erichson, *Polylobus* Solier, *Spanioda* Blackwelder, *Topikoglossa* Pace. **Pseudoperinthini:** *Austrointhus* Kistner & Pasteels, *Pseudoperinthus* Wasmann. **Tachyusini:** *Gnypeta* Thomson, *Thripsophaga* Cameron. **Termitohospitini:** *Hetairotermes* Cameron. **Termitonannini:** *Eutermitophila* Cameron, *Lauella* Mann. **Termitusini:** *Pseudotermitoecia* Kistner, *Termitoecia* Bernhauer. **Trichopseniini:** *Mastopsenius* Seevers, *Prorhinopsenius* Pasteels & Kistner.

Scaphidiinae: Scaphidiini: *Scaphidium* Olivier. **Scaphisomatini:** *Baeocera* Erichson, *Scaphicoma* Motschulsky, *Scaphisoma* Leach, *Scaphobaeocera* Csiki, *Scaphoxium* Löbl, *Toxidium* LeConte, *Xotidium* Löbl.

Piestinae: *Prognathoides* Steel.

Osoriinae: Eleusinini: *Eleusis* Laporte, *Zeoleusis* Steel. **Thoracophorini:** *Allotrochus* Fagel, *Clavilispinus* Bernhauer, *Espeson* Schaufuss, *Lispinus* Erichson, *Myrmelibia* Newton, *Nacaeus* Blackwelder, *Neolosus* Blackwelder, *Pseudespeson* Lecoq, *Thoracophorus* Motschulsky. **Osoriini:** *Arpagonus* Blackwelder, *Holotrochus* Erichson, *Mimogonellus* Fagel, *Mimogonus* Fauvel, *Oeophronistus* Blackburn, *Osorius* Latreille, *Typhlobledius* Lea. **Leptochirini:** *Plastus* Bernhauer.

Oxytelinae: Coprophilini: *Homalotrichus* Solier. **Euphaniini:** *Oxypius* Newton. **Blediini:** *Bledius* Samouelle. **Oxytelini:** *Anotylus* C.G. Thomson, *Blediotrogus* Sharp, *Carpelimus* Leach, *Oxytelus* Gravenhorst, *Sartallus* Sharp, *Teropalpus* Solier, *Thinobius* Kiesenwetter, *Thinodromus* Kraatz.

Megalopsidiinae: *Megalopinus* Eichelbaum.

Scydmaeninae: Cephenniini: *Cephennodes* Reitter (= *Coatesia* Lea, *Cephennomicrus* Reitter (= *Neseuthia* Scott. **Scydmaenini:** *Palaeoscydmaenus* Franz, *Scydmaenus* Latreille. **Cyrtoscydmini:** *Euconnus* C.G. Thomson, *Horaeomorphus* Schaufuss, *Maorinus* Franz, *Microscydmus* Saulcy & Croissandeau, *Psepharobius* King, *Sciacharis* Broun (= *Neuraphoconnus* Franz), *Stenichnus* C.G. Thomson, *Syndicus* Motschulsky. **Clidicini:** *Clidicus* Laporte. **Eutheiini:** *Paraneseuthia* Franz.

Steninae: *Stenus* Latreille, New genus.

Euaesthetinae: *Austroesthethus* Oke, *Chilioesthetus* Saiz, *Edaphus* Motschulsky, *Mesoaesthetus* Cameron, *Octavius* Fauvel, *Protopristus* Broun, *Tasmanosthetus* Puthz.

Leptotyphlinae: New genera.

Paederinae: Paederini: *Astenus* Dejean, *Cephalochetus* Kraatz, *Charichirus* Sharp, *Dibelonetes* Sahlberg, *Dicax* Fauvel, *Eupaederus* Scheerpeltz, *Hyperomma* Fauvel, *Lathrobium* Gravenhorst, *Lithocharis* Dejean, *Macrodicax* Lea, *Medon* Stephens, *Monocrypta* Casey, *Notobium* Solsky, *Ochthephilum* Stephens, *Ophiomedon* Sharp, *Paederus* Fabricius, *Pinobius* MacLeay, *Pseudomedon* Mulsant & Rey, *Rugilus* Leach, *Scopaeus* Erichson, *Scymbalium* Erichson, *Stilicoderus* Sharp, *Stilicopsis* Sachse, *Suniopsis* Fauvel, *Sunius* Stephens, *Thinocharis* Kraatz, *Tripectenopus* Lea (= *Scopaeodracus* Scheerpeltz). **Pinophilini:** *Araeocerus* Nordmann, *Oedichirus* Erichson, *Palaminus* Erichson, *Pinophilinus* Eichelbaum, *Pinophilus* Gravenhorst, *Procirrus* Latreille.

Staphylininae: Diochini: *Diochus* Erichson. **Xantholinini:** *Archaites* Bordoni, *Australinus* Bordoni, *Bathyllia* Bordoni, *Eachamia* Bordoni, *Enervia* Bordoni, *Grevillia* Bordoni, *Gyrohypnus* Leach, *Leptacinus* Erichson, *Metolinus* Cameron, *Mitomorphus* Kraatz, *Neohypnus* Coiffait & Sáiz, *Neoxantholinus* Cameron, *Notolinus* Casey, *Pachycorynus* Motschulsky, *Phacophallus* Coiffait, *Queenslandina* Bordoni, *Thyreocephalus* Guérin-Méneville, *Walesia* Bordoni, *Xantholinus* Serville, *Xanthophius* Motschulsky, *Zeteotomus* Jacquelin du Val. **Staphylinini:** *Actinus* Fauvel, *Acylophorus* Von Nordmann, *Anchocerus* Fauvel, *Antimerus* Fauvel, *Atanygnathus* Jacobson, *Australotarsius* Solodovnikov & Newton, *Belonuchus* Nordmann, *Bisnius* Stephens, *Cafius* Stephens, *Creophilus* Samouelle, *Ctenandropus* Cameron, *Gabrius* Stephens, *Gabronthus* Tottenham, *Hesperus* Fauvel, *Heterothops* Stephens, *Lonia* Strand, *Myotyphlus* Fauvel, *Mysolius* Fauvel, *Neobisnius* Ganglbauer, *Philonthus* Stephens, *Quediomimus* Cameron, *Quediopsis* Fauvel, *Quedius* Stephens, *Remus* Holme, *Sternotoxus* Bernhauer, *Tasgius* Stephens, *Tympanophorus* Nordmann.

SCARABAEIFORMIA: SCARABAEOIDEA

This is a distinctive and obviously monophyletic group adults of which are distinguished by: (1) a highly modified, burrowing prothorax with large procoxae (Fig. 17CD) (almost always with concealed trochantins and closed cavities, Fig. 44L) and usually dentate tibiae with only one spur; (2) hind wings with reduced venation and a strong radial bending zone coupled with a hinge; (3) a lamellate antennal club (Fig. 15H); (4) no metacoxal plates; (5) abdominal sternite II represented by a pair more or less concealed lateral sclerites (Figs 32C, 61K–L); (6) tergite VIII forming a pygidium and not concealed by tergite VII; and (7) four Malpighian tubules. Larvae are grub-like and usually C-shaped (Figs 33F, 61H), with well developed antennae and legs, no urogomphi, and usually cribriform spiracles. Trends in scarabaeoids include the reduction of adult spiracles on segment VIII and movement of terminal spiracles dorsally or ventrally, loss of the larval spiracular closing apparatus, reduction of the number of adult antennomeres, and increase in the number of larval antennomeres from three to four. The group appears to be primitively adapted for burrowing in soil, with secondary trends in other directions. Loss of the spiracular closing apparatus may have accompanied the development of cribriform spiracles, although in the family Trogidae, the two structures occur together. [Janssens 1949; Ritcher 1966; Hinton 1967b; Howden 1982; Lawrence and Newton 1982; Scholtz 1986a, 1990a; Scholtz *et al.* 1987a, 1987b, 1988, 1994; Browne and Scholtz 1994, 1995, 1996, 1998, 1999; Scholtz and Chown 1995; Allsopp 1995, 1999; Scholtz and Browne 1996; Jameson and Ratcliffe 2002a; Grebennikov 2004a; Scholtz and Grebennikov 2005a, 2005b; Smith 2006; Smith *et al.* 2006; Kohlmann 2006].

Scarabaeoid families not occurring in Australia include the Glaresidae (Holarctic, Neotropical, Afrotropical and Oriental), Pleocomidae (western North America), Diphyllostomatidae (California), Belohinidae (Madagascar), Ochodaeidae (Holarctic, Neotropical, Afrotropical and Oriental) and Glaphyridae (Holarctic). [Paulian 1958; Holloway 1972; Scholtz *et al.* 1987a, 1988; Carlson 2002a, 2002b; Hovore 2002; Jameson 2002a, Jameson and Ratcliffe 2002b].

17. GEOTRUPIDAE

Adult (Figs 3E, 60K–L, 68I)

Length ~5–25 mm. Body ~1.25–1.5 times as long as wide, sides of pronotum and elytra independently rounded; strongly convex; colour yellowish-brown to dark reddish-brown or black (metallic blue ventrally in *Geotrupes*); dorsal surfaces glabrous, but ventral surfaces, sides of body and legs hirsute.

Head prognathous, deeply inserted into prothorax, usually abruptly constricted immediately behind eyes forming very short temples and a short, broad neck (not in Geotrupes). Eyes well developed, usually strongly protuberant, deeply emarginate or completely divided by canthus, finely facetted, without interfacetal setae; ommatidium eucone in *Geotrupes*, exocone in Bolboceratinae. Antennal insertions concealed from above; subantennal grooves present. Frontoclypeal suture absent in Bolboceratinae, well developed and angulate in *Geotrupes*; males sometimes with median elevation on frons and with dorsally or anteriorly projecting horn on anterior portion of clypeus (Fig. 60L); anterior edge of clypeus truncate in Bolboceratinae, rounded in *Geotrupes*. Labrum strongly transverse and truncate to slightly emarginate. Antennae short, 11-segmented with strong, 3-segmented, compact, lamellate club; scape usually more than twice as long as wide, more than twice as long as pedicel, and slightly curved or asymmetrically expanded. Mandibles moderately elongate, with expanded, slightly raised and carinate outer edge and unidentate or weakly bidentate apex; with or without subapical tooth or retinaculum; molae well developed and subovate to strongly transverse; prostheca densely setose. Maxilla in *Geotrupes* with 2-segmented, apically expanded, densely setose galea and a densely setose lacinia bearing a weak, bidentate uncus; galea in Bolboceratinae subfalcate with fewer setae and lacinia with apical and mesal unci; palp 4-segmented with apical palpomere fusiform. Mentum subquadrate or slightly transverse and apically convex (emarginate in *Geotrupes*); ligula bilobed; labial palps 3-segmented with apical palpomere fusiform. Gular sutures separated; gula usually elongate, strongly convex and slightly to strongly produced posteriorly. Corpotentorium slender and arched; laminatentoria well developed and broadly joined at midline. Cervical sclerites well developed.

Pronotum ~0.42–0.55 times as long as wide, widest posteriorly, wider at base than combined elytral bases; sides strongly curved, not explanate. Lateral pronotal carinae complete, simple, with raised margin or bead; anterior angles slightly produced forward and rounded to subacute; posterior angles broadly rounded; posterior edge evenly rounded; disc usually simple, with various depressions and elevations (horns) in males of some Bolboceratinae (Fig. 60L), with paired deep pits near anterior edge in *Blackburnium*. Prosternum in front of coxae much shorter than shortest diameter of procoxal cavity, biconcave or slightly tumid; anterior edge sometimes produced to form chin piece. Prosternal process complete but sometimes interrupted at middle, distinctly expanded and truncate to angulate apically. Notosternal sutures complete. Hypomeron usually concave anteriorly forming pockets for the reception of the eyes when head retracted. Procoxae large with mesal portion projecting below prosternum, without concealed lateral extensions; trochantins concealed; endopleuron fused to notum dorsolaterally. Procoxal cavities strongly transverse, very narrowly separated or contiguous, closed externally by postcoxal hypomeral projections fitting into lateral cavities in prosternal process, internally open.

Scutellar shield well developed, sometimes abruptly elevated anteriorly, posteriorly broadly rounded to acute. Elytra ~0.75–0.95 times as long as combined width and 1.8–2.2 times as long as pronotum; sides moderately curved and apices conjointly rounded concealing all abdominal tergites or exposing part of pygidium only; disc with 10–15 puncture rows or striae and no scutellary striole; epipleura relatively narrow and complete or almost complete. Mesoventrite separated by complete sutures from mesanepisterna, which are widely separated at midline; anterior edge of mesoventrite usually on different plane than metaventrite, without paired procoxal rests, sometimes with median prosternal rest or broad collar; mesoventral cavity absent; discrimen absent. Mesocoxae strongly transverse and slightly oblique, with exposed trochantins. Mesocoxal cavities moderately widely separated in most Bolboceratinae, narrowly separated in *Eucanthus* and *Elephastomus*, contiguous in *Geotrupes*, partly closed laterally by mesepimeron. Meso-metathoracic joint within coxal cavities absent. Mesometaventral junction usually absent in Bolboceratinae due to fusion meso- and metaventrites; ventrites abutting or overlapping in *Elephastomus* and *Eucanthus*, not quite meeting in *Geotrupes*. Metaventrite strongly transverse and strongly convex, without postcoxal lines; transverse (metakatepisternal) suture short and curved towards posterior edge of ventrite; discrimen complete, extending to anterior edge of ventrite; exposed portion of metanepisternum relatively short and broad. Metacoxae strongly transverse, subcontiguous, extending laterally to meet elytra; plates absent. Metendosternite with long, slender stalk, long, transverse lateral arms and short to moderately long anterior process with narrowly separated tendons. Hind wing with long radial bending zone combined with distinct hinge at end of radial bar; apical field about half as long as total wing length, with as many as 4 radial extensions (RA_3, apically fused RA_4 and RP_1 and RP_2,); radial cell incomplete, without base; cross-vein r3 absent; r4 more or less complete; R-M loop forming acute angle; basal portion of RP relatively short and apical extension extending to wing margin; medial spur barely curved posteriorly, extending to wing margin; medial field with 4 free veins; MP_{3+4} absent, MP_3 and MP_4 not connected basally; wedge cell absent; AA_{3+4} forked or not; anal lobe highly reduced, separated by wide, deep embayment, AP_{3+4} not forked, sometimes absent. Legs rather stout, with expanded and flattened femora; trochanterofemoral joint strongly oblique with femur in contract with coxa; meso- and metatibiae in Bolboceratinae apically expanded and outer edge with at least one transversely oblique subapical ridge (apex less expanded and outer edge with several transverse ridges in *Geotrupes*); meso- and metatibial spurs well developed and usually asymmetrical; protibia apically expanded with one spur, its outer edge with a variable number of lobes or teeth. Tarsi 5-5-5; tarsomeres relatively long and slender, simple; pretarsal claws simple; empodium well developed and bisetose.

Abdomen with 7 free ventrites (sternites II–VIII), but sternite II divided by coxae into 2 small lateral sclerites; metacoxal cavities

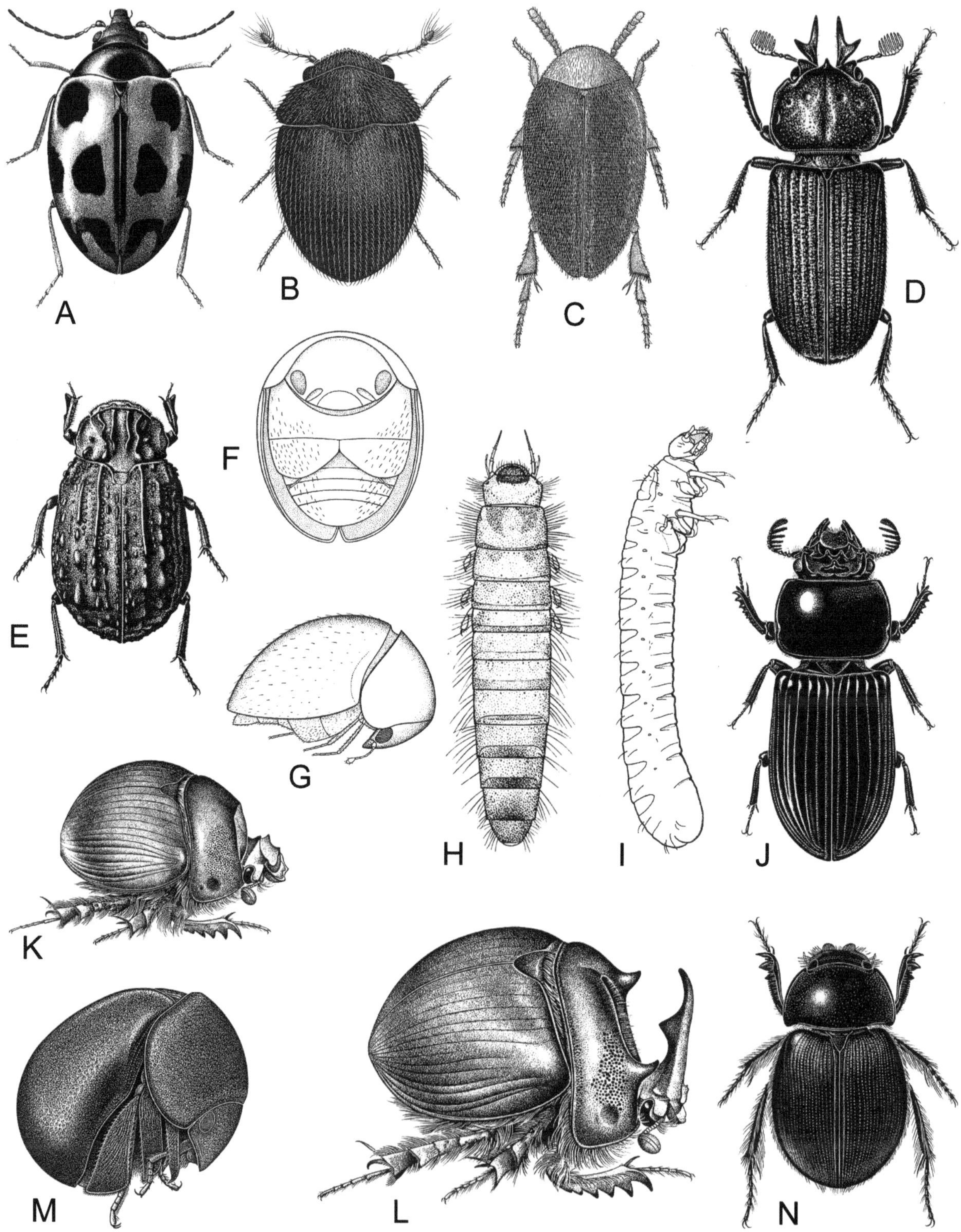

Fig. 60. A–E, Adult, dorsal: A, *Macrohelodes crassus* Blackburn (Scirtidae); B, *Acalyptomerus asiaticus* Crowson (Clambidae: Acalyptomerinae); C, *Eucinetus* sp. (Eucinetidae); D, *Syndesus cornutus* (Fabricius) (Lucanidae: Syndesinae); E, *Omorgus australasiae* (Erichson) (Trogidae). F, *Clambus simsoni* Blackburn (Clambidae), adult ventral; G, same, lateral. H–I, Larva: H, Genus ? sp. (Scirtidae), dorsal; I, *Aulacocyclus* sp. (Passalidae), lateral. J–N, adult: J, *Mastachilus questionis* (Kuwert) (Passalidae), dorsal; K–L, *Blackburnium cavicolle* (Macleay) (Geotrupidae: Bolboceratinae), female and male, lateral; M, *Cyphopisthes descarpentriesi* Paulian (Hybosoridae: Ceratocanthinae), lateral; N, *Phaeochrous emarginatus* (Laporte) (Hybosoridae: Hybosorinae), dorsal.

weakly defined and intercoxal process absent. Functional spiracles in *Geotrupes* on abdominal segments I–VIII, located mainly in pleural membrane, those in Bolboceratinae on segments I–VII, located in sternal notches. Abdominal tergites relatively lightly sclerotised, except for pygidium (tergite VIII), with patches of wing-folding asperities usually on tergites IV–VII. Anterior edge of sternite VIII in male without median strut. Anterior edge of segment IX with broadly rounded subgenital plate; tergite IX almost completely divided into two laterotergites (united by a slender strip of membrane); tergum X free. Aedeagus of modified trilobate type, with large, well sclerotised phallobase; parameres free, fused to phallobase and/or fused to one another, symmetrical or asymmetrical; penis mostly membranous, sometimes with paired struts; endophallus with or without armature. Sternite VIII in female without spiculum ventrale. Sternite IX in female divided into two parts; proctiger usually sclerotised. Ovipositor short and broad; paraprocts slightly shorter than coxites, without baculi; coxites undivided, setose, without styli. [Howden 1955, 1979, 1985; Howden and Cooper 1977; D'Hotman and Scholtz 1990a; Jameson 2002c; Scholtz and Grebennikov 2005b].

Larva

Body subcylindrical and strongly curved ventrally (C-shaped); relatively weakly sclerotised and lightly pigmented; vestiture in *Geotrupes* relatively sparse, consisting of long and short setae, that in Bolboceratinae a dense covering of very short hairs.

Head protracted and hypognathous (Fig. 42K), about as long as wide or slightly transverse, strongly rounded laterally and slightly flattened; posterior edge straight; epicranial stem moderately long; frontal arms V-shaped or U-shaped, with bases contiguous; median endocarina coincident with or extending anterior to epicranial stem. Stemmata absent. Frontoclypeal suture absent; clypeus sometimes asymmetrical; labium free, transverse, broadly rounded or weakly trilobed; tormae united mesally in *Geotrupes*. Antennae less than half as long as head width, 3-segmented; antennomere 2 in *Geotrupes* with a single, dome-like sensorium, with 2 to 4 or 5 short, conical sensoria in Bolboceratinae. Mandibles large and more or less asymmetrical, unidentate or weakly bidentate, with accessory ventral process; incisor edge usually with distinct retinaculum; mola well developed; prostheca absent. Ventral mouthparts slightly retracted, with well developed maxillary articulating area; cardines more or less oblique, divided; stipes as long as wide or slightly transverse, its dorsal surface with stridulatory teeth; galea and lacinia well separated, relatively sparsely setose, the former articulated or fixed, with terminal spine, the latter with bidentate uncus; palp 3-segmented with segment-like palpifer. Labium consisting of prementum, mentum and submentum; ligula more or less truncate; palps 2-segmented without distinct palpiger, widely separated. Hypopharyngeal bracon absent; hypopharynx with paired symmetrical or asymmetrical oncyli. Hypostomal rods moderately long and diverging or absent; ventral epicranial ridges absent. Gular region absent.

Thoracic segments more or less equal in length, without distinct tergal plates. Prothoracic venter without distinct sclerites. Legs short to moderately long, usually 4-segmented with trochanter absent and pretarsus claw-like but minute (legs 3-segmented and pretarsal claw absent in some Bolboceratinae; metathoracic leg reduced in *Geotrupes*, Fig. 42K); stridulatory organ in *Geotrupes* formed by pars stridens on posterior surface of the mesocoxa and a longitudinal row of teeth (plectrum) on the anterior edge of the metafemur; coxae moderately widely separated. Abdominal terga without sclerotised plates, terga I–VIII usually transversely divided forming 2 plicae. Tergum IX about as long as VIII, not extending onto ventral surface, without urogomphi; segment X visible from above, more or less cylindrical in Bolboceratinae, laterally expanded and obliquely flattened in Geotrupes; anal opening Y-shaped, surrounded by single dorsal lobe and paired ventral lobes. Functional spiracles usually cribriform and slightly reniform (biforous in *Eucanthus*), with spiracular scar located ventrally, those on segment VIII slightly reduced; closing apparatus absent. [Howden 1955; Ritcher 1966; Carlson 1991; Verdú *et al.* 1998; Jameson 2002c; Grebennikov and Scholtz 2004; Scholtz and Grebennikov 2005].

Classification and World Distribution

The family Geotrupidae is a widely distributed family containing ~65 genera and 500 species placed in four subfamilies. The group was considered by Browne and Scholtz (1995) to be polyphyletic based on a series of characters based on the axillary region of the wing; according to them Bolboceratinae belongs to a different lineage than the other three subfamilies used here. (Scholtz and Browne 1996). The subfamily Geotrupinae is primarily Holarctic in distribution, the genus *Lethrus* Scopoli (Lethrinae) occurs in eatern Europe and central Asia, Taurocerastinae includes two genera, *Taurocerastes* Philippi and *Frickius* Germain, from southern South America, and the Bolboceratinae is widely distributed but most abundant and diverse in arid regions. [Howden 1982; Howden and Peck 1987].

Australian Fauna

Australian geotrupids include the introduced European *Geotrupes spiniger* (Marsham) (Fig. 68I) and ten genera of Bolboceratinae. Australian geotrupines are usually stout, very strongly convex beetles, yellowish to reddish-brown in colour (except for *G. spiniger*, which is dark metallic blue), with a prognathous head and usually with conspicuous cephalic and pronotal horns in the male (Fig. 60L). The antennal club is relatively small and oval with closely adpressed lamellae. Geotrupid larvae differ from those of other scarabaeoids in having a very lightly pigmented head capsule and 3-segmented antennae. The *Geotrupes* larva has a reduced hind leg acting as a stridulatory plectrum and a complexly lobed anal region; and bolboceratine larvae tend to have all legs somewhat reduced, sometimes lacking a pretarsal claw. Adult geotrupids are often attracted to lights at night; they lay their eggs in deep burrows in the soil and provide the larvae with food, which may include fungi, decaying organic matter or dung. Houston and Bougher (2010) and Houston (2011) found that food items in the burrows of at least some species of *Blackbolbus* and *Bolborhachium* consisted of fruiting bodies of glomeralean mycorhizal fungi. Carne (1965) observed *Elephatomus meraldus* Carne burying large quantities of cattle dung in its oviposition burrows, but dung feeding is not known in other bolboceratines. *G. spiniger*, on the other hand, uses dung for larval provisioning and was introduced into Australia for the control of cattle dung. [Howden 1979, 1985; Howden and Cooper 1977; Howden and Peck 1987; Howden *et al.* 2007; Telnov 2011].

Australian Subfamilies and Genera

Bolboceratinae: *Australobolbus* Howden & Cooper, *Blackbolbus* Howden & Cooper, *Blackburnium* Boucomont, *Bolbobaineus* Howden & Cooper, *Bolboleaus* Howden & Cooper, *Bolborhachium* Boucomont, *Elephastomus* W. S. Macleay, *Eucanthus* Westwood, *Gilletinus* Boucomont, *Stenaspidius* Westwood.

Geotrupinae: *Geotrupes* Latreille.

18. PASSALIDAE

Adult (Figs 3G, 32C, 60J)

Length ~18–55 mm. Body ~2.2–2.5 times as long as wide, more or less parallel-sided, but with distinct constriction between prothorax and elytra, with narrow peduncle usually visible; moderately convex to strongly flattened; colour almost always black, elytra sometimes reddish-brown; dorsal surfaces glabrous and shiny; ventral surfaces and legs with or without dense, yellow hairs.

Head prognathous, not deeply inserted into prothorax, not abruptly constricted posteriorly. Eyes well developed, usually strongly protuberant, deeply emarginate, divided by elevated canthus, finely facetted, without interfacetal setae; ommatidium exocone. Antennal insertions concealed from above; subantennal grooves well developed, extending behind eyes. Frontoclypeal suture usually absent (weakly developed and incomplete mesally in *Aulacocyclus*); frontoclypeal region usually with various ridges and tubercles, sometimes with median, anteriorly curved horn; anterior edge of clypeus often dentate (simple in *Aulacocyclus*). Labrum well sclerotised, subquadrate to slightly transverse, truncate to slightly emarginate. Antennae moderately long, 10-segmented, capable of being anteriorly curled, with 3-segmented or 6-segmented club, the segments of which are uniformly pubescent and not capable of close apposition; scape more than twice as long as wide, more than twice as long as pedicel, and more or less curved. Mandibles moderately elongate, projecting beyond apex of labrum; outer edge with dorsal and ventral elevated carinae, the former enclosing sides of labrum and the latter enclosing ventral mouthparts; apex broadly bidentate; incisor edge with 1 or 2 retinacula and with basally articulated, broad, flattened and apically emarginate process; mola well developed but slightly sub-basal; prostheca absent. Maxilla with galea 1-segmented, falcate, sparsely clothed with stout setae; lacinia only slightly narrower than galea and with similar vestiture, with 1 or 2 unci; palp 4-segmented with apical palpomere fusiform. Mentum subquadrate or slightly transverse and deeply emarginate apically, completely enclosing prementum; ligula trilobed; labial palps 3-segmented with apical palpomere fusiform. Gular sutures well separated; gula about as long as wide. Corpotentorium moderately broad and strongly arched; anterior tentorial arms approaching one another and fused together at midline forming short laminatentorial bridge. Cervical sclerites well developed.

Pronotum ~0.7–0.85 times as long as wide, widest at about middle and as wide at base as combined elytral bases; sides subparallel, not explanate. Lateral pronotal carinae complete, simple, with raised margin or bead; anterior angles not produced forward; posterior angles broadly rounded; posterior edge straight or evenly rounded; disc simple, with or without median longitudinal groove. Prosternum in front of coxae distinctly shorter than shortest diameter of procoxal cavity, at least slightly tumid; anterior edge produced to form apically rounded chin piece. Prosternal process usually complete (but interrupted at middle in *Aulacocyclus*), narrow at middle, abruptly expanded before apex, then slightly to strongly narrowed; apex usually broadly rounded to truncate (acute in *Aulacocyclus*). Notosternal sutures complete. Hypomeron simple, without antennal cavities or crural impressions. Procoxae large, with mesal portion projecting below prosternum in *Aulacocyclus*, without or with very short concealed lateral extensions; trochantins concealed; endopleuron fused to notum. Procoxal cavities strongly transverse, usually very narrowly separated (subcontiguous in *Aulacocyclus*), broadly closed externally by postcoxal hypomeral projections abutting or overlapping prosternal process, internally open.

Scutellar shield well developed, not abruptly elevated anteriorly, posteriorly acute, located on peduncle and on a lower plane than main portion of elytra. Elytra ~1.5–1.75 times as long as combined width and 2.2–2.55 times as long as pronotum; humeri well developed; sides subparallel; apices broadly, conjointly rounded concealing all abdominal tergites; disc with 10 striae and no scutellary striole; epipleura very narrow but more or less complete. Mesoventrite separated by complete sutures from mesanepisterna, which are relatively narrowly separated at midline but not subcontiguous; anterior edge of mesoventrite on same plane as metaventrite, without paired procoxal rests, sometimes with median prosternal rest or broad collar formed with mesanepisterna; mesoventral cavity absent; discrimen absent. Mesocoxae more or less globular, with concealed trochantins. Mesocoxal cavities moderately widely separated, broadly closed laterally by meeting of meso- and metaventrites. Meso-metathoracic joint within coxal cavities absent. Mesometaventral junction absent, due to fusion of ventrites and loss of suture between them. Metaventrite slightly transverse and moderately convex, usually with vaguely defined postcoxal lines delimiting crural impressions; transverse (metakatepisternal) suture absent; discrimen absent; exposed portion of metanepisternum long and very slender; metepimeron slightly exposed. Metacoxae strongly transverse, subcontiguous, not extending laterally to meet elytra; plates absent. Metendosternite with short, broad stalk, long, transverse or slightly curved lateral arms and long, slender anterior process with narrowly separated tendons. Hind wing with long radial bending zone combined with distinct hinge at end of radial bar; apical field about one-fourth as long as total wing length, with as many as 4 radial extensions (RA_3, apically fused RA_4 and RP_1 and RP_2,); radial cell incomplete, without base; cross-vein r3 absent; r4 more or less complete; R-M loop absent; basal portion of RP absent, apical extension extending to wing margin; medial spur not curved posteriorly, extending to wing margin; medial field with 3 or 4 free veins; MP_{3+4} absent, MP_3 and MP_4 not connected basally; wedge cell absent; AA_{3+4} forked and AA_3 joining Cu; anal lobe well developed, without embayment, AP_{3+4} forked or not. Legs relatively long and slender; trochanterofemoral joint strongly oblique with femur in contract with coxa; meso- and metatibiae not expanded apically, outer edge with longitudinal row of stout setae; outerapical angle with 2 fixed spines; meso- and metatibial spurs well developed and asymmetrical; protibia apically expanded with one large, subapical spur, its outer edge with a variable number of lobes or teeth. Tarsi 5-5-5; tarsomeres relatively long and slender, simple; pretarsal claws simple; empodium well developed and bisetose.

Abdomen usually with 6 ventrites (sternites II–VII) (5 ventrites in *Aulacocyclus*), the last 3 of which are more or less connate; sternite II divided by coxae into 2 small lateral sclerites (but these concealed by metacoxae in *Aulacocyclus*); metacoxal cavities well defined with acute intercoxal process (a reduced intercoxal process on sternite II concealed between metacoxae). Functional spiracles on abdominal segments I–VII located in pleural membrane; a reduced pair on segment VIII. Abdominal tergites relatively lightly sclerotised (although VII and VII slightly darker than others), with patches of wing-folding asperities sometimes on tergite IV. Anterior edge of sternite VIII in male without median strut. Anterior edge of segment IX with broadly rounded subgenital plate; tergite IX almost completely divided into two laterotergites (united by a slender strip of membrane); tergum X membranous. Aedeagus of trilobate type, but reduced in size, with short broad phallobase, equally short and broad parameres more or less fused together and penis mostly membranous. Sternite VIII in female without spiculum ventrale. Sternite IX in female divided into two parts; proctiger membranous. Ovipositor very short and broad; paraprocts much shorter than coxites, without baculi; coxites undivided, without styli. [Tanner 1927; Dibb 1938; D'Hotman and Scholtz 1990a; Schuster 2002; Boucher 2005; Scholtz and Grebennikov 2005].

Larva (Fig. 60I)

Body subcylindrical and slightly curved ventrally; relatively weakly sclerotised, with lightly pigmented tergal plates; vestiture relatively sparse, consisting of long and short setae.

Head (Fig. 42J) more or less prognathous, distinctly transverse, strongly rounded laterally and slightly flattened; posterior edge straight; epicranial stem moderately long; frontal arms V-shaped or U-shaped, with bases contiguous; median endocarina Y-shaped, coincident with epicranial stem and bases of frontal arms. Stemmata absent. Frontoclypeal suture present; labium free, transverse, broadly rounded or somewhat angulate; tormae not united, without distinct mesal arms. Antennae less than a fifth as long as head width, 2-segmented, without sensorium. Mandibles large, symmetrical, tridentate, without accessory ventral process; incisor edge simple; mola well developed, sub-basal with basally projecting lobe bearing tuft of hairs; prostheca absent. Ventral mouthparts slightly retracted, with well developed maxillary articulating area; cardines slightly oblique, divided; stipes longer than wide, its dorsal surface with stridulatory teeth; galea and lacinia distinctly separated, relatively sparsely setose and falcate (each with terminal spine), lacinia with several stout setae along inner edge; palp 2-segmented with distinct, palpifer. Labium consisting of prementum, mentum and submentum; ligula about as long as first palpomere, broadly rounded; palps 2-segmented with distinct palpiger, widely separated. Hypopharyngeal bracon absent; hypopharynx with slender, curved sclerotised bar. Hypostomal rods moderately long and slightly diverging; ventral epicranial ridges absent. Gular region absent.

Prothorax slightly longer than meso- or metathorax. Prothoracic venter with well developed cervical region with paired lateral expansions. Prothoracic and mesothoracic legs moderately long, 4-segmented with trochanter absent and pretarsus claw-like and bisetose; metathoracic leg reduced (Figs 42J, 50N), 1-segmented with row of anterior teeth forming operating as plectrum of stridulatory organ, with pars stridens on posterior surface of mesocoxa; coxae moderately widely separated. Abdominal terga not transversely divided. Tergum IX about as long as VIII, not extending onto ventral surface, without urogomphi; segment X visible from above, slightly longer than IX, more or less cylindrical; anal opening Y-shaped, with paired dorsal lobes and single ventral lobe. Functional spiracles cribriform, reniform, with spiracular scar located anteriorly on thorax and posteriorly on abdomen; spiracles on segment VIII not enlarged; closing apparatus present. [Böving and Craighead 1931; Costa and Fonseca 1986; Costa *et al.* 1986; Carlson 1991; Schuster 2002; Scholtz and Grebennikov 2005].

Classification and World Distribution

Passalidae is a widespread but mainly tropical family including ~60 genera and 700 species. The subfamily Passalinae occurs on all continents, while Aulacocyclinae are restricted to South-east Asia and the Australian region.

Australian Fauna

Australian Passalidae occur primarily in the northern and eastern part of the continent, with one species, *Pharochilus rugiceps* (Hope and Westwood) extending into TAS. The **Aulacocyclinae** is represented by 11 species of *Aulacocyclus* Kaup, and the **Passalinae** includes members of two tribes: Leptaulacini with the genus *Leptaulax* Kaup, and Macrolinini with the genera *Analaches* Kuwert, *Austropassalus* Mjöberg, *Gonatas* Kaup, *Labienus* Kaup, *Mastachilus* Kaup, *Pharochilus* Kaup and *Protomocoelus* Zang. Passalids are relatively large, shining, black, somewhat flattened beetles, with a prognathous head, characteristically curved, pectinate antennae, and a narrow waist or pedicel between the large prothorax and the hind body; the scutellum is not visible and the elytra are conspicuously striate. Adults stridulate by rubbing a plectrum on each hind wing against a pars stridens on the dorsal surface of the abdomen (Reyes-Castillo and Jarman 1983). Larvae differ from those of other scarabaeoids by being less curved and in having the hind leg reduced to a stump, which produces a sound when rubbed against a file on the mesocoxa. Passalids live in rotten logs and exhibit a type of subsocial behaviour discussed in detail by Reyes-Castillo and Halffter (1983). [Boucher 2005].

19. TROGIDAE

Adult (Figs 3H, 60E)

Length ~5–30 (usually 7–20) mm. Body ~1.55–1.95 times as long as wide, oblong to oval with sides of prothorax and elytra slightly to moderately, independently rounded; moderately to strongly convex above but relatively flat below; colour brown to grey or black; dorsal surfaces dull, often granulate to tuberculate and sometimes encrusted with dirt; vestiture of short to very short, erect, bristles, usually yellow to brown or black in colour.

Head hypognathous, deeply inserted into prothorax, not abruptly constricted posteriorly; with low, curved occipital carina, which extends anteriorly on each side to meet supraocular carinae. Eyes well developed, usually strongly protuberant, shallowly emarginate, finely facetted, without interfacetal setae; ommatidium usually of exocone or duocone type (eucone in *Trox*). Antennal insertions concealed from above by frontal carina, which is continuous with supraocular carina on each side and extends to midline; subantennal grooves well developed but short, not extending below or behind eyes. Frontoclypeal suture very weak or incomplete mesally; frontoclypeal region abruptly declined anterior to frontal carina (reflexed area very short in *Trox*). Labrum well sclerotised, strongly transverse and slightly emarginate. Antennae very short, 10-segmented, with 3-segmented, lamellate club with pubescent segments capable of close apposition; scape usually more than twice as long as wide, more than twice as long as pedicel, more or less curved and clothed with stout bristles (scape much shorter and broader in *Trox*). Mandibles short and broad, unidentate; incisor edge sharply carinate; mola weakly developed and slightly sub-basal; prostheca a membranous lobe. Maxilla with galea short, 2-segmented, more or less truncate and clothed with stout setae; lacinia only slightly narrower than galea, with densely setose inner edge, apicolateral 2- or 3-dentate uncus and apicomesal unidentate uncus; palp 4-segmented with apical palpomere fusiform. Mentum subquadrate or slightly transverse, apically convex; ligula short and truncate or angulate; labial palps 3-segmented with bases concealed from below by mentum and apical palpomere fusiform. Gular sutures well separated; gula about as long as wide. Corpotentorium moderately broad and slightly arched; laminatentoria fused together at midline forming bridge. Cervical sclerites well developed.

Pronotum ~0.55–0.8 times as long as wide, widest at about middle but almost always sharply constricted posteriorly, so that base is distinctly narrower than combined elytral bases; sides usually moderately rounded (less so in *Trox*), sometimes explanate. Lateral pronotal carinae complete, simple, undulate or crenulate, without raised margin or bead; anterior angles produced forward and rounded to subacute; posterior angles usually right to broadly rounded (subacute in *Trox*); posterior edge rounded, undulate or with distinct mesal lobe; disc with various impressions and elevations, sometimes forming longitudinal costae. Prosternum in front of coxae distinctly shorter than shortest diameter of procoxal cavity, biconcave, oblique to almost vertical and on different plane than prosternal process. Prosternal process complete, anteriorly narrow to laminate, abruptly expanded and subtruncate apically. Notosternal sutures absent. Hypomeron with large anterior pockets for reception of head. Procoxae large, transverse, with mesal portion projecting slightly below prosternum, each with rounded ridge below apex and moderately long concealed lateral extension; trochantin concealed; endopleuron not fused to notum. Procoxal cavities moderately transverse, narrowly separated or subcontiguous, broadly closed externally by postcoxal hypomeral projections fitting into lateral cavities near apex of prosternal process, internally open.

Scutellar shield well developed, not abruptly elevated anteriorly, usually subapically expanded (hastate) (not in *Trox*), usually posteriorly acute (rounded in *Trox*), occasionally fused to elytra. Elytra ~1.1–1.32 times as long as combined width and 2–3.2 times as long as pronotum; humeri strong and abrupt to weakly developed and broadly rounded; sides slightly to strongly rounded; apices broadly, conjointly rounded concealing all abdominal tergites; disc usually without distinct puncture rows or striae, but usually with tubercles and/or longitudinal costae (with 10 distinct puncture rows and no scutellary striole in *Trox*); epipleura moderately broad and complete; elytra sometimes completely fused together. Mesoventrite separated by complete sutures from mesanepisterna, which are widely separated at midline; anterior edge of mesoventrite on same plane as metaventrite, without paired procoxal rests but with median prosternal rest or broad collar formed with mesanepisterna; mesoventral cavity absent; discrimen absent. Mesocoxae more or less globular, with concealed trochantins. Mesocoxal cavities (Fig. 44V) contiguous, broadly closed laterally by meeting of meso- and metaventrites. Meso-metathoracic joint within coxal cavities absent. Mesometaventral junction absent (mesoventral and metaventral processes not meeting). Metaventrite slightly to strongly transverse and moderately convex, without postcoxal lines; transverse (metakatepisternal) suture absent or located very close to posterior edge of ventrite; discrimen usually short to moderately long, sometimes absent; exposed portion of metanepisternum moderately long and narrow; metepimeron concealed. Metacoxae strongly transverse, subcontiguous, extending laterally to meet elytra; plates absent. Metendosternite usually with moderately long and narrow stalk, long, transverse lateral arms and moderately long, anterior process with narrowly separated tendons. Hind wing with long radial bending zone combined with distinct hinge at end of radial bar; apical field about half as long as total wing length, with 2 or 3 radial extensions; radial cell complete, shorter than wide, more or less oblique, with lumen completely filled by ptgerostigma; cross-vein r3 absent; r4 more or less complete; R-M loop forming a narrowly acute angle; basal portion of RP short, apical extension interrupted basally but extending to wing margin; medial spur not curved posteriorly, extending to wing margin; medial field with 4 free veins; MP_{3+4} absent, MP_3 and MP_4 not connected basally; wedge cell absent; AA_{3+4} forked but AA_3 not joining Cu; anal lobe well developed, with very small, notch-like embayment, AP_{3+4} not forked; hind wings sometimes absent. Legs relatively long and slender; trochanterofemoral joint strongly oblique with femur in contract with coxa; profemur expanded and flattened with carinate anterior edge (concealing tibiae, tarsi and prosternum in repose) and with finely striate area on inner surface; all tibiae more or less expanded apically; meso- and metatibiae in *Trox* with one or two teeth at outer apical angle and a longitudinal row of spines along outer edge, those in *Omorgus* without teeth but often with socketed spines; outer edge and outer apical angle of protibia with several large teeth; meso- and metatibial spurs well developed and more or less asymmetrical; protibia one large, apical spur. Tarsi 5-5-5; tarsomeres slender, simple; pretarsal claws simple; empodium absent or not apparent.

Abdomen usually with 5 connate ventrites (sternites III–VII), with ventrite 1 strongly inclined and much shorter than 2, and sternites 2–5 all on the same plane and separated by distinct grooves; lateral portions of sternite II completely concealed by coxae; metacoxal cavities well defined with acute intercoxal process. Functional spiracles usually on abdominal segments I–VIII (on I–VII in *Trox*) and located in pleural membrane. Abdominal tergites (including tergite VIII) relatively lightly sclerotised, with patches of wing-folding asperities on tergites V and VI. Anterior edge of sternite VIII in male without median strut. Anterior edge of segment IX with broadly rounded subgenital plate; tergite IX almost completely divided into two laterotergites;

tergum X membranous with paired struts. Aedeagus of trilobate type, usually symmetrical, with short broad phallobase, slender, simple parameres and broad, somewhat flattened penis without basal struts. Sternite VIII in female without spiculum ventrale. Sternite IX in female divided into two parts; proctiger membranous with lateral struts. Ovipositor very short and broad; paraprocts much shorter than coxites, without baculi; coxites undivided, with strongly oblique basal baculi, with or without apical styli. [Tanner 1927; Scholtz 1986a, 1986b, 1990b; D'Hotman and Scholtz 1990a; Jameson 2002b; Scholtz and Grebennikov 2005].

Larva (Fig. 34M)

Body subcylindrical and strongly curved ventrally (C-shaped); relatively weakly sclerotised and lightly pigmented, except for heavily pigmented head and weakly pigmented prothoracic plate; vestiture relatively dense, consisting of long and short setae and short spines.

Head hypognathous, distinctly transverse, strongly rounded laterally and slightly flattened; posterior edge straight; epicranial stem moderately long; frontal arms V-shaped or U-shaped, with bases contiguous; median endocarina coincident with epicranial stem. Stemmata one on each side. Frontoclypeal suture usually present; clypeus symmetrical; labium free, transverse, emarginate; tormae united mesally. Antennae about a third as long as head width, 3-segmented; antennomere 3 min; antennomere 2 with dome-like sensorium. Mandibles large, unidentate, with accessory ventral process; incisor edge usually with 1 or 2 teeth or lobes; mola well developed, somewhat asymmetrical, with small sub-basal tuft of hairs; prostheca absent. Ventral mouthparts slightly retracted, with well developed maxillary articulating area; cardines more or less oblique, internally divided; stipes elongate its dorsal surface with row of stridulatory teeth; galea and lacinia well separated, the former articulated, 2-segmented, with few setae and a terminal spine, the latter with bidentate uncus and stout setae along inner edge; palp 3-segmented with segment-like palpifer. Labium consisting of prementum, mentum and submentum; ligula weakly developed and truncate; palps 2-segmented without distinct palpiger, widely separated. Hypopharyngeal bracon absent; hypopharynx with slender undulate transverse bar with 2 pairs of distal projections. Hypostomal rods short and very slightly diverging; ventral epicranial ridges absent. Gular region absent.

Thoracic segments more or less equal in length, with distinct tergal plate on prothorax. Prothoracic venter without distinct sclerites; presternal area short. Legs moderately long, slender, 5-segmented with pretarsus claw-like and bisetose; stridulatory mechanism absent; coxae moderately widely separated. Abdominal terga without sclerotised plates, terga I–VI or VII transversely divided forming 3 plicae, each with at least one transverse row of short, spine-like setae, in addition to longer setae. Tergum IX about as long as VIII, not extending onto ventral surface, without urogomphi; segment X visible from above, more or less cylindrical and terminal, more or less evenly clothed with short and long, stiff hairs; anal opening Y-shaped, surrounded by single dorsal lobe and paired ventral lobes. Functional spiracles usually cribriform and slightly reniform with spiracular scar located antero ventrally on both thorax and abdomen (spiracles biforous in Trox); closing apparatus present. [Böving and Craighead 1931; Ritcher 1966; Baker 1968; Scholtz and Peck 1990; Carlson 1991; Scholtz 1991; Jameson 2002b; Scholtz and Grebennikov 2005b].

Classification and World Distribution

The Trogidae is a small, distinctive, cosmopolitan family containing four genera: *Afromorgus* Scholtz from the southern Palaearctic, Afrotropical and Oriental regions, *Omorgus* from the New World, Australasia and central to southern Africa, *Trox* from the Holarctic region and Africa, and the Neotropical *Polynoncus* Burmeister. [Scholtz 1990b].

Australian Fauna

The Australian trogid fauna includes an introduced species, *Trox scaber* (Linnaeus) and ~50 species of *Omorgus*. Trogids are robust, heavily sclerotised, dull and uniformly dark beetles, with a tuberculate upper surface, strongly declined head and broadly closed mesocoxal cavities. Larvae differ from those of other scarabaeoids in having a darkly pigmented head with stemmata, three antennomeres and abdominal terga with transverse folds and stiff spines. Adults and larvae apparently feed on dry animal remains, being one of the last in the succession of insects that invade carcasses; however, gut contents of several adults of the wingless species *Omorgus rotundulus* (Haaf) contained several insect parts, particularly of termites and ants (Houston *et al.* 2010). Trogid larvae live in vertical burrows in the soil beneath the carcass. Adults stridulate by rubbing an abdominal plectrum against a file along the internal edges of the elytra. The group appears to flourish in the more arid parts of Australia. [Scholtz 1986a, 1986b, 1991].

20. LUCANIDAE

Adult (Figs 2E, G, 60D, 68C–E)

Length ~6–60 (usually 10–40) mm. Body ~1.5–2.8 times as long as wide, oblong to oval with sides of prothorax and elytra subparallel or slightly to moderately, independently rounded; with distinct constriction between prothorax and elytra, so that narrow peduncle sometimes visible; slightly flattened to moderately convex; colour usually reddish-brown to black (metallic red to green or blue in Lampriminae); dorsal surfaces dull, often granulate to tuberculate and sometimes encrusted with dirt; dorsal surfaces subglabrous and shiny in *Syndesus* and many Lucaninae (e.g. *Figulus*, *Aegus*, *Lissotes*, *Prosopocoilus*), with a velvety appearance due to microsculpture in *Rhyssonotus*, or clothed with expanded setae or flattened scales (e.g. *Ceratognathus*, *Cacostomus*); ventral surfaces usually with similar vestiture, but densely clothed with longer, fine setae in *Ceratognathus* and Lampriminae.

Head prognathous, not deeply inserted into prothorax, not abruptly constricted posteriorly; without occipital carina. Eyes moderately well developed, strongly protuberant in *Syndesus*, deeply emarginate or completely divided into dorsal and ventral parts only in Lucaninae, finely facetted, without interfacetal setae; ommatidium usually of eucone type (acone at least in *Lissotes*). Antennal insertions concealed from above by frontal carina, which at least partly bisects eye in Lucaninae; subantennal grooves weakly developed or absent. Frontoclypeal suture absent; frontoclypeal region usually more or less abruptly declined anteriorly but not sharply carinate (frontal carinae not meeting at midline). Labrum usually more or less fused to clypeus (free in *Ceratognathus* and Lampriminae). Antennae moderately long, 10-segmented, usually with 3-segmented, asymmetrical, pectinate or flabellate club, with segments pubescent and not capable of close apposition (club 6-segmented in *Sphaenognathus*, 7-segmented in *Syndesus*); club segments only slightly expanded and pubescence only at apex in some Lucaninae (e.g. *Figulus*, *Lissapterus*); distinctly geniculate in Lucaninae, slightly so in Lampriminae; scape more than 3 times as long as wide, and more than 3 times as long as pedicel, usually more or less curved. Mandibles elongate, usually much larger in males, unidentate to tridentate and sometimes with additional dorsal or mesal teeth or ridges; mola and prostheca absent. Maxilla (Fig. 16G) with galea 1-segmented, apically rounded and densely setose; lacinia usually somewhat reduced, setose, with undivided uncus in Lampriminae, sometimes absent; palp 4-segmented with apical palpomere fusiform. Mentum about as long as wide to strongly transverse, apically convex; ligula undivided, sometimes highly reduced; labial palps 3-segmented with bases concealed from below by mentum in Lucaninae; apical palpomere fusiform. Gular sutures well separated; gula usually longer than wide. Corpotentorium moderately broad and arched; anterior tentorial arms not expanded. Cervical sclerites well developed.

Pronotum ~0.45–0.88 times as long as wide, usually widest at about middle or anteriorly; base usually at least as wide as combined elytral bases (often wider in males); sides subparallel to moderately rounded, not explanate. Lateral pronotal carinae complete, simple, crenulate or crenate, with or without raised margin or bead; anterior angles at least slightly produced forward and rounded to subacute; posterior angles obtuse to broadly rounded; posterior edge rounded, straight or biemarginate; disc usually simple, with various impressions in *Rhyssonotus*, with anterior median horn in male *Syndesus*. Prosternum (Fig. 44L) in front of coxae at least slightly shorter than shortest diameter of procoxal cavity, flat to biconcave or mesally tumid, without distinct chin piece. Prosternal process complete, at least slightly curved behind coxae in lateral view in *Ceratognathus* and most Lucaninae, abruptly expanded and subtruncate apically. Notosternal sutures present. Procoxae large, transverse, projecting below prosternum in *Ceratognathus*, Lampriminae and in a few Lucaninae, with densely setose patch on posterior surface, with very short concealed lateral extension; trochantin concealed; endopleuron not fused to notum. Procoxal cavities moderately to strongly transverse, moderately to very narrowly separated, broadly closed externally by postcoxal hypomeral projections fitting into lateral cavities near apex of prosternal process, internally open.

Scutellar shield well developed, abruptly elevated anteriorly or not, posteriorly broadly rounded to acute. Elytra ~1.25–1.65 times as long as combined width and 1.8–2.8 times as long as pronotum; humeri strongly developed; sides subparallel to slightly rounded; apices broadly, conjointly rounded concealing all abdominal tergites; disc usually without distinct puncture rows or striae (with 10 distinct puncture rows in *Syndesus* and *Figulus*; with 5 striae in *Aegus*); epipleura usually narrow but more or less complete. Mesoventrite separated by complete sutures from mesanepisterna, which are widely separated at midline; anterior edge of mesoventrite usually on different plane than metaventrite, without paired procoxal rests but with median prosternal rest or broad collar formed with mesanepisterna; mesoventral cavity absent; discrimen absent. Mesocoxae slightly to strongly transverse. sometimes slightly oblique, with exposed trochantins. Mesocoxal cavities narrowly to moderately widely separated, partly closed laterally by mesanepisterna and mesepimera. Meso-metathoracic joint within coxal cavities absent. Mesometaventral junction usually absent due to fusion of ventrites and loss of suture between them (ventrites abutting and suture present in *Ceratognathus*). Metaventrite slightly to strongly transverse and slightly convex, without postcoxal lines; transverse (metakatepisternal) suture absent; discrimen usually moderately long, sometimes absent; exposed portion of metanepisternum moderately long and narrow; metepimeron usually exposed. Metacoxae strongly transverse, subcontiguous, extending laterally to meet elytra; plates absent. Metendosternite usually with moderately long and narrow stalk, long, transverse lateral arms and long, anterior process with narrowly separated tendons. Hind wing with long radial bending zone combined with distinct hinge at end of radial bar; apical field ~0.4 times as long as total wing length, with 3 radial extensions; radial cell without base; cross-vein r3 absent; r4 complete or incomplete; R-M loop forming a narrowly acute angle; basal portion of RP moderately long, apical extension extending to wing margin; medial spur distinctly curved posteriorly, extending to wing margin; medial field with 3 or 4 free veins; MP_{3+4} absent, MP_3 and MP_4 not connected basally; wedge cell absent; AA_{3+4} forked and AA_3 usually joining Cu; anal lobe well developed, without embayment, AP_{3+4} not forked. Legs relatively long and slender; trochanterofemoral joint strongly oblique with femur in contract with coxa; profemur expanded and flattened, with densely setose patch on anterior surface; meso- and metatibiae usually slightly expanded apically, usually with one or two teeth and spines at outer apical angle and several spines along outer edge; protibia usually distinctly expanded apically with relatively large teeth along outer edge gradually increasing in size to outer apical angle; meso- and metatibial spurs well developed, paired and more or less symmetrical; protibia one large, apical spur (greatly expanded and flattened in males of *Lamprima*). Tarsi 5-5-5; tarsomeres slender, simple; pretarsal claws simple; empodium well developed and bisetose.

Abdomen almost always with 5 free ventrites (sternites III–VII) (sternite VIII often partly visible in *Syndesus*), with ventrite 1 shorter than 2; sternite II completely concealed by coxae, although sometimes more or less complete, with distinct intercoxal process; metacoxal cavities well defined with acute intercoxal process. Functional spiracles on abdominal segments I–VIII, located in pleural membrane. Abdominal tergites relatively lightly sclerotised, sometimes with the exception of pygidium, with patches of wing-folding asperities sometimes present on tergites V and VI. Anterior edge of sternite VIII in male without median strut. Anterior edge of segment IX with truncate to broadly or narrowly rounded subgenital plate; tergite IX (Fig. 31A) apically emarginate and usually divided at midline by suture (sometimes completely divided into laterotergites); tergum X evenly sclerotised. Aedeagus of trilobate type, symmetrical, with short broad phallobase, slender, simple parameres and broad, somewhat flattened penis with or without basal struts; endophallus permanently everted in Lucaninae. Sternite VIII in female without spiculum ventrale. Sternite IX in female divided into two parts; proctiger evenly sclerotised or sometimes membranous. Ovipositor short and broad; paraprocts shorter than coxites, with baculi; coxites undivided, usually setose, often with well sclerotised, apical styli (absent in some Lucaninae). [Holloway 1960, 1968, 1969, 1997, 1998, 2007; Howden and Lawrence 1974; Moore 1978a, 1994; D'Hotman and Scholtz 1990a; Bartolozzi 1996; Chalumeau and Brochier 1993, 1995; Reid 1999a; Ratcliffe 2002; Moore and Monteith 2004; Scholtz and Grebennikov 2005].

Larva (Fig. 34F)

Body subcylindrical and strongly curved ventrally (C-shaped); relatively weakly sclerotised and lightly pigmented; vestiture usually a mixture of long, fine setae and very short, spine-like setae. Head hypognathous and at least slightly transverse, strongly rounded laterally and slightly flattened; posterior edge straight; epicranial stem moderately long; frontal arms V-shaped or U-shaped, with bases contiguous; median endocarina coincident with or extending anterior to epicranial stem. Stemmata usually absent (occasionally 1 on each side). Frontoclypeal suture present; clypeus symmetrical; labium (Fig. 46O) free, slightly transverse, broadly rounded or truncate; tormae united mesally and with anteriorly projecting epitorma. Antennae usually about half as long as head width, 3- or 4-segmented, with a distinct antennifer; preapical antennomere with a single, dome-like sensorium. Mandibles (Fig. 41B) large, bidentate or tridentate, with accessory ventral process; incisor edge sometimes with teeth; molae well developed, asymmetrical; prostheca absent. Ventral mouthparts slightly retracted, with well developed maxillary articulating area; cardines more or less oblique, internally divided; stipes slightly longer than wide, its dorsal surface usually without stridulatory teeth; galea and lacinia well separated, relatively sparsely setose, the former articulated with terminal spine, the latter with unidentate uncus; palp 3-segmented with segment-like palpifer. Labium consisting of prementum, mentum and submentum; ligula absent; palps 2-segmented without distinct palpiger, widely separated. Hypopharyngeal bracon absent; hypopharynx with complex, asymmetrical armature. Hypostomal rods short and slightly diverging; ventral epicranial ridges absent. Gular region absent.

Thoracic segments more or less equal in length, without distinct tergal plates; protergum in *Syndesus* with pair of anteriorly projecting sclerotised lobes. Prothoracic venter without distinct presternal sclerite, but usually with paired lateral cervical sclerites. Legs moderately long, slender, 4-segmented, with pretarsus often somewhat reduced, but usually claw-like and bisetose or with more than 2 setae; stridulatory organ formed by pars stridens (Fig. 50K) on posterior surface of mesocoxa and plectrum (Fig. 50I) on anterior edge of the metatrochanter; coxae moderately widely separated. Abdominal terga without sclerotised plates, not transversely divided to form plicae; most abdominal terga sometimes with patches of very short, spine-like setae. Tergum IX usually about as long as VIII, not extending onto ventral surface, without urogomphi; segment X visible from above, more or less cylindrical; anal opening Y-shaped, surrounded by reduced dorsal lobe and large, paired ventral lobes (Fig. 53F), the latter often with an asetose anal pad; sternite X sometimes with paired patches of setae forming a raster. Functional spiracles cribriform and slightly reniform, with spiracular scar located anteriorly on thorax and anteroventrally on abdomen; posterior abdominal spiracles sometimes smaller in size; closing apparatus present. [Böving and Craighead 1931; Emden 1935; Ritcher 1966; Lawrence 1981; Costa *et al.* 1988; Carlson 1991; Onore 1994; Ratcliffe 2002; Moore and Monteith 2004; Scholtz and Grebennikov 2005].

Classification and World Distribution

The Lucanidae is a widespread family containing ~95 genera and 1250 species. The group may be divided into six subfamilies (Aesalinae, Nicaginae, Ceratognathinae, Syndesinae, Lampriminae and Lucaninae), the last four of which occur in Australia, but Holloway (1960, 1968, 1969, 1997, 1998, 2007) considered the first three of these to form a single subfamily, Aesalinae, while Howden and Lawrence (1974) united Nicaginae and Ceratognathinae. The Penichrolucaninae (Ratcliffe 1984) probably belongs in the subfamily Lucaninae, and according to Nikolajev (1999b), the two included genera belong to two different tribes.

Australian Fauna

The Australian lucanid fauna consists of *Ceratognathus*, *Syndesus*, three genera of Lampriminae and at least 12 genera of Lucaninae. Adults are characterised by the relatively thick antennal lamellae, long scape and prognathous head with well developed, often sexually dimorphic mandibles; they are usually black, brown or metallic in colour and usually glabrous but occasionally clothed with scales. Adults are mainly nocturnal and some, at least, are nectar feeders. The group is more or less restricted to the wetter parts of the continent. The genus *Syndesus* (Fig. 60D) (**Syndesinae**) differs from most lucanids in having a 7-segmented antennal club with lamellae longer than usual; the genus also occurs in New Caledonia and South America. *Ceratognathus* (Figs 2E, 68D) (**Ceratognathinae**) includes several smaller species clothed with scales and having three long lamellae in the antennal club; its closest relatives include *Holloceratognathus* Nikolajev and *Mitophyllus* Parry from New Zealand (Holloway 2007) and the South American *Hilophyllus* Paulsen and Mondaca (2006). The subfamily **Lampriminae** includes the brilliant metallic *Phalacroglathus muelleri* Macleay, the most attractive of Australian beetles, which occurs in the rainforests of northern QLD, and species of *Lamprima* (Figs 2G, 68C), which may vary from metallic green or bronze to blue, red or violet. The remaining Australian genera belong to the largest subfamily **Lucaninae** and appear to represent several tribes, although the tribal classification is in need of revision. *Rhyssonotus nebulosus* (Kirby) (Fig. 68E) is relatively common at lower elevations in eastern Australia. *Lissotes* includes several flightless species in TAS, and the genera *Sphaenognathus* and *Cacostomus* are related to the South American *Chaisognathus* Stephens, and *Casignetus* Macleay, respectively (Moore 1978a, 1994; Reid 1999a; Moore and Monteith 2004). Lucanid larvae differ from other scarabaeoids in having a stridulatory apparatus consisting of a pars stridens on the mesocoxa and plectrum on the metatrochanter, combined with lack of folds on the abdominal terga and the presence of a longitudinally oval pad on either side of the vertical anal opening (Lawrence 1981; Ritcher 1966). Larvae almost always feed in rotten wood; however, those of at least some *Sphaenognathus* are atypical in that they may live freely in the soil (Onore 1994). [Chalumeau and Brochier 1993, 1995; Bartolozzi 1996].

Australian Subfamilies and Genera

Ceratognathinae: *Ceratognathus* Westwood.
Syndesinae: *Syndesus* W. S. Macleay.
Lampriminae: *Homolamprima* Macleay, *Lamprima* Latreille, *Phalacrognathus* Macleay.
Lucaninae: *Aegus* W. S. Macelay, *Cacostomus* Newman, *Dorculus* Didier, *Eucarteria* Lea, *Figulus* W. S. Macleay, *Hoplogonus* Parry, *Lissapterus* Deyrolle, *Lissotes* Westwood, *Prosopocoilus* Hope & Westwood, *Pseudodorcus* Parry, *Rhyssonotus* W. S. Macleay, *Serrognathus* Motschulsky, *Sphaenognathus* Buquet.

21. HYBOSORIDAE

Adult (Figs 60M–N, 68J)

Length ~2.8–15 mm. Body ~1.45–1.9 times as long as wide, oblong to oval with sides of prothorax and elytra subparallel or slightly to moderately, independently rounded; with distinct constriction between prothorax and elytra, so that narrow peduncle sometimes visible; slightly to strongly convex; Ceratocanthinae and some *Liparochrus* capable of conglobation; colour usually reddish-brown to black; dorsal surfaces usually shiny subglabrous or relatively sparsely clothed with short, fine setae (relatively dull and clothed with short to very short, stout bristles in Ceratocanthinae); ventral surfaces and legs often more densely clothed with longer setae.

Head prognathous or apparently deflexed in conglobular forms, sometimes deeply inserted into prothorax, more or less flattened, not abruptly constricted posteriorly; without occipital carina. Eyes moderately well developed, not strongly protuberant, shallowly to very deeply emarginate (almost completely divided into dorsal and ventral parts in Ceratocanthinae), finely facetted, without interfacetal setae; ommatidium of eucone type. Antennal insertions concealed from above by frontal carina, which at least partly bisects eyes; subantennal grooves weakly developed. Frontoclypeal suture absent; frontoclypeal region usually more or less abruptly declined anteriorly and sharply carinate, so that frontal carinae meet at midline (not so in *Phaeochrous*). Labrum free, well sclerotised and at least partly exposed, transverse and apically truncate to slightly emarginate. Antennae short, usually 10-segmented (9-segmented in *Pterorthochaetes*), with 3-segmented, asymmetrical, cupuliform club, with segments capable of close apposition and first segment more or less glabrous and shiny; scape short and broad in Liparochrinae, elongate and slender in *Phaeochrous*, and elongate but abruptly angulate in Ceratocanthinae. Mandibles short and broad, laterally expanded and evenly rounded, unidentate, sometimes with additional subapical tooth; mola moderately well developed, usually obliquely transverse; prostheca well developed, with dense fringe of hairs. Maxilla with broad, densely setose galea and reduced lacinia with densely setose inner edge, usually with weak undivided uncus (not in Ceratocanthinae); palp 4-segmented with apical palpomere usually fusiform (slightly expanded apically in some *Liparochrus*). Mentum about as long as wide, usually truncate (emarginate in Ceratocanthinae); ligula deeply divided and densely setose; labial palps 3-segmented with well developed, exposed palpigers; apical palpomere fusiform. Gular sutures well separated; gula usually longer than wide. Corpotentorium slender; anterior tentorial arms expanded but laminatentoria not fused at midline. Cervical sclerites well developed.

Pronotum ~0.5–0.7 times as long as wide, usually widest posteriorly (widest at middle in Ceratocanthinae); base at least as wide as combined elytral bases; sides strongly rounded, more or less obliquely explanate in Ceratocanthinae. Lateral pronotal carinae complete, simple, with or without raised margin or bead; anterior angles at least slightly produced forward and rounded to subacute; posterior angles obtuse to broadly rounded; posterior edge rounded or biemarginate; disc simple. Prosternum in front of coxae distinctly shorter than shortest diameter of procoxal cavity, flattened to biconcave, often with short, broad, truncate to biemarginate chin piece. Prosternal process always complete, but interrupted (with concave section) in Ceratocanthinae, usually abruptly expanded and truncate at apex (expanded preapically and then narrowed, with acute apex in *Phaeochrous*). Notosternal sutures present. Procoxae large, transverse, projecting below prosternum, with moderately long concealed lateral extension; trochantin concealed; endopleuron fused to notum. Procoxal cavities transverse, narrowly separated to contiguous, broadly closed externally by postcoxal hypomeral projections fitting into lateral cavities near apex of prosternal process, internally open.

Scutellar shield well developed, not elevated anteriorly, posteriorly acute. Elytra ~0.95–1.35 times as long as combined width and 1.6–2.7 times as long as pronotum; humeri moderately well developed; sides moderately to strongly rounded; apices broadly, conjointly rounded concealing all abdominal tergites; disc usually with 11 or as many as 18 puncture rows or striae (sometimes fewer; punctation irregular in Ceratocanthinae); longitudinal costae present in some *Liparochrus*; epipleura narrow to moderately broad and more or less complete. Mesoventrite separated by complete sutures from mesanepisterna, which are widely separated at midline; anterior edge of mesoventrite on different plane than metaventrite, without paired procoxal rests but with median prosternal rest or broad collar formed with mesanepisterna; mesoventral cavity absent; discrimen absent. Mesocoxae transverse and slightly to strongly oblique, with concealed trochantins. Mesocoxal cavities usually very narrowly separated (contiguous in *Phaeochrous*), partly closed laterally by mesepimera. Meso-metathoracic joint within coxal cavities absent. Mesometaventral junction with ventrites abutting or overlapping (absent in *Phaeochrous*). Metaventrite strongly transverse and slightly convex, without postcoxal lines; transverse (metakatepisternal) suture absent; discrimen complete; exposed portion of metanepisternum usually moderately long and narrow (short and broad in Ceratocanthinae); metepimeron exposed or concealed. Metacoxae large, strongly transverse, subcontiguous, extending laterally to meet elytra; plates absent. Metendosternite usually with moderately long and narrow stalk, long, transverse lateral arms and short to moderately long anterior process with narrowly separated tendons (deeply emarginate anteriorly with tendons widely separated in Ceratocanthinae). Hind wing with long radial bending zone combined with distinct hinge at end of radial bar; apical field about half as long as total wing length, with 2 or 3 radial extensions; radial cell without base; cross-vein r3 absent; r4 incomplete or absent; R-M loop usually forming a narrowly acute angle and basal portion of RP moderately long (loop absent and basal portion of RP absent in Ceratocanthinae), apical extension of RP extending to wing margin; medial spur not curved posteriorly, extending to wing margin; medial field with 3 or 4 free veins; MP_{3+4} absent, MP_3 and MP_4 not connected basally or MP_4 absent; wedge cell absent; AA_{3+4} neither forked nor joining Cu; anal lobe reduced, without embayment, AP_{3+4} reduced or absent. Legs usually relatively long and slender; trochanterofemoral joint strongly oblique with femur in contract with coxa; profemur usually with sub-basal, more or less circular, densely setose patch on anterior surface (with dense longitudinal fringe of setae in Ceratocanthinae); meso- and metatibiae usually slightly expanded apically, with outer edge somewhat denticulate and lined with longitudinal rows of stout setae and with apical fringe of spines (mesotibiae slightly, evenly expanded and flattened and metatibiae strongly so in Ceratocanthinae); protibia usually distinctly expanded apically, usually with numerous small teeth and 2 or 3 larger teeth along outer edge (uniformly slender, with outer edge simple or with fine teeth and apex with 2 or 3 teeth in Ceratocanthinae); meso- and metatibial spurs well developed, paired and usually asymmetrical; protibia with one large, subapical spur. Tarsi 5-5-5; tarsomeres slender, simple; protarsi usually attached subapically, especially in Ceratocanthinae; pretarsal claws simple; empodium usually well developed and bisetose (absent in Ceratocanthinae).

Abdomen apparently with 6 free ventrites (sternites III–VIII) but sternites III–VII are capable of very limited movement in *Phaeochrous*; ventrite 1 usually about as long as or longer than 2; metacoxal cavities usually weakly defined, without intercoxal process.

Functional spiracles on abdominal segments I–VIII, those on I–VII located in pleural membrane, those on segment VIII located on tergite. Abdominal tergites relatively lightly sclerotised, with the exception of the pygidium (tergite VIII), with patches of wing-folding asperities usually present on tergites III–VI. Anterior edge of sternite VIII in male without median strut. Anterior edge of segment IX with narrowly rounded subgenital plate or short spiculum gastrale, sometimes with posteriorly projecting sternal lobe; tergite IX completely divided into laterotergites; tergum X usually evenly sclerotised. Aedeagus of trilobate type, usually asymmetrical, with phallobase much longer than freely-articulated parameres and penis usually short and weakly sclerotised, but with long basal struts basal; endophallus well developed. Sternite VIII in female without spiculum ventrale. Sternite IX in female divided into two parts; proctiger evenly sclerotised or sometimes membranous. Ovipositor short and broad; paraprocts slightly longer and located lateral to coxites, without baculi; coxites undivided, usually setose, without styli. [Paulian 1977, 1978, 1980; Allsopp 1982; D'Hotman and Scholtz 1990a; Howden and Gill 2000; Jameson 2002d, 2002e; Scholtz and Grebennikov 2005].

Larva

Body subcylindrical and strongly curved ventrally (C-shaped); relatively weakly sclerotised and lightly pigmented; vestiture usually a mixture of long and short, fine setae. Head hypognathous, at least slightly transverse, more or less symmetrical, slightly rounded laterally and slightly flattened; posterior edge not emarginate; epicranial stem moderately long; frontal arms V-shaped or U-shaped, with bases contiguous; median endocarina lying beneath epicranial stem and anteriorly between frontal arms. Stemmata absent. Frontoclypeal suture present in *Phaenognatha* (absent in *Cyphopisthes*); clypeus symmetrical; labium free, slightly transverse or as long as wide, symmetrical or asymmetrical; anterior edge serrate (Fig. 46L) or with several blunt projections (Fig. 46K); tormae not united mesally or absent. Antennae about half as long as head width, 4-segmented, with a distinct antennifer; preapical antennomere with a single, dome-like or broadly conical sensorium. Mandibles large, unidentate; with accessory ventral process; incisor edge with single subapical tooth; molae somewhat reduced, highly asymmetrical, with basal penicillus; prostheca absent. Ventral mouthparts very slightly retracted, with well developed maxillary articulating area; cardines more or less oblique; stipes slightly longer than wide, its dorsal surface with stridulatory teeth; galea and lacinia well separated, relatively sparsely setose, the former with terminal spine, the latter usually more or less subequal in length to galea and with unidentate uncus (much shorter, without uncus in *Cyphopisthes*); palp 3-segmented with well developed palpifer. Labium consisting of prementum, mentum and submentum; ligula absent; palps usually 2-segmented without distinct palpiger, widely separated (1-segmented in *Cyphopisthes*). Hypopharyngeal bracon absent; hypopharyngeal sclerome present. Hypostomal rods short and converging; ventral epicranial ridges absent. Gular region absent.

Thoracic segments more or less equal in length, without distinct tergal plates; prothoracic venter with paired lateral cervical sclerites. Legs moderately long, slender, 5-segmented, with pretarsus somewhat reduced; stridulatory organ in *Phaeochrous* formed by pars stridens on posterior surface of procoxa and plectrum on anterior edge of mesofemur and mesotibiotarsus; coxae moderately widely separated. Abdominal terga without sclerotised plates, anterior terga transversely divided to form 3 plicae, each with transverse rows or patches of short, spine-like setae. Tergum IX usually about as long as VIII, not extending onto ventral surface, without urogomphi; segment X visible from above, more or less cylindrical; anal opening transverse, facing dorsally in *Cyphopisthes*; sternite X sometimes with paired lines of setae forming a weak raster in *Phaeochrous*. Functional spiracles cribriform and ovate to slightly reniform, much larger on thorax; closing apparatus absent. [Gardner 1935b; Ritcher 1966; Costa *et al.* 1988; Grebennikov *et al.* 2002, 2004; Jameson 2002d, 2002e; Scholtz and Grebennikov 2005].

Classification and World Distribution

Hybosoridae is a worldwide family containing ~70 genera and 550 species. The group was divided into five subfamilies by Ocampo (2006a, 2006b): Anaidinae from the Neotropical region, the widely distributed Ceratocanthinae with two of its three tribes Neotropical in distribution, Pachyplectrinae from North America, eastern Europe, North Africa, Asia Minor and Central Asia, the widely distributed Hybosorinae, and Liparochrinae from Australia. A sixth subfamily was added by Nikolajev (1999a), who suggested that the scarabaeid genera *Orubesa* Reitter (Dynamopodinae) and *Thinorycter* Semenov-Tian-Shanskii & Reichardt (Thinorycterini) be added to this family. [Howden and Gill 2000; Howden 2001; Ocampo and Hawks 2006].

Australian Fauna

The Australian hybosorine fauna includes *Phaeochrous* Laporte (Fig. 60N) (**Hybosorinae**), *Liparochrus* Erichson (Fig. 68J) and *Antiochrus* Sharp (**Liparochrinae**), and the genera *Cyphopisthes* Gestro (Fig. 60M) and *Pterorthochaetes* Gestro (**Ceratocanthinae**). Hybosoridae are shining, dark brown to black beetles with sharply curved, acute mandibles, which are visible beyond the strongly transverse labrum, and 10-segmented antennae with a club resembling those in geotrupids. The ceratocanthine subgroup consists of small and highly convex, beetles, with a large head and pronotum, which can be strongly deflexed (as in Clambidae), so that mouthparts, abdomen and parts of the legs are entirely concealed, the body assuming a globular form; the scutellum is very large and the eyes completely divided. Hybosorid larvae are distinguished from those of other scarabaeoids by the 4-segmented antennae (sometimes with the fourth antennomere highly reduced and fused to the third) combined with a trilobed or multidentate labrum; some species have a stridulatory organ formed from the fore legs and mid legs. Phaeochrous and the liparochrines are known to feed on carrion, whereas the ceratocanthine species *Cyphopisthes descarpentriesi* Paulian has been found in the nests of the termite *Mastotermes darwiniensis* Froggatt. [Paulian 1977, 1978, 1980; Kistner 1982; Allsopp 1982, 1984; Iwata *et al.* 1992; Ocampo and Ballerio 2006].

22. SCARABAEIDAE

Adult (Figs 2F, H, 3A–D, F, 61A–F, I, 68F–H, K–L)

Length ~1.8–60 (usually 3–50) mm. Body ~1.3–2.4 times as long as wide, sides of prothorax and elytra subparallel or slightly to strongly, independently rounded, usually with distinct constriction or slight change in curvature between prothorax and elytra; slightly to strongly convex, not capable of conglobation; colour varied but usually yellowish-brown to black and rarely bicolored, but brightly coloured, bicolored or with metallic lustre in some Melolonthinae and Rutelinae and many Cetoniinae; dorsal surfaces usually shiny and subglabrous or sparsely clothed with short to long fine hairs; surface relatively dull in some Aphodiinae (e.g. *Odontolochus*) and Cetoniinae (e.g. *Protaetia*), clothed with scale-like setae in some Aphodiinae (e.g. *Euparotrix*, *Australoxenella*), Melolonthinae (e.g. *Lepidiota*), and Cetoniinae (*Microvalgus*), and densely clothed with decumbent hairs in some Melolonthinae; ventral surfaces and legs often more densely clothed with long hairs.

Head usually prognathous or slightly declined, but moderately to strongly declined in some Aphodiinae; usually not abruptly constricted posteriorly, or neck constriction, if present (e.g. some Scarabaeinae) concealed by pronotum; transverse occipital ridge present in some Aphodiinae and Scarabaeinae. Eyes usually well developed, strongly protuberant or not, usually deeply, narrowly emarginate, completely divided into dorsal and ventral parts in some Scarabaeinae, shallowly emarginate in most Aphodiinae; finely facetted, without interfacetal setae; ommatidium of eucone type. Antennal insertions almost always concealed from above by frontal carina, which often partly bisects eyes (insertions exposed in many Cetoniinae); subantennal grooves present or absent. Vertex or frontoclypeal region in males of *Phaenognatha* and some Scarabaeinae and Dynastinae modified to form elevated ridges or horns. Frontoclypeal suture usually well developed in Melolonthinae, Rutelinae and Dynastinae, often absent in Aphodiinae and Scarabaeinae and absent in *Phaenognatha* and Cetoniinae; anterior portion of clypeus usually more or less abruptly declined to form a sharply carinate transverse ridge, sometimes elevated, often extending laterally onto eye surface as an ocular canthus and forming one or more teeth or horns in males of *Phaenognatha* and some Scarabaeinae and Dynastinae. Labrum usually free (but often closely adpressed to clypeus and not freely movable), well sclerotised and at least partly exposed in *Saprus*, *Phaenognatha* and some other Melolonthinae (e.g. *Diphucephala*, *Phyllotocus*), Rutelinae (e.g. *Parastasia*) and Dynastinae (e.g. *Neodon*), sclerotised but concealed from above by clypeus (many Melolonthinae, Rutelinae and Dynastinae), or membranous and concealed (all Scarabaeinae, almost all Aphodiinae and Cetoniinae). Antennae shorter than head width, 10-segmented in most Rutelinae, Dynastinae and Cetoniinae and some Melononthinae, 9-segmented in *Phaenognatha*, almost all Aphodiinae and Scarabaeinae, and some Melolonthinae, Rutelinae and Dynastinae, 8-segmented in a few Aphodiinae (e.g. *Phycochus*), Scarabaeinae (e.g. *Sisyphus*), and Dynastinae (e.g. some *Cryptodus*) and several Melolonthinae (e.g. *Automolius*, *Heteronyx*, *Sericesthis*) and 7-segmented in a few Melolonthinae (e.g. *Nepytis*); club highly asymmetrical and lamellate, with segments capable of close apposition, usually 3-segmented, but with 5 or 6 segments in some Melolonthinae (e.g. *Telura*, *Rhopaea*, *Xylonichus*) and as many as 7 or 8 segments in a few melolonthines (e.g. *Allothnonius*, *Microrhopaea*, *Zietzia*); scape usually much longer than pedicel. Mandibles short and broad to moderately elongate, often broad at base, flattened apically and deeply inserted within head capsule, so that they are not visible in lateral view, usually unidentate and often more or less rounded apically; mola and prostheca usually well developed, but sometimes highly reduced or absent (e.g. *Phaenognatha*). Maxilla usually with well developed, broad galea and reduced, slender lacinia (both reduced in *Phaenognatha*); galea and lacinia often weakly sclerotised and densely setose, but usually heavily sclerotised and one or several teeth or spines in Melolonthinae, Rutelinae and Dynastinae; palp 4-segmented with apical palpomere cylindrical or fusiform. Mentum usually about as long as wide or slightly transverse, sometimes apically emarginate; prementum partly concealed by mentum; ligula usually emarginate or bilobed, lightly sclerotised, and densely setose; labial palp 3-segmented; apical palpomere cylidrical or fusiform. Gular sutures well separated; gula usually longer than wide and convex. Corpotentorium slender and arched. Cervical sclerites well developed.

Pronotum ~0.3–0.9 times as long as wide, usually widest posteriorly or at middle; base usually as wide as combined elytral bases (slightly wider in some Scarabaeinae, narrower in most Cetoniinae); sides slightly to strongly rounded. Lateral pronotal carinae almost always complete (incomplete or absent in a few Scarabaeinae), almost always simple (crenulate in *Rhopaea*), with or without raised margin or bead; anterior angles not to distinctly produced forward and rounded to subacute; posterior angles usually right to obtuse or broadly rounded (subacute in most Cetoniinae); posterior edge usually straight, rounded or biemarginate (distinctly triemarginate in most Cetoniinae, with distinct mesal lobe in *Lompatera*); disc usually simple (with several longitudinal costae in *Rhyparus*, several longitudinal grooves in *Australoxenella*); males of some Scarabaeinae and Dynastinae (Fig. 61B–D) with anterior excavations or horns. Prosternum in front of coxae distinctly shorter than shortest diameter of procoxal cavity, biconcave or mesally tumid, occa-

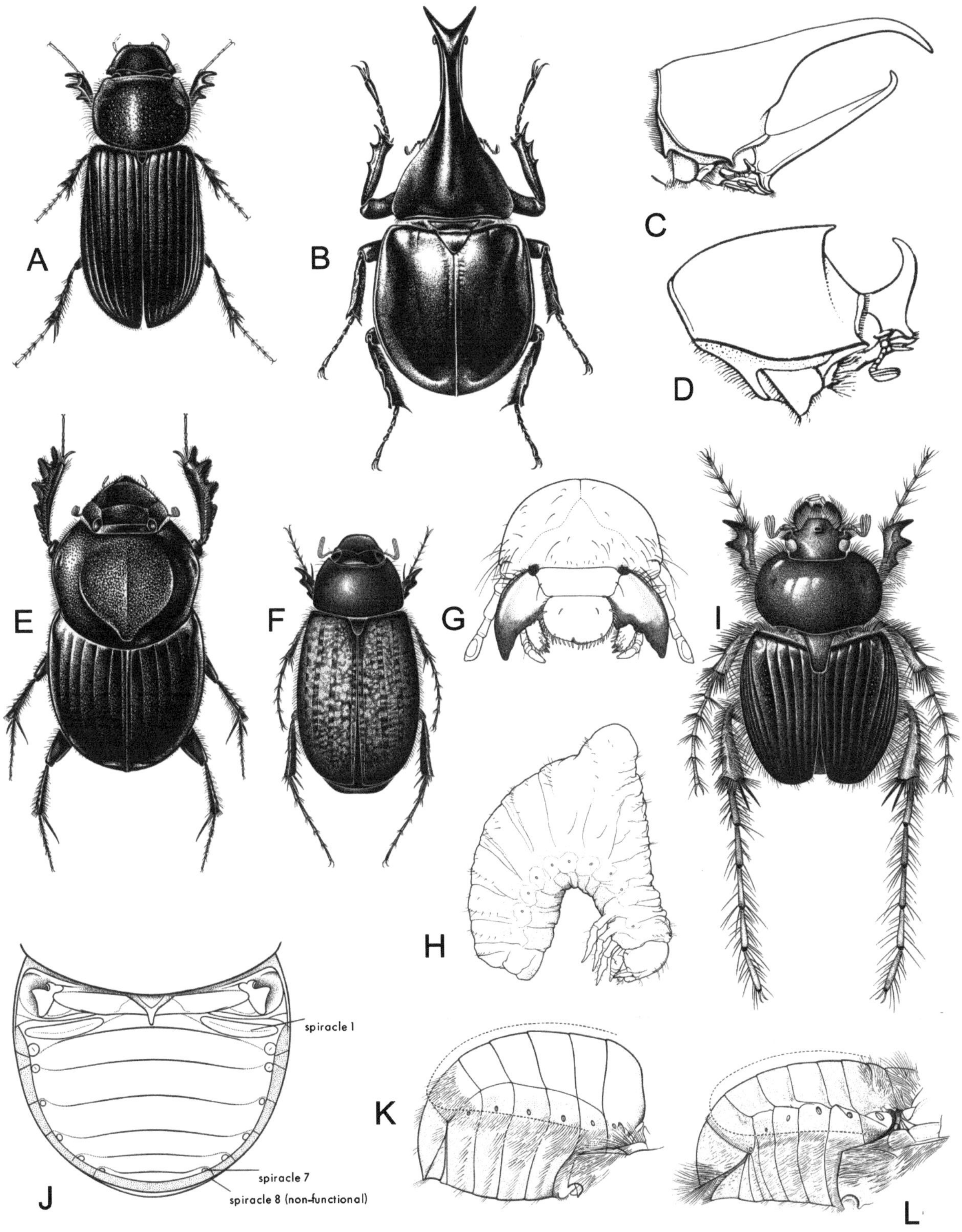

Fig. 61. A–F, Adult: A, *Acrossidius tasmaniae* (Hope) (Scarabaeidae: Aphodiinae), dorsal; B, *Xylotrupes ulysses australicus* (Thomson) (Scarabaeidae: Dynastinae), dorsal; C–D, same, larger and smaller males, head and prothorax, lateral; E, *Onthophagus declivis* Harold (Scarabaeidae: Scarabaeinae), dorsal; F, *Sericesthis geminata* Boisduval (Scarabaeidae: Melolonthinae), dorsal. G–H, Larva: G, *Anoplognathus montanus* Macleay (Scarabaeidae: Rutelinae), head, anterodorsal; H, *Onthophagus gazella* ((Fabricius), lateral. I, *Phaenognatha* sp. (Scarabaeidae: Melolonthinae), adult, dorsal. J–K, Scarabaeidae abdomen with elytra removed: J, *Onthophagus declivis* Harold (Scarabaeinae), dorsal; K, *Colpochila* sp. (Melolonthinae), lateral; L, *Anoplognathus* sp. (Rutelinae), lateral.

sionally with mesal projection; prosternum and hypomera combined forming cavities for antennae and fore legs in some Aphodiinae (e.g. *Gongrolophus*, *Riparus*) and Scarabaeinae (e.g. *Coptodactyla*, *Lepanus*). Prosternal process always complete, but sometimes interrupted at middle (with concave section), abruptly expanded and truncate at apex. Notosternal sutures almost always complete. Procoxae large, transverse, projecting below prosternum, with short concealed lateral extensions; trochantin concealed; endopleuron fused to notum. Procoxal cavities transverse, narrowly separated to contiguous, closed externally by postcoxal hypomeral projections fitting into lateral cavities near apex of prosternal process, internally open.

Scutellar shield usually well developed, not to abruptly elevated anteriorly, narrowly acute to broadly rounded posteriorly; shield absent (scutellum completely concealed) in some Aphodiinae (e.g. *Phycochus*, *Australoxenella*), almost all Scarabaeinae (shield exposed in *Euoniticellus*) and some Cetoniinae (e.g. *Lomaptera*). Elytra ~0.65–1.75 times as long as combined width and 0.96–2.8 times as long as pronotum; humeri moderately well developed; mesepimeron visible from above between pronotum and elytral humeri in almost all Cetoniinae (not *Microvalgus*); sides of elytra subparallel to strongly rounded; apices usually broadly, conjointly rounded (somewhat truncate or slightly acute in Cetoniinae, independently rounded in *Phaenognatha*), concealing all abdominal tergites (although vertically oriented pygidium may be slightly visible from above); disc often with irregular punctation, poorly defined puncture rows or with 8 or fewer distinct striae, but with 10 distinct puncture rows and no scutellary striole in most Aphodiinae and with 9 or 10 puncture rows or striae in *Phaenognatha*, some Scarabaeinae, Melolonthinae, Rutelinae and Dynastinae; longitudinal costae present in some Aphodiinae and Scarabaeinae; epipleura usually narrow and incomplete. Mesoventrite separated by complete sutures from mesanepisterna, which are widely separated at midline; anterior edge of mesoventrite on same plane as metaventrite (on different plane in Phaenognatha, some Melolonthinae and Rutelinae and all Cetoninae), without paired procoxal rests but with median prosternal rest or broad collar formed with mesanepisterna; mesoventral cavity absent; discrimen absent. Mesocoxae transverse and slightly to strongly oblique (longitudinally oriented in most Scarabaeinae), with trochantins exposed or concealed. Mesocoxal cavities usually subcontiguous to narrowly separated (broadly separated in Scarabaeinae, Cetoniinae and some Aphodiinae), usually partly closed laterally by mesepimera, but occasionally by both mesanepisterna and mesepimera. Mesometathoracic joint within coxal cavities absent. Mesometaventral junction with ventrites abutting and usually fused together (sometimes without suture between them). Metaventrite slightly to strongly transverse and slightly convex, without postcoxal lines; anterior edge strongly produced forward so that metaventral process conceals mesoventrite from below in most Cetoniinae and some Rutelinae; transverse (metakatepisternal) suture absent; discrimen complete to base of metaventral process; exposed portion of metanepisternum usually short to moderately long and subquadrate or slightly narrowed posteriorly (long and slender in most Aphodiinae, broader anteriorly but strongly narrowed posteriorly in Scarabaeinae); metepimeron usually exposed. Metacoxae large, strongly transverse, and slightly to moderately oblique, often as wide as or wider laterally than mesally, subcontiguous, extending laterally to meet elytra; plates absent. Metendosternite usually with long, narrow stalk, long, transverse lateral arms and moderately to very long anterior process with narrowly separated tendons. Hind wing with long radial bending zone combined with distinct hinge at end of radial bar; apical field about half as long as total wing length, with sclerotised anterior edge extending almost to apex, with 2 or 3 radial extensions; radial cell without base or absent; cross-vein r3 absent; r4 incomplete or absent; R-M loop forming a narrowly acute angle and basal portion of RP usually moderately long, apical extension of RP extending to wing margin; medial spur curved posteriorly and extending to wing margin; medial field with 2 or 3 free veins; MP represented by single detached vein; wedge cell absent; AA_{3+4} forked and joining Cu or not; anal lobe well developed, without or with shallow embayment, AP_{3+4} present. Legs relatively long and slender to moderately stout; trochanterofemoral joint strongly oblique with femur in contract with coxa; profemur with sub-basal, more or less circular, densely setose patch on anterior surface in Scarabaeinae and a similar but less well defined patch in at least some Aphodiinae; meso- and metatibiae usually slightly to strongly expanded apically, with a variety of modifications, such as stout, socketed spines, sometimes forming clusters or oblique or transverse rows, often with apical fringe of spines or with 2 or 3 apical teeth in Cetoniinae; protibia (Fig. 30H) usually distinctly flattened and expanded apically, with 3 large teeth along the outer edge (teeth occasionally reduced or absent); all legs apically widened and without large teeth in *Australoxenella*); meso- and metatibial spurs usually well developed and paired but often of different lengths (Fig. 16L), sometimes reduced to one spur, at least on metatibia; protibia with one large, usually subapical spur. Tarsi almost always 5-5-5 (4-4-4 in *Australoxenella*, 3-3-3 in *Phycochus*); tarsomeres slender, simple; pretarsal claws usually simple (toothed in some Scarabaeinae, Melolonthinae and Rutelinae; absent in *Phycochus* and some Scarabaeinae); empodium usually well developed and bisetose.

Abdomen (Fig. 61J–L) with 6 ventrites (sternites III–VIII); sternite II well developed or represented by lateral sclerotisations only, but always concealed by large metacoxae; ventrites 1–5 almost always connate (apparently free in *Phaenognatha*) and ventrite 6 (sternite VIII) freely movable; abdomen shorter than metaventrite in *Phaenognatha* and Scarabaeinae; ventrite 1 about as long as or shorter than 2; metacoxal cavities weakly to strongly defined and intercoxal process well developed and acute or sometimes reduced or absent. Sternites III–VII usually without lateral carinae delimiting laterosternites (present in Aphodiinae and Scarabaeinae). Abdominal tergites I–VI usually lightly sclerotised (more heavily sclerotised in Cetoniinae), while tergite VIII (pygidium) and sometimes tergite VII more heavily sclerotised; tergite and sternite VII in Melolonthinae, Rutelinae, Dynastinae and Cetoniinae fused together without suture between them. Functional spiracles on abdominal segments I–VII in Aphodiinae, Scarabaeinae and *Phaenognatha* located in pleural membrane, while those in other groups located on sternites and the last 1 to 3 visible below elytra in intact specimens. Anterior edge of sternite VIII in male without median strut. Anterior edge of segment IX with narrowly rounded subgenital plate often produced to form a broad to narrow spiculum gastrale; tergite IX completely divided into laterotergites; tergum X more or less membranous. Aedeagus of modified trilobate type, usually symmetrical, with phallobase usually longer than articulated parameres, which may be fused together at base; penis membranous, usually with long basal struts basal; endophallus well developed and often complex. Sternite VIII in female without spiculum ventrale. Sternite IX in female divided into two parts; proctiger usually membranous. Ovipositor short and broad; usually without baculi; coxites undivided, without styli. [D'Hotman and Scholtz 1990b; Browne and Scholtz 1998; Scholtz and Grebennikov 2005].

Larva (Figs 33F, 42L, 61H)

Body usually subcylindrical (wider near middle with dorsal hump-like protrusion in Scarabaeinae) and strongly curved ventrally (C-shaped); relatively weakly sclerotised and lightly pigmented, except for head capsule, which may be yellow to dark brown in colour; vestiture of long and/or short, fine setae, with short, stout, spine-like setae dorsally in most groups (not in Aphodiinae or Scarabaeinae). Head (Fig. 61G) hypognathous, slightly transverse, more or less symmetrical, slightly to strongly rounded laterally and slightly flattened; posterior edge not emarginate; epicranial stem moderately long; frontal arms V-shaped or U-shaped, with bases contiguous; median endocarina usually lying beneath epicranial stem and often extending anteriorly between frontal arms. Stemmata usually absent (single stemma on each side in most Dynastinae, some Cetoniinae and a few Melolonthinae). Fronto-clypeal suture present; labium free, slightly transverse or as long as wide, symmetrical or asymmetrical; anterior edge usually broadly rounded, sometimes weakly trilobed; tormae united mesally in Aphodiinae and Scarabaeinae, not in other groups; epipharynx complex and variable. Antennae moderately long, often more than half as long as head width, usually 4-segmented (5-segmented in some Aphodiinae), with a distinct antennifer; preapical antennomere with a single, dome-like or broadly conical sensorium in Aphodiinae and Scarabaeinae, without sensorium in other subfamilies; apical antennomere with one or more flat sensory areas in Aphodiinae, Melolonthinae, Rutelinae (Fig. 61G), Dynastinae and Cetoniinae. Mandibles large, unidentate or bidentate; with accessory ventral process; incisor edge with or without one or more subapical teeth; molae-developed, more or less asymmetrical, with or without basal penicillus; ventral surface lateral to mola with large, ovate stridulatory area (pars stridens) Rutelinae Dynastinae and Cetoniinae; prostheca absent. Ventral mouthparts not or very slightly retracted, with well developed maxillary articulating area; cardines more or less oblique; stipes usually slightly longer than wide, its dorsal surface sometimes with row or patch of stridulatory teeth; galea and lacinia well separated from one another in Aphodiinae and Scarabaeinae, but either fitting very tightly together or partly or completely fused in Melolonthinae, Rutelinae, Dynastinae and Cetoniinae; galea and lacinia both clothed with setae and/or spines, plus one or more terminal stout spines or unci; palp 3-segmented with well developed, segment-like palpifer. Labium consisting of prementum, mentum and submentum; ligula short and undivided or absent; palps usually 2-segmented (1-segmented in some Aphodiinae) without distinct palpiger, widely separated; submentum in *Cephalodesmius* with stridulatory pars stridens. Hypopharyngeal bracon absent; hypopharynx with complex, asymmetrical armature. Hypostomal rods usually moderately long and at least slightly diverging (short and converging in Scarabaeinae); ventral epicranial ridges absent. Gular region absent.

Thoracic segments more or less equal in length, usually without distinct tergal plates (occasionally with pair of small plates). Prothoracic venter without distinct presternal sclerite, but usually with paired lateral cervical sclerites. Legs moderately long, slender, usually 4-segmented (2-segmented in Scarabaeinae), usually clothed with long spine-like setae; with usually pretarsus well developed, claw-like, and bisetose (absent in Scarabaeinae, not claw-like and multisetose in Cetoniinae); coxae moderately widely separated; stridulatory organs absent on legs. Abdominal terga without sclerotised plates, first 6 or 7 terga (as well as meso and metathoracic terga) usually transversely divided to form 3 plicae, which are usually clothed with transverse patches of short, spine-like setae (terga not subdivided in Scarabaeinae; not spinose in Aphodiinae or Scarabaeinae). Segments IX and X (or sometimes VII to X) larger than anterior segments; IX subterminal, not extending onto ventral and without urogomphi; X terminal and visible from above, usually more or less cylindrical (somewhat flattened in Scarabaeinae); anal opening more or less transverse in Aphodiinae, Scarabaeinae, Rutelinae, Dynastinae and Cetoniinae, usually somewhat angulate or Y-shaped in Melolonthinae; sternum X with one or two anal pads in Aphodiinae or several pads in Scarabaeinae and often with a raster consisting of patches or rows of short, stout spines or palllidia (sternite X in *Cephalodesmius* with stridulatory plectrum which rubs against pars stridens on submentum). Functional spiracles cribriform and more or less reniform, with spiracular scar usually located posteriorly on prothorax and anteriorly or anteroventrally on abdomen (all ventrally in Scarabaeinae); closing apparatus absent. [Böving and Craighead 1931; Gardner 1935b; Jerath 1960; Ritcher 1966; McQuillan 1985; Kim and Lumaret 1988; Costa *et al.* 1988; Carlson 1991; Ratcliffe *et al.* 2002; Scholtz and Grebennikov 2005; Mico *et al.* 2008].

Classification and World Distribution

Scarabaeaidae is a very large, diverse and cosmopolitan family containing ~1600 genera and 27 000 species. As here understood, the family contains 12 subfamilies, six of which occur in Australia. Exotic subfamilies include Orphninae (Neotropical, southern Palaearctic and Afrotropical regions) (Schmidt 1913; Vulcano *et al.* 1966; Paulian 1976, 1984; Paulian and Lumaret 1982; Morón 1991), Allidiostomatinae (South America) (Schmidt 1913; Ruiz 1924), Aclopinae (South America) (Ocampo and Mondaca 2012), Phaenomeridinae (Afrotropical) (Ohaus 1933; D'Hotman and Scholtz 1990b), Pachypodinae (southern Europe, North Africa) (Crovetti 1969) and Euchirinae (Asia Minor, Central Asia, Oriental region) (Ohaus 1933; Marquet 1992; Young 1989). Smith (2006) also recognises Chironinae, Aegialiinae, Eremazinae, Aulonocneminae and Termitotroginae (included here under Aphodiinae), and considers Pachypodinae and Euchirinae as tribes of Melolonthinae. An additional subfamily Dynamopodinae has been mentioned under Hybosoridae.

Australian Fauna

The Australian scarab fauna includes members of six subfamilies. Scarabaeids are usually moderately to large in size, occasionally brightly coloured (few Melolonthinae, some Rutelinae, many Cetoniinae) or rarely small and clothed with scales (*Microvalgus*). With the exception of *Phaenognatha*, *Saprus* and certain Rutelinae, either the labrum or the mandibles are concealed from above by the clypeus, and in most cases both are concealed. Except in *Saprus*, exposed mandibles have non-opposable apices. The antennal club is variable, but not cupuliform. Scarab larvae (Fig. 33F) always have four or apparently five antennomeres, and the legs are never modified as stridulatory organs; they always live in concealed habitats, feeding on roots, dung or decaying vegetable matter. [McQuillan 1985].

Members of the **Aphodiinae** tend to be much smaller and more elongate than most scarabs, and with the exception of the genus *Saprus*, they have a membranous, concealed labrum. Of the 28 genera known from Australia, the largest are *Podotenus*, *Ataenius* and *Saprosites*. Most aphodiines are dung feeders, and some may be found in mammal nests ands burrows. *Acrossidius tasmaniae* (Hope) (Fig. 61A) is a phytophagous species, which is regarded as an important pasture pest; the larvae form vertical burrows from which they emerge at night to cut off leaves and drag them into their tunnels for consumption (Carne 1957a). Aphodiine larvae have a sclerotised head capsule and 5-segmented legs with a claw-like pretarsus. [Dellacasa 1988a, 1988b, 1989; Howden and Storey 1992, 2000; Stebnicka and Howden 1995, 1996; Storey and Howden 1996; Rakovič 1997; Stebnicka 1998, 2001, 2009].

The **Scarabaeinae** (dung beetles, Figs 1A–B, 61E) is a large subfamily, the adults of which, like those of most Aphodiinae, have modified, liquid-feeding mouthparts with membranous, complexly setose labrum-epipharynx and partly membranous, apically lamellate mandibles; they differ from aphodiines, however, in being more robust and rounded in form, with widely separated mesocoxae, an exposed pygidium and usually conspicuous sexual dimorphism. Their larvae have a distinctive, humped appearance and reduced legs. Scarabaaeine adults usually feed on the faeces of various animals and use this material as provisions for their larvae, which live in chambers or burrows in the ground, but details of feeding behaviour and nest construction are known for few Australian species (Bornemissza 1971a). Mycophagy is known in some *Onthophagus* species, including *O. dunningi* Harold (Bornemissza 1971b), and species of *Cephalodesmius* have been shown to form brood balls from dead leaves and other plant material, rather than dung (Monteith and Storey 1981). Although many scarabaeines build their brood chambers directly beneath the food source, some (ball-rollers) form a ball of dung, push this away from the dung mass, and form their brood chambers at some distance from the food source. The native Australian fauna comprises ~20 genera placed in three of the six recognised tribes: Canthonini (e.g. *Amphistomus*, *Cephalodesmius*, *Lepanus*, *Temnoplectron*), Coprini (e.g. *Coptodactyla*, *Demarziella*) and Onthophagini (*Onthophagus*), but species of *Copris* (Coprini), *Sisyphus* (Sisyphini), *Onitis* (Onitini) and *Euoniticellus* (Oniticellini) have been introduced from South Africa and the Mediterranean Region for the biological control of cattle dung. *Onthophagus* is the largest genus with almost 200 species, and a few are remarkable in that the adults have prehensile claws and cling to the hair around the anus of wallabies. [Matthews 1972, 1974, 1976; Halffter and Edmonds 1982; Matthews and Stebnicka 1986; Doube *et al.* 1991; Howden and Howden 1991; Moczek 1998; Moczek and Emlen 1999; Storey and Monteith 2000; Davis *et al.* 2002; Bell *et al.* 2004; Philips *et al.* 2004; Weir and Monteith 2010].

Adult **Melolonthinae** are usually called chafers and are often reddish-brown in colour, but may be black, or sometimes metallic or bicoloured. Members of the unusual tribe Phaenognathini, usually included in the subfamily Aclopinae (Ocampo and Mondaca 2012), differ from other melolonthines in having exposed, rounded and non-opposable mandibles. The group is represented in Australia by a few species of *Phaenognatha* (Fig. 61I) in northern QLD and the NT; adults are usually taken at light and nothing is known of the biology (Allsopp 1981, 1983). Melolonthine larvae, commonly known as white grubs, feed on roots and other organic matter and may cause serious damage to vegetables, wheat, sugarcane and pastures. Pupation occurs in a cell at some depth in the soil, and the adult after eclosion, remains in the cell until rain softens the soil. This synchronises adult emergence and flight. Most adults are crepuscular or nocturnal and many, including *Sericesthis geminata* Boisduval (Fig. 61F) and species of *Heteronyx*, *Automolus* and *Liparetrus*, feed on leaves and may be serious defoliators. Common diurnal forms include the metallic species of *Diphucephala* (Fig. 68K) and bicoloured species of *Phyllotocus*, which may swarm around flowering trees and shrubs. Some *Maechidius* occur in termite nests. The profusion of species of Melolonthinae may be explained by temporal isolation resulting from local rainfall in semidesert areas. Special features which might lead to the isolation of limited populations in time are: (1) the short adult life; (2) the fact that adults remain in pupal cells until released by rain; (3) the duration of the life cycle, which is often more than one year; (4) the fact that adults, though winged, fly only short distances; and (5) the fact that some species do not feed as adults. [Allsopp 1990a, 1990b; Britton 1957, 1978, 1980, 1986, 1987, 1988, 1990, 1995, 2000].

The **Rutelinae** (Figs 2F, 3C–D, 68H) are of stout build and are usually brightly coloured and metallic. Larvae are similar to those of melolonthines and also occur in the soil; those of some species of *Anoplognathus* cause damage to lawns and pastures, and adults of the same genus, commonly called Christmas beetles, often defoliate eucalypts. [Carne 1957b, 1958].

Dynastinae (Figs 3F, 61B–D) are stout bodied, mainly black or brown beetles, often exhibiting obvious sexual dimorphism, the males bearing horns or tubercles which are disproportionately large in larger specimens (allometry). Larvae feed on roots or decaying vegetable matter in soil or rotten logs, and the nocturnal adults do not feed on leaves. The subfamily includes several well known soil pests, including *Heteronychus arator* (Fabricius), introduced from South Africa. Species of the largest genus, *Cryptodus*, live in association with ants and termites. The subfamily is most abundant and diverse in the north and west of the continent, and includes some of our largest beetles, such as *Haploscapanes barbarossa* (Fabricius). [Carne 1957c; Endrödi 1985].

The subfamily **Cetoniinae** (Figs 2H, 68F–G, 68L) includes several somewhat flattened beetles, which have exposed mesepimera and pygidium and usually a strikingly patterned, metallic or 'enamelled' colouring. Adults are diurnal nectar feeders, and can fly rapidly by spreading their wings without the necessity of raising the elytra. Larvae occur in rotten wood or decaying vegetation and normally are able to move on their backs, even on the soil surface, by using their tergal folds, which are densely clothed with strong bristles. *Microvalgus* species differ from most members of the group in being less than 5 mm in length and clothed with scale-like setae; little is known of their habits but they may be associated with termites. Some of the common and well known cetoniines include *Eupoecila australasiae* Donovan (Fig. 2H) and species of

Diaphonia, *Chlorobapta* and *Lomaptera*. [Krikken 1984; Hiller 1990; Allard 1995; Antoine 2003; Hasenpusch and Moseneder 2010].

Australian Subfamilies, Tribes and Genera

Aphodiinae: Aegialiini: *Saprus* Blackburn. **Aphodiini:** *Acrossidius* Schmidt, *Aphodius* Illiger, *Candezeollus* Stebnicka & Howden, *Ozodius* Skelley, *Lorditomaeus* Peringuey, *Podotenus* Schmidt. **Eupariini:** *Airapus* Stebnicka & Howden, *Ataenius* Harold, *Australammoecius* Petrovitz, *Cnematoplatys* Schmidt, *Euparotrix* Stebnicka & Howden, *Parataenius* Balthasar, *Saprosites* Redtenbacher. **Odontolochini:** *Gongrolophus* Stebnicka & Howden, *Odontolochus* Schmidt. **Proctophanini:** *Australaphodius* Balthasar, *Harmogaster* Harold, *Proctophanes* Harold. **Psammodiini:** *Aphodopsammobius* Endrödi, *Leiopsammodius* Rakovic, *Neotrichiorhyssemus* Rakovic & Kral, *Phycochus* Broun, *Platytomus* Mulsant, *Psammorpha* Stebnicka, *Rhyssemus* Mulsant, *Tesarius* Rakovic. **Rhyparini:** *Rhyparus* Westwood. **Stereomerini:** *Australoxenella* Howden & Storey, *Daintreeola* Howden & Storey.

Scarabaeinae: Canthonini: *Amphistomus* Lansberge, *Aptenocanthon* Matthews, *Aulacopris* White, *Boletoscapter* Matthews, *Canthonosoma* W.J. Macleay, *Cephalodesmius* Westwood, *Coproecus* Reiche, *Diorygopyx* Matthews, *Labroma* Sharp, *Lepanus* Balthasar, *Mentophilus* Laporte, *Monoplistes* Lansberge, *Pseudignambia* Paulian & Pluot-Sigwalt, *Sauvagesinella* Paulian, *Temnoplectron* Westwood, *Tesserodon* Hope. **Coprini:** *Copris* Geoffroy, *Coptodactyla* Burmeister, *Demarziella* Balthasar, *Thyregis* Blackburn. **Oniticellini:** *Euoniticellus* Janssens, *Liatongus* Reitter. **Onitini:** *Bubas* Mulsant, *Onitis* Fabricius. **Onthophagini:** *Onthophagus* Latreille. **Sisyphini:** *Sisyphus* Latreille.

Melolonthinae: Automoliini: *Aplopsis* Blanchard, *Automolius* Britton, *Bryantella* Britton, *Maechidinus* Lea, *Phyllochlaenia* Blanchard. **Colymbomorphini:** *Chariochilus* Brenske, *Colymbomorpha* Blanchard, *Exostethus* Britton, *Nitorellus* Britton, *Phyllococerus* Waterhouse, *Xylonichus* Boisduval. **Comophorinini:** *Comophorina* Strand. **Diphucephalini:** *Cunderdinia* Lea, *Diphucephala* Dejean, *Watkinsia* Britton. **Heteronycini:** *Acheilo* Britton, *Callabonica* Blackburn, *Cubidens* Britton, *Eurychelus* Blanchard, *Heteronyx* Guérin-Méneville, *Neoheteronyx* Blackburn, *Nepytis* Erichson, *Odontotonyx* W. J. Macleay, *Proborhinus* Britton, *Pseudoheteronyx* Blackburn, *Webbella* Britton. **Liparetrini:** *Allara* Britton, *Anacanthodes* Britton, *Aneucomides* Blackburn, *Aphanesia* Britton, *Astibicola* Britton, *Biphyllocera* White, *Cheilo* Britton, *Colobostoma* Blanchard, *Colpochila* Erichson, *Dikellites* Britton, *Dysphanochila* Blackburn, *Engyopsina* Britton, *Frenchella* Blackburn, *Glossocheilifer* Blackburn, *Hadropechys* Britton, *Ictigaster* Britton, *Leonotus* Britton, *Liparetrus* Guérin-Méneville, *Macleayella* Britton, *Neso* Blackburn, *Nosphisthis* Blackburn, *Pachygastra* Germar, *Parasciton* Britton, *Paronyx* Britton, *Petinopus* Blackburn, *Scitaloides* Britton, *Sciton* Blackburn, *Stenochelyne* Britton, *Teluroides* Britton, *Xyridea* Britton. **Macrodactylini:** *Plectris* Saint-Fargeau & Serville. **Maechidiini:** *Epholcis* Waterhouse, *Harpechys* Britton, *Maechidius* W.S. Macleay, *Microcoenus* Britton, *Microthopus* Burmeister, *Termitophilus* Britton. **Melolonthini:** *Allothnonius* Britton, *Antitrogus* Burmeister, *Dermolepida* Arrow, *Holorhopaea* Britton, *Hypolepida* Britton, *Lepidiota* Kirby, *Megacoryne* Britton, *Megarhopaea* Britton, *Metatrogus* Britton, *Microrhopaea* Lea, *Nanorhopaea* Britton, *Othnonius* Olliff, *Pararhopaea* Blackburn, *Pseudholophylla* Blackburn, *Rhopaea* Erichson, *Zietzia* Blackburn. **Pachytrichini:** *Pachytricha* Hope. **Phaenognathini:** *Phaenognatha* Hope. **Phyllotocidiini:** *Phyllotocidium* Blackburn. **Scitalini:** *Byrrhomorpha* Blackburn, *Colpochelyne* Britton, *Colpochilodes* Blackburn, *Gnaphalopoda* Reiche, *Hadrops* Britton, *Homolotropus* W. J. Macleay, *Idanastes* Britton, *Ophropyx* Britton, *Phorine* Britton, *Protelura* Brittonm *Scitala* Erichson, *Sericesthis* Boisduval, *Synchilus* Britton, *Telura* Erichson, *Xyrine* Britton, *Xyroa* Britton, *Xyrodes* Britton. **Sericini:** *Adossa* Britton, *Ancylonyx* Britton, *Anthotocus* Britton, *Cheiragra* W. J. Macleay, *Cheirodontus* Britton, *Cheirora* Britton, *Neophyllotocus* Blackburn, *Opsitocus* Britton, *Phyllotocus* Fischer, *Sphaeroscelis* Burmeister. **Systellopini:** *Chilodiplus* Sharp, *Enamillus* Sharp, *Prochelyna* Erichson, *Sarothromerus* Blackburn, *Sphyrocallus* Sharp, *Systellopus* Sharp.

Rutelinae: Adoretini: *Adoretus* Laporte. **Anomalini:** *Anomala* Samouelle. **Anoplognathini:** *Amblochilus* Blanchard, *Amblyterus* W.S. Macleay, *Anoplognathus* Leach, *Anoplostethus* Brullé, *Bilobatus* Machatschke, *Calloodes* White, *Clilopocha* Lea, *Dungoorus* Carne, *Eosaulostomus* Carne, *Epichrysus* White, *Exochogenys* Carne, *Mesystoechus* Waterhouse, *Mimadoretus* Arrow, *Paraschizognathus* Ohaus, *Pseudoschizognathus* Ohaus, *Repsimus* W.S. Macleay, *Saulostomus* Waterhouse, *Schizognathus* Fischer, *Trioplognathus* Ohaus, *Wambo* Allsopp. **Rutelini:** *Parastasia* Westwood.

Dynastinae: Cyclocephalini: *Cyclocephala* Latreille. **Dynastini:** *Haploscapanes* Arrow, *Xylotrupes* Hope. **Oryctini:** *Oryctes* Illiger. **Oryctoderini:** *Oryctoderinus* Endrödi, *Oryctoderus* Boisduval. **Pentodontini:** *Adoryphorus* Blackburn, *Anomalomorpha* Arrow, *Cavonus* Sharp, *Cheiroplatys* Hope, *Corynophyllus* Hope, *Cryptoryctes* Carne, *Dasygnathus* W. S. Macleay, *Dipelicus* Hope, *Epironastes* Carne, *Heteronychus* Burmeister, *Hyphoryctes* Blackburn, *Metanastes* Arrow, *Neocnecus* Sharp, *Neocorynophyllus* Carne, *Neodasygnathus* Carne, *Neodon* Carne, *Neonastes* Carne, *Nephrodopus* Sharp, *Novapus* Sharp, *Orthocavonus* Carne, *Papuana* Arrow, *Pentodon* Hope, *Phylliocephala* Blackburn, *Phyllognathus* Eschscholtz, *Pimelopus* Erichson, *Pseudocavonus* Blackburn, *Pseudoryctes* Sharp, *Semanopterus* Hope, *Teinogenys* Sharp, *Temnorrhynchus* Hope, *Trissodon* Burmeister, *Xynedria* Blackburn. **Phileurini:** *Cryptodus* W.S. Macleay, *Eophileurus* Arrow.

Cetoniinae: Cetoniini: *Glycyphana* Burmeister, *Protaetia* Burmeister. **Schizorhinini:** *Aphanesthes* Kraatz, *Bisallardiana* Antoine, *Chalcopharis* Heller, *Chlorobapta* Kraatz, *Chondropyga* Kraatz, *Clithria* Burmeister, *Diaphonia* Newman, *Dilochrosis* Thomson, *Eupoecila* Burmeister, *Hemichnoodes* Kraatz, *Hemipharis* Burmeister, *Ischiopsopha* Gestro, *Lenosoma* Kraatz, *Lomaptera* Gory & Percheron, *Lyraphora* Kraatz, *Macrotina* Strand, *Metallesthes* Kraatz, *Micropoecila* Kraatz, *Mycterophallus* van de Poll, *Neoclithria* van de Poll, *Neorrhina* Thomson, *Phyllopodium* Schoch, *Poecilopharis* Kraatz, *Pseudoclithria* van der Poll, *Schizorhina* Kirby,

Schochidia Berg, *Stenopisthes* Moser, *Storeyus* Hasenpusch & Moeseneder, *Tapinoschema* Thomson, *Trichaulax* Kraatz. **Valgini:** *Microvalgus* Kraatz.

SCIRTIFORMIA: SCIRTOIDEA

Scirtiformia is equivalent to the Eucinetiformia of Crowson (1981), containing the families Scirtidae, Eucinetidae and Clambidae, with the addition of the more recently described Decliniidae. The group was included in an expanded Elateriformia by Lawrence *et al.* (1995a) and Lawrence and Newton (1995), but subsequent accumulation of both molecular (Caterino *et al.* 2002; Hunt *et al.* 2007; Maddison *et al.* 2009, Pons *et al.* 2010; Song *et al.* 2010); and morphological (Lawrence 1999, 2001; Lawrence and Yoshitomi 2007) evidence indicates that these families lie at the base of the polyphagan clade, possibly sister to the remaining Polyphaga (but see Derodontiformia below). Although the group is usually considered to be monophyletic, the inclusion of Clambidae was recently questioned by Lawrence *et al.* (2011). The only scirtoid family not occurring in Australia is Decliniidae, which is restricted to the eastern Palaearctic region (eastern Siberia and Japan) (Nikitsky *et al.* 1994, Lawrence *et al.* 1995a, Sakai and Satô 1996).

23. CLAMBIDAE

Adult (Fig. 60B, F–G)

Length ~0.75–1.6 mm. Body ~0.95–1.45 times as long as wide, usually globose, highly convex, and conglobate (somewhat flattened in *Acalyptomerus*); sides of prothorax and elytra usually strongly, continuously rounded, when at rest; colour yellowish-brown to black; upper surfaces usually smooth and shiny and either glabrous, sparsely clothed with scattered, long erect setae or more densely clothed with fine, recumbent hairs (surfaces somewhat uneven and clothed with short, erect setae in *Acalyptomerus*).

Head strongly declined, usually concealed from above in conglobate forms, usually strongly transverse and strongly flattened (about as long as wide and slightly flattened in *Acalyptomerus*), not abruptly constricted posteriorly, without transverse occipital ridge; frontal region not declined. Eyes entire and strongly protuberant (*Acalyptomerus*), entire but not protuberant (*Calyptomerus*), partly divided by canthus (*Sphaerothorax*) or entirely divided into dorsal and ventral sections (*Clambus*), coarsely facetted, without interfacetal setae; ommatidium of acone type. Antennal insertions exposed and widely separated; subantennal grooves well developed. Frontoclypeal suture distinct, usually strongly curved (straight in *Acalyptomerus*). Labrum free, transverse, concealed from above in Clambinae. Antennae slender, usually 10-segmented (9-segmented in *Acalyptomerus*) with 2-segmented club. Mandible short and broad, with slender, abruptly curved, unidentate or weakly bidentate apex, well developed membranous prostheca, articulated, sclerotised process and well developed mola. Maxilla with distinct, slender galea and lacinia; lacinia without uncus; apical palpomere cylindrical or fusiform. Ligula; apical labial palpomere fusiform. Subgenal ridges well developed, venter between ridges more or less concave; gular sutures well separated. Corpotentorium narrow. Cervical sclerites present.

Pronotum ~0.3–0.45 times as long as wide, usually widest at middle with sides strongly rounded and somewhat obliquely explanate (widest posteriorly with sides more or less straight in *Acalyptomerus*). Lateral pronotal carinae complete, usually simple, without raised margin or narrow bead (finely denticulate in *Acalyptomerus*); anterior angles usually slightly produced forward and broadly rounded (not in *Acalyptomerus*); posterior angles usually absent, with posterior edge evenly rounded and continuous with lateral edges (posterior angles somewhat acute and posterior edge biemarginate in *Acalyptomerus*); disc usually simple (with pair of broad impressions in Acalyptomerus). Prosternum in front of coxae highly reduced, much shorter than shortest diameter of weakly impressed coxal cavities, not produced anteriorly, flat to slightly concave. Prosternal process usually incomplete or absent (complete but very slender in *Calyptomerus*). Notosternal suture complete but very short. Hypomeron with antennal groove anteriorly and crural impression posteriorly in *Acalyptomerus*. Procoxae usually elongate, projecting well below prosternum, without concealed lateral extensions, coxal plates present in *Acalyptomerus*. Procoxal cavities strongly transverse, contiguous or very narrowly separated, externally broadly open, without narrow lateral extensions, internally broadly closed. Trochantin exposed. Promesothoracic interlocking mechanism weakly developed, without pronotal crenulations.

Scutellar shield abruptly elevated, anteriorly simple, posteriorly acute or obtusely angulate. Elytra ~0.75–1.2 times as long as wide and 1.9–5 times as long as pronotum, entire and conjointly rounded at apex, more or less apunctate in Clambinae, finely, irregularly punctate in *Calyptomerus* and with ~14 puncture rows mesally and irregular punctation laterally in *Acalyptomerus*, without scutellary striole; epipleura short or absent. Mesoventrite separated by complete or incomplete sutures from mesanepisterna, which are widely separated at midline; anterior edge of mesoventrite on different plane than metaventrite, without anterior paired procoxal rests, but with entire ventrite concave or biconcave to receive procoxae; discrimen and metaventral cavity absent. Mesocoxae projecting or not, with exposed trochantins; mesocoxal cavities narrowly separated or contiguous, slightly to strongly transverse, not or slightly oblique, partly closed laterally by mesanepisterna and mesepimera in *Acalyptomerus* or by mesepimera and metanepisterna in *Calyptomerus* and Clambinae; membranous joint present between meso- and metathoracic portions of coxal cavities. Mesoventral process narrowly rounded to acute and mesometaventral junction, if present, formed by meeting of ventrites. Metaventrite strongly transverse and slightly convex, without or occasionally with short discrimen, and without transverse (metakatepisternal) suture; paired postcoxal lines absent, but with single, curved, postcoxal line extending along posterior edges of mesocoxal cavities in *Calyptomerus* and with curved line at about middle of ventrite, which delimits a deep anterior impression in Clambinae. Exposed portion of metanepisternum relatively broad in *Calyptomerus*, narrow in *Acalyptomerus*, broad anteriorly and narrow posteriorly in Clambinae. Metacoxae more or less horizontal, usually contiguous (well separated in *Acalyptomerus*), extending laterally to meet elytra; plates well developed and complete, concealing most of first ventrite (Fig. 60F). Metendosternite with relatively short, broad stalk, well developed lateral arms, no laminae or ventrolateral processes and anterior tendons, if present, located on arms. Hind wing moderately short and broad, but narrowed at base in *Acalyptomerus* and Clambinae; apical field at least two-thirds of total wing length with two distinct radial extensions and with transverse folds meeting anterior wing margin at separate points; radial bending zone short and almost hinge-like; radial cell absent (RA_{1+2} and RA_{3+4} not recombining apically); cross-vein r3 absent and r4 sometimes incomplete or absent; R-M loop usually highly reduced or absent, but very broad with formation of oblongum cell in *Calyptomerus*; basal section of RP usually absent (short in *Calyptomerus*); medial spur moderately long, sometimes extending to wing margin; medial field highly reduced, without or with only one free vein; wedge cell absent; anal lobe usually reduced (absent in *Acalyptomerus*); anal embayment deep usually absent or weak (deep and notch-like in *Calyptomerus*); AP present and undivided in *Calyptomerus* and *Sphaerothorax*; wing more or less completely lined with short to long fringe hairs. Legs moderately long and very slender; tibiae without spines; tibial spurs paired. Tarsi usually 4-4-4 (3-3-3 in *Acalyptomerus*); tarsomeres slender and simple; claws simple; empodium absent.

Abdomen with usually five free ventrites (six in *Calyptomerus*); ventrite 1 with curved postcoxal lines in *Calyptomerus* and Clambinae, intercoxal process usually acute (broadly rounded in *Acalyptomerus*). Functional spiracles in Clambinae present on segments I–V and VIII, those in *Calyptomerus* and *Acalyptomerus* present on segments I–V only. Anterior edge of sternite VIII in male without median strut; tergite VIII without lateral struts. Anterior edge of sternite IX in male with median strut (spiculum gastrale); tergite IX usually simple and undivided (divided by longitudinal suture in *Calyptomerus*), posterior edge shallowly to deeply emarginate; tergite X well developed, free and more or less evenly sclerotised, without baculi. Aedeagus of trilobate type, symmetrical or asymmetrical; anterior edge without struts; parameres free or fused together and articulated with or fused to phallobase; penis without dorsal and ventral lobes, with or without anterior struts. Tergite VIII in female undivided and without median strut; tergite IX not divided into two parts, articulating apically with tergite X, which lacks paired baculi. Ovipositor usually very short and broad, slightly more elongate in Clambinae, without baculi; paraprocts slightly to distinctly longer than gonocoxites, which are undivided and bear terminal, palpiform styli in Clambinae (reduced to patches of setae or absent in *Calyptomerus* and *Acalyptomerus*). [Endrödy-Younga 1959, 1990; Crowson 1979; Young 2002b; Leschen 2005b].

Larva

Body elongate to broadly elliptical and slightly flattened, with broad thorax in *Acalyptomerus*, more or less lightly sclerotised and evenly pigmented, generally smooth or weakly granulate (with small, setiferous tubercles in *Acalyptomerus* and *Calyptomerus*); vestiture usually consisting of fine hairs or setae only.

Head prognathous to somewhat declined, protracted, transverse and somewhat flattened. Posterior edge of head capsule not or only slightly emarginate. Epicranial stem very short; frontal arms lyriform, contiguous at base; median endocarina absent. Stemmata 5 on each in *Calyptomerus*, 1 in *Acalyptomerus*, usually absent in Clambinae. Antennae very short to moderately long, 3-segmented, with conical or palpiform sensorium on antennomere 2 longer than apical antennomere in *Acalyptomerus* and *Calyptomerus*, not in Clambinae. Frontoclypeal suture usually present (absent in *Acalyptomerus*). Labrum free. Mandibles symmetrical, somewhat flattened, broad at base and narrow apically, unidentate to tridentate; incisor edge with 2 or 3 teeth or serrations or with an articulated, complex, subapical lobe; accessory ventral process present in *Acalyptomerus* and *Calyptomerus*, absent in Clambinae. Mola well developed, tuberculate or asperate; prostheca consisting of fixed hyaline process with acute or bifid apex. Ventral mouthparts retracted; maxillary articulating area present. Maxilla with fixed galea and lacinia or with apically cleft mala; cardo distinct, stipes longer than wide; maxillary palp usually 3-segmented (2-segmented in *Acalyptomerus*), palpifer present or absent. Labium consisting of prementum and postmentum. Ligula very short and undivided. Labial palps usually 2-segmented (1-segmented in *Acalyptomerus*), widely separated. Hypopharyngeal bracon and sclerome present. Hypostomal rods and ventral epicranial ridges absent. Gular sutures separate, gula wider than long, separated from labium by suture.

Thoracic segments about equal in length; lateral tergal processes present on thorax and most abdominal segments in *Acalyptomerus*. Prothoracic venter without distinct armature. Thoracic legs well developed, not spinose, 5-segmented, including pretarsus, which is acute, claw-like and usually bisetose, with one seta distal to the other; mesocoxae well separated. Abdomen more than twice as long as thorax; terga with irregular, transverse rows of asperities in *Acalyptomerus* and *Calyptomerus*. Abdominal segment IX usually about as long as VIII, somewhat shorter in *Acalyptomerus*, tergum IX completely dorsal, without paired urogomphi. Sternum IX simple. Segment X usually well developed, cylindrical and terminal (reduced, somewhat flattened in *Acalyptomerus*). Spiracles annular, those on segment VIII reduced in Clambinae. [Crowson 1979; Lawrence 1991b; Young 2002b; Leschen 2005b].

Classification and World Distribution

This is a worldwide family of minute beetles including ~150 described species. There are three subfamilies and five genera, four of which occur in Australia. As mentioned above, there is some doubt concerning the placement of this family. The hindwing in all clambids contains a distinct hinge on the medial bar (MP_{1+2}) and in *Calyptomerus*, there is a distinct oblongum cell; these two features are characteristic of Archostemata, Myxophaga and Adephaga and occur in no other polyphagan (Kukalová-Peck and Lawrence 1993; Lawrence *et al.* 2011).

Australian Fauna

The Australian fauna consists of ~20 species of minute, oval and convex beetles, usually with partly or completely divided eyes, large metacoxal plates, and capable of rolling into a ball (conglobation), with the relatively large head strongly deflexed so that it abuts the metaventrite. The subfamily **Acalyptomerinae** is represented in Australia by *Acalyptomerus asiaticus* Crowson (Fig. 60B) (the Australian record cited by Crowson 1979). This species differs from other clambids in having a smaller head, much narrower than the prothorax, with protuberant entire eyes, the elytral epipleura very broad at base but disappearing apically, and the body not capable of conglobation. **Calyptomerinae** are not native to Australia, but *Calyptomerus dubius* (Marsham) has been introduced into TAS. This species is similar to the Clambinae but has entire (although not protruding) eyes, closed procoxal cavities and no functional spiracles on abdominal segment VIII. **Clambinae** includes two genera: *Clambus* Fischer von Waldheim (Fig. 60F–G) and *Sphaerothorax* Endrödy-Younga. Larvae are similar to those of Eucinetidae, being more or less fusiform and lightly pigmented, without urogomphi. Clambids usually occur in leaf litter and hay stacks and feed on fungal spores. [Crowson 1979; Endrödy-Younga 1990].

24. EUCINETIDAE

Adult (Fig. 60C)

Length ~1.3–3.0 mm. Body ~1.25–1.9 times as long as wide, slightly to strongly convex and more or less fusiform (widest near anterior end of elytra and narrowed anteriorly and posteriorly), sides of prothorax and elytra slightly to strongly, continuously rounded, upper surfaces relatively smooth and shiny, but clothed with fine, decumbent hairs.

Head hypognathous, usually entirely concealed from above by pronotum, subquadrate to transverse, not abruptly constricted posteriorly, with fine transverse occipital ridge lying immediately in front of occipital foramen; frontal region not declined. Eyes more or less entire, nor or slightly protuberant, coarsely facetted, with or without interfacetal setae; ommatidium of eucone type. Antennal insertions exposed or slightly concealed from above and well separated. Frontoclypeal suture straight; clypeus very short. Labrum visible, free, slightly transverse, truncate to rounded. Antennae 11-segmented, filiform to weakly serrate; scape subequal to or shorter than pedicel. Mandible short and broad, somewhat flattened, unidentate; incisor edge two or more teeth or serrations; mola and prostheca well developed. Maxilla with distinct galea and lacinia; galea densely setose; lacinia usually with unidentate uncus; apical palpomere fusiform or slightly expanded and apically obliquely truncate. Ligula undivided, broad, apically emarginate or truncate; apical labial palpomere fusiform or short, broad and somewhat truncate. Subgenal ridges present immediately beneath eyes; venter between ridges more or less biconcave; gular sutures well separated. Corpotentorium narrow to moderately broad and straight. Cervical sclerites well developed.

Pronotum ~0.35–0.43 times as long as wide, widest posteriorly; sides slightly to strongly curved. Lateral pronotal carinae complete, simple, not usually visible for their entire lengths from above, without raised margin or narrow bead; anterior angles not or barely produced forward and broadly rounded; posterior angles broadly rounded to slightly acute; posterior edge evenly rounded; disc simple. Prosternum in front of coxae distinctly shorter than shortest diameter of coxal cavity, not produced anteriorly, flat to slightly convex. Prosternal process complete, slender, parallel-sided, not curved in lateral view, apex narrowly rounded. Notosternal suture complete but very short. Procoxae at least slightly projecting below prosternum, obliquely oriented, without concealed lateral extensions, each coxa with distinct carina which fits against mesanepisternum. Procoxal cavities strongly transverse, narrowly separated, externally broadly open, without narrow lateral extensions, internally open. Trochantin exposed.

Scutellar shield usually well developed (somewhat reduced in *Noteucinetus*), abruptly elevated or not, anteriorly simple, posteriorly acute or obtusely angulate. Elytra ~1.0–1.6 times as long as wide and 3.8–5.8 times as long as pronotum, entire and conjointly rounded at apex; disc irregularly punctate or with ~10 longitudinal bands of minute punctures, without scutellary striole, transversely strigulate in *Eucinetus*; ventral surface with longitudinal carina near lateral edge; epipleura narrow and incomplete. Mesoventrite separated by complete sutures from mesanepisterna, which are relatively narrowly separated at midline in *Eucinetus*, but more widely separated in *Noteucinetus*; anterior edge of mesoventrite on slightly different plane than metaventrite, with large, paired, strongly declined, procoxal rests extending laterally onto mesanepisterna; discrimen well developed but incomplete anteriorly; mesoventral cavity large but shallow. Mesocoxae not projecting, with exposed trochantins; mesocoxal cavities narrowly separated, strongly transverse and oblique, partly closed laterally by mesepimera; membranous joint present between meso- and metathoracic portions of coxal cavities. Mesoventral process usually apically divided, mesometaventral junction monocondylic. Metaventrite more or less Y-shaped, with slender posterior stem, projecting anteriorly as metaventral process, and with narrow, oblique lateral arms, flat, with moderately long discrimen, with or without short, transverse (katepisternal) suture; paired postcoxal lines absent. Exposed portion of metanepisternum relatively short and broad and usually concealed posteriorly by metacoxae. Metacoxae (Fig. 26K) large and strongly oblique, contiguous, extending laterally to meet elytra; plates large and complete, completely concealing metafemora when legs folded. Metendosternite greatly enlarged, with long, oblique arms strongly expanded at apex; laminae and ventrolateral processes absent; anterior process short to moderately long and emarginate; anterior tendons widely separated but not on arms. Hind wing relatively short and broad; apical field about a third of total wing length with a few vague sclerotisations, a single radial extension and transverse folds meeting anterior wing margin at separate points; radial bending zone long and relatively weak; radial cell short and broad with inner edge oblique, forming equilateral triangle with lumen moderately pigmented; cross-vein r3 short and strongly oblique, r4 usually complete, curved and lightly pigmented; R-M loop very broad; basal section of RP absent or very short; medial spur slightly curved and extending to wing margin; medial field usually with four free veins; wedge cell present but incomplete; MP_{3+4} with basal cross-vein but no spur; AA_4 running along anal fold; anal lobe moderately well developed, anal embayment deep and notch-like, AP undivided; hind wings absent in *Noteucinetus*. Fore legs relatively long and slender; mid and hind legs shorter and stouter; meso- and metatibiae gradually expanded apically, with ring of dark spines at apex and several dark spines along outer edge; tibial spurs usually double, metatibia with spurs of unequal length or with a single spur only. Tarsi 5-5-5 without ventral lobes; meso- and metatarsomeres 1–4 with ring of dark apical spines; claws simple; empodium absent or not apparent.

Abdomen with five ventrites, with basal two connate; ventrite 1 almost completely divided into two parts by metacoxae; postcoxal lines absent, intercoxal process narrowly acute. Functional spiracles on segments I–VIII, located in pleural membrane. Anterior edge of sternite VIII in male with median strut. Anterior edge of segment IX in male with median strut (spiculum gastrale); tergite IX deeply emarginate; tergite X well developed, free. Aedeagus of trilobate type, symmetrical, with phallobase reduced, without basal strut; parameres free or fused to phallobase, without apical hooks; penis without dorsal and ventral lobes, with or without anterior struts. Sternite VIII in female without spiculum ventrale; tergite IX deeply emarginate but not divided; tergite X (proctiger) well developed and free. Ovipositor moderately short and broad, relatively lightly sclerotised; paraprocts without baculi, located lateral to coxites, which are divided, with basal oblique baculi; styli well developed, apically attached. [Vit 1977, 1985; Young 2002; Leschen 2005a].

Larva

Body elongate to broadly elliptical, narrowing posteriorly, slightly flattened. Head, dorsal body surfaces and ventral abdominal surfaces relatively lightly sclerotised and evenly pigmented, generally smooth, but often with dorsal setiferous tubercles; vestiture consisting of fine hairs or setae only.

Head prognathous to somewhat declined, protracted, transverse and somewhat flattened. Posterior edge of head capsule not emarginate. Epicranial stem moderately long; frontal arms lyriform, contiguous at base; median endocarina absent. Stemmata 5 on each side. Antennae moderately long, 3-segmented, with antennomere 2 much longer than 1 and conical sensorium on antennomere 2 much shorter than antennomere 3; antennal insertions well separated from mandibular articulations. Frontoclypeal suture absent. Labrum partly fused to head capsule. Mandibles (Fig. 39C) symmetrical, bidentate. Accessory ventral process present. Incisor edge more or less serrate. Mola well developed, asperate; prostheca consisting of fixed, hyaline process, broadly angulate at apex. Ventral mouthparts retracted; maxillary articulating area well developed. Maxilla with cardines distinct, divided, transverse to slightly oblique, well separated from one another; stipes longer than wide; galea and lacinia fixed and closely associated, the former rounded and setose, the latter falcate, without distinct uncus; maxillary palp 3-segmented, with weak, shelf-like palpifer. Labium consisting of prementum, mentum and submentum; ligula short and broad, bilobed; labial palps 2-segmented, widely separated. Hypopharyngeal bracon present; hypopharyngeal sclerome consisting of transverse bar. Hypostomal rods and ventral epicranial ridges absent. Gular sutures separate, gula wider than long, separated from labium by suture.

Prothorax slightly larger than meso- or metathorax; all thoracic terga and most abdominal terga with weakly pigmented plates, some darkly pigmented areas and numerous dark, setiferous tubercles; abdominal terga I–VIII with well developed subspiracular lobes. Prothoracic venter short and simple. Thoracic legs moderately stout, 5-segmented, including pretarsus, which is acute, claw-like and bisetose; coxae well separated. Abdomen less than twice as long as thorax, gradually tapering posteriorly. Abdominal segment IX slightly longer than VIII, subcylindrical; tergum IX completely dorsal, without paired urogomphi. Sternum IX simple. Segment X terminal, with sclerotised tergite and short, paired ventral pygopods, without anal hooks. Spiracles annular, placed at ends of short spiracular tubes. [Böving and Craighead 1931; Lawrence 1991b; Young 2002a; Leschen 2005a].

Classification and World Distribution

Eucinetidae is a relatively small but widely distributed family with ten genera and ~50 species. The two largest genera are the widely distributed *Eucinetus* Germar and the Holarctic *Nycteus* Latreille. Species of the Holarctic genus *Tohlezkus* Vit and the Neotropical *Jentozkus* Vit have tubular, suctorial mouthparts (Vit 1977).

Australian Fauna

Several species of *Eucinetus* Germar occur in Australia, but none of them have yet been described. In addition, there are at least two undescribed species of *Noteucinetus* Bullians and Leschen, a genus presently known only from New Zealand and Chile (Bullians and Leschen 2004). Eucinetids are small, fusiform beetles, with short prothorax, narrow, deflexed head, oblique metacoxae with plates partly concealing the femora, and short legs with meso- and metatibiae and tarsi bearing fringes of short, dark spines. Larvae are more or less oblong and moderately lightly sclerotised, with setiferous tubercles dorsally, and broad head narrowed anteriorly. Feeding habits of Australian species are unknown, but some members of the family feed on slime mould spores or on the fruiting bodies of Basidiomycetes (Boletaceae, Coniophoraceae) (Wheeler and Hoebeke 1984).

25. SCIRTIDAE

Adult (Fig. 60A)

Length ~1.5–12 mm. Body broadly ovate to narrowly elongate, 1.0–2.5 times as long as wide, strongly convex to strongly flattened; sides of prothorax and elytra usually slightly to strongly, independently rounded, rarely continuously rounded (e.g. some *Scirtes*) or subparallel (e.g. some *Accolabass*); usually finely pubescent (subglabrous in *Macrohelodes*).

Head moderately to strongly declined (Fig. 15J), sometimes almost concealed from above by pronotum, subquadrate to strongly transverse, not abruptly constricted posteriorly, rarely with transverse occipital ridge lying immediately in front of occipital foramen; frontal region not to moderately, gradually declined. Eyes entire or shallowly emarginate, slightly to strongly protuberant, finely or coarsely facetted, without interfacetal setae; ommatidium of eucone or acone type. Antennal insertions exposed and widely separated, sometimes countersunk so that they lie in saucer-like fossae, which are continuous with subantennal grooves. Frontoclypeal suture absent, but clypeus usually distinctly divided, with membranous anteclypeus. Labrum at least partly visible, free, slightly to strongly transverse, truncate to slightly emarginate. Antennae 11-segmented, filiform to serrate. Mandible short and broad to moderately elongate, slightly to strongly and gradually to abruptly curved mesally, unidentate; incisor edge simple or with one or more teeth or retinacula; mola usually highly reduced or absent (well developed in *Macrohelodes*, *Pseudomicrocara*, *Heterocyphon* and *Peneveronatus*); prostheca absent. Maxilla with distinct galea and lacinia; galea densely setose; lacinia without uncus; apical palpomere fusiform. Ligula undivided to deeply emarginate or bilobed; apical labial palpomere fusiform, sometimes arising subapically in *Prionocyphon*, *Scirtes* and *Ora*. Subgenal ridges strongly developed, often extending well behind eyes and occasionally curved ventrally to meet transverse occipital ridge; venter between ridges more or less concave (Figs 14J, 44F); gular sutures well separated. Corpotentorium usually moderately broad and anterior tentorial arms not or only barely expanded. Cervical sclerites well developed.

Pronotum ~0.25–0.70 times as long as wide, usually widest posteriorly or at middle; sides slightly to strongly curved, rarely somewhat explanate. Lateral pronotal carinae complete, simple, usually visible for their entire lengths from above, with or without raised margin or narrow bead; anterior angles not to distinctly produced forward and broadly rounded to acute; anterior edge sometimes forming continuous curve with lateral edges (e.g. *Pseudomicrocara* and *Heterocyphon*); posterior angles usually right to broadly rounded, somewhat acute in *Scirtes* and *Ora*; posterior edge more or less straight, evenly rounded or biemarginate, with or without narrow raised margin or bead; disc usually simple, with pair of small, deep pits just in front of hind margin in *Heterocyphon*. Prosternum in front of coxae much shorter than shortest diameter of coxal cavity, not produced anteriorly, flat to slightly convex, but obliquely oriented in lateral view. Prosternal process complete, slender, in ventral view slightly expanded subapically, then narrowed to subacute apex, strongly projecting ventrally between coxae and in lateral view abruptly bent near apex and then flattened, strongly overlapping mesoventrite. Notosternal suture complete but very short. Procoxae not projecting below prosternum, but strongly obliquely oriented, without concealed lateral extensions, each coxa with distinct carina which fits against mesanepisternum. Procoxal cavities strongly transverse, narrowly separated, externally broadly open, without narrow lateral extensions, internally open. Trochantin broadly exposed and very narrow. Promesothoracic interlocking mechanism weakly developed, without pronotal crenulations.

Scutellar shield well developed, abruptly elevated, anteriorly simple, posteriorly acute or obtusely angulate. Elytra ~0.85–1.95 times as long as wide and 2.9–8.5 times as long as pronotum, entire and conjointly rounded at apex, apunctate or irregularly punctate, without scutellary striole; epipleura relatively narrow but more or less complete. Mesoventrite separated by complete sutures from mesanepisterna, which are widely separated at midline; anterior edge of mesoventrite on same plane as metaventrite, with paired, slightly to strongly oblique procoxal rests; discrimen usually present but anteriorly obscured by mesoventral cavity, which may be small to moderately large and relatively shallow. Mesocoxae not projecting, with exposed trochantins; mesocoxal cavities narrowly separated, slightly to strongly transverse, often more or less oblique, partly closed laterally by mesanepisterna and mesepimera; membranous joint present between meso- and metathoracic portions of coxal cavities. Mesoventral process apically divided and mesometaventral junction complex (monocondylic or dicondylic). Metaventrite flat to strongly convex, with short to very long discrimen, transverse (katepisternal) suture often more or less complete but located closed to posterior edge of ventrite; paired postcoxal lines absent, but usually with single, curved, postcoxal line extending along posterior edges of mesocoxal cavities and base of metaventral process and continuing laterally across anterior part of metanepisternum. Exposed portion of metanepisternum relatively short and broad (usually not more than 2.5 times as long as wide). Metacoxae more or less horizontal, contiguous or narrowly separated, extending laterally to meet elytra; plates usually complete but weaker laterally; reduced in *Scirtes* and *Ora*. Metendosternite with arms short to very long, laminae absent, ventrolateral processes usually strongly developed, anterior process short to moderately long, with tendons approximate or well separated but not on arms. Hind wing (Fig. 23D) relatively short and broad; apical field about a fourth of total wing length with a few vague sclerotisations and transverse folds meeting anterior wing margin at separate points; radial bending zone long and relatively weak; radial cell short and broad with inner edge oblique, forming equilateral triangle with lumen moderately pigmented; cross-vein r3 short and strongly oblique, r4 usually complete, curved and lightly pigmented; R-M loop very broad; basal section of RP absent or very short; medial spur short, curved and meeting wing margin where there is distinct embayment; medial field usually with three free veins; wedge cell absent; AA4 running along anal fold; anal lobe well developed, anal embayment deep and notch-like, AP often divided with posterior branch meeting basal wing margin. Metafemora greatly enlarged in *Scirtes* and *Ora*. Outer edges of tibiae with paired longitudinal carinae which may be more or less spinose. Tibial spurs double, glabrous or pubescent, hind pair of unequal length in *Scirtes* and *Ora*. Tarsi 5-5-5 with ventral lobe on penultimate tarsomere; claws simple; empodium absent or not apparent.

Abdomen with five ventrites, with basal two more or less connate; ventrite 1 without postcoxal lines, intercoxal process acute or narrowly rounded. Functional spiracles on segment VIII absent, those on I–VII located in pleural membrane. Anterior edge of sternite VIII in male with or without median strut; anterior edge of tergite VIII with paired lateral struts. Anterior edge of sternite IX in male often with median strut; tergite IX with paired lateral struts, which never unite with sternite to form subgenital plate or spiculum gastrale. Aedeagus highly variable, but basically a trilobate type with phallobase reduced or fused to bases of parameres, symmetrical or asymmetrical; anterior edge without struts; parameres free or fused; penis with or without dorsal and ventral lobes and without anterior struts. Ovipositor elongate and slender, lightly sclerotised, except for baculi; paraprocts much longer than gonocoxites, which are undivided and bear terminal, palpiform styli. Internal female tract complex, with glands sometimes opening near entrance of common oviduct, a long vagina, occasionally with sclerotizations, and a large and multilobed bursa, sometimes with a gland opening near its base. [Armstrong 1953; Nyholm 1972; Kitching and Allsopp 1987; Lawrence 2001, 2005a; Young 2002c; Watts 2004, 2007, 2009, 2010a, 2010b, 2011; Yoshitomi 2005; Lawrence and Yoshitomi 2007; Ruta and Yoshitomi 2010].

Larva (Fig. 60H)

Body elongate to broadly elliptical and slightly to strongly flattened. Head, dorsal body surfaces and ventral abdominal surfaces lightly to more or less heavily sclerotised and evenly pigmented, generally smooth; thoracic venter usually more lightly sclerotised; vestiture consisting of fine hairs or setae only.

Head prognathous, protracted, not concealed from above by prothorax, transverse and somewhat flattened. Posterior edge of head capsule not to moderately deeply emarginate. Epicranial stem very short or absent; frontal arms V-shaped or U-shaped, contiguous or distinctly separated at base. Stemmata as many as 5 on each side, but often tightly clustered, sometimes absent. Antennae at least 10-segmented, often with more than 20 antennomeres, with or without sensorium on antennomere 2. Clypeus highly reduced and frontoclypeal suture absent, incomplete or vaguely indicated. Labrum enlarged, free; epipharynx with complex armature. Mandibles (Fig. 48I) symmetrical, somewhat flattened, at least slightly curving mesally at apex, which is unidentate and acute or somewhat rounded. Accessory ventral process present. Incisor edge usually simple but lined with dendritic setae. Mola present, finely ridged; prostheca consisting of fixed process with comb-like mesal edge and sometimes a brush of hairs. Ventral mouthparts retracted; maxillary articulating area present but more or less concealed. Maxilla with fixed galea and lacinia; cardo distinct, undivided, transverse to slightly oblique, well separated from one another; stipes longer than wide; apex of galea rounded or truncate, bearing comb-hairs (Fig. 27H) or spatulate hairs; lacinia falciform, with or without uncus; maxillary palp 3-segmented or 4-segmented with reduced apical palpomere, the apical or subapical palpomere often with transverse rows of microspines; palpifer usually well developed. Labium consisting of prementum and postmentum, the latter large, subquadrate or trapezoidal, with blunt posterior edge, not divided longitudinally, laterally expanded to conceal maxillary articulating area. Ligula very short and broad or absent. Labial palps 2-segmented or 3-segmented with reduced apical palpomere, widely separated. Hypopharyngeal region complex, consisting of a series of combs, plates, and patches of simple and complex hairs. Hypostomal rods and ventral epicranial ridges absent. Gular sutures separate, gula wider than long, separated from labium by suture.

Thoracic segments about equal in length or with prothorax slightly larger; lateral tergal processes sometimes present on thorax. Prothoracic venter usually with anterolateral cervical sclerites but without other armature. Thoracic legs 5-segmented, including pretarsus, which is acute, claw-like and bisetose; mesocoxae large, more or less conical and projecting, narrowly to moderately widely separated; femur and tibiotarsus often clothed with stout setae or spines. Abdomen less than twice as long as thorax; terga sometimes extending laterally beyond edges of sterna. Abdominal apex with respiratory chamber formed from terga VIII and IX and enclosing pair of enlarged, posteriorly oriented, annular spiracles; thoracic and anterior abdominal spiracles reduced and non-functional or absent. Abdominal segment IX shorter than VIII, sometimes concealed from above by tergum VIII; tergum IX completely dorsal, without paired urogomphi. Sternum IX simple without hinged operculum. Segment X membranous and more or less fused to IX. Anal region posteriorly or terminally oriented, often with five or more, simple or branched, anal papillae or gills; anal hooks absent. [Böving & Craighead 1931; Beier 1952; Striganova 1961; Bertrand 1972; Costa *et al.* 1988; Hannappel & Paulus 1987, 1992, 1994; Kitching & Allsopp 1987; LeSage 1991; Lawrence *et al.* 1999a; Yoshitomi 2005; Lawrence 2005a; Zwick 2008, 2011; Ruta and Yoshitomi 2010].

Classification and World Distribution

Scirtidae are cosmopolitan in distribution but far more abundant and diverse in the cool temperate parts of the Northern and Southern Hemispheres. The family contains ~45 genera and 1000 species. There are three recognised subfamilies: Nipponocyphoninae (Japan), Stenocyphoninae (Chile and New Zealand) and Scirtinae including the remaining genera. [Lawrence 2001, 2005; Yoshitomi 2005; Lawrence and Yoshitomi 2007; Ruta *et al.* 2011].

Australian Fauna

The fauna currently includes 17 genera, but it is likely that more remain to be described. Scirtid adults are small to moderately large, oblong to ovoid beetles, usually somewhat flattened, with a large, strongly deflexed head, short and broad prothorax with large, projecting procoxae and reduced sternal region, moderately broad and complete elytral epipleura and lobes fourth tarsomeres. A pair of sharp genal ridges (Fig. 14J) rest against the procoxae when the head is fully deflexed. Metacoxal plates are present, except in *Scirtes* and the related genera *Ora* and *Exochomoscirtes*, which have enlarged metafemora. Male genitalia are complex and variable within the family (Nyholm 1972). Larvae are usually elongate and somewhat flattened (or sometimes short, broad and fusiform), with large head, multisegmented antennae (unique in beetle larvae), complex mouthparts with maxillary comb-hairs (Fig. 27H) and comb-like hypopharyngeal armature, a single pair of large spiracles at the abdominal apex and sometimes a set of five anal papillae. Adult scirtids are often found on vegetation near water, while the larvae usually occur in lentic habitats, such as ponds, marshes, bogs or tree holes (Kitching and Allsopp 1987), but have also been found in wet, rotten wood well away from water (Hannappel and Paulus 1992; Klausnitzer 2006). Larvae are filter-feeding detritivores (Beier 1952). They use the eighth spiracles to breathe at the water surface, while the anal papillae are thought to be osmoregulatory in function (Treherne 1954). The largest Australian genera are *Heterocyphon*, *Macrohelodes*, *Prionocyphon*, *Pseudomicrocara* and *Scirtes*. [Watts 2004, 2007, 2009, 2010a, 2010b, 2011; Zwick 2012, 2013].

Australian Subfamilies and Genera

Scirtinae: *Accolabass* Watts, *Chameloscyphon* Watts, *Cyphon* Paykull, *Daploeuros* Watts, *Dasyscyphon* Watts, *Eurycyphon* Watts, *Exochomoscirtes* Pic, *Heterocyphon* Armstrong, *Macrodascillus* Carter, *Macrohelodes* Blackburn, *Ora* Clark, *Pachycyphon* Zwick, *Peneveronatus* Armstrong, *Petrocyphon* Watts, *Prionocyphon* Redtenbacher, *Pseudomicrocara* Armstrong, *Scirtes* Illiger, *Spaniodascillus* Watts.

ELATERIFORMIA

The series is used here in the sense of Lawrence (1988b) and differs from that of Crowson (1955) with respect to the inclusion of Dascillidae. Larval similarities between scarabaeoids and Dascillidae are either plesiomorphic or associated with soil dwelling habits, and adult dascillids share many more features with elateriform taxa such as Buprestidae, Callirhipidae and Ptilodactylidae (Lawrence and Newton 1982), and the adult mouthparts of dascilloids were found to be similar to those of various Elateroidea (*sensu lato*) in the presence of an inframandibular antepipharynx and associated liquid-feeding (Anton and Beutel 2011). A general feature shared by many members of the series is the heterogeneous life cycle, with long-lived larvae and short-lived adults usually occupying different habitats. Surface-active adults often have a stream-lined body and complex promesothoracic interlocking device, or in its absence, a chemical defence system and aposematic coloration. The antennae are rarely clubbed and often serrate, and with the exception of derived, soft-bodied forms, the metacoxae are more or less excavate to receive the femora. [Beutel 1995b; Beutel and Leschen 2005e; Bocákova *et al.* 2007].

DASCILLOIDEA

Although Dascillidae and Rhipiceridae share several adult features, such as the type of mouthparts, promesothoracic interlocking mechanism, wing venation and folding and aedeagus, rhipicerid larvae differ from those of dascillids and resemble more derived elateriform larvae in having consolidated ventral mouthparts, biforous spiracles and no mandibular mola. [Crowson 1971a].

26. DASCILLIDAE

Adult (Fig. 62A)

Length 7.5–12 mm. Body ~2.6–2.75 times as long as wide, slightly flattened. Upper surfaces clothed with decumbent, whitish hairs, which are variously thickened forming vaguely indicated longitudinal lines.

Head about as long as wide, slightly flattened, not declined and not abruptly constricted posteriorly; transverse occipital ridge absent; vertex with vaguely indicated V-shaped ridge. Eyes large, entire, moderately strongly protuberant, finely facetted, without interfacetal setae; ommatidium of the exocone type. Antennal insertions exposed and moderately widely separated, not borne on raised tubercles. Subantennal groove absent or very weakly developed. Frontoclypeal suture distinctly impressed; clypeus short, strongly transverse and truncate. Antennae 11-segmented, serrate in male beginning on antennomere 3 (more or less filiform in female); scape less than twice as long as pedicel. Labrum visible, free, transverse, lightly pigmented and truncate. Mandible moderately elongate, apex abruptly curved mesally and unidentate, incisor edge with one or two teeth or retinacula; mola and prostheca absent. Maxilla with distinct galea and lacinia; galea lightly pigmented, densely pubescent and bilobed; lacinia lightly pigmented, without uncus; apical palpomere fusiform. Apical labial palpomere fusiform; ligula bilobed, lightly pigmented and setose. Subgenal ridges absent. Gular sutures widely separated. Corpotentorium very broad; anterior tentorial arms not expanded mesally. Cervical sclerites well developed.

Pronotum ~0.55–0.67 times as long as wide, distinctly narrower at base than combined elytral bases; widest posteriorly; sides almost straight, slightly narrowing anteriorly; lateral carinae complete, simple, with narrow raised margin or bead; anterior angles not produced forward; posterior angles slightly acute; posterior edge triemarginate and weakly crenulate; disc simple. Promesothoracic interlocking mechanism moderately well developed. Anterior portion of prosternum subequal to shortest diameter of procoxal cavity; slightly convex, anterior edge not produced forward, without mesal excavation. Prosternal process complete, slightly elevated or curved behind coxae and slightly, gradually expanded at apex, which is narrowly rounded. Notosternal suture complete. Procoxae not to strongly projecting below prosternum, without concealed lateral extensions. Procoxal cavities strongly transverse, narrowly separated, externally broadly open, without narrow lateral extensions, internally open. Postcoxal projection absent. Trochantin broadly exposed, but not abutting prosternum or hypomeron.

Scutellar shield well developed, abruptly elevated, anteriorly simple, posteriorly broadly rounded. Elytra ~2.1–2.3 times as long as wide and 4.55–4.85 times as long as pronotum; sides subparallel and apices conjointly rounded; disc with 12 puncture rows and no scutellary striole; epipleuron narrow and complete. Mesoventrite separated by complete sutures from mesepisterna, which are well separated from one another; anterior edge on same plane as metaventrite, with paired, slightly declined, procoxal rests; discrimen well developed; mesoventral cavity small, shallow and indistinct. Mesocoxae not projecting, with exposed trochantins. Mesocoxal cavities narrowly separated, slightly transverse and oblique, partly closed laterally by mesepimeron. Mesoventral process divided at apex and mesometaventral junction monocondylic. Metaventrite strongly convex, with discrimen complete to base of metaventral process and with complete transverse (katepisternal) suture; paired postcoxal lines absent. Exposed portion of metanepisternum ~2.5 times as long as wide; metepimeron exposed. Metacoxae contiguous, extending laterally to meet elytra, metacoxal plates well developed and more or less uniform. Metendosternite with long stalk, very short lateral arms and moderately long anterior process with approximate anterior tendons; laminae and ventrolateral processes absent. Hind wing (Fig. 24C) moderately elongate; radial bending zone moderately long; apical field about a fourth of total wing length, with 2 radial extensions; apical folds meeting anterior wing margin at separate points; radial cell ~2.5 times as long as wide with lumen distinctly pigmented; inner edge slightly oblique and inner posterobasal angle slightly obtuse; cross-vein r3 strongly oblique and meeting RP, but interrupted in two places by folds; r4 long, slightly curved and complete; R-M loop moderately broad but forming acute angle; basal portion of RP moderately long, apical extension strongly curved; medial field with five free veins, but MP_3 often detached at base; MP_{3+4} with basal cross-vein and spur; wedge cell well developed and apically acute; CuA_2 appearing to arise from MP_{3+4}; anal lobe well developed, anal notch absent; AP divided, with posterior branch meeting basal wing margin. Legs relatively long and slender; tibiae slightly widened at apex, outer edge simple; tibial spurs serrate, paired on all legs. Tarsi 5-5-5; membranous lobes present on tarsomeres 1–4 (Fig. 44X), at least distal ones divided. Claws simple, empodium absent.

Abdomen with five ventrites, the basal two connate; ventrite 1 about as long as 2; intercoxal process acute. Most tergites well sclerotised. Functional spiracles present on segments I–VIII, those on III–VII located on laterosternites, those on VIII in pleural membrane. Anterior edge of sternite VIII in male without median strut. Anterior edge of segment IX in male with broadly rounded subgenital plate; tergite IX deeply emarginate, X well developed, evenly sclerotised and free. Aedeagus of trilobate type, symmetrical; phallobase relatively large and about as long as wide, with median longitudinal endocarina, without anterior struts; parameres articulated, not outwardly hooked. Penis with longer dorsal lobe and shorter ventral lobe, anterior edge with short paired struts. Sternite VIII in female with long anterior strut (spiculum ventrale). Ovipositor narrowly elongate, lightly sclerotised, except for baculi; paraprocts more than twice as long as coxites, which are located distally, subdivided, with short basal baculum and well developed, terminal styli. [Kasap & Crowson 1975; Caveney 1986; Lawrence *et al.* 1999b; O'Keefe 2002a; Lawrence 2005b].

Larva

Body elongate, more or less parallel-sided, subcylindrical or slightly flattened, slightly curved ventrally. Surfaces lightly pigmented, smooth, clothed with longer and shorter, fine hairs, mainly laterally and ventrally.

Head (Fig. 45M) protracted and moderately declined (hypognathous), not concealed from above; large, transverse, strongly rounded laterally and slightly flattened; posterior edge not or only weakly emarginate. Epicranial stem very short or absent; frontal arms V-shaped, contiguous at base; median endocarina absent; paired endocarinae beneath frontal arms. Stemmata absent. Antennae 3- segmented but with antennomere 3 min, so that they appear to be 2-segmented; sensorium on preapical antennal segment dome-like. Frontoclypeal suture absent. Labrum partly fused to head capsule (suture absent mesally), asymmetrical; epipharynx complex, consisting of several sclerites. Mandibles (Fig. 41D) asymmetrical, robust, at least slightly curving mesally at apex, which is unidentate; accessory ventral process present; incisor edge of mandible with 2 or 3 heavily sclerotised scissorial teeth (retinacula); mola well developed, with numerous fine, longitudinal ridges; prostheca consisting of simple, flattened, membranous area and a small, articulated process. Ventral mouthparts slightly retracted; cardo distinct, sclerotised, externally divided, transverse to slightly oblique; cardines separated from each other by labium; stipes longer than wide; maxillary articulating area well developed and exposed; galea articulated, 1-segmented and falcate; lacinia fixed with bifid uncus at apex and several spine-like setae along inner edge; palp 3-segmented, without distinct palpifer. Labium consisting of prementum, mentum and submentum; ligula shorter than labial palp, bilobed; palps 2-segmented, without distinct palpiger, separated by more than width of first palpomere. Hypopharyngeal bracon absent; hypopharynx with complex armature, consisting of several sclerites and a pair of combs; hypostomal rods short and strongly diverging; ventral epicranial ridges absent; gular sutures well separated; gula wider than long.

Prothorax only slightly longer than meso- or metathorax. Thoracic terga and abdominal terga I to IX with well developed, yellowish plates; meso- and metaterga and abdominal terga I to VIII each with transverse carina near anterior edge; metatergum and abdominal terga I to IV or V with heavier, asperate carina near posterior edge. Thorax and abdomen without patches of asperities or lateral tergal processes. Prothoracic venter with lateral cervical sclerites. Thoracic legs moderately long, clothed with spine-like setae, 5-segmented including pretarsal claw, which bears 3 or 4 setae. Metathoracic leg about same size as mesothoracic leg; meso- and metathoracic legs not forming stridulatory organ. Coxae well separated. Abdomen about twice as long as thorax, with 10 visible segments, I–VIII more or less equal, IX slightly shorter, its tergum dorsal, bearing a pair of short, lightly pigmented, fixed urogomphi. Sternum IX exposed, simple, not enclosed by sternum VIII; segment X reduced, posteriorly oriented; anal opening transverse, with a pair of oval lobes beneath it. Spiracles cribriform, with spiracular scar anterior to sieve plate, large and reniform on thorax and first abdominal segment, becoming smaller and more semicircular posteriorly, located laterally just below edges of tergal plates. Spiracular closing apparatus absent. [Gahan 1908a; Verhoeff 1923; Böving 1929; Böving and Craighead 1931; Emden 1942b; Zocchi 1961; Lawrence 1991b; Lawrence *et al.* 1999a; O'Keefe 2002a; Grebennikov and Scholtz 2003; Lawrence 2005b].

Classification and World Distribution

The family Dascillidae includes ~15 genera and 80 species primarily from the Northern Hemisphere. There are two subfamilies: Dascillinae from western North America, the Greater Antilles, Eurasia, tropical India and Africa, south-east Asia, The East Indies and Australia, and Karumiinae from arid and semiarid parts of western North America, southern South America, Africa and Central Asia.

Australian Fauna

There are three species of dascilline Dascillidae in Australia: *Notodascillus brevicornis* (Macleay), *N. iviei* Jin *et al.* and *N. sublineatus* Carter (Fig. 62A). Adults are more or less parallel-sided, brown beetles clothed with decumbent grey or yellow pubescence, which usually forms narrow stripes on the elytra. They differ from the related Rhipiceridae in having serrate antennae, which are not borne on tubercles, and a strongly transverses pronotum with complete lateral carinae visible from above, and in lacking an empodium projecting between the pretarsal claws. Larvae are somewhat similar to those of Scarabaeoidea in being more or less C-shaped, with a large head, well developed legs, and complex epipharyngeal and hypopharyngeal armature, but they differ from them in having the labrum partly fused to the clypeus, the antennae appearing to be 2-segmented with the sensorium at the apex, and tergum IX with a pair of minute urogomphi. Dascillid larvae live on organic matter in the soil, and some Holarctic species are known to damage roots. [Jin *et al.* 2013].

27. RHIPICERIDAE

Adult (Figs 5C, 62D)

Length 10–25 mm. Body ~1.9–3 times as long as wide, slightly flattened to moderately convex. Colour brown to black, sometimes with legs red or red and black; vestiture of short decumbent, dark or translucent setae and longer, thickened, whitish setae, which may be recumbent or suberect and which usually form small whitish spots or irregular maculi on elytra but which may cover entire pronotal disc, ventral surfaces and sometimes legs.

Head longer than wide, prognathous or slightly declined, not abruptly constricted posteriorly; transverse occipital ridge absent; vertex sometimes with vaguely indicated longitudinal line. Eyes entire, somewhat protuberant, finely facetted, without interfacetal setae; ommatidium of the exocone type. Antennal insertions

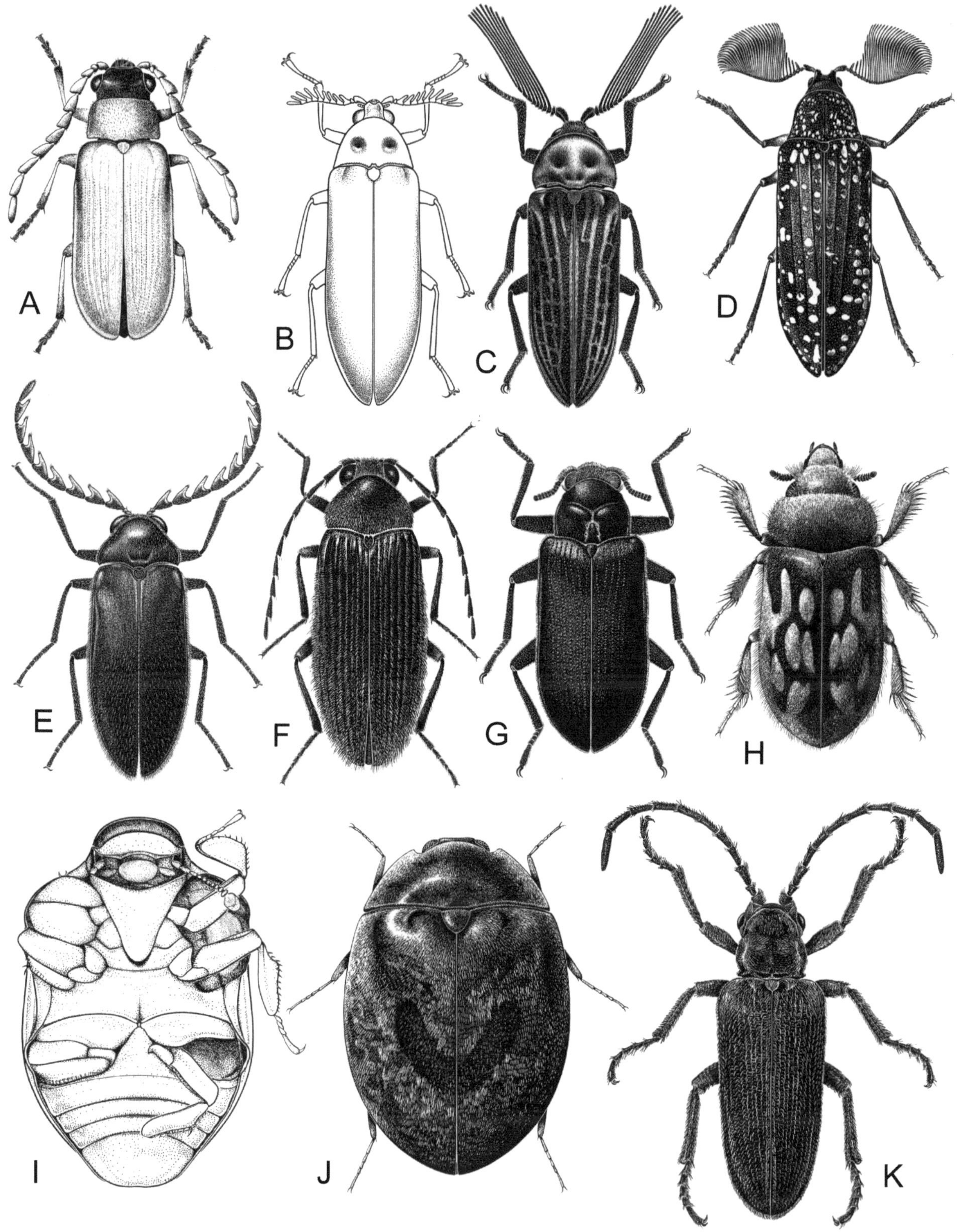

Fig. 62. A–H, Adult, dorsal. A, *Notodascillus sublineatus* Carter (Dascillidae); B, *Callirhipis cardwellensis* Blackburn (Callirhipidae), female; C, same, male; D, *Rhipicera femorata* Kirby (Rhipiceridae); E, *Austrolichas monteithi* Lawrence & Stribling (Ptilodactylidae: Cladotominae), male; F, *Byrrocryptus oblongus* (Carter) (Ptilodactylidae: Anchytarsinae); G, *Stetholus elongatus* Carter & Zeck (Elmidae: Larainae); H, *Heterocerus flindersi* Blackburn (Heteroceridae). I, *Microchaetes* sp. (Byrrhidae: Syncalyptinae), ventral. J–K, Adult, dorsal: J, *Sclerocyphon* sp. (Psephenidae); K, *Rhinorhipus tamborinensis* Lawrence (Rhinorhipidae).

concealed from above, but borne on strongly raised tubercles. Subantennal groove absent. Frontoclypeal suture absent; frontoclypeal region strongly, sharply declined between antennal tubercles; clypeus more or less vertical and apically truncate. Antennae flabellate in both sexes, with up to ~20–30 antennomeres in females and up to 40 in males; antennal rami beginning on antennomere 3; scape more than 3 times as long as pedicel, which is strongly transverse. Labrum visible, densely setose, separated from clypeus by distinct suture, but not movable, without tormae, about as long as wide with apex narrowly rounded. Mandible elongate and narrow, apex abruptly curved mesally and unidentate, incisor edge with two teeth or retinacula; mola and prostheca absent. Maxilla with reduced galea and lacinia; galea articulated, broadly rounded, lightly pigmented and densely pubescent; lacinia very short, lightly pigmented, without uncus; apical palpomere cylindrical or fusiform. Mentum trapezoidal, wider at base, truncate apically; ligula reduced, truncate, lightly pigmented and setose; apical labial palpomere cylindrical or fusiform. Subgenal ridges absent. Gular sutures well separated, but gula longer than wide. Corpotentorium broad; anterior tentorial arms not expanded mesally. Cervical sclerites well developed.

Pronotum ~0.6–0.97 times as long as wide, base as wide as or only slightly narrower than combined elytral bases; widest posteriorly; sides straight or only slightly curved, narrowing anteriorly; lateral carinae incomplete or vaguely indicated, simple, without narrow raised margin or bead; anterior angles not produced forward; posterior angles slightly acute; posterior edge triemarginate and weakly crenulate or denticulate; disc with weak median longitudinal groove and vaguely indicated lateral and basomesal impressions. Promesothoracic interlocking mechanism moderately well developed. Anterior portion of prosternum distinctly shorter than shortest diameter of procoxal cavity; slightly convex but strongly declined, anterior edge not produced forward, without mesal excavation. Prosternal process complete or almost complete, somewhat declined but not curved behind coxae, more or less parallel sides and narrowly rounded at apex. Notosternal suture complete. Procoxae conical and projecting below prosternum, without concealed lateral extensions. Procoxal cavities slightly to strongly transverse, narrowly separated, externally broadly open, without narrow lateral extensions, internally open. Postcoxal projection absent or short and broadly rounded. Trochantin broadly exposed, but not abutting prosternum or hypomeron.

Scutellar shield well developed, abruptly elevated, anteriorly simple, posteriorly broadly rounded. Elytra ~1.5–2.55 times as long as wide and 3.1–5.5 times as long as pronotum; sides subparallel or slightly curved and apices almost meeting but independently narrowly rounded; disc irregularly punctate, without scutellary striole, usually with 3 to 5 weak longitudinal costae, variously joined posteriorly; epipleuron narrow and complete. Mesoventrite separated by complete sutures from mesepisterna, which are well separated from one another; anterior edge on same plane as metaventrite, with paired, moderately declined, procoxal rests separated by carina; discrimen absent; mesoventral cavity small and shallow or absent. Mesocoxae conical and projecting, with exposed trochantins. Mesocoxal cavities moderately narrowly separated, slightly transverse, weakly delimited (without distinct posterior edge), partly closed laterally by mesanepisterna and mesepimera, with solid joint between meso- and metathoracic portions of coxal cavities. Mesoventral process not divided at apex, which is subtruncate; mesometaventral junction monocondylic. Metaventrite strongly convex, with discrimen complete to base of metaventral process and with complete transverse (katepisternal) suture; paired postcoxal lines absent. Exposed portion of metanepisternum about twice as long as wide; metepimeron exposed. Metacoxae contiguous, extending laterally to meet elytra, metacoxal plates well developed and narrowing laterally but usually complete. Metendosternite with long stalk, very short lateral arms and long anterior process with approximate anterior tendons; laminae and ventrolateral processes absent. Hind wing moderately elongate; radial bending zone moderately long; apical field about a fifth of total wing length, with 2 radial extensions; apical folds meeting anterior wing margin at separate points; radial cell ~2.5 times as long as wide; inner edge more or less straight and inner posterobasal angle right; cross-vein r3 short and longitudinally oriented; r4 long, slightly curved and complete; R-M loop moderately broad but forming acute angle; basal portion of RP long, apical extension strongly curved; medial field with five free veins; MP_{3+4} with basal cross-vein and spur; wedge cell somewhat reduced and apically acute; CuA_2 appearing to arise from MP_{3+4}; anal lobe well developed, anal notch absent; AP undivided. Legs relatively long and slender; tibiae slightly widened at apex, outer edge lined with spine-like setae; tibial spurs usually asymmetrical, paired on all legs. Tarsi 5-5-5; membranous lobes present on tarsomeres 1–4, at least distal ones divided. Claws simple, empodium well developed and multisetose.

Abdomen with five ventrites, the basal two connate; ventrite 1 about as long as 2, without postcoxal lines; intercoxal process weakly developed or absent. Most tergites well sclerotised, without patches of wing-folding spicules. Functional spiracles on segments I–VIII located in pleural membrane. Anterior edge of sternite VIII in male without median strut. Anterior edge of segment IX in male with broadly rounded subgenital plate; tergite IX shallowly emarginate, X well developed, evenly sclerotised and free. Aedeagus of trilobate type, symmetrical; phallobase strongly transverse, without anterior struts; parameres articulated, not outwardly hooked. Penis with shorter dorsal lobe and longer ventral lobe, anterior edge with short paired struts. Sternite VIII in female with long anterior strut (spiculum ventrale). Ovipositor narrowly elongate; proctiger lightly sclerotised, without baculi; paraprocts ~5 times as long as coxites, with distinct baculi; coxites distal to paraprocts, subdivided, lightly sclerotised, with oblique basal baculum and well developed, terminal styli. [Emden 1924a, 1927, 1930, 1931a, 1931b; Katovich 2002a; Lawrence 2005c].

Larva (late instar) (based on exotic genera Sandalus *Knoch and* Polymerius *Philippi)*

Body elongate, grub-like, with enlarged abdomen, widest behind middle and narrowing anteriorly and posteriorly, not to slightly curved ventrally. Surfaces very lightly pigmented, smooth, clothed with fine setae only.

Head protracted and prognathous, relatively small, slightly flattened; posterior edge not or only weakly emarginate. Epicranial suture absent or with moderately long epicranial stem and V-shaped frontal arms; median endocarina absent. Stemmata absent. Antennae very short and subconical, either 1-segmented or 3-segmented with small sensorium on antennomere 2. Frontoclypeal suture absent. Labrum completely fused to head capsule (suture absent). Mandibles symmetrical, flat and subtriangular, unidentate; accessory ventral process absent; incisor edge simple or with small tooth; mola absent or represented by a slight mesal expansion which is finely denticulate. Ventral mouthparts strongly retracted, with maxillary articulating area. Maxilla with cardo distinct or not clearly indicated and stipes at least slightly longer than wide; galea and lacinia fixed and narrowly or broadly rounded, the latter without uncus; palp 2- or 3-segmented. Labium consisting of prementum and postmentum or of single plate; ligula about as long as labial palps and bilobed or not; labial palps 1- or 2-segmented and widely separated. Hypopharyngeal bracon present; hypopharyngeal sclerome absent. Hypostomal rods and ventral epicranial ridges absent. Gular sutures widely separated; gula transverse.

Prothorax narrower than abdomen; thoracic terga and abdominal terga without patches or rows of asperities, lateral processes or transverse plicae. Thoracic legs highly reduced, 4-segmented; pretarsus claw-like with 1 seta; coxae widely separated. Abdomen almost 4 times as long as thorax, with 10 visible segments, segment IX much shorter than VIII, with or without pair of short, fixed, upturned urogomphi. Sternum IX exposed, simple, not enclosed by sternum VIII; segment X distinct, posteriorly or posteroventrally oriented, without paired pygopods. Spiracles biforous with tubes longitudinally aligned (absent on segment VIII in *Polymerius*). Spiracular closing apparatus absent. [Craighead 1921; Katovich 2002a; Solervicens 2005b; Lawrence 2005c].

Larva (first instar)

Minute (less than 1 mm in length), elongate, parallel-sided, flattened, lightly pigmented and clothed with posteriorly-directed setae. Antennae 2-segmented with sensorium on antennomere 1 longer than antennomere 2. Mandible flattened, with apex more or less rounded. Maxilla with single rounded mala, indistinctly cleft; maxillary and labial palps 1-segmented, long and tusk-like. Legs long and slender, 5-segmented; pretarsus long, with 1 seta. Paired egg bursters on abdominal terga III–VII. Tergum IX without urogomphi; X well developed, posteriorly oriented, with 4 lobes surrounding anus. Spiracles longitudinally oval (possibly biforous), present on mesothorax and abdominal segments I and VIII. Spiracular closing apparatus absent. [Dodge 1941; Elzinga 1977].

Classification and World Distribution

Rhipiceridae includes seven genera and ~100 species occurring in all major regions but not common in collections. The genus *Sandalus* Knoch is known from North to South America, China, Japan, India, south-east Asia and Africa. *Rhipicera* Latreille occurs in Brazil, New Caledonia and Australia. *Desmodocus* Guérin-Méneville, *Microrhipis* Guérin-Méneville and *Chamaerhipis* Latreille are African, *Arrhaphipterus* Schaum ranges from south-eastern Europe to Central Asia, and *Polymerius* Philippi is endemic to Chile.

Australian Fauna

Australian rhipicerids are placed in two subgenera of *Rhipicera*: *Oligorhipis* Guérin-Méneville, with one species, *R.* (*O.*) *vestita* (Laporte), and *Agathorhipis* Guérin-Méneville, with five species. Adults are easily recognised by the multisegmented flabellate antennae of the male (strongly pectinate in the female) and membranous lobes on tarsomeres 1–4. Unlike the closely related Dascillidae, they have weak lateral pronotal carinae, which are incomplete anteriorly and barely visible from above, antennae borne on raised tubercles, and a large, setose empodium. Late instar larvae of Australian species have never been studied, but the parasitic larva on a cicada nymph shown in Moulds (1990) is almost certainly a *Rhipicera* larva. Larvae of the North American *Sandalus niger* Knoch are lightly sclerotised ectoparasitoids of immature cicadas (Hemiptera: Cicadidae) (Craighead 1921). Minute, triungulin-like first instars very similar to those of *S. niger* described by Dodge (1941) and Elzinga (1977) have been seen from WA. [Emden 1931].

BUPRESTOIDEA

The phylogenetic relationships of the family Buprestidae have been the subject of some controversy, and the group is usually placed in a monotypic superfamily. Forbes (1926, 1942) noted similarities in the wing folding of schizopodine buprestids and members of the Dascilloidea. Crowson (1982) pointed out several features shared by Buprestidae and members of the dryopid complex (Dryopidae, Elmidae, Lutrochidae, Limnichidae, Heteroceridae). Lawrence (1988b) suggested that the family could be the sister group of the Byrrhoidea in the broad sense (**see** below), but an equally convincing case might be made for the derivation of buprestids directly from a dascilloid ancestor.

28. BUPRESTIDAE

Adult (Figs 4A–H, 69A–G)

Length 1.5–65 (usually 5–40) mm. Body ~1.5–3.5 times as long as wide, sides usually subparallel but converging posteriorly, occasionally strongly curved; slightly flattened to strongly convex. Colour reddish-orange to black, but often bicolored, forming light and dark pattern, or with metallic coloration. Upper surfaces often subglabrous, sometimes clothed with short, decumbent to erect, fine hairs, rarely with thickened or scale-like setae forming patches on elytra.

Head (Fig. 15G) slightly transverse, somewhat flattened, deeply inserted into prothorax, not declined at base but with vertex strongly declined, so that mouthparts are ventrally oriented, not abruptly constricted posteriorly; transverse occipital ridge absent; occipital region almost always with distinct, moderately long median endocarina. Eyes large, entire, vertically oval, and not protuberant, finely facetted, without interfacetal setae; ommatidium of the eucone type. Antennal insertions exposed, usually widely separated, but narrowly separated in Agrilinae, not borne on raised tubercles. Subantennal groove usually absent or weakly developed, but well developed in some Polycestinae (e.g. *Mastogenius*, *Xyroscelis*) and Agrilinae (e.g. *Cisseis*, *Aphanisticus*). Frontoclypeal suture absent; clypeus short and usually anteriorly emarginate (sometimes with membranous anteclypeus visible within emargination). Antennae 11-segmented, usually weakly to strongly serrate (rarely pectinate) beginning on antennomere 4. Labrum visible, free, small, transverse. Mandible small, short and broad; apex unidentate or bidentate; incisor edge sometimes with one or two teeth; mola and prostheca absent. Maxilla almost always with distinct, setose galea and lacinia, rarely with single lobe; lacinia without uncus; apical palpomere cylindrical or fusiform. Mentum quadrate to triangular; ligula undivided. Gular region short to moderately long; gular sutures usually widely separated, sometimes narrowly separated or absent. Corpotentorium slender and arched; anterior tentorial arms not expanded mesally. Cervical sclerites well developed.

Pronotum ~0.32–0.85 times as long as wide, not distinctly narrower at base than combined elytral bases; usually widest posteriorly or at middle; sides almost straight or weakly curved; lateral carinae usually complete or incomplete anteriorly, occasionally absent, simple, usually without narrow raised margin or bead; anterior angles usually not produced forward (produced and acute in Trachydini and *Aphanisticus*); posterior angles right to acute; posterior edge straight or biemarginate, sometimes weakly crenulate (e.g. *Castiarina*) or very finely, sharply crenulate (e.g. *Helferella*, *Paratrachys*, *Xyroscelis*); disc usually simple, sometimes with median groove (e.g. *Polycesta*, *Prospheres*), several impressions (e.g. *Nascio*, *Cyphogastra*, *Aphanisticus*) or curved, oblique costae (e.g. *Xyroscelis*); sublateral carinae present in Haplostethini and some Agrilinae (e.g. *Cisseis*, *Agrilus*). Promesothoracic interlocking mechanism moderately well developed. Anterior portion of prosternum slightly to much longer than shortest diameter of procoxal cavity; moderately to strongly convex, without mesal excavation; anterior edge produced forward to form short, broad chin piece in *Xyroscelis* and some Agrilinae. Prosternal process complete, not elevated or curved behind coxae, usually parallel-sided or gradually narrowed apically, but sometimes abruptly narrowed at apex, which may be acute or narrowly rounded to broadly rounded or truncate. Notosternal suture complete, with antennal groove in *Xyroscelis*; hypomeron with short antennal groove in *Aphanisticus*. Procoxae not projecting below prosternum, without concealed lateral extensions. Procoxal cavities subglobular or slightly transverse, moderately widely separated, externally broadly open, without narrow lateral extensions, internally open. Postcoxal projection absent. Trochantin broadly exposed, and more or less abutting prosternum and hypomeron.

Scutellar shield well developed or somewhat reduced, abruptly elevated, anteriorly simple, posteriorly acute to truncate. Elytra ~1.05–2.85 times as long as wide and 1.4–3.8 times as long as pronotum; sides usually subparallel anteriorly and converging posteriorly (more or less converging from base to apex in *Paratrachys* and Trachydini); apices sometimes conjointly rounded, but often independently rounded, acute or biacute; lateral edges serrate posteriorly in some Polycestinae (e.g. *Polycesta*, *Xyroscelis*), Chrysochroinae (e.g. *Cyphogastra*), Buprestinae (e.g. *Melobasis*, *Chrysobothris*) and Agrilinae (e.g. *Cisseis*); disc with irregular punctation or with 10 puncture rows, with or without scutellary striole; epipleuron gradually or abruptly narrowed and usually incomplete. Mesoventrite (Fig. 44Z) separated by complete sutures from mesepisterna, which are well separated from one another; anterior edge on same plane as or different plane than metaventrite, with paired, slightly to strongly declined or vertical, procoxal rests; discrimen more or less developed in *Polycesta*, *Xyroscelis* and *Prospheres*, but usually obliterated by mesoventral cavity, which is relatively large and deep, extending onto metaventrite in *Astraeus*, Haplostethini and members of all other subfamilies. Mesocoxae not projecting, with exposed trochantins. Mesocoxal cavities widely separated, subcircular or longitudinally oval, partly closed laterally by mesanepisterna and mesepimera. Mesoventral process not divided, more or less truncate or emarginate and either fused to or abutting metaventral process. Metaventrite (Fig. 44Z) slightly to strongly convex, with discrimen usually complete to apex of metaventral process and with moderately long, distinctly curved, transverse (katepisternal) suture, which crosses midline well in front of metacoxae; paired postcoxal lines absent. Exposed portion of metanepisternum usually ~2.5 times as long as wide (more elongate and slender in Haplostethini and Aphanisticini; metanepisternum concealed in *Paratrachys*); metepimeron exposed or concealed. Metacoxae contiguous, extending laterally to meet elytra and usually as broad laterally as mesally; metacoxal plates well developed and more or less uniform. Metendosternite with moderately long stalk, short lateral arms and short to moderately long anterior process with approximate anterior tendons; laminae and ventrolateral processes absent. Hind wing moderately elongate; radial bar extending well beyond end of radial cell with long, but weak bending zone; apical field relatively short, usually with 2 radial extensions; apical folds meeting anterior wing margin at separate points; radial cell usually narrowly elongate, as much as 10 times as long as wide, sometimes absent; inner edge slightly oblique and inner posterobasal angle slightly acute; cross-vein r3 either absent or fused basally with r4, which is usually short and straight; R-M loop forming acute angle; basal portion of RP very short or absent, apical extension usually present; median spur usually straight and extending to wing margin; medial field with 6 free veins and no wedge cell (when CuA_{3+4} extends to wing margin), 5 free veins with an apically acute wedge cell, or fewer veins in some smaller species; MP_{3+4} sometimes with cross-vein and basal spur, but often detached; anal lobe usually well developed with embayment weak or absent and AP undivided, lobe sometimes reduced, with deep, notch-like embayment in Haplosatethini. Legs usually relatively long and slender; femora flattened and expanded to receive tibiae in *Aphanisticus*; tibiae flattened and expanded to receive tarsi in *Xyroscelis*; outer edge simple; tibial spurs usually paired on all legs. Tarsi 5-5-5; membranous lobes present on at least some preapical tarsomeres. Claws simple, empodium not apparent.

Abdomen usually with five ventrites (sometimes with slender lateral portion of sternite II visible between lateral edge of metacoxa and elytron), the basal two of which are connate; ventrite 1 slightly to distinctly longer than 2; intercoxal process usually acute (broadly rounded or angulate in Trachyini); suture between ventrites 1 and 2 more or less obliterated in *Paratrachys*, Haplostethini, Agrilini, Aphanisticini and Trachyini. Paired small, oval impressions (cuticularia) frequently present at or near joint between ventrites 2 and 3 and less commonly between 1–2 and 3–4. All tergites usually heavily sclerotised, but occasionally very lightly so (e.g. Haplostethini), with as many as 5 pairs of cuticularia similar to those on ventrites. Functional spiracles present on segments I–VIII, those on II–VIII located on tergites or laterotergites, that on VIII sometimes reduced in size. Anterior edge of sternite VIII in male without median strut. Anterior edge of segment IX in male with emarginate subgenital plate; tergite IX either deeply emarginate and X well developed and free, or IX partly or completely fused to X, which may be membranous. Aedeagus of trilobate type, more or less symmetrical, with phallobase highly reduced and more or less fused to parameres, which form a flattened tube more or less enveloping penis. Sternite VIII in female without spiculum ventrale. Ovipositor usually moderately to very elongate, more or less flattened and lightly sclerotised, except for baculi; proctiger usually more or less evenly sclerotised, usually with 2 short, basolateral struts articulating with apex of paraproctal baculi; the latter, in turn, abruptly curved or angulate ventrally near anterior end and articulated with coxital baculi; coxites located mesoventrally, undivided and usually slightly expanded at apex, with 2 pairs of longitudinal baculi (sometimes with pair of short, transverse, basal baculi) and with short, palpiform, terminal or subterminal styli. Ovipositor short and broad in the buprestinae tribe Stigmoderini and in a few Polycestinae (e.g. *Paratrachys*, Haplostethini) and Agrilinae (e.g. *Germarica*), and in these groups, paraproctal baculi may be absent, coxital baculi transverse or absent, and styli reduced or absent. In Stigmoderini an additional pair of projections or dorsal valves are present between coxites and proctiger. [Good 1925; Barker 1986; Bellamy 1986, 1997, 2003; Bílý 2000b; Kubán *et al.* 2000; Jendek 2001; Volkovitsh 2001; Bellamy and Nelson 2002; Bellamy and Volkovitsh 2005].

Larva (Figs 5B, 34L, 65A)

Body elongate, usually more or less parallel-sided, except for thorax, which is much wider than abdomen (less elongate, widest anteriorly and gradually narrowing posteriorly in Trachyini and in other leaf-mining forms), subcylindrical to slightly flattened, except for thorax, which is usually distinctly flattened. Very lightly pigmented, except for mouth frame and mandibles, but often with yellow dorsal and ventral prothoracic plates and sometimes patches of small sclerites; with paired, sclerotised, acute processes on segment X in *Agrilus*; vestiture of short to moderately long suberect, fine hairs.

Head prognathous and deeply retracted into prothorax, with only short anterior portion visible; slightly rounded laterally and strongly flattened; posterior edge deeply emarginate dorsally and ventrally. Epicranial stem absent; frontal arms V-shaped, contiguous at base; median endocarina lying between frontal arms and extending to frontoclypeal suture; additional paired endocarinae lying beneath frontal arms. Stemmata usually absent (2 on each side in some Trachyini). Antennae highly reduced, either 3-segmented, with conical sensorium on antennomere 2 much longer than antennomere 3, or 2-segmented, with sensorium at apex of antennomere 2. Frontoclypeal suture present. Labrum free, lightly sclerotised, and broadly rounded to truncate. Mandibles robust, usually bidentate or tridentate, without accessory ventral process; incisor edge sometimes with additional teeth; mola and prostheca absent; mesal surface of mandibular base sometimes with brush of hairs in Agrilinae. Ventral mouthparts protracted; cardines more or less distinct, sclerotised or not, separated from each other by labium; stipes not or only slightly longer than wide; maxillary articulating area absent; maxilla with articulated, rounded or truncate mala; palp 2-segmented, without distinct palpifer. Labium consisting of prementum and postmentum or a single labial plate; ligula broadly rounded, undivided; labial palps 1-segmented and widely separated or absent. Hypopharyngeal bracon and sclerome absent; hypostomal rods long and diverging; ventral epicranial ridges absent; gular region short; sutures narrowly separated.

Prothorax (Fig. 45H) usually longer than meso- and metathorax combined (not so in leaf-mining groups) and moderately to strongly flattened; often with moderately sclerotised dorsal and ventral plates, usually with V- or Y-shaped dorsal endocarina and single median ventral endocarina, occasionally with single median endocarina both dorsally and ventrally (e.g. *Prospheres*, most Agrilinae) and sometimes with additional small sclerites (with sclerotised plates in some Trachyini); mesothorax strongly transverse and usually shorter than metathorax, which may be transverse or subquadrate, sometimes with ventrolateral ampullae (e.g. in *Neocuris*). Thoracic legs almost always absent (vestiges occasionally present). Abdominal segments I to IX more or less similar in length, often subdivided by transverse folds (strongly transverse, undivided and somewhat expanded laterally in leaf-mining forms), without patches or rows of asperities or lateral processes; tergum IX shorter than VIII, subterminal, without urogomphi; sternum IX well developed and simple; segment X usually well developed and posteriorly oriented with paired, oval lobes lying on each side of vertical anal opening (segment X somewhat reduced in leaf-mining forms); with posteriorly projecting, acute processes in *Agrilus*. Spiracles cribriform, usually reniform with spiracular scar anterior to sieve plate, but sometimes reduced and so resembling multiforous or uniforous type. Spiracular closing apparatus present. [Böving and Craighead 1931; Rees 1941; Hawkeswood 1985; Bílý 1972, 1986, 1989, 1997, 1999, 2000a; Volkovitsh and Hawkeswood 1987, 1990, 1993, 1999; Bellamy and Nelson 2002; Bílý and Volkovitsh 2003a, 2003b, 2005; Volkovitsh *et al.* 2003; Bellamy and Volkovitsh 2005; Bílý *et al.* 2008].

Classification and World Distribution

The Buprestidae is a cosmopolitan family with ~475 genera and 15 000 species. The subfamily classification varies among authors, and the present treatment recognises seven subfamilies, four of which occur in Australia. Exotic subfamilies include Schizopodinae, which is restricted to south-western North America and often considered as a separate family (Nelson and Bellamy 1991, 2002; Bellamy and Nelson 2002; Bellamy 2003) and Julodinae and Galbellinae, both from the Palaearctic, Afrotropical and Oriental regions. [Kolibáč 2000; Kubán *et al.* 2000; Volkovitsh 2001; Bellamy 2008a, 2008b, 2008c, 2008d, 2009].

Australian Fauna

The Australian buprestid fauna includes four of the subfamilies and about half of the tribes known for the world. Buprestid adults are heavily sclerotised, rigid beetles, usually elongate and brilliantly coloured, with a metallic sheen. The head is more or less globular, deflexed and deeply inserted into the prothorax, which is closely applied to the elytra, antennae short and usually serrate, eyes reniform and not very protuberant, mesoventrite with a large cavity for reception of the prosternal process, metaventrite with a distinct transverse (metakatepisternal) suture, tarsi with membranous lobes beneath, abdominal tergites heavily sclerotised, and first two ventrites fused together with suture between them indistinct or incomplete. Larvae (Figs 34L, 65A) of most species may be distinguished by the greatly enlarged and flattened prothorax, very short antennae, simple abdominal apex and cribriform spiracles; some Agrilinae, however, have paired processes on segment X and leaf-mining agriline larvae are small, flattened and widest at middle. Adults are active in hot weather and will readily fly in sunlight; they often occur in nectar-bearing flowers.

Larvae usually feed in wood (often in the phloem just beneath the bark) or in root systems of trees or shrubs, but some feed in herbaceous stems or galls or mine leaves.

The Australian **Polycestinae** include species of *Polycesta*, *Prospheres* and *Astraeus* (Fig. 69B), as well as some unusual genera such as *Helferella*, *Paratachys* and *Xyroscelis*. Among the **Chalcophorinae** are the large, tropical species of *Cyphogastra* and *Pseudotaenia*. **Buprestinae** is the largest group with several tribes and 35 genera. Included are the conifer-feeding *Diadoxus* and *Araucariana* and the two species of *Julodimorpha*, the largest of Australian buprestids; *J. bakewelli* (White) (Fig. 4G) is known for its habit of congregating around 'stubby' beer bottles, the tinted and pitted glass of which resemble the beetles' elytra (Gwynne and Rentz 1983). *Merimna atrata* (Hope) is known as the fire beetle, because of its habit of flying to burnt trees. The tribe Stigmoderini contains almost half of the described Australian buprestids, including species of *Calodema* (Fig. 69G), *Stigmodera* (Fig. 69F), *Metaxymorpha* (Fig. 69E), *Temognatha* (Fig. 69E), and *Castiarina* (Fig. 69C). Species of *Castiarina* are often aposematically coloured and form part of a mimicry ring with species of Lycidae and several other families; these and other members of the family contain bitter chemicals called buprestins (Brown *et al.* 1985; Moore and Brown 1989). Other common buprestine genera include *Bubastes*, *Nascio*, *Melobasis* and *Selagis* (Fig. 69D). **Agrilinae** is a large group of smaller species, some of which are gall makers or leaf miners. The largest genera are *Cisseis* and *Agrilus* (Fig. 69A). Also included are leaf miners in the genera *Habroloma* and *Trachys*. A species of *Habroloma* from India is known to have a close association with *Oecophylla* ants; larvae mine the leaves comprising the ant nest and adults are not killed by the ants but merely thrown out of the nest along with dead ants (Bílý *et al.* 2008). The South American agriline *Hylaeogena jureceki* Obenberger has recently been introduced into Australia for control of cat's claw creeper (Dhileepan *et al.* 2012). [Carnaby 1986; Gardner 1989; Curletti 2001; Bellamy 1986, 1997, 2002, 2003; Williams and Weir 1987, 1988; Bellamy and Weir 2008].

Australian Subfamilies, Tribes and Genera

Polycestinae: Haplostethini: *Helferella* Cobos, *Mastogenius* Solier. **Paratrachydini:** *Paratrachys* Cobos. **Polycestini:** *Polycesta* Solier, *Strigoptera* Dejean. **Xyroscelini:** *Xyroscelis* Saunders. **Prospherini:** *Prospheres* Saunders. **Astraeini:** *Astraeus* Laporte & Gory.

Chrysochroinae: Chrysochroini: *Chrysodema* Laporte & Gory, *Cyphogastra* Deyrolle, *Cyphogastrella* Théry, *Iridotaenia* Deyrolle, *Metataenia* Théry, *Mroczkowskia* Hołyn ski, *Paracupta* Deyrolle, *Pseudotaenia* Kerremans. **Poecilonotini:** *Melobasina* Kerremans.

Buprestinae: Maoraxiini: *Maoraxia* Obenberger. **Julodimorphini:** *Julodimorpha* Gemminger & Harold, *Nascio* Lporte & Gory, *Nascioides* Kerremans. **Epistomentini:** *Araucariana* Levey, *Balthasarella* Obenberger, *Cyrioides* Carter, *Diadoxus* Saunders, *Neobuprestis* Kerremans. **Buprestini:** *Buprestina* Obenbnerger. **Bubastini:** *Bubastes* Gory & Laporte, *Euryspilus* Lacordaire, *Microcastalia* Heller, *Notobubastes* Carter. **Stigmoderini:** *Calodema* Laporte, *Calotemognatha* Peterson, *Castiarina* Gory & Laporte, *Metaxymorpha* Parry, *Stigmodera* Eschscholtz, *Temognatha* Solier. **Pterobothrini:** *Neobubastes* Blackburn. **Curini:** *Anilara* Saunders, *Anthaxoschema* Obenberger, *Australorhipis* Bellamy, *Barakula* Peterson & Bellamy, *Neocuris* Saunders, *Notographus* Thomson, *Pseudanilara* Théry, *Selagis* Dejean, *Theryaxia* Carter, *Torresita* Gemminger & Harold. **Melobasini:** *Melobasis* Gory & Laporte. **Melanophilini:** *Merimna* Saunders. **Actenodini:** *Belionota* Eschscholtz. **Chrysobothrini:** *Chrysobothris* Eschscholtz.

Agrilinae: Coraebini: *Aaaba* Bellamy, *Ano* Bellamy, *Cisseis* Gory & Laporte, *Dinocephalia* Obenberger, *Ethonion* Kubán , *Hypocisseis* Thomson, *Meliboeithon* Obenberger, *Neospades* Blackburn, *Pachycisseis* Théry, *Paracephala* Saunders, *Sambus* Deyrolle, *Stanwatkinsius* Barker & Bellamy, *Synechocera* Deyrolle. **Agrilini:** *Agrilus* Curtis. **Aphanisticini:** *Anthaxomorphus* Deyrolle, *Aphanisticus* Latreille, *Endelus* Deyrolle, *Germarica* Blackburn. **Trachydini:** *Habroloma* Thomson, *Trachys* Fabricius. **Pachyschelini:** *Hylaeogena* Obenberger.

BYRRHOIDEA

Byrrhoidea, as used here, includes Byrrhoidea and Dryopoidea in the sense of Crowson (1960, 1973b, 1978, 1982), Callirhipidae from the Artematopoidea of Crowson (1973b), Cneoglossidae from the Cantharoidea of Crowson (1972b), and Byrrhoidea plus Psephenoidea of Lawrence (1988b). The group is variable and not easily characterised using either adult or larval features. Adults have a specialised type of wing folding (dryopoid type of Forbes 1926), which has been modified subsequently in large, elongate forms (Callirhipidae and Eulichadidae) to resemble that found in most elateroids (Kukalová-Peck and Lawrence 1993). The group may not be monophyletic; in analyses based on larvae (Beutel 1995b) or larvae and adults (Costa *et al.* 1999; Lawrence *et al.* 2011), the psephenoid and dryopoid taxa were in two or more different clades. Byrrhoid families not occurring in Australia include Dryopidae (widely distributed, except for Australia) (Kodada and Jäch 2005b), Lutrochidae (Nearctic and Neotropical) (Ide *et al.* 2005), Cneoglossidae (Neotropical) (Costa *et al.* 1999, 2005), Podabrocephalidae (southern India) (Lawrence *et al.* 2010d) and Eulichadidae (western North America and eastern Asia) (Ivie 2005; Hájek 2007).

29. BYRRHIDAE

Adult (Figs 62I, 69I–J)

Length 1.5–5.5 mm. Body ~1.25–1.5 times as long as wide, more or less ovate, with sides uniformly rounded or pronotum and elytra independently rounded, strongly convex dorsally and flat to slightly convex ventrally; colour sometimes uniformly reddish-brown to black, or metallic green, occasionally with red elytral tubercles or ridges; upper surfaces either glabrous or clothed with decumbent and/or erect setae, usually interspersed with erect bristles or sometimes with clusters of expanded, ribbed setae.

Head strongly declined at base, deeply inserted into prothorax and often not or barely visible from above, not abruptly constricted posteriorly. Occipital region with short median endocarina, with sharp transverse carina in *Microchaetes* (Fig. 44G). Eyes relatively small, not protuberant, entire or weakly emarginate, finely facetted, without interfacetal setae. Antennal insertions often slightly exposed in Byrrhinae, concealed in *Microchaetes*; subantennal groove short or absent. Frontoclypeal suture absent; frontoclypeal region usually with weak to strong transverse ridge. Labrum relatively large, well sclerotised and exposed, its base on a lower plane forming a distinct groove between main portion of sclerite and clypeus; anterior edge broadly rounded or truncate. Antennae short to moderately long, almost always 11-segmented (rarely 10-segmented), clavate or capitate with a weak club 2- to 5- or 6-segmented club. Mandible broad and subtriangular or wedge-like, tridentate or bidentate with an additional subapical tooth; with small but distinct mola and extensive prostheca in Microchaetes; mola and prostheca highly reduced or absent in Byrrhinae; mesal surface of mandibular base always with brush of hairs or setose lobe. Maxilla with short, setose galea and lacinia, the latter without uncus; apical palpomere fusiform. Mentum strongly transverse; ligula simple or bilobed; apical labial palpomeres fusiform. Subgenal ridges absent. Gular sutures well separated; gula transverse. Corpotentorium broad, straight. Cervical sclerites well developed.

Pronotum ~0.38–0.65 times as long as wide, widest posteriorly with sides slightly curved and converging anteriorly. Lateral pronotal carinae complete, with or without distinct bead; anterior angles not or slightly produced forward and acute; posterior angles right to acute; posterior edge straight to weakly biemarginate; disc strongly convex and simple. Prosternum in front of coxae slightly to distinctly shorter than shortest diameter of procoxal cavity, slightly convex, with paired lateral antennal grooves and a concave head rest above, with anteriorly diverging precoxal lines in *Microchaetes* forming mesal portions of crural impressions extending onto hypomera. Prosternal process complete, parallel-sided or narrowed apically and broadly to narrowly rounded at apex. Notosternal sutures complete, open anteriorly and associated with antennal grooves. Procoxae more or less cylindrical, with very short concealed lateral extensions, trochantin well developed and exposed. Procoxal cavities strongly transverse, moderately widely separated, broadly open externally and internally.

Scutellar shield small to minute or absent, abruptly elevated, anteriorly simple, posteriorly acute. Elytra ~0.9–1.2 times as long as combined width and 1.9–2.9 times as long as pronotum, complete and conjointly rounded apically, often with internal subapical interlocking flange; disc usually with irregular punctation, with 10 or 11 irregular striae or puncture rows in *Microchaetes*; epipleura usually broad at base and then abruptly narrowed and incomplete. Mesoventrite usually separated by complete sutures from mesanepisterna, which are very widely separated (rarely fused to mesanepisterna in Byrrhinae); anterior edge often on different plane than metaventrite, with a pair of large, strongly declined to vertical procoxal rests continued onto mesanepisterna and with shallow to deep, mesoventral cavity; discrimen usually present just behind cavity. Mesoventral process short, broad and apically truncate to emarginate. Mesocoxae subglobular to transverse, with or without short plates and with exposed trochantins. Mesocoxal cavities very widely separated, partly closed laterally by mesepimera sometimes in combination with mesanepisterna. Mesometaventral junction either dicondylic or a straight line, with meso- and metaventrites abutting. Metaventrite slightly convex, very short in wingless species; sometimes with pair of curved or recurved coxal lines; discrimen complete in *Microchaetes*, short or absent in Byrrhinae; transverse (metakatepisternal) suture absent in Byrrhinae, present but relatively short in *Microchaetes*; exposed portion of metanepisternum moderately elongate to very long and narrow or absent (metanepisternum completely concealed). Metacoxae usually contiguous or narrowly separated and extending laterally to meet elytra; strongly transverse with well developed, more or less complete coxal plates in *Microchaetes* and some Byrrhinae, slightly transverse and subconical, without plates in some wingless Byrrhinae. Metendosternite in *Mic*rochaetes and some Byrrhinae with moderately long stalk, short, broad lateral arms and well-developed anterior process with well separated anterior tendons (most wingless Byrrhinae with short or long stalk and long lateral arms, but no anterior process. Hind wing relatively long and narrow with long apical field either without sclerotisation or with vaguely indicated radial extensions; radial cell strongly oblique; cross-vein r3 absent; basal portion of RP short or absent; medial spur long and straight; medial field with two or three veins and no wedge cell; anal lobe without embayment; hind wings absent in almost all Byrrhinae. Femora more or less excavated to receive tibiae; tibiae slender and not apically expanded with simple outer edge in most Byrrhinae, distinctly expanded in *Microchaetes* and bearing socketed spines along outer edge in *Microchaetes* and some Byrrhinae; tibial spurs usually paired, reduced or absent in *Microchaetes*. Tarsi usually 5-5-5 (rarely 4-4-4); tarsomeres simple in *Microchaetes*, tarsomere 3 lobed and 4 reduced in Byrrhinae; pretarsal claws simple.

Abdomen with five ventrites, the first three of which are usually connate (two ventrites connate in *Microchaetes*); ventrite 1 shorter than 2, often with paired crural impressions, delimited by postcoxal lines in *Microchaetes*; intercoxal process usually acute (narrowly rounded in *Microchaetes*). Abdominal tergites lightly sclerotised or membranous. Functional spiracles on abdominal segments I–VIII. Anterior edge of segment IX in male with short pregenital plate but without strut-like spiculum gastrale; tergite IX emarginate and X free. Aedeagus of trilobate type; phallobase asymmetrical; parameres individually articulated or partly fused together (completely fused to form tube in *Microchaetes*); penis long and slender, with very short to moderately long basal struts. Sternite VIII in female without slender spiculum ventrale. Ovipositor short and broad; paraprocts short, usuallywith oblique baculi, much shorter than coxites, each of which is subdivided with distal lobe slender, lightly to moderately sclerotised and setose; gonostyli short to moderately long and apical, sometimes acute and spur-like. [Lea 1907, 1920; Wilson 1921; Paulus 1972, 1973; Pütz 2000; Johnson 2002a, 2005].

Larva (Figs 34N, 65B)

Body elongate and subcylindrical or oblong to ovate and slightly flattened, slightly curved ventrally (sometimes strongly curved near posterior end), usually moderately to heavily pigmented dorsally and lightly so ventrally (lightly pigmented except for head and tergum IX in at least some *Microchaetes*); surfaces smooth or somewhat rugulose, subglabrous above and setose below or clothed both above and below by long and short, erect setae.

Head protracted and moderately to strongly declined (hypognathous), large, globular; epicranial stem moderately long; frontal arms V-shaped, contiguous at base and forming narrowly acute angle; median endocarina often absent beneath epicranial stem; cuticular glands sometimes present on head. Stemmata usually six on each side, well separated, forming two groups of three. Frontoclypeal suture distinct. Labrum free. Antennae short, 3-segmented, with small conical sensorium on antennomere 2. Mandibles symmetrical, short and broad, unidentate or multidentate, without accessory ventral process; mola and prostheca absent; mesal surface of mandibular base with dense brush of hairs (Fig. 47J). Ventral mouthparts (Fig. 40F) strongly retracted forming maxillolabial complex, with well developed maxillary articulating area; cardines transverse and well separated; stipes elongate; galea articulated and palpiform, lacinia fixed and falciform or truncate; palps 3-segmented with segment-like palpifer. Labium consisting of prementum, mentum and submentum; ligula absent; labial palps 2-segmented and widely separated. Hypopharyngeal sclerome and hypostomal rods absent; ventral epicranial ridges well developed. Gular region very short, transverse.

Prothorax somewhat longer than meso- and metathorax but not as long as the two combined; thoracic terga and abdominal terga I–IX each with transverse row of small cuticular glands; additional larger paired glands occasionally present anterolaterally on meso- and metaterga and abdominal terga I–IX; prothoracic venter often with paired cervical sclerites and precoxalia, Legs well developed, relatively stout, often armed with short spines, 5-segmented, with claw-like pretarsus bearing two setae lying side by side; coxae large and strongly projecting, moderately to widely separated. Abdomen about twice to more than 3 times as long as thorax; tergal plates usually well developed (mostly lightly sclerotised in at least some *Microchaetes*), often distinctly punctate and setose, at least laterally and posteriorly, but without rows or patches of asperities or tubercles; with 1 to 3 pairs of laterotergites and often 1 or 2 pairs of laterosternites; tergum VIII sometimes with transverse carina or microasperate elevation near anterior end. Tergum IX subequal to or slightly shorter than VIII, usually not extending onto ventral surface (with slight ventral extension in *Microchaetes*), without paired urogomphi, forming a declined, circular disc (Fig. 44X) in *Microchaetes*; sternite IX exposed and simple; segment X usually posteriorly oriented (posteroventrally oriented in *Microchaetes*), with distinct tergum, sternum and paired pygopods, each sometimes with a hook-like sclerotisation. Spiracles biforous, without closing apparatus, located on or above laterotergites. [Böving and Craighead 1931; Emden 1958; Johnson 1987; Szpojda 1989; 2002a, 2005; Lawrence 1991b].

Classification and World Distribution

The family Byrrhidae includes 28 genera and ~300 described species occurring primarily in the temperate parts of the Northern and Southern Hemispheres (amphipolar distribution of Crowson 1980). The family divided into three subfamilies: Amphicyrtinae, with two species from north-western North America, Syncalyptinae and Byrrhinae. [Paulus 1972, 1973].

Australian Fauna

The Australian byrrhid fauna includes nine species of *Microchaetes* (Fig. 69I) (**Syncalyptinae**) and ~20 species of **Byrrhinae**, including *Morychus torrensensis* (Blackburn) and 20 species currently placed in the genus *Pedilophorus* Steffahny (Fig. 69J). Except for the species of *Microchaetes*, the Australian byrrhids belong to several undescribed genera. Byrrhids are short, stout, strongly convex beetles with a deflexed head, broad prosternal process fitting into a cavity on the mesoventrite, legs which may be capable of retraction into ventral cavities (Fig. 62I), metaventrite without a transverse suture, and tarsi which usually fit into grooves on the tibiae and bear a membranous appendage on tarsomere 3. Species of *Microchaetes* are dull black, tuberculate and clothed with stiff bristles or scales; larvae (Fig. 65B) are cylindrical and lightly sclerotised, with a concave plate on tergum IX. Those species currently placed in *Pedilophorus* are either glabrous or clothed with mixed setae and are usually metallic green. The most striking species is the Tasmanian *Pedilophorus gemmatus* Lea (Fig. 69J). Their larvae are variable but are commonly onisciform with a heavily sclerotised dorsal surface. Adult and larval byrrhids are usually found feeding on mosses and liverworts, but some may occur among the roots of higher plants. An undescribed species from Mt. Kosciusko, NSW, was found feeding on the rosettes of *Cardamine* (Brassicaceae) and apparently transmits yellow mosaic virus (TYMV-Cd) (Guy and Gibbs 1985). [Lea 1907, 1920; Wilson 1921].

30. LIMNICHIDAE

Adult (Figs 63D, 69H)

Length 1–3.2 mm. Body ~1.45–2.1 times as long as wide, usually widest at pronoto-elytral junction and narrowing anteriorly and posteriorly; sides slightly to strongly, usually uniformly rounded, moderately to strongly convex dorsally and slightly convex ventrally; colour uniformly reddish-brown to black; upper surfaces usually densely clothed with decumbent hairs, which may differ in colour and form a distinct pattern or a silvery sheen, often interspersed with erect hairs, or with erect hairs only or sparsely distributed decumbent hairs.

Head not to slightly declined at base but with vertex steeply sloping, so that mouthparts are either hypognathous or opisthognathous, deeply inserted into prothorax in Limnichinae, but produced forward and sometimes slightly elevated at base and abruptly declined from about the upper edges of the eyes in *Parathroscinus* Wooldridge and the Thaumastodinae, not abruptly constricted posteriorly. Occipital region usually with fine transverse ridge or carina in Limnichinae. Eyes usually well developed, not protuberant, entire, finely facetted, without interfacetal setae (deeply divided by ocular canthus in *Paralimnichus* Delève, somewhat reduced in *Hyphalus* Britton). Antennal insertions exposed or barely concealed; subantennal groove usually short or absent in *Hyphalus* and Limnichinae, well developed in *Parathroscinus* and Thaumastodinae. Frontoclypeal suture present; anterior edge of clypeus truncate or broadly rounded. Labrum usually relatively large, well sclerotised and exposed, its base on a lower plane forming a distinct groove between main portion of sclerite and clypeus; anterior edge broadly rounded. Antennae short to moderately long and slender, 11-segmented, filiform or with apical antennomeres slightly thickened, sometimes subserrate (*Parathroscinus*) or forming a weak club (3-segmented in *Hyphalus*, 4-segmented in Thaumastodinae). Mandible broad at

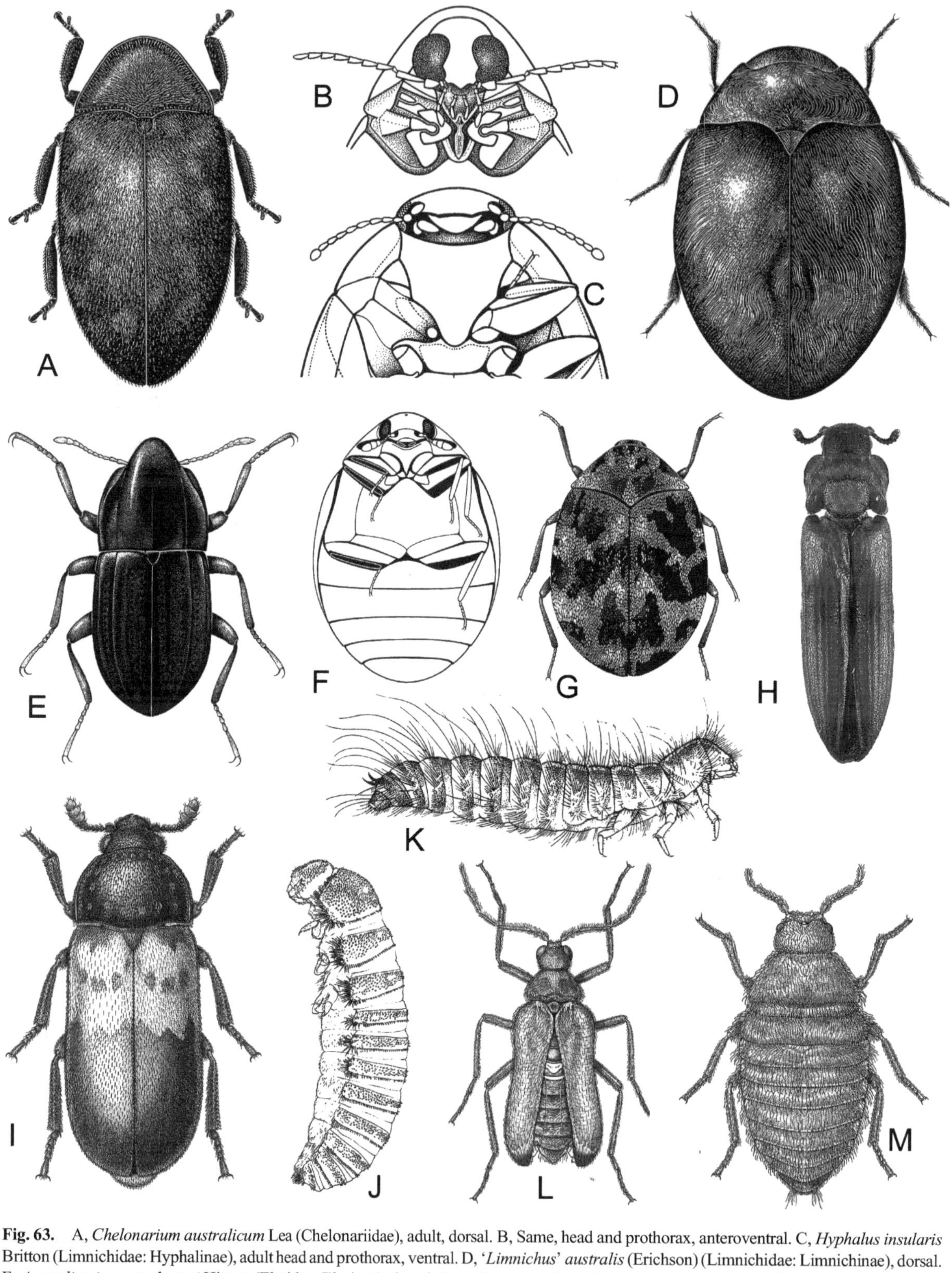

Fig. 63. A, *Chelonarium australicum* Lea (Chelonariidae), adult, dorsal. B, Same, head and prothorax, anteroventral. C, *Hyphalus insularis* Britton (Limnichidae: Hyphalinae), adult head and prothorax, ventral. D, '*Limnichus*' *australis* (Erichson) (Limnichidae: Limnichinae), dorsal. E, *Austrolimnius waterhousei* Hinton (Elmidae: Elminae), dorsal. F, *Anthrenus verbasci* (Linnaeus) (Dermestidae: Megatominae), ventral. G, same, dorsal. H, *Derbyana matthewsi* Lawrence and Ślipiński (Dermestidae: Dermestinae), adult, dorsal. I, *Dermestes lardarius* Linnaeus (Dermestidae: Dermestinae), adult, dorsal. J–K, Larva, lateral: J, *Anthrenocerus stigmacrophilus* Armstrong (Dermestidae: Megatominae); K, *Dermestes maculatus* Degeer. L, *Thylodrias contractus* Motschulsky (Dermestidae: Trinodinae), male, dorsal; M, same, female, dorsal. I, L, M from Busquet 1990, © Agriculture and Agri-Food Canada; J from Kisyelova and McHugh 2006, © T. Kisyelova and J.V. McHugh.

base and narrow at apex, usually bidentate with an additional subapical tooth (unidentate in *Paralimnichus*), with well developed mola and prostheca usually consisting of sclerotised, apically forked process attached to a fringed membrane (fringed membrane only in *Paralimnichus*). Maxilla in *Hyphalus* and Limnichinae with lacinia usually broader than galea and setose without uncus; in *Parathroscinus* and Thaumastodinae lacinia is narrower and more spinose with uncus; apical palpomere usually fusiform (cylindrical in *Paralimnichus*); mentum usually transverse (more elongate in Thaumastodinae); ligula truncate; apical labial palpomeres usually fusiform (cylindrical in *Paralimnichus* and Thaumastodinae, expanded and truncate in *Parathroscinus*); labial palps aciculate in Thaumastodinae. Subgenal ridges present in *Paralimnichus*. Gular sutures well separated; gula slightly to strongly transverse. Corpotentorium very narrow, slightly arched; laminatentoria fused at midline forming anterior bridge. Cervical sclerites well developed.

Pronotum ~0.32–0.65 times as long as wide, widest posteriorly with sides slightly to strongly curved and converging anteriorly. Lateral pronotal carinae complete, usually with distinct bead; anterior angles slightly to strongly produced forward and acute; posterior angles usually right to slightly acute; posterior edge weakly biemarginate in *Hyphalus*, triemarginate in many Limnichinae, with four distinct emarginations in *Byrrhinus* Motschulsky, well developed interlocking device; disc slightly convex and simple. Prosternum in front of coxae usually slightly to much longer than shortest diameter of procoxal cavity (shorter in Thaumastodinae), moderately convex, usually with truncate chin piece, above which is a concave head rest (absent in Thaumastodinae), with curved precoxal lines in *Paralimnichus* forming mesal portions of crural impressions extending onto hypomera. Prosternal process complete, parallel-sided or slightly narrowed apically and broadly rounded to truncate at apex (Fig. 63C). Notosternal sutures complete, usually open anteriorly and often associated with antennal grooves. Procoxae more or less cylindrical, without concealed lateral extensions, usually with short coxal plates; trochantin usually large, well sclerotised and abutting sternum and notum (reduced and concealed Thaumastodinae). Procoxal cavities strongly transverse, moderately widely separated, broadly open externally and internally.

Scutellar shield abruptly elevated, anteriorly simple, posteriorly subacute to broadly angulate. Elytra ~1.15–1.63 times as long as combined width and 2.4–4.6 times as long as pronotum, usually more or less attenuate near suture, forming an interlocking device with abdominal apex; an internal interlocking ridge sometimes present at anterior third; posterolateral edge minutely denticulate in *Babalimnichus*; disc usually with irregular punctation, sometimes forming indistinct rows in *Byrrhinus*; epipleura usually broad at base and then abruptly narrowed but more or less complete. Mesoventrite (Fig. 28V) separated by complete sutures from mesanepisterna, which are very widely separated; anterior edge often on different plane than metaventrite; at middle with a pair of short carinae flanked by large, paired procoxal rests, which are usually almost vertical (moderately declined in *Hyphalus*) and continued onto mesanepisterna, and delimiting a midline channel leading to a large, usually deep, mesoventral cavity (shallow in *Hyphalus*); discrimen usually present. Mesoventral process short and apically truncate. Mesocoxae globular to slightly transverse, with short plates and exposed trochantins. Mesocoxal cavities very widely separated, partly closed laterally by mesepimera sometimes in combination with mesanepisterna. Mesometaventral junction a straight line, with meso- and metaventrites abutting or slightly overlapping. Metaventrite slightly to moderately convex, that portion behind mesocoxal cavities distinctly shorter than cavities in *Hyphalus* and Thaumastodinae; discrimen usually complete at least to base of metaventral process and transverse (metakatepisternal) suture usually well developed (not in *Hyphalus*); with pair of crural impressions extending onto metanepisterna in Limnichinae; exposed portion of metanepisternum moderately elongate to very long and narrow. Metacoxae usually contiguous (narrowly separated in *Paralimnichus*, widely so in *Hyphalus*), usually extending laterally to meet elytra (not in *Hyphalus*), usually strongly transverse with well developed, more or less complete coxal plates; coxae greatly enlarged and oblique with plates more or less longitudinal and restricted to mesal portions of coxae in Thaumastodinae; plates reduced in *Hyphalus*. Metendosternite usually with moderately long, slender stalk, short or long lateral arms, and a short, apically emarginate anterior process with tendons widely separated on either side of emargination (stalk very short and anterior process absent in *Hyphalus*; stalk long, broad, and well sclerotised with moderately long, apically expanded arms in Thaumastodinae). Hind wing with apical field relatively long and usually without distinct sclerites (*Parathroscinus* with an anterior linear sclerite and a single radial extension); radial bar relatively thick apically and folding point located behind anterior wing margin near one end of linear sclerite; radial cell elongate with inner posterobasal angle acute or with basal edge absent; cross-vein r3 moderately long and longitudinal; r4 sometimes incomplete or absent; basal portion of RP moderately to very long; R-M loop moderately broad; medial spur usually extending almost to wing margin, where there is a weak medial embayment; medial field with two or three veins and no wedge cell; anal embayment absent. Legs slender; femora excavated to receive tibiae in Limnichinae; tibiae slender and not apically expanded; mesotibiae with a few short spines and metatibiae with several long spines in Thaumastodinae; tibial spurs absent. Tarsi usually 5-5-5 (4-4-4 in *Hyphalus* and Thaumastodinae); tarsomeres almost always simple; tarsomere 4 reduced and tarsomeres 1–3 with ventral, setose lobes in *Paralimnichus*; pretarsal claws simple; empodium usually not apparent.

Abdomen with five ventrites, the first three of which are usually connate (five ventrites connate in *Paralimnichus*); ventrite 1 usually slightly shorter than 2, with recurved postcoxal lines delimiting crural impressions in Limnichinae; intercoxal process usually acute (broad and truncate in *Hyphalus*). Abdominal tergites lightly sclerotised or membranous. Functional spiracles on abdominal segments I–VIII. Anterior edge of sternite VIII in male usually with median strut in *Hyphalus* and *Paralimnichus*. Anterior edge of segment IX in male with spiculum gastrale in some Limnichinae; tergite IX and X usually membranous. Aedeagus of trilobate type; phallobase asymmetrical; parameres individually articulated, partially fused together or completely fused; penis long and slender, usually with long struts. Sternite VIII in female with long, fixed spiculum ventrale. Ovipositor narrowly elongate; paraprocts with large, thickened baculi, much longer than coxites, each of which is undivided, heavily sclerotised and apically acute or somewhat hook-like and usually laterally compressed, without styli. [Spilman 1959, 1972; Britton 1971, 1973, 1977; Wooldridge 1983, 1984;

Satô 1994; Spangler 1995; Satô and Yoshitoma 1998; Hernando and Ribera 2000, 2003, 2004, 2005; Shepard 2002b; Yoshitomi 2006].

Larva

Body elongate, more or less parallel-sided, and subcylindrical, often distinctly curved ventrally, especially abdominal apex. moderately pigmented dorsally, especially head, protergum and tergum IX and brownish in colour, more lightly pigmented ventrally; surfaces smooth, densely clothed with long and short, erect setae.

Head prognathous and protracted, as wide as prothorax, slightly transverse, weakly rounded laterally; epicranial stem short; frontal arms V-shaped or U-shaped and contiguous at base; endocarina absent. Stemmata five on each side, well separated. Frontoclypeal suture distinct. Labrum free, relatively large, slightly transverse and slightly emarginate, with long, heavily sclerotised tormae. Antennae very short, 3-segmented, with large, dome-like or broadly conical sensorium on antennomere 2. Mandibles symmetrical, somewhat wedge-like with excavate mesal edge, tridentate with one or two subapical teeth; mola and prostheca absent; mesal surface of mandibular base usually simple, with articulated, pectinate process in *Hyphalus* and *Babalimnichus*. Ventral mouthparts strongly retracted forming maxillolabial complex, without maxillary articulating area (Fig. 38G); cardines approximate or contiguous, divided by internal ridge; stipes elongate, well separated from one another; galea articulated, lacinia fixed, both apically rounded or truncate. Labium consisting of prementum and postmentum, the latter elongate; ligula broad, truncate and at least as long as basal palpomere; labial palps 2-segmented and widely separated. Hypopharyngeal sclerome and hypostomal rods absent; ventral epicranial ridges well developed. Gular region moderately long but much more lightly sclerotised than head capsule on either side; gular sutures well separated; gula slightly longer than wide.

Prothorax much longer than meso- or metathorax but not as long as the two combined; venter without distinct armature. Legs well developed, 5-segmented, with claw-like pretarsus bearing 1 seta; coxae large and narrowly separated. Abdomen about twice as long as thorax; terga and sterna simple. Tergum IX shorter than VIII, not extending onto ventral surface, without paired urogomphi or other armature; sternite IX exposed, but reduced and strongly transverse, not enclosed by sternum VIII, segment X forming weakly sclerotised operculum, narrowly rounded apically and bearing a pair of curved teeth or hooks; anal opening usually concealed by operculum. Spiracles cribriform type, without closing apparatus, those on abdomen somewhat reduced. [Paulus 1970; Britton 1971; Brown 1991; Yoshitomi and Satô 2001; Shepard 2002b; Hernando and Ribera 2005].

Classification and World Distribution

The Limnichidae is a widely distributed family with ~400 species in 37 genera. There are four subfamilies, all of which are represented in Australia.

Australian Fauna

The Australian Limnichidae include a single species of **Cephalobyrrhinae**, *Parathroscinus ater* (Lea), two genera of **Thaumastodinae**, *Babalimnichus* Satô and *Pseudeucinetus* Heller, *Hyphalus insularis* Britton (Fig. 63 C) (**Hyphalinae**) and several **Limnichinae**, including *Paralimnichus castaneus* (Lea) and several species of *Byrrhinus* Motschulskly (Fig. 69H) and *Pelochares* Mulsant and Rey (Fig. 63D). The Australian species need revision and other genera may occur here. Limnichids are small, oval and moderately convex beetles, clothed with decumbent, erect or occasionally minute and pruinose vestiture. They resemble some Byrrhidae in having ventral cavities into which the legs can be folded, but they differ from byrrhids in being less convex and usually by having slender antennae, a longer prosternum which extends well in front of the procoxae, a distinct frontoclypeal suture, and usually a more or less complete metakatepisternal suture. *Paralimnichus* lacks the katepisternal suture and has lobed tarsi similar to those in byrrhids, but differs from them in having all five ventrites connate and the eyes deeply emarginate. The two thaumastodines are distinguished by having narrowly separated and dorsally placed eyes, large, oblique metacoxae, spinose metatibiae and 4-4-4 tarsi. The genus *Hyphalus* also has 4-4-4 tarsi and differs from other limnichids in having widely separated metacoxae and no metakatepisternal suture. Limnichid larvae are cylindrical and resemble some kinds of byrrhid larvae, but they are easily distinguished from them by having the maxillolabial complex consolidated, so that the cardines are contiguous, the maxillary articulating areas absent and the postmentum joined to the stipites for most of its length. Limnichids occur in riparian situations and large numbers may be attracted to lights at night. Larvae usually live in sand, mud and leaf litter, and apparently feed on decaying organic matter. The Australian fauna is relatively rich but few species have yet been described. Many northern species belong to the pan-tropical genus *Byrrhinus*. *Hyphalus insularis* was described from the interstices of intertidal coral slabs on the Great Barrier Reef, but the genus is also known from Lord Howe Island, Norfolk Island, the Cocos-Keeling Islands, the Seychelles, New Zealand, Taiwan and the Ryukyu Islands. *Hyphalus* larvae have anal gills for living submerged during high tides. [Spilman 1959, 1972; Britton 1971, 1973, 1977; Wooldridge 1983, 1984; Lawrence 1987a; Spangler 1995; Satô and Yoshitomi 1998; Spangler *et al.* 2001; Hernando and Ribera 2000, 2003, 2004, 2005, 2010; Yoshitomi 2006].

31. HETEROCERIDAE

Adult (Fig. 62H)

Length 2.2–5.3 mm. Body usually ~2 to 3 times as long as wide (as much as 3.5 times as long as wide in *Elythomerus*), more or less parallel-sided but constricted at pronoto-elytral junction, slightly flattened dorsally and convex ventrally; colour yellow to black, but usually with pattern of light spots or streaks on a dark background or, less commonly, with dark markings on a light background; upper surfaces usually densely clothed with decumbent hairs interspersed with erect hairs.

Head prognathous, deeply inserted into prothorax, transverse, not abruptly constricted posteriorly. Occipital region without transverse ridge or longitudinal endocarina. Eyes well developed, slightly protuberant, entire, finely facetted, without interfacetal setae; ommatidium of exocone type. Antennal insertions exposed; subantennal groove absent. Frontoclypeal suture distinct and not or only slightly curved; anterior edge of clypeus truncate or slightly concave, usually with pair of horn-like processes in male (not in *Elythomerus*). Labrum large and well sclerotised, its anterior edge strongly convex. Antennae short, 11-segmented, with last 7 antennomeres usually forming a subserrate club (club compact and not serrate in *Elythomerus*); scape usually distinctly widened apically and not much longer than pedicel (slightly widened and much longer than pedicel in *Elythomerus*). Mandible elongate and flattened, unidentate, usually with 2 subapical teeth on mesal edge and a single tooth on outer edge (absent in *Elythomerus*); outer edge near base with several spine-like setae; mola moderately well developed; prostheca usually long and fringed with seta-like processes (reduced in *Elythomerus*). Maxillae with galea 2-segmented and articulated, lacinia fixed and narrowly rounded or subacute, both lined internally with stout setae; apical palpomere more or less fusiform; mentum large, subtrapezoidal and apically emarginate; prementum elongate, ligula slightly emarginate; apical labial palpomeres fusiform. Subgenal ridges present. Gular sutures well separated; gula transverse. Corpotentorium slender and arched; laminatentoria fused at midline forming anterior bridge. Cervical sclerites well developed.

Pronotum ~0.5–0.6 times as long as wide, widest at middle, with sides slightly to strongly curved. Lateral pronotal carinae very weakly developed or absent, without distinct bead; anterior angles not produced forward; posterior angles broadly rounded; posterior edge weakly rounded, without interlocking device; disc slightly convex and simple. Prosternum in front of coxae distinctly longer than shortest diameter of procoxal cavity, moderately convex, with rounded chin piece, above which is a concave head rest. Prosternal process complete, slightly expanded apically, strongly curved dorsally behind coxae and truncate at apex. Notosternal sutures complete. Procoxae not projecting below prosternum, without concealed lateral extensions; trochantin exposed, moderately large, but not abutting sternum or notum. Procoxal cavities strongly transverse, moderately narrowly separated, broadly open externally and internally.

Scutellar shield abruptly elevated, anteriorly simple, posteriorly acute. Elytra ~1.53–1.63 times as long as combined width and 3.4–3.6 times as long as pronotum, with sides subparallel and apices conjointly rounded; an internal interlocking ridge present just anterior to middle; disc with irregular punctation; epipleura abruptly narrowed at anterior third but more or less complete. Mesoventrite separated by complete sutures from mesanepisterna, which are well separated; anterior edge on same plane as metaventrite; at middle with a pair of short carinae flanked by large, paired, moderately declined procoxal rests continuing onto mesanepisterna, and delimiting a midline channel leading to a moderately large but shallow mesoventral cavity; discrimen absent. Mesoventral process short and apically truncate or slightly concave. Mesocoxae slightly transverse, with exposed trochantins. Mesocoxal cavities usually very widely separated (narrowly separated in *Elythomerus*), partly closed laterally by mesepimera. Mesometaventral junction a slightly curved line, with meso- and metaventrites slightly overlapping. Metaventrite moderately convex; discrimen usually complete to base of metaventral process and transverse (metakatepisternal) suture well developed; with pair of curved postcoxal lines extending from middle of each coxal cavity to about middle of metanepisterna; exposed portion of metanepisternum moderately elongate. Metacoxae narrowly separated, extending laterally to meet elytra, transverse and obliquely oriented, with weak but more or less complete coxal plates. Metendosternite with moderately long, slender stalk, short lateral arms, and a short anterior process with tendons widely separated. Hind wing with short radial bending zone and relatively long apical field with one vaguely indicated radial extension and folding point located well behind anterior wing margin near one end of a transversely oblique, linear sclerite; radial cell elongate with partly pigmented lumen but no basal edge; cross-vein r3 moderately long and slightly oblique; r4 complete; basal portion of RP moderately long, with apical extension reaching wing margin; R-M loop moderately broad; medial spur short and ending well before wing margin; medial field with 2 or 3 free veins and no wedge cell; anal lobe somewhat reduced, without embayment and with undivided AP. Legs relatively slender; metafemora with anteroventral, longitudinal carina (stridulatory plectrum); tibiae more or less expanded subapically (more strongly so on fore leg) with long, slender, socketed spines along outer edge (Fig. 27B); tibial spurs paired, slender and unequal. Tarsi 4-4-4; tarsomeres slender and simple, without ventral lobes; pretarsal claws simple (reduced and hair-like in *Elythomerus*); empodium usually not apparent.

Abdomen with five ventrites, the first three of which are connate; ventrite 1 usually slightly shorter than 2, with pair of curved lines beginning at posterolateral edges of metacoxal cavities, usually ending near midline, and consisting of a series of short, fine transverse ridges (pars stridens); intercoxal process broad at base but apically acute. Abdominal tergites moderately well sclerotised, with patches of wing folding spicules on segments V–VII. Functional spiracles on abdominal segments I–VIII, those on I–VII in pleural membrane, those on VIII on tergite. Anterior edge of sternite VIII in male without median strut. Anterior edge of segment IX in male with narrowly acute subgenital plate, sometimes forming short spiculum gastrale; tergite X membranous. Aedeagus of modified trilobate type; phallobase more or less symmetrical, well sclerotised and anteriorly blunt, but with apically rounded anterior projection; parameres usually individually articulated to phallobase, except in *H. flindersi* Blackburn, where they are fused to phallobase; penis somewhat asymmetrical, often with anterior apical projection on one side only; base with single strut. Sternite VIII in female without spiculum ventrale. Ovipositor highly reduced and membranous, except for remnants of paraproctal baculi; styli absent. [Pacheco 1964; Charpentier 1965, 1968; Vanin *et al.* 1995, 2005; Katovich 2002b].

Larva

Body elongate, more or less parallel-sided or slightly wider anteriorly and narrowed towards abdominal apex, subcylindrical or slightly flattened, abdominal apex usually distinctly curved ventrally; moderately pigmented dorsally and yellow or yellowish-brown in colour, more lightly pigmented ventrally; surfaces smooth, clothed with long and short, erect setae.

Head prognathous and protracted, as wide as prothorax, slightly transverse, weakly rounded laterally; epicranial stem absent; frontal arms V-shaped or U-shaped and contiguous at base; endocarina absent. Stemmata five on each side, well separated. Frontoclypeal suture distinct. Labrum free, strongly transverse and subtruncate with long, heavily sclerotised tormae. Antennae very short, 3-segmented, with large, dome-like sensorium on antennomere 2. Mandibles (Fig. 41O) symmetrical, somewhat wedge-like with excavate mesal edge, broadly tridentate; mola and prostheca absent (mandible bidentate and more flattened with complex pseudomola in at least some exotic species). Ventral mouthparts (Fig. 38I) strongly retracted forming maxillolabial complex, without maxillary articulating areas; cardines approximate, subtriangular, divided by internal ridge; stipes elongate, well separated from one another and separated from labium by complete sutures; galea articulated, lacinia fixed, both apically rounded or truncate. Labium consisting of prementum and postmentum, the latter elongate and broadly rounded at base; ligula broadly rounded; labial palps 2-segmented and widely separated. Hypopharyngeal sclerome and hypostomal rods absent; ventral epicranial ridges well developed. Gular region transverse but more lightly sclerotised than head capsule on either side; gular sutures well separated.

Prothorax distinctly longer than meso- or metathorax but not as long as the two combined; terga with lightly sclerotised plates and meso- and metatergum with fine transverse, anterior carina; prothoracic venter with large, subtriangular presternite vaguely divided at midline, flanked by large precoxalia. Legs well developed, 5-segmented, with claw-like pretarsus bearing 1 seta; coxae large and narrowly separated. Abdomen about twice as long as thorax; terga I–IX with lightly sclerotised plates and first 8 with fine transverse, anterior carina; sterna simple and more lightly sclerotised. Tergum IX subequal in length to VIII, not extending onto ventral surface, without paired urogomphi or other armature; sternum IX exposed, short and strongly transverse; segment X posteriorly oriented, sternum X forming weakly sclerotised operculum bearing a pair of fine hooks at apex; anal opening transverse. Spiracles cribriform, without closing apparatus, those on abdomen with reduced sieve plate (Figs 38H, 53S). [Böving and Craighead 1931; Gwynn-Silvey 1935; Kaufmann and Stansly 1979; Lawrence 1991b; Vanin *et al.* 1995, 2005; Katovich 2002b].

Classification and World Distribution

The family Heteroceridae is a worldwide family with ~15 described genera and 300 species. There are two subfamilies: Elythomerinae with a single Australian genus and species, and Heterocerinae containing the remaining taxa. Most heterocerine genera were described by Pacheco (1964), but Miller (1988) questioned the validity of some of Pacheco's genera, and King *et al.* (2011) proposed that only the widely distributed *Heterocerus* Fabricus, the Holarctic *Auglyes* Schiödte, the Asian *Micilus* Mulsant and Rey and the New World *Tropicus* Pacheco be recognised.

Australian Fauna

The Australian fauna includes *Elythomerus elongatulus* Waterhouse (**Elythomerinae**), known from only two specimens collected at Rockhampton, QLD, and seven species of *Heterocerus* Fabricius (**Heterocerinae**). These are elongate, densely pubescent beetles with maculate elytra, prognathous head, large labrum, thick, usually subserrate antennae, flattened, spinose tibiae and simple tarsi. Larvae are elongate, widest at the prothorax, tapered posteriorly, and clothed with scattered dark, erect setae; the mandibles are flattened with a sub-basal, concave mola, the ventral mouthparts are consolidated (Fig. 38I), and segment X forms a conical pygopod. Hetrerocerids tunnel in mud at the edges of ponds and streams and apparently subsist on algae, diatoms and organic debris. Egg guarding has been reported for one North American species (Folkerts 1989). [Gwynn-Silvey 1935; Bro-Larsen 1936; Charpentier 1968; Kaufmann and Stansly 1989; Vanin *et al.* 2005].

32. ELMIDAE

Adult (Figs 62G, 63E)

Length 0.9–10.2 mm. Body ~1.75 to 2.65 times as long as wide, oblong or subovate to elongate, with pronotum and elytra usually independently curved (sides almost parallel in *Graphelmis*), slightly flattened to strongly convex; colour usually black, occasionally metallic, sometimes with red or yellow maculae or linear markings on elytra (with yellow and black pattern on pronotum and elytra in *Graphelmis*, with legs partly yellow in some Larainae); dorsal surfaces densely clothed with short, decumbent hairs in Larainae, subglabrous in Elminae; ventral surfaces clothed with hydrofuge hairs and appearing uniformly grey, sometimes with a silver sheen.

Head slightly to moderately declined, deeply inserted into prothorax, about as long as wide, not abruptly constricted posteriorly. Occipital region without transverse ridge or longitudinal endocarina. Eyes well developed, slightly to moderately protuberant, entire, finely facetted, without interfacetal setae, but surrounded by long, dark 'eyelash' setae in Larainae; ommatidium of exocone type. Antennal insertions exposed; subantennal groove absent. Frontoclypeal suture distinct, straight or slightly curved; clypeus transverse, usually with well developed membranous anteclypeus, anteriorly truncate. Labrum large and well sclerotised, sometimes laterally expanded, anterior edge strongly truncate to slightly emarginate. Antennae moderately short, rarely exceeding base of prothorax, 11-segmented, subserrate or clavate and pubescent in Larainae, filiform and glabrous in Elminae; scape elongate, much longer than pedicel in Larainae, not or only slightly longer in Elminae. Mandible moderately short, broad and flattened, bidentate, or obliquely tridentate, with well developed, asymmetrical molae and large, membranous prostheca. Maxillae with articulated, palpiform galea and broad, fixed lacinia with complexly fringed apex; maxillary palps usually 4-segmented (3-segmented in *Kingolus*); apical palpomere more or less fusiform or apically expanded; mentum strongly transverse and apically truncate; ligula laterally expanded, and apically truncate; apical labial palpomeres fusiform or apically expanded. Subgenal ridges absent or weakly developed. Submentum separated from pregular region, which lies between pregular sutures; tentorial pits located near posterior end of head, so that gula proper is reduced. Corpotentorium narrow and straight; laminatentoria well developed but not fused at midline. Cervical sclerites well developed.

Pronotum ~0.6–1.1 times as long as wide, usually near posterior end with sides straight to strongly curved or undulate and narrowing anteriorly (widest at middle with sides strongly rounded in some *Notriolus*). Lateral pronotal carinae usually sharply defined and complete but without distinct bead (weakly defined in *Potamophilinus* and incomplete anteriorly in *Stetholus*); anterior angles produced forward and acute in Kingolus and Notriolus; posterior angles right to acute; posterior edge usually triemarginate (not or weakly so in *Austrolimnius*), with moderately well developed interlocking device; disc sometimes weakly impressed forming anterior and posterior lobes, with shallow posterolateral depressions in some Larainae, with short, paired basal sublateral carinae in *Hydora*, *Ovolara* and *Kingolus*, with long sublateral carinae and a median longitudinal sulcus in *Austrolimnius*, and with 2 paramedian and 2 sublateral, longitudinal carinae in *Graphelmis*. Prosternum in front of coxae usually slightly shorter than shortest diameter of procoxal cavity, flattened to slightly convex, and with a short, broad, truncate chin-piece (much shorter without chin-piece in *Hydora*, *Stetholus* and *Potamophilinus*), with paired, longitudinal carinae in *Kingolus* and *Coxelmis*. Prosternal process complete, parallel-sided or posteriorly narrowed, not strongly curved dorsally behind coxae and subacute to broadly rounded or truncate at apex. Notosternal sutures complete. Procoxae projecting below prosternum, without or with short concealed extensions and with exposed trochantins in Larainae, not projecting, with more or less well developed concealed extensions and concealed trochantins in Elminae. Procoxal cavities strongly transverse in Larainae, slightly transverse to subcircular in Elminae, narrowly to widely separated, broadly open externally and internally.

Scutellar shield abruptly elevated, anteriorly simple, posteriorly acute to broadly angulate. Elytra ~1.3–2.1 times as long as combined width and 2.1–3.75 times as long as pronotum, with sides subparallel, curved, or sinuate and apically converging, with apices conjointly or independently rounded; disc usually with 10 puncture rows, but sometimes with reduced number or with punctures not seriate (e.g. *Coxelmis*), with or without scutellary striole; each elytron with 2 sublateral longitudinal carinae in *Austrolimnius*; lateral interlocking ridge present or absent; epipleura sometimes abruptly narrowed at anterior third, complete or incomplete. Mesoventrite separated by complete sutures from mesanepisterna, which are widely separated; anterior edge usually on same plane as metaventrite, with a pair of slightly to strongly declined procoxal rests continuing onto mesanepisterna; mesoventral cavity small and shallow to large and deep, sometimes delimited laterally by paired carinae; discrimen usually absent, sometimes short. Mesoventral process short and more or less emarginate. Mesocoxae subglobular or slightly transverse, sometimes slightly projecting, with exposed trochantins. Mesocoxal cavities usually widely to very widely separated, partly closed laterally by mesepimera. Mesometaventral junction dicondylic. Metaventrite moderately convex but flattened at middle; discrimen usually complete to base of metaventral process and transverse (metakatepisternal) suture well developed; with pair of straight postcoxal lines in *Kingolus*; exposed portion of metanepisternum narrowly elongate; metepimera concealed. Metacoxae moderately to very widely separated, transverse, oblique, extending laterally to meet elytra in Larainae, with coxal plates moderately well developed and complete in Larainae, but variously reduced in some Elminae. Metendosternite with moderately long, stalk, short to moderately long and transverse lateral arms, and a short to long anterior process with tendons approximate or widely separated. Hind wing without or with weak radial bending zone and relatively long apical field, with vaguely indicated radial extensions and folding point located just behind anterior wing margin; radial cell without basal edge; cross-vein r3 short, r4 complete; basal portion of RP moderately long; R-M loop moderately broad; medial spur extending to wing margin; medial field with as many as 5 free veins and small wedge cell in Larainae, with fewer free veins and no wedge cell in Elminae; anal lobe moderately well developed, without embayment and with undivided AP; wings absent some Elminae. Legs relatively long and slender, not expanded or flattened and without spines or other modifications; tibial spurs small and often not apparent. Tarsi 5-5-5, sometimes almost as long as tibiae; tarsomeres slender and simple, without ventral lobes; pretarsal claws usually simple, sometimes with basal tooth; empodium usually not apparent.

Abdomen usually with 5 ventrites (6 in *Potamophilinus*), the first three of which are connate; ventrite 1 usually about as long as 2; intercoxal process broad at base, apically angulate to broadly rounded, but in Elminae with pair of dorsal, subapical cavities for receiving paired posterior projections of metaventrite. Abdominal tergites usually lightly sclerotised or posterior tergites more heavily sclerotised, with patches of wing folding spicules on segments V–VII. Functional spiracles on abdominal segments I–VIII in Larainae, I–VII in Elminae, usually located in pleural membrane (sometimes in laterosternites of segments VI and VII). Anterior edge of sternite VIII in male usually with short median strut. Anterior edge of segment IX in male with narrowly acute, usually asymmetrical subgenital plate, sometimes forming short spiculum gastrale; tergite X usually well sclerotised but at least partly fused to IX, which is deeply emarginate. Aedeagus of trilobate type; phallobase usually asymmetrical, with apically rounded anterior projection in most Larainae; parameres usually individually articulated to phallobase (partly fused together in some Kingolus, almost completely fused to form tube surrounding penis in *Potamophilinus* and *Graphelmis*); penis with short basal struts. Sternite VIII in female with short to long spiculum ventrale. Ovipositor usually moderately elongate (only slightly longer than wide in *Potamophilinus*), symmetrical; proctiger usually membranous with lateral baculi; paraprocts subequal to or somewhat longer than coxites, with longitudinal baculi; coxites attached apically, each divided into short, broad basal lobe with transverse or oblique baculum and long narrow apical one, usually more heavily sclerotised; styli usually well developed, palpiform and apically attached, sometimes reduced and subapical or occasionally absent. [Carter and Zeck 1929; Hinton 1965, 1968; Brown 1981; Glaister 1999; Shepard 2002a; C iampor and Kodada 2004; Kodada and Jäch 2005].

Larva (Fig. 65H)

Body elongate, more or less parallel-sided or wider anteriorly and narrowed towards abdominal apex (occasionally broadly elliptical), subcylindrical to strongly flattened; moderately strongly pigmented dorsally, yellow to dark reddish-brown or black, sometimes with darker colour forming patterns, lightly to heavily pigmented ventrally; surfaces usually more or less granulate, often with smooth areas (signa), large setiferous tubercles, or asperities and with vestiture of simple and modified setae of various types.

Head prognathous to moderately declined and protracted or slightly retracted, as wide as to distinctly narrower than prothorax, about as long as wide, weakly rounded laterally and slightly flattened; epicranial stem short or absent; frontal arms V-shaped or U-shaped and contiguous at base; endocarina absent. Stemmata usually 5 on each side, but tightly clustered and often appearing to be a single stemma. Frontoclypeal suture present or absent. Labrum free, strongly transverse and subtruncate with long, heavily sclerotised tormae. Antennae moderately well developed but usually less than half as long as head width (longer in *Austrolimnius*), 3-segmented, with palpiform sensorium on antennomere 2 about as long as short apical antennomere (sensorium setiform and much longer than apical antennomere in *Austrolimnius*). Mandibles symmetrical, somewhat wedge-like with excavate mesal edge, broadly tridentate; mola absent; mesal surface of mandibular base with articulated, pubescent process. Ventral mouthparts strongly retracted forming maxillolabial complex, without maxillary articulating areas; cardines well separated, transverse, undivided; stipes elongate, well separated from one another and separated from labium by complete sutures; galea articulated, lacinia fixed, both apically rounded or truncate and setose; maxillary palp 3-segmented, sometimes with segment-like palpifer. Labium consisting of prementum and postmentum, the latter elongate, subtruncate at base, with transverse basal section lightly pigmented, somewhat set off from remainder of sclerite and possibly representing submentum; ligula transverse and broadly rounded or truncate; labial palps 2-segmented, with or without palpiger, widely separated. Hypopharyngeal sclerome and hypostomal rods absent; ventral epicranial ridges well developed. Gular region short to moderately long; gular sutures well separated or absent. Cervical region lightly sclerotised, usually with paired sclerites, sometimes forming V.

Prothorax slightly longer than meso- or metathorax; thoracic terga and abdominal terga I–IX with well sclerotised plates, occasionally with short lateral expansions (e.g. *Kingolus tinctus* Carter & Zeck) or with median dorsal projections (e.g. *Stetholus*, some Elminae); prothoracic venter usually with large precoxalia meeting at midline, smaller postcoxalia, slender laterotergite and small sternal sclerite between coxae (with posterior median sclerite in some Larainae); mesothorax and metathorax with anterior and posterior laterotergites. Legs short and stout, 5-segmented, with claw-like pretarsus bearing 1 seta; coxae large and relatively narrowly separated (at least on prothorax). Abdomen usually about twice as long as thorax; with distinct mediotergites and sternites, usually separated by slender laterotergites on segments I–IX in *Potamophilinus*, I–VII in *Hydora*, *Ovolara*, *Graphelmis*, and I–V or VI in most genera; tergum and sternum usually completely fused together on segments VI–IX, VII–IX or VIII and IX; segment IX usually at least twice as long as VIII (more than 3 times as long in *Potamophilinus*), extending into ventral surface, slightly concave to mesally tumid, without urogomphi, but with paired hook-like processes near anterior edge in *Graphelmis*; apex deeply forked in *Coxelmis*, with pair of long, slender processes in *Potamophilinus*); sternum IX exposed, short and strongly transverse; sternum X forming well sclerotised ventral operculum (Figs 42A, 51H, 65M) bearing a pair of hooks at apex and concealing anal opening and 3 tufts of retractable tracheal anal gills. Spiracles biforous, without closing apparatus, usually located on main body of mesotergum and abdominal terga I–VIII (last 2 to 4 pairs of abdominal spiracles reduced in *Kingolus*, all abdominal spiracles reduced in some *Austrolimnius*). [Böving and Craighead 1931; Glaister 1985, 1999; Spangler and Santiago-Fragoso 1992; Shepard 2002a; Kodada and Jäch 2005; Archangelsky and Manzo 2006].

Classification and World Distribution

Elmidae is a large, worldwide family with 145 genera and ~1400 species. There are two subfamilies, both of which occur in Australia.

Australian Fauna

The Australian elmid fauna consists of four genera of **Larainae** (*Hydora* Broun, *Ovolara* Broun, *Potamophilinus* Grouvelle and *Stetholus* Carter and Zeck, Fig. 62G) and five genera of **Elminae** (*Austrolimnius* Carter and Zeck, Fig. 63E, *Coxelmis* Carter and Zeck, *Graphelmis* Delève, *Kingolus* Carter and Zeck and *Notiolus* Carter and Zeck). Elmids are small to minute, darkly pigmented beetles, with moderately short, slender or slightly incrassate antennae, moderately to widely separated coxae, and long legs with simple tarsi and long pretarsal claws. Larvae (Fig. 65H) are onisciform to cylindrical, with a hinged operculum with paired hooks and with retractile anal gills. Adult Elminae are glabrous or sparsely pubescent above but have a ventral plastron; adult Larainae are more densely pubescent and usually lack a plastron. Most adult elmids and all larvae live under water, usually in lotic habitats with rocky bottoms, clear water and higher oxygen content, and feed on algae or on decaying wood or other vegetation. Larvae breathe through anal gills, but they possess functional spiracles in the last instar, when they leave the water to pupate. Most adults are capable of remaining under water for long periods by maintaining an air bubble. The laraine genera each contain one or a few species, whereas some elmine genera are speciose (~60 species in *Austrolimnius* and more than 20 each in *Kingolus* and *Notriolus*). Larvae of several Australian elmids have been reared (Glaister 1985) and are illustrated in Glaister (1999). According to McKie and Cranston (1998) xylophagous elmids are the principal modifiers of submerged wood, where their gouging of channels increases surface area, facilitating invasion by other macroinvertebrates and wood-decaying microorganisms and fungi. [Stride 1955; Hinton 1965, 1968; Brown 1981, 1987].

33. PSEPHENIDAE

Adult (Fig. 62J)

Length 2.2–5.3 mm. Body ~1.4–1.7 times as long as wide, broadly ovate, distinctly flattened. Colour yellowish-brown to dark brown or black, sometimes more or less variegated; densely clothed with decumbent setae, those on the upper surfaces fine and yellow or somewhat thickened and white, the latter usually forming a pattern on pronotum and elytra.

Head strongly declined, deeply inserted into prothorax but not concealed from above, strongly transverse, not abruptly constricted posteriorly; with transverse occipital carina immediately in front of occipital foramen. Vertex moderately, gradually declined. Eyes large, entire, finely facetted, without interfacetal setae; ommatidium of exocone type. Antennal insertions exposed from above, widely separated, not borne on raised tubercles. Subantennal groove absent. Frontoclypeal suture absent; frontoclypeal region not sharply declined, but clypeus strongly convex, its anterior edge more or less truncate. Labrum free, partly concealed, strongly transverse, truncate. Antennae 11-segmented, weakly serrate beginning on antennomere 3; stipes elongate, almost 3 times as long as pedicel. Mandible short, broad, flattened; apex weakly bidentate; incisor edge simple; mola and prostheca absent. Maxilla with galea biacuminate and setose, lacinia acuminate and setose, without uncus; apical palpomere expanded and truncate. Mentum transverse; ligula bilobed with each lobe biacuminate and setose; apical labial palpomere expanded and truncate. Subgenal ridges absent. Gular sutures widely separated; gula strongly transverse. Corpotentorium absent; anterior tentorial arms closely approaching one another but not mesally expanded. Cervical sclerites large and subtriangular.

Pronotum ~0.3–0.4 times as long as wide, widest posteriorly and as wide at base as combined elytral bases; sides strongly curved, apically narrowed; lateral carinae sharply defined and complete, simple, with raised margin or narrow bead; anterior angles produced forward and narrowly rounded; posterior angles obtuse; posterior edge triemarginate and crenulate; disc slightly convex and simple. Promesothoracic interlocking mechanism well developed. Prosternum in front of coxae (Fig. 19K) distinctly longer than shortest diameter of coxal cavity, slightly convex; anterior edge with broad, truncate chin piece. Prosternal process complete, gradually narrowed posteriorly, curved behind coxae; apex narrowly rounded. Notosternal suture complete. Procoxae not projecting below prosternum, without concealed, lateral extensions. Procoxal cavities strongly transverse, without narrow lateral extensions, narrowly separated, externally and internally open. Postcoxal projections moderately long and acute. Trochantins broadly exposed.

Scutellar shield (Fig. 16Q) well developed, abruptly elevated, anteriorly crenulate, posteriorly acute. Elytra ~1.12–1.37 times as long as wide and 3.45–4.45 times as long as pronotum; complete and conjointly rounded at apex, irregularly punctate, but sometimes with very weak striae laterally; scutellary striole absent; epipleura moderately broad and complete. Mesoventrite (Fig. 21F) separated by complete sutures from mesepisterna, which are well separated from one another; anterior edge on different plane than metaventrite, with paired, vertical, procoxal rests; discrimen complete to edge of mesoventral cavity, which is moderately large, extending to anterior edge of ventrite, narrow and deep. Mesocoxae not projecting, trochantins exposed. Mesocoxal cavities transverse and slightly oblique, widely separated, partly closed laterally by mesanepisterna and mesepimera; with membranous joint between meso- and metathoracic portions of coxal cavities. Mesoventral process apically divided and mesometaventral junction dicondylic. Metaventrite moderately convex, with discrimen relatively long, paired postcoxal lines absent and transverse (katepisternal) suture short but distinct. Exposed portion of metanepisternum about twice as long as wide; metepimeron exposed. Metacoxae contiguous, transversely oriented, extending laterally to meet elytra; plates weakly developed but compete. Metendosternite with long, slender stalk, short, broad lateral arms, no laminae, weakly developed ventrolateral processes and short, broad, emarginate anterior process with widely separated tendons. Hind wing moderately elongate; radial bending zone well developed, moderately long; apical field about a half of total wing length, with anterior and posterior radial extensions; apical folds meeting at pivot point well behind anterior wing margin; radial cell large, ~2.5 times as long as wide, without pigmented lumen; inner edge straight and strongly oblique and inner posterobasal angle narrowly acute; cross-vein r3 strongly oblique and meeting RP, but variously obscured at middle; r4 moderately short, straight and complete; R-M loop broadly rounded; basal portion of RP moderately long, apical extension obscured at base but extending to wing margin; medial spur straight, ending before wing margin which has a weak embayment; medial field with 3 or 4 free veins; MP_{3+4} with long basal cross-vein but no spur; wedge cell absent; anal lobe well developed, anal embayment absent; AP undivided. Legs relatively long and slender; tibiae not widened; tibial spurs paired but relatively short. Tarsi 5-5-5; tarsomeres simple; pretarsal claws simple; empodium reduced and not apparent.

Abdomen with five ventrites, the first three of which are connate; ventrite 1 not longer than 2, without curved postcoxal lines; intercoxal process acute; ventrites with sharp lateral projections and well defined laterosternitres; posterior edges of ventrites 2–4 usually distinctly denticulate; posterior edge of ventrite 3 in male at least sometimes with median cleft and dense patch of setae. Tergites relatively lightly sclerotised, with wing-folding spicule patches present on most segments. Functional spiracles present on segments I–VII, those on III–VII located in pleural membrane. Anterior edge of sternite VIII in male with short median lobe but no distinct strut. Anterior edge of segment IX in male with subacute or narrowly rounded subgenital plate; tergite IX partly fused to tergite X, which is evenly sclerotised. Aedeagus of trilobate type, symmetrical; phallobase about as long as wide, anteriorly truncate, without anterior struts; parameres articulated. Penis with dorsal and ventral lobes and short anterior struts. Sternite VIII without anterior strut (spiculum ventrale); anterior edge broadly truncate. Ovipositor moderately elongate, broad at base and apically narrowed, more or less depressed; paraprocts converging posteriorly, slightly longer than coxites, which are apically attached, moderately well sclerotised, subdivided into proximal and distal lobes more or less equal in length, the former with basal transverse baculi; styli apically attached, palpiform and 2-segmented. [Smith 1981; Davis 1986; Shepard 2002c; Lee *et al.* 2005a, 2007].

Larva (Fig. 65I)

Body broadly ovate and widest at middle or widest near anterior end and narrowing posteriorly, strongly flattened. Dorsal surfaces moderately heavily sclerotised, granulate, yellowish-brown to black and usually with pattern of light and dark pigment; ventral surfaces more lightly sclerotised; vestiture of fine hairs, irregularly distributed but densely placed along entire outer edges of thorax and abdomen.

Head protracted or slightly retracted and prognathous, but completely concealed from above by pronotum, about as long as wide and slightly flattened; sides very slightly curved; posterior edge not emarginate. Epicranial stem absent; frontal arms V-shaped, contiguous at base; median endocarina absent. Stemmata 6 on each side, but densely clustered so they may appear to be a single stemma. Antennae more than half as long as head width, 3-segmented; antennomere 3 much shorter than 2 and palpiform sensorium on antennomere 2 almost as long as antennomere 3. Frontoclypeal suture present. Labrum free, strongly transverse, truncate. Mandibles short and broad, symmetrical, with broad, tridentate apex and usually 1 subapical tooth; mesal surface broadly concave; accessory ventral process absent;

mola absent; mesal surface of mandibular base with plumose process. Ventral mouthparts strongly retracted; maxillary articulating areas present but concealed beneath laterally expanded postmentum. Cardines distinct, sclerotised, undivided, slightly oblique. Stipes longer than wide. Galea articulated, setose; lacinia subacute. Maxillary palp 3-segmented with segment-like palpifer. Labium consisting of prementum and postmentum; ligula shorter than labial palps, truncate, not bilobed; palps 2-segmented, without distinct palpiger, widely separated. Hypopharyngeal bracon, hypopharyngeal sclerome, hypostomal rods and ventral epicranial ridges absent; gular region moderately long; gular sutures absent.

Prothorax longer than meso- and metathorax combined, forming a large anteriorly evenly curved plate concealing head; meso- and metathorax and and abdominal segments I–VIII each with a pair of flattened, laterally or posterolaterally projecting processes or laminae; tergum IX forming posterior plate the edge of which is more or less continuous with the outer edges of laminae; all edges densely lined with hairs; all thoracic and abdominal terga covered with small cuticular beads and patches of hair-like processes. Prothoracic venter with large, trapezoidal undivided presternum extending posteriorly between procoxae. Legs well developed, slender, 5-segmented, including pretarsus, which is claw-like and unisetose; coxae widely separated. Abdominal terga I–VIII with well developed laterotergites beneath laminae and sterna I–VIII transverse and undivided; abdominal terga usually with 3 or 4 pairs of gin traps on adjoining edges of terga III–VI or VII (rarely with only 2 gin traps involving terga III–V). Tergum IX distinctly longer than VIII, extending onto ventral surface, its disc flattened with posterior edge usually broadly rounded (occasionally angulate or subtruncate); anterolateral corners of tergum in last instar each bearing a spiracular brush; sternum IX short and strongly transverse; segment X forming apically rounded, ventrally-hinged operculum without hooks, which conceals anal region. Spiracles absent or non-functional except in last instar, which has a single pair, each located dorsally at posterior edge of tergum VIII near base of lamina and opposite spiracular brush on tergum IX; spiracles biforous, without closing apparatus. [Bertrand and Watts 1965; Bertrand 1972; Smith 1981; Davis 1986; Shepard 2002c; Lee *et al.* 2005a, 2007].

Classification and World Distribution

The family Psephenidae includes 35 genera and ~275 species occurring on all continents, but absent from New Zealand. There are four subfamilies: Eubrianacinae (Holarctic, Afrotropical and Oriental regions), Psepheninae (New World, eastern Palaearctic and Oriental regions), Psephenoidinae (eastern Palaearctic and Oriental regions, New Guinea) and Eubriinae (North and Central America, Palaearctic, Oriental and Afrotropical regions, Australia and Chile). [Lee *et al.* 2005a].

Australian Fauna

A single genus of **Eubriinae**, *Sclerocyphon* Blackburn, with 13 described species, occurs in Australia, while its sister group *Tychepsephus* Waterhouse occurs in Chile. Adults (Fig. 62J) are broadly oval, flattened beetles clothed with short, dense pubescence forming a pattern which is superimposed on a maculate dorsal surface. The head is deflexed and only partly visible from above within the anterior pronotal emargination, and the bases of the pronotum, scutellum and elytra are all finely crenulate forming an interlocking device. The head is narrowed anteriorly forming a short rostrum, the mouthparts are usually concealed by the prosternal chin piece, the prosternal process fits into a mesoventral cavity, the metakatepisternal is distinct, and the tarsi are long and simple. Larvae (Fig. 65I) are broadly oval, strongly flattened and disc-like, with head and legs concealed from above, and the disc clothed with a pattern of granules and modified setae, lined with a marginal fringe of hairs, and provided with two to four pairs of defensive gin traps. The larval antennae are long, the labrum large, and the maxillae freely movable but partly concealed behind the expanded postmentum. Functional spiracles occur in the final instar only; they are placed on tergum VIII opposite a pair of spiracular brushes on tergum IX, which prevent silt from clogging the openings. Sternum IX has a ventrally-articulated operculum that conceals a chamber housing paired retractile anal gill tufts. Larval psephenids are commonly found adhering to rocks in streams throughout the wetter parts of eastern Australia, but one species has been recorded from the George Gill Ranges in the central part of the continent. Larvae feed on algae covering the rocks, and because of their unique, streamlined form, they are able to move across surfaces within high energy water flows and thus exploit resources unavailable to their competitors. Pupation takes place within the larval skin in concealed areas along stream banks. Adults are very cryptic, but they have been observed flying above streams on hot days. [Hinton 1955, 1966; Smith and Dartnall 1980; Smith 1981; Davis 1986].

34. PTILODACTYLIDAE

Adult (Fig. 62E–F)

Length 2.8–8 mm. Body ~2.1–2.55 times as long as wide, slightly flattened to strongly convex. Colour yellowish-brown to dark reddish-brown; surfaces clothed with fine, decumbent setae, which may form variegated pattern.

Head slightly to strongly declined, about as long as wide or slightly transverse, not abruptly constricted posteriorly; transverse occipital ridge weakly developed in *Byrrocryptus*. Frontal region not to moderately, gradually declined. Eyes entire, strongly protuberant, finely facetted, without interfacetal setae; ommatidium of exocone type. Antennal insertions exposed, narrowly to widely separated, not borne on raised tubercles. Subantennal groove absent. Frontoclypeal suture absent to distinctly impressed; frontal region with median longitudinal ridge in *Austrolichas*. Antennae 11-segmented, filiform or serrate with rami beginning on antennomere 3, or with articulated appendages beginning on antennomere 4. Labrum at least partly visible, free, subquadrate to strongly transverse. Mandible short and broad moderately to strongly, gradually or abruptly curved mesally; apex bidentate; incisor edge simple; mola and prostheca usually well developed (absent in *Austrolichas*). Maxilla with galea and lacinia usually short, broad and densely setose, the latter without uncus; galea acuminate and lacinia reduced in *Austrolichas*; apical palpomere cylindrical, fusiform, or somewhat expanded apically and securiform. Mentum strongly transverse; ligula usually bilobed (biacuminate in *Austrolichas*); apical labial palpomere cylindrical, fusiform or somewhat expanded apically. Subgenal ridges present or absent. Gular sutures widely separated. Corpotentorium very slender or absent; anterior tentorial arms expanded mesally forming laminatentoria, which may meet at midline. Cervical sclerites well developed.

Pronotum ~0.5–0.65 times as long as wide, widest posteriorly, and as wide as or slightly wider at base than combined elytral bases; sides strongly curved or somewhat sinuate, and apically narrowed; lateral carinae usually sharply defined posteriorly but incomplete anteriorly (complete in *Austrolichas*), often not visible for their entire lengths from above, simple, with or without raised margin or narrow bead; anterior angles not produced forward; posterior angles right to somewhat acute; posterior edge distinctly sinuate or triemarginate and crenulate; disc usually at least slightly elevated posteriorly (strongly elevated and anterior produced in *Austrolichas* and *Pherocladus*), with transverse posteromesal impression in *Austrolichas*. Promesothoracic interlocking mechanism well developed. Prosternum in front of coxae (Fig. 44W) slightly longer than to slightly shorter than shortest diameter of coxal cavity, flat to slightly convex; anterior edge not produced anteriorly, with broad mesal excavation in *Austrolichas*. Prosternal process complete, parallel-sided or gradually narrowed apically, not curved behind coxae; apex narrowly rounded or truncate. Notosternal suture complete. Procoxae often at least slightly projecting below prosternum, without or with short concealed, lateral extensions. Procoxal cavities slightly to strongly transverse, usually without narrow lateral extensions (but with sternal cowling overlapping postcoxal process in Ptilodactylinae), narrowly separated, externally broadly to narrowly open, internally open. Postcoxal projections short and acute to truncate or rounded. Trochantin usually exposed (concealed in Ptilodactylinae).

Scutellar shield well developed, abruptly elevated, anteriorly notched, posteriorly acute to broadly rounded. Elytra about1.65–2.1 times as long as wide and 3.8–4.9 times as long as pronotum; complete and conjointly rounded at apex, irregularly punctate or with 10 or 11 puncture rows or striae; scutellary striole absent; epipleura complete. Mesoventrite separated by complete sutures from mesepisterna, which are well separated from one another; anterior edge on same plane as metaventrite, with paired, slightly to strongly declined procoxal rests; discrimen usually more or less complete (absent in Ptilodactylinae); mesoventral cavity small and shallow or absent. Mesocoxae slightly to moderately strongly projecting; trochantins exposed. Mesocoxal cavities transverse and somewhat oblique, narrowly separated, partly closed laterally by mesanepisterna and mesepimera; without membranous joint between meso- and metathoracic portions of coxal cavities. Mesoventral process apically divided and mesometaventral junction dicondylic. Metaventrite moderately to strongly convex, with moderately long discrimen; paired postcoxal lines absent; transverse (katepisternal) suture short or absent. Exposed portion of metanepisternum moderately elongate; metepimeron exposed. Metacoxae contiguous, obliquely oriented, sometimes extending laterally to meet elytra; plates moderately well developed mesally but narrowing laterally. Metendosternite with long, slender stalk, short lateral arms, no laminae, weakly developed ventrolateral processes and relatively short anterior process with well separated tendons. Hind wing (Fig. 22A) moderately elongate; radial bending zone absent; apical field about a third of total wing length, with anterior and posterior oblique linear sclerites, probably representing radial extensions; apical folds meeting at pivot point well behind anterior wing margin; radial cell ~3.5 to 4 times as long as wide, without pigmented lumen; inner edge straight or curved, but strongly oblique and inner posterobasal angle acute to narrowly acute; cross-vein r3 usually moderately long and slightly oblique (very short but almost in contact with RP in Ptilodactylinae); r4 moderately long, straight and complete; R-M loop moderately broad but forming acute angle; basal portion of RP long, apical extension vaguely indicated; medial spur meeting wing margin, where there is slight embayment; medial field with 4 or 5 free veins; MP_{3+4} with basal cross-vein but no spur; wedge cell reduced and apically acute or absent; anal lobe well developed, anal embayment absent; AP undivided. Legs relatively long and slender; tibiae not widened at apex, sometimes lined with spine-like setae; tibial spurs paired on all legs. Tarsi 5-5-5 with simple tarsomeres in *Byrrocryptus* and *Austrolichas*, pseudotetramerous, with tarsomere 3 lobed beneath and tarsomere 4 reduced and more or less enclosed within lobe on 3, in Ptilodactylinae; claws usually simple (with tooth at base in Ptilodactylinae), empodium reduced and more or less concealed.

Abdomen with five ventrites, the first three of which are connate; ventrite 1 not longer than 2; intercoxal process acute; ventrites usually with sharp lateral projections and well defined laterosternitres. Tergites relatively lightly sclerotised, with wing-folding spicule patches present on most segments. Functional spiracles present on segments I–VII, those on III–VII located on laterosternites or pleural membrane. Anterior edge of sternite VIII in male usually without median strut (with strut in Ptilodactylinae). Anterior edge of segment IX in male usually with truncate or broadly rounded subgenital plate (with spiculum gastrale in Ptilodactylinae); tergite IX subtruncate to deeply emarginate, X well developed, evenly sclerotised and free. Aedeagus of trilobate type, symmetrical; phallobase relatively large, truncate or slightly emarginate at base, without anterior struts; parameres articulated, usually simple, sometimes complex in Ptilodactylinae. Penis longitudinally divided, without ventral lobe, with paired anterior struts. Sternite VIII in female with long anterior strut (spiculum ventrale) and often with pair of shorter anterolateral struts. Ovipositor (Fig. 30D) long and slender, more or less depressed, usually lightly sclerotised except for baculi (heavily sclerotised and blade-like in *Austrolichas*); paraprocts much longer than coxites, which are located distally and usually undivided and apically acute without styli (coxites fused to paraprocts in *Austrolichas* and styli present and terminal but reduced and button-like in *Byrrocryptus*). [Stribling 1985; Lawrence 1987a, 1988b, 2005d; Lawrence and Stribling 1992; Lawrence *et al.* 1999b; Ivie 2002a; Lee *et al.* 2005b].

Larva (Fig. 65J)

Body elongate, parallel-sided, cylindrical to slightly flattened, somewhat curved ventrally in *Ptilodactyla*. Surfaces lightly to moderately heavily sclerotised, usually smooth (distinctly granulate in *Austrolichas*); colour yellow to dark reddish-brown; vestiture usually of scattered, long, fine hairs, mainly laterally (with slender or broadly expanded pubescent hairs in *Austrolichas*).

Head protracted and prognathous; globular or slightly flattened; sides subparallel to slightly curved; posterior edge not to deeply emarginate. Epicranial stem absent; frontal arms V-shaped, usually contiguous at base (well separated in *Austrolichas*); median endocarina absent. Stemmata five or six on each side, in two vertical rows, but densely clustered so they may appear to be a single stemma. Antennae less than a third as long as head width in *Austrolichas* and *Ptilodactyla*, but about half as long in *Byrrocryptus*, 3-segmented with palpiform sensorium on antennomere 2 about as long as antennomere 3. Frontoclypeal suture weakly to strongly impressed; clypeus very short, with membranous anteclypeus. Labrum free, strongly transverse, usually more or less truncate. Mandibles short and broad to moderately elongate, symmetrical, with broad, tridentate apex and usually 2 subapical teeth; mesal surface broadly concave; accessory ventral process absent; mola absent; mesal surface of mandibular base simple in *Austrolichas*, with setose, articulated process in *Ptilodactyla* (Fig. 40B), and with two dense brushes of hairs in *Byrrocryptus* (Fig. 41N). Ventral mouthparts (Figs 38C, 48O, 65F) strongly retracted; maxillary articulating area well developed in *Byrrocryptus* and *Ptilodactyla*, absent in *Austrolichas*. Cardo distinct and sclerotised, undivided, slightly to strongly oblique; cardines separated from each other by labium. Stipes subquadrate to elongate. Maxilla with articulated, 1- or 2-segmented, setose galea and fixed, truncate or falciform, setose lacinia. Maxillary palp 3-segmented with segment-like palpifer. Labium consisting of prementum, mentum and submentum, but with submental sclerite reduced or not clearly delimited in *Byrrocryptus*; postmentum laterally expanded concealing maxillary articulating area in *Byrrocryptus* and *Ptilodactyla*; divided longitudinally into 3 parts in *Byrrocryptus*; ligula shorter than labial palps, rounded or truncate, not bilobed; palps 2-segmented, with or without distinct palpiger, separated by more than width of first palpomere. Hypopharyngeal bracon and hypopharyngeal sclerome absent; hypostomal rods absent; ventral epicranial ridges present; gular sutures widely separated in *Byrrocryptus*, narrowly separated in *Austrolichas*, fused together in *Ptilodactyla*.

Prothorax slightly longer than meso- or metathorax; thoracic and abdominal terga relatively smooth and even in *Byrrocryptus* and *Ptilodactyla* (often very lightly pigmented in *the latter*), somewhat uneven and finely tuberculate in Austrolilchas, without patches or rows of asperities, transverse plicae or lateral tergal processes. Thoracic terga extending ventrally, so that pleural and sternal regions are highly reduced in *Austrolichas*; prothoracic venter with distinct, undivided presternum well developed precoxalia more or less meeting at midline in *Austrolichas* and *Byrrocryptus*; presternum absent in *Ptilodactyla*. Legs moderately long in *Byrrocryptus* and *Ptilodactyla*, relatively short in *Austrolichas*, 5-segmented, including pretarsus, which is claw-like and unisetose; legs partly clothed with short, spine-like setae in *Austrolichas* and *Byrrocryptus*; coxae relatively narrowly separated or subcontiguous. Abdominal terga I–VII extending onto ventral surface, so that sterna are very narrow, becoming narrower posteriorly and almost absent on segment VII; laterotergites slender and elongate on segments I–V, absent on VI and VII. Segment VIII in *Byrrocryptus* and *Ptilodactyla* subterminal, cylindrical and similar to segment VII, with well developed sternum, with lateral pockets in *Ptilodactyla*; segment VIII in *Austrolichas* terminal and produced posteriorly to form conical siphon bearing spiracles at apex, with sternum highly reduced and concealed by segment VII. Tergum IX in *Byrrocryptus* and Ptilodactyla terminal, only slightly extending onto ventral surface, its disc somewhat flattened in *Byrrocyptus* and lined with semicircular ridge or carina, that in *Ptilodactyla* similar but strongly declined posteriorly, with carina weak or absent; sternum IX very narrow in *Byrroctyprus*. Tergum IX in *Austrolichas* short, transverse and entirely ventral to segment VIII; sternum IX forming a hinged operculum lacking hooks. Segment X in *Byrrocryptus* and *Ptilodactyla* with a pair of pygopods, each with several sclerotised hooks and two osmoregulatory papillae in *Byrrocryptus* (Figs 52S, 65L) or fewer hooks and no papillae in *Ptilodactyla*; segment X in *Austrolichas* membranous and concealed by operculum. Thoracic spiracles and abdominal spiracles I–VII small, functional, biforous and located laterally on terga in *Ptilodactyla*, *Austrolichas* and in last instar of *Byrrocryptus*. Spiracles on segment VIII similar in size to anterior spiracles and located laterally in *Ptilodactyla* and *Byrrocryptus*, enlarged and located at end of posterior siphon in *Austrolichas* (Fig. 65J); spiracles reduced and non-functional in early instars of *Byrrocryptus*; spiracular closing apparatus absent. [Böving 1929; Böving & Craighead 1931; Hudson 1934; Bertrand 1966, 1972; Britton 1970; Süss & Puppin 1976; LeSage & Harper 1976; Spangler 1983; Hayashi 1986; Stribling 1986; Costa *et al.* 1988; 1999; Lawrence 1991b, 2005d; Lawrence & Stribling 1992; Funk and Fenstermacher 2002; Ivie 2002a; Lee *et al.* 2005b].

Classification and World Distribution

The family Ptilodactylidae contains ~34 genera and more than 500 species; they are widely distributed but most abundant and diverse in tropical and subtropical regions. The family is presently divided into five subfamilies, but the classification is in need of revision. The subfamily Araeopidiinae includes the single species *Araeopidius monachus* (Leconte) from north-western North America, Aploglossinae with about ten Neotropical species of *Aploglossa* Guérin-Méneville, and three subfamilies, Anchytarsinae, Cladotominae and Ptilodactylinae, with representatives in most regions. [Stribling 1985; Lee *et al.* 2005b].

Australian Fauna

The Australian ptilodactylid fauna includes one species of **Cladotominae**, *Austrolichas monteithi* Lawrence and Stribling (Fig. 62E), four described species of *Byrrocryptus* Broun (Fig. 62F) (**Anchytarsinae**), which also occurs in New Zealand, and a few undescribed species of **Ptilodactylinae** belonging to the genera *Ptilodactyla* Illiger and *Pherocladus* Fairmaire. Adults are elongate and finely pubescent, with moderately long, filiform or serrate antennae, or sometimes with articulated appendages attached to antennomeres 4–10. The head is strongly deflexed; the pronotum has sharp lateral carinae, which are almost always incomplete anteriorly, and a strongly developed basal interlocking mechanism with crenulations; the scutellum is notched; the elytra are often striate, and the tarsi may be simple, provided with ventral brushes, or sometimes with a reduced 4th tarsomere and membranous appendage on tarsomere 3. Adults are not common but may be collected at lights. There are three larval types, all of which feed on decaying plant material. The aquatic larvae of *Byrrocryptus* species (Fig. 65C) are elongate and subcylindrical, with long antennae, a mandible bearing two basal brushes of hairs, movable maxillae with the articulating area partly concealed behind the postmentum, which is longitudinally divided into three parts, paired gular sutures and laterally placed abdominal spiracles. Tergum IX forms a sclerotised, concave plate and segment X has a pair of lobes, each bearing several hooks, and usually two osmoregulatory papillae. Ptilodactyline larvae, which occur in leaf litter and wet rotten wood, are more ventrally curved at the apex and have an articulated process at the mandibular base, an undivided postmentum, fused gular sutures, paired lateral defence glands on the mesothorax and segment VIII, no anal papillae and fewer anal hooks. The *Austrolichas* larva, which is also found in leaf litter, is clothed with complex setae (Fig. 27D), has simple mandibles, an undivided postmentum, very short antennae, eighth spiracles located at the end of a spiracular siphon, and sternite IX with a ventral operculum (Fig. 65J). [Lawrence 1987a, 1988b].

35. CHELONARIIDAE

Adult (Fig. 63A)

Length 4.8–5.5 mm. Body ~1.8–2.15 times as long as wide, ovoid and biconvex (lenticular), highly compacted. Colour dark reddish-brown; surfaces shiny but clothed with slender, yellow and slightly thicker, white decumbent setae, the latter completely covering scutellary shield and forming a pattern of spots on elytra.

Head strongly declined, deeply inserted into prothorax and concealed from above by pronotum; about as long as wide, not abruptly constricted posteriorly; transverse occipital ridge weakly developed and immediately in front of occipital foramen. Frontal region not to moderately, gradually declined. Eyes large, entire, strongly protuberant, finely facetted, without interfacetal setae; ommatidium of exocone type. Antennal insertions concealed from above, narrowly separated, not borne on raised tubercles. Subantennal groove absent. Frontoclypeal suture absent; frontal region sharply declined, forming a ridge between antennal insertions which is mesally produced forming a subacute process, which continues ventrally as a median, vertical carina separating a pair of frontoclypeal impressions extending laterally to antennal sockets; anterior edge of clypeus more or less truncate. Labrum free, concealed, subtruncate. Antennae 11-segmented, with first 3 antennomeres heavily pigmented and last 8 very lightly so; scape highly reduced and transverse, pedicel ~3 times as long as scape and slightly longer than wide, antennomere 3 2.5 times as long as 3 and 5 times as long as wide, antennomere 4 highly reduced, and remaining antennomeres slightly elongate and weakly serrate. Mandible (Fig. 28S) short, broad, flattened; apex rounded; incisor edge simple; mola and prostheca absent. Maxilla with galea biacuminate and setose, lacinia very short and setose, without uncus; apical palpomere securiform. Mentum reduced, transverse; ligula biacuminate and setose; apical labial palpomere securiform. Subgenal ridges present. Gular sutures well separated; gula longer than wide. Corpotentorium absent; anterior tentorial arms expanded mesally forming laminatentoria, which do not meet at midline. Cervical sclerites large and subtriangular.

Pronotum ~0.5–0.6 times as long as wide, widest posteriorly and as wide at base as combined elytral bases; sides strongly curved, apically narrowed and continuous with and on same plane as anterior edge (Fig. 28E); lateral carinae sharply defined and complete, simple, without raised margin or narrow bead; anterior angles not produced forward; posterior angles somewhat acute; posterior edge distinctly triemarginate and crenulate and transversely striolate; disc slightly concave and simple. Promesothoracic interlocking mechanism well developed. Prosternum in front of coxae (Fig. 63B) slightly shorter than shortest diameter of coxal cavity, slightly convex, but with deep mesal excavation into which the anterior portion of the head and the enlarged basal antennomeres fit when head is at rest. Prosternal process complete, more or less parallel-sided, and not curved behind coxae; apex narrowly rounded. Notosternal suture complete. Procoxae not projecting below prosternum, with very short concealed, lateral extensions. Procoxal cavities strongly transverse, without narrow lateral extensions, narrowly separated, externally broadly open, internally open. Postcoxal projections absent. Trochantins broadly exposed.

Scutellar shield well developed, abruptly elevated, anteriorly crenulate, posteriorly broadly rounded. Elytra ~1.43–1.55 times as long as wide and 3.35–3.9 times as long as pronotum; complete and conjointly rounded at apex, irregularly punctate; scutellary striole absent; epipleura narrow but complete. Mesoventrite separated by complete sutures from mesepisterna, which are well separated from one another; anterior edge on different plane than metaventrite, with paired, strongly declined; almost vertical, procoxal rests; discrimen absent; mesoventral cavity large, extending to anterior edge of ventrite, and deep. Mesocoxae not projecting, with short mesal plates; trochantins exposed. Mesocoxal cavities transverse and slightly oblique, widely separated, partly closed laterally by mesanepisterna and mesepimera; without membranous joint between meso- and metathoracic portions of coxal cavities. Mesoventral process apically broadly rounded and mesometaventral junction monocondylic. Metaventrite moderately convex, with discrimen complete to base of mesoventral process; paired postcoxal lines strongly curved, delimiting crural impressions; transverse (katepisternal) suture very short. Exposed portion of metanepisternum ~3 times as long as wide; metepimeron exposed. Metacoxae contiguous, transversely oriented, extending laterally to meet elytra; plates well developed and compete. Metendosternite with long, slender stalk, short, broad lateral arms, no laminae, weakly developed ventrolateral processes and long anterior process with narrowly separated tendons. Hind wing relatively short and broad; radial bending zone absent; apical field about a half of total wing length, with anterior and posterior oblique linear sclerites, and a vaguely indicated posterior radial extension; apical folds meeting at pivot point well behind anterior wing margin; radial cell large, ~4 times as long as wide, without pigmented lumen; inner edge straight and strongly oblique and inner posterobasal angle narrowly acute; cross-vein r3 absent; r4 moderately long, sinuate and complete; R-M loop moderately broad but forming acute angle; basal portion of RP long, apical extension vaguely indicated; medial spur meeting wing margin; medial field with 5 free veins; MP_{3+4} with basal cross-vein and spur; wedge cell absent; anal lobe well developed, anal embayment absent; AP divided at base. Legs relatively short and capable of being compacted, with tibia folded into groove on femur and tarsus folded against side of tibia; tibiae at least slightly widened at middle (Fig. 28O), protibia more strongly expanded with outer edge dentate; tibial spurs absent. Tarsi 5-5-5 but pseudotetramerous, with tarsomere 3 with membranous lobe beneath and tarsomere 4 reduced and more or less enclosed within lobe on 3; tarsomeres 1 and 2 with ventral setose pad; pretarsal claws toothed at base, empodium reduced and not apparent.

Abdomen with five ventrites, the first three of which are connate; ventrite 1 not longer than 2, with curved postcoxal lines delimiting crural impressions; intercoxal process acute; ventrites usually with sharp lateral projections and well defined laterosternitres. Tergites relatively lightly sclerotised, with wing-folding spicule patches present on most segments. Functional spiracles present on segments I–VII, those on III–VII located in pleural membrane. Anterior edge of sternite VIII in male with median strut. Anterior edge of segment IX in male with acute subgenital plate; tergite IX slightly emarginate, X well developed, evenly sclerotised and free. Aedeagus of trilobate type, symmetrical; phallobase more than twice as long as parameres and broadly rounded at base, without anterior struts; parameres articulated. Penis complex, with dorsal and ventral lobes and endophallic armature, with paired anterior struts. Sternite VIII in female with long, fixed anterior strut (spiculum ventrale) and paired anterolateral struts. Ovipositor long and slender, more or less depressed; paraprocts and proctiger lightly sclerotised except for baculi; paraprocts much longer than coxites, which are located distally, undivided, well sclerotised, acute, without styli. [Méquignon 1934, 1935; Paulus 1969; Ivie 2002b; Beutel and Leschen 2005f].

Larva

Body elongate, subcylindrical or slightly flattened, narrowing posteriorly. Surfaces moderately heavily sclerotised, granulate; colour reddish-brown with legs yellowish-brown; vestiture of long pubescent hairs, often arising in clusters or lines; surfaces usually coated with debris.

Head protracted and prognathous; globular or slightly flattened; sides slightly curved; posterior edge not emarginate. Epicranial stem moderately long; frontal arms V-shaped, contiguous at base; median endocarina absent. Stemmata probably 5 on each side, but densely clustered so they appear to be a single stemma. Antennae less than a fifth as long as head width, 3-segmented; antennomere 3 much longer and narrower than 2, and palpiform sensorium on antennomere 2 longer than antennomere 3. Frontoclypeal suture weakly impressed; clypeus very short, with membranous anteclypeus. Labrum free, strongly transverse, truncate. Mandibles short and broad, symmetrical, with broad, tridentate apex and usually 1 subapical tooth; mesal surface broadly concave; accessory ventral process absent; mola absent; mesal surface of mandibular base simple. Ventral mouthparts strongly retracted, forming compact maxillolabial complex without maxillary articulating areas. Cardines distinct, sclerotised, undivided, transverse and fused together forming a single plate. Stipes completely fused to postmentum, forming a single plate not divided by sutures. Galea articulated, 2-segmented and palpiform; lacinia absent. Maxillary palp 3-segmented with segment-like palpifer. Labium consisting of prementum and postmentum, but the latter not clearly separated from stipites; ligula shorter than labial palps, truncate, not bilobed; palps 2-segmented, without distinct palpiger, separated by more than width of first palpomere. Hypopharyngeal bracon and hypopharyngeal sclerome absent; hypostomal rods absent; ventral epicranial ridges present; gular region moderately long but gular sutures absent.

Prothorax distinctly longer than meso- or metathorax; thoracic terga and abdominal terga without patches or rows of asperities or transverse plicae, but with paramedian longitudinal rows tubercles; thoracic terga and abdominal terga I–VIII each with pair of small, flattened, apically rounded dorsolateral processes bearing a row of long, pubescent setae. Prothoracic venter with large, triangular, undivided presternum and well developed precoxalia meeting at midline. Legs relatively short and stout, 5-segmented, including pretarsus, which is claw-like and unisetose; legs partly clothed with short, expanded setae; coxae relatively narrowly separated. Abdominal terga I–VII extending ventrally, but sterna I–VII well developed, transverse and clearly separated from terga; sterna VII and VIII completely fused to terga forming complete ring. Tergum IX distinctly longer than VIII, not or only slightly extending onto ventral surface, its disc somewhat flattened, produced posteriorly and broadly rounded, with several short, rounded processes bearing rows of pubescent setae and more or less joined together to form a semicircular carina; sternum IX well developed and moderately long; segment X forming a circular, ventral, hinged operculum without hooks, which conceals anal region. Spiracles of a modified biforous type with the openings long and sinuous (Fig. 53Q), located laterally on short lateral protuberances, those on segment VIII similar in size to anterior spiracles and located laterally; spiracular closing apparatus absent. [Böving and Craighead 1931; Spangler 1980a, 1991; Costa *et al.* 1988; Lawrence *et al.* 1999a; Ivie 2002b; Beutel and Leschen 2005f].

Classification and World Distribution

Chelonariidae contains three genera and ~300 species occurring mainly in the tropical regions of the world. *Brounia thoracica* Sharp is a rare New Zealand endemic (Leschen and Early 2004), species of *Pseudochelonarium* Pic occur throughout South-east Asia and the East Indies to New Guinea and the Moluccas (Méquignon 1935, Paulus 1969), and *Chelonarium* Fabricius is a primarily a New World genus, with some species in Asia and Australia (Méquignon 1934; Beutel and Leschen 2005f).

Australian Fauna

The family is represented in Australia by a single species *Chelonarium australicum* Lea (Fig. 63A) from northern QLD. Adults are oval, lenticular, highly compact beetles with a strongly deflexed head not visible from above and legs capable of being retracted into cavities on the ventral surface; our species is clothed with decumbent setae, some of which are thickened and form oval patches on the elytra. The labrum and mandibles are highly reduced and concealed, the second and third antennomeres enlarged, the fourth minute and the last seven slender and weakly serrate, and the pronotum has a continuous carina extending from one posterior angle to the other and a basal interlocking mechanism with crenulations. Ventrally the prosternum is highly reduced and the mesoventrite has a large cavity for receiving the prosternal process (Fig. 63B); tibiae are flattened and provided with external spines, tarsomere 4 is reduced, 3 bears a membranous appendage, and the pretarsal claws are toothed. Chelonariid larvae are elongate, posteriorly narrowed and clothed with tubercles and long pubescent hairs. The head is globular with a long epicranial stem; mandibles are simple at the base and sutures are absent between the maxillae and labium. The abdominal apex has a ventrally hinged operculum and spiracles are of a unique, undulating type borne on lateral tubercles. Larvae are terrestrial and New World species have been recorded from the roots of epiphytes and from the nests of ants and termites (Lenko 1967; Spangler 1980a).

36. CALLIRHIPIDAE

Adult (Fig. 62B–C)

Length 9–23 mm. Body ~3–3.15 times as long as wide, slightly flattened above, moderately convex below. Colour reddish-brown to black; surfaces either clothed with fine, decumbent yellow setae or subglabrous.

Head about as long as wide or slightly transverse, strongly declined, not abruptly constricted posteriorly; transverse occipital ridge absent; vertex moderately steeply sloping. Eyes very large, entire, strongly protuberant, finely facetted, without interfacetal setae; ommatidium of the exocone type. Antennal insertions exposed, narrowly to moderately widely separated, and slightly to strongly raised. Subantennal groove absent. Frontoclypeal suture absent; frontoclypeus more or less abruptly declined (especially in *Callirhipis*); clypeus anteriorly truncate. Antennae 11-segmented, serrate in female, flabellate in male, with rami beginning on antennomere 3; scape more than 3 times as long as pedicel. Labrum free, exposed or partly concealed beneath clypeus, slightly to strongly transverse with apex truncate or convex, moderately heavily sclerotised with base on lower plane, so that distinct groove formed between body of labrum and clypeus. Mandible short, broad and strongly curved mesally, unidentate, but with dorsal, subapical tooth which fits over anterior labrum; mola and prostheca absent. Maxilla with galea and lacinia; galea short, simple and pubescent, the latter without uncus; apical palpomere fusiform. Mentum about as long as wide or slightly transverse and apically convex; ligula undivided and apically rounded; apical labial palpomere fusiform. Subgenal ridges absent. Gular sutures well separated; gula longer than wide. Corpotentorium very slender; anterior tentorial arms expanded mesally and almost meeting at midline. Cervical sclerites well developed.

Pronotum (Fig. 20C) ~0.63–0.7 times as long as wide, widest posteriorly and as wide at base as combined elytral bases; sides subsinuate and narrowing anteriorly; lateral carinae incomplete, weakly developed and simple, without narrow raised margin or bead; anterior angles not produced forward; posterior angles slightly acute; posterior edge triemarginate, simple in *Ennometes*, crenulate in *Callirhipis*); disc slightly convex, with pair of broad posterolateral impressions and a smaller posteromesal impression. Promesothoracic interlocking mechanism moderately well developed. Anterior portion of prosternum shorter than shortest diameter of procoxal cavity; slightly convex, anterior edge weakly produced forward and truncate in *Ennometes*, with mesal excavation in *Callirhipis*. Prosternal process complete, not elevated behind coxae, parallel-sided or slightly narrowed posteriorly and narrowly rounded or acute at apex. Notosternal suture complete. Procoxae slightly to strongly projecting below prosternum, without concealed lateral extensions. Procoxal cavities transverse, narrowly separated, externally broadly open, without narrow lateral extensions, internally open. Postcoxal projections short and acute. Trochantin broadly exposed, but not abutting prosternum or hypomeron.

Scutellar shield well developed, abruptly elevated, anteriorly simple in *Ennometes*, crenulate in *Callirhipis*, posteriorly subacute to broadly rounded. Elytra ~2.45–2.57 times as long as wide and 4.35–4.82 times as long as pronotum; sides subparallel and apices conjointly rounded; disc with 10 more or less distinct puncture rows and scutellary striole in Ennometes; punctation irregular in *Callirhipis*; each elytron with 3 weak costae in female *Callirhipis* or with series of anastomosing costae in male *Callirhipis*; epipleuron narrow and complete. Mesoventrite separated by complete sutures from mesepisterna, which are well separated from one another; anterior edge on same plane as metaventrite, with paired, declined, procoxal rests; discrimen well developed; mesoventral cavity small, shallow and indistinct; discrimen almost complete (weakly indicated in *Ennometes*). Mesocoxae slightly to strongly projecting, with exposed trochantins. Mesocoxal cavities narrowly separated, slightly transverse, partly closed laterally by mesanepisterna and mesepimera, without joint between meso- and metathoracic portions of coxal cavities. Mesoventral process divided at apex; mesometaventral junction dicondylic. Metaventrite moderately to strongly convex, with discrimen complete to base of metaventral process and with transverse (katepisternal) suture short or absent; paired postcoxal lines absent. Exposed portion of metanepisternum about times as long as wide; metepimeron exposed. Metacoxae contiguous, extending laterally to meet elytra, metacoxal plates weakly developed but more or less complete. Metendosternite with long stalk, short lateral arms, no laminae, well developed ventrolateral processes and long anterior process with approximate anterior tendons. Hind wing moderately elongate; radial bending zone absent; apical field about a fifth of total wing length, with anterior and posterior oblique linear sclerites, probably representing radial extensions; apical folds meeting at pivot point well behind anterior wing margin; radial cell almost 5 times as long as wide, without pigmented lumen; inner edge straight and inner posterobasal angle slightly obtuse; cross-vein r3 moderately long and oblique; r4 moderately long, straight and complete; R-M loop moderately broad but forming acute angle; basal portion of RP long, apical extension vaguely indicated; medial spur short and straight, meeting wing margin; medial field with five free veins; MP_{3+4} with basal cross-vein and spur; wedge cell well developed and apically acute; CuA_2 meeting MP_{3+4}; anal lobe well developed, anal embayment absent; AP undivided. Legs relatively long and slender; tibiae not widened at apex, outer edge simple; tibial spurs small, paired on all legs. Tarsi 5-5-5; membranous lobes absent (Fig. 44Y); claws simple, empodium well developed and multisetose (Fig. 27E).

Abdomen with five ventrites, the first three of which are connate; ventrite 1 not longer than 2; intercoxal process acute. Tergites relatively lightly sclerotised, without wing-folding spicule patches. Functional spiracles present on segments I–VIII, those on III–VIII located on laterosternites. Anterior edge of sternite VIII in male without median strut. Anterior edge of segment IX in male with truncate subgenital plate; tergite IX deeply emarginate, X well developed, evenly sclerotised and free. Aedeagus of trilobate type, symmetrical; phallobase relatively large, transverse, without anterior struts; parameres articulated, not outwardly hooked. Penis longitudinally divided, without ventral lobe but with narrow rod lying in ventral membrane, with short paired anterior struts. Sternite VIII in female with long anterior strut (spiculum ventrale) and pair of shorter anterolateral struts. Ovipositor long and moderately slender, heavily sclerotised, depressed; paraprocts ~4 times as long as coxites, which are located distally, undivided and apically acute; styli absent. [Lea 1914; Emden 1931b, 1932b, 1933; Lawrence 2005e; Young 2002d].

Larva

Body elongate, parallel-sided, cylindrical, not curved ventrally. Dorsal and ventral surfaces heavily sclerotised; colour dark reddish-brown to black or yellow with darker markings; surfaces generally smooth; vestiture of scattered, long, fine hairs, mainly laterally.

Head (Fig. 38D) protracted and prognathous or slightly declined, not concealed from above; globular; posterior edge not emarginate. Epicranial stem moderately long; frontal arms V-shaped, contiguous at base; median endocarina beneath epicranial stem. Stemmata absent. Antennae very short, 2-segmented with apical, dome-like, C-shaped sensorium. Frontoclypeal suture weakly impressed; clypeus very short, with slender membranous anteclypeus. Labrum free, strongly transverse, apically multilobed. Mandibles short and broad, robust and wedge-like, symmetrical, with broad, tridentate apex, with 1 or 2 subapical teeth; accessory ventral process absent; mesal surface broadly concave; mola and prostheca absent. Ventral mouthparts (Fig. 49I) strongly retracted and consolidated, forming maxillolabial complex; maxillary articulating area absent. cardines distinct, sclerotised, undivided, transverse and closely approximate, with membranous area between them, not separated by labium; stipites longer than wide and broadly contiguous basally; galea and lacinia fused to from articulated, apically truncate mala; palp 3-segmented, with segment-like palpifer. Labium consisting of prementum and postmentum, the latter closely associated with stipites, not longitudinally divided and posteriorly acute; ligula longer than labial palps, forming truncate sclerome; palps 2-segmented, without distinct palpiger, separated by more than width of first palpomere. Hypopharyngeal bracon present; hypopharyngeal sclerome absent; hypostomal rods absent; ventral epicranial ridges present; gular sutures well separated; gula wider than long.

Prothorax distinctly longer than meso- or metathorax, but not longer than two combined. Thoracic terga extending ventrally, so that pleural and sternal regions are highly reduced, without distinct ventral sclerites; abdominal terga I–IX extending ventrally to midline forming complete sclerotised rings. Legs very short, 5-segmented, including pretarsus, which is claw-like, without setae; coxae subcontiguous. Meso- and metatergum and abdominal tergum I each with two dorso-lateral pairs of longitudinally oval, sclerotised impressions subdivided by several transverse ridges; abdominal terga II–VII each with single pair of similar depressions and with deep cavity lined by sclerotised ridges; tergum VIII with a more narrowly elongate, oblique, subdivided impression. Tergum IX forming dorsally-articulated operculum (Fig. 52Q); sternite IX and segment X concealed and membranous. Spiracles biforous with spiracular scar located anterior to accessory openings, which are horizontal; spiracular closing apparatus absent. [Böving and Craighead 1931; Emden 1932a; Zimmerman 1942; Costa *et al.* 1988; Lawrence 1991b, 2005e; Young 2002d].

Classification and World Distribution

Callirhipidae contains ~175 described species in ten genera, occurring mainly in tropical parts of the Oriental, Australian and Neotropical regions, with a few species reaching the southern Palaearctic region, the monotypic genus *Zenoa* Say in North America, and *Callirhipis philiberti* Fairmaire in the Seychelles. [Hájek 2011].

Australian Fauna

The Australian callirhipid fauna includes two species of *Callirhipis* Latreille: *C. cardwellensis* Blackburn (Fig. 62B–C) and *C. leai* Pic, and two species of *Ennometes* Pascoe: *E. bifoveicollis* Lea and *E. ruficornis* (Gray). Callirhipids are large, elongate beetles with pectinate (female) or flabellate (male) antennae borne on a frontal prominence. The eyes are large and projecting, the prothorax lacks distinct lateral carinae but has a well developed basal interlocking mechanism, which may be crenulate, the elytra are ribbed or provided with window punctures, and the tarsi are simple, with a large multisetose empodium between the pretarsal claws. Larvae are heavily sclerotised and cylindrical, with a globular head (Fig. 38D), very short antennae, a solidly fused maxillolabial complex, a sclerotised ligular plate, short legs and a dorsally-hinged operculum at the abdominal apex. Callirhipids occur in the wet forests of northern NSW and QLD; larvae live in rotten logs and adults have been collected at lights. [Lea 1914].

ELATEROIDEA

This superfamily includes most of the Elateroidea and Cantharoidea in the traditional sense, plus the families Brachypsectridae, Artematopodidae and Rhinorhipidae, although the last family formed a clade with Rhipiceridae and/or Dascillidae in analyses by Lawrence *et al.* (1995, 2011). All elateroid adults lack a mandibular mola and metakatepisternal suture and have a low-grade type of wing folding (Hammond 1979) with a reduced apical field, and all but the doubtfully included Rhinorhipidae have four Malpighian tubules. Known larvae have a single large stemma on each side (sometimes absent) and mouthparts adapted for liquid feeding (mola absent, ventral mouthparts connate, buccal cavity blocked). The phylogeny of elateroids, the relationships of soft-bodied elateroids formerly included in Cantharoidea, and the evolution of neoteny and bioluminescence in this group have been the subjects of several recent works, including those of Branham and Wenzel (2001, 2003), Bocáková *et al.* (2007), Sagegami-Oba *et al.* (2007a, 2007b), Bocák *et al.* (2008), Vahtera *et al.* (2009) and Kundrata and Bocak (2011). The following elateroid families do not occur in Australia: Artematopodidae (Holarctic and Neotropical), Cerophytidae (Holarctic, Neotropical and Afrotropical), Plastoceridae (Asia Minor and South-east Asia), Drilidae (western Palaearctic and Afrotropical), Omalisidae (Europe), Telegeusidae and Phengodidae (North to South America), Rhagophthalmidae (Oriental), Omethidae (eastern Asia and North America). [Crowson 1972b; Muona

1995; Calder 1998; Johnson 2002b; Miller 2002b; O'Keefe 2002b; Ramsdale 2002a, 2010a; Young 2002e; Costa *et al.* 2003, 2010b; Lawrence 2005f, 2010b, 2010c; Branham 2010a; Bocák and Bocákova 2010a; Bocák *et al.* 2010; Costa and Zaragoza-Caballero 2010; Kawashima *et al.* 2010; Lawrence *et al.* 2010c, 2010d].

37. RHINORHIPIDAE

Adult (Fig. 62K)

Length 5–8.5 mm. Body ~3 times as long as wide; heavily sclerotised; slightly flattened above, moderately convex below; colour usually grey (brown in females); upper surfaces clothed with relatively stout, somewhat flattened, decumbent, pale setae.

Head strongly declined, abruptly constricted immediately behind eyes, but without distinct temples. Eyes large, protuberant, circular, entire, without interfacetal setae, ommatidia with exocone lenses (Fig. 29H); Antennal insertions obliquely raised and exposed from above, separated by slightly more than the diameter of a socket; subantennal groove absent. Frontoclypeal region deflexed; frontoclypeal suture absent; clypeus elongate, narrowly rounded at apex. Labrum membranous, concealed beneath clypeus. Antennae 11-segmented, filiform to weakly subserrate in both sexes, much longer in male. Mandible elongate, gradually curved and unidentate, with deep, setose cavity at base (concealed by clypeal lobe when mandibles closed); mola absent, prostheca reduced. Maxilla with reduced, membranous, densely setose galea and lacinia and no laciniar uncus; apical maxillary and labial palpomeres fusiform. Mentum apically biemarginate; ligula reduced, bilobed and setose. Corpotentorium and laminatentoria fused to form broad bridge. Cervical sclerites reduced.

Pronotum (Fig. 20H) about as long as wide, widest posteriorly and distinctly narrower than combined elytral bases, abruptly narrowed anteriorly forming rounded lobe partly concealing head; sides sinuate, with oblique, narrow, deep cavity at anterior fourth. Lateral pronotal carinae absent, anterior angles not produced; posterior angles broadly rounded; posterior edge trisinuate, without crenulations or interlocking device. Prosternum in front of coxae shorter than shortest diameter of procoxal cavity. Prosternal process complete, narrow, slightly curved, narrowly rounded at apex. Procoxae conical and projecting (Fig. 20H), without concealed lateral extensions; trochantin exposed, closely adpressed to sternum; endopleuron free. Procoxal cavities very slightly transverse, narrowly separated, broadly open externally and internally.

Scutellar shield abruptly elevated, posteriorly angulate. Elytra ~2 times as long as combined width and 3 times as long as pronotum, more or less parallel-sided, with deflected sutural flange; each elytron with 12 puncture rows and no scutellary striole; epipleura weakly developed. Mesoventrite without anterior procoxal rests, but body of ventrite with pair of longitudinally oblique carinae mesally delimiting shallow mesoventral cavity for receiving prosternal process and laterally delimiting paired procoxal rests, which continue onto the mesanepisterna. Mesocoxa conical and projecting, with broadly exposed trochantins. Mesocoxal cavities narrowly separated, partly closed laterally by mesanepisternum and mesepimeron, which are partly fused together. Meso- and metathoracic portions of coxal cavities joined by complete suture. Mesometaventral junction complex. Metaventrite with short discrimen and no transverse (metakatepisternal) suture; exposed portion of metanepisternum relatively short and broad Metacoxae enlarged and contiguous, extending laterally to meet elytra; plates weakly developed. Metendosternite with long, narrow stalk, most of which is fused to mesal internal edges of metacoxae, moderately long, curved lateral arms, well developed, expanded and plate-like ventrolateral processes, and anterior process moderately long and deeply divided, so that anterior tendons are on separate lobes. Hind wing with short apical field vaguely pigmented and with single oblique linear sclerite, but without distinct radial extensions; radial cell ~4 times as long as wide, with inner posterobasal angle more or less right; cross-vein r3 short, slightly oblique and fused at base with cross-vein r4; basal portion of RP well developed; medial spur extending to wing margin; medial field with five free veins; wedge cell well developed and apically acute; anal embayment absent, AP undivided. Outer edges of tibiae lined with small teeth and apices somewhat expanded (more so in hind legs of female). Tarsi 5-5-5; tarsomeres simple; pretarsal claws pectinate; empodium well developed, bisetose or trisetose (Fig. 27F).

Abdomen with five ventrites the first three of which are connate; intercoxal process acute; last ventrite (sternite VII) with narrow, deep groove bordering free edge. Laterosternites well developed, those on VII delimited by distinct ridge; parasternites narrow, absent on posterior half of segment VII; tergites all heavily pigmented. Functional spiracles on abdominal segments I–VIII. Anterior edge of sternite VIII in male with short, triangular apodeme. Anterior edge of segment IX in male broadly rounded, without spiculum gastrale; tergite IX emarginate; tergite X free. Aedeagus (Fig. 31B) symmetrical, trilobate, with short phallobase, long, narrow parameres and longitudinally divided penis with short basal struts and no ventral sclerotised lobe. Sternite VIII in female without median strut (spiculum ventrale) or anterolateral struts. Ovipositor (Fig. 30A) ~2.5 times as long as wide, relatively lightly sclerotised, except for proctigeral, paraproctal and transverse coxital baculi; paraprocts ~1.5 times as long as coxites, which are divided into broad basal and narrow apical lobes; styli well developed and terminal. [Lawrence 1988b, 2010a; Lawrence *et al.* 1999b].

Larva

Unknown.

Classification and World Distribution

The family contains the single Australian genus *Rhinorhipus* Lawrence.

Australian Fauna

Rhinorhipus tamborinensis Lawrence is known only from higher elevations in extreme southern Qld (Tamborine Mountain, Lamington National Park, Mount Glorious), but it is likely to occur in parts of northern NSW. Males have been collected, sometimes in large numbers, in early Spring on introduced vegetation bordering rainforest. Females were found in much smaller numbers in the same situations. When disturbed, the beetles feign death and drop to the ground. Specimens often have clay adhering to the legs, indicating that they may have emerged from the soil, and the expanded tibial apices, especially in the female, may support this hypothesis. Larvae are unknown. [Lawrence 1988b, 2010a].

38. BRACHYPSECTRIDAE

Adult (Fig. 70H)

Length ~4–8 mm. Body ~2–2.5 times as long as wide; lightly sclerotised; flattened; colour usually yellowish-brown; upper surfaces finely pubescent.

Head transverse, not declined, deeply inserted into prothorax. Eyes large, protuberant, circular, entire, without interfacetal setae. Antennal insertions slightly exposed from above, moderately widely separated; subantennal groove short and weakly developed. Frontoclypeal region deflexed; frontoclypeal suture absent; clypeus short, broad, apically emarginate. Labrum small, free, well sclerotised and apically rounded. Antennae 11-segmented, weakly to strongly pectinate or somewhat clavate. Mandible small, subtriangular and unidentate, without mola or prostheca. Maxilla with membranous, setose galea and lacinia and no laciniar uncus; apical maxillary and labial palpomeres fusiform. Mentum transverse, slightly emarginate; ligula short, slightly emarginate and setose. Corpotentorium absent. Cervical sclerites moderately well developed.

Pronotum ~0.6 times as long as wide, widest posteriorly, at least as wide as combined elytral bases; sides slightly curved. Lateral pronotal carinae complete, but with anterior third concealed from above; anterior angles not produced; posterior angles produced, acute and embracing elytral bases; posterior edge trisinuate, without crenulations but with moderately well developed interlocking device. Prosternum in front of coxae about twice as long as shortest diameter of procoxal cavity, with short, subtruncate chinpiece. Prosternal process complete, narrow, slightly curved, narrowly rounded at apex. Notosternal sutures complete and more or less open anteriorly. Procoxae without concealed lateral extensions; trochantin slender, partly concealed by expansion of prosternal cowling; endopleuron free. Procoxal cavities strongly transverse, narrowly separated, broadly open externally and internally.

Scutellar shield abruptly elevated, posteriorly broadly rounded. Elytra ~1.5–2 times as long as combined width and 3.75–5 times as long as pronotum, slightly wider posteriorly; each elytron with nine weakly impressed striae and no scutellary striole; epipleura complete. Mesoventrite with procoxal rests shallow and slightly inclined; mesoventral cavity moderately large, extending beyond anterior edges of coxal cavities; discrimen absent. Mesocoxae transverse, with exposed trochantins. Mesocoxal cavities narrowly separated, partly closed laterally by mesanepisternum and mesepimeron. Mesometaventral junction monocondylic. Metaventrite relatively long and flat with long discrimen and no transverse (metakatepisternal) suture; exposed portion of metanepisternum relatively narrow. Metacoxae strongly subcontiguous, extending laterally to meet elytra; plates narrow but more or less complete. Metendosternite with long, narrow stalk, short, broad lateral arms, weakly developed ventrolateral processes and anterior process broadly emarginate, so that anterior tendons are on widely separated lobes. Hind wing with short apical field bearing an anterior and posterior oblique linear sclerite; radial cell moderately large, elongate, with inner posterobasal angle right; cross-vein r3 moderately long and almost parallel to long wing axis, well separated from r4; basal portion of RP long; medial spur extending almost to wing margin; medial field with five free veins; MP_{3+4} with basal cross-vein and spur, apically forked; CuA_1 meeting MP_4; wedge cell absent; anal embayment absent, AP undivided. Tibiae simple, slender; spurs absent. Tarsi 5-5-5; tarsomere 4 somewhat reduced, with weak ventral lobe; pretarsal claws simple; empodium bisetose.

Abdomen with five free ventrites; intercoxal process acute. Functional spiracles on abdominal segments I–VIII. Anterior edge of sternite VIII in male bisinuate. Anterior edge of segment IX in male broadly rounded, without spiculum gastrale; tergite IX emarginate; tergite X free. Aedeagus symmetrical, trilobate, with short phallobase, slightly emarginate anteriorly, and long parameres, each with lateral hook at apex; penis undivided, with short, widely diverging basal struts. Ovipositor moderately elongate, lightly sclerotised, except for paraproctal and transverse coxital baculi; paraprocts slightly longer than coxites, which are divided into broad basal and narrow apical lobes; styli well developed and terminal. [Blair 1930; Lawrence *et al.* 1999b; Young 2002f; Costa *et al.* 2006; 2010a].

Larva (Figs 5G, 65K)

Body broadly ovate, strongly flattened and disc-like, with long, branched lateral tergal projections on thoracic and most abdominal segments; head and segment IX dark brown, remaining dorsal surfaces unevenly pigmented, varying from white to yellowish brown and covered with small setiferous tubercles, flattened pigmented areas and a median row of subconical projections lined with setiferous tubercles; ventral surfaces lightly pigmented; vestiture consisting of expanded, modified setae, which may form irregularly multisided, flattened scales with median stalk.

Head hyperprognathous, elongate, more or less parallel-sided and much narrower than prothorax; posterior edge broadly emarginate; epicranial stem short; frontal arms lyriform with bases contiguous; median endocarina absent. Stemmata on each side one, with well developed lens. Frontoclypeal suture absent or weakly developed. Labrum free, apically biemarginate, with four tubercles bearing stout setae. Antennae well developed, 3-segmented; antennomeres 2 much longer than 1, bearing highly reduced apical antennomere and palpiform sensorium. Mandibles (Fig. 40G) symmetrical, narrow and falcate, strongly curved, widely separated at base, and perforate with completely enclosed internal channel; without mola or prostheca. Ventral mouthparts (Fig. 48J) deeply retracted forming maxillolabial complex, without maxillary articulating area; cardines fused together forming single, median plate; stipes elongate; galea small, palpiform, articulated; lacinia fixed and falcate; 4-segmented, with palpomere 2 as long as 1, 3 and 4 combined. Labium consisting of prementum and postmentum, the latter completely fused to stipes on each side (separated by slight impression only); palps 2-segmented, approximate, with palpomere 1 much longer than 2; ligula absent. Hypopharyngeal sclerome absent. Hypostomal rods and ventral epicranial ridges absent. Gular region longer than maxillolabial complex; gular sutures fused.

Thoracic terga and abdominal terga I–VIII transverse, the former each with two pairs of lateral processes, the latter each with one pair. Prothoracic venter unpigmented, except for pair of sclerites on cervicosternum. Legs well developed, 5-segmented; pretarsus forming distinct claw bearing 2 setae lying side by side; coxae widely separated. Abdominal terga I–VIII ~1.5 times as long as thorax; terga I–VII each with median projection. Tergum IX forming long, narrow, heavily sclerotised plate, which is narrowed to subacute apex; sternum IX membranous and segment X not distinguishable. Spiracles reduced and biforous, with closing apparatus, all borne on short spiracular tubes; mesothoracic spiracles ventral, on spiracular sclerites; abdominal spiracles dorsal, on lateral portion of tergite. [Blair 1930; Böving and Craighead 1931; Lawrence 1991b; Lawrence *et al.* 1999a; Costa *et al.* 2006; 2010a].

Classification and World Distribution

The family Brachypsectridae contains the single genus *Brachypsectra* LeConte with species from the south-western part of North America, the Greater Antilles, western Iran, southern India, Singapore and Australia. [Costa *et al.* 2006, 2010a; Hájek 2010].

Australian Fauna

The Australian species of *Brachypsectra* has not yet been described but is apparently widely distributed in the arid parts of Australia. Larvae having been collected from the Flinders Range, SA, Windjana Gorge, WA and Diamantina National Park, western QLD, and a single female was reared from the last locality (Monteith and Turco 2011). Adults of exotic species are oblong, somewhat flattened and yellowish-brown, with the hind angles of the pronotum acute and carinate, the prosternum long forming a short chin piece, the antennae relatively short with apical antennomeres weakly pectinate in the male, the mandibles reduced, elytra weakly striate, tarsi simple and abdominal ventrites free. Larvae (Fig. 65K) are broadly ovate and flattened, with complex lateral processes on most segments and with segment IX produced posteriorly to form a flexible, tapered spine. The head is prognathous, with one pair of large stemmata, club-like antennae, free but reduced labrum, perforated, falcate mandibles, and strongly consolidated ventral mouthparts. Thoracic spiracles are biforous, but those on the abdomen are reduced. Brachypsectrid larvae occur under loose bark or in rock crevices and prey on various small invertebrates. It is thought that prey are attracted to an allomone produced by dorsal glands and when they walk over the flattened body of the larva, the head and tail spine are both raised, pinning the prey between them. The beetle then pierces the body of the prey and feeds by extraoral digestion. Pupation occurs in a loose cocoon composed silken threads. [Costa *et al.* 2006, 2010a].

39. EUCNEMIDAE

Adult (Fig. 64B–C)

Length 1.5–18 mm. Body ~2.4–3.6 times as long as wide, more or less parallel-sided, widest at pronoto-elytral junction and narrowing anteriorly and posteriorly; heavily sclerotised, or narrowed at pronoto-elytral junction, as in most Elateridae; slightly flattened to strongly convex above; ventral surfaces flat to convex; colour usually yellowish-brown to black and uniform, occasionally with elytra lighter than pronotum; upper surfaces often clothed with fine, decumbent hairs, but sometimes with somewhat flattened, yellow or white hairs forming a pattern.

Head strongly transverse, moderately to strongly declined and deeply inserted into prothorax, not abruptly constricted posteriorly. Occipital region sometimes with fine transverse ridge. Vertex sometimes with long median carina. Eyes small to large, flat to protuberant, more or less entire, finely facetted, without interfacetal setae. Antennal insertions exposed from above and narrowly to moderately widely separated; subantennal groove short or absent, sometimes containing pores or deep pits. Frontoclypeal region gradually declined; frontoclypeal suture absent; anterior edge of clypeus usually broadly rounded, concave or bisinuate. Labrum usually membranous and concealed beneath clypeus (Figs 14I, 19H); occasionally lightly sclerotised and partly visible (*Anischia*). Antennae 11-segmented, moniliform, serrate, pectinate, flabellate, clavate, or abruptly expanded to form 3- or 4-segmented club (*Anischia*); scape usually much longer than pedicel, which is attached subapically. Mandible elongate or short and broad, unidentate or bidentate, without mola or prostheca. Maxilla usually with articulated galea and apically narrowed lacinia without uncus; apical maxillary and labial palpomeres usually fusiform, sometimes apically widened; ligula truncate. Gular sutures often widely separated. Corpotentorium narrow and arched; laminatentoria sometimes fused at midline. Cervical sclerites well developed.

Pronotum transverse to elongate, usually widest at about middle; sides subparallel to strongly rounded. Lateral pronotal carinae usually entire, with distinct bead, sometimes incomplete anteriorly or weakly developed, without bead; anterior angles not produced forward; posterior angles produced and acute; posterior edge with well developed interlocking device, sometimes with paired cavities for receiving anterior elytral projections (*Anischia*); disc flattened to strongly convex, usually simple, with large gibbosities in *Galbites*, with sub-lateral basal carinae in *Anischia* or oblique anterior carinae in some Melasinae. Prosternum in front of coxae (Fig. 19H) more than twice as long as shortest diameter of procoxal cavity, with broad chin piece. Prosternal process complete, narrow, slightly curved, narrowly rounded at apex. Notosternal sutures complete to almost entirely obliterated. Hypomera with deep lateral antennal grooves in some Eucneminae and Macraulacinae, with antennal grooves extending across notosternal suture in Dirhagini. Procoxae globular, with very short concealed lateral extensions; trochantinopleuron reduced, concealed and fused to hypomeron. Procoxal cavities subcircular, moderately widely separated, broadly open externally and internally.

Scutellar shield abruptly elevated or not, anteriorly simple, posteriorly acute. Elytra ~1.7–3 times as long as combined width and 2.3–4.9 times as long as pronotum, more or less parallel-sided; each elytron usually with nine puncture rows or striae and no scutellary striole; epipleura narrow and complete. Mesoventrite with anterior procoxal rests horizontal; mesoventral cavity deep and extending well beyond anterior edges of mesocoxal cavities. Mesocoxae globular, with concealed trochantins. Mesocoxal cavities moderately widely separated, partly closed laterally by mesepimeron, which is completely fused to mesanepisternum. Mesometaventral junction complex. Metaventrite with or without discrimen and without transverse (metakatepisternal) suture, with two pairs of postcoxal lines in *Anischia*; exposed

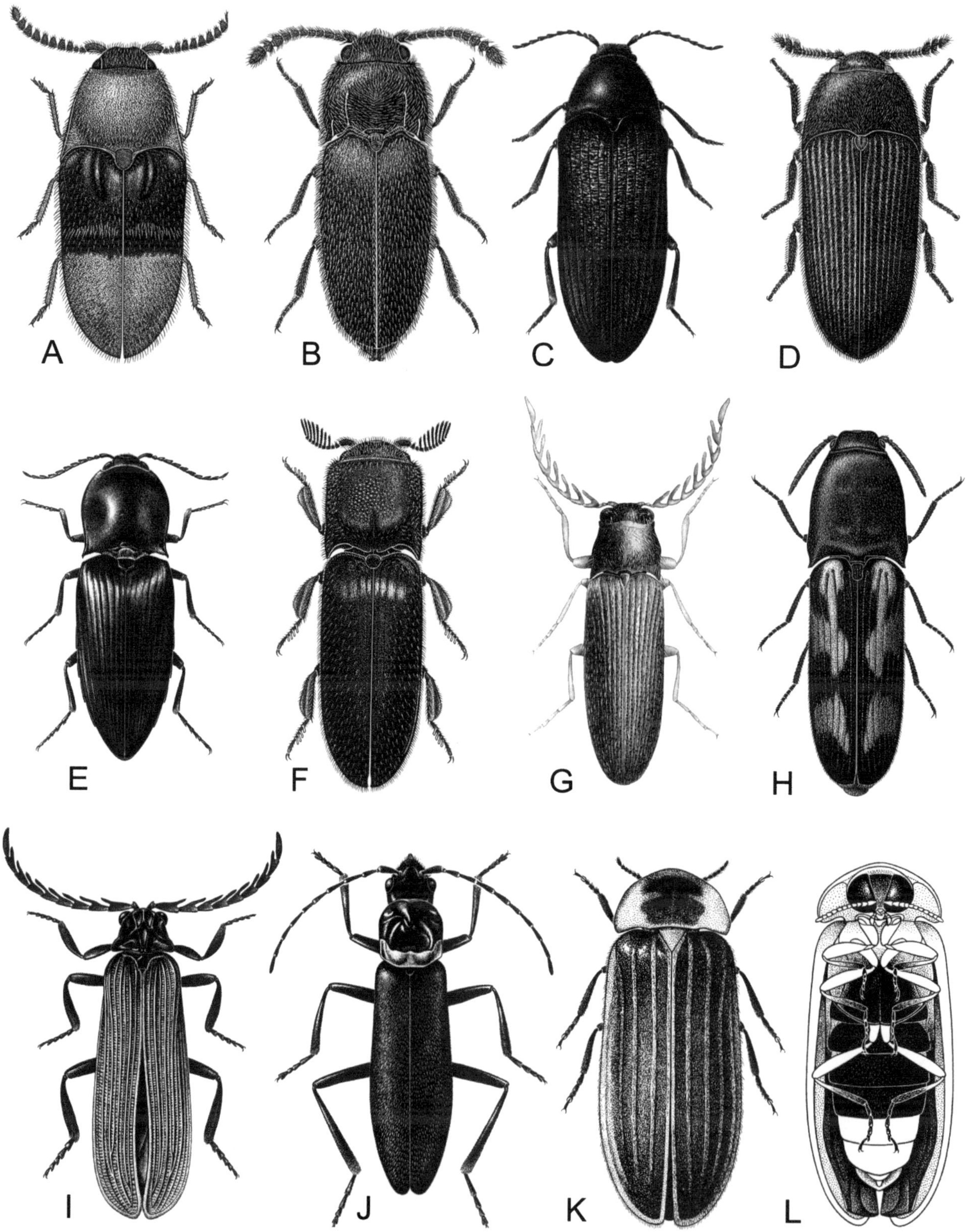

Fig. 64. A–J, Adult, dorsal: A, *Osslimus freyi* (Cobos) (Elateridae: Lissominae); B, *Anischia stupenda* Fleutiaux (Eucnemidae: Anischiinae); C, *Euryptychus porrosus* (Lea) (Eucnemidae: Macraulacinae); D, *Aulonothroscus* sp. (Throscidae); E–H, Elateridae: E, *Lanelater mastersi* (Agrypninae); F, *Cussolenis mutabilis* (Bonvouloir) (Thylacosterninae); G, *Austrelater macphersonensis* Calder (Lissominae); H, *Subprotelater* sp. (Subprotelaterinae). I, *Porrostoma rhipidius* (Macleay) (Lycidae), dorsal; J, *Chauliognathus lugubris* (Fabricius) (Cantharidae), dorsal. K, *Atyphella lychnus* Olliff (Lampyridae), dorsal; L, same, ventral.

portion of metanepisternum usually relatively narrow. Metacoxae usually slightly oblique, contiguous and extending laterally to meet elytra (widely separated and not elytra in *Anischia*); plates usually well developed and complete or narrowed laterally (absent in *Anischia*). Metendosternite usually with long, narrow stalk, short lateral arms and long, narrow anterior process with approximate tendons (stalk short and broad and lateral arms long and oblique with tendons near their apices in *Anischia*). Hind wing usually with short apical field (longer in *Anischia*), usually with anterior, posterior and median linear sclerites, but without distinct radial extensions; radial cell 1.5–5 times as long as wide, with inner posterobasal angle more or less right; cross-vein r3 usually almost longitudinal and sometimes absent, well separated from r4; basal portion of RP short to very long; R-M loop forming acute angle; medial spur usually well developed; medial field usually with five free veins (sometimes reduced to three or four); CuA joining MP_{3+4}; wedge cell absent or well developed with oblique apex; anal embayment absent or notch-like, AP undivided. Trochanters very long in *Anischia*. Tibiae usually slender, tibial spurs almost always reduced to one on fore legs. Tarsi 5-5-5; tarsomeres usually simple or with setose lobe on tarsomere 4; membranous lobes on 1–4 in Galbitini; male protarsomere 1 sometimes with sex combs; pretarsal claws usually simple, with basal setae in *Pseudomenes*; empodium usually not visible.

Abdomen almost always with five connate ventrites (basal three connate in *Anischia*); ventrite 1 with two postcoxal lines in *Anischia*; intercoxal process acute. Functional spiracles on abdominal segments I–VIII. Anterior edge of sternite VIII in male with lateral struts. Anterior edge of segment IX in male usually broadly rounded, without spiculum gastrale; tergites IX and X usually more or less fused together. Aedeagus of trilobate type but highly variable; symmetrical or asymmetrical, tubular to flattened, with parameres of variable shape and free or fused together and/or to penis. Sternite VIII in female usually with fixed or articulated spiculum ventrale. Ovipositor usually long, slender and relatively lightly sclerotised, except baculi; coxites divided or not; styli usually minute and subapical. [Cobos 1964; Muona 1991b, 1993, 2000, 2002, 2010; Lawrence *et al.* 1999b, 2007].

Larva (Fig. 35I)

Body elongate, usually parallel-sided (widest at metathorax in Melasini), usually strongly flattened (not so in Melasini and some Dirhagini), swollen and maggot-like in some Eucnemini; usually evenly lightly or heavily sclerotised; head and tergum IX often more heavily pigmented than remainder of body; vestiture consisting of a few macrosetae and numerous microsetae, usually with dorsal and/or ventral patches of dense spicules and microtrichia.

Head prognathous and protracted, usually forming a serrate, wedge-like plate (Fig. 42I), widest at base and tapering anteriorly with broadly rounded anterior edge and tapering from base to apex in lateral view (head in *Pseudomenes* with deep longitudinal grooves); posterior edge emarginate or not; epicranial suture and median endocarina absent. Stemmata almost always absent (one on each side in *Anischia*). Frontoclypeal suture absent. Labrum almost always fused to clypeus (free in *Pseudomenes*). Antennae very short, 2-segmented with large sensorium, located within deep cavities. Mandibles usually reduced and fused to head capsule, movable and opposable in *Anischia* and *Pseudomenes* (Fig. 42H), without mola or prostheca. Ventral mouthparts very deeply retracted forming maxillolabial complex, without maxillary articulating area; stipes, cardines and postmentum more or less fused together and to head capsule; apical lobes and palps reduced or absent. Hypopharyngeal sclerome, hypostomal rods and ventral epicranial ridges absent. Gular region highly reduced or absent.

Prothorax usually shorter and narrower than meso- or metathorax; thoracic terga and sterna often with variably-sized plates or T-shaped rods (Fig. 45J). Legs highly reduced, almost always reduced to small dark spots, with one short, broad segment bearing several setae in *Anischia* and *Pseudomenes*. Abdominal segments I–VIII more or less subequal in length, each divided into two pseudosegments in Melasini. Mesothorax, metathorax and abdominal segments I–VI often with variably placed plates of two types: those which are densely clothed with fine spicules and those which are smooth (areoles); thoracic and most abdominal terga in *Anischia* more heavily sclerotised than sterna, which have goblet-shaped spicule patches and no areoles. Tergum IX shorter or longer than previous ones, usually simple, furrowed or rugose, occasionally with urogomphi, extending onto ventral surface; sternite IX with two patches of asperities in *Anischia*; segment X ventral, elongate-oval, without asperities. Spiracles biforous, with closing apparatus, and surrounded by sclerotised ring. [Böving and Craighead 1931; Gardner 1935c; Mamaev 1976a; Burakowski 1991; Muona 2002, 2010; Lawrence *et al.* 1999a, 2007; Muona and Teräväinen 2008].

Classification and World Distribution

The family Eucnemidae are cosmopolitan in distribution but most abundant and diverse in tropical regions. The group has been divided into nine subfamilies and several tribes. The following subfamilies do not occur in Australia: Perothopinae (Nearctic), Phyllocerinae (North and Central America, Palaearctic and Afrotropical regions); Palaeoxeninae (south-western North America) and Phlegoninae (Neotropical). [Muona 1991a, 1991b, 2010].

Australian Fauna

Almost 50 genera of Eucnemidae occur in Australia and these represent five subfamilies. **Pseudomeninae** includes the single species *Pseudomenes bakewelli* (Bonvouloir), while **Anischiinae** includes three species of *Anischia*, which were once included in the family Cerophytidae based on the loss of the metacoxal plates. The remaining three subfamilies are each composed of several genera. Eucnemid adults are elongate, cylindrical to slightly flattened and compact, usually finely pubescent. They resemble Elateridae in the form of the prothorax, with acute posterior angles and a clicking device involving the prosternal spine and mesoventral cavity, but they differ from all elaterids in having the labrum concealed beneath the clypeus, all five ventrites connate and the antennae sometimes received in hypomeral grooves just beneath the lateral pronotal carinae or ventrally crossing or bordering the notosternal suture. The tarsi are usually simple, but in *Galbites* there are ventral membranous appendages on tarsomeres 2–4. Larvae care usually flattened, with no legs and a reduced head with minute antennae and non-opposable or immovable mandibles. There are two common larval types – one in which the body is parallel-sided with a sclerotised, wedge-like head, and another resembling a buprestid larva in having a greatly enlarged prothorax bearing rod-like sclerotisations; however, larvae are known for relatively few Australian groups and at least two other larval types occur in *Anischia* (Lawrence *et al.* 2007) and *Pseudomenes*, and additional ones have been described by Gardner (1935c). Most larvae occur in dead wood, but their food source and the mechanics of their locomotion through the woody substrate are still matters for speculation. They are liquid feeders and there is no evidence that they are predators; it is possible that they digest wood fibres extraorally, but they may also feed on slime mould plasmodium or some other nutrient-rich fluid within the wood. According to Muona and Teräväinen (2008) the areoles on the larval body are water excreting organs.

Australian Subfamilies, Tribes and Genera

Pseudomeninae: Pseudomenini: *Pseudomenes* Fleutiaux.
Anischiinae: *Anischia* Fleutiaux
Melasinae: Calyptocerini: *Calyptocerus* Guérin-Méneville. **Xylobiini:** *Agalba* Broun. **Neocharini:** *Neocharis* Sharp. **Epiphanini:** *Hylis* Des Gozis. **Dirhagini:** *Arrhipis* Bonvouloir, *Cafolus* Bonvouloir, *Entomophthalmus* Bonvouloir, *Farsus* Jacquelin Du Val, *Freyiola* Cobos, *Microrhagus* Dejean, *Porraulacus* Fleutiaux, *Prodirhagus* Fleutiaux. *Rhagomicrus* Fleutiaux.
Eucneminae: Phaenocerini: *Phaenocerus* Bonvouloir, *Pinaroo* Muona. **Proutianini:** *Myall* Muona, *Onya* Muona. **Dendrocharini:** *Dendrocharis* Guérin-Méneville, *Scopulifer* Fleutiaux. **Yangini:** *Yanga* Muona. **Entomosatopini:** *Entomosatopus* Bonvouloir. **Dyscharachthini:** *Dyscharachthis* Blackburn. **Mesogenini:** *Anabolodes* Strand, *Mesogenus* Bonvouloir, *Vitellius* Bonvouloir. **Eucnemini:** *Epipleurus* Fleutiaux, *Poecilochrus* Bonvouloir. **Galbitini:** *Galbites* Fleutiaux.
Macraulacinae: Jenibuntorini: *Jenibuntor* Muona. **Echtrogasterini:** *Echthrogaster* Blackburn, *Dictyeucnemis* Lea, *Hemiopsida* W.J. Macleay. **Euryptychini** *Euryptychus* Leconte. **Orodotini:** *Orodotes* Bonvouloir. **Macraulacini:** *Heterotaxis* Bonvouloir, *Cladidus* Fleutiaux, *Discaptothorax* Blackburn, *Galbodema* Laporte, Macroscython Fleutiaux, *Plesiofornax* Coquerel, *Fornax* Laporte, *Dromaeoloides* Cobos, *Dromaeolus* Kiesenwetter. **Nematodini:** *Neomathion* Fleutiaux, *Nematodes* Berthold, *Trigonopleurus* Bonvouloir.

40. THROSCIDAE

Adult (Fig. 64D)

Length 1.6–3.5 mm. Body ~2–2.6 times as long as wide, usually widest at pronoto-elytral junction and narrowing anteriorly and posteriorly (sides subparallel in *Potergus*); slightly flattened to moderately convex dorsally and ventrally; colour uniformly reddish-brown to dark brown; upper surfaces clothed with decumbent hairs, usually interspersed with erect hairs, sometimes with silvery sheen.

Head transverse, moderately to strongly declined and deeply inserted into prothorax, not abruptly constricted posteriorly. Occipital region usually with fine transverse ridge. Vertex with long median carina in *Potergus* Bonvouloir. Eyes well developed, not protuberant, entire to deeply emarginate, finely facetted, without interfacetal setae. Antennal insertions exposed and moderately widely separated, located in circular fossae in *Potergus*; subantennal groove well developed, sometimes containing pores or deep pits. Frontoclypeal region gradually declined; frontoclypeal suture absent; anterior edge of clypeus more or less truncate. Labrum small, transverse, exposed, its anterior edge convex. Antennae 11-segmented, usually with symmetrical to asymmetrical 3-segmented club (antennae filiform in *Potergus*); scape only slightly longer than pedicel. Mandible short, broad, somewhat flattened and unidentate, without mola or prostheca. Maxilla usually with short galea and lacinia, without uncus; apical palpomere usually expanded and truncate (fusiform in Potergus); ligula short, broad, cleft, lightly saclerotised; apical labial palpomeres apically widened or fusiform. Gula strongly transverse. Corpotentorium narrow. Cervical sclerites well developed.

Pronotum as long as wide or transverse, usually widest posteriorly with sides almost always slightly to strongly curved and converging anteriorly (subparallel in *Potergus*). Lateral pronotal carinae usually incomplete (complete but not entirely visible from above in *Potergus*), with or without distinct bead; anterior angles not produced forward; posterior angles produced and acute, fitting into cavities beneath elytral humeri; posterior edge biemarginate, with well developed interlocking device; disc slightly convex and simple. Prosternum in front of coxae longer than shortest diameter of procoxal cavity, slightly to strongly convex, with paired longitudinal carinae (except in *Potergus*) and with truncate or rounded chin piece. Prosternal process complete, parallel-sided, extending well behind coxae and rounded at apex. Notosternal sutures complete, widely open anteriorly forming antennal grooves, which curve laterally and extend almost to lateral edges of hypomera. Procoxae globular, without concealed lateral extensions; trochantinopleuron reduced, concealed and fused to hypomeron. Procoxal cavities subcircular, moderately widely separated, broadly open externally and internally.

Scutellar shield abruptly elevated, posteriorly subacute to broadly rounded. Elytra usually ~1.5–2.2 times as long as combined width and 2–4 times as long as pronotum (2.4–2.5 times as long as wide and 2.5–3 times as long as pronotum in *Potergus*), anteriorly produced and carinate, fitting into transverse grooves at base of pronotum; each elytron usually with 9 puncture rows and no scutellary striole; epipleura narrow and incomplete. Mesoventrite with anterior edge on different plane than metaventrite; at middle with raised area flanked by large, paired procoxal rests and continued posteriorly as diagonal slide leading to large, deep mesoventral cavity; discrimen absent. Mesocoxae globular, with concealed trochantins. Mesocoxal cavities widely separated, partly closed laterally by mesepimeron. Mesometaventral junction a straight line or absent due to fusion. Metaventrite usually without discrimen (discrimen long but weakly indicated in *Potergus*) and without transverse (metakatepisternal) suture; usually with pair of curved, deep grooves extending from mesocoxal cavities to posterolateral edges of metaventrite (except in Trixagus, which has a pair of curved postcoxal lines delimiting leg impressions); exposed portion of metanepisternum narrowly elongate and parallel-sided; anterior part of metanepisternum and metaventrite slightly impressed forming with mesepimeron a rest for mesofemur and mesotibia. Metacoxae contiguous, extending laterally to meet elytra; plates well developed and more or less complete. Metendosternite with moderately long stalk, long lateral arms, short to moderately long anterior process and approximate to well separated anterior tendons. Hind wing relatively short and broad; apical field well developed but without distinct sclerites; radial bar heavily pigmented; radial cell vaguely indicated, without basal edge; cross-vein r3 absent, r4 incomplete or absent; basal portion of RP usually short or absent (long in *Potergus*); R-M loop forming acute angle; medial spur slightly to strongly curved; medial field with three veins and no wedge cell; anal embayment deep and notch-like. Trochanterofemoral joint on fore legs almost transverse. Femora flattened and dorsally excavate to receive tibiae, which are relatively slender, flattened, widest at middle and usually sinuate towards apex; outer apical angle of protibia toothed in *Potergus*; tibial spurs absent. Tarsi 5-5-5; tarsomeres slender. 3 and 4 with slight ventral lobes, except in *Potergus*; empodium absent.

Abdomen with five connate ventrites; ventrite 1 not longer than 2, usually with paired transverse impressions for housing hind legs; with short, oblique postcoxal lines in *Potergus*; intercoxal process narrow and acute; ventrites 2 and 3 on each side usually with vague lateral impressions (with deep grooves for reception of metatarsi in *Potergus*). Abdominal tergites usually membranous (moderately pigmented in *Potergus*). Functional spiracles usually on abdominal segments I–VII (on I–VIII in *Potergus*). Anterior edge of sternite VIII in male usually without or with very short median strut (with long strut in *Potergus*). Anterior edge of segment IX in male usually broadly rounded, without spiculum gastrale; tergite IX usually emarginate and X well developed and free (sometimes with IX and X more or less fused together). Aedeagus of trilobate type, symmetrical; phallobase elongate, slightly flattened, with deep anterior emargination; parameres individually articulated, lancet-shaped, not laterally curved and without teeth; penis long and slender, flattened, with long anterior struts. Sternite VIII in female with long, fixed spiculum ventrale. Ovipositor narrowly elongate and lightly sclerotised; paraprocts with baculi, much longer than coxites, each of which has a transverse basal baculum and apical stylus. [Cobos 1960, 1961, 1967; Coffin 1993; Lawrence *et al.* 1999b; Johnson 2002c; Muona *et al.* 2010].

Larva

Body moderately elongate, widest at middle, slightly flattened, dorsally curved, evenly lightly pigmented and weakly sclerotised, except for portions of head, mandibles, prothoracic venter and pretarsal claws.

Head prognathous or slightly elevated and protracted, distinctly narrower than prothorax, anterior depressed and wedge-like; epicranial suture and median endocarina absent. Stemmata absent. Frontoclypeal suture absent. Labrum completely fused to head capsule. Antennae very short, located on raised articulating membrane, 2-segmented; antennomere 1 short and broad, 2 more than twice as long and much narrower; sensorium on antennomere 1 as long as or longer than antennomere 2, palpiform. Mandibles symmetrical, non-opposable and fused to head capsule, flattened and more or less triangular, without mola or prostheca. Ventral mouthparts strongly retracted forming maxillolabial complex, without maxillary articulating area; stipes longer than wide, not separated from cardines, well separated from one another; galea and lacinia combined to form subacute lobe but separated by longitudinal suture; Labium consisting of prementum and postmentum; prementum slightly elongate; ligula slender and about as long as labial palps, which are 1-segmented and separated by less than basal width. Hypopharyngeal sclerome, hypostomal rods, ventral epicranial ridges and gula absent. Ventral portion of head capsule supported by heavily sclerotised frame consisting of thick bars extending from mandibular insertions to to base of maxillolabial complex, then forking to form lateral and mesal branch, both of which join a posterior bar extending along base of head.

Prothorax about equal in length to meso- or metathorax; venter with pair of lateral, subtrapezoidal, slightly curved sclerites, extending from neck to procoxal articulations, and slender, longitudinal sternal rods converging posteriorly. Legs very short, ~0.15 times as long as mesothoracic width, 5-segmented, with claw-like pretarsus bearing 2 setae; fore legs slightly larger than other two pairs, with large, oval coxae separated by about two diameters, meso- and metacoxae smaller and separated by several diameters. Abdomen ~2.5 times as long as thorax; terga and sterna simple. Tergum IX shorter than VIII, not or only slightly extending onto ventral surface, surface simple or with pair of minute urogomphi; sternite IX simple; segment X posteroventrally oriented. Spiracles biforous, with closing apparatus. [Böving and Craighead 1931; Burakowski 1975b, 1991; Becker 1991; Lawrence *et al.* 1999a; Johnson 2002c; Muona *et al.* 2010].

Classification and World Distribution

Throscidae is a cosmopolitan family containing ~150 species in four genera: *Potergus* Bonvouloir, with a few species in the Oriental and Australian regions, *Pactopus* LeConte with one species in north-western North America, and the widely distributed *Aulonothroscus* Horn and *Trixagus* Kugelann. [Muona 1993a, 1993b].

Australian Fauna

The described Australian Throscidae includes *Potergus freyi* Cobos from northern QLD, *Trixagus leechi* (Cobos) from SA, and seven species of *Aulonothroscus*; it is likely, however, that others remain to be described. Throscids are oblong to elongate, slightly flattened or rarely (*Potergus*) subcylindrical, finely pubescent beetles with visible labrum, acute posterior pronotal angles, which fit closely against the elytral humeri, five connate ventrites and tarsi without membranous appendages (although tarsomere 4 is usually lobed below). The antenna usually has a distinct 3-segmented club, but in *Potergus* it is moniliform, with compact cylindrical segments. Like the Elateridae, throscids are able to jump by means of a clicking mechanism involving the prosternal process and mesoventral cavity. Larvae are lightly sclerotised and grub-like, with a small head and reduced legs; antennae are very short, mandibles flattened, rounded and fused to the head capsule, the prothorax has supporting rods and tergum IX has a pair of minute fixed urogomphi. The biology of Australian throscids is unknown, but larvae of the European *Trixagus dermestoides*

(Linnaeus) feed in the soil on ectomycorrhizal fungi on the roots of various trees, and North American throscid larvae have been collected in rotten wood and grass tufts.

41. ELATERIDAE

Adult (Figs 5D, 64A, E–H, 70A–F)

Length 2.5–60 (usually 4–45) mm . Body ~2–5 times as long as wide, usually more or less parallel-sided, with elytra slightly expanded basally and tapered apically and the prothorax often distinctly curved laterally, so that the narrowest point is at the pronoto-elytral junction; slightly flattened to moderately convex dorsally and ventrally; colour highly variable, although many species are brown to black in colour; upper surfaces glabrous or pubescent.

Head subquadrate to strongly transverse, prognathous to strongly declined, not abruptly constricted posteriorly. Eyes small to large, often protuberant, entire, finely facetted, without interfacetal setae. Antennal insertions exposed or concealed and moderately to widely separated; subantennal groove present or absent. Frontoclypeal region highly variable, not or only slightly declined to vertical or inflexed at apex, forming frontal carina (Fig. 14K); frontoclypeal suture absent. Labrum free, slightly to strongly transverse, moderately to heavily sclerotised. Antennae almost always 11-segmented (12-segmented in *Pseudotetralobus* and *Wardulupicola*), filiform, serrate, pectinate or flabellate, with rami beginning on antennomere 3 or 4; scape usually much longer than pedicel. Mandible short and broad to long and narrowly falcate, with apex unidentate or bidentate; mola almost always absent (present in *Melanotus*); mandibular base often with brush of hairs. Maxilla with galea and lacinia pubescent, without uncus, sometimes reduced; apical maxillary and labial palpomeres usually at least slightly expanded apically, sometimes fusiform; ligula absent or membranous. Subgenal ridges present or absent. Corpotentorium narrow or absent. Cervical sclerites well developed.

Pronotum ~0.5–1.25 times as long as wide, usually widest posteriorly; sides usually more or less sinuate, sometimes straight or curved anteriorly. Lateral pronotal carinae almost always complete (incomplete anteriorly or absent in some Cardiophorinae); anterior angles not or slightly produced forward; posterior angles strongly produced posteriorly or posterolaterally and more or less acute, often carinate, with carinae extending well onto disc in *Pseudotetralobus*; posterior edge distinctly sinuate or variously lobed to form well developed interlocking device; disc usually simple or with paired, shallow sub-basal impressions, with paired, basal, sublateral carinae in *Rivulicola* and the Cardiophorinae (Fig. 27G). Prosternum in front of coxae almost always much longer than shortest diameter of procoxal cavity, with broadly rounded chin piece (Figs 18B, 44M) (somewhat shorter with subtruncate chin piece in *Austrelater*); surface flat to moderately convex, with paired longitudinal carinae in *Drapetes* (Fig. 18C). Prosternal process complete, narrow, parallel-sided or slightly narrowed apically, slightly to strongly elevated behind and narrowly rounded at apex, extending well beyond procoxae and fitting into mesoventral cavity (occasionally shorter or flattened). Notosternal sutures usually complete and straight, occasionally curved or sinuate, sometimes anteriorly open (with gap between hypomeron and sternum), forming deep antennal cavity in *Agrypnus*, *Trieres* and *Lanelater*, forming broad internal antennal cavity above hypomeral wall in *Drapetes* (Fig. 18C), *Osslimus* and *Cussolenis* (Fig. 18I). Hypomera with deep lateral antennal grooves in *Subprotelater*. Procoxae globular, without or with very short concealed lateral extensions; trochantinopleuron reduced, concealed and fused to hypomeron. Procoxal cavities narrowly to moderately widely separated, open externally and internally.

Scutellar shield abruptly elevated or not, sometimes strongly oblique, usually anteriorly simple (notched in Cardiophorinae, Fig. 27G), posteriorly acute to broadly rounded or truncate. Elytra ~1.3–3.5 times as long as combined width and 1.5–5 times as long as pronotum, more or less parallel-sided or with sides converging posteriorly, complete and conjointly rounded at apex or independently rounded or acute; each elytron usually with 9 puncture rows or striae and no scutellary striole (sometimes irregularly punctate); epipleura complete, incomplete or absent. Mesoventrite (Fig. 17K) with anterior edge on same plane as or different plane than metaventrite; anterior procoxal rests horizontal to vertical or absent; mesoventral cavity almost always deep and extending well beyond anterior edges of mesocoxal cavities, with anterior boss and inclined slide forming part of clicking mechanism. Mesocoxae globular, with exposed or concealed trochantins. Mesocoxal cavities usually moderately to widely separated (narrowly separated in *Austrelater*), usually partly closed laterally by both mesanepisternum and mesepimeron (in addition to meso- and metaventrites), partly closed by mesepimeron only in *Cussolenis*, *Drapetes*, *Osslimus*, *Anchastus*, *Diadysis*, *Melanotus*, *Hapatesus*, *Crepidomenus* and *Glypheus*, and closed by meeting of meso- and metaventrites in *Agrypnus*, *Trieres*, *Rivulicola* and the Cardiophorinae. Mesometaventral junction consisting of a monocondylic or dicondylic joint, or of a straight or curved line formed by the abutment or fusion of meso- and metaventrites. Metaventrite moderately to very long and slightly to strongly convex, without transverse (metakatepisternal) suture and usually with a long discrimen (occasionally absent), exposed portion of metanepisternum moderately to very elongate. Metacoxae (Fig. 26C) contiguous or narrowly separated, horizontally or slightly obliquely oriented, extending laterally to meet elytra; plates usually well developed but narrowed laterally, uniformly broad in *Cussolenis*, *Drapetes* and *Osslimus*. Metendosternite usually with long, narrow stalk, short lateral arms and long anterior process with approximate tendons. Hind wing (Fig. 22B) usually with short apical field and one of three patterns of linear sclerites: a single anterior oblique sclerite, anterior and posterior oblique sclerites or three sclerites forming an epsilon figure), without distinct radial extensions; radial cell usually well developed and elongate with inner posterobasal angle right or slightly obtuse (cell sometimes shorter and broader or without basal edge); cross-vein r3 moderately long, horizontal and arising well away from r4, well separated from r4; basal portion of RP moderately to very; R-M loop usually forming acute angle; medial spur well developed; medial field usually with five free veins, no medial fleck and a well developed wedge cell with truncate apex (veins reduced to four or three in some smaller species and wedge cell absent in Agrypninae, *Subprotelater*, *Rivulicola* and Cardiophorinae; anal notch usually absent (present in, Lissomini, *Rivulicola* and Cardiophorinae). Trochanters very long in Lissomini; metatrochanter and metafemur greatly enlarged in *Patriciella*. Tibiae usually slender (widened apically in *Patriciella*, subapically in *Cussolenis*), outer edge and outer apical angle usually simple; tibial spurs occasionally single or absent. Tarsi 5-5-5; tarsomeres often simple, sometimes with ventral pubescent pads or membranous lamellae on preapical tarsomere or on more than one tarsomere; pretarsal claws usually simple, appendiculate in *Drapetes*, pectinate in *Melanotus*, with one or more setae near the base of each in Agrypninae; empodium usually well developed, exposed and bisetose (multisetose in *Pseudotetralobus*).

Abdomen five ventrites, the first four of which are connate (Fig. 26C); ventrite 1 with intercoxal process acute or narrowly rounded. Functional spiracles on abdominal segments I–VIII. Anterior edge of segment IX in male usually broadly rounded, without spiculum gastrale; tergite IX truncate to deeply emarginate or almost completely divided into two parts; tergite X usually well developed and free. Aedeagus of trilobate type, usually symmetrical, with parameres almost always individually articulated, with or without apicolateral hooks; penis without distinct ventral lobe, sometimes with rod-like sclerite, with short basal struts. Sternite VIII in female with long, fixed spiculum ventrale and shorter paired lateral struts. Ovipositor (Fig. 29B–C) long and slender, lightly to heavily sclerotised; paraprocts much longer than coxites, which may or may not be subdivided; styli usually present and subapical, sometimes absent. Internal tract variable and sometimes complex (Fig. 32D). [Crowson 1955, 1961b; Gur'yeva 1974; Dolin 1975b, 2000; Calder 1990b, 1992, 1996; Calder and von Hayek 1992; Calder *et al.* 1993; Lawrence *et al.* 1999b; Johnson 2002d; Costa *et al.* 2010c].

Larva (Figs 35B, F, 65D)

Body almost always elongate and parallel-sided and flattened to more or less cylindrical (broader and grub-like in *Pseudotetralobus* and *Cussolenis*), dorsal and ventral surfaces similarly darkly or lightly pigmented, surfaces lightly pigmented except for head capsule and tergum IX, dorsal surfaces more heavily pigmented than ventral ones or dorsal surfaces with dark and light pattern; surfaces generally smooth, but sometimes with tubercles, transverse carinae or patches of specialised setae; vestiture usually consisting of fine hairs or setae (densely clothed with long hairs in *Pseudotetralobus*).

Head prognathous and protracted, usually more or less flattened; epicranial stem usually absent or very short; frontal arms usually lyriform, V-shaped or U-shaped (absent in *Pseudotetralobus*), distinctly separated or contiguous at base, joined anteriorly by transverse line in Cardiophorinae; median endocarina absent. Stemmata usually absent (sometimes with one on each side). Frontoclypeal suture absent. Labrum fused to frontoclypeal plate (Figs 36D, 65P), which is often produced to form median nasale consisting of one to several teeth and a pair of adnasalia (deeply emarginate in Lissominae, Fig. 38F). Antennae moderately long, 3-segmented, with conical, palpiform or dome-like sensorium (occasionally with compound or multiple sensoria). Mandibles symmetrical, usually narrow and falcate, broad at base and narrowed at apex or stout and wedge-like (elongate and apically-cleft with lateral and mesal teeth in Cardiophorinae); incisor edge usually simple or with one or two heavily sclerotised retinacula (with bifurcate retinacula in *Austrelater* and Lissomini); mesal surface simple or with brush of hairs; mola absent. Ventral mouthparts deeply retracted forming maxillolabial complex, without maxillary articulating area; Cardines sometimes externally divided, sometimes approximate or fused together at midline; maxillary apex with articulated mala or with articulated galea and fixed lacinia; palp 4-segmented. Labium consisting of prementum and postmentum; posterior edge of postmentum truncate (Elaterinae, Denticollinae and Thylacosterninae) to narrowly rounded or acute (Agrypninae and Pityobiinae), extending posteriorly beyond stipital bases in Lissominae; ligula absent or shorter than labial palps, which are 2-segmented and narrowly to widely separated. Hypopharyngeal sclerome absent or consisting of transverse bar. Hypostomal rods absent; ventral epicranial ridges present. Gular sutures separate to partly or completely fused together.

Thoracic terga with or without plates, usually without patches or rows of asperities. Prothoracic venter usually with large subtriangular presternal plate, sometimes divided longitudinally, rarely with small additional posterior sclerite. Meso- and metathorax often without ventral plates, sometimes with reduced presternal plate, precoxalia, laterotergites, and small intersternites. Legs usually well developed, slender or stout, clothed with fine setae or short, stout spines; usually 5-segmented including claw-like pretarsus usually with two setae lying side by side; legs reduced and 4-segmented without claw setae in *Cussolenis*; coxae usually separated by less than a basal coxal diameter. Abdominal terga often simple with vestiture of fine hairs; anterior part of tergum with transverse carina abruptly curved laterally in many Denticollinae; transverse striated impressions on anterolateral portions of terga in many Elaterinae; patches of short, stout, specialised setae in Lissominae. Abdominal tergum IX shorter than to slightly longer than VIII, always extending onto ventral surface; often with pair of bifurcate urogomphi (Fig. 65O); urogomphi occasionally simple, minute and hook-like in Thylacosterninae; tergum simple and rounded to acute or slightly flattened with scalloped or tridentate apex in Elaterinae and Cardiophorinae. Sternum IX partly or entirely exposed, simple or sometimes U-shaped with posterolateral teeth, not enclosed by sternum VIII. Segment X often forming short, cylindrical pygopod, flanked by paired teeth or hooks in many Agrypninae, with pair of moderately long pygopods in Cardiophorinae. Spiracles biforous, without closing apparatus, with air tubes projecting posteriorly and with anterior spiracular scar. [Gardner 1936; Burakowski 1973; Dolin 1978a, 2000; Becker 1991; Costa *et al.* 1988, 1992, 2010c; Calder *et al.* 1993; Lawrence *et al.* 1999a; Johnson 2002d].

Classification and World Distribution

The Elateridae is a large cosmopolitan family with ~400 genera and 10 000 species. There are 17 subfamilies recognised, nine of which occur in Australia. This subfamily classification, however, was not supported by two recent molecular studies (Sagegami-Oba *et al.* 2007a; Kundrata and Bocak 2011). Exotic subfamilies include Campyloxeninae (Neotropical), Cebrioninae (relatively widely distributed), Eudicronychinae (Afrotropical), Hemiopinae (Oriental), Morostominae (Madagascar), Oxynopterinae (Oriental and Neotropical), Physodactylinae (Oriental, Afrotropical, Neo-

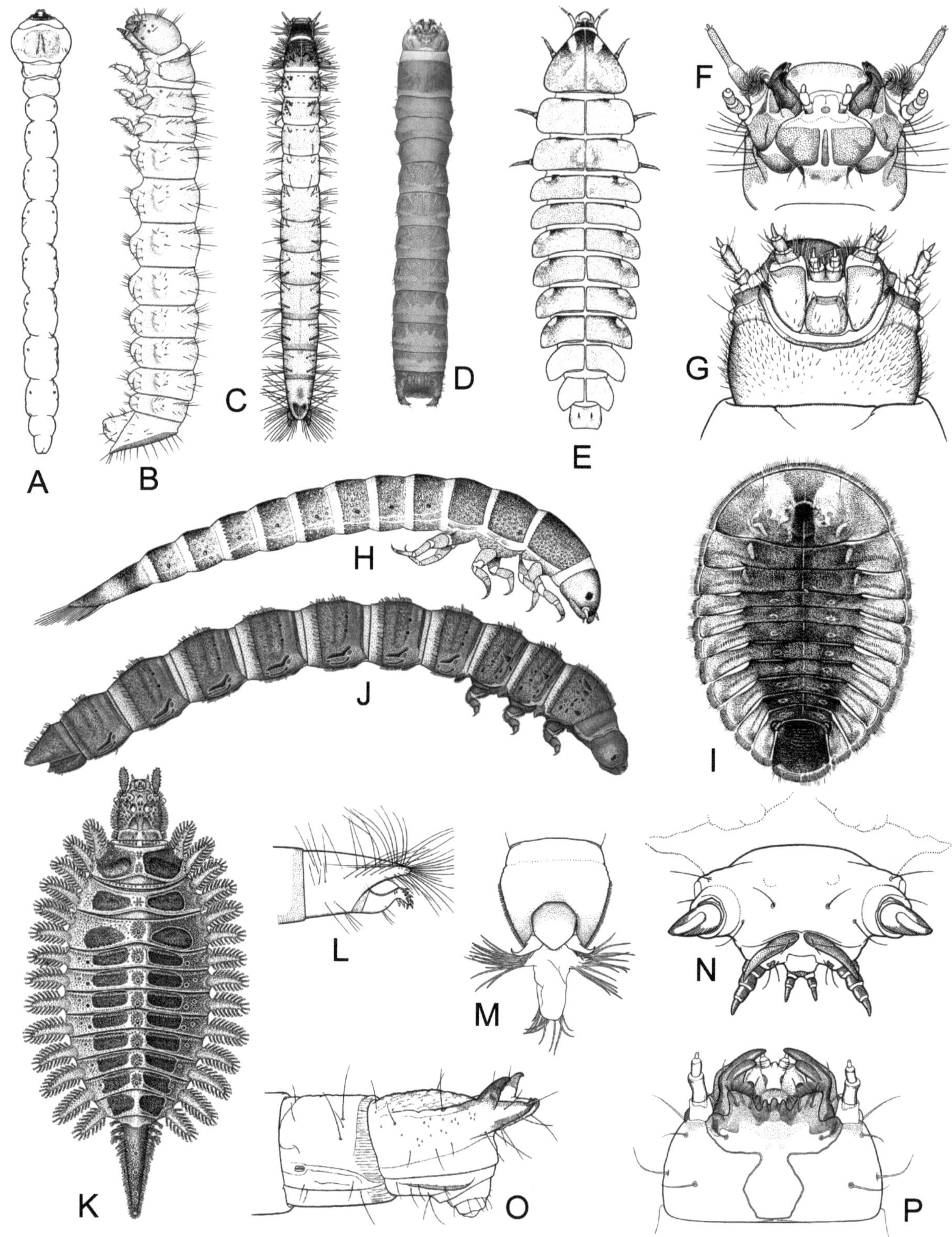

Fig. 65. A–E, Larva: A, Genus ? sp. (Buprestidae), dorsal; B, *Microchaetes* sp. (Byrrhidae: Syncalyptinae), lateral; C, *Byrrocryptus* sp. (Ptilodactylidae: Anchytarsinae), dorsal; D, *Austrelater macphersonensis* Calder (Ptilodactylidae: Cladotominae), dorsal; E, *Atyphella* sp. (Lampyridae), dorsal. F–G, Larval head, ventral: F, *Byrrocryptus* sp. (Ptilodactylidae); G, *Chauliognathus lugubris* (Fabricius) (Cantharidae). H–K, larva: H, *Austrolimnius waterhousei* Hinton (Elmidae: Elminae), lateral; I, *Sclerocyphon* sp. (Psephenidae), dorsal; J, *Austrolichas monteithi* Lawrence & Stribling (Ptilodactylidae: Cladotominae), lateral; K, *Brachypsectra* sp. (Brachypsectridae), dorsal. L–M, Larval abdominal apices: L, *Byrrocryptus* sp., lateral; M, *Notriolus* sp. (Elmidae: Elminae), ventral. N, Genus ? sp. (Lycidae), larval head, anterodorsal. O, *Hapatesus hirtus* Candèze (Elateridae: Dendrometrinae) larva, abdominal apex, lateral; P, same, head, dorsal.

tropical) and Semiotinae (Neotropical). [Costa *et al.* 2010c; Bouchard *et al.* 2011].

Australian Fauna

The Australian fauna is large and poorly understood, in spite of the relatively recent faunal work by Calder (1996). The largest Australian subfamilies are Agrypninae, Dendrometrinae and Elaterinae. **Agrypninae** includes species of *Pseudotetralobus* (Fig. 70E), the largest of the Australian elaterids, with flabellate antennae in the male and with densely setose, grub-like larvae which live in termite nests. Other common agrypnine genera are *Agrypnus* and *Conoderus*, which have soil-inhabiting larvae, *Agrypnus variabilis* (Candèze) is a major pest of sugarcane in QLD. Species of *Paracalais* (Fig. 70C) are predators of wood-boring insects. Australian **Dendrometrinae** include the genera *Hapatesus*, *Dicteniophorus*, *Elatichrosis* and *Glypheus*, as well as the large Bassian genus *Crepidomenus*. Common genera of **Elaterinae** include *Anilicus*, *Glyphochilus*, *Megapenthes* and *Melanoxanthus*. Species of *Paracardiophorus* (**Cardiophorinae**) are abundant, and their unusual, slender larvae are predators in soil. The subfamily The genus *Subprotelater* (Fig. 64H) (**Subprotelaterinae**), confined to northern Australia, is peculiar in having antennal grooves along the sides of the pronotal hypomera, as in some Eucnemidae; its affinities, however, are probably with the Agrypninae. **Pityobiinae** includes *Parablax* and a few related genera mainly southern in distribution. The **Negastriinae** is represented by a few species of *Rivulicola* (Fig. 70F), which occur in sandy areas and resemble some cardiophorines. Two groups often included in the family Throscidae but indistinguishable from elaterids on the basis of larval features are the **Thylacosterninae**, with the cylindrical and eucnemid-like *Cussolenis mutabilis* (Bonvouloir) (Fig. 64F) from northern QLD (Vahtera *et al.* 2009), and species of *Drapetes* and *Osslimus* (Fig. 64A) (**Lissominae**) which have compact throscid-like bodies; in both groups the adults have a pair of large prothoracic pockets which receive the antennae and tarsomeres with membranous lobes beneath. The larva of *Cussolenis* is more or less grub-like but with typical elaterid mouthparts, while the lissomine larvae are flattened and well sclerotised above, with patches of unique spines capable of moving forwards or backwards and mandibles with a complex retinaculum. The genus *Austrelater* (Fig. 64G) is unusual in that the larvae are of the lissomine type (Fig. 65G), while adults are typical elaterids resembling in some respects species of Protelaterini, a group confined to New Zealand and Chile (Calder *et al.* 1993; Lawrence and Arias 2009). Almost all adult elaterids have a characteristic elongate form, with acute hind pronotal angles and a clicking mechanism enabling them to jump by forcing the prosternal spine into the deep mesoventral cavity causing a sudden movement of the prothorax relative to the hind body (Evans 1972, 1973). The labrum in this group is always visible, the frontal region often has a sharp transverse ridge between the eyes, the antennae are almost always serrate, the pronotum is usually elongate and often produced forward to form a chin piece beneath the mouthparts, tarsi may be simple or variously provided with setal brushes or membranous lobes, and ventrite 5 is more or less free. Larvae are elongate and cylindrical to slightly flattened and either uniformly sclerotised and yellow to brown in colour, or more lightly sclerotised except for the head capsule, protergum and tergum IX. The labrum and clypeus are solidly fused forming a clypeolabrum, which is often lobed or toothed to form a nasale, stemmata are often reduced or absent, ventral mouthparts are consolidated to form a maxillolabial complex, segment X is reduced and ventrally located, and tergum IX is terminal with characteristic modifications (cylindrical and acute or rounded, forming concave plate or with simple or bifurcate urogomphi) in different groups. Elaterid larvae may be saprophagous, feeding in rotten wood, phytophagous, feeding mainly on roots of plants, or predaceous; in all cases, however, they are liquid feeders and practise extraoral digestion. Phytophagous larvae, known as wireworms, are pests of cereals, grasses and root crops. [Eidt 1959; Neboiss 1967; Gur'yeva 1969; Burakowski 1973; Dolin 1978b; Calder 1990b, 1992, 1996, 1998; Calder and von Hayek 1992; Calder *et al.* 1993].

Australian Subfamilies, Tribes and Genera

Lissominae: *Austrelater* Calder & Lawrence, *Drapetes* Dejean, *Osslimus* Calder.

Thylacosterninae: *Cussolenis* Fleutiaux.

Agrypninae: Agrypnini: *Agrypnus* Eschscholtz, *Lanelater* Arnett, *Trieres* Candèze. **Hemirhipini:** *Anthracalaus* Fairmaire, *Aphileus* Candèze, *Austrocalais* Neboiss, *Paracalais* Neboiss, *Tetrigus* Candèze. **Oophorini:** *Aeoloderma* Fleutiaux, *Conoderus* Eschscholtz, *Drasterius* Eschscholtz, *Heteroderes* Latreille, *Pseudaeolus* Candèze. **Laconini:** *Lacon* Castelnau. **Tetralobini:** *Pseudotetralobus* Schwarz. **Tribe?:** *Chrostus* Candèze, *Macromalocera* Hope.

Subprotelaterinae: *Subprotelater* Fleutiaux.

Dendrometrinae: *Arachnodima* Candèze, *Carterelater* Calder, *Corystelater* Calder, *Crepidomenus* Erichson, *Dicteniophorus* Candèze, *Doloporus* Candèze, *Drymelater* Calder, *Elatichrosis* Hyslop, *Enischnelater* Calder, *Glypheus* Candèze, *Hapatesus* Candèze, *Litotelater* Calder, *Microdesmes* Candèze, *Neboisselater* Calder, *Paracrepidomenus* Schwarz, *Rousia* Calder, *Stichotomus* Candèze, *Toorongus* Neboiss, *Tyiwiphila* Calder, *Wardulupicola* Calder.

Pityobiinae: *Parablax* Schwarz, *Parasaphes* Candèze, *Tasmanelater* Calder, *Wynarka* Calder. Xuthelater Calder.

Elaterinae: *Acroniopus* Erichson, *Anchastus* LeConte, *Anilicoides* Candèze, *Anilicus* Candèze, *Antoligostethus* Blackburn, *Ascesis* Candèze, *Augenotus* Gullan, *Diadysis* Candèze, *Glyphochilus* Candèze, *Idiomilta* Calder, *Lingana* Neboiss, *Megapenthes* Kiesenwetter, *Melanotus* Eschscholtz, *Melanoxanthus* Eschscholtz, *Nullarborica* Blackburn, *Ochosternus* Candèze, *Ophidius* Candèze, *Paranilicus* Candèze, *Rangsia* Calder, *Simodactylus* Candèze, *Yalganus* Neboiss.

Cardiophorinae: *Cardiotarsus* Eschscholtz, *Paracardiophorus* Schwarz, *Patriciella* Van Zwaluwenburg.

Negastriinae: *Rivulicola* Calder.

42. LYCIDAE

Adult (Figs 5E, 64I)

Length ~4.5–25 mm. Body ~2.6–4.5 times as long as wide; lightly sclerotised; more or less parallel-sided; somewhat flattened; colour sometimes uniformly black but often brightly coloured and aposematically patterned (usually black and red or yellow);

vestiture usually consisting of short, fine hairs, sometimes subglabrous.

Head small, much narrower than prothorax, transverse to elongate, prognathous to moderately declined, sometimes concealed from above by pronotum. Eyes moderately to very large, protuberant, circular and entire, without interfacetal setae. Antennal insertions exposed from above and narrowly separated; subantennal groove absent. Frontoclypeal region slightly to strongly deflexed, sometimes forming frontal carina; frontoclypeal suture absent. Labrum small, free, often apically rounded. Antennae 11-segmented, slightly serrate to flabellate, with rami usually beginning on antennomere 3; pedicel much smaller than scape or antennomere 3. Mandibles narrowly elongate, sometimes reduced, unidentate, without mola or prostheca. Maxilla with setose galea and lacinia or a single mala, without uncus; apical maxillary and labial palpomeres usually fusiform, sometimes slightly to strongly expanded apically. Mentum small, transverse; prementum usually elongate; ligula absent. Corpotentorium absent. Cervical sclerites well developed.

Pronotum ~0.45–0.80 times as long as wide, about as wide as combined elytral bases; usually widest posteriorly with sides converging anteriorly (sides subparallel in *Calochromus*). Lateral pronotal carinae complete, sometimes sharply defined or somewhat elevated, but without margin or bead; anterior angles not produced; posterior angles right to produced, subacute and slightly embracing elytral bases; posterior edge usually bisinuate, without crenulations or interlocking device; disc highly variable, often with longitudinal, oblique and transverse carinae defining cells of various sizes, sometimes with several shallow impressions. Prosternum in front of coxae shorter than shortest diameter of procoxal cavity, without chin-piece. Prosternal process incomplete and broadly rounded or absent. Procoxae projecting, without concealed lateral extensions; trochantin exposed; endopleuron free. Procoxal cavities slightly transverse, contiguous, broadly open externally and internally.

Scutellar shield not abruptly elevated, usually truncate or emarginate posteriorly. Elytra ~2.3–3.8 times as long as combined width and 4.5–8 times as long as pronotum (elytra much shorter and broader in female *Porrostoma apterus* (Lea)), subparallel or widened posteriorly, complete and usually independently rounded at apex incomplete and truncate, exposing several abdominal tergites in female *Porrostoma apterus*; each elytron with either nine or four longitudinal costae and numerous transverse costae forming deep transverse to subquadrate cells similar to window punctures; with weak longitudinal costae and no distinct transverse costae in *Calochromus*; scutellary striole absent; epipleura incomplete or absent. Mesoventrite transverse, anteriorly emarginate, procoxal rests, mesoventral cavity and discrimen absent. Mesocoxae projecting, with slender, exposed trochantins. Mesocoxal cavities barely impressed, without distinct posterior edge, contiguous, partly closed laterally by mesanepisternum and mesepimeron. Mesometaventral junction absent. Metaventrite relatively long and convex, with long discrimen and no transverse (metakatepisternal) suture; exposed portion of metanepisternum relatively narrow. Metacoxae subcontiguous, extending laterally to meet elytra; plates weakly developed or absent. Metendosternite with long, robust stalk, short, broad lateral arms and anterior process with anterior tendons narrowly separated. Hind wing with short apical field bearing two or three vaguely indicated sclerites; radial cell moderately large, elongate, with inner posterobasal angle obtuse; cross-vein r3 moderately long, almost parallel to long wing axis and well separated from r4 or absent; basal portion of RP very long; medial spur extending to wing margin; medial field usually with four or five free veins; MP_{3+4} with basal cross-vein but no spur; CuA_1 meeting MP_4; wedge cell absent; AA_4 short or absent; anal embayment shallow or absent, AP undivided. Tibiae simple, slender, slightly wider before apex; spurs almost always absent (present in some *Porrostoma*). Tarsi 5-5-5; tarsomeres 1–4 or 2–4 often with setose pads; pretarsal claws simple; empodium reduced.

Abdomen with eight free ventrites in male (seven in female); intercoxal process absent. Functional spiracles on abdominal segments I–VIII. Abdomen usually shorter and much narrower than elytra. Anterior edge of segment IX in male narrowly rounded to acute, without spiculum gastrale; tergite IX emarginate; tergite X free. Aedeagus trilobate, but often with phallobase asymmetrical and parameres absent; penis undivided, without basal struts. Sternite VIII in female usually without spiculum ventrale (well developed and long in *Calochromus*). Ovipositor usually moderately elongate (very long in Calochromus); segment X (proctiger) sclerotised and separated from IX, which is undivided; paraproctal baculi usually somewhat longer than coxites, which are undivided and well sclerotised, sometimes fused to coxites or connected by bridge; styli well developed and terminal. [Miller 2002a; Bocák 2002; Bocák and Bocákova 1990, 2008, 2010b].

Larva (Fig. 42C)

Body elongate and parallel-sided or widest at middle, slightly to strongly flattened; moderately to heavily sclerotised; dorsum smooth or with granulate or tuberculate surface and usually with broad, flat, lateral tergal projections (long and slender projections in some Metriorrhynchini); vestiture of scattered long setae.

Head prognathous and protracted, slightly to strongly transverse; epicranial suture and median endocarina absent. Stemmata one on each side. Frontoclypeal suture absent. Labrum fused to frontoclypeal plate, which lacks a distinct nasale or adnasalia. Antennae moderately long, with large basal articulatory membrane, 2-segmented, with strongly transverse and ring-like basal antennomere and long apical antennomere bearing a large, dome-like sensorium. Mandibles symmetrical, falcate, slightly curved to almost straight, approximate at base with apices diverging, deeply longitudinally cleft (Figs 42G, 43J, 65N). Ventral mouthparts deeply retracted forming maxillolabial complex, without maxillary articulating area; cardines usually distinct and undivided, well separated; stipites completely fused to postmentum without separating sutures; maxillary apex with mala partly fused to segment-like palpifer, sometimes reduced; palp 3-segmented. Labium consisting of prementum and postmentum; posterior edge of postmentum broadly rounded; ligula absent; labial palps 2-segmented and moderately widely separated. Hypopharyngeal sclerome absent. Hypostomal rods subparallel or slightly converging; ventral epicranial ridges absent. Gular region not clearly defined.

Thoracic terga with plates, sometimes with lateral and posterior projections. Prothoracic venter with large subtriangular, undivided presternal plate, precoxalia, laterally projecting spiracular sclerites and a sternal plate. Meso- and metathorax with precoxalia and a median sternal plate and. Legs well developed, 5-segmented including claw-like pretarsus bearing two setae lying side by side; coxae relatively large and moderately narrowly separated. Abdominal terga I–VIII simple or with two pairs of lateral tergal processes, one above the other, the ventral ones bearing spiracles. Abdominal tergum IX about as long as VIII, always extending onto ventral surface; with or without pair of simple. posteriorly projecting urogomphi. Sternum IX exposed, simple, not enclosed by sternum VIII; sometimes longitudinally divided. Segment X forming short, cylindrical pygopod, without teeth or hooks. Spiracles biforous, without closing apparatus and with spiracular scar well separated from paired openings. [Miller 1997, 2002a; Burakowski 1988; Costa *et al.* 1988; Lawrence *et al.* 1999a; Bocák and Matsuda 2003; Kazantsev 2005, 2006a; Bocák and Bocákova 2010b].

Classification and World Distribution

The family includes ~4600 described species in 160 genera distributed throughout the world, but most abundant and diverse in humid tropical regions and with no species native to New Zealand. The group has recently been divided into six subfamilies three of which occur in Australia. Exotic subfamilies include Libnetinae (eastern Palaearctic and Oriental), Ateliinae (eastern Palaearctic and Oriental) and Dexorinae (Afrotropical). [Bocák and Bocákova 1990, 2008, 2010b; Bocákova *et al.* 2007; Bocák *et al.* 2008].

Australian Fauna

Although most of Australian Lycidae belong to the large subfamily Lycinae in the broad sense of Bocák and Bocákova (2008), at least two Australian species are now known representing two primarily exotic subfamilies: *Autaphes australis* Kazantsev from the NT belongs to the Miniduliticolini (**Lypropaeinae**) and *Proterotaphes sempiternus* (Kazantsev) from northern NSW belongs to the Taphini (**Dictyopterinae**) (Kazantsev 2004a, 2004b, 2006b). The Australian **Lycinae** include two genera of Calochromini (*Calochromus* Guérin-Méneville and *Dumbrellia* Lea) and one genus of Platerodini (*Plateros* Bourgeois); the remaining genera (*Achras* Waterhouse, *Cladophorus* Guérin-Méneville, *Enylus* Waterhouse, *Metriorrhynchus* Gemminger and Harold, *Porrostoma* Castelnau (Fig. 64I), *Procautires* Kleine, *Stadenus* Waterhouse, *Synchonnus* Waterhouse, *Trichalus* Waterhouse, *Xylobanus* Waterhouse) are members of the tribe Metriorrhynchini. Lycids are elongate beetles with soft integument, flattened form, narrow, more or less dehiscent elytra bearing longitudinal ridges or rows of window punctures, and aposematic coloration involving various combinations of reddish-yellow and black. The head is more or less triangular to rostrate and partly concealed by the pronotum, antennae are long, thick and approximate or contiguous, the eyes are well separated and the maxillary palps are apically expanded. The surface of the pronotum is usually divided by ridges into cell-like compartments, femora and tibiae are more or less flattened, and tarsomeres 3 and 4 are lobed beneath. Lycid larvae are elongate, somewhat flattened, usually heavily sclerotised, sometimes tuberculate with segment IX varied in shape and segment X small and ventrally situated. The head is small with non-opposable, blade-like mandibles, which are longitudinally divided into two parts (Figs 42C, 43J, 47C), one of which is apparently derived from the clypeus; Kazantsev (2005, 2006a) illustrated lycid larval mouthparts in detail, but his interpretations and phylogenetic conclusions were criticised by Beutel *et al.* (2007). Adult lycids are often seen flying at dusk, resting on plant surfaces or sometimes in blossoms. They are distasteful to birds and often form part of a mimicry complex which also includes beetles in the families Buprestidae, Meloidae, Pyrochroidae, Oedemeridae, Cerambycidae and Belidae, as well as moths, flies and wasps (Figs 5E, 7D, 9A, 79A). Lycid larvae occur beneath bark or in leaf litter and soil, and they are thought to feed on soft, decaying plant material or possibly slime mould plasmodia or yeasts. [Lea 1909; Bocák 2002; Bocák and Matsuda 2003].

43. LAMPYRIDAE

Adult (Figs 5F, H, 64K–L)

Length ~3.8–18 mm. Body ~1.7–3.25 times as long as wide; relatively lightly sclerotised; elongate-oval and wider at middle or parallel-sided; somewhat flattened; colour often yellow to reddish-yellow or yellowish-brown; elytra sometimes entirely black or dark brown, reddish-yellow with black apices, yellowish-brown with slender yellow longitudinal costae, or black with reddish or yellow suture, lateral edges or both; vestiture consisting of short, fine hairs.

Head (Fig. 44D) transverse to elongate, prognathous to moderately declined, sometimes concealed from above by pronotum, abruptly narrowed immediately behind eyes, so that temples are absent and the neck, if present, consists of an incomplete ring bordering the occipital foramen. Eyes very large, protuberant, circular, and entire or slightly emarginate posteriorly, approximate ventrally, without interfacetal setae. Antennal insertions exposed from above and very narrowly separated; subantennal groove absent. Frontoclypeal region moderately deflexed but not carinate; frontoclypeal suture absent. Labrum small, free, apically rounded. Antennae relatively short, almost always 11-segmented (rarely with 9- or 10-segmented), filiform. Mandibles small, narrowly elongate and strongly curved, unidentate, without mola or prostheca, but setose at base. Maxilla with setose galea and reduced lacinia without uncus; apical maxillary and labial palpomeres usually fusiform, sometimes slightly expanded apically. Mentum small, transverse; ligula absent. Gular sutures narrowly separated. Corpotentorium absent. Cervical sclerites well developed.

Pronotum ~0.45–0.80 times as long as wide, about as wide as combined elytral bases; usually widest posteriorly with sides rounded and converging anteriorly; sides occasionally subparallel, rarely widest anteriorly (e.g. *Luciola nigra* Olivier), sometimes explanate. Lateral pronotal carinae complete but not sharply defined, without margin or bead, sometimes forming continuous curve with anterior edge; anterior angles not produced; posterior angles right, rounded or slightly acute but not embracing elytral bases; posterior edge straight to bisinuate, without crenulations or interlocking device; disc usually with broad, shallow sub-basal impressions and an incomplete longitudinal furrow. Procoxal cavities barely impressed. Prosternum in front of coxae much shorter than shortest diameter of cavity, without chin-piece. Prosternal process incomplete and broadly rounded. Procoxae projecting, without concealed lateral extensions; trochantin exposed; endopleuron free. Procoxal cavities transverse, contiguous, broadly open externally and internally.

Scutellar shield not abruptly elevated, broadly emarginate posteriorly. Elytra usually ~1.65–2.35 times as long as combined width and 2.7–5.3 times as long as pronotum (much shorter and broader in brachypterous females), subparallel or slightly wider at middle, conjointly rounded or sometimes independently rounded at apex, exposing several abdominal tergites in some females; punctation irregular, but elytra sometimes with four weak longitudinal costae; epipleura complete. Mesoventrite transverse, anteriorly emarginate; procoxal rests, mesoventral cavity and discrimen absent. Mesocoxae projecting, with slender, exposed trochantins. Mesocoxal cavities barely impressed, without distinct posterior edge, contiguous, partly closed laterally by mesanepisternum and mesepimeron. Mesometaventral junction absent. Metaventrite relatively long and convex, with long discrimen and no transverse (metakatepisternal) suture; exposed portion of metanepisternum moderately broad, with slightly curved mesal edge. Metacoxae subcontiguous, somewhat oblique, extending laterally to meet elytra; plates weakly developed. Metendosternite with long, narrow stalk, short, broad lateral arms and anterior process with anterior tendons narrowly separated. Hind wing with very short apical field bearing two vaguely indicated linear sclerites; radial cell moderately large, elongate, with inner posterobasal angle obtuse; cross-vein r3 short to long, oblique or almost parallel to long wing axis, well separated from r4; basal portion of RP very long; medial spur extending to wing margin; medial field with five free veins; MP_{3+4} with basal cross-vein but no spur; CuA1 meeting MP_{3+4} wedge cell absent; anal embayment absent or very shallow, AP undivided. Metafemora enlarged and swollen and metatibiae somewhat curved and subapically expanded in males of *Pyrophanes* Olivier; tibial spurs absent. Tarsi 5-5-5; tarsomere 4 with emarginate ventral membranous lobe; pretarsal claws simple; empodium reduced.

Abdomen with six free ventrites (sternites II–VII) in male and seven (II–VIII) in female; ventrite 1 shorter than 2, without distinct coxal cavities or intercoxal process. Light organs in male on segments VI and VII (those in female on VI only), usually entire (bipartite in *Pyrophanes*). Functional spiracles on abdominal segments I–VIII located on sternites. Anterior edge of segment IX in male narrowly rounded, without spiculum gastrale; tergite IX more or less truncate; tergite X free. Aedeagus trilobate; phallobase asymmetrical; parameres articulated to phallobase but more or less fused together at base and fused to base of penis, which is undivided, with short basal struts. Ovipositor moderately elongate; lightly sclerotised except for paraproctal and coxital baculi; paraprocts much longer than coxites, which are setose and divided with slender apical lobe and well developed terminal styli. [McDermott 1964; Ballantyne 1968, 1987a, 1987b; Ballantyne and McLean 1970; Crowson 1972b; Ballantyne and Lambkin 2000, 2009; Lloyd 2002; Branham 2010b].

Larva (Figs 65E)

Body elongate and parallel-sided or widest anterior to middle and narrowing posteriorly, slightly to strongly flattened; thoracic and most abdominal terga laterally expanded; dorsum moderately to heavily sclerotised, smooth or somewhat granulate vestiture of scattered long setae.

Head small, prognathous and retracted, usually concealed from above by prothorax; about as long as wide or elongate with sides slightly curved to almost straight; posterior edge deeply emarginate; epicranial stem short; frontal arms V- or U-shaped, without median endocarina, but with paired endocarinae beneath frontal arms. Stemmata one on each side, with well developed lens. Frontoclypeal suture absent. Labrum fused to frontoclypeal plate, which lacks a distinct nasale or adnasalia, but may be emarginate. Antennae moderately long, with large basal articulatory membrane, 3-segmented, with basal two antennomeres long and subequal and apical antennomere reduced and about equal in length to palpiform sensorium. Mandibles (Fig. 40K) symmetrical, falcate, curved, unidentate, with internal perforation, weak retinaculum and basal fringe of setae, without mola. Ventral mouthparts deeply retracted forming maxillolabial complex, without maxillary articulating area; cardines longitudinally oriented and subdivided, well separated; stipites elongate; galea 2-segmented and palpiform; lacinia reduced and setose; maxillary palp 3-segmented with large, segment-like palpifer (thus appearing 4-segmented). Labium consisting of prementum and postmentum; posterior edge of postmentum broadly truncate; ligula absent; labial palps 2-segmented and widely separated. Hypopharyngeal sclerome absent. Hypostomal rods subparallel; ventral epicranial ridges absent. Gular region not clearly defined and membranous.

Prothorax widest posteriorly with sides slightly rounded and converging anteriorly and anterior edge produced and narrowly rounded or slightly emarginate; thoracic terga with simple plate. Prothoracic venter with large subtriangular, undivided presternal plate and well developed precoxalia. Legs well developed, 5-segmented including claw-like pretarsus bearing two setae lying side by side; coxae relatively large and moderately narrowly separated. Tergum IX about as long as VIII, but distinctly narrower, extending onto ventral surface; without paired urogomphi. Sternum IX exposed, simple, not enclosed by sternum VIII. Segment X short and cylindrical, with several eversible, tubular holdfast organs covered with minute hooks. Spiracles biforous, without closing apparatus and with spiracular scar, located on laterotergites. [Gardner 1946; LaBella and Lloyd 1991; Lawrence *et al.* 1999a; Ballantyne and Lambkin 2000; Lloyd 2002; Branham 2010b].

Classification and World Distribution

The Lampyridae is a widespread family of ~85 genera 2000 species, with the greatest diversity in the more humid parts of the Neotropical and Oriental regions. The group is here considered to be divided into five subfamilies, only one of which occurs in Australia. The constitution of the family and its subfamily classification, however, have been the subject of controversy (Branham and Wenzel 2001, 2003; Bocákova *et al.* 2007; Branham 2010b; Lawrence *et al.* 2010d; Kundrata and Bocak 2011). Exotic subfamilies include Pterotinae (north-western North America), Ototretinae (including Ototretadrilinae) (western North America, eastern Palaearctic and Oriental regions), Psilocladinae (Holarctic, Neotropical and Oriental regions), Lampyrinae (including Amydetinae) (widely distributed) and Photurinae (New World). [McDermott 1964, 1966; Crowson 1972b; Jeng *et al.* 1998].

Australian Fauna

The Australian lampyrid fauna includes 26 species in six genera of the subfamily **Luciolinae**: *Aquilonia* Ballantyne & Lambkin, *Atyphella* Olliff, *Lloydiella* Ballantyne & Lambkin, *Luciola* Laporte, *Pteroptyx* Olivier and *Pyrophanes* Olivier. Adults are elongate, somewhat flattened, soft-bodied beetles with very large eyes in the male and with luminous organs (yellowish-white in colour) on the fifth (female) or fifth and sixth (male) ventrites. The head is partly or entirely concealed from above, the pronotum is often laterally explanate, the antennae are relatively short, the mandibles small, curved, falcate and perforate, and tarsomere 4 is lobed beneath. The colour of the pronotum is usually yellow and the elytra dark brown with yellow lateral and sutural edges, entirely yellow or with brown and yellow stripes. In most species the elytra are simple and complete, but in some females they may be shortened (Fig. 5H), and in *Pteroptyx* males they are apically deflexed. Larvae (Fig. 65E)) are elongate, tapered at both ends, somewhat flattened, dorsally sclerotised and often with laterally expanded thoracic and abdominal terga; the head is small and retractable head bearing curved, perforate mandibles. There is a luminous organ on segment VIII, tergum IX is terminal and segment X is reduced and bears a holdfast organ consisting of several eversible, asperate, tubular filaments. Lampyrids are restricted to the wetter areas along the eastern coast and are usually found in rainforests and mangrove flats. Males fly just after dusk and emit a series of controlled flashes from the light organs as part of the mating sequence; females also flash but have not been observed to fly with the males. Synchronised flashing, well known in some New Guinea species, has been observed in northern QLD. Adults are not known to feed, but larvae prey on small land snails, which they paralyse using secretions produced by a pair of acinose glands at the anterior end of the alimentary canal and injected through the perforate mandibles; the prey is digested extraorally and the liquified tissues are imbibed. Larvae and pupae are faintly luminous. The possible adaptive significance of larval luminescence has been discussed by Sivinsky (1981). [Ballantyne 1968, 1987a, 1987b; Ballantyne and McLean 1970; Crowson 1972b; Ballantyne and Lambkin 2000, 2009; Stanger-Hall *et al.* 2007].

44. CANTHARIDAE

Adult (Figs 6C, 64J, 70G)

Length ~2.5–15 mm. Body ~2.2–6.2 times as long as wide; relatively lightly sclerotised; more or less parallel-sided and somewhat flattened; colour uniformly black, reddish-yellow, yellow and black in various combinations, or with blue elytra, black head and black and yellow prothorax (some species of *Chauliognathus* Hentz); vestiture consisting of short, fine, decumbent hairs.

Head (Fig. 44B) transverse to elongate, prognathous, exposed from above, not abruptly narrowed behind. Eyes moderately well developed, somewhat protuberant, circular to transversely oval and not emarginate, without interfacetal setae. Antennal insertions exposed from above and widely separated; subantennal groove absent. Frontoclypeal region not or only slightly deflexed; frontoclypeal suture absent. Labrum usually fused to frontoclypeus, except in *Chauliognathus*, where it may be partly or completely divided at midline; epipharynx sometimes projecting anteriorly beyond edge of labrum. Antennae moderately long, 11-segmented and filiform to weakly serrate (with particular antennomeres enlarged and modified in males of some species). Mandibles narrowly elongate and strongly curved, unidentate (but with subapical, dorsal tooth in *Chauliognathus*); incisor edge sometimes finely serrate; mola absent; mesal surface of mandibular base without fringe or brush of hairs. Maxilla with setose galea and lacinia, the latter reduced, without uncus; apical maxillary and labial palpomeres either fusiform or expanded and securiform. Mentum elongate, transverse; ligula cleft, setose. Gular sutures narrowly separated or fused together. Corpotentorium absent. Cervical sclerites well developed.

Pronotum ~0.5–1.25 times as long as wide, usually about as wide as combined elytral bases (distinctly narrower in some *Chauliognathus*), usually widest at middle or anteriorly, with sides straight or weakly rounded. Lateral pronotal carinae complete, with or without distinct bead, sometimes slightly elevated; anterior angles not produced; posterior angles right or rounded, not embracing elytral bases; posterior edge more or less straight to bisinuate, without crenulations or interlocking device; disc simple. Procoxal cavities not or only barely impressed. Prosternum in front of coxae much shorter than shortest diameter of cavity, without chin-piece. Prosternal process absent. Procoxae projecting, without concealed lateral extensions; trochantin exposed; endopleuron free. Procoxal cavities transverse, contiguous, broadly open externally and internally.

Scutellar shield not abruptly elevated, broadly rounded to truncate posteriorly. Elytra usually ~1.8–2.75 times as long as combined width and 3.3–5.2 times as long as pronotum (much shorter and broader in brachypterous species of *Chauliognathus*), subparallel, conjointly rounded or independently rounded at apex, sometimes exposing one to several abdominal tergites; punctation irregular; epipleura narrow and incomplete or absent. Mesoventrite transverse, anteriorly emarginate; procoxal rests, mesoventral cavity and discrimen absent. Mesocoxae projecting, with slender, exposed trochantins. Mesocoxal cavities barely impressed, without distinct posterior edge, contiguous, partly closed laterally by mesanepisternum and mesepimeron. Mesometaventral junction absent. Metaventrite relatively long and convex, with long discrimen and no transverse (metakatepisternal) suture; exposed portion of metanepisternum moderately broad, with sinuate mesal edge. Metacoxae narrowly separated and somewhat oblique, extending laterally to meet elytra; plates weakly developed. Metendosternite with long, narrow stalk, short, broad lateral arms and short anterior process with anterior tendons moderately widely separated. Hind wing with short apical field bearing two linear sclerites (sometimes with vaguely indicated pigmented areas only); radial cell moderately large, elongate, with inner posterobasal angle obtuse; cross-vein r3 moderately long, almost parallel to long wing axis and close to r4, or sometimes absent; basal portion of RP very long; medial spur extending to wing margin; medial field with three free veins; MP_{3+4} with basal cross-vein but no spur; wedge cell absent; AA_4 absent; anal embayment absent or very shallow, AP undivided. Legs slender, not expanded; tibial spurs absent in *Chauliognathus*. Tarsi 5-5-5; tarsomere 4 shortened but with deeply emarginate ventral lobe; pretarsal claws simple; empodium reduced.

Abdomen (Fig. 28H) with seven free ventrites (sternites II–VIII) in females and some males, seven (II–IX) in most males; ventrite 1 shorter than 2, without distinct coxal cavities or intercoxal process. Tergites I–VIII with paired lateral glandular pores (Fig. 28R); intersegmental membranes sometimes with lateral eversible glands. Light organs absent. Functional spiracles on abdominal segments I–VIII located in pleural membrane at edges of tergites. Anterior edge of segment IX in male broadly rounded to subacute, without spiculum gastrale; tergite IX slightly emarginate and tergite X free, or tergites IX and X fused together. Aedeagus trilobate; highly asymmetrical in *Chauliognathus*; parameres well developed but highly varied in form, usually fused to phallobase, sometimes complexly lobed; penis undivided, without basal struts. Sternite VIII in female without spiculum ventrale. Ovipositor short and broad, well sclerotised; proctiger well developed and free; paraprocts evenly sclerotised, without baculi, longer than coxites, which are undivided, setose and located mesal to paraprocts; styli palpiform, well developed or highly reduced, terminal. [Wittmer 1955, 1964, 1966, 1969, 1971, 1972a, 1972b, 1979; Brancucci 1980, 1981a, 1981b, 1981c, 1981d, 1982, 1983, 1985; Ramsdale 2002b, 2010b].

Larva (Figs 6E, 35A)

Body elongate and more or less parallel-sided, slightly flattened; lightly sclerotised and either pale or with a dorsal pattern of darker pigment; surface covered with minute protuberances and appearing 'velvety'; vestiture almost always of very short, erect hairs (mixed with expanded frayed setae in some Dimorphocerinae).

Head well developed, prognathous and protracted, about as long as wide with sides slightly curved to almost straight; posterior edge weakly emarginate; epicranial suture and median endocarina absent. Stemmata one on each side, with well developed lens. Frontoclypeal suture absent. Labrum fused to frontoclypeal plate, the anterior edge finely denticulate (*Chauliognathus*) or with median tooth (and sometimes two lateral teeth). Antennae moderately long, with large basal articulatory membrane, 3-segmented, with basal two antennomeres long and subequal and apical antennomere reduced and slightly longer than palpiform sensorium. Mandibles well developed, falcate, symmetrical, curved and unidentate, with broadly open longitudinal channel and retinaculum; mola absent; mandibular base with brush of setae. Ventral mouthparts (Figs 48P, 65G) somewhat retracted forming maxillolabial complex, without maxillary articulating area; cardines longitudinally oriented and undivided, well separated; stipites elongate; galea articulated and palpiform (absent in *Chauliognathus*) and lacinia well developed and setose; maxillary palp 3-segmented, with or without segment-like palpifer. Labium consisting of prementum and postmentum; posterior edge of postmentum broadly truncate; ligula absent; labial palps 2-segmented and well separated. Hypopharyngeal bracon and sclerome absent. Hypostomal rods absent; ventral epicranial ridges well developed, curved and meeting at midline, forming subfacial sinus. Gular region well developed; gular sutures fused together or absent.

Thoracic terga and abdominal terga I–VIII each with paired, lateral glandular pores. Prothoracic venter with weakly defined, undivided presternal plate. Legs well developed, 5-segmented including claw-like pretarsus bearing four or more setae; coxae relatively large and narrowly to moderately widely separated. Abdominal terga I–VIII simple. Tergum IX distinctly shorter and narrower than VIII, without paired urogomphi. Sternum IX exposed, simple, not enclosed by sternum VIII. Segment X short and cylindrical, more or less terminal or posteroventrally oriented, with pair of short, broad membranous lobes. Spiracles reduced, biforous, without closing apparatus. [Böving and Craighead 1931; Šulc 1949; Crowson 1972b; Fitton 1975; Brancucci 1980; Costa *et al.* 1988; Ramsdale 2002b, 2010b].

Classification and World Distribution

Cantharidae is a cosmopolitan family containing ~140 genera and more than 5000 species. The group is usually divided into five subfamilies, three of which occur in Australia. The exotic subfamilies Cantharinae and Malthininae are occur primarily in the Holarctic and Oriental regions, the former extending into Mexico and the latter with a few Neotropical and Afrotropical representatives. The subfamiles Silinae and Chauliognathinae are widely distributed mainly in tropical and subtropical parts of the world and Dysmorphocerinae is a Southern Hemisphere group with one or two taxa extending into the Nearctic region. [Miskimen 1961a, 1961b; Magis and Wittmer 1974; Delkeskamp 1977, 1978].

Australian Fauna

There are at least three cantharid genera in Australia: the widely distributed *Chauliognathus* Hentz (**Chauliognathinae**) and *Heteromastix* Boheman (**Dysmorphocerinae**), and the northern *Sphaerarthrum* Waterhouse (**Silinae**). Cantharids are elongate, parallel-sided, soft-bodied beetles usually yellow, black or a combination of red and black or bluish-grey in colour. The head is relatively large, the labrum often inconspicuous, and the antennae moderately long and usually filiform with widely separated insertions; the pronotum is oval or subquadrate, the fourth tarsomere lobed beneath and the male genitalia usually large, asymmetrical and exposed. The elytra are usually complete, but they are abbreviated in *Chauliognathus apterus* (Olliff) from Lord Howe Island. Larvae are elongate and lightly sclerotised, although often darkly pigmented or patterned, with a characteristic surface of minute protuberances producing a velvety appearance; the vestiture consists of fine hairs only, except for a few Tasmanian species which are clothed with modified expanded setae. Stemmata are relatively large, the ventral mouthparts are slightly retracted, the gular region is well developed without or with fused gular sutures, and there are paired gland openings on all thoracic and most abdominal segments; segment X is reduced and terminal. Adult cantharids are active during daylight and are sometimes found in large numbers on flowers and vegetation; large masses of soldier beetles (*Cauliognathus* sp.) were reported by Zeck (1919). They are known to attack other insects, but they may feed as well on pollen, nectar and fresh foliage. Like the Lycidae, they produced defensive chemicals (Meinwald *et al.* 1968) and belong to mimicry complexes. Larvae are common in soil and litter, where they are usually general predators; some Holarctic species, however, are known to be phytophagous. [Lea 1909; Schultz 1994].

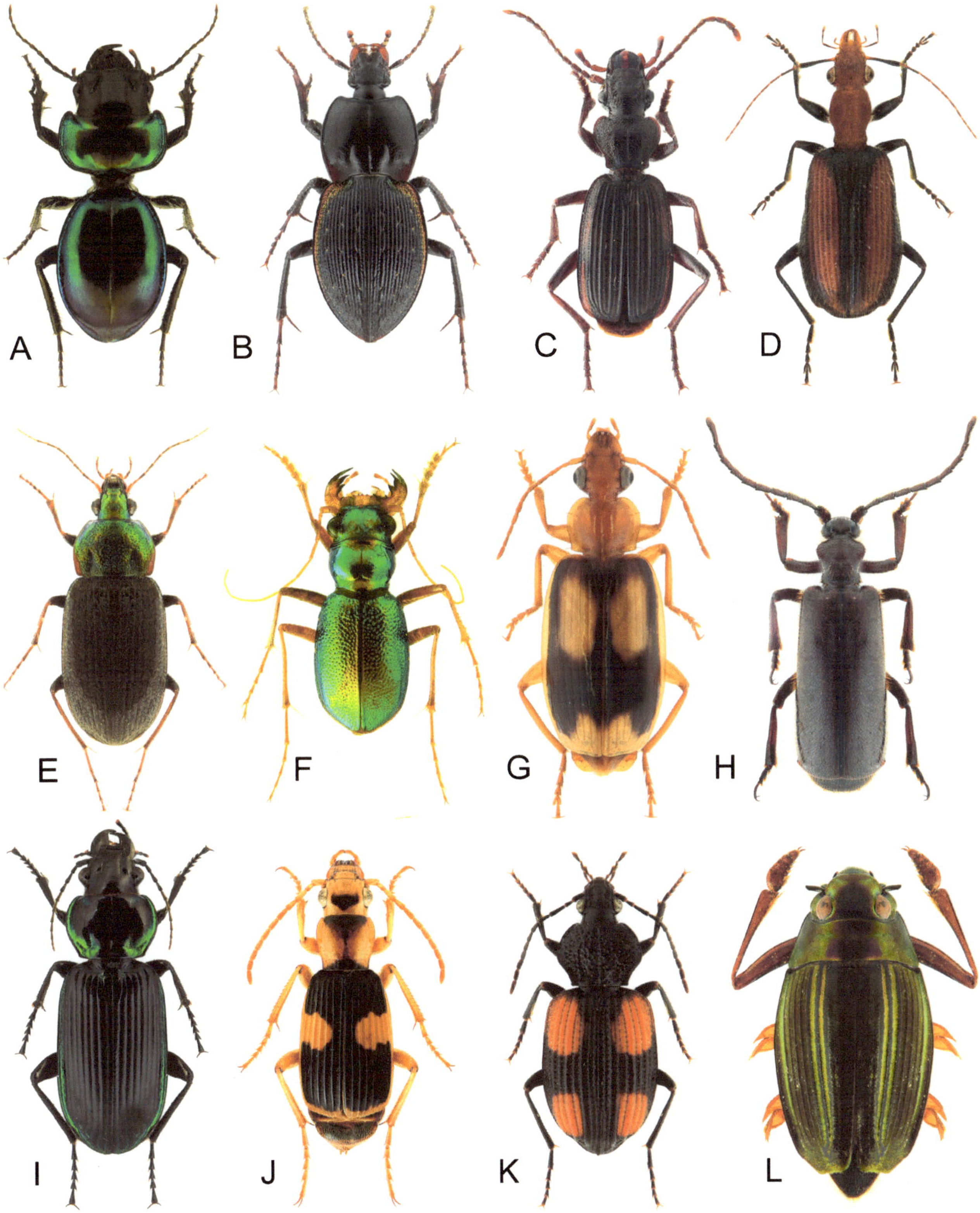

Fig. 66. Adult. A–K, Carabidae, dorsal: A, *Carenum elegans* Macleay (Scaritinae: Carenini); B, *Pamborus* sp. (Carabinae); C, *Helluo costatus* Bonelli (Harpalinae: Helluonini); D, *Dendrocellus smaragdinus* Chaudoir (Harpalinae: Dryptini); E, *Chlaenius australis* Dejean (Harpalinae: Chlaeniini); F, *Megacephala* sp. (Cicindelinae); G, *Trigonothops pacificus* (Erichson) (Harpalinae: Lebiini); H, *Megalopaussus amplipennis* Lea (Paussinae); I, *Catadromus lacordairei* Boisduval (Harpalinae: Morionini); J, *Pheropsophus* sp. (Brachininae); K, *Craspedophorus australasiae* (Chaudoir) (Harpalinae: Panagaeini). L, *Macrogyrus viridisulcatus* Mjöberg (Gyrinidae).

Fig. 67. Adult. A–H, Staphylinidae, dorsal: A, *Aleochara* sp. (Aleocharinae); B, *Paederus* sp. (Paederinae); C, *Plastus miles* Bernhauer (Osoriinae); D, *Antimerus* sp. (Staphylininae); E, *Ochthephilum mastersi* (Macleay) (Paederinae); F, *Scaphidium* sp. (Scaphidiinae); G, *Glypholoma rotundulum* Thayer & Newton (Glypholomatinae); H, *Creophilus erythrocephalus* (Fabricius) (Staphylininae). I–J, Leiodidae: I, *Dietta* sp. (Leiodinae); J, *Myrmecholeva ligulata* Lea (Camiarinae). K–L, Staphylinidae: Oxytelinae: K, *Sartallus* sp.; L, *Bledius* sp.

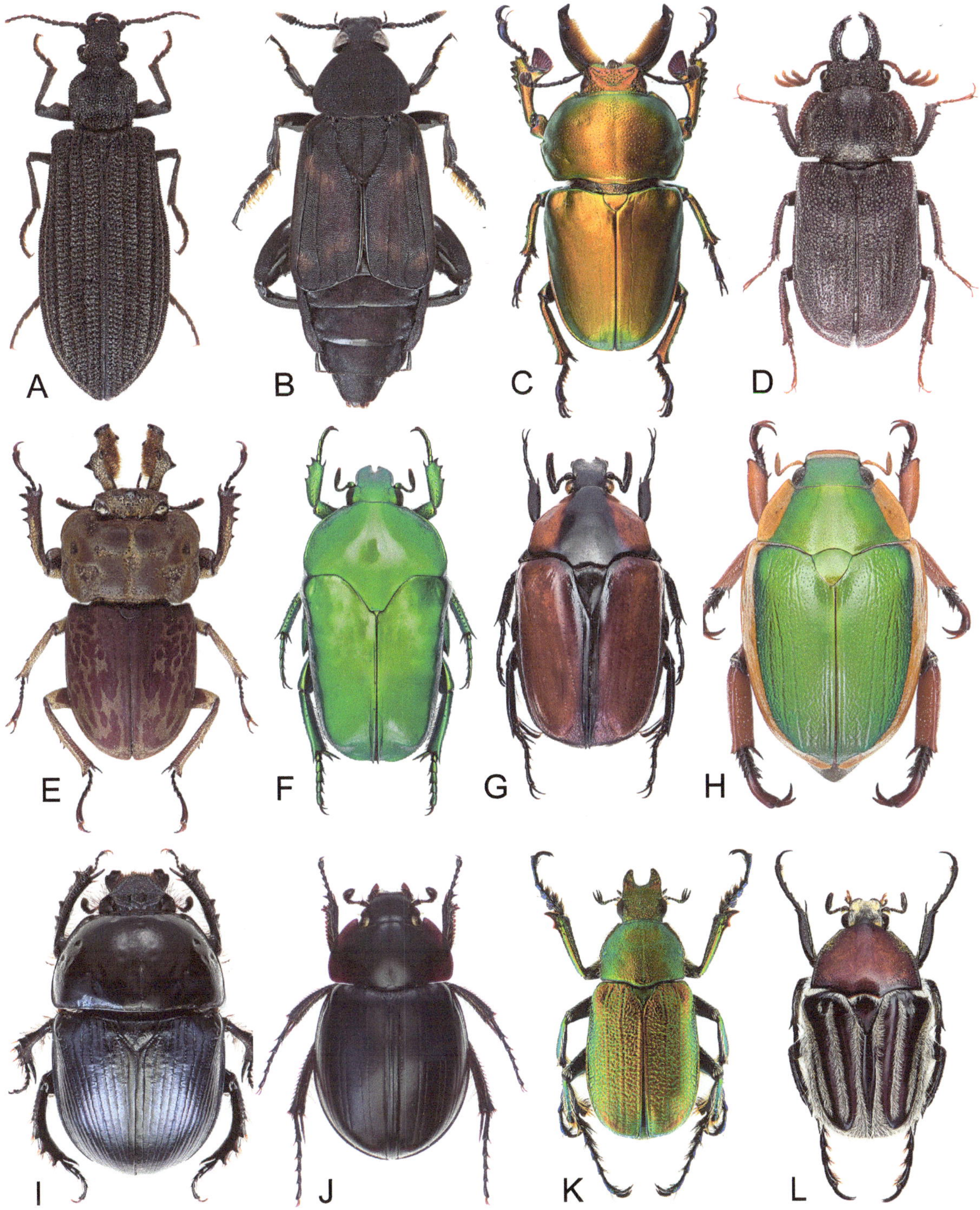

Fig. 68. Adult, dorsal. A, *Omma stanleyi* Newman (Ommatidae). B, *Diamesus osculans* (Vigors) (Silphidae). C–E, Lucanidae: C, *Lamprima* sp. (Lampriminae); D, *Ceratognathus bitumulatus* Carter (Ceratognathinae); E, *Rhyssonotus nebulosus* (Kirby) (Lucaninae). F–G, Scarabaeidae: Cetoniinae: F, *Ischiopsopha wallacei* (Thomson); G, *Dolichrosis bakewellii* (White). H, *Callodes grayianus* (White) (Scarabaeidae: Rutelinae). I, *Geotrupes spiniger* Marsham (Geotrupidae: Geotrupinae). J, *Liparochrus hackeri* Blackburn (Hybosoridae: Hybosorinae). K, *Diphucephala* sp. (Scarabaeidae: Melolonthinae). L, *Trichaulax philipsii* (Schreibers). (Scarabaeidae: Cetoniinae).

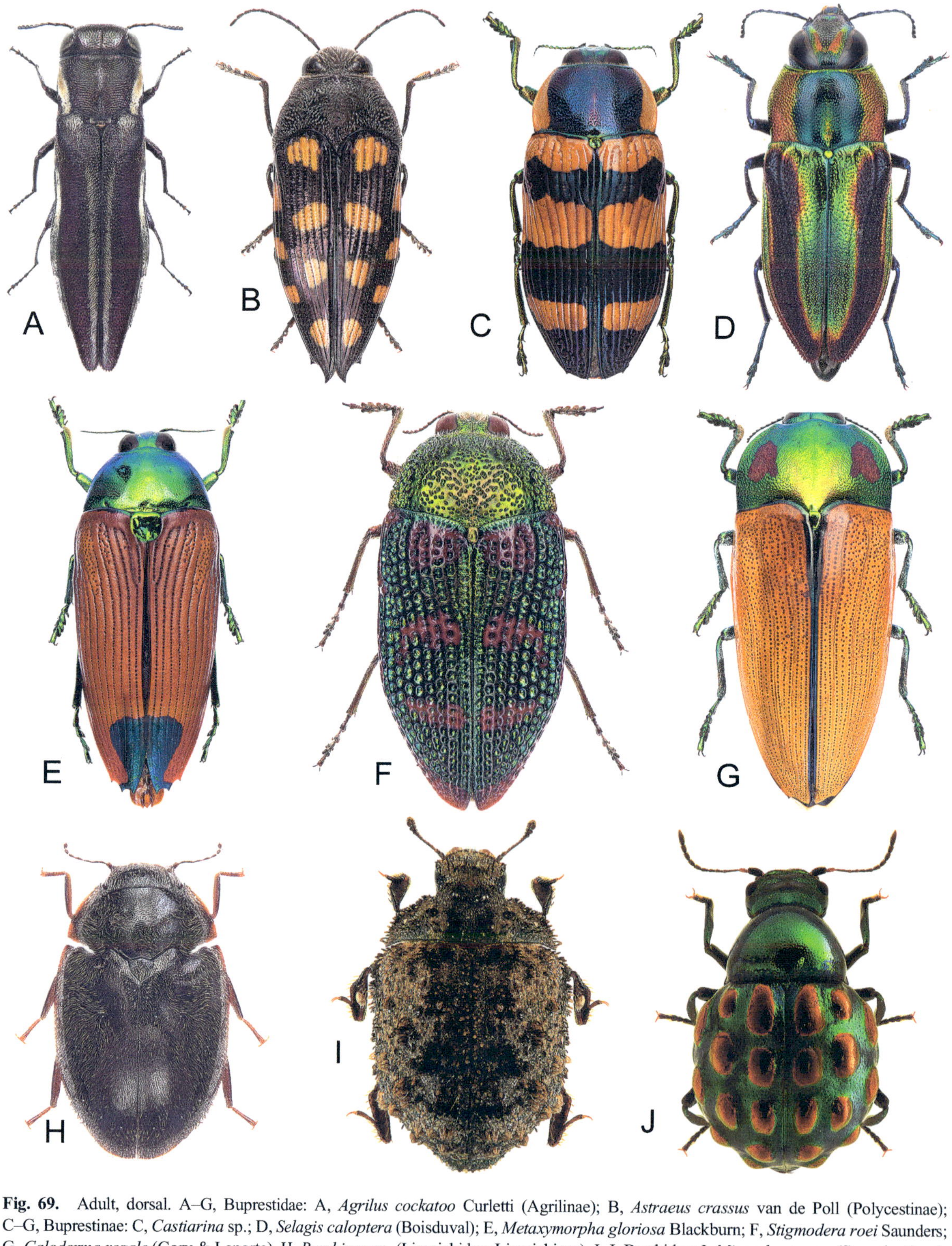

Fig. 69. Adult, dorsal. A–G, Buprestidae: A, *Agrilus cockatoo* Curletti (Agrilinae); B, *Astraeus crassus* van de Poll (Polycestinae); C–G, Buprestinae: C, *Castiarina* sp.; D, *Selagis caloptera* (Boisduval); E, *Metaxymorpha gloriosa* Blackburn; F, *Stigmodera roei* Saunders; G, *Caloderma regale* (Gory & Laporte). H, *Byrrhinus* sp. (Limnichidae: Limnichinae). I–J, Byrrhidae: I, *Microchaetes* sp. (Syncalyptinae); J, '*Pedilophorus*' *gemmatus* Lea (Byrrhinae).

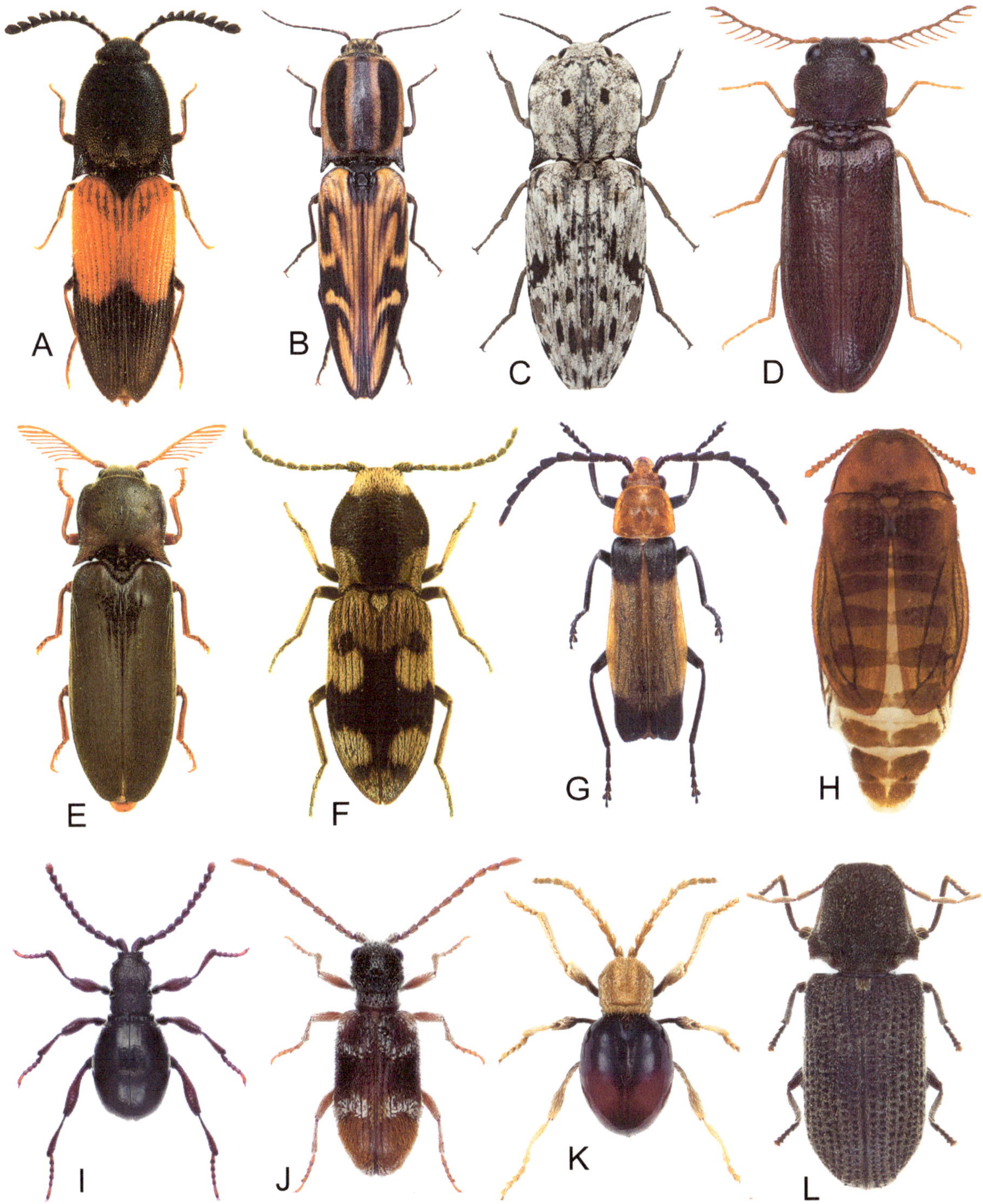

Fig. 70. Adult, dorsal: A–F, Elateridae: A, *Anilicus xanthomus* (Macleay) (Elaterinae); B, *Ophidius histrio* (Boisduval) (Elaterinae); C, *Paracalais gibboni* (Newman) (Agrypninae); D, *Wardulupicola brittoni* Calder (Dendrometrinae ?); E, *Tetralobus* sp. (Agrypninae); F, *Rivulicola variegatus* (Macleay) (Negastriinae). G, *Sphaerarthrum* sp. (Cantharidae). H, *Brachypsectra* sp. (Brachypsectridae). I–K, Ptinidae: Ptininae: I, *Diplocotes* sp.; J, *Ptinus* sp.; K, *Mezium* sp. L, *Hadrobregmus australis* Pic (Ptinidae: Anobiinae).

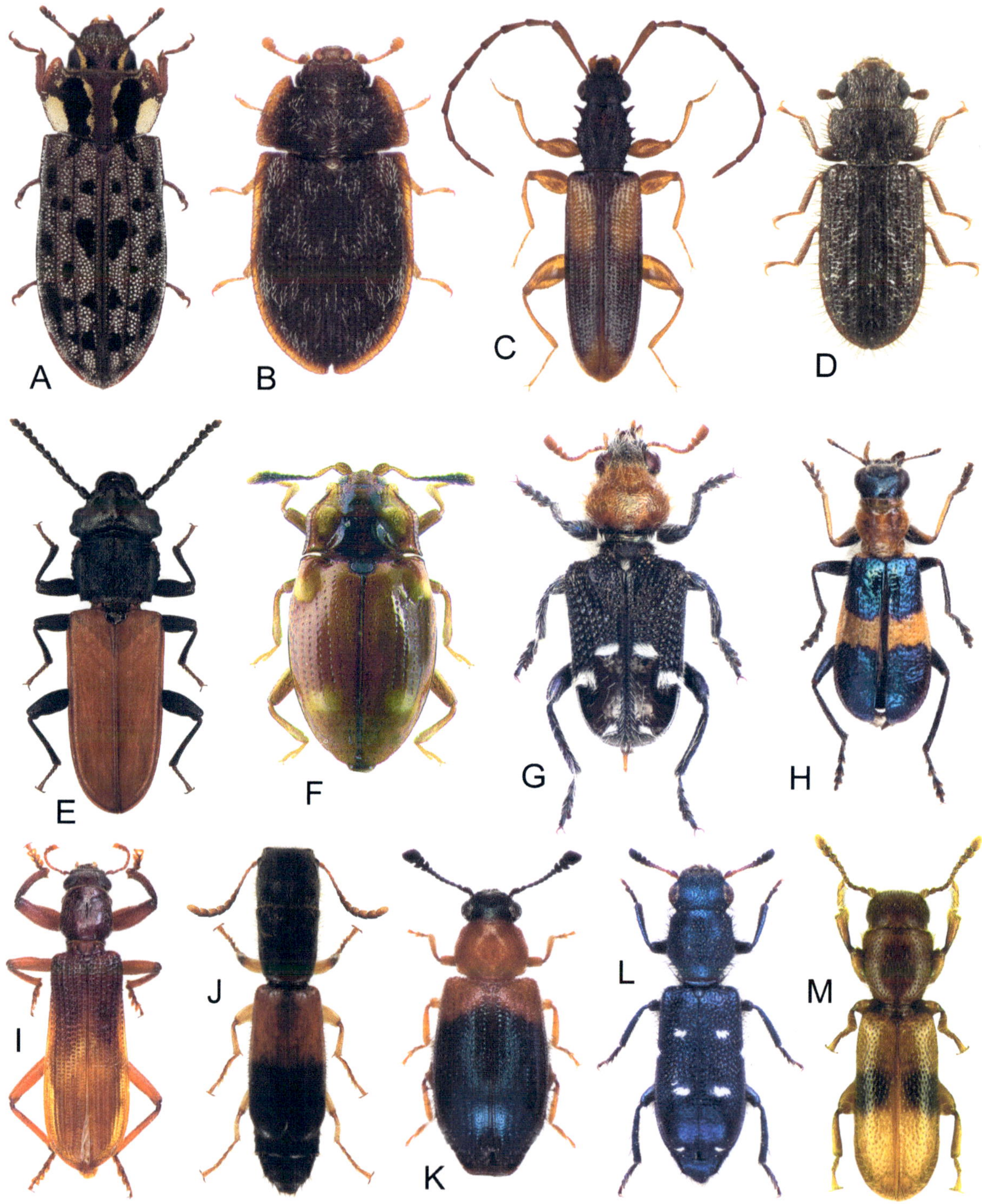

Fig. 71. Adult, dorsal. A–B, Trogossitidae: A, *Leperina decorata* (Erichson) (Trogossitinae); B, *Neaspis* sp. (Lophocaterinae). C, *Macrohyliota militaris* (Erichson) (Silvanidae: Brontinae). D, *Necrobiopsis tasmanica* Crowson (Trogossitidae: Trogossitinae). E, *Platisus* sp. (Cucujidae). F, *Periptyctus quadripunctatus* Tomaszewska & Slipinski (Corylophidae: Periptyctinae). G–L, Cleridae: G, *Zenithicola crassus* (Newman) (Clerinae); H, *Lemidia* sp. (Hydnocerinae); I, *Euatalis* sp. (Clerinae); J, *Cylidrus basalis* Macleay (Tillinae); K, *Necrobia ruficollis* Fabricius (Korynetinae); L, *Cleromorpha* sp. (Clerinae). M, *Isoclerus cipisek* Kolibác (Thanerocleridae).

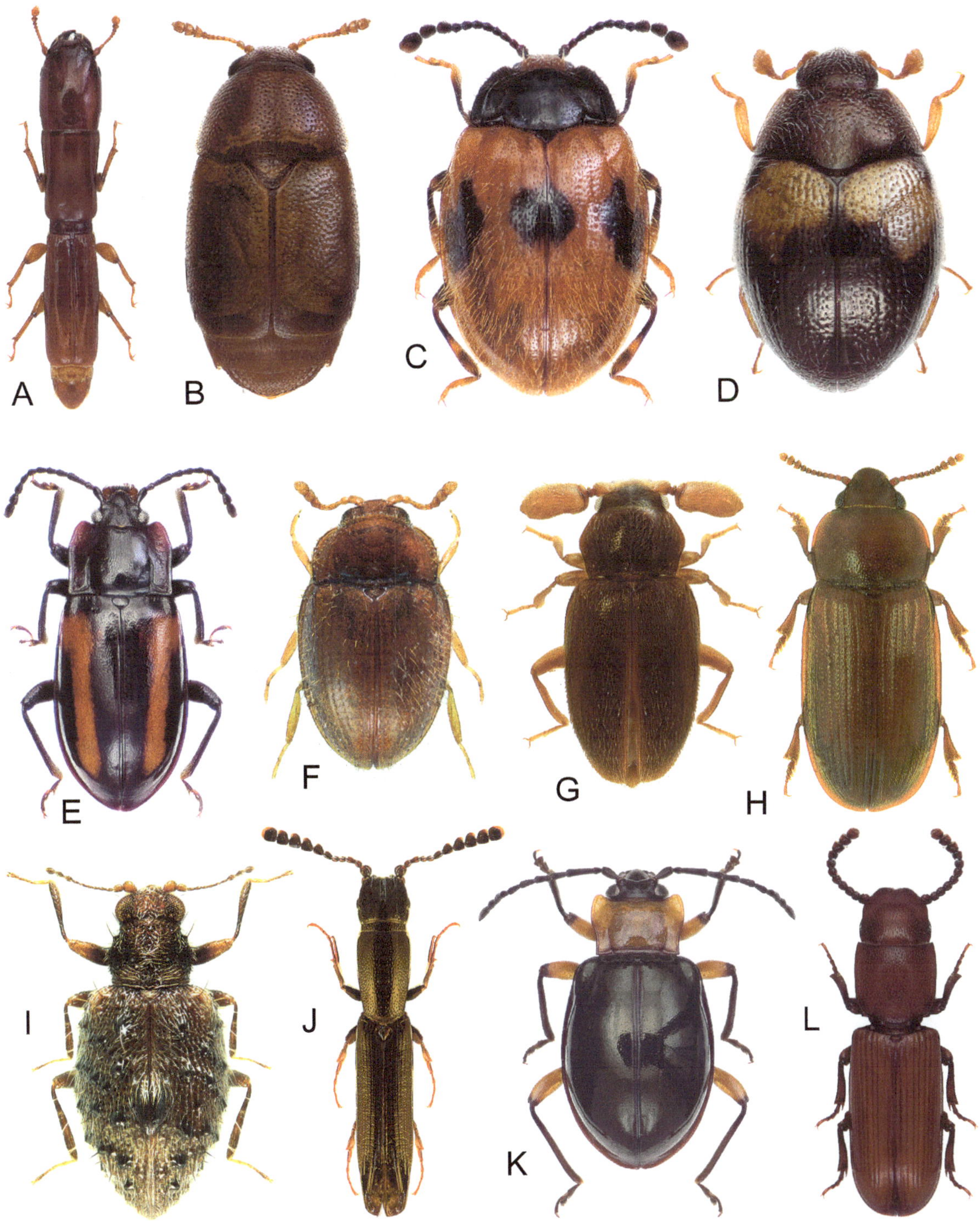

Fig. 72. Adult, dorsal. A, *Shoguna termitiformis* (Fairmaire) (Monotomidae). B, *Notobrachypterus* sp. (Kateretidae). C, *Stenotarsus commodus* Blackburn (Endomychidae: Stenotarsinae). D, *Hypodacnella kasiae* Slipinski (Cerylonidae: Euxestinae). E–G, Endomychidae: E, *Indalmus lineellus* (Chapuis) (Lycoperdininae); F, *Idiophyes brevis* Blackburn (Anamorphinae); G, *Trochoideus desjardinsi* Guérin–Méneville (Pleganophorinae). H, *Xenocryptus tenebroides* Arrow (Erotylidae: Xenoscelinae). I, *Bicava* sp. (Latridiidae: Corticariinae); J, *Ancistria retusa* (Fabricius) (Passandridae); K, *Encymon immaculatus* (Montrouzier) (Endomychidae: Lycoperdininae); L, *Aulonosoma tenebrioides* Motschulsky (Passandridae).

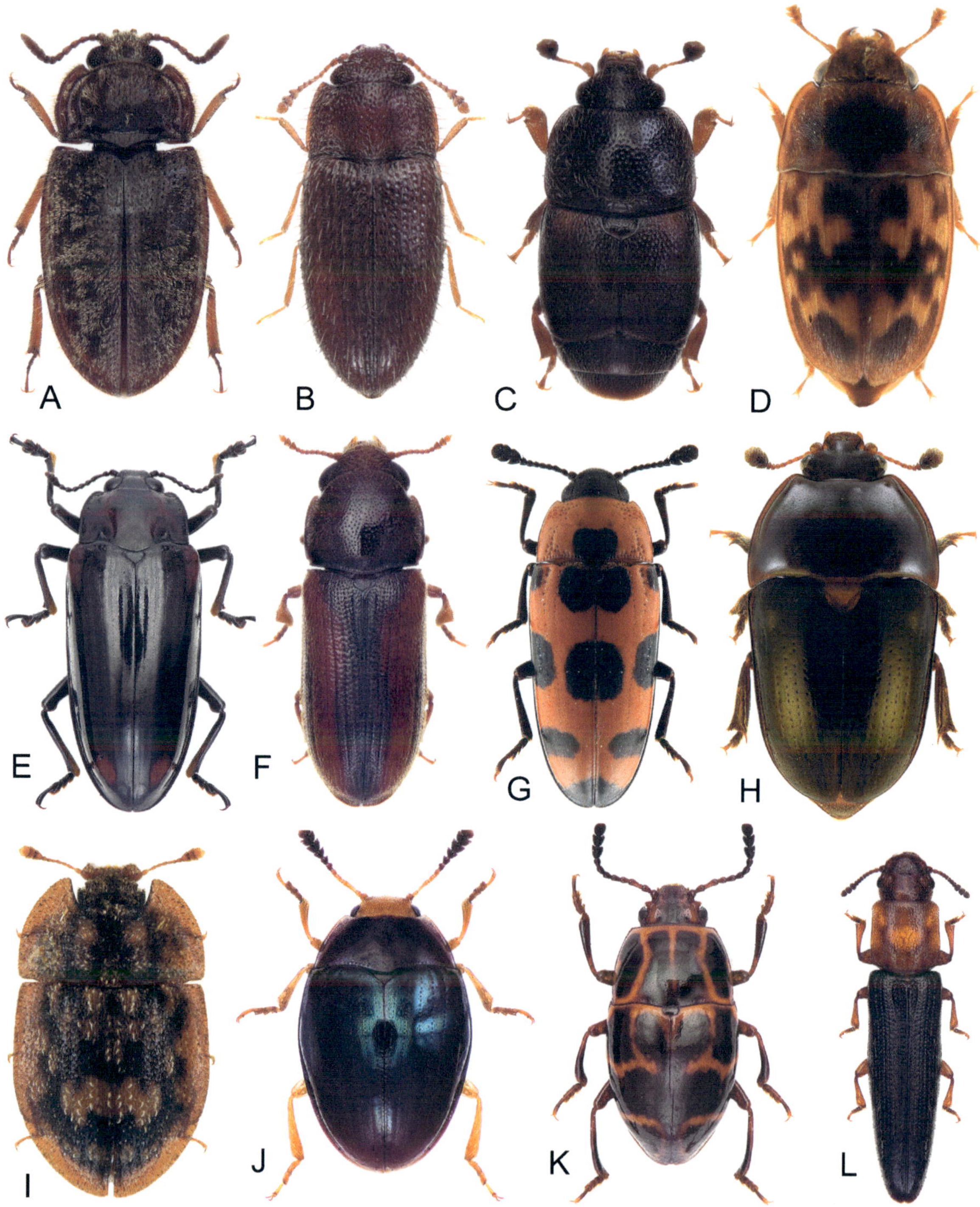

Fig. 73. Adult, dorsal. A–B, Biphyllidae: A, *Althaesia sericea* Lea; B, *Diplocoelus latus* Lea. C–D, Nitidulidae: C, *Carpophilus* sp. (Carpophilinae); D, *Cryptarcha* sp. (Cryptarchinae). E–G, Erotylidae: Erotylinae: E, *Encaustes hercules* (Lea); F, *Cnecosa* sp.; G, *Episcaphula australis* (Boisduval). H–I, Nitidulidae: Nitidulinae: H, *Gymnocychramus politus* Lea; I, *Macleayana vesca* (Olliff). J–K, Erotylidae: Erotylinae: J, *Spondotriplax* sp.; K, *Episcaphula rufolineata* Wilson; L, *Hapalips* sp. (Erotylidae: Languriinae).

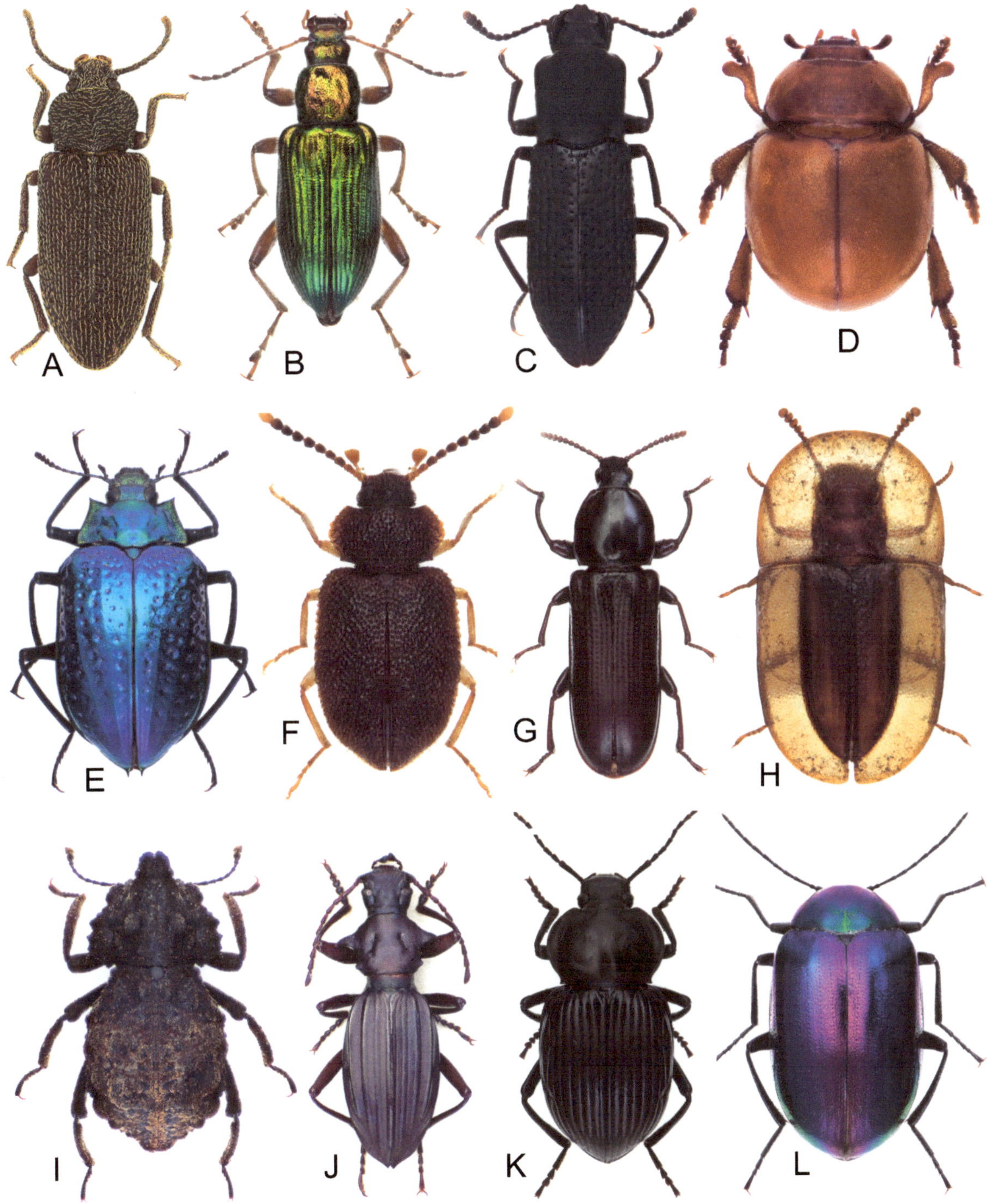

Fig. 74. Adult, dorsal. A–L, Tenebrionidae: A, *Exangeltus* sp. (Pimeliinae: Vacronini), from Matthews and Bouchard 2008, © Agriculture and Agri-Food Canada; B, *Aethyssius* sp. (Alleculinae); C, *Toxicum* sp. (Tenebrionidae: Toxicini); D, *Sphargeris physodes* Pascoe (Lagriinae: Chaerodini); E, *Cyphaleus mastersi* Pascoe (Tenebrioninae: Heleini); F, *Pseudolyprops punctatus* (Carter) (Lagriinae: Goniaderini); G, *Tanylypa morio* Pascoe (Zolodininae); H, *Cossyphus odewahnii* Pascoe (Lagriinae: Cossyphini); I, *Mychestes* sp. (Tenebrioninae: Toxicini); J, *Blepegenes* sp. (Lagriinae: Adeliini); K, *Adelium* sp. (Lagriinae: Adeliini); L, *Chalcopteroides* sp. (Tenebrionidae: Amarygmini).

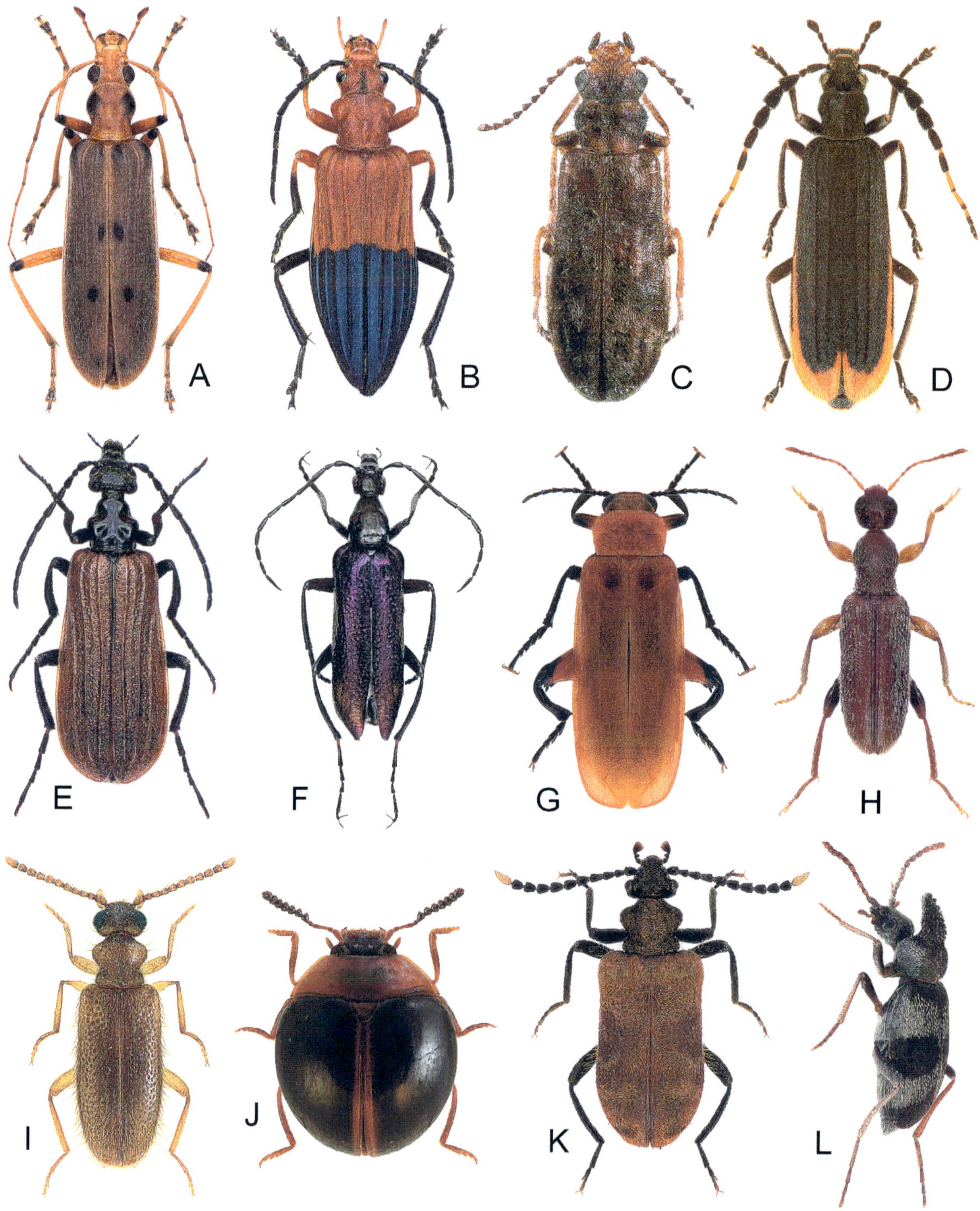

Fig. 75. Adult. A–K, dorsal: A–D, Oedemeridae: A, *Sessinia nigronotata* (Boheman); B, *Agasma semicrudum* Newman; C, *Dasytomima rachelae* Lawrence; D, *Pseudolycus* sp. E–G, Meloidae: E, *Palaestra* sp.; F, '*Zonitis*' sp.; G, *Horia mira* (Blackburn). H–I, Anthicidae: H, *Macratria* sp. (Macratriinae); I, *Pseudotomoderus* sp. (Tomoderinae). J, *Leiochrodes suturalis* Westwood (Tenebrionidae: Diaperinae). K–L, Anthicidae: K, *Lemodes coccinea* Boheman (Lemodinae); L, *Mecynotarsus kreusleri* King (Notoxinae), dorsolateral.

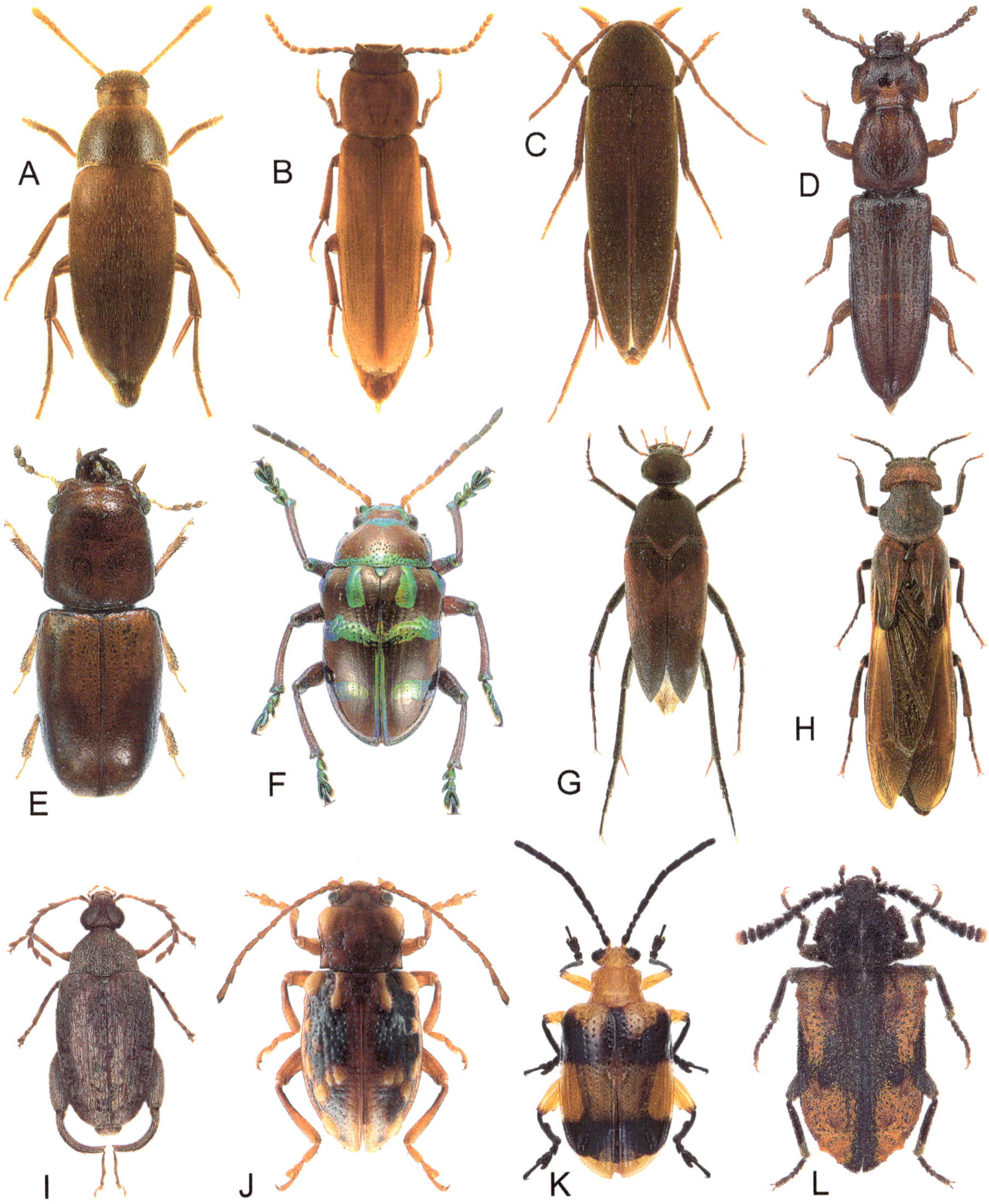

Fig. 76. Adult, dorsal. A–C, Melandryidae: A, *Orchesia* sp.; B, *Mystes planatus* Champion; C, *Talayra elongata* Macleay. D, *Dryocora cephalotes* (Waterhouse) (Prostomidae). E, *Octotemnus walkeri* Blair (Ciidae). F, *Spilopyra stirlingi* Lea (Chrysomelidae: Spilopyrinae), from Reid and Beatson 2010, © Australian Museum. G–H, Ripiphoridae: G, *Macrosiagon* sp. (Ripiphorinae); H, *Sitarida* sp. (Hemirhipidiinae). I, *Caryedon gonagra* (Fabricius) (Chrysomelidae: Bruchinae). J, *Chalcolampra pustulata* Baly (Chrysomelidae: Chrysomelinae); K, *Stethopachys formosa* Baly (Chrysomelidae: Criocerinae). L, *Ganyme sapphira* (Newman) (Ulodidae).

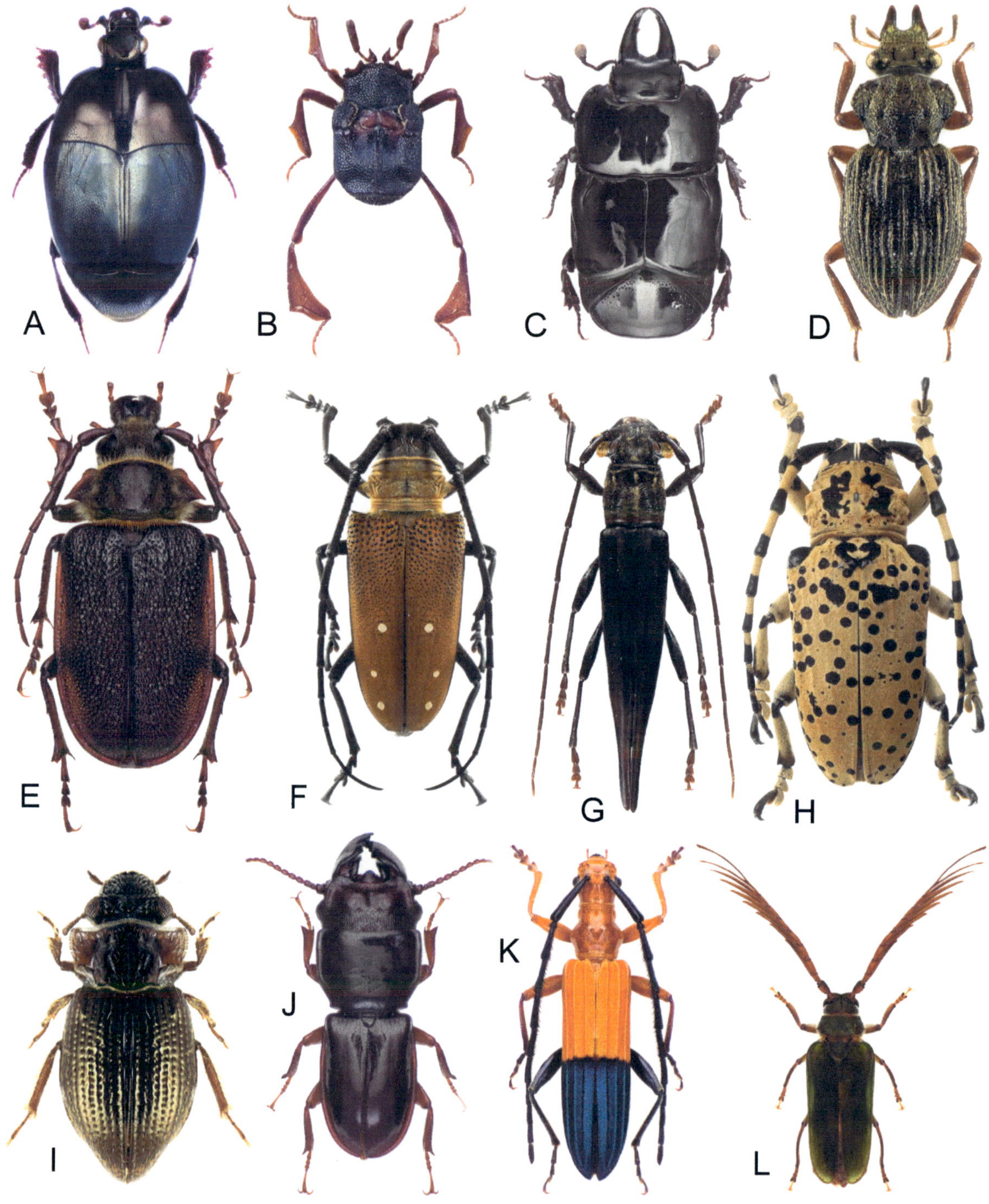

Fig. 77. Adult, dorsal. A–C, Histeridae): A, *Saprinus splendens* (Paykull) (Saprininae); B, *Chlamydopsis* sp. (Chlamydopsinae); C, *Hololepta* sp. (Histerinae). D, *Tympanogaster porchi* Perkins (Hydraenidae: Ochthebiinae). E, *Sclerocantha* sp. (Cerambycidae: Prioninae); F–H, Cerambycidae: Lamiinae: F, *Batocera* sp.; G, *Enichodes fichteli* (Schreibers); H, *Penthea pardalis* (Newman). I, *Gymnochthebius setosus* Perkins. (Hydraenidae: Ochthebiinae). J–K, Cerambycidae: J, *Storeyandra frenchi* (Blackburn) (Parandrinae); K, *Stenoderus ostricilla* Newman (Cerambycinae); L, *Rhipidocerus australasiae* Westwood (Prioninae).

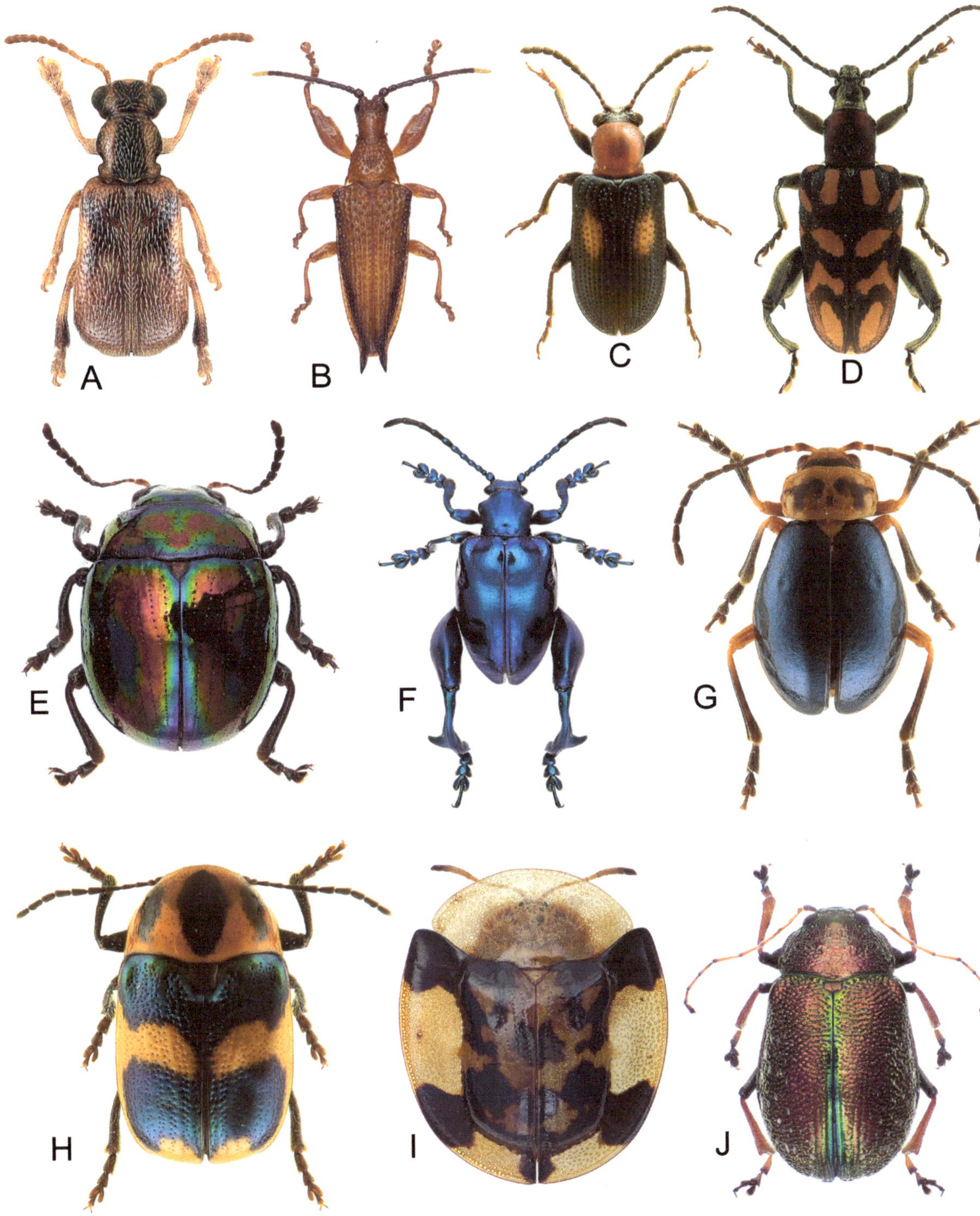

Fig. 78. Adult, dorsal. A, *Zeugophora vitinea* (Oke) (Megalopodidae: Zeugophorinae). B–J, Chrysomelidae: B, *Aproida balyi* Pascoe (Cassidinae); C, *Lema rufocincta* Clark (Criocerinae); D, *Mecynodera coxalgica* (Boisduval) (Sagrinae); E, *Callidemum* sp. (Chrysomelinae); F, *Sagra papuana* Jacoby (Sagrinae); G, *Ellopidia amplipennis* (Lea) (Galerucinae); H, *Aporocera parentheticus* (Suffrian) (Cryptocephalinae); I, *Aspidimorpha interrupta* (Fabricius) (Cassidinae); J, *Eudusella* sp. (Eumolpinae).

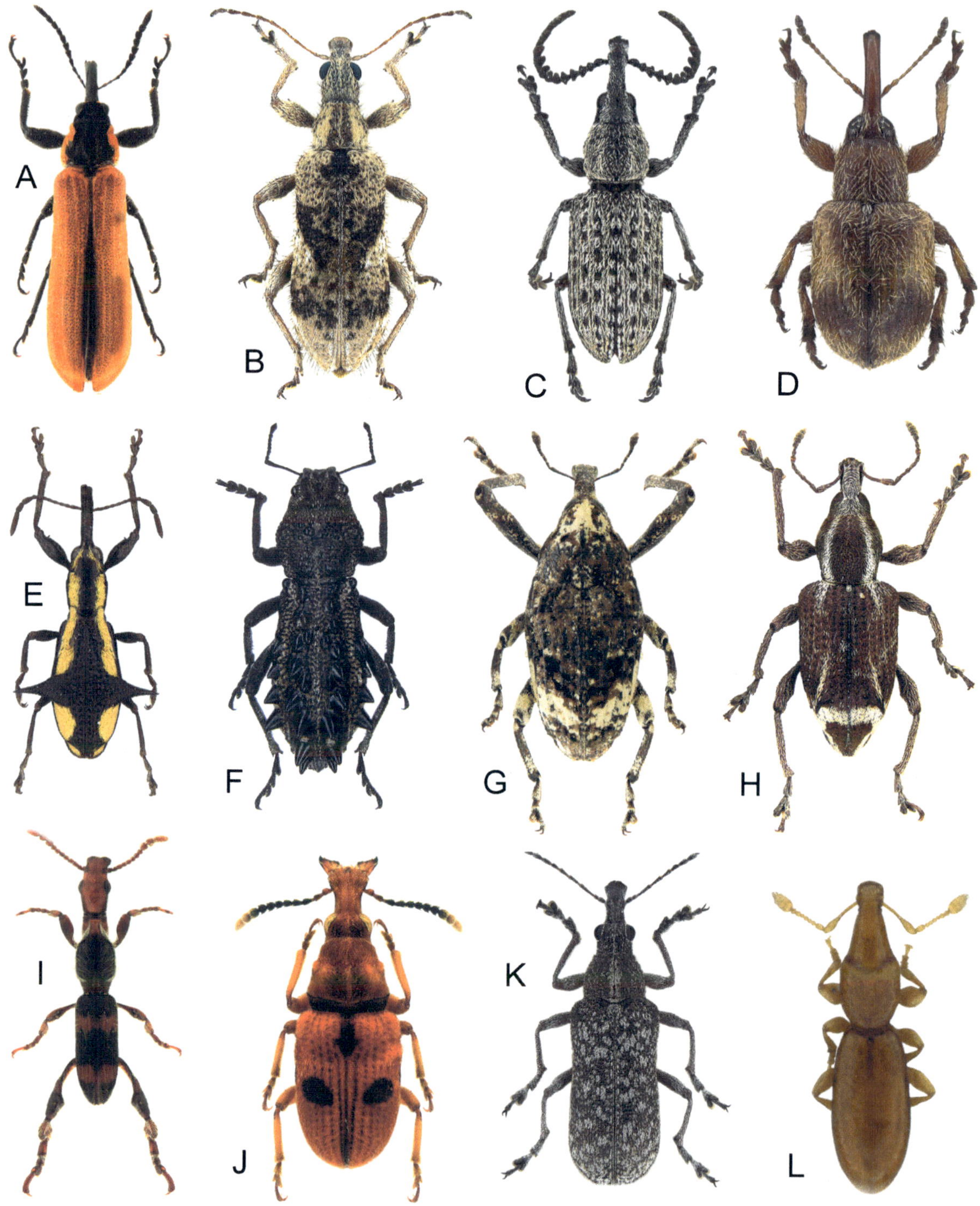

Fig. 79. Adult Curculionoidea, dorsal. A–B, Belidae: A, *Rhinotia haemoptera* Kirby; B, *Cyrotyphus blandus* (Faust). C, *Ctenaphides maculatus* (Pascoe) (male) (Brentidae: Eurhynchinae). D, *Car pini* Lea (Caridae). E, *Aporhina splendida* (Blackburn) (Brentidae: Eurhynchinae). F–H, Curculionidae: F, *Gagatophorus* sp. (Cyclominae); G, *Axionicus insignis* Pascoe (Molytinae: Cryptorhynchini); H, *Lycosura bispinosa* Pascoe (Subfamily ?: Hyperini). I, *Cyphagogus modiglianii* Senna (Brentidae: Brentinae). J, *Eupanteos* sp. (Anthribidae). K, *Pachyura australis* Hope (Belidae). L, *Myrtonymus* sp. (Curculionidae: Brachycerinae).

Fig. 80. Adult Curculionoidea, dorsal. A–C, Curculionidae: A, *Platynotocis* sp. (Curculioninae: Viticiini); B, *Pycnochirus fuscovarius* Klima (Curculioninae: Curculionini); C, *Aades cultratus* (Fabricius) (Cyclominae: Aterpini). D, *Austronanodes nigrovarius* Lea (Brentidae: Nanophyinae). E–K, Curculionidae: E, *Syarbis* sp. (Subfamily ?: Gonipterini); F, *Eurhamphus fasciculatus* Shuckard (Molytinae: Orthorhinini); G, *Leptopius areolatus* Blackburn (Entiminae: Leptopiini); H, *Cossonus varius* Lea (Cossininae); I, *Aonychus* sp. (Brachycerinae: Aonychini); J, *Haplonyx* sp. (Curculioninae: Cryptoplini); K, *Myllocerus* sp. (Entiminae: Cyphicerini). L, *Cylas formicarius* (Fabricius) (Brentidae: Brentinae: Cyladini).

DERODONTIFORMIA: DERODONTOIDEA

This series and superfamily includes three families, Derodontidae, Nosodendridae and Jacobsoniidae, placed by Crowson (1955, 1960) in his superfamily Dermestoidea and included within Bostrichiformia by Lawrence and Britton (1994). These families appear to lie at the base of the polyphagan lineage and may not form a monophyletic group. In the recent morphological analysis by Lawrence *et al.* (2011), Jacobsoniidae was sister to basal Staphylinoidea, while Derodontidae and Nosodendridae formed a clade with Scirtidae, Eucinetidae and Decliniidae. [Ge *et al.* 2007; Lawrence *et al.* 2010e].

45. DERODONTIDAE

Adult (Fig. 81C)

Length 1.8–1.95 mm. Body ~1.75–2.1 times as long as wide; moderately well sclerotised; slightly convex above, flattened below; colour of head and pronotum reddish-brown, elytra dark brown with suture reddish-brown, undersurfaces and legs yellowish-brown; upper surfaces clothed with short, fine decumbent hairs, which may be yellow or white in colour, the latter forming two transverse bands on elytra.

Head prognathous. Vertex with transverse band of large, glandular pores. Ocelli located close to inner edges of eyes, each ocellus closely associated anteriorly with three tubercles forming bridges over short, transverse and longitudinal canals. Eyes well developed, protuberant, circular, entire, coarsely facetted, without interfacetal setae; Antennal insertions exposed from above, well separated; subantennal groove short. Frontoclypeal region abruptly declined forming a weak frontal carina, in front of which is another pair of median tubercles flanking a short, transverse canal. Frontoclypeal suture absent; clypeus broadly rounded at apex. Labrum well developed, free and exposed. Antennae 11-segmented, with weak, loose, 3-segmented club. Mandible broad at base and narrowed apically, gradually curved mesally, bidentate, with well developed mola and prostheca. Maxilla with setose galea and lacinia, the latter bearing a bidentate uncus; apical maxillary and labial palpomeres fusiform. Mentum about as long as wide, apically truncate; ligula short, broadly rounded. Corpotentorium straight, moderately broad; laminatentoria expanded and meeting at midline. Cervical sclerites well developed.

Pronotum ~0.75–0.8 times as long as wide, widest at middle and distinctly narrower at base than combined elytral bases; sides distinctly curved and slightly explanate. Lateral pronotal carinae complete, with distinct bead, edges irregularly denticulate; anterior angles very slightly produced; posterior angles broadly right; posterior edge evenly rounded; disc with complex of longitudinal and oblique carinae delimiting cells or impressions. Prosternum in front of coxae slightly shorter than mid length of procoxal cavity. Prosternal process complete, narrow, not curved dorsally, abruptly expanded and truncate at apex. Procoxae conical and projecting, with short, concealed lateral extensions; trochantin concealed or barely exposed; endopleuron fused to notum. Procoxal cavities very slightly transverse, narrowly separated, narrowly closed externally, open internally.

Scutellar shield abruptly elevated, posteriorly broadly rounded. Elytra ~1.65–1.7 times as long as combined width and 2.7–2.9 times as long as pronotum, sides slightly diverging to middle, then converging to conjointly rounded apices; each elytron with 11 distinct punctate striae and no scutellary striole; striae incomplete or absent anteriorly, where the minute strial punctures are replaced by linearly arranged, large glandular pores (Fig. 28P); epipleura narrow and incomplete. Mesoventrite without anterior procoxal rests; broad prothoracic rest present, extending onto mesanapisterna; mesoventral cavity absent; mesoventral process carinate, apically truncate. Mesocoxae subglobular, with exposed trochantins. Mesocoxal cavities narrowly separated, partly closed laterally by mesepimeron and metanepisternum. Meso- and metathoracic portions of coxal cavities joined by complete membranous joint. Mesometaventral junction simple, with metaventrite overlapping mesoventrite. Metaventrite anteriorly with pair of short carinae extending posteriorly from base of metaventral process and with two glandular pores behind each mesocoxal cavity; discrimen moderately long and transverse (metakatepisternal) suture complete; exposed portion of metanepisternum relatively long and narrow; anterolateral portion with elytral interlocking device. Metacoxae approximate, extending laterally to meet elytra; plates well developed and complete. Metendosternite with short but narrow stalk, long, curved lateral arms, no laminae or ventrolateral processes, and anterior process moderately long and deeply divided, so that anterior tendons are on separate lobes. Hind wing with very long apical field containing two radial extensions and a distinct hinge joint at the apex of the radial bar; radial cell and cross-veins r3 and r4 absent; R-M loop forming narrowly acute angle and basal portion of RP very short, but apical extension extending to wing margin; medial spur well developed, slightly anteriorly curved; medial field with three free veins and no medial fleck; wedge cell absent; anal lobe absent; radial bar and entire hind wing margin lined with moderately long hairs. Femoral attachment of mid trochanter slightly oblique. Legs slender; tibiae simple; apical spurs small. Tarsi 5-5-5; tarsomeres simple; pretarsal claws simple; empodium bisetose.

Abdomen with five free ventrites; ventrite 1 not longer than 2, with short, straight, postcoxal line on either side of acute intercoxal process; tergites moderately pigmented. Functional spiracles on abdominal segments I–VIII. Anterior edge of sternite VIII in male with short strut (spiculum relictum). Anterior edge of segment IX in male broadly rounded, without spiculum gastrale; tergite IX emarginate and longitudinally divided by suture; tergite X free. Aedeagus symmetrical, trilobate, with basally rounded phallobase, short, blunt parameres and short, broad penis without paired basal struts. Sternite VIII in female with short spiculum ventrale. Ovipositor ~1.75 times as long as wide, relatively lightly sclerotised, without distinct baculi; paraprocts about as long as coxites, which are divided into broad basal and narrow apical lobes; styli well developed and terminal. [Crowson 1959; Lawrence and Hlavac 1979; Lawrence 1985a; Leschen 2002a; Leschen and Beutel 2010a].

Larva

Body elongate and more or less parallel-sided or slightly wider at middle, slightly flattened; moderately lightly pigmented but slightly darker above than below; dorsal surfaces granulate and tuberculate; vestiture of short, simple setae.

Head well developed, prognathous and protracted, transverse and slightly flattened, with sides distinctly rounded. Epicranial stem very short or absent; frontal arms lyriform and contiguous at base; median endocarina absent. Stemmata six on each side (sometimes only five). Frontoclypeal suture absent. Labrum large, free, broadly rounded at apex. Antennae ~0.25 times as long as head width, 3-segmented, with basal antennomere short and antennomeres 2 and 3 moderately long; sensorium on antennomere 2 about half as long as antennomere 3. Mandibles symmetrical, broad at base and narrow at apex, bidentate, with serrate cutting edge and accessory ventral process; mola well developed; prostheca a narrowly acute hyaline process. Ventral mouthparts retracted, with well developed maxillary articulating area; cardines transverse, well separated; stipites elongate; galea and lacinia fixed and more or less fused at base; galea narrowly rounded; lacinia falcate with two apical teeth; maxillary palp 3-segmented. Labium consisting of prementum, mentum and submentum; ligula broadly rounded; labial palps 2-segmented and well separated. Hypopharyngeal sclerome a transverse bar. Hypostomal rods absent. Ventral epicranial ridges present. Gular region very short.

Thoracic terga and abdominal terga I–VIII with dorsal and lateral processes. Legs moderately well developed, 5-segmented including claw-like pretarsus bearing two setae lying side by side; coxae moderately widely separated. Tergum IX slightly shorter than VIII, not extending onto ventral surface, with well developed, simple, subparallel or slightly converging paired urogomphi. Sternum IX exposed, simple, not enclosed by sternum VIII. Segment X short and cylindrical, at least slightly sclerotised dorsally and more or less terminal, without paired pygopods. Spiracles annular–biforous, all borne at ends of subconical spiracular tubes. [Crowson 1959; Lawrence and Hlavac 1979; Leschen 2002a; Zilahi-Balogh *et al.* 2006; Leschen and Beutel 2010a].

Classification and World Distribution

Derodontidae contains ~40 described species in four genera. The family is subdivided into four subfamilies: Peltasticinae, with the genus *Peltastica* Mannerheim from western North America and eastern Asia, Derodontinae, with the Holarctic genus *Derodontus*, and Laricobiinae, including the genus *Laricobius* Rosenhauer, occurring in the Holarctic and Oriental regions, and *Nothoderodontus* from New Zealand, Australia and Chile. [Crowson 1959; Lawrence and Hlavac 1979; Leschen 2011].

Australian Fauna

The family is represented in Australia by a single species, *Nothoderodontus darlingtoni* Lawrence (**Laricobiinae**), which occurs in the wet forests of TAS, VIC and the ACT. It is a distinctive beetle with paired ocelli and a series of canals and bridges on the head, a pronotum divided into several cells by longitudinal ridges and transverse elevations, and elytra with several deep pits at the bases of striae. The mesocoxal cavities are partly closed by the metanepisterna, the metaventrite has a distinct katepisternal suture, the metacoxal plates are well developed, and there is a pair of longitudinal carinae near the middle of the first ventrite. Larvae are elongate, slightly tapered posteriorly, with the dorsal surfaces granulate, tuberculate and usually with paired processes on all segments except X, which is cylindrical and posteroventrally oriented. The head is transverse with no epicranial stem and contiguous frontal arms and the spiracles are borne on tubular processes. The habits of the Australian species is unknown, but the New Zealand *N. gourlayi* Crowson feeds on the hyphae and conidia of sooty moulds (Ascomycetes: Capnodiaceae and related families). [Crowson 1959; Lawrence 1985a].

46. NOSODENDRIDAE

Adult (Fig. 81G–H)

Length 3.5–5.5 mm. Body ~1.4–1.6 times as long as wide; well sclerotised; dorsally convex and ventrally flattened; colour uniformly black; upper surfaces subglabrous.

Head prognathous, inserted well into prothorax. Eyes entire, not strongly protuberant, finely facetted, without interfacetal setae. Antennal insertions widely separated and concealed from above; subantennal grooves well developed, extending behind eyes. Frontoclypeal region not declined. Frontoclypeal suture absent; clypeus broadly rounded at apex. Labrum well developed, free, partly exposed and apically subtruncate. Antennae 11-segmented, with well developed, slightly asymmetrical, 3-segmented club. Mandible short and broad, gradually curved mesally, unidentate, with small, deep fovea near base of outer edge; molae large and asymmetrical; prostheca reduced to setal fringe. Maxilla with setose galea and lacinia, the latter bearing a tridentate uncus; maxillary palp relatively short, not exceeding apex of galea, more or less cylindrical. Mentum large, about as long as wide, broadly rounded apically; ligula bilobed; insertions of labial palps concealed from below by mentum; apical labial palpomere subcylindrical. Gular region well developed; gular sutures absent. Corpotentorium broad and straight; laminatentoria not expanded. Cervical sclerites relatively small.

Pronotum ~0.3–0.35 times as long as wide, widest at base and as wide as combined elytral bases; sides distinctly curved anteriorly. Lateral pronotal carinae complete, with distinct bead, edges irregularly denticulate; anterior angles produced and acute or narrowly rounded; posterior angles right or slightly obtuse; posterior edge evenly rounded; disc simple. Prosternum in front of coxae much shorter than mid length of procoxal cavity; surface of prosternum forming raised mesal shelf continuous with the prosternal process, which is complete, not curved dorsally, posteriorly narrowed and rounded at apex. Hypomeron with shallow anterior impression for receiving antennal club and a shallow crural impression posteriorly. Procoxae subcylindrical and somewhat excavate, without concealed lateral extensions; trochantin broadly exposed and more ot less abutting sternum and notum; endopleuron free. Procoxal cavities strongly transverse, well separated, broadly open externally and internally.

Scutellar shield abruptly elevated, with sharp edges, posteriorly acute. Elytra ~1.15–1.3 times as long as combined width and 4–5 times as long as pronotum, widest at anterior third and converging posteriorly to conjointly rounded apices; anterolateral corners almost right but humeri very weak; each elytron with 11 rows of very fine punctures and no scutellary striole or punctation irregular; epipleura moderately broad and somewhat oblique anteriorly, gradually narrowed posteriorly and incomplete. Mesoventrite with anterior edge on same plane as metaventrite, but mesal portion on different plane than lateral portions, which form with mesopleura impressions for receiving mesofemora; anterior procoxal rests almost vertical; mesoventral cavity relatively small and deep, not extending beyond anterior edges of mesocoxal cavities; mesoventral process flat, shallowly concave at apex. Mesocoxae with exposed trochantins. Mesocoxal cavities very widely separated, partly closed laterally by mesepimeron and metanepisternum. Meso- and metathoracic portions of coxal cavities joined by complete membranous joint. Mesometaventral junction forming broad, monocondylic joint. Metaventrite transverse, flattened; anterolateral portion with curved line delimiting part of crural impression; discrimen short and transverse (metakatepisternal) suture complete; exposed portion of metanepisternum long and narrow; anterior portion forming part of crural impression. Metacoxae narrowly separated, extending laterally almost as far as elytra, but separated from them by metepimeron; plates well developed and complete. Metendosternite with short stem slightly longer than wide, long, straight lateral arms, no laminae and short anterior process with widely separated. Hind wing (Fig. 29F) with moderately long apical field containing two vaguely indicated radial extensions; radial bar with long, weak bending zone; radial cell moderately well developed, about as long as wide, with basal edge oblique and slightly curved; cross-vein r3 short and strongly oblique, well separated from r4, which is complete and curved; R-M loop broadly rounded, basal portion of RP short and curved and apical extension extending to wing margin; MP_{1+2} with long, weak bending zone; medial spur short, ending well before wing margin; medial field with five free veins and no medial fleck; MP_{3+4} with sub-basal cross-vein and basal spur; CuA1 meeting MP_{3+4}; wedge cell moderately well developed and obliquely truncate; anal lobe well developed with notch-like embayment. Legs short, stout and modified for compaction; femur basally inflated, with linear groove for receiving tibia, which is also impressed for receiving tarsi; meso- and metatibiae slightly expanded at apex; protibiae strongly expanded with broad excavation at apical third; outer edges of all tibiae with short socketed spines. Tarsi 5-5-5; tarsomeres simple; pretarsal claws simple; empodium well developed and bisetose.

Abdomen with five free ventrites; ventrite 1 slightly longer than 2, with pair of crural impressions which extend onto ventrite 2; intercoxal process broadly rounded. Functional spiracles on abdominal segments I–VII. Anterior edge of sternite VIII in male with short strut (spiculum relictum). Anterior edge of segment IX in male with narrowly rounded subgenital plate (without spiculum gastrale); tergite IX deeply emarginate; tergite X free. Aedeagus symmetrical, trilobate, with basally rounded phallobase, simple, articulated parameres and penis with paired basal struts. Sternite VIII in female with long, slender spiculum ventrale. Ovipositor ~5 times as long as wide, relatively lightly sclerotised, with distinct proctigeral and paraproctal baculi; paraprocts longer than coxites, which are divided into broad basal lobe with obliquely transverse baculum and narrower apical lobe; styli well developed and terminal. [Reichardt 1976; Endrödy-Younga 1991b; Lawrence *et al.* 1999b; Háva 2000; Ivie 2002c; Leschen and Beutel 2010b].

Larva (Fig. 35J)

Body elongate and fusiform, somewhat wider at middle; upper surfaces distinctly sclerotised and pigmented, granulate and tuberculate; undersurfaces lighter; vestiture of dorsal surfaces consisting of very short, fine setae, but lateral portions head, thorax and abdomen with much longer setae, sometimes forming tufts.

Head (Fig. 38J) well developed, prognathous and protracted, transverse and somewhat flattened, with sides distinctly rounded. Epicranial stem absent; frontal arms U-shaped or weakly lyriform and widely separated at base; median endocarina absent. Stemmata five on each side, well separated. Frontoclypeal suture absent or only vaguely indicated. Labrum completely fused to clypeus; clypeolabrum produced forward with sides converging apically and apex slightly emarginate. Antennae ~0.30 times as long as head width, 3-segmented, inserted on raised antennifer; basal antennomere short and slightly transverse, 2 long and narrow with sensorium about half as long as antennomere 3, which is short and narrow. Mandibles (Fig. 41F) symmetrical, broad at base and narrow at apex, unidentate, with bifurcate retinaculum and accessory ventral process; mola well developed; prostheca consisting of dense fringe of hairs. Ventral mouthparts retracted, with well developed maxillary articulating area; stipites elongate; galea and lacinia fixed; galea blunt and apically setose; lacinia narrowed apically with tridentate uncus; maxillary palp 3-segmented with segment-like palpifer. Labium consisting of prementum, mentum and submentum; ligula bilobed; labial palps 2-segmented and widely separated. Hypopharyngeal sclerome a transverse bar. Hypostomal rods absent. Ventral epicranial ridges well developed. Gular region very short and membranous.

Protergum without distinct plate or rows of asperities, but more or less irregularly covered with small tubercles. Meso- and metaterga and abdominal terga I–VII each with anterior row of tubercles, in addition to irregularly arranged tubercles; abdominal terga I–VII also with small, lateral projections. Legs moderately well developed, 5-segmented including claw-like pretarsus bearing two minute setae lying side by side; coxae separated by one diameter or less. Tergum VIII ~4 times as long as VII, tapering posteriorly to form terminal process (Fig. 38L); lateral edges of tergum extending onto ventral surface and meeting at midline to form a median suture, incomplete posteriorly where the tergum articulates with paired spiracular sclerites and anteriorly where it meets the reduced sternum and encloses segments IX and X, which are reduced to a pair of ventrally oriented anal pads. Spiracles annular–biforous; thoracic spiracles ventral, not located on tubular process; abdominal spiracles on segments I–VII located on dorsolateral processes, those on VIII at end of terminal process. [Hayes and Chu 1946; Crowson 1959; Costa *et al.* 1986; Lawrence 1991b; Beutel 1996; Lawrence *et al.* 1999a; Ivie 2002c; Leschen and Beutel 2010b].

Classification and World Distribution

Nosodendridae contains the single genus *Nosodendron* Latreille with ~60 species of worldwide occurrence, including New Zealand. [Reichardt 1976; Endrödy-Younga 1991b; Háva 2000; Leschen and Beutel 2010b].

Australian Fauna

There are at least two described species of Australian *Nosodendron*: *N. australicum* Lea and *N. interruptum* Lea, both occurring in QLD. Nosodendrids are oval, convex, glabrous and shining, black beetles, capable of folding their legs into ventral cavities, as in some

Byrrhidae, but differing from Byrrhidae in having a prognathous head, well marked 3-segmented, tomentose antennal club and maxillae and prementum concealed by an enlarged mental plate. Larvae are elongate, fusiform, somewhat flattened and dorsally granulate and tuberculate, with tergum VIII forming a tapered, terminal process with spiracles at the apex. The frontal arms are distant at base, antennomere 3 is reduced, mandibles have complex armature and vestiture, and maxillae are provided with comb-hairs. Both larvae and adults live in slime fluxes, which develop when tree wounds are attacked by certain fungi, and the spiracular siphon enables the larva to obtain air when immersed in the fluid medium. [Osborne and Kulhavy 1975; Sokoloff 1959].

47. JACOBSONIIDAE

Adult (Fig. 81D)

Length 0.65–1.2 mm (*Derolathrus* Sharp) or 1.7–2.1 mm (*Sarothrias* Grouvelle). Body ~2.1–3 times as long as wide; lightly to heavily sclerotised, convex to slightly flattened; colour yellow-brown to reddish-brown in *Derolathrus*, black in *Sarothrias*; surfaces in *Derolathrus* glabrous or clothed with short, those in *Sarothrias* glabrous with a few scattered setae, but with a dense pruinosity (formed by densely packed micropapillae) forming a pattern on head and prothorax, in some elytral striae and extensively covering legs, antennomeres and much of ventral surface; apical rims of most antennomeres with additional stout, flattened setae.

Head prognathous, abruptly constricted behind eyes to form short temples in *Derolathrus*. Occipital with low transverse carina in *Sarothrias*. Eyes usually well developed, protuberant, circular, entire, coarsely facetted, without interfacetal setae; Antennal insertions concealed from above; subantennal groove absent. Frontoclypeal region in *Sarothrias* with transverse groove and anterior declination. Frontoclypeal suture absent; clypeus broadly rounded at apex. Labrum strongly transverse with broadly rounded anterior edge. Antennae 11-segmented, with weak, loose, 3-segmented club in *Sarothrias*, but most antennomeres laterally expanded, so that club segments are not clearly distinguishable; antennae in *Derolathrus* 10-segmented with 1-segmented club. Mandible short and broad, bidentate, with well developed mola and with prostheca. Maxilla with galea bearing an expanded flattened lobe composed of microtrichial fringes and lacinia with uncus; maxillary palps aciculate, 4-segmented, with palpomere 2 inflated and 3 and 4 much narrower and cylindrical in *Sarothrias*, 3-segmented, with 2 inflated and 3 narrow and cylindrical in *Derolathrus*. Mentum well developed, about as long as wide, apically convex; ligula bilobed; labial palps reduced, 3-segmented with apical palpomere cylindrical. Gulamentum with distinct transverse groove. Corpotentorium straight, broad or narrow; laminatentoria well developed in *Sarothrias*. Cervical sclerites absent.

Pronotum ~1.14–1.55 times as long as wide, widest at middle or anteriorly and not narrower at base than combined elytral bases; sides subparallel and slightly sinuate in *Sarothrias*, weakly curved in *Derolathrus*. Lateral pronotal carinae absent, anterior angles not produced; posterior angles right to rounded; posterior edge straight or weakly rounded; disc with complete median groove and a few weak anterolateral impressions filled with microvilli in *Sarothrias*, simple, with broad, shallow impression, incomplete median, longitudinal groove or several longitudinal grooves and carinae in *Derolathrus*. Prosternum in front of coxae slightly to much longer than mid length of procoxal cavity. Prosternal process complete, not curved dorsally, more or less parallel-sided, interrupted at middle in *Sarothrias*, narrowly rounded at apex. Notosternal sutures absent. Procoxae more or less globular, projecting below prosternum in *Sarothrias*, with long, concealed lateral extensions; trochantins concealed; endopleuron free. Procoxal cavities usually narrowly separated (contiguous in *Sarothrias*), open externally in *Derolathrus*, broadly closed externally in *Sarothrias* (with notal projections fused to prosternal process), open internally.

Scutellar shield absent. Elytra ~1.5–2.2 times as long as combined width and 1.75–3 times as long as pronotum, sides slightly diverging to middle, then converging to conjointly rounded apices; each elytron with nine striae (some incomplete or joined before apex) and no scutellary striole in *Sarothrias*, with sutural stria and one or more additional striae in *Derolathrus*; epipleura incomplete or absent. Mesoventrite partly or completely fused to mesanepisterna, without procoxal rests, with broad prothoracic rest in *Sarothrias*; mesoventral cavity absent. Mesoventral process with lateral projections forming secondary mesal articulations with coxae. Mesocoxae subglobular, slightly projecting in *Sarothrias*, with concealed trochantins. Mesocoxal cavities narrowly separated in *Derolathrus*, moderately widely separated in *Sarothrias*, laterally closed by meeting of mesoventrite and metaventrite. Mesometaventral junction usually simple; meso- and metaventrites fused together in *Sarothrias*. Metaventrite markedly elongate, at least 2.5 times as long as mesoventrite; more or less flattened and simple in *Derolathrus*, in *Sarothrias* with pair of long, slightly curved carinae which extend posteriorly from mesocoxal cavities to posterior end of ventrite, flanking a large, shallow impression and lined with strips of micropapillae; discrimen and transverse (metakatepisternal) suture absent; exposed portion of metanepisternum very long and narrow, almost absent in *Sarothrias*. Metacoxae slightly transverse or subglobular and separated by more than their width, plates absent. Metendosternite without stalk and with long, narrow lateral arms, laminae and anterior process absent and anterior tendons not apparent. Hind wing narrowly elongate, with apical field more than half of total wing length, usually with one or two vaguely indicated radial extensions, radial cell, cross-veins r3 and r4, R-M loop, basal portion of RP and medial spur absent; medial field highly reduced or absent, at most with single vein; anal lobe absent; hind wing margin lined with long hairs. Femoral attachment of mid trochanter transverse to slightly oblique. Legs slender; tibiae slightly expanded at apex in *Sarothrias*; tibial spurs short. Tarsi 3-3-3; tarsomeres simple; pretarsal claws simple; empodium not obvious.

Abdomen at least slightly shorter than metaventrite, with five free ventrites; ventrite 1 longer than 2, fused to metathorax and with large setose impression in *Sarothrias*; intercoxal process truncate. Functional spiracles on abdominal segments I–VIII located in pleural membrane or at edges of tergite. Anterior edge of sternite VIII in male with median strut (spiculum relictum) in *Derolathrus*. Anterior edge of segment IX in male with spiculum gastrale; tergite IX truncate; tergite X free. Aedeagus symmetrical, trilobate, with phallobase reduced and parameres fused together forming a tube containing penis, which may or may not have short basal struts. Sternite VIII in female with long spiculum ventrale in Derolathrus. Ovipositor ~2 times as long as wide in *Sarothrias*, somewhat shorter and broader in *Derolathrus*, relatively lightly sclerotised, without distinct baculi; coxites undivided; styli well developed and terminal in *Derolathrus*, absent in *Sarothrias*. [Crowson 1959; Löbl and Burckhardt 1988; Burckhardt and Löbl 1990; Lawrence *et al.* 1999b; Philips *et al.* 2002; Lawrence and Leschen 2010a].

Larva

Body elongate and more or less parallel-sided, subcylindrical, more heavily pigmented dorsally than ventrally; vestiture of simple setae.

Head prognathous and protracted, transverse and slightly flattened, with sides distinctly rounded. Epicranial stem absent; frontal arms lyriform and well separated at base; median endocarina absent. Stemmata absent. Frontoclypeal suture distinct. Labrum free. Antennae very short, 3-segmented; sensorium more than half as long as antennomere 3. Mandibles symmetrical, broad at base and narrow at apex, bidentate, without accessory ventral process; incisor edge serrate; mola well developed, finely ridged; prostheca an acute hyaline process. Ventral mouthparts retracted, with well developed maxillary articulating area; stipites elongate; galea and lacinia fixed and more or less fused at base; galea with multilobed or fringed process; lacinia with unidentate uncus; maxillary palp 3-segmented. Labium consisting of prementum and postmentum; ligula usually bilobed, shorter than labial palps, which are 2-segmented and well separated. Hypopharyngeal sclerome molar-like. Hypostomal rods and ventral epicranial ridges absent. Gula wider than long.

Thoracic and abdominal terga with at least weakly sclerotised plates. Legs moderately well developed, 5-segmented including claw-like pretarsus bearing one seta; mesocoxae moderately separated by less than one coxal diameter. Tergum IX slightly shorter than VIII, not extending onto ventral surface, with simple paired urogomphi. Sternum IX exposed, simple, not enclosed by sternum VIII. Segment X short and cylindrical, without paired pygopods. Spiracles annular, not placed at ends of spiracular tubes. [Crowson 1959; Sen Gupta 1979; Lawrence *et al.* 1999a; Lawrence and Leschen 2010a].

Classification and World Distribution

The family Jacobsoniidae currently includes *Saphophagus minutus* Sharp from New Zealand, 13 species of *Sarothrias* Grouvelle distributed from India and South-east Asia to the Seychelles, Fiji, New Caledonia, Indonesia, New Guinea and Australia, and *Derolathrus* with eight described species from Florida, Barbados, the West Indies, Hawaii, the Canary Islands, Madeira, Sri Lanka, Mauritius and Fiji and numerous undescribed species from Australia, New Zealand, South America and Africa. [Ślipiński 1986; Háva and Löbl 2005; Peck 2010].

Australian Fauna

The Australian jacobsoniid fauna consists of *Sarothrias lawrencei* Löbl and Burckhardt from northern QLD and several undescribed species of *Derolathrus* or related genera from Christmas Island, Norfolk Island, Lord Howe Island, QLD, SA and the ACT

Jacobsoniids are elongate to narrowly elongate beetles with the metaventrite as long as or longer than all five abdominal ventrites combined, maxillary palps aciculate, and the basal tarsomeres reduced, so that tarsi may appear to be 2-segmented. Larvae are elongate and subcylindrical and sclerotised dorsally, with short antennae and short, fixed urogomphi. *Sarothrias* (Fig. 81D) are particularly distinctive in having scale-like setae on the antennae and tracts of silvery tomentum covering much of the body surface. Derolathrines are minute (less than 1 mm) and yellowish-brown, with the head abruptly constricted behind the eyes to form short temples and the antennae 10- or 11-ssegmented with a 1- or 2-segmented club. Habits are unknown, but *Sarothrias* have been collected in leaf litter and rotten wood, while *Derolathrus* have been found in leaf litter extractions, rotten palm wood, bat guano and fungal fruiting bodies.

BOSTRICHIFORMIA: BOSTRICHOIDEA

The series Bostrichiformia, as defined by Crowson (1955), was a paraphyletic assemblage of polyphagan families belonging to the cucujiform lineage (Crowson 1960, Lawrence and Newton 1982) but lacking the apomorphic features of the series Cucujiformia. The term is used here to include only the Bostrichoidea, as delimited by Lawrence and Newton (1982) and including the families Dermestidae, Endecatomidae, Bostrichidae and Ptinidae. Dermestidae was included in this group based on the presence of a modified type of cryptonephridism (Saini 1964) in which the Malpighian tubules are attached to the hind gut in a single bundle, an aedeagal structure somewhat intermediate between the trilobate type and that found in cucujiforms, and lack of a basal mandibular mola in the larva. Ivie (1985, 2002c) added Nosodendridae (redefined to include the exotic dermestid genus *Orphilus* Erichson) to this group, assuming a secondary loss of cryptonephridism in that family. Endecatomidae, the only bostrichoid family not occurring in Australia, is often placed in Bostrichidae (Ivie 2002d; Lawrence 2010d; Liu and Schönitzer 2011). [Lawrence *et al.* 2010f].

48. DERMESTIDAE

Adult (Figs 6D, 63F–I, L–M)

Length 1–13 mm. Body ~1.4–3.5 times as long as wide; almost always broadly ovate to oblong (elongate and more or less parallel-sided in *Derbyana*); well sclerotised; moderately to strongly convex dorsally and slightly convex to flattened ventrally; colour often uniformly black or brown, sometimes with lighter colour forming patterns; upper surfaces usually densely clothed with fine, decumbent hairs or setae which are thickened and flattened or scale-like and which may also may form a pattern; most Trinodinae with stiff, erect, dehiscent bristles.

Head prognathous, not to strongly declined, abruptly constricted behind eyes to form neck and temples in *Orphilodes*. Occipital region relatively short, with transverse occipital ridge in *Orphilodes*; usually without median endocarina and with median ocellus (Fig. 28F) (median endocarina present in *Dermestes* Linnaeus and *Derbyana* and ocellus absent in *Dermestes*, *Derbyana* and *Trichelodes*). Eyes entire, strongly protuberant or not, finely or coarsely facetted, without interfacetal setae. Antennal insertions widely to narrowly separated and exposed from above; subantennal grooves usually well developed (weakly developed or absent in *Dermestes*, *Anthrenus* and *Thylodrias*), extending behind eyes in *Orphilodes*. Frontoclypeal region strongly deflexed anteriorly in *Trichelodes*. Frontoclypeal suture usually absent (weakly impressed in Dermestinae); clypeal apex straight to slightly concave. Labrum slightly to strongly transverse. Antennae usually 11-segmented (9-segmented in *Thylodrias*), usually clavate or capitate with 1- to 6-segmented club (filiform in *Thylodrias*, pectinate in some species of *Trogoderma*). Mandibles short and broad, unidentate or bidentate; mola absent except in *Orphilodes*. Maxilla with distinct galea and lacinia, the latter with spine-like uncus in Dermestinae and most Trinodinae; apical maxillary and labial palpomeres cylindrical to fusiform; ligula not to deeply emarginate. Corpotentorium usually narrow to moderately broad (absent in most Trinodinae). Cervical sclerites present.

Pronotum ~0.3–1.2 times as long as wide, almost always widest at base and as wide as combined elytral bases with sides moderately to strongly curved and converging anteriorly (sides straight and subparallel in *Derbyana* and straight and converging in most Trinodinae). Lateral pronotal carinae usually complete (incomplete or absent in some Trinodinae), simple, with or without distinct bead; anterior angles usually not produced (produced and broadly rounded in Derbyana); posterior angles usually broadly rounded to right (somewhat acute in *Attagenus* and *Tichelodes*); posterior edge relatively straight to deeply bisinuate, forming mesal lobe; disc with or without basal impressions, with sublateral carinae in *Evorinea* Beal and *Trichelodes*). Prosternum in front of coxae (Fig. 13C) usually slightly to much shorter than mid length of procoxal cavity (somewhat longer in *Thylodrias*), surface usually flat to slightly convex, rarely concave; anterior edge forming short chin piece in some Megatominae and Trinodinae. Prosternal process usually complete and narrowly to broadly rounded at apex (incomplete or absent in Dermestinae and most Trinodinae; apically expanded to form secondary coxal articulation in *Orphilodes*, Fig. 20L). Notosternal sutures absent in *Thylodrias*. Hypomera sometimes with broad shallow impressions or discrete antennal cavities. Procoxae usually not or only slightly projecting below prosternum (strongly projecting in *Thylodrias*), without concealed lateral extensions; trochantin almost always at least partly exposed. Procoxal cavities slightly to strongly transverse and contiguous or narrowly separated (widely separated in *Evorinea*), broadly open externally, internally closed in *Dermestes*, *Attegenus* and *Orphilodes*.

Scutellar shield usually abruptly elevated (not elevated in *Derbyana* and male *Thylodrias*), posteriorly acute to rounded or truncate, scutellum absent in female *Thylodrias*. Elytra ~1–2.75 times as long as combined width and 1.7–5.35 times as long as pronotum, usually widest at middle and converging posteriorly to conjointly rounded apices, independently rounded and separated by gap in *Thylodrias* males; irregularly punctate without scutellary striole; epipleura usually incomplete or absent; elytra absent in female *Thylodrias*. Mesoventrite usually with anterior edge on same plane as metaventrite (on different plane in *Orphilodes*), anterior procoxal rests almost always well developed and moderately to strongly inclined or vertical (absent in *Thylodrias*); mesoventral cavity usually well developed, absent in Dermestinae, *Orphilodes* and Trinodinae. Mesoventral process flat or somewhat elevated, broadly rounded, truncate or concave at apex. Mesocoxae usually not projecting (strongly projecting in *Thylodrias*), with exposed trochantins. Mesocoxal cavities usually circular to slightly transverse, narrowly to widely separated and partly closed laterally by mesanepisternum and mesepimeron (cavities strongly oblique and barely impressed in *Thylodrias*). Mesometaventral junction simple or complex; secondary mesocoxal articulation formed by metaventral projection and coxal impression in Dermestinae (Fig. 25D) or by mesocoxal projection and metaventral impression in *Orphilodes*. Metaventrite with discrimen usually moderately to very long (absent in *Thylodrias*) and transverse (metakatepisternal) suture absent; exposed portion of metanepisternum short and broad to moderately elongate. Metacoxae usually distinctly transverse, contiguous to narrowly separated, and not extending laterally to meet elytra (widely separated in Evorinea and *Thylodrias*, subconical in *Thylodrias*); plates usually well developed and complete (Fig. 63F) (weakly developed laterally in Dermestinae, weakly developed or absent in *Thylodrias*). Metendosternite with moderately to very long lateral arms and no laminae; ventrolateral processes sometimes present; anterior process short with widely separated tendons or absent. Hind wing usually well developed apical field and at least one radial extension; radial bar with or without weak bending zone; radial usually small and less than twice as long as wide, with inner posterobasal angle acute; cross-vein r3 usually absent, moderately strongly oblique in *Orphilodes* and *Attagenus*, slightly oblique in Dermestinae; R-M loop broadly rounded, basal portion of RP short to very long with apical extension usually extending to wing margin; MP_{1+2} with or without weak bending zone; medial spur usually extending to wing margin; medial field with six free veins in Attagenus (Fig. 29K), five in Dermestinae, four in *Orphilodes* and some Megatominae and three of fewer in Trinodinae and most Megatominae; wedge usually absent (well developed in Dermestinae); anal embayment usually deep and notch-like (absent in Dermestinae); wings absent in *Thylodrias* females. Femoral attachment of mid trochanter usually transverse or slightly oblique (more strongly oblique in *Attagenus*). Outer edges of tibiae often lined with spines or teeth (simple in most Trinodinae and Megatominae); tibial spurs reduced or absent in Trinodinae and some Megatominae. Tarsi 5-5-5; tarsomeres simple; pretarsal claws simple.

Abdomen usually with five free ventrites (first two ventrites connate in Megatominae;, eight free ventrites in female *Thylodrias*); intercoxal process acute to broadly rounded; post-coxal lines sometimes present in Megatominae; sternite III with paired oblique grooves arising from deep pits in Dermestinae, with medial setal patch in *Thylodrias* males; sternites V and VI with setal patches in males of some *Dermestes*. Functional spiracles on abdominal segments I–VIII. Anterior edge of sternite VIII in male with median strut (spiculum relictum) in some Megatominae. Anterior edge of segment IX in male with broadly to narrowly rounded subgenital plate, without spiculum gastrale; tergite IX slightly to moderately emarginate and free from tergite X in *Orphilodes*, Dermestinae and some Megatominae, partly or entirely fused to X in Attageninae, Trinodinae and most Megatominae. Aedeagus symmetrical, trilobate, but usually with phallobase overlapping parameres, which are free from one another except at base, where they may be fused together and to base of penis (Fig. 31C–D), with additional transverse bridge joining parameres ventrally (Fig. 29G) in *Orphilodes*, Trinodinae and

Megatominae; anterior edge of penis with or without paired struts. Sternite VIII in female with slender spiculum ventrale. Ovipositor usually moderately elongate and lightly sclerotised except for proctigeral, paraproctal and transverse coxital baculi; paraprocts usually slightly longer than coxites, which are divided into broad basal and narrow apical lobes; styli usually well developed and terminal. [Armstrong 1945; Hinton 1945a; Peacock 1978, 1993; Bousquet 1990; Lawrence *et al.* 1999b; Roach 2000; Vijay Veer and Rao 2000; Kingsolver 2002a; Beal 2003; Háva 2004; Lawrence and Ślipiński 2005, 2010a].

Larva (Fig. 63J–K)

Body elongate to oblong or ovate, parallel-sided or narrowed posteriorly; dorsal surfaces usually more heavily pigmented than ventral ones.

Vestiture usually including one or more types of modified setae: barbed spicisetae (Fig. 46G) in *Dermestes*, Trinodinae and Megatominae, ribbed, scale-like setae in *Attagenus*, hastisetae (Fig. 46H), with apical recurved barbs, in Megatominae.

Head moderately to strongly declined (hypognathous), globular to slightly flattened with rounded sides. Epicranial stem moderately long; frontal arms V-or U-shaped with bases contiguous; median endocarina usually absent (short in *Dermestes*). Stemmata six on each side in Dermestes, Megatominae and most Trinodinae, four in *Attagenus*, three in *Thylodrias* and one in *Orphilodes*. Frontoclypeal suture usually present. Labrum free. Antennae very short to moderately long, 3-segmented, sensorium palpiform and almost or quite as long as apical antennomere in *Orphilodes* and Trinodinae. Mandibles symmetrical, broad and stout or more or less wedge-like, usually unidentate (bidentate in most Trinodinae, tridentate in *Dermestes*), without accessory ventral process; lightly pigmented basally and more heavily so apically in Attagenus, Trinodinae and Megatominae; incisor edge simple; mesal surface of mandibular base often simple, with fixed, rigid hyaline process in *Dermestes*, *Attagenus* and some Megatominae, with basal penicillus in *Dermestes*, *Attagenus* and *Orphilus*; mola usually absent, reduced in *Orphilus*. Ventral mouthparts retracted, with reduced maxillary articulating area; stipites elongate; galea and lacinia fixed; galea rounded or truncate and setose; lacinia falciform; maxillary palp 3- or 4-segmented. Labium consisting of prementum, mentum and submentum; ligula absent; labial palps 2-segmented, usually widely separated. Hypopharyngeal sclerome absent. Hypostomal rods and ventral epicranial ridges absent. Gular region very short.

Thoracic terga with distinct plates, protergum more or less vertically explanate, with deep pit on each side in *Orphilodes*. Legs moderately well developed, 5-segmented including claw-like pretarsus bearing two minute setae lying side by side; coxae narrowly to widely separated. Segment IX excluding appendages shorter than VIII, not forming articulated plate, usually not or only slightly extending onto ventral surface (completely ventral in Trinodinae), usually without urogomphi (with acute, posteroventrally curved urogomphi in *Dermestes*). Sternite IX usually somewhat reduced and simple (well developed in *Dermestes*); segment X usually reduced and membranous, without paired pygopods (well developed, sclerotised and cylindrical in *Dermestes*); anal region posteriorly to ventrally oriented. Spiracles annular, those on segment VIII enlarged in *Orphilodes* and Trinodinae. [Böving and Craighead 1931; Rees 1943; Korschefsky 1944; Hinton 1945a; Beal 1985, 1991; Lawrence *et al.* 1999a; Vijay Veer and Rao 2000; Kiselyova and McHugh 2006; Kiselyova 2008; Lawrence and Ślipiński 2005, 2010a].

Classification and World Distribution

The Dermestidae includes ~1000 species in 50 genera distributed throughout the world but most abundant and diverse in arid and semiarid regions. The family is divided into five subfamilies, four of which occur in Australia. The exotic subfamily Thorictinae occur primarily in southern Europe, Asia Minor, Central Asia and Africa; *Thorictodes heydeni* Reitter is more widely distributed but has not yet been recorded from Australia (Emden 1924b; Anderson 1949; John and Andreae 1967). Dermestinae and Attageninae are almost entirely exotic (**see** below) with their greatest diversity in the more arid parts of the Holarctic and Afrotropical regions. [Mroczkowski 1968; Mroczkowski and Ślipiński 1997; Zhantiev 2000; Háva 2003; Lawrence and Ślipiński 2005; Kiselyova and McHugh 2006].

Australian Fauna

The Australian fauna includes a single native species of **Dermestinae**, *Derbyana matthewsi* Lawrence and Ślipiński (Fig. 63H), which is known from a single specimen collected at Derby, WA in 1892. All species of *Dermestes* (Fig. 63I) and most *Attagenus* species (**Attageninae**) in Australia have been introduced from elsewhere. The subfamily **Orphilinae** includes the endemic genus *Orphilodes*, which has also been collected in Sabah. **Trinodinae** includes the introduced *Thylodrias contractus* Motschulsky (Fig. 63L–M) plus two endemic genera, and the remaining Australian dermestids are placed in **Megatominae**. Dermestids are oblong to broadly ovate beetles almost always with a deflexed head and clothed with erect or decumbent hairs or scales, sometimes forming a pattern. Except for the Dermestinae, Thorictinae and *Trichelodes delicatula* Carter, all dermestids have a median ocellus on the head (Leschen and Beutel 2004). Larvae are elongate and cylindrical to ovate and fusiform, with distinct sclerotised terga, and are densely clothed with long hairs, some of which may be barbed (spicisetae) or spear-headed (hastisetae), serving as a defence mechanism by detaching and entangling potential predators. (Nutting and Spangler 1969). The lacinia terminates in a sclerotised spur or spine, the mandible usually bears a membranous lobe and setal brush at the base, and the urogomphi are absent, except in *Dermestes* (and in the exotic genus *Orphilus* Erichson and the Thorictinae). Dermestids usually feed as larvae on dry material of animal origin, such as insect remains, dry carcasses, hair or feathers, and they may occur in the nests of mammals, birds or social insects. Adults of some species (*Anthrenus*, *Trogoderma*) are commonly found on flowers and feed on nectar or pollen, but others feed on the larval food source or not at all. The family includes several cosmopolitan stored products pests, such as *Dermestes* species, which attack hides, furs and skins, as well as some foodstuffs, species of *Anthrenus* and *Attagenus*, which may damage carpets, blankets or other woollen fabrics, and various dermestids, particularly *Anthrenus verbasci* (Linnaeus) (Fig. 63G), which damage dried insect specimens. Only *Trogoderma granarium* Everts is known to feed normally on cereal products, but it has not yet become established in Australia. Two other introduced pests known to attack insect collections have been found in the ACT: the parthenogenetic *Reesa vespulae* (Milliron), which has also been introduced into New Zealand (Waller and Watt 1979), and the unusual *Thylodrias contractus*, females of which are larviform. Megatomines of note include *Anthrenocerus australis* (Hope), which has been reported as a household pest, the myrmecophilus *Myrmeanthrenus frontalis* Armstrong, and *Thaumaglossa nigricans* (Macleay), which lives in mantid egg cases. [Armstrong 1945; Hinton 1945a; Mroczkowski 1968; Peacock 1978; Lawrence and Ślipiński 2005].

Australian Subfamilies and Genera

Dermestinae: *Dermestes* Linnaeus, *Derbyana* Lawrence and Ślipiński.

Attageninae: *Attagenus* Latreille.

Orphilinae: *Orphilodes* Lawrence and Ślipiński.

Trinodinae:, *Evorinea* Beal, *Thylodrias* Motschulsky, *Trichelodes* Carter.

Megatominae: *Adelaidia* Blackburn, *Anthrenocerus* Arrow, *Anthrenus* Müller, *Myrmeanthrenus* Armstrong, *Neoanthrenus* Armstrong, *Orphinus* Motschulsky, *Reesa* Beal, *Thaumaglossa* Redtenbacher, *Trogoderma* Berthold.

49. BOSTRICHIDAE

Adult (Figs 6B, 81B, E, J)

Length 2–20 mm. Body ~2.45–3.65 times as long as wide; more or less parallel-sided and usually cylindrical (somewhat flattened in Lyctinae), well sclerotised; colour usually black, reddish-brown or a combination of the two, upper surfaces almost always either glabrous or clothed with fine hairs, with erect, scale-like setae in *Minthea rugicollis* (Walker).

Head moderately to strongly declined in Dinoderinae and Bostrichinae, sometimes abruptly constricted immediately behind eyes to form very short temples. Occipital region moderately long, posterior edge above occipital foramen slightly to strongly produced posteriorly and usually angulate to subacute, with median endocarina. Eyes strongly protuberant, entire, more or less circular or slightly oval and finely facetted without interfacetal setae. Antennal insertions concealed from above; subantennal grooves absent. Frontoclypeal region not or only slightly declined. Frontoclypeal suture present; anterior edge of clypeus truncate to emarginate. Labrum free and visible, usually transverse and subtruncate (convex in some Dinoderinae, emarginate in some Lyctinae). Antennae usually 10-segmented (11-segmented in Lyctinae, 9-segmented in some Bostrichinae), usually with loose and often asymmetrical, 3-segmented club (with compact, 2-segmented club in Lyctinae). Mandibles short and broad, usually unidentate or bidentate, mola and prostheca usually well developed, sometimes reduced. Maxilla with distinct galea and lacinia, the latter without uncus; palps small, apical maxillary and labial palpomeres cylindrical to fusiform; ligula usually undivided. Gular sutures narrowly separated or partly fused together. Corpotentorium usually broad (narrow in Dinoderinae and some Lyctinae). Cervical sclerites absent.

Pronotum ~0.7–1.2 times as long as wide, usually widest at middle (widest anteriorly in Lyctinae) and as wide as combined elytral bases, with sides straight to weakly curved. Lateral pronotal carinae usually incomplete or absent (complete in most Lyctinae), simple, with or without distinct bead; anterior angles usually not produced (rarely produced and acute in Lyctinae); posterior angles right to broadly rounded; posterior edge straight to broadly rounded; disc without carinae or basal impressions. Prosternum in front of coxae usually shorter than mid length of procoxal cavity (longer in some Lyctinae), usually flat to moderately convex, rarely concave. Prosternal process usually incomplete and narrowly rounded to acute at apex (complete, flattened, apically expanded and truncate in Lyctinae). Notosternal sutures present or absent. Procoxae usually projecting well below prosternum, without or with very short concealed extensions (not projecting and with long concealed extensions in some Lyctinae); trochantins concealed. Procoxal cavities more or less circular and usually contiguous (narrowly to widely separated in Lyctinae), usually open externally (narrowly to broadly closed in Lyctinae), without narrow lateral extensions; internally closed.

Scutellar shield usually abruptly elevated, posteriorly broadly rounded to truncate. Elytra ~1.75–2.7 times as long as combined width and 1.9–3.25 times as long as pronotum, parallel-sided; usually irregularly punctate (with 10 or 11 puncture rows and a scutellary striole in some Dinoderinae and Lyctinae; elytral apex sometimes with distinct, concave declivity; epipleura usually incomplete. Mesanepisterna may be narrowly separated or contiguous at midline in Bostrichinae; anterior edge of mesoventrite usually on different plane than metaventrite (on same plane in Lyctinae); without anterior procoxal rests, with broad prothoracic rest forming part of peduncle in Lyctinae (in Dinoderinae and Bostrichinae peduncle formed entirely by mesanepisterna). Mesoventral process narrowly to broadly rounded or truncate at apex. Mesocoxae with concealed trochantins. Mesocoxal cavities circular, usually narrowly separated (contiguous in Dinoderinae, moderately to widely separated in Lyctinae), usually closed laterally by meeting of meso- and metaventrite (occasionally partly closed by mesepimeron). Mesometaventral junction usually complex (a straight line in some Lyctinae); Metaventrite with discrimen usually moderately to very long (short or absent in some Dinoderinae and Lyctinae) and transverse (metakatepisternal) suture absent; exposed portion of metanepisternum moderately to very elongate. Metacoxae usually contiguous to narrowly separated (widely separated in some Lyctinae), extending laterally to meet elytra in some Lyctinae; plates absent. Metendosternite with moderately to very long lateral arms and no laminae; anterior process short to moderately long with narrowly to widely separated tendons. Hind wing with moderately long apical field and distinct extension of RP; radial bar usually strongly diverging from wing margin, with well developed bending zone; radial usually small and less than twice as long as wide (lumen sometimes highly reduced or absent), with inner edge oblique and inner posterobasal angle right or acute; cross-vein r3 usually moderately long, strongly oblique and fused at base with r4, which is long, sinuate or angulate and complete; R-M loop broadly rounded, basal portion of RP usually short; MP1+2 with or without weak bending zone; medial field usually with four free veins (three or fewer in Dinoderinae and Lyctinae); wedge usually well developed; anal embayment deep and notch-like in Dinoderinae and Lyctinae. Femoral attachment of mid trochanter usually transverse or slightly oblique (more strongly oblique in some Lyctinae). Outer edges of protibiae and mesotibiae often lined with spines in Dinoderinae and Bostrichinae and outer apical angles of same tibiae usually produced to form tooth or spine in these groups and in Lyctinae; tibial spurs usually double, but protibia with single large curved spur in Dinoderinae, Lyctinae and Bostrichinae. Tarsi 5-5-5; tarsomeres usually simple, sometimes with weak tarsal lobes; pretarsal claws simple; empodium often well developed.

Abdomen almost always with five free ventrites (first two ventrites rarely connate in Lyctinae); ventrite 1 much longer than 2 in some Lyctinae; intercoxal process usually acute to narrowly rounded (truncate to slightly emarginate in some Lyctinae); sternites VI and VII highly modified in females of some Bostrichinae. Functional spiracles on abdominal segments I–VIII. Anterior edge of sternite VIII in male with median strut (spiculum relictum) in some Lyctinae. Anterior edge of segment IX in male with truncate to subacute subgenital plate, without spiculum gastrale; tergite IX more or less fused to X. Aedeagus symmetrical, of a modified trilobate type; phallobase often anteriorly emarginate or with paired struts, posteriorly overlapping parameres, which may be free from or fused with phallobase, usually joined together near base and with anteriorly projecting parameral struts attached to base of penis; anterior edge of penis with paired struts. Sternite VIII in female with or without spiculum ventrale. Ovipositor usually very long and slender, lightly sclerotised except for paraproctal baculi; paraprocts usually much longer than coxites, which are undivided and setose; styli well developed and subapical; ovipositor shorter and broader in some Bostrichinae and may be heavily sclerotised without styli. [Lesne 1896, 1897, 1901, 1906, 1909, 1911, 1921; Crowson 1961a; Vrydagh 1958; Lawrence *et al.* 1999b; Ivie 2002d; Lawrence 2010e].

Larva (Figs 34A, 81K)

Body oblong to elongate, parallel-sided or with broad thorax and narrower abdomen, subcylindrical or slightly flattened, usually somewhat C-shaped; very lightly pigmented except for mouthframe and mandibles; surfaces smooth and vestiture of short, fine hairs.

Head prognathous and deeply retracted within prothorax, but not completely concealed from above, at least slightly longer than wide and slightly flattened. Epicranial suture absent; median endocarina long. Stemmata absent. Frontoclypeal suture present. Labrum free. Antennae short to moderately long, either 3-segmented with palpiform sensorium on antennomere 2 shorter than to longer than antennomere 3, or with antennomeres 2 and 3 partly or entirely fused together, so that sensorium is at middle of apical antennomere. Mandibles symmetrical, broad, stout and often more or less wedge-like, unidentate or bidentate, without accessory ventral process; mesal surface of mandibular base simple in Bostrichinae, with sub-basal mola accompanied by membranous prostheca and basal brush of hairs in Dinoderinae and Lyctinae. Ventral mouthparts retracted, with reduced maxillary articulating area; stipites longer than wide; galea rounded and setose; lacinia falciform; maxillary palps usually 3-segmented (2-segmented in Dinoderinae, Lyctinae and some Bostrichinae). Labium usually consisting of prementum, mentum and submentum (mentum and submentum fused in Dinoderinae); ligula absent or short and simple; labial palps usually 2-segmented (1-segmented in Dinoderinae and Lyctinae), widely separated. Hypopharyngeal sclerome absent. Hypostomal rods very short or absent; ventral epicranial ridges absent. Gular region absent; labium contiguous with thoracic membrane.

Prothorax sometimes longer than meso- and metathorax combined; terga without distinct plates; prothorax laterally with narrow, oblique rod on each side in most Bostrichinae. Legs moderately well developed, 5-segmented including claw-like pretarsus, usually without setae (with two setae lying side by side in Dinoderinae); coxae widely separated. Abdominal terga with two to four transverse folds. Segment IX excluding appendages usually as long as or longer than VIII, not extending onto ventral surface, without urogomphi or median process. Sternite IX exposed, simple. Segment X somewhat reduced, with longitudinally oval pads on either side of groove (Fig. 52V); without paired pygopods; anal region posteriorly (terminally) oriented. Spiracles annular, those on segment VIII enlarged in Lyctinae. [Böving and Craighead 1931; Gardner 1933b; Pringle 1938; Crowson 1961a; Lawrence 1991b; Lawrence *et al.* 1999a; Ivie 2002d; Lawrence 2010e; Liu and Schönitzer 2011].

Classification and World Distribution

Bostrichidae is a widely distributed family with ~90 genera and 600 species. The group is divided into seven subfamilies, three of which occur in Australia. Exotic subfamilies include Dysidinae (Oriental and Neotropical regions), Polycaoninae (Holarctic and South Africa), Euderiinae (New Zealand) and Psoinae (Holarctic and South Africa). Within the major subfamily Bostrichinae, the following tribes do not occur in Australia: Dinapatini (South-western North America), Apatini (Holarctic, Neotropical, Afrotropical, Oriental) and Chileniini (southern South America). [Crowson 1961a; Borowski and Węgrzynowicz 2007; Ivie 2010; Lawrence 2010e].

Australian Fauna

The endemic Australian bostrichid fauna consists mostly of **Bostrichinae**, with three genera of Bostrichini and nine genera of Xyloperthini. Our largest bostrichine is *Bostrychopsis jesuita* (Fabricius) (Fig. 81J), which is widely distributed across the northern part of the continent and normally attacks wattles and eucalypts. *Mesoxylion*, *Xylobosca*, *Xylopsocus* and related genera include most of our smaller species, many of which have concave elytral apices with which the males block tunnel entrances against predators and parasitoids, while the female is laying eggs. In *Xylobosca bispinosa* (Macleay) males have in addition to the concave declivity a pair of diverging spines, which are lacking in the female. The tribe Synoxylonini is represented by two cosmopolitan species of *Synoxylon* Lesne. The subfamily **Dinoderinae** includes three genera, most species of which are cosmopolitan or introduced. The **Lyctinae** also includes several cosmopolitan or introduced species, especially in the genera *Lyctus* (Fig. 81B), *Minthea* and *Trogoxylon*, plus endemic species, such as *Tristaria grouvellei* Reitter, *Lyctodon bostrychoides* Lesne, *Lyctoxylon dentatum* Pascoe and *Minthea acanthacollis* (Carter and Zeck). The last species is known to attack a mistletoe in the genus *Loranthus*. Bostrichids are elongate, cylindrical to somewhat flattened, black or brown beetles, with the head visible from above (Lyctinae) or concealed from above by the pronotum, which is distorted so that the anterior foramen is ventral in position (Dinoderinae and Bostrichinae). The antennal club is usually relatively loose, the eyes circular and strongly projecting laterally, the lateral pronotal carinae weak or absent, the metacoxae contiguous to widely separated without coxal plates, and the tarsomeres relatively slender and not lobed beneath. In the more cylindrical, burrowing forms, the head, pronotum and elytral apices may be tuberculate, toothed or otherwise modified, and the protibiae may be dentate; these resemble scolytine weevils, which differ in having geniculate antennae with a more compact club, elongate, less protruding eyes, and lobed tarsi with a reduced fourth tarsomere. Larvae are C-shaped and very lightly sclerotised, with enlarged thoracic region and strongly retracted head. Most bostrichids bore into moribund or freshly felled trees, dead and dry sapwood, or other kinds of dry plant material containing starches and sugars. They cannot produce cellulases and are thus restricted to sapwood, which the larvae reduce to a soft powder, earning them the name powderpost beetles. There are several pest species in Australia, most of which are introduced. *Lyctus brunneus* (Stephens) (Fig. 81B) attacks structural timber containing sapwood (Peters *et al.* 2002); *Dinoderus minutus* (Fabricius) commonly infests bamboos, cane furniture or basketwork; and *Rhyzopertha dominica* (Fabricius) (Fig. 81E) is our most serious pest of stored grain. [Aitken 1975; Vrydagh 1958; 1959a, 1959b].

Australian Subfamilies, Tribes and Genera

Bostrichinae: Bostrichini: *Bostrychoplites* Lesne, *Bostrychopsis* Lesne, *Calophorus* Lesne, *Heterobostrychus* Lesne. **Synoxylonini**: *Sinoxylon* Duftschmidt. **Xyloperthini:** *Mesoxylion* Vrydagh, *Xylobosca* Lesne, *Xylodectes* Lesne, *Xylodeleis* Lesne, *Xylogenes* Lesne, *Xylopertha* Guérin-Méneville, *Xylopsocus* Lesne, *Xylothrips* Lesne, *Xylotillus* Lesne.

Dinoderinae: *Dinoderus* Stephens, *Rhyzopertha* Stephens, *Stephanopachys* Waterhouse.

Lyctinae: *Lyctodon* Lesne, *Lyctoxylon* Reitter, *Lyctus* Fabricius, *Minthea* Pascoe, *Tristaria* Reitter, *Trogoxylon* LeConte.

50. PTINIDAE

Adult (Figs 6A, 70I–L, 81A, F, I, L)

Length 1–8.5 mm. Body ~1.4–3.65 times as long as wide; more or less parallel-sided and cylindrical to oblong or ovate, moderately to strongly convex; sides of body strongly rounded or not, sometimes capable of conglobation, with head and legs fitting into thoracic excavations; colour usually brown to black; upper surfaces clothed with fine recumbent and/or erect hairs or scales, sometimes forming patterns, occasionally subglabrous.

Head usually deflexed and inserted well into prothorax, sometimes concealed from above by pronotum. Occipital region often with median endocarina, transverse carina present in some Dorcatominae and Mesocoelopodinae; Eyes small to large, entire, more or less circular to slightly emarginate (occasionally deeply divided by canthus), with or without interfacetal setae. Antennal insertions usually exposed, often widely separated (usually narrowly separated in Ptininae), sometimes located within fossae; subantennal grooves usually absent. Frontoclypeal region not or only slightly declined. Frontoclypeal suture present or absent; anterior edge of clypeus straight, convex or shallowly emarginate. Labrum free and visible, usually transverse. Antennae usually 11-segmented (rarely 3-, 6- or 10-segmented), usually with loose and often asymmetrical, 3-segmented club (club sometimes 1- or 2-segmented), but sometimes filiform, serrate, pectinate or rarely flabellate. Mandibles short and broad, usually bidentate without distinct mola or prostheca. Maxilla usually with distinct galea and lacinia, the latter without uncus; apical maxillary and labial palpomeres usually fusiform (sometimes apically expanded and truncate or emarginate; ligula undivided, cleft to emarginate. Gular sutures widely to narrowly separated or absent. Corpotentorium narrow. Cervical sclerites absent.

Pronotum usually 0.3–1.1 times as long as wide (more elongate in a few Ptininae), usually widest at base, which is wide as combined elytral bases, and with sides strongly curved ant anteriorly converging (sometimes widest at middle or anteriorly, with sides straight to weakly curved, and with base distinctly narrower than combined elytral bases). Lateral pronotal carinae complete, incomplete or absent, simple or finely denticulate, with or without distinct bead; anterior angles usually not produced (sometimes produced and broadly rounded); posterior angles obtuse, right or moderately acute; posterior edge straight to broadly rounded; disc usually simple (disc and sides often variously modified in myrmecophilous species). Prosternum in front of coxae at least slightly shorter than mid length of procoxal cavity and usually much shorter, flat or slightly convex to concave or biconcave, sometimes excavated to receive head. Prosternal process complete to absent, flat, concave or slightly elevated behind coxae, with apex acute, rounded or truncate. Notosternal sutures present or absent. Hypomeron sometimes with anterior head impression and/or posterior crural impression. Procoxae more or less globular to strongly projecting below prosternum, without or with very short concealed extensions; trochantins exposed or concealed. Procoxal cavities circular to transverse and contiguous to well separated, open externally, internally open or closed by slender bar.

Scutellar shield abruptly elevated or not, posteriorly acute to broadly rounded, truncate or emarginate, sometimes absent. Elytra ~1–2.6 times as long as combined width and 1.6–4.5 times as long as pronotum, often parallel-sided but sometimes widest at base or at middle and moderately to strongly rounded; often irregularly punctate, sometimes with ten puncture rows or punctate striae, with or without scutellary striole; epipleura incomplete or absent. Mesoventrite usually separated by complete sutures from mesanepisterna, which are well separated at midline (sometimes fused to mesanepisterna), anterior edge sometimes on different plane than metaventrite; with or without anterior procoxal rests. Mesoventral cavity small to large, shallow to deep or sometimes absent. Mesoventral process acute to broadly rounded or truncate at apex. Mesocoxae projecting or not, with concealed trochantins. Mesocoxal cavities more or less circular, nearly contiguous to widely separated, closed laterally by meeting of meso- and metaventrite. Metaventrite flat to strongly convex; discrimen long to short or absent; transverse (metakatepisternal) suture absent; exposed portion of metanepisternum short and broad to narrowly elongate or sometimes absent. Metacoxae strongly transverse to subtriangular, narrowly to widely separated and usually extending laterally to meet elytra; plates well developed and complete, well developed mesally and weak laterally, weakly developed or absent. Metendosternite with moderately to very long lateral arms and no laminae; anterior process short to moderately long with narrowly to widely separated tendons. Hind wing with moderately long apical field and vaguely indicated extension of RP; radial bar usually strongly diverging from wing margin, with well developed bending zone; radial usually absent; basal portion of RP usually short; medial field usually with three or fewer free veins, rarely with wedge cell; notch-like anal embayment sometimes present. Femoral attachment of mid trochanter transverse or slightly oblique. Femora sometimes inflated; tibiae slender; spurs often reduced or absent. Tarsi 5-5-5; tarsomeres simple; pretarsal claws simple.

Abdomen almost always with five ventrites (four in *Gibbium*); first two ventrites usually connate (first three connate in Ptininae; rarely all connate, sutures sometimes nearly obsolete; ventrite 1 not much longer than 2; intercoxal process acute to broadly rounded or angulate, sometimes absent; recurved postcoxal lines present in Dorcatominae and Mesocoelopodinae. Functional spiracles usually on abdominal segments I–VIII. Anterior edge of sternite VIII in male with median strut. Anterior edge of segment IX in male without spiculum gastrale, except in Ptininae and Dryophilinae; tergite IX more or less fused to X or X membranous. Aedeagus of a modified trilobate type; phallobase posteriorly overlapping parameres, which may be symmetrical or asymmetrical, free from or fused with phallobase, usually joined together near base and with anteriorly projecting parameral struts attached to base of penis; anterior edge of penis usually without paired struts. Sternite VIII in female with spiculum ventrale. Ovipositor usually moderately elongate, lightly sclerotised except for paraproctal and coxital baculi; paraprocts usually longer than coxites, which are divided; styli well developed and subapical. [Lea 1924; Hinton 1941; Lawrence and Reichardt 1969; Ford 1970; Español 1970, 1976, 1977a, 1977b, 1979, 1982; Bellés 1985; Bellés and Lawrence 1990; Lawrence *et al.* 1999b; Philips 2000, 2002; Philips and Bell 2010].

Larva (Fig. 81M)

Body elongate, parallel-sided, subcylindrical (occasionally with enlarged thorax), strongly curved and C-shaped; very lightly pigmented except for mouth-frame and mandibles; surfaces smooth and vestiture of short to long, fine hairs.

Head protracted and prognathous, about as long as wide and somewhat flattened. Epicranial stem long; frontal arms indistinct to absent, sometimes V-shaped; median endocarina coincident with epicranial stem. Stemmata absent or one on each side. Frontoclypeal suture present. Labrum free. Antennae very short, either 2-segmented with sensorium on antennomere 1 longer than antennomere 2, or 1-segmented with sensorium at the apex. Mandibles symmetrical, short, stout, unidentate to tridentate, without accessory ventral process; mesal surface of mandibular base simple or with brush of hairs. Ventral mouthparts retracted, with maxillary articulating area reduced or absent; stipites longer than wide; galea truncate or rounded and setose; lacinia reduced, rounded to falciform, sometimes with spine or spur; maxillary palps usually 3-segmented (occasionally 2- or 4-segmented). Labium usually consisting of prementum, mentum and submentum; ligula short and broad or absent; labial palps 2-segmented, widely separated. Hypopharyngeal sclerome absent. Hypostomal rods and ventral epicranial ridges absent. Gular region absent; labium contiguous with thoracic membrane.

Prothorax without distinct tergal plates. Legs usually moderately well developed, 5-segmented including claw-like pretarsus, usually with paired setae (sometimes with one seta or numerous hairs); coxae widely separated. Abdominal terga almost always with two transverse folds. Segment IX excluding appendages usually as long as or longer than VIII, not extending onto ventral surface, without urogomphi or median process. Sternite IX exposed, simple. Segment X somewhat reduced, with longitudinally oval pads on either side of groove (Fig. 52V); without paired pygopods; anal region posteriorly (terminally) oriented. Spiracles annular or annular–uniforous. [Böving and Craighead 1931; Parkin 1933; Hall and Howe 1953; Böving 1954; Lawrence and de Viedma 1991; Lawrence 1991b; Lawrence *et al.* 1999a; Philips 2002; Philips and Bell 2010].

Classification and World Distribution

The family Ptinidae is a cosmopolitan family with ~2500 species in 230 genera. The group is commonly divided into ten subfamilies, nine of which occur in Australia; however, Bellés (1985) and Philips and Bell (2010) include *Gibbium*, *Mezium* and related genera into a separate subfamily, Gibbiinae (**see** Ptininae below). The exotic subfamily Alvarenganiellinae includes the genus *Dasytanobium* Pic with a species in Brazil and another in Hawaii. [White 1974; Bellés 1982; Philips 2000].

Australian Fauna

The Australian ptinid fauna contains ~200 described species in 40 genera, but the classification is in need of revision and several species remain to be described. The subfamilies **Eucradinae, Ernobiinae** and **Ptilininae** are all poorly represented, the first by *Ptilineurus marmoratus* (Reitter) introduced into NSW from the Oriental region and *Clada posticum* Gahan, occurring only on Christmas Island, the second by the introduced species *Ernobius mollis* (Linnaeus) and *Xestobium rufovillosum* (DeGeer) and the third by the introduced *Ptilinus pectinicornis* (Linnaeaus). **Dryophilinae** includes a large endemic genus *Dryophilodes* containing ~50 species. The **Ptininae** is well represented by both native and introduced species; among the autochthonous forms are several species of *Ptinus* (Fig. 70J), *Pitnus* (*Octopitnus*) *australiae* (Lea) from SA and WA (Bellés 1992), two species of *Sphaericus* (*Leasphaericus*) from the North-west Cape Peninsula and Barron Island, WA (Bellés 1998), two species of *Neoptinus* (Fig. 81I) from North QLD and Christmas Island (Bellés and Lawrence 1990), and at least 35 myrmecophilous species of *Diplocotes* (Fig. 70I) and related genera (Lawrence and Reichardt 1969). Cosmopolitan or introduced ptinines include species of *Gibbium* (Fig. 81A) and *Mezium* (Fig. 70K), *Sphaericus gibboides* (Boieldieu), *Ptinus fur* (Linnaeus), *P. tectus* Boieldieu, *Niptus hololeucus* (Faldermann) and *Trigonogenius globulum* Solier. The subfamily **Anobiinae** includes the introduced *Anobium punctatum* (Degeer) and *Stegobium paniceum* (Linnaeus) (Fig. 81F, L), as well as *Tasmanobium mimicum* Lea and several native *Hadrobregmus* (Fig. 70L). **Xyletininae** includes the introduced *Lasioderma serricorne* (Fabricius) and native *Leanobium*, *Pronus* and related genera. **Dorcatominae** and **Mesocoelopodinae** include several smaller and more compact species. *Calymmaderus*, *Methemus* and *Caenocara* are placed in Dorcatominae, along with several species currently placed in *Dorcatoma* Herbst but probably belonging elsewhere. Within Mesocoelopodinae are species of *Deltocryptus* and *Secretipes*, plus *Stagetomorphus lanigera* (Olliff), which has a strongly flabellate antennal club.

Anobiids are convex and elongate and cylindrical to broadly ovate and globose beetles, variously clothed with decumbent or erect hairs or scales and only occasionally glabrous above. The head is strongly deflexed and usually concealed from above, the prosternum is reduced and sometimes deeply excavate forming a cavity which continues onto the mesoventrite, trochanters are squarely attached to the bases of the femora, and antennae are varied but usually filiform, serrate or with a large, loose, elongate club. The metacoxae are usually contiguous and excavate to receive the femora (with distinct coxal plates) except in Ptininae, where they are widely separated and not excavate (lacking coxal plates); most ptinines also have approximate antennal insertions and no lateral pronotal carina, but this is not the case in the genus *Neoptinus*. Ptinid larvae are C-shaped and lightly sclerotised, with an exerted, hypognathous head, very small antennae, and a pair of narrowly oval lobes on segment X beneath the anus. Most larval Ptinidae bore into the wood or bark of dead trees, and a few have become pests of furniture or timber in Australia. The most serious of these pests is the furniture beetle, *Anobium punctatum* (Degeer). *Ernobius mollis* attacks pine timber which still has some bark attached, and the native species *Calymmaderus incisus* Lea attacks softwoods, but has also been reported from *Eucalyptus* wood (Hockey 1986). Most species of *Dryphiloides*, our largest genus, feed on the woody fruits of *Eucalyptus* and other Myrtaceae (Andersen and New 1987), but at least one is known to attack *Araucaria* seed cones. The two cosmopolitan species *Stegobium paniceum* and *Lasioderma serricorne* are important stored products pests, while some members of the Dorcatominae feed on woody or fibrous fungal fruiting bodies. Members of the Ptininae feed mainly on dry materials of plant or animal origin and are common inhabitants of the nests of birds, mammals or social insects. *Ptinus exulans* Erichson lives on the arthropod remains in spider nests (Hickman 1974), the exotic *Ptinus californicus* Pic is a depredator in bee nests (Linsley and MacSwain 1942), and the introduced Ptininae mentioned above are pests of stored products. The species of *Diplocotes*, *Polyplocotes*, *Enasiba* and *Ectrephes* are specialised symphiles in the nests of ants, especially those of *Iridomyrmex* species (Lawrence and Reichardt 1969).

Australian Subfamilies, Tribes and Genera

Eucradinae: *Clada* Pascoe, *Ptilineurus* Reitter.

Ptininae: *Diplocotes* Westwood, *Ectrephes* Pascoe, *Enasiba* Olliff, *Gibbium* Scopoli, *Mezium* Curtis, *Neoptinus* Gahan, *Niptus* Boieldieu, *Pitnus* (*Octopitnus*) Bellés, *Ptinus* Linnaeus, *Sphaericus* (*Leasphaericus*) Bellés, *Sphaericus* Wollaston, *Trigonogenius* Solier.

Dryophilinae: *Dryophilodes* Blackburn.

Ernobiinae: *Ernobius* Thomson, *Xestobium* Motschulsky.

Anobiinae: *Anobium* Fabricius, *Gastrallus* Jacquelin du Val, *Hadrobregmus* Thomson, *Priobium* Motschulsky, *Stegobium* Motschulsky, *Tasmanobium* Lea.

Ptilininae: *Ptilinus* Müller.

Xyletininae: *Deroptilinus* Lea, *Lasioderma* Stephens, *Leanobium* Español, *Pronus* Lea, *Pseudopronus* Español.

Dorcatominae: *Caenocara* Thomson, *Calymmaderus* Solier, *Dorcatoma* Herbst ?, *Methemus* Broun, *Mirosternomorphus* Español, *Mysticephala* Ford, *Serianotus* Ford.

Mesocoelopodinae: *Deltocryptus* Lea, *Secretipes* Lea, *Stagetomorphus* Pic.

CUCUJIFORMIA

The monophyly of this group appears to be well founded, being based on several adult features, such as the presence of acone eyes with open rhabdomes (Wachmann 1977; Caveney 1986), cryptonephridial Malpighian tubules of the 'normal' type (not joining the hind gut in a single bundle), absence of plates on the metacoxae, absence of functional spiracles on segment VIII, reduction of the pregenital segments (IX and X) and the presence of the hylecoetoid type of metendosternite (with well developed laminae and approximate anterior tendons) and cucujoid type of aedeagus (with a sheath or ring-type of tegmen surrounding the penis) or one derived from this type. Larvae retain most of the plesiomorphic polyphagan features, but usually have the galea and lacinia fused to form a mala. Of the six included superfamilies, there is little doubt about the monophyly of Chrysomeloidea and Curculionoidea, but the limits of and relationships among the remaining four are far from clear. [Leschen and Ślipiński 2010a; Lawrence *et al.* 2011].

LYMEXYLOIDEA

This group was considered by Crowson (1955) to be an isolated lineage in which the specialised, wood-boring larva retains a separate galea and lacinia, and the aedeagus resembles to some extent that found in Bostrichoidea. Crowson (1981) placed the Strepsiptera in this superfamily based on relatively weak evidence reviewed by Lawrence and Newton (1982) and Wheeler (1986). In recent analyses Lymexylidae formed a clade within Tenebrionoidea (Lekaničová 2009; Lawrence *et al.* 2011) or with Tenebrionoidea and Cleroidea (Hunt *et al.* 2007). [Lawrence 2010f].

51. LYMEXYLIDAE

Adult (Figs 8B, 82A, G)

Length 6–35 mm. Body ~4.5–12 times as long as wide; parallel-sided and slightly flattened; lightly sclerotised; colour yellowish-brown to black, with metallic blue hind wings in *Arractocetus* Kurosawa, upper surfaces usually clothed with short, fine hairs.

Head prognathous, abruptly constricted behind eyes in Melittommatinae to form short temples (very short and hardly visible in *Melittomma*). Occipital region moderately long. Eyes usually large to very large and strongly protuberant, sometimes almost meeting dorsally, at least shallowly emarginate, finely to coarsely facetted with short to long interfacetal setae. Antennal insertions concealed from above by frontal ridges which encroach upon eye; subantennal grooves absent. Frontoclypeal suture usually present, sometimes weakly impressed; anterior edge of clypeus convex to slightly concave. Labrum small, free, slightly to strongly transverse; anterior edge slightly to strongly convex. Antennae 11-segmented, filiform, serrate or more or less clavate, with antennomeres gradually expanded towards middle and then narrowed towards apex. Mandibles short and broad, unidentate or bidentate, mola absent; prostheca present or absent. Maxilla with short, simple, setose galea and lacinia, the latter without uncus, sometimes reduced; apical maxillary palpomere cylindrical or slightly expanded; complex palp organ attached to palpomere 3 in males of *Melittomma* and in males and females of Atractocerinae. Ligula short and broad to elongate and narrow, apically rounded or emarginate; apical labial palpomere cylindrical or slightly expanded. Corpotentorium broad or narrow. Cervical sclerites present, sometimes very large.

Pronotum ~0.95–1.25 times as long as wide, usually widest at middle and as wide as combined elytral bases, with sides straight to weakly curved. Lateral pronotal carinae complete and simple, with distinct bead in Melittommatinae, weakly developed and not sharply delimited in Atractocerinae; anterior angles not produced; posterior angles more or less right; posterior edge straight to bisinuate; disc with median longitudinal groove in Atractocerinae. Prosternum in front of coxae (Fig. 44Q) shorter than to slightly longer than mid length of procoxal cavity, flat, without paired lines or carinae. Prosternal process incomplete and broadly rounded or absent. Hypomera simple, highly reduced in Atractocerinae. Procoxae projecting well below prosternum, without concealed extensions; trochantins exposed. Procoxal cavities very weakly impressed, slightly transverse and contiguous, broadly open externally and internally.

Scutellar shield not abruptly elevated, sometimes deeply notched anteriorly, broadly rounded to truncate or emarginate posteriorly. Elytra ~3–4.7 times as long as combined width and 2.8–5 times as long as pronotum in Melittommatinae (1.4–2 times as long as wide and 1.5–2 times as long as pronotum in Atractocerinae), parallel-sided and more or less complete with conjointly or independently rounded apices, exposing one or two abdominal tergites in Melittommatinae, very short, independently rounded and widely separated at base, exposing metatergum, abdomen and hind wings in Atractocerinae; irregularly punctate, without scutellary striole; epipleura very narrow and incomplete or absent. Mesoventrite (Fig. 44S) separated by complete sutures from mesanepisterna, which are narrowly separated at midline in Melittommatinae and broadly joined at midline in Atractocerinae; anterior edge without procoxal rests. Mesoventral process very short and acute to broadly rounded at apex. Mesocoxae conical and projecting, with exposed trochantins. Mesocoxal cavities weakly impressed, obliquely oriented and contiguous, partly closed laterally by mesanepisternum and mesepimeron. Mesometaventral junction absent. Metaventrite elongate, without discrimen, postcoxal lines or transverse (metakatepisternal) suture; exposed portion of metanepisternum moderately to very elongate. Metacoxae transverse but often with mesal portion strongly projecting, more or less obliquely oriented, contiguous, extending laterally to meet elytra or sides of body; plates weakly developed or absent. Metendosternite with long stalk and long lateral arms; laminae well developed in Melittommatinae, absent in Atractocerinae; anterior process absent and anterior tendons on lateral arms or absent. Hind wing with very short apical field containing anterior and posterior radial extensions at least vaguely indicated, without transverse folds in Atractocerinae; radial bar extending slightly beyond radial cell in Melittommatinae, continuing to wing apex in Atractocerinae; radial well developed in *Australymexylon*, somewhat reduced in *Melittomma* (absent in Atractocerinae), with inner edge obliquely curved and inner posterobasal angle obtuse; cross-vein r3 absent, r4 long and complete; R-M loop narrowly rounded and basal portion of RP moderately long in Melittommatinae, both absent in Atractocerinae; medial spur complete to wing margin; medial field with four free veins and no medial fleck; wedge well developed in Melittommatinae (very narrow in *Australymexylon*), absent in Atractocerinae; anal lobe well developed, without embayment. Legs long and slender, greatly elongated in some Atractocerinae. Tibiae simple; tibial spurs paired on all legs in *Australymexylon* and *Melittomma*, absent in Atractocerinae. Tarsi 5-5-5; tarsomeres simple; pretarsal claws simple; empodium well developed, enlarged and multisetose in Atractocerinae.

Abdomen with five free ventrites (III–VII) in *Australemyxylon*, six (II–VII) in *Melittomma* and female Atractocerinae and seven (II–VIII) in male Atractocerinae; ventrite 1 not much longer than 2, intercoxal process absent; Functional spiracles on abdominal segments I–VII. Anterior edge of segment IX in male without spiculum gastrale; tergite IX truncate to deeply emarginate; tergite X free or partly fused to IX. Aedeagus of cucujiform type, symmetrical; anterior edge of tegmen without struts, its sclerotised edges more or less joined anteriorly and fused to base of penis; parameres individually articulated to phallobase or fused to phallobase but free from one another; anterior edge of penis without or with short struts. Ovipositor usually long and narrow, with indistinctly divided, lightly sclerotised, setose coxites and well developed apical or slightly subapical styli, shorter and broader with more heavily sclerotised coxites and reduced styli in some Atractocerinae. [Wheeler 1986; Lawrence *et al.* 1999b; Young 2002g; Paulus 2004; Lawrence 2010f].

Larva (Fig. 82B)

Body narrowly elongate and more or less cylindrical, with enlarged prothorax; dorsal and ventral surfaces lightly pigmented or heavily pigmented on tergum IX and sometimes one or two preceding terga (most terga with transverse sclerites in early instars); surfaces smooth and vestiture of scattered setae.

Head hypognathous, more or less globular. Epicranial stem moderately long with underlying endocarina; frontal arms V-shaped with contiguous bases. Stemmata usually absent (five on each side in first instars). Frontoclypeal suture present in first instars only. Labrum free. Antennae minute, 3-segmented, with palpiform sensorium shorter than apical antennomere. Mandibles broad, stout and more or less wedge-like, unidentate, without accessory ventral process; incisor edge simple' mola well developed, with numerous fine transverse ridges; prostheca absent. Ventral mouthparts retracted, with maxillary articulating area; cardines divided; stipites longer than wide; mala rounded and at least partly divided into galea and lacinia,clothed with setae and spines; inner surface usually with sclerotised peg; maxillary palps 3-segmented with distinct palpifer. Labium consisting of prementum, mentum and submentum; ligula simple, shorter or longer than labial palps, which are 2-segmented and separated by less than width of first palpomere. Hypopharyngeal sclerome subtriangular and convex. Hypostomal rods very short or absent; ventral epicranial ridges weakly developed or absent. Gula wider than long.

Prothorax longer than meso- or metathorax, sometimes longer than two combined; protergum expanded and often produced anteriorly over head; usually with patches of small dark granules or spicules laterally or over much of surface. Meso- and metaterga with small lateral patches of asperities in *Australymexylon* and some Atractocerinae. Prothoracic venter with greatly enlarged presternum, without armature. Legs moderately well developed, 5-segmented including claw-like pretarsus bearing two or several setae; coxae widely separated. Abdomen in late instars more than twice as long as thorax; some posterior terga armed with transverse carinae in *Melittomma*; most terga simple or with weak, transverse ampullae in Atractocerinae. Segment IX excluding appendages usually somewhat longer than VIII, tergum extending onto ventral surface, without urogomphi, forming circular, concave plate lined with large blunt teeth in Melittommatinae and first-instar Atractocerinae, forming a lightly pigmented cylindrical, blunt process in late instar Atractocerinae. Sternite IX exposed, reduced, distinctly transverse, sometimes forming with segment X a pygopod-like ventral projection; anal region ventrally oriented. Spiracles annular. [Britton 1974a; Wheeler 1986; Korolev 1987; Lawrence and Britton 1991; Lawrence *et al.* 1999a; Young 2002g; Lawrence 2010f; Casari and Teixeira 2011].

Classification and World Distribution

The Lymexylidae is a widely distributed family with ~65 species in nine genera. The group has been divided into four subfamilies, two of which occur in Australia. The exotic subfamilies Hylecoetinae (*Elateroides* Schaeffer) and Lymexylinae (*Lymexylon* Fabricius) are Holarctic and Palaearctic–Oriental, respectively, while both Melittommatinae and Atractocerinae are mainly Pantropical in distribution, with *Melittomma* extending into temperate North America and the atractocerine genera *Urtea* Paulus occurring in Greece and *Homaloxylon* Kurosawa in northern India, Nepal and China. [Kurosawa 1985; Wheeler 1986; Paulus 2004; Lawrence 2010f].

Australian Fauna

The family is represented in Australia by two genera of **Melittommatinae**, the endemic *Australymexylon* Wheeler (Fig. 82G) and the Pantropical *Melittomma* Murray, and three or four genera of **Atractocerinae (see** below). Lymexylids are narrowly elongate, finely pubescent beetles with well developed eyes, sometimes very large and dorsally contiguous, long legs with projecting coxae, metanepisterna meeting at the midline and apically rounded elytra exposing the abdominal apex (Melittommatinae) or most of the abdomen (Atractocerinae). A branched sensory organ arises from the third maxillary palpomere in males of *Melittomma* and males and females of Atractocerinae. Larvae are very long, narrow and cylindrical, with a globular head, very short antennae, enlarged and dorsally humped prothorax and a terminal tergum IX, which forms a lightly sclerotised, rounded and asperate protuberance (Atractocerinae) a heavily sclerotised, concave plate lined with teeth (Melittommatinae). Lymexylid larvae bore into hard wood of weakened or dead trees and feed on ambrosia fungi of the genus *Ascoidea* (Ascomycetes: Endomycetales), which grows on their tunnel walls. Adult females carry the fungal spores in pouches near the end of the ovipositor and deposit them in a slimy matrix with the eggs. The first instar larvae then transport the spores into the wood on their bodies (Francke-Grosmann 1967). In Australia lymexylids damage commercial timber because of the darkly stained 'pinholes' they form within the wood (Clark 1925). *Australymexylon* and *Melittomma* each contain two species distributed along the eastern coast of Australia. The Atractocerinae are widely distributed, but there is some question about the generic placement of some of them. *Atractocerus crassicornis* Clark (Fig. 82A) from northern Australia and *Raractocetus kreuslerae* (Pascoe) appear to be correctly placed, but the placement of the south-eastern species *tasmaniensis* Lea and *victoriensis* Blackburn in the Chilean genus *Fusicornis* Philippi by Paulus (2004) requires reassessment. The genus *Arractocetus* Kurosawa has not been recorded from Australia, but a species with the distinctive head shape of this genus and with metallic blue wings has been seen from North QLD.

CLEROIDEA

Some basal members of this superfamily are difficult to distinguish from Cucujoidea, and recent analyses based on both morphological and molecular data do not support a monophyletic Cleroidea (Leschen *et al.* 2005; Hunt *et al.* 2007; Lawrence *et al.* 2011). Typical cleroids have a distinctive adult habitus with projecting procoxae and five, often lobed, tarsomeres, and larvae have only one pretarsal seta, no mandibular mola and usually a pedunculate seta on the mala. Although phloiophilids and some trogossitids are mycophagous, most other cleroids are predacious as larvae and pollen-feeding as adults. . Relationships among all cleroid families were discussed by Leschen (2010a) and among those taxa belonging to the melyrid lineage by Bocákova *et al.* (2011). In the latter paper, a cladistic analysis based on molecular data did not support a family Melyridae as recognized here, since the exotic melyrid subfamily Rhadalinae (based on *Aplocnemus* Stephens and *Microjulistus* Reitter) weas found to be sister to a clade comprising Mauroniscidae, Prionoceridae and the melyrid subfamilies Melyrinae, Dasytinae and Malachiinae. Exotic families include Phloiophilidae (Europe) and Chaetosomatidae (New Zealand and Madagascar), Metaxinidae (New Zealand), Prionoceridae (Palaearctic, Oriental and Afrotropical), Mauroniscidae (Nearctic and Neotropical) and Rhadalidae, if that group is recognised at the family level. [Crowson 1964b; Ekis and Menier 1980; Menier and Ekis 1982; Majer 1994b, 1995; Kolibáč 1999, 2004b, 2008; Leschen 2010a, 2010b; Kolibáč and Leschen 2010b; Lawrence and Leschen 2010d; Lawrence 2010g].

52. TROGOSSITIDAE

Adult (Figs 8E, 71A–B, D, 82D, H–I)

Length 1–18 (usually 3–15) mm. Body ~1.2–3.3 times as long as wide; elongate to broadly ovate and usually slightly to strongly flattened (broadly ovate and strongly convex in Rentoniini); well sclerotised; colour yellowish-brown to black, occasionally bicolored, upper surfaces usually clothed with decumbent and/or erect hairs or stout bristles, with erect, scale-like setae or with flattened scales, often forming a pattern (more or less glabrous in *Egolia*, *Temnoscheila* and *Tenebroides*).

Head usually prognathous (strongly deflexed in Rentoniini). Occipital region moderately long, rarely with transverse carina. Eyes small to moderately large, slightly to moderately protuberant, not or very shallowly emarginate, with or without interfacetal setae. Antennal insertions exposed or concealed from above by frontal ridges. Frontoclypeal suture present in some Lophocaterinae; anterior edge of clypeus usually truncate to emarginate. Labrum free, often slightly emarginate. Antennae 8- to 11-segmented, with a distinct 1- to 3-segmented club, somewhat flattened and asymmetrical in Trogossitinae. Mandibles unidentate or bidentate, with or without mola and prostheca. Maxilla galea and lacinia, the latter with unidentate to tridentate uncus; apical maxillary and labial palpomeres usually cylindrical or fusiform. Ligula shallowly or deeply emarginate. Gular sutures widely or narrowly separated. Corpotentorium present or absent. Cervical sclerites present.

Pronotum ~0.3–1.1 times as long as wide, almost always slightly to strongly transverse and somewhat narrower at base than combined elytral bases in most Trogossitinae; sides straight to strongly rounded, sometimes horizontally explanate (obliquely, almost vertically explanate in Rentoniini). Lateral pronotal carinae complete, with or without distinct bead, its edge usually simple, sometimes crenulate or denticulate; anterior angles usually slightly to strongly produced and broadly rounded to acute (not produced in *Egolia* or Rentoniini); posterior angles rounded to right; posterior edge straight to slightly rounded; disc almost always simple (irregularly impressed in *Egolia*). Prosternum in front of coxae (Fig. 44H) usually slightly longer than mid length of procoxal cavity (much shorter in Rentoniini), flat, without paired lines or carinae. Prosternal process usually complete, apically expanded and broadly rounded, angulate or truncate (incomplete and narrowly rounded or acute in Rentoniini). Procoxae without concealed extensions, with distinct procoxal plates in Rentoniini; trochantins exposed. Procoxal cavities slightly to strongly transverse, narrowly to moderately widely separated, broadly closed externally and open internally in *Parapeltis* and most Trogossitinae (internally closed by slender bar in *Larinotus*), externally and internally open in Lophocaterinae, and broadly open externally and broadly closed internally in Rentoniini.

Scutellar shield usually not abruptly elevated, narrowly to broadly rounded posteriorly (abruptly elevated in *Larinotus* and Rentoniini, absent in *Parapeltis*). Elytra ~0.8–2.4 times as long as combined width and 1.9–4.2 times as long as pronotum, usually more or less parallel-sided and complete with conjointly rounded apices, widest basally and apically narrowed in Rentoniini; punctures usually forming 10 or 11 regular rows without scutellary striole (irregular in Rentoniini; regular or irregular but often with longitudinal continuous or interrupted costae in *Leperina*); with subapical sutural interlocking mechanism in Trogossitinae; epipleura complete or incomplete. Mesoventrite with anterior edge on same plane as metaventrite (except in Rentoniini); usually with a broad prothoracic rest or prepectus (with a median prosternal rest in Lophocaterinae; with paired, vertical procoxal rests in Rentoniini). Mesoventral process moderately long and narrowly rounded to truncate at apex. Mesocoxae not conical and projecting, with exposed trochantins. Mesocoxal cavities usually circular to slightly transverse (strongly transverse in Rentoniini), narrowly to moderately widely separated, usually closed laterally partly by mesepimeron (closed by meeting of meso- and metaventrite in Egoliini). Mesometaventral junction usually monocondylic. Metaventrite more or less flattened, often with discrimen and transverse (metakatepisternal) suture; postcoxal lines present in *Larinotus* and *Parapeltis*; exposed portion of metanepisternum usually moderately elongate (short and broad in *Parapeltis*). Metacoxae contiguous or narrowly separated and extending laterally to meet elytra (except in *Parapeltis* and Rentoniini); plates absent. Metendosternite with short to long stalk, long lateral arms; laminae well developed in *Egolia*, usually very slender or reduced; anterior process short and broad (absent in *Parapeltis* and some Rentoniini) and anterior tendons widely separated or sometimes absent. Hind wing with moderately well developed apical field containing one or more distinct sclerites; radial bar with distinct bending zone; radial cell as long as wide or shorter than wide, sometimes highly reduced or absent; cross-vein r3 moderately well developed and closely associated with r4 or absent; R-M loop broadly rounded and basal portion of RP moderately long, short or absent; medial field often with four free veins, a wedge cell and no medial fleck (often reduced in smaller species); anal lobe well developed, embayment often notch-like; hind wings absent in *Parapeltis* and some Rentoniini). Legs relatively short. Tibiae sometimes with row of spines along outer edge and with tibial apex bearing one or two hooked spurs. Tarsi almost always 5-5-5, often with tarsomere 1 short and somewhat enclosed within tibial apex or partly fused to 2; tarsomeres simple; pretarsal claws simple; empodium well developed and bisetose.

Abdomen almost always with five or six free ventrites (III–VII or VIII) and ventrite 1 not much longer than 2 (ventrite 1 longer than rest combined and ventrites 1–4 partly or completely fused together in some Rentoniini); intercoxal process acute to narrowly rounded. Functional spiracles on abdominal segments I–VII. Anterior edge of sternite VIII in male sometimes with median strut. Anterior edge of segment IX in male with or without spiculum gastrale; tergites IX and X often partly or completely fused together. Aedeagus of cucujiform type, symmetrical; anterior edge of tegmen usually with ventral strut and two opposing dorsal struts (Fig. 31F–G); parameres individually usually fused to phallobase but free from one another; anterior edge of penis with paired struts. Ovipositor usually moderately elongate and lightly sclerotised except for baculi, coxites with well developed apical styli. [Crowson 1964b, 1966a, 1970; Ślipiński 1992; Lawrence *et al.* 1999b; Leschen 2002b; Kolibáč 2005, 2006, 2007, 2009, 2010; Kolibáč and Leschen 2010a].

Larva (Fig. 35L)

Body elongate, slightly flattened; dorsal and ventral surfaces lightly pigmented, except for well sclerotised head capsule, tergum IX and sometimes one or more thoracic terga; surfaces smooth and vestiture usually of scattered setae.

Head prognathous and protracted; posterior edge slightly to moderately emarginate. Epicranial stem absent to moderately long; frontal arms V-shaped with contiguous bases. Median endocarina underlying epicranial stem, if present, and extending to clypeal region. Stemmata usually five on each side, but often only two.

Frontoclypeal suture almost always absent (apparently present in Rentoniini). Labrum free. Antennae short, 3-segmented, with palpiform sensorium shorter than apical antennomere. Mandibles usually unidentate or bidentate, without accessory ventral process; incisor edge often with one or two teeth (serrate in Rentoniini); mola absent; mesal surface of mandibular base usually with simple or complex hyaline process (lacinia mobilis) or brush of hairs. Ventral mouthparts retracted, without maxillary articulating area; stipites longer than wide; mala rounded or truncate; with large pedunculate seta in Egoliini, Gymnochilini and Trogossitini; maxillary palps usually 3-segmented (2-segmented in Rentoniini), with or without distinct palpifer. Labium usually consisting of prementum, mentum and submentum (prementum and postmentum in *Parapeltis*); ligula present or absent; labial palps usually 2-segmented (1-segmented in Rentoniini), usually narrowly separated. Hypopharyngeal sclerome absent. Hypostomal rods usually moderately to very long and subparallel; ventral epicranial ridges present or absent. Gula often longer than wide (strongly transverse in *Larinotus*).

Protergum sometimes with single large plate, and meso- and metaterga sometimes with paired plates. Prothoracic venter sometimes with well developed sternal plate. Legs moderately well developed, 5-segmented including claw-like pretarsus bearing one seta; coxae widely separated. Abdomen more than twice as long as thorax; terga without patches or rows of asperities; abdominal ampullae present or absent. Segment IX excluding appendages shorter than or subequal to VIII, tergum almost always extending onto ventral surface (not so in *Larinotus*); dorsal surface usually with sclerotised plate (divided into anterior and posterior plates in Lophocaterinae, Fig. 52K); posterior edge almost always with pair of well developed, posteriorly oriented urogomphi upturned at apex (urogomphi very small and hook-like in *Larinotus*, joined together on median process in Rentoniini); with median pit between urogomphi in *Parapeltis*; with median, tooth-like process in Lophocaterinae. Sternite IX exposed, simple; anal region usually ventrally oriented (posteriorly oriented in *Larinotus*). Spiracles usually annular–biforous (annular–cribriform in *Parapeltis*, Fig. 43H). [Crowson 1964b, 1966a, 1970; Mamaev 1976b; Ślipiński 1992; Lawrence *et al.* 1999a; Leschen 2002b; Kolibáč 2005, 2006; Kolibáč and Leschen 2010a; Kolibáč and Zaitzev 2010].

Classification and World Distribution

The family Trogossitidae includes ~50 genera and 600 species distributed throughout the world, including some Pacific islands. The subfamily classification has varied over time (Crowson 1964b, 1970; Ślipiński 1992; Lawrence and Newton 1995; Kolibáč 2006), but three subfamilies are recognised here, following Kolibáč and Zaitzev (2010): Peltinae, Lophocaterinae and Trogossitinae. All three subfamilies occur in Australia, but within Peltinae, the following tribes are exotic: Peltini (Holarctic), Protopeltini (New Zealand) and Decamerini (Neotropical).

Australian Fauna

The Australian **Peltinae** are comprised of the genus *Parapeltis* Ślipiński (Colydiopeltini), *Parentonium* Crowson (Rentoniini), and at least one undescribed genus of Rentoniini. The subfamily **Lophocaterinae** includes the introduced *Lophocateres pusillus* (Klug) and several species belonging to the genera *Ancyrona* Reitter, *Neaspis* Pascoe (Fig. 71B) and *Peltonyxa* Reitter. **Trogossitinae** includes four tribes. The Egoliini includes *Egolia variegata* Erichson (Fig. 82H) and *Necrobiopsis tasmanica* Crowson (Fig. 71D) from TAS, with a second species in Chile (Arias *et al.* 2009), and *Paracalanthis binnaburrense* Crowson from southern QLD. The related tribe Larinotini includes *Larinotus umbilicatus* Carter and Zeck (Fig. 82I) and an undescribed species, both from QLD. Gymnochilini includes *Leperina* (Fig. 71A) with ~12 described Australian species and a few others from New Guinea and New Caledonia and the primarily exotic genus *Phanodesta* with the species *P. pudica* (Olliff) from Lord Howe Island. The tribe Trogossitini is represented by two introduced species only: *Tenebroides mauritanica* (Linnaeus) and *Temnoscheila virescens* (Fabricius). Trogossitids may be broadly ovate to elongate and usually more or less flattened, with a prognathous head, short, clubbed antennae, and a vestiture of short or long hairs, bristles or scales. *Larinotus* is unusual in being strongly convex with a deflexed head, *Egolia* is subcylindrical and glabrous and both larinotines and egoliines have a 1- or 2-segmented compact antennal club. Members of the peltine tribe Rentoniini are also unusual in being minute, globose and clambid-like. Although trogossitid tarsi are almost always 5-segmented, the first tarsomere may be reduced and partly concealed by the tibial apex. Larvae are usually elongate, parallel-sided, slightly to strongly flattened and lightly sclerotised except for the head and usually additional sclerites on the thorax and tergum IX; the mandible usually has one or several hyaline processes at the base and the ventral mouthparts are consolidated, without articulating areas; hypostomal rods are usually long and parallel; and urogomphi are usually well developed but minute in *Larinotus*. Most trogossitids are predaceous as larvae and adults, but *Larinotus umbilicatus* feeds on fungal fruiting bodies and some lophocaterines may also be mycophagous. Many species occur under bark, but Rentoniini and some Lophocaterinae occur in leaf litter. Our most common trogossitines in the genus *Leperina* are clothed with scales or bristles forming a pattern. The introduced *Tenebroides mauritanicus* is a cosmopolitan stored products pest; although basically predaceous, both adults and larvae will feed directly on grain and seeds. Another trogossitine, *Temnoscheila virescens*, was introduced from North America to control the bark beetle *Ips grandicollis* (Eichhoff). [Aitken 1975; Crowson 1964b, 1966a, 1970; Ślipiński 1992; Leschen and Lackner 2013].

Australian Subfamilies, Tribes and Genera

Peltinae: Colydiopeltini: *Parapeltis* Ślipiński; **Rentoniini:** New genus, *Parentonium* Crowson.

Lophocaterinae: *Ancyrona* Reitter, *Neaspis* Pascoe, *Peltonyxa* Reitter.

Trogossitinae: Egoliini: *Egolia* Erichson, *Necrobiopsis* Crowson, *Paracalanthis* Crowson; **Larinotini:** *Larinotus* Carter & Zeck; **Gymnochilini:** *Leperina* Erichson, *Phanodesta* Reitter; **Trogossitini:** *Temnoscheila* Westwood, *Tenebroides* Piller & Mitterpacher.

53. THANEROCLERIDAE

Adult (Fig. 71M)

Length 1.7–8 mm. Body ~2.5–3.2 times as long as wide; elongate and cylindrical to slightly flattened; colour reddish-brown or a combination of bright red and black, forming a pattern on elytra; upper surfaces dull or shiny, densely clothed with short and long, erect and suberect hairs.

Head prognathous. Occipital region long. Eyes small to moderately large, slightly protuberant, not or very shallowly emarginate, with very short interfacetal setae. Antennal insertions concealed; subantennal grooves absent. Frontoclypeal suture absent, anterior edge of clypeus distinctly emarginate. Labrum small, free, transverse, slightly emarginate. Antennae 11-segmented, with distinct 3-segmented club. Mandibles unidentate; incisor edge with one or two teeth; mola and prostheca absent. Maxilla setose galea and lacinia, the latter without uncus; apical maxillary palpomeres fusiform. Ligula deeply emarginate; apical labial palpomeres more or less expanded and securiform. Gular sutures narrowly separated. Corpotentorium present. Cervical sclerites well developed.

Pronotum ~0.85–1.12 times as long as wide, widest anteriorly, and somewhat narrower at base than combined elytral bases; sides straight or weakly rounded. Lateral pronotal carinae weak, simple or crenulate, always incomplete posteriorly where there is a distinct collar; anterior angles not produced; posterior angles rounded; posterior edge straight to slightly rounded; disc usually with three distinct impressions, except in *Isoclerus parallelus* (Lewis). Prosternum in front of coxae at least slightly longer than mid length of procoxal cavity, flat, without paired lines or carinae. Prosternal process complete, apically expanded and truncate. Procoxae with long concealed extensions; trochantins concealed. Procoxal cavities circular, narrowly separated, broadly closed externally and open internally.

Scutellar shield abruptly elevated, broadly rounded posteriorly. Elytra ~1.8–2.2.25 times as long as combined width and 2.15–3.1 times as long as pronotum, parallel-sided; punctures irregular without scutellary striole; epipleura incomplete. Mesoventrite with anterior edge on different plane than metaventrite, with broad prothoracic rest or prepectus. Mesoventral process long and narrowly rounded at apex. Mesocoxae not conical and projecting, with concealed trochantins. Mesocoxal cavities circular, narrowly separated, closed laterally by meeting of meso- and metaventrite. Mesometaventral junction monocondylic. Metaventrite slightly convex; discrimen and transverse (metakatepisternal) suture absent; exposed portion of metanepisternum narrowly elongate. Metacoxae strongly transverse, narrowly separated, extending laterally to meet elytra; plates absent. Metendosternite with long, slender stalk, long lateral arms; weak laminae, and short, broad, emarginate anterior process, with anterior tendons on either side of emargination. Hind wing with well developed apical field containing distinct post-radial sclerite; radial cell reduced or absent; basal portion of RP very short or absent; medial field often with at most four free veins, wedge cell absent; anal embayment notch-like. Legs relatively short; femora somewhat inflated, especially profemora. Tibiae without marginal spines. Tarsi 5-5-5 in *Thaneroclerus* Lefebvre, 5-4-4 in *Isoclerus* Lewis; tarsomeres usually simple; pretarsal claws simple; empodium well developed and bisetose.

Abdomen with five or six free ventrites; ventrite 1 not much longer than 2; intercoxal process acute to narrowly rounded. Functional spiracles on abdominal segments I–VII. Anterior edge of segment IX in male with spiculum gastrale; tergites IX and X often partly or completely fused together. Aedeagus of cucujiform type, symmetrical; anterior edge of tegmen with two struts; parameres fused to phallobase and to one another; anterior edge of penis with paired struts. Ovipositor narrowly elongate and lightly sclerotised except for baculi, coxites with well developed apical styli. [Kolibáč 1992, 1998b; Kolibáč and Leschen 2010c].

Larva

Body elongate, parallel-sided and cylindrical to slightly flattened; dorsal and ventral surfaces lightly sclerotised, except for well sclerotised head capsule and tergal plates on thorax and a portion of tergum IX, unpigmented or somewhat reddish; surfaces smooth and vestiture usually of scattered long setae.

Head prognathous and protracted, elongate, parallel-sided and somewhat flattened; posterior edge moderately emarginate. Epicranial stem moderately long; frontal arms V-shaped with contiguous bases. Median endocarina absent or extending anterior to epicranial stem.

Stemmata usually five on each side. Frontoclypeal suture absent. Labrum free. Antennae short, 3-segmented, with palpiform sensorium shorter than apical antennomere. Mandibles unidentate, without accessory ventral process; incisor edge with or without subapical tooth; mola absent; mesal surface of mandibular base with simple hyaline process (lacinia mobilis). Ventral mouthparts protracted, without maxillary articulating area; stipites transverse; mala rounded; with pedunculate seta; maxillary palps usually 3-segmented with distinct palpifer. Labium consisting of prementum and postmentum; ligula rounded; labial palps 2-segmented and moderately narrowly separated. Hypopharyngeal sclerome absent. Hypostomal rods and ventral epicranial ridges absent. Gula distinctly longer than wide, sometimes bearing a large tubercle.

Protergum with single large plate, meso- and metaterga sometimes with paired plates. Prothoracic venter with well developed sternal plate. Legs moderately well developed, 5-segmented including claw-like pretarsus bearing one seta; coxae widely separated. Segment IX excluding appendages shorter than VIII, tergum extending onto ventral surface; dorsal surface with small sclerotised plate; posterior edge with pair of very small urogomphi in *Thaneroclerus*. Sternite IX exposed, simple; anal region posteroventrally oriented. Spiracles annular–biforous. [Böving and Champlain 1922; Crowson 1964b; Foster 1976; Kolibáč 1992, 1999, 2004b; Lawrence *et al.* 1999a; Kolibáč and Leschen 2010c].

Classification and World Distribution

Thanerocleridae is a small group with ~30 described species in seven genera. There are two subfamilies: Zenodosinae with a single Nearctic genus and species and the widely distributed Thaneroclerinae.

Australian Fauna

The Australian thaneroclerine fauna consists of the introduced *Thaneroclerus buquet* (Lefèvre) and four species of *Isoclerus*

Lewis (Fig. 71M). Thaneroclerines resemble members of the related family Cleridae, but our species may be distinguished by their small size, globular procoxae with externally closed coxal cavities, laterally closed mesocoxal cavities, compact protarsi with four basal tarsomeres expanded, mandibles with a deep basal notch, and tegmen without a basal strut. Larvae resemble some clerid larvae, in having an elongate, parallel-sided head capsule, with protracted mouthparts and a narrowly elongate gula, but there is a moderately long epicranial stem, no distinct endocarina and urogomphi on tergum IX rudimentary or absent. *Thaneroclerus buquet* is a predator of various stored products insects, such as the ptinid *Lasioderma serricorne*, while the species of *Isoclerus* are usually found in woody bracket fungi, where they feed on Ciidae or small Tenebrionidae. [Crowson 1964b; Kolibáč 1992, 1998b; Kolibáč and Leschen 2010c].

54. CLERIDAE

Adult (Figs 8F–G, 71G-L, 82J)

Length 2.4–44 (usually 5–25) mm. Body ~2–5 times as long as wide; usually elongate and more or less cylindrical, slender and elongate in Tillinae and some Hydnocerinae; colour variable, often brightly coloured or metallic; upper surfaces usually densely clothed with erect and suberect hairs, sometimes combined with decumbent hairs or tufts of thickened setae, but never with scales.

Head prognathous. Eyes usually moderately large and slightly to moderately protuberant (flattened in *Cylidrus*; strongly protuberant in Hydnocerinae), very shallowly to deeply emarginate, with or without interfacetal setae. Antennal insertions exposed or concealed; subantennal grooves absent. Frontoclypeal suture present or absent. Labrum free, usually slightly emarginate. Antennae almost always 11-segmented (10-segmented in *Monophylla* Spinola, 9-segmented in *Sedlacekvia* Winkler), filiform, serrate, pectinate, clavate or capitate; club usually 3-segmented and often asymmetrical (6-segmented in *Cylidrus* Latreille, 1-segmented in *Sedlacekvia* and *Monophylla*). Mandibles unidentate; incisor edge with one or two teeth; mola absent; prostheca well developed to absent. Maxilla with setose galea and lacinia, the latter without uncus; apical maxillary palpomeres cylindrical or fusiform to securiform. Ligula deeply emarginate; apical labial palpomeres cylindrical or fusiform to securiform. Gular sutures strongly convergent. Corpotentorium present. Cervical sclerites present.

Pronotum ~0.5–1.5 times as long as wide, widest at middle or anteriorly, and usually narrower at base than combined elytral bases (not so in flightless forms); often constricted anteriorly and posteriorly to form collars; sides usually strongly rounded, rarely almost straight. Lateral pronotal carinae absent in Tillinae, Hydnocerinae and Clerinae, almost always complete in Korynetinae (weakly developed in *Tarsostenus* Spinola), often not visible from above, rarely with bead, their edges usually simple (denticulate in *Crobenia* Blackburn); anterior angles not produced; posterior angles usually right; posterior edge straight to slightly rounded; disc usually simple, sometimes with median groove or impression. Prosternum in front of coxae (Fig. 18L) shorter than to somewhat longer than mid length of procoxal cavity, without paired lines or carinae. Prosternal process usually more or less complete, parallel-sided and narrowly rounded or acute at apex or apically expanded and truncate. Procoxae projecting, with or without long concealed extensions; trochantins usually exposed. Procoxal cavities transverse to circular, narrowly separated, broadly open to broadly closed externally and usually open internally (broadly closed in *Cylidrus*).

Scutellar shield abruptly elevated or not, broadly rounded to truncate posteriorly. Elytra ~1.45–3.65 times as long as combined width and 1.4–5.3 times as long as pronotum, parallel-sided and almost always complete with conjointly rounded apices (occasionally exposing one or two abdominal tergites); punctures irregular or forming 10 puncture rows, without scutellary striole; epipleura usually narrow and incomplete, sometimes complete or absent. Mesoventrite with anterior edge almost always on same plane as metaventrite (on different plane in *Zenithicola* Spinoloa); with paired, slightly inclined procoxal rests or with broad prothoracic rest or prepectus incorporating mesanepisterna. Mesoventral process long and narrowly rounded at apex. Mesocoxae usually projecting, with exposed trochantins (concealed in *Cylidrus*). Mesocoxal cavities circular to slightly transverse, narrowly separated, usually partly closed laterally by mesepimeron (occasionally by both mesanepisternum and mesepimeron), closed laterally by meeting of meso- and metaventrite in *Cylidrus*. Mesometaventral junction usually monocondylic. Metaventrite with discrimen very long to short or absent; transverse (metakatepisternal) suture absent; postcoxal lines absent; exposed portion of metanepisternum narrowly elongate. Metacoxae narrowly separated, extending laterally to meet elytra; plates absent. Metendosternite usually with long, slender stalk, long lateral arms; reduced, slender laminae; anterior process short to moderately long, with anterior tendons narrowly to widely separated, or sometimes absent. Hind wing with well developed apical field containing distinct post-radial sclerite; radial cell well developed and slightly elongate to reduced or absent; cross-vein r3 horizontal or oblique, sometimes fused at base to r4 or absent; basal portion of RP usually short; medial field at most with four free veins, wedge cell present or absent; anal embayment usually weakly developed or absent, sometimes notch-like. Legs well developed; femoral attachment of mid trochanter slightly to strongly oblique; femora not clavate. Tibiae without marginal spines; tibial apex with or without spurs. Tarsi always 5-5-5; tarsomere 1 small and concealed from above by 2 in Clerinae, some Hydnocerinae and Korynetinae, as large as 2 in Tillinae and some Korynetinae; tarsomere 4 reduced and more or less concealed within base of 3 in Korynetinae; tarsomeres 1–4 always with membranous lobes; pretarsal claws simple or with one or two teeth; empodium well developed and bisetose.

Abdomen usually with six free ventrites; ventrite 1 not much longer than 2; intercoxal process acute to narrowly rounded. Functional spiracles on abdominal segments I–VII. Anterior edge of segment IX in male with spiculum gastrale; tergites IX and X often partly or completely fused together. Aedeagus of cucujiform type, symmetrical; anterior edge of tegmen with single strut; parameres fused to phallobase and to one another; anterior edge of penis with paired struts. Ovipositor narrowly elongate and lightly sclerotised except for baculi, coxites with well developed apical styli. [Crowson 1964b; Kolibáč 1987, 1989a, 1997, 1998a, 1998c, 1999, 2003, 2004a; 2010; Lawrence *et al.* 1999b; Opitz 2002].

Larva (Figs 34H, 82C)

Body elongate, usually parallel-sided and cylindrical to slightly flattened; dorsal and ventral surfaces lightly sclerotised, except for well sclerotised head capsule, thoracic plates, legs, tergum IX and urogomphi; usually pink to white in colour, but occasionally mottled with violet; surfaces smooth and sparsely or densely clothed with long or short hairs.

Head prognathous and protracted, parallel-sided or with sides more rounded, usually somewhat flattened, but occasionally more globular; posterior edge slightly emarginate. Epicranial stem usually short or absent; frontal arms V-shaped or U-shaped, usually with contiguous bases (well separated in Hydnocerinae). Median endocarina almost always present between frontal arms. Stemmata usually five on each side, occasionally fewer or none. Frontoclypeal suture absent. Labrum free. Antennae short to moderately long, 3-segmented, with palpiform sensorium shorter than apical antennomere. Mandibles unidentate, without accessory ventral process; incisor edge simple or with one or two subapical teeth; mola absent; mesal surface of mandibular base with lacinia mobilis consisting of two acute and spine-like hyaline processes. Ventral mouthparts protracted, without maxillary articulating area; stipites transverse; mala rounded; with pedunculate seta; maxillary palps usually 3-segmented with distinct palpifer. Labium consisting of prementum and postmentum; ligula rounded; labial palps 2-segmented and moderately narrowly separated. Hypopharyngeal sclerome absent. Hypostomal rods short and subparallel or absent; ventral epicranial ridges absent. Gula distinctly longer than wide.

Protergum usually with single large plate and meso- and metaterga with paired plates (all plates absent in some larvae). Prothoracic venter sometimes with one or more sternal plates. Legs moderately well developed, 5-segmented including claw-like pretarsus bearing one seta; coxae widely separated. Abdomen sometimes with tergal ampullae. Segment IX excluding appendages shorter than VIII, tergum extending onto ventral surface; dorsal surface with or without sclerotised plate; posterior edge usually with pair of well developed urogomphi (sometimes reduced and hook-like). Sternite IX exposed, simple; anal region posteroventrally oriented. Spiracles annular or annular–biforous. [Gardner 1937; Crowson 1964b; Moore 1973; Mamaev 1977; Kolibáč 1989b, 1997, 2010; Lawrence *et al.* 1999a].

Classification and World Distribution

Cleridae is a cosmopolitan family containing ~200 genera and 3500 species. There are four subfamilies, all of which occur in Australia. Tillinae are most abundant in the Oriental and Afrotropical regions, with a few genera extending into the Holarctic and Australian regions. Hydnocerinae, Clerinae and Korynetinae are all worldwide in distribution.

Australian Fauna

Clerids are elongate, usually parallel-sided beetles, clothed with erect or erect and decumbent hairs and often either metallic in colour or patterned with red or yellow. The head is prognathous and the eyes often slightly to strongly emarginate; antennae are relatively short and labial palps are often enlarged and securiform. The pronotum usually lacks lateral carinae, the prosternum is well developed in front of the coxae, one or more tarsomeres are lobed below or bear membranous appendages, tarsomere 1 is sometimes reduced and concealed within the tibial apex and in Korynetinae tarsomere 4 is reduced and enclosed within the base of tarsomere 3. Larvae are elongate, cylindrical to somewhat flattened and lightly sclerotised except for head, one of more tergal plates and tergum IX, but cuticle often pigmented with yellow, pink, blue or brown. The larval head is well sclerotised dorsally and ventrally, with the epicranial stem absent, a median endocarina usually present, a long, slender gula, and the mandible with a hyaline process at the base. Both larvae and adults of most Cleridae prey upon other insects, especially those associated with bark or wood. Adults are very active and usually diurnal; they are often found on flowers, foliage and tree trunks. There are only two native Australian species of **Tillinae**, both in the genus *Cylidrus*. Other tillines, including *Cylidrus cyaneus* Fabricius, *C. wallacei* Thomson, *Cylidroctenus chalybaeus* Wester, *Monophylla terminata* (Say) and *Tilloidea notata* (Klug) are adventive. *Cylidrus nigrinus* White and *C. basalis* Macleay (Fig. 71J) were observed entering the tunnels of two species of xyloperthine Bostrichidae boring into freshly cut branches of *Eucalyptus mannifera maculosa* in the ACT and their larvae were later found in the tunnels; the exotic *Cylidrus fasciatus* Laporte is a predator of *Synoxylon senegalense* Karsch (Peake 1953). The subfamily **Hydnocerinae** includes the large genus *Lemidia* (Fig. 71H) and the related *Allelidea*; these are relatively small, brightly coloured beetles with a slender prothorax and protuberant eyes. Adults of *Lemidia subaenea* Gorham have been observed feeding of psyllids (Hemiptera: Psyllidae) on the surfaces of *Acacia* leaves (New 1978a), and *Lemidia* larvae, recognisable by the head with well separated frontal arms, are probably the most common clerids in leaf litter samples. **Clerinae** and **Korynetinae** are both large groups, the former almost always having the first tarsomere reduced and more or less concealed (not so in *Eleale*) and the latter with the fourth tarsomere reduced and enclosed within the lobe on tarsomere 3. Little is known of the biology of our larger Clerinae, such as *Eunatalis titana* (Thompson) and *Trogodendron fasciculatum* (Schreibers), but the latter is a predator of wood-boring Cerambycidae. The unusual clerine genus *Sedlacekvia* was unable to be placed in a subfamily by its author (Winkler 1989). *Zenithicola crassus* (Newman) (Fig. 71G) is a predator in termite nests and its larva is clothed with very long hairs like those of the termite-feeding elaterid larva of the genus *Pseudotetralobus* (Moore 1973). The Korynetinae is defined here in the broadest sense following Kolibáč (1997, 2003) and including the subfamilies Enopliinae, Epiphloeinae and Tarsosteninae recognised in most general works (Lawrence and Newton 1995; Solervicens 2005a), as well as the additional subfamilies Peloniinae and Neorthopleurinae recognised by Opitz (2010). Some korynetines, such as *Paratillus carus* (Newman) and *Tarsostenus univittatus* (Rossi) invade the tunnels of various Bostrichidae and a few scolytine Curculionidae. The introduced *Necrobia ruficollis* (Fabricius) (Fig. 71K) and *N. rufipes* (Degeer) occur in carrion and in stored products of animal origin; the latter, however, is also common in copra cargoes, where it feeds directly on the copra, as well as on other insects. [Beeson 1926; Froggatt 1927; Crowson 1964b; Aitken

1975; Winkler 1982; Kolibáč 1987, 1989a, 1989b, 1998a, 1998c; Gerstmeier 1998a, 1998b, 2002; Solervicens 2007; Bartlett 2009a, 2009b, 2009c, 2009d].

Australian Subfamilies, Tribes and Genera

Tillinae: *Cylidrus* Latreille, *Monophylla* Spinola, *Tillus* Olivier.
Hydnocerinae: *Allelidea* Waterhouse, *Lemidia* Spinola (=*Blaesiophthalmus* Schenkling).
Clerinae: *Astigmus* Kuwert, *Cleromorpha* Gorham, *Eleale* Newman, *Natalis* Laporte, *Neoscrobiger* Blackburn, *Odontophlogistus* Elston, *Olesterus* Spinola, *Omadius* Castelnau, *Opilo* Latreille, *Orthrius* Gorham, *Phlogistomorpha* Hintz, *Phlogistus* Gorham, *Sedlacekvia* Winkler, *Scrobiger* Spinola, *Stigmatium* Gray, *Thanasimus* Latreille, *Trogodendron* Spinola, *Zenithicola* Spinola.
Korynetinae: *Anthicoclerus* Schenkling, *Apopylus* Kolibáč, *Apteropilo* Lea, *Blackburniella* Chapin, *Cregya* LeConte, *Crobenia* Blackburn, *Necrobia* Olivier, *Opetiopalpus* Spinola, *Pylus* Newman, *Tarsostenodes* Blackburn, *Tarsostenus* Spinola, *Teneromimus* Gahan, *Teneropsis* Chapin, *Tenerus* Laporte, *Thriocera* Gorham, *Thriocerodes* Wolcott & Dybas.

55. ACANTHOCNEMIDAE

Adult (Fig. 82F)

Length 2.5–5.5 mm. Body ~2.3–2.55 times as long as wide; more or less parallel-sided, slightly flattened; colour dark brown to black; upper surfaces densely clothed with stiff, dark suberect bristles.

Head prognathous. Eyes protuberant, not emarginate, without interfacetal setae. Antennal insertions exposed; subantennal grooves absent. Frontoclypeal suture indistinct; anterior edge of clypeus truncate. Labrum free, strongly transverse, subtruncate. Antennae 11-segmented with 3-segmented club. Mandibles bidentate; incisor edge with broad tooth or retinaculum; mola absent; prostheca a large brush of hairs; mandibular base with large, membranous, setose process. Maxilla with moderately broad, setose galea and lacinia, the latter without uncus; apical maxillary palpomeres weakly fusiform, almost cylindrical. Ligula distinctly emarginate; apical labial palpomeres weakly fusiform. Gula longer than wide. Corpotentorium absent. Cervical sclerites present.

Pronotum ~0.7–0.75 times as long as wide, widest at about middle, base not or only slightly narrower than elytral bases; sides slightly curved and slightly explanate. Lateral pronotal carinae complete with narrow bead; anterior angles not produced; posterior angles rounded; disc simple. Prosternum in front of coxae (Fig. 44I) longer than mid length of procoxal cavity, flat to slightly convex. Prosternal process complete, narrowed apically, not elevated or curved behind coxae, its apex narrowly acute. Hypomera each with a heat-sensing organ (Fig. 26G) consisting of a broad, shallow cavity partly closed by a circular flap. Procoxae projecting, without concealed extensions; trochantins broadly exposed. Procoxal cavities strongly transverse, narrowly separated, externally broadly open and internally open.

Scutellar shield abruptly elevated, posteriorly truncate. Elytra ~1.75–1.87 times as long as combined width and 2.85–3.2 times as long as pronotum; punctures irregular, without scutellary striole; epipleura narrow and incomplete. Mesoventrite without paired procoxal rests. Mesoventral process long and subacute at apex. Mesocoxae with exposed trochantins. Mesocoxal cavities slightly transverse, narrowly separated, partly closed laterally by mesepimeron; meso- and metathoracic portions of mesocoxal cavity joined by membranous suture. Mesometaventral junction simple, with ventrites meeting at single point. Metaventrite with long discrimen; transverse (metakatepisternal) suture more or less complete and postcoxal lines absent; exposed portion of metanepisternum moderately elongate. Metacoxae contiguous, extending laterally to meet elytra; plates absent. Metendosternite with long stalk, long lateral arms and well developed laminae; anterior process very short with anterior tendons widely separated. Hind wing with well developed apical field containing distinct post-radial sclerite and vaguely indicated anterior and posterior radial extensions; radial cell short and broad with inner posterobasal angle acute; cross-vein r3 short, oblique, fused at base to r4; basal portion of RP very short; medial spur straight, extending almost to wing margin where there is a slight embayment; medial field with four free veins and no medial fleck; wedge cell well developed, obliquely truncate at apex; anal embayment deep and notch-like. Legs moderately long and slender; femoral attachment of mid trochanter slightly oblique. Tibiae outwardly lined with stiff spines; spurs paired on all legs. Tarsi always 5-5-5 in female, 4-5-5 in male; tarsal lobes absent; pretarsal claws simple; empodium small but exposed between claws.

Abdomen usually with five free ventrites in male, six in female; ventrite 1 not much longer than 2; intercoxal process acute. Functional spiracles on abdominal segments I–VII. Anterior edge of sternite VIII in male with median strut. Anterior edge of segment IX in male acute but without spiculum gastrale; tergites IX and X fused together. Aedeagus of cucujiform type, symmetrical; anterior edge of tegmen with ventral strut and two opposing dorsal struts; parameres fused to phallobase and to one another; anterior edge of penis with paired struts. Sternite VIII in female with spiculum ventrale. Ovipositor ~4.5 times as long as wide, lightly sclerotised except for baculi, paraprocts ~1.8 times as long as coxites, each of which is subdivided, with transverse basal baculum and elongate distal lobe; styli well developed, and apical. [Champion 1922; Lawrence *et al.* 1999b; Lawrence and Leschen 2010c].

Larva

Body elongate and more or less parallel-sided, lightly pigmented, except for head and tergum IX; vestiture of scattered setae.

Head prognathous and protracted, somewhat flattened; posterior edge distinctly emarginate. Epicranial stem absent or very short; frontal arms V-shaped and contiguous at base. Median endocarina present between frontal arms. Stemmata five on each side. Frontoclypeal suture absent. Labrum free. Antennae moderately long, 3-segmented, with broadly conical sensorium much shorter than apical antennomere. Mandibles bidentate, without accessory ventral process; incisor edge with one or two subapical teeth; mola absent; mesal surface of mandibular base with lacinia mobilis consisting of five or six acute and spine-like hyaline processes. Ventral mouthparts retracted, without maxillary articulating area; stipites longer than wide; mala truncate, with pedunculate seta; maxillary palps 3-segmented. Labium consisting of prementum and postmentum; ligula simple, not bilobed; labial palps 2-segmented and separated by more than width of first palpomere. Hypopharyngeal sclerome absent. Hypostomal rods long and subparallel; ventral epicranial ridges absent. Gula longer than wide.

Protergum usually with single large plate and meso- and metaterga with paired plates. Prothoracic venter with elongate, subquadrate sternal plate. Legs short, 5-segmented including claw-like pretarsus bearing one seta; coxae widely separated. Segment IX excluding appendages shorter than VIII, tergum extending onto ventral surface; dorsal surface lightly sclerotised; posterior edge with pair of well developed urogomphi. Sternite IX exposed, simple; anal region posteroventrally oriented. Spiracles annular-biforous. [Lawrence and Leschen 2010c].

Classification and World Distribution

Acanthocnemidae are native to Australia but have been introduced into southern Europe, Africa, India, Thailand, Myanmar and New Caledonia. [Lawrence and Leschen 2010c].

Australian Fauna

Acanthocnemus nigricans (Hope) (= *Dasytes fuscipennis* Hope) is widely distributed throughout Australia, from the Cape York Peninsula and Kakadu National Park in the north to TAS and the southern part of WA. These are elongate, flattened, black beetles, clothed with erect hairs and similar to some dasytine Melyridae, from which they differ in having a distinct antennal club and a pair of partly covered, shallow hypomeral cavities which have been shown to be highly sensitive infrared receptor organs called microbolometers serving to attract both sexes to sites of recent fires (Schmitz and Schmitz 2002; Schmitz *et al.* 2002; Kreiss *et al.* 2005). Adults are known to be attracted to log dumps and campfires; larvae have not yet been collected but those reared from eggs laid in the laboratory fed on cooked meat and boiled eggs. [Lawrence and Leschen 2010c].

56. PHYCOSECIDAE

Adult (Figs 8H, 82E)

Length 1.5–3.4 mm. Body ~1.9–2.25 times as long as wide; with sides of pronotum and elytra rounded and body narrowest at pronoto-elytral junction; moderately convex to slightly flattened; brown to black in colour; upper surfaces densely clothed with white suberect or adpressed scales.

Head prognathous. Occipital region relatively long. Eyes small, entire, protuberant and coarsely facetted, with fine interfacetal setae. Antennal insertions barely exposed from above; subantennal grooves short. Frontoclypeal suture absent; anterior edge of clypeus emarginate. Labrum free, strongly transverse, its apex slightly concave. Antennae almost 10-segmented, with 1-segmented club. Mandibles short, broad, bidentate; incisor edge with one tooth; mola absent; prostheca reduced. Maxilla with moderately broad, setose galea and lacinia, the latter without uncus; apical maxillary palpomere fusiform. Ligula shallowly emarginate; apical labial palpomere fusiform. Genae with anteriorly projecting processes. Gula longer than wide. Corpotentorium absent. Cervical sclerites present.

Pronotum ~0.85–1.0 times as long as wide, widest just behind middle, anteriorly produced forward to partly cover head. Lateral pronotal carinae complete, simple, without raised margin or bead; anterior angles produced forward and narrowly rounded; posterior angles rounded; posterior edge evenly rounded; disc simple. Prosternum in front of coxae about as long as mid length of procoxal cavity, flat to slightly convex. Prosternal process complete, distinctly curved, abruptly expanded apically and truncate. Procoxae with long concealed extensions; trochantins concealed. Procoxal cavities slightly transverse, narrowly separated, broadly closed externally, with narrow lateral extension; open internally.

Scutellar shield absent. Elytra ~1.22–1.4 times as long as combined width and 1.55–1.95 times as long as pronotum, broadly curved laterally; punctures irregular, without scutellary striole; epipleura usually narrow and complete. Mesoventrite at least partly fused to mesanepisterna, which are usually widely separated at midline; anterior edge without paired procoxal rests, with broad prothoracic rest or prepectus incorporating mesanepisterna. Mesoventral process moderately long and narrowly rounded at apex. Mesocoxae with trochantins concealed. Mesocoxal cavities circular to slightly transverse, narrowly separated, closed laterally by meeting of meso- and metaventrite. Mesometaventral junction simple. Metaventrite shorter than first abdominal ventrite, with short discrimen; exposed portion of metanepisternum short and broad. Metacoxae somewhat transverse, narrowly separated, not extending laterally to meet elytra; plates absent. Metendosternite with long lateral arms, no laminae; anterior process and anterior tendons apparently absent. Hind wing absent. Legs slender; femoral attachment of mid trochanter slightly oblique. Tibiae with marginal and apical marginal spines. Tarsi 4-4-4; tarsomeres without ventral lobes; pretarsal claws simple; empodium small.

Abdomen with five free ventrites; ventrite 1 not much longer than 2; intercoxal process acute. Functional spiracles on abdominal segments I–VII. Anterior edge of segment IX in male without spiculum gastrale; tergites IX and X fused together. Aedeagus of cucujiform type, symmetrical; anterior edge of tegmen with single strut; parameres fused to phallobase; anterior edge of penis without paired struts. Sternite VIII in female with short spiculum ventrale. Ovipositor short and broad, lightly sclerotised except for baculi, paraprocts shorter than coxites, which are located mesal to them, each divided into broad basal and narrow apical lobe with small, button-like apical stylusi. [Crowson 1964b; Majer 1994b; Lawrence *et al.* 1999b; Leschen and Beutel 2010c].

Larva

Body elongate, slightly flattened, with metathorax and abdominal segments distinctly rounded laterally; lightly sclerotised and unpigmented, except for head capsule, protergal plate and tergum IX; surfaces smooth and sparsely clothed with setae.

Head prognathous and protracted, almost parallel-sided and somewhat flattened. Epicranial stem moderately long; frontal arms V-shaped with contiguous bases. Median endocarina absent.

Stemmata six on each side, an anterior curved row of four and two more above and behind. Frontoclypeal suture absent. Labrum free. Antennae very short, 3-segmented, with broadly conical sensorium almost as long as apical antennomere. Mandibles bidentate, without accessory ventral process; incisor edge with single tooth; mola absent; mesal surface of mandibular base with lacinia mobilis consisting of a plumose basal appendage. Ventral mouthparts retracted, without maxillary articulating area; stipites elongate; mala rounded; with row of strong spines and pedunculate seta; maxillary palps 3-segmented with distinct palpifer. Labium consisting of prementum and postmentum; ligula rounded; labial palps 2-segmented and moderately widely separated. Hypopharyngeal sclerome absent. Hypostomal rods moderately long and slightly converging; ventral epicranial ridges absent. Gular region short and broad, much more lightly pigmented than rest of head capsule (ventrally open head capsule).

Protergum with small medial plate; meso- and metaterga without plates; without rows of asperities or carinae. Prothoracic venter sometimes with one or more sternal plates. Legs moderately well developed, 5-segmented including claw-like pretarsus bearing one seta; coxae widely separated. Segment IX excluding appendages shorter than VIII, tergum extending onto ventral surface; dorsal surface lightly scledrotised; posterior edge with pair of well developed, subparallel, posteriorly projecting urogomphi. Sternite IX exposed, simple; anal region posteroventrally oriented. Spiracles annular. [Crowson 1964b; Lawrence *et al.* 1999a; Beutel and Pollock 2000; Leschen and Beutel 2010c].

Classification and World Distribution

Phycosecidae contains the single genus *Phycosecis* Pascoe, with species known from New Zealand, Australia, New Caledonia and Vanuatu.

Australian Fauna

There are three Australian species of *Phycosecis*: *P. litoralis* (Pascoe), which is distributed along the southern coast from WA to NSW, *P. ammophila* Lea from WA and *P. hilli* Lea from northern QLD. Phycosecids are small, ovate, convex beetles with a prognathous head, partly covered by a semicircular projection of the pronotum and a dorsal vestiture of whitish scales or scale-like setae. The anterior pronotal angles are produced forward, the metaventrite is very short and the hind wings are absent. Larvae are similar to those of Melyridae, except for the short epicranial stem and presence of six stemmata. Larvae and adults live among sand dunes along coastlines and are scavengers on dead fishes and birds. Crowson (1964b) found insect parts in the larval gut of the New Zealand species *P. limbata* (Fabricius), but *P. litoralis* has been reared through an entire generation in a closed container with fish, lettuce and bran, and larvae were usually clustered beneath pieces of rotting fish.

57. MELYRIDAE

Adult (Figs 8A, C–D, 83A–B, I)

Length 1.5–10 mm. Body ~1.75–5.55 times as long as wide; moderately convex to more or less flattened; colour usually uniformly black, brown or yellow in Dasytinae and often bicolored with a combination of red or orange with blue or black in Malachiinae; upper surfaces usually clothed with fine, decumbent hairs, often mixed with stiff, dark, erect bristles. Prothorax, metathorax and/or abdomen with eversible vesicles in Malachiinae.

Head prognathous. Eyes not to strongly protuberant, entire or shallowly emarginate, with or without interfacetal setae. Antennal insertions exposed, moderately narrowly to widely separated; subantennal grooves absent. Frontal region sometimes excavate in males. Frontoclypeal suture usually distinct. Labrum free, usually transverse to subquadrate. Antennae 11-segmented (apparently 10-segmented in *Dicranolaius* Champion and a few related genera, in which the pedicel is more or less embedded within the apex of the scape), usually filiform or serrate, but pectinate in *Flabellolaius* Wittmer and *Balanophorus* Macleay. Mandibles moderately elongate, unidentate or bidentate, with incisor edge sometimes denticulate; mola and prostheca reduced or absent. Maxilla with galea densely setose and lacinia without uncus; apical maxillary palpomeres fusiform. Ligula undivided or bilobed; apical labial palpomeres fusiform. Subgenal ridges absent. Gula longer than wide. Corpotentorium narrow or absent. Cervical sclerites usually weakly developed.

Pronotum ~0.55–1.5 times as long as wide, usually widest at middle or near base (widest apically in some Carphurini); base usually not or only slightly narrower than combined elytral bases (distinctly narrower in some Carphurini); sides straight to strongly curved or occasionally sinuate, not explanate. Lateral pronotal carinae usually complete and simple, with of without narrow bead; anterior angles not produced; posterior angles rounded or right; disc usually simple (excavate in males of *Thoraxocarphurus* Wittmer). Prosternum in front of coxae shorter than to longer than mid length of procoxal cavity, flat to slightly convex. Prosternal process incomplete or absent, usually narrowed with subacute apex. Procoxae projecting, without concealed extensions; trochantins broadly exposed. Procoxal cavities slightly to strongly transverse, contiguous, externally broadly open and internally open.

Scutellar shield not abruptly elevated, posteriorly narrowly rounded to truncate. Elytra ~0.45–1.8 times as long as combined width and 0.5–5.35 times as long as pronotum, parallel-sided and complete with conjointly rounded apices in Dasytinae; often truncate, exposing several abdominal tergites in Carphurini; punctation irregular, without scutellary striole; epipleura narrow and incomplete or absent. Mesoventrite with or without paired procoxal rests. Mesoventral process complete or incomplete, with narrowly rounded or subacute at apex. Mesocoxae sometimes conical and projecting, with exposed trochantins. Mesocoxal cavities slightly transverse, narrowly separated or contiguous, partly closed laterally by mesanepisternum and mesepimeron; meso- and metathoracic portions of mesocoxal cavity solidly fused. Mesometaventral junction absent or simple, with ventrites meeting at single point. Metaventrite with moderately to very long discrimen; postcoxal lines absent; exposed portion of metanepisternum usually moderately to very elongate (shorter and broader in some Malachiinae). Metacoxae contiguous, usually extending laterally to meet elytra; plates absent. Metendosternite with long stalk and long lateral arms; laminae sometimes present; anterior tendons widely separated on lateral arms. Hind wing with short to moderately long apical field containing one or two post-radial sclerites and sometimes vaguely indicated radial extensions; radial cell usually short and broad with inner posterobasal angle acute, sometimes absent; cross-vein r3 short or absent, sometimes fused at base to r4; basal portion of RP short or absent; medial spur usually sinuate and complete; medial field usually with three free veins (sometimes four) and no medial fleck; wedge cell usually absent; anal embayment deep and notch-like. Legs moderately long and slender. Tibiae without spines, except at apex; metatibiae unusually long in some Malachiinae; spurs reduced or absent. Tarsi 5-5-5 in both sexes; one or more tarsal lobes present in some Malachiinae; black apical sex comb present in males of all Carphurini and at least some males of *Dicranolaius*; pretarsal claws usually dentate or with membranous appendages; empodium moderately well developed and bisetose.

Abdomen usually with six free ventrites (seven in some Carphurini); ventrite 1 not much longer than 2; intercoxal process acute or absent. Functional spiracles on abdominal segments I–VII. Anterior edge of sternite VIII in male with median strut in some Dasytinae. Anterior edge of segment IX in male with spiculum gastrale; tergites IX and X fused together. Aedeagus of cucujiform type, symmetrical; anterior edge of tegmen occasionally with single strut; parameres fused to phallobase and usually short or not apparent; anterior edge of penis usually without struts. Sternite VIII in female with long, articulated spiculum ventrale. Ovipositor moderately elongate, lightly sclerotised except for baculi, paraprocts distinctly longer than coxites, each of which is subdivided into basal and apical sections, the former with oblique basal baculum; styli well developed, and apical. [Crowson 1964b; Majer 1987, 1994b; 2002; Mayor 2002; Lawrence and Leschen 2010e].

Larva (Fig. 34I)

Body elongate, more or less parallel-sided or with head and tergum IX somewhat narrower than anterior trunk segments, slightly flattened. Dorsal surfaces lightly sclerotised, except for head and usually tergum IX, but sometimes with reddish pigmentation similar to that in larval Cleridae, often with one or more pigmented areas on thoracic terga or rarely on trunk segments. Surfaces smooth and vestiture usually consisting of sparsely or densely distributed long setae.

Head (Fig. 43B) prognathous and protracted, flattened, slightly transverse to elongate, often parallel-sided. Epicranial stem moderately long; frontal arms V-shaped and contiguous at base. Median endocarina absent. Stemmata varying from one to five on each side (usually five in Dasytinae and four in Malachiinae). Frontoclypeal suture absent. Labrum free. Antennae short to moderately long, 3-segmented, with broadly conical sensorium shorter than apical antennomere. Mandibles symmetrical, bidentate, without accessory ventral process; incisor edge usually with one subapical tooth; mola absent; mesal surface of mandibular base with lacinia mobilis which may be subdivided and ciliate. Ventral mouthparts retracted, without maxillary articulating area; cardines divided or not; stipites longer than wide; mala rounded, not cleft, with pedunculate seta; maxillary palps 3-segmented. Labium consisting of prementum and postmentum; ligula simple or absent; labial palps 2-segmented and narrowly to moderately widely separated. Hypopharyngeal sclerome, hypostomal rods and ventral epicranial ridges absent. Gular region well developed but very lightly pigmented (ventrally 'open' head capsule).

Protergum sometimes with one or more pigmented areas. Prothoracic venter without plates or maculae. Legs moderately long, 5-segmented including claw-like pretarsus bearing one seta; coxae widely separated. Abdomen with paired glandular orifices. Segment IX excluding appendages usually as long as or slightly longer than VIII, tergum extending onto ventral surface; dorsal surface often with sclerotised plate; posterior edge almost always with pair of well developed, posteriorly projecting or slightly upturned urogomphi, usually simple but with accessory processes in some Malachiinae (urogomphi absent in an unassociated dasytine larva from the ACT). Sternite IX exposed, simple; anal region posteroventrally oriented. Spiracles annular. [Böving and Craighead 1931; Crowson 1964b; Fiori 1971; Costa *et al.* 1988; Constantin 1990; Constantin and Menier 1990; Foster and Lawrence 1991; Mayor 2002; Asano and Kojima 2009; Asano and Yoshitomi 2011; Lawrence and Leschen 2010e].

Classification and World Distribution

Melyridae is a large, cosmopolitan family with ~300 genera and 6000 species. The group may be divided into four subfamilies, two of which occur in Australia; however, Majer (1994a, 1994b, 2002) and Bocákova *et al.* (2011) treat this group as four or five families. The exotic subfamily Rhadalinae occurs in north-western North America, Mexico and Central America, Eurasia, the Afrotropical Region, the Seychelles and Mascarenes, and the Indo-Australian Archipelago (Peacock 1987), and is considered by Bocákova *et al.* (2011) to be a separate family. The Melyrinae occur from the eastern United States to Chile and Argentina, and in southern Europe, Asia Minor, Central Asia, India and Africa. Dasytinae and Malachiinae are both very widely distributed, but appear to be most abundant and diverse in arid and semiarid regions.

Australian Fauna

The Australian melyrid fauna consists of a relatively small number of **Dasytinae**, which are either undescribed or incorrectly placed in the genus *Dasytes* Paykull (Fig. 83B), and a larger number of **Malachiinae**, which belong to three tribes: Malachiini with *Dicranolaius* Champion (Fig. 83A), *Flabellolaius* Wittmer, *Laius* Guérin-Méneville, *Simoderus* Abeille de Perrin and *Troglolaius* Wittmer, Lemphini with the single genus *Hypattalus* Blackburn, and Carphurini, including *Balanophorus* Macleay (Fig. 83I), *Carphurus* Erichson, *Helcogaster* Boheman, *Neocarphurus* Lea and *Thoraxocarphurus* Wittmer. Melyrids are oblong to elongate, somewhat flattened, soft-bodied beetles, clothed with decumbent hairs and scattered erect setae, and either uniformly black or grey (most Dasytinae) or bicoloured with red or yellow and blue or black (Malachiinae). The labrum is well sclerotised, the tarsal claws are almost always either toothed or provided with a pair of fleshy appendages, the elytra in Carphurini are very short, exposing several abdominal tergites, and there are often eversible vesicles on the prothorax, mesothorax and abdomen. Males of *Carphurus*, *Neocarphurus* and *Helcogaster*, and males and females of *Balanophorus* have the first protarsomere enlarged and provided with a black comb. Those malachiines with complete elytra may have 11-segmented antennae as in *Hypattalus* or apparently 10-segmented (with the pedicel reduced and concealed within the apex of the scape). In males of *Dicranolaius* and *Flabellolaius* Wittmer, the scape and antennomere 3 are enlarged and distorted (Wittmer 1952). Males of various Malachiinae also have excitory glands on various parts of the body (Matthes 1962). Larvae are elongate, slightly flattened, and lightly sclerotised except for the head, thorax and tergum IX, and sometimes red or pink in colour (like some clerid larvae). The larval head capsule has a long epicranial stem, no median endocarina, and lightly sclerotised gular and paragular regions, so that it is often said to be 'open' behind the ventral mouthparts; as in most cleroids, there is a mesal appendage at the base of the mandible. Adult melyrids are often found on flowers and several them feed on pollen (Schicha 1974; Hawkeswood 1987b), but Crowson (1964b) found insect fragments in the guts of adults belonging to several species. Malachiine adults contain toxins, some of which have been shown to be incorporated and utilised by certain poisonous frogs and birds (Dumbacher *et al.* (2004). Larvae are usually predacious, but may also be scavengers (Fiori 1971); larvae of *Dicranolaius villosus* (Lea) are known to feed on locust egg pods (Farrow 1974). [Lea 1909; Wittmer 1952, 1954, 1955, 1957, 1959; Evers 1997].

CUCUJOIDEA

Members of the superfamily are usually distinguished from those of Tenebrionoidea by the presence of a cucujoid or ring-type of aedeagus in the adult male and a more or less cylindrical pygopod--like abdominal segment X in the larva. In the adults, the trochanterofemoral joint is rarely strongly oblique (Bothrideridae is a notable exception), females never have 5-5-4 tarsi, and there may be more than four free veins in the medial field of the hind wing. In the larva, the mandible usually has a distinct prostheca, the mala is often falciform, and there may be six pairs of stemmata. Cucujoids are often difficult to distinguish from basal Cleroidea, and it has been demonstrated in various analyses based on both morphological and/or molecular data (Beutel and Pollock 2000; Beutel and Ślipiński 2001; Hunt *et al.* 2007; Leschen and Ślipiński 2010a, 2010b; Lawrence *et al.* 2011) that none of the three superfamilies is monophyletic. The family concepts and the order in which the families are covered are based primarily on the work of Leschen *et al.* (2005), with a few recent additions, such as the families Cyclaxyridae (Gimmel *et al.* 2009) and Akalyptoischiidae (Lord *et al.* 2010).

The following cucujoid families do not occur in Australia: Byturidae (mainly Holarctic and Oriental), Helotidae (Oriental and Afrotropical); Agapythidae (New Zealand); Cyclaxyridae (New Zealand); Smicripidae (Nearctic and Neotropical); Alexiidae (western Palaearctic) and Akalyptoischiidae (North America). [Sen Gupta and Crowson 1969a; Springer and Goodrich 1983, 1986, 1990, 1994; Goodrich and Springer 1988, 1995; Ślipiński and Pakaluk 1992; Malloch *et al.* 2001; Goodrich 2002a; Price 2002; Leschen *et al.* 2005, 2010b; Robertson *et al.* 2008; Cline 2010; Cline *et al.* 2010; Lawrence *et al.* 2010g; Leschen and Lawrence 2010a, Ślipiński and Tomaszewska 2010b].

58. BOGANIIDAE

Adult (Fig. 85F, J)

Length 1.5–4 mm. Body ~1.8–2.33 times as long as wide; more or less parallel-sided or with sides slightly curved, slightly flattened; colour yellow to reddish-brown or black; upper surfaces subglabrous in *Paracucujus* Sen Gupta and Crowson and in *Athertonium parvum* Crowson, clothed with decumbent or suberect hairs in all species of *Boganium* Sen Gupta and Crowson and in an undescribed *Athertonium* species.

Head (Fig. 14G) prognathous, not abruptly constricted behind eyes. Eyes protuberant, not emarginate, coarsely facetted, without or with very short interfacetal setae. Antennal insertions exposed; subantennal grooves absent. Frontoclypeal suture distinctly impressed, straight; base of clypeus slightly impressed laterally to receive mandibular tubercles, its anterior edge usually truncate (with two small teeth in *Paracucujus*). Labrum lightly sclerotised and more or less concealed beneath clypeus, its apex emarginate. Antennae 11-segmented with weak, 3-segmented club (barely distinguishable in *Paracucujus*). Mandibles (Fig. 16E) usually moderately elongate and unidentate; dorsally with setose cavity partly covered by mesally directed tubercle; incisor edge with two small subapical teeth in *Boganium*, with two dorsal subapical teeth and an additional ventral one in Paracucujini; mola well developed and tuberculate; prostheca a short setal fringe. Maxilla with elongate, setose galea and very short lacinia with undivided uncus; maxillary palp long, its apical palpomere cylindrical or fusiform. Ligula more or less bilobed; apical labial palpomeres cylindrical or fusiform. Corpotentorium very narrow, sometimes absent. Cervical sclerites reduced.

Pronotum ~0.45–0.65 times as long as wide, widest at about middle, sides moderately to strongly curved, not explanate. Lateral pronotal carinae complete, with raised margin or narrow bead; anterior angles not or very slightly produced and rounded, with prominent callosities containing gland openings in *Boganium* or gland openings only in *Athertonium*; posterior angles obtuse or right, with gland openings in *Athertonium*; posterior edge weakly bisinuate; disc simple. Prosternum in front of coxae (Fig. 20K) as long as or slightly longer than mid length of procoxal cavity, flat to moderately convex. Prosternal process complete, slightly, gradually expanded apically, flat or slightly elevated behind coxae, its apex truncate to emarginate. Procoxae without concealed extensions; trochantins exposed. Procoxal cavities strongly transverse, narrowly separated, externally broadly open, without narrow lateral extensions, internally open.

Scutellar shield not abruptly elevated, laterally expanded and rounded, posteriorly broadly rounded or obtusely angulate. Elytra ~1.38–1.8 times as long as combined width and 2.7–3.8 times as long as pronotum; suture not deflected near apex; disc irregularly punctate in *Boganium* and *Athertonium*, with 10 puncture rows and scutellary striole in *Paracucujus*; epipleura incomplete. Mesoventrite with slightly declined paired procoxal rests. Mesoventral process long and truncate to emarginate at apex. Mesocoxae with exposed trochantins. Mesocoxal cavities subcircular, moderately widely separated, partly closed laterally by mesepimera; meso- and metathoracic portions of mesocoxal cavity joined by membranous suture. Mesometaventral junction monocondylic. Metaventrite with moderately to very long discrimen; exposed portion of metanepisternum moderately elongate. Metacoxae contiguous or narrowly separated, extending laterally to meet elytra; plates absent. Metendosternite with long stalk, long lateral arms, narrow laminae, and short anterior process with anterior tendons narrowly separated. Hind wing with long apical field containing small, dark sclerite just beyond radial cell and two veinal remnants almost reaching wing margin; radial cell well developed, elongate, with inner posterobasal angle acute; cross-vein r3 short and oblique; basal portion of RP short to moderately long; medial spur slightly sinuate, not reaching wing margin; medial field with four free veins and no medial fleck; wedge cell absent; anal embayment deep and notch-like. Tibial apices widened and spinose at apex in *Athertonium*. Tarsi 5-5-5 in both sexes; penultimate tarsomere reduced and one or more preceding ones lobed beneath; pretarsal claws usually simple (toothed in *Athertonium*).

Abdomen with five free ventrites; ventrite 1 not much longer than 2; intercoxal process acute. Functional spiracles on abdominal segments I–VII, those on VII at lateral edges of tergite. Anterior edge of sternite VIII in male with median strut. Anterior edge of segment IX in male more or less acute or with paired struts not meeting at midline, without spiculum gastrale; posterior edge of sternite IX with truncate mesal lobe and pair of slender lateral lobes; tergite IX and X fused together or X membranous. Aedeagus of cucujiform type, asymmetrical, lying on one side when retracted; anterior edge of tegmen without struts; parameres usually not apparent, partly fused together but articulated with phallobase in *Paracucujus*; penis laterally compressed, strongly curved, its anterior edge with paired, slender, articulated struts. Ovipositor moderately long, lightly sclerotised except for baculi, paraprocts subequal in length to coxites, each of which is subdivided into broad basal and slender apical lobes; styli moderately long and subapical. [Sen Gupta and Crowson 1966; Lawrence *et al.* 1999b; Lawrence and Ślipiński 2010b].

Larva (Fig. 84A)

Body elongate and more or less parallel-sided or slightly wider at middle and tapering at both ends; head, thoracic terga and abdominal terga I–IX with moderately heavily pigmented plates in *Paracucujus* and first instars of *Boganium*; ventral surfaces and dorsal surfaces in late instars of *Boganium* lightly pigmented); vestiture of long and short, scattered setae, many of which are apically expanded.

Head prognathous and protracted, narrower than thorax, somewhat flattened. Epicranial stem absent; frontal arms at least slightly lyriform and contiguous at base. Median endocarina absent. Stemmata six on each side. Frontoclypeal suture distinct; clypeus large and subquadrate or apically expanded. Labrum completely fused to clypeus. Antennae moderately long, 3-segmented, with conical sensorium much shorter than apical antennomere. Mandibles unidentate, with accessory ventral process; incisor edge with or without one short tooth; mola well developed, tuberculate; prostheca a rigid hyaline process with broad base and acute apex. Ventral mouthparts (Fig. 39I) retracted, with maxillary articulating area; stipites about as long as wide; mala articulated, falciform, and cleft forming two subacute lobes (partly fused galea and lacinia); maxillary palps 3-segmented. Labium consisting of prementum, mentum and submentum; ligula simple, rounded, shorter than labial palps, which are 2-segmented and separated by more than width of first palpomere. Hypopharyngeal sclerome complex, including triangular mesal lobe flanked by two anteriorly projecting process. Hypostomal rods very short and diverging; ventral epicranial ridges absent. Gula wider than long.

Thoracic and abdominal terga usually with distinct sclerotised plates. Legs well developed, 5-segmented including claw-like pretarsus bearing two setae lying side by side; mesocoxae separated by more than one basal coxal diameter. Segment IX excluding appendages subequal in length to VIII, tergum not extending onto ventral surface; with short, posteriorly oriented urogomphi in *Paracucujus* (absent in *Boganium*); sternum IX exposed, simple, not enclosed by sternum VIII; segment X visible from above, with paired pygopods, which are simple and longer than wide; anal region posteriorly oriented. Spiracles annular–biforous. [Lawrence *et al.* 1999a; Lawrence and Ślipiński 2010b].

Classification and World Distribution

Boganiidae is a small family with five genera, nine described species and several more to be described. There are two subfamilies: Boganiinae with *Afroboganium* Endrödy-Younga & Crowson from southern Africa and the Australian *Boganium* Sen Gupta & Crowson, and Paracucujinae with *Metacucujus encephalartii* Endrödy-Younga and the Australian genera *Paracucujus* Sen Gupta and Crowson and *Athertonium* Crowson. [Endrödy-Younga and Crowson 1986; Crowson 1990; Endrödy-Younga 1991a; Lawrence and Ślipiński 2010b].

Australian Fauna

Paracucujus rostratus Sen Gupta and Crowson (Fig. 85J) is known only from the southern part of WA, while species of *Boganium* (85F) occur across the southern part of the continent . The only described species of *Athertonium* occurs in northern QLD, but an undescribed species is known from northern NSW. [Crowson 1990; Lawrence and Ślipiński 2010b].

59. PROTOCUCUJIDAE

Adult (Fig. 83G–H)

Length 3.5–5.8 mm. Body ~2.1–2.5 times as long as wide; more or less parallel-sided, slightly flattened to moderately convex; colour reddish-brown to black, sometimes with yellow areas on elytra; vestiture of fine, recumbent setae, often with thickened silvery setae forming small patches on elytra.

Head prognathous, not abruptly constricted behind eyes. Eyes prominent, not emarginate, moderately coarsely facetted, with very fine interfacetal setae. Antennal insertions concealed; subantennal grooves present. Frontoclypeal suture weakly or strongly impressed and strongly curved; base of clypeus without lateral impressions, its anterior edge truncate. Labrum strongly transverse, its apex subtruncate. Antennae 11-segmented with loose, 3-segmented club. Mandible ~1.5 times as long as wide, bidentate, without dorsal tubercle or cavity; incisor edge with a small subapical tooth; mola well developed and transversely ridged, with setose sub-basal lobe; prostheca consisting of tuft of seta and narrow membrane. Maxilla with broad, setose galea and narrow lacinia without uncus but with several apical spines or setae; apical maxillary palpomere fusiform. Mentum strongly transverse; ligula more or less bilobed; apical labial palpomere fusiform. Corpotentorium moderately broad, straight. Cervical sclerites well developed.

Pronotum ~0.65–0.75 times as long as wide, widest anteriorly or at about middle and almost as wide at base as combined elytral bases; sides weakly curved, slightly explanate. Lateral pronotal carinae complete, simple or slightly crenulate; anterior angles not or very slightly produced and rounded; posterior angles obtuse or right; posterior edge straight; disc simple and even or with two or several impressions. Prosternum in front of coxae 1.2–1.5 times as long as mid length of procoxal cavity, flat to moderately convex. Prosternal process complete, flat or weakly curved behind coxae, gradually expanded posteriorly, with truncate apex. Procoxae without concealed extensions; trochantins exposed. Procoxal cavities slightly transverse, narrowly separated, externally broadly closed, with narrow lateral extensions, internally open.

Scutellar shield not abruptly elevated, laterally expanded and rounded, posteriorly broadly rounded or obtusely angulate. Elytra ~1.5–2 times as long as combined width and 2.7–3.2 times as long as pronotum, with well developed humeri, usually complete with conjointly rounded apices (apices independently rounded and separated by gap in *Ericmodes costatus* Ślipiński and Pakaluk); suture not deflected near apex; disc slightly convex, with 10 puncture rows and scutellary striole (somewhat broken up by tubercles and costae in *E. costatus*; epipleura narrow but complete. Mesoventrite with anterior edge on different plane than metaventrite, without paired procoxal rests. Mesoventral process long and emarginate at apex. Mesocoxae with exposed trochantins. Mesocoxal cavities subcircular, narrowly separated, partly closed laterally by mesanepisterna and mesepimera; meso- and metathoracic portions of mesocoxal cavity joined by incomplete suture. Mesometaventral junction monocondylic. Metaventrite with long discrimen; postcoxal lines absent; exposed portion of metanepisternum long and narrow. Metacoxae narrowly separated; plates absent. Metendosternite with moderately long, broad stalk, long lateral arms, narrow laminae, and short anterior process with anterior tendons narrowly separated. Hind wing with long apical field containing a small sclerite just beyond radial cell and two veinal remnants almost reaching wing margin; radial cell short and broad, with inner posterobasal angle acute; cross-vein r3 strongly oblique and fused basally with r4, which bears an apical spur; basal portion of RP moderately long; medial spur strongly curved, not reaching wing margin, where there is an embayment; medial field with five free veins and a divided medial fleck; wedge cell well developed, apically acute; anal embayment deep and notch-like. Tibial apices not apically widened. Tarsi 5-5-5 in female, 5-5-4 in male; penultimate tarsomere reduced and the three preceding ones more or less lobed beneath; pretarsal claws simple; empodium well developed and bisetose.

Abdomen with five free ventrites; ventrite 1 not much longer than 2; intercoxal process acute. Functional spiracles on abdominal segments I–VII, those on VII at lateral edges of tergite. Sternite VIII in male reduced, without median strut. Anterior edge of segment IX in male with short spiculum gastrale; posterior edge of sternite IX without mesal lobe; tergite IX and X fused together and membranous. Aedeagus of cucujiform type, asymmetrical, lying on one side when retracted; anterior edge of tegmen with median strut; parameres fused together and to tegmen; penis not subdivided, anterior edge with pair of broad, apically expanded struts; endophallus complex. Ovipositor about twice as long as wide, lightly sclerotised except for baculi, paraprocts slightly shorter than coxites, each of which is subdivided into basal and apical lobes; styli long and apical. [Crowson 1955; Ślipiński 1998; Lawrence *et al.* 1999b; Ślipiński and Beutel 2010].

Larva (Fig. 84N)

Body elongate and more or less parallel-sided or slightly wider at middle and tapering at both ends; colour yellowish-white, slightly darker dorsally than ventrally with only mandibles and tips of urogomphi heavily pigmented; dorsal surfaces more or less tuberculate and vestiture of long and short, scattered setae.

Head prognathous and protracted, slightly transverse and slightly flattened. Epicranial stem short; frontal arms lyriform and contiguous at base. Median endocarina absent. Stemmata five on each side. Frontoclypeal suture absent; clypeus moderately long and trapezoidal. Labrum free, transverse. Antennae moderately long, 3-segmented, with conical sensorium about as long as apical antennomere. Mandibles usually tridentate, with accessory ventral process; incisor edge simple; mola well developed, tuberculate; prostheca an acute hyaline process, somewhat reduced in later instars and acute apex. Ventral mouthparts retracted, with maxillary articulating area; cardines divided by internal ridge; stipites longer than wide; mala fixed, falciform; maxillary palps 3-segmented, without distinct palpifer. Labium consisting of prementum, mentum and submentum; ligula simple, rounded, shorter than labial palps, which are 2-segmented and separated by more than width of first palpomere. Hypopharyngeal sclerome a transverse bar. Hypostomal rods moderately long and diverging; ventral epicranial ridges absent. Gular region short and lightly sclerotised.

Thoracic and abdominal terga with numerous granules, transverse rows of tubercles, and transverse carinae, all much more well developed posteriorly. Legs well developed, 5-segmented including claw-like pretarsus bearing two setae lying side by side; coxae moderately widely separated. Segment IX excluding appendages subequal in length to VIII, tergum extending onto ventral surface; with moderately long, upturned urogomphi bearing accessory tubercles near the base and with several pregomphi on tergal disc. Sternum IX exposed, simple, not enclosed by sternum VIII; segment X not visible from above, posterioventrally oriented and forming cylindrical pygopod, but with tergum slightly darker. Spiracles annular–biforous, all placed at the ends of spiracular tubes. [Ślipiński 1998; Lawrence *et al.* 1999a; Beutel and Ślipiński 2001; Ślipiński and Beutel 2010].

Classification and World Distribution

Protocucujidae contains a single genus, *Ericmodes* Reitter with three Australian species and four from southern South America.

Australian Fauna

Ericmodes australis Grouvelle (Fig. 83H) is known from SA, VIC and TAS, *E. costatus* Ślipiński and Pakaluk in Ślipiński (1998) (Fig. 83G) from eastern NSW and *E. lawrencei* Ślipiński and Pakaluk in Ślipiński from north-eastern QLD. Protocucujids are elongate, somewhat flattened beetles clothed with stout, recumbent setae. The head has a strongly curved frontoclypeal suture resembling that in Sphindidae and a weak antennal club, the pronotum is somewhat explanate laterally, tarsi are densely setose below and the elytra have deep window-punctures and a series of weak, longitudinal elevations or sometimes sharp, incomplete costae. Larvae have a tuberculate upper surface and a distinctive tergum IX with several pregomphi, in addition to the urogomphi.

Little is known about the biology of these beetles. Tillyard (1926) recorded *E. australis* from *Uromycladium* galls (Basidiomycetes: Uredinales) on *Acacia*, but this has not been confirmed. Larvae of *E. australis* were found in an aquatic sample, where they may have fallen from the forest canopy or from riparian vegetation, and those of *E. costatus* were found on tree fern fronds.

60. SPHINDIDAE

Adult (Fig. 83J–K)

Length 1.2–2.4 mm. Body ~1.37–2.1 times as long as wide; broadly ovate to more or less parallel-sided, slightly flattened to moderately convex; colour red with dark markings on elytra (*Notosphindus* McHugh and Wheeler) or reddish-brown to black, sometimes with red spots on elytral humeri (*Aspidiphorus* Latreille); vestiture of fine, decumbent and suberect setae (*Aspidiphorus*) or minute hairs not or barely longer than an elytral puncture (*Notosphindus*).

Head (Fig. 44E) prognathous, not abruptly constricted behind eyes. Dorsal posterior edge of head capsule deeply bisinuate in *Aspidiphorus*, less so in *Notosphindus*. Eyes prominent, not emarginate, moderately coarsely facetted, without interfacetal setae. Antennal insertions concealed; subantennal grooves well developed, extending at least slightly behind eye. Frontal region not deflexed, with a series of diverging longitudinal grooves on each side bordering eye in *Notosphindus*. Frontoclypeal suture strongly impressed and strongly curved; base of clypeus with paired lateral impressions into which mandibular tubercles fit; anterior edge of clypeus broadly rounded. Labrum free, strongly transverse, its apex truncate. Antennae 10-segmented with large, compact, densely setose, 3-segmented club. Mandible unidentate or bidentate, with dorsal tubercle and setose cavity; incisor edge with one or two small subapical teeth; mola well developed; prostheca reduced to fringe of hairs. Maxilla with slender, setose galea and lacinia, the latter with one or two stout apical spines; apical maxillary palpomere fusiform. Ligula slightly emarginate; apical labial palpomere fusiform. Corpotentorium narrow, straight or arched. Cervical sclerites absent.

Pronotum ~0.42–0.77 times as long as wide, widest at about middle, as wide at base as combined elytral bases; sides weakly curved. Lateral pronotal carinae complete, with distinct margin or bead and simple edge; anterior angles not produced in *Notosphindus*, produced and broadly rounded in *Aspidiphorus*; posterior angles rounded; posterior edge evenly rounded or bisinuate; disc moderately convex, simple. Prosternum in front of coxae shorter than mid length of procoxal cavity, flat to moderately convex, with paired longitudinal carinae in *Aspidiphorus*. Prosternal process complete, straight or weakly curved behind coxae, gradually expanded posteriorly, with truncate apex . Procoxae without concealed extensions; trochantins concealed or slightly exposed. Procoxal cavities strongly transverse, moderately widely separated, externally broadly open in *Aspidiphorus*, narrowly open in *Notosphindus*, without narrow lateral extensions, internally open in *Notosphindus*, closed by narrow bar in *Aspidiphorus*.

Scutellar shield not abruptly elevated, anteriorly simple, posteriorly broadly rounded or obtusely angulate. Elytra ~1.02–1.47 times as long as combined width and 2.25–3.1 times as long as pronotum, with well developed humeri; suture not deflected near apex; disc with 10 puncture rows and scutellary striole; epipleura narrow and incomplete. Mesoventrite with paired procoxal rests. Mesoventral cavity absent in *Notosphindus*, very shallow in *Aspidiphorus*. Mesoventral process short and truncate or emarginate at apex. Mesocoxae with exposed trochantins. Mesocoxal cavities slightly transverse, very widely separated, partly closed laterally by mesepimera. Mesometaventral junction monocondylic in *Notosphindus*, simple with meso- and metaventrites slightly overlapping in *Aspidiphorus*. Metaventrite with long discrimen; postcoxal lines absent; exposed portion of metanepisternum long and narrow. Metacoxae strongly transverse, narrowly separated, extending laterally to meet elytra; plates absent. Metendosternite with short, broad stalk, long lateral arms, narrow laminae, and short anterior process with anterior tendons widely separated or without anterior process and anterior tendons on bases of lateral arms. Hind wing with long apical field sometimes with vaguely indicated radial veinal remnant; radial cell in *Notosphindus* short and broad, with inner posterobasal angle acute (cell absent in *Aspidiphorus*); cross-vein r3 absent and r4 incomplete; basal portion of RP moderately long; R-M loop rounded; medial spur ending well before wing margin; medial field with two or three free veins and no medial fleck; wedge cell absent; anal embayment deep and notch-like. Tibial apices apically widened and outer edges with row of small spines in *Notosphindus*, more slender, widest at middle, without spines in *Aspidiphorus*; spurs absent. Tarsi 5-5-5 in female, 5-5-4 in male; tarsomeres simple; pretarsal claws simple.

Abdomen with five free ventrites; ventrites not in same plane; ventrite 1 not much longer than 2; intercoxal process acute in *Notosphindus*, truncate in *Aspidiphorus*; anterior edges of ventrites 2–5 each with transverse row of large punctures. Pygidium (tergite VII) in *Aspidiphorus* with sharply defined median longitudinal groove. Functional spiracles on abdominal segments I–V. Anterior edge of segment IX in male without spiculum gastrale; posterior edge of sternite IX without mesal lobe; tergite IX and X fused together and membranous. Aedeagus of cucujiform type, symmetrical, lying on one side when retracted; anterior edge of tegmen without median strut; parameres fused together and to tegmen; penis not subdivided, anterior edge with pair of broad struts. Ovipositor slightly longer than wide, lightly sclerotised except for baculi, paraprocts about as long as coxites, each of which is subdivided into basal and apical lobes; styli palpiform, but small and subapical. [Sen Gupta and Crowson 1979; Burakowski and Ślipiński 1987; McHugh and Wheeler 1991; McHugh 1993, 2002; Lawrence *et al.* 1999b; Forrester and McHugh 2010].

Larva

Body elongate and more or less parallel-sided or slightly wider at middle and tapering at both ends; body pale or with vaguely indicated pigment patches on some terga; head darker but with pattern of pale areas; dorsal surfaces smooth but with irregular patches of setiferous spicules; vestiture of moderately to very long setae.

Head prognathous and protracted, transverse, slightly flattened with sides rounded. Epicranial stem short; frontal arms lyriform and contiguous at base. Median endocarina absent. Stemmata six on each side. Frontoclypeal suture absent; clypeus well developed, with parallel sides. Labrum at least partly fused to clypeus, strongly transverse. Antennae moderately to very long, 3-segmented, with conical sensorium shorter than apical antennomere. Mandibles bidentate in *Notosphindus*, unidentate in *Aspidiphorus*, with accessory ventral process; incisor edge simple in *Notosphindus*, serrate in *Aspidiphorus*; mola well developed, tuberculate; prostheca an acute hyaline process. Ventral mouthparts retracted, with maxillary articulating area; stipites longer than wide; mala fixed, truncate or rounded; maxillary palps 3-segmented. Labium consisting of prementum, mentum and submentum; ligula simple, rounded, shorter than labial palps, which are 2-segmented and separated by more than width of first palpomere. Hypopharyngeal sclerome a transverse bar. Hypostomal rods short to moderately long, diverging; ventral epicranial ridges absent. Gular region short and lightly sclerotised.

Thoracic and abdominal terga simple. Legs well developed, 5-segmented including claw-like pretarsus bearing one seta in *Notosphindus*, one long and one short in *Aspidiphorus*; coxae well separated. Abdomen less than twice as long as thorax. Segment IX shorter than VIII, tergum, not extending onto ventral surface; without urogomphi. Sternum IX exposed, simple, not enclosed by sternum VIII; segment X visible from above, posteriorly oriented; anal opening transverse, sometimes surrounded by several lobes. Spiracles annular, placed at the ends of short spiracular tubes in *Notosphindus*. [Sen Gupta and Crowson 1979; Burakowski and Ślipiński 1987; Lawrence *et al.* 1999a; Chiao and McHugh 2000; Beutel and Ślipiński 2001; McHugh 2002; McHugh and Kiselyova 2003; Forrester and McHugh 2010].

Classification and World Distribution

The Sphindidae is a widely distributed family containing nine genera and ~70 species. There are four subfamilies, three of which are of limited distribution and contain a single genus: Protosphininae (*Protosphindus* Sen Gupta and Crowson, Chile), Odontosphindinae (*Odontosphindus* LeConte, North America and Europe) and Sphindiphorinae (*Sphindiphorus* Sen Gupta and Crowson, South Africa). The remaining genera are placed in Sphindinae. [Sen Gupta and Crowson 1979; McHugh 1993, 2000; Freeman *et al.* 2003; Forrester and McHugh 2010].

Australian Fauna

The Australian sphindid fauna includes *Notosphindus slateri* McHugh (Fig. 83K) and Wheeler from VIC and TAS and five species of *Aspidiphorus* Macleay (Fig. 83J), a widely distributed Old World genus. Sphindids are oblong to broadly ovate, convex beetles, finely pubescent or subglabrous (*Notosphindus*) with a large, abrupt, tomentose antennal club. The frontoclypeal suture is characteristically arched (Fig. 44E), the mandible has a dorsal tubercle and setose cavity, the elytral punctation is seriate, tibial spurs are usually absent, tarsi are simple and the pygidium in *Aspidiphorus* is longitudinally grooved. Larvae are elongate, tapered posteriorly and lightly sclerotised, except for head and weak tergal plates on either prothorax or abdominal segments III–IX. The antennae are usually more than half as long as the head width. Both adult and larval Sphindidae feed on spores of slime moulds (Myxomycetes).

61. BIPHYLLIDAE

Adult (Figs 73A–B, 86H)

Length usually 1.4–3 mm (as much as 8.5 mm in *Althaesia*). Body ~1.5–2.9 times as long as wide; usually ovate to oblong with rounded sides, occasionally more or less parallel-sided, slightly flattened to moderately convex; colour yellowish-brown to reddish-brown or black, sometimes bicolored or with pattern on elytra; vestiture of shorter, recumbent and longer suberect setae.

Head prognathous, not abruptly constricted behind eyes. Occipital region with low transverse carina. Eyes prominent, not emarginate, coarsely facetted, with short to long interfacetal setae. Antennal insertions concealed; subantennal grooves present, extending behind eyes. Frontoclypeal suture absent; clypeus well developed, with parallel sides and subtruncate apex, without lateral impressions at base. Labrum strongly transverse, its apex subtruncate. Antennae 11-segmented, usually with 3-segmented club (2-segmented in *Biphyllus* Dejean). Mandible bidentate, without dorsal tubercle or cavity; incisor edge simple; mola well developed, transversely ridged, prostheca well developed. Maxilla with elongate, apically rounded, setose galea and somewhat narrower setose lacinia without uncus; apical maxillary palpomere fusiform. Mentum trapezoidal, apically biemarginate; ligula simple; apical labial palpomere securiform. Head ventrally with pair of setose tubular invaginations opening laterally into bases of subantennal grooves. Corpotentorium moderately broad, straight. Cervical sclerites well developed.

Pronotum ~0.45–0.72 times as long as wide, usually widest at or behind middle and almost as wide at base as combined elytral bases, with sides weakly curved (widest anteriorly and sides straight in *Diplocoelus angustatus* Lea). Lateral pronotal carinae complete, usually simple (sometimes serrulate or denticulate); anterior angles often produced forward and narrowly rounded or acute; posterior angles more or less right; posterior edge usually bisinuate; disc sometimes with one or two pairs of longitudinal, sublateral carinae and in *Althaesia* with two longitudinal sulci joined posteriorly by a sub-basal transverse groove. Prosternum in front of coxae slightly shorter than mid length of procoxal cavity, flat to moderately convex. Prosternal process complete, flat or weakly curved behind coxae, more or less parallel-sided with apex broadly rounded or truncate. Procoxae without or with very short concealed extensions; trochantins concealed or barely exposed. Procoxal cavities transverse, moderately narrowly separated, externally broadly closed, with very narrow lateral extensions, internally broadly closed.

Scutellar shield not abruptly elevated, laterally expanded, posteriorly broadly rounded or obtusely angulate. Elytra ~1.2–2.2 times as long as combined width and 2.6–3.2 times as long as pronotum, with well developed humeri; suture not deflected near apex; disc with 10 puncture rows and scutellary striole; epipleura complete. Mesoventrite with paired, slightly inclined procoxal rests. Mesoventral process truncate at apex. Mesocoxae with exposed trochantins. Mesocoxal cavities subcircular, moderately broadly separated, partly closed laterally by mesepimera. Mesometaventral junction monocondylic. Metaventrite with long discrimen and usually two pairs of postcoxal lines, one straight and another posteriorly curved and meeting the lateral edges of the ventrite; exposed portion of metanepisternum moderately long and narrow. Metacoxae strongly transverse, well separated, not extending laterally to meet elytra; plates absent. Metendosternite with moderately long, broad stalk, long lateral arms, narrow laminae, and short anterior process with anterior tendons widely separated on either side of anterior emargination. Hind wing with small linear sclerite just beyond radial cell and usually with at least one veinal remnant; radial cell short and broad, with inner posterobasal angle acute; cross-vein r3 strongly oblique, well separated basally from r4, which bears an apical spur; basal portion of RP short or absent; medial spur reaching wing margin, where there is an embayment; medial field with four free veins, the anterior one extending into medial fleck; wedge cell present or absent; anal embayment deep and notch-like. Legs well developed; trochanterofemoral joint strongly oblique with base of femur in contact with coxa. Protibial apices sometimes widened; spurs paired on all legs. Tarsi 5-5-5 in both sexes; tarsomeres 1 and 4 somewhat reduced, tarsomeres 1–3 or 2–3 usually with setose lobes beneath; pretarsal claws simple; empodium weakly developed, bisetose.

Abdomen (Fig. 25H) with five free ventrites; ventrite 1 not much longer than 2, with two pairs of straight postcoxal lines; intercoxal process narrowly to broadly rounded. Functional spiracles on abdominal segments I–VII. Sternite VIII in male with or without median strut. Anterior edge of segment IX in male with subacute subgenital plate or with spiculum gastrale; posterior edge of sternite IX with mesal lobe; tergite IX deeply emarginate and segment X free. Aedeagus of cucujiform type, symmetrical; anterior edge of tegmen with single median strut and opposing paired struts; parameres fused together and to phallobase and partly fused to one another; penis not subdivided, anterior edge with paired struts. Sternite VIII in female with short spiculum ventrale. Ovipositor moderately to very elongate, lightly sclerotised except for baculi, paraprocts subequal in length to coxites, each of which is subdivided into basal and apical lobes, the latter sometimes long and slender; styli long and apical. [Lea 1922b; Nikitsky 1983; Goodrich and Springer 1992; Goodrich 2002b; Cline and Shockley 2010].

Larva (Fig. 84G)

Body elongate, parallel-sided and more or less cylindrical; colour yellowish brown; dorsal surfaces lightly pigmented and finely granulate; ventral surfaces unpigmented; vestiture of scattered, simple setae.

Head prognathous and protracted, slightly transverse and slightly flattened. Epicranial stem very short or absent; frontal arms lyriform and contiguous at base. Median endocarina absent. Stemmata six on each side. Frontoclypeal suture absent. Labrum partly fused to head capsule. Antennae well developed, ~0.25 times as long as head width, 3-segmented, with conical sensorium about half as long as apical antennomere. Mandibles (Figs 27J, 48E) symmetrical, bidentate, with accessory ventral process; incisor edge sometimes serrate; mola well developed, transversely ridged with asetose hyaline lobe at base; prostheca a brush of complex, comb-like hairs. Ventral mouthparts retracted, with well developed maxillary articulating area; stipites elongate; mala falciform, with row of spines along inner edge; maxillary palps 3-segmented. Labium consisting of prementum, mentum and submentum; ligula simple, transverse, truncate; labial palps 2-segmented and widely separated. Hypopharyngeal sclerome a transverse bar. Hypostomal rods short and diverging; ventral epicranial ridges well developed. Gular region short and lightly sclerotised.

Mesotergum, metatergum and abdominal terga I–VIII each with anterior transverse carina. Legs well developed, 5-segmented including claw-like pretarsus bearing one seta; coxae very narrowly separated. Segment IX slightly shorter than VIII, tergum not extending onto ventral surface; without urogomphi; sternum IX exposed, simple, not enclosed by sternum VIII; segment X visible from above, posteriorly oriented and subcylindrical, with tergum pigmented as in previous segments. Spiracles annular–biforous, those on segment VIII placed at the ends of short spiracular tubes at posterior edge of tergum. [Nikitsky 1983; Lawrence *et al.* 1999a; Goodrich 2002b; Cline and Shockley 2010].

Classification and World Distribution

Biphyllidae is a widely distributed family containing six genera and ~200 species.

Australian Fauna

The Australian biphyllid fauna includes three species each in the genera *Althaesia* Pascoe (Fig. 73A) and *Biphyllus* Dejean, and 13 described species of *Diplocoelus* Guérin-Méneville (Figs 73B, 85H, L). Biphyllids are oblong to elongate, somewhat flattened, reddish-brown to black or occasionally bicoloured beetles, densely clothed with erect and decumbent hairs. The eyes are very coarsely facetted, the pregular region has a pair of laterally-opening, setose cavities, the lateral pronotal margins are sometimes finely crenulate or serrulate, the pronotal disc has a pair of longitudinal, sublateral carinae and occasionally a pair of basal grooves or foveae, femoral lines are present on ventrite 1 and occasionally the metaventrite, and the penultimate tarsomere is reduced while the preceding ones have lobes beneath. Larvae (Fig. 84G) are elongate, cylindrical and dorsally sclerotised, with a complex mandibular prostheca bearing comb-hairs, a falcate mala, posterodorsally displaced spiracles on segment VIII, tergum IX without or with minute urogomphi and segment X terminal. Biphyllids occur under bark or in leaf bases, especially where fermentation is occurring, but they also feed on pyrenomycetous fungi (Ascomycetes), such as *Daldinia* and *Hypoxylon*. [Lea 1922b].

62. EROTYLIDAE

Adult (Figs 73E–G, J–L, 86G, K)

Length usually 1.4–25 (usually 2–8.5) mm . Body ~1.5–5 times as long as wide; usually ovate to elongate, parallel-sided, laterally evenly rounded or with sides of prothorax and elytra independently rounded, strongly flattened to moderately convex; colour yellowish-brown to reddish-brown or black, occasionally metallic blue; usually bicolored with pattern of red and black or yellow and black in Erotylinae; vestiture often consisting of shorter, and/or longer, recumbent to erect fine setae (setae thickened and decumbent in *Leucohimatium* Rosenhauer), surfaces subglabrous in many Languriinae and Erotylinae. Glandular pores present on many parts of the body, especially along the lateral pronotal carinae (Drilling *et al.* 2010).

Head prognathous, constricted behind eyes to form short temples in *Protoloberus* Leschen. Occipital region with low transverse carina in *Leucohimatium* and *Toramus* Grouvelle, with median stridulatory file in *Leucohimatium* (Fig. 14E), *Protoloberus* Leschen and *Cryptophilus* Reitter and with paired stridulatory files in many Erotylinae (Figs 14B, F). Eyes usually well developed, not emarginate, finely to coarsely facetted, with or without interfacetal setae. Antennal insertions concealed or exposed; subantennal grooves relatively short or absent. Frontoclypeal suture present or absent; clypeus without lateral impressions at base. Labrum strongly transverse, its apex subtruncate. Antennae 11-segmented, almost always with 3-segmented club (4-segmented in *Spondylotriplax* Crotch, *Anadastus* Gorham and *Caenolanguria* Gorham); club flattened in *Anadastus, Caenolanguria* and several Erotylinae; terminal antennomere reduced so that the club may appear 2-segmented in *Cathartocryptus*). Mandible usually short and broad, usually bidentate, without dorsal tubercles or cavity; incisor edge usually simple or with single tooth; mola well developed, prostheca occasionally reduced. Ventral mouthparts retracted and enclosed by lateral expansion of mandible in some Erotylinae. Maxilla with well developed, setose galea and lacinia with or without unidentate to tridentate uncus; apical maxillary palpomere usually fusiform (slightly expanded and truncate to securiform or cultriform in some Erotylinae). Mentum trapezoidal and often transverse, apically biemarginate; ligula simple or emarginate; apical labial palpomere usually fusiform (sometimes slightly to strongly expanded). Head ventrally without setose invaginations. Corpotentorium narrow, straight. Cervical sclerites present.

Pronotum ~0.45–1.1 times as long as wide, usually widest at middle or at base (widest anteriorly in *Leucohimatium*, some species of *Hapalips* Reitter and a few other taxa), almost always as wide at base as combined elytral bases (distinctly narrower in *Protoloberus*), with sides straight to strongly curved or occasionally sinuate. Lateral pronotal carinae complete, with distinct raised margin or bead, usually simple (crenulate in *Protoloberus* and some species of *Thallis* Erichson and *Cnecosa* Pascoe); anterior angles produced forward in some Erotylinae, with anterolateral callosities in *Leucohimatium*; posterior angles usually obtuse or right; posterior edge usually straight, distinctly bisinuate most Erotylinae; disc usually simple or with very small pair of basal pits (with two distinct pits joined by transverse groove in *Loberus* LeConte. Prosternum in front of coxae (Fig. 19L) as long as or longer than mid length of procoxal cavity, flat to strongly convex, with paired lines in some Erotylinae. Prosternal process complete, flat, curved behind coxae, or elevated but not curved, more or less parallel-sided or gradually to abruptly expanded, with apex broadly rounded to truncate or emarginate. Notosternal sutures almost always complete (absent in *Leucohimatium*). Procoxae with short to moderately long concealed extensions; trochantins concealed. Procoxal cavities slightly transverse to circular, moderately widely separated, broadly open externally in *Loberus*, *Microlanguria*, *Toramus* and the Languriini, narrowly open in *Xenocryptus*, *Leucohimatium*, *Protoloberus* and *Hapalips*, and broadly closed in Erotylinae and Cryptophilini, sometimes with very narrow lateral extensions in *Hapalips* and *Loberus*, internally closed by slender bar in *Xenocryptus*, *Leucohimatium*, *Protoloberus*, Cryptophilini and *Toramus*.

Scutellar shield usually abruptly elevated, anteriorly simple, posteriorly broadly rounded, obtusely angulate or truncate. Elytra ~1.0–3.5 times as long as combined width and 2.0–5.5 times as long as pronotum, almost always with well developed humeri; suture not deflected near apex; punctation usually seriate, with or without scutellary striole, sometimes irregular; epipleura complete (Fig. 26H). Mesoventrite (Fig. 21E) usually with paired slightly inclined procoxal rests. Mesoventral process truncate at apex. Mesocoxae with concealed trochantins. Mesocoxal cavities subcircular, moderately to widely separated, closed laterally by meeting of meso- and metaventrite. Mesometaventral junction simple (abutting or overlapping ventrites), monocondylic or dicondylic. Metaventrite with well developed discrimen, occasionally with postcoxal lines; exposed portion of metanepisternum moderately to very long and narrow. Metacoxae contiguous to widely separated, not extending laterally to meet elytra; plates absent. Metendosternite with moderately long, broad stalk and long lateral arms; laminae broad to narrow; anterior process usually short with anterior tendons widely separated to narrowly separated (anterior process absent and anterior tendons located on lateral arms in Cryptophilinae). Hind wing with small linear sclerite just beyond radial cell and usually with at least one veinal remnant; radial cell slightly elongate to short and broad, often with inner edge curved and inner posterobasal angle right to obtuse (cell absent in Cryptophilinae); cross-vein r3 slightly to strongly oblique and well separated basally from r4, which lacks an apical spur; basal portion of RP usually moderately long; medial field with four or fewer free veins and a medial fleck; wedge cell present or absent; anal embayment deep and notch-like or absent. Protibial apices sometimes widened; spurs usually paired. Tarsi 5-5-5 in both sexes; tarsomeres without ventral setose lobes or with lobes on antepenultimate tarsomere only or on two or three basal tarsomeres; pretarsal claws almost always simple (toothed in some *Loberus* and *Toramus*); empodium weakly developed, bisetose.

Abdomen usually with five free ventrites (ventrites 1 and 2 connate in some Erotylinae); ventrite 1 not much longer than 2; postcoxal lines present (Fig. 25G) or absent; intercoxal process acute to broadly rounded or subtruncate. Functional spiracles usually on abdominal segments I–VII (on segments I–V in *Toramus*). Sternite VIII in male usually without median strut. Anterior edge of segment IX in male usually with narrowly rounded to truncate subgenital plate, sometimes more or less acute, but not forming a single strut or spiculum gastrale; posterior edge of sternite IX with distinct mesal lobe; tergite IX usually completely subdivided and segment X often free. Aedeagus of cucujiform type, usually symmetrical; anterior edge of tegmen without struts; parameres individually articulated to phallobase; penis not subdivided, anterior edge with single or paired struts. Ovipositor short and broad to moderately elongate, lightly sclerotised except for baculi or with more heavily and evenly sclerotised coxites; paraprocts subequal in length to coxites, which are subdivided, narrower apically and usually located apical to paraprocts (located mesal to paraprocts in *Cathartocryptus*); styli well developed and apical, reduced and subapical or absent. [Lea 1922b; Sen Gupta 1967, 1968a, 1968b, 1969; Chûjô 1969, 1987; Sen Gupta and Crowson 1969b, 1971; Lawrence 1988a; McHugh *et al.* 1997;

Lawrence *et al.* 1999b; Węgrzynowicz 2002; Skelley and McHugh 2002; Leschen and Skelley 2002b; Leschen 2003; Skelley and Leschen 2007; Leschen *et al.* 2010b].

Larva (Figs 34E, 84I, K)

Body elongate, parallel-sided and more or less cylindrical to fusiform, narrowed at either end and slightly flattened; dorsal surfaces lightly or heavily pigmented and often with fine asperities, granules and/or setiferous tubercles (smooth in *Leucohimatium* and Languriini); ventral surfaces unpigmented; vestiture almost always of sparse to dense, simple setae (apically expanded setae in *Cryptophilus*).

Head prognathous to hypognathous and protracted, transverse and slightly flattened to globular. Ecdysial lines often obscure or absent, especially in later instars; epicranial stem usually short or absent (long in some Dacnini); frontal arms often lyriform and contiguous at base (well separated at base in *Xenocryptus* and *Hapalips*). Median endocarina usually absent (present in *Leucohimatium* and some Dacnini). Stemmata usually six on each side (sometimes five, two or none). Frontoclypeal suture present or absent. Labrum free. Antennae short to very long, 3-segmented, with conical sensorium about half as long as to almost as long as apical antennomere (longer than apical antennomere in *Toramus*). Mandibles symmetrical, bidentate or tridentate, with accessory ventral process in Cryptophilinae, Dacnini and *Hapalips*; incisor edge usually with subapical teeth; mola usually well developed (weakly developed or replaced with basal membranous lobe in some Dacnini, Fig. 41M); prostheca (Fig. 41L) usually a rigid hyaline process obtusely angulate at apex. Ventral mouthparts retracted, with well developed maxillary articulating area; stipites elongate; mala falciform to broadly rounded or truncate, sometimes deeply cleft, setose or spinose, inner edge sometimes with one to several small teeth; maxillary palps 3-segmented. Labium consisting of prementum, mentum and submentum; ligula simple, transverse, truncate; labial palps 2-segmented and narrowly to widely separated. Hypopharyngeal sclerome usually molar-like or a transverse bar (absent in some Dacnini). Hypostomal rods usually long and diverging, occasionally absent; ventral epicranial ridges present or absent. Gular region transverse.

Thoracic and abdominal terga with or without sclerotised plates, often with rows of setiferous tubercles and/or patches or rows of asperities or granules; lateral tergal processes sometimes present (e.g. *Cryptophilus*). Legs well developed, 5-segmented including claw-like pretarsus bearing two setae usually lying side by side (one distal to the other in *Toramus*); coxae well separated. Segment IX shorter than or as long as VIII, tergum extending onto ventral surface; almost always with paired urogomphi, which (absent in *Microsternus*, which has single median process, Fig. 84I); urogomphi usually with accessory processes and with pregomphi (simple in Languriini); sternum IX exposed, simple, not enclosed by sternum VIII; segment X not visible from above, posteroventrally or ventrally oriented. Spiracles almost always annular-biforous, sometimes placed at the ends of short spiracular tubes. [Böving and Craighead 1931; Rymer Roberts 1939, 1958; Sen Gupta 1969; Sen Gupta and Crowson 1969b, 1971; Hayashi and Takanaka 1965; Hayashi 1986; Lawrence 1991b; Lawrence *et al.* 1999a; Carlton *et al.* 2000; Skelley and McHugh 2002; Leschen *et al.* 2010b].

Classification and World Distribution

The family Erotylidae contains ~260 genera and more than 3500 species, and is worldwide in distribution but more common in forested areas and absent from oceanic islands (except for a few Cryptophilinae and some stored products pests). The group is divided into six subfamilies, three of which are further subdivided into tribes. All subfamilies occur in Australia, but the languriine tribe Thalisellini (Neotropical), cryptophiline tribes Empocryptini (Neotropical) and erotyline tribe Erotylini (Neotropical) do not occur here. [Sen Gupta 1968a, 1968b; Chûjô 1988, 1989, 1990; Leschen and Wegrzynowicz 1998; Wegrzynowicz 2000, 2002; Robertson *et al.* 2004; Skelley and Leschen 2007; Leschen and Buckley 2007; Leschen *et al.* 2010b].

Australian Fauna

The Australian fauna includes two species of **Xenoscelinae**: *Xenocryptus tenebroides* Arrow (Fig. 72H), from the southern part of WA, with a sister species in South Africa (Wegrzynowicz 2000), and the endemic *Protoloberus singularis* (Blackburn) (Fig. 86L) from NSW and QLD (Leschen 2003). The **Pharaxonothinae** is represented by *Leucohimatium arundinaceum* (Forskål), and the **Loberinae** by six species of *Loberus*. **Languriinae** includes the genus *Hapalips* (Hapilipsini), with *H. investigatus* Leschen and Wegrzynowicz from Lord Howe Island and several undescribed species (Fig. 73L) from the mainland, and three genera of Languriini. **Cryptophilinae** includes one species of *Toramus* Grouvelle (Toramini) from Christmas Island and the cryptophiline genera *Cryptophilus* (Fig. 86K) and *Cathartocryptus* (Fig. 86G). The **Erotylinae** is represented by two species of the tritomine genus *Spondotriplax* (Fig. 73J), which is widely throughout the eastern Palaearctic, Oriental and Australian regions and probably includes *Hedista* Weise (Arrow 1944), the megalodacnine genus *Episcaphula* (Fig. 73G, K), three genera of Encaustini, including *Encaustes hercules* (Lea) (Fig. 73E) and four endemic dacnine genera, including *Cnecosa* (Fig. 73F) and *Thallis* (Fig. 86C), plus the adventive *Dacne picta* Crotch. Erotylids are ovate to narrowly elongate and convex to moderately flattened, small, setose or larger, glabrous beetles, which may be uniformly yellowish-brown or reddish-brown but are more often brightly coloured and patterned with red or yellow and black or occasionally metallic blue. The antennal club is well developed and usually 3-segmented (4-segmented and flattened in Tritomini and most Languriini), occipital stridulatory file or files are present in many species, the apical maxillary and labial palpomeres expanded in some Erotylinae, lateral pronotal carinae are usually simple (rarely crenulate or undulate and with anterior callosities in *Leucohimatium*), tarsi are lobed and setose beneath, the elytral epipleura are complete, and ventrite 1 is only slightly longer than 2. Larvae elongate and cylindrical to oblong and slightly flattened, sometimes with lateral thoracic and abdominal processes (Fig. 84K) usually with a more or less granulate or tuberculate upper surface (not in Languriini or in *Leucohimatium*) and tergum IX usually with paired urogomphi plus accessory processes or pregomphi; the mandible usually has a well developed mola and a broad-based prostheca and the maxillary mala may be falcate or truncate. The larva of *Microsternus* species (Fig. 84I) is unusual in having a median spine on tergum IX. The feeding habits of Erotylidae are varied. *Leucohimatium arundinaceum* is known to occur in stored products where it probably feeds on moulds; larvae and adults were collected in WA in the spore masses of smuts (Basidiomycetes: Ustilaginales) growing on beach grass (*Spinifex sericeus*). *Xenocryptus tenebroides* larvae and adults feed on the male cones of *Macrozamia* (Cycadaceae) in WA, while some *Hapalips* species are associated with *Cycas* in northern Australia. *Loberus* species probably feed on decaying vegetation; the North American *Loberus impressus* (LeConte) were associated with fungi growing on seed capsules of *Iris hexagona* (Iridaceae) (Carlton *et al.* 2000; Van Zandt *et al.* 2003). *Cryptophilus* species occur in leaf litter and mouldy vegetation, and *C. integer* (Heer) is a minor stored products pest. Species of *Toramus* also feed on decaying vegetable matter; adults and larvae of a Neotropical species (misidentified as *Loberoschema* Reitter) were found in rotten flowers of *Pseudobombax* (Bombacaceae) (Lawrence 1991b). Little is known about the biology of Australian Languriini, but at least some exotic species are stem borers in Asteraceae and Fabaceae (Lawrence 1991b; Leschen and Skelley 2002b). Most Erotylinae feed on the fruiting bodies of Agaricales and Aphyllophorales, feeding either internally (*Microsternus*) or on the surface (*Episcaphula*); a species of *Neothallis*, however, was observed feeding on spores of *Scleroderma* sp. (Sclerodermataceae) and a species of *Cnecosa* was collected in the spore mass of the unusual desert agaric, *Podaxis pistillaris*. [Lea 1922b; Aitken 1975; Lawrence 1988a; Sasaji 1989].

Australian Subfamilies, Tribes and Genera

Xenoscelinae: *Xenocryptus* Arrow; *Protoloberus* Leschen.
Pharaxonothinae: *Leucohimatium* Rosenhauer.
Loberinae: *Loberus* LeConte.
Languriinae: Hapalipsini: *Hapalips* Reitter. **Languriini**: *Anadastus* Gorham, *Caenolanguria* Gorham, *Microlanguria* Lewis.
Cryptophilinae: Cryptophilini: *Cryptophilus* Reitter, *Cathartocryptus* Sharp (including *Xenoscelinus* Grouvelle); **Toramini**: *Toramus* sp. (Christmas Island).
Eroytlinae: Dacnini: *Cnecosa* Pascoe, *Dacne* Latreille, *Hoplepiscapha* Lea, *Microsternus* Lewis, *Neothallis* Fauvel, *Thallis* Erichson. **Encaustini:** *Aulacochilus* Lacordaire, *Encaustes* Lacordaire, *Micrencaustes* Crotch. **Megalodacnini:** *Episcaphula* Crotch. **Tritomini:** *Spondotriplax* Crotch (=? *Hedista* Weise).

63. MONOTOMIDAE

Adult (Figs 72A, 83F, 87Q)

Length usually 1.4–5 mm. Body ~2.33–5 times as long as wide; parallel-sided and subcylindrical to flattened; colour yellow to reddish-brown or black, sometimes bicolored; usually subglabrous or sparsely clothed with short, fine, erect setae (setae decumbent and sometimes thickened in *Monotoma* Herbst.

Head prognathous, usually abruptly constricted behind eyes to form temples (not or very weakly constricted in *Shoguna* Lewis). Eyes entire, moderately prominent (except in *Shoguna*), finely facetted, without or with very short interfacetal setae. Antennal insertions exposed and widely separated; subantennal grooves absent or very short. Frontoclypeal suture absent; clypeus well developed, usually with broadly rounded apex (emarginate in *Shoguna*), without lateral impressions at base. Labrum more or less fused to clypeus and concealed from above. Antennae short, 10-segmented, with distinct 1- or 2-segmented club. Mandible unidentate, without dorsal tubercle or cavity; incisor edge simple or with one or more subapical teeth; mola well developed, prostheca reduced. Maxilla with elongate, slender, finger-like galea and broad setose lacinia, without uncus; apical palpomere fusiform. Mentum often elongate, ligula simple; apical labial palpomere securiform; maxillary and labial palps with palpomere 2 greatly enlarged in *Monotoma*. Corpotentorium narrow, straight, without median process. Cervical sclerites highly reduced.

Pronotum ~0.7–1.85 times as long as wide, usually about as wide at base as combined elytral bases (distinctly narrower in some *Monotoma*), with sides straight or very weakly curved. Lateral pronotal carinae more or less complete, usually crenulate or denticulate (smooth in *Shoguna*); anterior angles slightly produced in *Monotoma*; posterior angles right or rounded; posterior edge simple; disc simple in *Shoguna*, with punctures sometimes unevenly distributed and forming longitudinal rows in Europini, with paired sub-basal impressions in *Monotoma*. Prosternum in front of coxae more than twice as long as procoxal cavity, usually flat to slightly convex (strongly convex in *Shoguna*). Prosternal process complete, flat, gradually expanded posteriorly and truncate at apex. Notosternal sutures incomplete or absent; hypomera in *Shoguna* broadly, shallowly impressed to receive fore legs. Procoxae with long concealed extensions; trochantins concealed. Procoxal cavities circular, broadly separated, externally broadly closed, without narrow lateral extensions, internally closed.

Scutellar shield not abruptly elevated, posteriorly broadly to narrowly rounded. Elytra ~1.5–2.7 times as long as combined width and 1.4–2.25 times as long as pronotum, with well developed humeri, parallel-sided, with truncate apices, exposing tergite VII (female) or tergites VII and VIII (male); suture not deflected near apex; disc usually with 10 puncture rows and no scutellary striole; rows forming impressed striae in some *Shoguna*; epipleura well developed but ending at truncation. Mesoventrite with broad prothoracic rest or prepectus. Mesoventral process emarginate at apex. Mesocoxae with concealed trochantins. Mesocoxal cavities subcircular, broadly separated, partly closed laterally by mesepimera. Mesometaventral junction monocondylic. Metaventrite usually slightly convex (strongly so in *Shoguna*) with long discrimen; exposed portion of metanepisternum very long and narrow. Metacoxae usually widely separated (narrowly so in *Shoguna*), plates absent. Metendosternite usually without distinct stalk, with long lateral arms, no laminae, no anterior process, and anterior tendons located on arms (stalk long and narrow and anterior tendon approximate and located on short anterior process in *Shoguna*). Hind wing with long to very long apical field containing no sclerites or vein remnants; radial cell and cross-vein r3 absent; cross-vein r4 often incomplete or absent; basal portion of RP and medial spur absent; medial field with one free vein and usually with medial fleck (absent in *Shoguna*); anal lobe often reduced, embayment absent. Legs relatively short; trochanterofemoral joint slightly oblique; tibiae sometimes widened apically; outer apical angle in *Shoguna* with several stout teeth. Tarsi in *Monotoma* and Europini 5-5-5 in female and 5-5-4 in male, with tarsomere 1 min and often difficult to see (tarsi 5-5-5 in both sexes in *Shoguna*); tarsomeres not lobed; pretarsal claws simple.

Abdomen with five free ventrites; ventrite 1 as long as next two combined, with postcoxal lines in Europini; with small setose pocket in males of *Mimemodes*; intercoxal process broadly to narrowly rounded. Tergite VII well sclerotised. Functional spiracles usually on abdominal segments I–VI (on segments I–VII in *Shoguna*). Tergite VIII in male forming exposed genital capsule; sternite VIII in male reduced. Segment IX in male highly reduced, without spiculum gastrale. Aedeagus of cucujiform type, symmetrical; anterior edge of tegmen broadly rounded; parameres fused together and to phallobase and apparently absent; penis short, broad and flattened with long paired struts. Ovipositor moderately short and broad, lightly sclerotised except for baculi, paraprocts shorter than coxites, each of which is subdivided into basal and apical lobes; styli reduced and usually subapical. [Kuschel 1979; Nikitsky 1986a; Sen Gupta 1988; Sen Gupta and Pal 1995; Otero *et al.* 1996; Lawrence *et al.* 1999b; Bousquet and Laplante 2000; Bousquet 2002, 2010].

Larva (Fig. 84L)

Body elongate, parallel-sided or somewhat fusiform and slightly to moderately flattened; dorsal surfaces lightly pigmented and usually granulate or tuberculate (unpigmented and almost smooth in *Shoguna*); ventral surfaces unpigmented; vestiture of scattered, simple setae.

Head prognathous and protracted, slightly transverse and slightly flattened, abruptly constricted posteriorly in *Monotoma*. Epicranial stem absent; frontal arms U-shaped or slightly lyriform and widely separated at base. Median endocarina absent. Stemmata usually two on each side (absent in *Shoguna*). Frontoclypeal suture absent. Labrum usually free (fused to head capsule in *Shoguna*). Antennae short, 3-segmented, with conical sensorium more than half as long as apical antennomere. Mandibles (Fig. 27I) symmetrical, bidentate, with accessory ventral process; incisor edge simple or serrate; mola well developed, asperate or tuberculate; prostheca a narrowly acute hyaline process. Ventral mouthparts retracted, with well developed maxillary articulating area; stipites elongate; mala falciform, with row of setae along inner edge; maxillary palps 3-segmented. Labium consisting of prementum, mentum and submentum; ligula short and broad; labial palps 1-segmented and widely separated. Hypopharyngeal sclerome a transverse bar. Hypostomal rods usually absent (short and diverging in *Shoguna*); ventral epicranial ridges usually well developed (absent in *Shoguna*). Gular region short.

All thoracic terga and abdominal terga I–VIII with complex lateral processes in some *Monotoma*. Legs well developed, 5-segmented including claw-like pretarsus bearing two setae lying side by side; coxae moderately to widely separated. Abdominal terga often with rows of tubercles or asperities. Segment IX slightly shorter than VIII, tergum usually extending onto ventral surface and with short, subparallel, posteriorly directed urogomphi (with accessory processes and pregomphi in some *Monotoma*), tergum not extending onto ventral surface and urogomphi minute and dorsally oriented in *Shoguna*; sternum IX exposed, simple, not enclosed by sternum VIII; segment X subcylindrical, usually not visible from above and posteroventrally oriented (visible from above and posteriorly oriented in *Shoguna*). Spiracles annular–biforous, sometimes placed at the ends of short spiracular tubes. [Lawrence *et al.* 1999a; De Marzo 2000; Bousquet 2002, 2010].

Classification and World Distribution

Monotomidae is a cosmopolitan family of 33 genera and ~250 species. The group is usually divided two subfamilies: Rhizophaginae with the single Holarctic and Oriental genus *Rhizophagus* Herbst and Monotominae with the remaining genera. Within the Monotominae, four tribes are recognised, three of which occur in Australia; the tribe Lenacini includes only the New Zealand genus *Lenax* Sharp. [Sen Gupta 1988; Bousquet 2002, 2010].

Australian Fauna

The species of **Monotominae** inhabiting Australia are primarily northern in distribution. Species of the genus *Mimemodes* Reitter (Europini) (Fig. 83F) occur in northern NSW and QLD, while *Monotomopsis monotomoides* Grouvelle, another europine, is restricted to northern QLD, as are the species of *Shoguna* Lewis (Thionini) (Fig. 72A). The adventive species of the widespread genus *Monotoma* Herbst (Monotomini) (Fig. 87Q) are more widely distributed. Monotomids are elongate, parallel-sided, subcylindrical to flattened beetles, which may be subglabrous or clothed with decumbent hairs and have truncate elytra exposing the pygidium. Except in *Shoguna*, the head is abruptly narrowed posteriorly forming distinct temples behind the eyes, and the pronotum is either crenulate or dentate laterally. Larvae are elongate, slightly flattened, slightly tapered posteriorly and lightly pigmented above but with tubercles on the dorsal surfaces and sometimes with lateral tergal processes on the thorax and abdomen (Fig. 84L). The mala of monotomids is falciform, the mandibular prostheca is narrow and sometimes serrate, and the urogomphi on tergum IX are weakly developed. The introduced species *Monotoma longicollis* Gyllenhal, *M. picipes* Herbst and *M. spinicollis* Aubé occur in the more populated areas, where they are common inhabitants of grass piles or compost heaps. *Shoguna* species are very distinctive, subcylindrical beetles, which have a long, narrow head and prothorax and apparently live in the tunnels of ambrosia beetles (Schedl 1962). Species of *Mimemodes* have a greatly expanded head in the male. Monotomids all appear to be mycophagous, feeding on moulds and other less conspicuous fungi. [Kuschel 1979].

64. HOBARTIIDAE

Adult (Fig. 86J)

Length usually 2–2.9 mm. Body ~1.9–2.05 times as long as wide; ovate to oblong and convex; dorsal surfaces moderately shiny, yellowish-brown to reddish-brown or dark-brown, with pronotum darker than elytra and elytra sometimes bicolored; vestiture of decumbent and erect hairs.

Head slightly declined, transverse, narrowing anteriorly, not abruptly constricted posteriorly. Eyes entire, prominent, finely facetted, without interfacetal setae. Antennal insertions exposed and widely separated; subantennal grooves absent. Frontoclypeal suture sharply defined, straight; base of clypeus with paired lateral impressions into which mandibular tubercles fit; anterior edge of clypeus truncate. Labrum free, strongly transverse. Antennae 11-segmented, with distinct 3-segmented club in *Hobartius* Sen Gupta and Crowson, but with antennomere 7 also enlarged in *Hydnobiodes* Sen Gupta and Crowson, so that the club appears to be 5-segmented with the second club segment reduced (as in many Leiodidae). Mandible bidentate, with distinct dorsal tubercle but without associated cavity; incisor edge simple; mola and prostheca well developed. Maxilla with broad, setose galea and slender lacinia, with unidentate uncus; apical palpomere fusiform. Mentum transverse and trapezoidal, ligula truncate; apical labial palpomere securiform in *Hydnobioides*, slightly expanded and truncate in *Hobartius*. Corpotentorium moderately broad, without median process. Cervical sclerites well developed.

Pronotum ~0.5–0.65 times as long as wide, widest at base, which is as wide as combined elytral bases; sides very weakly curved. Lateral pronotal carinae complete, the edges weakly crenulate to denticulate; anterior angles slightly produced and rounded; posterior angles obtuse; posterior edge rounded or weakly bisinuate; disc simple. Prosternum in front of coxae about as long as procoxal cavity, slightly convex. Prosternal process complete, moderately curved, parallel-sided or weakly expanded apically, broadly rounded at apex. Procoxae with very short concealed extensions; trochantins exposed. Procoxal cavities transverse, narrowly separated, externally broadly open (but with short postcoxal projections acute), without narrow lateral extensions, internally open.

Scutellar shield more or less abruptly elevated, laterally expanded and posteriorly broadly rounded. Elytra ~1.3–1.6 times as long as combined width and 2.6–3.3 times as long as pronotum, with well developed humeri, sides weakly rounded; suture not deflected near apex; disc irregularly punctate, without scutellary striole; epipleura narrow and incomplete. Mesoventrite with slightly inclined procoxal rests. Mesoventral cavity small and shallow. Mesoventral process emarginate at apex. Mesocoxae with exposed trochantins. Mesocoxal cavities slightly transverse, moderately broadly separated, partly closed laterally by mesanepisterna and mesepimera. Mesometaventral junction monocondylic. Metaventrite flat, with long discrimen; exposed portion of metanepisternum moderately elongate. Metacoxae narrowly separated; plates absent. Metendosternite with broad stalk, long lateral arms, narrow laminae, very short anterior process, and anterior tendons approximate. Hind wing with long apical field containing anterior and posterior veinal remnants; radial cell reduced, elongate, sometimes without lumen; cross-vein r3 short and oblique, sometimes fused to r4, which may be incomplete; basal portion of RP moderately long; medial spur not extending to wing margin; medial field with one or two free veins and no medial fleck; wedge cell absent; anal lobe moderately well developed; embayment notch-like. Tibiae slightly widened apically in *Hobartius*, more strongly so in *Hydnobioides*, which also has spines on outer edge and at apex. Tarsi 5-5-5 females and 5-5-4 in males of *Hobartius*, 5-5-5 in both sexes in *Hydnobioides*; tarsomeres without ventral lobes; pretarsal claws simple.

Abdomen with five free ventrites; ventrite 1 almost as long as next two combined; intercoxal process acute. Functional spiracles on abdominal segments I–VII. Anterior edge of segment IX in male with spiculum gastrale; tergite IX and segment X membranous. Aedeagus of cucujiform type, symmetrical, turned on side when retracted; anterior edge of tegmen with single strut (*Hobartius*) or paired, contiguous struts (*Hydnobioides*); parameres articulated and separate or basally fused together; penis flattened and basally truncate, without paired struts. Ovipositor moderately elongate, lightly sclerotised except for baculi, paraprocts shorter than coxites, each of which is subdivided into basal and apical lobes; styli well developed and apical. [Sen Gupta and Crowson 1966, 1969a; Lawrence *et al.* 1999b; Tomaszewska and Ślipiński 1995, 2010].

Larva (Fig. 84M)

Body elongate, fusiform and somewhat flattened; dorsal surfaces lightly pigmented, except for mandibles and tips of urogomphi, bearing simple or complex setiferous tubercles.

Head prognathous and protracted, slightly transverse and slightly flattened; dorsal surface with irregular patterns of granules. Epicranial stem absent; frontal arms lyriform, widely separated at base. Median endocarina absent. Stemmata usually five on each side. Frontoclypeal suture absent. Labrum large, free, rounded anteriorly. Antennae moderately long, 3-segmented, with conical sensorium less than half as long as apical antennomere. Mandibles symmetrical, bidentate, with accessory ventral process; incisor edge serrate in *Hobartius*, simple in *Hydnobioides*; mola well developed, transversely ridged; prostheca an acute hyaline process. Ventral mouthparts retracted, with well developed maxillary articulating area; stipites elongate; mala falciform in *Hobartius*, truncate in *Hydnobioides*, with two or three teeth at inner apex; maxillary palps 3-segmented. Labium consisting of prementum, mentum and submentum; ligula short and broad; labial palps 1-segmented and widely separated. Hypopharyngeal sclerome molar-like. Hypostomal rods long and diverging; ventral epicranial ridges absent. Gular region short.

Tergal sclerites not clearly delimited, with asperities and setiferous tubercles. Legs short, 5-segmented including claw-like pretarsus bearing two setae; coxae moderately widely separated. Abdominal terga with prominent, complex tubercles in *Hobartius* or small conical tubercles in *Hydnobioides*. Segment IX shorter than VIII, tergum not extending onto ventral surface, with short, simple, subparallel or slightly convergent, posteriorly directed urogomphi upturned at apex; sternum IX exposed, simple, not enclosed by sternum VIII; segment X subcylindrical, visible from above and posteriorly oriented. Spiracles annular–biforous, placed at the ends of short spiracular tubes. [Sen Gupta and Crowson 1969a; Lawrence 1991b; Lawrence *et al.* 1999a; Tomaszewska and Ślipiński 1995, 2010].

Classification and World Distribution

The family Hobartiidae includes six species in two genera. *Hobartius chilensis* Tomaszewska & Ślipiński occurs in Chile and Argentina and the remaining members of the family are Australian.

Australian Fauna

The Australian Hobartiidae include *Hobartius eucalypti* (Blackburn) from TAS and Victoria, *H. newtonorum* Tomaszewska &

Ślipiński from VIC and NSW, *H. niger* Tomaszewska & Ślipiński from the ACT and NSW, *Hydnobioides pubescens* Sen Gupta & Crowson, from northern NSW and southern QLD and *H. lawrencei* Tomaszewska & Ślipiński from southern NSW. Hobartiids are oblong, pubescent beetles with a strong antennal club (in *Hydnobioides* of the leiodid type with antennomere 8 smaller than either 7 or 9), a well developed, curved frontoclypeal suture, a visible labrum, dorsal mandibular tubercles (but not cavities), and contiguous elytral apices, without an expanded sutural flange. Larvae are elongate, tapered posteriorly and relatively lightly sclerotised, with a tuberculate dorsal surface, truncate or falcate mala, and 1-segmented labial palps. Adults and larvae are known to feed on the softer fruiting bodies of some Basidiomycetes, but adults of the South American species were attracted in large numbers to carrion traps. [Tomaszewska and Ślipiński 1995, 2010].

65. CRYPTOPHAGIDAE

Adult (Fig. 86E–F)

Length 0.85–2.7 mm. Body ~1.5–2.95 times as long as wide; usually more or less parallel-sided (sometimes ovoid or with prothorax and elytra independently rounded laterally), strongly convex to strongly flattened; colour yellowish-brown to reddish-brown or dark-brown, elytra sometimes bicolored; vestiture usually of decumbent or decumbent and erect hairs, rarely subglabrous.

Head slightly declined, transverse, usually not abruptly constricted posteriorly (short temples present in some Cryptophaginae). Occipital region with or without transverse carina. Eyes large or small, sometimes highly reduced, entire, coarsely or finely facetted, with or without interfacetal setae. Antennal insertions exposed and widely to very narrowly separated; subantennal grooves usually absent (present in *Ephistemus* and *Microatomaria*). Frontoclypeal suture present in most Atomariinae; base of clypeus without paired lateral impressions. Labrum free, strongly transverse. Antennae 11-segmented, usually with distinct 3-segmented club (club 2-segmented in *Agnetaria* Bruce and *Microatomaria* Leschen, weakly developed in *Anitamaria* Leschen). Mandible bidentate; incisor edge simple or with single tooth; mola and prostheca usually well developed. Maxilla (Fig. 16C) with well developed galea and lacinia, the latter with or without uncus; apical palpomere cylindrical. Mentum transverse to quadrate, ligula undivided; apical labial palpomere cylindrical. Corpotentorium narrow, sometimes arched, without median process. Cervical sclerites present in most Cryptopphaginae.

Pronotum ~0.45–0.90 times as long as wide, usually as wide or almost as side as combined elytral bases; sides straight or moderately to strongly curved. Lateral pronotal carinae usually complete, with or without margin or bead (incomplete in Cryptafricini and some Atomariini), edge usually simple, serrate in *Henoticus*, with one or more lateral teeth plus a prominent glandular callosity at anterior end in *Cryptophagus* (Figs 18J–K); anterior angles almost never produced forward (produced and acute in *Ephistemus*); posterior angles usually right to obtuse (rounded in some Atomariinae); posterior edge straight or weakly bisinuate; disc usually simple (paired sub-basal impressions or transverse basal impression sometimes present). Prosternum in front of coxae about as long as to distinctly longer than procoxal cavity, slightly to strongly convex, occasionally with paired lines. Prosternal process complete, usually flat, parallel-sided to weakly expanded apically, and broadly rounded or truncate at apex. Notosternal sutures present or absent. Procoxae with short concealed extensions; trochantins concealed. Procoxal cavities slightly to strongly transverse, moderately widely separated, externally open, without narrow lateral extensions; usually internally open (closed by slender bar in some Cryptophaginae).

Scutellar shield abruptly elevated, posteriorly broadly rounded or obtusely angulate. Elytra ~1.05–1.90 times as long as combined width and 1.90–3.60 times as long as pronotum, with well developed humeri, sides parallel to weakly rounded; suture deflected near apex (Fig. 26I); disc irregularly punctate, without scutellary striole; epipleura narrow and incomplete. Anterior edge of mesoventrite usually on same plane as metaventrite (on different plane in *Ephistemus*), with horizontal to strongly inclined procoxal rests. Mesoventral cavity, if present small to large and very shallow. Mesoventral process truncate or slightly emarginate at apex. Mesocoxae with concealed trochantins. Mesocoxal cavities subovate (distinctly oblique in Cryptafricini), widely separated, closed laterally by meeting of meso- and metaventrite. Mesometaventral junction simple or dicondylic. Metaventrite usually with discrimen and without postcoxal lines; exposed portion of metanepisternum moderately to very long and narrow. Metacoxae narrowly to widely separated, not extending laterally to meet elytra; plates absent. Metendosternite with short, broad stalk, long lateral arms and narrow laminae; anterior process very short or absent and anterior tendons usually approximate. Hind wing with long apical field usually with vaguely indicated anterior and posterior veinal remnants; radial cell reduced or absent; cross-vein r3 usually absent and r4 incomplete; R-M loop acute and basal portion of RP short to moderately long; medial spur often extending to wing margin; medial field with five or fewer free and no medial fleck; wedge cell absent; anal lobe reduced; embayment notch-like; posterior edge with long fringe hairs in some small species. Legs usually long and slender; trochanterofemoral joint transverse to somewhat oblique; tibiae not apically expanded; tibial spurs single in some Atomariinae. Tarsi 5-5-5 in males and females of Atomariinae, 5-5-5 in female and 5-5-4 in males of Cryptophaginae; tarsomeres without or with weak ventral lobes; pretarsal claws simple; empodium bisetose.

Abdomen with five free ventrites; ventrite 1 almost always much longer than 2 (only slightly longer in Cryptafricini), recurved postcoxal lines present in *Ephistemus*; intercoxal process narrowly to broadly rounded or truncate. Functional spiracles usually present on abdominal segments I–VII (on I–VI in *Henoticus*, I–V in most Atomariinae and 1–IV in Cryptafriciini). Anterior edge of segment IX in male usually with spiculum gastrale; tergite IX and segment X membranous. Aedeagus of cucujiform type, usually more or less symmetrical, anterior edge of tegmen with or without single strut; parameres articulated and separate or fused together and to phallobase; penis flattened and basally rounded truncate, without paired struts; endophallus often with asperities or sclerites. Sternite VIII in female usually with spiculum ventrale (absent in Cryptafricini). Ovipositor usually ~1.5 times as long as wide (reduced in Cryptafricini), lightly sclerotised except for baculi, paraprocts subequal to coxites, each of which is subdivided into broader basal and narrower apical lobe; styli well developed and apical. [Hinton 1945; Evans 1961a, 1961c; Bousquet 1989; Leschen 1996, 2010c Lawrence *et al.* 1999b; Leschen and Skelley 2002a].

Larva (Fig. 84B)

Body elongate, parallel-sided, cylindrical to distinctly flattened; dorsal surfaces lightly to moderately pigmented, smooth or finely asperate, granulate or tuberculate; vestiture of sparsely distributed setae.

Head prognathous and protracted, slightly transverse and slightly flattened. Epicranial stem absent; frontal arms lyriform and contiguous at base. Median endocarina absent. Stemmata either absent or forming a small tight cluster (four in *Henoticus* and at least some *Cryptophagus*, sometimes two or one). Frontoclypeal suture absent. Labrum free. Antennae usually moderately long (very short in *Cryptogasterus*), 3-segmented, with conical sensorium less than half as long as apical antennomere. Mandibles symmetrical, broad at base and narrow at apex, bidentate, with accessory ventral process; incisor edge usually serrate; mola well developed, asperate or tuberculate; prostheca a hyaline process, which is usually broad at base and narrowly acute or bidentate and more heavily pigmented apically. Ventral mouthparts retracted, with well developed maxillary articulating area; stipites elongate; mala falciform, not cleft; maxillary palps 3-segmented. Labium consisting of prementum, mentum and submentum; ligula simple; labial palps 1- or 2-segmented and widely separated. Hypopharyngeal sclerome molar-like or a transverse bar. Hypostomal rods moderately long and diverging; ventral epicranial ridges absent. Gula wider than long.

Thoracic and abdominal terga without sclerotised plates and usually without lateral processes. Legs well developed, 5-segmented including claw-like pretarsus bearing two setae lying side by side; coxae moderately to widely separated. Segment IX slightly shorter than or subequal to VIII, tergum sometimes extending onto ventral surface and with short, subparallel, posteriorly directed, simple urogomphi upturned at apex; urogomphi absent in *Cryptogasterus* and some Atomariini (e.g. *Atomaria lewisi* Reitter, *Microatomaria*); sternum IX exposed, simple, not enclosed by sternum VIII; segment X subcylindrical, posteriorly or postoventrally oriented and often visible from above. Spiracles annular in Atomariinae, annular–biforous in Cryptophaginae, rarely placed at the ends of short spiracular tubes. [Hinton 1945; Evans 1961b; Lawrence 1991b; Lawrence *et al.* 1999a; De Marzo 2000; Leschen and Skelley 2002a; Leschen 2010c].

Classification and World Distribution

Cryptophagidae is a worldwide family with 60 recognised genera and ~600 described species; the group displays an amphipolar distribution (Crowson 1980) with the highest diversity in the temperate parts of both Northern and Southern Hemispheres. There are two subfamilies and six tribes, four of which occur in Australia. The cryptophagine tribe Caenoscelini is mainly Holarctic, with a few species at higher altitudes in tropical Asia and Central America. The atomariine tribe Hypocoprini occurs in the Holarctic region and in Baja California and Chile. [Leschen and Lawrence 1991; Leschen 1996, 2001, 2010c; Lyubarsky 1998; Leschen and Skelley 2002a].

Australian Fauna

The Australian cryptophagid fauna includes members of both Cryptophaginae and Atomariinae. Among the **Cryptophaginae**, the tribe Cryptophagini is represented by a few itroduced species of *Cryptophagus* Herbst and *Henoticus* Thomson. The Notogean tribe Cryptosomatulini includes all of the endemic species in the genera *Agnetaria* Bruce, *Chiliotis* Reitter and *Micrambina* Reitter, plus a few species misplaced in the genus *Cryptophagus* (Fig. 86E–F) and several undescribed taxa. In the subfamily **Atomariinae**, members of the tribe Atomariini includes endemic species in the genera *Micratomaria* Leschen and *Chilatomaria* Leschen plus introduced species of *Atomaria* Stephens and *Ephistemus* Stephens. The tribe Cryptogasterini includes the endemic genus *Cryptogasterus* Leschen. Cryptophagids are usually oblong to elongate and slightly flattened beetles, usually red or brown in colour and clothed with erect and decumbent hairs (rarely globose and/or glabrous). The antennal insertions are exposed, lateral and well separated or more or less approximate; the pronotum is subquadrate with distinct lateral carinae or more rounded, sometimes with paired glandular calllosities at the anterior angles; the prosternal process is moderately broad and overlapping the mesoventrite; the elytra have apically widened sutural flanges and incomplete epipleura; the trochanters are more or less elongate; and ventrite 1 is much longer than 2. Cryptophagid larvae are elongate and subcylindrical to flattened and lightly sclerotised, with short and straight or sharply curved urogomphi (absent in an introduced *Atomaria*). The prostheca is usually serrate and the mala falcate. Cryptophagids feed on spores and hyphae of moulds and other fungi usually occurring under bark, in leaf litter, or in nests of various kinds. The adventive species of *Cryptophagus*, *Henoticus*, *Atomaria* and *Ephistemus* may occur in foodstuffs and some are common in grass cuttings and compost heaps. The Australian fauna, which is large and poorly studied, includes several genera with New Zealand and South American affinities. [Evans 1961b; Aitken 1975; Leschen 2001].

66. PRIASILPHIDAE

Adult (Fig. 85A)

Length 3.3–5.5 mm. Body ~1.7–2.1 times as long as wide; oblong with sides subparallel or weakly rounded, slightly to moderately convex; colour dark-brown or black; surfaces more or less smooth, but with short carinae or tubercles on elytra and somewhat shiny but usually covered with exudate and debris; vestiture of stout, decumbent bristles evenly distributed or forming clusters.

Head not or slightly declined, about as long as wide, not abruptly constricted posteriorly. Eyes relatively small but prominent, entire, coarsely facetted, without interfacetal setae. Antennal insertions slightly concealed from above; subantennal grooves present. Frontoclypeal suture absent; clypeus extending well in front of antennal insertions, with subparallel sides and truncate apex. Labrum free, strongly transverse. Antennae 11-segmented, with distinct 3-segmented club. Mandible bidentate or unidentate, without dorsal tubercle or cavity; incisor edge simple; mola well developed and transversely ridged; prostheca well developed. Maxilla with broad, setose galea and narrow lacinia with bifid uncus; apical palpomere fusiform. Mentum strongly transverse, ligula laterally expanded and apically emarginate; apical labial palpomere fusiform. Genae somewhat inflated and anteriorly projecting. Corpotentorium moderately broad and straight, without median process. Cervical sclerites small and more or less contiguous with head capsule.

Pronotum ~0.5–0.65 times as long as wide, widest at about middle and almost as wide at base as combined elytral bases; sides strongly curved and explanate. Lateral pronotal carinae complete, without margin or bead, edge broadly undulate; anterior angles produced and narrowly to broadly rounded; posterior angles more or less right; posterior edge straight; disc simple or with very weak anterior and posterior lateral impressions. Prosternum in front of coxae (Figs 18E, 44K) about as long as procoxal cavity, moderately convex, without paired lines. Prosternal process complete, slightly curved, with subparallel sides and truncate apex. Procoxae without concealed extensions; trochantins exposed. Procoxal cavities strongly transverse, widely separated, externally narrowly open, without narrow lateral extensions; internally open.

Scutellar shield small, strongly transverse, abruptly elevated, posteriorly obtusely angulate. Elytra ~1.23–1.36 times as long as combined width and 2.05–2.6 times as long as pronotum, with well developed humeri, sides weakly rounded; suture not deflected near apex; disc irregularly punctate, without scutellary striole; epipleura incomplete. Mesoventrite with strongly inclined procoxal rests. Mesoventral process short and truncate. Mesocoxae with exposed trochantins. Mesocoxal cavities transverse, widely separated, partly closed laterally by mesepimera. Mesometaventral junction a straight line with metaventrite overlapping mesoventrite or with metaventrite fitting into a very broad and shallow cavity (monocondylic). Metaventrite short and slightly convex, without discrimen or postcoxal lines; exposed portion of metanepisternum moderately elongate. Metacoxae widely separated; plates absent. Metendosternite with short, broad stalk, long lateral arms and narrow laminae; anterior process absent and anterior tendons located on lateral arms. Hind wings absent. Legs usually long and slender; trochanterofemoral joint strongly oblique with femur separate from coxa; tibiae not apically expanded, edges usually minutely denticulate; spurs paired on all legs. Tarsi 5-5-5 in females and 5-5-4 in males; tarsomeres without ventral lobes; pretarsal claws simple; empodium not apparent.

Abdomen with five ventrites, the first two of which are connate; ventrite 1 about as long as 2; intercoxal process truncate. Functional spiracles present on abdominal segments I–VII. Anterior edge of segment IX in male with spiculum gastrale; posterior edge with articulated lobe; tergite IX completely divided; segment X membranous with lateral baculi. Aedeagus of cucujiform type, symmetrical and inverted; anterior edge of tegmen subacute; parameres articulated and separate; penis divided into basal and apical lobes, broadly rounded anteriorly, with median carina, without paired struts; endophallus with sclerites. Ovipositor moderately elongate, lightly sclerotised except for baculi, paraprocts subequal to coxites, each of which is subdivided into broad basal and narrow apical lobe; styli well developed and apical. [Crowson 1973; Lawrence *et al.* 1999b; Leschen *et al.* 2005; Leschen and Lawrence 2010b].

Larva (Fig. 84O)

Body moderately elongate, somewhat elliptical, widest near anterior end and narrowing posteriorly, flattened, with log lateral processes on all thoracic terga and abdominal terga I–VIII; dorsal surfaces moderately pigmented, smooth and finely granulate, with setiferous tubercles forming straight and curved rows and patches.

Head prognathous and protracted, strongly transverse and slightly flattened. Epicranial stem absent; frontal arms lyriform and well separated at base. Median endocarina absent. Stemmata six on each side, three on a single projection, and three on individual projections. Frontoclypeal suture absent. Labrum free. Antennae moderately long, 3-segmented, with conical sensorium almost as long as reduced apical antennomere. Mandibles symmetrical, broad at base and narrow at apex, unidentate, with accessory ventral process; incisor edge simple; mola well developed, transversely ridged; prostheca a slender, apically acute hyaline process. Ventral mouthparts retracted, with well developed maxillary articulating area; stipites elongate; mala falciform, not cleft; maxillary palps 3-segmented. Labium consisting of prementum, mentum and submentum; ligula simple; labial palps 2-segmented and widely separated. Hypopharyngeal sclerome a transverse bar. Hypostomal rods moderately diverging; ventral epicranial ridges present. Gula wider than long.

Prothorax with two long, posteriorly curved, lateral processes on each side; meso- and metatergum each with one similar process on each side. Legs well developed, 5-segmented including claw-like pretarsus bearing two setae lying side by side; coxae moderately widely separated. Abdomen less than twice as long as thorax; tergum I with long, straight spiracle-bearing process on each side; terga II–VIII each with a long, curved spiracle-bearing process on each side. Segment IX slightly shorter than VIII, tergum barely extending onto ventral surface, with moderately long, subparallel, posteriorly directed, each with numerous accessory processes; sternum IX exposed, simple, not enclosed by sternum VIII; segment X sclerotised at least ventrally but not or only slightly visible from above; anus posteroventrally oriented. Spiracles annular–biforous, all placed at the ends of long spiracular tubes. [Lawrence *et al.* 1999a; Leschen *et al.* 2005; Leschen and Lawrence 2010b].

Classification and World Distribution

The family Priasilphidae includes *Chileosilpha elguetai* Leschen *et al.* from the Valdivian region of Chile, *Priasilpha* Broun with eight species in New Zealand and the Subantarctic Islands, and the Australian genus *Priastichus* Crowson. [Leschen *et al.* 2005; Leschen and Michaux 2005; Leschen and Lawrence 2010b].

Australian Fauna

The three species of *Priastichus*. *P. tasmanica* Crowson, *P. crowsoni* Leschen *et al.* and *P. megathorax* Leschen *et al.* (Fig. 85A), occur only in TAS. Priastichids are oblong, somewhat flattened beetles, wingless beetles with a laterally explanate pronotum, widely separated coxae, and a dull, tuberculate upper surface bearing clusters of coarse, decumbent setae and often caked with mud and debris. Larvae are flattened with complex lateral processes on thorax and abdomen with the abdominal spiracles located at the ends of these processes. Both adults and larvae occur in leaf litter. [Crowson 1973; Leschen *et al.* 2005].

67. PHLOEOSTICHIDAE

Adult (Figs 83C, 87O)

Length 2.4–6.5 mm. Body ~2.5–3.5 times as long as wide; almost always parallel-sided (*Hymaea succinifera* Pascoe with sides of prothorax and elytra independently weakly rounded), moderately convex to slightly flattened; colour yellowish-brown (*Bunyastichus* Leschen *et al.*) or reddish-brown to black; elytra with yellow maculae in *Hymaea* Pascoe; surfaces more or less smooth, but usually with short carinae or tubercles usually on elytra; usually shiny (with a dull lustre in *Rhopalobrachium* Boheman); vestiture of scattered, erect setae in *Rhopalobrachium* and *Hymaea*; *Bunyastichus* more densely clothed with decumbent and suberect yellow hairs.

Head not or slightly declined (but abruptly bent ventrally at about middle in *Hymaea*), at least as long as wide, abruptly constricted posteriorly to form distinct temples in *Hymaea* (except *H. succinifera*). Eyes strongly protuberant, entire, coarsely facetted, without interfacetal setae. Antennal insertions usually exposed from above (barely concealed in *Bunyastichus*); subantennal grooves absent. Frontoclypeal suture indistinct and curved (absent in *Bunyastichus*); clypeus extending well in front of antennal insertions, with distinct basolateral emarginations to receive mandibular tubercles; anterior edge truncate. Labrum free, strongly transverse. Antennae 11-segmented, with 3-segmented club (club segments not clearly distinguished from funicular antennomeres in *Rhopalobrachium*). Mandible usually bidentate (tridentate in *Bunyastichus*), with dorsal tubercle (weakly developed in *Bunyastichus*) and with glabrous cavity in *Hymaea* and *Rhopalobrachium*; incisor edge simple or with a rounded or angulate process; mola well developed and transversely ridged; prostheca somewhat reduced. Maxilla with broad, setose galea and narrow lacinia with bifid uncus; apical palpomere fusiform. Mentum subtrapezoidal; ligula shallowly to moderately deeply emarginate; apical labial palpomere fusiform. Genae sometimes laterally projecting. Pregular region with slight to distinct transverse impression. Corpotentorium short, narrow to moderately broad, without median process. Cervical sclerites reduced and adpressed to head capsule.

Pronotum ~0.9–1.4 times as long as wide, widest at about middle and usually distinctly narrower at base than combined elytral bases (not so in *H. succinifera*); sides slightly to strongly curved (irregularly sinuate in *Rhopalobrachium*). Lateral pronotal carinae complete and coarsely dentate in *Bunyastichus*, absent in *Hymaea* and *Rhopalobrachium*; anterior angles not produced forward; posterior angles right to rounded; posterior edge straight; disc simple in *Bunyastichus*, with tubercles and/or impressions in *Rhopalobrachium* and *Hymaea*. Prosternum in front of coxae longer than procoxal cavity, flat to strongly convex. Prosternal process complete, at least slightly curved, parallel-sided to slightly or strongly expanded at apex, which is broadly rounded or truncate. Notosternal sutures incomplete in *Rhopalobrachium*, absent in *H. succinifera*. Procoxae without or with very short concealed extensions; trochantins exposed. Procoxal cavities strongly transverse in *Bunyastichus*, slightly transverse in *Hymaea* and *Rhopalobrachium*, narrowly to widely separated, externally narrowly open in *Hymaea* and *Rhopalobrachium* (narrowly closed in *Bunyastichus*), with narrow lateral extensions (except in *Bunyastichus*); internally open.

Scutellar shield abruptly elevated, posteriorly obtusely angulate. Elytra usually ~1.85–2.85 times as long as combined width (1.55–1.65 times in *H. succinifera*) and 2.5–3 times as long as pronotum (1.7–1.8 in *H. succinifera*), usually with well developed humeri, subparallel sides and suture deflected near apex (humeri absent, sides rounded and suture not deflected in *H. succinifera*); disc usually with ten distinct puncture rows and a scutellary striole (irregularly punctate in *Rhopalobrachium*); epipleura narrow but almost complete (except in *H. succinifera*). Mesoventrite with slightly to strongly inclined procoxal rests. Mesocoxae with exposed trochantins. Mesocoxal cavities narrowly to moderately widely separated, partly closed laterally by mesepimera. Mesometaventral junction monocondylic. Metaventrite usually slightly transverse to elongate, with moderately to very long discrimen (strongly transverse without discrimen in *H. succinifera*); postcoxal lines absent; exposed portion of metanepisternum very long and narrow. Metacoxae usually narrowly separated (widely separated in *H. succinifera*); plates absent. Metendosternite usually with moderately long, broad stalk, long lateral arms and broad or narrow laminae; anterior process short and anterior tendons more or less approximate (laminae not apparent and anterior tendons on lateral arms in *H. succinifera*). Hind wings usually with long apical field and at least vaguely indicated radial extensions; radial cell incomplete basally; cross-veins r3 absent; R-M loop forming an acute angle; medial spur long; medial field with five free veins and no medial fleck; wedge cell absent; anal notch distinct; hind wings absent in *H. succinifera.* Legs relatively long and slender; femora somewhat inflated in *Rhopalobrachium*; tibiae not apically expanded. Tarsi 5-5-5 in females and 5-5-4 in males; tarsomeres 2 and 3 with ventral lobes and tarsomere 4 highly reduced in *Bunyastichus*; pretarsal claws simple.

Abdomen usually with five free ventrites (ventrites 1–3 connate in *H. succinifera*); ventrite 1 not much longer than 2; intercoxal process usually acute (broadly rounded in *H. succinifera*). Functional spiracles present on abdominal segments I–VII. Anterior edge of segment IX in male usually with spiculum gastrale (with narrowly acute subgenital plate in *Rhopalobrachium*); posterior edge with articulated lobe; tergites IX and X fused together and membranous. Aedeagus of cucujiform type, symmetrical; anterior edge of tegmen usually subacute (forming short strut in *Bunyastichus*); parameres articulated and separate; penis not subdivided, usually narrow and parallel-sided anteriorly, without paired struts; endophallus usually with rod-like sclerite. Ovipositor moderately elongate, lightly sclerotised except for baculi and apical portions of coxites, paraprocts subequal to or shorter than coxites, each of which is subdivided into broad basal and narrow apical lobe; styli well developed and apical. [Crowson 1973; Sen Gupta and Crowson 1966, 1969a; Lawrence 1995; Lawrence *et al.* 1999b; Leschen *et al.* 2005; Lawrence and Ślipiński 2010c].

Larva

Body elongate, parallel-sided, flattened, lightly pigmented except for mandibles, urogomphi and sometimes part or all of tergum IX; vestiture of scattered setae.

Head prognathous and protracted, strongly transverse and slightly flattened; as wide as or wider than prothorax. Epicranial stem absent; frontal arms lyriform and well separated at base.

Median endocarina absent. Stemmata six on each side. Frontoclypeal suture absent. Labrum partly or completely fused to clypeus. Antennae moderately long, 3-segmented, with dome-like sensorium shorter than apical antennomere. Mandibles symmetrical, broad at base and narrow at apex, tridentate, with accessory ventral process; incisor edge simple or with one or two subapical teeth; mola well developed with transversely ridges or rows of tubercles; prostheca a slender, apically acute hyaline process. Ventral mouthparts retracted, with well developed maxillary articulating area; cardines divided; stipites elongate; mala falciform or narrowly truncate with acute inner angle, not cleft; maxillary palps 3-segmented. Labium consisting of prementum, mentum and submentum; ligula broadly rounded or truncate; labial palps 2-segmented and widely separated. Hypopharyngeal sclerome a transverse bar. Hypostomal rods moderately long and diverging; ventral epicranial ridges absent. Gula wider than long.

Legs well developed, 5-segmented including claw-like pretarsus bearing two setae lying side by side; coxae widely separated. Abdomen with terga and sterna I–VII each with pair of curved rows of asperities. Segment IX as long as or slightly shorter than VIII, tergum extending onto ventral surface, not forming articulated plate; with long, simple, posteriorly or posterodorsally directed, widely separated urogomphi; sternum IX exposed, simple, not enclosed by sternum VIII; segment X not visible from above, without paired pygopods; anal region ventrally or posteroventrally oriented. Spiracles annular–biforous. [Lawrence *et al.* 1999a; Leschen *et al.* 2005; Lawrence and Ślipiński 2010c].

Classification and World Distribution

The family Phloeostichidae contains four genera. *Phloeostichus denticollis* Redtenbacher occurs from eastern Europe to Siberia, *Rhopalobrachium* species occur in Chile, Argentina and Australia, while *Hymaea* Pascoe and *Bunyastichus* Leschen *et al.* are Australian genera. [Lawrence and Ślipiński 2010c].

Australian Fauna

The genus *Hymaea* (Fig. 83C) extends from TAS to southern QLD, *Rhopalobrachium crowsoni* Lawrence (Fig. 87O) is known from northern NSW and southern QLD, and *Bunyastichus monteithi* Leschen *et al.* is recorded from northern QLD. Most Australian phloeostichids are narrowly elongate, parallel-sided and fully-winged, with a more or less tuberculate upper surface, a relatively sparse clothing of erect hairs, mandibles with dorsal tubercles and cavities, no lateral pronotal carinae, narrowly open procoxal cavities, long, somewhat clavate femora, simple tarsi and an expanded sutural flange on the elytra. In the Tasmanian *Hymaea succinifera* Pascoe, however, the body is not parallel-sided, the hind wings are absent and the sutural flange is not expanded, while in *Bunyastichus*, the procoxal cavities are narrowly closed, the mandibular cavity is absent, the lateral pronotal carinae are complete and more or less dentate and two tarsomeres are ventrally lobed. Phloeostichid larvae are elongate, parallel-sided and flattened, with widely separated urogomphi and curved rows of tergal and sternal asperities on the first six abdominal segments. It is likely that both adults and larvae feed on rotten cambium or other plant material. [Lawrence 1995; Leschen *et al.* 2005; Lawrence and Ślipiński 2010c].

68. SILVANIDAE

Adult (Figs 71C, 85D, H)

Length 1.2–15 mm. Body ~1.9–4.5 times as long as wide; almost always parallel-sided, sometimes with sides of pronotum and elytra weakly, independently rounded, moderately convex to strongly flattened; colour yellowish-brown to reddish-brown, dark brown or black; elytra red or yellow and black in *Macrohyliota militaris* (Erichson), many Telephanini and a few Silvaninae; surfaces usually more or less smooth and shiny (dull in *Silvanus* Latreille, dull and carinate in *Nepharis* Laporte); vestiture of decumbent or decumbent and erect hairs (subglabrous in *Nepharis* and *Nepharinus* Grouvelle).

Head not or slightly declined, usually slightly to distinctly longer than, usually abruptly constricted posteriorly to form distinct temples, which may be very short to moderately long (constriction more gradual or absent in Telephanini). Occipital region occasionally with low transverse carina or furrow. Eyes weakly to strongly protuberant, entire, coarsely facetted, without or with very short interfacetal setae. Antennal insertions usually concealed from above (barely exposed in some Brontini); subantennal grooves usually absent or very short (well developed and long in *Nepharis* and *Nepharinus*). Frontoclypeal suture usually absent (present in Telephanini); clypeus without basolateral emarginations; anterior edge almost always broadly rounded to truncate (distinctly emarginate in *Nepharis* and *Nepharinus*). Labrum free, usually transverse. Antennae in Silvaninae usually 11-segmented with weak but distinct 3-segmented club (9-segmented with 2-segmented club in *Nepharis* and *Nepharinus*, club 4-segmented in *Nausibius*); antennae in Brontinae filiform or with a barely perceptible club; antennal scape longer than head width and much longer than scape in Brontini. Mandible short and broad, moderately to strongly curved, abruptly so in Brontini, usually bidentate, with dorsal setose cavities in Brontinaae and small remnants of these in most Silvaninae; incisor edge often with single tooth; mola usually well developed and transversely ridged, with a membranous, setose sub-molar lobe and slender setose prostheca (mandible in *Nepharis* reduced and lightly sclerotised with reduced mola and prostheca). Maxilla with broadly rounded setose galea and narrow, setose lacinia, the latter sometimes with bifid uncus; apical palpomere usually cylindrical to fusiform (expanded and securiform in *Psammoecus*). Mentum transverse, broadly emarginate; ligula simple or laterally expanded and broadly emarginate apically; apical labial palpomere usually cylindrical or fusiform (expanded in *Psammoecus*). Corpotentorium narrow to moderately broad and straight, often with median process (not in Telephanini). Cervical sclerites absent or reduced and adpressed to head capsule.

Pronotum ~0.6–2.5 times as long as wide, usually widest at about middle (widest anteriorly in some Telephanini and Silvaninae), distinctly narrower at base than combined elytral bases in Brontinae (except for *Brontoliota* Thomas) and some Silvaninae; sides straight to slightly curved. Lateral pronotal carinae usually more or less complete, although sometimes interrupted or not visible from above, with or without raised margin or bead, edges almost always dentate, denticulate, tuberculate, crenate or crenulate; anterior angles rarely produced forward (e.g. *Silvanoprus* Reitter), but often produced laterally; posterior angles usually right to obtuse; posterior edge straight; disc usually simple, with weak longitudinal ridges in some Brontini and sharp longitudinal ridges in *Oryzaephilus* Ganglbauer and *Nepharis*. Prosternum in front of coxae (Figs 18G, 44J) much longer than procoxal cavity, flat to slightly convex. Prosternal process complete, flat to slightly curved, parallel-sided, gradually expanded, with sinuate sides or abruptly expanded at apex, which is truncate. Notosternal sutures often incomplete or absent. Procoxae with long concealed extensions; trochantins concealed. Procoxal cavities circular moderately to widely separated, usually broadly closed externally (broadly or narrowly open in Brontini), without narrow lateral extensions; internally open.

Scutellar shield abruptly elevated, posteriorly broadly rounded to truncate. Elytra ~1.3–3 times as long as combined width and 1.8–3.9 times as long as pronotum, usually with well developed humeri (except in *Brontoliota*); suture not deflected near apex; apices usually conjointly rounded (independently acute in most *Brontoliota*); disc usually with ten distinct puncture rows; scutellary striole present or absent; epipleura well developed and more or less complete in Silvaninae and *Brontoliota*, incomplete in most Brontini and in Telephanini. Mesoventrite in Brontini separated by complete sutures from mesanepisterna, which are widely separated at midline (mesoventrite fused to mesanepisterna in Telephanini and Silvaninae); anterior edge without procoxal rests but usually with broad prothoracic rest or prepectus. Mesocoxae with concealed trochantins. Mesocoxal cavities narrowly to widely separated, partly closed laterally by mesepimera. Mesometaventral junction a straight line. Metaventrite slightly transverse and slightly convex, with well developed discrimen; exposed portion of metanepisternum very long and narrow. Metacoxae narrowly separated; plates absent. Metendosternite with moderately long, broad stalk, long lateral arms, slender laminae and short anterior process with narrowly to moderately widely anterior tendons in most Brontini, usually more reduced without laminae or anterior process in most taxa. Hind wings with relatively short apical field bearing a post-radial sclerite and at least vaguely indicated radial extensions; radial cell incomplete basally or absent; cross-veins r3 absent; R-M loop broadly rounded or acute; medial spur short; medial field with five or fewer free veins and no medial fleck; wedge cell absent; anal notch distinct; hind wings absent in *Brontoliota*. Legs relatively long and slender; femora somewhat inflated in some Brontini; tibiae not apically expanded. Tarsi 5-5-5 in both sexes; tarsomere 1 much smaller than 2 in Brontinae; tarsomeres 2 and 3 lobed beneath and tarsomere 4 highly reduced in Telephanini and some Silvaninae; pretarsal claws simple.

Abdomen usually with five free ventrites (ventrites 1 and 2 connate in some Silvaninae); ventrite 1 not much longer than 2 (except in *Psammoecus*), with or without postcoxal lines; intercoxal process usually acute or narrowly rounded, sometimes broadly rounded. Functional spiracles present on abdominal segments I–VII. Anterior edge of segment IX in male with or without spiculum gastrale; posterior edge without articulated lobe; tergites IX and X fused together and membranous. Aedeagus of cucujiform type, more or less symmetrical, inverted (Brontinae) or not (Silvaninae); anterior edge of tegmen broad and truncate, narrowly rounded or with short strut; parameres articulated or fused to phallobase and separate; penis not subdivided, anteriorly with single broad strut; endophallus usually with well developed armature. Ovipositor moderately elongate, lightly sclerotised except for baculi and apical portions of coxites, paraprocts shorter than coxites, each of which is subdivided into broader proximal and narrower distal lobe; styli well developed and apical. [Arrow 1901; Halstead 1973, 1980, 1993; Thomas 1984, 2002a, 2004; Pal 1985; Pal *et al.* 1985; Sen Gupta and Pal 1996; Lawrence *et al.* 1999b; Leschen *et al.* 2005; Thomas and Nearns 2008; Thomas and Leschen 2010a].

Larva (Fig. 34J)

Body elongate, parallel-sided, slightly to strongly flattened, lightly to moderately pigmented, except for heavily pigmented mandibles and tips of urogomphi; vestiture of scattered setae.

Head prognathous and protracted, slightly to strongly transverse and slightly to strongly flattened; as wide as or wider than prothorax in Brontinae. Epicranial stem absent; frontal arms lyriform and contiguous at base. Median endocarina absent. Stemmata six on each side. Frontoclypeal suture absent. Labrum free in Telephanini and Silvaninae, partly fused to clypeus in Brontini. Antennae moderately to very long, 3-segmented, with antennomere 3 almost as long as 2 and much longer than conical sensorium (Brontinae) or highly reduced and only slightly longer than sensorium, so that antennae appear to be 2-segmented. Mandibles (Fig. 41K) symmetrical, broad at base and narrow at apex, bidentate or tridentate, with accessory ventral process; incisor edge simple or with one or two subapical teeth; mola well developed, tuberculate or asperate, usually with brush of hairs or spines at base; prostheca a hyaline process that is slender and apically acute or with broad base and narrowly acute apex. Ventral mouthparts retracted, with well developed maxillary articulating area; stipites elongate; mala falciform, not cleft; maxillary palps 3-segmented. Labium consisting of prementum, mentum and submentum; ligula simple; labial palps 2-segmented and widely separated. Hypopharyngeal sclerome a transverse bar or molar-like. Hypostomal rods moderately long and diverging; ventral epicranial ridges absent. Gula wider than long.

Prothorax with or without sclerotised plates. Legs well developed, 5-segmented including claw-like pretarsus bearing one or two setae; coxae widely separated. Abdomen with posterolaterally projecting lobes in some Brontinae. Segment IX without appendages much shorter than VIII, tergum not extending onto ventral surface, not forming articulated plate; with long, slender, whip-like, posterodorsally directed urogomphi in Brontini (Fig. 52E) and *Cryptamorpha*; without urogomphi in *Psammoecus* and Silvaninae (Fig. 52U); sternum IX concealed by sternum VIII; segment X visible from above, usually forming cylindrical pygopod, without paired pygopods; anal region posteriorly or terminally oriented. Spiracles annular, usually not placed at the ends of spiracular tubes (on very short tubes in some Brontini). [Böving 1921; Böving and Craighead 1931; Thomas 1988, 2002a; Lawrence 1991b; Lawrence *et al.* 1999a; Thomas and Leschen 2010a].

Classification and World Distribution

Silvanidae is a worldwide family including ~60 described genera and 500 species. There are two subfamilies: Brontinae and Silvaninae, both of which occur in Australia.

Australian Fauna

The Australian **Brontinae** are placed in two tribes: Telephanini with the genera *Cryptamorpha* (Fig. 85H) and *Psammoecus* and Brontini with several endemic genera, such as *Brontoliota* and *Macrohyliota* (Fig. 71C). The **Silvaninae** contains several adventive species, such as *Ahasverus advena* (Waltl), *Cathartus quadricollis* (Guérin-Méneville), *Monanus concinnula* (Walker), *Nausibius clavicornis* (Kugelann), *Oryzaephilus mercator* (Fauvel), *O. surinamensis* (Linnaeus) and *Silvanus unidentatus* (Olivier) (Fig. 85D) plus endemic species in genera such as *Austronausibius* and *Nepharis*. Silvanids are usually elongate, parallel-sided, more or less flattened, pubescent beetles with a prognathous head abruptly constricted to form temples behind the eyes and a pronotum with projecting anterior angles and/or dentate or crenulate lateral carinae. The procoxal cavities are broadly closed externally and the tarsi lobed below in Silvaninae, but not in Brontinae, which have a very long antennal scape. Larvae are elongate, parallel-sided, subcylindrical to flattened and lightly pigmented, with long antennae and legs, a reduced ninth segment, bearing a pair of slender urogomphi in Brontinae, and a well developed, terminal tenth segment. Silvanids are common in leaf litter and under bark, where they feed primarily on dead plant material and fungi. *Cryptamorpha desjardinsi* (Guérin-Méneville) and several silvanine species noted above are cosmopolitan pests of stored foodstuffs. Members of the Brontini are likely to be found under loose bark, while species of *Psammoecus* and native *Cryptamorpha* species, many of which are wingless, are common litter inhabitants. Three wingless brontines in the genus *Brontoliota* Thomas (2004) occur in the rainforests of north-eastern QLD, a fourth species has been described from the Valdivian region of Chile (Thomas 2010) and the related genus *Brontopriscus* Sharp occurs in New Zealand. The silvanine genera *Nepharis* and *Nepharinus* include symphiles found in ant nests (particularly of *Iridomyrmex* and *Crematogaster* species). [Lea 1905, 1910; Halstead 1973, 1980].

Australian Subfamilies, Tribes and Genera

Brontinae: Brontini: *Australodendrophagus* Thomas, *Australohyliota* Thomas, *Brontoliota* Thomas, *Macrohyliota* Thomas; **Telephanini:** *Cryptamorpha* Wollaston, *Psammoecus* Latreille.

Silvaninae: *Ahasverus* Gozis, *Austronausibius* Halstead, *Cathartus* Reiche, *Monanus* Sharp, *Nausibius* Redtenbacher, *Neosilvanus* Grouvelle, *Nepharinus* Grouvelle, *Nepharis* Laporte, *Oryzaephilus* Ganglbauer, *Silvanolomus* Reitter, *Silvanoprus* Reitter, *Silvanus* Latreille.

69. CUCUJIDAE

Adult (Fig. 71E)

Length 2.5–18 mm. Body ~2.6–2.95 times as long as wide; parallel-sided, strongly flattened; colour yellowish-brown to black; surfaces usually more or less smooth, shiny or with dull sheen and subglabrous.

Head large, as wide as or wider than thorax, flattened, abruptly constricted posteriorly to form distinct temples in *Platisus* Erichson. Occipital region with distinct transverse furrow. Eyes weakly to strongly protuberant, entire, coarsely facetted, with very short interfacetal setae. Antennal insertions barely exposed or concealed from above; subantennal grooves absent or very short. Frontoclypeal suture absent; clypeus with sides converging anteriorly, without basolateral emarginations; anterior edge truncate or emarginate. Labrum free but reduced and strongly transverse, concealed by clypeus in *Platisus*. Antennae 11-segmented and filiform or somewhat moniliform in *Platisus*, with weak 3-segmented club in *Pediacus*. Mandible short and broad, strongly, abruptly curved, broadly bidentate, with broad, subapical, dorsal impression in *Pediacus*; incisor edge with single tooth; *Pediacus* with well developed mola and prostheca and membranous, setose sub-molar lobe; mola and prostheca highly reduced in *Platisus*. Maxilla with broadly rounded setose galea and slightly narrower, setose lacinia, the latter with undivided uncus in *Platisus* or several stout spines in *Pediacus*; apical palpomere fusiform. Mentum transverse, broadly emarginate or not, concealing palpal articulations in some *Platisus*; ligula usually membranous and simple to broadly emarginate apically (well sclerotised in some *Platisus*); apical labial palpomere fusiform. Transverse pregular groove present in some *Platisus*. Corpotentorium narrow to moderately broad and straight, without median process. Cervical sclerites reduced and adpressed to head capsule.

Pronotum ~0.75–0.85 times as long as wide, widest at about middle or anterior to middle, distinctly narrower at base than combined elytral bases; sides straight to slightly curved. Lateral pronotal carinae complete, with raised margin or bead, edges simple or coarsely denticulate; anterior angles not or barely produced forward; posterior angles more or less right; posterior edge straight; disc with small paired basal impressions. Prosternum in front of coxae much longer than procoxal cavity, flat to slightly convex. Prosternal process complete, flat, parallel-sided or very gradually expanded; apex broadly rounded or angulate. Procoxae without concealed extensions; trochantins exposed. Procoxal cavities slightly transverse, widely separated, broadly open externally, with narrow or somewhat broader lateral extensions; internally open.

Scutellar shield not abruptly elevated, posteriorly broadly rounded to obtusely angulate. Elytra ~1.95–2.25 times as long as combined width and 2.9–3.35 times as long as pronotum, with well developed humeri; suture not deflected near apex; disc with main portion flat but with longitudinal apically curved ridge delimiting an outer portion which is more or less vertical; punctation irregular and scutellary striole absent; epipleura broad and complete. Mesoventrite with small, slender procoxal rests in *Pediacus*, with weakly indicated broadly triangular prosternal rest in *Platisus*. Mesocoxae with exposed trochantins. Mesocoxal cavities moderately to widely separated, partly closed laterally by mesepimera. Mesometaventral junction a straight line. Metaventrite elongate, with well developed discrimen; exposed portion of metanepisternum long and narrow. Metacoxae narrowly separated; plates absent. Metendosternite with moderately long, broad stalk, long lateral arms, broad laminae and very short anterior process bearing moderately widely separated anterior tendons. Hind wings with relatively short apical field bearing one or two post-radial sclerites and at least a posterior radial extension; radial cell in *Platisus* slightly elongate with right or obtuse inner posterobasal angle, that in *Pediacus* without basal edge; cross-veins r3 moderately long and longitudinally oriented; R-M loop acute; medial spur short; medial field with five free veins and no medial fleck; wedge cell absent; anal notch distinct. Legs relatively long and slender; femora somewhat inflated; tibiae not apically expanded. Tarsi 5-5-5 in females, 5-5-4 in males; tarsomeres 1–4 subequal in length and not lobed beneath; pretarsal claws simple; empodium in *Platisus* well developed and bisetose.

Abdomen with five free ventrites; ventrite 1 not much longer than 2, intercoxal process acute or narrowly rounded. Functional spiracles present on abdominal segments I–VII. Anterior edge of segment IX in male with spiculum gastrale; posterior edge with articulated lobe; tergite IX completely divided into laterotergites; segment X free, lightly sclerotised with lateral baculi. Aedeagus of cucujiform type, more or less symmetrical, inverted in *Pediacus*, lying on side in *Platisus*; anterior edge of tegmen broad and truncate, narrowly rounded in *Pediacus*, with short strut in *Platisus*; parameres articulated to phallobase and separate; penis subdivided into basal and apical sections, without basal strut; endophallus usually with well developed armature. Ovipositor moderately elongate, lightly sclerotised except for baculi and apical portions of coxites, paraprocts about as long as coxites, each of which is subdivided into proximal and distal lobes; styli well developed and apical. [Lawrence *et al.* 1999b; Watt *et al.* 2001; Thomas 2002c; Leschen *et al.* 2005; Thomas and Leschen 2010b].

Larva (Fig. 84D)

Body elongate, parallel-sided, strongly flattened, lightly pigmented, except for head and segments VIII and IX; vestiture of scattered setae.

Head prognathous and protracted, transverse and flattened; as wide as or wider than prothorax. Epicranial stem short to moderately long; frontal arms lyriform and contiguous at base. Median endocarina absent. Stemmata six on each side. Frontoclypeal suture absent. Labrum free in *Pediacus*, partly fused to clypeus in *Platisus*. Antennae moderately long, 3-segmented, with antennomere 3 almost as long as 2 and much longer than conical sensorium. Mandibles symmetrical, broad at base and narrow at apex, tridentate, without accessory ventral process; incisor edge with one or two subapical teeth; mola well developed, although narrow, and asperate in *Pediacus* (Figs 39D, G), replaced in *Platisus* by thin lobe lined with hairs; prostheca a hyaline process that is slender and apically acute (highly reduced or absent in *Platisus*). Ventral mouthparts slightly retracted, with well developed maxillary articulating area; stipites as wide as or wider than long; mala falciform, not cleft; maxillary palps 3-segmented. Labium consisting of prementum, mentum and submentum; ligula simple; labial palps 2-segmented and widely separated. Hypopharyngeal sclerome a transverse bar. Hypostomal rods short and slightly diverging in *Pediacus*, absent in *Platisus*; ventral epicranial ridges absent. Gula wider than long.

Prothorax with or without sclerotised plates. Legs well developed, 5-segmented including claw-like pretarsus bearing two setae lying side by side; coxae widely separated. Segment VIII elongate and much longer than VII; tergum IX shorter, extending onto ventral surface and forming articulated plate; with median, posteriorly projecting process that is forked at apex; sternum IX partly enclosed by sternum VIII, which also bears a pair of posteriorly projecting teeth; segment X not visible from above, reduced and membranous, without paired pygopods; anal region ventrally oriented. Spiracles annular–biforous, all placed at the ends of spiracular tubes. [Lawrence 1991b; Lawrence *et al.* 1999a; Watt *et al.* 2001; Thomas 2002c; Thomas and Leschen 2010b].

Classification and World Distribution

The family Cucujidae includes ~50 described species in four genera, two of which occur in Australia. *Cucujus* Fabricius and *Pediacus* Shuckard are primarily Holarctic in distribution, although the latter extends into Central America and northern Australia. The genus *Palaestes* Perty is restricted to the Neotropical region, occurring at higher altitudes in Central and South America, and *Platisus* species are known from Australia and New Zealand. [Watt *et al.* 2001; Thomas and Leschen 2010b].

Australian Fauna

The Australian cucujid fauna consists of four species of *Platisus* Erichson (Fig. 71E) and an undescribed species of *Pediacus* from northern QLD. Adults are elongate, parallel-sided, distinctly flattened and glabrous, with large heads, usually wider than the prothorax, relatively small eyes, distinct lateral pronotal carinae, and reduced labrum; the antennae are moniliform in *Platisus*, but have a 3-segmented club in *Pediacus*. Larvae (Fig. 84D) are strongly flattened with a hinged, forked process (tergum IX) at the abdominal apex; the head is much broader than the thorax, the antennae are moderately long with a small sensorium, ventral mouthparts are only slightly retracted, the mandibular mola is well developed (*Pediacus*) or reduced (*Platisus*), the prostheca is slender and acute or absent, and sternum IX and segment X are reduced and partly enclosed by sternum VIII. Both adults and larvae occur under bark and at least those of *Platisus* are predaceous. [Olliff 1885; Watt *et al.* 2001].

70. MYRABOLIIDAE

Adult (Fig. 83E)

Length 1.8–3.65 mm. Body ~2.7–3.8 times as long as wide; parallel-sided, distinctly flattened; colour yellowish-brown to dark brown, rarely with bicolored elytra; upper surfaces more or less shiny; vestiture of decumbent and/or suberect hairs.

Head slightly longer than wide, slightly narrowed anteriorly, not constricted posteriorly. Occipital region with median stridulatory file. Eyes strongly protuberant, entire, coarsely facetted, with short interfacetal setae. Antennal insertions barely concealed from above; subantennal grooves well developed but not extending behind eyes. Frontoclypeal suture absent; clypeus with sides converging anteriorly, without basolateral emarginations; anterior edge truncate. Labrum free, strongly transverse, with truncate apex. Antennae short, 11-segmented, with weak 3-segmented club. Mandible short and broad, bidentate, without dorsal setose cavity; incisor edge simple or with single tooth; mola well developed, asperate, with membranous, setose sub-molar lobe; prostheca consisting of slender membranous lobe and short setal fringe. Maxilla with moderately broad, setose galea and narrower lacinia with bidentate uncus; apical palpomere cylindrical or fusiform. Mentum strongly transverse, broadly emarginate; ligula simple; apical labial palpomere fusiform. Genae with anteriorly projecting, subacute processes. Pregular area moderately long, without transverse groove. Corpotentorium narrow, without median process. Cervical sclerites reduced and adpressed to head capsule.

Pronotum ~0.7–0.95 times as long as wide, widest at about middle, not narrower at base than combined elytral bases; sides straight to slightly curved. Lateral pronotal carinae complete, with raised margin or bead, edges finely crenulate or denticulate; anterior angles not produced forward; posterior angles more or less right; posterior edge slightly, evenly rounded; disc without sublateral carinae or paired basal impressions. Prosternum in front of coxae much longer than procoxal cavity, flat to slightly convex, without paired lines or carinae; anterior edge finely denticulate. Prosternal process complete, flat, gradually but distinctly expanded posteriorly; with truncate apex. Procoxae with long concealed extensions; trochantins concealed or very slightly exposed. Procoxal cavities more or less circular, moderately to widely separated, externally narrowly open, with narrow lateral extensions, internally open.

Scutellar shield abruptly elevated, posteriorly obtusely angulate. Elytra ~1.65–2.0 times as long as combined width and 2.15–2.6 times as long as pronotum, with well developed humeri; suture not deflected near apex; disc with ten rows of larger punctures, alternating with ten rows of smaller punctures and often ten weak longitudinal carinae, with or without scutellary striole; epipleura narrow and complete. Mesoventrite with paired, horizontal or slightly declined procoxal rests. Mesocoxae with concealed trochantins. Mesocoxal cavities moderately to widely separated, closed laterally by ventrites. Mesometaventral junction monocondylic. Metaventrite moderately long, flat, with moderately well developed discrimen; exposed portion of metanepisternum very long and narrow. Metacoxae transverse, narrowly separated, not extending laterally to meet elytra; plates absent. Metendosternite with short, broad stalk, long lateral arms, narrow laminae and very short anterior process bearing moderately narrowly separated anterior tendons. Hind wings with long apical field bearing one post-radial sclerite; radial cell without basal edge; cross-veins r3 absent, r4 incomplete or absent; R-M loop acute; medial spur short; medial field with four free veins and no medial fleck; wedge cell absent; anal notch distinct. Legs relatively long and slender; femora somewhat inflated; tibiae not apically expanded. Tarsi 5-5-5 in females, 5-5-4 in males; tarsomeres not lobed beneath; pretarsal claws simple.

Abdomen (Fig. 26J) with five free ventrites; ventrite 1 much longer than 2, usually with paired punctate impressions in male; intercoxal process acute. Functional spiracles present on abdominal segments I–VII. Anterior edge of segment IX in male with spiculum gastrale; posterior edge with articulated lobe; tergites IX and X membranous. Aedeagus of cucujiform type, symmetrical, uninverted; anterior edge of tegmen broadly rounded or truncate, without struts; parameres short, articulated to phallobase and separate; penis not divided, widest at base, with short, paired basal struts; endophallus with well developed armature. Ovipositor moderately elongate, lightly sclerotised except for baculi and apical portions of coxites, paraprocts slightly shorter than coxites, each of which is subdivided into broad proximal and much narrower distal lobes; styli well developed and apical. [Leschen *et al.* 2005; Tomaszewska and Ślipiński 2008; Ślipiński *et al.* 2010a; Szawaryn and Tomaszewska 2011].

Larva

Body elongate, parallel-sided, flattened, lightly pigmented, except for head and posteromesal portion of tergum IX, which are yellow, and mandibles and urogomphi, which are more darkly pigmented; vestiture of scattered setae.

Head prognathous and protracted, transverse, flattened. Epicranial stem absent; frontal arms lyriform and contiguous at base. Median endocarina absent. Stemmata two on each side. Frontoclypeal suture absent. Labrum free. Antennae moderately long, 3-segmented, with antennomeres 2 and 3 subequal in length and conical sensorium long but shorter than apical antennomere. Mandibles symmetrical, broad at base and narrow at apex, tridentate, without accessory ventral process; mola absent; mesal surface of mandibular base with three blunt, hyaline processes and acute lobe distal to these. Ventral mouthparts somewhat retracted, with well developed maxillary articulating area; stipites about as long as wide; mala weakly falciform, not cleft; maxillary palps 3-segmented. Labium consisting of prementum, mentum and submentum; ligula simple; labial palps 2-segmented and moderately widely separated. Hypopharyngeal sclerome absent. Hypostomal rods moderately long and diverging; ventral epicranial ridges absent. Gula wider than long.

Legs well developed, 5-segmented including claw-like pretarsus bearing two setae lying side by side; coxae widely separated. Tergum IX excluding appendages much longer than VIII, extending onto ventral surface, not forming articulated plate; with short, approximate and more or less parallel, apically upturned urogomphi; sternum IX simple, not enclosed by sternum VIII; segment X not visible from above, more or less circular, without paired pygopods; anal region ventrally oriented. Spiracles annular-biforous. [Lawrence *et al.* 1999a; Tomaszewska and Ślipiński 2008; Ślipiński *et al.* 2010a].

Classification and World Distribution

The family Myraboliidae contains a single genus, which is known only from Australia. [Tomaszewska and Ślipiński 2008; Ślipiński *et al.* 2010a].

Australian Fauna

The species of *Myrabolia* Reitter are distributed primarily in south-eastern Australia and TAS, but *M. leai* (Grouvelle) extends into WA, *M. micra* Tomaszewska and Ślipiński is known only from WA, and *M. brevicornis* (Erichson) extends to the Kuranda area of northern QLD (Tomaszewska and Ślipiński 2008; Szawaryn and Tomaszewska 2011). Myraboliid adults tend to resemble some silvanine Silvanidae in being elongate, parallel-sided, flattened, and pubescent, with a 3-segmented antennal club, crenulate or denticulate lateral pronotal carinae; however, the tarsi are simple, the first abdominal ventrite is more than twice as long as the second, usually with lateral pore plates in the male, and there is a stridulatory file on the vertex. Myraboliid larvae are elongate, somewhat flattened and lightly pigmented, except for the head and tergum IX and with upturned, approximate and more or less parallel urogomphi; the head is prognathous with two stemmata on each side, moderately long antennae, no epicranial stem or endocarinae, an unusual mandible with no mola but with three blunt, hyaline processes and an acute process distal to these, weakly falcate mala and diverging hypostomal rods. Myraboliid adults and larvae have been found under the bark of living eucalypts, but details of their feeding habits are unknown.

71. CAVOGNATHIDAE

Adult (Fig. 86I)

Length 2.4–3.3 mm. Body ~2.2–2.5 times as long as wide; parallel-sided, somewhat flattened; colour uniformly reddish-brown, upper surfaces slightly shiny; vestiture of moderately long, recumbent hairs.

Head transverse, not constricted posteriorly. Occipital region without or with very weak transverse ridge. Eyes strongly protuberant, entire, coarsely facetted, with a few short interfacetal setae. Antennal insertions concealed from above; subantennal grooves absent. Frontoclypeal suture absent; frontoclypeal region with pair of pits (*T. australis*) or with U-shaped groove (*T. pullivora*); clypeus extending well in front of antennal insertions, without basolateral emarginations; anterior edge subtruncate. Labrum free, strongly transverse, with truncate apex. Antennae short, 11-segmented, with 3-segmented club; antennomeres 5 and 7 wider than neighbouring antennomeres, especially in males. Mandible slightly longer than wide, bidentate, without dorsal tubercle or setose cavity, but outer edge with outlets of internal vesicles; incisor edge with single subapical tooth; mola reduced; prostheca well developed. Maxilla with setose galea slightly wider than lacinia, which bears unidentate uncus; apical palpomere fusiform. Mentum transverse; ligula transverse, sclerotised and undivided; apical labial palpomere fusiform. Genae with anteriorly projecting, subacute processes. Pregular area with longitudinal, curved grooves ending in small pits below maxillary articulations. Corpotentorium narrow, without median process. Cervical sclerites reduced and adpressed to head capsule.

Pronotum ~0.63–0.75 times as long as wide, widest at about middle, slightly narrower at base than combined elytral bases; sides slightly curved. Lateral pronotal carinae weakly defined and incomplete, without raised margin or bead; anterior angles not produced forward; posterior angles more or less rounded; posterior edge evenly curved; disc simple. Prosternum in front of coxae not or slightly longer than procoxal cavity, slightly convex. Prosternal process complete, flat, weakly expanded posteriorly; with broadly rounded apex. Notosternal sutures incomplete. Procoxae with short concealed extensions; trochantins concealed. Procoxal cavities slightly transverse, narrowly separated, externally narrowly open, with very narrow lateral extensions, internally open.

Scutellar shield not abruptly elevated, posteriorly obtusely angulate. Elytra ~1.65–2.0 times as long as combined width and 2.75–3.25 times as long as pronotum, with well developed humeri; suture deflected near apex; disc densely, irregularly punctate, without scutellary striole; epipleura narrow and complete. Mesoventrite without paired procoxal rests; with weak prosternal rest. Mesocoxae with slightly exposed trochantins. Mesocoxal cavities moderately separated, partly closed laterally by mesepimera. Mesometaventral junction dicondylic. Metaventrite elongate, slightly convex, with well developed discrimen; exposed portion of metanepisternum very long and narrow. Metacoxae narrowly separated; plates absent. Metendosternite with short, broad stalk, long lateral arms, narrow laminae and very short anterior process bearing narrowly separated anterior tendons. Hind wings with short apical field bearing one post-radial sclerite; radial cell without basal edge; cross-veins r3 absent, r4 incomplete or absent; R-M loop acute; medial spur short; medial field with four free veins and no medial fleck; wedge cell absent; anal notch distinct. Legs relatively slender; tibiae sometimes apically expanded and acute. Tarsi 5-5-5 in both sexes; tarsomeres not lobed beneath, but tarsomeres 1–3 on male fore legs and mid legs expanded and spongy beneath; pretarsal claws simple.

Abdomen with five free ventrites; ventrite 1 not much longer than 2, intercoxal process acute. Functional spiracles present on abdominal segments I–VII. Anterior edge of segment IX in male with spiculum gastrale; posterior edge without articulated lobe; tergites IX and X membranous. Aedeagus of cucujiform type, symmetrical, partly or completely inverted; anterior edge of tegmen with short broad strut; parameres short, articulated to phallobase and separate; penis not distinctly divided, parallel-sided, its anterior edge broadly rounded; endophallus with armature. Ovipositor moderately long and broad, lightly sclerotised except for baculi and apical portions of coxites, paraprocts slightly shorter than coxites, each of which is subdivided into broad proximal and much narrower distal lobes; styli well developed and apical. [Crowson 1964a, 1973a; Sen Gupta and Crowson 1966, 1969a; Watt 1980; Leschen *et al.* 2005; Ślipiński and Tomaszewska 2010a, 2010d].

Larva (Fig. 84E)

Body elongate, parallel-sided, slightly flattened, lightly pigmented, except for head, thoracic and abdominal terga and urogomphi; vestiture of scattered long setae.

Head prognathous and protracted, transverse, slightly flattened, rounded laterally. Epicranial stem absent; frontal arms lyriform and well separated at base. Median endocarina absent. Stemmata six on each side. Frontoclypeal suture absent. Labrum free. Antennae moderately long, 3-segmented, with antennomeres 3 ~0.33 times as long as 2 and conical sensorium minute. Mandibles symmetrical, broad at base and narrow at apex, tridentate, without accessory ventral process; mola absent, replaced by two processes: one lightly sclerotised and acute and the other hyaline and setose. Ventral mouthparts somewhat retracted, with moderately well developed maxillary articulating area; stipites longer than wide; mala falciform, not cleft, with several teeth along inner margin; maxillary palps 3-segmented. Labium consisting of prementum, mentum and submentum; ligula simple; labial palps 2-segmented and narrowly separated. Hypopharyngeal sclerome absent. Hypostomal rods moderately long and diverging; ventral epicranial ridges absent. Gula wider than long.

Legs well developed, 5-segmented including claw-like pretarsus bearing two setae with one distal to the other; coxae moderately widely separated. Tergum IX excluding appendages as long as VIII, not extending onto ventral surface, not forming articulated plate; with short, subparallel, posteriorly oriented urogomphi; sternum IX simple, not enclosed by sternum VIII; segment X visible from above, more or less cylindrical, without paired pygopods; anal region posteriorly or posteroventrally oriented. Spiracles annular–bifforous. [Crowson 1964a; Sen Gupta and

Crowson 1966; Watt 1980; Lawrence 1991b; Lawrence *et al.* 1999a; Ślipiński and Tomaszewska 2010a, 2010d].

Classification and World Distribution

The family Cavognathidae contains the single genus *Taphropiestes* Reitter with nine described species occurring in Chile, Argentina, Venezuela, New Zealand and Australia. [Ślipiński and Tomaszewska 2010a, 2010d].

Australian Fauna

Two species of *Taphropiestes* occur in Australia: the widely distributed *T. australis* Ślipiński and Tomaszewska, and *T. pullivora* (Crowson), which occurs in the eastern part of the continent. Cavognathids are elongate, slightly flattened, pubescent beetles with apically widened sutural flange on the elytra and a pair of frontal cavities sometimes joined by a strongly curved impression; the prothorax is simple, with rounded angles and indistinct lateral carinae. Larvae are elongate and parallel-sided, with tergal plates on the thorax and most abdominal segments and with short, posteriorly projecting urogomphi; mandibles lack a mola and have a sclerotised, acute process and hyaline, setose process at the base of the mesal edge and the mala is falciform. Adult and larval cavognathids have been found in the nests or sifted from nest debris of various bird species, but the nature of the host association is not well understood.

In Australia, cavognathids have been associated with the Australian magpie (*Cracticus tibicen* (Latham) and the weebill (*Smicrornis brevirostris* (Gould), but several other hosts have been recorded from New Zealand bird species. It is likely that these beetles are nest scavengers, although Crowson (1964a) reported the occurrence of *T. pullivora* on magpie nestlings. [Chisolm 1952; Watt 1980; Ślipiński and Tomaszewska 2010d].

72. LAMINGTONIIDAE

Adult (Fig. 86A–B)

Length 2.5–4 mm. Body ~2.1–2.4 times as long as wide; broadly elongate, more or less parallel-sided, slightly to strongly flattened; colour of head and pronotum reddish-brown or black, elytra yellow with black or dark brown pattern or reddish-brown with broad black area along suture, upper surfaces shiny and subglabrous.

Head slightly transverse, somewhat flattened, not constricted posteriorly, distinctly narrowed anteriorly. Occipital foramen with pair of deep notches dorsally. Eyes moderately large, slightly protuberant, entire, finely facetted, without interfacetal setae. Antennal insertions exposed from above, laterally placed and located in saucer-like depression; subantennal grooves absent. Frontoclypeal suture absent; frontal carina on each side slightly elevated and bordered mesally by groove, which extends forward and ends at base of clypeus, forming a small notch on either side of clypeus, which is anteriorly truncate. Labrum free, transverse, with broadly rounded apex. Antennae 11-segmented, with distinct, 3-segmented club; scape and pedicel more or less globular. Mandible abruptly curved, tridentate, with rounded incisor lobe and without dorsal tubercle or cavity; mola absent; mesal surface of mandibular base with dense fringe of hairs. Maxilla with broad, rounded and setose galea and lacinia, the latter without uncus; apical palpomere subconical. Mentum strongly transverse; ligula simple; apical labial palpomere subconical. Genae with short, anteriorly projecting, subtruncate processes. Pregular area broadly concave. Corpotentorium narrow and slightly arched, without median process. Cervical sclerites small, transverse and adpressed to head capsule.

Pronotum ~0.6–0.7 times as long as wide, widest posteriorly or at middle, not narrower at base than combined elytral bases; sides weakly curved. Lateral pronotal carinae well defined and complete, with raised margin or bead; edges simple; anterior angles slightly produced forward; posterior angles more or less right; posterior edge evenly curved or slightly sinuate; disc with paired, shallow, prebasal impressions. Prosternum in front of coxae longer than procoxal cavity, slightly convex. Prosternal process complete, flat, laterally margined, slightly but more or less abruptly expanded posteriorly; with truncate apex. Procoxae without concealed extensions; trochantins at least partly exposed. Procoxal cavities slightly transverse, moderately widely separated, externally broadly open, with very narrow lateral extensions, internally open.

Scutellar shield abruptly elevated, posteriorly broadly rounded. Elytra ~1.55–1.85 times as long as combined width and 2.9–3.6 times as long as pronotum, with well developed humeri; disc with ten puncture rows and no scutellary striole; epipleura narrow and incomplete. Mesoventrite with paired, strongly inclined procoxal rests. Mesocoxae with exposed trochantins. Mesocoxal cavities moderately widely separated, partly closed laterally by mesepimera. Mesometaventral junction an anteriorly curved line with metaventrite overlapping mesoventrite. Metaventrite moderately long, flattened, with discrimen very short or absent; exposed portion of metanepisternum moderately long and narrow. Metacoxae narrowly separated; plates absent. Metendosternite with short, broad stalk, long lateral arms, narrow laminae and short anterior process bearing narrowly separated anterior tendons. Hind wings with moderately long apical field bearing a post-radial sclerite and at least one vaguely indicated radial extension; radial cell elongate with inner posterobasal angle right; cross-veins r3 absent, r4 complete; R-M loop acute; basal portion of RP very short; medial spur short; medial field with four free veins and no medial fleck; wedge cell absent; anal notch distinct. Legs relatively slender; tibiae very slightly expanded apically expanded. Tarsi 5-5-5 in both sexes; tarsomeres subequal and not lobed beneath; pretarsal claws simple.

Abdomen with five free ventrites; ventrite 1 not much longer than 2, intercoxal process acute. Functional spiracles present on abdominal segments I–VII. Anterior edge of segment IX in male with spiculum gastrale; posterior edge without articulated lobe; tergites IX and X membranous. Aedeagus of cucujiform type, symmetrical, anterior edge of tegmen with very short acute strut; parameres short, articulated to phallobase and separate; penis not distinctly divided, parallel-sided, its anterior edge subtruncate or notched; endophallus with distinct rod. Ovipositor moderately elongate, lightly sclerotised except for baculi and apical portions of coxites, paraprocts slightly shorter than coxites, each of which is subdivided into broad proximal and much narrower, heavily sclerotised distal lobe; styli well developed and slightly subapical. [Sen Gupta and Crowson 1969b; Lawrence *et al.* 1999b; Lawrence and Leschen 2003, 2010f].

Larva (Fig. 84C)

Body elongate, parallel-sided or slightly wider behind middle; slightly flattened; lightly pigmented, except for head and posterior portion of tergum IX; vestiture of scattered long setae.

Head prognathous and protracted, transverse, flattened, strongly rounded laterally; posterior edge distinctly emarginate. Epicranial stem very short or absent; frontal arms lyriform, contiguous at base, usually joined anteriorly by transverse line. Median endocarina absent, but paired endocarinae beneath bases of frontal arms. Stemmata five or six on each side. Frontoclypeal suture absent. Labrum free. Antennae short, 3-segmented, with conical sensorium subequal in length to antennomere 3. Mandibles symmetrical, broad at base and narrower apically; broadly tridentate, without accessory ventral process; mola absent; mesal surface of mandibular base with cluster of acute, hyaline processes. Ventral mouthparts strongly protracted, without maxillary articulating area; stipites short and broad; mala rounded, not cleft; maxillary palps 3-segmented. Labium consisting of prementum, mentum and submentum; ligula absent; labial palps 2-segmented and widely separated. Hypopharyngeal sclerome absent. Hypostomal rods long and diverging; ventral epicranial ridges absent. Gula longer than wide and lightly pigmented.

Presternum large and triangular, with median, longitudinal endocarina. Legs well developed, 5-segmented including claw-like pretarsus bearing two setae with one distal to the other; coxae very widely separated. Abdomen with dorsal and ventral, asperate ambulatory warts. Tergum IX slightly longer than and somewhat narrower than VIII, extending onto ventral surface, not forming articulated plate; with short, upturned, acute urogomphi; sternum IX simple, not enclosed by sternum VIII; segment X not visible from above, more or less cylindrical, without paired pygopods; anal region posteroventrally oriented. Spiracles annular–uniforous, with long accessory tubes, those on segment VIII placed at the ends of posteriorly oriented acute processes. [Lawrence *et al.* 1999a; Lawrence and Leschen 2003, 2010f].

Classification and World Distribution

Lamingtoniidae is an endemic Australian family with a single genus *Lamingtonium* Sen Gupta & Crowson and three described species.

Australian Fauna

Lamingtonium loebli Lawrence & Leschen and *L. thayerae* Lawrence & Leschen (Fig. 86B) are known from the cool temperate forests of VIC and TAS and *L. binnaburrense* Sen Gupta & Crowson (Fig. 86A) from the subtropical rainforests of northern NSW and southern QLD. Lamingtoniids are oblong, slightly flattened, subglabrous and usually bicoloured beetles with sharply defined supra-antennal ridges, a distinct frontoclypeal prominence, a small, transverse, more or less concealed labrum, and distinctive mandibles without a mola but with a dense fringe of hairs at the base. Larvae are elongate, somewhat flattened and lightly pigmented, with paired endocarinae on the head, long, diverging hypostomal rods, an acute, hyaline lobe at the base of the mandible, short urogomphi bearing accessory tubercles, and the eighth spiracles located at the ends of long tubes. Lamingtoniids feed as both adults and larvae in the softer fruiting bodies of some polypore fungi.

73. PASSANDRIDAE

Adult (Figs 72J, L, 85I)

Length 3–25 mm. Body ~2.4–7.5 times as long as wide; elongate, more or less parallel-sided, flattened to subcylindrical; colour usually reddish-brown to black; upper surfaces subglabrous.

Head large, as long as or longer than wide, about as wide as and not retracted into prothorax, not or slightly declined, weakly constricted posteriorly in *Ancistria* Erichson, with longitudinal grooves in *Ancistria* and *Passandra* Dalman and additional transverse groove in *Passandra*. Eyes small to moderately large, not protuberant, entire, finely facetted, without interfacetal setae. Antennal insertions usually concealed from above, but frontal ridges in *Ancistria* distinctly elevated so that insertions are partly exposed; subantennal grooves weakly to strongly developed. Frontoclypeal suture absent; anterior edge of clypeus with a short narrowly rounded to slightly emarginate projecting lobe. Labrum free, small, inclined and not visible from above. Antennae stout, 11-segmented, stout and more or less moniliform in *Passandra*, but with apical antennomeres irregularly expanded, with asymmetrical, 6-segmented club in *Ancistria* and with weak, 3-segmented club in *Aulonosoma* Motschulsky; scape more or less globular and pedicel transverse. Mandible unidentate; incisor lobe usually with one or two weak subapical teeth; dorsal surface sometimes with oblique ridge fitting against edge of clypeus; mola absent; prostheca a fringe of hairs. Maxilla with broad, rounded, setose galea and similar but much shorter lacinia without uncus; apical palpomere fusiform. Mentum strongly transverse; ligula bilobed; apical labial palpomere fusiform. Genae forming distinct, anteriorly projecting processes, which conceal maxillae (Fig. 28A). Gular sutures approximate or partly fused; pregular area broadly concave, with pair of setose pores in *Passandra*. Corpotentorium absent. Cervical sclerites reduced or absent.

Pronotum ~0.85–2.45 times as long as wide, widest anteriorly, not or only slightly narrower at base than combined elytral bases; sides straight to very weakly curved. Lateral pronotal carinae complete with raised margin or bead in *Passandra* and *Aulonosoma*, incomplete anteriorly in *Ancistria*; edges simple; anterior angles slightly produced and acute in *Passandra*; posterior angles more or less right or obtuse; posterior edge evenly curved; disc with paired, sublateral carinae in *Passandra*. Prosternum in front of coxae much longer than procoxal cavity, moderately convex. Notosternal suture incomplete anteriorly in *Passandra* and *Ancistria*. Prosternal process complete, distinctly curved behind coxae, strongly and abruptly expanded apically with truncate or biemarginate apex. Procoxae with short concealed extensions in *Passandra* and *Ancistria* (longer in *Aulonosoma*); trochantins concealed. Procoxal cavities circular, widely separated, externally narrowly open in *Passandra* and *Ancistria*, narrowly closed in *Aulonosoma*, without narrow lateral extensions, internally open in *Passandra* and *Aulonosoma*, closed in *Ancistria*.

Scutellar shield more or less abruptly elevated, posteriorly broadly rounded to subacute. Elytra ~1.7–5 times as long as combined width and 2.15–2.85 times as long as pronotum, with well developed humeri; disc with nine or ten puncture rows and no scutellary striole in *Aulonosoma* and *Ancistria* and impunctate with three striae in *Passandra*; posterior portion of disc in *Ancistria* with a distinct declivity flanked on each side by two sharp costae; epipleura complete. Mesoventrite flat to moderately convex, with or without weakly developed, horizontal procoxal rests; anterior edge sometimes on slightly different plane than metaventrite. Mesocoxae with concealed trochantins in *Passandra*, exposed trochantins in *Aulonosoma* and *Ancistria*. Mesocoxal cavities moderately widely separated, closed laterally by ventrites in *Passandra*, partly closed laterally by mesepimera in *Aulonosoma* and *Ancistria*. Mesometaventral junction an anteriorly curved line with metaventrite overlapping mesoventrite. Metaventrite long, moderately to strongly convex, usually with long discrimen (somewhat shorter in *Aulonosoma*) exposed portion of metanepisternum moderately to very long and narrow. Metacoxae narrowly separated; plates absent. Metendosternite in *Aulonosoma* and *Passandra* with long, narrow stalk, long lateral arms, broad laminae and short anterior process bearing narrowly separated anterior tendons (reduced in *Ancistria*). Hind wings in *Passandra* and *Aulonosoma* with relatively short apical field bearing one or two post-radial sclerites and one distinct radial extension (no distinct sclerotizations in *Ancistria*); radial cell incomplete, with no basal edge; cross-vein r3 absent, r4 complete; R-M loop usually broad (acute in *Ancistria*); basal portion of RP usually very short (long in *Passandra*); medial spur short or long; medial embayment present; medial field with three to five free veins and no medial fleck; wedge cell absent; anal lobe well developed and anal notch distinct in *Passandra* and *Aulonosoma*; anal lobe highly reduced in *Ancistria*. Legs stout, femora swollen; tibiae with sharp external edges and with outer apical angle produced and acute; tibial spurs highly unequal, with one long and hooked. Tarsi 5-5-5 in both sexes; tarsomeres not lobed beneath; pretarsal claws simple.

Abdomen with five free ventrites; ventrite 1 not much longer than 2, intercoxal process acute; last ventrite with preapical groove in *Passandra*. Functional spiracles present on abdominal segments I–VII. Anterior edge of segment IX in male narrowly rounded to truncate; posterior edge without articulated lobe; tergites IX and X membranous. Aedeagus (Fig. 31K) of cucujiform type, symmetrical, inverted; anterior edge of tegmen broadly rounded; parameres articulated to phallobase and separate; penis not distinctly divided, with long, slender anterior strut; endophallus with distinct rod. Ovipositor very long and slender, with extremely elongated paraprocts; coxites not distinctly divided; styli usually apical (subapical in *Ancistria*). [Ślipiński 1983, 1987, 1989; Lawrence *et al.* 1999b; Burckhardt and Ślipiński 1991, 1995, 2003, 2010; Lawrence *et al.* 1999b; Thomas 2002b].

Larva

Mature larva physogastric, lightly pigmented, with vestiture of scattered, simple setae. First instar (Fig. 84J) moderately flattened and distinctly pigmented dorsally and ventrally, with vestiture of long spines. Head prognathous and protracted, transverse. Epicranial suture absent in mature larva; epicranial stem absent and frontal arms U-shaped and well separated at base in first instar. Median endocarina absent. Stemmata absent in mature larva; five or six on each side in first instar. Frontoclypeal suture absent. Labrum completely fused to head capsule. Antennae very short and 2-segmented in mature larva; short and 3-segmented with conical sensorium about half as long as antennomere 3 in first instar. Mandibles symmetrical, apically tridentate in mature larva (falcate and unidentate with groove along inner edge in first instar); mola and prostheca absent; mesal edge of mandibular base in mature larva with single or multiple hyaline processes. Ventral mouthparts strongly protracted in mature larva (retracted in first instar), without maxillary articulating area; cardo absent or indistinct in mature larva, distinct and somewhat elongate in first instar; mala obtuse; maxillary palps usually 2-segmented in mature larva (1-segmented in *Ancistria*), 3-segmented in first instar. Labrum free to base; ligula absent; labial palps usually 2-segmented (1-segmented in mature larva of *Ancistria*). Hypostomal rods moderately long and diverging. Thoracic and abdominal terga without plates in mature larva; with sclerotised plates in first instar. Legs highly reduced and not distinctly articulated in mature larva; moderately well developed, widely separated with unisetose, dentate pretarsal claw in first instar. Tergum IX with short, fixed urogomphi in mature larvae of *Ancistria*, without urogomphi in those of *Passandra*; with six long caudal filaments in first instars. Segment X posteriorly oriented and visible from above in mature larvae; ventrally oriented in first instars. Spiracles annular, not placed at the ends of spiracular tubes in mature larvae; annular–biforous and placed at ends of short spiracular tubes in first instars. [Gravely 1916; Böving and Craighead 1931; Hayashi 1994; Lawrence 1991b; Lawrence *et al.* 1999a; Burckhardt and Ślipiński 2010].

Classification and World Distribution

The family Passandridae is a predominantly tropical group comprising 109 described species in nine genera, three of which occur in Australia.

Australian Fauna

The genus *Passandra* Dalman includes three Australian species: the widely distributed *P. heros* (Fabricius) (Fig. 85I), *P. trigemina* (Newman) from northern Australia, and *P. marginata* Grouvelle. The genus *Ancistria* Erichson is represented by the widely distributed *A. retusum* (Fabricius) (Fig. 72J). *Aulonosoma tenebrioides* Motschulsky (Fig. 72L) is known from Asia, New Guinea and north-western Australia. *Passandra* adults are relatively large, elongate, parallel-sided, moderately flattened and glabrous, with a longitudinally grooved head, reduced and concealed labrum, moniliform antennae and well developed, unequal tibial spurs. *Ancistria retusum* is similar in most respects, but is more narrowly elongate and subcylindrical, with particularly long tibial spurs and subserrate antennae. *A. tenebrioides* is a smaller beetle with a simple head and a 3-segmented antennal club. Passandrid larvae are of two types: a triungulinoid first instar, which is flattened and clothed with simple, long, conical spines, falcate mandibles and six long caudal filaments on tergum IX, and an ectoparasitic mature larva, which is physogastric, with reduced antennae and palps, tridentate mandibles, and a ninth tergum with or without short, fixed urogomphi. Passandrid larvae are ectoparasites of wood-inhabiting insect larvae or pupae. *Aulonosoma tenebrioides* inhabits bamboo and rattan canes, where it feeds on larvae of lyctine and bostrichine Bostrichidae. [Aitken 1975; Lawrence 1991b; Burckhardt and Ślipiński 2010].

74. PHALACRIDAE

Adult (Fig. 85C, E)

Length 1–4.5 mm. Body ~1.15–1.6 times as long as wide; broadly ovate to circular, almost always moderately to strongly convex dorsally (somewhat flattened in *Platyphalacrus*) and flat to somewhat concave ventrally; colour often black, brown or yellow, sometimes with dark and light colour pattern, upper surfaces shiny and subglabrous, often highly irridescent.

Head slightly to moderately declined, not concealed from above, strongly transverse and somewhat flattened, not abruptly constricted posteriorly (gradually narrowed in Phaenocephalinae). Eyes moderately large, not protuberant, entire, finely facetted, without interfacetal setae. Antennal insertions barely exposed or concealed from above; subantennal grooves absent. Frontoclypeal suture absent. Labrum free, transverse, with broadly rounded to truncate apex. Antennae 11-segmented, with large, well defined, elongate and often somewhat asymmetrical, 3-segmented club; condyle of scape displaced distally. Mandible usually unidentate or bidentate (rarely tridentate), without dorsal tubercle or cavity; distinctly curved (except in *Phalacrinus*); incisor edge with or without subapical tooth; mola and prostheca usually well developed (mola reduced in *Phalacrinus*). Maxilla with broad setose galea and narrow lacinia, the latter with bidentate uncus; apical palpomere fusiform. Mentum strongly transverse; ligula simple or bilobed; apical labial palpomere fusiform. Gular sutures very short or absent. Corpotentorium absent. Cervical sclerites absent.

Pronotum ~0.26–0.45 times as long as wide, widest posteriorly, not narrower at base than combined elytral bases; sides strongly curved and obliquely explanate. Lateral pronotal carinae complete, with or without raised margin or bead; edges simple; anterior angles produced and broadly to narrowly rounded or acute; posterior angles right to slightly acute, distinctly overlapping elytral humeri; posterior edge straight to weakly sinuate; disc simple. Prosternum in front of coxae shorter than procoxal cavity, flat to slightly convex. Prosternal process complete, parallel-sided or gradually expanded apically and gradually or abruptly curved behind coxae, sometimes extending posteriorly to meet metaventral process; broadly rounded or truncate at apex. Procoxae with very short to moderately long concealed extensions; trochantins partly exposed or concealed. Procoxal cavities slightly transverse to circular, narrowly to moderately widely separated, externally broadly open, with or without narrow lateral extensions, internally closed.

Scutellar shield abruptly elevated, posteriorly acute to broadly rounded. Elytra ~0.9–1.22 times as long as combined width and 2.6–3.85 times as long as pronotum, with well developed humeri; disc usually with nine puncture rows and no scutellary striole, but puncture rows sometimes very fine or absent or striae reduced to one or two near suture; epipleura complete or incomplete, usually broadly, obliquely to vertically explanate. Mesoventrite usually with anterior edge on different plane than metaventrite (not so in Phaenocephalinae), paired procoxal rests highly reduced or absent; mesoventrite sometimes almost or completely concealed by the approximate or abutting prosternal and metaventral processes. Mesocoxae with concealed trochantins. Mesocoxal cavities moderately widely separated, closed laterally by meeting of meso- and metaventrites. Mesometaventral junction usually an anteriorly curved line (straight line in Phaenocephalinae) with mesoventrite overlapping or abutting metaventrite (or with junction absent due to fusion). Metaventrite strongly transverse, flattened, sometimes with curved or angulate postcoxal lines; discrimen short to very long; exposed portion of metanepisternum moderately to very long and narrow. Metacoxae narrowly separated; plates absent. Metendosternite with short, broad stalk or without stalk, with long lateral arms and narrow laminae, and with very short anterior process bearing widely separated anterior tendons or without anterior process and anterior tendons on lateral arms. Hind wings with moderately long apical field bearing one or two post-radial sclerites and at least one vaguely indicated radial extension; radial cell and cross-vein r3 absent; r4 often incomplete; R-M loop broad; basal portion of RP short to long, with or without apical extension; medial spur short; medial field with two or three free veins and no medial fleck; wedge cell absent; anal notch distinct. Femora often inflated; tibiae sometimes expanded apically and spinose at apex; spurs very short to long, equal or unequal, sometimes reduced to one on fore leg. Tarsi usually 5-5-5 in both sexes with penultimate tarsomere reduced and at least antepenultimate lobed beneath (5-5-4 in males of *Olibrus* and *Acylomus*, 5-5-4 in both sexes of *Stilbus*, 5-5-4 or 4-5-4 in both sexes of *Augasmus*, and 4-4-4 in both sexes of Phaenocephalinae with tarsomeres 1–3 distinctly lobed beneath); pretarsal claws toothed.

Abdomen with five free ventrites; ventrite 1 not much longer than 2, intercoxal process acute or narrowly rounded. Functional spiracles usually present on abdominal segments I–VII (present on I–VI in *Phalacrus*). Anterior edge of segment IX in male with or without spiculum gastrale; posterior edge without articulated lobe; tergites IX and X membranous. Aedeagus of cucujiform type, sometimes asymmetrical, anterior edge of tegmen with or without short, paired ventral struts and sometimes with median, dorsal strut; parameres articulated or fused to phallobase and separate or fused together; penis broad and flattened, its anterior edge without struts; endophallus often with sclerites or spicules. Ovipositor short and broad to moderately elongate, usually with proctigeral and paraproctal baculi; coxites usually subdivided, apex heavily sclerotised and complex in *Phalacrus*; styli apical or subapical. [Lea 1932; Thompson and Marshall 1980; Pakaluk 1991; Lawrence *et al.* 1999b, 2010h; Steiner 2002; Gimmel 2013].

Larva

Body usually moderately to very elongate and more or less parallel-sided to slightly wider at middle (thorax widest and abdomen gradually narrowing posteriorly in *Phalacrinus*), slightly to strongly flattened; usually lightly pigmented, except for head and all or a portion of tergum IX, or with pigmented tergal plates on most segments; vestiture usually of scattered simple setae (setae very long and apically expanded in *Phalacrinus*).

Head usually prognathous and protracted, transverse, flattened and strongly rounded laterally (hypognathous, about as long as wide and slightly rounded laterally in *Phalacrinus*); posterior edge not to moderately emarginate. Epicranial stem usually very short or absent; frontal arms lyriform, contiguous at base (epicranial suture not apparent in *Phalacrinus*). Median endocarina present or absent; paired endocarinae absent. Stemmata five or six on each side.

Frontoclypeal suture present or absent. Labrum usually free (partly or completely fused to head capsule in *Phalacrus*). Antennae usually short to moderately long (very long and well separated from mandibular articulation in *Phalacrinus*), 3-segmented, with sensorium usually on preapical antennomere (on basal antennomere in *Phalacrus*, Fig. 46P), shorter or longer than apical antennomere. Mandibles symmetrical, usually broad at base and narrow at apex (more elongate and not mesally curved in *Phalacrinus*)); bidentate or tridentate, occasionally with accessory ventral process; incisor edge simple, with one or two subapical teeth, or serrate; mesal edge of mandibular base usually with large, extensive mola, with a smaller, sub-basal mola, which is finely or coarsely tuberculate, or with two or more hyaline processes (simple in *Phalacrinus*). Ventral mouthparts (Fig. 49A) strongly protracted, without maxillary articulating area; cardo not distinct (separated from stipes but fused to labium in *Phalacrinus*); stipites usually wider than long; mala rounded or truncate, not cleft; maxillary palps 3-segmented. Labium consisting of prementum and postmentum, the latter rarely more or less fused with maxillae; ligula simple, not bilobed, shorter than labial palps, which are 2-segmented and moderately widely separated. Hypopharyngeal sclerome absent or consisting of transverse bar. Hypostomal rods almost always long and diverging (absent in *Phalacrinus*); ventral epicranial ridges absent. Gula usually longer than wide (absent in *Phalacrinus*).

Legs moderately well developed, 5-segmented including claw-like pretarsus bearing one seta; coxae moderately widely separated. Abdominal terga and sterna sometimes with rows of asperities forming longitudinally oriented, open or closed ovals on either side of midline; thoracic and abdominal terga with setiferous tubercles in *Phalacrinus*. Tergum IX excluding appendages shorter than VIII, extending onto ventral surface and sometimes forming articulated plate (Fig. 52B); with paired dorsally or posterodorsally oriented urogomphi (very long, diverging and apically setose in *Phalacrinus*). Sternum IX simple, not enclosed by sternum VIII, usually concealed (Fig. 52B). Segment X without paired pygopods, usually concealed; anal region posteriorly or posteroventrally oriented. Spiracles usually annular–uniforous or annular–biforous (annular in *Phalacrinus*), sometimes placed at ends of spiracular tubes on segments I–VIII or on I and VIII only. [Thompson and Marshall 1980; Steiner 1984, 2002; Steiner and Singh 1987; Lawrence 1991b; Lawrence *et al.* 1999a, 2010h].

Classification and World Distribution

The Phalacridae is a widely distributed family with 34 described genera and ~650 species. There are two subfamilies – Phaenocephalinae, with a few genera in the Oriental, Afrotropical and Australian regions, and Phalacrinae with the remaining genera. [Gimmel 2009, 2013].

Australian Fauna

The Australian fauna includes 12 genera and ~100 species. The subfamily **Phaenocephalinae** by the relatively common and widespread Australian genus *Phalacrinus* and a few undescribed species of the primarily Oriental genus *Phaenocephalus*. The most common and speciose genera of **Phalacrinae** are *Litochrus* and *Austroporus*, each containing ~25 described Australian species. The cosmopolitan genus *Phalacrus* is represented by three described species, one of which, *P. uniformis* (Blackburn), has been introduced into New Zealand (Thompson and Marshall 1980). *Platyphalacrus lawrencei* Gimmel is known only from the southern tip of WA. The remaining genera include the widely distributed *Acylomus*, *Stilbus* and *Olibrus*, the Oriental *Augasmus* and *Sveculus*, and the primarily New World *Litochropus*, each of which includes one or more, mainly undescribed Australian species.

Almost all phalacrids are broadly ovate, strongly convex, glabrous and shiny beetles, with the lateral portions of the prothorax and elytra broadly, obliquely explanate and deflexed; *Platyphalacrus* differs in being somewhat flattened. The head is short, broad and deeply inserted into an anterior pronotal emargination, the antennae are relatively long with a large club, the tibiae have a conspicuous apical comb, the first three tarsomeres are lobed and the pretarsal claws toothed or appendiculate. Larvae of almost all phalacrids are elongate, subcylindrical to flattened and lightly pigmented, except for the head, protergal plate, apex of segment VIII and the urogomphi or the entire tergum IX, which may form an articulated plate. The head is broad and somewhat flattened, with short antennae and usually with a median endocarina; the mala is blunt; hypostomal rods are long and divergent, the eight spiracles are located dorsally near the posterior end of the tergum, and sternum IX and segment X are usually concealed. Larvae of *Phalacrus* are unusual in lacking an endocarina and in having the antennal sensorium on the first, rather than the second, antennomere, and the presumed larva of *Phalacrinus* is unusual in several respects, including the shape (widest anteriorly and narrowing posteriorly), vestiture of long, apically expanded setae, hypognathous head, longer antennae, more elongate mandibles with simple base, absence of hypostomal rods, and long, diverging, apically setose urogomphi. Phalacridae are primarily mycophagous and most feed on spores of various kinds; however, larvae of *Litochrus major* Blackburn have been collected in rotten stringy bark and in rotten wood at the base of a living eucalypt. *Phalacrus uniformis* (Blackburn) is known to feed on spores produced on the galls of *Uromycladium* (Basidiomycetes: Uredinales), *Phalacrinus* adults (Fig. 85E) have been collected in numbers by beating dead eucalypt leaves, and *Platyphalacrus* have been collected in the male cones of cycads. [Lea 1932; Thompson and Marshall 1980; Steiner 1984; Gimmel 2013].

Australian Subfamilies, Tribes and Genera

Phaenocephalinae: *Phaenocephalus* Wollaston, *Phalacrinus* Blackburn.

Phalacrinae: Acylomus Sharp, *Augasmus* Motschulsky, *Austroporus* Gimmel, *Litochropus* Casey, *Litochrus* Erichson, *Olibrus* Erichson, *Phalacrus* Paykull, *Platyphalacrus* Gimmel, *Stilbus* Seidlitz, *Sveculus* Gimmel.

75. PROPALTICIDAE

Adult (Fig. 85B)

Length 1.2–1.5 mm. Body ~1.25–1.65 times as long as wide; broadly ovate to circular with sides evenly curved; strongly flattened; colour yellowish-brown to reddish-brown, upper surfaces either dull or with a slight sheen; vestiture either if fine, decumbent setae or thickened decumbent setae that may form a pattern.

Head not or slightly declined, strongly transverse and somewhat flattened. Occipital region with low transverse carina. Eyes large, entire, coarsely facetted, with very short and fine interfacetal setae. Antennal insertions exposed from above; subantennal grooves absent. Frontoclypeal suture absent. Labrum free, transverse, with subtruncate to broadly rounded apex. Antennae short, 11-segmented, with narrow, elongate, 3-segmented club; condyle of scape displaced distally. Mandible short, broad, tridentate, without dorsal tubercle or cavity; incisor edge simple; mola and prostheca well developed. Maxilla with broad setose galea and slender lacinia with bidentate uncus; apical palpomere narrowly fusiform. Mentum strongly transverse; ligula simple; apical labial palpomere narrowly fusiform. Gular sutures very short or absent. Corpotentorium narrow, strongly arched. Cervical sclerites absent.

Pronotum ~0.45–0.55 times as long as wide, widest posteriorly, not narrower at base than combined elytral bases; sides moderately to strongly curved. Lateral pronotal carinae complete, with or without narrow margin or bead; edges simple; anterior angles produced and more or less acute; posterior angles right to obtuse, more or less abutting elytral humeri; posterior edge straight to weakly sinuate; disc with long median endocarina. Prosternum in front of coxae much longer than procoxal cavity, flat to slightly convex. Prosternal process complete, strongly and abruptly expanded apically, flat, with apex truncate. Procoxae with long, more or less oblique, concealed lateral extensions; trochantins concealed. Procoxal cavities slightly transverse, externally narrowly open, without narrow lateral extensions, internally open.

Scutellar shield strongly transverse, not abruptly elevated, posteriorly obtusely angulate. Elytra ~0.75–1.2 times as long as combined width and 1.65–2.2 times as long as pronotum, with well developed humeri; disc irregularly punctate, with eight puncture rows and no scutellary striole, or with three longitudinal carinae; epipleura incomplete. Mesoventrite without paired procoxal rests. Mesocoxae with concealed trochantins. Mesocoxal cavities widely separated, closed laterally by meeting of meso- and metaventrites. Mesometaventral junction a straight line. Metaventrite short and broad, flattened, with long discrimen; exposed portion of metanepisternum moderately elongate. Metacoxae widely separated; plates absent. Metendosternite without stalk and with widely separated, long lateral arms; laminae and anterior process absent; anterior tendons on lateral arms or not apparent. Hind wings with long apical field and reduced venation; radial cell, cross-vein r3 and R-M loop absent; medial field with single; anal lobe absent. Legs slender; protibia longer than mesotibia or metatibia, armed with single large spur, which may be broad and apically dentate; spurs on mesotibiae and metatibiae inconspicuous or absent. Tarsi usually 5-5-5 in both sexes; tarsomeres simple. penultimate one not reduced; pretarsal claws simple.

Abdomen with five free ventrites; ventrite 1 not much longer than 2, intercoxal process broad and truncate. Functional spiracles present on abdominal segments I–VII (those on VII reduced). Anterior edge of segment IX in male with spiculum gastrale; tergites IX and X fused and membranous. Aedeagus of cucujiform type, symmetrical, anterior edge of tegmen broadly truncate; parameres partly or entirely fused together but articulated to phallobase; anterior edge of penis with single strut; endophallus with long, slender plate. Ovipositor relatively short and broad, widest at base and narrowed posteriorly; paraprocts slightly longer than coxites, each of which is subdivided, with distal lobe narrow, palpiform and heavily sclerotised; styli long, cylindrical and apical. [John 1960; Lawrence *et al.* 1999b; Lawrence 2010h].

Larva

Body somewhat elongate, slightly flattened; lightly pigmented.

Head prognathous and protracted. Epicranial stem absent; frontal arms parallel and widely separated at base. Median endocarina absent; paired endocarinae present beneath frontal arms. Stemmata two on each side. Frontoclypeal suture absent. Labrum free. Antennae very short, 3-segmented, with antennomeres 1 and 2 strongly transverse and 3 long and narrow, with moderately long sensorium. Mandibles broad at base and narrow at apex, bidentate; incisor edge with single tooth; mesal edge of mandibular base with several hyaline processes. Ventral mouthparts strongly protracted; mala blunt; maxillary palps 3-segmented. Labial palps 2-segmented. Hypopharyngeal sclerome absent. Hypostomal rods long and diverging.

Legs well developed, 5-segmented including claw-like pretarsus bearing two setae; coxae widely separated. Abdominal segment VIII slightly shorter than VII. Segment IX unusually short, with pair of articulations at base; urogomphi absent. Spiracles annular. [Lawrence 2010h].

Classification and World Distribution

Propalticidae is a small family of ~40 species in two genera. *Propalticus* Sharp is distributed throughout the warmer parts of the Old World, from Japan and India to Australia, New Caledonia, several Pacific island groups, Madagascar and West Africa. The genus *Slipinskogenia* Gimmel occurs only in tropical Africa. [John 1960; Crowson and Sen Gupta 1969; Lawrence 2010h; Gimmel 2011].

Australian Fauna

There are two described species of Propalticidae from the Australian mainland, *Propalticus simplex* Sen Gupta & Crowson from northern NSW and southern QLD and *P. doddi* John from northern QLD, and two more species have been seen from Christmas Island. Propalticids are minute, broadly ovate and strongly flattened beetles, clothed with short, decumbent, somewhat thickened setae and with a large head and eyes, a median endocarina on the pronotum, a strongly transverse scutellum, an enlarged, serrate tibial spur on each fore leg, and very widely separated procoxae. The presumed larva is slightly elongate, slightly flattened and lightly pigmented, with paired endocarinae beneath the well separated frontal arms, very short antennae, mandibles with several hyaline processes at the base, protracted ventral mouthparts, long diverging hypostomal rods, and a reduced, but basally articulated tergum IX without urogomphi. Adult propalticids have be collected on palm trunks covered with algae and on or under bark of standing trees or logs. The presumed larva was collected under bark.

76. LAEMOPHLOEIDAE

Adult (Figs 85K, 87R)

Length 1.5–3.5 mm. Body ~1.75–3.3 times as long as wide; elongate, moderately to strongly flattened; colour yellow to reddish-brown or black; elytra sometimes with dark and light colour pattern, upper surfaces glabrous or clothed with short, fine, decumbent hairs.

Head (Fig. 28B) not declined, not abruptly constricted posteriorly, usually transverse, but sometimes longer than wide, often with supra-antennal and superocular ridges and often with long median endocarina; occipital region sometimes with transverse ridge or groove. Eyes small to moderately large, protuberant or not, entire, coarsely facetted, with very short, fine interfacetal setae. Antennal insertions exposed or concealed from above; subantennal grooves absent or short and weakly developed. Frontoclypeal suture present or absent; clypeus short to moderately long with its apex truncate, emarginate or with three emarginations. Labrum free, large, apically rounded to deeply emarginate. Antennae 11-segmented, filiform or with weak, 3-segmented club, often quite long; scape enlarged and distorted in some males. Mandible large, stout to elongated, bidentate, without dorsal tubercle or cavity; incisor edge usually with subapical tooth; mola and prostheca usually well developed. Maxilla with broad setose galea and narrow lacinia, the latter with bidentate uncus; apical palpomere fusiform. Mentum strongly transverse; ligula sclerotised, slightly emarginate; apical labial palpomere fusiform. Genae sometimes produced forward. Gular sutures sometimes narrowly separated or absent. Corpotentorium narrow and arched. Cervical sclerites absent.

Pronotum ~0.55–1.2 times as long as wide, usually widest anteriorly and not narrower at base than combined elytral bases (rarely widest at middle or distinctly narrower than elytra); sides straight to slightly curved. Lateral pronotal carinae complete, with raised margin or bead; edges simple; anterior angles sometimes slightly produced and narrowly rounded or acute; posterior angles more or less right; posterior edge straight; disc with pair of sublateral carinae extending from posterior to anterior edge. Prosternum in front of coxae much longer than procoxal cavity, flat to slightly convex. Prosternal process complete, flat, gradually to abruptly expanded apically and truncate at apex. Notosternal sutures absent. Procoxae with moderately long, more or less oblique, concealed lateral extensions; trochantins concealed. Procoxal cavities circular, widely separated, externally open or closed, without narrow lateral extensions, internally open.

Scutellar shield abruptly elevated, posteriorly broadly rounded or obtusely angulate. Elytra ~1.2–2.3 times as long as combined width and 2.15–3.2 times as long as pronotum, with well developed humeri; disc impunctate, irregularly punctate or with eight or nine puncture rows and no scutellary striole, sometimes with longitudinal costae; epipleura complete or incomplete. Mesoventrite without paired procoxal rests, with broad prothoracic rest. Mesocoxae with concealed trochantins. Mesocoxal cavities widely separated, usually partly closed laterally by mesepimeron (closed by meeting of meso- and metaventrites in *Lathropus* Erichson). Mesometaventral junction a straight line with mesoventrite overlapping or abutting metaventrite. Metaventrite slightly transverse to elongate, more or less flattened, with long discrimen; exposed portion of metanepisternum moderately to very long and narrow. Metacoxae moderately to widely separated; plates absent. Metendosternite without stalk, with long lateral arms, no laminae or anterior process and anterior tendons on lateral arms or not apparent. Hind wings with long apical field without sclerites and with or without vaguely indicated radial extension; radial cell, cross-veins r3 and r4 absent; R-M loop acute; basal portion of RP short absent; medial field usually with one or two free veins and no medial fleck; wedge cell absent; anal lobe highly reduced or absent. Legs relatively long and slender; trochanterofemoral joint usually transverse or slightly oblique; tibial spurs usually paired, small and subequal. Tarsi usually 5-5-5 in female and 5-5-4 in male (rarely 5-5-5 in both sexes); tarsomeres simple, with first one reduced; pretarsal claws simple.

Abdomen with five free ventrites; ventrite 1 not much longer than 2, post-coxal lines sometimes present; intercoxal process broadly rounded to truncate. Functional spiracles usually present on abdominal segments I–VII. Anterior edge of segment IX in male without spiculum gastrale; tergites IX and X membranous. Aedeagus of cucujiform type, symmetrical, inverted; anterior edge of tegmen without struts; parameres fused together and to phallobase; penis with long anterior strut; endophallus sometimes complex. Ovipositor moderately elongate; paraprocts subequal to coxites, which are subdivided with apical portion narrow; styli apical, subapical or absent. [Lefkovitch 1959, 1962; Lawrence *et al.* 1999b; Thomas 2002d; Thomas and Leschen 2010c].

Larva (Fig. 34K)

Body elongate and parallel-sided; strongly flattened; lightly pigmented, except for head and abdominal segment VIII and tergum IX; vestiture usually of scattered simple setae.

Head usually prognathous and protracted, transverse, flattened and strongly rounded laterally; posterior edge distinctly emarginate. Epicranial stem absent; frontal arms lyriform or U-shaped, well separated at base. Median endocarina absent; paired endocarinae present beneath bases of frontal arms. Stemmata five on each side. Frontoclypeal suture absent. Labrum free, with long, widely separated tormae. Antennae short to moderately long, 3-segmented, with conical sensorium on preapical antennomere usually much shorter than apical antennomere (almost as long as preapical antennomere in *Lathropus*). Mandibles symmetrical, broad at base and narrow at apex; tridentate, without accessory ventral process; incisor edge usually with one or more subapical teeth; mesal edge of mandibular base with well developed mola in *Placonotus*, reduced, sub-basal mola in *Cryptolestes*, and an acute sclerotised, sub-basal process plus two or more hyaline processes in *Laemophloeus*; prostheca absent. Ventral mouthparts strongly protracted, without maxillary articulating area; cardo absent; stipites wider than long; mala obtuse, not cleft; maxillary palps 3-segmented. Labium consisting of prementum and postmentum; ligula simple, shorter than labial palps, which are 2-segmented and moderately widely separated. Hypopharyngeal sclerome present or absent. Hypostomal rods long and more or less parallel, extending to posterior edge of head; ventral epicranial ridges absent. Gula usually longer than wide.

Legs moderately well developed, 5-segmented including claw-like pretarsus bearing one seta; coxae moderately widely separated. Abdominal terga and sterna I–VI or VII often with curved, oblique rows of asperities on either side of midline. Tergum VIII distinctly longer than VII; tergum IX excluding appendages much shorter than VIII, extending onto ventral surface and forming articulated plate (Fig. 52A); with paired posteriorly projecting and slightly upturned urogomphi. Sternum IX and segment X membranous and enclosed within a deep emargination of sternum VIII. Anal region ventrally oriented. Spiracles usually annular–biforous or annular–uniforous, those on segment VII facing posterolaterally. [Thomas 1988, 2002d; Lawrence 1991b; Lawrence *et al.* 1999a; Thomas and Leschen 2010c].

Classification and World Distribution

Laemophloeidae is widely distributed with its greatest diversity in tropical regions. The family contains ~450 species in 37 described genera.

Australian Fauna

Since has been no major revision of Australian Laemophloeidae, the generic placement of several our species is still in doubt. The Australian fauna includes *Microlaemus piceicollis* (Lea) and *M. brightensis* (Blackburn), *Microbrontes blackburni* (Grouvelle), several species of *Placonotus* Macleay and *Cryptolestes* Ganglbauer (Fig. 87R) (including cosmopolitan pests, such as *C. ferrugineus* (Stephens)), and several *Laemophloeus* species (Fig. 85K), which may eventually be moved to other genera. Laemophloeids are small, elongate, parallel-sided, usually flattened and often glabrous beetles, which never have a large or abrupt antennal club and may have a modified scape in the male; the pronotum has a pair of longitudinal, submarginal carinae, and the elytra are often carinate. Larvae are narrowly elongate, parallel-sided, flattened and lightly sclerotised, except for the head and segments VIII and IX. The head is flattened, with moderately long antennae, paired dorsal endocarinae and long, parallel hypostomal rods, the mandible is varied, with a basal or sub-basal mola or a group of mesobasal hyaline processes, the mala is blunt, tergum IX forms an articulated plate and sternum IX and segment X are enclosed within an emargination of sternum VIII (Fig. K165). Laemophloeids usually feed on moulds or on the spores and stromaata of various other Ascomycetes, but *Cryptolestes ferrugineus* and some related species are important pests of stored grains. [Thomas and Leschen 2010c].

77. TASMOSALPINGIDAE

Adult (Fig. 85G)

Length 1.2–2.2 mm. Body ~1.95–2.05 times as long as wide; widest at posterior third; slightly flattened; colour dark-brown to black; elytra usually with yellow maculae; upper surfaces shiny, clothed with very short, fine, decumbent hairs.

Head not or slightly declined, but anterior portion bent ventrally; not abruptly constricted posteriorly, slightly longer than wide, somewhat flattened; occipital region with median stridulatory file. Eyes small, protuberant, entire, coarsely facetted, without interfacetal setae. Antennal insertions exposed from above; subantennal grooves present. Frontoclypeal suture absent; clypeus extending well in front of antennal insertions, forming with genae a short, broad, flattened, muzzle, truncate at apex. Labrum strongly transverse, its apex subtruncate to slightly convex. Antennae 11-segmented with weak, 3-segmented club. Mandible slightly longer than wide, flattened, tridentate; incisor edge with single subapical tooth; mesal edge of mandibular base with membranous setose process and narrow fringe of setae. Maxilla with galea only slightly wider than lacinia, both setose and the latter without uncus; apical palpomere narrowly fusiform. Mentum slightly transverse; ligula sclerotised, undivided, truncate; apical labial palpomere narrowly fusiform. Genae projecting anteriorly and acute. Pregular region concave, in male with curved, transverse groove in the centre of which is a patch of glandular hairs. Corpotentorium narrow. Cervical sclerites reduced and contiguous with head capsule.

Pronotum ~0.78–0.85 times as long as wide, widest at middle, with base distinctly narrower than combined elytral bases; sides moderately curved anteriorly and sinuate posteriorly. Lateral pronotal carinae complete, but deflected posteriorly by pair of large, deep, laterally opening setose cavities, with raised margin or bead; edges finely crenulate; anterior angles not produced; posterior angles obtuse or right; posterior edge straight; disc simple. Prosternum in front of coxae about as long as procoxal cavity, slightly convex. Prosternal process complete, flat, abruptly expanded apically and emarginate at apex. Procoxae without or with very short, concealed lateral extensions; trochantins at least partly exposed. Procoxal cavities slightly transverse, moderately widely separated, broadly open externally, without narrow lateral extensions, internally closed by slender bar.

Scutellar shield abruptly elevated, posteriorly broadly rounded. Elytra ~1.34–1.4 times as long as combined width and 2.05–2.2 times as long as pronotum, with weak humeri and sides moderately rounded; elytral suture deflected near apex; disc irregularly punctate without scutellary striole; epipleura incomplete. Mesoventrite with paired, horizontal or slightly declined procoxal rests. Mesocoxae with exposed trochantins. Mesocoxal cavities moderately widely separated, partly closed laterally by mesepimeron. Mesometaventral junction an anteriorly curved line with mesoventrite overlapping metaventrite. Metaventrite strongly transverse and slightly convex, without discrimen; exposed portion of metanepisternum very long and narrow. Metacoxae moderately widely separated; plates absent. Metendosternite with short, broad stalk, long lateral arms, narrow laminae, a short anterior process and anterior tendons widely separated. Hind wings with very short apical field without sclerites and without radial cell, cross-veins r3 or r4 or R-M loop; medial spur short but medial embayment well developed; medial field with one free veins and no medial fleck; wedge cell absent; anal lobe highly reduced or absent. Legs slender; tibial spurs well developed, paired. Tarsi usually 5-5-5 in female and 5-5-4 in male; tarsomeres simple; pretarsal claws simple.

Abdomen with five free ventrites; ventrite 1 not much longer than 2; intercoxal process broadly rounded. Functional spiracles present on abdominal segments I–V. Anterior edge of segment IX in male with spiculum gastrale; tergites IX and X membranous. Aedeagus of cucujiform type, symmetrical; anterior edge of tegmen broadly angulate, without struts; parameres short and broad, fused to phallobase but free from one another; penis with paired basal struts; endophallus usually with long, rod-like sclerite. Ovipositor with paraprocts subequal to coxites, which are subdivided with apical portion narrow and more heavily pigmented; styli apical. [Leschen *et al.* 2005; Lawrence and Ślipiński 2010d].

Larva

Body elongate, more or less parallel-sided and somewhat flattened; lightly pigmented, except for head; vestiture of long, fine setae.

Head usually prognathous and protracted, transverse, flattened; posterior edge deeply emarginate with pair of deep notches. Epicranial stem absent; frontal arms lyriform, well separated at base, joined anteriorly by transverse line. Median endocarina absent; paired endocarinae lying beneath frontal arms. Stemmata five on each side. Frontoclypeal suture present or absent. Labrum free. Antennae moderately long, 3-segmented, with sensorium on antennomere 2 longer than antennomere 3. Mandibles symmetrical, tridentate, without accessory ventral process; incisor edge with one subapical tooth; mesal edge of mandibular base with three hyaline processes. Ventral mouthparts protracted, without maxillary articulating area; cardo distinct; stipites wider than long; mala blunt, not cleft; maxillary palps 3-segmented. Labial palps 2-segmented and moderately widely separated. Hypopharyngeal sclerome absent. Hypostomal rods diverging; ventral epicranial ridges absent. Gular region long.

Prothorax with ventral median longitudinal endocarina. Legs moderately widely developed, 5-segmented including claw-like pretarsus bearing one seta; coxae moderately widely separated. Tergum IX simple, extending onto ventral surface, without urogomphi; sternum IX simple; segment X concealed from above; anal region ventrally oriented. Spiracles annular not placed at ends of spiracular tubes. [Leschen *et al.* 2005; Lawrence and Ślipiński 2010d].

Classification and World Distribution

Tasmosalpingidae contains the single Australian genus *Tasmosalpingus* Lea.

Australian Fauna

Tasmosalpingus quadrispilotus Lea and *T. promiscuus* Lea were both described from TAS, but a specimen of the latter has been seen from Victoria. These small beetles resemble members of the salpingid subfamily Salpinginae, but they differ from salpingines in several respects, including the 5-5-5- tarsi in the female, the median stridulatory file on the head and the pair of posterolateral setose cavities on the pronotum. Larvae are elongate, parallel-sided, flattened, and lightly pigmented except for the head; the frontal arms are lyriform and well separated at base, with underlying paired endocarinae, mandibles are tridentate, with three hyaline processes at the base, the prothorax has a ventral longitudinal, endocarina and tergum IX is simple. Adult tasmosalpingids have been collected in Malaise traps and a putative larva was collected by pyrethrum fogging the bark of *Phyllocladus* (Podocarpaceae). The gut of a dissected male was found to be packed with fungal hyphae and spores possibly of a hyphomycete.

78. KATERETIDAE

Adult (Fig. 72B)

Length usually 1.5–4 mm. Body ~1.55–2.1 times as long as wide; oblong and moderately convex; yellowish-brown to black; dorsal surfaces dull to moderately shiny with vestiture of short, decumbent hairs.

Head slightly declined, transverse, not abruptly constricted posteriorly. Eyes entire, moderately prominent, finely facetted, without interfacetal setae. Antennal insertions exposed and widely separated; subantennal grooves absent. Frontoclypeal suture straight, incomplete laterally; base of clypeus without paired lateral impressions; anterior edge of clypeus truncate or emarginate. Labrum free, transverse, apically emarginate. Antennae 11-segmented, with weak, loose, 3-segmented club. Mandible bidentate, without dorsal tubercle or cavity; incisor edge simple; mola coarsely tuberculate; prostheca moderately well developed. Maxilla with galea and lacinia both slender, the latter with unidentate uncus; apical palpomere subconical or fusiform. Mentum transverse, trapezoidal and apically emarginate, ligula truncate; apical labial palpomere slightly expanded, more or less ovate. Corpotentorium slender and arched, without median process. Cervical sclerites well developed.

Pronotum ~0.57–1.05 times as long as wide, usually widest near base, which is as wide as combined elytral bases; sides usually weakly curved (subparallel in *Jelinekiella* Kirejtshuk). Lateral pronotal carinae complete, with raised margin or bead, the edges simple; anterior angles not or barely produced and broadly rounded; posterior angles broadly rounded; posterior edge evenly rounded; disc simple. Prosternum in front of coxae slightly shorter than procoxal cavity, slightly convex. Prosternal process complete, strongly curved and abruptly raised behind coxae, strongly expanded at apex, which is truncate. Procoxae without concealed lateral extensions; trochantins exposed. Procoxal cavities strongly transverse, narrowly separated, externally broadly closed, without narrow lateral extensions, internally broadly closed.

Scutellar shield large, not abruptly elevated, posteriorly narrowly rounded. Elytra ~0.8–1.04 times as long as combined width and 1.35–2.25 times as long as pronotum, with well developed humeri, sides usually weakly rounded (subparallel in *Jelinekiella*); apices independently rounded or truncate, exposing pygidium (tergite VII) and a portion of tergite VI; suture not deflected near apex; disc irregularly punctate, without scutellary striole; epipleura sharply defined but incomplete. Mesoventrite without paired procoxal rests; with broad prothoracic rest or prepectus. Mesoventral process truncate at apex. Mesocoxae with exposed trochantins. Mesocoxal cavities transverse, moderately broadly separated, partly closed laterally by mesanepisterna and mesepimera. Mesometaventral junction a straight line, with meso- and metaventrites abutting or slightly overlapping. Metaventrite flat, with long discrimen; exposed portion of metanepisternum moderately elongate. Metacoxae widely separated; plates absent. Metendosternite with short, broad stalk, long lateral arms, no laminae, short anterior process, and approximate anterior tendons. Hind wing with very long apical field containing a longitudinal post-radial sclerite and at least one vaguely indicated radial extension; radial cell, cross-veins r3 and r4, R-M loop and medial spur absent; medial field with one free veins and no medial fleck; wedge cell absent; anal embayment notch-like. Tibiae strongly widened apically, with row of short, stout apical spines. Tarsi 5-5-5 in both sexes; tarsomeres 1–3 with strong ventral lobes; tarsomere 4 reduced but not enclosed within lobe of preceding tarsomere; pretarsal claws simple or toothed.

Abdomen with five free ventrites; ventrite 1 subequal to 2; intercoxal process broadly rounded. Functional spiracles on abdominal segments I–VI located on tergites. Most tergites well sclerotised. Tergite VIII in male exposed; sternite VIII reduced and membranous. Anterior edge of segment IX in male with spiculum gastrale; tergite IX and segment X membranous. Aedeagus of cucujiform type, symmetrical; anterior edge of tegmen broadly rounded; parameres well developed, articulated to phallobase and separate; penis somewhat asymmetrical, with slender, articulated basal strut and endophallic sclerites. Ovipositor elongate, broadest at base and narrowing apically, well sclerotised; paraprocts not clearly separated from coxites, which are undivided and moderately heavily sclerotised; styli highly reduced and slightly subapical. [Audisio 1993; Lawrence *et al.* 1999b; Habeck 2002a; Jelinek and Cline 2010].

Larva

Body elongate, parallel-sided to slightly fusiform and slightly flattened; dorsal surfaces lightly pigmented or with weakly to moderately pigmented tergal plates on thorax and abdominal segments I–IX; surfaces smooth and vestiture of short, simple setae.

Head protracted and prognathous or slightly declined, transverse and slightly flattened. Epicranial stem absent; frontal arms lyriform, narrowly separated at contiguous at base. Median endocarina absent. Stemmata four on each side. Frontoclypeal suture absent. Labrum free, strongly transverse and broadly rounded anteriorly. Antennae moderately short, 3-segmented, with conical sensorium at least as long as reduced apical antennomere. Mandibles symmetrical, unidentate, without accessory ventral process; incisor edge simple; mola well developed, coarsely tuberculate; prostheca absent. Ventral mouthparts moderately retracted, without maxillary articulating area; cardines only slightly oblique, stipites shorter than wide; mala apically rounded, not setose; maxillary palps 3-segmented. Labium consisting of prementum and postmentum; ligula well developed, rounded; labial palps 1-segmented and narrowly separated. Hypopharyngeal sclerome a transverse bar. Hypostomal rods long and diverging; ventral epicranial ridges absent. Gular region slightly transverse.

Tergal sclerites sometimes not clearly delimited. Legs relatively stout, 5-segmented including short, claw-like pretarsus bearing one seta and a broad, membranous lobe; coxae moderately widely separated. Segment IX slightly shorter than VIII, tergum extending slightly onto ventral surface, without urogomphi; sternum IX exposed, simple, not enclosed by sternum VIII; segment X partly visible from above, posteroventrally oriented, with pair of distinct pygopods. Spiracles annular–biforous. [Hayashi 1978; Audisio 1993; Lawrence 1999b; Lawrence *et al.* 1999a; Habeck 2002a; Jelinek and Cline 2010].

Classification and World Distribution

The family Kateretidae includes 14 genera and ~95 species distributed throughout the temperate and subtropical areas of both Northern and Southern Hemispheres, with the major diversity occurring in more arid regions. Two genera occur in Australia.

Australian Fauna

The Australian kateretid fauna includes *Jelinekiella aterrima* (Kirejtshuk) from WA and ten species of *Notobrachypterus*, widely distributed across the southern part of the continent. Kateretids resemble small carpophiline Nitidulidae in general form and in the truncate elytra exposing the pygidium and part of the propygidium; they differ from nitidulids, however, in having a weak antennal club, a distinct frontoclypeal suture and a maxillary galea. Larvae differ from those of nitidulids in having the frontal arms contiguous, the prostheca reduced, the cardines only slightly oblique and the urogomphi absent. All kateretids appear to be pollen feeders in flowers of a variety of angiosperms. *Notobrachypterus* larvae have been collected in wax flower heads (*Chamelaucium uncinatum*, Myrtaceae). [Blackburn 1892; Kirejtshuk 1986b, 2000].

79. NITIDULIDAE

Adult (Figs 6H, 73C–D, H–I, 83D, 87A–B)

Length 0.8–11 (usually 1.5–5) mm. Body ~1.2–5.4 times as long as wide; shape highly variable: body usually oblong to ovate and moderately convex to flattened, sometimes elongate to narrowly elongate and parallel-sided (many Cillaeinae), or strongly convex to globose (many Nitidulinae, *Cybocephalus* Erichson); colour typically uniformly yellowish-brown to red, reddish-brown or black, with elytra occasionally bicolored; dorsal surfaces usually either subglabrous (*Cybocephalus*, Cychramptodini, Lawrencerosini, many Cryptarchinae and Cyllodini and a few Cillaeinae) or clothed with fine decumbent and/or erect hairs, occasionally with stout, decumbent or erect bristles, rarely with irridescence caused by a diffraction grating.

Head prognathous to strongly declined, usually transverse, often constricted immediately behind eyes or gradually narrowed posteriorly, but without clearly delimited temples. Occipital region often with low transverse occipital ridge or carina, with median stridulatory file in *Cryptarcha* Shuckard. Eyes entire or weakly emarginate, usually moderately to strongly prominent, finely or coarsely facetted, with or without interfacetal setae. Antennal insertions widely separated and almost always concealed or barely exposed (clearly exposed in *Cybocephalus* Erichson); subantennal grooves usually well developed, often extending behind the eyes and sometimes convergent. Frontoclypeal suture usually weakly impressed or absent; base of clypeus without paired lateral impressions; anterior edge of clypeus truncate to emarginate. Labrum usually free, transverse and apically bilobed; fused to clypeus in Cryptarchinae. Antennae almost always 11-segmented with a strong, compact, 3-segmented club (antennae 10-segmented in *Cychramptodes* and *Lawrencerosus*, club 4- or 5-segmented in Lawrencerosini; club weak with last antennomere reduced in *Cybocephalus*); scape usually short and broad. Mandible usually unidentate, sometimes minutely bidentate, without dorsal tubercle or cavity; incisor edge usually with well developed subapical tooth or retinaculum; mola usually well developed, with transverse rows of small teeth (absent in *Cybocephalus*, Cychramptodini and Lawrencerosini); prostheca usually moderately well developed and setose, sometimes consisting of a narrow fringe of setae or small setal brush. Maxilla (Fig. 16B) with galea absent and lacinia broad, rounded and setose, without uncus; apical palpomere usually subconical or fusiform. Mentum transverse, trapezoidal and apically emarginate, ligula simple, broadly emarginate or bilobed; apical labial palpomere more or less fusiform. Corpotentorium slender, with or without median process. Cervical sclerites present or absent.

Pronotum ~0.35–1.15 times as long as wide, usually widest near base or at middle (widest anteriorly in *Lawrencerosus* Kirejtshuk) and as wide at base as combined elytral bases; sides very weakly to strongly curved, horizontally explanate in some Nitidulinae (e.g. *Soronia* Erichson), obliquely or vertically explanate in *Cybocephalus*, Cychramptodini and some Cyllodini); lateral pronotal carinae complete, with or without raised margin or bead, the edges simple; anterior angles produced or not and broadly rounded to narrowly rounded or acute; posterior angles broadly rounded to slightly acute; posterior edge evenly rounded to distinctly bisinuate; disc simple. Notosternal sutures absent or indistinct. Prosternum in front of coxae usually slightly shorter to slightly longer than procoxal cavity, occasionally much shorter (e.g. *Cybocephalus*, Cychramptodini) or distinctly longer (e.g. Cyllaeinae), flattened or slightly convex to strongly tumid or carinate. Prosternal process complete, either flat, strongly curved behind coxae or abruptly elevated and truncate behind coxae forming vertical wall, not to gradually or abruptly expanded apically, with narrowly rounded to truncate apex. Procoxae without concealed lateral extensions; trochantins exposed. Procoxal cavities strongly transverse, narrowly to moderately widely separated, externally usually narrowly to broadly closed (externally open in Amphicrossinae, Cryptarchinae and Cybocephalinae), without narrow lateral extensions, internally closed by slender bar or broadly closed.

Scutellar shield large, not abruptly elevated, posteriorly broadly rounded to obtusely angulate. Elytra ~0.68–1.9 times as long as combined width and 1.35–2.75 times as long as pronotum, sides strongly rounded to subparallel; apices complete and conjointly rounded or independently rounded to squarely truncate, exposing all or part of pygidium (tergite VII) and sometimes part of tergite VI (some Epuraeinae, Carpophilinae and Nitidulinae), or exposing two or three tergites (Cillaeinae); suture not deflected near apex; disc often irregularly punctate, but with ~10 puncture rows, usually alternating with rows of smaller punctures and no scutellary striole in Cillaeinae and number of Nitidulini; epipleura often sharply defined but incomplete. Mesoventrite with anterior edge on the same plane as the metaventrite in Carpophilinae, Meligethinae, Cillaeinae and Cryptarchinae, on a slightly more dorsal plane in Epuraeinae and many Nitidulinae and on a much higher plane in some Nitidulinae and in Cybocephalinae, without anterior paired procoxal rests; usually with a median, subtriangular prosternal rest or a broad prothoracic rest or prepectus, which continues onto the mesanepisterna (in *Cybocephalus* almost the entire surface of the mesoventrite is occupied by paired impressions separated by a low carina and acting as housings for the procoxae. Mesoventral process usually truncate or emarginate at apex. Mesocoxae with exposed trochantins. Mesocoxal cavities usually transverse and moderately broadly separated (narrowly separated in Cychramptodini and Lawrencerosini), partly closed laterally by mesanepisterna and mesepimera. Mesometaventral junction a straight or anteriorly curved line, with meso- and metaventrites abutting or slightly overlapping. Metaventrite flat to slightly convex, almost always with long discrimen (absent in Cybocephalinae); postcoxal lines often present, which are tangential to the posterior edges of the mesocoxal cavities and curve posterolaterally to meet the lateral edges of the metaventrite, enclosing two areas usually referred to as an axillary spaces; exposed portion of metanepisternum moderately elongate. Metacoxae narrowly to moderately widely separated; plates absent. Metendosternite with short, broad stalk, long lateral arms, laminae very narrow or usually absent, anterior process short to moderately long or absent and anterior tendons well separated, sometimes on lateral arms. Hind wing with very long apical field usually containing two post-radial sclerites and two radial extensions; radial cell reduced or absent; cross-vein r3 absent; R-M loop forming narrowly acute angle; medial field with three or fewer free veins and no medial fleck; wedge cell absent; anal embayment notch-like. Outer edge of protibia often with finely crenulate or denticulate ridge (more coarsely dentate in *Pria* Stephens) and outer edges of meso- and metatibiae often with two rows of socketed spines. Tarsi almost always 5-5-5 in both sexes (4-4-4 in *Cybocephalus*); tarsomeres 1–3 usually with setose ventral lobes beneath and often greatly expanded; tarsomere 4 reduced but not enclosed within lobe of preceding tarsomere; pretarsal claws usually simple (toothed in *Trimenus* Murray and most species of *Aethina* Erichson.

Abdomen with five free ventrites; ventrite 1 usually subequal to or slightly longer than 2 (much longer in Carpophilinae); intercoxal process subacute to broadly rounded or truncate, with or without recurved or somewhat angulate postcoxal lines; ventrite 5 (sternite VII) forming hypopygidium, which may be apically concave or emarginate in male. Functional spiracles usually on abdominal segments I–VI (on segments I–V in *Cybocephalus*). Tergite VII well sclerotised and often enlarged forming pygidium; preceding two or three tergites also well sclerotised in some groups. Tergite VIII in male exposed, sometimes visible ventrally; sternite VIII reduced and membranous. Anterior edge of segment IX in male with spiculum gastrale; tergite IX divided into two parts and segment X membranous. Aedeagus of cucujiform type, symmetrical; anterior edge of tegmen with or without short to moderately long strut; parameres fused to phallobase but more or less separated from one another in Carpophilinae and Epuraeinae, mostly fused together forming an emarginate tegmen in Meligethinae and completely fused together in other groups; penis usually with single anterior strut and with endophallic sclerites. Ovipositor moderately elongate, wide at base and gradually but distinctly narrowing apically, lightly sclerotised except for baculi and sometimes coxital apices, paraprocts shorter than coxites, each of which is usually subdivided into basal and apical lobes; coxital apices sometimes acute or with recurved teeth; styli often well developed and apical or subapical, sometimes reduced or absent. [Kirejtshuk 1986a, 1986b, 1986c, 1986d, 1987, 1990, 1996, 1998a, 2004, 2006a, 2006b, 2008, 2010, 2011a, 2011b; Kirejtshuk and Lawrence 1990, 1992a, 1992b, 1999; Kirejtshuk and Leschen 1998; Audisio 1993; Jelinek 1999; Lawrence *et al.* 1999b; Kirejtshuk and Kvamme 2001, 2002; Habeck 2002b; Cline 2008; Audisio *et al.* 2009; Jelinek *et al.* 2010].

Larva (Figs 84F, H)

Body elongate, parallel-sided to fusiform, slightly to strongly flattened; usually not or slightly, ventrally curved (dorsally curved in *Pocadius*; usually not or slightly, ventrally curved (dorsally curved in *Pocadius* Erichson); dorsal surfaces weakly to moderately sclerotised, sometimes with head, tergum IX or VIII and IX only pigmented, or with tergal plates on all body segments;

surfaces often smooth but sometimes granulate or with patches of asperities or setiferous tubercles; rarely coated with slime produced by cuticular glands (few Nitidulinae); vestiture of short, simple or modified setae.

Head protracted and prognathous, transverse and somewhat flattened. Epicranial stem usually absent and frontal arms usually lyriform and well separated at base (epicranial stem present in *Pocadius* and frontal arms contiguous at base in *Cybocephalus*). Median endocarina absent. Stemmata two to four on each side. Frontoclypeal suture absent. Labrum usually free (fused to clypeus in Meligethinae and *Cybocephalus*), strongly transverse and usually broadly rounded or truncate anteriorly (emarginate in *Cybocephalus*). Antennae moderately long, 3-segmented, with conical sensorium usually less than half as long as antennomere 3. Mandibles symmetrical, broadly triangular at base, with strongly projecting ventral condyle, usually bidentate with serrate or complexly lobed incisor edge or edges (serrations continue on outer surface of mandibular apex in some Cyllodini), unidentate in Meligethinae and *Cybocephalus*; without accessory ventral process; mola almost always well developed with transverse rows of asperities (reduced in *Cybocephalus*); prostheca almost always complex, consisting of a series of fringed processes (Fig. 39A) or clusters of spines (a simple hyaline lobe in Meligethinae, absent in *Cybocephalus*). Ventral mouthparts (Fig. 49F) very strongly retracted, with the hypostomal cavity narrowing posteriorly and extending to the posterior edge of the head; maxillary articulating area usually reduced or absent; cardines strongly oblique, almost longitudinal; stipites slightly longer than wide; mala usually apically rounded and setose (Fig. 38E) with a bidentate uncus located on inner edge well away from apex (mala without or with few setae and without uncus in Meligethinae and *Cybocephalus*); maxillary palps almost always 3-segmented, with distinct palpiger (palpiger indistinct in Meligethinae; palp 2-segmented with distinct palpiger in *Cybocephalus*). Labium usually consisting of prementum, mentum and submentum (prementum and postmentum in *Cybocephalus*); ligula simple and rounded or absent; labial palps 1-segmented and narrowly to moderately widely separated. Hypopharyngeal sclerome a transverse bar or molar-like structure. Hypostomal rods usually absent (moderately long and diverging in *Cybocephalus*); ventral epicranial ridges absent. Gular region very short or absent.

Tergal sclerites distinct, indistinct or absent. Legs relatively short and broad, 5-segmented including short, claw-like pretarsus, which is sometimes enlarged at base and usually bears two setae (one seta in *Pocadius*; one of two setae modified as adhesive process in Meligethinae and in some *Cybocephalus*); coxae moderately widely separated. Abdominal segments I–VIII with lateral lobes or processes in *Amphicrossus*, *Cryptarcha* and some Nitidulinae). Segment IX without appendages slightly shorter than VIII, tergum usually extending slightly onto ventral surface, posterior edge usually with simple, bifid or branched urogomphi and often with additional pregomphi anterior to their bases (Fig. 52M) (urogomphi absent in Meligethinae and *Cybocephalus*); sternum IX exposed, simple, not enclosed by sternum VIII; segment X sometimes partly visible from above, pygopod-like and posteroventrally oriented. Spiracles annular-biforous, situated at apex of short to long spiracular tubes in Cryptarchinae, Carpophilinae, Amphicrossinae and several Nitidulinae. [Hayashi 1978; Audisio 1993; Lawrence 1991b; Kirejtshuk *et al.* 1997; Carlton and Leschen 2007; Jelinek *et al.* 2010].

Classification and World Distribution

Nitidulidae is a large, cosmopolitan family with ~350 genera and 44500 species. Ten subfamilies are recognised, eight of which occur in Australia. Non-Australian subfamilies include Calonecrinae (South-east Asia) and Maynipeplinae (Africa). [Kirejtshuk 1998b; Jelinek *et al.* 2010].

Australian Fauna

The Australian nitidulid fauna is large and diverse, but at least two subfamilies are poorly represented here, considering their diversity in the Holarctic and Afrotropical regions. **Carpophilinae** includes the widely distributed *Urophorus humeralis* (Fabricius) and several species of Carpophilus (Fig. 73C), and the only species of **Meligethinae** is *Pria pumila* Cooper. **Amphicrossinae** includes a few species of *Amphicrossus*, and **Cybocephalinae** is represented by three species of *Cybocephalus*. The **Epuraeinae** is moderately well represented, with ten genera; included are the widely distributed fruit beetles *Epuraea* (*Haptoncus*) *ocularis* (Fairmaire) and *E.* (*H.*) *luteolus* (Erichson), several endemic *Epuraea* species, several other genera of Epuraeini and four genera of Taenioncini. **Cillaeinae** and **Cryptarchinae** are both moderately well represented, including several species of *Brachypeplus* (Fig. 83D) and *Cryptarcha* (Fig. 73D), respectively, plus several endemic or Indo-Australian genera. The largest and most diverse subfamily is the **Nitidulinae**, which has been divided into five tribes and several 'complexes' within the nominate tribe Nitidulini. The classification of this subfamily has been confused, and the only tribe based on a cladistic analysis is the Cyllodini, which includes the *Cyllodes* and *Oxycnemus* groups of genera (Leschen 1999). Two other tribes that appear to be monophyletic are the Cychramptodini and Lawrencerosini, both based on highly modified adults that are adapted to survive a close association with aggressive ant species; cychramptodines are predators of scale insects (Coccoidea) tended by ants and lawrencerosines are apparently symphiles in ant nests. These tribes, as well as the Cychramini and the exotic Mystropini, are recognised in the recent work on nitidulid genera (Kirejtshuk 2008) and are used in the list below. The Nitidulini, however, is certainly a paraphyletic assemblage and the currently recognised generic complexes, genera and subgenera are still in a state of flux.

Nitidulid adults are highly varied in form, but are usually oblong to ovate (less commonly elongate and parallel-sided), strongly convex to flattened, usually brown or black in colour, but occasionally bicoloured, and glabrous or pubescent. The antennal club is almost always distinct, the frontoclypeal suture is almost always absent, the labrum often emarginate or bilobed (sometimes fused to clypeus), the maxillary galea is always absent and the palp not expanded apically, metacoxae are well separated, tibiae are often expanded and serrate or spinose, the first four tarsomeres often have expanded, setose lobes, and the elytra are sometimes truncate, sometimes exposing one to several tergites. *Cybocephalus* is a minute, clambid-like form with only four tarsomeres. Nitidulid larvae are oblong to elongate and subcylindrical to flattened, with dorsal surfaces often granulate or tuberculate and urogomphi (absent in *Cybocephalus*) often accompanied by a second pair of projections (pregomphi). Most larvae have the frontal arms widely separated at base (contiguous at base in *Pocadius* and *Cybocephalus*) and the ventral mouthparts deeply recessed, with longitudinally oblique cardines and a blunt mala. The mandibles are usually distinctive in having a complex prostheca with fringes of comb-hairs, but this is absent in *Cybocephalus*. Many nitidulid larvae have all spiracles (Fig. 84F) or just the eighth pair (Fig. 52M) located at the ends of tubular processes.

Nitidulids have a wide variety of habits and habitats and those in Australia may be found in leaf litter (*Stelidota*, *Thalycrodes*), rotten fruits (*Lasiodites*, *Carpophilus*), flowers of palms and *Pandanus* (*Amystrops*), dicotyledonous flowers (*Idaethina*), seeds of Proteaceae and Sterculiaceae (*Australaethina*), rotten cactus (introduced *Camptodes*), mushrooms (*Cychramus*, *Pallodes*), puffballs (*Pocadius*), stinkhorns (*Gymnocychramus*, Fig. 73H), male cones of cycads (*Circopes*) or *Araucaria* (*Testudorea*), fermenting bark or tree wounds (*Brachypeplus*, *Cryptarcha*, *Amphicrossus*), or preying on scale coccoid or aleyrodoid Hemiptera (*Cychramptodes*, *Cybocephalus*). Several introduced species, such as *Urophorus humeralis*, *Carpophilus dimidiatus* (Fabricius), *C. hemipterus* (Linnaeus), *C. marginellus* Motschulsky and *C. mutilatus* Erichson are pests of stored dried fruits, but may also attack ripe fruits in orchards. Another introduced species, *Aethina tumida* Murray, has recently become a serious pest of bee hives (Menier and Jouan 2003; Stedman 2006; Trend 2007). *Cychramptodes murrayi* Reitter (Fig. 87A–B) may be seen feeding on the honeydew produced by the large females of *Cryptes baccatus* (Hemiptera: Coccoidea); they are protected from the tending ants by their ability to conceal all moving parts beneath the explanate and deflexed pronotum and elytral epipleura, and their larvae feed internally on the female and her brood (Kirejtshuk and Lawrence 1992). [Lawrence 1988a; 1991b; Gough and Hamacek 1989; Leschen and Marris 1995; Leschen 1999; Kirejtshuk and Kvamme 2001].

Australian Subfamilies, Tribes and Genera

Epuraeinae: Epuraeini: *Amystrops* Grouvelle, *Ceratomedia* Kirejtshuk, *Epuraea* (*Epuraea*) Erichson, *Epuraea* (*Blackburnaea*) Kirejtshuk & Kvamme, *Epuraea* (*Ceratomedia*) Kirejtshuk, *Epuraea* (*Haptoncus*) Murray, *Mystronoma* Kirejtshuk, *Stauromenus* Kirejtshuk & Kvamme, *Trimenus* Murray. **Taenioncini:** *Csiromenus* Kirejtshuk & Kvamme, *Eutaenioncus* Kirejtshuk & Kvamme, *Taeniolinus* Kirejtshuk, *Taenioncus* Kirejtshuk.

Carpophilinae: *Carpophilus* Stephens, *Urophorus* Murray.

Amphicrossinae: *Amphicrossus* Erichson.

Meligethinae: *Pria* Stephens.

Nitidulinae: Nitidulini: *Aethina* (*Circopes*) Reitter, *Aethina* (*Cleidorura*) Kirejtshuk & Lawrence, *Aethina* (*Idaethina*) Gemminger & Harold, *Australycra* Kirejtshuk & Lawrence, *Gaulodes* Erichson, *Kryzhanovskiella* Kirejtshuk, *Macleayania* Kirejtshuk, *Neopocadius* (*Australaethina*) Kirejtshuk & Lawrence, *Nitidula* Fabricius, *Omosita* Erichson, *Ostomarcha* Kirejtshuk, *Perilopsis* (*Testudorea*) Kirejtshuk, *Phenolia* (*Aethinodes*) Blackburn, *Phenolia* (*Lasiodites*) Jelinek, *Phenolia* (*Plesiothina*) Kirejtshuk, *Pocadius* Erichson, *Prometopia* Erichson, *Rixerodes* Kirejtshuk & Lawrence, *Soronia* Erichson, *Stelidota* Erichson, *Thalycrodes* Blackburn. **Cychramini:** *Axychramus* Kirejtshuk, *Ceratochramus* Kirejtshuk, *Cychramus* Kugelann, *Xenostrongylus* Wollaston. **Cyllodini:** *Camptodes* Erichson, *Cyllodes* Erichson, *Gymnocychramus* Lea, *Neopallodes* Reitter, *Pallodes* Erichson. **Cychramptodini:** *Cychramptodes* Reitter, *Cylindroramus* Kirejtshuk & Lawrence, *Gonoglypha* Kirejtshuk, *Miskoramus* Kirejtshuk & Lawrence. **Lawrencerosini:** *Koryaga* Kirejtshuk, *Koryaginus* Kirejtshuk, *Krakingus* Kirejtshuk, *Lawrencerosus* Kirejtshuk.

Cillaeinae: *Brachypeplus* Erichson, *Brittonema* Kirejtshuk, *Cillaeus* Laporte, *Ithyphenes* Murray, *Onicotis* Murray.

Cryptarchinae: Cryptarchini: *Cryptarcha* Shuckard, *Kaszabena* Kirejtshuk. **Platyarchini:** *Amlearcha* Kirejtshuk, *Platyarcha* Kirejtshuk.

Cybocephalinae: *Cybocephalus* Erichson.

80. BOTHRIDERIDAE

Adult (Fig. 87G, I–J)

Length 1.4–12 mm. Body usually ~1.85–4.5 times as long as wide (somewhat shorter and broader in *Dastarcus* Walker and *Xylariophilus* Pal and Lawrence), cylindrical to slightly flattened; colour uniformly yellow to red or black, with elytra bicolored in *Xylariophilus*; dorsal surfaces usually either glabrous or clothed with very short, fine hairs (vestiture of long, fine hairs in *Teredolaemus* Sharp and *Xylariophilus* or with scale-like setae in *Dastarcus*.

Head prognathous or slightly declined, about as long as wide, usually not constricted behind eyes (with a week postocular impression in *Deretaphrus* Newman. Eyes entire (absent in *Anommatus* Wesmael), usually strongly protuberant, finely or coarsely facetted, without or with very short, fine interfacetal setae. Antennal insertions widely separated and exposed from above; subantennal grooves usually well developed (weakly developed or absent in *Teredolaemus* and *Xylariophilus*). Frontoclypeal suture distinct and acuate. Labrum free, transverse, usually with long, subparallel outer tormal arms; anterior edge truncate. Antennae almost always 11-segmented with a strong, compact, 2-segmented club (club 3-segmented, weaker and less compact in *Deretaphrus* and *Sysolus* Grouvelle). Mandible usually unidentate or obliquely bidentate with well developed subapical tooth; mola and prostheca usually well developed, both reduced in *Deretaphrus* and mola replaced by fleshy lobe in *Dastarcus*. Maxilla with galea moderately broad lacinia broad and setose; lacinia narrower with unidentate or bidentate uncus; apical palpomere fusiform. Ligula usually simple and truncate or slightly expanded; apical labial palpomere fusiform. Gular sutures strongly convergent or anteriorly fused in most Bothriderinae. Corpotentorium usually slender, with (*Teredolaemus* and *Xylariophilus*) or without median process; sometimes absent. Cervical sclerites highly reduced or absent.

Pronotum ~0.7–1.5 times as long as wide, almost always widest anteriorly (not so in Dastarcus) and usually as wide at base as combined elytral bases (distinctly narrower in *Deretaphrus* and several Bothriderinae); sides straight, weakly rounded or sinuate, not explanate; lateral pronotal carinae usually complete (incomplete anteriorly in Deretaphrus), with or without raised margin or bead, the edges simple; anterior angles not produced or produced and acute; posterior angles right or oblique to slightly acute; posterior edge straight or biemarginate; disc simple in *Anommatus*, *Xylariophilus*, *Teredolaemus* and *Sosylus* Erichson, with median groove in *Asosylus* Grouvelle and *Deretaphrus*, with two to four carinae in *Machlotes* Pascoe and *Erotylathris* Motschulsky, with two pairs of undulating, longitudinal grooves in *Dastarcus*, and with various curved impressions or channels in other bothriderine genera. Prosternum in front of coxae at least twice as long as procoxal cavity, flattened or slightly convex. Prosternal process almost always complete (but interrupted at middle in *Sosylus*), flat or slightly concave, parallel-sided or weakly to strongly expanded apically, with narrowly rounded to truncate apex. Procoxae with long concealed lateral extensions; trochantins concealed. Procoxal cavities circular, partly confluent in *Sosylus*, narrowly separated in *Anommatus*, *Teredolaemus*, *Xylariophilus*, *Deretaphrus* and *Sysolus*, moderately to very widely separated in most Bothriderinae (although sometimes with transverse groove on prosternal process), narrowly open externally in *Anommatus*, *Teredolaemus* and *Xylariophilus*, narrowly to broadly closed in Bothriderinae, without narrow lateral extensions, usually internally open.

Scutellar shield often small (absent in *Anommatus*), abruptly elevated, posteriorly broadly rounded to obtusely angulate. Elytra ~1.3–3.35 times as long as combined width and 1.6–3.2 times as long as pronotum, sides subparallel; disc often with ten puncture rows and no scutellary striole, but puncture rows sometimes partly effaced or fewer in number and elytra with variously developed longitudinal costae; apical declivity present in *Sosylus* and *Sysolus*; epipleura narrow and incomplete; elytral apex often forming interlocking device with last ventrite. Mesoventrite with anterior edge usually on same plane as metaventrite (on slightly higher plane in *Deretaphrus*); procoxal rests absent; usually with a broad prothoracic rest or prepectus, which continues onto the mesanepisterna. Mesoventral process usually broadly rounded to truncate (emarginate in *Deretaphrus*). Mesocoxae almost always with concealed trochantins (exposed in *Sysolus*). Mesocoxal cavities circular, narrowly to broadly separated, almost always closed laterally by meeting of meso- and metaventrites (very narrowly open in *Xylariophilus* and *Sysolus*). Mesometaventral junction either monocondylic (*Anommatus*, *Xylariophilus*, *Teredolaemus*, *Deretaphrus*) or a straight line with meso- and metaventrites abutting or slightly overlapping. Metaventrite flat to moderately convex, usually with short to moderately long discrimen; straight or slightly curved postcoxal lines present in *Pseudobothrideres* Grouvelle and *Erotylathris* Motschulsky; exposed portion of mesanepisternum very long and narrow. Metacoxae strongly transverse and narrowly separated to slightly transverse and widely separated; plates absent. Metendosternite usually with very short, broad stalk (longer and narrower in *Teredolaemus* and *Deretaphrus*), long lateral arms, well developed or very narrow laminae and anterior process either short or absent and anterior tendons widely separated. Hind wing with relatively short apical field usually containing one or two post-radial sclerites and sometimes one radial extension; radial cell usually incomplete or absent; complete and slightly elongate in *Deretaphrus*, with inner posterobasal angle right; cross-vein r3 usually absent (very short in *Deretaphrus*, moderately long in *Teredolaemus* and *Sysolus*); R-M loop relatively broad and basal portion of RP long in *Deretraphrus* and *Dastarcus* (R-M loop narrowly acute and base of RP short or absent in *Teredolaemus* and *Sysolus*); medial spur usually moderately long and medial embayment usually present; medial field with five free veins in *Deretaphrus*, three or four free veins in *Teredolaemus*, *Sysolus* and most Bothriderinae, and one in *Xylariophilus*; medial fleck present in *Teredolaemus*, *Xylariophilus* and *Sysolus*; wedge cell absent; anal embayment notch-like; wings absent in *Anommatus*. Legs well developed; trochanterofemoral joint strongly oblique with tibiae in contact with coxae, somewhat internalised and not visible from below in Bothriderinae. Tibial apices often somewhat widened and outer apical angle of protibia usually produced and acute; tibial spurs paired and often unequal in length. Tarsi almost always 4-4-4 (3-3-3 in *Anommatus*); tarsomeres simple; pretarsal claws simple, and empodium reduced.

Abdomen with five free ventrites; ventrite 1 longer than 2, sometimes with postcoxal lines in Bothriderinae; intercoxal process subacute to truncate. Functional spiracles on abdominal segments I–VII. Sternites VIII in male with short anterior strut in *Teredolaemus* and *Xylariophilus*. Segment IX in male usually produced to form narrowly acute, truncate or slightly expanded subgenital plate (with paired anterior struts in *Deretaphrus*); tergite IX usually divided into two parts by segment X, which may be membranous or lightly sclerotised with lateral baculi (laterotergites meet at midline in *Sosylus* in front of segment X, which is surrounded by heavily sclerotised rim). Aedeagus of cucujiform type, symmetrical; anterior edge of tegmen usually with long slender strut (short and acute in *Dastarcus*, absent in *Deretaphrus*); parameres well developed, articulated and setose in *Xylariophilus* and *Teredolaemus*, free with basal, hook-like processes in *Deretaphrus*, articulated but fused together in *Dastarcus*, *Sosylus* and Bothriderinae; penis long and slender, with short to long paired anterior struts. Sternite VIII in female with long spiculum ventrale. Ovipositor moderately elongate to very long and narrow, lightly sclerotised except for baculi; paraprocts usually subequal to coxites (much longer than coxites in *Deretaphrus* and *Dastarcus*); coxites subdivided with relatively long apical or subapical styli. [Carter and Zeck 1937; Pal and Lawrence 1986; Ślipiński *et al.* 1989, 2010b; Lawrence *et al.* 1999b; Philips and Ivie 2002].

Larva

Free-living larvae (*Xylariophilus*, *Teredolaemus*, *Anommatus*) elongate and more or less parallel-sided, slightly flattened; dorsal surfaces granulate, asperate or tuberculate, with vestiture consisting of simple, apically expanded or scale-like setae. Ectoparasitic larvae physogastric, lightly pigmented, with vestiture of scattered, simple setae. Triungulinid larvae moderately flattened and distinctly pigmented dorsally and ventrally, with vestiture of sparse simple setae or stout dorsal and ventral spines (*Sosylus*).

Head prognathous and protracted, usually transverse (slightly elongate in *Bothrideres* triungulinid). Epicranial suture often absent, short in free-living *Anommatus* and *Teredolaemus*; frontal arms lyriform in free-living *Anommatus* and *Teredolaemus*; ectoparasitic larvae sometimes with no epicranial stem, but with straight, widely separated frontal. Median endocarina absent. Stemmata absent in *Anommatus* and in ectoparasitic larvae; six on each side in *Teredolaemus* and *Xylariophilus*, one or two (*Dastarcus*) triungulinids. Frontoclypeal suture very weak or absent. Labrum free in free-living larvae and usually ectoparasitic larvae, completely fused to head capsule in triungulinid. Antennae relatively short, 3-segmented in free-living larvae, with conical sensorium slightly longer than apical antennomere, 1- or 2-segmented and separated from mandibular articulation by broad strip of cuticle in ectoparasitic larva, with sensorium inserted on antennomere 1; antennae in triungulinid short and 2-segmented with antennomere 1 very short and antennomere 2 elongate and longer than sensorium. Mandibles symmetrical, broad at base and tridentate in free-living larvae, with accessory ventral process (reduced in *Anommatus*), well developed, tuberculate mola and prostheca reduced or absent; those in ectoparasitic larvae broad at base and unidentate or bidentate, without accessory ventral process mola or prostheca (base of mandible with slender, acute hyaline lobe in *Deretaphrus*); those in triungulinid narrow and sickle-shaped. Ventral mouthparts retracted in free-living and ectoparasitic larvae, protracted in triungulinid; maxillary articulating area usually present in free-living and ectoparasitic larvae (indistinct in *Sosylus*) and absent in triungulinid; free-living larvae with cardo transverse, stipes about as long as wide, mala truncate (*Anommatus* and *Xylariophilus*) or falciform (*Teredolaemus*) and palp 3-segmented; ectoparasitic larva of *Deretaphrus* with cardo transverse, stipes about as long as wide, mala falciform and palp 2-segmented; cardo and stipes indistinct and mala and palps absent or highly reduced in *Sosylus*; in cardo and stipes distinct in triungulinid of *Sosylus* or fused with labrum in that of *Dastarcus*; mala broadly rounded in triungulinid *Sosylus* or strongly reduced in *Dastarcus*; palps in triungulinid long and 2- or 3-segmented. Labium in free-living larvae with prementum, mentum and submentum and with ligula shorter than labial palps, which are 2-segmented and narrowly separated; most ectoparasitic larvae with ligula usually longer than short, 2-segmented palps; labium in ectoparasitic larva of *Sosylus* not subdivided and palps and ligula absent; ligula indistinct and labial palps 2-segmented and approximate in triungulinids. Hypopharyngeal sclerome present and hypostomal rods long and diverging in free-living larvae; both absent in ectoparasitic larvae and their triungulinids; ventral epicranial ridges absent.

Legs well developed and 5-segmented in free-living larvae, ectoparasitic larvae of *Deretaphrus* and all triungulinid larvae; those of other ectoparasitic larvae highly reduced; pretarsus, if present, with one seta. Abdominal terga in free-living larvae sometimes extending laterally to form short processes; terga without sclerotised plates but with transverse rows of scale-like setae or asperities, especially prominent on terga V–IX in *Anommatus*. Tergum IX in free-living larvae complex, with median elevation (*Xylariophilus*), a pair of upturned urogomphi with accessory processes, and additional pairs of processes anterior to urogomphi; in ectoparasitic larvae tergum IX short, without or with short, simple, upturned urogomphi; tergum IX somewhat longer in triungulinid, without urogomphi but with pair of long-terminal setae; sternum IX simple. Segment X in free-living larvae oval and ventrally oriented, in ectoparasitic larvae oval and posteroventrally oriented, and in triungulinid long and posteriorly oriented. Spiracles annular–biforous in *Xylariophilus* and *Teredolaemus*, with accessory openings much longer than peritreme and placed at the sides of spiracular tubes, annular in *Anommatus* and all ectoparasitic forms, with heavily sclerotised peritreme in ectoparasitic larvae, absent on thorax and segment VIII in *Sosylus*, placed at the ends of short spiracular tubes in *Dastarcus* triungulinid. [Böving and Craighead 1931; Piel 1938; Sen Gupta and Crowson 1971; Dajoz 1968, 1977; Sen Gupta and Crowson 1973; Pal and Lawrence 1986; Lawrence 1991b; Lawrence *et al.* 1999a; Ślipiński *et al.* 2010b].

Classification and World Distribution

The family Bothrideridae occurs in all major zoogeographic regions and comprises ~400 species in 38 genera. Thee are four recognised subfamilies, all of which occur in Australia. [Ślipiński *et al.* 2010b].

Australian Fauna

The subfamily **Xylariophilinae** contains the singe genus *Xylariophilus*, with three described species: *X. comatus* Pal & Lawrence from north-eastern India, *X. bicoloripennis* Pal & Lawrence from southern QLD, and *X. honoratus* Pal & Lawrence from northern QLD. Undescribed species have also been seen from Malaysia, Vanuatu and Papua New Guinea. The closely related subfamily **Teredinae** is represented by *Teredolaemus leai* (Grouvelle) (Fig. 87J), which occurs in TAS, VIC and NSW, undescribed *Teredolaemus*, and an undescribed species of *Sysolus* Grouvelle from northern QLD. The introduced *Anommatus duodecimstriatus* (Müller) (Fig. 87I), which has been collected in TAS and on Norfolk Island, is our only representative of the primarily Palearactic **Anommatinae**. Within the subfamily **Bothriderinae**, the genus *Deretaphrus* contains several species distributed across the southern part of the continent, and the only non-Australian species in the genus, *D. oregonensis* Horn, occurs in north-western North America. Also in the subfamily are species of *Ascetoderes* (Fig. 87G) and several other genera mentioned below.

Bothriderids are oblong to narrowly elongate, subcylindrical to flattened beetles, usually subglabrous, but sometimes clothed with decumbent and erect hairs or rarely (*Dastarcus*) scale-like setae. Adults are superficially similar to some colydiine Zopheridae, from which they differ in having exposed antennal insertions and a cucujoid or ring-type of aedeagus. Antennae are relatively short, usually with a compact 1- or 2-segmented club (weakly 3-segmented in *Deretaphrus*), the frontoclypeal suture is usually present (sometimes indistinct), tibiae have expanded and spinose apices and sometimes enlarged and unequal tibial spurs, the trochanterofemoral articulation is strongly oblique, so that a portion of the femur is in contact with the coxa, and in most species the trochanter is reduced and more or less concealed by the femoral base. Larvae of Xylariophilinae, Teredinae and Anommatinae are typically cucujoid with a granulate or tuberculate upper surface and well developed mandibular molae and urogomphi; while those of Bothriderinae have no molae and minute or no urogomphi, and they are either lightly sclerotised, physogastric ectoparasitoids or minute, fusiform triungulinoids. *Anommatus duodecimstriatus* occurs in soil and leaf litter and is probably saprophagous. *Xylariophilus* species feed on the fruiting bodies of pyrenomycetous fungi in rainforests of QLD (Pal and Lawrence 1986), and *Teredolaemus* are known to inhabit the tunnels of ambrosia beetles (Curculionidae: Platypodinae) where they probably feed on the yeasts cultivated by their hosts (Lawrence 1985b). Bothriderinae feed on larvae and pupae of various wood-boring insects, particularly Coleoptera (Buprestidae, Bostrichidae, Ptinidae, Cerambycidae, Curculionidae) but also Hymenoptera (Xiphydriidae, xylocopine Apidae) (Craighead 1920; Piel 1938). These predacious or ectoparasitic forms construct silken cocoons or waxen chambers for pupation (Burke 1919). Species of *Sosylus* appear to be restricted to the tunnels of ambrosia beetles and are known to have triungulinoid first instars and ectoparasitic late instars (Browne 1962; Roberts 1980).

Australian Subfamilies, Tribes and Genera

Anommatinae: *Anommatus* Wesmael.
Xylariophilinae: *Xylariophilus* Pal & Lawrence
Teredinae: *Teredolaemus* Sharp, *Sysolus* Grouvelle.
Bothriderinae: *Ascetoderes* Pope, *Asosylus* Grouvelle, *Dastarcus* Walker, *Deretaphrus* Newman, *Erotylathris* Motschulsky, *Leptoglyphus* Sharp, *Machlotes* Pascoe, *Pseudobothrideres* Grouvelle, *Sosylus* Erichson.

81. CERYLONIDAE

Adult (Figs 72D, 87C–F)

Length 0.9–3 mm . Body usually ~1.25–2.5 times as long as wide, broadly ovate to elongate and strongly convex to distinctly flattened; colour uniformly yellow to red or black, with elytra bicolored in some *Hypodacnella* Ślipiński and *Australiorylon* Ślipiński; dorsal surfaces either glabrous or clothed with short and/or long fine hairs.

Head globular, moderately declined, often deeply retracted into prothorax, without postocular constriction. Occipital region with fine transverse carina in *Murmidius*. Eyes well developed or somewhat reduced, subcircular to distinctly higher than wide, moderately protuberant, finely or coarsely facetted, with short, fine interfacetal setae. Antennal insertions widely separated and exposed from above; subantennal grooves well developed in Euxestinae, Murmidiinae and Ostomopsinae, short or weakly developed in Ceryloninae. Frontoclypeal suture distinct in Euxestinae, Murmidiinae and Ostomopsinae, absent in Ceryloninae. Labrum free, strongly transverse to elongate and apically narrowed' often with long, subparallel outer tormal arms and with or without labral rods; apex truncate, rounded or acute. Antennae usually 10-segmented with a 1-segmented club (11-segmented with a 3-segmented club in *Protoxestus* Sen Gupta and Crowson, 11-segmented with a 2-segmented club in *Lawrenciella* Ślipiński, *Australiorylon* Ślipiński and some species of *Philothermus* Aubé, 9-segmented with a 2-segmented club in *Cautomus* Sharp, and 8-segmented with a 1-segmented club in *Mychocerus* Erichson). Mandible usually tridentate (bidentate or drawn out to form piercing blade in Ceryloninae), usually with basal mola and fringed prostheca (poorly developed or obsolete in *Cautomus*). Maxilla usually with brush-like galea and slender lacinia, the latter with unidentate uncus in Euxestinae (maxillary lobes styliform in *Cautomus*, Fig. 16A); apical palpomere usually cylindrical or fusiform (aciculate, much narrower than preapical palpomere in Ceryloninae). Mentum trapezoidal, rarely elongate; ligula bilobed in Murmidiinae, reduced in Euxestinae and Ostomopsinae, well developed and membranous in Ceryloninae (elongate in *Cautomus*); apical labial palpomere aciculate in Ceryloninae. Gular sutures widely separated. Corpotentorium slender, with median process. Cervical sclerites absent.

Pronotum ~0.35–0.9 times as long as wide, widest posteriorly or at middle and as wide at base as combined elytral bases; sides weakly rounded or subparallel; lateral pronotal carinae complete, usually with raised margin or bead, the edges usually simple (Fig. 19D) (finely denticulate in *Ostomopsis* and *Cautomus*); anterior angles usually not produced (produced and acute in *Ostomopsis*, produced and broadly rounded in *Cerylonopsis* Heinze); posterior angles more or less right; posterior edge often biemarginate or with distinct mesal lobe; disc usually simple (with paired anterolateral cavities for receiving antennal club in *Murmidius* (Fig. 19F); with paired, shallow, sublateral impressions in *Lawrenciella*). Prosternum more than twice as procoxal cavity (Fig. 26B), flattened or slightly convex; with paired longitudinal lines and strongly produced anterior edge concealing head from below in *Murmidius*. Anterior portion of hypomera in *Mychocerus* with deep antennal cavities (Fig. 19G). Prosternal process complete, flat, parallel-sided or weakly to strongly expanded apically, with broadly rounded, truncate or slightly emarginate apex. Procoxae with long concealed lateral extensions; trochantins concealed. Procoxal cavities circular, moderately to very widely separated, externally open in *Murmidius*, *Ostomopsis*, *Mychocerus* and *Lawrenciella*, externally closed in Euxestinae and other Ceryloninae, internally open.

Scutellar shield usually well developed, more or less abruptly elevated and broadly rounded to obtusely angulate posteriorly (narrow and acute in *Murmidius* and *Lawrenciella*). Elytra ~0.9–1.9 times as long as combined width and 1.9–3.5 times as long as pronotum, sides usually slightly to strongly rounded (subparallel in *Metacerylon* Grouvelle); disc usually with eight or nine puncture rows and no scutellary striole (punctures irregular or rows not clearly indicated in some *Hypodacnella*), each with four longitudinal costae in *Lawrenciella*; suture slightly deflected near apex in *Ostomopsis*; epipleura complete or incomplete, with small cavity near apex in *Metacerylon*. Mesoventrite with anterior edge usually on same plane as metaventrite (on higher plane in *Murmidius*); procoxal rests usually absent (weakly developed and more or less horizontal in Ostomopsis and *Lawrenciella*), sometimes with a prothoracic rest or prepectus. Mesoventral process broadly rounded to truncate. Mesocoxae with concealed trochantins. Mesocoxal cavities circular, moderately to very broadly separated, closed laterally by meeting of meso- and metaventrites. Mesometaventral junction a straight line with meso- and metaventrites abutting or slightly overlapping. Metaventrite flat to slightly convex, usually without or with very short discrimen (very long in *Murmidius*); curved postcoxal lines sometimes present (Fig. 26B) present; exposed portion of mesanepisternum very long and narrow or absent (concealed by elytra). Metacoxae slightly to strongly transverse and widely separated; plates absent. Metendosternite with very short and broad stalk and long lateral arms, with or without reduced, narrow laminae, without or with very short anterior process and anterior tendons well separated but either on anterior process or at bases of lateral arms. Hind wing with long apical field usually without sclerites or radial extensions (post-radial sclerite and vaguely indicated RP extensions present in *Metacerylon* and *Hypodacnella*); radial cell, cross-vein r3 and basal portion of RP absent; medial spur moderately long and medial embayment present in *Murmidius* and most Euxestinae; medial field usually with one or two free veins (none in *Ostomopsis*, *Murmidius* and most *Ceryloninae*) and with divided medial fleck in Murmidius and most Euxestinae (not Metacerylon); wedge cell present only in *Protoxestus* Sen Gupta and Crowson); anal lobe absent in *Murmidius*, *Ostomopsis* and Ceryloninae; anal embayment notch-like in Euxestinae. Legs relatively short; trochanters elongate and trochanterofemoral joint slightly oblique in most Ceryloninae, joint weakly to strongly oblique in other groups. Tibiae usually not or only weakly expanded apically; outer apical angle of protibia produced and acute in some *Cerylonopsis*, spinose in *Metacerylon*; tibial spurs often reduced or absent. Tarsi usually 4-4-4 (3-3-3 in *Ostomopsis*, *Cerylonopsis and Cautomus*); tarsomeres simple; pretarsal claws simple.

Abdomen with five free ventrites; ventrite 1 usually much longer than 2, with straight (Fig. 87D) or curved (Fig. 26B) postcoxal lines; intercoxal process truncate. Ventrite 5 (sternite VII) inflexed with hind edge crenulate (Fig. 44AA) in *Murmidius*, *Ostomopsis* and Ceryloninae, forming an interlocking mechanism with corresponding crenulations beneath elytral apices. Functional spiracles on abdominal segments I–VII in Euxestinae and I–V in remaining groups. Tergite VII usually moderately to heavily sclerotised. Sternite VIII in male with short anterior strut in Euxestinae and *Ostomopsis*. Anterior edge of segment IX in male usually produced to form a spiculum gastrale (forming narrowly rounded subgenital plate in *Murmidius*); tergite IX divided into two parts and segment X membranous. Aedeagus of cucujiform type, symmetrical or asymmetrical; anterior edge of tegmen with long strut in *Murmidius* and *Ostomopsis*; parameres weakly develolped, fused to phallobase and sometimes absent; penis long and slender, with paired anterior struts in *Ostomopsis*. Sternite VII in female with spiculum ventrale in *Ostomopsis* and Ceryloninae (absent in *Murmidius* and Euxestinae). Ovipositor usually moderately elongate, lightly sclerotised except for baculi; paraprocts usually shorter than coxites, which are subdivided with relatively long apical styli (reduced without styli in *Murmidius*). [Besuchet 1972; Sen Gupta and Crowson 1973; Ślipiński 1984, 1988, 1990; Lawrence *et al.* 1999b; Thomas 2002e; Ślipiński and Lawrence 2010a].

Larva (Figs 88A, F, J)

Body elongate, parallel-sided to somewhat onisciform and slightly flattened in Euxestinae and Ceryloninae, broadly ovate and strongly flattened in *Murmidius*; dorsal surfaces more or less lightly pigmented, but often either tuberculate or densely granulate and asperate; vestiture of long and short simple setae borne on elongate tubercles in *Hypodacnella*, long clubbed or apically expanded setae in *Euxestus*, barbed setae in *Murmidius* and scale-like setae in at least some Ceryloninae.

Head usually protracted and prognathous or slightly declined (hypognathous in Ceryloninae), transverse and slightly flattened, completely concealed by prothorax in *Murmidius* and at least partly so in Ceryloninae. Epicranial suture absent. Median endocarina present in *Murmidius*. Stemmata two on each side in *Hypodacnella*, absent in *Euxestus*, *Murmidius* and Ceryloninae. Frontoclypeal suture absent. Labrum free in Euxestinae and *Murmidius*, fused to clypeus, elongate and apically acute in Ceryloninae. Antennae moderately long, 3-segmented, with sensorium longer than apical antennomere. Mandibles symmetrical, bidentate, with blunt subapical tooth in *Euxestus*, tridentate in *Hypodacnella* and *Murmidius*, and stylet-like and enclosed in tubular beak in Ceryloninae; with accessory ventral process, except in Ceryloninae; mola well developed, with transverse rows of asperities in Euxestinae and *Murmidius*, absent in Ceryloninae; prostheca hyaline with broad base in *Murmidius*, absent in Euxestinae and Ceryloninae. Ventral mouthparts retracted, with well developed maxillary articulating area in Euxestinae and *Murmidius*; cardines transverse in Euxestinae, elongate in *Murmidius*; stipites long and narrow in Euxestinae, short and trapezoidal in *Murmidius*; neither cardinae nor stipes clearly delineated in Ceryloninae; mala somewhat falciform in Euxestinae, blunt in *Murmidius* and styliform in Ceryloninae; maxillary palps 3-segmented. Labium consisting of prementum and postmentum; ligula well developed, rounded; labial palps 2-segmented, widely separated in Euxestinae and *Murmidius*, enclosed within tubular beak in Ceryloninae (Fig. 46M). Hypopharyngeal sclerome present in Euxestinae and *Murmidius*. Hypostomal rods short and subparallel in *Murmidius*, absent in other taxa; ventral epicranial ridges absent. Gular region transverse.

Thorax with long, subdivided lateral processes in *Hypodacnella*, with lateral and dorsal setiferous tubercles in *Euxestus*, with short lateral lobes in most taxa. Legs relatively stout, 5-segmented including short, claw-like pretarsus bearing one seta; coxae moderately widely separated. Most abdominal segments with lateral lobes similar to those on thorax, those in *Murmidius* with paired glandular openings. Tergum IX shorter than or slightly longer than VIII, extending slightly onto ventral surface, with long, diverging urogomphi in Euxestinae (simple in *Euxestus* and with numerous accessory processes in *Hypodacnella*), without urogomphi in *Murmidius* and Ceryloninae; sternum IX exposed, simple, not enclosed by sternum VIII; segment X concealed from above, ventrally oriented and usually pygopod-like. Spiracles annular-biforous in Euxestinae (on long spiracular tubes in *Hypodacnella*), annular in *Murmidius* and Ceryloninae. [Böving and Craighead 1931; Halstead 1968; Sen Gupta and Crowson 1973; Ślipiński 1984, 1990; Lawrence 1991b; Lawrence *et al.* 1999a; Thomas 2002e; Ślipiński and Lawrence 2010a].

Classification and World Distribution

The family Cerylonidae includes 52 genera and ~450 species occurring in most regions but most abundant and diverse in tropical and subtropical forests. There are five subfamilies, four of which occur in Australia. The exotic subfamily Loebliorylorinae contains the single Oriental genus *Loebliorylon* Ślipiński. [Ślipiński 2003; Ślipiński and Lawrence 2010a].

Australian Fauna

The Australian **Murmidiinae** consists of the single adventive species *Murmidius ovalis* (Beck), and the **Ostomopsinae** includes *Ostomopsis solitaria* Scott, which occurs on Christmas Island but has not yet been found on the Australian mainland. The subfamily **Euxestinae** includes the genera *Euxestus* Wollaston, *Hypodacnella* Ślipiński (Fig. 72D), *Metacerylon* Grouvelle (Fig. 87C–D) and *Protoxestus* Sen Gupta & Crowson, while **Ceryloninae** includes *Australiorylon* Ślipiński, *Cautomus* Sharp (Fig. 87F), *Cerylonopsis* Heinze (Fig. 87E), *Lawrenciella* Ślipiński, *Mychocerus* Erichson and *Philothermus* Aubé. Cerylonids are small, broadly ovate to narrowly elongate, slightly to strongly flattened beetles, red to brown or black in colour (occasionally bicoloured) and subglabrous or sparsely clothed with erect hairs. Antennae vary but always have a compact club with rarely more than 2 segments, metacoxae are almost always widely separated, and femoral lines are often present on metaventrite and/or first abdominal ventrite. In Ceryloninae the maxillary palps are aciculate, the frontoclypeal suture absent, the mandibles and maxillae narrow and often styliform, and the fifth ventrite with a finely crenulate edge. In some cerylonines (e.g. *Cautomus mirabilis* (Oke), Fig. 87F) the labrum is acute and forms part of a sucking tube (Besuchet 1972). Larvae are elongate to broadly ovate and disc-like (*Murmidius*, Fig. 88A) with lateral tergal processes on all thoracic and most abdominal segments and usually with a granulate or tuberculate dorsal surface, clothed with variously modified setae. Cerylonids are commonly found under bark of rotten logs and in decaying leaf litter, where they probably feed on fungal hyphae and spores. *Murmidius ovalis* occurs in stored food products which have been contaminated with moulds or yeasts; it is not considered to be a pest. [Halstead 1968; Sen Gupta and Crowson 1973; Pal and Lawrence 1986; Ślipiński 1988, 1990; Ślipiński and Lawrence 2010a].

82. DISCOLOMATIDAE

Adult (Fig. 87K)

Length 1.3–2 mm. Body usually ~1.25–1.33 times as long as wide, broadly ovate and more or less flattened; colour reddish-brown to black, with elytra bicolored in some species; dorsal surfaces either glabrous or clothed with decumbent or suberect hairs.

Head moderately declined, deeply retracted into prothorax and often concealed from above by pronotum, sometimes abruptly constricted behind eyes to form distinct temples. Eyes moderately well developed, elliptical, moderately protuberant, coarsely facetted, without interfacetal setae. Antennal insertions widely separated and exposed from above; subantennal grooves well developed, extending slightly behind eyes. Frontoclypeal suture distinct, straight to slightly curved; clypeal apex truncate to emarginate. Labrum free, strongly transverse, broadly rounded apically. Antennae 9-segmented with a 1-segmented club. Mandible broad at base and narrow at apex, bidentate; incisor area with one subapical tooth very close to apex; mola well developed and transversely ridged; prostheca well developed, mostly membranous with distinct fringe but with sclerotised, tridentate or quadridentate apex on left mandible (reduced on right mandible). Maxilla with setose galea and slightly narrower, setose lacinia, the latter without uncus; second palpomere inflated, third reduced and apical one somewhat conical. Mentum elongate; ligula simple; apical labial palpomere conical. Gular sutures very short and widely separated. Corpotentorium absent. Cervical sclerites absent.

Pronotum ~0.3–0.6 times as long as wide, widest posteriorly and as wide at base as combined elytral bases; sides strongly curved, narrowly explanate; lateral pronotal carinae complete, without raised margin or bead, the edges simple; anterior angles not produced; posterior angles acute; disc simple, with two laterally opening glandular pores on each side, one at each anterior angle and one on each side a short distance in front of posterior angle (Fig. 26F). Notosternal sutures absent. Prosternum slightly longer than procoxal cavity, flattened or slightly convex. Prosternal process complete, flat or slightly curved, strongly expanded apically, with emarginate apex. Procoxae with long concealed lateral extensions; trochantins concealed. Procoxal cavities circular, very widely separated, externally closed, internally open.

Scutellar shield abruptly elevated, triangular and apically acute. Elytra ~1.0–1.1 times as long as combined width and 3.75–3.85 times as long as pronotum, sides strongly rounded; disc irregularly punctate, without scutellary striole, lateral edge with six glandular pores; epipleura broad, sharply defined and complete. Mesoventrite (Fig. 26E) not separated by sutures from mesanepisterna, which are fused to mesepimera, anterior edge of ventrite on higher plane than metaventrite; procoxal rests absent, but with broad prothoracic rest or prepectus continuing onto mesanepisterna; main portion of ventrite with slightly produced and broadly rounded anterior edge, which abuts apex of prosternal process. Mesoventral process truncate. Mesocoxae with long concealed lateral extensions; trochantins concealed or absent. Mesocoxal cavities circular,very widely separated, least partly closed laterally by mesepimeron or fused mesepimeron and mesanepisternum. Mesometaventral junction a straight line with meso- and metaventrites abutting. Metaventrite flat to slightly convex, without discrimen or postcoxal lines; exposed portion of mesanepisternum very long and narrow or absent (concealed by elytra). Metacoxae strongly transverse but mostly concealed by metaventrite and appearing globular (Fig. 26E), widely separated; plates absent. Metendosternite without stalk, with long lateral arms, without laminae or anterior process and with anterior tendons not apparent. Hind wing with long apical field containing two radial extensions; radial cell, cross-veins r3 and r4, R-M loop and basal portion of RP absent; medial spur very short; medial field without veins or medial fleck; anal lobe highly reduced, its embayment deep and notch-like. Legs relatively long and slender; tibiae not expanded apically; tibial spurs reduced or absent. Tarsi 3-3-3; tarsomeres simple; pretarsal claws simple; empodium not apparent.

Abdomen with five free ventrites; ventrite 1 much longer than 2, with transverse row of glandular pores; intercoxal process slightly emarginate. Functional spiracles on abdominal segments I–V. Tergite VII heavily sclerotised, with median groove forming an interlocking mechanism with elytra. Sternite VIII and segments IX and X in male highly reduced and membranous, without spiculum gastrale. Aedeagus of cucujiform type, symmetrical, lying on side in repose; anterior edge of tegmen with short strut; parameres fused to phallobase and to each other or apparently absent; penis slender, curved, without anterior struts. Sternite VII in female reduced, without spiculum ventrale. Ovipositor reduced and membranous, without styli. [Lea 1922b; John 1959; Lawrence *et al.* 1999b; Cline and Ślipiński 2010].

Larva (Fig. 88B)

Body broadly ovate, flattened and disk-like; dorsal surfaces moderately sclerotised with dense fine granules; vestiture of short, simple setae, expanded or frayed setae and pubescent setae, the latter forming clusters.

Head protracted and prognathous, completely concealed by prothorax, slightly transverse and slightly flattened. Epicranial stem absent; frontal arms lyriform, contiguous at base. Median endocarina absent. Stemmata three on each side. Frontoclypeal suture present. Labrum free, almost as long as wide and broadly rounded at apex. Antennae about as long as head width, separated from mandibular articulation by distinct strip of cuticle, 2-segmented, with setiferous sensorium at apex of apical antennomere. Mandibles symmetrical, broad at base, tridentate, without accessory ventral process; incisor edge simple; mola well developed, tuberculate; prostheca a broadly angulate hyaline lobe. Ventral mouthparts deeply retracted, with maxillary articulating area; cardines transverse, stipites elongate, mala narrowly rounded; maxillary palps 2-segmented, with well developed palpifer. Labium consisting of prementum and postmentum; ligula simple, rounded; labial palps 2-segmented, with segment-like palpiger, narrowly separated. Hypopharyngeal sclerome large, subquadrate. Hypostomal rods weakly indicated, short and converging; ventral epicranial ridges absent. Gular region very short.

Thoracic terga each with two rounded processes on each side clothed with pubescent setae; protergum trapezoidal, much longer than meso- or metatergum. Legs well developed, 5-segmented including short, claw-like pretarsus bearing one seta; tibiotarsus with rows of short spines; coxae moderately widely separated. Abdominal terga I–VII each with one rounded, setose lateral lobe on each side; terga VIII and IX more or less fused together forming posterior plate which is bordered by four rounded processes, but without urogomphi; sternum IX exposed, simple, not enclosed by sternum VIII; segment X concealed from above, ventrally oriented, with pair of distinct pygopods. Spiracles annular. [Fukuda 1969; Lawrence *et al.* 1999a; Cline and Ślipiński 2010].

Classification and World Distribution

The Discolomatidae is a circumtropical group most diverse in the Afrotropical region and containing ~400 species in 16 genera. There are five currently recognised subfamilies, one of which occurs in Australia. Exotic groups include Notiophyginae (Neotropical, Oriental and Afrotropical regions), Discolomatinae (Neotropical and Afrotropical), Cephalophaninae (South-east Asia, East Indies) and Pondonatinae (Afrotropical region). [John 1959; Cline and Ślipiński 2010].

Australian Fauna

Australian discolomatids all belong to the *Aphanocephalus* Wollaston (**Aphanocephalinae**), a genus widely distributed in the Old World. Species are known from various parts of Australia, but primarily in the north. These are broadly ovate and flattened, subglabrous or pubescent beetles, with more or less explanate side margins bearing glandular pores on both prothorax and elytra; antennae have an enlarged and apically expanded terminal segment forming a club, the frontoclypeal suture is present, ventrite 1 is much longer than 2, and all coxae are externally small and globose but with a long internal extension. Larvae are broadly ovate, flattened and disc-like, with the head more or less concealed from above with long antennae and the vestiture consisting of club-like setae around the edges and star-shaped scales on the dorsal surfaces. Discolomatids may be collected in leaf litter or under bark, and they are known to feed on the fruiting bodies of various polypore fungi. [Lea 1922b; John 1959; Fukuda 1969; Cline and Ślipiński 2010].

83. ENDOMYCHIDAE

Adult (Figs 6F, 72C, E–G, K, 89D–E, G)

Length 1–7.5 mm. Body ~1–2.55 times as long as wide, usually broadly ovate to oblong, but occasionally elongate and parallel-sided (some *Holoparamecus* Curtis), strongly convex to slightly flattened; colour yellow to reddish-brown or black, sometimes bicolored with pronotum lighter than elytra or with a dark and light pattern on elytra; dorsal surfaces either glabrous or clothed with decumbent or decumbent and suberect hairs.

Head prognathous or slightly declined, often deeply retracted into prothorax but always visible from above, not abruptly constricted behind eyes; with broad median stridulatory file in Lycoperdininae. Eyes usually well developed, elliptical, moderately protuberant, coarsely facetted, with or without fine interfacetal setae (somewhat reduced in *Eidoreus* Sharp and some Anamorphinae; consisting of a few facets only in *Cholovocerida* Belon). Antennal insertions almost always widely separated and exposed from above (concealed in Merophysiinae); subantennal grooves almost always absent (extending behind eyes in *Cholovocerida*). Frontoclypeal suture distinct, usually straight to slightly curved (abruptly arched and angulate in *Cholovocerida*); clypeal apex usually truncate. Labrum free, transverse. Antennae usually moderately long and often 11-segmented (10-segmented in *Idiophyes* Blackburn, *Eidoreus* and *Holoparamecus*, 9-segmented in *Geoendomychus*, 8-segmented in *Cholovocerida* and 4-segmented in *Trochoideus* Westwood) with a well developed 3-segmented club (2-segmented in *Eidoreus* and *Holoparamecus*, 1-segmented in *Cholovocerida* and *Trochoideus*). Mandible broad at base and narrow at apex, bidentate; incisor area usually with one subapical tooth very close to apex; mola well developed and transversely ridged, with setose basal lobe; prostheca well developed, membranous with distinct fringe (with sclerotised, tridentate lobe at apex on left mandible in *Cholovocera*). Maxilla with setose galea and narrower, setose lacinia, the latter without distinct uncus; apical palpomere more or less conical or fusiform. Mentum transverse; ligula simple; apical labial palpomere usually expanded and truncate. Gular sutures usually widely separated. Corpotentorium usually slender and arched (absent in Merophysiinae), without median process; laminatentoria often fused at midline. Cervical sclerites usually moderately well developed but slender.

Pronotum ~0.3–0.9 times as long as wide, usually widest posteriorly or at middle and as wide at base as combined elytral bases (widest anteriorly and narrower at base than elytral bases in *Holoparamecus*, *Trochoideus* and *Encymon* Gerstaecker); sides evenly curved, or sinuate; lateral pronotal carinae complete, usually with raised margin or bead, the edges almost always simple (coarsely crenulate or denticulate in *Mycetaea* Stephens, *Daulis* Erichson and *Daulotypus* Lea); anterior angles almost always produced and narrowly to broadly rounded (not so in *Erotendomychus* Lea, *Holoparamecus* and *Trochoideus*); posterior angles more or less right; disc often with pair of sub-basal pits and/or short to long sublateral carinae. Prosternum almost always slightly shorter than to slightly longer than procoxal cavity (much longer in *Holoparamecus*), flattened to moderately convex. Notosternal sutures usually complete; anterior part of each hypomeron in *Eidoreus* and *Cholovocerida* concave forming antennal cavity partly concealed from below by prosternum. Prosternal process almost always complete, flat or slightly curved, parallel-sided to gradually expanded apically with narrowly rounded to truncate or slightly emarginate apex (incomplete with subacute apex in *Erotendomychus*). Procoxae with long concealed lateral extensions; trochantins concealed. Procoxal cavities circular, very narrowly to moderately broadly separated, externally open, internally closed.

Scutellar shield abruptly elevated and apically narrowly rounded to broadly rounded or truncate. Elytra ~0.9–1.75 times as long as combined width and 1.9–4.5 times as long as pronotum, sides weakly to strongly rounded; disc usually irregularly punctate, without scutellary striole (occasionally with eight or nine puncture rows); epipleura sometimes broad, but usually incomplete. Mesoventrite usually separated by sutures from mesanepisterna, which are well separated from one another (partly or completely fused to ventrite in *Cholovocerida* and *Trochoideus*), anterior edge of ventrite on higher plane than metaventrite in *Erotendomychus*, *Cholovocerida* and *Eidoreus*; procoxal rests usually weakly declined (more strongly so in *Erotendomychus*, *Cholovocerida* and *Eidoreus*); body of mesoventrite produced anteriorly to form rounded process extending beneath procoxae in *Erotendomychus*. Mesoventral process usually truncate, sometimes slightly emarginate. Mesocoxae with trochantins exposed or concealed. Mesocoxal cavities circular, moderately to very widely separated, usually partly closed laterally by mesepimeron (closed by meeting of ventrites in Anamophinae, Merophysiinae and *Trochoideus*). Mesometaventral junction usually a straight line with meso- and metaventrites abutting or overlapping (junction monocondylic in *Trochoideus*). Metaventrite flat to slightly convex, with discrimen long, short or absent; curved postcoxal lines present in *Eidoreus*; postcoxal pits present in *Encymon* and *Stenotarsus* Perty; exposed portion of mesanepisternum moderately to very long and narrow. Metacoxae widely separated, usually strongly transverse (slightly transverse in Merophysiinae); plates absent. Metendosternite without or with short, broad stalk, with long lateral arms, without or with very slender laminae and with anterior tendons usually on lateral arms (moderately close together near midline in *Holoparamecus*). Hind wing with long apical field without distinct sclerites or veins; radial cell and cross-vein r3 absent, cross-vein r4 usually incomplete; R-M loop forming acute angle; basal portion of RP short or absent; medial embayment often present; medial field without or with one free vein; medial fleck almost always present; anal lobe often highly reduced, its embayment, if present notch-like. Legs relatively long; trochanterofemoral joint often oblique, trochanters elongate in *Holoparamecus*; tibiae not or slightly expanded apically; tibial spurs reduced or absent. Tarsi usually 4-4-4 (3-3-3 in *Merophysiinae*, *Idiophyes* and *Geoendomychus*); tarsomeres often simple, but tarsomeres 1 and 2 lobed below and 3 very short and partly concealed by 2 (pseudotrimerous) in *Stenotarsus* and Lycoperdininae; pretarsal claws simple; empodium usually not apparent.

Abdomen with five or six free ventrites; ventrite 1 usually distinctly longer than 2, with curved postcoxal lines in *Cholovocerida* and *Eidoreus*; intercoxal process truncate to slightly emarginate. Functional spiracles on abdominal segments I–V. Anterior edge of segment IX usually rounded to acute, rarely produced to form spiculum gastrale (e.g. *Cholovocerida*, *Erotendomychus*, *Mycetaea*). Aedeagus of cucujiform type, symmetrical, almost always lying on side in repose; tegmen often reduced with short tegminal plate, short, fused parameres and no anterior strut (with long, articulated strut in *Eidoreus*); tegmen absent in *Erotendomychus*; penis usually slender and curved, without anterior struts. Sternite VII in female without spiculum ventrale. Ovipositor short and broad, with undivided coxites usually lying mesal to paraprocts; styli short and terminal or absent. [Lea 1922b; Strohecker 1953; Sasaji 1978, 1990; Lawrence *et al.* 1999b; Tomaszewska 2000, 2004a, 2004b, 2005, 2010a; Skelley and Leschen 2002].

Larva (Fig. 88E, G, K)

Body ovate to elongate and convex to slightly flattened; dorsal surfaces lightly to moderately sclerotised and smooth or granulate or tuberculate; vestiture of simple and/or expanded or frayed setae.

Head protracted or slightly retracted and usually hypognathous, slightly to strongly transverse, sometimes widest posteriorly and narrowed anteriorly, slightly flattened. Epicranial stem usually absent (short in *Trochoideus* and some Lycoperininae); frontal arms often weakly developed or absent; if present, U-shaped, V-shaped or weakly lyriform and contiguous at base. Median endocarina absent. Stemmata absent in Merophysiinae and most Anamorphinae, two on each side in *Eidoreus*, *Mycetaea* and *Trochoideus*, and four in Stenotarsus and Lycoperdininae. Frontoclypeal suture present or absent. Labrum free, truncate to shallowly emarginate at apex. Antennae short to very long and often well separated from mandibular articulations, 3-segmented, with segment 2 longest, 3 reduced and sensorium at least as long as apical antennomere. Mandibles broad, usually triangular, with apex unidentate to multidentate (reduced or entirely absent in Anamorphinae), usually without accessory ventral process; mola usually well developed and transversely ridges; prostheca, if present, simple or complex. Ventral mouthparts retracted, with maxillary articulating area; cardines transverse, stipites elongate, mala usually obtuse, but falciform in Anamorphinae and *Trochoideus*; maxillary palps 3-segmented, with well developed palpifer. Labium consisting of prementum and postmentum; ligula simple, rounded; labial palps 2-segmented, with well developed palpiger, usually widely separated. Hypopharyngeal sclerome a transverse bar or molar-like. Hypostomal rods usually subparallel or diverging, absent in Stenotarsus; ventral epicranial ridges absent. Gular region very short.

Protergum usually almost as long as meso- and metaterga combined; terga with or without sclerotised plates, sometimes with well developed lateral processes. Legs usually long and slender, 5-segmented including short, claw-like pretarsus bearing one seta; tibial apex with spatulate setae in some Anamorphinae; coxae moderately widely separated. Abdominal terga I–VIII on each side sometimes with one or two lateral tergal processes, which may or may not be dehiscent. Tergum IX short, extending onto ventral surface, with a pair of short, acute urogomphi in *Mycetaea*; sternum IX exposed, simple, not enclosed by sternum VIII; segment X concealed from above, ventrally or posteroventrally oriented. Spiracles annular. [Kemner 1924; Böving and Craighead 1931; Lawrence 1991b; McHugh and Pakaluk 1997; Lawrence *et al.* 1999a; Burakowski and Ślipiński 2000; Tomaszewska 2010a].

Classification and World Distribution

The family Endomychidae comprises ~1800 species in 130 genera and is widely distributed but most diverse in tropical and subtropical regions. The group is currently divided into twelve subfamilies, seven of which occur in Australia. Exotic subfamilies include Leiestinae (Holarctic region), Xenomycetinae (north-western North America), Danascelinae (northern North America and Pakistan), Endomychinae (Holarctic and Orriental regions) and Epipocinae (New World). [Shockley *et al.* 2009; Tomaszewska 2010a].

Australian Fauna

The Australian endomychid fauna includes one to a few genera in each of seven subfamilies. **Merophysiinae** is represented primarily by widely distributed adventive species, such as *Cholovocerida maderae* (Wollaston), *Holoparamecus caularum* (Aubé); *H. depressus* Curtis and *H. kunzei* (Aubé), but native *Holoparamecus* species (Fig. 89D) are also present. Within the **Pleganophorinae**, the widely distributed *Trochoideus desjardinsi* Guérin-Méneville (Fig. 72G) is common on Christmas Island, but two species of this genus (possibly quarantine interceptions) have been seen from Darwin and Broome. Australian records for the subfamilies **Eupsilobiinae** and **Mycetaeinae** are also based on the occurrence of adventive species *Eidoreus minutus* Sharp from Christmas Island and *Mycetaea subterranea* (Fabricius) (Fig. 89G) from various mainland localities. the genus *Stenotarsus* Perty (**Stenotarsinae**) (Figs 72C, 89E) includes several endemic Australian species, and the **Anamorphinae** includes species of *Idiophyes* Blackburn (Fig. 72F), *Geoendomychus* Lea and *Erotendomychus* Lea. The Australian **Lycoperdininae** includes species of *Daulis* Erichson, *Daulotypus* Lea, *Encymon* Gerstaecker (Fig. 72K) and *Indalmus* Gerstaecker (Fig. 72E). Endomychids are broadly ovate to elongate and globose to moderately flattened beetles, glabrous or clothed with decumbent and erect hairs, and often brightly coloured with red or yellow and black. The frontoclypeal suture is always present in this group, the antennae are moderately long, usually with a relatively loose club, the apical maxillary palpomeres are fusiform, although the labial palps may be apically securiform, tarsomere 2 is commonly lobed, extending below the reduced third tarsomere, pretarsal claws are simple, and femoral lines are lacking on the first ventrite. Larvae are usually oblong to ovate (rarely elongate), slightly to strongly flattened, usually with lateral tergal processes on all thoracic and most abdominal segments and only rarely with short urogomphi. In *Encymon immaculatus* (Montrouzier) the lateral processes are deciduous, breaking off when grasped by a predator. Larval antennae are often long (with antennomere 2 long and 1 and 3 reduced) and their sockets are well separated from the mandibular articulations; the head is often broad at base and narrowed anteriorly, and the mala is obtuse. Endomychids feed on a wide variety of fungi, but little is known about feeding habits of most Australian species. *Trochoideus desjardinsi* has been recorded from termite fungus gardens in Java (Kemner 1924), but on Christmas Island both adults and larvae of this species were observed feeding on fruiting bodies of Auriculariaceae and Tremellaceae. Both *Mycetaea subterranea* and various Holoparamecus species are introduced contaminants of mouldy food stores. [Lea 1922b; Tomaszewska 2010a].

84. COCCINELLIDAE

Adult (Figs 6G, 87L–N)

Length 0.9–12 mm. Body usually ~1.15–1.75 times as long as wide, broadly ovate to oblong, dorsally strongly convex to distinctly flattened and almost always ventrally flat or slightly convex; colour often yellow to red or black, sometimes with elytra and less commonly pronotum bicolored, occasionally metallic blue, green or violet; dorsal surfaces either subglabrous or clothed with short to moderately long setae, stiff hairs or a combination of both.

Head small, usually flattened and subquadrate to distinctly transverse (distinctly rostrate in Microweiseini), usually prognathous (moderately hypognathous in Serangiini and Chilocorini, strongly so in Microweiseini), often deeply retracted into prothorax and sometimes partly or completely concealed from above, without postocular constriction. Eyes well developed and usually finely facetted (coarsely facetted in many Coccidulini), often slightly emarginate, deeply divided by ocular canthus in Telsimiini and some Chilocorini), with or without interfacetal setae (prominent in *Rodolia*). Antennal insertions usually exposed and located near inner edges of eyes, concealed by expanded clypeus in Telsimiini and Chilocorini, well separated from eyes in Epilachnini and some Coccinellini, and located well in front of eyes in Microweiseini and Serangiini; subantennal grooves usually present, often extending to hind edges of eyes. Frontoclypeal suture always absent. Labrum free, transverse and usually about as wide as clypeus (much narrower in Telsimiini and Chilocorini), tormae poorly developed and labral rods absent. Antennae variable in length but often short, usually 11-segmented (10-segmented in *Pharellus*, *Scymnomorphus*, *Hong*, some *Sticholotis*, *Bucolellus*, *Exochomus*, *Trichorcus*, *Cryptolaemus*, *Scymnodes*, some *Scymnus*, *Lucailla*, *Iracilda*, *Poorani* and some *Epilachna*, 9-segmented in some *Serangium*, some *Orcus* and *Hypoceras*, 8-segmented in *Parapharellus*, some *Serangium*, *Brumoides*, *Chilocorus*, some *Orcus* and *Rodolia*, and 6- or 7-segmented in Telsimiini) with a relatively weak, 3-segmented club (2-segmented in *Scymnomorphus*, 1-segmented in *Serangium* and *Andrzej* and 4-segmented in *Chaetolotis*, *Telsimia*, *Cyrema*, *Dichaina*, *Hypoceras* and *Poorani*). Mandible more or less triangular, usually bidentate (unidentate in various groups, multidentate in *Epilachna*); incisor edge in *Illeis* with denticulate incisor edge; mola usually more or less concave with a sharp distal tooth and hyaline prostheca (mola and prostheca absent in Microweiseini and Serangiini, replaced by membranous lobe in *Epilachna*). Maxilla usually with broad, apically setose galea and much smaller lacinia (strongly reduced in Microweiseinae); apical palpomere varying from narrow and apically acute to strongly securiform. Mentum usually transverse and distinctly narrowed posteriorly (parallel-sided cor broadest at base in *Epilachna*); ligula usually short and reduced (distinctly expanded in many Coccinellini); labial palp 2-segmented in *Rodolia*, apical palpomere more or less fusiform. Gular sutures widely separated and usually long (reduced in Serangiini). Corpotentorium slender, simple. Cervical sclerites well developed to absent.

Pronotum ~0.30–0.75 times as long as wide, almost always widest posteriorly, occasionally at middle and as usually wide at base as combined elytral bases (distinctly narrower in some Chilocorini, Coccinellini and Epilachnini); sides usually strongly rounded, occasionally weakly rounded and rarely subparallel, sometimes obliquely explanate; lateral pronotal carinae complete, with or without raised margin or bead, the edges simple; anterior angles usually not produced (sometimes produced and broadly rounded); posterior angles broadly rounded or absent; posterior edge usually straight or weakly rounded; disc usually simple (with short carinae near anteror angles in most Microweiseini). Prosternum in front of coxae usually about as long as or slightly longer than shortest diameter of coxal cavity (much shorter in *Rodolia* and Sticholotini), flattened or slightly convex; sometimes with paired longitudinal lines; with strongly produced anterior edge forming chin-piece in *Serangium* and *Bucolus*. Notosternal suture usually complete; hypomera often modified to receive antennae and/or legs in repose. Prosternal process complete, parallel-sided or weakly expanded apically, with broadly rounded or truncate apex. Procoxae with long concealed lateral extensions; trochantins concealed. Procoxal cavities circular to slightly transverse, usually moderately widely separated (narrowly so in some Microweiseini), always externally open and internally closed.

Scutellar shield abruptly elevated and acute to broadly rounded or truncate posteriorly. Elytra ~0.9–1.25 times as long as combined width and 2.4–5.4 times as long as pronotum, sides almost always weakly to strongly rounded; disc with irregular punctation and no scutellary striole; epipleura usually complete, often wide and sometimes oblique, with clearly delimited cavities for receiving meso- and metafemora in *Serangium* and less well defined cavities in Chilocorini. Mesoventrite with anterior edge usually on higher plane than metaventrite, with distinct, strongly oblique to vertical procoxal rests and a small mesoventral cavity. Mesoventral process broadly rounded to truncate. Mesocoxae with concealed trochantins. Mesocoxal cavities circular, moderately to very widely separated, partly closed laterally by mesepimera. Mesometaventral junction a straight or curved line with meso- and metaventrites abutting or slightly overlapping. Metaventrite flat to slightly convex, usually with discrimen of varying length; curved postcoxal lines almost always present; exposed portion of mesanepisternum moderately elongate to very long and narrow. Metacoxae slightly to strongly transverse and widely separated; plates absent. Metendosternite with stalk short and broad to very short or absent; lateral arms long, without or with reduced, narrow laminae, without or with very short anterior process and anterior tendons widely separated and sometimes well out on lateral arms. Hind wing (Fig. 29I) with long apical field without post-radial sclerite and with or without radial extensions; radial cell, cross-vein r3 and often r4 absent; R-M loop forming acute angle; medial spur usually long; medial field without or with one or two free veins and no medial fleck; wedge cell absent; anal lobe present or absent, its embayment notch-like. Legs relatively short; trochanterofemoral joint oblique. Femora flattened and usually broad. Tibiae usually not or only weakly expanded apically; tibial spurs often reduced or absent. Tarsi almost always 4-4-4 (3-3-3 in Telsimiini, *Cryptolaemus* and *Stethorus*); tarsomeres 1 and 2 large and lobed beneath and 3 very small and simple (pseudotrimerous); pretarsal claws usually appendiculate, sometimes simple.

Abdomen with five or six ventrites (sometimes with six in male only), the first two of which are usually solidly fused; ventrite 1 often much longer than 2, almost always with curved postcoxal lines (Fig. 25E) (lines absent or hardly separable from edges of coxal cavities in *Rhynchortalia* and *Nat*); intercoxal process truncate to slightly emarginate. Functional spiracles on abdominal segments I–V. Tergite VII usually moderately to heavily sclerotised. Anterior edge of segment IX in male usually produced to form a spiculum gastrale (strongly reduced in Telsimiini); tergite IX distinct. Aedeagus (Fig. 31H) of a complex cucujiform type, located on left side when retracted; usually symmetrical (not so in Microweiseini and Serangiini); anterior edge of tegmen with tegminal strut (trabes) and posterior edge with penis guide (median lobe); parameres usually well develolped and articulated; penis (sipho) long, slender and curved, usually enlarged at base, without endophallic sclerites. Ovipositor very short to moderately long and lightly sclerotised, usually without distinct baculi; styli usually well developed and terminal. [Kamiya 1960; Sasaji 1967, 1968b, 1971; Gordon 1985; Kovář 1996a; Vandenberg 2002; Ślipiński 2007; Ślipiński and Tomaszewska 2010c].

Larva (Fig, 88L)

Body usually oblong to elongate, parallel-sided to fusiform and slightly flattened (broadly ovate, strongly flattened and disc-like in *Scymnodes*); dorsal surfaces sometimes covered with dense, waxy exudate, usually in combination with soft, lightly pigmented integument; often either granulate or spinose, often with distinct sclerotised plates sometimes accompanied by complex cuticular armature in larvae without waxy exudate; vestiture usually of long and short simple setae (expanded, scale-like setae in *Scymnodes*).

Head protracted and prognathous, usually transverse and slightly flattened, sometimes concealed from above by pronotum (e.g. *Scymnodes bellus* Pope & Lawrence). Epicranial stem present or absent; frontal arms usually lyriform (V-shaped in *Epilachna*), contiguous at base. Median endocarina absent. Stemmata usually three on each side forming a triangle (two distinct and one with reduced lens in *Stethorus*). Frontoclypeal suture usually absent (present in *Epilachna*). Labrum usually free. Antennae usually very short (longer in *Epilachna*), usually 3-segmented (2-segmented in many Coccinellini, 1-segmented in *Stethorus* and some Chilocorini), with antennomere 3 reduced and shorter than conical sensorium; antennae usually separated from mandibular articulation by a distance greater than diameter of antennal socket. Mandibles symmetrical, broad at base and narrowed apically, usually unidentate or bidentate (multidentate in *Epilachna*, Fig. 39E), incisor edge in *Illeis* forming pectinate, scoop-like structure; accessory ventral process absent; mola usually reduced forming a tooth-like structure bearing microtrichia (Fig. 48K) (absent in Microweiseinae and *Epilachna*); prostheca absent. Ventral mouthparts retracted, without maxillary articulating area; cardines indistinct, fused to stipites; mala rounded to falcate, without uncus but often with a specialised setiferous process called a stylus; maxillary palps almost always 3-segmented (2-segmented in *Rodolia*). Labium usually consisting of a single sclerite; ligula short and simple; labial palps usually 2-segmented. Hypopharyngeal sclerome usually absent (sometimes consisting of transverse bar). Hypostomal rods short and subparallel or weakly convergent; ventral epicranial ridges absent. Gular region transverse.

Prothorax longer than meso- or metathorax, sometimes longer than the two combined. Thoracic and abdominal terga sometimes laterally expanded forming processes of various kinds. Legs well developed (in Coccinellini forelegs usually longer than remaining pairs), 5-segmented including pretarsus, which is claw-like with basal tooth bearing single, simple seta; apex of tibiotarsus usually with specialised clubbed or expanded and flattened setae; coxae moderately to widely separated. Abdominal terga often with transverse row of six setose protuberances (Fig. 51C) or sometimes branched processes; terga or intersegmental membranes often with glandular openings in Chilocorini, Diomini and Sticholotidini. Tergum IX slightly longer and narrower than VIII, extending onto ventral surface, without urogomphi; sometimes forming evenly rounded, flattened plate; sternum IX exposed, simple, not enclosed by sternum VIII; segment X concealed from above, ventrally or posteroventrally oriented and pygopod-like. Spiracles annular, those on abdomen located dorsolaterally. [Kapur 1950; Kamiya 1965; Sasaji 1968a; Gordon and Vandenberg 1995; Vandenberg 2002; Ślipiński 2007; Ślipiński and Tomaszewska 2010c].

Classification and World Distribution

The family Coccinellidae is a widely distributed family with ~6000 species in 360 genera. In the most recent classification (Ślipiński 2007; Seago *et al.* 2011; Escalona and Ślipiński 2012), there are two subfamilies: Microweiseinae with four tribes, three of which occur in Australia, and Coccinellinae with 17 tribes, nine of which extend into Australia. Non-Australian groups include the microweiseine tribe Carinodulini (Mesoamerica, South-east Asia and Africa) and the coccinelline tribes Aspidimerini (India and Japan to East Indies), Monocorynini (Thailand, Malaysia and Indonesia), Platynaspidini (Eurasia, Africa, Madagascar), Cephaloscymnini (New World), Argentipilosini (Brazil), Plotinini (Oriental), Shirozuellini (Oriental) and Limnichopharini (Oriental). [Kovář 1996b].

Australian Fauna

The subfamily **Microweiseinae** is represented by two tribes: Serangiini containing a single genus, and Microweiseini with four genera. There are nine tribes of Coccinellinae occurring in Australia, the largest being Coccidulini (including Scymnini) with 20 genera and Coccinellini with 15. Monotypic tribes include Noviini (*Rodolia*) and Epilachnini (*Epilachna*), the tribes Sticholotidini, Telsimiini and Scymnillini each contain two genera, Diomini three, and Chilocorini six. Some of the largest coccinelline genera are *Rhyzobius* (51 spp.), *Diomus* (45), *Scymnus* (21), *Telsimia* (15), *Rodolia* (13) and *Scymnodes* (10). Almost all Australian coccinellids are broadly ovate, moderately to strongly convex beetles with pseudotrimerous tarsi (4-segmented with 3 reduced and 2 strongly lobed beneath it). The frontoclypeal suture is absent, antennae are usually short and weakly clubbed, the apical maxillary palpomere is almost always large and securiform, and there are curved femoral lines on the first abdominal ventrite and sometimes the metaventrite. Many larger species are glabrous and brightly patterned with red or yellow and black or blue, but most of the smaller species are more uniformly coloured and finely pubescent. Larvae may be elongate (Fig. 88L), but are often oblong and sometimes broadly ovate and slightly to strongly flattened, usually with a transverse row of six tubercles or prominences on most abdominal segments and often covered with a waxy exudate; some forms are spinose above and may be aposematically coloured like some adults. Larval antennae are very short, the mandibles often have a sub-basal lobe, which probably represents a reduced mola, the mala is obtuse and usually bears a stylus, there are paired glandular openings on the thorax and abdomen, and the tibiotarsus usually has two to several expanded setae exending beneath the pretarsus. The pupa is obtext and partly enclosed within the larval skin, which is attached to the substrate by the anal end.

Adults and larvae of most species are predators of aphids, mealybugs, scales or other small insects and mites. *Epilachna* species, however, are phytophagous, damaging the foliage of Solanaceae and Cucurbitaceae, including vegetable crops like pumpkin, cucumber, tomato and potato, and *Illeis galbula* (Mulsant) feeds on powdery mildews (Ascomycetes: Erysiphales). Among the carnivorous species, several have been used with notable success as biological control agents. *Rodolia cardinalis* (Mulsant), for instance, saved the California citrus industry from destruction by the cottony cushion scale (*Icerya purchasi* Maskell), while other Australian species, such as *Cryptolaemus montrouzieri* Mulsant, *Rhyzobius ventralis* (Erichson) and *R. forestieri* (Mulsant) were employed in Hawaii and California for the control of several scale insects (Hemiptera: Coccoidea). Species of *Stethorus* are predators of the two-spotted mite (*Tetranychus urticae*) (Britton and Lee 1972). The unusual, flattened and scale-covered larva of *Scymnodes bellus* Pope & Lawrence is known to feed on small ants under eucalypt bark (Pope and Lawrence 1990). Some coccinellids, when alarmed, feign death and discharge droplets of yellow haemolymph, which is toxic to vertebrates, from the femorotibial articulations. [Hagen 1962; Bielawski 1963; Pope 1979, 1981, 1989; Richards 1980, 1981, 1983, 1985; Anderson 1981; Pope and Lawrence 1990; Richards and Goletsos 1991; Li 1993; Hodek and Honek 1996; Kovár 1996b; Ślipiński 2004, 2007; Ślipiński *et al.* 2005; Ślipiński and Tomaszewska 2005, 2010c; Ślipiński and Giorgi 2006; Ślipiński and Burckhardt 2006; Poorani *et al.* 2008; Giorgi *et al.* 2009; Poorani and Ślipiński 2009; Łączyński and Tomaszewska 2009; Pang and Ślipiński 2009, 2010; Escalona and Ślipiński 2010; Tomaszewska 2010b; Seago *et al.* 2011; Tomaszewska and Ślipiński 2011; Escalona and Ślipiński 2012].

Australian Subfamilies, Tribes and Genera

Microweiseinae: Microweiseini: *Pharellus* Sicard, *Parapharellus* Chazeau, *Scymnomorphus* Weise, *Hong* Ślipiński. **Serangiini:** *Serangium* Pope.

Coccinellinae: Sticholotidini: *Chaetolotis* Ślipiński, *Sticholotis* Crotch. **Telsimiini:** *Hypocyrema* Blackburn, *Telsimia* Casey. **Scymnillini:** *Bucolellus* Blackburn, *Cyrema* Blackburn. **Chilocorini:** *Brumoides* Chapin, *Chilocorus* Leach, *Exochomus* Redtenbacher, *Halmus* Mulsant, *Orcus* Mulsant, *Trichorcus* Blackburn. **Diomini:** *Andrzej* Ślipiński, *Dichaina* Weise, *Diomus* Mulsant. **Coccidulini:** *Apolinus* Pope & Lawrence, *Bucolus* Mulsant, *Cryptolaemus* Mulsant, *Erithionyx* Blackburn, *Guillermo* Ślipiński, *Hypoceras* Chapuis, *Iracilda* Ślipiński, *Lucailla* Ślipiński, *Poorani* Ślipiński, *Rhyzobius*

Stephens, *Robert* Ślipiński, *Rodatus* Mulsant, *Scymnodes* Blackburn, *Scymnus* Kugelann, *Stethorus* Weise, *Wioletta* Ślipiński. **Noviini:** *Rodolia* Mulsant. **Coccinellini:** *Adalia* Mulsant, Antineda Iablokoff-Khnzorian, Archegleis Iablokoff-Khnzorian, *Australoneda* Iablokoff-Khnzorian, *Cleobora* Mulsant, *Coccinella* Linnaeaus, *Coelophora* Mulsant, *Harmonia* Mulsant, *Hippodamia* Chevrolat in Dejean, *Illeis* Mulsant, *Menochilus* Timberlake, *Micraspis* Chevrolat in Dejean, *Oenopia* Mulsant, *Phrynocaria* Timberlake, *Synona* Pope. **Epilachnini:** *Epilachna* Chevrolat in Dejean.

85. CORYLOPHIDAE

Adult (Fig. 71F, 89A–C, 89F)

Length usually 0.6–2.2 mm (up to 3.9 mm in Periptyctinae). Body usually ~1.1–1.9 times as long as wide (up to 2.4 times in *Priamima* and *Pseudostanus*), sides often evenly curved (more or less parallel-sided in *Priamima*, *Pseudostanus* and some species of *Arthrolips* Wollaston), dorsally strongly convex to distinctly flattened; colour yellow to brown or black, sometimes bicolored; dorsal surfaces either subglabrous or clothed with fine, recumbent hairs.

Head small, subquadrate to elongate; usually concealed from above by pronotum (exposed in Periptyctinae, *Priamima*, *Pseudostanus* and *Orthoperus*). Eyes almost always prominent and moderately to coarsely facetted, entire, without interfacetal setae. Antennal insertions exposed and narrowly to widely separated; subantennal grooves present or absent. Frontoclypeal suture absent. Labrum free and at least partly visible, usually slightly to strongly transverse (narrowly elongate and apically acute in *Gloeosoma*). Antennae relatively long, usually 11-segmented (10-segmented in some species of *Arthrolips*, *Sericoderus* and *Gloeosoma*, 9-segmented in *Orthoperus*) with long scape and pedicel and large, loose club, which is 3-segmented or 5-segmented but with the second club segment much smaller than the first or third one; antenna capable of folding at apex of pedicel and at midpoint, appearing doubly geniculate; antennal club with sensory vesicles on at least first two segments. Mandible usually short and broad to moderately elongate (narrowly elongate and endognathous in *Gloeosoma*), unidentate to multidentate; mola well developed in *Priamima*, *Pseudostanus*, *Holopsis* and *Orthoperus* but reduced or absent in Periptyctinae and most Corylophinae; prostheca variable. Maxilla with single lobe or mala, which is slender with spinose apex; apical maxillary and labial palpomeres cylindrical to fusiform, usually smaller than preapical palpomere in Corylophinae. Mentum usually quadrate, narrow in Periptyctinae; ligula often large and undivided. Gular sutures usually widely separated (partly confluent in *Gloeosoma*). Corpotentorium narrow or absent. Cervical sclerites present or absent.

Pronotum usually ~0.3–0.6 times as long as wide (as much as 0.7 or 0.8 in Periptyctinae, *Priamima* and *Pseudostanus*), usually widest posteriorly with sides strongly curved (widest anteriorly or at middle in *Pseudostanus*, *Priamima* and some *Periptyctus* and sides straight basally or angulate in Periptyctinae), base not or only slightly narrower than combined elytral bases; lateral pronotal carinae complete, with or without raised margin or bead, the edges simple; anterior angles usually not produced forward (produced and broadly to narrowly rounded in some Periptyctinae); posterior angles obtuse or right in *Holopsis*, *Arthrolips* and *Orthoperus*, moderately to strongly acute in Periptyctinae, *Corylophus*, *Sericoderus* and *Gloeosoma*; posterior edge straight, rounded or bisinuate; disc usually simple (with long sublateral carinae in *Periptyctus*). Prosternum in front of coxae usually shorter than shortest diameter of coxal cavity (as long as or longer than this in Peripytctinae, *Priamima*, *Pseudostanus*, and some *Arthrolips*), sometimes with paired longitudinal carinae defining a subquadrate, triangular or narrow plate (intercoxal area); with strongly produced anterior edge forming chin-piece in Periptyctinae. Notosternal suture usually incomplete or absent in Corylophinae. Prosternal process complete, usually more or less parallel-sided or weakly expanded apically. Procoxae not projecting and with long concealed lateral extensions in Periptyctinae, *Priamima*, *Pseudostanus*, *Sericoderus* and *Arthrolips*, strongly projecting without concealed extensions in *Holopsis*, *Corylophus*, *Orthoperus* and *Gloeoporus*; trochantins concealed. Procoxal cavities slightly to strongly transverse, narrowly to moderately widely separated, almost always externally closed (open in *Holopsis*)), internally closed.

Scutellar shield usually well developed (reduced in *Orthoperus*), abruptly elevated and acute to broadly rounded or truncate posteriorly. Elytra usually 0.8–1.05 times as long as combined width (up to 1.25 times in *Arthrolips*, 1.4 times in Periptyctinae and *Priamima* and 1.65 times in *Pseudostanus*) and usually 1.45–3.25 times as long as pronotum (as much as 4.15 times in *Holopsis*); sides always weakly to strongly rounded and conjointly rounded at apex (truncate exposing most or all of pygidium in *Pseudostanus*, *Priamima*, *Sericoderus* and some *Arthrolips*); disc with irregular punctation and no scutellary striole; epipleura incomplete, usually narrow and horizontal (wide and inflexed in *Sericoderus*). Mesoventrite on higher plane than metaventrite in *Orthoperus*, *Holopsis* and *Gloeosoma*; anterior edge with distinct, strongly oblique to vertical procoxal rests in *Holopsis*, *Corylophus* and *Gloeosoma*, with median prosternal rest in Periptyctinae. Mesoventral cavity almost always absent (small and shallow in *Corylophus*). Mesoventral process more or less truncate. Mesocoxae with concealed trochantins. Mesocoxal cavities usually circular to slightly transverse (strongly transverse in *Orthoperus*), moderately to very widely separated, usually closed laterally by meeting of meso- and metaventrites (partly closed by mesepimera in *Orthoperus*). Mesometaventral junction a straight or slightly curved line with meso- and metaventrites abutting or slightly overlapping. Metaventrite usually short and broad, without discrimen (short discrimen in *Priamima* and some Periptyctinae); curved postcoxal lines present in *Holopsis* and *Orthoperus*; exposed portion of mesanepisternum moderately elongate to very long and narrow. Metacoxae slightly to strongly transverse and widely separated; plates absent. Metendosternite without stalk, laminae or anterior process; anterior tendons located on long lateral arms. Hind wing relatively narrow with long, simple apical field; radial cell and R-M loop absent; medial field without or with one free vein and with medial fleck in Periptyctinae and *Holopsis*; wedge cell absent; anal lobe absent; long fringe setae restricted to posterior margin; wing slender and strap-like with medial fleck at apex in some Periptyctinae, shortened and stub-like in some Corylophinae. Trochanterofemoral joint moderately to strongly oblique. Femora flattened and usually broad. Tibiae usually simple, not strongly widened, with outer edge simple, often sexually dimorphic; tibial spurs single or absent. Tarsi 4-4-4 with tarsomeres 1 and 2 large with one or both lobed beneath and 3 smaller and simple; pretarsal claws simple.

Abdomen usually with six free ventrites (five in *Periptyctus* and *Weirus*); ventrite 1 usually much longer than 2, with straight postcoxal lines in Priamima and curved ones in Holopsis; intercoxal process truncate. Functional spiracles usually on abdominal segments I–VII (on I–V in *Orthoperus*). Anterior edge of segment IX in male without spiculum gastrale; tergites IX and X fused. Aedeagus of cucujiform type; anterior edge of tegmen without strut; parameres fused together but articulated with phallobase; penis large, distinctly curved and lying on its side, without basal struts, with endophallic sclerites. Ovipositor very short to moderately long and lightly sclerotised, usually without distinct baculi; styli well developed and terminal. [Paulian 1950, 1962; Pakaluk 1985; Pakaluk and Lawrence 1986; Bowestead 1999; Bowestead *et al.* 2001; Ślipiński *et al.* 2001, 2009, 2010c; Bowestead and Leschen 2002; Tomaszewska and Ślipiński 2002].

Larva (Fig. 88C–D, H)

Body usually oblong to elongate, more or less parallel-sided and slightly flattened (widest anteriorly and gradually narrowed posteriorly in Periptyctinae; broadly ovate, strongly flattened and disc-like in *Holopsis* and *Gloeosoma*); dorsal surfaces usually more or less heavily sclerotised and granulate, in contrast to lightly pigmented ventral surfaces in Periptyctinae, *Holopsis* and *Gloeosoma*; in most other groups more lightly pigmented dorsally, sometimes with pigmented areas on protergum and/or tergum IX; vestiture of simple setae, often mixed with modified setae.

Head protracted and prognathous, more or less concealed from above by pronotum in Periptyctinae, *Holopsis* and *Gloeosoma*, usually about as long as wide and slightly flattened, strongly transverse in *Gloeosoma*, more elongate in *Holopsis*. Epicranial stem absent; frontal arms, when present, not joined at base, with paired endocarinae beneath them in Periptyctinae and at least some *Arthrolips*. Stemmata usually two on each side (one in *Pakalukodes* and *Arthrolips*). Frontoclypeal suture absent. Labrum usually free (fused to head capsule in Periptyctinae, *Holopsis* and *Gloeosoma*). Antennae usually moderately long (very short in *Orthopterus*, more than half as long as head width in Periptyctinae, *Holopsis* and *Gloeosoma*), well separated from mandibular articulations, usually 3-segmented (2-segmented in *Sericoderus* and *Glososoma*); sensorium often longer than antennomere 3, sometimes absent. Mandibles symmetrical, usually broad at base and narrowed apically, usually bidentate or tridentate (unidentate in *Holopsis* and *Gloeosoma*), more or less endognathous in Periptyctinae and *Gloeosoma*, without accessory ventral process; mola usually well developed (rudimentary in Periptyctinae and *Holopsis*, absent in *Gloeosoma*); prostheca usually absent (consisting of two hyaline processes in *Arthrolips*). Ventral mouthparts usually protracted (retracted in Periptyctinae and *Gloeosoma*) without maxillary articulating area; cardines indistinct, usually more or less fused to stipites, which are strongly transverse (somewhat longer in Periptyctinae); mala usually rounded to truncate (falciform in *Gloeosoma*); maxillary palps usually 3-segmented (2-segmented in Periptyctinae, *Orthoperus* and *Gloeosoma*) with relatively long apical palpomere. Labium consisting of a single sclerite; labial palps usually 2-segmented (1-segmented in *Holopsis* and *Orthoperus*). Hypopharyngeal sclerome usually absent (sometimes consisting of transverse bar). Hypostomal rods and ventral epicranial ridges absent. Gular sutures separate; gula usually longer than wide (shortened in *Gloeosoma*); gular region sometimes with paired endocarinae.

Prothorax longer than meso- or metathorax, but not as long as the two combined. Thoracic and most abdominal terga more or less laterally expanded forming rounded laterotergal lobes in most Corylophinae, more strongly expanded to form distinct lateral projections in Periptyctinae or expanded and flattened forming continuous to slightly overlapping plates lined with dense fringes of modified setae in *Holopsis* and *Gloeosoma*. Legs well developed, 5-segmented including pretarsus, which is claw-like bearing single, simple seta; coxae moderately to widely separated. Paired dorsal glandular openings present on abdominal segments I–VII in Aenigmaticini (Fig. 42E), *Orthoperus* and *Arthrolips* and on segments I and VIII (Fig. 45L) in *Holopsis*, *Sericoderus* and *Gloeosoma*; openings in Arthrolips located at ends of short tubes. Tergum IX slightly longer and narrower than VIII, extending onto ventral surface, usually forming evenly rounded, flattened plate without urogomphi, but with paired rounded projections in Periptyctinae; sternum IX exposed, simple, not enclosed by sternum VIII; segment X concealed from above, ventrally oriented and pygopod-like. Spiracles annular, those on abdomen located dorsally or dorsolaterally. [Böving and Craighead 1931; Chandler 1983; Pakaluk 1985; Costa *et al.* 1988; McHugh and Pakaluk 1997; Ślipiński *et al.* 2001, 2009, 2010c; Bowestead and Leschen 2002; Polilov and Beutel 2010].

Classification and World Distribution

The family Corylophidae comprises ~400 species in 26 genera and occurs in all major regions of the world. There are two subfamilies, the Periptyctinae with three Australian genera and Corylophinae, which is divided into eleven tribes, seven of which occur in Australia. The exotic tribes are Aenigmaticini (North and Central America), Stanini (New Zealand), Cleidostethini (East Africa) and Teplinini (Mediterranean region). [Bowestead 1999; Bowestead *et al.* 2001; Ślipiński *et al.* 2001, 2009, 2010c; Robertson *et al.* 2013].

Australian Fauna

Among the **Periptyctinae**, species of *Periptyctus* Blackburn (Fig. 71F) are distributed throughout the eastern part of Australia from TAS to Cape York Peninsula, while *Pakalukodes bimaculatus* Ślipiński *et al.* (Fig. 89F) and *Weirus tozer* Ślipiński *et al.* occur in northern QLD. The Australian **Corylophinae** comprise seven genera, each representing a different tribe, plus *Pseudostanus* Robertson *et al.*, whose tribal affinities are uncertain. The most unusual, more or less elongate adults with exposed heads belong to *Priamima rubicunda* Macleay (Foadiini) (Fig. 89C) and *Pseudostanus*. The remaining Australian corylophines are moderately to very short and broad and either somewhat flattened, like *Serico-*

derus and *Arthrolips* (Fig. 89B) or subcircular and dorsally convex, like *Holopsis* (Fig. 89A), *Corylophus*, *Orthoperus* and *Gloeosoma*, with the head more or less concealed from above by the pronotum (slightly exposed in *Orthoperus*). Corylophines are minute beetles, with a small head and relatively long antennae with a large club. Species of *Orthoperus*, *Holopsis*, *Corylophus* and *Gloeosoma* are glabrous, while *Priamima*, *Pseudostanus*, *Sericoderus* and *Arthrolips* are almost always pubescent. In *Sericoderus*, the posterior pronotal angles extend posteriorly to embrace the elytral humeri, and in *Gloeosoma* the labrum is acute and the mandibles and maxillae stylet-like. Larvae are oblong to broadly ovate, more or less flattened, clothed with variously modified setae and usually with paired gland openings on abdominal segments I–VII or I and VIII. Antennae are moderately to very long, the head may be strongly transverse to elongate, the mouthparts are sometimes stylet-like and endognathous, and the pretarsal setae are usually clavate. Species of *Sericoderus* and *Orthoperus* are commonly associated with grass cuttings or mouldy hay, and those of *Periptyctus*, *Holopsis* and *Gloeosoma* have been taken in leaf litter samples. Some Parmulini and Sericoderini have been collected feeding on surface spore fields produced by certain Ascomycetes. Corylophids have also been collected under bark, in rotten wood and on the pore surfaces of bracket fungi. As in Coccinellidae, the pupa is obtect and partly enclosed within the larval skin. [Pakaluk and Lawrence 1986; Ślipiński *et al.* 2001, 2009, Ślipiński *et al.* 2010c; Tomaszewska and Ślipiński 2002].

Australian Subfamilies and Genera

Periptyctinae: *Pakalukodes* Ślipiński *et al.*, *Periptyctus* Blackburn, *Weirus* Ślipiński *et al.*

Corylophinae: Foadiini: *Priamima* Pakaluk & Lawrence; **Peltinodini**: *Holopsis* Broun; **Orthoperini**: *Orthoperus* Stephens; **Parmulini**: *Arthrolips* Wollaston; **Corylophini**: *Corylophus* Stephens; **Sericoderini**: *Sericoderus* Stephens; **Rypobiini**: *Gloeosoma* Wollaston; **Tribe ?:** *Pseudostanus* Robertson *et al.*

86. LATRIDIIDAE

Adult (Figs 72I, 87H, P)

Length usually 0.9–2.3 mm. Body usually ~1.7–2.5 times as long as wide; sides of prothorax and elytra slightly independently curved or more or less parallel, sometimes with sides of prothorax straight or sinuate and those of elytra curved; strongly convex to distinctly flattened; colour yellowish-brown to reddish-brown or black, rarely bicolored; upper surfaces in Latridiinae subglabrous, sometimes partly covered with waxy exudate, those in Corticariinae usually clothed with short fine, recumbent hairs (occasionally with long decumbent and erect hairs).

Head prognathous, not concealed from above, usually constricted behind eyes to form temples of various lengths. Eyes usually well developed, entire and finely to coarsely facetted, without interfacetal setae (reduced to a small number of facets in *Dienerella*, *Eufallia* and *Lithostygnus*. Antennal insertions exposed and widely separated; subantennal grooves usually present, extending well behind eyes in some Latridiinae. Frontoclypeal suture present; clypeus on lower plane than frons in Latridiinae. Labrum free and at least visible, short and truncate to large and laterally expanded. Antennae relatively long, usually 11-segmented (10-segmented in *Migneauxia*) with large, globuar scape and usually a weak, loose club (very rarely strong and compact), usually with 3-segments, but 2-segmented in *Lithostygnus*, *Cartodere* and *Dienerella filum* (Aubé). Mandible usually unpigmented apically (more evenly pigmented in *Dienerella*), unidentate or narrowly bidentate; mola well developed; prostheca well developed to highly reduced. Maxilla with lacinia reduced or apparently absent; apical maxillary and labial palpomeres cylindrical to fusiform, but often much narrower than preapical palpomere (aciculate). Mentum trapezoidal, sometimes fused with prementum; ligula undivided. Gular sutures short and widely separated or absent. Corpotentorium slender, arched, with median process in Corticariinae; sometimes absent. Cervical sclerites absent.

Pronotum usually ~0.45–1.3 times as long as wide, widest anteriorly or at middle; sides evenly arcuate, sinuate or emarginate, base usually distinctly narrower than combined elytral bases (not so in *Corticaria*, *Dienerella*, *Eufallia* and *Lithostygnus*); lateral pronotal carinae usually complete (incomplete or absent in a few Corticariinae), with or without raised margin or bead, the edges simple in Latridiinae, denticulate in some Corticariinae; anterior angles usually not produced forward (produced and broadly in *Lithostygnus*); posterior angles obtuse or right; posterior edge straight or slightly rounded; disc in Corticariinae either simple, with shallow medial impression or paired impressions, that in Latridiinae sometimes with paired longitudinal carinae or transverse groove. Prosternum in front of coxae usually slightly shorter than to slightly longer than shortest diameter of coxal cavity, usually flat or slightly convex, with paired setose impressions in some species of *Corticaria*; with median carina in some *Enicmus*. Notosternal suture incomplete or absent. Prosternal process incomplete and apically acute or complete, parallel-sided or weakly expanded apically and narrowly rounded to truncate at apex, concave, flat or carinate, but not strongly curved behind coxae(a concave process in some Corticariinae is more or less interrupted at middle, so that the coxae are almost contiguous). Procoxae often projecting below prosternum, with short to long concealed lateral extensions; trochantins concealed. Procoxal cavities circular to slightly transverse, subcontiguous to moderately widely separated, narrowly open to narrowly closed externally by meeting of postcoxal projections with prosternal process in Corticariinae, broadly closed by meeting of the two postcoxal projections in Latridiinae, internally closed.

Scutellar shield small or absent, abruptly elevated or not and usually broadly rounded to truncate posteriorly. Elytra usually 1.3–1.95 times as long as combined width and 2.2–4.8 times as long as pronotum; sides usually weakly to strongly rounded and conjointly rounded at apex; suture may be fused in flightless species; disc in Latridiinae usually with eight distinct puncture rows and no scutellary striole; in Corticariinae major puncture rows are not as distinct and are often interspersed with rows of finer punctures and punctation may be irregular; epipleura narrow and more or less complete in Latridiinae, usually incomplete or absent in Corticariinae. Mesoventrite fused to mesanepisterna in *Eufallia*; anterior edge with broad prothoracic rest or prepectus. Mesoventral process narrowly rounded to truncate. Mesocoxae sometimes projecting, with concealed trochantins. Mesocoxal cavities circular, narrowly to moderately widely separated, closed laterally by meeting of meso- and metaventrites. Mesometaventral junction a straight or slightly curved line with meso- and metaventrites abutting or slightly overlapping (junction absent due to complete fusion of meso- and metaventrite in *Eufallia*. Metaventrite with very long to short discrimen (absent in *Eufallia*); postcoxal lines absent, but deep postcoxal fovea sometimes present in Lathridiinae; exposed portion of mesanepisternum moderately elongate to very long and narrow or absent. Metacoxae usually strongly transverse (slightly transverse in *Eufallia*), moderately to very widely separated; plates absent. Metendosternite without or with very short and broad stalk and long lateral arms; laminae highly reduced and slender; anterior tendons located on lateral arms. Hind wing with long apical field lacking distinct sclerites or veinal remnants; radial cell and R-M loop absent; medial field without or with one free vein and with medial fleck; wedge cell absent; anal lobe absent; long fringe setae on posterior margin; wings sometimes absent. Legs relatively long and slender; trochanterofemoral joint sometimes strongly oblique with femur in contact with coxa; joint almost transverse and trochanter elongate and narrow in *Eufallia*. Tibiae usually simple, not strongly widened, with outer edge simple. Tarsi 3-3-3 with tarsomeres usually simple; pretarsal claws almost always simple (appendiculate in some species of *Bicava*).

Abdomen usually with five free ventrites in Latridiinae, with six free ventrites in both sexes or in male only in Corticariinae; ventrite 1 distinctly longer than 2, without postcoxal lines, sometimes with deep cavities; intercoxal process broadly rounded to truncate. Functional spiracles usually on abdominal segments I–VII. Anterior edge of segment IX in male without spiculum gastrale; tergite IX emarginate and X free in at least some Latridiinae and Corticariinae. Aedeagus of cucujiform type; anterior edge of tegmen with broad to slender plate or strut; parameres fused together and to phallobase, apparently absent; penis usually with single basal strut. Sternite VIII in female without spiculum ventrale. Ovipositor short and broad, lightly sclerotised, without baculi; coxites undivided and styli absent. [Watt 1969; Johnson 1975; Lawrence *et al.* 1999b; Andrews 2002; Hartley and McHugh 2010].

Larva (Fig. 88I)

Body elongate to oblong, more or less parallel-sided, slightly wider at middle or widest anteriorly and gradually narrowed posteriorly, slightly flattened; dorsal surfaces usually lightly pigmented; vestiture usually of scattered simple setae, mixed with apically expanded setae in Corticariinae.

Head protracted and prognathous, slightly transverse and slightly flattened. Epicranial suture often absent, when present, epicranial stem short and frontal arms U- or V-shaped. Stemmata usually three to five on each side, sometimes fewer or stemmata absent. Frontoclypeal suture absent. Labrum free. Antennae short to moderately long, narrowly separated from mandibular articulations, 3-segmented, with antennomeres 2 and 3 subequal and much longer than 1, or antennomere 2 longer than either 1 or 3; sensorium shorter than to longer than antennomere 3. Mandibles symmetrical, usually lightly sclerotised except for base or base and apex; outer edge or apex with two long setae; apex broadly rounded or with one to four teeth; mola well developed and tuberculate; prostheca usually consisting of broadly rounded, hyaline lobe. Ventral mouthparts retracted, with maxillary articulating area in Corticariinae, protracted, without articulating area in Latridiinae; cardines sometimes indistinct or fused to stipites, which are subquadrate or transverse in Latridiinae or elongate in Corticariinae; mala broadly to narrowly rounded, sometimes with apical spine; maxillary palps 3-segmented. Labium consisting of prementum and postmentum or a single sclerite; ligula simple or absent; labial palps 1- or 2-segmented and widely separated. Hypopharyngeal sclerome usually consisting of transverse bar. Hypostomal rods strongly diverging; ventral epicranial ridges absent. Gular region moderately to very long; gular sutures separate or absent.

Thoracic and most abdominal terga without lateral expansions or distinct tergal plates. Prothorax usually narrower and longer than meso- or metathorax. Legs well developed, 5-segmented including pretarsus, which is claw-like bearing single, simple seta; coxae moderately to widely separated. Paired setiferous tubercles sometimes present in terga I–VIII. Tergum IX slightly shorter and narrower than VIII, not extending onto ventral surface, without urogomphi; sternum IX exposed, simple, not enclosed by sternum VIII; segment X exposed from above, posteriorly oriented. Spiracles annular. [Böving and Craighead 1931; Hinton 1945a; Hammad 1953; Chandler 1983; Lawrence 1991b; Lawrence *et al.* 1999a; Andrews 2002; Hartley and McHugh 2010].

Classification and World Distribution

Latridiidae is a cosmopolitan family with 29 recognised genera and ~1100 species. The group is consistently divided into two subfamilies, both of which occur in Australia.

Australian Fauna

Australian species of both **Latridiinae** and **Corticariinae** contain a mixture of endemic groups and adventive species that are often cosmopolitan in distribution and associated with imported food products. Most of the endemic latridiines are in the genera *Aridius*, *Enicmus* and *Lithostygnus* and endemic corticariines are primarily in the genus *Bicava* (Fig. 72I). Latridiids are minute beetles with the pronotum usually much narrower than the elytral bases, elytral punctures more or less seriate, clypeus laterally expanded in front of the antennal insertions and antennal scape usually large and more or less globular. Latridiinae are subglabrous, with the clypeus on a lower plane than the frons, separated procoxae and usually with carinae on the pronotum or elytra and a waxy exudate covering various portions of the body. Corticariinae are usually setose with contiguous procoxae, and no carinae or waxy exudate. Larvae are elongate and usually tapered posteriorly, lightly sclerotised except for the head, and with a vestiture of long, fine setae or shorter, frayed setae. Antennae are relatively long, mandibles often partly membranous and mala obtuse. Adults and larvae feed on the spores on a variety of fungi and are commonly taken from dead vegetation and leaf litter. Some *Enicmus* species feed on the spores of Myxomycetes, while other latridiids specialise in moulds or other ascomycetous fungi. *Cortinicara hirtalis* (Broun) (Fig. 87H) is often beaten from foliage. Several Australian species, such as *Latridius minutus* (Linnaeus), *Cartodere constricta* (Gyllenhal), *Dienerella* spp., *Adistemia watsoni* (Wollaston), *Corticaria japonica* Reitter, *C. elongata* (Gyllenhal) and *C. fenestralis* (Linnaeus), have been introduced in ballast or mouldy food stores, while others like

Aridius nodifer (Westwood) (Fig. 87P), *A. bifasciatus* (Reitter) and *A. australicus* Belon are probably endemic taxa that have been introduced elsewhere. [Aitken 1975; Hinton 1945a; Watt 1969; Johnson 1975; Hartley and McHugh 2010].

Australian Subfamilies and Genera

Latridiinae: *Adistemia* Fall, *Aridius* Motschulsky, *Cartodere* Thomson, *Dienerella* Reitter, *Enicmus* Thomson, *Eufallia* Muttkowski, *Latridius* Herbst, *Lithostygnus* Broun,

Corticariinae: *Bicava* Belon, *Cortinicara* Johnson, *Corticaria* Marsham, *Migneauxia* Jacquelin du Val.

TENEBRIONOIDEA

Basic features of adult Tenebrionoidea, as here delimited, include 5-5-4 tarsi in both sexes (with reduction to 4-4-4 in several lineages), strongly oblique trochanterofemoral joint with the femur in contact with the coxa, reduction in the number of wing veins in the medial field (between MP_{1+2} and the anal fold) to 4 or fewer, aedeagus of the tenebrionoid type, in which the incomplete tegmen lies above the penis (orientation reversed in some groups), and a tendency for the procoxae to be conical and projecting or partly concealed by the procoxal cowling and their trochantins to be reduced and/or concealed. Larvae have fewer than six stemmata on each side, a blunt mala, no mandibular prostheca, and a transverse or somewhat rectangular segment X that does not form a cylindrical pygopod and is often concealed from above.

This phylogeny of Tenebrionoidea has received much less attention than that of other superfamilies, and its constitution is still somewhat in doubt. Although Crowson (1966b) and Lawrence (1977) proposed possible evolutionary scenarios, the only morphological cladistic analyses of Beutel and Friedrich (2005) was based on a small sample of larvae but the larger tenebrionoid datasets of Hunt *et al.* (2007) and Lekanicová and Bocak (2009) recovered a monophyletic Tenebrionoidea only when including Lymexylidae. Several tenebrionoid families, as here constituted, may not be monophyletic, and three subfamilies of Anthicidae (Afreminae, Ischaliinae and Lagrioidinae) were treated as Tenebrionoidea *incertae sedis* by Lawrence *et al.* (2010m). Tenebrionoid families not occurring in Australia include Pterogeniidae (India and Philippines to New Guinea and Solomon Is.) (Lawrence 1977, 2010j; Burckhardt and Löbl 1992; Löbl and Leschen 1994; Löbl 2005), Tetratomidae (Holarctic, Neotropical, Afrotropical and Oriental) (Nikitsky 1991, 1998, 2004a, 2004b; Young and Pollock 2002; Lawrence and Leschen 2010h; Pollock 2012), Chalcodryidae (New Zealand) (Watt 1974a; Lawrence and Leschen 2010i), Trachelostenidae (Chile) (Lawrence and Ślipiński 2010f), Synchroidae (Holarctic, Oriental) (Young 2002k; Ślipiński and Lawrence 2010c), Stenotrachelidae (Holarctic) (Young 2002l; Lawrence and Ślipiński 2010g), Trictenotomidae (eastern Palaearctic and Oriental) (Pollock and Telnov 2010). [Lawrence *et al.* 2010i].

87. MYCETOPHAGIDAE

Adult (Fig. 90G, K–L)

Length usually 2.2–3.6 (0.9–1.2 in *Nesolathrus* Scott) mm. Body usually ~1.79–2.3 times as long as wide (as much as 2.54 time in *Nesolathrus*); sides usually slightly rounded (more or less parallel in *Nesolathrus*), slightly flattened to moderately convex; colour yellowish-brown to dark brown, sometimes bicolored with pattern on elytra; vestiture of usually of dense, short to moderately long, recumbent or recumbent and suberect hairs (hairs minute in *Nesolathrus*).

Head prognathous to slightly declined, abruptly constricted behind eyes to form neck and temples only in *Nesolathrus*. Eyes prominent, not or slightly emarginate and coarsely facetted, without or with very short interfacetal setae. Antennal insertions at least slightly concealed from above; subantennal grooves absent. Frontoclypeal suture distinct and more or less curved; clypeus transverse, subtrapezoidal and broadly rounded or truncate at apex. Labrum free. Antennae 11-segmented, almost always with weak to moderately strong, relatively loosely-jointed, 3-segmented club (4-segmented in *Mycetophagus quadriguttatus* Müller). Mandible bidentate, with well developed mola and prostheca. Maxilla with moderately broad, setose galea and narrow lacinia without uncus; apical maxillary palpomere cylindrical or fusiform to sightly expanded and truncate. Mentum transverse; ligula simple and truncate; apical labial palpomere cylindrical to fusiform. Corpotentorium narrow and arched. Cervical sclerites usually moderately well developed.

Pronotum ~0.4–0.6 times as long as wide, usually widest at or behind middle and as wide at base as combined elytral bases, sides usually weakly curved and anteriorly converging (subparallel in *Nesolathrus*). Lateral pronotal carinae complete, usually simple (minutely denticulate in *Nesolathrus* and *Mycetophagus quadriguttatus*); anterior angles not produced forward; posterior angles more or less right in *Typhaea* Stephens, *Nesolathrus* and *M. punctatus,* subacute and partly overlapping elytral humeri in *Litargus* Erichson and *Litargops* Reitter; posterior edge straight or bisinuate; disc with pair of small sub-basal pits or grooves at the attachments of internal apodemes. Prosternum in front of coxae not or only slightly longer than mid length of procoxal cavity in *Typhaea* and *M. quadriguttatus*, much longer in *Litargus*, *Litargops* and *Nesolathrus*, flat to moderately convex, with shallow, setose impression near anterolateral corner in males of *Nesolathrus.* Prosternal process complete, weakly curved behind coxae, more or less parallel-sided or gradually expanded with apex narrowly rounded to truncate. Procoxae without or with very short concealed extensions; trochantins concealed or barely exposed. Procoxal cavities slightly transverse to more or less circular, moderately narrowly separated, externally broadly open, with or without narrow lateral extensions, usually internally closed (internally open in *Nesolathrus*).

Scutellar shield usually not abruptly elevated, posteriorly broadly rounded to obtusely angulate or truncate. Elytra ~1.35–1.7 times as long as combined width and 2.9–3.95 times as long as pronotum; disc usually with 10 or 11 puncture rows and no scutellary striole, punctation sometimes irregular; epipleura incomplete. Mesoventrite usually separated by complete sutures from mesepisterna, which are narrowly separated at midline so that ventrite is strongly narrowed anteriorly (mesoventrite and mesanepisterna fused in *Nesolathrus*), anterior edge without paired procoxal rests, but with median prosternal rest. Mesoventral process narrowly rounded to slightly emarginate at apex. Mesocoxae with exposed trochantins. Mesocoxal cavities subcircular, moderately narrowly separated, partly closed laterally by mesepimera. Mesometaventral junction monocondylic. Metaventrite with short to long discrimen and no postcoxal lines; exposed portion of metanepisternum usually 3 or 4 times as long as wide, much narrower in *Nesolathrus*. Metacoxae strongly transverse and narrowly separated. Metendosternite usually with long stalk, long lateral arms, short, broad laminae, and no anterior process, with anterior tendons on lateral arms (stalk and laminae absent in *Nesolathrus*). Hind wing usually with one or two small post-radial sclerites, an oblique linear sclerite just apical to and partly contiguous with cross-vein r4, and vaguely indicated radial extensions; radial cell short and broad to slightly elongate, with inner posterobasal angle obtuse; cross-vein r3 absent; basal portion of RP moderately long; medial spur reaching wing margin, where there is a slight embayment; medial field always with weakly divided medial fleck, with four free veins in *Mycetophagus*, three in *Litargus*, *Litargops* and *Typhaea* and none in *Nesolathrus*; elongate wedge cell only in *Mycetophagus*; anal lobe usually well developed, without embayment, lobe absent and hind wing margin with long fringe hairs in *Nesolathrus*. Legs well developed; trochanterofemoral joint oblique, sometimes with base of femur in contact with coxa. Protibial apices not strongly widened, with simple outer edges; spurs paired on all legs, sometimes serrate or pectinate. Tarsi almost always 4-4-4 in female and 3-4-4 in male (3-3-3 in both sexes in *Nesolathrus*); tarsomeres simple, 1 usually much longer than 2 and sometimes as long as 2 and 3 combined; pretarsal claws simple; empodium usually not apparent.

Abdomen with five free ventrites; ventrite 1 not much longer than 2; intercoxal process acute or narrowly rounded. Functional spiracles on abdominal segments I–VII, those on VI and VII somewhat reduced. Sternite VIII in male with short anterior strut. Anterior edge of segment IX in male with spiculum gastrale; tergite IX divided into two parts and segment X membranous. Aedeagus of uninverted tenebrionoid type, with phallobase and parameres lying dorsal to penis; parameres free or fused together but articulated to phallobase; penis with short, paired anterior struts. Sternite VIII in female with long spiculum ventrale. Ovipositor moderately elongate, lightly sclerotised except for baculi; proctiger elongate with distinct baculi; paraprocts slightly longer than coxites, each of which is subdivided into three lobes and uniformly clothed with fine hairs; distal coxital lobe as long as basal ones combined; styli well developed, sclerotised and subterminal. [Scott 1922; Hinton 1945a; Miyatake 1957; Leschen and Lawrence 1991; Nikitsky 1993; Lawrence *et al.* 1999b; Young 2002h; Lawrence and Leschen 2010g].

Larva (Fig. 90F)

Body elongate, more or less parallel-sided and slightly flattened; dorsal surfaces usually more heavily pigmented than ventral ones; surfaces generally smooth; vestiture of fine setae only.

Head prognathous and protracted, slightly transverse and slightly flattened. Epicranial stem very short or absent; frontal arms lyriform and contiguous at base. Median endocarina absent. Stemmata four or five on each side. Frontoclypeal suture absent. Labrum free. Antennae well developed, 3-segmented, with sensorium conical, palpiform or dome-like and shorter than apical antennomere. Mandibles (Fig. 39F) asymmetrical, broad at base and narrow at apex, bidentate; incisor edge simple or with one or two subapical teeth; mola well developed. Ventral mouthparts retracted, with well developed maxillary articulating area; stipites elongate; mala rounded or truncate, not cleft; maxillary palps 3-segmented. Labium consisting of prementum, mentum and submentum; ligula simple, shorter than labial palps, which are 2-segmented and narrowly to moderately widely separated. Hypopharyngeal sclerome a transverse bar. Hypostomal rods moderately long, subparallel or diverging; ventral epicranial ridges absent. Gular region moderately long; gular sutures well separated.

Mesotergum, metatergum and abdominal terga I–IX often with well defined tergal plate and sometimes with fine, anterior transverse carina. Legs well developed, 5-segmented including claw-like pretarsus bearing two seta; coxae narrowly to moderately widely separated. Segment IX distinctly shorter than VIII, tergum extending onto ventral surface, with pair of posteriorly-oriented, apically upturned, simple urogomphi; sternum IX exposed, simple, not enclosed by sternum VIII; segment X not visible from above, posteroventrally oriented and transverse, sometimes with tergum and sternum equally pigmented. Spiracles annular-biforous, or annular–uniforous or annular. [Böving and Craighead 1931; Hinton 1945a; Hayashi 1971; Wallage and Rose 1982; Lawrence 1991b; Lawrence *et al.* 1999a; De Marzo 2000; Young 2002h; Lawrence and Leschen 2010g].

Classification and World Distribution

The family Mycetophagidae is widely distributed (with a few cosmopolitan species), but its greatest diversity is in the temperate parts of the Northern Hemisphere (especially Eurasia) with secondary areas of divergence in New Zealand and Chile. The ~130 described species in 18 genera are currently placed in three subfamilies, two of which occur in Australia. The exotic subfamily Esarcinae occurs in southern Europe and North Africa. [Leschen and Lawrence 1991; Lawrence and Leschen 2010g].

Australian Fauna

The Australian mycetophagid fauna is relatively depauperate; the **Mycetophaginae** is represented by two endemic species of *Litargops* Reitter (Fig. 90G), plus the adventive species *Litargus balteatus* LeConte (Fig. 90K), *Mycetophagus quadriguttatus* Müller and *Typhaea stercorea* (Linnaeus) (Fig. 90L), and the **Bergininae** by *Nesolathrus typicus* Scott. Mycetophagids are oblong to ovate, somewhat flattened, densely pubescent beetles, which are uniformly brown or yellow, or bicoloured with a yellow and black pattern on the elytra. The eyes are relatively large and coarsely facetted, the frontoclypeal suture is distinct, the tibial spurs are often well developed and serrate, and the male protarsi have three, instead of four tarsomeres (except in *Nesolathrus*, where all tarsi are 3-segmented). Larvae are elongate, parallel-sided, slightly flattened, and moderately sclerotised dorsally (except in *Nesolathrus*), with well developed legs and slightly curved, simple urogomphi. Adults and larvae are primarily mycophagous and may be found in rotten wood or associated with fungal fruiting bodies. The introduced *Litargus balteatus* and *Typhaea stercorea* usually occur in stored foodstuffs affected by mould, but larvae and adults of the former species (as *Triphyllus minor* Lea) were observed feeding on *Claviceps paspali* (ergot) on *Paspalum dilatatum* (Wallage and Rose 1982). [Aitken 1975; Hinton 1945a; Lawrence 1987b; Lawrence and Leschen 2010g].

88. ARCHEOCRYPTICIDAE

Adult (Fig. 89H–I)

Length 1.35–4.5 mm. Body usually ~1.44–1.82 times as long as wide; sides slightly to strongly, more or less uniformly, rounded, slightly flattened to moderately convex; colour yellowish-brown to dark brown or black, rarely vaguely bicolored; vestiture of dense, short to very short, recumbent hairs, occasionally forming a marbled pattern.

Head prognathous to moderately declined, not abruptly constricted behind eyes; posterior edge of above occipital foramen with short median endocarina flanked by a pair of obliquely transverse endocarinae. Eyes slightly prominent, usually slightly emarginate anteriorly and flattened posteriorly, the vertical diameter exceeding the horizontal, coarsely facetted, without or with very short, fine interfacetal setae. Antennal insertions concealed from above; subantennal grooves weakly developed or absent. Frontoclypeal suture usually distinct and more or less curved; clypeus short, broad and anteriorly rounded. Labrum free, with exposed basal membrane. Antennae 11-segmented, usually with weak, loose, 3-segmented club (5- or 7-segmented in *Enneboeopsis*, 2-segmented in *Gondwanenneboeus* and *Nothenneboeus*). Mandible short and broad, bidentate, with well developed, transversely ridged mola and membranous, apically pubescent prostheca. Maxilla with galea and lacinia setose, the latter much narrower than the former, with terminal uncus; apical maxillary palpomere almost always expanded apically and subtriangular or securiform (fusiform in *Falsoplatydema*). Mentum subquadrate; ligula simple and truncate; apical labial palpomere fusiform. Corpotentorium narrow and strongly arched. Cervical sclerites absent.

Pronotum ~0.35–0.6 times as long as wide, always widest posteriorly and as wide at base as combined elytral bases, sides slightly to strongly curved and anteriorly converging, somewhat explanate in *Nothenneboeus*. Lateral pronotal carinae complete, simple, with variously developed raised margin or bead; anterior angles usually weakly produced and broadly rounded; posterior angles obtuse or right; posterior edge bisinuate; disc usually simple without sub-basal pits (in *Wattianus* disc with pair of long sublateral ridges and adjacent grooves beginning at posterior angles and converging anteriorly to meet the anterior edge of disc well mesal to anterior angles). Prosternum in front of coxae shorter than mid length of procoxal cavity, flat to strongly convex. Prosternal process complete, extending well behind coxae, but not dorsally curved (Fig. 20A) strongly and abruptly expanded at apex, which narrowly to broadly rounded. Procoxae without or with short concealed lateral extensions; trochantins concealed or slightly exposed. Cryptopleuron fused to notum at junction of notosternal suture and procoxal cavity. Procoxal cavities slightly transverse to subcircular, moderately widely separated, externally broadly closed, with very narrow lateral extensions, usually internally closed by slender bar.

Scutellar shield not abruptly elevated, posteriorly broadly rounded to obtusely angulate. Elytra ~1.0–1.35 times as long as combined width and 1.8–3.9 times as long as pronotum; disc irregularly punctate or with nine or ten distinct puncture rows and a scutellary striole; epipleura complete. Mesoventrite on same plane as metaventrite; anterior edge with large, subtriangular prosternal rest partly divided at middle by weak carina. Mesoventral process truncate to emarginate at apex. Mesocoxae with exposed trochantins. Mesocoxal cavities subcircular, moderately to widely separated, partly closed laterally by mesanepisterna and mesepimera. Mesometaventral junction monocondylic. Metaventrite transverse, slightly convex, with short to moderately long discrimen and no postcoxal lines; exposed portion of metanepisternum ~3 times as long as wide and anteriorly widened. Metacoxae contiguous. Metendosternite with moderately to very long stalk and long lateral arms; laminae well developed, reduced or absent; anterior process short or absent and anterior tendons located on lateral arms or not apparent. Hind wing usually with moderately long apical field containing two or three small post-radial sclerites but no distinct radial extensions (highly reduced in *Nothenneboeus*); radial cell short and broad with inner posterobasal angle acute; cross-vein r3 absent; basal portion of RP moderately long; medial spur strongly curved and ending well before wing margin; medial field with four free veins and a divided medial fleck; wedge cell elongate and apically acute; CuA_{1+2} absent, so that CuA_2 and CuA_1 appear as a single bent veins attached to MP_{3+4}; anal lobe usually well developed, without embayment (absent in *Nothenneboeus*). Legs well developed; trochanterofemoral joint strongly oblique, with base of femur in contact with coxa. Tibiae usually not strongly widened, but outer edges often lined with stout spines and outer apical angles sometimes produced and spinose; spurs small and simple, paired on all legs. Tarsi almost always 5-5-4 in both sexes (4-4-4 in *Nothenneboeus carinatus* Lawrence); first two or three tarsomeres densely setose beneath; penultimate tarsomere sometimes highly reduced; pretarsal claws simple and empodium not apparent.

Abdomen with five ventrites, the first two of which are connate; ventrite 1 not much longer than 2; intercoxal process acute. Functional spiracles on abdominal segments I–VII. Sternite VIII in male with short anterior strut. Anterior edge of segment IX in male with long spiculum gastrale; tergite IX and segment X more or less membranous. Aedeagus of uninverted tenebrionoid type, with phallobase dorsal to penis; parameres partly fused together forming articulated apicale, which is always much shorter than phallobase; penis with well developed anterior struts; ejaculatory duct with scleotised seminal pump. Sternite VIII in female with long spiculum ventrale. Ovipositor moderately elongate, with paraprocts subequal in length to coxites, each of which is divided into three lobes, with long, narrow, apical styli. [Kaszab 1981, 1984; Merkl 1988; Lawrence 1994b; Lawrence *et al.* 1999b; Lawrence 2010i].

Larva (Fig. 90C)

Body elongate, more or less parallel-sided and slightly flattened; dorsal and ventral surfaces lightly pigmented, smooth, clothed with scattered, erect setae.

Head prognathous and protracted, slightly transverse and slightly flattened. Epicranial stem short; frontal arms lyriform and contiguous at base. Median endocarina absent. Stemmata five on each side. Frontoclypeal suture distinct. Labrum free, slightly transverse, apically rounded. Antennae moderately long, 3-segmented, with conical sensorium shorter than apical antennomere. Mandibles (Fig. 41H) asymmetrical, broad at base and narrow at apex, bidentate; incisor edge simple or with one or two subapical teeth; mola well developed, without hyaline lobe or brush at base. Ventral mouthparts retracted, with well developed maxillary articulating area; mala truncate, not cleft; maxillary palps 3-segmented. Labium consisting of prementum, mentum and submentum; ligula absent; labial palps 2-segmented and narrowly separated. Hypopharyngeal sclerome molar-like. Hypostomal rods short and subparallel; ventral epicranial ridges absent. Gular region transverse; gular sutures well separated.

Thoracic and abdominal terga without patches or rows of asperities or carinae. Legs moderately long, not spinose; 5-segmented including claw-like pretarsus bearing two setae lying in row, one distal to the other; coxae separated by one or two basal diameters. Tergum IX as long as or slightly longer than VIII, not extending onto ventral surface or forming articulated plate, with pair of posteriorly-oriented, simple urogomphi, which are not or barely darker than tergum and not upturned at acute apex (Figs 52H, 90C); sternum IX exposed, simple, not enclosed by sternum VIII; segment X visible from above, posteriorly oriented and transverse, with tergum and sternum equally pigmented. Spiracles annular–biforous, not placed at ends of spiracular tubes, those on segment VIII not enlarged. [Lawrence 1991b; Lawrence *et al.* 1999a; Young 2002i; Lawrence 2010i].

Classification and World Distribution

Archeocrypticidae are widely distributed in most parts of the world but rare in the Holarctic region and most diverse in Australia. The group currently comprises 53 species in ten genera, nine of which occur in Australia.

Australian Fauna

The Australian archeocrypticid fauna includes nine genera and several species. *Archeocrypticus topali* Kaszab is restricted to TAS, but the same species is known from New Zealand and Chile. *Enneboeus* Waterhouse is the largest genus with species occurring from TAS to QLD, but the genus also extends throughout the New World from the USA A. to Chile. *Wattianus* Kaszab (Fig. 89H) occurs in northern NSW and southern QLD, *Gondwanenneboeus* occurs from south-western NSW and the ACT to WA, and *Falsoplatydema* Kaszab includes a south-eastern and south-western species. The genus *Australenneboeus* Kaszab occurs in the central and northern part of the continent, *Nothenneboeus* Lawrence (Fig. 89I) contains a north-eastern and north-western species, and *Enneboeopsis* Champion a eastern and south-western species. The genus *Sivacrypticus greensladei* Kaszab is known only from SA, but several other species occur throughout the Oriental and Afrotropical regions. Archeocrypticids are more or less ovate, slightly flattened, finely pubescent, black or brown beetles. The frontoclypeal suture is distinct, the apical maxillary palpomere is expanded and truncate, the lateral pronotal carinae are well developed, the prosternal process is abruptly expanded apically and the tibiae are more or less spinose. Larvae are elongate, parallel-sided, slightly flattened and lightly pigmented, with well separated, simple, acute, posteriorly-projecting urogomphi. Adult and larval archeocrypticids may be found in leaf litter or in the softer fruiting bodies of some polypores, such as *Piptoporus*. [Kaszab 1984; Merkl 1988; Lawrence 1994b].

89. CIIDAE

Adult

(Figs 76E, 89J). Length 0.9–4.5 mm (usually 2,5 mm or less). Body usually ~1.35–2.65 times as long as wide; sides usually subparallel to slightly rounded, occasionally strongly, uniformly rounded; usually slightly flattened to strongly, rarely strongly flattened; colour yellow to red, reddish-brown, dark brown or black, occasionally with head and pronotum a lighter colour than elytra, rarely with dark and light pattern on elytra or elytra and pronotum; dorsal surfaces either subglabrous or with vestiture of short and/or long fine hairs or stout bristles; ventral surfaces with or without fine, decumbent setae.

Head slightly to strongly declined, usually more or less globular, not abruptly constricted behind eyes. Eyes entire, somewhat prominent, coarsely facetted, with or without interfacetal setae. Antennal insertions exposed or concealed; subantennal grooves well developed. Vertex flat to strongly concave or biconcave, occasionally with small setose fovea in male. Frontoclypeal suture distinct and strongly curved; frontoclypeal region in males often raised to form one or two plates or horns or two or four teeth or tubercles. Labrum free, visible, slightly transverse to distinctly longer than wide; apex subtruncate to strongly convex or narrowly rounded, occasionally asymmetrical. Antennae usually 10-segmented with strong, loosely joined, 3-segmented club with four large multi-pronged sensoria (sensillifers) on each club segment (antennae 9-segmented in *Neoennearthron*, Australian species of *Hadreule* and *Ceracis*, and some species of *Orthocis*, 8-segmented in *Ocotemnus*; club 2-segmented in *Hadreule* and *Dimerapterocis*; club more compact, with more than four sensillifers per segment in *Scolytocis* and *Xylographella*). Mandible short and broad, bidentate, with well developed, mola and no prostheca. Maxilla with galea broad and setose and lacinia shortened, more or less fused to stipes, without uncus; apical maxillary palpomere cylindrical to conical or fusiform. Mentum short and subquadrate; prementum often elongate, with or without ligula; apical labial palpomere usually cylindrical to fusiform (expanded and subtriangular in *Scolytocis* and *Xylographella*). Corpotentorium very narrow and strongly arched. Cervical sclerites absent.

Pronotum ~0.55–0.95 times as long as wide, usually widest at middle or posteriorly (slightly wider anteriorly in *Hadreule*), as wide at base as combined elytral bases, sides straight to strongly curved. Lateral pronotal carinae complete, simple, with or without raised margin or bead; anterior angles not to strongly produced forward and broadly rounded to narrowly rounded or acute; posterior angles obtuse or rounded; posterior edge straight or evenly rounded; anterior edge usually at least slightly produced and rounded and in some males strongly produced to form a flattened plate or paired teeth or horns; disc simple, without paired sub-basal pits or impressions, sometimes anteriorly impressed in males. Prosternum in front of coxae much shorter than to much longer than mid length of procoxal cavity, concave or biconcave to strongly convex or carinate. Prosternal process complete, incomplete or absent, narrowed apically or parallel-sided, flat, concave or only slightly elevated behind coxae; not laterally expanded at apex, which is finely acute to broadly rounded or truncate. Procoxae without concealed lateral extensions; trochantins concealed. Cryptopleuron fused to notum at junction of notosternal suture and procoxal cavity. Procoxal cavities slightly transverse, contiguous to moderately widely separated, externally narrowly open or closed, without narrow lateral extensions, internally open.

Scutellar shield well developed to reduced or absent, abruptly elevated, posteriorly broadly rounded to obtusely angulate. Elytra ~0.85–1.80 times as long as combined width and 1.2–2.3 times as long as pronotum; disc usually irregularly punctate, sometimes with indistinctly defined puncture, without scutellary striole; with weak elytral declivity in *Scolytocis* and *Xylographella*; epipleura complete or incomplete. Mesoventrite partly or completely fused to mesanepisterna; anterior edge on same plane as metaventrite, without paired procoxal rests, usually with broad prothoracic rest or prepectus continuing onto mesanepisterna. Mesoventral process acute to narrowly rounded at apex. Mesocoxae with concealed trochantins. Mesocoxal cavities circular to slightly transverse, contiguous to narrowly separated, partly closed laterally by mesepimera. Mesometaventral junction either absent or simple, with ventrites abutting or overlapping. Metaventrite flat to strongly convex, with discrimen moderately long to absent; postcoxal lines absent; exposed portion of metanepisternum long and narrow. Metacoxae contiguous or narrowly separated. Metendosternite without stalk; with widely separated and diverging lateral arms, no laminae or anterior process and anterior process arising near apices of lateral arms. Hind wing with long apical field usually containing one or two vaguely indicated radial extensions; radial cell, cross-veins r3 and r4 and R-M loop absent; medial spur distinctly curved and usually extending to wing margin, where there is a distinct medial embayment; medial field usually with one free vein and a divided medial fleck; anal lobe weakly developed or absent. Legs short and stout; trochanterofemoral joint strongly oblique, but with base of femur separated from coxa. Tibiae often strongly widened at apex or at middle, with outer edges sometimes serrate or with socketed spines; outer apical angle of protibia often produced forming one or more teeth or a rounded lobe bearing socketed spines; spurs inconspicuous or absent. Tarsi almost always 4-4-4, with tarsomere 4 as long as first three combined; tarsomeres simple; pretarsal claws simple.

Abdomen with five free ventrites; ventrite 1 not much longer than 2, often with median setose patch or fovea in male; intercoxal process acute. Functional spiracles on abdominal segments I–VII. Sternite VIII in male with or without short anterior strut. Anterior edge of segment IX in male usually forming broadly rounded to acute subgenital plate (with spiculum gastrale in *Scolytocis* and *Xylographella*); tergite IX and segment X more or less membranous. Aedeagus of inverted tenebrionoid type, with phallobase ventral to penis; phallobase without basal strut; parameres partly or completely fused together forming apicale, which is longer than and overlaps apex of phallobase; penis without anterior struts. Sternite VIII in female with long spiculum ventrale. Ovipositor usually moderately elongate, lightly sclerotised, with proctigeral and paraproctal baculi, and paraprocts longer than wide and longer than coxites, which are subdivided with apical styli; in some genera (e.g. *Ceracis*, *Xylographus*, *Octotemnus*, *Scolytocis* and *Xylographella*), paraprocts are not longer than wide and subequal in length to coxites, and styli are reduced or absent. [Blair 1940; Lawrence *et al.* 1999b; Thayer and Lawrence 2002; Lawrence and Lopes-Andrade 2010].

Larva (Fig. 90B)

Body elongate, more or less parallel-sided and cylindrical and slightly curved ventrally; dorsal and ventral surfaces lightly pigmented, except for mouth frame and tips of urogomphi (rarely with light tergal plates on most segments); smooth with vestiture of scattered, long, simple setae.

Head protracted and hypognathous, globular. Epicranial stem long; frontal arms lyriform and contiguous at base. Median endocarina present beneath epicranial stem. Stemmata usually three to five on each side (sometimes two, one or none). Frontoclypeal suture distinct. Labrum free. Antennae (Fig. 46Q) very short, 2-segmented, with narrowly conical sensorium longer than apical antennomere and arising near its base and a long seta arising at the antennal apex. Mandibles large, robust and asymmetrical, bidentate, without true mola (sometimes with sub-basal pseudomola); with acute, hyaline on one mandible only. Ventral mouthparts retracted, with reduced, slender maxillary articulating area; cardo transverse; stipes elongate; mala (or galea) rounded or truncate, not cleft; lacinia represented by short, truncate, subapical lobe on dorsal surface (Fig. 49G); maxillary palps 3-segmented. Labium consisting of prementum, mentum and submentum; ligula short or absent; labial palps 2-segmented and narrowly separated. Hypopharyngeal sclerome absent. Hypostomal rods absent; ventral epicranial ridges present. Gular region transverse; gular sutures well separated.

Prothorax only slightly longer than meso- or metathorax; terga without patches or rows of asperities or carinae. Legs moderately short and broad, 5-segmented including claw-like pretarsus bearing two seta; coxae narrowly separated. Abdomen more than twice as long as thorax; terga and sterna without patches or rows of asperities. Tergum IX slightly longer than VIII, extending onto ventral surface, not forming articulated plate, usually with pair of upturned urogomphi; sternum IX exposed, simple, not enclosed by sternum VIII; segment X not visible from above, posteroventrally oriented and transverse. Spiracles annular or annular–uniforous, not placed at ends of spiracular tubes, those on segment VIII not enlarged. [Entwistle 1955; Lawrence 1974, 1991b; Costa *et al.* 1988; Rühm and Milewski 1996; Lawrence *et al.* 1999a; Thayer and Lawrence 2002; Lawrence and Lopes-Andrade 2010].

Classification and World Distribution

Ciidae is a widely distributed family occurring on all continents and most island groups and comprising ~650 described species currently placed in 43 genera. There are two subfamilies: Sphindociinae with a single species *Sphindocis denticollis* Fall from coastal California, and Ciinae containing the remainder of the genera.

Australian Fauna

Although there are only 17 species described from Australia, placed in four genera, at least 120 Australian species are known to the senior author, several more genera are known to occur here (**see** below), and several new genera remain to be described. Within the subfamily **Ciinae**, the tribe Xylographellini (Lopes-Andrade 2008a) is represented by one species each of *Xylographella* Miyatake and *Scolytocis* Blair, both from northern QLD. Orophiini, as presently defined, is represented by several species of *Xylographus* Mellié, including *X. bynoei* Blair, and four species of *Octotemnus* Mellié (Kawanabe 2002), including *O. dilutipes* (Blackburn) (Fig. 76E). The remainder of the species belong to Ciini and include one species each of *Dichodontus* Kawanabe (Kawanabe 1994a), *Dimerapterocis* Scott (1926) (Christmas Island), *Hadreule* Thomson and *Neoennearthron* Miyatake (Kawanabe 1994b), at least five species of *Ceracis* Mellié and ten of *Orthocis* Casey, including *O. leanus* (Blackburn), and as many as 80 species of *Cis* Latreille. Ciidae may be elongate and subcylindrical or moderately flattened to ovate and globose beetle, which may be subglabrous or clothed decumbent and/or erect hairs or bristles. The head is somewhat deflexed, usually with a distinct frontoclypeal ridge, which may be produced in the male to form a plate or a pair of tubercles or horns. The antennal club is large and usually loose, with well developed multi-pronged sensoria, the maxillae have reduced lobes and fusiform apical palpomeres, the pronotum is large, sometimes with an anterior plate or paired horns in the male, the tibiae lack distinct spurs and may have spines along the outer edge, the protibia often has an acute tooth or comb of spines at the apex, and ventrite 1 in the male often bears a median setose fovea or patch of glandular setae, which is thought to produce a pheromone (Lopes-Andrade *et al.* 2003). Ciid larvae are elongate, subcylindrical and lightly sclerotised with varied armature on tergum IX, but usually paired urogomphi, and a pygopod-like segment X. The antennae are very short and 2-segmented, with a long sensorium arising from near the base of the terminal antennomere, the mandible often has an acute, hyaline process at the base, and the maxilla bears a dorsal, subapical, laciniar lobe. Most Ciidae feed on the hyphae comprising the more durable frtuiting bodies of various Badisiomycetes, particularly Polyporaceae, Ganodermataceae and Hymenochaetaceae, but some feed in rotten vines or branches and others specialise on the gelatinous fruiting bodies of Auriculariaceae. Patterns of host use in this group have been studied by several workers (Paviour-Smith 1960a; Lawrence 1973; Orledge and Reynolds 2005). A few species, such as the introduced *Cis chinensis* Lawrence, are known to damage commercial dried fungi (Lawrence 1991c; Lopes-Andrade 2008b). A relatively common Australian species, *Cis bilamellatus* Wood, has been introduced into Britain and is now generally distributed throughout England, Wales, southern Scotland, Ireland, the Channel Islands and north-western France (Paviour-Smith 1960b; Orledge *et al.* 2010). [Blair 1940; Lawrence 1971; Lawrence and Lopes-Andrade 2010].

90. MELANDRYIDAE

Adult (Figs 76A–C, 93A)

Length 1.2–15 mm. Body usually ~3.7–4.6 times as long as wide; usually widest near anterior end and narrowed posteriorly; sides subparallel to slightly rounded; usually slightly flattened to moderately convex (strongly flattened in *Mystes*); colour yellow to dark brown or black, with dark and light pattern on elytra in *Dircaea*, *Megapselaphus* and some species of *Orchesia*; dorsal surfaces clothed with very short or moderately long, fine, recumbent hairs (with groups of short spines along posterior and sutural edges of elytra in *Ctenoplectron*).

Head slightly to strongly declined, sometimes concealed from above in *Orchesia*, *Phloeotrya* and *Dircaea*), usually more or less globular, not abruptly constricted behind eyes. Eyes usually not or only slightly emarginate (distinctly emarginate in *Orchesia* and *Mystes*), not prominent, finely or coarsely facetted, without interfacetal setae. Antennal insertions usually slightly concealed (exposed in *Orchesia*); subantennal grooves absent. Vertex usually strongly, evenly declined (not so in *Mystes*, in which frontal region is abruptly declined and broadly concave). Frontoclypeal suture absent; anterior edge of clypeus with distinct hyaline anteclypeus. Labrum free, visible, transverse; apex broadly rounded to truncate. Antennae 11-segmented, usually moderately long and filiform or serrate (shorter and incrassate in *Orchesia*). Mandible short and broad, bidentate, with well developed, mola and prostheca. Maxilla with galea and lacinia apically rounded and setose and lacinia without uncus; apical maxillary palpomere expanded and securiform to cultriform and palpomeres 2 and 3 sometimes similarly modified. Mentum slightly transverse; ligula simple; apical labial palpomere usually cylindrical to fusiform or slightly expanded and truncate. Corpotentorium narrow and arched or sometimes absent. Cervical sclerites absent.

Pronotum ~0.50–0.95 times as long as wide, widest at middle or posteriorly, as wide at base as combined elytral bases, sides slightly curved. Lateral pronotal carinae usually incomplete (complete in some *Orchesia*), simple, with or without raised margin or bead; anterior angles not produced forward; posterior angles obtuse to right or slightly acute and sometimes embracing elytral humeri; posterior edge at least slightly biemarginate; disc simple or with vaguely indicated, shallow sub-basal impressions. Prosternum in front of coxae usually slightly shorter than mid length of procoxal cavity (much shorter in *Orchesia*, much longer in *Mystes*), usually flattened or slightly convex (concave in *Phloiotrya* and *Dircaea*). Prosternal process usually incomplete, narrowed apically and acute (complete and narrowly rounded in *Orchesia*). Procoxae occasionally projecting below prosternum, without or with very short concealed lateral extensions; trochantins concealed. Cryptopleuron fused to notum at junction of notosternal suture and procoxal cavity. Procoxal cavities slightly transverse, contiguous or narrowly separated, externally broadly open, without narrow lateral extensions, internally closed.

Scutellar shield well developed, not abruptly elevated, posteriorly broadly rounded to obtusely angulate or truncate. Elytra ~2.0–3.5 times as long as combined width and 3.4–4.6 times as long as pronotum, usually complete with apices conjointly rounded concealing all tergites, sometimes independently rounded exposing apex of tergite VII; disc irregularly punctate, without scutellary striole; epipleura incomplete or absent. Mesoventrite sometimes partly or completely fused to mesanepisterna, which may be narrowly separated at midline; anterior edge usually on same plane as metaventrite, without paired procoxal rests, sometimes with median prosternal rest; body of ventrite in *Orchesia* with median longitudinal carina separating a pair of large, shallow impressions, which serve as procoxal rests. Mesoventral process acute to narrowly rounded at apex. Mesocoxae with concealed trochantins. Mesocoxal cavities slightly to strongly transverse, and oblique, contiguous to narrowly separated, partly closed laterally by mesanepisterna and mesepimera. Mesometaventral junction either absent or simple, with ventrites abutting or overlapping.

Metaventrite slightly to strongly convex, sometimes strongly tumid or carinate, with discrimen moderately long; postcoxal lines absent; exposed portion of metanepisternum long and narrow. Metacoxae often enlarged, contiguous or narrowly separated. Metendosternite with long, narrow stalk; long, narrowly diverging lateral arms, usually small, broad laminae (absent in *Orchesia*) and no anterior process; anterior tendons arising on lateral arms. Hind wing with short apical field without distinct radial extensions but with distinct, oblique, linear post-radial sclerite in *Orchesia*; radial cell incomplete (without base); cross-vein r3 absent and r4 sometimes incomplete; R-M loop forming acute angle, with basal portion of RP very short or absent; medial spur extending to wing margin; medial field usually with three or four free veins and a vaguely defined, undivided medial fleck; wedge cell well developed or reduced; anal lobe well developed, without embayment. Legs well developed; fore legs usually shorter than mid or hind legs; trochanterofemoral joint usually strongly oblique with base of femur in contact with coxa. Outer edge of protibia usually with longitudinal row of spines (absent in *Phloeotrya*); outer edges of meso- and metatibiae with longitudinal spine rows in *Orchesia*, *Phloeotrya* and, with numerous short, obliquely transverse spine rows in *Megapselaphus*, *Ctenoplectron*, *Talayra* Champion and *Mystes*. Tarsi 5-5-4; penultimate protarsomere bilobed and penultimate mesotarsomere with small lobe (except in *Mystes*); penultimate metatarsomere lobed in *Orchesia*, *Dircaea* and *Phloiotrya*; pretarsal claws simple; empodium usually not apparent.

Abdomen usually with five free ventrites (ventrites 1 and 2 appear to be connate in Orchesia); ventrite 1 not much longer than 2; intercoxal process acute or narrowly rounded. Functional spiracles on abdominal segments I–VII. Anterior edge of segment IX in male usually forming narrowly rounded to acute subgenital plate, but without spiculum gastrale; tergite IX usually completely divided and segment X free or partly fused to IX. Aedeagus of uninverted tenebrionoid type, with phallobase dorsal to penis; parameres partly or completely fused together forming apicale, which is articulated to the posterior end of the phallobase; apicale undivided or cleft in Serropalpini or with a broad central lobe and a pair of slender, fixed lateral lobes in *Orchesia*, *Dircaea* and *Phloiotrya*; penis slender, usually lightly sclerotised except for lateral struts and with paired anterior struts. Sternite VIII in female with spiculum ventrale. Ovipositor moderately to very elongate, lightly sclerotised, with proctigeral and paraproctal baculi; paraprocts slightly to much longer than coxites, which are subdivided with apical or subapical styli. [Champion 1916a, 1916b; Crowson 1966b; Nikitsky *et al.* 1998; Pollock 2002a; Nikitsky and Pollock 2010].

Larva (Figs 34B–C, 91E)

Body elongate (very long and slender in *Mystes*), more or less parallel-sided and cylindrical or slightly flattened; lightly pigmented, except for head, tips of urogomphi, if present, sometimes with protergum and tergum VIII lightly pigmented and tergum IX heavily so; dorsal surfaces smooth, with vestiture of scattered, long, simple setae.

Head protracted and hypognathous, subglobular to slightly flattened. Epicranial stem short to moderately long; frontal arms V-shaped and contiguous at base. Median endocarina Y-shaped, located beneath epicranial stem and extending beneath or between bases of frontal arms (Fig. 46C). Stemmata usually absent (five or two on each side in *Orchesia*). Frontoclypeal suture absent. Labrum free. Antennae relatively short, 3-segmented, with antennomeres more or less equal in length, conical sensorium usually shorter than apical antennomere (longer in *Orchesia*). Mandibles large, robust and more or less symmetrical, bidentate or unidentate, sometimes with single subapical tooth; mola and prostheca absent. Ventral mouthparts retracted, with well developed maxillary articulating area; cardo transverse; stipes elongate; mala rounded or truncate, not cleft; maxillary palps 3-segmented. Labium consisting of prementum and postmentum, the latter narrowed posteriorly; ligula rounded, longer than labial palps, which are usually 1-segmented (2-segmented in *Orchesia*) and narrowly to very narrowly separated. Hypopharyngeal sclerome absent. Hypostomal rods short to very short and diverging or subparallel; ventral epicranial ridges absent. Gular region slightly transverse; gular sutures well separated.

Prothorax usually much longer than meso- or metathorax (not in *Orchesia*); protergum sometimes with large patch of asperities. Presternum usually well defined, sometimes partly divided by longitudinal line or endocarina. Legs moderately well developed, 5-segmented including claw-like pretarsus bearing one or two setae; coxae widely separated. Abdomen more than twice as long as thorax (often more than 4 times as long); terga I–V often with pair of dorsal, asperity-bearing ampullae (not in *Orchesia*) and segments I–VII often with pair of ventrolaterally projecting ampullae. Tergum IX usually slightly to distinctly shorter than VIII, extending onto ventral surface, and either simple or bearing a pair of urogomphi, which may be short and hook-like or moderately long, basally contiguous, apically diverging; sternum IX simple. Segment X usually not visible from above, posteroventrally oriented and transverse. Spiracles usually annular (annular–biforous in *Orchesia*), those on segment VIII not enlarged. [Hayashi 1975; Lawrence 1991b; Lawrence *et al.* 1999a; Nikitsky 2002; Nikitsky and Pollock 2010].

Classification and World Distribution

The widely distributed family Melandryidae comprises ~420 species in 60 genera. There are two subfamilies: Melandryinae with genera in all major regions and Osphyinae occurring in the Holarctic, Oriental and Neotropical regions.

Australian Fauna

The Australian melandryid fauna contains members of three tribes of **Melandryinae**: Orchesiini with the single genus *Orchesia* Latreille (Fig. 76A), Dircaeini, with the genera *Dircaea* Fabricius and *Phloiotrya* Stephens. and the Serropalpini with the genera *Ctenoplectron* Redtenbacher, *Megapselaphus* Montrouzier, *Mystes* Champion (Fig. 76B) and *Talayra* Champion (Figs 76C, 93A). Melandryids are elongate beetles that are usually somewhat wedge-shaped (tapering posteriorly and more convex ventrally than dorsally), with a vestiture of fine, decumbent hairs. The head is not constricted posteriorly and is deeply inserted into the prothorax, the eyes are vertical and at least slightly emarginate, the antennal insertions are exposed, the maxillary palps are variously modified, often with the apical palpomere strongly expanded and securiform or cultriform, the lateral pronotal carinae are incomplete anteriorly and the meso- and metatibiae often bear several short combs, which may also be present on the tarsomeres. Larvae are elongate, usually subcylindrical, and lightly pigmented, except for urogomphi, which may be minute or absent. The head has a long median endocarina, antennae are very short, stemmata are never well developed, mandibles are stout and symmetrical, and malae are simple and obtuse. Melandryid larvae occur in dead wood or in the durable fruiting bodies of some Basidiomycetes. Adults are often active on surfaces at night, and those of *Orchesia* species are capable of jumping. *Orchesia* is the largest Australian genus. The most unusual species is *Mystes planatus* Champion, which has strongly flattened adults and very long slender larvae (Fig. 91E), which resemble those of some Lymexylidae. [Champion 1916a, 1916b; Crowson 1966b; Nikitsky and Pollock 2010].

91. MORDELLIDAE

Adult (Figs 7C, 93B)

Length 1.5–15 mm. Body usually ~2.3–4.7 times as long as wide; wedge-shaped, widest at base of prothorax, slightly to strongly narrowed anteriorly and posteriorly; slightly flattened above and convex below; colour yellowish-brown to dark brown or black, sometimes with dark and light pattern on elytra and pronotum; dorsal surfaces clothed with fine, recumbent hairs, sometimes forming pattern.

Head large, transverse, flattened and strongly declined, with mouthparts directed posteriorly and usually extending to anterior edge of metaventrite at rest, abruptly constricted behind eyes to form short temples and a narrow neck, which is always concealed by pronotum. Occipital region with transverse ridge or carina that abuts anterior edge of pronotum. Eyes not protuberant, entire or weakly emarginate, lying just above subgenal ridges, finely or coarsely facetted, with or without interfacetal setae. Antennal insertions concealed from above; subantennal grooves absent or weakly developed. Frontoclypeal suture absent; anterior edge of clypeus straight or convex. Labrum free, transverse, its apex subtruncate to slightly convex. Antennae relatively short, not extending behind prothorax, 11-segmented, filiform, serrate or incrassate. Mandible short and broad, unidentate or bidentate; mola and prostheca well developed or reduced. Maxilla with galea and lacinia setose, often with expanded setae; lacinia without uncus; apical maxillary palpomere slightly expanded and truncate to broadly triangular, securiform or cultriform. Mentum transverse; ligula deeply emarginate or bilobed; apical labial palpomere fusiform or slightly expanded and truncate. Ventral surface of head more or less concave between sharp subgenal ridges. Corpotentorium narrow and arched. Cervical sclerites well developed.

Pronotum ~0.45–0.87 times as long as wide, widest posteriorly, as wide as or slightly wider than combined elytral bases, sides moderately to strongly curved and obliquely explanate. Lateral pronotal carinae complete, simple, without raised margin or bead; anterior angles not produced forward; posterior angles rounded or right, more or less embracing elytral humeri; posterior edge strongly biemarginate to form broad mesal lobe; disc simple, without sub-basal impressions. Prosternum in front of coxae much shorter than mid length of procoxal cavity, concave or biconcave. Prosternal process complete or incomplete, parallel-sided or narrowed apically and acute or narrowly rounded at apex. Procoxae projecting below prosternum, without or with very short concealed lateral extensions; trochantins partly exposed. Procoxal cavities strongly transverse, contiguous or narrowly separated, externally broadly open, without narrow lateral extensions, internally closed.

Scutellar shield well developed, not abruptly elevated, posteriorly narrowly rounded or acute to broadly rounded, truncate or emarginate. Elytra ~1.55–2.53 times as long as combined width and 2.7–3.75 times as long as pronotum, with apices independently rounded exposing all or most of tergite VII; disc irregularly punctate, without scutellary striole; epipleura incomplete or absent. Mesoventrite separated by complete sutures from mesanepisterna, which are contiguous or broadly joined at midline; anterior edge on much higher plane than metaventrite, without paired procoxal rests; posterior portion of mesoventrite almost vertical. Mesoventral process broadly rounded at apex. Mesocoxae with exposed trochantins. Mesocoxal cavities slightly to strongly transverse, vertically oblique, moderately to widely separated, partly closed laterally by mesepimera. Mesometaventral junction simple, with ventrites abutting or overlapping. Metaventrite strongly convex with very long but weakly impressed discrimen; anterior edge with transverse, bisinuate carina joining posterior edges of mesocoxal cavities; exposed portion of metanepisternum short and broad to moderately long and narrow. Metacoxae very large, vertically oblique, contiguous and immobile, without plates, extending laterally and dorsally beneath elytra. Metendosternite large with long, narrow stalk; long, narrowly diverging lateral arms, and no laminae or anterior process; anterior tendons arising on lateral arms or not apparent. Hind wing with moderately long apical field with a complex of sclerites and radial extensions; radial cell elongate, apically truncate with inner postobasal angle right; cross-vein r3 long and oblique; r4 complete; R-M loop relatively narrow and basal portion of RP moderately long; medial spur extending to wing margin; medial field usually with four free veins and no medial fleck; wedge cell absent; anal lobe well developed, without embayment. Legs well developed; hind legs longer and stouter than fore or mid legs; trochanterofemoral joint strongly oblique with base of femur in contact with coxa. Metatibiae with apical and subapical comb-like ridges; tibial spurs pubescent, those on hind leg enlarged and usually unequal. Tarsi 5-5-4; penultimate protarsomere sometimes lobed beneath; pretarsal claws more or less serrate with bristle-like ventral appendages; empodium not apparent.

Abdomen distinctly narrowed posteriorly, with five free ventrites; ventrite 1 distinctly longer than 2; intercoxal process broadly angulate or absent. Ventrite 5 (sternite VII) usually elongate but shorter than tergite VII, which extends well beyond elytral apices and is usually narrowly acute and spine-like. Posterior edge of sternite VIII in male sometimes deeply emarginate. Anterior edge of segment IX in male without spiculum gastrale; tergite IX almost completely divided into two parts and segment X free or partly fused to IX. Aedeagus of inverted tenebrionoid type, with phallobase ventral to penis; parameres partly or completely fused together forming apicale, which is articulated to the posterior end of the phallobase and is sometimes highly asymmetrical; penis long and very slender, without anterior struts. Sternite VIII in female with spiculum ventrale. Ovipositor elongate, lightly sclerotised, with proctigeral and paraproctal baculi; paraprocts much longer than coxites, which are subdivided with small, subapical styli. [Lea 1917b; Ermisch 1950, 1962; Franciscolo 1957, 1965, 1967; Lu *et al.* 1997; Lawrence *et al.* 1999b; Jackman and Lu 2002; Lawrence and Ślipiński 2010e].

Larva (Fig. 91D)

Body elongate, more or less parallel-sided and cylindrical or slightly flattened; lightly pigmented; dorsal surfaces smooth, with vestiture of scattered, long, simple setae.

Head (Fig. 42D) protracted and hypognathous, subglobular. Epicranial suture absent. Median endocarina long. Stemmata usually absent (sometimes one to three on each side). Frontoclypeal suture distinct. Labrum free. Antennae very short, 1- to 3-segmented; conical sensorium apical, or if preapical, then shorter than or longer than apical antennomere. Mandibles (Fig. 41P) stout, robust and more or less symmetrical and wedge-like, unidentate with one or two subapical teeth; mola and prostheca absent. Ventral mouthparts retracted, with well developed maxillary articulating area; cardo transverse; stipes elongate; mala rounded or truncate, not cleft; maxillary palps 2- or 3-segmented. Labium consisting of prementum and postmentum; ligula simple, shorter than to longer than labial palps, which are 2-segmented and narrowly separated. Hypopharyngeal sclerome absent. Hypostomal rods absent or very short; ventral epicranial ridges absent. Gular region absent (labium contiguous with cervical membrane).

Prothorax usually distinctly longer than meso- or metathorax; protergum sometimes with patches of asperities. Legs (Figs 42D, 50C) short, 3- or 4-segmented with pretarsus rounded or truncate; coxae widely separated. Abdomen more than twice as long as thorax; terga I–VI sometimes with pairs of weak ampullae. Tergum IX as long as or slightly longer than VII, extending onto ventral surface, and bearing either a pair of small, acute urogomphi or with a median process, which is sclerotised and truncate or emarginate at apex (Fig. 52W); sternum IX simple. Segment X not visible from above, posteroventrally oriented, without paired pygopods. Spiracles annular, those on segment VIII not enlarged. [Lawrence 1991b; Odnosum 1991; Lawrence *et al.* 1999a; Jackman and Lu 2002; Lawrence and Ślipiński 2010e].

Classification and World Distribution

The cosmopolitan family Mordellidae comprises ~1500 species and 85 genera. There are two subfamilies: Ctenidiinae, with a single South African species *Ctenidia mordelloides* Laporte, and Mordellinae containing the remaining genera, which are placed in five tribes, two of which occur in Australia. Exotic tribes include Reynoldsiellini (Venezuela), Stenaliini (southern Europe to southern Africa, Kamtshatka and South-east Asia) and Conaliini (Holarctic, Neotropical, Sumatra, Africa). [Ermisch 1950; Franciscolo 1957, 1965, 1967; Lawrence and Ślipiński 2010e].

Australian Fauna

Australian Mordellidae belong to the two largest tribes of **Mordellinae**: Mordellini and Mordellistenini. Adults are distinctive wedge-shaped, laterally compressed and posteriorly tapered beetles, with tergite VII produced forming an acute process extending well behind the elytral apices. The body is clothed with fine, decumbent hairs, the head is strongly deflexed, concealing the prosternum, procoxae and mesocoxae, and abruptly constricted behind the eyes to form a narrow neck, which is concealed by the abutment of the head against the pronotum. Lateral pronotal carinae are complete, metacoxae are very large, tibiae and tarsi often bear combs of spines, and pretarsal claws are serrate or pectinate, each with a bristle-like appendage beneath. Larvae are elongate, subcylindrical and very lightly pigmented, with a globular head, long endocarina, robust mandibles, short legs, which are indistinctly segmented without claws and tergum IX with a median terminal spine or a pair of minute urogomphi. Adults are often found on flowers and foliage, and when disturbed they usually tumble about by rapidly moving their hind legs. Larvae usually bore through dead wood or mine the stems of various herbs, but a few species occur in fungus fruiting bodies. Most Australian species belong to the genera *Hoshihananomia*, *Mordella* or *Mordellistena*, but it is possible that some of those in the last two genera are misplaced. [Lea 1917b; Ermisch 1962].

Australian Subfamilies, Tribes and Genera

Mordellinae: Mordellini: *Austromordella* Ermisch, *Glipa* LeConte, *Hoshihananomia* Kôno, *Mordella* Linnaeus, *Mordellaria* Ermisch, *Opthalmoglipa* Franciscolo, *Plesitomoxia* Ermisch, *Tomoxia* Costa, *Tomoxioda* Ermisch, *Trichotomoxia* Franciscolo. **Mordellistenini:** *Glipostenoda* Ermisch, *Mordellistena* Costa, *Mordellistenoda* Ermisch, *Paramordellistena* Ermisch.

92. RIPIPHORIDAE

Adult (Figs 7F–G, 76G–H, 93I, K–L)

Length 3.5–38 (usually 5–20) mm. Body usually ~2.3–4.8 times as long as wide; usually wedge-shaped, widest at base of prothorax, slightly to strongly narrowed anteriorly and posteriorly;' sometimes almost parallel-sided; usually slightly flattened above and convex below; colour yellowish-brown to reddish-brown, dark brown or black, sometimes with pronotum darker than elytra or with elytra or pronotum bicolored; dorsal surfaces usually clothed with fine, recumbent hairs, sometimes forming pattern, less commonly subglabrous.

Head large, subglobular to elongate and strongly declined, with mouthparts directed ventrally or posteriorly at rest; almost always abruptly constricted behind eyes to form short to long temples and a narrow neck, which may or may not be concealed by pronotum (head gradually narrowed behind eyes in Ripidiinae). Occipital region usually with transverse ridge (absent in Ripidiinae), which may be rounded, often extending well above the anterior edge of pronotum or sharply carinate and abutting the anterior edge of pronotum and concealing the neck from above. Eyes not protuberant, usually slightly to moderately deeply emarginate (very deeply emarginate and almost completely subdivided in *Ptilophorus* Dejean and some Hemirhipidiinae; entire in Ripiphorinae), moderately to coarsely facetted, usually with convex lenses (very finely facetted with flat lenses in Ripiphorinae); interfacetal setae present or absent, very prominent Ripidiinae, in which the eyes are almost contiguous.

Antennal insertions usually exposed (at least slightly concealed in Ripidiinae, *Ptilophorus* and *Macrosiagon* Hentz); subantennal grooves absent. Frontoclypeal suture absent; anterior edge of clypeus usually truncate or convex. Labrum almost always free and exposed (partly concealed beneath clypeus in *Ptilophorus* and *Ripiphorus* Bosc; absent in Ripidiinae), usually transverse to subquadrate (elongate in some *Macrosiagon*), its anterior edge truncate to emarginate. Antennae almost always 11-segmented (3-segmented in female *Rhipidioides* Riek), usually serrate, unipectinate or uniflabellate (bipectinate or biflabellate in *Ripiphorus*, filiform or clavate in female Ripidiini). Mandibles short and broad to elongate and slender (absent in Ripidiinae), unidentate, with or without subapical tooth on incisor edge; mola usually absent (present in *Trigonodera* Dejean); prostheca present or absent. Maxilla usually with galea and lacinia setose, the latter without uncus (lacinia sometimes reduced, absent or fused to galea, both lobes absent in Ripidiinae; maxillary palps reduced or fused into single lobe in Ripidiinae, apical palpomere cylindrical to fusiform. Mentum transverse or highly reduced; ligula undivided to deeply emarginate or bilobed; apical labial palpomere cylindrical to fusiform. Gular sutures widely separated to confluent. Corpotentorium absent. Cervical sclerites usually well developed.

Pronotum usually ~0.4–1.45 times as long as wide (0.25–0.5 in female Ripidiinae), widest at posterior end and narrowed anteriorly, as wide at base as combined elytral bases, sides straight to slightly curved or somewhat sinuate. Lateral pronotal carinae incomplete, simple, without raised margin or bead, or absent; anterior angles not produced forward; posterior angles often acute and more or less embracing elytral humeri (right to oblique or rounded in Ripiphorinae, Ripidiinae and some Hemirhipidiinae); posterior edge usually rounded or sinuate (produced to form mesal lobe in *Macrosiagon*); disc usually simple, occasionally with shallow sub-basal impressions. Prosternum in front of coxae shorter than mid length of procoxal cavity, usually strongly oblique and flat to concave. Prosternal process usually complete or almost complete, slender and narrowly rounded or acute at apex (absent in Hemirhipidiinae, Ripidiinae and *Ripiphorus*). Procoxae usually conical and ventrally projecting (long and narrow in Ripidiinae), without concealed lateral extensions; trochantins partly exposed or concealed, often reduced or apparently absent. Procoxal cavities transverse, contiguous or narrowly separated, externally broadly open, without narrow lateral extensions, very shallow and only vaguely indicated in Hemirhipidiinae, Ripidiinae and *Ripiphorus*, internally closed by slender bar (which may appear to be an external closure).

Scutellar shield well developed, not abruptly elevated, posteriorly usually rounded, truncate or emarginate (acute and concealed by pronotum in *Macrosiagon*). Elytra ~0.5–3.3 times as long as combined width and 0.9–5.0 times as long as pronotum, with apices more or less conjointly rounded concealing all abdominal tergites or exposing only part of tergite VII in Ptilophorinae, independently acuminate, exposing most of tergite VII in Macrosiagon and some Pelecotominae, and variously shortened with apices independently rounded or truncate, exposing much of hind wing, abdomen and often pterothorax in Hemirhipidiinae, Ripidiinae and *Ripiphorus*; disc irregularly punctate, without scutellary striole, with longitudinal costae in Pelecotominae; epipleura incomplete or absent; elytra absent in female Ripidiinae. Mesoventrite fused to mesanepisterna in Ripidiinae and *Macrosiagon*; anterior edge with shallow, paired procoxal rests in *Macrosiagon*; body of ventrite with longitudinal carina Trigonodera and *Euctenia* Gerstaecker. Mesoventral process narrowly rounded to truncate at apex. Mesocoxae often conical and projecting, with exposed trochantins. Mesocoxal cavities slightly to strongly transverse, sometimes oblique, contiguous or narrowly separated, partly closed laterally by mesanepisterna and mesepimera. Mesometaventral junction, if present, simple, with ventrites abutting or overlapping. Metaventrite usually strongly convex with discrimen moderately long to absent; exposed portion of metanepisternum short and broad to moderately long and narrow; metepimera broadly exposed in Hemirhipidiinae and Rhipiphorinae. Metacoxae almost always narrowly separated or contiguous (widely separated in female Ripidiinae), sometimes moderately large and strongly oblique, extending laterally to meet elytra or sides of body (except in Ripiphorinae). Metendosternite usually with long, narrow stem; long, narrowly diverging lateral arms, no laminae, and a short to moderately long and narrow anterior process with approximate anterior tendons. Hind wing usually moderately long (shorter and broader in Ripidiinae); apical field short, with a complex of sclerites and one or two radial extensions in Ptilophorinae, vaguely indicated anterior and posterior oblique sclerites in Pelecotominae and Hemirhipidiinae, without sclerotizations in Ripidiinae and Ripiphorinae; radial cell in Ptilophorinae and Pelecotominae elongate, with apex truncate, inner posterobasal angle right or acute, and cross-vein r3 long and longitudinal or oblique; radial cell in *Macrosiagon* narrowly elongate with both ends acute and r3 absent; cell without basal edge or absent in Hemirhipidiinae, Ripidiinae and *Ripiphorus*; basal portion of RP moderately long in Ptilophorinae Pelecotominae, Hemirhipidiinae and *Macrosiagon*, short or absent in Ripidiinae and *Ripiphorus*; medial spur extending to wing margin; medial field with four free veins in Ptilophorinae, Pelecotominae and some Hemirhipidiinae (reduced to three or fewer in other groups); wedge cell apically acute in Ptilophorinae, absent in other groups; anal lobe well developed, without embayment; wings not folded beneath elytra in Hemirhipidiinae, Ripidiinae and *Ripiphorus*. Legs moderately long and slender; trochanterofemoral joint usually strongly oblique with base of femur in abutting coxa (femur well separated from coxa in Ripidiinae and joint slightly oblique in Hemirhipidiinae). Protibia shorter than protarsus in Pelecotominae and *Macrosiagon*; tibial spurs 2-2-2 in Ptilophorinae and *Ripiphorus*, 0-0-1 in *Nephrites* Shuckard and *Heteromeroxylon* Pic, 1-1-1 in *Sitarida* White, 0–0-0 in Ripidiinae and variable in Pelecotominae and *Macrosiagon*. Tarsi almost always 5-5-4 in both sexes (4-4-4 in females of *Rhipidioides*); tarsomeres simple; pretarsal claws simple in *Sitarida* and Ripidiinae, bifid in *Macrosiagon*, toothed in *Nephrites* and *Heteromeroxylon*, and pectinate in *Ripiphorus*, most Ptilophorinae and some Pelecotominae, without bristle-like ventral appendages; empodium well developed and multisetose in *Trigonodera*, *Euctenia* and Ripiphorinae.

Abdomen usually with five or six free ventrites (six or seven in ripidiine males, eight or nine in ripidiine females); ventrite 1 distinctly longer than 2 only in *Macrosiagon*; intercoxal very slender and more or less concealed or absent. Ventrite 5 (sternite VII usually elongate but shorter than tergite VII, which extends well beyond elytral apices and is usually narrowly acute and spine-like. Anterior edge of segment IX in male usually with narrowly acute subgenital plate or with spiculum gastrale; tergite IX almost completely divided into two parts and segment X free in Ptilophorinae, Pelecotominae, Hemirhipidiinae and *Macrosiagon*; IX and X more or less fused and membranous in Ripidiinae and *Ripiphorus*. Aedeagus of tenebrionoid type, almost always asymmetrical; phallobase sometimes with anterior strut on one side only; parameres usually articulated, much shorter than phallobase and more or less fused together forming apicale or fused to phallobase and apparently absent; penis long and very slender, usually without anterior struts. Sternite VIII in female with or without spiculum ventrale. Ovipositor long and slender, telescopic, lightly pigmented except for baculi; paraprocts much longer than coxites, which may be divided into three lobes, the basal one with an oblique baculum, the apical two setose; styli either small, lightly pigmented and subapical or absent. [Riek 1955, 1973; Besuchet 1956a; Selander 1957; Callan 1981; Švácha 1994; Tôyama and Ohbayashi 1994; Falin 2002, 2004; Batelka 2006; Lawrence *et al.* 2010j].

Larva (Fig. 93J)

Known larvae of Pelecotominae, Ripidiinae and Ripiphorinae exhibit hypermetamorphosis of varying complexity. Body of first instar Pelecotominae elongate, relatively narrow and parallel-sided, slightly flattened; lightly pigmented; dorsal surfaces smooth, with vestiture of scattered, long, simple setae. First instars of Ripidiinae and Ripiphorinae of the triungulinid type, more or less boat-shaped, tapering posteriorly and more heavily pigmented. Intemediate instars usually very lightly pigmented. Final instar in Pelecotominae moderately elongate, wider at middle and slightly curved, with lateral locomotory protuberances, lightly pigmented except for sclerotised urogomphi, and those of Ripidiinae slightly flattened and very lightly pigmented. Final instar Ripiphorinae subcylindrical or slightly flattened and distinctly curved, with large conical protuberances on thorax and most abdominal segments. Vestiture of short, fine setae, except in first instars of Ripidiinae and Ripiphorinae, which posteriorly directed setae and/or spines.

Head of first instars protracted and prognathous, elongate in Pelecotominae, short and broad, widest posteriorly with truncate hind edge in Ripidiinae and Ripiphorinae. Intermediate and final instars of Ripidiinae and Ripiphorinae with hypognathous head, those of Pelecotominae prognathous. Epicranial suture almost always absent (indistinct with long stem and U-shaped frontal arms in final instar of Pelecotominae. Median endocarina usually absent (present beneath epicranial stem in final instar of Pelecotominae). Stemmata absent in intermediate and final instars of Pelecotominae and Ripiphorinae, two on each side in first instar of Pelecotominae, five in first instar of Ripiphorinae, and four in all instars of Ripidiinae. Frontoclypeal suture usually absent (indistinct in final instar of Pelecotominae). Labrum free in first instars and in final instar of Pelecotominae (fused to clypeus in all intermediate instars and in final instars of Ripidiinae and Ripiphorinae). Antennae moderately long in first instars of Pelecotominae, inserted close to mandibular articulations and 3-segmented, with sensorium as long as elongate antennomere 3; antennae shorter and located well away from mandibular articulations in first instars of Ripiphorinae and Ripidiinae, 3-segmented with short, palpiform sensorium in the former, 2-segmented with long, setiform sensorium in the latter; antennae short, 3-segmented, with sensorium about as long as antennomere 3 in intermediate and final instars of Pelecotominae, short and unsegmented, sometimes dome-like in intermediate and final instars of Ripidiinae and Ripiphorinae. Mandibles in first instar Pelecotominae broad at base, strongly curved and falcate apically, with serrate incisor edge and closed canal, those in first instars of Ripidiinae and Ripiphorinae similarly curved but without closed canal; mandibles of intermediate and final instars of Pelecotominae short, robust and unidentate, those of same instars in Ripidiinae absent, and those in Ripiphorinae more or less flattened and exodont; mola and prostheca always absent. Ventral mouthparts retracted in first instar Pelecotominae and Ripiphorinae, protracted in Ripidiiane, rarely with maxillary articulating area; cardo fused to stipes (or ventral mouthparts vestigial); mala simple and rounded or absent; maxillary palps 3-segmented in first instar Ripiphorinae and final instar Pelecotominae and Ripidiinae, 2-segmented in first instar Pelecotominae and Ripidiinae, unsegmented or absent in intermediate instars and in final instar Ripiphorinae. Labium often reduced or absent; well developed with widely separated 1-segmented palps in final instar Pelecotominae, consisting of small papillae in first instars of Ripidiinae and Ripiphorinae.

Legs well developed and 5-segmented in first instars, with spatulate seta at apex of femur in Pelecotominae and with extremely long tibiotarsus and fleshy pulvillus on pretarsus in Ripidiinae; legs short and not distinctly segmented or absent in all intermediate instars and in final instar Ripiphorinae; legs in final instar Pelectominae and Ripidiinae well developed and 5-segmented, with pretarsus more or less dome-like without setae in the former and claw-like with two setae in the latter. Abdomen in late instar Pelecotominae with distinct lateral projections, that of Ripiphorinae with transverse rows of large conical projections. Tergum IX usually not extending onto ventral surface and with short, tooth-like lateral urogomphi only in later instars of Pelecotominae; with long-terminal setae in first instars of Ripiphorinae. Segment X usually more or less terminal. Spiracles, if present, small and annular, those in first instar Pelecotominae on abdominal segments I and VIII, those in first instar *Macrosiagon* on mesothorax and abdominal segments I–VIII, those in first instar Ripiphorus on VIII only and those in Ripidiinae on VII and VIII; spiracles on mesothorax and abdominal segments I–VIII in intermediate and final instars of Pelecotominae, on mesothorax and segments I–VII in final instar Ripiphorinae, and mesothorax and segments I–VI in Ripidiinae; spiracles absent in intermediate instar Ripidiinae. [Linsley *et al.* 1952; Besuchet 1956; Selander 1991a; Švácha 1994; Lawrence *et al.* 1999a; Falin 2002, 2004; Lawrence *et al.* 2010j].

Classification and World Distribution

Ripiphoridae is a widely distributed group including 30 genera and more than 400 species. The group has been divided into five subfamilies, all of which occur in Australia.

Australian Fauna

The largest number of Australian ripiphorids, and those which are most likely to be encountered in general collecting belong to the genera *Euctenia* Gerstaecker, *Ptilophorus* Dejean and *Trigonodera* Dejean (Figs 7G, 93I) in the subfamily **Ptilophorinae**. Also relatively common are species of *Macrosiagon* Hentz (Fig. 76G) in the subfamily **Ripiphorinae**, but the other ripiphorine genus, *Ripiphorus* Bosc, is known only from an apparently undescribed species from NT. There are three Australian genera of **Hemirhipidiinae**: *Heteromeroxylon* Pic, *Nephrites* Shuckard and *Sitarida* White (Fig. 76H), while subfamily **Pelecotominae** is represented by a specimen from northern NSW, which has not been placed to genus. **Ripidiinae** are fairly well represented by several species of *Rhipidioides* Riek (Fig. 93K–L), three species of *Riekella* Selander (Fig. 7F), two of *Neonephrites* Riek and one each of *Neorhipidius* Riek and *Paranephrites* Riek. Ripiphorids are elongate beetles, often laterally compressed and posteriorly tapered, as in Mordellidae, but without the abdominal spine. The body is often subglabrous or clothed with decumbent hairs, the head may be strongly deflexed and abruptly constricted behind the eyes to form a neck, which may or may not be visible, the eyes are slightly to strongly emarginate and sometimes very large, the elytra may be entire and slightly dehiscent, exposing the abdominal apex, or highly reduced exposing most of the abdomen. Females of Ripidiinae (Figs 7F, 93K) lack both elytra and hind wings and have reduced mouthparts. Known larvae are usually very lightly sclerotised parasitoids, but in Ripiphorinae and Ripidiinae the first instars are minute, well sclerotised triungulinids (Fig. 93J). The Ptilophorinae include the least specialised forms, in which adults have complete elytra and minor sexual dimorphism and the larvae are probably predators or ectoparasitoids of wood-boring larvae. In the Hemirhipidiinae the antennal dimorphism is slight, but the elytra are abbreviated; larvae of *Sitarida scabriceps* Lea was reared from a beetle-infested *Acacia* log. In Ripiphorinae the male antennae are biflabellate, while those of the female are pectinate, and the elytra are either dehiscent and more or less acute at apex (*Macrosiagon*) or very short (*Ripiphorus*). Ripiphorinae lay their eggs in flowers, on the undersides of leaves, or in the soil, and the triungulinids attach themselves to solitary bees (*Ripiphorus*) or wasps of the families Vespidae, Sphecidae, Tiphiidae or Scoliidae (*Macrosiagon*). These minute larvae enter the nest with the female and eventually bore into a hymenopteran larva. This endoparasitic phase is followed by an ectoparasitic one, in which the larva leaves the body of the host and feeds externally. Ripidiine triungulins attach themselves to cockroach nymphs, pierce the cuticle and feed externally for a short time and then bore into the host to become endoparasitic. *Riekella* are parasitoids of Blattidae, *Paranephrites xenus* Riek has been reared from Blaberidae, and all other genera, including *Rhipidioides*, have been associated with Blatellidae. [Hudson 1934; Linsley *et al.* 1952; Riek 1955, 1973; Besuchet 1956a; Selander 1957; Callan 1981; Tôyama and Ohbayashi 1994; Falin 2002, 2004; Batelka 2006; Lawrence *et al.* 2010j].

93. ZOPHERIDAE

Adult (Fig. 92B–L)

Length usually 1.1–13.5 mm (25–34 mm in *Zopherosis* White). Body ~1.4–6.8 times as long as wide; usually parallel-sided, sides of pronotum and elytra sometimes independently rounded (sides uniformly rounded only in *Monomma*), strongly convex to strongly flattened; colour yellowish-brown to dark brown or black, rarely vaguely bicolored; upper surfaces glabrous or clothed with hairs, bristles, expanded scale-like setae, the latter occasionally forming a pattern; occasionally covered with hardened exudate mixed with debris or rarely with algal growth.

Head prognathous, sometimes deeply imbedded in prothorax, abruptly constricted behind eyes in Latometini, *Rhopalocerus* and a few Synchitini. Eyes not to strongly protuberant, usually entire (distinctly emarginate in *Pseudendestes*, very deeply emarginate or completely divided in *Docalis* and *Cotulades*, strongly narrowed and vertical in *Zopherosis*, almost meeting dorsally in *Monomma*), usually finely facetted, with interfacetal setae in some Synchitini. Antennal insertions concealed from above; subantennal grooves often weakly developed or absent (well developed in *Nematidium* and in some Synchitini, where they may extend well behind eyes). Frontal region not deflexed, with supraorbital carinae in *Pseudendestes* and *Munaria*. Frontoclypeal suture absent or incomplete. Labrum free, usually transverse and often more or less concealed beneath clypeus (subquadrate or longer than wide in *Zopherosis* and *Nematidium*). Antennae usually 11-segmented, with a 2- or 3-segmented club (antennae 10-segmented in some Synchitini and sometimes more or less clavate or with club segments not easily distinguishable; club 1-segmented in some Synchitini and in some species of *Pycnomerus*). Mandible short and broad, unidentate or bidentate; incisor edge simple or with single tooth; mola well developed, with or without prostheca. Maxilla with setose galea and lacinia with or without uncus; apical maxillary palpomere cylindrical to subtriangular. Ligula undivided; apical labial palpomere usually cylindrical or fusiform; labial palps absent in *Ablabus*, *Pristoderus* and related Notogean Synchitini. Submentum with setose pit in males of Pycnomerini, Latometini and *Monomma*. Gular sutures widely separated. Corpotentorium broad or narrow. Cervical sclerites absent.

Pronotum ~0.4–2.3 times as long as wide, usually widest at middle or posteriorly (widest anteriorly in *Pseudendestes*), usually as wide at base as combined elytral bases (distinctly narrower in some Latometini and Pycnomerini), sides straight to strongly curved, horizontally explanate or variously lobed in various Synchitini (e.g. *Tentablabus*, some species of *Ablabus* and *Pristoderus*). Lateral pronotal carinae almost always complete (weakly developed in *Nematidium*, absent in *Rhopalocerus*, *Cotulades* and some species of *Enhypnon*), simple to coarsely dentate or crenulate, with or without raised margin or bead; anterior angles and posterior angles variable; posterior edge straight, evenly rounded or sinuate; disc highly variable, sometimes tuberculate or with longitudinal carinae or grooves. Prosternum in front of coxae (Fig. 20B) as long as or longer than mid length of procoxal cavity (much longer in *Nematidium* and Gempylodini), flat to strongly convex. Prosternal process complete, parallel-sided, gradually expanded or strongly and abruptly expanded apically, not dorsally curved behind coxae and usually broadly rounded to truncate, emarginate or biemarginate at apex. Notosternal sutures often absent in Colydiinae; hypomera with antennal grooves or cavities in *Monomma*, *Zopherosis* and some Synchitini (e.g. *Epistranus*, *Faecula*, *Norix*) and with impressions for housing tibiae and tarsi in *Orthocerodes*. Procoxae with long concealed lateral extensions; trochantins concealed; cryptopleuron fused to notum at junction of notosternal suture and procoxal cavity. Procoxal cavities more or less circular, externally closed in *Zopherosis*, *Pycnomerus*, *Docalis*, *Nematidium*, *Munaria*, *Pseudendestes* and in the synch-

tine genera *Acostonotus*, *Antilissus*, *Enhypnon*, *Epistranodes*, *Faecula*, *Lascobitoma*, *Lasconotus*, *Namunaria*, *Synagathis*, *Tentablabus* and *Todima*, open in *Rhopalocerus*, *Latometus*, *Orthocerodes*, *Cotulades*, *Monomma* and the remaining Australian genera of Synchitini), narrowly to widely separated, internally open.

Scutellar shield usually abruptly elevated and acute to truncate (sometimes reduced or absent). Elytra ~1.0–4.8 times as long as combined width and 1.5–3.5 times as long as pronotum, with or without humeri; disc irregularly punctate or with eight to eleven distinct puncture rows or striae and with or without scutellary striole; apices meeting or almost meeting at suture, which is deflected in Latometini; epipleura complete, incomplete or absent. Mesoventrite with anterior edge usually on same plane as metaventrite (on different plane in *Zopherosis* and *Monomma*), with or without procoxal rests (with large, strongly declined rests and relatively deep mesoventral cavity in *Monomma*). Mesoventral process narrowly to broadly rounded, truncate or emarginate. Mesocoxae with concealed trochantins. Mesocoxal cavities subcircular, moderately to widely separated, closed laterally by meeting of ventrites. Mesometaventral junction simple or complex (monocondylic or dicondylic). Metaventrite with discrimen moderately long to short or absent, without postcoxal lines; exposed portion of metanepisternum usually moderately to very elongate. Metacoxae contiguous to widely separated. Metendosternite with stalk moderately long to short and broad or absent and with long lateral arms; laminae usually absent, sometimes reduced but broadly rounded or very slender; anterior process short or absent and anterior tendons narrowly to widely separated and on anterior process or lateral arms. Hind wings, if present, with moderately long apical field sometimes with one vaguely indicated radial extension and one or two small transverse post-radial sclerites; radial cell short and broad to elongate with inner posterobasal angle right or obtuse, cell sometimes reduced or absent; cross-vein r3 absent; basal portion of RP usually moderately long; medial spur extending to wing margin, where there is a distinct medial embayment; medial field usually with four free veins and a subdivided medial fleck (absent in Pycnomerini and *Monomma*); wedge cell usually absent, sometimes long and narrow; anal lobe well developed, with embayment either notch-like or absent; hind wings absent in *Zopherosis*, *Cotulades* and several Synchitini. Trochanterofemoral joint usually strongly oblique, with base of femur in contact with coxa. Tibiae usually simple, outer edge with longitudinal carina in *Monomma*; outer apical angle produced forming tooth or spine in *Pycnomerus*, Gempylodini and *Nematidium*; inner apical angle with curved spine in *Nematidium*. Tarsi 5-5-4 in *Zopherosis*, *Monomma*, *Latometus*, *Docalis* and *Cotulades*, 4-4-4 in *Nematidium*, Gempylodini, *Rhopalocerus* and most Synchitini, and 3-3-3 (or apparently 3-3-3 with partial fusion of first two tarsomeres) in the synchitine genera *Antilissus*, *Cebia*, *Endeitoma*, *Neotrichus* and *Tentablabus*; tarsomeres usually simple and similar in form (tarsomeres 1 and 2 slightly enlarged in *Todima*); pretarsal claws simple and empodium not apparent.

Abdomen with five ventrites, all of which are similarly articulated or with first three of four connate; ventrite 1 sometimes distinctly longer than 2; intercoxal process acute to broadly rounded, truncate or emarginate. Anterior edge of segment IX in male with or without spiculum gastrale; tergite IX almost completely divided into two parts or completely fused to X. Aedeagus of tenebrionoid type, symmetrical and almost always inverted, with tegmen ventral to penis (tegmen dorsal in some Colydiinae); parameres partly or entirely fused together forming articulated apicale, without accessory lobes; penis without or with single or double anterior struts; ejaculatory duct with scleotised seminal pump. Sternite VIII in female with long spiculum ventrale (Fig. 29D). Ovipositor (Fig. 29A) usually moderately to very long and slender, usually lightly sclerotised except for baculi; paraprocts distinctly longer than coxites, each of which is subdivided and apically palpiform, with well developed, apical or subapical styli (coxites moderately sclerotised in *Pycnomerus*, highly modified forming heavily sclerotised blade-like processes without styli in *Pseudendestes* and *Nematidium*). [Carter and Zeck 1937; Freude 1955; Doyen and Lawrence 1979; Lawrence 1980; Ślipiński 1987b; Ślipiński and Lawrence 1997, 1999, 2010b; Lawrence *et al.* 1999b; Ivie 2002e, 2002f, 2002g; Foley and Ivie 2008].

Larva (Fig. 91H–I)

Body elongate, more or less parallel-sided and subcylindrical to slightly flattened; dorsal surfaces usually lightly pigmented, except for head and tergum IX, generally smooth; vestiture of fine hairs or setae only.

Head prognathous and protracted, as long as wide or transverse, slightly flattened. Epicranial stem usually very short or absent and frontal arms usually lyriform (stem longer and arms V-shaped or U-shaped in Pycnomerini and possibly *Zopherosis*); bases always contiguous. Median endocarina usually absent (possibly present in *Zopherosis*). Stemmata on each side usually five, sometimes three or none. Frontoclypeal suture almost always absent (possibly present in *Zopherosis*). Labrum free. Antennae usually well developed, 3-segmented, with conical or palpiform sensorium shorter than antennomere 3. Mandibles symmetrical, broad at base and narrow at apex or broad, stout and wedge-like, usually bidentate; incisor edge often with one or two subapical teeth or serrate; mola usually well developed, tuberculate (Fig. 41G) or transversely ridged (41J) (reduced or absent in *Monomma* and replaced with several hyaline processes, Fig. 41C). Ventral mouthparts retracted, with well developed maxillary articulating area; cardines more or less transverse and undivided or divided by internal ridge; mala rounded or truncate, usually cleft or distinctly emarginate; inner apical angle often with one to several small teeth; maxillary palps 3-segmented. Labium consisting of prementum, mentum and submentum; ligula usually short and simple; labial palps almost always 2-segmented (1-segmented in *Nematidium* and Gempylodini) and narrowly to widely separated. Hypopharyngeal sclerome usually consisting of molar-like sclerite or transverse bar (absent in *Monomma*). Hypostomal rods usually moderately long and diverging (occasionally short and converging; absent in Zopherini); ventral epicranial ridges absent. Gular region usually transverse (longer than wide in *Nematidium*); gular sutures well separated.

Thoracic and abdominal terga often with patches or single transverse rows of asperities (with double rows forming open or closed ovals in *Monomma*), sometimes with one or more sclerotised plates, without lateral tergal processes. Legs well developed, slender, 5-segmented including claw-like pretarsus bearing two setae; coxae narrowly to widely separated. Tergum IX shorter than to slightly longer than VIII, extending onto ventral surface, not forming articulated plate, almost always with pair of simple urogomphi, often with pit between them (urogomphi minute in *Rhopalocerus*); sternum IX exposed, simple, not enclosed by sternum VIII; segment X not visible from above; anal region posteroventrally oriented. Spiracles annular–biforous, not placed at ends of spiracular tubes, those on segment VIII not enlarged. [Fletcher 1916; Doyen and Lawrence 1979; Lawrence 1980, 1999b; Ślipiński and Lawrence 1999; Lawrence *et al.* 1999a; Ivie 2002e, 2002f, 2002g; Ślipiński and Lawrence 2010b].

Classification and World Distribution

Zopheridae is a cosmopolitan family comprising 190 genera and ~1700 species. There are two subfamilies: Zopherinae and Colydiinae, the former with six tribes, four of which occur in Australia, and the latter with nine tribes, four of which occur in Australia. Exotic tribes include Usechini and Phellopsini (Holarctic), Colydiini (Holarctic, Neotropical), Acropini and Adimerini (Neotropical), Rhagoderini (Nearctic) and Orthocerini (Palaearctic).

Australian Fauna

The Australian fauna of **Zopherinae** includes members of the tribes Monommatini, Latometini, Pycnomerini and Zopherini. The presence of Monommatini is based on the single record of *Monomma australe* Thomson from Australia with no specific locality; since no other specimens of this or any other monommatine have been recorded from Australia, it is likely that the type was mislabelled (Freude 1955). Zopherini is represented by *Zopherosis georgei* White (Fig. 92L), a large and distinctive 'iron-clad beetle' occurring in northern NSW and southern QLD. Latometini is a Notogean group with the genera *Latometus* (Fig. 92G) and *Orthocerodes* from south-eastern Australia and the Chilean genus *Notorthocerus* Ślipiński and Lawrence. The Pycnomerini includes the cosmopolitan genus *Pycnomerus* and the endemic genera *Docalis* (Fig. 92K) and *Cotulades* (Fig. 92H); all three genera are widely distributed in eastern Australia. The **Colydiinae** include a species of *Nematidium* (Nematidiini) (Fig. 92C) and *Rhopalocerus* (Rhopalocerini), two Gempylodini (*Munaria* and *Pseudendestes*, Fig. 92I), and at least 30 genera of Synchitini (Fig. 92B, D–F, J). Zopherids are narrowly elongate to oblong or ovate, strongly convex to flattened beetles, usually brown or black in colour, which may be glabrous but are usually clothed with decumbent and/or erect hairs, bristles or scales. Antennal insertions are almost always concealed (exposed in *Nematidium*), antennae with a 1- to 3-segment club, lateral pronotal carinae almost always complete and often crenulate or dentate, prosternal process not curved dorsally at apex, tarsomeres usually simple, and abdominal ventrites equally articulated or with first three or four connate and the apical ones freely movable. Larvae are usually elongate and subcylindrical to slightly flattened, and lightly pigmented, except for head and sometimes tergum IX or tips of urogomphi, which may have a pit between them (Fig. 91H); the epicranial stem may be short or long, the mandibles usually with a distinct mola, the mala rounded or truncate and usually cleft and provided with one or more teeth at the inner apical angle, the hypostomal rods are usually long and diverging, and the thoracic and abdominal terga may bear transverse rows or patches of asperities (Fig. 91I). Zopherids are usually associated with wood or fungi. Larvae may be found under bark or in rotten wood and adults on the surfaces of wood or fungal fruiting bodies. *Pycnomerus* species are often associated with white rotten wood. Several members of the Colydiinae, such as species of *Nematidium* and *Pseudendestes*, are predators of platypodine weevils (Roberts 1977; Lawrence 1980). Some Synchitini may feed on lichens or pyrenomycetous fungi, and *Todima* (Fig. 92E) may feed on the rotting flower stalks of the grass tree (*Xanthorrhoea*). The exotic *Monomma brunneum* Thomson is known to feed on rotten stems of papaya (Fletcher 1916). [Carter and Zeck 1937; Lawrence 1980; Ślipiński and Lawrence 1997, 1999, 2010b; Turco *et al.* 2012].

Australian Subfamilies, Tribes and Genera

Colydiinae: Nematidiini: *Nematidium* Erichson. **Gempylodini:** *Munaria* Reitter, *Pseudendestes* Lawrence. **Rhopalocerini:** *Rhopalocerus* Redtenbacher. **Synchitini:** *Ablabus* Broun, *Acolophoides* Ślipiński & Lawrence, *Acolophus* Sharp, *Acostonotus* Ślipiński & Lawrence, *Antilissus* Sharp, *Bitoma* Herbst, *Bolcocius* Dajoz, *Bupala* Pascoe, *Caprodes* Pascoe, *Cebia* Pascoe, *Cerchanotus* Erichson, *Cicablabus* Ślipiński & Lawrence, *Ciconissus* Broun, *Colobicones* Grouvelle, *Colobicus* Latreille, *Endeitoma* Sharp, *Enhypnon* Carter, *Epistranodes* Ślipiński & Lawrence, *Epistranus* Sharp, *Faecula* Ślipiński & Lawrence, *Lascobitoma* Ślipiński & Lawrence, *Lasconotus* Erichson, *Lobomesa* Ślipiński & Lawrence, *Namunaria* Reitter, *Neotrichus* Sharp, *Norix* Broun, *Phorminx* Carter & Zeck, *Pristoderus* Hope, *Synagathis* Carter & Zeck, *Synchita* Hellwig, *Tentablabus* Ślipiński & Lawrence, *Todima* Grouvelle.

Zopherinae: Latometini: *Latometus* Erichson, *Orthocerodes* Ślipiński & Lawrence. **Pycnomerini:** *Cotulades* Pascoe, *Docalis* Pascoe, *Pycnomerus* Erichson. **Monommatini:** *Monomma* Klug. **Zopherini:** *Zopherosis* White.

94. ULODIDAE

Adult (Figs 76L, 92A)

Length 3.5–9 mm. Body usually ~1.93–2.67 times as long as wide; more or less parallel-sided and slightly flattened; colour yellowish-brown to dark brown or black; vestiture usually of short, recumbent, thickened setae, usually forming a pattern, combined with erect thickened setae in *Ulodes* Erichson or dark erect bristles in *Phaennis* Champion (fine recumbent setae only in *Meryx* Latreille).

Head prognathous, often deeply embedded within prothorax, not abruptly constricted posteriorly. Eyes slightly to moderately prominent, usually slightly emarginate anteriorly and flattened posteriorly, the vertical diameter exceeding the horizontal, finely to coarsely facetted, with fine interfacetal setae in *Phaennis*. Antennal insertions almost always exposed from above (concealed by cephalic horns in *Notocerastes* Carter; subantennal grooves absent. Frontoclypeal suture absent. Labrum free, visible, transverse to subquadrate, with exposed basal membrane. Antennae almost always 11-segmented with a weak to moderately strong 3-segmented club (10-segmented with a 2-segmented club in an undescribed genus from QLD). Mandible short and broad, bidentate; incisor edge with distinctive truncate process in *Meryx*; mola and prostheca well developed. Maxilla with broad, setose galea and narrow lacinia with terminal uncus (except in *Meryx*); apical maxillary palpomere short and broad, fusiform or slightly expanded and truncate. Mentum slightly transverse; ligula simple and truncate or emarginate; apical labial palpomere fusiform. Submentum without setose pit. Corpotentorium narrow. Cervical sclerites absent.

Pronotum ~0.52–0.88 times as long as wide, usually widest anteriorly or at middle and distinctly narrower at base than combined elytral bases (widest posteriorly and as wide at base as elytral bases in *Dipsaconia* Pascoe), sides slightly curved or sinuate, with sharp lateral tooth in *Ganyme* Pascoe. Lateral pronotal carinae complete, simple, without raised margin or bead; anterior angles produced and acute to narrowly rounded in *Notocerastes*, *Ulodes* and *Ganyme*; posterior angles right or slightly acute; posterior edge usually straight (bisinuate in *Ulodes*); disc usually simple or with slight elevations or tubercles (with paired sub-basal pits in *Phaennis* and *Notocerastes*). Prosternum in front of coxae (Fig. 91I) as long as or slightly longer than mid length of procoxal cavity, flat to slightly convex. Prosternal process complete, not or only slightly curved behind coxae, strongly and abruptly expanded at apex (Fig. 44P), which is truncate to slightly emarginate. Notosternal sutures present or absent. Procoxae with short or moderately long concealed lateral extensions; trochantins reduced. Cryptopleuron fused to notum at junction of notosternal suture and procoxal cavity. Procoxal cavities subcircular to slightly transverse, narrowly to moderately widely separated, externally broadly closed, with notal projections broadly overlapping lateral projections of prosternal process, with very narrow lateral extensions in *Meryx*, internally open.

Scutellar shield abruptly elevated, posteriorly subacute to broadly rounded to obtusely angulate. Elytra ~1.48–2.1 times as long as combined width and 2.7–3.48 times as long as pronotum, completely and conjointly rounded at apex; disc usually with ten distinct puncture rows and a scutellary striole (irregularly punctate in *Meryx*, with irregular pattern of short carinae, sometimes enclosing hexagonal or pentagonal cells in *M. rugosa* Latreille), epipleura complete. Mesoventrite separated by complete sutures from mesanepisterna, which are fairly narrowly separated from one another; anterior edge on same plane as metaventrite, without paired procoxal rests, with median prosternal rest (except in *Meryx*). Mesoventral process truncate to emarginate at apex. Mesocoxae with exposed trochantins. Mesocoxal cavities subcircular, narrowly to moderately widely separated, partly closed laterally by mesepimera. Mesometaventral junction monocondylic or dicondylic. Metaventrite subquadrate or longer than wide, slightly convex, with moderately long discrimen and with transverse (metakatepisternal) suture distinct but not crossing midline; exposed portion of metanepisternum elongate and narrow. Metacoxae usually contiguous or narrowly separated (more widely separated in *Ganyme*) and extending laterally to meet elytra. Metendosternite with long stalk and long lateral arms, weakly developed laminae (absent in *Meryx*), no anterior process and anterior tendons located on lateral arms. Hind wing with moderately long apical field containing one or two small post-radial sclerites and a vaguely defined radial extension; radial cell short and broad with inner posterobasal angle more or less right; cross-vein r3 absent or very short; basal portion of RP moderately long; medial spur strongly curved and extending to wing margin, where there is a distinct medial embayment; medial field with four free veins and a vaguely defined or undivided medial fleck; wedge cell elongate and apically acute or narrowly truncate; anal lobe well developed, without embayment. Legs well developed; trochanterofemoral joint strongly oblique, with base of femur in contact with coxa. Tibiae not strongly widened; spurs distinct, paired on all legs. Tarsi usually 5-5-4 (4-4-4 in *Meryx*); tarsomeres simple and similar in form; pretarsal claws simple and empodium bisetose.

Abdomen with five ventrites, the first three of which are connate; ventrite 1 not much longer than 2; intercoxal process usually subacute (broadly rounded in *Ganyme*); posterolateral flanges sometimes present on ventrites 3 and 4. Functional spiracles on abdominal segments I–VII. Sternite VIII in male with or without median strut. Anterior edge of segment IX in male with acute subgenital plate or spiculum gastrale; tergite IX and segment X fused together and more or less membranous. Aedeagus of uninverted tenebrionoid type, with phallobase dorsal to penis; parameres partly fused together forming articulated apicale, which is slightly shorter than phallobase; penis without anterior struts. Sternite VIII in female with long spiculum ventrale. Ovipositor long and slender, with paraprocts distinctly longer than coxites, each of which is subdivided, apically slender, with well developed, apical or subapical styli. [Watt 1974b; Doyen and Lawrence 1979; Lawrence 1994a; Lawrence *et al.* 1999b; Leschen and Ślipiński 2010c].

Larva (Fig. 91J)

Body elongate, more or less parallel-sided and subcylindrical to slightly flattened; dorsal surfaces moderately darkly pigmented or lightly pigmented except for head and abdominal apex, usually more or less tuberculate or asperate, but smooth in *Meryx*, except for tergum IX, which is covered in tubercles, vestiture of scattered, erect setae.

Head prognathous and protracted, slightly transverse and slightly flattened. Epicranial stem short to moderately long; frontal arms lyriform and contiguous at base. Median endocarina usually present beneath epicranial stem (not in *Meryx*). Stemmata five on each side. Frontoclypeal suture present or absent. Labrum free. Antennae relatively short, 3-segmented, with conical sensorium shorter than apical antennomere. Mandibles asymmetrical, broad at base and narrow at apex, bidentate; incisor edge with or without subapical tooth; mola well developed but with simple surface, without hyaline lobe at base. Ventral mouthparts retracted, with well developed maxillary articulating area; cardo divided by internal ridge; stipes longer than wide; mala truncate, more or less cleft; maxillary palps 3-segmented. Labium consisting of prementum, mentum and submentum; ligula simple, shorter than labial palps, which are 2-segmented and narrowly separated. Hypopharyngeal sclerome molar-like. Hypostomal rods short and diverging; ventral epicranial ridges absent. Gular region transverse; gular sutures well separated.

Thoracic and most abdominal terga with moderately strongly pigmented smooth tergal plates in *Meryx* or lightly pigmented plates covered with tubercles in *Dipsaconia*. Legs slender, not spinose; 5-segmented including claw-like pretarsus bearing two setae lying side by side; coxae separated by one basal diameter. Tergum IX as long as or slightly longer than VIII, extending onto ventral surface, strongly tuberculate, with pair of moderately long, posteriorly-oriented urogomphi, slightly upturned at apex in *Meryx* or strongly upturned, short urogomphi, a pair of pregomphi and a pair of short, posteriorly projecting teeth in *Dipsaconia*; sternum IX exposed, simple, not enclosed by sternum VIII; segment X not visible from above, posteroventrally oriented and slightly transverse. Spiracles annular–biforous, not placed at ends of spiracular tubes, those on segment VIII not enlarged. [Lawrence 1994a; Leschen and Ślipiński 2010c].

Classification and World Distribution

Ulodidae is a Notogean family with six Australian genera, five from New Zealand, one from New Caledonia and two from Chile, but with only ~30 described species. [Watt 1974b; Leschen and Rhode 2002].

Australian Fauna

The Australian Ulodidae include *Phaennis fasciculata* Champion, two species each in the genera *Dipsaconia* Pascoe, *Meryx* Latreille (Fig. 92A), *Notocerastes* Carter and *Ulodes* Erichson, and three species of *Ganyme* Pascoe (Fig. 76L). Most species occur in south-eastern Australia and TAS, but *Ganyme demarzi* (Kulzer) is known from the south-western part of the continent. Australian ulodids are elongate, more or less parallel-sided, slightly flattened beetles with the prothorax distinctly narrower than the combined elytral bases (except in *Dipsaconia*) and with vestiture varying from more or less uniform, decumbent setae (*Meryx rugosa* Latreille) to decumbent setae of different thicknesses and/or colour forming a pattern (*Meryx aequalis* Blackburn, *Notocerastes*), sometimes combined with dark, erect setae (*Phaennis fasciculata*), or short erect bristles of different colours (*Ganyme*) sometimes combined with scales (*Dipsaconia*, *Ulodes*). The antennal insertions are exposed, the antennae have a weak 3-segmented club, the prosternal process is abruptly expanded apically closing the coxal cavities, and there are always three connate ventrites. Meryx differs from other ulodids in having 4-4-4 tarsi. Larvae are elongate, more or less parallel-sided and often lightly pigmented (with pigmented tergal plates on all trunk segments in *Meryx*), with well developed urogomphi sometimes accompanied by accessory processes or pregomphi. The frontal arms are lyriform and the frontoclypeal suture is usually absent (present in *Dipsaconia*), the mandibles have well developed molae and the mala is truncate with or without a cleft and short uncus. Adults and larvae of most Ulodidae feed on the soft fruiting bodies of certain Basidiomycetes, such as *Piptoporus* (Polyporaceae) and *Pleurotus* (Tricholomataceae).

95. PROMECHEILIDAE

Adult (Fig. 93C)

Length 4.5–18 mm. Body usually ~1.85–3.05 times as long as wide; more or less parallel-sided and slightly flattened or with laterally, independently rounded and moderately convex prothorax and elytra; colour uniformly dark brown in *Melytra* Pascoe or bicolored, dark reddish-brown with yellow sides or irregular yellow maculae on elytra; upper surfaces subglabrous, with very short, fine, recumbent setae, with a few scattered erect setae.

Head prognathous, not abruptly constricted posteriorly. Eyes not protuberant, shallowly emarginate in *Sirrhas* Champion, finely facetted, without interfacetal setae. Antennal insertions exposed from above; subantennal grooves absent. Frontoclypeal suture weakly impressed or absent; anterior edge of clypeus membranous. Labrum free, visible, strongly transverse and subtruncate. Antennae moderately long, 11-segmented with a very weak 3-segmented club, so that they may appear filiform. Mandible short and broad, bidentate; incisor edge simple or with single tooth; mola and prostheca well developed. Maxilla with broad, setose galea and narrow lacinia with or without terminal uncus; apical maxillary palpomere slightly expanded and truncate to securiform. Mentum transverse; ligula simple and truncate; apical labial palpomere fusiform. Corpotentorium narrow. Cervical sclerites absent.

Pronotum ~0.75–1.05 times as long as wide, widest at middle, distinctly narrower at base than combined elytral bases in *Sirrhas* (not so in *Melytra*), sides weakly curved or slightly sinuate. Lateral pronotal carinae complete, simple, with raised margin or bead in *Sirrhas*, absent in Melytra; anterior angles not produced; posterior angles obtuse or rounded; posterior edge straight or weakly sinuate; disc simple but with weak impressions in *Sirrhas*. Prosternum in front of coxae slightly shorter than mid length of procoxal cavity, flat to slightly convex. Prosternal process complete, not curved behind coxae, strongly and abruptly expanded at apex, which is truncate. Notosternal sutures complete in *Sirrhas*, absent in *Melytra*. Procoxae projecting well below prosternum, with short concealed lateral extensions; trochantins concealed. Cryptopleuron not fused to notum. Procoxal cavities slightly transverse and narrowly separated in Sirrhas, subcircular and more widely separated in *Melytra*, externally closed, without narrow lateral extensions, internally open.

Scutellar shield not abruptly elevated, posteriorly acute in *Melytra*, broadly rounded in *Sirrhas*. Elytra ~1.3–2.5 times as long as combined width and 2.2–4.3 times as long as pronotum, with well developed humeri in *Sirrhas*, completely and conjointly rounded at apex; disc with eight puncture rows and a scutellary striole in Melytra, irregularly punctate in *Sirrhas*; epipleura complete in *Sirrhas*, incomplete in *Melytra*. Mesoventrite in *Sirrhas* separated by complete sutures from mesanepisterna are narrowly separated from one another (partly fused to mesanepisterna in *Melytra*); anterior edge of mesoventrite on same plane as metaventrite, without paired procoxal rests. Mesoventral process truncate to emarginate at apex. Mesocoxae with exposed trochantins. Mesocoxal cavities subcircular, narrowly separated, partly closed laterally by mesepimera. Mesometaventral junction monocondylic. Metaventrite moderately long with well developed discrimen in *Sirrhas*, very short without discrimen in *Melytra*; exposed portion of metanepisternum elongate and narrow (shorter and broader in *Melytra*). Metacoxae narrowly separated, not extending laterally to meet elytra. Metendosternite in *Sirrhas* with long stalk and long lateral arms, well developed laminae, no anterior process and anterior tendons located on lateral arms; stalk shorter and laminae absent in *Melytra*. Hind wing in *Sirrhas* with vaguely defined radial extensions; radial cell short and broad with inner posterobasal angle more or less right; cross-vein r3 absent; basal portion of RP moderately long; medial spur not extending to wing margin, where there is a distinct medial embayment; medial field with four free veins and a vaguely defined medial fleck; wedge cell elongate and apically acute; anal lobe well developed, without embayment; hind wings absent in *Melytra*. Legs moderately to very long and slender; trochanterofemoral joint strongly oblique, with base of femur separate from or in contact with coxa. Tibiae not strongly widened; spurs distinct, paired on all legs. Tarsi usually 5-5-4; tarsomeres without ventral lobes, but clothed beneath with fine hairs; pretarsal claws simple.

Abdomen with five ventrites, the first three of which are connate; ventrite 1 not much longer than 2; intercoxal process subacute to broadly rounded. Functional spiracles on abdominal segments I–VII. Anterior edge of segment IX in male with spiculum gastrale; tergite IX and segment X fused together and more or less membranous. Aedeagus of uninverted tenebrionoid type, with phallobase dorsal to penis; parameres partly fused together forming articulated apicale, which is distinctly shorter than phallobase; penis without anterior struts. Sternite VIII in female with long spiculum ventrale. Ovipositor moderately elongate, lightly sclerotised except for baculi; paraprocts distinctly longer than coxites, each of which is divided into three parts, the proximal two short and the distal one longer and narrower, often setose apically, with slender, cylindrical, apical or subapical styli. [Watt 1967, 1974a; Doyen *et al.* 1990; Lawrence 1994a; Lawrence *et al.* 2010k].

Larva (Fig. 93D)

Body elongate, more or less parallel-sided, subcylindrical to slightly flattened; dorsal surfaces more or less heavily pigmented, with distinct dark and light colour pattern; surfaces smooth or very finely granulate, clothed with scattered, erect setae.

Head prognathous and protracted, slightly transverse and slightly flattened. Epicranial stem moderately long; frontal arms lyriform and contiguous at base. Median endocarina absent. Stemmata five on each side. Frontoclypeal suture present. Labrum free. Antennae moderately long, 3-segmented, with antennomere 2 slightly to much longer than 1, and 3 short and slender; sensorium shorter than apical antennomere. Mandibles broad and stout, more or less wedge-like, symmetrical or slightly asymmetrical, tridentate or bidentate with well developed subapical tooth; mola absent or reduced to a few small hyaline teeth. Ventral mouthparts retracted, with well developed maxillary articulating area; cardo externally divided; stipes longer than wide; mala truncate, more or less cleft, its inner apical angle with one to several small teeth; maxillary palps 3-segmented. Labium consisting of prementum, mentum and submentum; ligula simple, shorter than labial palps, which are 2-segmented and narrowly separated. Hypopharyngeal sclerome consisting of transverse bar. Hypostomal rods absent; ventral epicranial ridges absent. Gular region slightly transverse; gular sutures well separated.

Thoracic and most abdominal terga with moderately strongly pigmented, bicolored, smooth or weakly tuberculate tergal plates, a fine transverse carina along anterior edges of meso- and metaterga and abdominal terga I–IX. Legs moderately long and slender, 5-segmented including claw-like pretarsus bearing two setae lying side by side; coxae separated by one to two basal diameters. Tergum IX excluding appendage as long as or longer than VIII, extending onto ventral surface, strongly tuberculate, with pair of posteriorly-directed, upturned, bifurcate urogomphi and four or six pregomphi; sternum IX exposed, simple, not enclosed by sternum VIII; segment X not visible from above but posteriorly oriented and transverse with a sclerotised tergal plate. Spiracles annular–cribriform, not placed at ends of spiracular tubes, those on segment VIII not or only slightly enlarged. [Lawrence 1994a; Lawrence *et al.* 1999a; Lawrence *et al.* 2010k].

Classification and World Distribution

Promecheilidae comprises nine genera and 18 described species from southern Chile and Argentina, South Georgia, the Falkland Islands and TAS.

Australian Fauna

The Australian promecheilids include three Tasmanian species: *Sirrhas limbatus* Champion, *S. variegatus* Lawrence (Fig. 93C) and *Melytra ovata* Pascoe. *Sirrhas* species are elongate and parallel-sided, with the elytra distinctly broader at base than the prothorax and bicoloured with a pattern of reddish-brown and yellow; the antennae have a very weak club, often barely distinguishable from the funicle, the antennal insertions are exposed, the apical maxillary palpomere is large and securiform, the prosternal process is slender and abruptly expanded apically closing the cavities, and the tarsi are not lobed beneath. *Melytra* differs in being wingless, with ovate elytra and a very short metaventrite. Larvae are elongate and subcylindrical, with distinctive yellow and brown markings dorsally; the epicranial stem and antennae are long, the mola is reduced to a few patches of small asperities, the mala is obtuse and cleft with an uncus at the inner apical angle, the armature of tergum IX is complex with urogomphi bearing one or two basal processes, as well as pregomphal tubercles, and the spiracles are annular–cribriform.

96. TENEBRIONIDAE

Adult (Figs 7A–B, H, 74A–L, 75J, 90J, M–N)

Length usually 1.5–36 (usually 2.5–25) mm. Body ~0.9–4.6 times as long as wide; usually moderately to strongly convex, occasionally strongly flattened (e.g. *Cossyphus*, Heleina, *Platycotylus*); colour highly variable but often dark brown to black, sometimes yellow to light brown, bicolored (e.g. various Diaperinae) or metallic (e.g. some Cyphaleina, Amarygmini, Cnodalonini); upper surfaces glabrous or clothed with hairs, bristles or scales.

Head (Fig. 44A) prognathous to strongly declined, rarely abruptly constricted behind eyes to form neck and temples. Eyes usually vertically oriented and emarginate anteriorly or anteriorly and posteriorly, occasionally oval to circular (e.g. Phrenapatinae, Gnathidiini), sometimes completely divided by canthus (e.g. *Mesomorphus*), rarely reduced (Cnemeplatiini) or absent (*Eulea*, *Typhluloma*), finely to coarsely facetted, with or without interfacetal setae. Antennal insertions almost always concealed from above (exposed in Leiochrinini and some Lagriini); subantennal grooves usually absent (present in Phrenapatinae and Bolitophagini). Frontoclypeal suture absent to distinctly impressed. Labrum free, usually transverse, sometimes subquadrate to elongate, with or without exposed basal membrane. Antennae usually 11-segmented (10-segmented in *Triplehornia*, *Archaeo-*

glenes, *Eutermicola* and *Menimus*; 9-segmented in *Doyenia*, 8-segmented in *Kershawia*, 6-segmented in *Tyrtaeus* and 3-segmented in *Euclarkia*), and usually filiform, moniliform or clavate, occasionally capitate with a distinct club of one (*Kershawia*, *Euclarkia*, *Tyrtaeus*), two (e.g. *Archaeoglenes*, *Dioedus*, *Tagalinus*, *Orcopagia*, *Ilyxerus*, *Doyenia*), three (e.g. *Archeophthora*, *Pseudophthora*, *Hyocis*), four (e.g. *Toxicum*, some species of *Menimus*, *Lyphia*) or more than four segments; apical antennomeres often bearing compound (stellate) sensoria (Alleculinae, Diaperinae, Stenochiinae, Alphitobiini, Ulomini and Amarygmini), with placoid sensilla in Ulomini. Mandibles unidentate to tridentate or rounded at apex, almost always with well developed, simple or transversely ridged mola and distinct prostheca. Maxilla with or without laciniar uncus; apical maxillary and labial palpomeres cylindrical to subtriangular or securiform. Ligula usually small and undivided (sometimes emarginate or bilobed). Gular sutures widely to narrowly separated or confluent. Corpotentorium almost always narrow, sometimes absent. Cervical sclerites absent.

Pronotum ~0.2–1.5 times as long as wide, sides straight to strongly curved or sinuate, sometimes horizontally explanate (e.g. Cossyphini, Heleina), rarely obliquely explanate (Leiochrinini), usually as wide at base as combined elytral bases (distinctly narrower in a few Adeliini, Lagriini, Alleculinae and Stenochiinae). Lateral pronotal carinae almost always complete and usually simple (crenulate, dentate or various lobed in a few Adeliini, Goniaderini, Bolitophagini and Cnodalonini); anterior angles rounded or angulate to produced and rounded to strongly acute; posterior angles rounded to acute; disc often simple, sometimes with paired basal impressions, with longitudinal carinae in *Kershawia*, *Euclarkia* and *Eutermicola*, variously tuberculate in some Toxicini and Bolitophagini and a few Adeliini). Prosternum in front of coxae shorter than to longer than mid length of procoxal cavity, concave to strongly convex or occasionally carinate. Prosternal process complete (interrupted in Lagriini), parallel-sided or gradually expanded apically, and almost always dorsally curved behind coxae (Fig. 25C) (flat in *Leaus*, concave in Lagriina) and narrowly rounded to truncate or emarginate at apex. Notosternal sutures present or absent; hypomera with anteriorly opening antennal cavities in *Archaeoglenes*. Procoxae projecting below sternum only in Lagriina, with short to long concealed lateral extensions; trochantins concealed; cryptopleuron fused to notum at junction of notosternal suture and procoxal cavity. Procoxal cavities slightly transverse to circular, narrowly to widely separated (contiguous in Lagriina), almost always externally closed by postcoxal extensions of hypomera (Fig. 44N) (externally open in *Tanylypa*), almost always internally closed (internally open in *Tanylypa*, Toxicini, *Tenebrio*, Heleini (except *Edylius*), Triboliini, Leiochrinini and Scaphidemini), with narrow lateral extensions in *Triplehornia* and Leiochrinini.

Scutellar shield well developed to absent, abruptly elevated or not, posteriorly acute to truncate. Elytra ~0.3–3.5 times as long as combined width and 1.25–6.5 times as long as pronotum, with or without humeri; disc usually irregularly punctate or with nine distinct puncture rows or striae (ten rows or striae in various Adeliini, Statirina, *Tanylypa* and *Toxicum* and multiple rows in *Exangeltus* and *Lixionica*), usually with scutellary striole or with sutural stria distinctly diverging at base; epipleura complete, incomplete or absent. Mesoventrite on same plane as or different plane (Fig. 25A) than metaventrite; anterior edge without procoxal rests, sometimes with median prosternal rest or broad prothoracic rest. Mesoventral process narrowly to broadly rounded, truncate or emarginate. Mesocoxae subglobular with exposed or concealed trochantins. Mesocoxal cavities subcircular, narrowly to widely separated, usually partly closed laterally by mesepimeron (closed by meeting of meso- and metaventrites (Fig. 25L) in various groups, including most Lupropini, *Cossyphus*, *Archaeoglenes*, Cnemeplatiini, *Salax*, most Palorini, Alphitobiini, *Tribolium*, *Lyphia*, *Leichenum*, and several genera of Diaperinae); trochantins exposed or concealed. Mesometaventral junction simple, with abutting or overlapping ventrites or complex (monocondylic or dicondylic). Metaventrite with discrimen moderately long to short or absent; exposed portion of metanepisternum short and broad to long and narrow. Metacoxae contiguous to widely separated. Metendosternite usually with long stalk and long lateral arms; laminae usually absent, sometimes reduced; anterior process absent and anterior tendons on lateral arms. Hind wings well developed, highly reduced or absent; apical field short to moderately long, often without or with vague sclerotisations only, sometimes with one or more post-radial sclerites and a radial extension; radial cell short and broad to elongate with inner posterobasal angle right or obtuse to acute; cross-vein r3 often moderately long and closely associated with or basally fused with cross-vein r4; basal portion of RP usually moderately to very long, sometimes reduced or absent; medial field usually with four free veins and no medial fleck (medial fleck present *Tanylypa*, Toxicini, some Tenebrionini, Heleini, Titaenini and Heterocheirina; three or fewer free veins in many smaller species); wedge cell often large and apically acute or obliquely truncate, sometimes absent; anal lobe usually well developed without embayment, sometimes with anal notch. Legs moderately to very elongate; trochanterofemoral joint almost always strongly oblique, with base of femur in contact with coxa. Tibiae sometimes widened at or near apex; spurs simple, sometimes very small or apparently absent; outer apical angle of protibia sometimes with one or more outwardly facing teeth or spines or rounded lobe or process. Tarsi usually 5-5-4 (4-4-4 in *Kershawia*, *Euclarkia*, *Archaeoglenes*, *Eutermicola*, *Byrsax egenus* Pascoe, *Rhipidandrus*, *Tyrtaeus*, *Myrmechixenus*, *Csiro* (*Millstreamia*), *Brittona* and *Magela*); tarsomeres usually simple, penultimate tarsomere sometimes lobed beneath (Fig. 44BB); pretarsal claws usually simple (pectinate in almost all Alleculinae); empodium present or absent.

Abdomen (Figs 21B, 25K), with five ventrites, the first three of which are connate; ventrite 1 not much longer than 2, with postcoxal lines only in *Kershawia* and *Euclarkia*; intercoxal process acute to broadly rounded, truncate or emarginate (rarely absent). Membrane between sternites V and VI and VI and VII usually visible and articulation points joining these sternites lateral; membrane not visible and articulation points submedial in Belopini, *Cossyphus*, *Tanylypa*, all Pimeliinae, except *Trachyderma*); membrane not visible but articulation points lateral in Toxicini. Paired defensive glands usually with reservoirs opening between sternites VII and VIII (reservoirs opening between sternites VIII and IX in Adeliini; glands absent in Belopini, Cossyphini, Chaerodini, Phrenapatinae, *Tanylypa*, and Pimeliinae). Anterior edge of sternite VIII in male with or without median strut. Anterior edge of segment IX in male with or without spiculum gastrale; tergite IX and segment X in male more or less reduced and often membranous. Aedeagus usually of uninverted tenebrionoid type, with dorsal tegmen (inverted in *Tanylypa*, Pimeliinae and Palorini); parameres parallel, not diverging, without long apical setae, partly or entirely fused together forming apicale, which lacks acccessory lobes and is almost always articulated to phallobase, usually with a pair of internal sclerotised rods (alae) extending basally and enveloping penis; penis with or without anterior struts. Sternite VIII in female with short to long spiculum ventrale. Ovipositor usually moderately elongate and lightly sclerotised except for baculi (occasionally short and broad or heavily sclerotised and blade-like); paraproctal baculi usually elongate, sometimes strongly oblique or transverse; coxite often consisting of four parts, the basal one short and often subtriangular, with an oblique or transverse baculum articulating with the paraproctal baculum, and the distal one often narrow and palpiform; styli apical or subapical, rarely absent. [Doyen 1966, 1984, 1985, 1989; Watt 1974b, 1992; Doyen and Lawrence 1979; Tschinkel and Doyen 1980; Doyen and Tschinkel 1982; Matthews 1986, 1992b, 1993, 2003; Doyen *et al.* 1990; Matthews and Lawrence 1992, 2005; Bouchard and Yeates 2001; Iwan 2001, 2004; Aalbu *et al.* 2002; Banaszkiewicz 2006; Matthews and Bouchard 2008; Matthews and Monteith 2010; Matthews *et al.* 2010].

Larva (Fig. 91A–C, F, K)

Body usually elongate, more or less parallel-sided and subcylindrical to slightly flattened (rarely broadly ovate or strongly flattened); dorsal surfaces usually more or less evenly sclerotised and yellow to dark brown in colour (occasionally more or less membranous or with light and dark pigment pattern); generally smooth, sometimes with large, shallow punctures mainly on head and abdominal apex; vestiture usually of scattered, fine hairs or setae, sometimes densely hirsute.

Head (Fig. 36C) almost always prognathous and protracted (strongly declined in Leiochrinini), as long as wide or transverse, globular to slightly flattened. Epicranial stem usually moderately long (rarely absent); frontal arms usually V-shaped or U-shaped (rarely lyriform), contiguous at base, rarely joined anteriorly by transverse line. Median endocarina absent or occasionally lying beneath epicranial stem. Stemmata on each side five or fewer. Frons with median glandular sieve plate in *Corticeus*, Gnathidiini, *Spiloscapha* and Leiochrinini. Frontoclypeal suture always distinct. Labrum free. Antennae usually well developed, 3-segmented, with dome-like or C-shaped sensorium much shorter than antennomere 3; antennae 2-segmented in Lagriinae (Fig. 91B) and Gnathidiini and 1-segmented in Leiochrinini; sensorium occasionally conical or palpiform (e.g. Bolitophagini); antennal insertion usually separated from mandibular articulation by membrane or narrow strip of cuticle (more widely separated in Leiochrinini), sometimes concealed beneath frontal ridge. Mandibles (Fig. 40E) more or less asymmetrical, sometimes strongly so, usually broad at base and narrow at apex, which is bidentate or tridentate; incisor edge simple or with one or two subapical teeth; mola usually well developed, simple, irregularly ridged or with fine transverse ridges (reduced in *Spiloscapha*). Ventral mouthparts (Fig. 36A) retracted, with well developed maxillary articulating area; cardines more or less transverse and undivided or divided by internal ridge; mala rounded or truncate, rarely with one or more teeth at inner apical angle; maxillary palps 3-segmented. Labium consisting of prementum, mentum and submentum; ligula present or absent; labial palps almost always 2-segmented (1-segmented in Leiochrinini) and narrowly to widely separated. Hypopharyngeal sclerome almost always consisting of molar-like sclerite (Figs 38B, 50B) (absent in *Spiloscapha*). Hypostomal rods almost absent; ventral epicranial ridges present in *Spiloscapha*. Gular region usually transverse; gular sutures well separated.

Thoracic terga without patches or rows of asperities, transverse plicae or lateral processes. Paired sieve plates present on protergum and abdominal terga I–VII in *Corticeus* and on all thoracic terga and abdominal terga I–VIII in Gnathidiini, *Spiloscapha* and Leiochrinini. Prothoracic venter usually without distinct plates, often more heavily sclerotised laterally; postcoxalia greatly enlarged and meeting or fused at midline in some Pimeliinae. Legs well developed, slender to moderately stout, 5-segmented including claw-like pretarsus bearing two setae lying side by side; coxae usually separated; fore legs in some Pimeliinae greatly enlarged with pretarsus subdivided into lightly sclerotised base and heavily sclerotised, spade-like apex. Abdominal terga usually simple, without rows or patches of asperities. Median gland opening between terga VIII and IX in *Corticeus* and Gnathidiini. Tergum VIII with posterior protuberances forming pinching organ with tergum IX in some Cnodalonini. Tergum IX excluding appendages shorter than to slightly longer than VIII, usually extending onto ventral surface, not forming articulated plate, usually either without urogomphi (tergum IX posteriorly rounded, subacute or concave with sclerotised rim) or with pair of simple, acute urogomphi with no pit between them (bifurcate urogomphi in *Strongylium*); sternum IX exposed, simple, not enclosed by sternum VIII; segment X often reduced and concealed or with distinct pair of pygopods; anal region usually posteroventrally oriented. Spiracles annular or annular–multiforous, with peritreme sometimes crenulate, not placed at ends of spiracular tubes, those on segment VIII not enlarged. [St. George 1924, 1926; Böving and Craighead 1931; Doyen and Lawrence 1979; Costa *et al.* 1988; Lawrence and Spilman 1991; Lawrence *et al.* 1999a; Vidal and Guerrero 2007; Matthews *et al.* 2010].

Classification and World Distribution

The cosmopolitan family Tenebrionidae includes ~2300 genera and 20 000 species. The group is currently divided into nine subfamilies, eight of which occur in Australia. The exotic subfamily Nilioninae is restricted to the Neotropical region. [Bouchard *et al.* 2005b; Matthews *et al.* 2010].

Australian Fauna

Although most tenebrionid subfamilies occur in Australia, the large and diverse **Pimeliinae** is very poorly represented here. Of the 39 pimeliine tribes, only five extend into Australia, and at least two of these are based on apparent introductions: *Trachyderma* sp. (Pimeliini) from North Africa and *Salax* sp. (Tribolocarini) from Chile. The only endemic pimeliine genera are *Thorictosoma* and *Wattiana* from the Cnemeplatiini and *Exangeltus* (Fig. 74B) and *Lixionica* from the Vacronini. The subfamily **Zolodininae** includes only two species: the New Zealand *Zolodinus zealandicus* Blanchard and the Tasmanian *Tanylypa morio* Pascoe (Fig. 74G). **Lagriinae** are well represented with at least some members of seven of the nine tribes and a large and highly diverse fauna of the Notogean tribe Adeliini (Fig. 74J–K). Other lagriines of note include species of *Ecnolagria* (Fig. 94F), *Acerogria* (Fig. 7A) and related members of the Lagriini, the strongly flattened *Cossyphus odewahnii* Pascoe (Fig. 74H) of the Afro-Asian tribe Cossyphini, *Pseudolyprops punctatus* (Carter) (Fig. 74F) belonging to the pantropical Goniaderini, and the beach-dwelling *Sphargeris physodes* Pascoe (Chaerodini) (Fig. 74D) related to the New Zealand genus *Chaerodes* White. **Phrenapatinae** is represented by a few genera of Penetini and the unusual and widely distributed genus *Archaeoglenes*. About half of the tribes of **Tenebrioninae** occur in Australia, but some tribes like Blaptini, Melanimini and Pedinini are represented by single introduced (*Blaps polychresta* Forskål) or widely distributed (*Cheirodes sardous* Gené and *Leichenum canaliculatum* (Fabricius)) species, while Opatrini (Fig. 94H) is poorly represented as compared to the rich Palaearctic and Afrotropical faunas. The major tenebrionine tribe is Heleini with 38 endemic genera. The group was formerly placed in four different subfamilies because of the variation in adult body form; included are the strongly flattened pie-dish beetles of the genus *Helea* (Fig. 94I) and the more convex and often ridged species of *Nyctozoilus*, both common in the dry interior, as well as brightly coloured, metallic species of *Cyphaleus* (Fig. 74E) more common in forested areas. An unusual tenebrionine is *Leaus tasmanicus* Matthews & Lawrence (Fig. 94G), a Tasmanian species placed in the tribe Titaenini by Matthews and Bouchard 2008) but very similar in both larval and adult features to the Chilean species of *Trachelostenus* Solier (Trachelostenidae). Other tenebrionine tribes include the Toxicini (Fig. 74C, I), Ulomini (Fig. 74A) and Amarygmini (Fig. 74L). The subfamilies **Alleculinae** (Figs 74B, 94K) and **Stenochiinae** (Fig. 94J) are each represented by ~20 genera and numerous species, but the former is widely distributed throughout Australia, while the latter is most diverse in the tropical and subtropical parts of eastern Australia. All tribes of **Diaperinae** occur in Australia, although *Phaleria bimaculata* (Linnaeus) (Phaleriini) and *Myrmechixenus vaporariorum* Guérin-Méneville (Myrmechixenini) are introduced. Included are fungivorous groups, such as Scaphidemini and Diaperini, the small and mainly xeric species belonging to Hyociini, and the unusual, coccinellid-like algal grazers belonging to the tribe Leiochrinini (Fig. 75J).

Tenebrionids are highly variable in form, but usually black or brown and subglabrous, with concealed antennal insertions and usually stout, moniliform or incrassate antennae. The eyes are usually at least slightly emarginate, the lateral pronotal carinae are usually simple and complete, the prosternal process is usually moderately broad and strongly curved dorsally behind the coxae (except in Leaus) and neither strongly or abruptly expanded apically, and the junctions of the first three ventrites are different from those separating the last two, the basal ventrites being strongly connate and the apical ones more or less movable, often with exposed membrane. Externally open procoxal cavities occur only in *Tanylypa*, but internally open ones are known in some Tenebrioninae. Although almost all species have a 5-5-4 tarsal formula, reduction to 4-4-4 has occurred at least seven times in Australian tenebrionids (usually in small and relatively rare species). Most members of the family are also characterised by having paired glandular reservoirs, associated with a chemical defence system, located usually at the apex of the fifth ventrite (sternite VII) or at the apex of sternite VIII in Adeliini. In some Adeliini, these reservoirs are completely everted when the animal is disturbed. Most Lagriinae and Alleculinae differ from other tenebrionids in having a distinctly lobed penultimate tarsomere, and the latter are further distinguished by their pectinate pretarsal claws. Tenebrionid larvae are almost always elongate (ovate in Leiochrinini) and subcylindrical to slightly flattened. They may be very lightly sclerotised, moderately sclerotised and yellow in colour, with or without darker head and urogomphi, or uniformly heavily sclerotised and darkly pigmented, and the vestiture consists of scattered setae, except in Lagriini (Fig. 91A), whose larvae are densely setose. Some are called false wireworms because of their resemblance to larvae of Elateridae, but they have a very different type of head, which is less flattened and with a distinct clypeus and labrum. A combination of frontoclypeal suture, simple and rounded mala and no hypostomal rods will distinguish tenebrionid larvae from those of most related groups.

Tenebrionids are primarily scavengers, feeding on a variety of dead material of plant or fungal (rarely animal) origin, and they have been able to survive better than most beetles in arid habitats throughout the world. Many species (mainly Lagriini, Adeliini, Heleini and Opatrini) occur under rocks and in leaf litter, with the larvae living in the soil or scavenging on the ground at night; among these are several psammophilous species (*Sphargeris*, *Trachyscelis*, Heterocheirini, Hyociini, some Opatrini) occurring on dunes or near accumulations of debris along ocean beaches or on sandy river banks. Several tenebrionid larvae (some Adeliini, Phrenapatinae, *Tanylypa*, many Heleini, Ulomini, some Amarygmini, Alleculinae, Gnathidiini and Stenochiinae) feed in rotten wood; adults may also feed on the wood (Ulomini, Gnathidiini) but more often they occur on surfaces at night feeding on lichens, fungi or plant material. Some larval forms may also graze the surface at night; these are known in *Titaena*, *Amarygmus* and the Leiochrinini. Members of the Toxicini and Bolitophagini and most Diaperini feed as both larvae and adults on fungal fruiting bodies (particularly Polyporaceae), while other tenebrionids may occur in caves (*Brises*) or animal nests (*Ulomoides*). Some subterranean forms will feed on the roots of plants, including commercial crops; these false wireworms occur in several genera, including *Isopteron*, *Saragus*, *Pterohelaeus* and *Gonocephalum* (Allsopp 1980). Several species (known as mealworms and flour beetles) have become serious pests of stored products; these include *Alphitobius diaperinus* (Panzer), *A. laevigatus* (Fabricius), *Gnatocerus cornutus* (Fabricius) (Fig. 90N), *Latheticus oryzae* Waterhouse, *Palorus subdepressus* (Wollaston), *Tenebrio mol*itor Linnaeus (Fig. 90J), *T. obscurus* Fabricius, *Tribolium castaneum* (Herbst) and *T. confusum* Jacquelin DuVal (Fig. 90M). *Alphitobius diaperinus* may play a beneficial role in the control of fly populations in poultry houses; the beetlers modify the dung on which they feed, rendering it unsuitable for fly breeding (Wallace *et al.* 1985) and may also prey on the fly maggots and puparia (Despins *et al.* 1988). The Egyptian beetle (*Blaps polychresta* Forskäl) was accidentally introduced from North Africa and has become relatively common in SA around human habitations; however, it is not considered a pest species. [Carter 1926; Watt 1974b, 1992; Kaszab 1978, 1982a, 1982b, 1987, 1988; Lawrence and Medvedev 1982; Medvedev and

Lawrence 1983, 1984, 1986, 1987; Merkl 1986, 1987; Doyen *et al.* 1990; Matthews 1992, 1993, 2000; Matthews and Lawrence 1992, 2005; Merkl and Kompantsev 1996; Matthews and Bouchard 2008; Matthews and Monteith 2010; Mathews *et al.* 2010].

Australian Subfamilies, Tribes and Genera

Lagriinae: Belopini: *Doyenia* Matthews and Lawrence, *Euclarkia* Lea, *Eulea* Carter, *Kershawia* Lea. **Cossyphini:** *Cossyphus* Olivier. **Adeliini:** *Adelium* Kirby, *Adelodemus* Haag-Rutenberg, *Apasis* Pascoe, *Bellendum* Matthews, *Blepegenes* Pascoe, *Bluops* Carter, *Bolusculus* Matthews, *Brycopia* Pascoe, *Cardiothorax* Motschulsky, *Coripera* Pascoe, *Daedrosis* Bates, *Diaspirus* Matthews, *Dicyrtodes* Matthews, *Diemenoma* Matthews, *Dorrigonum* Matthews, *Epomidus* Matthews, *Isopteron* Hope, *Leptogastrus* Macleay, *Licinoma* Pascoe, *Monteithium* Matthews, *Nolicima* Matthews, *Nototrintus* Carter, *Seirotrana* Pascoe, *Yarranum* Matthews. **Goniaderini:** *Pseudolyprops* Fairmaire. **Lupropini:** *Iscanus* Fauvel, *Lorelus* Sharp, *Mesotretis* Bates, *Micropedinus* Lewis. **Lagriini: Lagriina:** *Acerogria* Borchmann, *Casnonidea* Fairmaire, *Ecnolagria* Borchmann, *Lagria* Fabricius, *Metriolagria* Merkl, *Sora* Walker, *Stenolagria* Merkl, *Xenolagria* Merkl. **Chaerodini:** *Sphargeris* Pascoe.

Phrenapatinae: Archaeoglenini: *Archaeoglenes* Broun. **Penetini:** *Archeophthora* Kaszab, *Dioedus* LeConte, *Pseudophthora* Kaszab, *Scolytocaulus* Fairmaire, *Tagalinus* Kaszab.

Zolodininae: *Tanylypa* Pascoe.

Pimeliinae: Cnemeplatiini: *Thorictosoma* Lea, *Wattiana* Matthews and Lawrence. **Vacronini:** *Exangeltus* Blackburn, *Lixionica* Blackburn. **Pimeliini:** *Trachyderma* Latreille. **Tentyriini:** *Stenosida* Solier. **Tribolocarini:** *Salax* Guérin-Méneville.

Tenebrioninae: Palorini: *Austropalorus* Halstead, *Eutermicola* Lea, *Palorus* Mulsant, *Platycotylus* Olliff, *Pseudeba* Blackburn, *Ulomina* Baudi di Selve. **Toxicini:** *Ilyxerus* Pascoe, *Mychestes* Pascoe, *Orcopagia* Pascoe, *Toxicum* Latreille. **Bolitophagini:** *Byrsax* Pascoe, *Rhipidandrus* LeConte. **Tenebrionini:** *Paratoxicum* Champion, *Tenebrio* Linnaeus. **Heleini:** *Agasthenes* Bates, *Aglypta* Gebien, *Amarygmimus* Bates, *Amphianax* Bates, *Asphalus* Pascoe, *Atoreuma* Gebien, *Bassianus* Matthews and Doyen, *Bolbophanes* Carter, *Boreosaragus* Matthews, *Brises* Pascoe, *Byallius* Pascoe, *Campanotiphilus* Lea, *Celibe* Boisduval, *Cillibus* Matthews, *Cyphaleus* Hope, *Dysarchus* Pascoe, *Edylius* Champion, *Emcephalus* Kirby, *Helea* Latreille, *Hemicyclus* Westwood, *Lepispilus* Westwood, *Meneristes* Pascoe, *Mithippia* Pascoe, *Mitrothorax* Carter, *Nyctozoilus* Guérin-Méneville, *Olisthaena* Erichson, *Onotrichus* Carter, *Ospidus* Pascoe, *Paraphanes* Macleay, *Phanechloros* Matthews and Bouchard, *Platyphanes* Westwood, *Prophanes* Westwood, *Pterohelaeus* Brême, *Saragus* Erichson, *Sloanea* Carter,*Styrus* Bates, *Sympetes* Pascoe, *Trichosaragus* Blackburn. **Titaenini:** *Leaus* Matthews and Lawrence, *Titaena* Erichson. **Alphitobiini:** *Alphitobius* Stephens, *Diaclina* Jacquelin du Val. **Triboliini:** *Latheticus* Waterhouse, *Lyphia* Mulsant and Rey, *Tribolium* Macleay. **Amarygmini:** *Amarygmus* Dalman, *Axynaon* Blackburn, *Chalcopteroides* Strand, *Trichamarygmus* Carter. **Ulomini:** *Achthosus* Pascoe, *Cenoscelis* Wollaston, *Typhluloma* Lea, *Uloma* Dejean. **Blaptini:** *Blaps* Fabricius. **Melanimini:** *Cheirodes* Gené. **Pedinini:** *Leichenum* Dejean. **Opatrini:** *Caediomorpha* Blackburn, *Caedius* Blanchard, *Diphyrrhynchus* Fairmaire, *Gonocephalum* Solier, *Heterocheira* Dejean, *Mesomorphus* Miedel, *Scymena* Pascoe, Sobas Pascoe.

Alleculinae: *Aethyssius* Pascoe, *Atoichus* Carter, *Barycistela* Blackburn, *Dimorphochilus* Borchmann, *Euomma* Boheman, *Hemicistela* Blackburn, *Homotrysis* Pascoe, *Jophon* Champion, *Melaps* Carter, *Metistete* Pascoe, *Neocistela* Borchmann, *Nocar* Blackburn, *Notocistela* Carter, *Nypsius* Champion, *Ommatophorus* Macleay, *Otys* Champion, *Pemanoa* Buck, *Scaletomerus* Blackburn, *Simarus* Borchmann, *Tanychilus* Newman, *Taxes* Champion.

Diaperinae: Crypticini: *Ellipsodes* Wollaston. **Trachyscelini:** *Trachyscelis* Latreille. **Phaleriini:** *Phaleria* Latreille. **Hyociini:** *Brittona* Medvedev and Lawrence, *Csiro* Medvedev and Lawrence, *Hyocis* Pascoe, *Megela* Medvedev and Lawrence, *Parahyocis* Kaszab, *Uptona* Medvedev and Lawrence. **Ectychini:** *Ectyche* Pascoe, *Micrectyche* Bates. **Hypophlaeini:** *Corticeus* Piller and Mitterpacher. **Myrmechixenini:** *Myrmechixenus* Chevolat. **Gnathidiini:** *Tyrtaeus* Champion, *Menimus* Sharp. **Tribe ?:** *Triplehornia* Matthews and Lawrence. **Leiochrinini:** *Derispia* Lewis, *Leiochrinus* Westwood, *Leiochrodes* Westwood. **Scaphidemini:** *Spiloscapha* Bates. **Diaperini:** *Adelina* Dejean, *Alphitophagus* Stephens, *Ceropria* Laporte and Brullé, *Gnatocerus* Thunerg, *Palembomimus* Matthews and Lawrence, *Pentaphyllus* Dejean, *Platydema* Laporte and Brullé, *Sciophagus* Sharp, *Ulomoides* Blackburn.

Stenochiinae: Cnodalonini: *Androsus* Gebien, *Apterotheca* Gebien, *Bradymerus* Perroud, *Campolene* Pascoe, *Chariotheca* Pascoe, *Cuemus* Bouchard, *Derosphaerus* Thomson, *Espites* Pascoe, *Hydissus* Pascoe, *Hypaulax* Bates, *Kaszaba* Matthews and Doyen, *Metisopus* Bates, *Neotheca* Carter, *Oectosis* Pascie, *Omolipus* Pascoe, *Promethis* Pascoe, *Scotoderus* Perroud, *Tetragonomenes* Chevrolat, *Zophophilus* Fairmaire. **Stenochiini:** *Eutherama* Carter, *Strongylium* Kirby.

97. PROSTOMIDAE

Adult (Figs 76D, 93E)

Length 5–12 mm. Body usually ~3.2–3.5 times as long as wide; subparallel and moderately flattened; colour uniformly yellowish-brown to dark reddish-brown; surfaces subglabrous.

Head prognathous, abruptly constricted behind eyes in *Dryocora* Pascoe forming broad neck and long temples. Eyes small, round and finely facetted, without interfacetal setae. Antennal insertions slightly concealed from above; subantennal grooves absent. Frontoclypeal suture distinctly impressed in *Prostomis*, weakly so in *Dryocora*; clypeus broad at base and slightly to strongly narrowed apically, with apex truncate to emarginate. Labrum free, strongly rounded in *Prostomis*, strongly transverse and truncate in *Dryocora*. Antennae 11-segmented, with weak, 3-segmented club. Mandible broad at base, elongate, projecting, weakly curved and obliquely bidentate or tridentate with multidentate incisor edge in *Prostomis*, shorter, strongly, abruptly curved with single rounded process on incisor edge in *Dryocora*; mola well developed and transversely ridged; prostheca well developed or reduced. Maxilla with sparsely setose galea and spinose lacinia without uncus, lobes slender and elongate in *Prostomis*, shorter and broader in *Dryocora*; apical maxillary palpomere elongate and fusiform. Mentum strongly transverse; ligula simple, short and truncate in *Dryocora*, elongate and narrowly rounded in *Prostomis*; apical labial palpomere fusiform. Subgenal ridges absent. Genae in *Prostomis* strongly produced anteriorly forming long, subacute horns, which are usually as long as the mandibles. Gular sutures well separated. Corpotentorium absent. Cervical sclerites absent.

Pronotum ~0.93–1.05 times as long as wide, widest at middle (*Prostomis*) or posteriorly (*Dryocora*) and as wide at base as combined elytral bases, sides very slightly to moderately curved, anteriorly constricted in *Dryocora*. Lateral pronotal carinae incomplete and not visible from above or absent, simple; anterior angles not produced forward; posterior angles obtuse; posterior edge slightly curved; disc simple. Prosternum in front of coxae more than twice as long as mid length of procoxal cavity, flat to slightly convex. Prosternal process complete, extending well behind coxae, not dorsally curved, strongly and abruptly expanded at apex, which is truncate. Notosternal sutures weakly impressed by more or less complete. Procoxae with long concealed lateral extensions; trochantins concealed. Procoxal cavities circular, very widely separated, externally broadly closed, without narrow lateral extensions, internally closed by slender bar.

Scutellar shield not abruptly elevated, posteriorly obtusely angulate to truncate. Elytra ~2.2–2.55 times as long as combined width and 2.15–2.8 times as long as pronotum; disc with weak to moderately strong post-humeral ridges delimiting steeply sloping sides, with five to seven regular puncture rows on mesal flattened portion of disc and irregularly punctate laterally; scutellary striole absent; epipleura narrow and complete or almost so. Mesoventrite on same plane as metaventrite; anterior edge without paired procoxal rests. Mesoventral process very broadly rounded or truncate at apex. Mesocoxae with concealed trochantins. Mesocoxal cavities subcircular, very widely separated, closed laterally by meeting of mesoventrite and metaventrite. Mesometaventral junction simple. Metaventrite elongate, flattened, with discrimen about half as long as ventrite; exposed portion of metanepisternum elongate and narrow. Metacoxae narrowly separated, extending laterally to meet elytra. Metendosternite with very short, broad stalk and long lateral arms; laminae and anterior process absent, anterior tendons located on lateral arms or not apparent. Hind wing usually with short apical field lacking sclerites or radial extensions; radial cell highly reduced or absent; cross-vein r3 absent; basal portion of RP very short or absent; medial spur strongly curved and extending almost to wing margin, where there is a medial embayment; medial field with two or three free veins and a medial fleck; wedge cell absent; anal lobe without embayment. Legs short and robust; trochanterofemoral joint strongly oblique, with base of femur in contact with coxa; femora slender basally and expanded apically. Tibiae widened apically with spinose outer apical angle, spurs large and stout, paired on all legs. Tarsi 4-4-4; tarsomeres without ventral lobes; pretarsal claws simple and empodium small, bisetose.

Abdomen with five ventrites, the first two of which are connate; ventrite 1 not much longer than 2; intercoxal process acute. Anterior edge of segment IX in male without spiculum gastrale, posterior edge with median projection; tergite IX and segment X more or less fused together. Aedeagus of inverted tenebrionoid type, small and lightly sclerotised; parameres in *Prostomis* fused together forming articulated apicale, which is divided into two lateral lobes and a median lobe or penis guide (parameres fused together and to phallobase in *Dryocora*); penis well sclerotised in Dryocora, membranous in *Prostomis*, without paired basal struts. Sternite VIII in female without or with very short spiculum ventrale. Ovipositor highly reduced, consisting of a pair of weakly sclerotised coxites and no styli. [Schawaller 1993, 2003; Lawrence *et al.* 1999b; Young 2002j; Schunger *et al.* 2003; Seago and Beutel 2010].

Larva (Fig. 35K)

Body elongate, more or less parallel-sided and flattened; dorsal and ventral surfaces lightly pigmented, white or cream in colour, with antennae and mouthparts somewhat darker; smooth, clothed with scattered, erect setae.

Head prognathous and protracted, strongly transverse, wider than thorax, strongly flattened and distinctly asymmetrical. Epicranial stem very short or absent; frontal arms lyriform and contiguous at base. Median endocarina absent. Stemmata absent. Frontoclypeal suture absent. Labrum small, free, very slightly transverse, asymmetrical, broadly rounded apically. Antennae moderately long, 3-segmented, with conical sensorium about half as long as apical antennomere. Mandibles, elongate, broad at base and narrow at apex, tridentate; incisor edge simple; molae well developed, highly asymmetrical, prostheca absent. Ventral mouthparts slightly retracted, with reduced maxillary articulating area; stipes short and broad; mala rounded, not cleft; maxillary palps 3-segmented. Labium consisting of prementum, mentum and submentum; ligula about as long as first palpomere; labial palps 2-segmented and moderately widely separated. Hypopharyngeal sclerome a transverse bar. Hypostomal rods moderately long and diverging; ventral epicranial ridges absent. Gular region slightly longer than wide; gular sutures well separated.

Thoracic and abdominal terga without patches or rows of asperities, carinae or ampullae. Legs fairly short and broad, 5-segmented including claw-like pretarsus bearing two setae; coxae very widely separated. Tergum IX about half as long as VIII, extending onto ventral surface and forming articulated plate, with pair of short, posteriorly-oriented, simple urogomphi, which are not or barely darker than tergum and only slightly upturned at apex; sternum IX exposed, partly enclosed by sternum VIII, apically lined with row of small teeth (Fig. 43I); segment X not visible from above, ventrally oriented and transverse. Spiracles annular-biforous, those on thorax placed at ends of lateral spiracular tubes, those on segment VIII not enlarged. [Young 1991, 2002j; Lawrence *et al.* 1999a; Schunger *et al.* 2003; Seago and Beutel 2010].

Classification and World Distribution

The family Prostomidae includes two genera: *Prostomis* Latreille with 28 species from north-western North America, South Africa, the Palaearctic and Oriental regions, New Guinea, Australia, New Caledonia, Fiji and Samoa, and *Dryocora* Pascoe with a species from New Zealand and one from Australia.

Australian Fauna

The Australian prostomid fauna consists of *Dryocora cephalotes* (Waterhouse) from TAS, VIC and the ACT and five species of *Prostomis* distributed along the eastern coast from TAS to northern QLD. Prostomids are elongate, more or less parallel-sided, flattened, reddish-brown, subglabrous beetles, with the head either wider than the prothorax (*Dryocora*) or elongate with large projecting mandibles and short to long, anteriorly projecting genal processes (*Prostomis*); the eyes are small, the frontoclypeal suture distinct, lateral pronotal carinae absent and procoxae and mesocoxae widely separated. Larvae are elongate, parallel-sided, flattened and lightly pigmented, with a broad, asymmetrical head, mandibles with large, asymmetrical, transversely ridged molae, a rounded and broadly cleft mala, a pair of lightly pigmented, slightly upturned urogomphi on tergum IX, and an apical row of fine asperities on sternum IX. Adult and larval prostomids are found in wet, brown rotten wood, and their presence is marked by the mud-like consistency of the substrate. [Schawaller 1993].

98. OEDEMERIDAE

Adult (Figs 7D, 75A–D, 93H)

Length 2.5–26 (usually 7–18) mm. Body usually ~2.4–4.4 times as long as wide; subparallel and moderately convex to slightly flattened; upper surfaces relatively smooth and weakly sclerotised; often bicolored, sometimes with a pattern on pronotum and/or elytra, sometimes uniformly yellowish-brown to black, occasionally metallic; vestiture usually consisting of decumbent fine hairs (mixed with erect bristles in *Dasytomima* Lawrence).

Head prognathous, not abruptly constricted posteriorly. Eyes not to moderately strongly protuberant, finely to coarsely facetted, without interfacetal setae (except in *Dasytomima*). Antennal insertions almost always exposed from above (slightly concealed ion *Dasytomima*); subantennal grooves absent. Frontoclypeal suture usually absent, weakly impressed in *Agasma* Newman, deeply so in *Dasytomima*). Frontoclypeal region produced to form short, flat rostrum in many Oedemerinae; anterior edge of clypeus truncate to convex with a distinct hyaline anteclypeus. Labrum free, strongly transverse to subquadrate or longer than wide, often expanded apically and truncate to emarginate. Antennae almost always 11-segmented (12-segmented in *Nacerdes* Dejean, sometimes apparently 12-segmented due to constriction in terminal antennomere, e.g. *Pseudolycus* Guérin-Méneville), usually filiform, but serrate in *Dasytomima*, with antennomeres 3–7 enlarged in *Pseudolycus* or 5–8 irregularly enlarged in male of *Dohrnia miranda* Newman). Mandibles short and broad to moderately elongate, unidentate or bidentate, with incisor edge simple or with single subapical tooth; mola and prostheca usually well developed. Maxilla with distinct galea and lacinia, both densely clothed with long setae, which may be expanded apically, the latter without uncus; apical maxillary palpomere usually expanded and truncate to securiform or cultriform, occasionally fusiform or only weakly expanded. Mentum usually strongly transverse; ligula usually emarginate or bilobed; apical labial palpomere fusiform to strongly expanded apically. Subgenal ridges absent. Gular sutures well separated, but gula often longer than wide. Corpotentorium narrow. Cervical sclerites present, but sometimes highly reduced.

Pronotum ~0.7–2.3 times as long as wide, usually widest at anterior third and narrowed posteriorly; sides usually sinuate; base distinctly narrower than combined elytral bases. Lateral pronotal carinae almost always absent; anterior angles not produced forward; posterior angles usually broadly rounded, sometimes obtuse or right; posterior edge more or less straight or evenly rounded, with very broad margin in *Agasma* and *Nacerdes*; disc usually simple, with vague impressions or a median longitudinal groove in some Oedemerinae. Prosternum in front of coxae as long as or longer than mid length of procoxal cavity, flat to moderately convex. Prosternal process almost incomplete or barely extending behind coxae, not dorsally curved, with apex acute to broadly rounded. Notosternal sutures complete. Procoxae projecting well below prosternum, without or with short concealed lateral extensions; trochantins usually concealed (sometimes partly exposed). Procoxal cavities slightly transverse to circular, contiguous or very narrowly separated, externally broadly open, with or without narrow lateral extensions, internally closed by slender bar.

Scutellar shield not abruptly elevated, posteriorly acute to broadly rounded or truncate. Elytra ~2.3–4.3 times as long as combined width and 3.1–5.6 times as long as pronotum, with conjointly rounded or independently rounded apices; disc irregularly punctate without scutellary striole, almost always with three or more weak longitudinal costae (absent in *Dasytomima*); epipleura very narrow and incomplete or absent. Mesoventrite separated by complete sutures from mesanepisterna, which are almost always well separated at midline (almost meeting at midline in *Dasytomima*); anterior edge without paired procoxal rests (procoxal rests on mesanepisterna in Dasytomima). Mesoventral process usually incomplete and apically acute (complete and narrowly rounded in *Dasytomima*). Mesocoxae usually conical and projecting with exposed trochantins. Mesocoxal cavities subcircular to slightly transverse, oblique to strongly oblique, almost always contiguous (narrowly separated in *Dasytomima*), partly closed laterally by mesepimera or both mesepimera and mesanepisterna. Mesometaventral junction almost always absent. Metaventrite with discrimen moderately to very long; exposed portion of metanepisternum elongate to very long and narrow. Metacoxae contiguous or narrowly separated, horizontal or oblique, extending laterally to meet elytra. Metendosternite with long stalk, long lateral arms; well developed laminae and short anterior process, which is bilobed with anterior tendons at end of each lobe. Hind wing usually with short apical field lacking radial extensions but usually with single anterior oblique linear sclerite (longer without sclerite in *Dasytomima*); radial cell usually elongate with inner posterobasal angle right or acute (sometimes with base missing); cross-vein r3 fused to r4 at base and often apparently absent; usually with basal portion of RP moderately long, R-M loop relatively narrow and medial spur straight and extending to wing margin where there is no medial embayment (in *Dasytomima*, basal portion of RP very short, R-M loop broader, medial spur curved and medial embayment present); medial field with four free veins and usually no medial fleck (present in *Dasytomima*); wedge cell well developed and apically acute; anal lobe without embayment. Legs long and slender; trochanterofemoral joint strongly oblique, with base of femur in contact with coxa or not. Tibiae not widened apically; spurs usually paired on all legs (protibia with single spur in *Nacerdes*). Tarsi 5-5-4 in both sexes, with penultimate tarsomere distinctly lobed beneath and preceding ones sometimes slightly expanded; special adhesive setae usually present on penultimate tarsomere (present on pro- and mesotarsomeres 2–4 and metatarsomeres 3 and 4 in *Agasma*); pretarsal claws usually simple or slightly expanded (sometimes with distinct basal tooth); empodium not apparent.

Abdomen with five or six ventrites, the first two of which are connate; ventrite 1 not much longer than 2; intercoxal process acute or narrowly rounded. Ventrites 3 and 4 in female *Dasytomima* with small, median patches of glandular hairs. Ventrite 5 (sternite VII) sometimes emarginate in males. Posterior edge of sternite VIII in male often deeply emarginate or bilobed. Anterior edge of segment IX in male usually with spiculum gastrale (subacute without strut in *Dasytomima*), posterior edge almost always with slender, hyaline, setose tegmenite; tergite IX deeply emarginate or almost completely divided into two parts; tergite X usually well developed, free, well sclerotised to completely membranous or fused to tergite IX. Aedeagus of tenebrionoid type, symmetrical; parameres fused to phallobase but partly free from one another (forming apically cleft tegmen) or entirely fused together and to phallobase, forming undivided tegmen; anterior edge of penis without paired struts, often distinctly curved, divided into dorsal and ventral lobes in *Dasytomima*, with sub-basal supporting sclerite in many Oedemerinae. Sternite VIII in female with long, basally articulated, spiculum ventrale. Ovipositor long and relatively slender, almost always lightly sclerotised except for baculi; paraprocts usually much longer than coxites, each of which is vaguely subdivided and setose, with transversely oblique basal baculum; styli well developed and terminal. [Lea 1917a, 1917b; Arnett 1948, 1951; Hudson 1975; Svihla 1986; Vásquez 1993, 1996; Lawrence *et al.* 1999b; Kriska 2002; Lawrence 2005g; Lawrence and Ślipiński 2010h].

Larva

Body elongate, more or less parallel-sided and subcylindrical; dorsal and ventral surfaces lightly pigmented, white or cream in colour, except for buccal region and mandibles; surfaces smooth, clothed with scattered, erect setae.

Head prognathous and protracted, slightly transverse, slightly flattened and often asymmetrical; sides weakly to strongly rounded. Epicranial stem moderately long; frontal arms V-shaped or U-shaped and contiguous at base. Median endocarina Y-shaped, coincident with stem and arms. Stemmata usually absent, sometimes two or five on each side. Frontoclypeal suture absent. Labrum free. Antennae well developed, 3-segmented, with apical antennomere reduced; sensorium conical or palpiform, shorter than to longer than apical antennomere. Mandibles broad at base and narrow at apex, bidentate or tridentate; incisor edge simple or with one or more subapical teeth; molae well developed, highly asymmetrical, with numerous, fine transverse ridges. Ventral mouthparts retracted, with well developed maxillary articulating area; cardo transverse to slightly oblique, sometimes divided; stipes longer than wide; mala more or less rounded, setose, often with distinct tooth along inner edge; maxillary palps 3-segmented. Labium consisting of prementum, mentum and submentum; ligula well developed, reduced or absent, simple; labial palps 2-segmented, usually narrowly separated. Hypopharyngeal sclerome molar-like, usually with long, columnar prehypopharynx. Hypostomal rods and ventral epicranial ridges absent. Gular region wider than long; gular sutures well separated.

Thorax relatively short and somewhat wider than abdomen; prothorax distinctly longer than meso- or metathorax. Thoracic terga usually with asperity-bearing ampullae on all segments. Prothoracic venter without special armature. Legs relatively short, 5-segmented including claw-like pretarsus bearing two setae; coxae moderately to very widely separated. Abdomen more than twice as long as thorax. Terga I and II or I–III with asperity-bearing ampullae; sterna with asperity-bearing prolegs on segments III and IV or II–IV. Tergum IX shorter than VIII, completely dorsal, not forming articulated plate, without urogomphi; sternum IX well developed, simple, not enclosed by sternum VIII; segment X usually not visible from above, posteriorly or posteroventrally oriented, without paired pygopods, with transverse anal opening. Spiracles annular or annular–multiforous, not placed at ends of lateral spiracular tubes, those on segment VIII not enlarged. [Duffy 1952; Rozen 1958, 1960; Lawrence 1991b; Lawrence *et al.* 1999a; Kriska 2002; Lawrence and Ślipiński 2010h].

Classification and World Distribution

Oedemeridae is a widely distributed family with ~80 genera and 1500 species. The group is currently divided into three subfamilies (Lawrence 2005): Calopodinae with the Holarctic genus *Calopus* Fabricius and *Sparedrus* Dejean from the Holarctic, Oriental and Neotropical regions, Polypriinae with the New World genus *Polypria* Chevrolat and the Australian genus *Dasytomima* Lawrence, and Oedemerinae with the remaining genera. The Oedmerinae is further subdivided into five tribes, three of which occur in Australia. The tribe Stenostomini is restricted to the Mediterranean region and the Oedemerini occurs in the Palaearctic and Oriental regions.

Australian Fauna

The subfamily **Polypriinae** is represented in Australia by the species *Dasytomima rachelae* Lawrence (Fig. 75C), a very small species distributed from TAS to southern QLD. Of the three tribes of **Oedemerinae**, Nacerdini is represented by the adventive species *Nacerdes melanura* (Linnaeus), Ditylini by the endemic *Agasma semicrudum* Newman (Fig. 75B), and the Asclerini by the genera *Asclera* Stephens, *Copidita* LeConte, *Danerces* Westwood, *Dohrnia* Newman, *Pseudolycus* Guérin-Méneville (Fig. 75D), *Selenopalpus* White and *Sessinia* Pascoe (Fig. 75A). Since a modern revision of the Australian Oedemeridae has not yet been attempted, it is likely that some generic placements will change and new genera and species will be described. Oedemerids are elongate, parallel-sided, soft-bodied beetles, often bicoloured and clothed with short, decumbent hairs, very rarely mixed with erect bristles. The head is usually somewhat produced forward to form a muzzle, the antennal insertions are located dorsally between the eyes, the lateral pronotal carinae are absent, the apical maxillary palpomere is more or less expanded and truncate (excavate in some males), and the penultimate tarsomere is lobed beneath. Adults occur in flowers or on foliage and may be beaten from vegetation by day or collected at lights at night; at least some adults feed on pollen that is stored in the large crop, but fungal spores and even pieces of cuticle are sometimes found in the gut. Some oedemerids are known to contain cantharidin, and several are aposematically coloured and belong to mimicry complexes (Moore and Brown 1989); species of *Pseudolycus* are often black and orange, mimicking various Lycidae. Oedemerid larvae are elongate, subcylindrical and very lightly pigmented, with a large, often asymmetrical head, and asperity-bearing prolegs on abdominal sterna III and IV or II–IV. The antennae are well developed, the epicranial stem long, the frontal arms V-shaped, the mandibles large and asymmetrical with transversely ridged molae and the ligula well developed, with a prehypopharyngeal lobe between it and the hypopharynx. Urogomphi are always absent in Australian species. Larvae feed on dead wood, and they are most common in that which has undergone white rot; those of the cosmopolitan *Nacerdes melanura* damage wharves and other wooden marine structures. [Lea 1917a, 1917b; Hudson 1975; Švihla 1986; Lawrence 2005g; Lawrence and Ślipiński 2010e].

99. MELOIDAE

Adult (Figs 7E, 75E–G)

Length 7–25 mm. Body usually ~2.2–3.2 times as long as wide; subparallel and moderately convex; upper surfaces relatively smooth and weakly sclerotised; colour uniformly yellowish-brown, bright red in Horiini, black with orange elytra (mimicking Lycidae) in *Palaestra* and *Palaestrida*, often metallic blue, green or violet, sometimes combined with red head and prothorax; surfaces usually glabrous, sometimes with very short, fine hairs.

Head deflexed, hypognathous, abruptly constricted posteriorly to form narrow neck and well defined temples, usually wider than anterior edge of prothorax (extremely enlarged in *Synhoria*). Eyes small to moderately large, usually not protuberant, entire or emarginate, finely facetted, without interfacetal setae. Antennal insertions exposed from above; subantennal grooves absent. Frontoclypeal suture usually absent or very weakly impressed (distinct in Horiini). Frontoclypeal region sometimes produced to form short, flat rostrum; anterior edge of clypeus truncate, with distinct hyaline anteclypeus. Labrum free, usually subquadrate to slightly elongate (strongly transverse in Horiini). Antennae 11-segmented, moderately long and filiform or moniliform. Mandibles short and broad to moderately elongate, unidentate, with incisor edge simple; mola and prostheca usually well developed (prostheca reduced and mola absent in Horiini). Maxilla with distinct galea and lacinia, but with the latter much shorter than the former, both densely clothed with long setae, which may be expanded apically, the latter without uncus; apical maxillary palpomere more or less fusiform. Mentum and prementum fused together; ligula usually emarginate or bilobed (simple in Horiini); apical labial palpomere fusiform. Subgenal ridges absent. Gular sutures well separated; gula longer than wide. Corpotentorium narrow. Cervical sclerites well developed.

Pronotum ~0.4–1.25 times as long as wide, usually widest posteriorly and gradually or abruptly narrowed anteriorly, with sides curved or sinuate (widest at middle and sides straight or very slightly curved in Horiini); base distinctly narrower than combined elytral bases. Lateral pronotal carinae absent; anterior angles not produced forward; posterior angles right to broadly rounded; posterior edge more or less straight or evenly rounded; disc usually simple, often with median longitudinal groove, with two pairs of impressions in *Palaestra*. Prosternum in front of coxae usually longer than mid length of procoxal cavity (much shorter in Horiini), flat to slightly convex. Prosternal process incomplete with apex obtusely angulate or absent. Notosternal sutures present or absent. Procoxae projecting well below prosternum, without concealed lateral extensions; trochantins exposed. Procoxal cavities slightly transverse to circular, contiguous, externally broadly open, without narrow lateral extensions, internally open.

Scutellar shield not abruptly elevated, posteriorly narrowly to broadly rounded. Elytra ~1.7–2.55 times as long as combined width and 2.9–4.9 times as long as pronotum, with conjointly rounded apices; disc irregularly punctate without scutellary striole, with four weak longitudinal costae in *Horia* and *Palaestra*; epipleura very narrow and incomplete or absent. Mesoventrite separated by complete sutures from mesanepisterna, which are broadly joined at midline; anterior edge without paired procoxal rests. Mesoventral process incomplete and apically acute. Mesocoxae more or less conical and projecting, with exposed trochantins. Mesocoxal cavities transverse and more or less oblique, contiguous, partly closed laterally by mesepimera and mesanepisterna. Mesometaventral junction absent. Metaventrite moderately convex, with well developed discrimen; exposed portion of metanepisternum long and narrow. Metacoxae contiguous or narrowly separated, horizontal or slightly oblique, extending laterally to meet elytra. Metendosternite with long stalk, long lateral arms; well developed laminae and no anterior process; anterior tendons widely separated near bases of lateral arms. Hind wing with short apical field lacking radial extensions but usually with anterior and posterior oblique linear sclerites; radial cell without with inner edge; cross-vein r3 absent; basal portion of RP moderately long, R-M loop relatively broad; medial spur straight and extending to wing margin where there may be a medial embayment; medial field with four free veins and no medial fleck; wedge cell usually absent or incomplete (apically acute in *Horia*); anal lobe without embayment. Legs usually long and slender; trochanterofemoral joint strongly oblique, with base of femur in contact with coxa or not; metafemora greatly enlarged and toothed in males of *Horia*. Tibiae usually not widened apically; spurs usually paired on all legs; metatibiae in males of *Horia* with long, acute spine at inner apical angle. Tarsi 5-5-4 in both sexes, tarsomere not lobed beneath; pretarsal claws pectinate and with ventral blade-like appendage; empodium small and bisetose.

Abdomen with six free ventrites; ventrite 1 not much longer than 2; intercoxal process absent. Ventrite 5 (sternite VII) sometimes emarginate in males. Posterior edge of sternite VIII in male often deeply emarginate or bilobed. Anterior edge of segment IX in male with spiculum gastrale, posterior edge without tegmenite; tergite IX deeply emarginate or almost completely divided into two parts; tergite X usually well developed, free, well sclerotised to completely membranous or fused to tergite IX. Aedeagus of tenebrionoid type, symmetrical; parameres fused together and articulated to phallobase, forming apicale; anterior edge of penis without paired struts. Sternite VIII in female without spiculum ventrale. Ovipositor short and broad, lightly sclerotised, without baculi; paraprocts subequal in length to coxites, each of which is undivided and setose; styli small, cylindrical, terminal. [Gahan

1908b; Bugnion 1910; Blair 1920; Cros 1924, 1927; Betrem 1932; Morgan 1968; Kaszab 1969; Gupta 1971; Berrioz-Ortiz 1985; Pinto and Bologna 2002; Bologna and Pinto 2001, 2002; Pinto and Bologna 2002; Bologna *et al.* 2008, 2010].

Larva

Larvae of Nemognathinae exhibit hypermetamorphosis. First instars of the triungulinid type, more or less boat-shaped, widest near anterior end and tapering anteriorly and posteriorly, moderately heavily pigmented and clothed with short posteriorly directed setae and a few very long setae. Intemediate and late instars usually very lightly pigmented and more grub-like.

Head of first instar protracted and prognathous, widest posteriorly; intermediate and final instars with hypognathous head. Epicranial suture and median endocarina absent in first instar and stemmata two on each side; later instars sometimes with epicranial suture and with or without single stemma on each side. Frontoclypeal suture absent and labrum fused to clypeus in first instar; labrum free in some later instars. Antennae in first instar 3-segmented without sensorium and with very long seta at apex of antennomere 3; antennae reduced in later instars, sometimes only 1-segmented. Mandibles in first instar transversely ridged, vertically oriented; those in later instars curving mesally and unidentate; mola and prostheca always absent. Ventral mouthparts more or less retracted in first instar, without maxillary articulating area; stipes short and broad; mala simple and rounded, and maxillary palps 3-segmented.

Legs well developed, slender, 5-segmented in first instar, with pretarsal claw bearing two setae (Fig. 51A); legs shorter and more robust in intermediate instars, reduced and bud-like in coarctate instar. Abdomen in first instar only slightly longer than thorax. Tergum IX not extending onto ventral surface, without urogomphi or long setae; segment X more or less terminal. Spiracles small and annular, on thorax and abdominal segments I–VIII, those on VIII borne on short spiracular tubes. [Bugnion 1910; Brues 1924; Cros 1924, 1927, 1938; MacSwain 1956; Kifune 1961; Bianchi 1962; Selander 1991b; Bologna and Laurenzi 1994; Pinto and Bologna 2002; Bologna *et al.* 2010; Di Giulio *et al.* 2010].

Classification and World Distribution

The cosmopolitan family Meloidae comprises ~3000 species in 120 genera. The group is currently divided into three or four subfamilies, one of which occurs in Australia. The exotic subfamilies include Eleticinae (Oriental, Afrotropical and Neotropical regions), Meloinae (widely distributed but absent from the Australian region) and Tetraonycinae (Neotropical and southern Nearctic regions); the last group, however, is sometimes considered to be a tribe within the Meloinae. [Bologna *et al.* 2010; Bologna and Di Giulio 2011].

Australian Fauna

The Australian meloid fauna consists of a few genera of Nemognathinae. The tribe Horiini is represented by two genera: *Horia* Fabricius with the species *H. mira* (Blackburn) (Fig. 75G) and *Synhoria* Kolbe with the species S. *maxillosa* (Fabricius), while the Nemognathini includes a few species of *Palaestrida* White and *Palaestra* Laporte (Fig. 75E), plus several species (Figs 75F, 94A) now placed in the genus *Zonitis* Fabricius but currently being reclassified by M. A. Bologna, J. D. Pinto and F. Turco. Meloids are elongate, moderately convex, soft-bodied beetles, often bicoloured with red or yellow and black or blue, but sometimes uniformly red or metallic blue or violet. The body is glabrous or clothed with short, decumbent hairs, the head is strongly constricted behind the eyes to form a narrow neck, the prothorax is usually narrowed anteriorly and lacks lateral carinae, and the pretarsal claws are pectinate with a blade-like process beneath each claw. Larvae are of three types: (1) a minute, fusiform, well sclerotised triungulin with long legs; (2) a grub-like, ectoparasitic larva, lightly sclerotised, with short legs and a hypognathous head; and (3) a coarctate form with reduced mouthparts and appendages. The evolution of parasitism and hypermetaboly in the family has been discussed by Bologna and Di Giulio (2011). All Australian meoids are parasitoids of bees (Hymenoptera: Apoidea). Eggs are laid on flowers and the active triungulins attach themselves to bees in order to gain access to the nest. Later instar consume both bee larva and provisions of honey and nectar. The coarctate larva is a resting stage, which gives rise to another grub-like prepupal form. The large species of Horiini parasitise carpenter bees of the apid subfamily Xylocopinae. [Brues 1924; Wilson 1920].

100. MYCTERIDAE

Adult (Fig. 93F–G)

Length 2.5–5.5 mm. Body usually ~2.7–4.85 times as long as wide; subparallel or widest at posterior third and narrowing anteriorly, usually moderately convex to slightly flattened (strongly flattened in *Hemipeplus*; surfaces relatively smooth and weakly to moderately well sclerotised; usually uniformly yellow to brown or black in colour; vestiture consisting of fine decumbent hairs.

Head prognathous, not abruptly constricted posteriorly. Eyes moderately strongly protuberant, not emarginate, coarsely facetted, usually with very short, fine interfacetal setae. Antennal insertions exposed from above in *Hemipeplus*, concealed from above by frontal ridges in Eurypinae; subantennal grooves weakly developed or absent. Frontoclypeal suture usually very weakly impressed and slightly curved. Anterior edge of clypeus truncate to convex with or without a distinct hyaline anteclypeus. Labrum free, slightly to strongly transverse. Antennae 11-segmented, relatively short, filiform or weakly serrate in Eurypinae, moniliform in *Hemipeplus*. Mandibles short and broad to moderately elongate, bidentate; incisor edge simple or sometimes serrate; mola well developed, with simple surface; prostheca usually long and slender. Maxilla with galea and lacinia short, broad and setose, the latter without uncus; apical maxillary palpomere usually expanded and truncate. Mentum strongly transverse; ligula truncate to slightly emarginate; apical labial palpomere fusiform. Subgenal ridges absent. Gular sutures well separated; gula longer than wide. Corpotentorium very narrow and arched. Cervical sclerites present, but reduced.

Pronotum ~0.75–0.95 times as long as wide, usually widest anteriorly (rarely at middle); sides very weakly rounded or sinuate; base usually distinctly narrower than combined elytral bases. Lateral pronotal carinae absent or vaguely indicated and incomplete; anterior angles not produced forward; posterior angles usually obtuse or right; posterior edge more or less straight;

disc always with pair of small sub-basal impressions, usually with transverse impression between them. Prosternum in front of coxae (Fig. 44R) distinctly longer than mid length of procoxal cavity, flat to slightly convex. Prosternal process incomplete, not dorsally curved behind coxae, with apex narrowly rounded or acute. Notosternal sutures usually complete. Procoxae not or slightly projecting below prosternum, with short concealed lateral extensions; trochantins concealed. Procoxal cavities subcircular, contiguous or very narrowly separated, externally broadly to narrowly open, with narrow lateral extensions, internally closed by slender bar.

Scutellar shield abruptly elevated, posteriorly broadly rounded to truncate. Elytra ~1.9–3.6 times as long as combined width and 3.6–5.2 times as long as pronotum; sides subparallel or weakly rounded and widest behind middle, with conjointly rounded apices; disc almost always irregularly punctate (rarely with ten puncture rows and scutellary striole); epipleura very narrow and incomplete. Mesoventrite separated by complete sutures from mesanepisterna, which are almost always narrowly separated or meeting at midline (more widely separated in *Hemipeplus*); anterior edge without paired procoxal rests, with broad prothoracic rest or prepectus in *Hemipeplus*. Mesoventral process complete and apically narrowly rounded or truncate. Mesocoxae not projecting, with concealed trochantins. Mesocoxal cavities subcircular, narrowly separated, closed laterally by meeting of meso- and metaventrites. Mesometaventral junction usually monocondylic (simple with ventrites abutting in *Hemipeplus*). Metaventrite relatively long, flattened to moderately convex with long discrimen; exposed portion of metanepisternum moderately to very long and narrow. Metacoxae contiguous or narrowly separated, horizontal, not extending laterally to meet elytra. Metendosternite with moderately long stalk, long lateral arms and well developed laminae; anterior process very short and usually emarginate, bearing widely separated anterior tendons. Hind wing with short apical field lacking radial extensions but usually with two post-radial sclerites (not in *Hemipeplus*); radial cell usually slightly elongate with inner posterobasal angle acute (cell absent in *Hemipeplus*); cross-vein r3 present or absent and r4 usually complete; basal portion of RP moderately long; medial spur extending to wing margin; medial field usually with four free veins (three in *Hemipeplus*) and a vaguely defined medial fleck; wedge cell usually narrow and apically acute and anal lobe well developed, without embayment (wedge cell and anal lobe absent in *Hemipeplus*). Legs short and more or less uniform; trochanterofemoral joint strongly oblique, with base of femur in contact with coxa. Tibiae not widened apically; spurs small and subequal, paired on all legs. Tarsi 5-5-4 in both sexes, with penultimate tarsomere distinctly lobed beneath and preceding ones also expanded and densely setose in *Hemipeplus*; pretarsal claws usually simple or slightly expanded (sometimes with distinct basal tooth); empodium not apparent.

Abdomen with five ventrites, the first two or four of which are connate; ventrite 1 not much longer than 2; intercoxal process acute or broadly angulate. Ventrite 2 in male Eurypinae with relatively large median patch of glandular hairs. Anterior edge of segment IX in male broadly rounded (Eurypinae) or somewhat truncate (*Hemipeplus*), without spiculum gastrale; tergite IX and X membranous. Aedeagus of tenebrionoid type, symmetrical; parameres partly fused together forming apicale, which is usually articulated to phallobase (fused to phallobase in *Hemipeplus*); penis broad at base and narrow at apex (Eurypinae) or long and slender (*Hemipeplus*), without paired struts. Sternite VIII in female with long, basally articulated, spiculum ventrale. Ovipositor long and slender, lightly sclerotised except for baculi; paraprocts much longer than coxites, each of which is subdivided and setose; proximal lobe with transversely oblique basal baculum, distal lobe narrow, cylindrical and sclerotised in Eurypinae; styli well developed and terminal. [Champion 1916b; Crowson and de Viedma 1964; Pollock 1999, 2002b, 2010a].

Larva

Body elongate, parallel-sided, distinctly flattened; dorsal and ventral surfaces lightly pigmented, white or cream in colour, except for buccal region, mandibles and tergum IX; surfaces smooth, clothed with short and long, scattered, erect setae.

Head prognathous and protracted, slightly transverse, slightly flattened, with sides evenly rounded. Epicranial stem short or absent; frontal arms lyriform and contiguous at base. Median endocarina usually absent (extending between frontal arms in *Hemipeplus*). Stemmata five on each side. Frontoclypeal suture absent. Labrum free. Antennae moderately long, 3-segmented, with antennomere 3 only slightly shorter than 2; sensorium usually dome-like (conical in *Hemipeplus*). Mandibles broad at base and narrow at apex, bidentate or tridentate; incisor edge with one or more subapical teeth; molae in Eurypinae (Fig. 41E) somewhat reduced, sub-basal, with surface simple or tuberculate, those in *Hemipeplus* well developed, asymmetrical and transversely ridged, prostheca absent. Ventral mouthparts slightly retracted, with well developed maxillary articulating area; cardo strongly oblique; stipes about as long as wide; mala more or less rounded, cleft, with or without inner apical tooth; maxillary palps 3-segmented. Labium consisting of prementum, mentum and submentum; ligula well developed, simple, longer than labial palps, which are 1-segmented and widely separated. Hypopharyngeal sclerome consisting of transverse bar. Hypostomal rods long and diverging. Gular region longer than wide; gular sutures well separated.

Thoracic segments about equal in length and width; terga without patches or rows of asperities or ampullae. Prothoracic venter without special armature. Legs moderately long, 5-segmented including claw-like pretarsus bearing two setae; coxae very widely separated. Abdomen more than twice as long as thorax, with posterior segments sometimes broader than anterior ones; terga and sterna II–VI in Eurypinae with paired longitudinal rows of asperities that are usually more or less curved (Fig. 43G). Tergum VIII much longer than VII; tergum IX much shorter than VIII, extending onto ventral surface and forming a broad, articulated plate (divided at midline in *Hemipeplus*), with a pair of posteriorly-projecting urogomphi, slightly upturned at apex, rarely with median, posteriorly-projecting process between urogomphi; sternum IX (Fig. 52L) U-shaped, partly enclosed by sternum VIII, with one to three teeth on each side posteriorly; segment X not visible from above, ventrally oriented and almost completely enclosed by sternite IX, without paired pygopods. Spiracles annular–multiforous, with a perimeter of small accessory openings or annular–cribriform (Fig. 53G) with cluster of small openings on one side only; thoracic spiracles placed at ends of short spiracular tubes, spiracles on segment VIII not enlarged. [Pollock *et al.* 2000; Pollock 1999, 2002b, 2010a; Thomas and Woodruff 2006].

Classification and World Distribution

The family Mycteridae occurs mainly in tropical and subtropical areas and includes 30 genera and ~160 species. There are three subfamilies, two of which occur in Australia. The exotic subfamily Mycterinae occurs in North America, Eurasia and South Africa.

Australian Fauna

The Australian mycterid fauna includes *Hemipeplus australicus* Arrow (Fig. 93G) from the **Hemipeplinae** and *Loboglossa australica* Champion and several species of *Trichosalpingus* Blackburn (Fig. 93F) from the **Eurypinae**. Eurypine mycterids are elongate, widest at the posterior third and narrowing anteriorly, moderately convex to slightly flattened and yellow to brown in colour, and clothed with small to minute, decumbent hairs. The eyes are prominent and not emarginate, the lateral pronotal carinae are absent, the mesanepisterna meet at midline, at least the penultimate tarsomere is lobed beneath, and there is a densely pubescent secretory patch on ventrite 2 in the male. *Hemipeplus australicus* differs from the rest of the Australian mycterids in being parallel-sided and strongly flattened, with the mesanepisterna separated at midline and the male secretory patch absent. Larvae are elongate, parallel-sided, flattened and lightly pigmented, with a broad head and articulated terminal plate formed by tergum IX; the head lacks a median endocarina (except in Hemipeplus), the mandibular molae are usually reduced and sub-basal (not in Hemipeplus), the legs are short and widely separated, there are paired rows of asperities forming incomplete rings on terga and sterna II–VI, and sternum IX forms a distinctive U-shaped sclerite enclosing segment X. Mycterid larvae occur under the bark of trees or logs or in the leaf axils of various monocotyledonous plants. *Hemipeplus australicus* breeds in the leaf axils of *Lomandra banksi* (Xanthorrhoeaceae) in northern QLD. [Champion 1916b; Pollock 2010a].

101. BORIDAE

Adult (Fig. 94D)

Length 3.7–12 mm. Body usually ~2.65–2.95 times as long as wide; subparallel and moderately convex; surfaces smooth and well sclerotised, uniformly brown to black and subglabrous.

Head prognathous, slightly longer than wide, not abruptly constricted posteriorly. Eyes strongly protuberant, not emarginate, coarsely facetted, without interfacetal setae. Antennal insertions concealed from above by frontal ridges; subantennal grooves absent. Frontoclypeal suture absent; anterior edge of clypeus slightly concave. Labrum free, strongly transverse and apically truncate. Antennae relatively short, 11-segmented, somewhat moniliform but with very weak 3-segmented club, antennomeres 9 and 10 with distal sensory patches enclosed in apical cavities and antennomere 11 apically acute, more or less surrounded by a dense row of flattened, adpressed sensilla. Mandibles moderately elongate, unidentate; incisor edge with single subapical tooth; mola highly reduced, surrounded by pubescent membranous lobes; prostheca well developed and densely setose. Maxilla with galea and lacinia short, broad and densely, setose, the latter without uncus; apical maxillary palpomere relatively small, slightly expanded and truncate. Mentum strongly transverse; ligula distinctly emarginate and setose; apical labial palpomere slightly expanded and truncate. Subgenal ridges absent. Gular sutures well separated; gula about as long as wide. Corpotentorium moderately broad and arched. Cervical sclerites absent.

Pronotum ~0.8–0.97 times as long as wide, widest anteriorly; sides weakly sinuate; base distinctly narrower than combined elytral bases. Lateral pronotal carinae complete but very fine and not visible from above (Fig. 19B); anterior angles not produced forward; posterior angles obtuse or rounded; posterior edge more or less straight; disc with pair of shallow posterolateral impressions. Prosternum in front of coxae slightly longer than mid length of procoxal cavity, moderately convex. Prosternal process incomplete, not dorsally curved behind coxae, with apex acute. Notosternal sutures complete. Procoxae conical and projecting below prosternum, with long concealed lateral extensions; trochantins concealed. Procoxal cavities subcircular, contiguous, externally narrowly open, without narrow lateral extensions, internally open.

Scutellar shield not abruptly elevated, posteriorly narrowly rounded. Elytra ~1.9–2.25 times as long as combined width and 2.7–3.4 times as long as pronotum; sides subparallel with conjointly rounded apices; disc irregularly punctate; epipleura very narrow and almost complete. Mesoventrite separated by complete sutures from mesanepisterna, which are well separated at midline, but not widely so; anterior edge without paired procoxal rests. Mesoventral process complete and apically truncate. Mesocoxae not projecting, with concealed trochantins. Mesocoxal cavities obliquely transverse, relatively narrowly separated, partly closed laterally by mesepimera. Mesometaventral junction simple with mesoventrite overlapping metaventrite. Metaventrite convex, with relatively long, discrimen; exposed portion of metanepisternum moderately long and narrow. Metacoxae narrowly separated, not extending laterally to meet elytra. Metendosternite with moderately long stalk, long lateral arms and well developed laminae; anterior process absent; anterior tendons widely separated, near bases of lateral arms. Hind wing with short apical field lacking distinct sclerotisations; radial cell small, short and broad, with inner posterobasal angle obtuse; cross-vein r3 absent; basal portion of RP moderately long and R-M loop broad; medial spur extending to wing margin; medial field with four free veins and no medial fleck; wedge cell narrow and apically acute; anal lobe well developed, without embayment. Legs slender; trochanterofemoral joint strongly oblique, with base of femur in contact with coxa. All trochanters and femora in males with small setose fovea on inner side; protibiae slightly widened apically; meso- and metatibiae spiculate; spurs well developed and subequal, paired on all legs. Tarsi 5-5-4 in both sexes; tarsomeres not lobed beneath; pretarsal claws simple; empodium bisetose.

Abdomen with five ventrites, the first two of which are connate; ventrite 1 not much longer than 2; intercoxal process acute. Anterior edge of segment VIII in male with short strut. Anterior edge of segment IX in male with very slender, acute subgenital plate; tergite IX completely divided and tergite X free and partly sclerotised. Aedeagus of tenebrionoid type, symmetrical; parameres fused together forming apicale, which is distinctly longer than and articulated to phallobase, bearing a pair of long, setose, articulated accessory lobes; penis somewhat flattened, with short paired anterior struts. Sternite VIII in female with long, basally articulated, spiculum ventrale. Ovipositor moderately long and slender, lightly sclerotised except for baculi; paraprocts much longer than coxites, each of which is vaguely subdivided and setose; proximal lobe with transversely oblique basal baculum, distal lobe not narrowed, lightly sclerotised; styli short, sclerotised and subterminal. [Lawrence and Pollock 1994; Pollock 2002c, 2010b].

Larva (based on exotic genera Boros *Herbst and* Lecontia *Champion)*

Body elongate, parallel-sided, slightly to strongly flattened; dorsal and ventral surfaces moderately lightly pigmented, brownish-yellow in colour, with head and tergum IX darker; surfaces smooth, clothed with short and long, scattered, erect setae.

Head prognathous and protracted, slightly transverse, flattened, with sides strongly rounded. Epicranial stem short; frontal arms lyriform and contiguous at base. Median endocarina absent. Stemmata on each side 5 in *Boros* (absent in *Lecontia*). Frontoclypeal suture absent. Labrum free. Antennae about half as long as head width in *Boros* (one-fifth as long in *Lecontia*), 3-segmented, with antennomeres 1 and 2 subequal in length, 3 about one-third as long as 2, and sensorium short and broadly conical or dome-like. Mandibles broad at base and narrow at apex, broadly tridentate; incisor edge with two subapical teeth; molae asymmetrical, well developed and transversely ridged in *Boros*, somewhat reduced, sub-basal and weakly tuberculate in *Lecontia*. Ventral mouthparts retracted, with well developed maxillary articulating area; cardo slightly oblique; stipes ~1.5 times as long as wide; mala truncate and cleft, with tridentate uncus at inner apical angle and inner edge spinose; maxillary palps 3-segmented. Labium consisting of prementum, mentum and submentum; ligula well developed, rounded, longer than basal palpomere; labial palps 2-segmented and widely separated. Hypopharyngeal sclerome well developed, transversely quadrate. Hypostomal rods moderately long and diverging. Gular region transverse; gular sutures well separated; gula about as long as wide.

Thorax elongate and parallel-sided; thoracic terga more or less equal in length; protergum subquadrate with distinct median ecdysial line (with pair of small, tooth-like processes on either side of its posterior end in *Lecontia*); sides of meso- and metathorax weakly rounded; anterior end of mesotergum in *Boros* with paired transverse carinae, each ending in a posteriorly-directed tooth on either side of midline; anterior end of metatergum with pair of fine, transverse, sinuate carinae not reaching midline. Prothoracic venter with pair of lateral precoxalia and a large, subquadrate or trapezoidal median sclerite extending posteriorly between coxae. Legs relatively short and stout, widely separated, 5-segmented with claw-like pretarsus bearing 2 setae lying side by side. Abdomen at least 3 times as long as thorax; terga and sterna without patches or rows of asperities or ampullae, but I–VII each with several fine longitudinal carinae in *Lecontia*; terga I–VIII each with pair of fine, transverse, sinuate carinae similar to those on thorax. Tergum VIII 1.3–1.5 times as long as VII; tergum IX much shorter than VIII and strongly transverse, extending onto ventral surface and forming a broad, articulated plate, longitudinally divided in *Boros*, with a pair posteriorly-projecting urogomphi, widely separated in *Boros*, more narrowly so in *Lecontia*; a broadly rounded median plate lying between urogomphi in *Boros* or slightly below them in *Lecontia*; a pair of posteriorly opening pits lying on either wide of median plate in *Boros* or just above plate in *Lecontia*; sternum IX short and U-shaped, partly enclosed by sternum VIII, with 1 to 3 sclerites on each side posteriorly (at ends of U); segment X ventrally oriented, slightly transverse or as long as wide and enclosed anteriorly and laterally by sternite IX, without paired pygopods; anal opening transverse. Thoracic and abdominal spiracles annular–biforous, not placed at ends of spiracular tubes; accessory chambers very small, so that spiracles may appear to be annular. [Böving and Craighead 1931; Young 1985b, 1991; Pollock 2002c, 2010b].

Classification and World Distribution

Boridae is a small family containing three genera and four species. The subfamily Borinae includes the Nearctic *Lecontia discicollis* (Leconte) and two species of the Holarctic genus *Boros* Herbst, while the Synercticinae includes only *Synercticus heteromerus* Pascoe from the Australian region.

Australian Fauna

Synercticus heteromerus is distributed throughout eastern Australia from central NSW to the Cape York Peninsula and also occurs in New Guinea. These are elongate, moderately convex beetles, dark brown to black in colour, subglabrous and shiny. The eyes are prominent and not emarginate, the head is gradually narrowed behind the eyes, the antennal insertions are partly concealed beneath frontal ridges, the antennal club is small with deep circular channels filled with sensilla at the apices of the first two segments and with the terminal segment surrounded by a row of flattened, adpressed sensilla, the lateral pronotal carinae are complete but very fine and not visible for their entire lengths from above and the tarsomeres lack ventral lobes and have spinose setae at their apices. Males have a distinctive type of sex patch located on each trochanter and femur. Larvae of *Synercticus* are unknown, but those of the North Temperate genera *Boros* and *Lecontia* are elongate, parallel-sided, strongly flattened and lightly sclerotised, with tergum IX forming a terminal, articulated plate and sternum IX forming a U-shaped structure, as in Mycteridae. Borid larvae differ from those of mycterids in lacking the curved rows of asperities on abdominal terga and sterna and in having well developed mandibular molae. [Lawrence and Pollock 1994].

102. PYTHIDAE

Adult (Fig. 94B)

Length 7–11.5 mm. Body usually ~2.9–3.1 times as long as wide; subparallel and moderately convex dorsally and ventrally; surfaces smooth and shiny, but with raised carinae and tubercles in *Anaplopus tuberculatus*; colour yellowish-brown to dark reddish-brown or black, with yellow raised areas on elytra and with some black tubercles in *A. tuberculatus*; vestiture of moderately long, decumbent hairs of varying density, dense patches forming a pattern in *A. cleroides*.

Head prognathous, slightly longer than wide, not abruptly constricted posteriorly. Eyes strongly protuberant, not emarginate, finely facetted, without interfacetal setae. Antennal insertions slightly concealed from above by raised frontal ridges; subantennal grooves absent. Frontoclypeal suture weakly impressed; anterior edge of clypeus truncate, with distinct anteclypeus. Labrum free, strongly transverse and apically truncate. Antennae moderately long, 11-segmented, filiform. Mandibles moderately elongate, bidentate; incisor edge with one or two subapical teeth; mola well developed, transversely ridged; prostheca well developed and densely setose. Maxilla with broad galea and narrower lacinia, both densely, setose, the latter without uncus; apical maxillary palpomere more or less securiform. Mentum slightly transverse, without setose pit in male; ligula truncate; apical labial palpomere expanded and truncate. Subgenal ridges absent. Gular sutures well separated; gula about as long as wide. Corpotentorium narrow and arched. Cervical sclerites absent.

Pronotum ~0.85–1.05 times as long as wide, widest anteriorly; sides more or less sinuate (with two lateral tubercles in *A. tuberculatus*); base distinctly narrower than combined elytral bases. Lateral pronotal carinae absent; anterior angles not produced forward; posterior angles rounded; posterior edge more or less straight; disc simple in *A. cleroides*, with pair of anterolateral impressions and two longitudinal rows of tubercles in *A. tuberculatus*. Prosternum in front of coxae about as long as mid length of procoxal cavity, moderately convex. Prosternal process incomplete, not dorsally curved behind coxae, with acute apex. Notosternal sutures incomplete. Procoxae conical and projecting below prosternum, with long concealed lateral extensions; trochantins concealed. Procoxal cavities subcircular, very narrowly separated, externally broadly open, without narrow lateral extensions, internally open.

Scutellar shield not abruptly elevated, posteriorly broadly rounded. Elytra ~2.2–2.4 times as long as combined width and 3.1–3.6 times as long as pronotum; sides subparallel with conjointly rounded apices; disc irregularly punctate, simple in *A. cleroides*, with several pairs of short longitudinal carinae and numerous tubercles in *A. tuberculatus*; epipleura very narrow and more or less complete. Mesoventrite separated by complete sutures from mesanepisterna, which are very narrowly separated at midline; anterior edge without paired procoxal rests. Mesoventral process complete and apically narrowly rounded. Mesocoxae not projecting, with exposed trochantins. Mesocoxal cavities subcircular, narrowly separated, partly closed laterally by mesepimera. Mesometaventral junction monocondylic. Metaventrite convex, with relatively long, discrimen; exposed portion of metanepisternum long and narrow. Metacoxae narrowly separated, extending laterally to meet elytra. Metendosternite with long stalk, long lateral arms and well developed laminae; anterior process absent; anterior tendons widely separated, near bases of lateral arms. Hind wing with moderately long apical field with two or three vague sclerotisations; radial cell small, short and broad, with inner posterobasal angle right; cross-vein r3 absent; basal portion of RP moderately long and R-M loop broad; medial spur extending almost to wing margin, where there is a slight medial embayment; medial field with four free veins and a vaguely indicate, undivided medial fleck; wedge cell narrow and apically acute; anal lobe well developed, without embayment. Legs slender; trochanterofemoral joint strongly oblique, with base of femur in contact with coxa. Tibiae not widened apically; spurs small and subequal, paired on all legs. Tarsi 5-5-4 in both sexes; tarsomeres not lobed beneath, but ventrally setose; pretarsal claws simple; empodium bisetose.

Abdomen with five free ventrites; ventrite 1 not much longer than 2; intercoxal process acute. Anterior edge of segment IX in male with very slender, subacute subgenital plate; tergite IX completely divided and tergite X free, membranous with paired baculi. Aedeagus of tenebrionoid type, symmetrical; parameres fused together forming apicale, which is articulated and subequal in length to phallobase, bearing a pair of long, apically setose, articulated accessory lobes; penis without paired anterior struts. Sternite VIII in female with long, basally articulated, spiculum ventrale. Ovipositor moderately long and slender, lightly sclerotised except for baculi; paraprocts much longer than coxites, each of which is subdivided and setose; proximal lobe with transversely oblique basal baculum, distal lobe not narrowed, lightly sclerotised; styli short, sclerotised and subterminal. [Lawrence 1987b; Pollock and Lawrence 1995; Pollock 2002d, 2010c].

Larva

Body elongate, parallel-sided, slightly flattened; dorsal and ventral surfaces moderately lightly pigmented, yellow in colour, with head and parts of tergum IX darker; surfaces smooth, clothed with short and long, scattered, erect setae.

Head prognathous and protracted, slightly transverse, flattened, with sides evenly rounded. Epicranial stem very short or absent; frontal arms lyriform and contiguous at base. Median endocarina absent. Stemmata five on each side. Frontoclypeal suture absent. Labrum free. Antennae moderately long, 3-segmented, with antennomere 3 about half as long as 2; sensorium short, conical. Mandibles broad at base and narrow at apex, bidentate; incisor edge with one or two subapical teeth; molae well developed, asymmetrical, transversely ridged. Ventral mouthparts retracted, with well developed maxillary articulating area; cardo slightly oblique; stipes longer than wide; mala more or less truncate, cleft, with tridentate uncus at inner apical angle; maxillary palps 3-segmented. Labium consisting of prementum, mentum and submentum; ligula well developed, rounded, about as long as labial palps, which are 2-segmented and widely separated. Hypopharyngeal sclerome consisting of transverse bar. Hypostomal rods short and slightly diverging; ventral epicranial ridges present. Gular region transverse; gular sutures well separated.

Protergum slightly longer than meso- or metatergum; anterior edge of mesotergum, metatergum and abdominal terga I–IX each with transverse carina, without patches or rows of asperities or ampullae. Prothoracic venter without special armature. Legs relatively short and armed with short spines, 5-segmented including claw-like pretarsus bearing two setae; coxae widely separated. Abdomen about three times as long as thorax, terga and sterna without with rows or patches of asperities. Tergum VIII not much longer than VII; tergum IX (Fig. 43K) slightly shorter than VIII, extending onto ventral surface but not forming articulated plate, posterior portion of tergum steeply sloping, with pair of narrowly separated, short and broad, quadrifid urogomphi, with two lobes mesally projecting, one posteriorly projecting and another dorsally curved; with row of short pregomphi anterior to urogomphi and a few other tubercles; sternum IX (Fig. 43L) transverse, slightly enclosed by sternum VIII, with two slightly curved rows of eight to ten asperities at base; segment X not visible from above, ventrally oriented, transverse, more or less membranous, without paired pygopods. Thoracic spiracles annular, abdominal spiracles annular–biforous but with very short accessory chambers, not placed at ends of spiracular tubes, spiracles on segment VIII not enlarged. [Pollock and Lawrence 1995; Pollock 2002d, 2010c].

Classification and World Distribution

The family Pythidae comprises 24 species in seven genera, one of which occurs in Australia. The genus *Pytho* Latreille is Holarctic in distribution, *Priognathus* LeConte, *Sphalma* Horn and *Trimitomerus* Horn are Nearctic, *Osphyoplesius* occurs in the southern part of the Palaearctic region, *Ischyomius* Chevrolat is a Neotropical genus, and *Anaplopus* Blackburn occurs in Australia. [Young 1978; Pollock 1998, 2010c].

Australian Fauna

The Australian pythid fauna consists of *Anaplopus tuberculatus* Blackburn from south-eastern QLD and *A. cleroides* Pollock & Lawrence from northern QLD. Adults are elongate, moderately convex, yellowish-brown to black beetles, usually with an irregular upper surface bearing tubercles and ridges plus black and yellow, shiny raised areas on the elytra (surfaces more even with a single raised shiny ridge on each elytron in *A. cleroides*); the vestiture is of moderately long erect and decumbent hairs. Eyes are prominent and not emarginate, the head is gradually narrowed behind the eyes, the frontal ridges are strongly raised, so that the antennal insertions are exposed from above, the last six antennomeres are evenly clothed with fine hairs, maxillary palps are relatively small, slightly to strongly expanded and truncate apically, lateral pronotal carinae are absent, and tarsomeres 2 to 4 have no ventral lobes but are densely, evenly clothed with fine pubescence below. Larvae are elongate, parallel-sided, moderately flattened and more or less uniformly yellow. The head is broad, the antennae well developed, the legs widely separated, tergum IX bears a pair of complexly forked urogomphi in front of which are several tubercles, and sternum IX has a basal, doubly curved row of asperities. Larvae usually occur under the bark of rotten logs, but they have also been taken in rotten epiphytes. [Lawrence 1987b; Pollock and Lawrence 1995].

103. PYROCHROIDAE

Adult (Fig. 95I, K)

Length 2.8–14 mm. Body usually ~2.3–3.2 times as long as wide; subparallel and slightly to moderately flattened; surfaces often smooth and shiny, sometimes densely, minutely punctate; colour usually red or yellow and black, occasionally uniformly black; upper surfaces subglabrous or clothed with short to moderately long, erect or erect and inclined fine setae.

Head prognathous, not abruptly constricted posteriorly, but more gradually narrowed to form broad neck. Eyes strongly protuberant, not emarginate, finely facetted, without interfacetal setae. Antennal insertions exposed; subantennal grooves absent. Frontoclypeal suture weakly to moderately impressed; anterior edge of clypeus truncate, with distinct anteclypeus. Labrum free, strongly transverse and apically truncate. Antennae short to moderately long, 11-segmented, filiform, moniliform, serrate, or unipectinate. Mandibles relatively small, short and broad, bidentate; incisor edge simple or with one subapical tooth; mola and prostheca usually well developed. Maxilla with short, broad, setose galea and slightly narrower lacinia without uncus; apical maxillary palpomere securiform or cultriform. Mentum strongly transverse, without setose pit in male; ligula undivided; apical labial palpomere fusiform. Subgenal ridges absent. Gular sutures well separated; gula about as long as wide. Corpotentorium narrow and arched. Cervical sclerites highly reduced or absent.

Pronotum ~0.63–0.92 times as long as wide, widest at middle; sides slightly to strongly rounded (with lateral tubercles in *Binburrum*); base distinctly narrower than combined elytral bases. Lateral pronotal carinae absent; anterior angles not produced forward; posterior angles right to rounded; posterior edge more or less straight; disc simple in *Temnopalpus and Paromarteon*, with pair of basolateral impressions in *Binburrum*, with weak longitudinal costae and various impressions in *Morpholycus*. Prosternum in front of coxae distinctly longer than mid length of procoxal cavity, slightly convex. Prosternal process complete or incomplete, not curved dorsally behind coxae, with acute apex. Procoxae conical and projecting below prosternum, with long concealed lateral extensions; trochantins slightly exposed. Procoxal cavities slightly transverse, very narrowly separated, externally broadly open, with narrow lateral extensions, internally open.

Scutellar shield not abruptly elevated, posteriorly broadly rounded or obtusely angulate. Elytra ~1.92–3.18 times as long as combined width and 4.0–6.7 times as long as pronotum; sides subparallel with conjointly rounded apices; disc irregularly punctate, usually simple, with four longitudinal costae *Morpholycus costipennis* Lea; epipleura usually very narrow and incomplete (complete in some *Morpholycus*). Mesoventrite (Fig. 44T) separated by complete sutures from mesanepisterna, which are very narrowly separated at midline; anterior edge without paired procoxal rests. Mesoventral process complete and apically narrowly rounded or acute. Mesocoxae not or only slightly projecting, with exposed trochantins. Mesocoxal cavities somewhat oblique, narrowly separated, partly closed laterally by mesepimera. Mesometaventral junction simple. Metaventrite convex, with relatively long, discrimen; exposed portion of metanepisternum long and narrow. Metacoxae narrowly separated, extending laterally to meet elytra. Metendosternite with long stalk, long lateral arms and well developed laminae; anterior process absent; anterior tendons widely separated on lateral arms. Hind wing with short apical field containing vague sclerotisations only; radial cell slightly elongate with inner posterobasal angle right or obtuse, base sometimes absent; cross-vein r3 absent; basal portion of RP very short and R-M loop often acute; medial spur straight and extending almost to wing margin; medial field with four free veins and a vaguely indicate, undivided medial fleck; wedge cell narrow and apically acute or absent; anal lobe well developed, with or without shallow embayment. Legs long and slender; trochanterofemoral joint strongly oblique, with base of femur in contact with coxa. Tibiae not widened apically; spurs small and subequal, paired on all legs. Tarsi 5-5-4 in both sexes; penultimate tarsomere lobed beneath and setose, antepenultimate sometimes lobed as well;

pretarsal claws slightly or distinctly dentate; empodium not obvious.

Abdomen with five or six free ventrites; ventrite 1 not much longer than 2; intercoxal process acute. Anterior edge of segment IX in male with very slender, subacute subgenital plate or spiculum gastrale; tergite IX completely divided and tergite X free, sclerotised or membranous. Aedeagus (Fig. 31E) of tenebrionoid type, symmetrical; parameres fused together forming apicale, which is articulated and longer than or subequal in length to phallobase, apically cleft in *Paromarteon,* bearing a pair of articulated accessory lobes, which may be inserted near the base in *Paromarteon* and *Binburrum* or near the apex in *Temnopalpus* and *Morpholycus*; penis with relatively short paired anterior struts. Sternite VIII in female with long, basally articulated, spiculum ventrale. Ovipositor moderately long and slender, lightly sclerotised except for baculi; paraprocts much longer than coxites, each of which is subdivided and setose; proximal lobe with transversely oblique basal baculum, distal lobe not narrowed, lightly sclerotised; styli short, sclerotised and terminal. [Nikitsky 1986b; Watt 1987; Young and Pollock 1991, 2010; Pollock 1992, 1994, 1995; Lawrence *et al.* 1999b; Young 2002m].

Larva (Fig. 91L)

Body elongate, parallel-sided, strongly flattened; dorsal and ventral surfaces moderately lightly pigmented, yellow in colour, with head and tergum IX darker; surfaces smooth, clothed with short and long, scattered, erect setae.

Head prognathous and protracted, slightly transverse and about as wide as thorax, flattened, with sides strongly rounded. Epicranial stem very short or absent; frontal arms lyriform and contiguous at base. Median endocarina absent. Stemmata two on each side. Frontoclypeal suture absent. Labrum free. Antennae moderately long, 3-segmented, with antennomere 3 not much shorter than 2; sensorium dome-like. Mandibles broad at base and narrow at apex, tridentate; incisor edge usually with one subapical tooth; molae well developed, highly asymmetrical, transversely ridged. Ventral mouthparts retracted, with well developed maxillary articulating area; cardo undivided; stipes longer than wide; mala more or less truncate, cleft, with unidentate uncus at inner apical angle; maxillary palps 3-segmented. Labium consisting of prementum, mentum and submentum; ligula well developed, rounded, about as long as labial palps, which are 2-segmented and widely separated. Hypopharyngeal sclerome molar-like. Hypostomal rods moderately long and strongly diverging; ventral epicranial ridges absent. Gular region transverse; gular sutures well separated.

Thoracic segments about equal in length; protergum parallel-sided or anteriorly widened, meso- and metaterga laterally rounded; thoracic terga without transverse carinae, or patches or rows of asperities. Prothoracic venter usually with cervicosternum divided longitudinally into three parts and large, undivided, triangular presternum. Legs relatively short, 5-segmented including claw-like pretarsus bearing two setae; coxae very widely separated. Abdomen about three times as long as thorax, terga and sterna without with rows or patches of asperities. Tergum VIII slightly longer than VII; tergum IX (Fig. 44M) about half as long as VIII, extending onto ventral surface and forming strongly transverse, articulated plate, posterior portion of tergum with pair of widely separated, short and broad, simple or bifid urogomphi, with two inter-urogomphal pits between them; sternum IX (Fig. 44N) slightly transverse, slightly enclosed by sternum VIII, with continuous straight row of asperities at base; segment X not visible from above, ventrally oriented, transverse, more or less membranous, without paired pygopods. Thoracic spiracles annular; abdominal spiracles annular with crenulate preitreme, not placed at ends of spiracular tubes, spiracles on segment VIII not enlarged. [Lawrence *et al.* 1999a; Young and Pollock 1991, 2010].

Classification and World Distribution

The family Pyrochroidae is a widespread family with ~185 species in 30 genera and its greatest diversity in the temperate parts of the Northern and Southern Hemispheres. There are five recognised subfamilies, one of which occurs in Australia. Tydessinae includes the single genus *Tydessa* Peacock from western North America and eastern Asia. Agnathinae includes two Palaearctic species of *Agnathus* Germar and a few species of *Cononotus* LeConte in south-western North America. The subfamily Pedilinae includes *Anisotria shooki* Young from north-western North America and several species of *Pedilus* Fischer von Waldheim from North America and Eurasia. Pyrochroinae is the largest subfamily, which is Holarctic in distribution but most abundant and diverse in eastern Asia. The subfamily Pilipalpinae is restricted to the Southern Hemisphere, with species in New Zealand, Australia and southern South America. [Doyen 1979; Nikitsky 1986b; Watt 1987; Pollock 1992, 1994, 1995; Young and Pollock 2010].

Australian Fauna

The Australian members of the subfamily **Pilipalpinae** include ~20 species in the genera *Binburrum* Pollock, *Morpholycus* Lea (Fig. 95I), *Paromarteon* Blackburn and *Temnopalpus* Blackburn (Fig. 95K), which occur in the eastern part of the continent from TAS to northern QLD. Pilipalpines are elongate, moderately convex to somewhat flattened, sometimes black, but usually bicoloured with various combinations of red, orange or yellow and black, and usually clothed with decumbent and erect hairs (occasionally subglabrous with scattered erect hairs). The eyes are prominent, not emarginate, the head is gradually narrowed behind the eyes, the antennal insertions are exposed, the maxillary palps are relatively small, slightly to strongly expanded apically and sometimes cultriform, lateral pronotal carinae are absent, at least the penultimate and antepenultimate tarsomeres are lobed beneath and all but the terminal tarsomere densely pubescent ventrally. Larvae (Fig. 91L) are elongate, parallel-sided and flattened, usually more or less yellow in colour, except for the darker head and tergum IX, occasionally with all terga more heavily pigmented than the sterna. The head is broad, the antennae are relatively long, the legs are widely separated, tergum IX forms an articulated plate bearing paired, posteriorly projecting urogomphi sometimes forked or with accessory processes at the base, the tergal plate with a pair of pits and sometimes a median process between the urogomphi, and sternum IX with a basal row of asperities which is either straight or only slightly curved. Larvae occur under the bark of dead limbs, standing dead trees or rotten logs. Species of *Morpholycus* may mimic the orange and black species of *Porrostoma* Laporte (Lycidae), while *Temnopalpus* species may be involved in a mimicry complex with black and red species of *Heteromastix* Boheman (Cantharidae). [Pollock 1995].

104. SALPINGIDAE

Adult (Fig. 95A–B, D-G)

Length usually 1.2–5 mm (up to 11.5 mm in *Tretothorax*). Body usually ~1.5–3.5 times as long as wide (up to 7 times in *Diagrypnodes*); subparallel and slightly to strongly flattened; surfaces usually smooth and shiny; colour yellow to dark brown or black, occasionally bicolored with light and dark pattern on elytra; upper surfaces usually subglabrous (densely clothed with fine, decumbent hairs in *Elacatis*, with short, thickened, decumbent setae in *Tretothorax*).

Head prognathous, often elongate, not abruptly constricted posteriorly. Eyes not to strongly protuberant, not emarginate, large and vertically oval in *Elacatis*, finely to coarsely facetted, without interfacetal setae, absent in *Aglenus*. Vertex strongly declined in *Elacatis*. Antennal insertions concealed from above, except in *Diagrypnodes*; subantennal grooves short or absent. Frontoclypeal suture usually weakly impressed or absent (distinct in *Euryplatus*). Frontoclypeal region produced to form rostrum in *Tretothorax* and some Salpinginae. Labrum free, strongly transverse and truncate or slightly convex (rarely subquadrate or longer than wide). Antennae usually relatively short (much longer in *Diagrypnodes*), 11-segmented with 3-segmented club in *Aglenus*, *Elacatis*, Prostominiinae and Salpinginae, 10-segmented and moniliform with a 1-segmented club in *Tretothorax*, and more or less filiform in *Euryplatus* and *Diagrypnodes*; scape much longer than pedicel in *Diagrypnodes*. Mandibles short and broad, bidentate; incisor edge usually with one or more subapical teeth; mola and prostheca usually well developed (reduced in some Salpinginae); mesal surface of mandibular base with elongate pubescent process in *Euryplatus*. Maxilla usually with densely setose galea (sparsely setose and palpiform in *Euryplatus*) and narrower lacinia with or without uncus; apical maxillary palpomere usually cylindrical or fusiform (slightly expanded and truncate in some Salpinginae). Mentum strongly transverse; ligula usually simple and undivided (slightly emarginate in *Diagrypnodes*); apical labial palpomere cylindrical or fusiform. Subgenal ridges absent. Gular sutures widely to narrowly separated. Corpotentorium usually absent (narrow in *Elacatis* and Prostominiinae). Cervical sclerites absent.

Pronotum ~0.55–1.6 times as long as wide, usually widest anteriorly; sides more or less straight to strongly rounded, sometimes sinuate (with lateral teeth or tubercles in some Salpinginae); base distinctly narrower than combined elytral bases in *Tretothorax*, *Euryplatus*, *Diagrypnodes* and most Salpinginae. Lateral pronotal carinae complete, incomplete or absent, simple or occasionally dentate; anterior angles not produced forward; posterior angles right to rounded; posterior edge usually more or less straight; disc usually simple, with median longitudinal groove in *Tretothorax*. Prosternum in front of coxae distinctly longer than mid length of procoxal cavity, flat to moderately convex. Prosternal process complete or incomplete, narrowed apically, parallel-sided or expanded at apex, not curved dorsally behind coxae, with apex acute to broadly rounded, truncate or emarginate. Procoxae sometimes projectng below prosternum, with or without long, concealed lateral extensions; trochantins concealed. Procoxal cavities slightly transverse to circular, contiguous to widely separated, externally broadly to narrowly open, without narrow lateral extensions, internally closed.

Scutellar shield abruptly elevated only in *Tretothorax*, posteriorly acute to broadly rounded; absent in *Aglenus*. Elytra ~0.95–2.7 times as long as combined width and 1.5–3.25 times as long as pronotum; sides usually subparallel and complete with conjointly rounded apices (apices independently rounded or truncate, exposing several abdominal tergites in *Euryplatus* and *Diagrypnodes*); disc irregularly punctate or with ten puncture rows and a scutellary striole; epipleura incomplete or absent. Mesoventrite separated by complete sutures from or partly fused with mesanepisterna, which may or may not meet at midline; anterior edge without paired procoxal rests. Mesoventral process complete and apically narrowly rounded to truncate or emarginate. Mesocoxae not or only slightly projecting, with concealed trochantins. Mesocoxal cavities circular to slightly transverse, narrowly to widely separated, closed laterally by meeting of ventrites. Mesometaventral junction simple or complex. Metaventrite elongate to short and broad, slightly to strongly flattened, usually with discrimen short or absent; exposed portion of metanepisternum moderately elongate to very long and narrow. Metacoxae usually contiguous or narrowly separated (widely separated in *Tretothorax* and *Euryplatus*), somewhat oblique in *Euryplatus*, not extending laterally to meet elytra. Metendosternite with long stalk and long lateral arms (stalk short and broad in *Aglenus*, arms short in *Diagrypnodes*); laminae well developed in *Elacatis*, *Tretothorax* and Salpinginae, absent in Prostominiinae, *Euryplatus* and *Diagrypnodes*; anterior process absent; anterior tendons widely separated on lateral arms. Hind wing usually with long apical field and no distinct sclerotisations or veinal remnants (shorter in *Tretothorax*, with anterior and posterior oblique linear sclerites); radial cell usually either absent or elongate with inner posterobasal angle right or obtuse (shorter and broader with angler acute in *Tretothorax*); cross-vein r3 either absent or fused at base with r4; basal portion of RP usually short or absent; medial spur distinctly curved and ending well before wing margin; medial field usually with two or four free veins and a medial fleck (Inopeplinae without or with a single vein and no medial fleck); wedge cell usually absent (narrow and apically acute in *Tretothorax*); anal lobe well developed without embayment or reduced. Legs usually long and slender; trochanterofemoral usually joint strongly oblique with base of femur in contact with coxa; femur not abutting coxa and femora somewhat inflated in Inopeplinae. Tibiae usually not widened apically; spurs small and subequal. Tarsi usually 5-5-4 in both sexes (4-4-4 in *Ocholissa* Pascoe and *Aglenus*); tarsomeres not lobed beneath; pretarsal claws simple.

Abdomen with almost always with five free ventrites (first two ventrites connate in *Elacatis*; ventrite 1 not much longer than 2; intercoxal process usually acute or narrowly rounded (broadly rounded in *Aglenus*, *Tretothorax* and Inopeplinae). Anterior edge of segment IX in male with very slender, subacute subgenital plate or spiculum gastrale; tergite IX completely fused to segment X, which may be membranous. Aedeagus of tenebrionoid type, symmetrical; parameres partly or entirely fused together forming apicale, which is articulated to phallobase and usually bears a pair of articulated accessory lobes (absent in *Elacatis*). Sternite VIII in female with long, basally articulated, spiculum ventrale (forked at base in *Elacatis*). Ovipositor elongate and slender, lightly sclerotised except for baculi; paraprocts much longer than coxites, each of which is subdivided with terminal styli. [Blair 1919, 1925; Watt 1967; Lawrence *et al.* 1999b, 2010l; Pollock 2002e].

Larva (Figs 35D–E, 91G)

Body elongate, parallel-sided, slightly to strongly flattened; dorsal and ventral surfaces usually lightly pigmented, except for head capsule and tergum IX; surfaces smooth, clothed with short and long, scattered, erect setae.

Head prognathous and protracted, slightly transverse, flattened, with sides strongly rounded; posterior edge slightly to moderately deeply emarginate. Epicranial stem very short or absent; frontal arms lyriform and contiguous at base. Median endocarina absent; paired endocarinae present beneath frontal arms in *Euryplatus* and the Salpinginae. Stemmata on each side five, two or none. Frontoclypeal suture absent. Labrum free. Antennae moderately long, 3-segmented; sensorium conical or palpiform and shorter than antennomere 3. Mandibles symmetrical or asymmetrical, broad at base and narrow at apex, bidentate or tridentate; incisor edge usually with one or two subapical teeth; mola well developed in *Elacatis*, *Aglenus*, *Ocholissa* and *Diagrypnodes*, somewhat reduced and sub-basal in *Euryplatus* and *Prostominia*, highly reduced or absent in Salpinginae (Fig. 39B). Ventral mouthparts retracted, with well developed maxillary articulating area; cardo undivided; stipes longer than wide; mala usually simple (cleft in *Elacatis*, Fig. 43D, and *Ocholissa*), rounded or truncate (Fig. 38K); maxillary palps 3-segmented. Labium consisting of prementum, mentum and submentum; ligula usually absent (well developed in *Elacatis* and Prostominiinae), labial palps 2-segmented and narrowly to widely separated. Hypopharyngeal sclerome molar-like in *Aglenus*, *Diagrypnodes*, *Elacatis* and Prostominiinae, a transverse bar in *Euryplatus*, absent in Salpinginae. Hypostomal rods usually moderately to very long and subparallel or very slightly arcuate (distinctly diverging in *Elacatis*); ventral epicranial ridges present in *Elacatis*. Gular region almost always transverse (rarely longer than wide); gular sutures well separated.

Thoracic segments about equal in length; thoracic terga without transverse carinae, or patches or rows of asperities. Prothoracic venter without armature. Legs well developed, 5-segmented including claw-like pretarsus bearing two setae; coxae usually very widely separated (less so in *Elacatis*). Abdomen more than twice as long as thorax, terga and sterna without with rows or patches of asperities. Tergum IX (Fig. 43E, excluding appendages, shorter than VIII, extending onto ventral surface but not forming articulated plate, with paired urogomphi, which are almost always either bifurcate or with accessory spines or tubercles (simple in some Salpinginae), without interurogomphal pits; sternum IX not or only slightly enclosed by sternum VIII, with continuous straight or slightly curved row of asperities at base in *Diagrypnodes*, a doubly curved row interrupted at middle in *Elacatis*; two to six asperities on each side in *Euryplatus* (Fig. 43F) and Prostominiinae, and one on each side in *Aglenus* and Salpinginae (Fig. 53D); segment X not visible from above, ventrally or posteroventrally oriented, transverse, usually more or less membranous, without paired pygopods. Thoracic spiracles usually annular-biforous (annular in *Aglenus*, *Diagrypnodes* and most Salpinginae), placed at ends of spiracular tubes in *Euryplatus* and Prostominiinae. Abdominal spiracles usually annular (annular–biforous in *Elacatis* and Prostominiinae), not placed at ends of spiracular tubes, spiracles on segment VIII not enlarged. [Hayashi 1969; Howden and Howden 1981; Young 1991; Lawrence *et al.* 1999a; Pollock 2002e; Lawrence *et al.* 2010l].

Classification and World Distribution

The family Salpingidae contains ~300 species in 35 genera and is widely distributed with most of its diversity in the cooler parts of the Northern and Southern Hemispheres. There are seven recognised subfamilies, six of which occur in Australia. The exotic Aegialitinae are known from islands and coastal fringes bordering the northern and southern ends of the Pacific Ocean (Spilman 1967; Doyen 1976; Zerche 2004).

Australian Fauna

Three subfamilies are each represented in Australia by a single species. *Tretothorax cleistostoma* Lea (Fig. 95F) is a member of the **Dacoderinae**, to which the New World genera *Dacoderus* LeConte and *Myrmecoderus* Aalbu *et al.* also belong (Aalbu *et al.* 2005), the subfamily **Agleninae** is represented by the adventive species *Aglenus brunneus* (Gyllenhal), and the **Othniinae** includes *Elacatis delusa* Pascoe (Fig. 95B), which also occurs in Papua New Guinea and Borneo (Brooks 1965). The subfamily **Inopeplinae** is represented in Australia by three species of *Euryplatus* Motschulsky: *E. angulicollis* (Lea), *E. dimidiatus* (Waterhouse) (Fig. 95E) and *E. olliffi* (Van de Poll). The **Prostominiinae** in Australia comprise several genera, including two species of *Ocholissa* Pascoe (Fig. 95D) one each of *Prostominia* Reitter and *Szekessya* Kaszab, an undescribed species of *Diagrypnodes* Waterhouse, which is probably more closely related to the prostominiines than to the inopeplines, where it is currently placed, and an undescribed genus (Fig. 95G) which occurs in rotten palm fronds. The largest group is the **Salpinginae**, but the generic classification is still in a state of flux. Some of our described salpingine species are currently placed in the genera *Lissodema* Curtis, *Neosalpingus* Blackburn (Fig. 95B) and *Platysalpingus* Blair, while a group of species related to *Lissodema frigidum* Blabkburn and once associated with the invalid generic name *Orphanotrophium* Spilman (1954), require a new generic name. Salpingids are elongate, slightly to strongly flattened beetles, which are subglabrous or clothed with scattered setae (except in *Elacatis*, which is densely clothed with decumbent hairs). The head is prognathous, sometimes rostrate and never abruptly narrowed behind the eyes; lateral pronotal carinae may be complete, incomplete or occasionally absent, and simple, undulate or dentate; the elytra are abbreviated, exposing several abdominal tergites in *Euryplatus* and *Diagrypnodes*; tarsi are almost always simple and the tarsal formula 4-4-4 occurs only in *Aglenus* and *Ocholissa*. Larvae are elongate, parallel-sided, slightly to strongly flattened, and moderately to lightly pigmented, except for head and abdominal apex. The epicranial stem is absent and the frontal arms have a relatively narrow angle between them and usually have paired endocarinae beneath them. The hypostomal rods are moderately to very long and subparallel, sternum IX usually has one to several teeth on each side at the base (sometimes forming a basal row interrupted at middle), and the armature on tergum IX is usually complex, with accessory processes on the urogomphi and/or additional spines or processes in front of or between them. Adult salpingids may be found on foliage or flowers and larvae commonly occur under bark or in dead twigs, stems or vines, where they feed on bark or phloem, and the European salpingine *Colposis mutilatus* (Beck) apparently feeds on fruiting bodies of *Cucurbitothis* (Ascomycetes: Pleosporales) (Franz 1955). At least some members of the Prostominiinae feed on dead plant tissue of various Arecaceae and Araceae (Escalona 2008). *Tretothorax cleistostoma* from North QLD is a myrmecophile, and *Aglenus brunneus* occurs in decaying vegetation and has been found in caves. [Blair 1919, 1925; Watt 1967; Lawrence 1977a; Howden and Howden 1981; Sasaji 1988].

105. ANTHICIDAE

Adult (Figs 75H–I, K–L, 95C, H, J)

Length usually 1.2–10 mm. Body usually ~1.9–3.8 times as long as wide; subparallel or with pronotum and elytra independently curved, moderately convex to somewhat flattened; surfaces usually smooth and shiny; colour often yellow to reddish-brown, dark brown or black, often bicolored or with light and dark pattern on elytra, brightly or metallically coloured in some Lemodinae; upper surfaces subglabrous to densely setose, often with distinct, erect tactile setae in rows and randomly distributed, shorter decumbent setae; some taxa with patches of thickened, decumbent setae forming a pattern.

Head prognathous or somewhat deflexed, almost always abruptly constricted posteriorly to form narrow neck and temples of varying length (not so in *Lagrioida*). Occipital region sometimes with a short but sharp transverse ridge (some Eurygeniinae), a well developed but somewhat rounded transverse ridge (e.g. *Chileanthicus*) or a median endocarina (e.g. *Omonadus*). Eyes usually large and prominent, not or only barely emarginate, finely to coarsely facetted, with or without interfacetal setae. Antennal insertions usually exposed from above (concealed in *Lagrioida* and *Ictistygna*); subantennal grooves absent. Frontoclypeal suture usually weakly impressed or absent (distinctly impressed in some Lemodinae and Anthicinae). Labrum free, usually transverse and subtruncate. Antennae short to relatively long, 11-segmented, usually filiform to moniliform or slightly clavate, with weak, short 3-segmented club in *Lagrioida*, with relatively long, narrow, 3-segmented club in *Macratria* and *Steropes*, with terminal antennomere elongate in males of some Eurygeniinae. Mandibles short and broad, usually bidentate, somewhat scoop-like in Eurygeniinae; incisor edge usually simple; mola and prostheca or somewhat reduced. Maxilla with galea and lacinia usually short and setose, the latter usually narrower, without uncus; apical maxillary palpomere varying from fusiform to strongly expanded and subtriangular, securiform or cultriform. Mentum transverse; ligula usually simple and undivided; apical labial palpomere cylindrical or fusiform to expanded and truncate. Subgenal ridges absent. Gular sutures widely separated; gula usually longer than wide. Corpotentorium narrow. Cervical sclerites well developed.

Pronotum ~0.75–2 times as long as wide, usually widest at or near anterior end (widest at middle in some Eurygeniinae and Notoxinae); sides slightly to strongly rounded, sinuate or with constriction near posterior end or at posterior third (rarely with lateral tubercles); base distinctly narrower than combined elytral bases, except in taxa with reduced humeri (e.g. *Tomoderus* and *Anthelephila*). Lateral pronotal carinae absent; anterior angles not produced forward; posterior angles right to rounded; posterior edge more or less straight; anterior edge of prothorax produced forward to form a horn covering most of head in Notoxinae, with a narrow marginal bead in Lemodinae, Steropinae and Tomoderinae or enlarged to form a ring-like collar of even thickness in Anthicinae and Macratriinae and expanded dorsally to form a broad flange in Eurygeniinae (absent in *Lagrioida*); disc usually with slender transverse antebasal sulcus continuing anteroventrally onto hypomeron (absent in *Lagrioida*, Tomoderinae and Lemodinae). Hypomeron in most taxa with a pit or shallow impression (Fig. 25F), which may project internally to form an apodeme. Prosternum in front of coxae (Fig. 18D) shorter than to longer than mid length of procoxal cavity, flat to slightly convex. Prosternal process incomplete, narrowed apically, not curved dorsally behind coxae, with apex acute to broadly angulate. Procoxae conical and projectng well below prosternum, with short to long concealed lateral extensions; trochantins concealed. Procoxal cavities more or less circular, almost always contiguous (rarely very narrowly separated), almost always broadly open externally (closed by meeting of postcoxal projections of hypomera in *Ictistygna* and *Egestriomima*, without narrow lateral extensions, almost always broadly closed internally (open in *Steropes*).

Scutellar shield not abruptly elevated, posteriorly acute to broadly rounded, truncate or emarginate. Elytra ~1.6–2.9 times as long as combined width and 1.9–4.2 times as long as pronotum; sides usually subparallel and complete with conjointly rounded apices, sometimes exposing tip of abdomen; humeri usually well developed (reduced in some Tomoderinae and Anthicinae); disc almost always irregularly punctate (with ten puncture rows becoming irregular in places in some *Macratria* and *Lemodinae*); epipleura narrow and incomplete or absent. Mesoventrite usually separated by complete sutures from mesanepisterna, which either meet at midline or are broadly joined at midline (partly fused to mesanepisterna in *Macratria*); anterior edge without paired procoxal rests (procoxal rests often present at anterior edges of mesanepisterna); lateral edges of mesoventrite usually straight (broadly expanded and rounded laterally in some Anthicinae). Mesoventral process complete and apically narrowly rounded to truncate or emarginate. Mesocoxae more or less conical and not to slightly projecting, with exposed trochantins. Mesocoxal cavities slightly to strongly obliquely transverse, narrowly separated, partly closed laterally by mesepimera or by both mesepimera and mesanepisterna. Mesometaventral junction usually monocondylic. Metaventrite convex, usually with distinct discrimen (weakly developed or absent in Lemodinae, *Ictistygna* and *Tomoderus*; exposed portion of metanepisternum moderately elongate. Metacoxae usually contiguous or narrowly separated (widely separated in Tomoderinae), usually extending laterally to meet elytra, with short internal apodeme near middle. Metendosternite usually with long stalk and long lateral arms, well developed laminae and anterior tendons widely separated on lateral arms. Hind wing usually with short to long apical field, usually without veinal remnants but sometimes with vague sclerotisations (with distinct post-radial sclerite in *Lemodes*, with RP remnant in *Lagrioida*); radial cell usually either absent or slightly elongate with inner posterobasal angle right or obtuse; cross-vein r3 usually absent (fused at base with r4 in *Lagrioida*); basal portion of RP short or absent; medial spur usually ending well before wing margin (extending to margin in Eurygeniinae); medial field with three or four free veins in larger species and with medial fleck (divided by MP_{3+4} in *Lemodes*); wedge cell moderately broad in *Lagrioida*, narrow and apically acute in *Macratria* and Eurygeniinae, absent in *Lemodes* and Anthicinae; anal lobe well developed, often without embayment (with notch-like embayment in most Anthicinae and Notoxinae). Legs relatively long and slender; trochanterofemoral joint slightly to moderately oblique with base of femur separated from coxa. Tibiae not or only slightly widened apically; spurs usually paired and subequal on all legs, pubescent in *Macratria*, indistinct or absent in Tomoderinae. Tarsi 5-5-4 in both sexes; penultimate tarsomere usually lobed beneath, often bilobed; pretarsal claws usually simple (with blunt or sharp basal tooth in most Eurygeniinae and *Macratria*).

Abdomen almost always with five free ventrites (first two ventrites connate in *Lagrioida*); ventrite 1 usually slightly to much longer than 2, with setose fovea on each side in *Sapintus*; intercoxal process acute to narrowly rounded or angulate. Anterior edge of segment IX in male with long spiculum gastrale; tergite IX completely fused to segment X, which may be membranous. Aedeagus of tenebrionoid type, symmetrical; parameres partly or entirely fused together forming undivided apicale, which is articulated to phallobase in Eurygeniinae and Anthicinae (apicale deeply divided in *Macratria* and *Steropes*, very deeply divided and mesally membranous, so that parameres appear to be separate and articulated in Notoxinae) or fused to phallobase in Lemodinae and Tomoderinae; penis with or without paired basal struts. Sternite VIII in female usually with long spiculum ventrale. Ovipositor usually elongate and slender (shorter in Notoxinae), lightly sclerotised except for baculi; coxites setose, subdivided or not, styli small, terminal. [Demarzo 1992, 1996; Hemp and Dettner 1997; Chandler 2002a, 2010].

Larva

Body elongate, parallel-sided, slightly to strongly flattened; dorsal and ventral surfaces usually lightly pigmented and yellowish-white in colour, except for head capsule and apices of urogomphi (when present) (*Lemodes* and *Lemodinus* with most terga pigmented); surfaces smooth, clothed with short and long, scattered, erect setae.

Head prognathous and protracted, slightly transverse, with sides slightly to strongly rounded. Epicranial stem short in Lagrioidinae, Lemodinae, Anthicinae and Notoxinae, absent in Eurygeniinae and Tomoderinae; frontal arms lyriform and contiguous at base. Median endocarina lying beneath epicranial stem and extending between frontal arms in Anthicinae and Notoxinae, V-shaped and lying beneath frontal arms in known Eurygeniinae; absent in *Lagrioida*, Lemodinae and Tomoderinae. Stemmata on each side five in *Lagrioida*, two in Lemodinae, one in Anthicinae, Notoxinae and Tomoderinae; stemmata absent in known Eurygeniinae. Frontoclypeal suture absent. Labrum free. Antennae short to moderately long, 3-segmented; sensorium dome-like or broadly conical and shorter than antennomere 3. Mandibles at least slightly asymmetrical, broad at base and narrow at apex, bidentate or tridentate; incisor edge usually with one or more subapical teeth; mola well developed and transversely ridged in *Lagrioida*, but usually reduced and sub-basal with a basal brush of hairs. Ventral mouthparts retracted, with narrowly oval maxillary articulating area; cardo undivided; stipes longer than wide; mala rounded or truncate, sometimes with one or more teeth or spines at inner apical angle; maxillary palps 3-segmented. Labium consisting of prementum, mentum and submentum; ligula usually well developed; labial palps 2-segmented and usually narrowly separated. Hypopharyngeal sclerome molar-like or a transverse bar. Hypostomal rods usually moderately long and diverging (slightly converging in Anthicinae and Notoxinae, apparently absent in known Eurygeniinae); ventral epicranial ridges absent. Gular region transverse; gular sutures well separated.

Prothorax not or only slightly longer than meso- or metathorax; thoracic terga without transverse carinae, or patches or rows of asperities. Prothoracic venter without armature. Legs well developed, 5-segmented including claw-like pretarsus bearing two setae, usually lying side by side (one distal to the other in Lemodinae and Tomoderinae); coxae narrowly to moderately widely separated. Abdomen more than twice as long as thorax, terga and sterna without with rows or patches of asperities. Tergum IX, excluding appendages, not to much shorter than VIII, extending onto ventral surface but not forming articulated plate, usually with paired, simple urogomphi (forked in known Eurygeniinae, absent in Tomoderinae), which may bear two or more small tubercles in *Lagrioida*, Anthicinae and Notoxinae, without interurogomphal pits; sternum IX not or only slightly enclosed by sternum VIII, usually without basal asperities (with one pair of asperities in *Lagrioida*), with two apicomesal teeth in known Eurygeniinae (Fig. 53C); segment X not visible from above, posteroventrally oriented, transverse, without paired pygopods. Thoracic and abdominal spiracles usually annular, not placed at ends of spiracular tubes (thoracic and first abdominal spiracles annular–biforous in *Lagrioida*); spiracles on segment VIII not enlarged. [Costa *et al.* 1995; Chandler 2002a, 2010; Hemp and Dettner 2003].

Classification and World Distribution

The cosmopolitan family Anthicidae comprises ~3000 species in 101 genera. There are eleven generally recognised subfamilies (Lawrence and Newton 1995), eight of which occur in Australia; but three of these, the southern African Afreminae, the Nearctic, eastern Palaearctic and Oriental Ischaliinae and the Notogean Lagrioidinae, are excluded by many workers and were considered as Tenebrionoidea *incertae sedis* by Lawrence *et al.* (2010m). In this work the Lagrioidinae is treated as a very distinctive group of Anthicidae. Another subfamily not occurring in Australia is the Copobaeninae, which occurs only in Chile. [Levey 1985; Young 1985a; Nikitsky and Egorov 1992; Nikitsky 1996; Chandler 1999; Chandler *et al.* 2004].

Australian Fauna

The subfamily **Lagrioidinae** includes the single species *Lagrioida australis* Champion, which is widely distributed on beaches along the east coast; other species of the genus are known from New Zealand, Chile and Brazil. Adults differ from all other Australian anthicids in lacking a narrow neck and temples and in having the first two ventrites connate and from almost all others in having concealed antennal insertions. The genus *Macratria* (Fig. 75H) (**Macratriinae**) includes several Australian species, but one of them, *M. davidsonae* Armstrong, may belong to *Steropes* Steven (**Steropinae**), a Palaearctic genus extending into South-east Asia (Telnov 2006a) but not yet recorded from Australia. The subfamily **Eurygeniinae** is well represented by six genera, two of which (*Ictistygna* and *Egestriomima*) differ from other members of the family in having externally closed procoxal cavities. Little is known of eurygeniine biology, and the only known larva is that of a Nearctic genus. **Lemodinae** includes species of *Lemodes* (Fig. 75K), adults of which are brightly coloured (red or red and blue) and pubescent, the unusual *Lemodinus tumidipennis* (Blair) and several species of *Trichananca*, some of which are wingless. The subfamily **Tomoderinae** include the litter-inhabiting *Tomoderus* and *Pseudotomoderus* (Fig. 75I), which occur in northern rainforest and southern eucalypt woodland, respectively. **Notoxinae** includes species of *Mecynotarsus* (Fig. 75L) and *Notoxus* with an anteriorly-projecting horn on the pronotum used to burrow through sand (Hashimoto and Hayashi 2012), and **Anthicinae** includes the great majority of genera and species, most of which are small in size. Several anthicine genera occur in this fauna, including *Anthicus* (Fig. 95H), *Anthelephila*, *Anthicomorphus*, *Chileanthicus*, *Omonadus* and *Sapintus*. Anthicids are elongate beetles, which are usually small in size and often distinctly narrowed between prothorax and elytra so they resemble ants; they may be black, brown, yellow or red in colour and are sometimes bicolored and almost always clothed with decumbent and/or erect hairs. The antennal club is weak or absent, the head abruptly narrowed behind the eyes to form a neck and the and the antennal insertions exposed (except in Lagrioidinae), the eyes not or only slightly emarginate, the apical maxillary palpomere slightly to strongly expanded apically and often securiform or cultriform, the pronotum without lateral carinae and almost always constricted at the middle, near the base or at base and apex, the mesanepisterna meet or almost meet at the midline, the penultimate tarsomere is lobed beneath, and the first two abdominal ventrites are free (except in Lagrioidinae). Larvae are elongate, parallel-sided and usually lightly pigmented (with a dark and light pattern dorsally in *Lemodes*). The antennal sensorium is usually broad and dome-like, there is almost always a single stemma on each side (five in Lagrioidinae), a median endocarina occurs in at least some Anthicinae, the mandibular molae are usually reduced and sub-basal, the mala is usually simple (cleft in Lagrioidinae), sternite IX is almost always simple (with a pair of basal teeth in Lagrioidinae), and urogomphi may be absent (Tomoderinae), simple (Lemodinae), forked (at least one exotic eurygeniine) or with accessory processes near the base (Lagrioidinae, Anthicinae, Notoxinae). Anthicids are usually found in decaying vegetation, fungus-infested logs and leaf litter, and some may be common in beach drift or sand dunes. they are often said to be general scavengers, and some exotic Anthicinae have been reported to feed on insect egg masses or fly puparia. *Lagrioida* larvae may be found among the roots of dune plants (Costa *et al.* 1995). *Lemodinus* feeds under the bark of rotten logs. *Lemodes* larvae were found in cells in wood, where they had been deposited by a predatory wasp; the bicoloured pattern on the dorsal surface of the larva suggests that it feeds on surfaces, possibly on fungal fruiting bodies. The species of *Trichananca* are common in leaf litter, as are those of *Tomoderus* and *Pseudotomoderus*. Some of the larger genera of Australian Anthicinae are *Mecynotarsus*, *Chileanthicus*, *Sapintus* and *Anthicus*, but as mentioned above, some species are currently misplaced in the last genus. *Omonadus floralis* is a cosmopolitan species that occurs in foodstuffs.

At least some anthicids are known to be attracted to the cantharidin or to meloid beetles that produce this chemical (Young 1984). Cantharidin is usually obtained by males, especially those having secretory glands concentrating this chemical in canals at the elytral apices (Schütz and Dettner 1992). Cantharidin is apparently a feeding deterrent to predators, and when females feed on male secretions they probably incorporate the cantharidin into their bodies and pass it onto their eggs and larvae, as has been documented for certain species of Nearctic Pyrochroidae Eisner et al. 1996a, 1996b). In some anthicid adults, mesothoracic glands secrete toxic iridoids through a median pore at the anterior end of the mesoventrite; this chemical appears to act as an effective feeding deterrent for ants (Hemp and Dettner 1997). [Lea 1922a; Armstrong 1948; Abdullah 1969; Lawrence 1977b, 1982; Young 1978, 1985a; Werner and Chandler 1995; Kejval 2003, 2009; Telnov 2003, 2006b, 2007, 2009a, 2009b, 2011b, 2011c; Uhmann 2007; Lawrence *et al.* 2010m].

Australian Subfamilies, Tribes and Genera

Lagrioidinae: *Lagrioida* Fairmaire & Germain.
Macratriinae: *Macratria* Newman.
Steropinae: *Steropes* Steven ?
Eurygeniinae: Eurygeniini: *Diacalla* Pascoe, *Egestria* Pascoe, *Egestrina* Champion, *Macratriomima* Champion; **Ictistygnini**: *Egestriomima* Champion, *Ictistygna* Pascoe.
Lemodinae: *Lemodes* Boheman, *Lemodinus* Blair, *Trichananca* Blackburn.
Tomoderinae: *Tomoderus* LaFerté-Sénectère, *Pseudotomoderus* Pic.
Notoxinae: *Mecynotarsus* LaFerté-Sénectère, *Notoxus* Geoffroy.
Anthicinae: Anthicini: *Acanthinus* LaFerté-Sénectère, *Anthicomorphus* Lewis, *Anthicus* Paykull, *Floydwernerius* Telnov, *Micranthicus* Champion, *Omonadus* Mulsant & Rey, *Pseudocyclodinus* Telnov, *Pseudoleptaleus* Pic, *Sapintus* Casey, *Stricticollis* Marseul; **Endomiini**: *Endomia* Laporte; **Formicomini**: *Anthelephila* Hope, *Chileanthicus* Werner.

106. ADERIDAE

Adult (Fig. 90H–I)

Length usually 1.3–3.5 mm (up to 7 mm in *Megaxenus*). Body usually ~1.8–3.1 times as long as wide; subparallel or rarely with pronotum and elytra independently curved, moderately convex to somewhat flattened; colour often yellow or red to dark brown or black, often bicolored or with light and dark pattern on elytra; upper surfaces clothed with fine, suberect to decumbent hairs, sometimes combined with fine, pruinose pubescence.

Head prognathous to moderately strongly declined, usually abruptly constricted posteriorly to form neck and temples of varying length (not so in *Megaxenus*, *Syzetonellus* or *Syzetoninus*). Occipital region sometimes with a transverse ridge, which may abut anterior edge of prothorax concealing neck. Eyes usually large and prominent, usually moderately to very deeply emarginate (not in *Syzetonellus* and *Scraptogetus*), sometimes approximate or almost contiguous dorsally, finely to coarsely facetted, with short to moderately long interfacetal setae. Antennal insertions usually exposed from above; subantennal grooves absent. Frontoclypeal suture usually distinctly impressed. Labrum free, usually slightly to strongly transverse. Antennae 11-segmented, usually filiform, moniliform, serrate or clavate, rarely pectinate or plumose from antennomere 6. Mandibles usually bidentate with well developed mola and prostheca (unidentate without mola and with reduced prostheca in *Megaxenus*). Maxilla with galea and lacinia short and setose, the latter without uncus; apical maxillary palpomere slightly expanded and truncate to securiform. Mentum transverse; ligula small, undivided; apical labial palpomere slightly expanded to securiform. Subgenal ridges absent. Gular sutures moderately widely separated. Corpotentorium narrow. Cervical sclerites present.

Pronotum ~0.6–1.1 times as long as wide, parallel-sided or widest at middle or anteriorly; sides straight to slightly rounded or sinuate; base distinctly narrower than combined elytral bases. Lateral pronotal carinae absent; anterior angles not produced forward; posterior angles right to rounded; posterior edge more or less straight; disc usually simple, rarely with single or paired shallow impressions. Prosternum in front of coxae usually shorter than mid length of procoxal cavity (slightly longer in *Scraptogetus*), flat to moderately convex. Prosternal process complete or incomplete, narrowed apically, not curved dorsally behind coxae, with apex acute. Procoxae conical and projectng well below prosternum, with short concealed lateral extensions; trochantins reduced and concealed. Procoxal cavities slightly transverse to circular, contiguous or narrowly separated, externally broadly open, without narrow lateral extensions, internally open.

Scutellar shield not abruptly elevated, posteriorly truncate to slightly emarginate. Elytra ~1.3–2.5 times as long as combined width and 2.1–4.5 times as long as pronotum; sides usually subparallel and complete with conjointly rounded apices; disc irregularly punctate without scutellary striole; epipleura incomplete or absent. Mesoventrite usually separated by complete sutures from mesanepisterna, which are approximate or contiguous at midline (sutures often abruptly curved mesally near anterior end, so that mesanepisterna may appear more widely separated); anterior edge without paired procoxal rests (procoxal rests often present at anterior edges of mesanepisterna). Mesoventral process complete and apically truncate or emarginate. Mesocoxae not or only slightly projecting, with exposed trochantins. Mesocoxal cavities slightly to strongly obliquely transverse, narrowly to moderately widely separated, partly closed laterally by mesepimera or by both mesepimera and mesanepisterna. Mesometaventral junction monocondylic. Metaventrite convex, with discrimen long, short or absent; exposed portion of metanepisternum usually moderately to very elongate. Metacoxae usually contiguous or narrowly separated and not extending laterally to meet elytra. Metendosternite usually with long stalk and long lateral arms; laminae reduced or absent (except in *Scraptogetus*); anterior process usually present bearing approximate tendons (anterior process absent and tendons located on arms in *Scraptogetus*). Hind wing usually with moderately long apical field, sometimes with veinal remnants, occasionally with post-radial sclerite); radial cell often absent (slightly elongate with subacute inner posterobasal angle in *Scraptogetus*); cross-vein r3 usually absent (present in *Scraptogetus*); basal portion of RP short or absent; medial spur usually short (extending to wing margin in *Megaxenus*); medial field with three or fewer free veins and with medial fleck; wedge cell absent; anal lobe more or less reduced, with notch-like embayment. Legs moderately long and slender; trochanterofemoral joint slightly to moderately oblique with base of femur separated from coxa. Metafemora in males often with small or large pubescent patch beneath. Tibiae not or only slightly widened apically; spurs weakly developed or absent. Tarsi 5-5-4 in both sexes; penultimate tarsomere reduced and antepenultimate one lobed beneath; pretarsal claws simple; empodium concealed or absent.

Abdomen with five ventrites, the first two of which are connate; ventrite 1 not much longer than 2; intercoxal process acute to moderately broadly rounded or angulate; suture sometimes partly or completely obliterated, so that there appears to be four ventrites with the first one greatly enlarged. Sternite IX in male with or without spiculum gastrale; tergites IX and X fused together and membranous. Aedeagus of inverted tenebrionoid type, symmetrical or asymmetrical; parameres fused together forming apicale, which is much longer than phallobase and sometimes bears accessory lobes; penis with or without paired anterior struts. Sternite VIII in female usually with spiculum ventrale. Ovipositor short and broad to moderately long and lightly sclerotised except for baculi; coxites subdivided or not, usually with acute, mesodistal projections; styli laterally placed and sometimes fused to coxites or absent. [Baguena Corella 1948, 1962; Lawrence *et al.* 1990; Chandler 2002b; Gompel and Barrau 2002; Lawrence and Ślipiński 2010i].

Larva (Fig. 90A, E)

Body elongate, parallel-sided, slightly flattened or slightly wider at middle; dorsal and ventral surfaces very lightly pigmented, except for mandibles and apices of urogomphi (when present); surfaces smooth, usually clothed with short and long, scattered, erect setae (more densely clothed with very short, fine hairs in *Megaxenus*).

Head prognathous and protracted, slightly transverse, flattened, with rounded sides. Epicranial stem short; frontal arms lyriform and contiguous at base. Median endocarina almost always absent (lying beneath epicranial stem in *Megaxenus*). Stemmata absent. Frontoclypeal suture absent or vaguely indicated. Labrum free, transverse. Antennae usually well developed (short in *Megaxenus*), 3-segmented; sensorium broadly conical and shorter than antennomere 3. Mandibles symmetrical, broad at base and narrowing apically, usually tridentate; incisor edge simple or with one or two subapical teeth; mola moderately well developed but sub-basal or reduced to narrow, flattened process, basal to which is a hyaline lobe or brush of hairs (mandible unidentate without mola in *Megaxenus*). Ventral mouthparts retracted, with maxillary articulating area; cardo undivided; stipes longer than wide; mala rounded or truncate; maxillary palps 3-segmented. Labium consisting of prementum, mentum and submentum; ligula shorter than labial palps, which are 2-segmented and moderately widely separated. Hypopharyngeal sclerome absent. Hypostomal rods short and diverging or absent; ventral epicranial ridges absent. Gular region transverse; gular sutures well separated.

Prothorax not or only slightly longer than meso- or metathorax; thoracic terga without transverse carinae, or patches or rows of asperities. Prothoracic venter without armature. Legs well developed, 5-segmented including pretarsus, which is usually slender and acute with two setae lying in row, one distal to the other; coxae widely separated. Abdomen about twice as long as thorax, terga and sterna without with rows or patches of asperities. Tergum IX, excluding appendages, as long as or slightly longer than VIII, extending onto ventral surface but not forming articulated plate, usually with paired, simple urogomphi, which are broad at base and narrowly acute apically, without interurogomphal pit (urogomphi absent in *Megaxenus*, which has small, lightly pigmented, median appendage); sternum IX exposed, simple, not or only slightly enclosed by sternum VIII; segment X posteroventrally oriented, transverse, without paired pygopods. Thoracic and abdominal spiracles annular, not placed at ends of spiracular tubes; spiracles on segment VIII not enlarged. [Lawrence *et al.* 1990; Chandler 2002b; Lawrence and Ślipiński 2010i].

Classification and World Distribution

The family Aderidae includes ~50 genera and 1000 species distributed throughout all major regions of the world. The group was subdivided into several tribes and subtribes by Baguena Corella (1948, 1962), but the classification has not been followed by most authors (Israelson 1971; Werner 1990; Gompel and Barrau 2002; Nardi 2007; Lawrence and Ślipiński 2010i).

Australian Fauna

The great majority of described Australian Aderidae are presently included in the *Aderus* Stephens (as *Xylophilus* Latreille), but none of them belong to that Holarctic genus, and the generic classification of the family is badly in need of revision. Some apparently endemic genera are *Syzetoninus* Blackburn, *Syzetonellus* Blackburn and *Pseudananca* Blackburn. The genus *Syzeton* Blackburn (Fig. 90I) was originally based on Australian species, but is apparently more widely distributed. Undescribed species of the New Zealand genus *Scraptogetus* Broun are known from TAS. The most unusual Australian aderid is *Megaxenus termitophilus* Lawrence (Fig. 90H), which occurs in northern QLD with related species in New Guinea and the Philippines and may be as much as 7 mm in length. Aderids are elongate, convex to slightly flattened beetles, almost always less than 3.5 mm in length (except *Megaxenus*), usually resembling smaller Anthicidae in having a narrow pronotum without lateral carinae, securiform apical maxillary palpomeres and the head is usually abruptly constricted behind the eyes; they differ from anthicids, however, in having larger, more coarsely facetted eyes, internally open procoxal cavities, the first two ventrites connate or solidly fused, the antepenultimate tarsomere lobed beneath and the penultimate one reduced. Many species have bicoloured elytra, and the vestiture consists of erect and/or decumbent hairs, the latter sometimes forming a pattern. In males of some species, the metafemora bear setose secretory organs. Larvae are almost always narrowly elongate, parallel-sided, somewhat flattened and lightly pigmented, except for the tips of the urogomphi; the mandible almost always has a hyaline lobe at the base of the mola, there is no median endocarina on the head, and there is no armature at the base of sternum IX. Adult Aderidae are usually collected on foliage or in Malaise or flight intercept traps, while larvae have been collected in rotten wood, under bark or in leaf litter. *Megaxenus termitophilus* inhabits the nests of termites of the genus *Microcerotermes* Silvestri and the large larvae mimic neotenic queens and are capable of entering into trophallactic behaviour with worker termites by stroking them with the fore legs feeding on the salivary secretions produced. [Lea 1917a; 1917b; Watt 1987; Lawrence *et al.* 1990; Lawrence and Ślipiński 2010i].

107. SCRAPTIIDAE

Adult (Fig. 94C)

Length usually 1.3–6 mm. Body usually ~2–3.6 times as long as wide; subparallel or widest at pronoto-elytral junction and narrowing anteriorly and posteriorly; somewhat flattened; colour yellow to dark brown or black, occasionally with red head and prothorax and black elytra or with light and dark pattern on elytra; upper surfaces clothed with short to moderately long, fine, suberect or decumbent hairs.

Head slightly to strongly declined, abruptly constricted posteriorly to form moderately broad neck and well developed temples. Occipital region with broadly rounded to sharp transverse ridge, which may abut anterior edge of prothorax concealing neck. Eyes relatively large and but not protuberant, moderately to very deeply emarginate, finely to coarsely facetted, with short to moderately long interfacetal setae. Antennal insertions exposed from above; subantennal grooves absent. Frontoclypeal suture distinctly impressed. Labrum free, transverse, broadly rounded to truncate. Antennae 11-segmented, usually filiform or suberrate (pectinate from antennomeres 5 to 10 in males of *Xylophilostenus*). Mandibles short and broad, bidentate with well developed mola and prostheca. Maxilla with setose galea and lacinia, the latter without uncus; apical maxillary palpomere apically expanded and securiform or cultriform. Mentum transverse; ligula deeply emarginate or bilobed; apical labial palpomere securiform. Subgenal ridges absent. Gular sutures well separated; gula sometimes longer than wide. Corpotentorium absent. Cervical sclerites well developed.

Pronotum ~0.48–1.07 times as long as wide, widest at base or at middle; sides slightly to strongly rounded; base not or only slightly narrower than combined elytral bases. Lateral pronotal carinae incomplete, simple, without raised margin or bead; anterior angles not produced forward; posterior angles right; posterior edge straight or weakly bisinuate; disc sometimes with paired shallow impressions. Prosternum in front of coxae shorter than mid length of procoxal cavity, flat to slightly convex. Prosternal process incomplete, narrowed apically, not curved dorsally behind coxae, with apex acute or broadly angulate. Notosternal sutures incomplete or absent. Procoxae conical and projectng well below prosternum, without or with short concealed lateral extensions; trochantins at least partly exposed. Procoxal cavities slightly to strongly transverse, contiguous, externally broadly open, without narrow lateral extensions, internally closed.

Scutellar shield not abruptly elevated, posteriorly acute to broadly rounded. Elytra ~1.55–3.1 times as long as combined width and 2.6–6.4 times as long as pronotum; sides subparallel or converging apically and complete with conjointly rounded apices or independently rounded and exposing two or three abdominal tergites; disc irregularly punctate without scutellary striole; epipleura incomplete or absent. Mesoventrite either separated by complete sutures from mesanepisterna, which are contiguous at midline or completely fused to mesanepisterna; anterior edge without paired procoxal rests. Mesoventral process incomplete and apically acute. Mesocoxae strongly projecting, with exposed trochantins. Mesocoxal cavities strongly obliquely transverse, contiguous or very narrowly separated, partly closed laterally by mesepimera. Mesometaventral junction absent. Metaventrite convex with long discrimen; exposed portion of metanepisternum very long and narrow. Metacoxae contiguous, obliquely oriented, extending laterally to meet elytra. Metendosternite with long stalk and long lateral arms; laminae and anterior process absent; anterior tendons located on arms. Hind wing usually with moderately long apical field, usually with post-radial sclerite and at least one radial extension; radial cell short and broad, sometimes incomplete; cross-vein r3 absent; basal portion of RP absent; medial spur straight and moderately long; medial field with three or four free veins and no medial fleck; wedge cell absent; anal lobe well developed or reduced, without or with slight embayment. Legs long and slender; trochanterofemoral joint strongly oblique with base of femur in contact with coxa. Tibiae not widened apically; spurs pubescent (Fig. 16I). Tarsi 5-5-4 in both sexes; penultimate tarsomere lobed beneath; pretarsal claws simple; empodium concealed or absent.

Abdomen with five free ventrites; ventrite 1 not much longer than 2; intercoxal process acute. Sternite IX in male with spiculum gastrale; tergites IX and X fused together and membranous. Aedeagus of uninverted tenebrionoid type, symmetrical; parameres completely fused together forming apicale, which is articulated to phallobase, without accessory lobes; penis with short paired anterior struts. Sternite VIII in female with basally articulated spiculum ventrale. Ovipositor long, slender and lightly sclerotised except for baculi; paraprocts distinctly longer than coxites, which are indistinctly subdivided and setose; styli small and terminal. [Franciscolo 1964, 1972; Watt 1987; Lawrence 1987b; Vanin *et al.* 1996; Pollock 2002f; Lawrence and Ślipiński 2010j].

Larva (Fig. 90D)

Body elongate, parallel-sided, slightly flattened; dorsal and ventral surfaces very lightly pigmented, except for mandibles; surfaces smooth, densely clothed with short and long fine hairs.

Head prognathous and protracted, slightly transverse, flattened, with slightly rounded sides. Epicranial stem very short; frontal arms lyriform and contiguous at base. Median endocarina absent. Stemmata absent or one on each side. Frontoclypeal suture vaguely indicated. Labrum free, transverse. Antennae well developed, 3-segmented, with antennomere 3 much shorter than 2 and sensorium large and dome-like. Mandibles asymmetrical, broad at base and narrow at apex, bidentate; incisor edge with several subapical teeth; mola moderately well developed, transversely ridged. Ventral mouthparts retracted, with maxillary articulating area; cardo undivided; stipes longer than wide; mala slender and truncate; maxillary palps 3-segmented. Labium consisting of prementum, mentum and submentum; ligula simple, shorter than labial palps, which are 2-segmented and moderately widely separated. Hypopharyngeal sclerome consisting of transverse bar. Hypostomal rods absent; ventral epicranial ridges absent. Gular region transverse; gular sutures well separated.

Thorax long and slender with segments more or less equal; thoracic terga without transverse carinae, or patches or rows of asperities. Prothoracic venter without armature. Legs well developed, 5-segmented including pretarsus, which is usually slender and acute with two setae lying in row, one distal to the other; coxae widely separated. Abdomen less than twice as long as thorax, terga and sterna without with rows or patches of asperities. Tergum IX, excluding appendages, shorter than VIII, not extending onto ventral surface or forming articulated plate, without urogomphi, with large, rounded, setose, dehiscent appendage; sternum IX simple, not enclosed by sternum VIII; segment X posteriorly oriented, transverse, without paired pygopods. Thoracic and abdominal spiracles annular, not placed at ends of spiracular tubes; spiracles on segment VIII not enlarged. [Böving and Craighead 1931; Watt 1987; Young 1991; Švácha 1995; Vanin *et al.* 1996; Pollock 2002f; Lawrence and Ślipiński 2010j].

Classification and World Distribution

The family Scraptiidae comprises ~400 species in 33 genera and cosmopolitan in distribution. The group is usually divided into two subfamilies: the very widely distributed Scraptiinae and the Anaspidinae, which occurs throughout the Holarctic and Afrotropical regions, with some extensions into the Neotropical and Oriental regions. [Franciscolo 1972; Lawrence and Ślipiński 2010j].

Australian Fauna

All Australian Scraptiidae are members of the subfamily **Scraptiinae** and the majority of these are now placed in the genus *Scraptia* Latreille (Fig. 94C). *Xylophilostenus octophyllus* Lea is unusual in that the last eight antennomeres bear long, flabellate processes. Scraptiids are oblong to narrowly elongate, parallel-sided to fusiform, soft-bodied and uniformly pubescent beetles with relatively long legs and antennae. The head is strongly deflexed and abruptly constricted behind the eyes forming a narrow neck, the frontoclypeal suture is distinct, the eyes are coarsely facetted and deeply emarginate, the apical maxillary palpomere is securiform, the lateral pronotal carinae are complete or incomplete anteriorly, tibial spurs are well developed and pubescent, and the penultimate tarsomere is lobed beneath. Larvae are elongate, parallel-sided, somewhat flattened and lightly pigmented, with an oblong, setose, deciduous process (often broken off) at the apex of tergum IX. The head lacks a median endocarina, the antennal sensorium is dome-like and the ligula moderately long. Adult scraptiids are usually collected on flowers or foliage and the larvae have been found under bark and in rotten wood and leaf litter. [Lawrence 1987b; Watt 1987].

CHRYSOMELOIDEA

Chrysomeloids are similar to curculionoids in having pseudotetramerous tarsi, a reduced cucujiform type of aedeagus and similar internal reproductive organs (Kasap and Crowson 1977; Mann and Crowson 1983) in the adult and lacking a distinct mandibular mola and gular region in the larva; they differ from weevils, however, in having well developed mouthparts but no rostrum or distinct antennal club in the adult and lacking a hypopharyngeal bracon in the larva. In most Twentieth Century classifications, including that of Crowson (1955, 1967) there were three generally recognised families: Cerambycidae, Bruchidae and Chrysomelidae. Chen (1985) recognised six chrysomeloid families: Megalopodidae (including Orsodacninae and Zeugophorinae), Cerambycidae, Crioceridae, Chrysomelidae, Eumolpidae and Hispidae. Megalopodidae and Orsodacnidae were treated as families by Lawrence and Newton (1995) and the family Bruchidae downgraded to a chrysomelid subfamily. Megalopodidae was expanded to include a new group Palophaginae by Kuschel and May (1990) and both Megalopodide and Orsodacnidae were recognised by Clark and Riley (2002a, 2002b). Bruchinae was retained as a family by Kingsolver (2002b) and this has been followed by some recent workers (Yus-Ramos *et al.* 2007); Mann and Crowson (1981) summarised evidence for the close relationship of Bruchinae and Sagrinae. The cerambycid subfamily Disteniinae was removed from the family by Linsley (1961), and Švácha and Danilevsky (1987) and Švácha *et al.* (1997) recognised the families Oxyeltidae, Disteniidae and Vesperidae (including the subfamilies Anoplodermatinae and Philinae). Although the chrysomelid subfamilies Cryptocephalinae (= Camptosomata), Galerucinae (including Alticinae) and Cassidinae (including Hispinae) are generally recognised as monophyletic groups, the constitution and systematic position of several smaller groups differ in early classifications of Monrós (1960), Medvedev (1971) and Mann and Crowson (1981). Comparative studies based on various morphological features have been published by Mann and Crowson (1983) (internal male reproductive organs), Schmitt (1992) (stridulatory devices), Suzuki (1994) (hind wings) and Samuelson (1996) (elytra-body binding sites). Reid (1995) produced a cladistic analysis of Chrysomeloidea based on adult and larval morphology, and Schmitt (1996) presented an overview of chrysomelid phylogeny. A DNA-based cladogram for Chrysomeloidea and Curculionoidea was published by Marvaldi *et al.* (2008), and Gómez-Zurita *et al.* (2005, 2007) produced a molecular cladogram for Eumolpinae and for Chrysomelidae as a whole. The following chrysomeloid families do not occur in Australia: Oxypeltidae (Chile), Vesperidae (Palaearctic, Neotropical, Oriental, Afrotropical), Disteniidae (Nearctic, Neotropical, eastern Palaearctic, Oriental, Afrotropical) and Orsodacnidae (Holarctic and Neotropical). [Švácha and Danilevsky 1987; Švácha *et al.* 1997; Santos-Silva and Hovore 2007].

108. CERAMBYCIDAE

Adult (Figs 9A–H, 77E–H, J–L, 97A–B)

Length ~2.3–90 (usually 8–50) mm; smallest species in Lamiinae and largest in Prioninae). Body ~1.6–8.0 times as long as wide; usually more or less parallel-sided, strongly convex to slightly flattened; variable in colour, often reddish-brown to dark brown in Parandrinae and Prioninae, with notable exceptions in *Rhipidocerus* (yellowish-green) and *Phaolus* (metallic blue or green), sometimes brightly bicolored in Cerambycinae and Lamiinae; body surfaces glabrous or clothed with distinct hairs, which may be thickened and form patterns but are not scale-like.

Head more or less prognathous, rarely strongly constricted posteriorly, occasionally constricted immediately behind eyes to form broad neck (e.g. in *Enicodes*); transverse occipital ridge absent. Frontal region not or only slightly deflexed in most subfamilies, but strongly deflexed from behind eyes and more or less vertical in most Lamiinae (Fig. 15F) (not Tmesisternini) and a few Cerambycinae (Fig. 15D), often with median longitudinal groove marking a more or less deep internal endocarina, which may continue into occipital region in some Prioninae and Lamiinae. Eyes not to strongly protuberant, oval to vertically elongate, not to deeply emarginate (Fig. 14D), rarely divided into upper and lower parts (e.g. in *Enicodes*, *Tmesisternus*), finely to very coarsely facetted. Antennal insertions almost always exposed (concealed in Parandrinae) and usually widely separated, almost always supported by raised tubercles; subantennal groove absent. Frontoclypeal suture distinctly impressed to absent, almost straight to strongly angulate; anterior edge of clypeus usually straight to convex, sometimes concave to shallowly emarginate. Frontoclypeal region produced anteriorly to form short to moderately long muzzle or rostrum in a few Cerambycinae (e.g. *Rhinophthalmus*). Labrum usually more or less visible and free (fused to clypeus in Parandrinae and some Prioninae), strongly transverse to distinctly longer than wide, of variable shape. Antennae almost always 11-segmented, (12-segmented in males of the prionine genus *Phaolus* and in both sexes of some lamiine genera, such as *Phelipara* and *Pseudocalamobius*), usually filiform or serrate and moderately to very long (often longer than body); usually distinctly longer in male than in female; occasionally moniliform, pectinate or flabellate (e.g. *Rhipidocerus*, *Enneaphyllus*), capitate with 3-segmented club in *Telocera*. Scape usually much longer than wide; pedicel usually very short, not or slightly longer than wide. Mandible short and broad to moderately elongate, apex usually unidentate, seldom bidentate or chisel-like, enlarged and strongly bent downwards in males of *Australodon*; incisor edge of mandible simple or with one or more teeth; mola and prostheca absent, but inner basal margin may bear a field or row of microtrichia arising from desclerotised cuticle. Maxilla almost always with distinct, setose galea and lacinia, latter strongly reduced or virtually absent in Parandrinae and most Prioninae; lacinia without uncus; apical maxillary palpomere fusiform to triangular or securiform. Ligula membranous or sclerotised, usually emarginate or bilobed, sometimes undivided or strongly reduced; apical labial palpomere fusiform to expanded apically. Gular sutures widely separated or absent. Corpotentorium broad to narrow or rudimentary. Cervical sclerites present or absent.

Pronotum ~0.35–2.0 times as long as wide, usually widest at middle; sides often slightly rounded, not explanate, occasionally with one or more lateral teeth or spines; base not to distinctly narrower than basal elytral width. Lateral pronotal carinae usually absent (Fig. 19C), complete and simple in Parandrinae, complete or incomplete and often serrate or dentate in most Prioninae and some Tmesisternini; anterior pronotal angles rarely produced forward; posterior angles broadly rounded to obtuse or right;

posterior edge usually more or less straight or evenly rounded, sometimes distinctly sinuate or variously lobed; disc occasionally with paired basal impressions, median longitudinal groove, or paired and/or unpaired tubercles. Prosternum in front of coxae shorter than to longer than shortest diameter of coxal cavity, prosternal process usually complete, narrowed apically, parallel-sided, gradually expanded or strongly and abruptly expanded at apex (Fig. 20E), latter usually narrowly rounded or truncate (sometimes emarginate). Notosternal sutures complete, incomplete or absent. Procoxae projecting well below prosternum in some Cerambycinae and Lamiinae, without or with short concealed lateral extension; trochantins often at least partly exposed (completely concealed in several Cerambycinae and Lamiinae). Procoxal cavities strongly transverse to circular, contiguous to widely separated, externally usually broadly open to narrowly closed (broadly closed in some Lamiinae), with or without narrow lateral extensions, internally more or less closed, except in Parandrinae and Prioninae.

Scutellar shield visible, usually not abruptly elevated, anteriorly simple, posteriorly acute to rounded or truncate or occasionally emarginate; mesoscutum usually with stridulatory file (Fig. 16O) (absent in Parandrinae, Prioninae and some Lamiinae). Elytra ~0.8–5.5 times as long as combined width and 1.25–7.3 times as long as pronotum; punctation irregular or with 8–10 puncture rows or striae, with or without scutellary striole; apices meeting at suture or independently rounded or acute; epipleura complete, incomplete or absent. Elytra in Cerambycinae sometimes short and truncate (e.g. *Hesthesis*) or apically narrowed (e.g. *Enchoptera*, *Macrones*), exposing several abdominal tergites. Mesoventrite almost always separated by complete sutures from mesanepisterna, which are distinctly separated at midline; anterior edge occasionally on different plane than metaventrite, with or without paired procoxal rests. Mesocoxae rarely slightly conical and projecting (e.g. *Macrones*, *Enchoptera*); trochantins exposed or concealed. Mesocoxal cavities narrowly to widely separated, circular to strongly transverse, not or slightly oblique, usually laterally open (partly closed by mesopleuron), but closed by meeting of meso- and metaventrites in some Cerambycinae and Lamiinae. Mesometaventral junction simple or complex; accessory mesocoxal articulation present in some Lamiinae (e.g. *Sybra*). Metaventrite with discrimen very long to absent; postcoxal lines absent; exposed portion of metanepisternum short and broad to very long and narrow. Metacoxae contiguous to widely separated, horizontally or obliquely oriented, extending laterally to meet elytra or not; plates absent. Metendosternite with lateral arms moderately to very long, laminae well developed, reduced or absent, anterior process short or absent and anterior tendons close together or widely separated. Hind wing usually with moderately long apical field with two more or less complete radial extensions and with a dark sclerite apical to radial cell and a subtriangular sclerite crossing r4; radial cell often well developed and more or less elongate (sometimes short and broad or without base), cross-vein r3 usually slightly to strongly oblique, sometimes absent; cross-vein r4 with or without apical spur; basal portion of RP long to very short; medial field usually with 4 or 5 free veins (sometimes with 3 or rarely fewer) and never with medial fleck or binding patch; wedge cell well developed in almost all Prioninae; absent in other subfamilies; hind wings lacking transverse folds in some Cerambycinae with shortened elytra; wings rarely reduced or absent. Legs usually moderately to very long and slender. Trochanterofemoral joint transverse to strongly oblique, occasionally with base of femur abutting coxa. Tibiae strongly widened and often toothed at apex and sometimes with teeth along outer tibial apex in Parandrinae and some Prioninae; tibial spurs usually 2-2-2, well developed, rarely reduced to 1 or 0 on some pairs of legs. Protibia in Lamiinae with oblique band of dense setae usually within pubescent groove (antenna cleaner) on inner face and mesotibia usually with similar structure on outer face. Tarsi almost always 5-5-5 in both sexes and usually pseudotetramerous with highly reduced penultimate tarsomere partly concealed at base of ventral lobe on antepenultimate one (4-4-4 in some Lamiinae where reduced penultimate tarsomere has fused with the terminal one); penultimate tarsomere relatively large and tarsal lobes absent in Parandrinae; tarsomeres 1–3 or at least 2 and 3 usually with dense pilosity beneath; pretarsal claws usually simple; empodium exposed (with one to several setae) or concealed.

Abdomen with 5 free ventrites; ventrite 1 usually not much longer than 2, without postcoxal lines; intercoxal process acute to broadly rounded or angulate; ventrite 2 with paired sex patches in males of some Laminae. Functional spiracles present on abdominal segments I–VII, located in pleural membrane. Anterior edge of sternite VIII in male with median strut. Males with anterior edge of sternite VIII almost always bearing median strut, anterior edge of sternite IX with spiculum gastrale; tergites IX and X fused together and usually membranous. Aedeagus cucujiform, symmetrical; anterior edge of tegmen without or with single strut; parameres usually fused to phallobase but free from one another, rarely fused together or virtually absent; anterior edge of penis almost always with paired struts; internal sac may bear distinctive sclerotised structures such as asperities, paired or unpaired sclerites, or a longitudinal sclerotised rod. Sternite VIII in female with anterior strut (spiculum ventrale). Ovipositor usually long and flexible with well developed apical or subapical styli (small in Lamiinae), secondarily shortened and modified particularly in some Cerambycinae ovipositing on host surface and often combined with ventral abdominal brushes used for covering eggs with debris; ovipositor well sclerotised in some groups (e.g. Parandrinae), with styli lateral or sometimes absent. One or two pairs of glandular reservoirs often present at base of ovipositor. [Thomson 1864, 1865; Lacordaire 1868, 1869, 1872; Gahan 1906; Lameere 1919; Saalas 1936; Gressitt 1951, 1956, 1959, 1984; Linsley 1961; Hayashi 1961; Harde 1966; Gressitt *et al.* 1970; Villiers 1978; Fragoso 1985; Saito 1989a, 1989b, 1990, 1991, 1992, 1993a, 1993b; Napp 1994; Švácha *et al.* 1997; Matthews 1997; Turnbow and Thomas 2002; Santos-Silva *et al.* 2010].

Larva (Figs 35G, 97I)

Body elongate and more or less parallel-sided or slightly wider anteriorly, subcylindrical or slightly flattened, weakly sclerotised and lightly pigmented, except for mouth frame and mandibles; prothorax sometimes weakly pigmented and tergum IX rarely so; vestiture usually of abundant short setae, sometimes with large areas of microscopic spines.

Head prognathous and usually deeply retracted (less retracted in some Lamiinae), moderately well sclerotised although lightly pigmented; posterior edge of head capsule not or only slightly emarginate. Epicranial stem absent or concealed beneath a longitudinal depression housing dorsal head retractor muscles (Fig. 45G); frontal arms V-shaped or U-shaped, usually extending to antennal insertions, with bases contiguous or narrowly separated by anterior end of muscle housing; entire posterior cranial region cemented to prothoracic pocket in some Lamiinae. Median endocarina long, sometimes extending to clypeus. Stemmata each side basically six, but usually only 3 in vertical row and often absent in late instars. Frontoclypeal suture usually present and associated with a sclerotised internal transverse bar (not clearly indicated in Prioninae, where postclypeus appears to be fused to frons forming a two-layered sclerotised ridge (epistomal margin), which is often carinate and/or toothed. Clypeus (or anteclypeus) usually broad, trapezoidal, lightly sclerotised, and occupying most of the area between dorsal mandibular articulations (except in Cerambycinae where the base is much narrower). Labrum free, usually transverse and broadly rounded apically (more or less elongate and narrowly rounded apically in Parandrinae and some Prioninae, with paired basal emarginations resulting in a cordate shape). Antennae moderately long to minute, often capable of deep retraction due to large articulating membrane; almost always 3-segmented, but with terminal antennomere reduced, occasionally 2-segmented; sensorium usually more or less conical, but flattened and dome-like in Parandrinae and Prioninae. Mandibles symmetrical and heavily sclerotised, almost always with unidentate apex, which may be subacute or obliquely truncate, but broadly rounded and more or less perpendicular to plane of movement in Cerambycinae; incisor edge concave to somewhat flattened or angulate, often with three keels or ridges converging towards apex, but with finely striate pseudomola in Parandrinae (Fig. 38A) and some Prioninae; mesal surface of mandibular base simple, without mola or other modifications. Ventral mouthparts not or only slightly retracted (Fig. 49J); maxillary articulating area well developed, but fused with cardines and submentum in Lamiinae; cardines usually well developed (but not distinguishable in Lamiinae), not divided; stipes shorter than broad; maxillary palp usually 3-segmented with well developed, partly articulated palpifer (reduced and fixed in Lamiinae); mala broadly triangular to slender and finger-like. Labium usually consisting of prementum, mentum and submentum; ligula usually well developed, simple and apically rounded; labial palps 2-segmented and usually well separated. Hypopharyngeal bracon and sclerome absent. Hypostomal rods usually moderately long and subparallel or anteriorly converging (very short in some Lamiinae); ventral epicranial ridges absent. Gular region short but moderately well developed in Cerambycinae, Lamiinae and *Arhopalus*, but highly reduced and sometimes concealed by post-gular lobe in Parandrinae and Prioninae; gular sutures well separated or absent. Tentorial bridge usually slender or rudimentary, but in Prioninae, Parandrinae (Fig. 49J) and to a lesser extent Cerambycinae moderately broad and ventrally located, so that occipital foramen appears to be divided into two parts.

Prothorax distinctly longer and usually wider than meso- or metathorax and often as long as meso- and metathroax combined; protergum without distinct plate but often somewhat sclerotised, at least anteriorly, and with sublateral longitudinal furrows; main anterior portion of proventer usually undivided or vaguely divided into an anteromesal and posterolateral regions, but posterior portion with distinct strongly transverse region bearing fore legs in most Cerambycinae or with separate posterolateral, leg-bearing sclerites in Parandrinae and Prioninae. Legs always very short and widely separated, 5-segmented in Parandrinae, Prioninae, *Arhopalus* and many Cerambycinae, but reduced or absent in some Cerambycinae and absent in all Lamiinae; coxae flat and poorly defined, never with sclerotised condyle, and pleuron undivided, without pleural sulcus; pretarsus acute, sometimes with sclerotised apex, almost always without setae.

Abdomen usually ~5 times as long as thorax, with segments I–VII often bearing dorsal and ventral ambulatory ampullae, with or without special cuticular modifications, but number of ampullae reduced in some groups. Segment IX usually as long as or somewhat shorter than VIII, with tergum not extending onto ventral surface and almost always simple (posterior end with minute, paired urogomphi in *Arhopalus* and with median sclerotised tubercle or spine in *Uracanthus* and *Ancita);* segment X terminal or posteroventral, sometimes fused to IX; anal opening usually Y-shaped (transverse in some Lamiinae). Thoracic and abdominal spiracles annular or annular–multiforous (annular-biforous in some first instars). [Craighead 1923; Gardner 1931a; Duffy 1953, 1963, 1968; Dumbleton 1957; Svácha and Danilevsky 1987, 1988, 1989; Švácha *et al.* 1997; Švácha 2001; Turnbow and Thomas 2002].

Classification and World Distribution

Cerambycidae is a cosmopolitan family comprised of ~25 000 species in over 5000 genera. The group is currently divided into seven subfamilies, four of which occur in Australia. The exotic subfamily Lepturinae is most abundant and diverse in the Holarctic region with southern incursions into Central and South America, East Africa and the Oriental region, while Necydalinae is confined to the Holarctic region and Dorcasominae is Palaearctic and Afrotropical, with an extensive radiation in Madagascar. [Linsley 1962a, 1962b, 1963, 1964; Linsley and Chemsak 1984, 1995, 1977; Monné and Bezark 2009; Bousquet *et al.* 2009].

Australian Fauna

The subfamily **Spondylidinae** is represented in Australia by two introduced Palaearctic species of the tribe Asemini: *Arhopalus syriacus* (Reitter), which has become established in NSW, and *A. rusticus* (Linnaeus), which has been recorded only from Melbourne (Wang and Leschen 2003). **Parandrinae** (Fig. 77J), represented by two genera and three species, are unusual in lacking the pseudotetramerous tarsi characteristic of other Australian Chrysomeloidea. **Prioninae** include several large, dark brown species with lateral pronotal carinae toothed as in *Sclerocantha* (Fig. 77E) or serrate as in *Paroplites.* Two unusual species are *Rhipidocerus australasiae* Westwood (Fig. 77L)), which is light green with flabellate antennae in the male, and *Phaolus metallicus* (Newman), which is metallic blue or green. Larvae of *Paroplites australis* (Erichson) may tunnel through old *Banksia* trees, while those of *Eurynassa australis* Boisduval and *Cnemoplites* species, usually referred to as witchetty grubs, were consumed by aboriginal Australians.

The **Cerambycinae** are the largest and most diverse group, containing more than 40 tribes and ~150 genera. Perhaps the best known species is *Phoracantha semipunctata* (Fabricius), a widely distributed eucalypt borer, which has been accidentally introduced into New Zealand, South Africa, South America, Europe, the

Middle East, and North America, where it is a serious pest of planted *Eucalyptus* (Carne and Taylor 1978; Scriven *et al.* 1986). *Hylotrupes bajulus* (Linnaeus), the European house borer, has been accidentally introduced into Australia on several occasions. In most instances establishment has been prevented (Howick and Carr 1971); however, the species has recently become established in Perth (http://www.ehb.wa.gov.au). *Hesthesis* species have shortened elytra, frequent flowers, and are striking mimics of wasps both in colour and behaviour; species of *Macrones* resemble narrow-waisted, sphecid wasps; and *Eroschema* species belong to a lycid mimicry complex. Other notable cerambycines include *Uracanthus triangularis* Hope, which attacks various shrubs, *U. cryptophagus* Olliff, a pest of cultivated *Citrus*, *U. cupressiformis* Rondonuwu and Austin, which damages ornamental cypress, and *Diotimana undulata* (Pascoe), which attacks *Araucaria cunninghamii* in NSW and QLD and also occurs on Norfolk Island, where its host is *A. heterophylla*. The larva of *Bardistus cibarius* Newman, called the bardee, was another aboriginal food source. The recently described *Australodon nearnsi* Escalona and Ślipiński from southern QLD and northern NSW was tentatively placed in Cerambycinae, but cannot be placed in a recognised tribe and could not be excluded from either Lepturinae or Dorcasominae, neither of which are known from Australia.

Among the **Lamiinae**, the dark-coloured, flightless species of *Microtragus* and *Athemistus* closely resemble amycterine weevils. *Batocera wallacei* Thomson is the largest Australian beetle; *B. boisduvali* (Hope) and the silvery grey *Rosenbergia megacephala* van der Poll both attack fig trees in QLD. Other common lamiines include *Ancita* species, which tunnel into sapwood of *Acacia* branches, and species of *Platyomopsis*, *Rhytiphora* and *Penthea* (Fig. 77H), many of which are also *Acacia* feeders. The Asian long-horned beetle, *Anoplophora glabripennis* (Motschulsky) has been intercepted by AQIS Victoria in timber pallets from China and is considered a possible risk to fruit trees in Australia (Lingafelter and Hoebeke 2002).

Cerambycids are usually elongate, subcylindrical to somewhat flattened beetles, usually pubescent, almost always with antennae more than two-thirds of the body length and inserted on prominences, so that they can be directed backwards over the body. The eyes are usually emarginate, often deeply so, the lateral pronotal carinae are usually absent (not so in Parandrinae and Prioninae), there are often lateral tubercles or spines on the prothorax, the mesonotum is provided with a stridulatory file, except in Parandrinae and Prioninae, tibial spurs are usually well developed, tarsomeres without ventral lobes only in Parandrinae, and pretarsal claws simple. Larvae are elongate, subcylindrical to somewhat flattened, lightly pigmented and subglabrous or clothed with fine setae or patches of spinules. The head is deeply retracted into the prothorax and excavated posteriorly forming a recess for housing retractor muscles, the antennae are short, the mandibles are large and stout, the prothorax is usually more or less enlarged, the legs are small and widely separated and may be vestigial or absent, the first six or seven abdominal segments may bear fleshy ampullae which aid in locomoation within wood, and tergum IX is very rarely provided with armature. Adult cerambycids are active fliers and may be diurnal or nocturnal. Many feed on flowers, foliage or bark, and some are attracted to sugar baits. Larvae are phytophagous and usually feed internally on bark, phloem, sapwood or heartwood of a variety of trees and shrubs. Some species attack herbs, others are root feeders or gall formers, and a few may feed on seeds or cones. [Froggatt 1894; 1923, 1927; Linsley 1959; Edwards 1961a, 1961b; Duffy 1963; McKeown 1947; Williams 1985; Rondonuwu and Austin 1988; Wang 1993a, 1993b, 1995a, 1995b, 2000; Webb 1994; Wang *et al.* 1994, 1999; Matthews 1997; Calder 2000; Martin 2000; Monné 2005; Thongphak and Wang 2007, 2008; Bousquet *et al.* 2009; Santos-Silva *et al.* 2010; Escalona and Ślipiński 2011; Vives 2012].

Australian Subfamilies, Tribes and Genera

Parandrinae: *Papuandra* Santos-Silva *et al.*, *Storeyandra* Santo-Silva *et al.*

Prioninae: Callipogonini: *Cacodacnus* Thomson, *Elaptus* Pascoe, *Sceleocantha* Newman, *Toxeutes* Newman. **Macrotomini:** *Agrianome* Thomson, *Analophus* Waterhouse, *Anomophysis* Quentin & Villiers, *Archetypus* Thomson, *Brephilydia* Pascoe,*Cnemoplites* Newman, *Eurynassa* Thomson. *Paroplites* Lameere, *Pseudoplites* Lameere, *Teispes* Thomson, *Utra* Jordan, *Xixuthrus* Thomson. **Meroscelisini:** *Aesa* Lameere, *Eboraphyllus* Mckeown, *Enneaphyllus* Waterhouse, *Howea* Olliff, *Phaolus* Pascoe, *Rhipidocerus* Westwood.

Spondylidinae: Asemini: *Arhopalus* Audinet-Serville.

Cerambycinae: Achrysonini: *Saporaea* Thomson. **Aphanasiini:** *Aphanasium* Thomson, *Aphanosperma* Britton, *Citriphaga* Lea, *Myrsellus* McKeown.

Aphneopini: *Aphneope* Pascoe, *Zoedia* Pascoe. **Bimiini:** *Bimia* White, *Akiptera* Saunders, *Proagapete* McKeown. **Callicromatini:** *Chloridolum* Thomson. **Callidiini:** *Callidium* Fabricius. **Callidiopini:** *Adrium* Pascoe, *Bethelium* Pascoe, *Callidiopis* Thomson, *Ceresium* Newman, *Coccothorax* Aurivillius, *Demelius* Waterhouse, *Didymocantha* Newman, *Ectinope* Pascoe, *Examnes* Pascoe, *Neobethelium* Blackburn, *Nerida* McKeown, *Notoceresium* Blackburn, *Oemona* Newman, *Paphora* Pascoe, *Porithea* Pascoe, *Porithodes* Aurivillius, *Psylacrida* Thomson, *Sidellus* McKeown, *Tethionea* Pascoe, *Thephantes* Pascoe, *Wahn* McKeown, *Yorkeica* Blackburn. **Cerambycini:** *Aeolesthes* Gahan, *Cerambyx* Linnaeus, *Pachydissus* Newman. **Certallini:** *Pytheus* Newman, *Titurius* Pascoe, *Brachytria* Newman, *Pempsamacra* Newman, *Omophaena* Pascoe, *Obrida* White. **Clytini:** *Xylotrechus* Chevrolat, *Clytus* Laicharting, *Chlorophorus* Chevrolat, *Demonax* Thomson. **Coptommatini:** *Navomorpha* Thomson. **Deilini:** *Schizopleurus* Lacordaire, *Telocera* White, *Eburophora* White. **Distichocerini:** *Distichocera* Kirby, *Paradistichocera* Van de Poll. **Eroschemini:** *Eroschema* Pascoe, *Chaodalis* Pascoe. **Glaucytini:** *Graphicoglaucytes* Breuning, *Scituloglaucytes* Breuning. **Gracilini:** *Gracilia* Audinet-Serville. **Hesperophanini:** *Phacodes* Newman, *Gastrophacodes* Lea, *Oebarina* Pascoe, Aesiotyche Pascoe. **Hesthesini:** *Hesthesis* Newman. **Heteropsini:** *Aridaeus* Thomson, *Cremys* Pascoe. **Macronini:** *Macrones* Newman, *Enchoptera* Saunders, *Brachopsis* Saunders, *Oroderes* Saunders. **Molorchini:** *Epania* Pascoe, *Molorchus* Fabricius, *Merionoeda* Pascoe, *Earinis* Pascoe, *Mecynopus* Erichson, *Omotes* Newman. **Neostenini:** *Neostenus* Pascoe, *Maltheba* Pascoe, *Aposites* Pascoe, *Mystrosa* Pascoe. **Obriini:** *Cleistimum* Thomson. **Phalotini:** *Phalota*

Pascoe, *Nenenia* Pascoe, *Maulia* Blackburn, *Xystoena* Pascoe. **Phlyctaenodini:** *Bardistus* Newman, *Tricheops* Newman, *Diotimana* Hawkins, *Phlyctaenodes* Newman, *Ambeodontus* Lacordaire. **Phoracanthini:** *Allotisis* Pascoe, *Atesta* Pascoe, *Coleocoptus* Aurivillius, *Coptocercus* Hope, *Epithora* Pascoe, *Paratesta* Wang, *Phoracantha* Newman, *Phytrocaria* Wang, *Skeletodes* Newman, *Steata* Wang, *Thoris* Pascoe. **Pseudocephalini:** *Cyclocranium* Van de Poll, *Pseudocephalus* Newman, *Formicomimus* Aurivillius. **Psilomorphini:** *Psilomorpha* Saunders, *Ischnauchen* Scambler, *Arcucornus* Scambler. **Stenoderini:** *Stenoderus* Dejean, *Syllitus* Pascoe, *Demomisis* Pascoe, *Simocrysa* Pascoe. **Rhagiomorphini:** *Rhagiomorpha* Newman, *Tritocosmia* Newman, *Tropis* Newman. **Rhopalophorini:** *Amphirhoe* Newman. **Sestyrini:** *Sophron* Newman. **Spintheriini:** *Anastetha* Pascoe. **Strongylurini:** *Aprosictus* Pascoe, *Bebius* Pascoe, *Exaeretiformis* McKeown, *Lygesis* Pascoe, *Opsidota* Pascoe, *Oxymagis* Pascoe, *Phantissus* McKeown, *Piesarthrius* Hope, *Strongylurus* Hope, *Tya* McKeown. **Tessarommatini:** *Tessaromma* Newman. **Tillomorphini:** *Tilloforma* McKeown, *Ochyra* Pascoe, *Obriomorpha* Aurivillius, *Homaemota* Pascoe, Acrogenoides McKeown. **Trachyderini:** *Ischnotes* Newman, *Purpuricenus* Germar. **Tragocerini:** *Tragocerus* Latreille. **Trichomesiini:** *Trichomesia* Pascoe. **Tropocalymmatini:** *Tropocalymma* Thomson. **Typhocesini:** *Taphos* Pascoe, *Typhocesis* Pascoe, *Hemesthocera* Newman. **Uracanthini:** *Scolecobrotus* Hope, *Uracanthus* Hope, *Neouracanthus* McKeown, *Emenica* Pascoe, Aethiora Pascoe, *Rhinophthalmus* Thomson. **Xystrocerini:** *Xystrocera* Audinet-Serville, *Nungena* McKeown. **Tribe ?:** *Australodon* Escalona & Ślipiński, *Monoplia* Newman, *Syllitosimilis* McKeown.

Lamiinae: Acanthocinini: *Acanista* Pascoe, *Australiorondonia* Breuning, *Australoleiopus* Breuning, *Didymocentrotus* McKeown, *Elongatocontoderus* Breuning, *Exocentrus* Mulsant, *Lagocheirus* Dejean, *Polyacanthia* Montrouzier, *Sciades* Pascoe, *Striomiaena* Breuning, *Subexocentrus* Breuning. **Agapanthiini:** *Phelipara* Pascoe, *Pseudocalamobius* Kraatz. **Ancitini:** *Ancita* Thomson. **Agniini:** *Acalolepta* Pascoe, *Callipyrga* Newman, *Pelargoderus* Audinet-Serville, *Periaptodes* Pascoe, *Potemnemus* Thomson. **Apomecynini:** *Apomecyna* Latreille, *Atimura* Pascoe, *Diasybra* Breuning, *Epilystoides* Breuning, *Eunidia* Erichson, *Ichthyodes* Newman, *Mimofalsoropica* Breuning, *Mimosybra* Breuning, *Mycerinopsis* Thomson, *Neosybra* Breuning, *Parahathlia* Breuning, *Phaeapate* Pascoe, *Ropica* Pascoe, *Sybra* Pascoe, *Zorilispe* Pascoe. **Astathini:** *Momisis* Pascoe. **Batocerini:** *Batocera* Castelnau, *Rosenbergia* Ritsema. **Cyrtinini:** *Brachyrhabdus* Aurivillius, *Cyrtillus* Aurivillius. **Dorcaschematini**: *Olenecamptus* Chevrolat. **Enicodini:** *Enicodes* Thomson. **Gnomini:** *Gnoma* Fabricius. **Gyaritini:** *Itheum* Pascoe. **Homonoeini:** *Mulciber* Thomson, *Notomulciber* Blackburn. **Obereini:** *Oberea* Mulsant. **Parmenini:** *Athemistus* Pascoe, *Austrosomatidia* McKeown, *Blaxotes* McKeown, *Ceraegidion* Boisduval, *Corestetha* Pascoe, *Dorcadida* White, *Echthistatus* Pascoe, *Elasmotena* McKeown, *Falsoceraegidion* Breuning, *Hoplathemistus* Aurivillius, *Mesathemistus* Breuning, *Mesolita* Pascoe, *Microtragoides* Breuning, *Microtragus* Thomson, *Parathemistus* Breuning, *Parmenomorpha* Blackburn, *Somatidia* Thomson, *Spinhoplathemistus* Breuning, *Spinopotemnemus* Breuning, *Xylotoles* Newman. **Pogonocherini:** *Cristhybolasius* Breuning, *Parahybolasius* Breuning. **Pteropliini:** *Cairnsia* Blackburn, *Cobria* Pascoe, *Corrhenes* Pascoe, *Corrhenispia* Breuning, *Corrhenodes* Breuning, *Depsages* Pascoe, *Eczemotes* Pascoe, *Etaxalus* Pascoe, *Hathliodes* Pascoe, *Ischioplites* Thomson, *Menyllus* Pascoe, *Mesiphiastus* Breuning, *Mimosaperdopsis* Breuning, *Notocorrhenes* Breuning, *Paradaxata* Breuning, *Pararhytiphora* Breuning, *Penthea* Castelnau, *Platycranium* Aurivillius, *Prosoplus* Blanchard, *Pterolophia* Newman, *Rhytiphora* Audinet-Serville, *Similosodus* Mckeown. *Symphyletes* Newman, *Trichohathliodes* Breuning, *Trichopenthea* Breuning, *Trichoprosoplus* Breuning, *Xiphohathlia* Breuning, *Xiphotheopsis* Breuning, *Zaeeropsis* Breuning, *Zygrita* Thomson. **Rhodopinini:** *Allomicrus* Gahan, *Amblymora* Pascoe, *Amblymoropsis* Breuning, *Australothelais* Breuning, *Bucynthia* Pascoe, *Cyocyphax* Thomson, *Demonassa* Thomson, *Dysthaeta* Pascoe, *Essisus* Pascoe, *Illaena* Erichson, *Meton* Pascoe, *Microlamia* Bates, *Mimotemnosternus* Breuning, *Mimoxylotoles* Breuning, *Oricopis* Pascoe, *Parapteridotelus* Breuning, *Paressisus* Aurivillius, *Pentacosmia* Newman, *Probatodes* Thomson, *Pseudancita* Breuning, *Setosophroniella* Breuning, *Spinostenellipsis* Breuning, *Stenellipsis* Bates, *Temnolamia* Breuning, *Tetroreopsis* Breuning, *Thyada* Pascoe, *Tuberothelais* Breuning, *Velora* Thomson. **Saperdini:** *Glenea* Newman. **Tmesisternini:** *Falsapolia* Breuning, *Temnosternopsis* Breuning, *Temnosternus* White, *Tmesisternus* Latreille, *Trigonoptera* Perroud. **Zygocerini:** *Disternopsis* Breuning, *Mimozygocera* Breuning, *Paradisterna* Breuning, *Pseudodisterna* Breuning, *Trichozygocera* Breuning, *Zygocera* Erichson, *Zygoceropsis* Breuning.

109. MEGALOPODIDAE

Adults (Figs 78A, 96A)

Length 2.7–7.5 mm. Body ~2.2–2.8 times as long as wide and more or less parallel-sided, slightly flattened, yellow to black in color, but often bicolored, with distinct pattern of yellow and black or red and black markings on elytra and pronotum in some *Zeugophora*, yellow with raised dark-brown or black areas in Palophaginae; usually clothed with fine, decumbent and suberect hairs (with patches of thickened light setae in some *Zeugophora*). Head not or slightly declined, strongly but gradually constricted posteriorly to form neck, but without distinct temples, in *Zeugophora*; occipital region without transverse ridge or stridulatory file. Frontal region without median groove or endocarina. Eyes strongly protuberant, shallowly emarginate in Palophaginae, more deeply so in *Zeugophora*, coarsely facetted in Palophaginae, finely so in *Zeugophora*, without interfacetal setae. Antennal insertions exposed from above; subantennal grooves absent. Frontal region without median groove. Frontoclypeal suture distinctly impressed, distinctly curved or angulate; clypeus anteriorly truncate. Labrum free, exposed, slightly to strongly transverse, subtruncate to slightly convex. Antennae 11-segmented, moderately long and filiform in Palophaginae, somewhat shorter and slightly clavate in *Zeugophora*; scape elongate but less than 3 times as long as pedicel. Mandible usually moderately elongate and gradually curved mesally, unidentate in Palophaginae, bidentate in

Zeugophora; with dorsal setose cavity in Palophaginae; incisor edge simple or with weak tooth; mola well developed and transversely ridged in Palophaginae, absent in *Zeugophora*; prostheca somewhat reduced, setose. Maxilla with setose galea and lacinia, the latter without uncus; apical maxillary palpomere cylindrical to fusiform. Mentum strongly transverse and apically emarginate; ligula distinctly bilobed in Palophaginae, truncate in *Zeugophora*; apical labial palpomere cylindrical to fusiform. Gular sutures widely separated. Corpotentorium moderately broad, straight; anterior tentorial arms not expanded mesally; cervical sclerites moderately well developed.

Pronotum ~0.7–0.93 times as long as wide, widest at about middle in Palophaginae or anterior to middle in *Zeugophora*, with base usually distinctly narrower than combined elytral bases; sides slightly curved or sinuate, not explanate, with single prominence or tubercle at about middle or slightly anterior to middle in Palophaginae, with sub-basal tubercle armed with 2 long, stiff setae in *Zeugophora*; lateral pronotal carinae absent in *Zeugophora*, more or less complete, simple and with distinct bead in Palophaginae, but obscured at middle by lateral tubercle; anterior angles not produced; posterior angles usually obtuse or right; posterior edge straight or evenly curved, finely margined in Palophaginae; disc with anterior transverse submarginal grooves in *Zeugophora*; with several longitudinal elevations in Palophaginae; paired basal impressions and median longitudinal groove absent. Prosternum in front of coxae about as long as shortest diameter of coxal cavity in Palophaginae, much shorter in Zeugophora, flat to moderately convex, without carinae. Prosternal process more or less complete, extending behind coxae in Palophaginae, flattened, strongly but gradually expanded apically and broadly rounded at apex in Palophaginae, subparallel and subacute at apex in *Zeugophora*, slightly curved and convex in Palophaginae, somewhat concave in *Zeugophora*. Notosternal sutures usually complete, sometimes obscured anteriorly. Procoxae projecting well below prosternum, without concealed lateral extensions; trochantins exposed. Procoxal cavities strongly transverse, narrowly to very narrowly separated, externally closed by meeting of postcoxal (hypomeral) processes behind prosternal process in *Zeugophora* or above apex of prosternal process in Palophaginae, without narrow lateral extensions, internally narrowly to broadly closed.

Scutellar shield not abruptly elevated, anteriorly simple, posteriorly narrowly rounded to truncate; mesoscutum with stridulatory file, divided in Palophaginae, undivided in *Zeugophora*. Elytra ~1.6–2.25 times as long as wide and 2.55–4.15 times as long as pronotum, irregularly punctate; elytral apices meeting or almost meeting at the suture, which is not deflected near apex; epipleura narrow and incomplete. Mesoventrite usually separated by complete sutures from mesanepisterna, which are distinctly separated at midline; anterior edge on same plane as metaventrite, without paired procoxal rests. Mesoventral cavity absent. Mesocoxae not conical and projecting, with exposed trochantins. Mesocoxal cavities narrowly separated, subcircular, not or slightly oblique, partly closed laterally by mesepimeron. Mesoventral process narrowly rounded or notched at apex; mesometaventral junction usually monocondylic. Metaventrite convex; transverse (metakatepisternal) suture long but close to posterior edge of ventrite; discrimen moderately to very long; postcoxal lines absent; exposed portion of metanepisternum moderately elongate in *Zeugophora*, narrowly elongate in Palophaginae; metepimeron usually at least slightly exposed. Metacoxae contiguous or narrowly separated, extending laterally to meet elytra in Palophaginae; plates absent. Metendosternite with long stalk, long lateral arms and well developed laminae; anterior process very short or barely indicated, with moderately widely separated anterior tendons. Hind wing with long apical field containing distinct sclerite just beyond radial cell and anterior and posterior remnants of RP; radial cell well developed, usually at least slightly longer than wide (not in *Zeugophora*), usually with apical protrusion, base more or less straight, inner posterobasal angle right to slightly acute; cross-vein r3 strongly oblique, arising well away from r4; basal portion of RP moderately long in *Palophagus*, short in *Zeugophora* and absent in *Cucujopsis*, apical extension extending to wing margin; R-M loop moderately broad; medial spur usually ending before wing margin, where there may be a weak embayment; medial field with 5 free veins in Palophaginae, 2 in *Zeugophora*; medial fleck present in Palophaginae, not in Zeugophorinae; MP_{3+4} with basal cross-vein but no spur in Palophaginae, without base in *Zeugophora*; wedge cell absent in *Cucujopsis* and *Zeugophora*, obliquely truncate at apex in *Palophagus*); anal lobe usually well developed with undivided AP (reduced in *Zeugophora*); anal embayment absent or very weak. Legs moderately long and slender; trochanterofemoral joint strongly oblique with base of femur separate from coxa; metafemora weakly inflated with small, mesal tooth near apex in *Cucujopsis*; tibial spurs paired on all legs. Tarsi pseudotetramerous, 5-5-5 in both sexes with tarsomere 4 reduced and 3 lobed beneath; pretarsal claws simple in Palophaginae, toothed in *Zeugophora*; empodium not apparent.

Abdomen with 5 free ventrites; ventrite 1 at its shortest point distinctly shorter than 2, without postcoxal lines; intercoxal process acute; ventrite 5 usually as long as preceding two ventrites combined, often apically emarginate in female. Abdominal terga usually lightly to moderately pigmented, except for pygidium (tergite VII); functional spiracles on segments I–VII, those on I–VI usually in pleural membrane, those on VII at lateral edges of tergite. Tergite VIII completely concealed beneath tergite VII; sternite VIII in male reduced and lightly sclerotised, with short median strut. Anterior edge of segment IX in male with distinct spiculum gastrale; tergite X free and somewhat sclerotised in Palophaginae, membranous in *Zeugophora*. Aedeagus cucujiform, often lightly asymmetrical; anterior edge of tegmen with short to long strut; parameres short, broad, separate and articulated in Palophaginae, fused together and to phallobase in *Zeugophora*; penis somewhat flattened and curved, its anterior edge with long, paired, slender struts, with long flagellum in *Zeugophora*. Segment VIII in female usually more or less membranous, with paired longitudinal sclerites and a long, basally articulated spiculum ventrale. Ovipositor reduced and enclosed in thickened membranous sheath; paraprocts fused to coxites and gonostyli absent. Internal tract with complexly curved, sclerotised capsular spermatheca. [Jacoby and Clavareau 2005; Crowson 1946; Monrós 1959; Reid 1989; Kuschel and May 1990, 1996a, 1996b; Schmitt 1994].

Larvae (Fig. 97E–F)

Body elongate and more or less parallel-sided or slightly wider anteriorly and narrowing towards apex, slightly flattened in Palophaginae, more strongly so in *Zeugophora*; all surfaces lightly pigmented, except for head capsule and prothorax; vestiture consisting of scattered long and short, simple setae. Head prognathous, slightly to moderately retracted; posterior edge of head capsule distinctly emarginate; epicranial stem absent; frontal arms V-shaped or somewhat lyriform, contiguous at base in Palophaginae, slightly separated at base in *Zeugophora*; median endocarina extending to clypeus. Stemmata on each side six in Palophaginae, two in *Zeugophora*. Antennae short, 3-segmented; sensorium about as long as apical antennomere in Palophaginae, shorter in *Zeugophora*, conical. Frontoclypeal suture distinct and labrum free in *Zeugophora*, suture absent and labrum fused to head capsule in Palophaginae. Mandibles more or less symmetrical, broad at base and narrow at apex, unidentate or weakly bidentate; accessory ventral process absent; incisor edge with one or two subapical teeth in Palophaginae, serrate in *Zeugophora*; mola reduced and subbasal in Palophaginae, absent in *Zeugophora*; prostheca absent. Ventral mouthparts retracted; maxillary articulating area present; cardo slightly to strongly oblique, undivided; stipes about as long as wide; mala fixed or articulated, apically rounded and setose; maxillary palp 3-segmented with weakly developed palpifer. Labium consisting of prementum, mentum and submentum (Palophaginae) or prementum and postmentum (*Zeugophora*); ligula short, broadly rounded and simple in Palophaginae, absent in *Zeugophora*; labial palps 2-segmented, moderately widely separated in Palophaginae, contiguous in *Zeugophora*. Hypopharyngeal bracon and hypopharyngeal sclerome absent. Hypostomal rods moderately long and slightly to strongly diverging; ventral epicranial ridges absent. Gular region absent.

Prothoracic tergum not or slightly longer than meso- or metaterga, with distinct yellowish plate. Legs in Palophaginae very short and widely separated, 4-segmented, with claw-like pretarsus lacking pretarsal setae; legs absent in *Zeugophora*. Abdomen more than twice as long as thorax; abdominal terga not extending laterally beyond edges of sterna; mesothorax, metathorax and abdominal segments I–VIII in *Zeugophora* with dorsal and ventral transverse patches of asperities; most abdominal terga and sterna in Palophaginae each with transverse groove or ambulatory ampulla; segment IX usually subequal in length to VIII; completely dorsal, without urogomphi; sternum 9 simple; segment X terminal, without pygopods; anal region posteriorly oriented. Anterior abdominal spiracles annular–biforous in Palophaginae, annular in *Zeugophora*, not placed at ends of spiracular tubes, those on segment VIII about same size as others on abdomen, facing laterally. [Medvedev and Zaitzev 1978; Hayashi 1986; Kuschel and May 1990, 1996a, 1996b; Lee 1990].

Classification and World Distribution

The family Megalopodidae primarily pantropical in distribution and includes ~450 species in 24 genera. There are three subfamilies, two of which occur in Australia. The exotic subfamily Megalopodinae occurs in the Neotropical, Afrotropical and Oriental regions. [Monrós 1947; Kuschel and May 1990; Schmitt 1992].

Australian Fauna

The Australian megalopodids include three species of **Palophaginae**, *Cucujopsis setifer* Crowson from northern QLD, *Palophagus bunyae* Kuschel (Fig. 96A) from southern QLD and *P. australiensis* Kuschel from northern NSW, and at least three species of *Zeugophora* (**Zeugophorinae**) (Fig. 78A), which occur along the east coast from VIC to northern QLD. Australian megalopodids are elongate, slightly flattened, yellow, black or bicoloured, pubescent beetles with prominent emarginate eyes, moderately long and filiform (Palophaginae) or shorter, incrassate (Zeugophorinae) antennae, the prothorax distinctly narrower at base than the combined elytral bases, and a laterally constricted prothorax and toothed pretarsal claws in Zeugophorinae. Both adults and larvae of Palophagus are known to feed on the pollen of Araucariaceae, and it is likely that *Cucujopsis setifer* has similar habits. Larvae of *Zeugophora* (*Pedrilia*) species have been recorded as leaf miners of *Santalum* (Santalaceae) in India and *Euonymus* and *Celastrus* (Celastraceae) in Japan and Australia, respectively. [Oke 1932; Crowson 1946; Medvedev and Zaitzev 1978; Reid 1989; Kuschel and May 1990].

110. CHRYSOMELIDAE

Adults (Figs 10A–H, 76F, I–K, 78C–J, 96C–E, G, I, K–L)

Length 1.2–32 (usually 1.8–16) mm. Body ~1.0 (Cassidini, Notasacanthini) to 4.6 (Cryptonychini, Eurispini) times as long as wide; more or less parallel-sided to broadly ovate with evenly rounded sides; usually moderately to strongly convex (somewhat flattened in several groups of Cassidinae); colour highly variable, yellow to black, but often either bicolored, or with metallic coloration; upper surfaces often glabrous or with scattered fine setae, densely clothed with fine, decumbent hairs or thickened hairs, sometimes forming a pattern, in Bruchinae and some Sagrinae, Galerucinae, Cryptocephalinae and Eumolpinae, rarely with scale-like setae (e.g. in *Pachnephorus*); ventral surfaces covered with silvery plastron in *Donacia*. Head not or only slightly declined at base, but frontal region usually strongly declined in Cassidinae (Fig. 15E) and Cryptocephalinae; usually not or only gradually constricted behind eyes, but with weak temples and a broad neck in *Donacia*, Criocerinae, Spilopyrinae and some Sagrinae; occipital region rarely with transverse ridge (e.g. in Chlamisini) or stridulatory file (one or two files present in some Cassidinae). Frontal region sometimes with median groove and/or diagonal grooves forming an X (Figs 15C, 96F). Eyes usually strongly protuberant and not or shallowly emarginate (deeply emarginate in some Bruchinae, Sagrinae, Criocerinae, Cryptocephalinae and Spilopyrinae), without interfacetal setae. Antennal insertions usually exposed (concealed in some Cassidinae), widely separated as in Chrysomelinae (Fig. 96B) to very narrowly separated as in Galerucinae (Fig. 96C, K) and Cassidinae (Figs 15A, 96I, L); subantennal grooves present in *Clavicornaltica*, most Cryptocephalinae and Eumolpinae. Frontoclypeal suture present or absent. Labrum free and usually exposed (concealed by clypeus in some Cassidinae), slightly to strongly transverse and usually subtruncate to emarginate. Antennae almost always 11-segmented (10-segmented in *Leucispa* and *Psylliodes*, 9-segmented in *Nonarthra*, 8-segmented in *Octotoma* and *Uroplata*), usually moderately long, but rarely extending beyond elytral apices and often much shorter in Cassidinae, usually filiform, moniliform or incrassate (clavate), sometimes serrate (Clytrini, Chlamisini) or pectinate (some Bruchinae), and rarely capitate, with a distinct club of four or more segments (e.g. *Johannica*, *Nonarthra*, *Clavicornaltica* and some Cassidinae); scape rarely more than three times as long as pedicel. Mandible usually relatively short and broad, without dorsal setose cavity; apex unidentate to multidentate; incisor edge usually simple but sometimes with one or more teeth; mola and prostheca usually absent, but both present in Bruchinae, Sagrinae, *Donacia* and Sagrinae, and in some Galerucinae (including *Clavicornaltica*); mesal surface of mandibular base usually with well developed setose process in Chrysomelinae and Galerucinae. Maxilla with setose galea and lacinia, the latter without uncus; apical maxillary palpomere usually cylindrical to fusiform, but occasionally expanded and truncate or securiform (e.g. *Cheiloxena*, some Chrysomelinae). Mentum strongly transverse and apically emarginate; ligula usually undivided, but more or less bilobed in Bruchinae and Sagrinae; apical labial palpomere almost always cylindrical to fusiform. Gular sutures widely separated. Corpotentorium usually absent, present but narrow in Bruchinae, Sagrinae, *Donacia*, and some members of other subfamilies; cervical sclerites absent.

Pronotum ~0.18–1.4 times as long as wide, the most elongate in Criocerinae and *Diaphanops*, with base distinctly narrower than combined elytral bases in Sagrinae (Fig. 78D, F), *Donacia*, Criocerinae (Figs 76K, 96F) and some members of most other subfamilies; sides straight to strongly curved or sinuate, explanate in some Cassidinae; lateral pronotal carinae complete, incomplete or absent, with loss of carinae occurring in some Sagrinae, *Donacia* and Criocerinae, a few Cassidinae (e.g. *Aproida*, *Eurispa*, *Promecotheca*), *Cheiloxena*, and a few Eumolpinae; anterior angles usually not produced forward; posterior angles usually obtuse or right; posterior edge usually straight or evenly curved; disc with or without paired basal impressions and median longitudinal groove. Prosternum in front of coxae shorter than to longer than shortest diameter of coxal cavity, flat to moderately convex, rarely with paired carinae. Prosternal process almost always complete, gradually to abruptly expanded and apically broadly rounded to truncate, but sometimes narrowed posteriorly and subacute apically. Notosternal sutures usually complete, sometimes obscured anteriorly. Procoxae sometimes projecting well below prosternum, without concealed lateral extensions; trochantins often at least partly exposed, but completely concealed in some groups . Procoxal cavities usually slightly to strongly transverse (more or less circular in some Sagrinae, Cassidinae and Eumolpinae, Fig. 19J), narrowly to moderately widely separated (contiguous in some Bruchinae and Galerucinae), usually externally closed (open in *Leasia* and in some Bruchinae, Chrysomelinae and Galerucinae), with or without narrow lateral extensions, and internally closed.

Scutellar shield abruptly elevated or not, anteriorly simple, posteriorly acute to truncate or emarginate; mesoscutum without stridulatory file. Elytra ~0.7–3.75 times as long as wide and 1.25–5.65 times as long as pronotum, irregularly punctate or with distinct puncture rows (impressed striae in some Bruchinae and Cryptocephalinae), usually 10 in number, but reduced to 9 or fewer in various taxa, with or without scutellary striole; elytral apices usually meeting or almost meeting at the suture, but independently rounded in Bruchinae, some Galerucinae and Cryptocephalinae (independently acute in *Aproida*, exposing pygidium in Bruchinae and almost all Cryptocephalinae, exposing additional tergites in some Bruchinae and a few Galeruicinae; epipleura complete or incomplete. Mesoventrite separated by complete sutures from mesanepisterna, which are distinctly separated at midline; anterior edge on same plane as or different plane than metaventrite, usually with broadly transverse prothoracic rest (with paired procoxal rests in some Chrysomelinae and Galerucinae and in *Leasia*). Mesoventral cavity absent. Mesocoxae not conical and projecting, with exposed or concealed trochantins. Mesocoxal cavities narrowly to widely separated, usually subcircular to slightly transverse and not to slightly oblique, usually open laterally (partly closed by mesepimeron and often mesanepisternum) (closed by meeting of ventrites in some Cassidinae and Eumolpinae). Mesoventral process narrowly to broadly rounded or notched at apex, forming secondary mesocoxal articulations in some Eumolpinae (Fig. 25B); mesometaventral junction often monocondylic. Metaventrite convex; transverse (metakatepisternal) suture absent or very close to posterior edge of ventrite; discrimen usually moderately long; postcoxal lines rarely present; exposed portion of metanepisternum usually moderately elongate, short and broad in Bruchinae and some Sagrinae and Cryptocephalinae, very long and narrow in many Cassidinae and in some Crysomelinae, Galerucinae and Eumolpinae. Metacoxae contiguous to widely separated, not extending laterally to meet elytra; plates absent. Metendosternite usually with long stalk, long lateral arms and reduced or no laminae (laminae well developed in Sagrinae, Criocerinae and Spilopyrinae); anterior process usually short or absent (moderately long in most Cryptocephalinae); anterior tendons usually widely separated. Hind wing with short to moderately long apical field containing anterior and posterior remnants of RP and at least one distinct sclerite just beyond radial cell; radial cell usually well developed but reduced in several groups; inner posterobasal angle usually acute; cross-vein r3 usually strongly oblique and arising close to or fused at base with r4 (r3 well removed from r4 in some Chrysomelinae); basal portion of RP usually moderately long; R-M loop moderately broad; medial spur often ending before wing margin, where there may be a weak embayment; medial field usually with three free veins (four in most Bruchinae, Sagrinae, Spilopyrinae and Eumolpinae; reduced to two or one in various groups); medial fleck present in several Eumolpinae and a few Spilopyrinae, Chrysomelinae and Cassidinae; MP_{3+4} usually without with basal cross-vein (present in some Chrysomelinae and Eumolpinae); wedge cell often absent, but present and sometimes well developed in a few Bruchinae and some Sagrinae, Cassidinae, Spilopyrinae and Eumolpinae; anal lobe usually well developed with undivided AP; anal embayment rarely present. Legs short and stout to moderately long and slender; trochanterofemoral joint usually strongly oblique with base of femur separate from coxa (abutting coxa in some Sagrinae and in *Cheiloxena*); metafemora moderately to strongly inflated with one or more teeth along the inner edge in Bruchinae, Sagrinae, *Donacia* and some Criocerinae; inflated metatibiae also occur in members of the galerucine tribe Alticini, in connection with a jumping mechanism involving a metefemoral spring; tibiae often somewhat expanded at apex, which may be densely setose or bear a sharp ridge densely lined with setae; meso- and metatibiae with subapical setose emarginations in some Eumolpinae (Fig. 16J); tibial spurs usually absent (paired spurs present in *Carpophagus*, Criocerinae, and *Spilopyra*, and single spurs in *Donacia* and some Galerucinae). Tarsi usually pseudotetramerous (Fig. 96H, J), 5-5-5 in both sexes with tarsomere 4 reduced and more or less concealed at base of lobe on preceding tarsomere; tarsi 4-4-4 in Cassidinae with lobe on tarsomere 3 and tarsomere 4c absent; ventral lobes bearing patches of adhesive setae present on tarsomere 3 and sometimes on 1–3; setae bifid in Bruchinae, Sagrinae, *Donacia*, Criocerinae and Cassidinae; pretarsal claws simple, toothed or bifid; empodium not apparent.

Abdomen usually with 5 free ventrites; ventrites 1 and 2 connate in Cassidinae and Spilopyrinae; ventrites 4 and 5 connate in Cryptocephalinae; ventrite 1 without postcoxal lines; intercoxal process acute to broadly rounded or truncate; ventrite 5 in female with distinct circular impression in Cryptocephalinae. Abdominal terga usually lightly to moderately pigmented, except for pygidium (tergite VII); functional spiracles usually on segments I–VII, those on I–VI usually in pleural membrane, those on VII at lateral edges of tergite. Tergite VIII more or less concealed beneath tergite VII; sternite VIII in male usually reduced and lightly sclerotised, sometimes with median strut (spiculum relictum). Anterior edge of segment IX in male with or without distinct spiculum gastrale; tergite X usually more or less membranous. Aedeagus cucujiform; anterior edge of tegmen usually with short to long, ventral strut, which may be laterally compressed or bear a median endocarina; posterodorsal portion of tegmen (cap piece) well developed with fixed parameres (usually partly or entirely fused together) in Bruchinae, Sagrinae and *Donacia*, but usually membranous with no trace of parameres; penis subcylindrical to somewhat flattened and curved, its anterior edge rarely with short, paired struts. Segment VIII in female usually with fixed spiculum ventrale (articulated in Eumolpinae, absent in Chrysomelinae and Cryptocephalinae). Ovipositor short and broad to moderately elongate; proctiger often membranous, sometimes with lateral sclerotisations or baculi; paraprocts either lateral, dorsal or basal to coxites, with or without baculi and sometimes partly fused to coxites, which are usually undivided with styli reduced to a patch of setae or absent (subdivided with distinct styli in Sagrinae, Spilopyrinae and at least some Eumolpinae); vaginal pouches multilobed in Sagrinae, simple and elongate in Spilopyrinae and Eumolpinae; usually with curved, sclerotised, capsular spermatheca; rectum with kotpresse consisting of sclerotised plates in Crypocephalinae or setal patches in Spilopyrinae. [Lea 1904b; Crowson 1946; Kimoto and Gressitt 1982; Borowiec 1987; Reid 1990a, 1990b, 1991a, 1991c, 1992a, 1992b, 1993, 1994, 1995, 1999b, 2000b, 2001, 2003, 2006; Matthews and Reid 2002; Kingsolver 2002b; Riley *et al.* 2002, Chaboo 2007; Reid *et al.* 2009a, 2009b; Reid and Beatson 2010a, 2010b, 2010c, 2010d, 2011].

Larvae (Figs 33G, 97C–D, G–H)

Body usually elongate and more or less parallel-sided, with thorax expanded in Bruchinae and Sagrinae or abdomen expanded in Criocerinae and Chrysomelinae, broadly ovate in some Cassidinae, usually subcylindrical to slightly flattened, but strongly flattened in most Cassidinae and some leaf-mining Galerucinae; body strongly curved ventrally (C-shaped) in Bruchinae, Sagrinae, *Donacia* and Cryptocephalinae; all surfaces more or less lightly sclerotised, except for head capsule, mouth-frame and mandibles, often a protergal plate, sometimes scattered small setiferous sclerites with or without projecting scoli, and occasionally tergum IX (some Galerucinae) or terga VIII and IX (Spilopyrinae, some Cassidinae); vestiture usually consisting of long and short, simple setae. Head usually protracted and hypognathous, but prognathous and more or less retracted in Bruchinae, Sagrinae, *Donacia*, some Cassidinae and a few Galerucinae; posterior edge of head capsule distinctly emarginate in some Cassidinae and Galerucinae; epicranial stem usually present, moderately long in Bruchinae and Sagrinae; frontal arms V-shaped and usually contiguous at base (well separated at base in some Cassidinae and Gallerucinae; median endocarina present or absent, sometimes extending to clypeus. Stemmata on each side six in *Donacia*, Criocerinae, Cassidinae and Cryptocephalinae, five in *Donacia* and Spilopyrinae, two at most but usually one or none in Eumolpinae, one in Galerucinae and none in Sagrinae or Bruchinae. Antennae short, 3-segmented in Bruchinae, Sagrinae, *Donacia*, Criocerinae, Cassidinae, Chrysomelinae and Cryptocephalinae, 2-segmented in Eumolpinae and 1-segmented in Galerucinae; conical sensorium either apical or preapical and about as long as reduced 3rd antennomere. Frontoclypeal suture present or absent; labrum usually free, but fused to heads capsule in Cryptocephalinae. Mandibles usually bidentate, with a variable number of subapical teeth, but broadly palmate with three or four more or less equal teeth in Criocerinae, Cassidinae, Chrysomelinae (Figs 40D, 41I), Galerucinae and Spilopyrinae; mesal surface of mandibular base usually simple, but occasionally with penicillus; mola never present. Ventral mouthparts slightly to strongly retracted; maxillary articulating area often absent; cardo slightly to strongly oblique, undivided; stipes short and broad to moderately elongate; apex almost always consisting of fixed, apically rounded and setose mala (rarely with galea and lacinia, rounded in a few Galerucinae or blade-like or styliform in *Donacia*); maxillary palp usually 3-segmented with or without distinct palpifer (2-segmented in some Bruchinae, Sagrinae and Cassidinae, 1-segmented in some Bruchinae and a few Cassidinae). Labium usually consisting of prementum and postmentum (occasionally with mentum and submentum separated); ligula usually undivided, occasionally absent; labial palps usually 2-segmented (1-segmented in *Donacia*, Criocerinae, Eumolpinae, most Cassidinae and a few Bruchinae, absent in most Bruchinae), usually well separated. Hypopharyngeal bracon and hypopharyngeal sclerome absent. Hypostomal rods often present; ventral epicranial ridges absent. Gular region absent.

Prothoracic tergum not to slightly longer than meso- or metaterga, often with distinct plate. Thoracic and abdominal terga with lateral processes in some Cassidinae and Galerucinae. Legs usually moderately long and stout, but slender in Cryptocephalinae and reduced or absent in Bruchinae, Sagrinae, and some Cassidinae and Galerucinae; usually moderately to widely separated, usually 5-segmented, but often with trochanter partly fused to femur, 4-segmented in *Donacia* and several Cassidinae, 1- or 2-segmented in Uroplatini and absent in Promecothecini and some Bruchinae; tibiotarsus with paronychial appendix in Criocerinae, Cassidinae, Chrysomelinae, Galerucinae and Spilopyrinae; pretarsus long and slender in Bruchinae, Sagrinae, Cryptocephalinae and Eumolpinae, but shorter and broad at base in *Donacia*, Criocerinae, Cassidinae, Chrysomelinae, Galerucinae and Spilopyrinae; pretarsal setae one or none. Thoracic egg bursters present in first-instars of Chrysomelinae, Galerucinae, Spilopyrinae, Cryptocephalinae and Eumolpinae. Abdomen usually more than twice as long as thorax (less so in many Cassidinae, but much longer in some

Galerucinae); abdominal terga often transversely divided forming two or three plicae; most sterna in Eumolpinae with ventral or ventrolateral lobes and with patches of relatively long setae. Abdominal egg bursters present on segments I–III in first-instars of Sagrinae and Spilopyrinae, I–II in those of Chrysomelinae and Eumolpinae, and I only in Bruchinae, Criocerinae and Cryptocephalinae. Tergum IX usually shorter than VIII, simple, and not extending ventrally, so that segment X is terminal; tergum well developed, somewhat sclerotised and extending ventrally, so that segment X is ventral in Spilopyrinae, most Galerucinae and Cassidinae, and tergum IX more or less fused to VIII in some Cassidinae; tergum IX or terga VIII and IX well sclerotised and sometimes forming circular, flattened or concave plate in Spilopyrinae, with minute paired urogomphi in some Galerucinae, posteriorly-projecting urogomphi in some Cassidinae (e.g. *Eurispa*, *Brontispa*) long, approximate, dorsally-projecting urogomphi (forming slender caudal processes on which exuviae from previous instars are attached) in some Cassidinae and a single, narrow, apically acute process in *Aproida*; segment X usually complexly lobed in Eumolpinae; anal region usually posteriorly oriented, but ventrally oriented in Spilopyrinae, some Cassidinae and most Galerucinae. Thoracic and anterior abdominal spiracles usually annular (annular–biforous in many first-instars, annular-uniforous in some Bruchinae, annular–multiforous in Hispini, annular–cribriform in some Cryptocephalinae and non-functional in Donaciinae), placed at ends of spiracular tubes in some Cassidinae; spiracles on segment VIII usually about the same size as others on abdomen and facing laterally, but dorsally placed and facing posteriorly in Spilopyrinae and many Cassidinae, incorporated into a pair of spine-like processes (Fig. 52R) in *Donacia*, or reduced in Cassidini. [Böving and Craighead 1931; Maulik 1931, 1932, 1941; Varley 1939; Gressitt 1957, 1963; Monteith 1970; Takizawa 1971, 1972, 1976; Kasap and Crowson 1976; Medvedev and Zaitzev 1978; Houston 1982; LeSage 1982; Costa *et al.* 1988; Lawson 1991; Lee 1991, 1992; Pfaffenberger 1991; Reid 1991b, 1993, 1995, 2000b; Hawkeswood and Takizawa 1997; Riley *et al.* 2002; Chaboo and Nguyen 2004; Chaboo 2007; Reid and Beatson 2010b, 2010c, 2011].

Classification and World Distribution

The family Chrysomelidae is a large and cosmopolitan family containing ~32 500 species in 2114 genera. With the exclusion of Megalopodidae and Orsodacninae, the family may be divided into twelve subfamilies, ten of which occur in Australia. The exotic subfamily Synetinae (Edwards 1953; Lee 1990a) is Holarctic in distribution, while Lamprosomatinae (Achard 1914; Monrós 1956) occurs in the Neotropical and Oriental regions, and in southern Africa, Europe, Japan, and New Caledonia. [Seeno and Wilcox 1982; Reid 1995; Rilet *et al.* 2003].

Australian Fauna

Chrysomelidae are extremely varied in adult body form, but they are often subglabrous and brightly coloured. Antennae are variable, but usually filiform, very rarely enlarged apically to form an elongate club of four or more segments, never inserted on prominences and usually not extending beyond the middle of the body. The head often has grooves or prominences, the labrum is visible, maxillae and labium are well developed with flexible palps, the head may be produced to form a short rostrum, which is very rarely slightly longer than wide, the lateral pronotal carinae are often complete, but may be incomplete or absent, distinct tibial spurs are not common (except in Galerucinae), the first three tarsomeres have ventral pads of simple, spatulate or bifid setae (Fig. 96H), the fourth tarsomere is either minute and concealed at the base of the preceding one or fused with the terminal tarsomere, pretarsal claws are usually cleft or appendiculate, and basal abdominal ventrites are rarely solidly fused as in most weevils. Larvae are also highly variable but usually lightly sclerotised, except for the head and sometimes numerous small dorsal plates or protuberances, and usually with well developed legs. The head is often hypognathous, with a median endocarina, mandibles usually have three or more apical teeth, the labrum is free except in Cryptocephalinae, and urogomphi are rarely present on tergum IX.

All Chrysomelidae are apparently phytophagous, the adults feeding externally and the larvae externally or internally on a wide variety of higher plant tissues comprising roots, foliage, herbaceous stems, growing tips, leaves, flowers, pollen, fruits and seeds. Except for larvae of Cryptocephalini, which specialise on dead leaves, most chrysomelids feed on living tissue, and few, if any, occur in wood.

The subfamily **Donaciinae** is represented in Australia by a single species, *Donacia* (*Cyphogaster*) *australasiae* Blackburn, which occurs in northern Australia and New Guinea (Reid 1993). Species of *Donacia* are associated with water lilies (Nymphaceae), sacred lotus (Nelumbonaceae) and marsh worts (Menyanthaceae); larvae have the eighth spiracles located in sharp spines, which can pierce plant roots or stems to obtain air (Varley 1939). **Sagrinae** have their greatest diversity in Australia, being represented elsewhere only by the genus *Sagra* in Africa and Asia, *Rhagiosoma* in Madagascar and *Atalasis* in Argentina (Monrós 1960). *Sagra papuana* Jacoby (Fig. 78F) occurs in northern QLD, while the remaining genera are widely distributed across the southern and western part of the continent. Larvae of *Sagra* are known to feed in herbaceous stems of various plants, including wild yams (Discoreaceae), the larvae of *Mecynodera coxalgica* (Boisduval) (Fig. 79D) were found in prepupal cocoons in soil at the base of a *Lomandra* (Xanthorrheaceae) and first instars of an unknown sagrine from WA were found in a multi-chambered mud egg case attached to a shrub. Sagrines are highly varied in form: *Carpophagus* species resemble giant seed weevils (Bruchinae), *Polyoptilus* are elongate with long antennae and resemble some Cerambycidae, *Diaphanops* (Fig. 10E) have a weevil-like rostrum, and *Megamerus* includes the largest species (more than 30 mm in length). Sagrines are noted for having enlarged metafemora bearing teeth or spines; however, they cannot jump. The Australian fauna of **Bruchinae** is not rich, and most endemic species belong to the genus *Bruchidius* (Fig. 96E), with the primarily Asian *Caryedon gonagra* (Fabricius) (Fig. 76I) extending into Qld (Delobel *et al.* 2003). Bruchinae have enlarged and spinose metafemora as in Sagrinae, but they are smaller and more compact beetles with projecting procoxae, a declivous head, and an exposed pygidium. Unlike sagrines, they are specialised seed-feeders as larvae. Introduced bruchines include *Bruchus pisorum*, a cosmopolitan pest of living beans and peas, and *Acanthoscelides obtectus* and several *Callosobruchus* species, which damage stored beans. Two other *Acanthoscelides* have been introduced from Mexico to control the giant sensitive plant. The **Criocerinae** are also poorly represented in Australia with four genera and 18 species. Criocerinae resemble small sagrines in having the deep head grooves and a narrow prothorax without lateral carinae, but they differ in having conical procoxae with concealed trochantins and a shorter basal abdominal ventrite. Most Criocerinae feed externally on monocotyledonous plants, such as Liliaceae, Dioscoraceae and Smilacaceae, with *Oulema* species specialising on Poaceae; however, some *Lema* species (Fig. 78C) feed on dicots, such as Solanaceae (Schmitt 1988), and some *Lilioceris* have been recorded feeding on Cycadaceae (Hawkeswood 1992; Forster and Machin 1994). *Stethopachys formosa* Baly (Fig. 76K) has become a pest of commercial orchids (Gough *et al.* 1994; larvae are covered in slime and pupation occurs in a cocoon composed of foam-like pellets produced by air bubbles being introduced into the slime, which then dries. The accidentally introduced criocerines, *Lema daturaphila* Koegen & Goeden from North America and *L. bilineata* Germar from South America are potential pests of various solanaceous plants (Koegen and Goeden 1970; Franzmann 1978; Stevens *et al.* 2010). The European *Oulema rufocyanea* (Suffrian) has been intercepted by AQIS but is not established here. An undescribed species of *Crioceris* from South Africa has been considered as a possible candidate for the control of *Asparagus asparagoides* in Australia (Witt and Edwards 2002).

The subfamily **Cassidinae,** used here in the broad sense suggested by Crowson (1955), Chen (1985), Borowiec (1995), Reid (1995), Chaboo (2007) and others, is relatively well represented by more than 20 genera representing eight tribes. Typical cassidines or tortoise beetles are characterised by their broadly ovate shape and by having elytra and prothorax broadly explanate, the latter concealing the head; however species of *Notosacantha* have a similar body form with an exposed head. A common Australian tortoise beetle genus is *Aspidimorpha* (Fig. 78I), the species of which have a strong preference for *Ipomoea* (Convolvulaceae) (Hawkeswood 1982). *Aproida balyi* is an unusual species that feeds externally on the leaves of *Eustrephus* (Philesiaceae); adults are green in colour (Fig. 10D), matching the colour of the leaves, but as in many other chrysomelids, the colour fades in dead specimens (Fig. 78B). Pupation takes place on the leaf surface and the pupa is suspended from the leaf by the withered larval skin (Monteith 1970; Samuelson 1989). Several endemic Cassidinae with the hispine body type mine the leaves of *Pandanus* (*Promecotheca*) or grasses (*Hispellinus*, Fig. 96L); others occur in leaf bases or among developing leaflets of palms (*Brontispa*), sedges and grasses (*Eurispa*) or other monocotyledonous plants. *Notosacantha dorsalis* is a miner in *Acacia* leaves. The exotic leaf-miners *Octotoma scabripennis* and *Uroplata* species were introduced for the control of the dicotyledonous weed *Lantana camara* (Verbenaceae).

Chrysomelinae is the largest subfamily in Australia with 46 genera (42 native) and ~750 species. Most Chrysomelinae have a short, broad, convex body, widely separated antennal insertions (unlike either Donaciinae or Galerucinae), well developed lateral pronotal carinae and transverse procoxae with exposed trochantins; there is also a broad, membranous anteclypeus, which is lacking in Eumolpinae and Cryptocephalinae. Chrysomeline adults and larvae feed externally on foliage and the latter are often armed with defence glands. The most important Australian group is the Paropsina, including several large and common genera, such as *Paropsis*, *Paropsisterna*, *Trachymela* and *Peltoschema*. The larval defence glands, which are located on tergum VIII, secrete hydrogen cyanide (Moore 1967a). Adult paropsines often have brilliant, species-specific colour patterns (Fig. 10B), but these quickly fade in dead specimens. Many paropsines are serious defoliators of eucalypts. *Paropsisterna bimaculata* (Olivier) is a notorious forest pest in TAS, *Paropsis charybdis* Stål has been introduced into New Zealand, where it causes severe damage to gum plantations, and *Trachymela tincticollis* (Blackburn) has become a pest in South Africa (Carne and Taylor 1978; De Little 1979; Tribe and Cillié 1985). Species of *Peltoschema* often specialise on the flowers of *Acacia* (New 1981). *Novacastria nothofagi* Selman feeds exclusively on young leaves of *Nothofagus moorei* in northern NSW (Selman and Lowman 1983). Adults and larvae of *Lamprolina aeneipennis* (Boisduval) (Figs AP13D–E) attack the leaves of *Pittosporum*, while the related *Johannica gemellata* (Westwood) feeds on the *vine Pandorea pandorana* (Reid 1991b). The European *Chrysolina hyperici* and *C. quadrigemina* were introduced into Australia to control St. John's wort, the North American *Zygogramma bicolorata* was introduced for ragweed (*Ambrosia artemisifolia*) control and another American chrysomeline, *Calligrapha pantherina*, was introduced to control a tropical pasture weed (*Sida* species). Other chrysomeline genera include *Callidemum* (Fig. 78E), *Calodema*, *Chalcolampra* and *Phyllocharis*. **Galerucinae** (Figs 78G, 96C, K) is used here in the broad sense to include those forms previously included in Alticinae and known as flea beetles because of their ability to jump using the enlarged metafemora (Furth 1982; Furth *et al.* 1983). Alticine larvae are not separable from those of other galerucines, and transitional forms occur among the adults as well (Lingafelter and Konstantinov 2000). The unusual endemic genus *Microdonacia* was placed in Eumolpinae by Monrós (1958) but moved to Alticinae by Jolivet (1968) and retained there by Reid (1992a). The group represented in Australia by 65 genera, but most diverse in tropical and subtropical forests. Galerucinae are usually more elongate and softer-bodied than chrysomelines, as well as having approximate antennal insertions and enlarged metafemora in some genera. The genus *Clavicornaltica* from tropical Queensland is unusual in having strongly clavate antennae. Galerucines have varied habits, and larvae may feed externally on foliage but often occur in soil and feed on roots. *Monolepta australis* attacks citrus fruit, maize and stone fruit, while adults of *Aulacophora hilaris* and *A. abdominalis* feed on foliage and flowers of cucumber, melon and pumpkin. Flea beetles often cause 'short-hole' damage to leaves. *Halticorcus platycerii* specialises on staghorn and elkhorn ferns. The elm leaf beetle, *Pyrrhalta luteola*, was accidentally introduced into VIC and is not a serious pest of elm (*Ulmus* species). Two European species of *Longitarsus* have been introduced to control exotic weed pests: *L. albineus* for heliotrope (*Heliotropium europaeum*) and *L. jacobeae* for ragwort (*Senecio jacobaea*). The South American species *Agasicles hygrophila* was imported to control alligator weed (*Alternanthera philoxeroides*).

Spilopyrinae (Reid 2000) includes the Australian genera *Allsortsia*, *Richmondia* and *Cheiloxena*, *Spilopyra* (Fig. 76F) and *Macrolema* from Australia and New Guinea, *Hornius* Fairmaire and *Stenomela* Erichson from Chile and *Bohumiljania* from New Caledonia. Spilopyrinae and Eumolpinae usually differ from chrysomelines in having the head more strongly declined with a broadly emarginate postclypeus without or with a reduced, centrally located anteclypeus, the prothorax distinctly narrower at base than the combined elytral bases and usually the procoxae globular with concealed trochantins. Spilopyrine adults resemble those of Eumolpinae and spilopyrine genera were formerly placed in four different eumolpine tribes. Reid (1995) was the first to place these genera in a single tribe, and in a later work (Reid 2000) he elevated the group to subfamily level. In his revised subfamily key, spilopyrines were distinguished from eumolpines by having short temples behind the eyes, broad metendosternal laminae, hind wing with a distinct wedge cell, meso- and metatibiae with two spurs; spatulate adhesive tarsal setae of the papillate type, ventrites 3 and 4 fused together (at least in all Old World species), and female rectum with a kotpresse. Larvae of Spilopyrinae feed externally but apparently shelter within cavities in the plant tissue and are usually protected by a well sclerotised circular plate formed from tergum IX or terga VIII and IX. **Eumolpinae** contains 26 Australian genera occurring throughout the continent in a variety of habitats with the largest numbers of species in genera such as *Rhyparida*, *Edusella* (Fig. 78J), *Geloptera* and *Eboo*. Eumolpinae adults are often shiny and metallic but sometimes dull in colour and occasionally with scale-like setae. Some of the larger genera are *Edusella*, *Geloptera*, *Eboo* and *Rhyparida*, the last of which contains economically important pests of sugarcane and other crops. Eumolpine larvae always live in the soil and feed on roots; they are invariable lightly sclerotised and often have ventral ampullae or appendages to assist movement through the soil and may have a series of lobes surrounding the anal region.

The **Cryptocephalinae** (Figs v78H, 96D) contains fewer genera but some of them (e.g. *Aporocera*, *Cadmus*, *Ditropidus*) contain large numbers of species. Chlamisini are represented by one or two native and one introduced species of *Chlamisus* in QLD (Reid 1991a), Clytrini by two species of *Aetheomorpha* in northern Australia, Pachybrachini by two introduced species of *Metallactus*, and Cryptocephalini by about eleven genera and more than 500 species, which are particularly abundant in open eucalypt forests (Reid 1990b). Within the Cryptocephalinae, *Chlamisus mimosae* Karren was introduced into Australia for the control of *Mimosa pigra. Adults* of Cryptocephalini feed primarily on foliage of *Eucalyptus* and *Acacia* and their larvae (Fig. 97D) are distinctive among chrysomelids in feeding on dead leaves accumulated on the ground and living within a close-fitting portable case (Fig. 97C) constructed of faeces and debris; pupation occurs within this case. The genus *Leasia* from WA has a greatly enlarged head in the male and was removed from the family by Monrós (1951); but according to Crowson (1955) and Reid (1990b), the genus has all the features of the subfamily and apparently occupies a basal position in the Cryptocephalini, along with *Atenesus*, *Semelvillea* and *Platycolaspis* (Reid 1990b, 1991c, 1994). [Lea 1904b, 1915, 1920; Maulik 1931, 1932; Gressitt 1957; New 1981; Reid 1990a, 1990b, 1991a, 1991c, 1992a, 1992b, 1994, 1999b, 2000b, 2001, 2003; Monteith 1991; Konstantinov 1995; Matthews and Reid 2002; Konstantinov and Duckett 2005a, 2005b; Reid *et al.* 2009a, 2009b; Reid and Beatson 2010a, 2010b, 2010c, 2010d, 2013].

Australian Subfamilies, Tribes and Genera

Sagrinae: *Ametalla* Hope, *Carpophagus* W.S. Macleay, *Coolgardica* Blackburn, *Diaphanops* Schönherr, *Duboulaia* Baly, *Mecynodera* Hope, *Megamerus* W.S. Macleay, *Neodiaphanops* Blackburn, *Polyoptilus* Germar, *Pseudotoxotus* Blackburn, *Sagra* Fabricius.

Bruchinae: Amblycerini: *Spermophagus* Schoenherr. **Bruchini:** *Acanthoscelides* Schilsky, *Algarobius* Bridwell, *Bruchidius* Schilsky, *Bruchus* Linnaeus, *Callosobruchus* Pic, *Mimosestes* Bridwell, *Parasulcobruchus* Anton, *Penthobruchus* Kingsolver. **Pachymerini:** *Buburra* Reid & Beatson, *Caryedon* Schoenherr, *Caryotrypes* Decelle.

Donaciinae: *Donacia* Fabricius.

Criocerinae: Criocerini: *Crioceris* Muller, *Lilioceris* Reitter. **Lemini:** *Lema* Fabricius. *Stethopachys* Baly.

Cassidinae: Aproidini: *Aproida* Pascoe. **Eurispini:** *Eurispa* Baly, *Leucispa* Chapuis. **Cryptonychini:** *Brontispa* Sharp, *Ceratispa* Gestro, *Plesispa* Chapuis. **Coelaenomenoderini:** *Heterrhachispa* Gressitt.. **Promecothecini:** *Promecotheca* Blanchard. **Hispini:** *Dactylispa* Weise, *Hispellinus* Weise. **Uroplatini:** *Octotoma* Dejean, *Uroplata* Chevrolat. **Notosacanthini:** *Notosacantha* Chevrolat. **Cassidini:** *Aspidimorpha* Hope, *Austropsecadia* Hincks, *Cassida* Linnaeus, *Charidotella* Weise, *Emdenia* Spaeth, *Laccoptera* Boheman, *Meroscalsis* Spaeth, *Thlaspidula* Spaeth.

Chrysomelinae: *Acernoides* Jacoby, *Alfius* Reid, *Aeratocerus* Blackburn, *Callidemum* Blanchard, *Calligrapha* Chevrolat, *Calomela* Hope, *Canobolas* Reid *et al.*, *Chalcolampra* Blanchard, *Chalcomela* Baly, *Chrysolina* Motschulsky, *Cyclonoda* Baly, *Deuterocampta* Chevrolat, *Diacosma* Weise, *Dicranosterna* Motschulsky, *Ethomela* Lea, *Eugastromela* Lea, *Eulina* Baly, *Ewanius* Reid, *Faex* Weise, *Geomela* Lea, *Gibbiomela* Daccordi, *Grammicomela* Lea, *Hysmatodon* Reid, *Johannica* Blackburn, *Lamprolina* Baly, *Novacastria* Selman, *Oomela* Lea, *Palaeomela* Daccordi, *Paropsides* Motschulsky, *Paropsimorpha* Lhoste, *Paropsis* Olivier, *Paropsisterna* Motschulsky, *Peltoschema* Reitter, *Philhydronopa* Weise, *Phola* Weise, *Phyllocharis* Dalman, *Plagiodera* Chevrolat, *Platymela* Baly, *Poropteromela* Lea, *Promechus* Boisduval, *Pterodunga* Daccordi, *Rhaebosterna* Weise, *Sphaerotritoma* Arrow, *Strumatophyma* Baly, *Tinosis* Weise, *Trachymela* Weise, *Zygogramma* Chevrolat.

Galerucinae: *Adoxia* Broun, *Altica* Geoffroy, *Amphimela* Chapuis, *Anoides* Weise, *Aphthona* Chevrolat, *Argopistes* Motschulsky, *Arsipoda* Erichson, Atysa Baly, *Aulacophora* Chevrolat, *Bellaltica* Reid, *Candezea* Chapuis, *Cassena* Weise, *Chabria* Jacoby, *Chaetocnema* Stephens, *Clavicornaltica* Scherer, *Crepidodera* Chevrolat, *Cydippa* Chapuis, *Docema* Waterhouse, *Ellopidia* Hincks, *Epitrix* Foudras, *Galerucella* Crotch, *Goweria* Lea, *Halticorcus* Lea, *Hoplosaenidea* Laboissière, *Hoplostines* Blackburn, *Hyphaltica* Blackburn, *Hyphalticoda* Oke, *Licyllus* Jacoby,*Longitarsus* Berthold, *Luperodes* Motschulsky, *Luperomorpha* Weise, *Malacorhinus* Jacoby, *Manobia* Jacoby, *Menippus* Clark, *Microdonacia* Blackburn, *Monolepta* Chevrolat, *Monomacra* Chevrolat, *Neodera* Duvivier, *Neorupilia* Blackburn, *Nisotra* Baly, *Nonarthra* Baly, *Nototrichaspis* Hincks, *Oedionychus* Berthold, *Oides* Weber, *Opisthopygme* Blackburn, *Orthaltica* Crotch, *Paradibolia* Baly, *Pedethma* Weise, *Pepila* Weise, *Phyllotreta* Chevrolat, *Platycepha* Baly, *Podontia* Dalman, *Polysastra* Shute, *Poneridia* Weise, *Prasyptera* Baly, *Psylliodes* Berthold, *Pyrrhalta* Joannis, *Rupilia* Clark, *Sastra* Baly, *Sittacella* Weise,

Sphaeroderma Stephens, *Sutrea* Baly, *Synodita* Chapuis, *Trirhabda* LeConte, *Xenidea* Baly, *Xuthea* Baly.

Spilopyrinae: *Allsortsia* Reid & Beatson, *Cheiloxena* Baly, *Macrolema* Baly, *Richmondia* Jacoby, *Spilopyra* Baly.

Eumolpinae: Typophorini: *Basilepta* Baly, *Cleorina* Lefevre, *Deretrichia* Weise, *Pagria* Lefevre, *Rhyparida* Baly. **Colasposomini:** *Colasposoma* Laporte. **Eumolpini:** *Abirus* Chapuis, *Agetinus* Lefevre, *Alittus* Chapuis, *Chrysolampra* Baly, *Cleptor* Lefevre, *Colaspoides* Laporte, *Cudnellia* Blackburn, *Edusella* Chapuis, *Edusoides* Blackburn, *Eucolaspinus* Lea, *Geloptera* Baly, *Hypoderes* Lefevre, *Ocnida* Lefevre, *Rhinobolus* Blackburn. **Adoxini:** *Demotina* Baly, *Eboo* Reid, *Neocles* Chapuis, *Neocloides* Jacoby, *Pseudocolaspis* Laporte, *Scelodonta* Westwood.

Cryptocephalinae: Chlamisini: *Chlamisus* Rafinesque, **Clytrini:** *Aetheomorpha* Lacordaire. **Pachybrachini:** *Metallactus* Suffrian. **Cryptocephalini:** *Aporocera* Saunders, *Aprionota* Maulik, *Atenesus* Weise, *Cadmus* Erichson, *Diachus* LeConte, *Ditropidella* Reid, *Ditropidus* Erichson, *Leasia* Jacoby, *Melatia* Reid, *Platycolaspis* Jacoby, *Semelvillea* Reid.

CURCULIONOIDEA

The Curculionoidea, previously known as the Rhynchophora, comprises the weevils, adults of which are generally characterised by the elongation of the anterior part of the head to form a rostrum and the accompanying mouthpart modifications, such as fusion of the labrum to the clypeus, reduction of mandible size, fusion of galea and lacinia, formation of short, rigid palps (Ting 1936). Exceptions occur in some basal groups, such as Nemonychidae and Anthribidae, or as reversals in some derived groups within Curculionidae. The possession of pseudotetramerous tarsi (secondarily reduced in some groups) and a reduced type of cucujiform aedeagus in the adult and loss of a mandibular mola in the larva are shared with members of the Chrysomeloidea, but the mouthpart modifications of the adult combined with the presence of a hypopharyngeal bracon in the larva distinguish weevils from chrysomeloids. Adult weevils usually have a more or less rigid body, with posteriorly closed procoxal cavities, mesocoxal cavities laterally closed by the meeting of the meso- and metaventrite, globular procoxae with concealed trochantins, and connation of two or more ventrites in almost all taxa. The antennae are almost always distinctly clubbed and usually have a long scape that may fit into a groove (scrobe) at the side of the rostrum. Larval weevils are almost always legless, but minute legs occur in some Nemonychidae and many Anthribidae and Brentidae, while well developed legs are present in the Palaearctic nemonychid genus *Nemonyx* Redtenbacher and in the Australian genus *Car* (Caridae). Phytophagous habits (including pollen-feeding in Nemonychidae and oxocorynine Belidae) are almost universal in the superfamily, notable exceptions being the mycophagous habits of some Anthribidae, coprophagy in *Tentegia* (see Curculionidae), and entomophagy in the Neotropical *Ludovix fasciatus* (Gyllenhal) (Zwolfer and Bennett 1969).

The classification of Curculionoidea has received considerable attention in the past 20 years beginning with three general works by Thompson (1992), Zimmerman (1991, 1992, 1993, 1994a, 1994b) and Kuschel (1995), the last one containing the first cladistic analysis of the group, and several papers dealing with particular character systems (Calder 1989, 1990a, internal anatomy; May 1993, 1994, larvae; Lyal 1995, ventral head structures; Howden 1995, ovipositors; Zherikhin and Gratshev 1995, the hind wing; Lyal and King 1996, elytro-tergal striculation; Morimoto and Kojima 2003, adult and larval head; Lyal *et al.* 2006, sclerolepidia). This was followed by several more cladistic analyses of Curculionoidea or particular subgroups by Marvaldi 1997, Curculionidae and some related groups, mainly larvae; Marvaldi and Morrone 2000, Curculionoidea, adults and larvae; Marvaldi *et al.* 2002, Curculionoidea, adults, larvae, 18S rDNA; Marvaldi 2005, oxycorynine Belidae, mainly larvae; Marvaldi *et al.* 2006, Belidae, adults and larvae; Marvaldi *et al.* 2009, Curculionoidea and Chrysomeloidea, 18S and 28S rDNA; McKenna *et al.* 2009, Curculionoidea, DNA sequence data (four nuclear and two mitochrondrial genes); and Kuschel and Leschen 2010, Nemonychidae, adults and larvae). The classification of weevil families and subfamilies has varies as a result of these conflicting sets of evidence, but there has usually been general agreement at least at the family level (Kuschel 1995; Morrone 1997; Alonso-Zarazaga and Lyal 1999, 2002; Oberprieler *et al.* 2007; Oberprieler 2010). Relationships within the family Curculionidae are still not well understood, and further morphological and molecular work is in progress. [Kasap and Crowson 1977; Wibmer and O'Brien 1986; Alonso-Zarazaga and Lyal 1999, 2002, 2009].

111. NEMONYCHIDAE

Adults (Fig. 98A)

Length 2.0–4.3 mm. Body ~2.2–2.8 times as long as wide and more or less parallel-sided, slightly flattened, yellow to dark brown in color, but often bicolored, with pattern of dark markings on elytra and occasionally pronotum; clothed with fine, decumbent and hairs, usually with scattered erect hairs. Head not or slightly declined at base, but vertex sometimes strongly declined, so that rostrum extends ventrally (e.g. in *Basiliorhinus*); head behind eyes not constricted posteriorly to form neck; occipital region without transverse ridge or stridulatory file. Frontal region without median groove or endocarina. Eyes slightly to strongly protuberant, not emarginate, finely or coarsely facetted, without or with short, fine interfacetal setae. Antennal insertions concealed or barely exposed from above, usually located close to apex of rostrum (located near middle of rostrum in *Basiliorhinus* and *Basiliogeus*); subantennal grooves (scrobes) well developed or reduced. Frontoclypeal suture absent; clypeal region produced forward to form rostrum, which may be only slightly longer than wide (as in *Bunyaeus*) but is usually much longer than wide (as much as 9 times as long as wide in *Basiliogeus*). Labrum free, exposed, on same plane as and fitting tightly against clypeus, slightly transverse, broadly rounded but sometimes finely denticulate at apex. Antennae 11-segmented, moderately long, with weak, loose, 3-segmented club; scape elongate but not much longer than pedicel. Mandible usually moderately elongate, slender and distinctly curved mesally, unidentate; incisor edge simple or with single tooth on one mandible; molae reduced and sometimes asymmetrical; reduced, setose prostheca sometimes present on one mandible; inner edge of mandibular base with moderately short pubescent process. Maxilla with short, setose galea and lacinia, the latter without

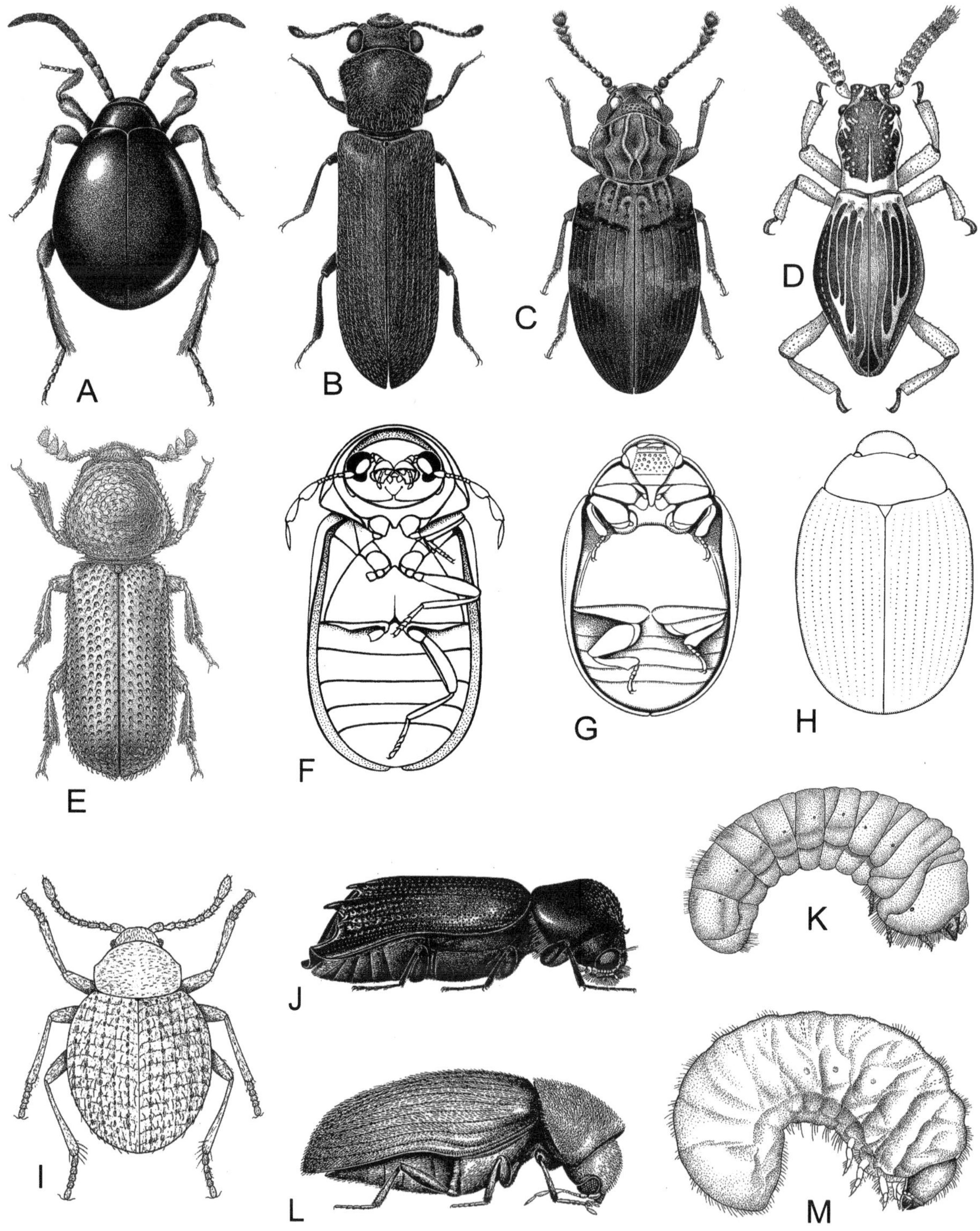

Fig. 81. A–E, Adult, dorsal. A, *Gibbium psylloides* (Czenpinski) (Ptinidae: Ptininae); B, *Lyctus brunneus* (Stephens) (Bostrichidae: Lyctinae); C, *Nothoderodontus darlingtoni* Lawrence (Derodontidae); D, *Sarothrias lawrencei* Lőbl & Burckhardt (Jacobsoniidae); E, *Rhyzopertha dominica* (Fabricius) (Bostrichidae: Dinoderinae), from Busquet 1990, © Agriculture and Agri-Food Canada. F, *Stegobium paniceum* (Linnaeus) (Ptinidae: Anobiinae), adult, ventral. G–H, *Nosodendron australicum* Lea (Nosodendridae): G, adult, ventral; H, same, dorsal. I, *Neoptinus marginicollis* (Bellés & Lawrence) (Ptinidae: Ptininae), adult, dorsal. J, *Mesoxylion cylindricus* (Macleay) (Bostrichidae: Bostrichinae), adult, lateral. K, *Dinoderus minutus* (Fabricius) (Bostrichidae: Dinoderinae), larva, lateral, from Costa *et al.* 1988, © Universidade de São Paulo, Brazil. L, *Stegobium paniceum* (Linnaeus) (Ptinidae: Anobiinae), adult, lateral. M, Same, larva, lateral.

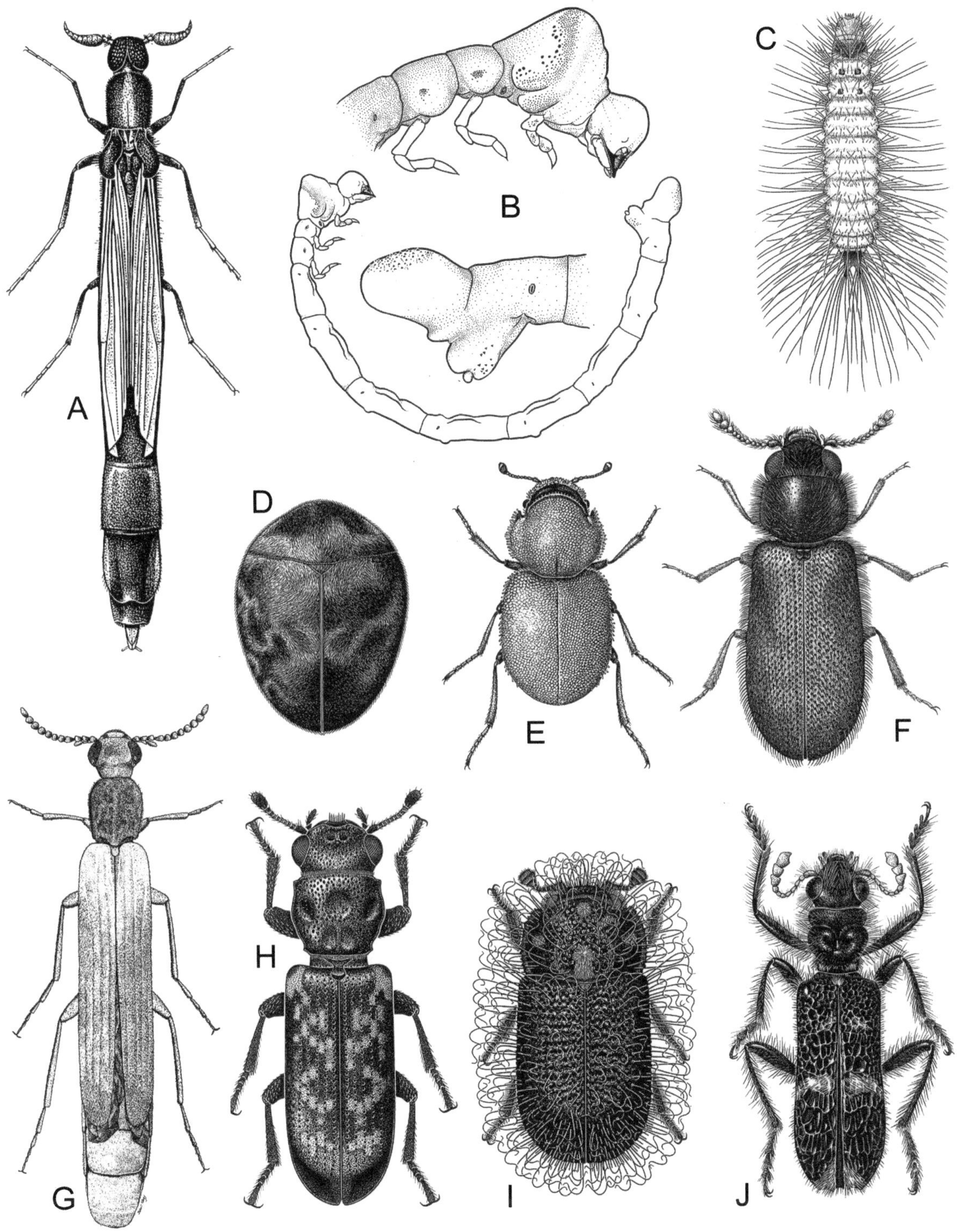

Fig. 82. A–B, Lymexylidae (Atractocerinae): A, *Atractocerus crassicornis* Clark, adult, dorsal; B, *Atractocerus* sp., larva, lateral. C, *Lemidia* sp. (Cleridae: Hydnocerinae), larva, dorsal. D–J, Adult, dorsal: D, *Parentonium australicum* Crowson. (Trogossitidae: Peltinae: Rentoniini); E, *Phycosecis litoralis* Pascoe (Phycosecidae); F, *Acanthocnemis nigricans* (Hope) (Acanthocnemidae); G, *Australymexylon fuscipennis* (Lea). (Lymexylidae: Melittommatinae); H, *Egolia variegata* Erichson (Trogossitidae: Trogossitinae: Egoliini); I, *Larinotus umbilcatus* Carter & Zeck (Trogossitidae: Trogossitinae: Larinotini); J, *Scrobiger splendidus* (Newman) (Cleridae: Clerinae).

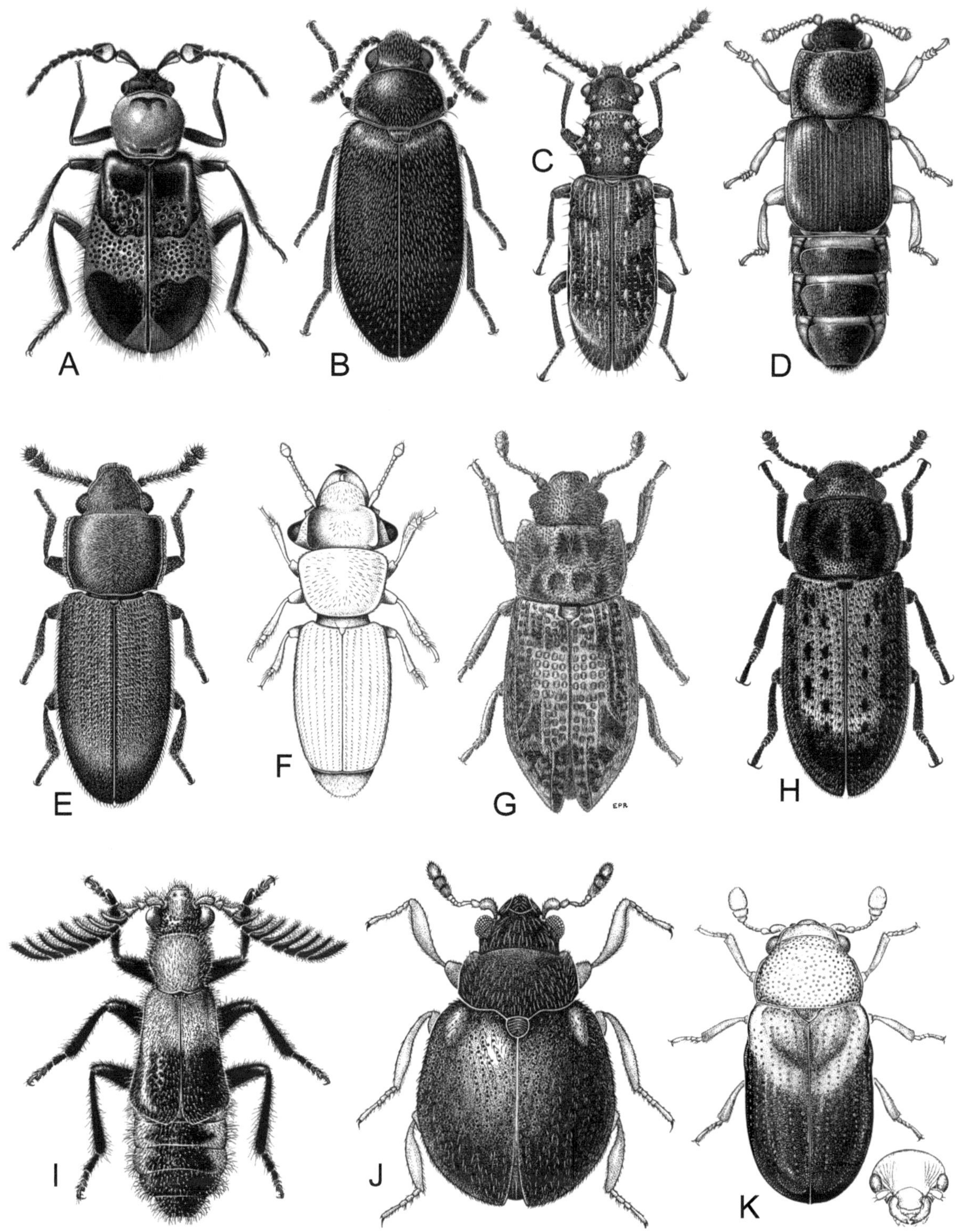

Fig. 83. Adult, dorsal. A, *Dicranolaius bellulus* (Guérin–Méneville) (Melyridae: Malachinae). B, '*Dasytes*' sp. (Melyridae: Dasytinae). C, *Hymaea magna* Sen Gupta & Crowson (Phloeostichidae). D, *Brachypeplus planus* Erichson (Nitidulidae: Cillaeinae). E, *Myrabolia grouvelliana* Reitter (Myraboliidae). F, *Mimemodes laticeps* (Macleay) (Monotomidae). G–H, Protocucujidae: G, *Ericmodes costatus* Slipinski & Pakaluk; H, *E. australis* Grouvelle. I, *Balanophorus mastersi* Macleay (Melyridae: Malachiinae). J–K, Sphindidae: J, *Aspidiphorus humeralis* Blackburn; K, *Notosphindus slateri* McHugh & Wheeler (with detail of head, anterodorsal).

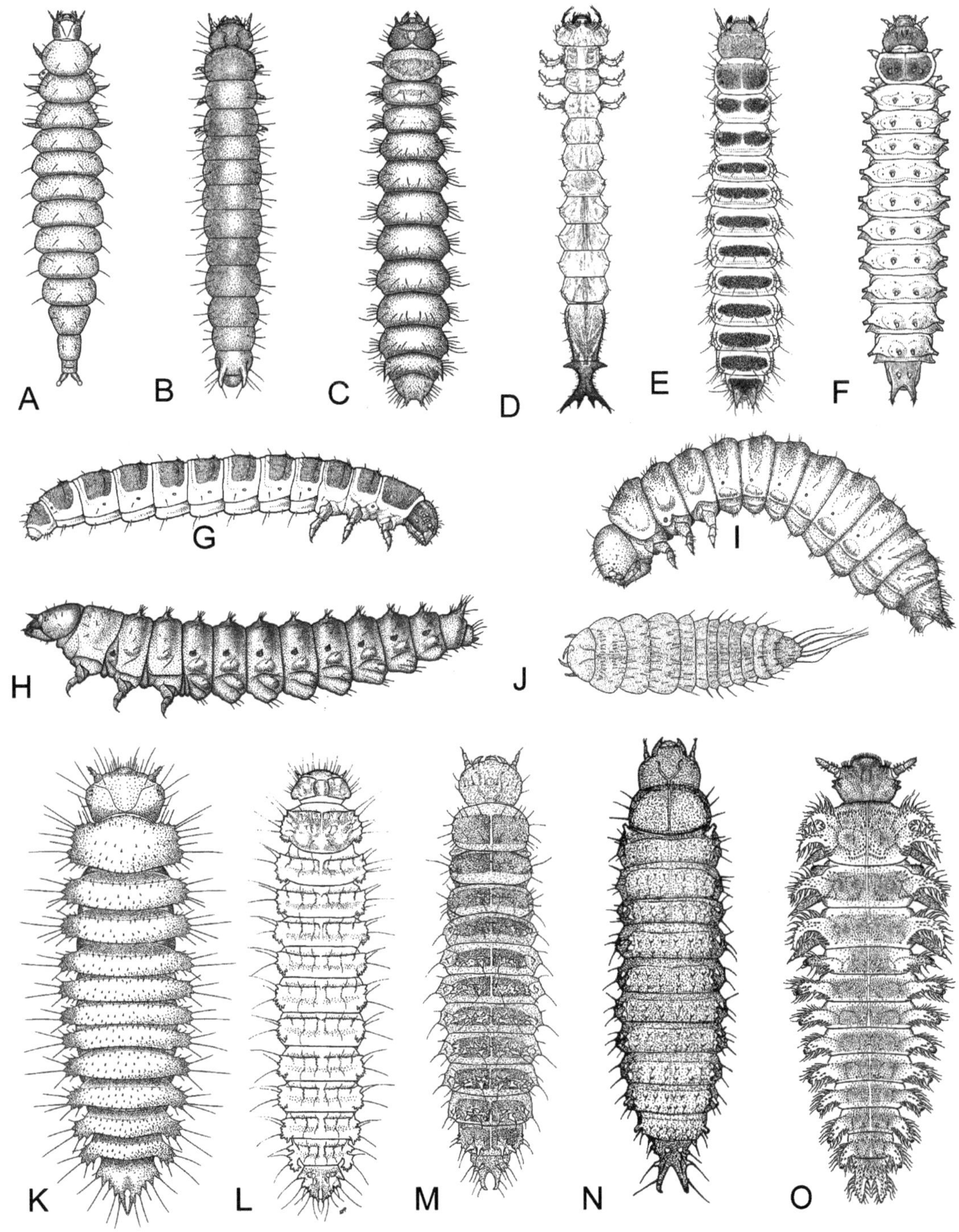

Fig. 84. A–F, Larva, dorsal: A, *Boganium* sp. (Boganiidae); B, *Atomaria* sp. (Cryptophagidae: Atomariinae); C, *Lamingtonium binnaburrense* Sen Gupta & Crowson (Lamingtoniidae); D, *Platisus moerosus* (Pascoe) (Cucujidae); E, *Taphropiestes* sp. (Cavognathidae); F, *Phenolia* (*Aethinodes*) sp. (Nitidulidae: Nitidulinae). G–I, larva, lateral: G, *Diplocoelus fasciatus* (Macleay) (Biphyllidae); H, *Pocadius kirejtshuki* Cline (Nitidulidae: Nitidulinae); I, *Microsternus* sp. (Erotylidae: Erotylinae). J–O, Larva, dorsal: J, *Ancistria* sp. (Passandridae); K, *Cryptophilus* sp. (Erotylidae: Cryptophilinae); L, *Monotoma americana* Aubé (Monotomidae); M, *Hobartius eucalypti* (Blackburn) (Hobartiidae); N, *Ericmodes* sp. (Protocucujidae); O, *Priastichus megathorax* Leschen *et al.* (Priasilphidae).

uncus; maxillary palps moderately well developed, 4-segmented, apical palpomere fusiform. Mentum slightly transverse, apically emarginate, fused at base; ligula bilobed; labial palp 3-segmented, apical palpomere fusiform. Gular sutures short and widely separated. Corpotentorium absent; anterior tentorial arms not expanded mesally; cervical sclerites reduced and transverse.

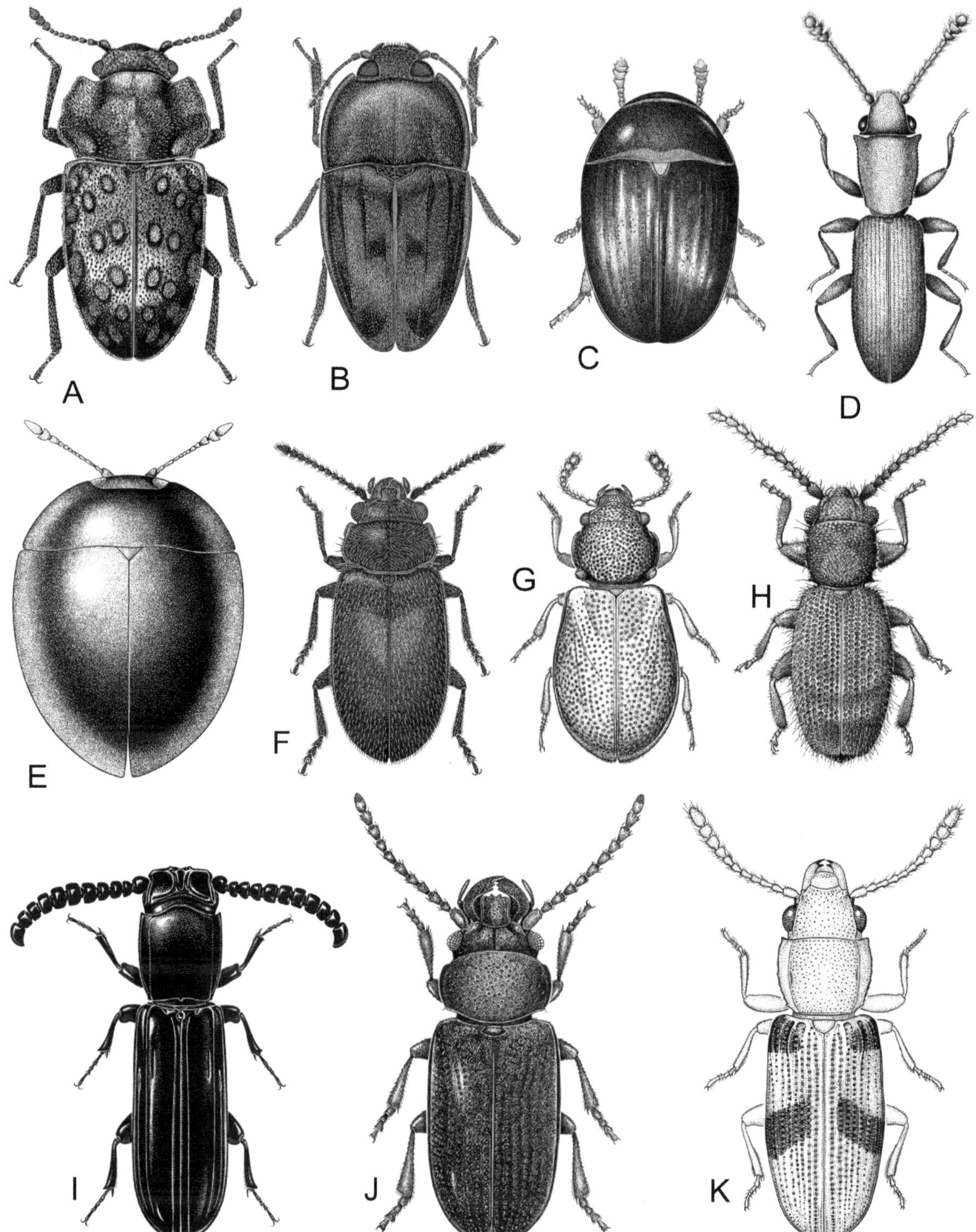

Fig. 85. Adult, dorsal: A, *Priastichus megathorax* Leschen *et al.* (Priasilphidae); B, *Propalticus simplex* Crowson & Sen Gupta (Propalticidae); C, *Litochrus major* Blackburn (Phalacridae: Phalacrinae); D, *Silvanus unidentatus* (Olivier) (Silvanidae: Silvaninae); E, *Phalacrinus rotundus* Blackburn (Phalacridae: Phaenocephalinae); F, *Boganium* sp. (Boganiidae); G, *Tasmosalpingus* sp. (Tasmosalpingidae); H, *Cryptamorpha* sp. (Silvanidae: Brontinae); I, *Passandra heros* (Fabricius) (Passandridae); J, *Paracucujus rostratus* Sen Gupta & Crowson (Boganiidae); K, *Laemophloeus amabilis* Olliff (Laemophloeidae).

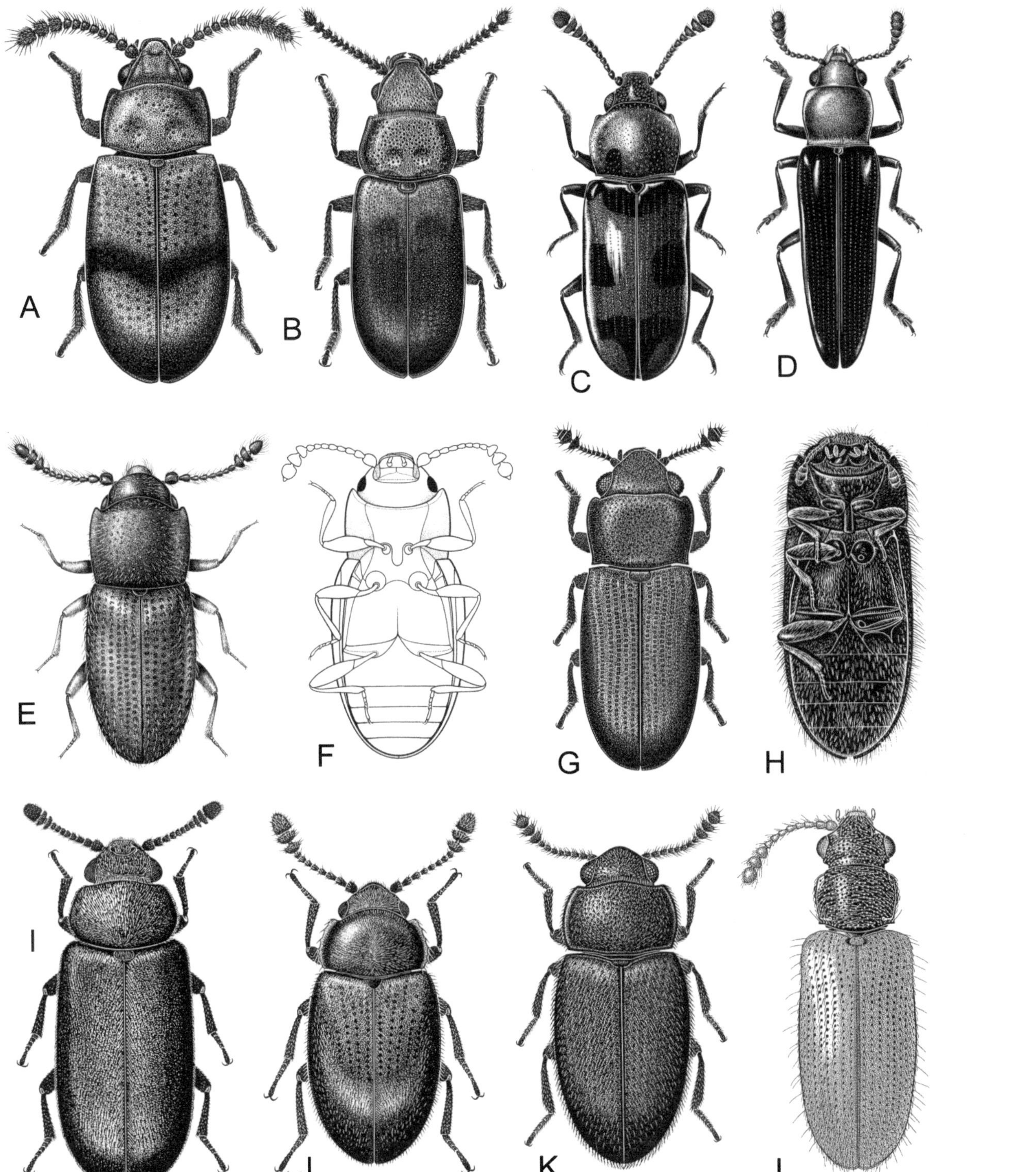

Fig. 86. Adult. A–B, Lamingtoniidae, dorsal: A, *Lamingtonium binnaburrense* Sen Gupta & Crowson; B, *L. thayerae* Lawrence & Leschen. C–D, Erotylidae, dorsal: C, *Thallis compta* Erichson (Erotylinae); D, *Caenolanguria vulgaris* (Harold) (Erotylidae: Languriinae). E, '*Cryptophagus*' *gibbipennis* Blackburn (Cryptophagidae: Cryptophaginae), dorsal; F, same, ventral. G, *Cathartocryptus* sp. (Erotylidae: Cryptophilinae), dorsal. H, *Diplocoelus punctatus* Lea (Biphyllidae), ventral. I, *Taphropiestes pullivora* (Crowson) (Cavognathidae), dorsal. J, *Hobartius* sp. (Hobartiidae), dorsal. K, *Cryptophilus* sp. (Erotylidae: Cryptophilinae), dorsal. L, *Protoloberus singularis* (Blackburn) (Erotylidae: Xenoscelinae), dorsal, from Leschen 2003; © Landcare Research, New Zealand.

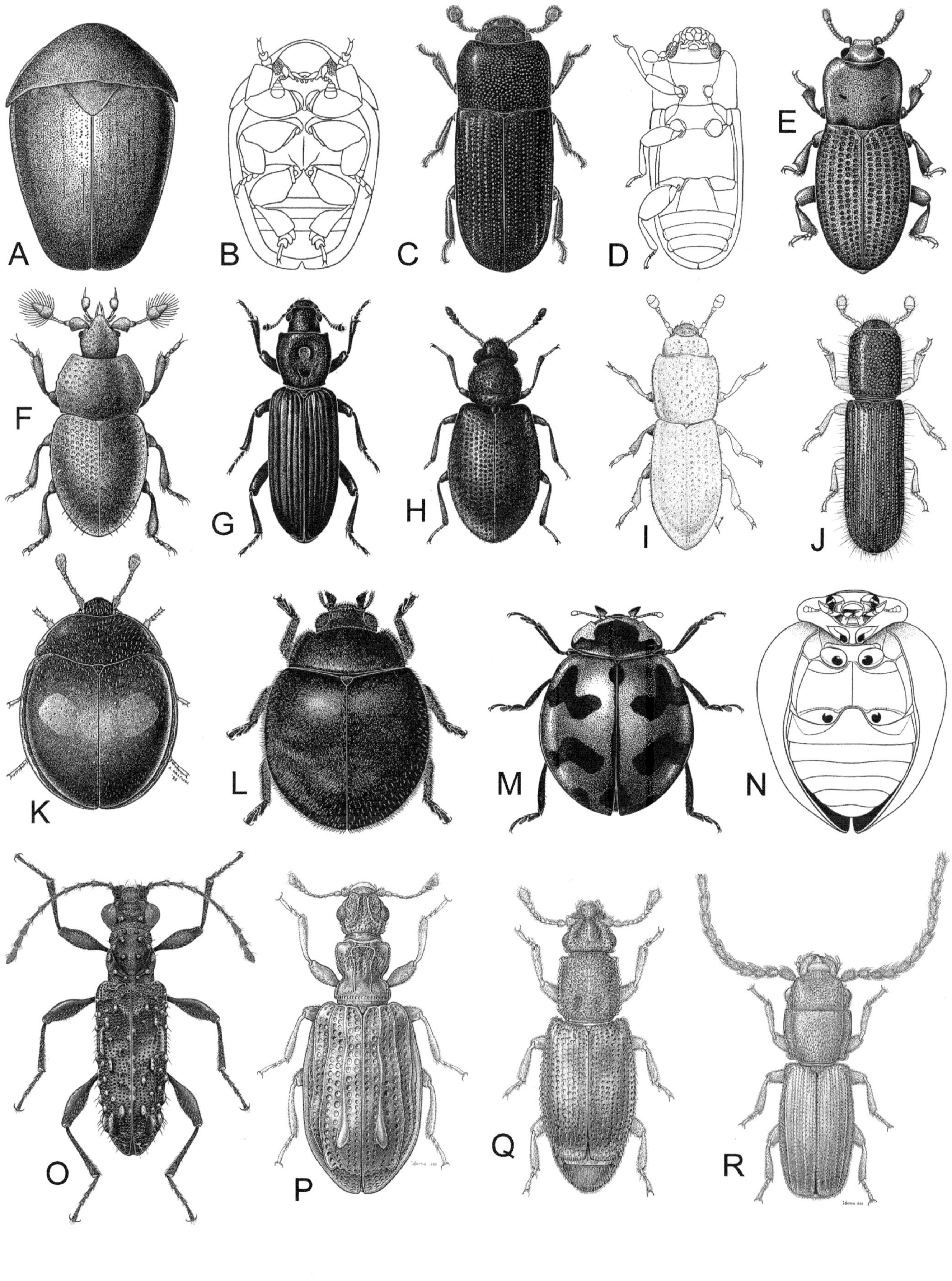
A
B
C
D
E
F
G
H
I
J
K
L
M
N
O
P
Q
R

Pronotum ~0.8–1.15 times as long as wide, widest at about middle; base narrower than combined elytral bases; sides slightly curved, not explanate, lateral pronotal carinae absent; anterior angles not produced; posterior angles right or rounded; posterior edge straight or evenly curved, finely margined; disc simple, without paired basal impressions or median longitudinal groove. Prosternum in front of coxae about as long as to somewhat longer than shortest diameter of coxal cavity, flat to slightly convex, without carinae. Prosternal process complete but interrupted at middle (intercoxal process strongly concave), the apex expanded and truncate or broadly angulate. Notosternal sutures incomplete. Procoxae projecting below prosternum, with short concealed lateral extensions; trochantins concealed. Procoxal cavities subcircular, contiguous, externally broadly closed, without narrow lateral extensions, internally broadly closed; postcoxal (hypomeral) processes meeting disjunct apex of prosternal process (sternellum).

Scutellar shield not abruptly elevated, anteriorly simple, posteriorly broadly rounded; mesoscutum with stridulatory file, which may be divided or undivided and sometimes asymmetrical. Elytra ~1.58–1.95 times as long as wide and 2.3–2.75 times as long as pronotum, irregularly punctate or with 11 puncture rows and scutellary striole; elytral apices meeting or almost meeting at the suture and exposing no abdominal tergites; epipleura narrow and incomplete. Mesoventrite separated by complete sutures from mesanepisterna, which are distinctly separated at midline; anterior edge on same plane as metaventrite, without paired procoxal rests, but with broadly transverse prothoracic rest. Mesoventral cavity absent. Mesocoxae not conical and projecting, with concealed trochantins. Mesocoxal cavities narrowly separated, subcircular, not or slightly oblique, closed laterally by meeting of mesoventrite and metaventrite. Mesoventral process narrowly rounded at apex; mesometaventral junction monocondylic. Metaventrite slightly convex; transverse (metakatepisternal) suture long but close to posterior edge of ventrite; discrimen about half as long as ventrite excluding anterior intercoxal process; postcoxal lines absent; exposed portion of metanepisternum moderately elongate; metepimeron slightly exposed. Metacoxae narrowly separated, not extending laterally to meet elytra; plates absent. Metendosternite with long stalk, long lateral arms, well developed but slender laminae and very short anterior process bearing widely separated anterior tendons. Hind wing with apical field about half as long as total wing length, containing distinct sclerite just beyond radial cell and vaguely indicated remnants of RP; radial cell moderately well developed, slightly longer than wide with base more or less straight and inner posterobasal angle right; cross-vein r3 short and moderately oblique, arising well away from r4, which is weakly sclerotised but more or less complete; basal portion of RP moderately long; R-M loop forming acute angle; medial spur ending before wing margin; medial field with 3 or 4 free veins and no medial fleck; MP_{3+4} without base; wedge cell absent; anal lobe moderately well developed with notch-like embayment. Legs moderately long and slender; trochanterofemoral joint strongly oblique with base of femur separate from coxa; femora slightly inflated; tibial spurs 2-2-2 in Mecomacerini, 2-2-1 in Rhinorhynchini. Tarsi pseudotetramerous, 5-5-5 in both sexes with tarsomere 4 reduced and 3 distinctly bilobed; pretarsal claws bifid or with basal tooth or lamina; empodium not apparent.

Abdomen with 5 free ventrites; ventrite 1 at its shortest point subequal in length to 2, without postcoxal lines; intercoxal process acute or narrowly rounded; apex of ventrite 5 broadly rounded or slightly emarginate. Abdominal terga lightly pigmented, except for tergite VII, which bears two patches of wing-folding asperities; functional spiracles on segments I–VII, those on I–VI in pleural membrane, those on VII at lateral edges of tergite. Tergite VIII in male partly exposed and sclerotised; sternite VIII reduced and lightly sclerotised, with short median strut. Anterior edge of segment IX in male with distinct spiculum gastrale; tergite X membranous. Aedeagus cucujiform, usually symmetrical; anterior edge of tegmen with moderately long strut, sometimes expanded at apex; parameres fused to phallobase and partly or entirely fused together. Penis flattened with dorsal and ventral lobes (tectum and pedon) and with paired anterior struts; endophallus with complex armature. Segment VIII in female more or less membranous, with long, fixed spiculum ventrale. Ovipositor elongate and lightly sclerotised, except for baculi; paraprocts usually subequal to or slightly shorter than coxites, with longitudinal baculi; coxites located distal to paraprocts, not or weakly divided, with narrow distal lobes bearing well developed, terminal styli. Internal tract with strongly curved, sclerotised capsular spermatheca and spermathecal gland attached at or near its base. [Kuschel 1959, 1983, 1989, 1993, 1994, 1995, 2003; Crowson 1985; Thompson 1992; Kuschel and May 1997; Anderson 2002a; Kuschel and Leschen 2010].

Larvae

Body elongate and more or less parallel-sided or slightly wider at middle and narrowing anteriorly and posteriorly, somewhat flattened; all surfaces lightly pigmented, except for head capsule and prothorax; vestiture consisting of scattered long and short, simple setae. Head prognathous or slightly declined, not retracted; posterior edge of head capsule not emarginate; epicranial stem moderately long; frontal arms V-shaped, contiguous at base; median endocarina extending almost to clypeal region. Stemmata on each side usually 4 (one stemma with lens and as

Fig. 87. Adult. A–B, *Cychramptodes murrayi* Reitter (Nitidulidae: Nitidulinae): A, dorsal; B, ventral. C–D, *Metacerylon minutum* (Carter) (Cerylonidae: Euxestinae): C, dorsal; D, ventral. E–F, Cerylonidae: Ceryloninae, dorsal: E, *Cerylonopsis alienigenus* (Blackburn); F, *Cautomus mirabilis* (Oke). G, *Ascetoderes aequinus* (Pascoe) (Bothrideridae: Bothriderinae), dorsal. H, *Cortinicara hirtalis* (Broun) (Latridiidae, Corticariinae), dorsal. I, *Anommatus duodecimstriatus* (Müller) (Bothrideridae: Anommatinae), dorsal. J, *Teredolaemus leae* (Grouvelle) (Bothrideridae: Teredinae), dorsal. K, *Aphanocephalus bimaculatus* Lea (Discolomatidae), dorsal. L–N, Coccinellidae: L, *Scymnodes bellus* Pope & Lawrence, dorsal; M, *Coccinella transversalis* Fabricius, dorsal; N, *Epilachna guttatopustulata* (Fabricius), ventral. O, *Rhopalobrachium crowsoni* Lawrence (Phloeostichidae), dorsal. P, *Aridius nodifer* (Westwood) (Latridiidae: Latridiinae), dorsal. Q, *Monotoma picipes Herbst* (Monotomidae), dorsal). R, *Cryptolestes turcicus* (Grouvelle) (Laemophloeidae), dorsal. P-R from Busquet 1990, © Agriculture and Agri-Food Canada.

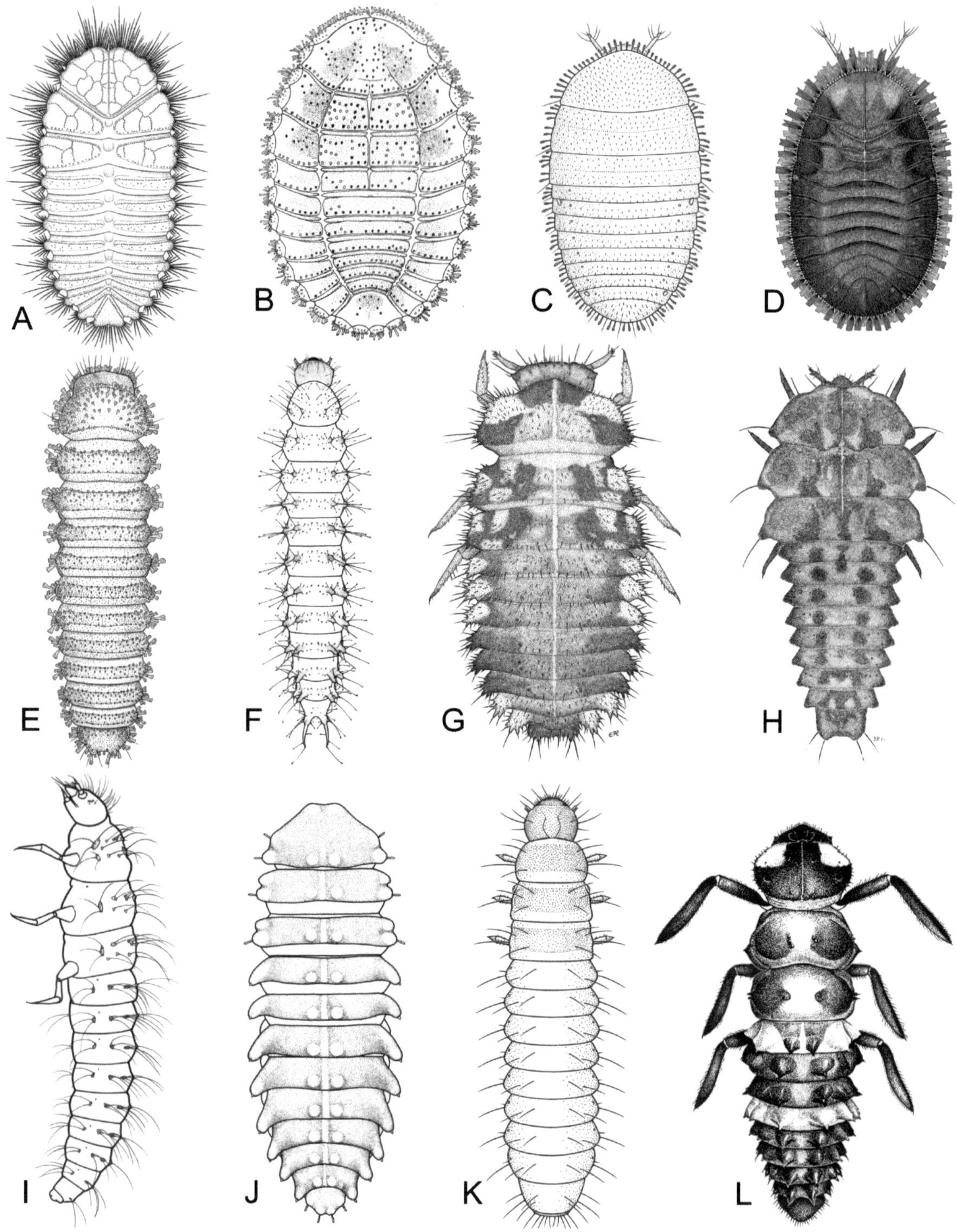

Fig. 88. Larva. A–H, Dorsal: A, *Murmidius ovalis* (Beck) (Cerylonidae: Murmidiinae); B, *Aphanocephalus hemisphericus* Wollaston (Discolomatidae), from Fukuda 1969; C, *Holopsis* sp. (Corylophidae: Corylophinae); D, *Gloeosoma* sp. (Corylophidae: Corylophinae); E, *Mycetaea subterranea* (Fabricius) (Endomychidae: Mycetaeinae); F, *Euxestus erithacus* Chevrolat (Cerylonidae: Euxestinae); G, *Stenotarsus commodus* Blackburn (Endomychidae: Stenotarsinae); H, *Periptyctus eximius* Arrow (Corylophidae: Periptyctinae). I, *Latridius minutus* (Linnaeus) (Latridiidae: Latridiinae), lateral. J, *Mychocerus* sp. (Cerylonidae: Ceryloninae), dorsal. K, *Holoparamecus* sp. (Endomychidae: Merophysiinae), dorsal. L, *Coccinella transversalis* Fabricius (Coccinellidae: Coccinellinae), dorsal. G, H from McHugh and Pakaluk 1997.

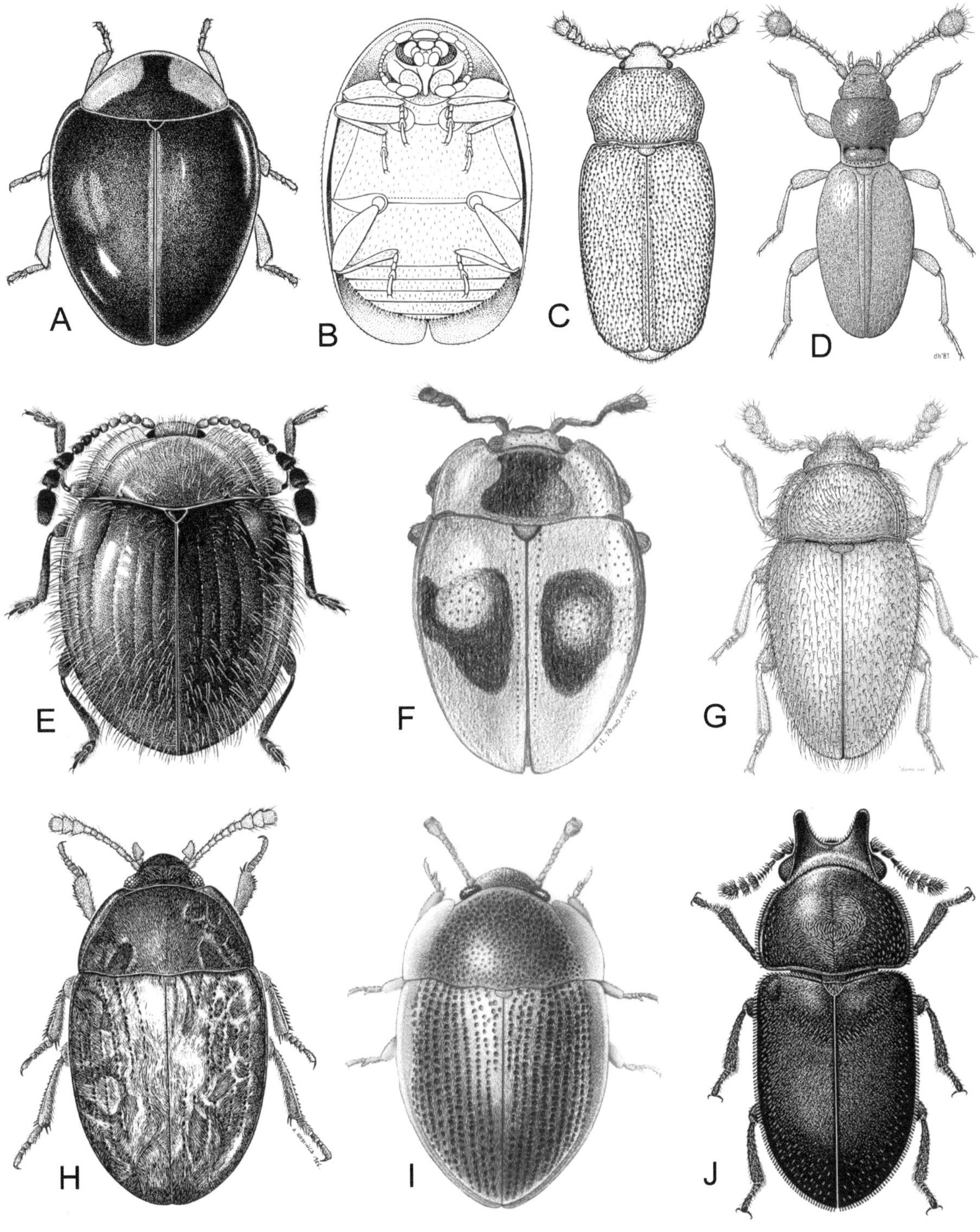

Fig. 89. Adult. A–C, Corylophidae: Corylophinae: A, *Holopsis* sp. (Peltinodini), dorsal; B, *Arthrolips pulchellus* Lea (Parmulini), ventral. C, *Priamima rubicunda* (Macleay) (Foadiini), dorsal. D–J, Dorsal: D, *Holoparamecus tenuis* Reitter (Endomychidae: Merophysiinae); E, *Stenotarsus pisoniae* Lea (Endomychidae: Stenotarsinae); F, *Pakalukodes bimaculatus* Ślipiński *et al.* (Corylophidae: Periptyctinae); G, *Mycetaea subterranea* (Fabricius) (Endomychidae: Mycetaeinae); H, *Wattianus queenslandicus* Kaszab (Archeocrypticidae); I, *Nothenneboeus punctatus* Lawrence (Archeocrypticidae); J, *Cis cervus* Blair (Ciidae). D, G from Watt *et al.* 1997; © Landcare Research, New Zealand.

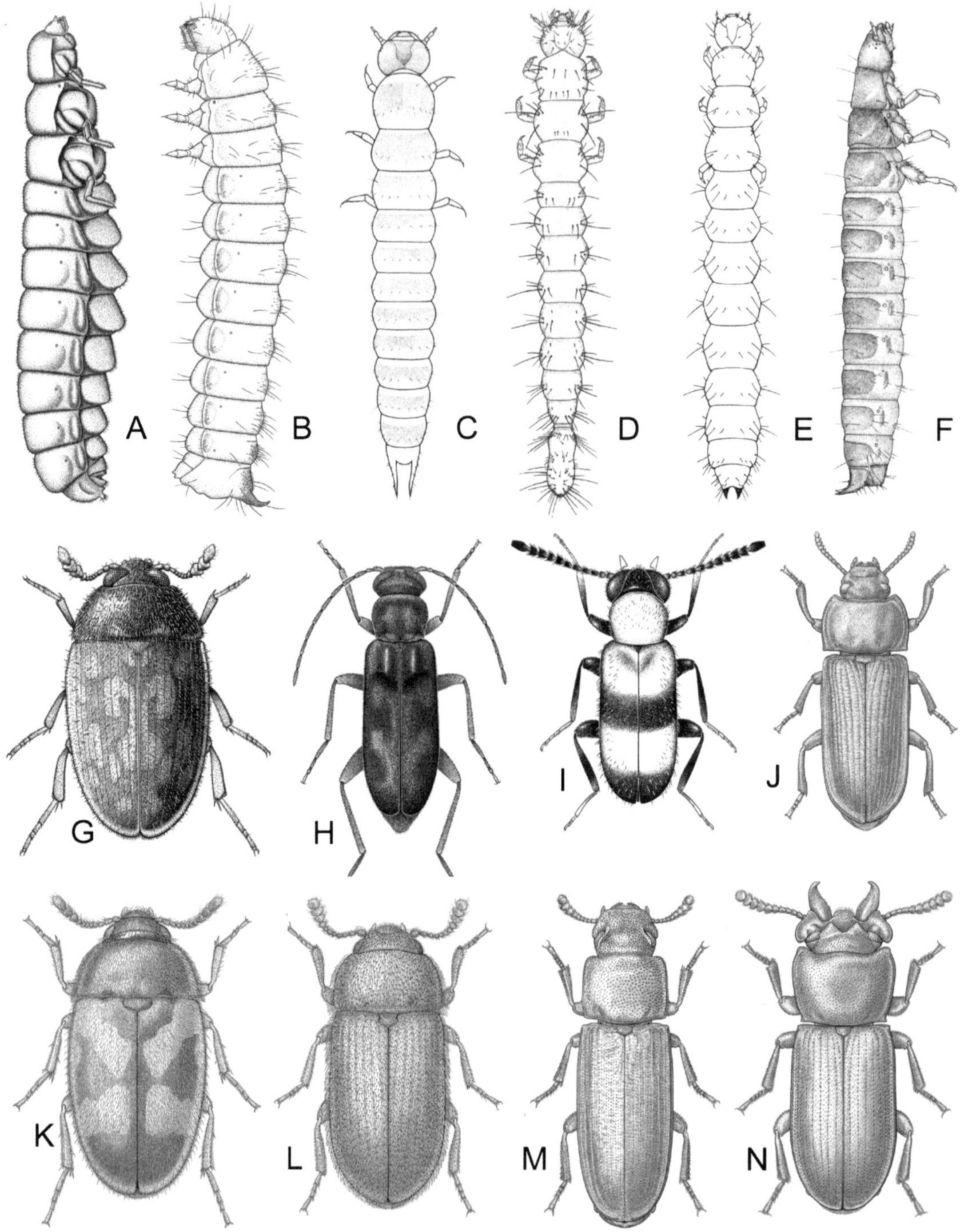

Fig. 90. A–F, larva: A, *Megaxenus termitophilus* Lawrence (Aderidae), lateral; B, *Cis* sp. (Ciidae), lateral; C, *Enneboeus* sp. (Archeocrypticidae), dorsal. D, *Scraptia* sp. (Scraptiidae), dorsal; E, Genus ? sp. (Aderidae), dorsal. F, *Mycetophagus* sp. (Mycetophagidae), lateral. G–N, adult, dorsal: G, *Litargops multipunctatus* (Lea) (Mycetophagidae); H, *Megaxenus termitophilus* Lawrence (Aderidae); I, *Syzeton abnormis* (King) (Aderidae); J, *Tenebrio molitor* Linnaeus (Tenebrionidae: Tenebrioninae); K, *Litargus balteatus* LeConte (Mycetophagidae); L, *Typhaea stercorea* (Linnaeus) (Mycetophagidae); M, *Tribolium confusum* Jacquelin DuVal (Tenebrionidae: Tenebrioninae); N, *Gnatocerus cornutus* (Fabricius) (Tenebrionidae: Diaperinae). J-N from Busquet 1990, © Agriculture and Agri-Food Canada.

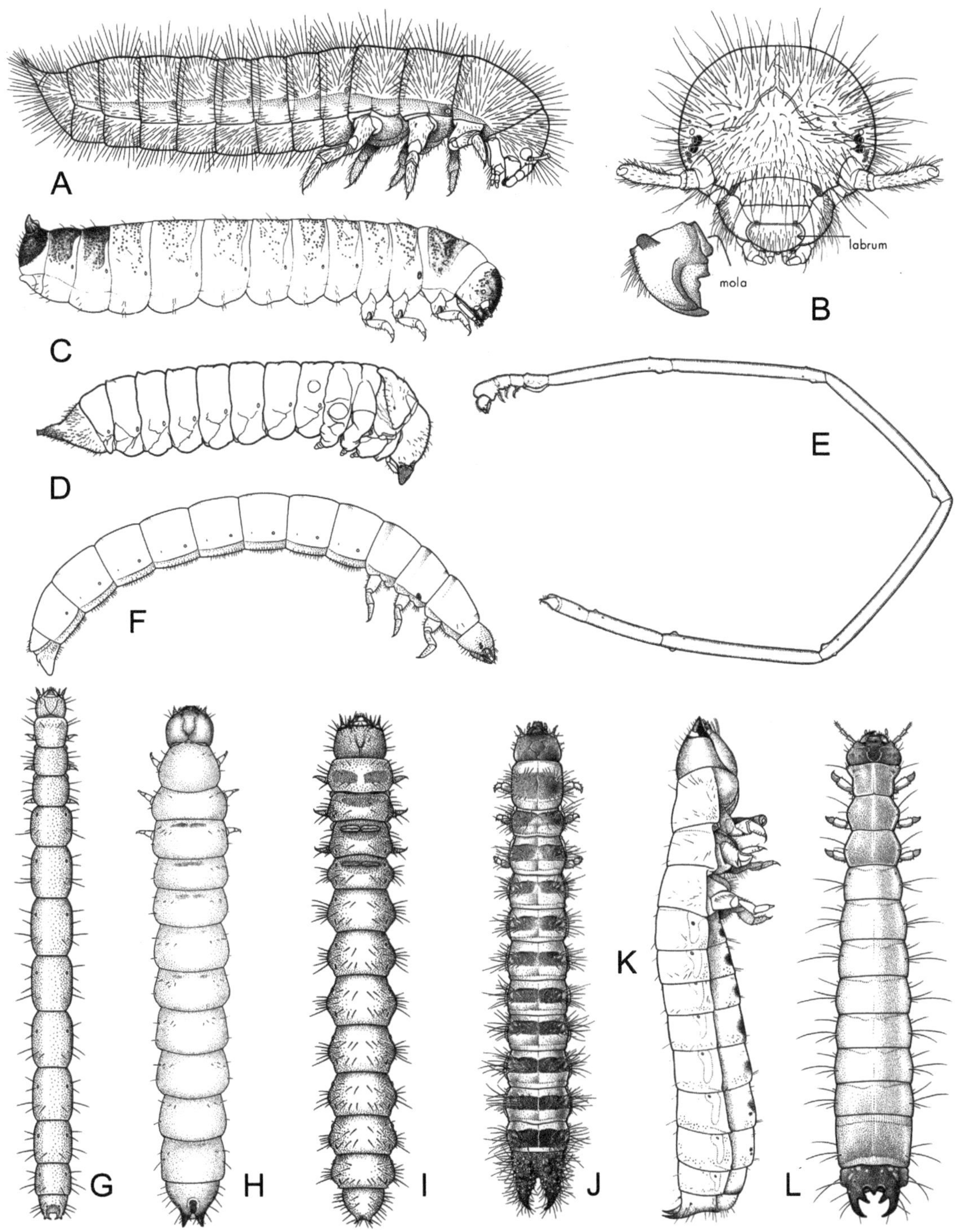

Fig. 91. Larva. A, *Ecnolagria grandis* (Gyllenhal) (Tenebrionidae: Lagriinae), lateral. B, Same, head, anterior and left mandible, ventral. C–F, Lateral: C, *Orcopagia monstrosa* Pascoe (Tenebrionidae: Tenebrioninae); D, Genus ? sp. (Mordellidae); E, *Mystes planatus* Champion (Melandryidae); F, *Scymena amphibia* Pascoe (Tenebrionidae: Tenebrioninae). G–J, Dorsal: G, Genus ? sp. (Salpingidae: Prostominiinae?); H, *Pycnomerus* sp. (Zopheridae: Zopherinae); I, *Cotulades brindabellae* Slipinski & Lawrence (Zopheridae: Zopherinae); J, *Meryx rugosa* Latreille (Ulodidae). K, *Promethis nigra* Blessig (Tenebrionidae: Stenochiinae), lateral. L, *Morpholycus costipennis* Lea (Pyrochroidae), dorsal.

many as 3 additional eye spots). Antennae very short, 1-segmented, with terminal, conical sensorium. Frontoclypeal suture absent; labrum free, but without tormae or labral rods. Mandibles more or less symmetrical, short and broad, obliquely bidentate; accessory ventral process absent; incisor edge with or without one subapical tooth, mola reduced and sub-basal; prostheca absent.

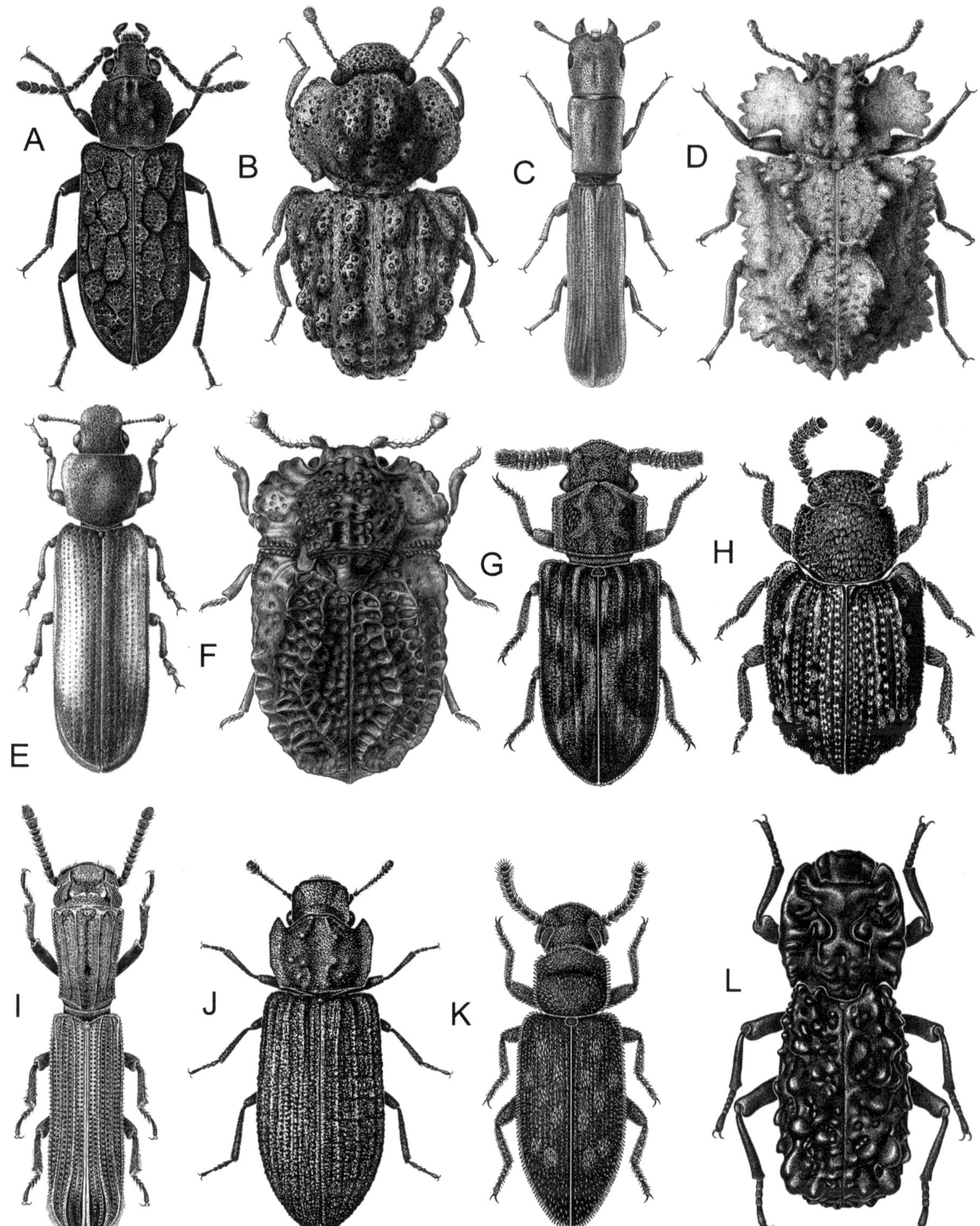

Fig. 92. Adult, dorsal. A, *Meryx rugosa* Latreille (Ulodidae); B–F, Zopheridae: Colydiinae: B, *Norix* sp.; C, *Nematidium poggi* Pal & Slipinski; D, *Pristoderus* sp.; E, *Todima fulvicincta* Elston; F, *Faecula cristata* Slipinski & Lawrence. G–H, Zopheridae: Zopherinae: G, *Latometus differens* Carter; H, *Cotulades brindabellae* Slipinski & Lawrence. I–J, Zopheridae: Colydiinae: I, *Pseudendestes australis* Lawrence; J, *Pristoderus elongatus* (Blackburn). K–L, Zopheridae: Zopherinae: K, *Docalis funerosus* (Hope); L, *Zopherosis georgei* White.

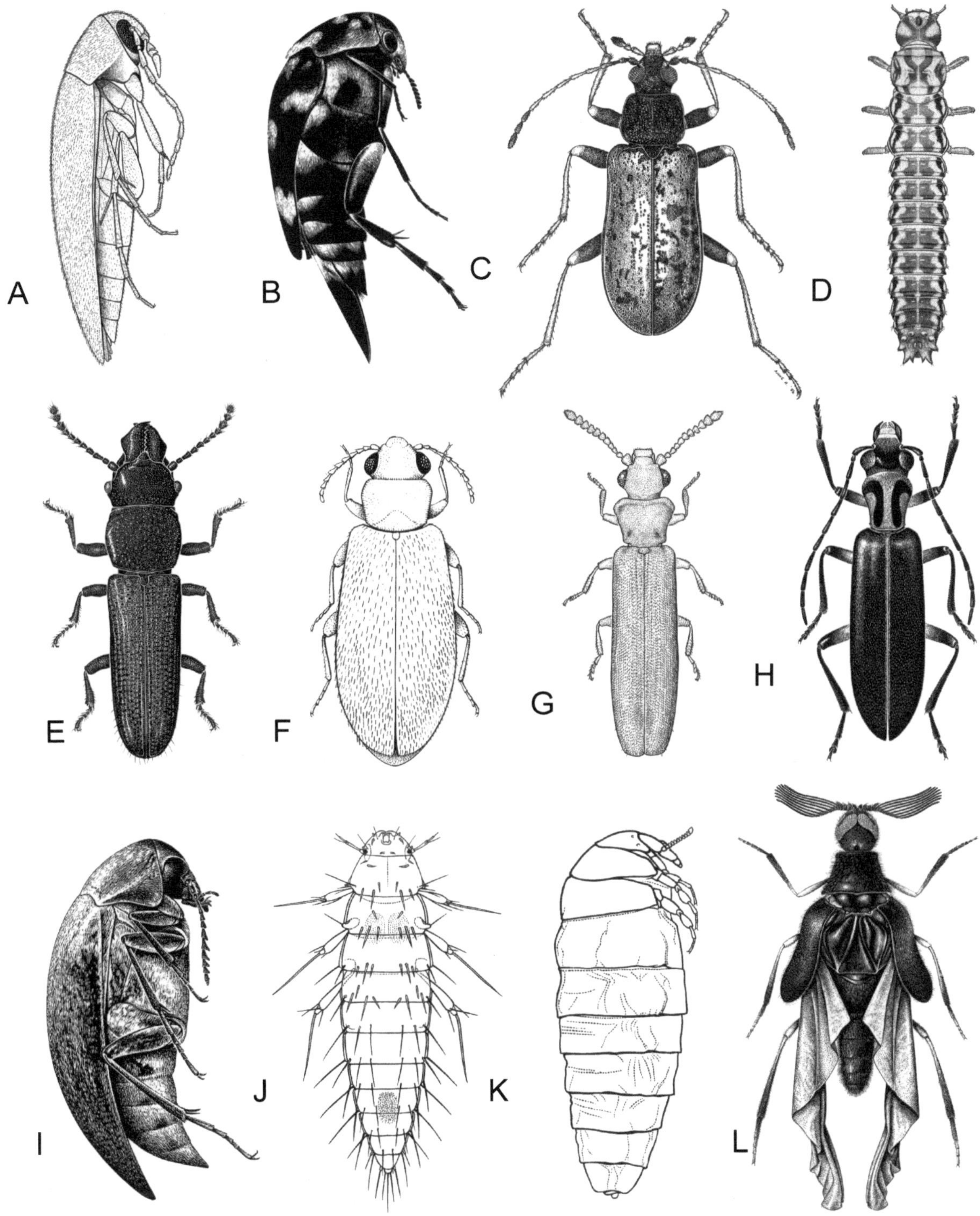

Fig. 93. A, *Talayra elongata* (Macleay) (Melandryidae), adult, lateral. B, *Hoshihananomia leucosticta* (Germar) (Mordellidae), adult, lateral. C–D, *Sirrhas variegatus* Lawrence (Promecheilidae): C, adult, dorsal; D, larva, dorsal. E–H, Adult, dorsal: E, *Prostomis atkinsoni* Waterhouse (Prostomidae); F, *Trichosalpingus brunneus* Blackburn (Mycteridae: Eurypinae); G, *Hemipeplus australicus* Arrow (Mycteridae: Hemipeplinae); H, 'Copidita' mira (Lea) (Oedemeridae). I–L, Ripiphoridae: I, *Trigonodera marmorata* (Macleay), adult, lateral; J, *Rhipidioides helenae* Riek, first instar larva (triungulinid), dorsal; K, *Riekella australis* (Riek), adult female, lateral; L, *Rhipidioides rubricatus* Riek, adult male, dorsal.

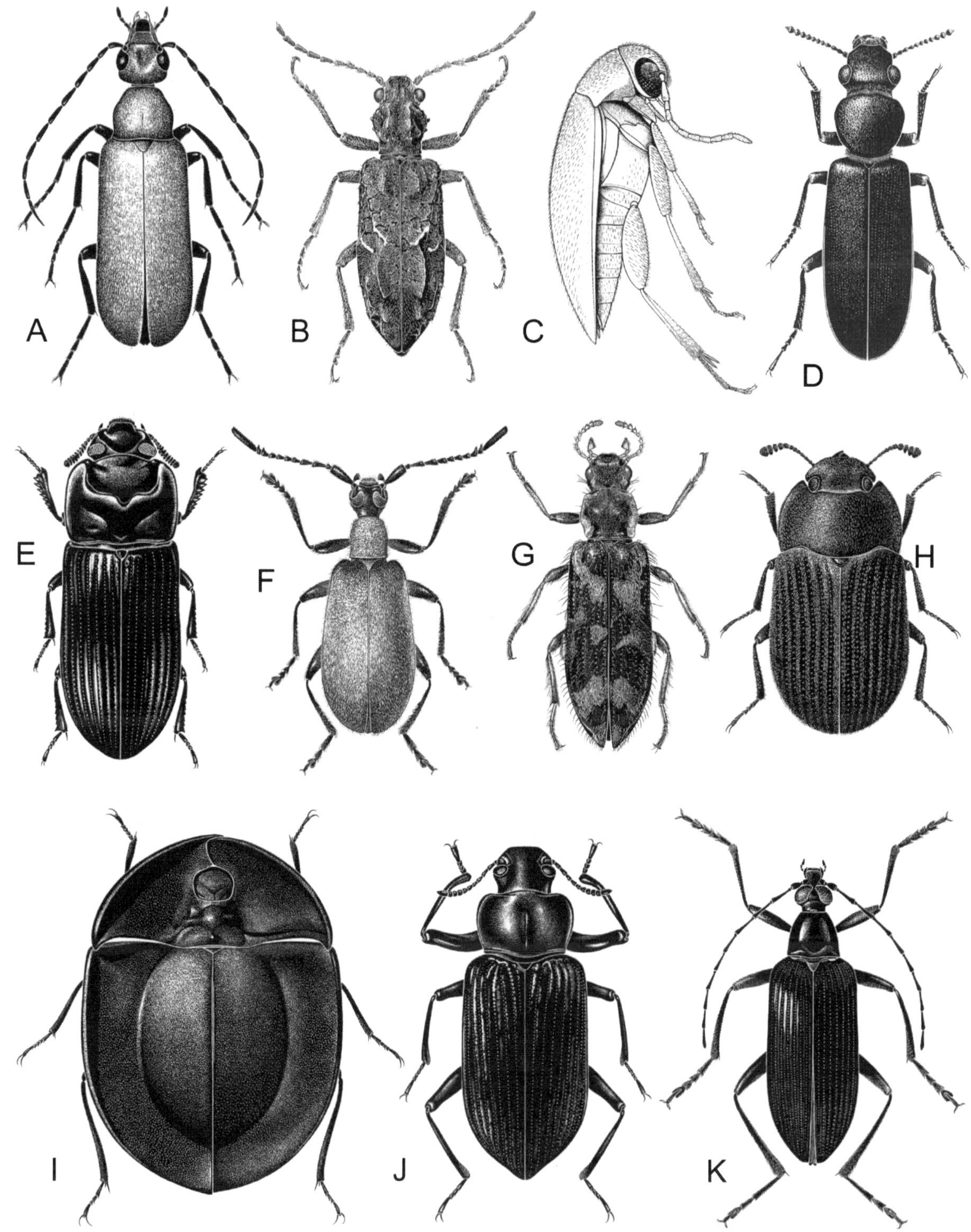

Fig. 94. Adult. A, '*Zonitis*' *lutea* Macleay (Meloidae), dorsal. B, *Anaplopus tuberculatus* Blackburn (Pythidae), dorsal. C, *Scraptia australis* Champion (Scraptiidae), lateral. D–K, Dorsal: D, *Synercticus heteromerus* Pascoe (Boridae); E, *Achthosus westwoodi* Newman (Tenebrionidae: Tenebrioninae); F, *Ecnolagria grandis* (Gyllenhal) (Tenebrionidae: Lagriinae); G–I, Tenebrionidae: Tenebrioninae: G, *Leaus tasmanicus* Matthews & Lawrence; H, *Gonocephalum meyricki* Blackburn; I, *Helea waitei* (Lea). J, *Promethis nigra* Blessig (Tenebrionidae: Stenochiinae); K, *Tanychilus striatus* Newman (Tenebrionidae: Alleculinae).

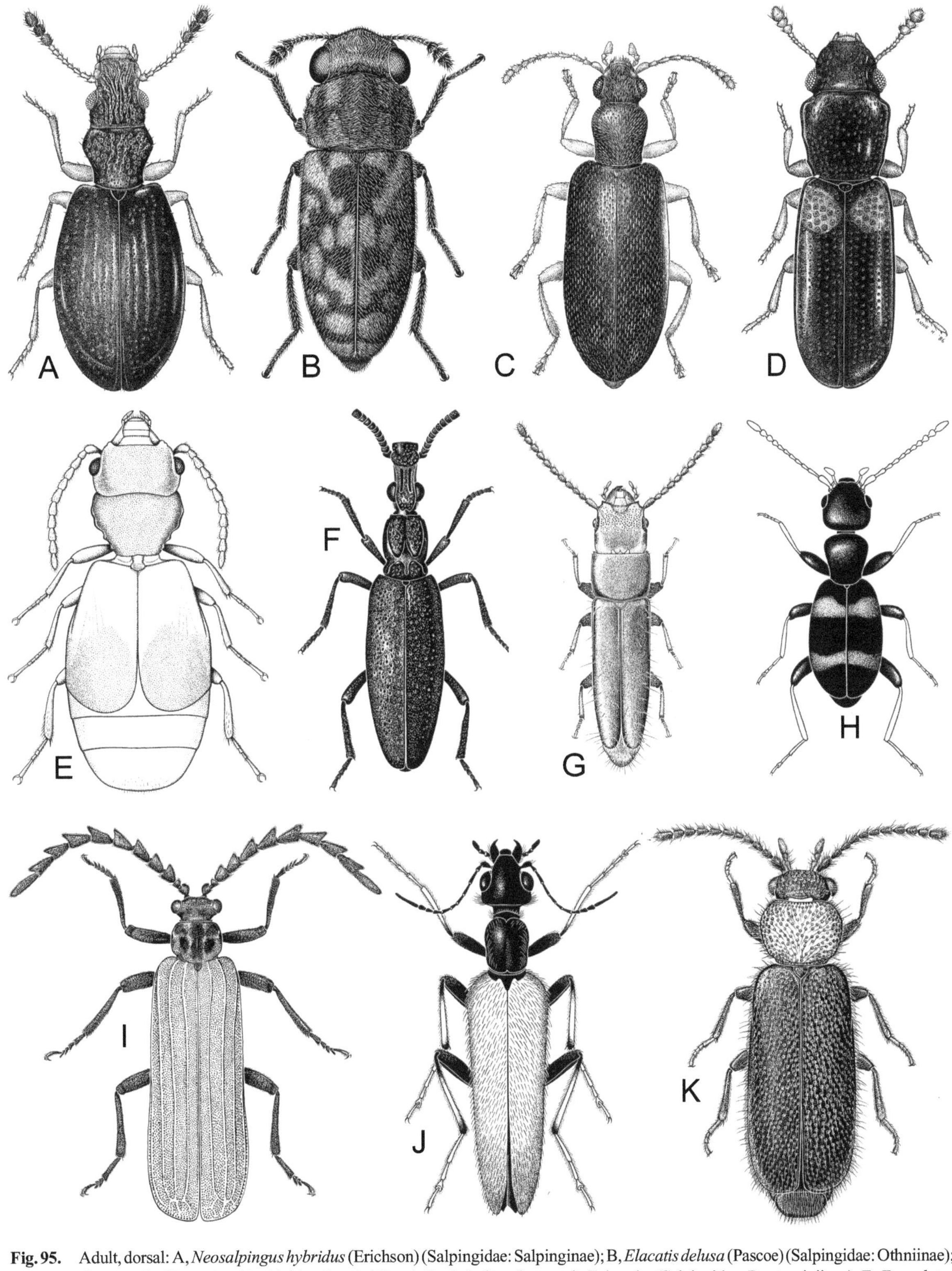

Fig. 95. Adult, dorsal: A, *Neosalpingus hybridus* (Erichson) (Salpingidae: Salpinginae); B, *Elacatis delusa* (Pascoe) (Salpingidae: Othniinae); C, *Lagrioida australis* Champion (Anthicidae: Lagrioidinae); D, *Ocholissa humeralis* Fairmaire (Salpingidae: Prostominiinae); E, *Euryplatus dimidiatus* (Waterhouse) (Salpingidae: Inopeplinae); F, *Tretothorax cleistostoma* Lea (Salpingidae: Dacoderinae); G, Genus ? sp. (Salpingidae: Prostominiinae); H, *Anthicus australis* King (Anthicidae: Anthicinae); I, *Morpholycus costipennis* Lea (Pyrochroidae); J, *Egestria taeniata* Pascoe (Anthicidae: Eurygeniinae); K, *Temnopalpus bicolor* Blackburn (Pyrochroidae).

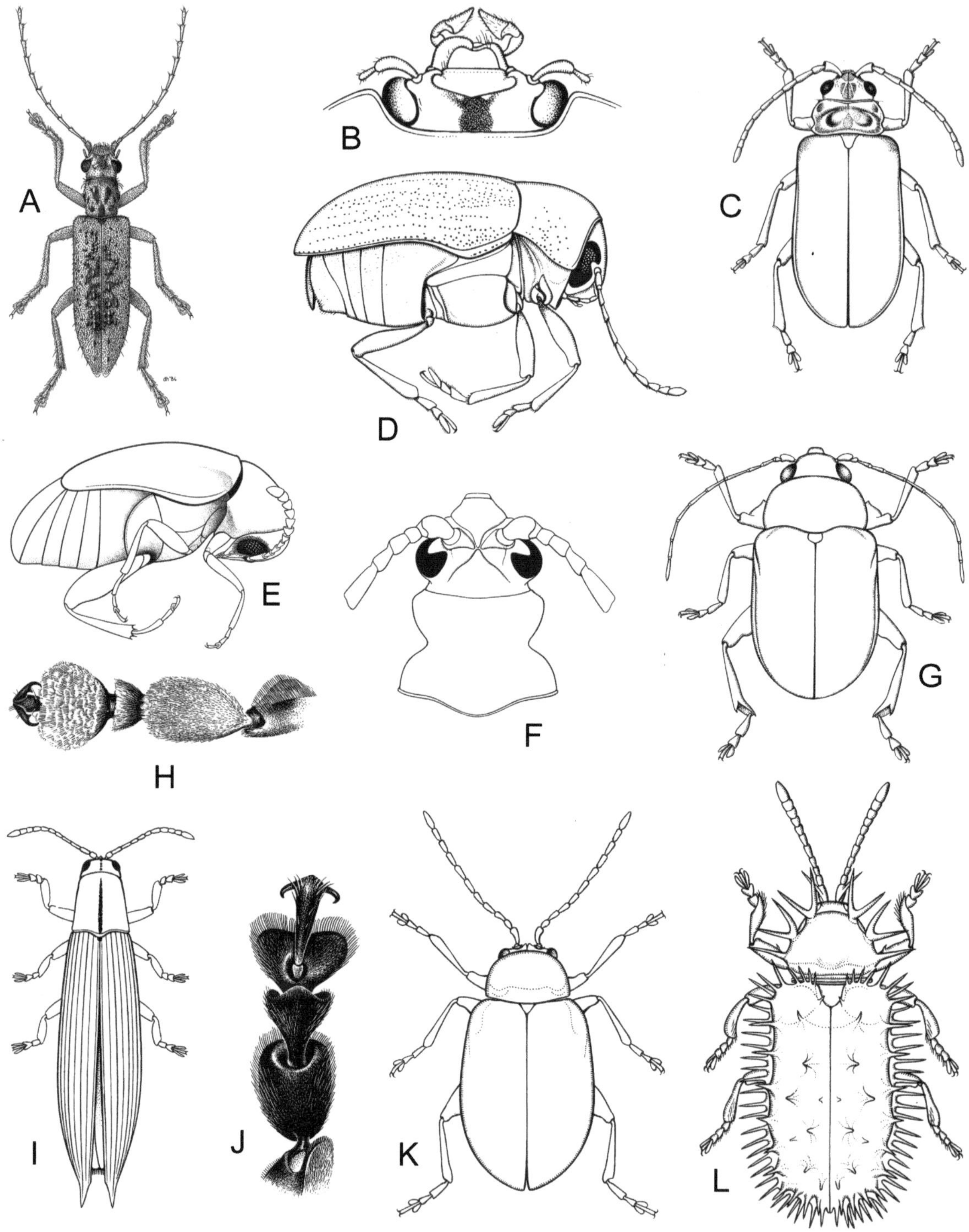

Fig. 96. A, *Palophagus bunyae* Kuschel (Megalopodidae: Palophaginae), adult, dorsal, from Kuschel and May 1990. B–L, Adult Chrysomelidae: B, *Sterromela trimaculata* (Chapuis) (Chrysomelinae), head, dorsal; C, *Poneridia australis* (Boheman) (Galerucinae), dorsal; D, *Aporocera haematodes* (Boisduval) (Cryptocephalinae), lateral; E, *Bruchidius lyndhurstensis* (Blackburn) (Bruchinaqe), lateral; F, *Stethopachys formosa* Baly (Criocerinae), head and prothorax, dorsal; G, *Edusella abdominalis* Lea (Eumolpinae), dorsal; H, *Sterromela trimaculata*, tarsus, ventral; I, *Eurispa howitti* Baly (Cassidinae), dorsal; J, *Sterromela trimaculata*, tarsus, dorsal; K, *Altica pagana* Blackburn (Galerucinae), dorsal; L, *Hispellinus multispinosus* (Germar) (Cassidinae), dorsal.

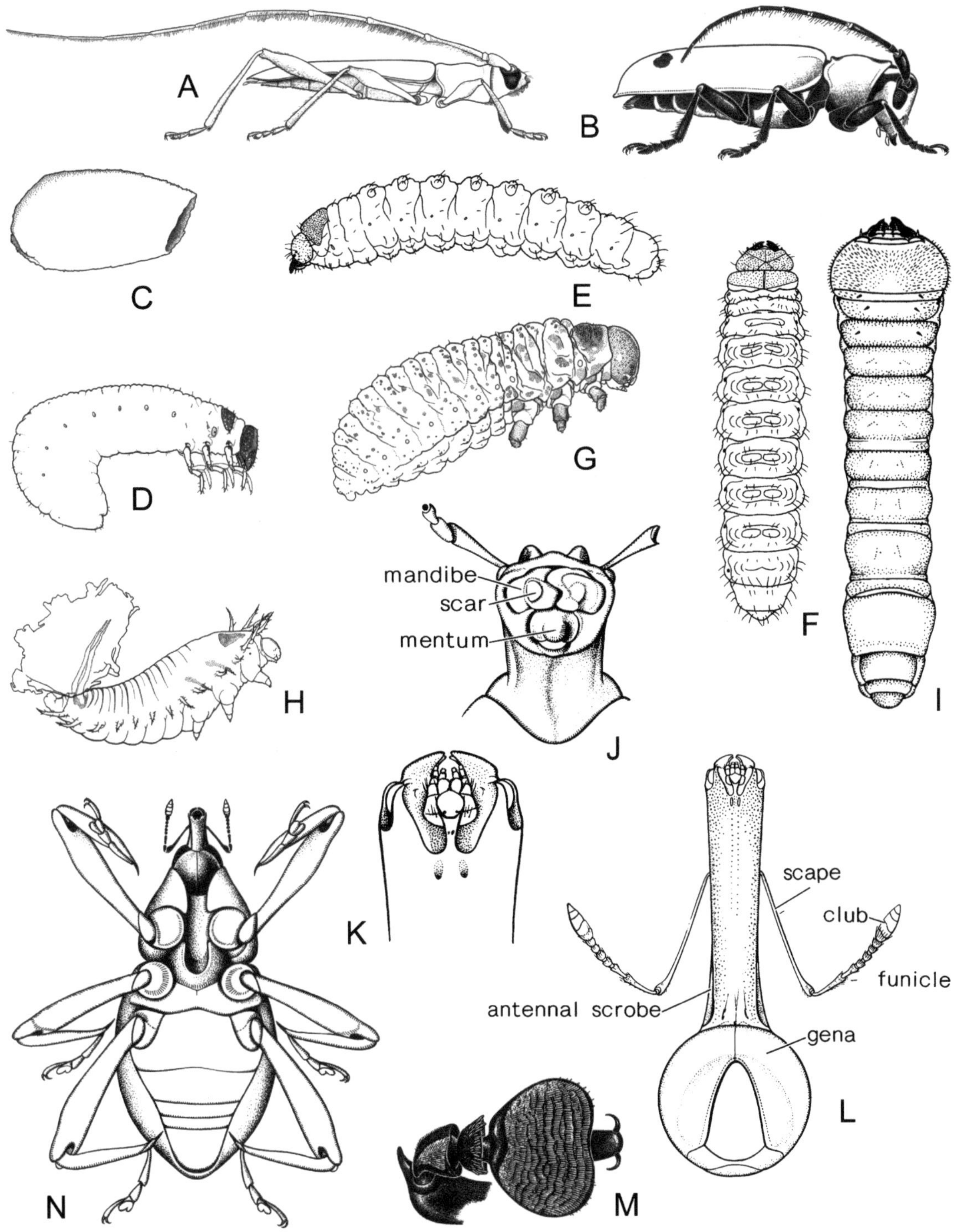

Fig. 97. A–B, Adult Cerambycidae, lateral: A, *Epithora dorsalis* (Macleay) (Cerambycinae); B, *Zygrita diva* Thomson (Lamiinae). C–I, Larva: C–D, Genus ? sp. (Chrysomelidae: Cryptocephalinae): C, larval case; D, lateral. E–F, *Palophagus bunyae* Kuschel (Megalopodidae: Palophaginae): E, lateral; F, dorsal, both from Kuschel and May 1990. G, *Lilioceris nigripes* (Fabricius) (Chrysomelidae: Criocerinae), lateral. H, *Aspidimorpha* sp. (Chrysomelidae: Cassidinae), lateral. I, *Phoracantha* sp. (Cerambycidae: Cerambycinae), ventral. J–N, Curculionidae: J, *Leptopius* sp. (Entiminae: Leptopiini), adult, rostral apex; K–L, *Orthorhinus cylindrirostris* (Fabricius) (Molytinae: Orthorhinini): K, adult rostral apex; L, head, ventral; M, *Diathetes morio* Pascoe (Dryophthorinae: Rhynchophorini), adult tarsus, ventral; N, *Paleticus subereus* (Pascoe) (Molytinae: Cryptorhynchini), adult, ventral.

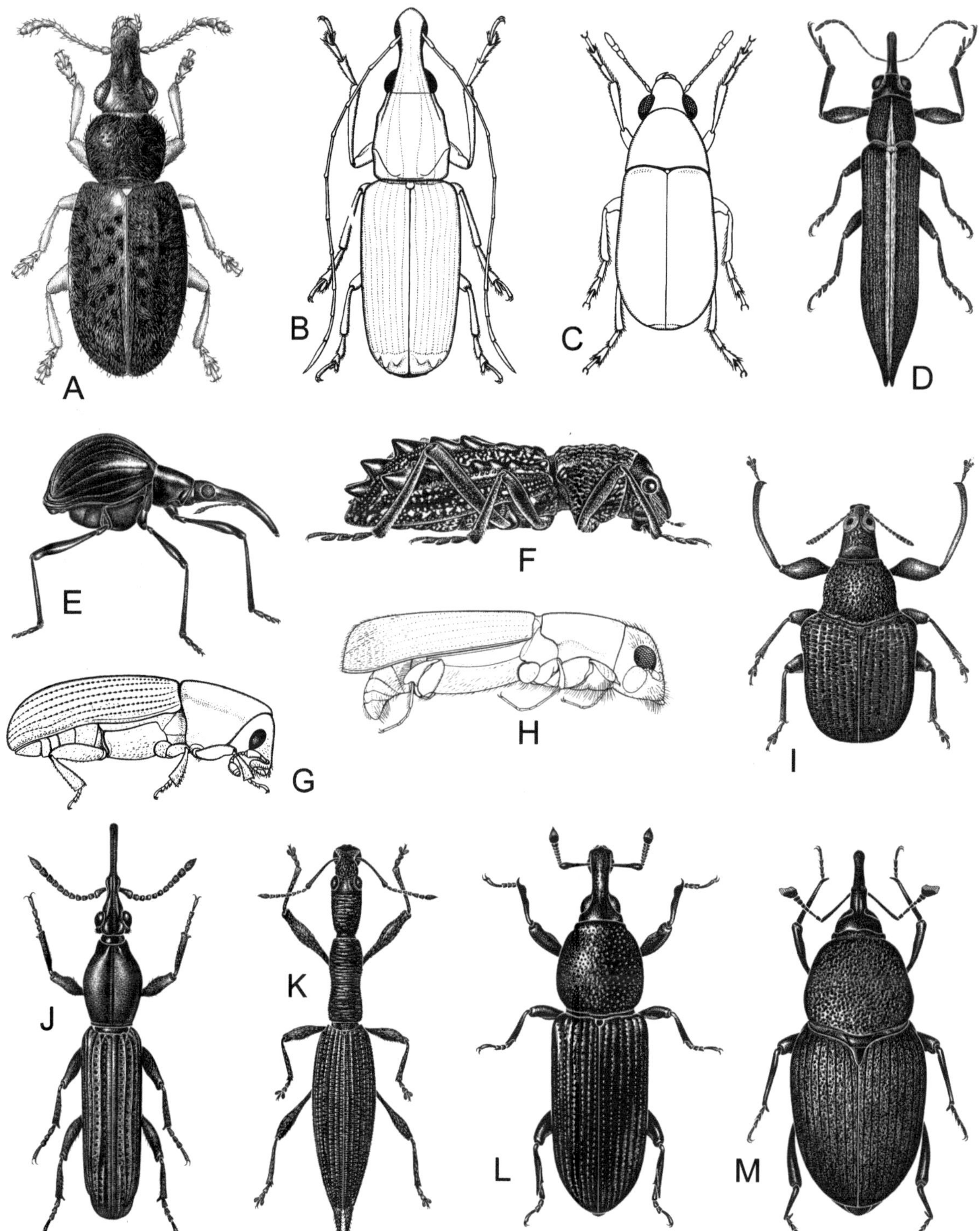

Fig. 98. Curculionoidea. A–D, Adult, dorsal: A, *Bunyaeus monteithi* Kuschel (Nemonychidae); B, *Ancylotropis waterhousei* Jekel (Anthribidae: Anthribinae); C, *Euparius lateripictus* (Jordan) (Anthribidae: Anthribinae); D, *Rhinotia suturalis* (Boisduval) (Belidae). E–H, Adult, lateral: E, '*Apion*' *convexipenne* Lea (Brentidae: Apioninae); F, *Gagatophorus boisduvalii* (Boisduval) (Curculionidae: Cyclominae); G, *Hylastes ater* (Paykull) (Curculionidae: Scolytinae); H, *Platypus subgranosus* Schedl (Curculionidae: Platypodinae). I–M, Adult, dorsal: I, *Euops falcata* (Guérin-Méneville) (Attelabidae: Attelabinae). J, *Euschizus internatus* (Pascoe) (Brentidae: Brentinae); K, *Rhadinosomus lacordairei* Pascoe (Curculionidae: Cyclominae); L. *Cossonus simsoni* Lea (Curculionidae: Cossoninae); M, *Trigonotarsus rugosus* (Boisduval) (Curculionidae: Dryophthorinae).

Ventral mouthparts retracted; maxillary articulating area present; cardo moderately strongly oblique, undivided; stipes about as long as wide; mala fixed, apically truncate with a few setae; inner apical angle with slight tooth; maxillary palp 3-segmented without distinct palpifer. Labium consisting of prementum and postmentum; ligula absent; labial palps 2-segmented, separated by distance equal to width of first palpomere. Hypopharyngeal bracon and sclerome present. Hypostomal rods short and slightly diverging; ventral epicranial ridges absent. Gular region very short.

Prothoracic tergum about as long as meso- or metatergum, with or without lightly pigmented plate. Legs absent. Abdomen more than twice as long as thorax; abdominal terga not extending laterally beyond edges of sterna, terga and sterna without patches or rows of asperities; meso- and metaterga and abdominal terga I–VIII each with transverse ambulatory ampulla; segment IX subequal to or slightly shorter than VIII; completely dorsal, without urogomphi; sternum 9 simple; segment X much shorter than IX, terminal, without pygopods; anal region posteriorly oriented; anal opening T-shaped. Spiracles annular–biforous with circular peritreme and moderately long accessory chambers. [WH Anderson 1947; May 1993, 1994; RS Anderson 2002a; Marvaldi 2003; Kuschel and Leschen 2010].

Classification and World Distribution

The family Nemonychidae is primarily an amphipolar group containing 75 species in 25 genera. There are three subfamilies: the Palaearctic Nemonychinae, the Holarctic Cimberidinae and the Rhinorhynchinae with three tribes, Mecomacerini from Australasia and southern South America, Rhynchitomacerini with two genera from Chile, and Rhinorhynchini from south-western North America, Central America, Venezuela, Chile, Argentina, Australia, New Guinea and New Zealand. [Kuschel and Leschen 2010].

Australian Fauna

Australian nemonychids belong to two tribes of **Rhinorhynchinae**: Mecomacerini with the genera *Aragomacer* Kuschel, *Bunyaeus* Kuschel (Fig. 98A), *Eutactobius* Kuschel, *Notomacer* Kuschel, and Rhinorhynchini with the genera *Basiliogeus* Kuschel, *Basiliorhinus* Kuschel, *Pagomacer* Kuschel, *Zimmiellus* Kuschel. Nemonychids are elongate and slightly flattened beetles clothed with decumbent hairs and usually scattered erect hairs. The rostrum is elongate and flattened to subcylindrical, the antennae are straight with a short scape and loose, weak club with insertions at the middle or near the apex of the rostrum, the labrum is free and on the same level as the clypeus, the mandibles are usually acute and opposable (exodont in *Zimmiellus*), palps are well developed and flexible, the pronotum lacks lateral carinae, there are one or two stridulatory files on the mesonotum, all tibiae have one or two spurs, pretarsal claws are toothed or bifid, ventrites are freely articulated and the pygidium is more or less concealed by the elytra. Larvae are elongate, slightly to moderately curved ventrally, slightly flattened and lightly pigmented, except for head capsule and protergal plate, and with a pair of well developed transverse folds on thoracic and most abdominal terga. The head is moderately deflexed, the epicranial stem short with a long median endocarina extending well in front of it. Both adult and larval nemonychids are pollen feeders, and Australian species have all been recorded from male cones of *Araucaria* and *Agathis* (Araucariaceae). All species occur in eastern Australia from northern NSW to northern QLD. [Kuschel 1994, Kuschel and Leschen 2010].

112. ANTHRIBIDAE

Adults (Figs 79J, 98B-C)

Length 1.5–20 mm. Body ~1.45–2.65 times as long as wide and usually either with elytra parallel-sided and pronotum at least slightly rounded, or with sides of pronotum and elytra independently rounded, rarely with sides evenly rounded (e.g. *Cisanthribus*), slightly flattened to strongly convex, usually yellowish-brown to dark brown or black (occasionally red), but often bicolored, with pattern of dark markings on elytra and occasionally pronotum; almost always densely clothed with decumbent, thickened setae usually forming a pattern, but sometimes with suberect or erect and usually fine hairs, rarely subglabrous (e.g. *Cisanthribus*). Head not to moderately strongly declined at base, or with vertex strongly declined, so that rostrum extends ventrally (e.g. in *Hucus*, *Litocerus*, *Mucronianus*). Head behind eyes not constricted posteriorly to form neck; occipital region without transverse ridge or stridulatory file, with short to moderately long occipital endocarina. Eyes slightly to strongly protuberant, usually well separated, but sometimes almost contiguous (e.g. in *Litocerus* and *Mucronianus*), not to deeply emarginate and finely or coarsely facetted, without interfacetal setae. Antennal insertions exposed from above and located between the eyes in Choraginae but more or less concealed from above and located laterally in front of eyes in Anthribinae, subantennal grooves (scrobes) well developed and sulciform or reduced and foveiform. Frontoclypeal suture absent; frontoclypeal region produced anteriorly to form a rostrum, which may be very short and strongly transverse to elongate and more than twice as long as wide, always more or less flattened, often with one or more longitudinal carinae dorsally; clypeal apex usually truncate to slightly emarginate. Labrum free, exposed, much narrower than clypeus, slightly transverse, broadly rounded, without tormae. Antennae 11-segmented, short (not reaching beyond base of pronotum) to very long, sometimes extending well beyond elytral apices, especially in males (e.g. in *Xenocerus*, *Ancylotropis*, *Exillis* and *Mecocerinopsis*), non-geniculate, very slender to moderately stout, usually with relatively weak, loose, 3-segmented club (appearing 4-segmented in *Eucorynus*), occasionally stronger and/or more compact; scape elongate but not much longer than pedicel. Mandible short and broad, unidentate or bidentate; incisor edge with or without tooth; molae well developed or reduced; prostheca absent; inner edge of mandibular base with short to moderately long, pubescent process. Maxilla with slender, articulated, setose galea and slightly wider setose lacinia, the latter with or without uncus; maxillary palps moderately well developed, 4-segmented, apical palpomere fusiform. Mentum slightly transverse, apically emarginate, more or less fused at base; ligula bilobed; labial palp 3-segmented, apical palpomere fusiform. Subgenal ridges absent. Gular region usually concave or transversely grooved; gular sutures absent. Corpotentorium absent; anterior tentorial arms not expanded mesally; cervical sclerites absent.

Pronotum ~0.40–1.0 times as long as wide, usually widest posteriorly or at middle; base not or only slightly narrower than combined elytral bases; sides usually slightly curved, not explanate, lateral pronotal carinae (Fig. 19A) almost always incomplete, not extending anteriorly beyond middle of pronotum, occasionally complete almost to anterior edge (e.g. in *Basitropis* and *Dendropemon*), often finely crenulate, occasionally highly reduced or absent; anterior angles not produced; posterior angles usually right or obtuse, occasionally acute; posterior edge straight, evenly curved or occasionally weakly sinuate, usually weakly margined; disc without paired basal impressions or median longitudinal groove, but almost always with a distinct transverse carina, which is finely crenulate, evenly rounded or sinuate, occasionally interrupted at middle and often anteriorly curved at each end or attached to lateral carina, if present, and located just anterior to posterior edge (sub-basal) or well in front of posterior edge (antebasal); surface of pronotum posterior to antebasal carina flat or slightly declined, that behind sub-basal carina steeply declined or vertical, so that carina abuts anterior edges of elytra. Prosternum in front of coxae shorter than to distinctly longer than shortest diameter of coxal cavity, flat to slightly convex, not carinate. Prosternal process complete but interrupted at middle (intercoxal process strongly concave or apparently divided by transverse suture), the apex expanded and truncate and more or less fused to postcoxal (notal) projections (Fig. 19E). Notosternal sutures probably absent, but a suture sometimes extending dorsolaterally from the lateral edge of the procoxal cavity and ending at the lateral edge of the pronotal disc (but never extending to anterior edge of prothorax). Procoxae not or slightly projecting below prosternum, with long concealed lateral extensions; trochantins concealed. Procoxal cavities subcircular, contiguous to moderately widely separated but joined by transverse suture, externally broadly closed, without narrow lateral extensions, internally broadly closed; postcoxal (hypomeral) processes fused to disjunct apex of prosternal process.

Scutellar shield well developed, highly reduced or absent, abruptly elevated, anteriorly simple, posteriorly acute to broadly rounded or truncate; mesoscutum without stridulatory file. Elytra ~0.85–2.0 times as long as wide and 1.75–3.0 times as long as pronotum; anterior edges often sharp or with raised carina; humeri well developed or not; sides usually subparallel to weakly rounded and conjointly rounded to slightly truncate at apex, but often exposing strongly declined or vertical pygidium; disc usually 10 distinct puncture rows and a scutellary striole, sometimes with various elevations or tubercle, with posteriorly directed horns in *Commista*; apical end of elytral suture with internal interlocking ridge; epipleura usually narrow and incomplete. Mesoventrite partly fused to or separated by complete sutures from mesanepisterna, which are distinctly separated at midline; anterior edge on different plane than metaventrite, without paired procoxal rests, but with broadly transverse prothoracic rest. Mesoventral cavity absent. Mesocoxae not conical and projecting, with concealed trochantins. Mesocoxal cavities moderately to very widely separated, subcircular, not or slightly oblique, closed laterally by meeting of mesoventrite and metaventrite. Mesoventral process narrowly rounded to broadly rounded or truncate at apex; mesometaventral junction usually formed by an abutting or overlapping of ventrites. Metaventrite slightly convex; transverse (metakatepisternal) suture long but close to posterior edge of ventrite; discrimen usually complete to base of anterior intercoxal process, occasionally short or absent; postcoxal lines usually absent, occasionally arched and strongly recurved; metaventrite reduced in wingless forms; exposed portion of metanepisternum moderately elongate; metepimeron slightly exposed. Metacoxae narrowly to widely separated, not extending laterally to meet elytra; plates absent. Metendosternite usually with long stalk, long lateral arms, well developed but slender laminae and very short or barely indicated anterior process bearing narrowly separated anterior tendons. Hind wing with apical field at least half as long as total wing length, containing distinct sclerite just beyond radial cell and two RP remnants; radial cell small, short and broad or absent; inner posterobasal angle right; cross-vein r3 moderately long and oblique to absent, sometimes fused basally with r4, which may have an apical spur; basal portion of RP short to moderately long; R-M loop relatively narrow; medial spur usually extending almost to wing margin; medial field with 3 or fewer free veins and no medial fleck; MP_{3+4} usually without base; wedge cell absent; anal lobe usually well developed without or with weak embayment; wings sometimes absent. Legs short and stout to moderately long and slender; trochanterofemoral joint strongly oblique with base of femur sometimes meeting coxa; femora not to slightly inflated; tibial spurs absent. Tarsi pseudotetramerous, 5-5-5 in both sexes with tarsomere 4 reduced and 3 distinctly bilobed; pretarsal claws usually toothed or cleft; empodium not apparent.

Abdomen with 5 ventrites, with the first 4 or all 5 connate; ventrite 1 at its shortest point subequal in length to 2, usually without postcoxal lines (occasionally with recurved lines just behind edges of coxal cavities); intercoxal process acute to broadly rounded or rarely truncate; ventrites 2, 3 and 4 with large, paired sex patches in *Dendropemon*. Abdominal terga usually lightly pigmented, except for tergite VII (pygidium), which bears a basal median notch interlocking with elytral apices; tergites III–VI in *Araecerus* with moderately sclerotised plates and numerous rounded microsclerites; patches wing-folding asperities sometimes present on tergites V and VI; functional spiracles on segments I–VII, those on I–VI in pleural membrane, those on VII at lateral edges of tergite (sometimes reduced). Tergite VIII in male partly concealed beneath pygidium and relatively lightly sclerotised; sternite VIII somewhat reduced, without median strut. Anterior edge of segment IX in male with distinct spiculum gastrale; tergite X membranous. Aedeagus cucujiform, usually symmetrical; anterior edge of tegmen with moderately long strut; parameres fused together and sometimes to phallobase. Penis flattened and usually curved, with dorsal and ventral lobes (tectum and pedon) and with long, paired anterior struts; endophallus with complex armature. Segment VIII in female more or less membranous, with short, fixed spiculum ventrale. Ovipositor elongate and lightly sclerotised, except for baculi and distal coxital lobes; proctiger elongate, with paired baculi; paraprocts more or less subequal in length to coxites and lateral to them, with longitudinal baculi; coxites divided into elongate basal lobe with sclerotised longitudinal baculum, and short, broad, distal lobe, which is heavily sclerotised with laterally directed teeth and subterminal, sclerotised styli. Internal tract with strongly curved, sclerotised capsular spermatheca. [Valentine 1960, 1999, 2002; Morimoto 1972, 1978, 1979, 1980, 1981; Holloway 1982, 1984, 1985; Thompson 1992; Zimmerman 1994a].

Larvae

Body elongate, usually more or less parallel-sided and circular in cross-section, sometimes with enlarged prothorax; all surfaces lightly pigmented, except for head capsule or mouth-frame and mandibles; vestiture consisting of scattered fine, simple setae, sometimes more abundant on thorax. Head hypognathous, protracted and more or less circular in outline or partially retracted, posterolaterally compressed and narrowed behind; posterior edge not emarginate, occasionally produced and acute; postoccipital condyles absent; ecdysial lines often obscure; epicranial stem long; frontal arms V-shaped, contiguous at base; median endocarina extending anteriorly between frontal arms. Stemmata on each side 3 or fewer (one stemma with lens and 1 or 2 additional eye spots). Antennae very short, 1-segmented, with terminal, conical sensorium. Frontoclypeal suture distinct or indistinct; labrum free, with 4 pairs of primary setae and lateral tormae. Mandibles short and broad, obliquely bidentate; accessory ventral process absent; incisor edge usually with subapical tooth, mola usually well developed (sometimes reduced); prostheca absent. Ventral mouthparts retracted; maxillary articulating area present; cardo slightly to strongly oblique, undivided; stipes longer than wide; usually with fixed, rounded, setose mala and sclerotised laciniar spine (absent in *Euciodes* and *Phloeobius*); maxillary palp 2- or 3-segmented, occasionally with distinct palpifer. Labium consisting of prementum and postmentum; ligula absent; labial palps usually 2-segmented (1-segmented in *Araecerus*), narrowly separated. Hypopharyngeal bracon present; hypopharyngeal sclerome usually present (absent in *Euciodes*). Hypostomal rods usually short and subparallel or converging; ventral epicranial ridges absent. Gular region very short or absent. Tentorial bridge divided on either side of midline.

Prothoracic tergum usually about as long as meso- or metatergum (as long as the two combined in *Phloeobius*), without pigmented plate. Legs usually very short and 1- or 2-segmented, without claw-like pretarsus (legs absent in *Euciodes*). Abdomen more than twice as long as thorax (more than 4 times as long in *Euciodes*); abdominal terga not extending laterally beyond edges of sterna, terga and sterna without patches or rows of asperities; most abdominal terga with 2 to 4 plicae; terga I–VI each with transverse ambulatory ampulla in *Euciodes*; segment IX usually shorter than VIII; completely dorsal, without urogomphi; sternum 9 simple; segment X much shorter than IX, terminal, without pygopods; anal region posteriorly oriented; anal opening X-shaped (4-lobed). Spiracles annular–biforous, annular–uniforous, annular–multiforous or annular; thoracic spiracles intersegmental or on mesothorax. [Anderson 1947; May 1993, 1994; Valentine 2002; Marvaldi 2003].

Classification and World Distribution

The family Anthribidae is a cosmopolitan family comprising ~3000 species in 365 genera. The group is divided into three subfamilies, two of which occur in Australia. The exotic subfamily Urodontinae, which occurs in Europe, Central Asia, Asia Minor, northern Africa and southern Africa, is sometimes excluded from the family (Crowson 1984b; Thompson 1992; Louw 1993). [Alonso-Zarazaga and Lyal 1999, 2002].

Australian Fauna

The Australian anthribid fauna includes 35 genera in 16 tribes of **Anthribinae**, (Fig. 98B–C) plus six genera not currently placed in a tribe, and six genera of **Choraginae**, placed in three tribes. In addition there are at least 28 new genera and many new species to be described (Zimmerman 1991, 1992, 1994a). Anthribids are moderately elongate to oblong or ovate, slightly to strongly convex beetles, usually clothed with black, brown and/or white decumbent hairs or scales forming a pattern. The rostrum is very short to moderately long and more or less flattened, the antennae are straight, very long in some males, with a club that may be indistinct, the labrum is free and either on a different plane than the clypeus or separated from it by membrane, the mandibles are well developed and more or less acute, the palps are well developed and flexible, gular sutures are absent, the pronotum usually has a transverse, sub-basal carina joined to the lateral carinae, which may be very short to complete, the legs lack tibial spurs, the pretarsal claws are usually toothed or cleft, and the pygidium is exposed beyond the edges of the elytra. Pheromone-producing setiferous sex patches are sometimes found in males on the abdominal ventrites or occasionally in other locations (Holloway 1985). Larvae are short, broad, more or less C-shaped and very lightly pigmented, with antennae minute and legs highly reduced without claws or absent. The frontoclypeal suture is present, mandibles are robust, the mala has a thorn-like lacinia near the middle of the inner edge, and most abdominal terga have two transverse folds. Most anthribid larvae feed in dead wood, but some are associated with the fruiting bodies of certain fungi (Ascomycetes: Pyrenomycetes) and others feed on seeds of various kinds. *Araecerus coffeae* (Fabricius) is a cosmopolitan pest of coffee, cocoa beans and spices, and *A. palmaris* (Pascoe) is unusual in that the larva develops in dried fruits, but Australian specimens are commonly found in fungus galls (*Uromycladium* spp., Uredinales). As noted by Zimmerman (1994a) the Australian anthribid fauna is dominated by invaders from Indonesia; however, some of the smaller, southern forms appears to be related to New Zealand taxa. One of the more unusual taxa is a species of *Cisanthribus* from Lord Howe Island, relatives of which are known from New Caledonia, Samoa and the Society Islands; these are minute, ovate, glabrous and wingless forms. [Holloway 1982].

Australian Subfamilies, Tribes and Genera

Anthribinae: Anthribini: *Streneoderma* Blackburn. **Basitropini:** *Basitropis* Jekel, *Plesiobasis* Jordan. **Cappadocini:** *Eczesaris* Pascoe, *Ethneca* Pascoe. **Corrhecerini:** *Apatenia* Pascoe, *Cacephatus* Blackburn. *Commista* Jordan, *Enspondus* Blackburn, *Phaulimia* Pascoe. **Cratoparini:** *Euparius* Schoenherr. **Discotenini:** *Ancylotropis* Jekel, *Genethila* Pascoe. **Eceloner-ini:** *Dendropemon* Schoenherr, *Eucorynus* Schoenherr. **Mauiini:** *Erichsonocis* Zimmerman, *Mauia* Blackburn, *Protaedus* Pascoe. **Ozotomerini:** *Ozotomerus* Perroud. **Platytsomini:** *Euciodes* Pascoe, *Exillis* Pascoe, *Phloeobius* Schönherr, *Platystomos* Schneider. **Proscoporhinini:** *Antribisomus* Perroud. **Sintorini:** *Epargemus* Blackburn, *Eupanteos* Jordan, *Telala* Jordan. **Stenocerini:** *Helmoreus* Holloway. **Tropiderini:** *Hucus* Pascoe, *Litocerus* Schönherr, *Mecocerinopis* Zimmerman, *Mucronianus* Jordan, *Notoecia* Blackburn. **Xenocerini:** *Xenocerus* Schönherr. **Zygaenodini:** *Xynotropis* Blackburn. **Tribe ?:** *Atropideres* Zimmerman. *Entromus* Blackburn, *Howeanthribus* Zimmerman, *Leaoan-*

thribus Zimmerman, *Taburnus* Zimmerman, *Teratanthribus* Zimmerman.

Choraginae: Choragini: *Araeocerodes* Blackburn. **Araecerini:** *Araecerus* Schönherr, *Deropygus* Sharp, *Misthosima* Pascoe, *Stenorhis* Jordan. **Cisanthribini:** *Cisanthribus* Zimmerman.

113. BELIDAE

Adults (Figs 11B, 79A–B, K, 98D)

Length 5–23 mm. Body ~1.9–6.4 times as long as wide, with elytra parallel-sided or slightly wider near apex and pronotum slightly to strongly rounded laterally, slightly flattened to moderately convex, usually yellowish-brown to reddish-brown, dark brown or black, occasionally bicolored on elytra or pronotum and elytra; almost always densely clothed with decumbent, thickened setae, variously coloured and often forming a pattern, sometimes interspersed with suberect finer setae, thickened setae sometimes limited to small patches. Head not to slightly declined at base. Head behind eyes not constricted posteriorly to form neck; occipital region without transverse ridge or stridulatory file, with moderately long occipital endocarina. Eyes strongly protuberant, well separated, entire and finely facetted, without of with few very short, fine interfacetal setae. Antennal insertions concealed from above and located in front of eyes, usually well out on rostrum; subantennal grooves (scrobes) well developed and sulciform. Frontoclypeal suture absent; frontoclypeal region produced anteriorly to form a rostrum, which may be very short (barely indicated in *Arhinobelus*) to elongate, up to 8 times as long as wide, and slightly flattened to subcylindrical, not or only slightly widened apically and without distinct longitudinal dorsal carinae; clypeal apex often notched. Labrum absent (completely fused to clypeus). Antennae 11-segmented, non-geniculate, moderately long, slender, and incrassate or with weak, 3- to 5-segmented club; scape at least slightly elongate and slightly to much longer than pedicel. Mandible very short and broad, bidentate or tridentate; incisor edge, without subapical teeth, mola or prostheca; inner edge of mandibular base with moderately long, pubescent process. Maxilla with short, broad, fixed, setose or spinose galea and lacinia, the latter without uncus; maxillary palps moderately very short, 4-segmented, apical palpomere cylindrical. Mentum elongate, apically truncate or emarginate, fused at base to gulamentum; ligula undivided; labial palp short, 3-segmented, apical palpomere cylindrical. Subgenal ridges absent. Pregular usually very long and narrow; gular sutures separated; gula longer than wide. Corpotentorium absent; anterior tentorial arms not expanded mesally; cervical sclerites absent.

Pronotum ~0.75–1.1 times as long as wide, widest behind middle, sometimes constricted anteriorly; base not to distinctly narrower than combined elytral bases; sides usually slightly curved or sinuate, not explanate, lateral pronotal carinae absent; anterior angles not produced; posterior angles rounded to slightly acute; posterior edge slightly to strongly sinuate or biemarginate, usually weakly margined, but with distinct ventral interlocking ridge; disc sometimes with median longitudinal groove and endocarina, and/or with paired, broad basal impressions into which anterior elytral projections fit. Prosternum in front of coxae slightly to much shorter than shortest diameter of coxal cavity, flat to somewhat declined, not carinate. Prosternal process complete but interrupted at middle (intercoxal process strongly concave in middle), the apex slightly expanded and then narrowed to form an apically acute or angulate, apparently disconnected piece usually referred to as a sternellum. Notosternal sutures incomplete, extending laterally from edge of procoxal cavity and abruptly curved anteriorly, but never reaching anterior edge of prothorax. Procoxae projecting well below prosternum, with short concealed lateral extensions; trochantins concealed pr barely exposed. Procoxal cavities subcircular, contiguous or narrowly separated, externally broadly closed, with or without narrow lateral extensions, internally broadly closed; postcoxal (hypomeral) processes meeting prosternal process or meeting one another behind prosternal process.

Scutellar shield well developed, abruptly elevated, anteriorly simple, posteriorly acute to broadly rounded or truncate; mesoscutum without stridulatory file. Elytra ~1.5–5.3 times as long as wide and 3.3–5.4 times as long as pronotum; anterior edges often sharp or with raised carinae; humeri well developed; sides usually subparallel, sometimes diverging to posterior third, usually conjointly broadly to narrowly rounded or subacute, with independently acute apices in some *Rhinotia*; abdominal tergites always concealed; disc irregularly punctate, without scutellary striole, with protuberances on either side of scutellum in *Cyrotyphus*, *Pachybelus* and *Isacantha*; lateral portion with internal interlocking ridge just anterior to middle; epipleura very narrow but sometimes complete. Mesoventrite separated by complete sutures from mesanepisterna, which are distinctly separated at midline; anterior edge on at least slightly different plane than metaventrite, without paired procoxal rests, but with broadly transverse prothoracic rest. Mesoventral cavity absent. Mesocoxae not projecting, with exposed trochantins. Mesocoxal cavities narrowly separated, subcircular, not or slightly oblique, closed laterally by meeting of mesoventrite and metaventrite or partly closed by mesepimeron. Mesoventral process broadly rounded or truncate at apex; mesometaventral junction monocondylic. Metaventrite moderately convex; transverse (metakatepisternal) suture long but not crossing midline; discrimen usually complete to base of anterior intercoxal process; postcoxal lines absent; exposed portion of metanepisternum moderately elongate; metepimeron slightly exposed. Metacoxae narrowly separated, not extending laterally to meet elytra; plates absent. Metendosternite with long stalk, long lateral arms, well developed laminae and very short anterior process bearing widely separated anterior tendons. Hind wing with apical field about a third as long as total wing length, containing distinct sclerite just beyond radial cell and two RP remnants; radial cell slightly elongate, with inner posterobasal angle right or obtuse; cross-vein r3 usually moderately long and oblique, sometimes absent; r4 without apical spur; basal portion of RP moderately long; R-M loop relatively broad; medial spur straight, extending to wing margin, where there is a distinct embayment; medial field usually with 5 free veins and no medial fleck; MP_{3+4} usually with basal cross-vein; wedge cell absent; anal lobe usually well developed without embayment. Legs usually short and stout, with the fore legs longer than the other two pairs; trochanterofemoral joint oblique but with base of femur usually separated from coxa; all tibiae, but especially fore tibia with row of teeth along inner edge; tibial spurs usually 2-2-2, but protibia often with single spur combined with mucro at internal apical angle. Tarsi pseudotetramerous, 5-5-5 in both sexes with tarsomere 4 reduced and 3 distinctly bilobed; pretarsal claws simple; empodium not apparent.

Abdomen with 5 free ventrites; ventrite 1 at its shortest point subequal in length to 2, without postcoxal lines; intercoxal process acute; ventrite 5 (sternite VII) emarginate exposing a portion of sternite VIII in some male *Rhinotia*. Abdominal tergites II–VI as darkly pigmented as tergite VII, which bears a pair of wing-folding patches; functional spiracles on segments I–VII, all in pleural membrane. Tergite VIII in male usually concealed beneath tergite VII, exposed in some *Rhinotia*; sternite VIII with short median strut. Anterior edge of segment IX in male with distinct spiculum gastrale; tergite X membranous. Aedeagus cucujiform, symmetrical; anterior edge of tegmen with moderately long strut; parameres partly or entirely fused together and fused to phallobase; penis with dorsal and ventral lobes. Penis flattened with dorsal and ventral lobes (tectum and pedon) and with paired anterior struts; endophallus with complex armature. Segment VIII in female with sclerotised tergite and sternite and long, fixed spiculum ventrale. Ovipositor moderately elongate and lightly sclerotised, except for baculi, distal coxital lobes and styli; proctiger with paired baculi; paraprocts more or less subequal in length to coxites and lateral to them, with longitudinal baculi; coxites divided into elongate basal lobe with sclerotised longitudinal baculum, and short distal lobe, which is heavily sclerotised with terminal sclerotised styli. Internal tract with slender, sclerotised capsular spermatheca. [Kuschel 1959, 1995; Vanin 1976; Thompson 1992; Anderson 2002b, 2005; Kuschel 2003; Kuschel and Leschen 2003; Marvaldi *et al.* 2006].

Larvae

Body usually elongate, more or less parallel-sided, circular in cross-section and ventrally curved or C-shaped, shorter, broader and less curved in *Brachybelus*; all surfaces lightly pigmented, except for mouth-frame and mandibles and sometimes posterior portion of protergum; vestiture consisting of simple setae, which may be dense. Head prognathous and strongly retracted, at least slightly elongate and somewhat flattened; posterior edge deeply emarginate in *Brachybelus* and *Sphinctobelus*, not in *Cyrotyphus* or *Rhinotia*, postoccipital condyles absent; ecdysial lines often obscure; epicranial stem long; frontal arms V-shaped, contiguous at base; endocarina V-shaped. Stemmata 1–6 on each side (3 vestigial eye spots). Antennae very short, usually 2-segmented, with conical sensorium on basal antennomere as long as or longer than apical antennomere (1-segmented with terminal sensorium in *Brachybelus*). Frontoclypeal suture present; labrum free, with 4 pairs of primary setae; tormae and epipharyngeal rods absent. Mandibles short and broad, unidentate or bidentate, with or without additional tooth on cutting edge; mola and prostheca absent. Ventral mouthparts retracted; maxillary articulating area present; cardo slightly oblique, undivided; stipes longer than wide; usually with fixed, rounded, setose mala, usually cleft at apex, without laciniar spine; maxillary palp 3-segmented, with distinct palpifer. Labium consisting of prementum and postmentum; ligula rounded, longer than basal palpomere; labial palps 2-segmented and widely separated. Hypopharyngeal bracon present; hypopharyngeal sclerome absent. Hypostomal rods and ventral epicranial ridges absent. Gular region absent. Tentorial bridge wide, not divided on either side of midline.

Prothoracic tergum usually longer than meso- and metatergum combined, usually with posterior sclerotised elevation that may be longitudinally striate or asperate. Legs absent. Abdomen more than twice as long as thorax; abdominal terga not extending laterally beyond edges of sterna, terga and sterna without patches or rows of asperities; most abdominal terga with 2 or 3 plicae, sometimes with transverse ambulatory ampullae; segment IX shorter than VIII; completely dorsal, without urogomphi; sternum 9 simple; segment X much shorter than IX, terminal, without pygopods; anal region posteriorly or posteroventrally oriented; anal opening Y-shaped (3-lobed). Spiracles annular-biforous or annular; thoracic spiracles on mesothorax. [Emden 1938; WH Anderson 1941; May 1993, 1994; RS Anderson 2002b; Marvaldi 2003, 2005].

Classification and World Distribution

The family Belidae includes ~350 species in 37 genera. The subfamily Belinae occurs in Australia, New Zealand, New Guinea, the Solomon Islands and Moluccas, and in Brazil, Argentina, Paraguay and Chile, while the Oxycoryninae occurs from southern Florida and Cuba to Argentina and Chile, and in South Africa, North Africa, the Canary Islands, South-east Asia, Indonesia, the Philippines, New Zealand, New Caledonia, Hawaii and some other Pacific island groups. [Vanin 1976; Kuschel and Leschen 2003; Marvaldi 2005; Anderson 2005].

Australian Fauna

All Australian belids belong to the subfamily **Belinae** and are placed in three tribes: Agnesiotidini, Pachyurini and Belini (Kuschel and Leschen 2003). Belids are elongate, usually more or less parallel-sided beetles, clothed with decumbent hairs (Figs 11B, 98D). The rostrum is usually moderately long, but may be relatively short in some Pachyurini (e.g. *Agathobelus* and *Apagobelus*) and almost absent in *Arhinobelus*. The antennae are straight, without or with a very weak, 4-segmented club (with intermediate antennomeres enlarged in *Rhinotiodes*) inserted at the middle or near the base of the rostrum, the labrum is not visible, the palps are more or less rigid and the gular sutures short and separate. The pronotum lacks lateral carinae, the tibiae usually have a row of small granules along the outer edge, the protibiae often have teeth on the inner edge, and the pygidium is concealed. Larvae are broad, slightly flattened, ventrally curved, lightly pigmented and pubescent, with an enlarged, strongly declivous prothorax. The head is elongate with an indistinct median endocarina, the frons sometimes bears a median spine, labral rods are absent; the mala is simple and rounded, the maxillary palpifer is well developed, the protergum often has a sclerotised keeled plate on the posterior half, and the abdominal terga have two indistinct transverse folds. Belid larvae are known to bore into wood and are sometimes relatively common in the branches and twigs of acacias. Adults fly actively during the day and some species of *Rhinotia* are mimics of Lycidae (Fig. 79A). [Zimmerman 1994a].

Australian Subfamily, Tribes and Genera

Belinae: Agnesiotidini: *Agnesiotis* Pascoe, *Cyrotyphus* Pascoe; **Pachyurini:** *Agathobelus* Zimmerman, *Apagobelus* Zimmerman, *Arhinobelus* Zimmerman, *Basiliobelus* Zimmerman, *Brachybelus* Zimmerman, *Habrobelus* Zimmerman, *Hadrobelus*

Zimmerman, *Macrobelus* Lea, *Pachybelus* Zimmerman, *Pachyura* Hope, *Sphinctobelus* Zimmerman; **Belini:** *Araiobelus* Zimmerman, *Isacantha* Hope, *Isacanthodes* Zimmerman, *Rhinotia* Kirby, *Rhinotiodes* Zimmerman, *Stenobelus* Zimmerman.

114. ATTELABIDAE

Adults (Figs 12C, 98I)

Length 1.3–7 mm. Body ~1.65–2.5 times as long as wide with sides of elytra subparallel or weakly rounded and pronotum independently rounded and usually narrowed anteriorly, moderately convex, yellowish-brown to red, dark brown or black in color, rarely bicolored; subglabrous (*Euops*) or clothed with fine, decumbent hairs (*Auletobius*). Head not to moderately declined at base, not constricted posteriorly to form neck; occipital region without transverse ridge or stridulatory file, with moderately long median endocarina. Frontal region abruptly declined in *Euops*, without median groove or endocarina. Eyes moderately to strongly protuberant, widely separated in *Auletobius*, subcontiguous dorsally in *Euops*, not emarginate, without interfacetal setae. Antennal insertions barely concealed from above, located relatively close to base of rostrum, laterally in *Auletobius*, dorsally on either side of a median elevation in *Euops*; subantennal grooves (scrobes) short and weakly defined. Frontoclypeal suture absent; clypeal region produced forward to form rostrum that is slightly longer than wide, sharply declined, somewhat flattened and distinctly expanded apically in *Euops* or more than 3 times as long as wide and not or only slightly declined or expanded in *Auletobius*. Labrum completely fused to clypeus. Antennae 11-segmented, moderately long with weak 3-segmented club; scape elongate, only slightly longer than pedicel. Mandible short and broad to slightly elongate, exodont in some *Auletobius*, with both mesally and laterally directed teeth at apex, non-exodont and unidentate in *Euops*; incisor edge with 1 or more subapical teeth; mola weakly developed and sub-basal, prostheca absent; inner edge of mandibular base with long pubescent process. Maxilla with short, broad, densely setose galea and narrower or reduced lacinia bearing a few spines; maxillary palps short and inflexible, extending only slightly beyond malar apex, 4-segmented, apical palpomere cylindrical. Mentum elongate and fused at base in *Auletobius*, shorter, broader and basally articulated in *Euops*; ligula simple; labial palp 3-segmented in *Auletobius*, absent in *Euops*. Gular sutures fused together. Corpotentorium absent; cervical sclerites absent.

Pronotum ~0.8–1.05 times as long as wide, widest at base in *Euops*, at about middle in *Auletobius*, base distinctly narrower than combined elytral bases; sides at least slightly curved, lateral pronotal carinae absent; anterior angles not produced; posterior angles rounded in *Auletobius*, right to obtuse in *Euops*; posterior edge straight or weakly sinuate; disc with weak, sub-basal transverse groove (obscured mesally in *Auletobius*), with narrow anterior collar in *Euops*. Prosternum in front of coxae about half as long as shortest diameter of coxal cavity and flat to slightly convex in *Euops*, much shorter and concave in *Auletobius*. Prosternal process but broadly interrupted at middle with anterior section short and acute and posterior section fused to postcoxal processes in *Euops*, with anterior section absent and posterior section forming a minute process (sternellum) lying between postcoxal processes in *Auletobius*. Notosternal sutures transverse and incomplete. Procoxae projecting below prosternum, with short concealed lateral extensions; trochantins concealed. Procoxal cavities subcircular, contiguous, externally broadly closed, postcoxal (hypomeral) processes meeting disjunct apex of prosternal process in *Auletobius* or fused with prosternal process in *Euops*, without narrow lateral extensions, internally broadly closed.

Scutellar shield not abruptly elevated, anteriorly simple, posteriorly broadly rounded. Elytra ~1.15–1.7 times as long as wide and1.7–2.5 times as long as pronotum, weakly rounded laterally with elytral apices meeting or almost meeting at the suture and exposing no abdominal tergites in *Auletobius*, subparallel laterally and broadly truncate exposing pygidium in *Euops*; disc irregularly punctate in *Auletobius*, with 9 puncture rows and a scutellary striole in *Euops*; epipleura narrow to very narrow and incomplete; with internal longitudinal interlocking ridge laterally at about middle in *Euops*. Mesoventrite separated by complete sutures from mesanepisterna, which are distinctly separated at midline in Euops, partly fused to mesanepisterna in *Auletobius*; anterior edge on different plane than metaventrite, without paired procoxal rests, but usually narrow transverse prothoracic rest continuing onto mesanepisterna. Mesoventral cavity absent. Mesocoxae not conical and projecting, with concealed trochantins. Mesocoxal cavities moderately narrowly separated in *Euops*, contiguous in *Auletobius*, subcircular, not or slightly oblique, closed laterally by meeting of mesoventrite and metaventrite. Mesoventral process emarginate at apex and mesometaventral junction dicondylic in *Euops*, process absent in *Auletobius*. Metaventrite moderately convex; transverse (metakatepisternal) suture long but not crossing midline; discrimen complete to base anterior intercoxal process; postcoxal lines absent; exposed portion of metanepisternum ~3 times as long as wide in *Auletobius*, less than twice as long as wide in *Euops*; metepimeron exposed. Metacoxae narrowly separated, not extending laterally to meet elytra; plates absent. Metendosternite with long stalk, long lateral arms, short, broad laminae and very short anterior process bearing narrowly to widely separated anterior tendons. Hind wing with apical field more than half as long as total wing length, containing distinct sclerite just beyond radial cell and remnants of RP; radial cell moderately well developed, slightly longer than wide with base more or less straight and inner posterobasal angle acute; cross-vein r3 moderately long and oblique, arising close to base of r4, which may be incomplete; basal portion of RP moderately long, its apical portion extending to wing margin; R-M loop moderately broad; medial spur sinuate and extending almost to wing margin; medial field with 3 free veins in *Euops*, 1 or 2 in *Auletobius*, and no medial fleck; MP_{3+4} without base; wedge cell absent; anal lobe moderately well developed or somewhat reduced, without embayment. Legs slender in *Auletobius* with femora not or only slightly inflated and tibial apices not expanded, with fringe of spines and paired spurs; legs in *Euops* with femora distinctly inflated, especially in male, in which fore legs are much longer than posterior pairs, meso- and metafemora may bear tufts of setae and protibiae are very long, slender, curved and lined with teeth internally; tibial apex with distinct uncus and no spurs. Tarsi pseudotetramerous, 5-5-5 in both sexes with tarsomere 4 reduced and 3 distinctly bilobed; tarsomeres 1–3, but especially 3, densely setose beneath; pretarsal claws in *Euops* simple and basally connate, those in *Auletobius* cleft and free at base; empodium not apparent.

Abdomen with 5 ventrites, the first 2 (male *Auletobius*), 3 (female *Auletobius*) or 4 (*Euops*) of which are connate; ventrite 1 at its shortest point not longer than 2, without postcoxal lines; intercoxal process acute in Euops, broadly rounded in *Auletobius*. Abdominal tergites in *Auletobius* relatively lightly pigmented, except for more or less horizontal pygidium (VII), with patches of wing-folding asperities on tergites V–VI and functional spiracles located on I–VI in pleural membrane; tergites moderately heavily sclerotised and accompanied by laterotergites in *Euops*, with pygidium heavily sclerotised and vertical, patches of wing-folding patches on IV–VII and spiracles on located on tergites. Tergite VIII in male more or less concealed beneath VII but well sclerotised; sternite VIII without median strut. Anterior edge of segment IX in male with distinct spiculum gastrale; tergite X membranous. Aedeagus cucujiform, symmetrical; anterior edge of tegmen with long strut, slightly expanded and truncate at apex; body of tegmen elongate; parameres fused together and to phallobase; penis with tectum, pedon and paired basal struts; endophallus with complex sclerites. Segment VIII in female more or less membranous, with long, fixed spiculum ventrale. Ovipositor reduced, with proctiger and paraprocts membranous and not clearly defined, without baculi; coxites undivided, without basal baculi in Auletobius, not clearly delimited in *Euops*; styli always absent. Internal tract with complex, well sclerotise, curved spermathecal capsule. [Thompson 1992; Zimmerman 1994a; Hamilton 2002].

Larvae

Body moderately elongate but robust, subcylindrical or widest at middle and ventrally curved (C-shaped); all surfaces lightly pigmented, except for mandibles and mouth frame; vestiture consisting of scattered to moderately abundant, short, simple setae. Head prognathous and retracted, longer than wide; posterior edge of head capsule not emarginate, posteriorly produced to form narrow, flattened extension; epicranial stem long; frontal arms V-shaped, usually indistinct, contiguous at base; median endocarina extending between frontal arms but not reaching clypeal region. Stemmata on each side variable, as many as 6, but usually indistinct eye spots or absent. Antennae very short, 1-segmented, with terminal, conical sensorium. Frontoclypeal suture present; labrum free, with 4 pairs of setae in *Auletobius*, 3 in Euops; epipharynx with very long and slender epipharyngeal rods. Mandibles short and broad, obliquely bidentate; accessory ventral process absent; mola and prostheca absent. Ventral mouthparts retracted; maxillary articulating area absent; cardo slightly oblique, undivided; stipes elongate; galea and lacinia present in *Auletobius*, apically rounded, with few stout spines at apex of each; *Euops* with undivided mala bearing distinct lacinial spine at inner apical angle; maxillary palp 2-segmented with distinct palpifer. Labium consisting of prementum and postmentum in *Auletobius*, a single sclerite (prementum fused to postmentum) in *Euops*; premental sclerite absent, ligula simple, longer than basal palpomere; labial palps 2-segmented, separated by about width of basal palpomere in *Euops*, more widely so in *Auletobius*. Hypopharyngeal bracon narrow in *Auletobius*, absent in *Euops*; sclerome absent. Hypostomal rods moderately long, diverging; ventral epicranial ridges absent. Gular region absent.

Prothoracic tergum subequal in length to meso- or metatergum, without or with very lightly pigmented plate. Legs absent. Abdomen more than twice as long as thorax; abdominal terga not extending laterally beyond edges of sterna, terga and sterna without patches or rows of asperities; abdominal terga I–VIII each with transverse groove forming 2 plicae, but posterior plica (postdorsum) subdivided laterally in *Auletobius*; abdominal terga strongly humped in middle in *Euops*; segment IX distinctly shorter than VIII; completely dorsal, without urogomphi; sternum 9 simple; segment X reduced and more or less fused to IX, without pygopods; anal region posteriorly or posteroventrally oriented; anal opening bordered by 4 lobes in *Euops*, transverse in *Auletobius*. Spiracles annular–biforous with very long accessory chambers; thoracic spiracle located on prothorax. [Gardner 1934; Emden 1938; Lee and Morimoto 1988a, 1988b; May 1993; Hamilton 2002].

Classification and World Distribution

Attelabidae comprises ~2000 species in 120 genera and is a widespread but most abundant and diverse in tropical regions. The group is divided into two subfamilies, both of which occur Australia.

Australian Fauna

The Australian attelabid fauna consists of the genera *Auletobius* Desbrochers (**Rhynchitinae**: Auletini) with ~50 species and *Euops* Schoenherr (**Attelabinae**: Euopini) (Figs 12C, 98I) with ~25, plus a single unnamed species of *Involvulus* Schrank (Rhynchitinae: Rhyunchitini) from North QLD (Zimmerman 1992, 1994a). Attelabids are oblong to slightly elongate beetles, narrowed anteriorly and subglabrous or clothed with decumbent and erect hairs. The rostrum in short, flattened and apically expanded (*Euops*) or long and narrow, the antennae are straight and inserted dorsally at the base of the rostrum, gular sutures are fused, the labrum is not visible, maxillary palps are well developed but labial palps are absent in *Euops*, the prothorax is much narrower than elytral bases and without lateral carinae, pretarsal claws are connate at base, cleft in *Auletobius*, the pygidium is partly or completely exposed in Euops, and the first three or four ventrites are solidly fused. Larvae are relatively short and broad, strongly curved, somewhat narrowed at each end and very lightly pigmented. The head is strongly retracted in some Rhynchitinae but only slightly so in Attelabinae, labral rods are present, most abdominal terga have two transverse folds (sometimes indistinct), and some Rhynchitinae have bands of tergal asperities. Rhynchitinae have varies habits, some laying their eggs in terminal shoots, flower buds or fruits, and others mining leaves; habits of the Australian *Auletobius* are unknown, but exotic Auletini breed in acorns of *Quercus* and have been associated with *Cistus* and *Tamarix*. Attelabinae are leaf rollers, the female laying eggs in cut portions of leaves, which she curls forming a roll in which the larvae develop. Females of *Euops* species have a series of brushes on the abdomen associated with mycetomes, from which fungal spores are rubbed onto the surface of the leaf roll; the fungi then breakdown the plant tissue making it more acceptable as food for the enclosed weevil larva (Sakurai 1985; Sawada and Morimoto 1985). [Hoffmann 1958; Zimmerman 1994a].

115. CARIDAE

Adults (Fig. 79D)

Length 2.0–5.5 mm. Body ~1.55–1.9 times as long as wide, with pronotum and elytra independently rounded, moderately convex, yellow to dark brown in color, sometimes bicolored, with pattern of dark markings on elytra and occasionally pronotum; clothed with decumbent hairs, some of which may be thickened, forming a pattern, and usually with scattered, fine erect hairs. Head not or slightly declined at base, head behind eyes not constricted posteriorly to form neck; occipital region without transverse ridge or stridulatory file. Frontal region without median groove or endocarina. Eyes not strongly protuberant, not emarginate, finely facetted, without interfacetal setae. Antennal insertions concealed from above, located relatively close to base of rostrum, arising ventrally close to the midline in *Car*; subantennal grooves (scrobes) short and weakly defined. Frontoclypeal suture absent; clypeal region produced forward to form rostrum that is much longer than wide, slender, subcircular and distinctly curved. Labrum completely fused to clypeus. Antennae 11-segmented, moderately long, with very weak, loose, 3- to 5-segmented club; scape elongate, slightly expanded apically and about twice as long as pedicel. Mandible moderately elongate, exodont in *Car* (Fig. 18A), with both mesally and laterally directed teeth at apex (not exodont in *Carodes*); incisor edge with single, large tooth; mola and prostheca absent; inner edge of mandibular base with long pubescent process. Maxilla with short, broad, densely setose mala; maxillary palps 3-segmented, extending only slightly beyond malar apex, apical palpomere cylindrical. Mentum elongate, fused at base; ligula simple, about as long as first 2 palpomeres; labial palp 3-segmented, separated by less than width of first palpomere, apical palpomere cylindrical. Gular sutures fused together. Corpotentorium absent; cervical sclerites absent.

Pronotum ~0.80–0.87 times as long as wide, widest at base, which is distinctly narrower than combined elytral bases; sides slightly curved, lateral pronotal carinae absent; anterior angles not produced; posterior angles rounded; posterior edge straight or weakly sinuate, not margined; disc simple, without paired basal impressions or median longitudinal groove. Prosternum in front of coxae about a fourth as long as shortest diameter of coxal cavity, flat to slightly convex, without carinae. Prosternal process more or less complete but broadly interrupted at middle (intercoxal process strongly concave), the apex expanded and truncate. Notosternal sutures transverse and incomplete. Procoxae projecting below prosternum, with short concealed lateral extensions; trochantins concealed. Procoxal cavities subcircular, contiguous, externally broadly closed, without narrow lateral extensions; postcoxal (hypomeral) processes meeting behind disjunct apex of prosternal process; cavities broadly closed internally.

Scutellar shield abruptly elevated, anteriorly simple, posteriorly broadly rounded. Elytra ~1.03–1.37 times as long as wide and 1.93–2.63 times as long as pronotum, weakly rounded laterally, with elytral apices meeting or almost meeting at the suture and exposing no abdominal tergites; disc with 10 puncture rows and no scutellary striole; epipleura very narrow and apparently absent; with internal longitudinal interlocking ridge laterally at about middle. Mesoventrite separated by complete sutures from mesanepisterna, which are distinctly separated at midline; anterior edge on different plane than metaventrite, without paired procoxal rests, but with narrow transverse prothoracic rest continuing onto mesanepisterna. Mesoventral cavity absent. Mesocoxae not conical and projecting, with concealed trochantins. Mesocoxal cavities narrowly separated, subcircular, not or slightly oblique, closed laterally by meeting of mesoventrite and metaventrite. Mesoventral process subtruncate at apex; mesometaventral junction monocondylic. Metaventrite slightly convex; transverse (metakatepisternal) suture absent; discrimen complete to base anterior intercoxal process; postcoxal lines absent; exposed portion of metanepisternum ~3 times as long as wide; metepimeron concealed. Metacoxae narrowly separated, not extending laterally to meet elytra; plates absent. Metendosternite with long stalk, long lateral arms, well developed but slender laminae and very short anterior process bearing widely separated anterior tendons. Hind wing with apical field more than half as long as total wing length, containing distinct sclerite just beyond radial cell and vaguely indicated remnants of RP; radial cell moderately well developed, slightly longer than wide with base more or less straight and inner posterobasal angle right; cross-vein r3 moderately long and strongly oblique, arising close to base of r4, which is weakly sclerotised but more or less complete; basal portion of RP moderately long, its apical portion extending to wing margin; R-M loop forming acute angle; medial spur sinuate and extending almost to wing margin; medial field with 3 free veins in *Car condensatus* Blackburn (1 in *C. pini* Lea) and no medial fleck; MP_{3+4} without base; wedge cell absent; anal lobe moderately well developed with shallow embayment in *C. condensatus*, highly reduced in *C. pini*. Legs short and stout; femora slightly inflated; outer edges of meso- and metatibiae in *Carodes* with beaded carina; tibial spurs 2-2-2, inner spur in *Carodes* curved and non-articulated; meso- and metatibiae slightly expanded at apex, with fringe of spines. Tarsi pseudotetramerous, 5-5-5 in both sexes with tarsomere 4 reduced and 3 distinctly bilobed; tarsomeres 1–3 densely setose beneath; pretarsal claws with slight basal tooth bearing single seta; empodium not apparent.

Abdomen with 5 free ventrites; ventrite 1 at its shortest point ~1.5 times as long as 2, without postcoxal lines; intercoxal process acute; apex of ventrite 5 broadly rounded. Abdominal tergites lightly pigmented, except for VII and VIII, with patches of wing-folding asperities on tergites IV–VII; functional spiracles on segments I–V located in pleural membrane, those on VI reduced and those on VII absent. Tergite VIII in male more or less concealed beneath VII but well sclerotised; sternite VIII moderately sclerotised, subdivided in *Carodes*, with short articulated median strut. Anterior edge of segment IX in male with distinct spiculum gastrale; tergite X membranous. Aedeagus cucujiform, symmetrical; anterior edge of tegmen with long strut, slightly expanded and truncate at apex; body of tegmen elongate; parameres short, broad, separate and articulated; penis with tectum, pedon and paired basal struts; endophallus with flagellum. Segment VIII in female more or less membranous, with long, fixed spiculum ventrale, slightly expanded at apex. Ovipositor slightly elongate and lightly sclerotised; proctiger and paraprocts membranous and not clearly defined, without baculi; coxites undivided, without basal baculi; styli moderately well developed, slender, terminal. Internal tract with complex, heavily sclerotised spermathecal complex. [Voss, 1932, 1965; Kuschel 1992, 1995; Thompson 1992; Zimmerman 1994a].

Larvae

Body moderately elongate but robust, somewhat flattened and ventrally curved (C-shaped); all surfaces lightly pigmented, except for head capsule and prothorax; vestiture consisting of scattered long and short, simple setae. Head strongly declined (hypognathous), not retracted; posterior edge of head capsule not emarginate; epicranial stem long; frontal arms V-shaped, contiguous at base; median endocarina extending between frontal arms but not reaching clypeal region. Stemmata on each side 4, all with distinct lenses. Antennae very short, 1-segmented, with terminal, conical sensorium. Frontoclypeal suture present; labrum free, with 4 pairs of setae; epipharynx with relatively short, broad epipharyngeal rods. Mandibles more or less symmetrical, short and broad, obliquely bidentate; accessory ventral process absent; incisor edge with 2 subapical teeth, mola and prostheca absent. Ventral mouthparts retracted; maxillary articulating area absent; cardo slightly oblique, undivided; stipes elongate; mala fixed, apically rounded with row of stout setae along inner edge; maxillary palp 2-segmented without distinct palpifer. Labium consisting of prementum and postmentum; premental sclerite with acute anterior extension and bifurcate posterior extension; ligula absent; labial palps 1-segmented and widely separated. Hypopharyngeal bracon present; sclerome absent. Hypostomal rods moderately long, subparallel; ventral epicranial ridges absent. Gular region absent.

Prothoracic tergum about as long as meso- or metatergum, with or without lightly pigmented plate. Legs moderately well developed, 3-segmented, including claw-like pretarsus bearing single seta. Abdomen more than twice as long as thorax; abdominal terga not extending laterally beyond edges of sterna, terga and sterna without patches or rows of asperities; abdominal terga I–VIII each with transverse groove forming 2 plicae; segment IX distinctly shorter than VIII; completely dorsal, without urogomphi; sternum 9 simple; segment X much shorter than IX, terminal, without pygopods; anal region posteriorly oriented; anal opening bordered by 4 lobes. Spiracles minute, annular–biforous with circular peritreme and moderately large accessory chambers; thoracic spiracle located on mesothorax. [May 1994; Oberprieler 2000].

Classification and World Distribution

The Caridae is a small family containing four genera: *Caenominurus* Voss and *Chilecar* Kuschel from southern South America and *Car* Blackburn and *Carodes* Zimmerman from Australia (Kuschel 1992; Zimmerman 1994a). The placement of the Jurassic fossil *Eccoptarthrus crassipes* Arnoldi (1977) and the Recent Caridae in the family Belidae and the resulting use of the older family-group name Eccoptarthrinae by Zherikhin and Gratshev (1995) has been criticised by Oberprieler *et al.* (2007) and is not followed here.

Australian Fauna

The genus *Car* (Fig. 79D) includes three described species distributed across the southern part of the continent and *Carodes* is based on the single species *C. revelatus* Zimmerman from the Atherton Tableland in northern QLD. Carids are oblong to slightly elongate beetles, narrowed anteriorly, with the base of the prothorax much narrowed than the combined elytral bases and with a vestiture of decumbent and erect hairs. The rostrum is long and narrow, the antennae straight, inserted ventrally near the base of the rostrum, with a very weak, loose club, mandibles are exodont in *Car* but smoothly rounded externally in *Carodes*, the labrum is not visible, maxillary and labial palps are well developed, the prothorax lacks lateral carinae and all ventrites are separated by deep grooves. Larvae are moderately short and broad, not strongly narrowed at either end, frontal arms do not extend to the mandibular articulations, there are four stemmata on each side, labral rods are present, abdominal terga have two transverse folds, legs are well developed and 4-segmented, including a claw-like pretarsus. Species of *Car* are associated with trees of the genus *Callitris* (Cupressaceae) and larvae are known to feed within the female cones (May 1994). The related South American species in the genera *Chilecar* and *Caenominurus* are also associated with Cupressaceae (*Pilgerodendron* and *Austrocedrus*, respectively).

116. BRENTIDAE

Adults (Figs 11A, 12D, 79C, E, I, 80D, L, 98E, J)

Length 1.5–40 mm. Body ~3–8 (Brentinae), 2.3–3.3 (Eurhynchinae) or 1.6–2.5 (Apioninae) times as long as wide, usually with pronotum and elytra independently rounded or with elytra more or less parallel-sided in most Brentinae, moderately convex, yellowish brown to red, dark brown or black in color, rarely metallic blue (e.g. some *Aporhina*), bicolored with red head and prothorax and black elytra in *Cylas* or with pattern of yellow markings on a dark background in some Brentinae, clothed with decumbent hairs, some of which may be thickened, forming a pattern, in most Eurhynchinae, *Austronanodes*, *Rhinorhynchidius* and some Apionini, but usually glabrous or with scattered, fine hairs in Brentinae, Myrmacicelini and Rhadinocybebini. Head prognathous, not declined at base, head behind eyes usually not sharply constricted posteriorly to form distinct temples and a narrow neck, but often slightly constricted and then widened behind eyes in Brentinae, so that a broad neck region is formed; occipital region without transverse ridge or stridulatory file, with median occipital endocarina and short frontal groove in *Eurhynchus*. Eyes not to strongly protuberant, not emarginate, finely or coarsely facetted, without interfacetal setae. Antennal insertions concealed from above, located at various positions on rostrum; subantennal grooves (scrobes) usually short (long in *Austronanodes*). Frontoclypeal suture absent; frontoclypeal region produced forward to form rostrum that may be shorter than wide and flattened in some Brentinae (e.g. *Cyphagogus*), but usually much longer than wide, slender and subcircular or widened apically, sexually dimorphic in some Brentinae. Labrum completely fused to clypeus. Antennae almost always 11-segmented (10-segmented in *Cylas*, 9-segmented in *Austronanodes* and the brentine genus *Neosebus*), moderately long, filiform, moniliform, incrassate, rarely pectinate (male *Ctenaphides*) or capitate with a relatively weak club, which is almost always 3-segmented (1-segmented in *Cylas*); scape elongate but rarely more than twice as long as pedicel (more than 4 times as long in *Austronanodes*, which is the only brentid with geniculate antennae). Mandible usually short and broad, occasionally slightly elongate, unidentate to obliquely tridentate or bidentate with large subapical tooth on incisor edge; mola and prostheca absent; inner edge of mandibular base with long pubescent process. Maxilla with short and broad to elongate, setose mala; maxillary palps 3-segmented in Eurhynchinae, Brentinae and Nanophyinae, 2-segmented in Apioninae, extending only slightly beyond malar apex and usually with distinct palpifer; apical palpomere cylindrical. Mentum usually elongate, fused at base; prementum usually elongate; ligula simple or absent; labial palps 1- or 2-segmented and more or less concealed within pits in Eurhynchinae and Brentinae, 2-segmented and exposed in Nanophyinae and 1-segmented and exposed in Apioninae; apical palpomere cylindrical. Gular sutures fused together. Corpotentorium absent; cervical sclerites absent.

Pronotum ~1.2–2.4 (Brentinae), 0.95–1.1 (Eurhynchinae) or 0.5–1.6 (Apioninae and Nanophyinae) times as long as wide, usually widest at about middle (widest at base in *Austronanodes* and some Apionini); base distinctly narrower than combined elytral bases in Eurhynchinae and some Brentinae and Apioninae; sides slightly to strongly curved, lateral pronotal carinae absent; anterior angles not produced; posterior angles rounded; posterior edge usually straight or weakly rounded (biemarginate in Eurhynchinae); disc without paired basal impressions, with median longitudinal groove in some Brentinae, constricted at base and sometimes with posterior margin or collar in some Brentinae, *Myrmacicelus* and some Rhadinocybebini, with anterior lateral impressions for housing profemora in Cyphagogini (Brentinae) . Prosternum in front of coxae much shorter than shortest diameter of coxal cavity in *Austronanodes*, Apioninae and Eurhynchinae, about as long in *Cylas* and much longer in most Brentinae, usually flat to slightly convex, strongly concave in *Austronanodes* and *Rhinorhynchidius*. Prosternal process usually more or less complete but broadly interrupted at middle (intercoxal process strongly concave), the apex usually represented by a disjunct sclerite (sternellum) (Fig. 18F), often very small, which may be completely fused to the postcoxal processes (concavity in some Brentinae reduced to transverse suture at middle of prosternal process). Notosternal sutures incomplete or absent. Procoxae (Fig. 44O) usually projecting below prosternum (not so in some Brentinae), with short concealed lateral extensions; trochantins concealed. Procoxal cavities subcircular, usually contiguous (well separated in some Brentinae), externally broadly closed, without narrow lateral extensions; postcoxal (hypomeral) processes meeting behind disjunct apex of prosternal process or fused together and sometimes to prosternal process; cavities broadly closed internally.

Scutellar shield often reduced and sometimes absent, usually abruptly elevated, anteriorly simple, posteriorly acute to broadly rounded. Elytra ~2.1–5.6 (Brentinae), 1.75–2.4 (Eurhynchinae) or 1.15–1.45 (Apioninae) times as long as wide and 1.3–2.65 (Brentinae), 2.9–3.35 (Eurhynchinae) or 1.7–3.45 (Apioninae) times as long as pronotum; sides subparallel (most Brentinae) to weakly rounded laterally, with elytral apices usually conjointly rounded, exposing no abdominal tergites (apices sometimes truncate or independently rounded or acute); disc sometimes with elevated ridges or spines in Eurhynchinae, often with 9 or 10 puncture rows or striae and no scutellary striole; epipleura very narrow or apparently absent; sometimes with internal longitudinal interlocking ridge laterally at about middle. Mesoventrite in Eurhynchinae and *Austronanodes* separated by complete sutures from mesanepisterna, which are distinctly separated at midline, partly or completely fused to mesanepisterna in Brentinae and Apioninae; anterior edge usually on same or slightly different plane than metaventrite, without paired procoxal rests, but usually with narrow transverse prothoracic rest continuing onto mesanepisterna; sides of mesothorax deeply emarginate in *Cylas*. Mesoventral cavity absent. Mesocoxae usually not conical and projecting (distinctly projecting in some Apioninae), with concealed trochantins. Mesocoxal cavities usually narrowly separated (contiguous in *Rhinorhynchidius*, widely separated in *Austronanodes*), subcircular, not or slightly oblique, closed laterally by meeting of mesoventrite and metaventrite. Mesoventral process usually narrowly rounded to subtruncate at apex (process absent in *Rhinorhynchidius*); mesometaventral junction, if present, often monocondylic (absent due to complete fusion of meso- and metaventrite in *Austronanodes*). Metaventrite moderately to strongly convex; transverse (metakatepisternal) suture absent or not extending across midline; discrimen short or absent; postcoxal lines absent; exposed portion of metanepisternum moderately to extremely long and narrow; metepimeron usually concealed. Metacoxae narrowly to moderately widely separated, sometimes extending laterally to meet elytra; plates absent. Metendosternite with long stalk, long lateral arms, well developed, broad laminae in Eurhynchinae, but reduced or no laminae in most groups, and very short anterior process bearing narrowly separated anterior tendons. Hind wing with apical field more than half as long as total wing length, with 2 remnants of RP; radial cell usually highly reduced or absent (moderately well developed in Eurhynchinae, with inner posterobasal angle acute); cross-vein r3 usually absent (joined basally with r4 in Eurhynchinae); r4 short or absent; basal portion of RP short or absent; R-M loop forming narrowly acute angle; medial spur sinuate and apically curved; medial field usually with 3 or fewer free veins (4 in Eurhynchinae) and no medial fleck; MP_{3+4} without base; wedge cell absent; anal lobe moderately well developed without embayment, reduced with notch-like embayment in Apioninae and some Brentinae. Legs usually moderately long and slender, but short and stout in some Brentinae (e.g. some Cyphagogini); trochanters elongate in Nanophyinae and most Apioninae (very short in Eurhynchinae and Brentinae); trochanterofemoral joint not or only slightly oblique in Nanophyinae and most Apioninae (strongly oblique in Eurhynchinae and Brentinae); femora usually not or only slightly inflated (strongly inflated in some Brentinae); tibiae usually not apically expanded or spinose, very short and stout in some Brentinae; protibia with subapical antenna cleaner in some Brentinae; tibial spurs 2-2-2 in Eurhynchinae and most Brentinae, reduced to 1-2-2 or 1-1-1 in various brentines and absent in *Cylas*, *Austronanodes* and Apioninae. Tarsi usually pseudotetramerous, 5-5-5 in both sexes with tarsomere 4 reduced and 3 distinctly bilobed, but tarsomere 3 not or barely lobed in Cyphagogini and tarsomere 4 is highly reduced or possibly absent in some members of that group; tarsomeres 1–3 or 3 only usually densely setose beneath (not in Cyphagogini); pretarsal claws simple and free in Eurhynchinae and most Brentinae, simple and connate in *Cylas* and *Austronanodes*, and toothed in most Apioninae; empodium not apparent.

Abdomen with 5 ventrites, the first 2 of which are connate; ventrite 1 at its shortest point subequal in length to 2, more or less equal in length to 5 and longer than 3 and 4 combined; ventrites 1 and 2 on different plane than 3–5, clearly separated by suture in Eurhynchinae, but with suture incomplete or vaguely indicated in most groups, without postcoxal lines; intercoxal process narrowly to broadly rounded; apex of ventrite 5 more or less truncate. Abdominal tergites I–VI or VII lightly to moderately pigmented, VIII and sometimes VII more heavily pigmented, usually with patches of wing-folding asperities on tergites IV–VII; functional spiracles on segments I–VII in Eurhynchinae and Brentinae (on I–VI in Nanophyinae and Apioninae), anterior spiracles located in pleural membrane, those on segment VII on tergite. Tergite VIII in male usually heavily sclerotised, forming pygidium; sternite VIII reduced, with short median strut in Eurhynchinae (Fig. 28L). Anterior edge of segment IX in male with distinct spiculum gastrale; tergite X membranous. Aedeagus (Fig. 31I–J) cucujiform, symmetrical; anterior edge of tegmen with long strut; body of tegmen elongate; parameres short and broad to moderately long and narrow, separate and articulated, at least partly separated but fused to phallobase or fused together and to phallobase; penis with tectum, pedon and paired basal struts; endophallus often with sclerites and flagellum. Tergite VIII in female sclerotised laterally in Eurhynchinae, forming heavily sclerotised toothed process in at least some Brentinae, more or less membranous in most groups; sternite VIII with long, fixed spiculum ventrale, with sclerotised posterior process in Eurhynchinae. Ovipositor slightly elongate and lightly sclerotised; proctiger and paraprocts membranous without baculi, except in Eurhynchinae; coxites undivided, with longitudinal baculi; styli moderately usually well developed, slender and terminal or subterminal. Internal tract with heavily sclerotised spermathecal capsule. [Morimoto 1962, 1976; Kissinger 1968; Alonso-Zarazaga 1990; Thompson 1992; Zimmerman 1993, 1994b; Kuschel 1995; Wanat 2001; Anderson and Kissinger 2002; Oberprieler 2000, 2004b].

Larvae

Body elongate and cylindrical to shorter, broader and wider at middle, ventrally curved (C-shaped); all surfaces lightly pigmented, usually with the exception of head capsule and protergum; vestiture consisting of scattered long and short, simple setae. Head strongly declined (hypognathous), not retracted; posterior edge of head capsule not emarginate; epicranial stem long; frontal arms V-shaped, contiguous at base, extending to mandibular membrane; median endocarina located beneath epicranial stem in most Brentinae, extending between frontal arms in *Cylas*, Eurhynchinae, Nanophyinae and Apioninae. Stemmata on each side 3 in Eurhynchinae, 2 in *Cylas*, 1 or 2 in Apioninae, 1 in *Austronanodes* and 0 in most Brentinae. Antennae very short, 1-segmented, with terminal, conical sensorium. Frontoclypeal suture present; labrum free, with 3 pairs of setae; epipharynx with relatively short and broad to longer and narrow epipharyngeal rods. Mandibles more or less symmetrical, short and broad, obliquely bidentate; accessory ventral process absent; incisor edge with or without subapical teeth, mola and prostheca absent. Ventral mouthparts retracted; maxillary articulating area absent; cardo slightly oblique, undivided; stipes elongate; mala fixed, apically rounded with row of stout setae along inner edge; maxillary palp 2-segmented without distinct palpifer. Labium consisting of prementum and postmentum; premental sclerite with anterior extension Eurhynchinae and *Austronanodes*); ligula simple and rounded or absent; labial palps usually 1-segmented and widely separated (2-segmented in Eurhynchinae and *Cylas*). Hypopharyngeal bracon present; sclerome absent. Hypostomal rods short to moderately long and diverging; ventral epicranial ridges present in Eurhynchinae. Gular region absent.

Prothoracic tergum about as long as meso- or metatergum, with or without pigmented plate. Legs very short, distinctly 2-segmented, without terminal claw in Eurhynchinae and some Apioninae, *Austronanodes* and some Brentinae. Abdomen usually more than twice as long as thorax (less so in some Apioninae); abdominal terga not extending laterally beyond edges of sterna, terga with patches asperities in Eurhynchinae and some Brentinae; abdominal terga I–VIII each with transverse groove forming 2 plicae in Apioninae, *Austronanodes* and *Cylas*, with 3 or 4 folds in Eurhynchinae and most Brentinae; segment IX distinctly shorter than VIII; completely dorsal, without urogomphi; sternum 9 simple; segment X much shorter than IX, terminal, without pygopods; anal region posteriorly or post-oventrally oriented; anal opening bordered by 4 lobes in Eurhynchinae, *Austronanodes* and *Cylas*, 3 lobes in at least some Brentinae and 2 lobes (with transverse anal opening) in Apioninae. Spiracles usually annular–biforous, annular in some Brentinae; thoracic spiracle usually located on prothorax; spiracles absent on segment VIII in Apioninae. [Gardner 1935d; Emden 1938; May 1993, 1994; Anderson and Kissinger 2002; Oberprieler 2000, 2004b].

Classification and World Distribution

Brentidae is a widely distributed family comprising ~4000? species in 475 genera. We recognise six subfamilies, four of which occur in Australia; however, several authors treat most or all of these groups as families. The exotic subfamily Ithycerinae contains the single North American genus *Ithycerus* Schoenherr (Sanborne 1981) and the Microcerinae contains a few Afrotropical genera (Louw 2004). Within the subfamily Apioninae as here delimited, the following tribes (sometimes raised to subfamily level) do not occur in Australia: Cybebini (Madagascar), Antliarhinini (South Africa) and Tanaini (Africa). [Alonso-Zarazaga 1990; Anderson 2002c; Oberprieler 2004a].

Australian Fauna

The subfamily **Eurhynchinae** is an Australian endemic group containing three genera: *Eurhynchus* (Fig. 12D), *Ctenaphides* (Fig. 79C) and *Aporhina* (Fig. 79E), the last of which extends into New Guinea. Larvae of *Eurhynchus laevior* (Kirby) tunnel into the woody branches of *Persoonia* (Proteaceae), while those of *Aporhina australis* (Heller) develop in twigs of *Litsea* (Lauraceae) (Oberprieler 2000). **Nanophyinae** contains several genera in the Holarctic, Oriental and Afrotropical regions, but in Australia it is represented by a few species of the endemic genus *Austronanodes* (Fig. 80D), which feed on aquatic plants, such as *Ludwigia* (Onagraceae) and *Nymphoides* (Gentianaceae). Brentinae and Apioninae are both well represented in Australia, with ~20 genera each. Typical **Brentinae** are often narrowly elongate (Fig. 98J) with a slender rostrum often exhibiting sexual dimorphism, with unusual forms including *Cylas formicarius* (Fabricius) (Fig. 80L) and members of the Cyphagogini (Fig. 79I). Most Brentinae are borers in dead wood, but Cyphagogini are known to enter the tunnels of other wood-borers, such as scolytine and platypodine ambrosia beetles, ejecting the host and feeding on the ambrosia fungi, and some Brentini are myrmecophilous. *Cylas formicarius* is unusual for the group in that it feeds in living stems and tubers of sweet potato and related Convolvulaceae. **Apioninae** contains several small species, usually having a distinctive shape shown in Fig. 98E. Almost all Apioninae feed on stems, galls, fruits and seeds of a variety of angiosperms, but *Rhinorhynchidius* species feed on the foliage of *Callitris* (Cupressaceae) forming characteristic deformities referred to by Zimmerman as 'pineapple-top' galls. [Zimmerman 1993, 1994b; Oberprieler 2004b; Wanat 2001].

Australian Subfamilies, Tribes and Genera

Eurhynchinae: *Aporhina* Boisduval, *Ctenaphides* Pascoe, *Eurhynchus* Kirby.

Nanophyinae: *Austronanodes* Zimmerman.

Brentinae: Brentini: *Ankleineella* Zimmerman, *Baryrhynchus* Lacordaire, *Cordus* Schönherr, *Ectocemus* Pascoe, *Kleineella* Strand, *Orychodes* Pascoe, *Schizoeupsalis* Kleine. **Cyphagogini:** *Allaeometrus* Senna, *Ancylobrentus* Damoiseau, *Callipareius* Senna, *Catagogus* Kleine, *Cerobates* Schönherr, *Cyphagogus* Parry, *Hoplopisthius* Senna, *Isomorphus* Kleine, *Mesoderes* Senna, *Neosebus* Senna. **Trachelizini:** *Araiorrhinus* Senna, *Euschizus* Kleine, *Higonius* Lewis, *Hormocerus* Schoenherr, *Ithystenus* Pasacoe, *Mesetia* Blackburn, *Microtrachelizus* Senna, *Miolispa* Pascoe, *Stenobrentus* Damoiseau, *Trachelizus* Schoenherr, *Tracheloschizus* Damoiseau, *Uropteroides* Kleine. **Cyladini:** *Cylas* Latreille.

Apioninae: Apionini: *Apion* Herbst, *Coelocephalapion* Wagner, *Conapion* Motschulsky, *Exapion* Bedel, *Flavopodapion* Korotyaev, *Perapion* Wagner, *Piezotrachelus* Schoenherr, *Pseudapion* Schilsky, *Pseudopiezotrachelus* Wagner, *Pseudorhinapion* Voss, *Rhopalapion* Schilsky, *Sterculapion* Rheinheimer. **Myrmacicelini:** *Myrmacicelus* Chevrolat. **Rhadinocybebini:** *Alissapion* Wanat, *Allonotapion* Wanat, *Coelapion* Wanat, *Ganoapion* Zimmerman, *Lissapion* Zimmerman, *Micronotapion* Wanat, *Myrmecapion* Wanat, *Nanomyrmacyba* Wanat, *Notapion* Zimmerman. **Rhinorhynchidiini:** *Rhinorhynchidius* Voss.

117. CURCULIONIDAE

Adults (Figs 11C–H, 12A–B, E–H, 79F–H, L, 80A–C, E–K, 98F–H, K–M)

Length 0.8–60 (usually 1.5–40) mm. Body ~1.2–5.7 times as long as wide, the most elongate forms in Cossoninae and Platypodinae, usually with pronotum and elytra independently rounded or with elytra more or less parallel-sided, usually slightly flattened to strongly convex (strongly flattened in a few Cossoninae), colour usually brown to black (rarely bicolored or metallic); vestiture usually consisting of fine hairs to stout setae or scales, which vary from white to brown in colour but may be metallic green or blue (e.g. *Myllocerus*, *Pantorhytes*, *Chrysolopus*) and often form a pattern, rarely subglabrous (e.g. some Cossoninae). Head usually more or less prognathous, not declined at base (except in Cryptorhynchini), although frontal region and rostrum variably declined; head behind eyes never abruptly constricted posteriorly to form temples and neck; occipital region without transverse ridge or stridulatory file, with median occipital endocarina most Platypodinae and Scolytinae. Eyes usually not or only slightly protuberant, entire and finely or coarsely facetted, without interfacetal setae; eyes more prominent in some taxa (e.g. some Aterpini and Entiminae), enlarged and dorsally approximate or contiguous in others (e.g. *Mecopus*, *Phaenomerus*, *Rhynchaenus*), shallowly to deeply emarginate in some Scolytinae, rarely absent (e.g. *Myrtonymus*). Antennal insertions almost always concealed from above (barely exposed in some Platypodinae), located at various positions on rostrum; subantennal grooves (scrobes) usually well developed and long (shorter and weakly developed in Platypodinae and Scolytinae). Frontoclypeal suture always absent; frontoclypeal region produced forward to form rostrum, which is usually narrowly elongate and circular in cross-section to somewhat flattened; rostrum short and broad in Amycterini, *Rhadinosomus*, various Entiminae and some Cossoninae and virtually absent in most Scolytinae and Platypodinae. Labrum completely fused to frontoclypeus. Antennae (Fig. 97L) usually 11-segmented with an elongate scape (usually more than 3 times as long as pedicel), a 7-segmented funicle and a compact, 3-segmented club, or apparently 12-segmented when the funicle is 7-segmented and the club apparently divided into four parts; funicle often 5-segmented in Cossoninae, 6-segmented in Amycterini, Stenopelmini and most Rhynchophorinae, 4-segmented in *Dryophthorus*, *Stenommatus*, *Dryophthoroides* and most Platypodinae, 3-segmented in *Notoplatypus*, and 3- to 6-segmented in some Scolytinae; club always 1-segmented (without subdivisions) in Dryophthorinae and Platypodinae. Mandibles usually short and broad, highly reduced in those taxa with long, narrow rostrum, but sometimes massive in those with short rostrum (e.g. Amycterini), unidentate or more or less truncate to bidentate or rarely tridentate; mola and prostheca absent; inner edge of mandibular base usually with long pubescent process (Fig. 28N) (short or absent in Platypodinae and Scolytinae). Mandibles with dehiscent cusp (or scar) (Fig. 97J) in most Entiminae. Maxilla with short and broad to moderately elongate, setose or spinose mala; maxillary palps usually 3-segmented, extending only slightly beyond malar apex (Fig. 27K) and usually with distinct palpifer; apical palpomere cylindrical. Mentum usually elongate, fused at base; prementum variable; ligula present or absent; labial palps usually 2-segmented with distinct palpiger, narrowly separated; apical palpomere cylindrical. Gular sutures fused together (Fig. 97L). Corpotentorium narrow, incomplete or absent; cervical sclerites absent.

Pronotum ~0.45–2.65 times as long as wide; base not to distinctly narrower than combined elytral bases; sides straight to strongly curved or sinuate; lateral pronotal carinae almost always absent (weakly defined and incomplete in Platypodinae and some Scolytinae where they are associated with crural impressions); anterior angles not produced forward; posterior angles rounded to obtuse or right; posterior edge usually straight or weakly rounded (rarely sinuate or lobed); disc usually simple, without basal impressions, median longitudinal groove or sublateral carinae, sometimes with anterior collar. Prosternum in front of coxae shorter than to longer than shortest diameter of coxal cavity, usually flat to slightly convex or slightly concave, sometimes strongly concave forming a channel (rostral canal) into which rostrum fits (e.g. in *Aonychus*, Cryptorhynchini, Fig. 97N). Prosternal process more or less complete but often broadly interrupted at middle (intercoxal process strongly concave), the apex represented by a disjunct sclerite (usually referred to as a sternellum), which is often very small and which may be completely fused to the postcoxal processes (concavity sometimes reduced to transverse groove at middle of prosternal process). Notosternal sutures usually incomplete or absent. Procoxae more or less globular but usually projecting at least slightly below prosternum, with short or occasionally long concealed lateral extensions; trochantins always concealed. Procoxal cavities subcircular and usually contiguous, but well separated in those groups with broad prosternal process and transverse groove (e.g. some Cossoninae, Fig. 20F) or with rostral canal, externally broadly closed, without narrow lateral extensions; postcoxal (hypomeral) processes meeting behind disjunct apex of prosternal process or fused together and sometimes to prosternal process (Fig. 20F); cavities broadly closed internally.

Scutellar shield often reduced and sometimes absent, usually abruptly elevated, anteriorly simple, posteriorly acute to broadly rounded. Elytra ~0.7–4.2 times as long as wide and 1.15–4.6 times as long as pronotum; sides subparallel to broadly rounded laterally, with elytral apices usually conjointly rounded, exposing no abdominal tergites, but occasionally independently rounded (e.g. *Ectemnorhinus*), independently acute (e.g. *Methypora*, *Rhadionosomus*) or truncate exposing the pygidium from above (e.g. *Sitophilus*, Mesoptiliini); underside with subapical stridulatory file in Platypodinae, Listroderini, Hylobiini, Orthorhinini, Cossoninae, Scolytinae, Ceutorhynchini and a few Curculioninae, or an apical file in *some* Amycterini; disc usually simple, occasionally tuberculate, often with 9 or 10 puncture rows or striae and no scutellary striole; epipleura very narrow or apparently absent. Mesoventrite usually separated by complete sutures from mesanepisterna, which are distinctly separated at midline, but partly or completely fused to mesanepisterna in various groups; anterior edge usually on same plane than metaventrite (on a different plane in those taxa with a mesothoracic extension of the rostral groove, e.g. Cryptorhynchini), without paired procoxal rests, but usually with narrow transverse prothoracic rest continuing onto mesanepisterna. Mesoventral cavity absent, but mesoventrite sometimes deeply concave forming continuation of rostral canal. Mesocoxae usually not or only slightly projecting, with concealed trochantins. Mesocoxal cavities narrowly to very widely separated, subcircular and closed laterally by meeting of mesoventrite and metaventrite. Mesoventral process narrowly rounded to broadly truncate; mesometaventral junction simple or complex, often a straight or curved line, occasionally absent due to fusion of ventrites. Metaventrite moderately to strongly convex; transverse (metakatepisternal) suture absent or not extending across midline; discrimen long to short or absent; postcoxal lines absent; exposed portion of metanepisternum usually long and narrow, with modified scales (sclerolepidia) lining inner edge in various Molytinae, Cossoninae, Scolytinae, Conoderinae and Curculioninae; metepimeron usually concealed. Metacoxae usually moderately to widely separated (narrowly separated in Platypodinae and some Scolytinae), sometimes extending laterally to meet elytra; plates absent. Metendosternite usually with long stalk, long lateral arms, reduced or no laminae and anterior tendons located on arms. Hind wing with apical field more than half as long as total wing length, with 2 remnants of RP; radial cell usually highly reduced or absent; cross-vein r3 usually absent and r4 short or absent; basal portion of RP short or absent; R-M loop forming narrowly acute angle; medial spur often sinuate; medial field with 3 or fewer free veins and no medial fleck; MP_{3+4} without base; wedge cell absent; anal lobe moderately well developed without embayment, occasionally reduced with notch-like embayment (e.g. some Scolytinae). Legs short and stout to long and slender; trochanterofemoral joint usually strongly oblique, sometimes with the femur abutting the coxa; femora often somewhat inflated; metafemora enlarged and internally dentate in some groups (e.g. Campyloscelini, Eugnomini, Derelomini and Rhamphini) and all femora dentate in *Acicnemis* (Trachodini); tibiae often expanded at apex to form an ovate corbel, which is often obliquely oriented and usually associated with a row of spines, an inwardly curved spine (uncus, Fig, 27A) arising from the outer apical angle or a tooth-like process (mucro) arising from the inner apical angle (rarely with both uncus and mucro); tibiae expanded subapically with outer edge lined with teeth or socketed spines in most Scolytinae (Fig. 27C) (not *Scolytus*), some Platypodinae and a few Cossoninae; tibial spurs almost always reduced or absent. Tarsi usually pseudotetramerous, 5-5-5 in both sexes with tarsomere 4 reduced and 3 distinctly lobed beneath, but basal three tarsomeres not or only weakly lobed in some groups (e.g. Dryophthorini, Platypodinae, Scolytinae, Cossoninae), tarsomere 4 only slightly reduced in Platypodinae, tarsomere 4 absent in *Myrtonymus*, and tarsomere 5 highly reduced in *Aonychus*; pretarsal claws usually simple, sometimes connate, toothed or cleft, absent in a few groups (e.g. *Aonychus*). empodium not apparent.

Abdomen with 5 ventrites, the first two of which are almost always connate (Fig. 28J) (not in Platypodini); ventrites 1 and 2 combined usually much longer than 3 and 4 combined (not so in Platypodinae); ventrite 1 without postcoxal lines; intercoxal process usually broadly rounded to truncate, without postcoxal lines; intercoxal process broadly rounded to truncate (narrowly rounded to acute in some Platypodinae and Scolytinae); apex of ventrite 5 usually broadly rounded to truncate. Abdominal tergites I–VI or VII lightly to moderately pigmented, VIII and sometimes VII more heavily so, with patches of wing-folding usually asperities on tergites IV–VII; functional spiracles usually on segments I–VII (but those on segment VII sometimes reduced and non-functional or absent); non-functional spiracles occasionally present on segment VIII (Thompson 1992, p. 840); anterior spiracles located in pleural membrane, those on segment VII sometimes on tergite. Tergite VIII in male usually heavily sclerotised, forming pygidium, which may or may not be exposed beyond elytral apices (tergite VIII concealed beneath VII in males of Dryophthorinae, Platypodinae, some Brachycerinae and a few Scolytinae); sternite VIII reduced, with median strut short or absent. Anterior edge of segment IX in male with distinct spiculum gastrale; tergite X membranous. Aedeagus cucujiform, with highly reduced, ring-like tegmen bearing long anterior strut; parameres absent; penis without tectum and pedon, usually with paired basal struts; endophallus often with sclerites and flagellum. Tergite VII in female terminal and sclerotised; tergite VIII lightly sclerotised and concealed beneath VIII; sternite VIII with long, fixed spiculum ventrale. Ovipositor highly variable, short and broad to narrowly elongate; coxites undivided; styli moderately well developed to reduced or absent. Internal tract usually with heavily sclerotised spermathecal capsule. [Zimmerman 1991, 1992,

1993, Thompson 1992; Howden 1995; Kuschel 1995; Lyal 1995; Lyal and King 1996; Anderson 2002d; Marvaldi *et al.* 2002; Lyal *et al.* 2006; Oberprieler *et al.* 2007; Oberprieler 2010].

Larvae (Fig. 33E)

Body elongate and cylindrical to shorter, broader and wider at middle, usually slightly to strongly curved ventrally (C-shaped). ventrally flattened in Gonipterini; all surfaces lightly pigmented, usually with the exception of mouth frame and mandibles or entire head capsule, sometimes protergum and rarely tergum IX; vestiture usually consisting of scattered long and short, simple setae, occasionally dense. Head almost always protracted and strongly declined (hypognathous), rarely retracted; posterior edge of head capsule not emarginate; epicranial stem long; frontal arms V-shaped, contiguous at base, not extending to mandibular membrane; median endocarina (Fig. 43A) located beneath epicranial stem and usually extending forward between frontal arms. Postoccipital condyles usually present. Stemmata on each side sometimes 1 or 2, often absent. Antennae (Fig. 40C) very short, 1-segmented, with terminal, conical sensorium. Frontoclypeal suture present or absent; labrum free, with 3 pairs of setae (fewer in Platypodinae); epipharynx with relatively short and broad to longer and narrow epipharyngeal rods. Mandibles (Fig. 40C) more or less symmetrical, short and broad, obliquely bidentate; accessory ventral process absent; incisor edge with or without subapical teeth, mola and prostheca absent, dorsal pseudomola present in Platypodinae and some Scolytinae. Ventral mouthparts retracted; maxillary articulating area absent; cardo slightly oblique, undivided; stipes short and broad to moderately elongate; mala fixed, apically rounded with row of stout setae along inner edge; maxillary palp usually 2-segmented without distinct palpifer (1-segmented in Platypodinae). Labium consisting of prementum and postmentum; prementum often with trident-shaped premental sclerite; ligula simple and rounded or absent; labial palps usually 2-segmented and well separated (1-segmented in Platypodinae and Bagoini). Hypopharyngeal bracon present; sclerome absent. Hypostomal rods present or absent; ventral epicranial ridges absent. Gular region absent.

Thoracic segments usually more or less equal in length and not much longer than anterior abdominal segments, but thorax greatly enlarged in a few groups (e.g. *Strongylorhinus*, *Neolaemosaccus*); prothoracic tergum usually about as long as meso- or metatergum, with or without pigmented plate, with a transverse row of setiferous rings in Platypodinae. Upper surfaces of thorax and abdomen usually covered with dark tubercles or conical protuberances in Gonipterini. Legs always absent. Abdomen usually more than twice as long as thorax; abdominal terga not extending laterally beyond edges of sterna, terga without patches or rows of asperities; abdominal terga I–VIII each with at least two transverse grooves forming three plicae (three grooves and four plicae in some Cossoninae). In most groups, segment VIII is subequal in length to VII, segment IX shorter than VIII, simple and subterminal, and segment X shorter than or more or less fused to IX and terminal; in most Brachycerinae segment VIII is enlarged and modified, forming a tergal disc or plate, while segments IX and X are ventral. Spiracles annular–biforous, annular–uniforous or annular; thoracic spiracles usually located on prothorax or pro-mesothoracic junction; abdominal spiracles usually lateral, those on segments II–VII in *Cyrtobagous*. I–VII in *Echinocnemus* and VIII in *Neochetina* associated with blade-like processes; spiracles on segments I and VIII vestigial in *Cyrtobagous*; those on VIII dorsally placed in *Bagous*, *Echinocnemus* and *Neohydronomus*. [Gardner 1932, 1934, 1938; Emden 1938, 1952; Browne 1972; May 1993, 1994; Kuschel 1995; Anderson 2002d; Marvaldi *et al.* 2002; Oberprieler *et al.* 2007].

Classification and World Distribution

Curculionidae is a very large, cosmopolitan family containing ~53 000 described species in 4800 genera. There is no general agreement on limits of the family or the number and constitution of the included subfamilies and tribes. The classification used here is based in large part on that in Oberprieler *et al.* (2007), with several the families and subfamilies listed by Alonso-Zarazaga and Lyal (1999) downgraded to subfamilies and tribes. The family may be divided into ten subfamilies, all of which occur in Australia. [Oberprieler 1993].

Australian Fauna

The subfamily **Dryophthorinae** (often referred to as Rhynchophorinae) is represented in Australia mainly by species belonging to widely distributed, usually Old World or Indo-Pacific genera, which have dispersed into Australia or been introduced through commerce. Of the four tribes occurring here, Dryophthorini includes a few species of *Dryophthorus* distributed along the east coast and an undescribed *Stenommatus* from northern QLD, Stromboscerini an undescribed *Dryophthoroides* from northern QLD, Orthognathini the widely distributed *Sipalinus gigas* (Fabricius) introduced into QLD, and Rhynchophorini several genera, mostly with northern affinities. A notable exception is the autochthonous *Trigonotarsus rugosus* (Boisduval) (Fig. 98M), which occurs on species of grass trees (Xanthorrhoeaceae: *Xanthorrhoea*). Dryophthorinae differ from all other weevils in having a single-segmented, shiny antennal club with a finely densely pubescent (spongiform) apex, possibly representing antennomere 8 with the entire club compressed within its apex, the prementum concealed from below by the postmentum, and the pretarsal claws separated by dorsal and ventral dermal lobes. *Rhabdoscelus obscurus* (Boisduval) and *Cosmopolites sordidus* (Germar) are pests of sugarcane and bananas, respectively, while *Sitophilus granarius* (Linnaeus), *S. oryzae* (Linnaeus) and *S. zeamais* Motschulsky are imported stored products pests. The subfamily **Platypodinae** is represented by the tesserocerine genus *Notoplatypus* and several genera of Platypodini (Fig. 98H). The inclusion of the genus *Diapus* in Tesserocerini by Wood (1993) and Alonso-Zarazaga and Lyal (2009) is not followed, since that genus lacks the characters listed by Thompson (1992) for the tribe (as Platytarsulini) and its inclusion by Wood was based on the presence of a maxillary lacinia, a plesiomorphy that seems to vary in both tribes. Wood's concept of Platypodidae included taxa now known to belong to other subfamilies (e.g. *Protohylastes* Wood in Molytinae: Psepholacini, *Protoplatypus* Wood in Cossoninae; Kuschel *et al.* 2000). Platypodines differ from other weevil groups in several respects including the lack of a rostrum (present but short in *Notoplatypus*), reduced funicle and 1-segmented, flattened, partly or entirely pubescent antennal club, short, spiculate or ridged tibiae and very long, slender tarsi without ventral lobes. The first abdominal ventrite is also much shorter than the second one in most members of the group (but not in *Notoplatypus*). The systematic position of Platypodinae is still an unresolved problem; Wood (1973, 1978, 1986, 1993) claimed that neither Platypodinae nor Scolytinae belonged within the family Curculionidae, Marvaldi (1997) and Oberprieler *et al.* (2007) placed the platypodines near Dryophthorinae, and Kuschel *et al.* (2000) and Farrell *et al.* (2001) considered the group to be derived from within the scolytine clade. Platypodines are ambrosia beetles, culturing and feeding on yeast-like fungi in a manner similar to that of some groups of Scolytinae (**see** below). Platypodine colonies are usually long-lived, and at least one species, *Austroplatypus incompertus* Schedl, is reported to be eusocial in that mature colonies are monogynous and exhibit cooperative brood care, division into reproductive and sterile castes, and overlap of two or more generations (Browne 1971; Kent and Simpson 1992, 2010). The subfamily **Brachycerinae** is used here in a broad sense including, not only the exotic Brachycerini, Desmidophorini and Ocladini (Morimoto and Kojima 2006), but the primarily aquatic weevils belonging to Erirhinini and Stenopelmini, as well as the blind and wingless Raymonionymini (Fig. 79L). Almost all members of the group differ from other true weevils (but not Dryophthorinae) in having the more plesiomorphic type of penis with a distinct pedon and tectum. A very unusual member of the group is the genus *Aonychus* (Fig. 80I), which has a rostral canal like Cryptorhynchini and an apparently 3-segmented tarsus (with pretarsus and tarsomeres 4 and 5 missing). The aquatic groups often have a ventral plastron and various respiratory adaptations in the larva. Included among them is *Cyrtobagous salviniae* Calder & Sands, which was introduced from Brazil to control salvinia (Calder and Sands 1985). The next two subfamilies, Cyclominae and Entiminae, are very large and well represented in Australia. Included among them are virtually all of the so-called broad-nosed weevils (with a short rostrum and ectophytic, subterranean larvae), which were grouped by Kuschel (1995) in an expanded Brachycerinae, but the Cyclominae now includes some groups (e.g. Aterpini) characterised by having a long, slender rostrum and endophytic larvae. The largest group of **Cyclominae** is the Amycterini, which are large, robust, ground-dwelling weevils (Figs 79F, 98F), which have a very short rostrum, very large, chisel-like mandibles and often an upper surface covered with tubercles or spines. Males of the genus *Amycterus* are unusual in having a pair of large forceps derived from sternite VIII. Amycterines are confined to Australia and feed on a variety of monocotyledonous plants, especially Poaceae and Liliaceae, the larvae living in the soil and feeding on underground stems, crowns, tubers and rhizomes, (Howden 1986). Among the Aterpini are the diamond beetle, *Chrysolopus spectabilis* (Fabricius), a spectacular green or blue species, whose larvae may destroy young *Acacia* trees, *Rhadinosomus* species (Fig. 98K), which are extremely elongate and narrow with each elytron terminating in an acute process, and *Strongylorhinus ochraceus* Schönherr, whose larvae live gregariously in large galls on eucalypts. Also in this subfamily are species of *Aphela* (Notiomimetini), which live on beaches associated with dune vegetation, and the imported Listroderini *Listroderes difficilis* Germain, the vegetable weevil, and *Listronotus bonariensis* (Kuschel), the Argentine stem weevil. The subfamily **Entiminae** is equivalent to the section Adelognatha of early classification and most of its members are characterised by the presence of a deciduous process or cusp on the mandible of the newly emerged adult, which is soon shed leaving a distinguishing scar, and the concealment of all of most of the maxillae by the large prementum. The largest of the endemic Australian tribes is the Leptopini, but the subfamily also includes the brightly coloured clown weevils (*Pantorhytes*) of New Guinea, one species of which extends into northern QLD, and several introduced pests, such as *Otiorhynchus* species, *Asynonychus cervinus* (Boheman) (Fuller's rose weevil), *Naupactus leucoloma* (Boheman) (whitefringed weevil), *Phlyctinus callosus* Schoenherr (garden weevil), and *Sitona discoideus* Gyllehnal. Species of *Leptopius* (Fig. 80G), called wattle pigs, are very common and have become pests of sugarcane and fruit trees. The fossil remains of their large pupal cases are often encountered in SA and WA (Lea 1925). The **Molytinae** is also a very large group, particularly with the inclusion of numerous genera formerly placed in Cryptorhynchinae, which are typically rather long-legged weevils with a long, backwardly directed rostrum fitting into a deep channel formed between the procoxae and often terminating in a receptacle on the mesoventrite (Fig. 97N). Species of the cryptorhynchine genus *Tentegia* is unusual in that their larvae subsist on dung pellets of wallabies and kangaroos, which are placed by the adult beetles in excavations beneath logs (Wassell 1966). Other notable Australian cryptorhynchs are the mango seed weevil, *Sternochetus mangiferae* (Fabricius) and *Axionicus insignis* Pascoe (Fig. 79G), which breeds in the seeds of kurrajong (*Brachychiton populneum*). Members of the tribe Psepholacini are unusual in having a much shorter rostrum and resembling bark beetles (Scolytinae) in general appearance. Included in the molytine tribe Orthorhinini are two of our most impressive weevils, *Orthorhinus cylindrirostris*, the elephant beetle, and *Eurhamphus fasciculatus* (Fig. 80F), a large and rare species the larvae of which tunnel into the wood of hoop pine (*Araucaria cunninghamii*). *Tranes* and related genera have large eyes, almost meeting beneath the head and are associated with the cones of cycads and flower stalks of *Xanthorrhoea*, while members of the Phrynixini sometimes feed on fern fronds and have many relatives in New Zealand. Species in the tribe Mesoptiliini are distinguished by having truncate elytra exposing a horizontal pygidium. The **Cossoninae** (Figs 80H, 998L) is a large subfamily worldwide, but is primarily tropical in distribution. They are relatively small, elongate weevils, with a short to moderately long rostrum, widely separated procoxae and usually a large curved uncus at the tibial apex; larvae and adults may be found beneath the bark of rotten logs or in rotten palm fronds. Many Australian species remain to be described. The **Scolytinae** (Fig. 98G) is a well known group often given family rank, mainly because of the extremely short rostrum and associated differences in the tentorium and mandibular articulations (Wood 1978, 1986). Scolytinae are usually referred to as bark beetles because the majority of them construct galleries in the phloem layer beneath the bark of living or dead tree trunks, branches or twigs and have characteristic gallery patterns resulting from the tunnelling of their larvae away from the oviposition site in the brood gallery; some species, like *Coccotrypes dactyliperda* (Fabricius), however, are seed feeders. Species of *Xyleborus* and related genera (Xyleborini) bore into sapwood and heartwood and feed on yeast-like ambrosia fungi (Ascomycetes: Endomycetales) that grow on the tunnel walls and are transported and cultured by the beetles; these, are often called ambrosia beetles, shot-hole borers or pinhole borers. Xyleborines often have complex life cycles involving inbreeding and haplodiploidy (Jordal *et al.* 2000). An interesting group of Oriental scolytines occurring in New Guinea and possibly northern Australia are members of the genus *Ozopemon* Hagedorn (Dryocoetini), the only known beetles with larviform males (Jordal *et al.* 2002). Several Northern Hemisphere Scolytinae have been accidentally introduced into Australia. These include *Scolytus multistriatus* (Marsham), which transmits Dutch elm disease, *Ips grandicollis* (Eichhoff), which damages pine plantations (Morgan 1967), *Hylastes ater* (Paykull) and *Hylurgus ligniperda* (Fabricius). The subfamily **Conoderinae** (including Baridini and Ceutorhynchini) includes several unusual types of weevils that are characterised by having a small exposed pygidium and an ascending mesanepisternum, which is visible from above between the prothorax and elytra. Baridini is a large and highly diverse tribe throughout tropical regions, particularly in the New World (Davis 2009, 2010), but is relatively poorly represented in Australia. Species of *Mecopus* (Conoderini) have long legs and very large, contiguous eyes. Weevils in the genera *Phaenomerus* Schoenherr and *Tomicoproctus* Faust (Campylosceliini) are unusual and quite different looking beetles that have been found as both larvae and adults in the tunnels of a platypodine weevil, *Crossotarsus armipennis* Lea (May 1997). It is thought that the *Tomicoproctus* larvae are cleptobiotic, living off the ambrosia fungi, while those of *Phaenomerus* are probably scavengers. The remaining weevil genera are placed in the subfamily **Curculioninae**, although the tribal placements of many Australian genera are still uncertain. The genus *Melanterius* includes a large number of species that attack the seed pods of *Acacia*. The tribe Ramphini includes several species of small jumping weevils with greatly enlarged metafemora and leaf-mining larvae; the related genus *Geochus* (Geochini) known from Lord Howe Island and New Zealand are minute, wingless weevils lacking pretarsal claws and their larvae have been reported to mine dead leaves (May 1992). *Curculio* is a widespread genus of nut weevils, and the Australian *C. fraudator* Zimmerman has been reared from the fruits of *Ficus macrophylla*. Species of *Elleschodes* are apparently involved in the fertilisation of *Eupomatia* (Eupomataceae) (Hamilton 1897) and larvae develop in the decaying flowers.

Three well known tribes are of uncertain relationships and have not been placed in one of the subfamilies noted above. **Bagoini** is usually included in the Brachyceinae, but unlike other members of that group, the pedon and tectum are entirely fused together. **Gonipterini** is sometimes tentatively included in Cyclominae, but the position of the group is still in doubt. Species of *Gonipterus* (Fig. 12A) and *Oxyops* are unusual in having extrenal, slug-like, foliage-feeding larvae; *G. scutellatus* (Gyllenhal) was accidentally introduced into New Zealand, Africa and South America, where it has become a pest of eucalyps. **Hyperini** is sometimes included in Entiminae, but may belong elsewhere. [Browne 1971; Zimmerman 1991, 1992, 1993; Alonso-Zarazaga and Lyal 2009].

Australian Subfamilies, Tribes and Genera

Dryophthorinae: Dryophthorini: *Dryophthorus* Germar, *Stenommatus* Wollaston. **Orthognathini:** *Sipalinus* Marshall. **Rhynchophorini:** *Cosmopolites* Chevrolat, *Diathetes* Pascoe, *Diocalandra* Faust, *Iphthimorhinus* Roelofs, *Polytus* Faust, *Rhabdoscelus*, *Sitophilus* Schönherr, *Sphenophorus* Schönherr, *Trigonotarsus* Schönherr. **Stromboscerini:** *Dryophthoroides* Roelofs.

Platypodinae: **Tesserocerini**: *Notoplatypus* Lea. **Platypodini**: *Austroplatypus* Browne. *Baiocis* Browne, *Carchesiopygus* Schedl, *Crossotarsus* Chapuis, *Diapus* Chapuis, *Dinoplatypus* Wood, *Platypus* Herbst, *Treptoplatypus* Schedl.

Brachycerinae: Aonychusini: *Aonychus* Schönherr. **Erirhinini:** *Bagoopsis* Faust, *Echinocnemus* Schönherr, *Tadius* Pascoe. **Myrtonymini**: *Myrtonymus* Kuschel. **Stenopelmini:** *Athor* Broun, *Baeosomus* Broun, *Cyrtobagous* Hustache, *Neochetina* Hustache, *Neohydronomus* Hustache, *Niphobolus* Blackburn.

Cyclominae: Amycterini: *Acantholophus* Boisduval, *Acherres* Pascoe, *Achorostoma* Baker & Thompson, *Aedriodes* Pascoe, *Alexirhea* Pascoe, *Amorphorhinus* Lacordaire, *Amycterus* Schönherr, *Anascoptes* Pascoe, *Antalaurinus* Zimmerman, *Atychoria* Pascoe, *Brachyrothus* Marshall, *Chriotyphus* Pascoe, *Cubicorhynchus* Lacordaire, *Cucullothorax* Ferguson, *Dialeptopus* Pascoe, *Dicherotropis* Ferguson, *Ennothus* Pascoe, *Euomella* Ferguson, *Euomus* Schönherr, *Gagatophorus* Jekel, *Hyborrhinus* Marshall, *Hypotomops* Baker & Thompson, *Lataurinus* Ferguson, *Melanegis* Pascoe, *Molochtus* Pascoe, *Myotrotus* Pascoe, *Mythites* Schönherr, *Neohyborrhynchus* Ferguson, *Notonophes* Sloane, *Oditesus* Pascoe, *Ophthalamycterus* Ferguson, *Parahyborrhynchus* Ferguson, *Pseudonotonophes* Ferguson, *Sclerorinus* Macleay, *Sclerorrhinella* Ferguson, *Sosytelus* Pascoe, *Talaurinellus* Zimmerman, *Talaurinus* Macleay, *Tetralophus* Waterhouse. **Aterpini:** *Aades* Schönherr, *Acalonoma* Pascoe, *Aesiotes* Pascoe, *Alphitopis* Pascoe, *Anomocis* Lea, *Aoplocnemis* Schönherr, *Aromagis* Pascoe, *Atelicus* Waterhouse, *Aterpodes* Zimmerman, *Brendamaya* Alonso-Zarazaga & Lyal, *Cechides* Pascoe, *Chrysolopus* Germar, *Chrysophoracis* Lea, *Dixoncis* Oke, *Euthyphasis* Pascoe, *Iphisaxus* Pascoe, *Kershawcis* Lea, *Nemestra* Pascoe, *Oenopus* Marshall, *Ophthalmorychus* Blackburn, *Pelororhinus* Schönherr, *Peltochirus* Lacordaire, *Rhadinosomus* Schönherr, *Rhinaria* Kirby, *Rhinoplethes* Pascoe, *Strongylorhinus* Schönherr, *Syagrius* Pascoe. **Listroderini:** *Anorthorhinus* Blackburn, *Listroderes* Schönherr, *Listronotus* Jekel, *Methypora* Pascoe, *Steriphus* Erichson. **Notiomimetini:** *Aphela* Pascoe, *Mandalotina* Oke, *Neosyagrius* Lea, *Notiomimetes* Wollaston, *Psaldus* Pascoe, *Tasmanica* Lea, *Wollastonicis* Lea. **Rhythirrinini:** *Aparete* Pascoe, *Ethemaia* Pascoe, *Medicasta* Pascoe, *Ophryota* Pascoe, *Phrenozemia* Pascoe.

Entiminae: Celeuthetini: *Apirocalus* Pascoe, *Celeuthetes Schönherr, Idiopsodes* Thompson, *Oribius* Marshall. **Cyphicerini:** *Atmesia* Pascoe, *Drymophoetus* Marshall, *Ecmyllocerus* Zimmerman, *Euphalia* Pascoe, *Hackeria* Lea, *Heteroptochus* Faust, *Matesia* Lea, *Myllocerus* Schönherr, *Phlyda* Pascoe, *Platytrachelus* Schönherr, *Sematia* Zimmerman, *Synomus* Pascoe, *Syrphax* Marshall, *Titinia* Pascoe. **Ectemnorhinini:** *Bothrometopus* Jeannel, *Ectemnorhinus* G.R. Waterhouse. **Embrithini:** *Afrophloeus* Borovec & Oberprieler. **Leptopiini:** *Amisallus* Schönherr, *Amnemus* Marshall, *Asceparnus* Schönherr, *Baryopadus* Pascoe, *Basedowia* Lea, *Bothynorhynchus* Schönherr, *Catasarcus* Schönherr, *Catastygnus* Pascoe, *Centyres* Pascoe, *Chaodius* Pascoe, *Decienus* Pascoe, *Ecrizothis* Blackburn, *Enchymus* Pascoe, *Esmelina* Pascoe, *Esso-*

lithna Pascoe, *Evadomorpha* Blackburn, *Evas* Pascoe, *Gastrocis* Lea, *Geosomus* Faust, *Hadrorhinus* Schönherr, *Howeocis* Lea, *Leptopius* Oke, *Lipothyrea* Pascoe, *Lysizone* Pascoe, *Mandalotus* Erichson, *Merimnetes* Schönherr, *Neomerimnetes* Blackburn, *Ochrometa* Pascoe, *Ocynoma* Pascoe, *Onesorus* Pascoe, *Oophthalmus* Marshall, *Pascoellus* Oke, *Peripagis* Pascoe, *Perperus Schönherr, Platypterocis* Lea, *Polyphrades* Schönherr, *Prosayleus* Schönherr, *Prostomus* Schönherr, *Prypnus* Schönherr, *Psapharus* Schönherr, *Rhinoscapha* Montrouzier, *Rhyparophilus* Schönherr, *Scotasmus* Schönherr, *Stenocorynus* Schönherr, *Synaptonyx* G.R. Waterhouse, *Telenica* Pascoe, *Timareta* Pascoe, *Uroleptops* Lea, *Xynaea* Pascoe, *Xyneella* Zimmerman, *Zymaus* Pascoe. **Naupactini:** *Asynonychus* Crotch, *Atrichonotus* Buchanan, *Eurymetopus* Schönherr, *Naupactus* Dejean. **Oosomini:** *Phlyctinus* Schönherr. **Otiorhynchini:** *Otiorhynchus* Germar. **Ottistirini:** *Eutinophaea* Pascoe, *Maleuterpes* Blackburn. **Pachyrhynchini** *Pantorhytes* Faust. **Sitonini:** *Sitona* Germar. **Tanymecini:** *Homoeotrachelus* Faust, *Tanymecus* Germar. **Tribe ?:** *Dermatodes* Schönherr.

Molytinae: **Aedemonini:** *Aristoxenus* Zimmerman, *Cincius* Zimmerman, *Damippus* Zimmerman, *Derbyiella* Lea, *Diatassa* Zimmerman, *Dystropicus* Pascoe, *Epirus* Zimmerman, *Magius* Zimmerman, *Mechistocerus* Fauvel, *Parendymia* Kirsch, *Rhadinomerus* Faust, *Roecus* Zimmerman. **Camptorhinini:** *Camptorrhinus* Schönherr. **Colobodini:** *Deretiosus* Pascoe, *Eprias* Heller, *Perrhaebius* Pascoe. **Cryptorhynchini:** *Acalles* Schönherr, *Achopera* Pascoe, *Achoperinus* Lea, *Acidinus* Zimmerman, *Acrotychreus* Lea, *Aethreus* Pascoe, *Agathicis* Lea, *Agenopus* Pascoe, *Alatidotasia* Lea, *Albius* Zimmerman, *Amorphocis* Lea, *Amleanus* Marshall, *Ampagia* Pascoe, *Ampagiosoma* Lea, *Anchithyrus* Pascoe, *Ancocis* Lea, *Anepigraphocis* Lea, *Anilaus* Pascoe, *Anoplocis* Lea, *Asytesta* Pascoe, *Athanasius* Zimmerman, *Athyreocis* Lea, *Australacalles* Rheinheimer, *Austrectopsis* Lea, *Axionicus* Pascoe, *Baeodontocis* Lea, *Bepharus* Pascoe, *Blepiarda* Pascoe, *Bleptocis* Lea, *Bothynacrum* Lea, *Brachyphyes* Marshall, *Cairnsicis* Lea, *Cardiopterocis* Lea, *Cedilaus* Lea, *Chaetectetorus* Schönherr, *Chimades* Pascoe, *Coelosternus* Sahlberg, *Coptocelis* Chevrolat, *Cratomerocis* Lea, *Critomerus* Lea, *Cyphoderocis* Lea, *Decilaus* Pascoe, *Diaphorocis* Lea, *Drassicus* Pascoe, *Dysopirhinus* Roelofs, *Ecildaus* Lea, *Elaeagna* Pascoe, *Elytroteinus* Marshall, *Embaphiodes* Pascoe, *Emydica* Pascoe, *Enteles* Schönherr, *Ephrycinus* Lea, *Ephrycus* Pascoe, *Episodiocis* Lea, *Ethocis* Lea, *Eucalyptocis* Lea, *Eufaustia* Lea, *Euoropsis* Pascoe, *Eurycis* Lea, *Eurymetopocis* Lea, *Euryporopterus* Lea, *Eutyrhinus* Dejean, *Evaniocis* Lea, *Exithioides* Lea, *Exithius* Pascoe, *Gasterocercus* Laporte & Brulle, *Genuacalles* Rheinheimer, *Glochinorrhinus* G.R. Waterhouse, *Glyptoporopterus* Lea, *Griffithia* Lea, *Gygaeus* Pascoe, *Gymnocis* Lea, *Gymnoporopterus* Lea, *Hexymus* Pascoe, *Hoplidotasia* Lea, *Hoplodecilaus* Lea, *Hybomorphus* Saunders & Jekel, *Hyparinus* Pascoe, *Hyperiosoma* Lea, *Hypsophorus* Dejean, *Idiopterocis* Lea, *Imaliodes* Pascoe, *Imalithus* Pascoe, *Isax* Pascoe, *Leptidotasia* Zimmerman, *Leucothyreocis* Lea, *Lophocheirus* Marshall, *Mallus* Marshall, *Menectetorus* Faust, *Meniomorpha* Lea, *Menios* Pascoe, *Mesalcus* Zimmerman, *Metacymia* Pascoe, *Methidrysis* Pascoe, *Metraniomorpha* Lea, *Metyrculus* Lea, *Metyrus* Pascoe, *Microcryptorrhynchus* Lea, *Microporopterus* Lea, *Miotus* Pascoe, *Mitrastethus* Redtenbacher, *Mormosintes* Pascoe, *Myrtesis* Pascoe, *Nechyrus* Pascoe, *Neodecilaus* Lea, *Neomystocis* Lea, *Neoporopterus* Faust, *Niconotus* Pascoe, *Nothotragopus* Zimmerman, *Notocalviceps* Lea, *Notocryptorrhynchus* Lea, *Nototragopus* Heller, *Odosyllis* Pascoe, *Oemethylus* Pascoe, *Omydaus* Pascoe, *Onidistus* Pascoe, *Ophrythyreocis* Lea, *Orochlesis* Pascoe, *Orphanistes* Pascoe, *Orthoporopterus* Lea, *Osaces* Pascoe, *Ouroporopterus* Lea, *Pachyporopterus* Lea, *Paleticus* Pascoe, *Paletonidistus* Lea, *Parasalcus* Zimmerman, *Paratituacia* Lea, *Perissops* Pascoe, *Pezichus* Waterhouse, *Phloeoglymma* Pascoe, *Plagiocorynus* Waterhouse, *Platyporopterus* Lea, *Platytenes* Pascoe, *Poropterculus* Lea, *Poropterellus* Lea, *Poropterinus* Lea, *Poropteroides* Lea, *Poropterus* Schönherr, *Pseudometyrus* Lea, *Pseudomydaus* Lea, *Pseudonidistus* Lea, *Pseudoporopterus* Lea, *Pseudotepperia* Lea, *Pteroporopterus* Lea, *Rhaebocnemocis* Lea, *Riboseris* Heller, *Roptoporus* Lea, *Salcus* Pascoe, *Scleropoides* Lea, *Scolyphrus* Pascoe, *Scotinocis* Lea, *Sophronocis* Lea, *Stenoporopterus* Lea, *Sternochetus* Pierce, *Sybulus* Pascoe, *Sympediosoma* Lea, *Tamphilus* Zimmerman, *Tapinocis* Lea, *Tapinosomus* Faust, *Tentegia* Pascoe, *Tepalicus* Lea, *Tepperia* Lea, *Terporopus* Lea, *Tetengia* Lea, *Tituacus* Marshall, *Tragopus* Schönherr, *Trigonopterus* Fauvel, *Triptocis* Lea, *Tropidotasia* Lea, *Tychreus* Pascoe, *Tylocis* Lea, *Tyrtaeosellus* Lea, *Tyrtaeosus* Pascoe, *Zenoporopterus* Lea. **Hylobiini:** *Aclees* Schönherr, *Lepyrus* Germar, *Hylobius* Germar. **Lixini:** Schoenherr, *Cleonis* Dejean, *Larinus* Dejean, *Lixus* Fabricius, *Rhinocyllus* Germar. **Mecysolobini:** *Sternuchopsis* Heller. **Mesoptiliini:** *Aethemagdalis* Zimmerman, *Allolaemosaccus* Zimmerman, *Atopomagdalis* Zimmerman, *Laemosaccellus* Zimmerman, *Melaemosaccus* Zimmerman, *Neolaemosaccus* Hustache, *Notomagdalis* Zimmerman, *Zimmermanianthus* Alonso-Zarazaga & Lyal. **Molytini:** *Howeotranes* Zimmerman, *Melanotranes* Zimmerman, *Miltotranes* Zimmerman, *Paratranes* Zimmerman, *Tranes* Schönherr. **Orthorhinini:** *Cylloramphus* Erichson, *Eurhamphus* Schuckard, *Ilacuris* Pascoe, *Homorthorhinus* Voss, *Orthorhinus* Schönherr, *Paraorthorhinus* Kuschel. **Phrynixini:** *Cisolea* Oke, *Crossixus* Kuschel, *Dinichus* Pascoe, *Lophixus* Kuschel, *Mallixus* Kuschel, *Sclerixus* Kuschel. **Psepholacini:** *Calpurnius* Zimmerman, *Catocalephe* Blackburn, *Celidaus* Lea, *Dactylipalpus* Chapuis, *Neozeneudes* Lea, *Oreda* White, *Protohylastes* Wood, *Psepholacipus* Lea, *Psepholasoma* Lea, *Psepholax* White, *Pseudotherebus* Lea, *Sympiezoscelus* G. R. Waterhouse, *Therebiosoma* Lea, *Therebus* Pascoe, *Wiburdia* Lea, *Zeneudes* Pascoe. **Trachodini:** *Acicnemis* Lacordaire. **Trypetidini:** *Arecophaga* Broun. **Tribe ?:** *Demyrsus* Pascoe, *Exeiratus* Broun, *Imbilius* Marshall, *Opsittis* Pascoe, *Peribleptus* Schönherr.

Cossoninae: *Allopentarthrum* Kuschel, *Aphanocorynes* Wollaston, *Conlonia* Lea, *Coptocorynus* Marshall, *Cossonideus* Wollaston, *Cossonus* Clairville, *Dryotribus* Horn, *Halorhynchus* Wollaston, *Hexarthroides* Lea, *Homalotrogus* Wollaston, *Hoplocossonus* Lea, *Omeretes* Marshall, *Orthotemnus* Wollaston, *Oxydema* Wollaston, *Macrorhyncolus* Wollaston, *Mastersinella* Lea, *Mesites* Schönherr, *Microcossonus* Wollaston, *Pentamimus* Wollaston, *Pentarthrocis* Lea, *Pentarthrum* Wollaston, *Phloeophagosoma* Wollaston, *Pselactus* Broun, *Rhyn-*

colus Germar, *Stereoborus* Wollaston, *Stereoderus* Wollaston, *Stereomimetes* Wollaston, *Xenocnema* Wollaston, *Xenotrupis* Wollaston.

Scolytinae: Corthylini: *Pityophthorus* Eichhoff. **Cryphalini:** *Cosmoderes* Eichhoff, *Cryphalus* Erichson, *Hypocryphalus* Hopkins, *Hypothenemus* Westwood, *Scolytogenes* Eichhoff. **Diamerini:** *Acacicis* Lea, *Diamerus* Erichson. **Dryocoetini:** *Coccotrypes* Eichhoff, *Cyrtogenius* Strohmeyer, *Dryocoetes* Eichhoff, *Dryocoetiops* Schedl, *Peridryocoetes* Wood. **Hylastini:** *Hylastes* Erichson. **Hylesinini:** *Dascylipalpus* Chapuis, *Ficicis* Lea, *Hylesinus* Fabricius. **Hylurgini:** *Chaetoptelius* Fuchs, *Hylurdrectonus* Schedl, *Hylurgus* Latreille, *Pachycotes* Sharp, *Xylechinus* Chapuis. **Hypoborini:** *Zygophloeus* Schedl. **Ipini:** *Ips* DeGeer. **Micracidini:** *Afromicracis* Schedl. **Phloeosinini:** *Hyleops* Schedl, *Phloeosinopsioides* Schedl, *Phloeosinus* Chapuis. **Phloeotribini:** *Aricerus* Blandford, *Phloeotribus* Latreille. **Scolytini:** *Scolytus* Geoffroy. **Xyleborini:** Amasa Lea, *Ambrosiodmus* Hopkins, *Cnestus* Sampson, *Coptodryas* Hopkins, *Cryptoxyleborus* Wood & Bright, *Cyclorhipidion* Hagedorn, *Eccoptopterus* Motschulsky, *Euwallacea* Hopkins, *Microperus* Wood, *Xyleborinus* Reitter, *Xyleborus* Eichhoff, *Xylosandrus* Reitter. **Xyloctonini:** *Scolytomimus* Blandford.

Conoderinae: **Baridini:** *Acythopeus* Pascoe, *Apinocis* Lea, *Baris* Germar, *Gymnobaris* Lea, *Orchidophilus* Buchanan, *Ipsichora* Pascoe, *Linogeraeus* Casey, *Liturgus* Schönherr, *Myctides* Pascoe, *Nyella* Oke, *Solenobaris* Lea. **Ceutorhynchini:** *Hypurus* Rey, *Indohypurus* Korotyaev, *Mecysmoderes* Schoenherr, *Mogulones* Reitter, *Rhinoncus* Schönherr, *Trichosirocalus* Colonnelli. **Campyloscelini:** *Phaenomerus* Schönherr, *Tomicoproctus* Faust. **Conoderini:** *Agametis* Pascoe, *Mecopus* Schönherr, *Lamitema* Lea, *Lobotrachelus* Schönherr, *Metialma* Pascoe, *Othippia* Pascoe, *Temialma* Lea.

Curculioninae: **Acalyptini:** *Amorphoidea* Motschulsky, *Eudela* Pascoe, *Eudelodes* Zimmerman, *Orsophagus* Roelofs, *Epamoebus* Blackburn. **Cleogonini:** *Achelocis* Lea, *Arthriticosoma* Lea, *Chalcodermus* Dejean, *Cycloporopterus* Lea, *Euthebus* Pascoe, *Hybophorus* Waterhouse, *Leucomelacis* Lea, *Lybaeba* Pascoe, *Melanterius* Erichson, *Moechius* Pascoe, *Neomelanterius* Lea, *Teutheria* Pascoe. **Cranopoeini:** *Cranopoeus* Marshall, *Cratoscelocis* Lea, *Ennaeus* Kuschel, *Fergusoniella* Zimmerman. **Cryptoplini:** *Cryptoplus* Erichson, *Haplonyx* Schönherr, *Sigastus* Pascoe, *Zeopus* Pascoe. **Curculionini:** *Curculio* Latreille, *Pycnochirus* Berg. **Eugnomini:** *Ancyttalia* Zimmerman, *Meripherinus* Lea. *Meriphus* Erichson, *Myossita* Pascoe, *Orpha* Pascoe. **Geochini:** *Geochus* Broun. **Rhamphini:** *Rhamphus* Clairville, *Rhynchaenus* Clairville. **Smicronychini:** *Smicronyx* Schoenherr. **Storeini:** *Emplesis* Pascoe, *Imathia* Pascoe, *Pseudostoreus* Lea, *Rhinidotasia* Lea. *Storeus* Schönherr. **Tychiini:** *Sibinia* Germar. **Viticiini:** *Eupholocis* Lea, *Peliocis* Lea. **Tribe ?:** *Abethas* Zimmerman, *Agestra* Pascoe, *Anarciarthrum* Blackburn, *Byrsodes* Marshall, *Cassythicola* Lea, *Cenchrena* Pascoe, *Cisowhitea* Lea, *Cydmaea* Pascoe, *Diapelmus* Erichson, *Dicomada* Pascoe, *Dyschoenium* Blackburn, *Elleschodes* Blackburn, *Ellescus* Dejean, *Empira* Pascoe, *Empolis* Blackburn, *Encosmia* Blackburn,*Eniopea* Pascoe, *Epacticus* Blackburn, *Eristinus* Lea, *Erytenna* Pascoe, *Ethadomorpha* Blackburn, *Euhackeria* Lea, *Gerynassa* Pascoe, *Glaucopela* Pascoe, *Hibberticola* Lea, *Ixamine* Pascoe, *Lexithia* Pascoe, *Meripherellus* Lea, *Mesembrinocis* Lea, *Micraonychus* Lea, *Microberosiris* Lea, *Minia* Pascoe, *Misophrice* Pascoe, *Misophricoides* Rheinheimer, *Myllorhinus* Boisduval, *Nedyleda* Pascoe, *Notionomus* Erichson, *Ochrophoebe* Pascoe, *Olanaea* Pascoe, *Olbiodorus* Blackburn, *Orichora* Pascoe, *Paryzeta* Pascoe, *Phaeodica* Pascoe, *Phaunaeus* Lea, *Placorrhinus* Marshall, *Plaesiorhinus* Blackburn, *Sellechus* Lea, *Symbothynus* Blackburn, *Thaumastophasis* Wollaston, *Thechia* Pascoe, *Themelia* Blackburn.

Subfamily Uncertain: Bagoini: *Bagous* Germar. **Hyperini:** *Agriochaeta* Pascoe, *Eurychirus* Waterhouse, *Hypera* Germar, *Lycosura* Pascoe, *Nothyperus* Marshall, *Xeda* Pascoe. **Gonipterini:** *Bryachus* Pascoe. *Gonipterus* Schönherr, *Iptergonus* Lea, *Oxyops* Schönherr, *Pantoreites* Pascoe, *Prophaesia* Pascoe, *Syarbis* Pascoe.

REFERENCES

Aalbu RL, Triplehorn CA, Campbell JM, Brown KW, Somerby RE, Thomas DB (2002) 106. Tenebrionidae Latreille 1802. In *American Beetles. Volume 2. Polyphaga: Scarabaeoidea through Curculionoidea.* (Eds RH Arnett Jr, MC Thomas, PE Skelley and JH Frank) pp. 463–509. CRC Press, Boca Raton, Florida.

Aalbu RL, Andrews FG, Pollock DA (2005) Revision of Western Hemisphere Dacoderinae with a phylogenetic analysis of world taxa (Coleoptera: Salpingidae). *Coleopterists Society Monographs* **59**, 1–40. doi:10.1649/0010-065X(2005)59[1:ROWHDW]2.0.CO;2

Abdullah M (1969) The natural classification of the family Anthicidae with some ecological and ethological observations. *Deutsche Entomologische Zeitschrift (N.F.)* **16**, 323–366. doi:10.1002/mmnd.19690160404

Achard J (1914) *Genera Insectorum de P. Wytsman. Fascicule 159. Coleoptera Phytophaga Fam. Chrysomelidae Subfam. Lamprosominae.* 13pp., 2 pl. P. Wytsman, Brussels.

Acorn JH, Ball GE (1991) The mandibles of some adult ground beetles: structure, function, and the evolution of herbivory (Coleoptera: Carabidae). *Canadian Journal of Zoology* **69**(3), 638–650. doi:10.1139/z91-094

Aitken AD (1975) *Insect Travellers. Volume I. Coleoptera.* H.M.S.O. (MAFF Tech. Bull. 31), London.

Akre AD, Hill WB (1973) Behavior of *Adranes taylori*, a myrmecophilous beetle associated with *Lasius sitkaensis* in the Pacific Northwest (Coleoptera: Pselaphidae; Hymenoptera: Formicidae). *Journal of the Kansas Entomological Society* **46**, 526–536.

Alarie Y (1991) Primary setae and pores on the cephalic capsule and head appendages of larval Hydroporinae (Coleoptera: Dytiscidae: Hydroporinae). *Canadian Journal of Zoology* **69**, 2255–2265. doi:10.1139/z91-316

Alarie Y, Nilsson AN, Hendrich L, Watts CHS, Balke M (2000) Larval morphology of four genera of Laccophilinae (Coleoptera: Adephaga: Dytiscidae) with an analysis of their phylogenetic relationships. *Insect Systematics & Evolution* **31**, 121–164. doi:10.1163/187631200X00354

Alarie Y, Watts CHS, Nilsson AN (2001) Larval morphology of the tribe Matini (Coleoptera: Dytiscidae, Colymbetinae): descriptions of *Batrachomatus daemeli, Matus bicarinatus*, and *Allomatus nannup* and phylogenetic relationships. *Canadian Entomologist* **133**, 165–196. doi:10.4039/Ent133165-2

Alarie Y, Archangelsky M, Nilsson AN, Watts CHS (2002) Larval morphology of genus *Lancetes* (Coleoptera: Adephaga: Dytiscidae): the hypothesis of sister-group relationship with the subfamily Dytiscinae revisited. *Canadian Entomologist* **134**, 467–501. doi:10.4039/Ent134467-4

Alarie Y, Beutel RG, Watts CHS (2004) Larval morphology of three species of Hygrobiidae (Coleoptera: Adephaga: Dytiscoidea) with phylogenetic considerations. *European Journal of Entomology* **101**, 293–311.

Alexeev AV (1993) Jurassic and Lower Cretaceous Buprestidae (Coleoptera) from Eurasia. *Paleontological Journal* **27**, (1A), 9–34.

Alexeev AV (1995) New taxa of metallic wood-boring beetles (Coleoptera, Buprestidae) from the Mesozoic of Russia, Kazakhstan, and Mongolia. *Paleontologicheskii Zhurnal* **1995**(4), 310–317 (in Russian; translation in *Paleontological Journal* **30**(3), 310–317, 1996).

Allard V (1995) *Les Coléoptères du Monde. Vol. 23. Schizorhini 1.* 152 pp. Sciences Nat, Venette, France.

Allsopp PG (1980) Identification of false wireworms (Coleoptera: Tenebrionidae) from southern Queensland and northern New South Wales. *Journal of the Australian Entomological Society* **18**, 277–286. doi:10.1111/j.1440-6055.1979.tb00852.x

Allsopp PG (1981) Revision of the Australian species of *Phaenognatha* Hope (Coleoptera: Scarabaeidae: Aclopinae). *Journal of the Australian Entomological Society* **20**, 185–195. doi:10.1111/j.1440-6055.1981.tb01028.x

Allsopp PG (1982) Corrections to Paulian (1980) "Révision taxonomique des *Liparochrus* Erichson, genre Australo-Mélanésien de Coléoptères Hybosoridae, Scarabaeoidea. *Journal of the Australian Entomological Society* **21**, 199–200. doi:10.1111/j.1440-6055.1982.tb01794.x

Allsopp PG (1983) *Neophaenognatha,* a new genus for the Neotropical species of *Phaenognatha* Hope (Scarabaeidae: Aclopinae) with the description of *N. capella* n. sp. and designation of lectotypes. *Coleopterists Bulletin* **37**, 208–211.

Allsopp PG (1984) Checklist of the Hybosorinae (Coleoptera: Scarabaeidae). *Coleopterists Bulletin* **38**, 105–117.

Allsopp PG (1990a) Revision of the Systellopini (Coleoptera: Scarabaeidae: Melolonthinae). *Invertebrate Taxonomy* **3**, 197–227. doi:10.1071/IT9890197

Allsopp PG (1990b) Revision of the Pachytrichini (Coleoptera: Scarabaeidae: Melolonthinae). *Invertebrate Taxonomy* **4**, 753–761. doi:10.1071/IT9900753

Allsopp PG (1995) Biogeography of the Australian Dynastinae, Rutelinae, Scarabaeinae, Melolonthini, Scitalini and Geotrupidae (Coloeoptera: Scarabaeoidea). *Journal of Biogeography* **22**, 31–48. doi:10.2307/2846071

Allsopp PG (1999) How localized are the distributions of Australian scarabs (Coleoptera, Scarabaeoidea)? *Diversity & Distributions* **5**, 143–149. doi:10.1046/j.1472-4642.1999.00050.x

Alonso-Zarazaga MA (1990) Revision of the supraspecific taxa in the Palaearctic Apionidae Schoenherr, 1823 (Coleoptera, Curculionoidea). 2. Subfamily Apioninae Schoenherr, 1823: introduction, keys and descriptions. *Graellsia* **46**, 19–156.

Alonso-Zarazaga MA, Lyal CHC (1999) *A World Catalogue of Families and Genera of Curculionoidea (Insecta: Coleoptera) (excepting Scolytidae and Platypodidae).* Entomopraxis, Barcelona.

Alonso-Zarazaga MA, Lyal CHC (2002) Addenda and corrigenda to 'A World Catalogue of Families and Genera of Curculionoidea (Insecta: Coleoptera)'. *Zootaxa* **63**, 1–37.

Alonso-Zarazaga MA, Lyal CHC (2009) A catalogue of family and genus group names in Scolytinae and Platypodinae with nomenclatural remarks (Coleoptera: Curculionidae). *Zootaxa* **2258**, 1–134.

Andersen AN, New TR (1987) Insect inhabitants of fruits *of Leptospermum, Eucalyptus* and *Casuarina* in South-Eastern Australia. *Australian Journal of Zoology* **35**, 327–336. doi:10.1071/ZO9870327

Anderson WH (1936) A comparative study of the labium of coleopterous larvae. *Smithsonian Miscellaneous Collections* **95** (13), 1–29, 8 pls.

Anderson WH (1941) On some larvae of the genus *Proterhinus* (Coleoptera: Aglycyderidae). *Proceedings of the Hawaiian Entomological Society* **11**(1), 25–35.

Anderson WH (1947) Larvae of some genera of Anthribidae (Coleoptera). *Annals of the Entomological Society of America* **40**, 489–517, pls. 1–4.

Anderson WH (1949) Larval description and transfer of *Thaumaphrastus karanisensis* from Colydiidae to a new subfamily of Dermestidae (Coleoptera). *Bulletin of the Brooklyn Entomological Society* **44**, 121–127.

Anderson WH (1952) Larvae of some genera of Cossoninae (Coleoptera: Curculionidae). *Annals of the Entomological Society of America* **45**, 281–309, 22 figs.

Anderson JME (1976) Aquatic Hydrophilidae (Coleoptera). The biology of some Australian species with descriptions of immature stages reared in the laboratory. *Journal of the Australian Entomological Society* **15**, 219–228. doi:10.1111/j.1440-6055.1976.tb01696.x

Anderson JME (1981) Biology and distribution of *Scymnodes lividigaster* (Mulsant) and *Leptothea galbula* (Mulsant) Australian ladybirds (Coleoptera: Coccinellidae). *Proceedings of the Linnaean Society of New South Wales* **105**, 1–15.

Anderson RS (2002a) 125. Nemonychidae Bedel 1882. In *American Beetles. Volume 2. Polyphaga: Scarabaeoidea through Curculionoidea.* (Eds RH Arnett Jr, MC Thomas, PE Skelley and JH Frank) pp. 692–694. CRC Press, Boca Raton, Florida.

Anderson RS (2002b) 127. Belidae Schönherr 1826. In *American Beetles. Volume 2. Polyphaga: Scarabaeoidea through Curculionoidea.* (Eds RH

Arnett Jr, MC Thomas, PE Skelley and JH Frank) pp. 701–702. CRC Press, Boca Raton, Florida.

Anderson RS (2002c) 130. Ithyceridae Schönherr 1823. In *American Beetles. Volume 2. Polyphaga: Scarabaeoidea through Curculionoidea.* (Eds RH Arnett Jr, MC Thomas, PE Skelley and JH Frank) pp. 720–721. CRC Press, Boca Raton, Florida.

Anderson RS (2002d) 131. Curculionidae Latreille 1802. In *American Beetles. Volume 2. Polyphaga: Scarabaeoidea through Curculionoidea.* (Eds RH Arnett Jr, MC Thomas, PE Skelley and JH Frank) pp. 722–815. CRC Press, Boca Raton, Florida.

Anderson RS (2005) New Oxycoryninae from Central and South America: phylogenetic and biogeographical implications (Coleoptera: Belidae). *Systematic Entomology* **30**, 644–652. doi:10.1111/j.1365-3113.2005.00304.x

Anderson RS, Kissinger DG (2002) 129. Brentidae Billberg 1820. In *American Beetles. Volume 2. Polyphaga: Scarabaeoidea through Curculionoidea.* (Eds RH Arnett Jr, MC Thomas, PE Skelley and JH Frank) pp. 711–719. CRC Press, Boca Raton, Florida.

Andrews FG (2002) 95. Latridiidae Erichson 1842. In *American Beetles. Volume 2. Polyphaga: Scarabaeoidea through Curculionoidea.* (Eds RH Arnett Jr, MC Thomas, PE Skelley and JH Frank) pp. 395–398. CRC Press, Boca Raton, Florida.

Aneshansley DJ, Jones TH, Alsop D, Meinwald J, Eisner T (1983) Thermal concomitants and biochemistry of the explosive discharge mechanism of some little known bombardier beetles. *Experientia* **39**, 366–368. doi:10.1007/BF01963125

Antoine P (2003) *Bisallardiana* nom. nouveau de replacement pour Allardiana Rigout, 1994 (Coleoptera, Cetoniidae). *Revue Française d'Entomologie (N. S.)* **25**(3), 112

Anton E, Beutel RG (2011) The adult head morphology of *Dascillus* (L.) (Dasacilloidea: Dascillidae) and *Glaresis* Erichson (Scarabaeoidea: Glaresidae) and its phylogenetic implications. *Arthropod Systematics & Phylogeny* **70**(1), 3–42.

Archangelsky M (1994) Description of the preimaginal stages of *Dactylosternum cacti* (Coleoptera: Hydrophilidae: Sphaeridiinae). *Entomologica Scandinavica* **25**, 121–128. doi:10.1163/187631294X00243

Archangelsky M (1996) A bibliographic compilation on the immature stages of Hydrophiloidea (Insecta, Coleoptera). *Entomologica Basiliensia* **19**, 563–673.

Archangelsky M (1997) Studies on the biology, ecology, and systematics of the immature stages of New World Hydrophiloidea (Coleoptera: Staphyliniformia). *Bulletin of the Ohio Biological Survey* (N.S.) **12**, ix + 207 pp.

Archangelsky M (1998) Phylogeny of Hydrophiloidea (Coleoptera: Staphyliniformia) using characters from adult and preimaginal stages. *Systematic Entomology* **23**, 9–24. doi:10.1046/j.1365-3113.1998.00039.x

Archangelsky M (1999) Adaptations of immature stages of Sphaeridiinae (Staphyliniformia, Hydrophiloidea: Hydrophilidae) and state of knowledge of preimaginal Hydrophilidae. *Coleopterists Bulletin* **53**(1), 64–79.

Archangelsky M (2004) Higher-level phylogeny of Hydrophilinae (Coleoptera: Hydrophilidae) based on larval, pupal and adult characters. *Systematic Entomology* **29**, 188–214. doi:10.1111/j.0307-6970.2004.00237.x

Archangelsky M, Manzo V (2006) The larva of *Hydora annectens* Spangler and Brown (Coleoptera: Elmidae: Larainae) and a key to New World Laraine larvae. *Zootaxa* **1204**, 41–52.

Archangelsky M, Beutel RG, Komarek A (2005) 10. Hydrophiloidea. Introduction, Phylogeny. 10.1. Hydrophilidae. In *Handbuch der Zoologie/Handbook of Zoology. Band/Volume IV Arthropoda: Insecta Teilband/Part 38. Coleoptera, Beetles. Volume 1. Morphology and Systematics (Archostemata, Adephaga, Myxophaga, Polyphaga partim).* (Eds RG Beutel and RAB Leschen) pp. 157–183. W. DeGruyter, Berlin.

Arias ET, Richardson BJ, Elgueta M (2008) The canopy beetle faunas of Gondwanan element trees in Chilean temperate rain forests. *Journal of Biogeography* **35**(5), 914–925. doi:10.1111/j.1365-2699.2007.01837.x

Arias ET, Ślipiński A, Lawrence JF, Elgueta M (2009) A review of the Chilean Egoliini (Coleoptera: Trogossitidae) with description of a new species of *Necrobiopsis* Crowson. *Zootaxa* **2170**, 37–45.

Armstrong JWT (1945) On Australian Dermestidae. Part iv. *Proceedings of the Linnean Society of New South Wales* **70**, 47–52.

Armstrong JWT (1948) On Australian Coleoptera. Part I. *Proceedings of the Linnean Society of New South Wales* **72**, 292–298.

Armstrong JWT (1953) On Australian Helodidae (Coleoptera). I. *Proceedings of the Linnean Society of New South Wales* **78**, 19–32.

Arndt E (1993) Phylogenetische Untersuchungen larvalmorphologischer Merkmale der Carabidae (Insecta: Coleoptera). *Stuttgarter Beitrage zur Naturkunde. Serie A, Biologie* **488**, 1–56.

Arndt E (1998) Phylogenetic investigations of Carabidae (Coleoptera) using larval characters. In *Phylogeny and Classification of Caraboidea. Proceedings of a Symposium (28 August, 1996, Florence, Italy), XX International Congress of Entomology.* (Eds GE Ball, A Casale and A. Vigna Taglianti) pp. 171–190. Museo Regionale di Scienze Naturali, Torino.

Arndt E, Beutel RG (1995) Larval morphology of *Systolosoma* Solier and *Trachypachus* Motschulsky (Coleoptera: Trachypachidae) with phylogenetic consideration. *Entomologica Scandinavica* **26**, 439–446. doi:10.1163/187631295X00099

Arndt E, Beutel RG, Will K (2005) 7.8. Carabidae Latreille, 1802. In *Handbuch der Zoologie/Handbook of Zoology. Band/Volume IV Arthropoda: Insecta Teilband/Part 38. Coleoptera, Beetles. Volume 1. Morphology and Systematics (Archostemata, Adephaga, Myxophaga, Polyphaga partim).* (Eds RG Beutel and RAB Leschen) pp. 119–146. W. DeGruyter, Berlin.

Arnett RH Jr (1948) A preliminary key to the genera of Oedemeridae of the world. *Coleopterists Bulletin* **2**, 13–14.

Arnett RH Jr (1951) A revision of the Nearctic Oedemeridae (Coleoptera). *American Midland Naturalist* **45**, 257–391. doi:10.2307/2421732

Arnett RH Jr (1983) *Checklist of the Beetles of North and Central America and the West Indies. Vols. 1–10.* Flora and Fauna Publications, Gainesville, Florida.

Arnett RH Jr, Ivie MA (2000) 5. Rhysodidae Laporte, 1840. In *American Beetles. Volume 1. American Beetles. Archostemata, Myxophaga, Adephaga, Polyphaga: Staphyliniformia.* (Eds RH Arnett Jr and MC Thomas) pp. 30–31. CRC Press, Boca Raton, Florida.

Arnett RH Jr, Thomas MC (eds) (2000) *American Beetles. Volume 1. American Beetles. Archostemata, Myxophaga, Adephaga, Polyphaga: Staphyliniformia.* xvi + 443 pp. CRC Press, Boca Raton, Florida.

Arnett RH Jr, Thomas MC, Skelley PE, Frank JH (2002) *American Beetles. Volume 2. Polyphaga: Scarabaeoidea through Curculionoidea.* xvi + 861 pp. CRC Press, Boca Raton, Florida.

Arnoldi LV (1977) Mesozoic Coleoptera. Rhynchophora. *Trudy Paleontologicheskogo Institut Akademia Nauk SSSR* **161**: 142–176 (in Russian; translation 1992, Smithsonian Institution Libraries and National Science Foundation, Washington, D. C., pp.191–241).

Arnoldi LV, Byzova YuB (1964) Family Curculionidae – Weevils. In *Determination Key to Soil-dwelling Insect Larvae.* (Ed. MS Ghilarov) pp. 537–573. "Nauka", Moscow.

Arrow GJ (1901) The genus *Hyliota*, of the coleopterous family Cucujidae, with descriptions of new forms and a list of the described species. *Transactions of the Entomological Society of London* **1901**(4), 593–601.

Arrow GJ (1944) On erotylid beetles belonging to *Spondylotriplax* and some allied genera, with descriptions of a few new species (Coleoptera). *Proceedings of the Royal Entomological Society of London (B)* **13**(5–6), 53–57.

Asano A, Kojima H (2009) On the mature larvae of the genus *Laius* (Coleoptera, Malachiidae) from Japan. Japanese. *Journal of Systematic Entomology* **15** (2), 481–486.

Asano A, Yoshitomi H (2011) The larva of *Carphurus pengaronae* Champion (Coleoptera, Malachiidae, Carphurinae). *Journal of Systematic Entomology* **17**(2), 225–229.

Ashe JS (1982) Construction of pupal cells by larvae of Aleocharinae (Coleoptera: Staphylinidae). *Coleopterists Bulletin* **35**, 341–343.

Ashe JS (1984) Major features of the evolution of relationships between gyrophaenine staphylinid beetles (Coleoptera: Staphylinidae: Aleocharinae) and fresh mushrooms. In *Fungus–Insect Relationships: Perspectives In Ecology And Evolution* (Eds Q Wheeler and M Blackwell) pp. 227–255 Columbia University Press, New York.

Ashe JS (2003) First record and four new species of *Adelarthra* Cameron 1920 from Australia and a related new genus and species *Notiomerinx zealandica*, from New Zealand (Coleoptera, Staphylinidae, Aleocharinae). In *Systematics of Coleoptera: Papers Celebrating the Retirement of Ivan* Löbl. (Eds G Cuccodoro and RAB Leschen) pp. 617–636. International Associated Publishers (Memoirs on Entomology International Volume 17), Gainesville, Florida.

Atkins MD (1958) Observations on the flight, wing movements and wing structure of male *Priacma serrata* (Lec.) (Coleoptera: Cupedidae). *Canadian Entomologist* **90**, 339–347. doi:10.4039/Ent90339-6

Atkins MD (1963) The Cupedidae of the world. *Canadian Entomologist* **95**, 140–162. doi:10.4039/Ent95140-2

Audisio P (1993) *Fauna d'Italia. 32. Coleoptera Nitidulidae – Kateretidae.* Edizioni Calderini, Bologna.

Audisio P, Cline AR, De Biase A, Antonini G, Mancini E, Trizzino M, Costantini L, Strika S, Lamanna F, Cerretti P (2009) Preliminary re-examination of genus-level taxonomy of the pollen beetle subfamily Meligethinae (Coleoptera: Nitidulidae). *Acta Entomologica Musei Nationalis Pragae* **49**(2), 341–504.

Austin A, Yeates DK, Cassis G, Fletcher MJ, La Salle J, Lawrence JF, McQuillan PB, Mound LA, Bickel DJ, Gullan PJ, Hales DF, Taylor GS (2004) Insects "Down Under" – Diversity, endemism and evolution of the Australian insect fauna: examples from selected orders. *Australian Journal of Entomology* **43**, 216–234. doi:10.1111/j.1326-6756.2004.00448.x

Baehr M (1979) Vergleichende Untersuchungen am Skelett und an der Coxalmuskulatur des Prothorax der Coleoptera. *Zoologica, Stuttgart* **44**(130), 1–76.

Baehr M (1984a) *Pogonus nigrescens* sp. n. from North Queensland (Coleoptera: Carabidae). *Journal of the Australian Entomological Society* **23**, 169–171. doi:10.1111/j.1440-6055.1984.tb01937.x

Baehr M (1984b) Revision der australischen Zuphiinae. 1. Gattung *Acrogenys* Macleay (Insecta, Coleoptera, Carabidae). *Spixiana* **7**, 115–134.

Baehr M (1985a) Revision of the Australian Zuphiinae. 3. The genus *Pseudaptinus* Castelnau (Insecta, Coleoptera, Carabidae). *Spixiana* **8**, 33–57.

Baehr M (1985b) Revision of the Australian Zuphiinae. 4. The genus *Parazuphium* Jeannel (Insecta, Coleoptera, Carabidae). *Spixiana* **8**, 295–321.

Baehr M (1986a) Review of the Australian species of genus *Tachyta* Kirby (Coleoptera, Carabidae, Bembidiinae). *Entomofauna* **7**, 305–314.

Baehr M (1986b) Revision of the Australian ground-beetle genus *Porocara* Sloane (Coleoptera: Carabidae: Odacanthinae). *Australian Journal of Zoology* **34**, 717–731. doi:10.1071/ZO9860717

Baehr M (1986c) Revision of the Australian Zuphiinae. 5. The genus *Zuphium* Latreille (Insecta, Coleoptera, Carabidae). *Spixiana* **9**, 1–23.

Baehr M (1986d) Revision of the Australian Zuphiinae. 6. The genus *Planetes* Macleay. Supplement to the other genera (Insecta, Coleoptera, Carabidae). *Spixiana* **9**, 151–168.

Baehr M (1987a) The Australian species of the carabid genus *Perileptus* (Coleoptera: Carabidae: Trechinae). *Invertebrate Taxonomy* **1**, 1–16. doi:10.1071/IT9870001

Baehr M (1987b) Revision of the Australian dromiine ground beetles, formerly placed in the genus *Microlestes* Schmidt-Göbel (Coleoptera, Carabidae, Lebiinae. *Entomologische Arbeiten aus dem Museum G. Frey* **35/36**, 21–65.

Baehr M (1988a) A review of the Australian tachyine beetles of the subgenera *Tachyura* Motschoulsky and *Sphaerotachys* Müller, with special regard to the tropical fauna (Insecta, Coleoptera, Carabidae, Bembidiinae). *Spixiana* **10**, 225–269.

Baehr M (1988b) Revision of the Australian Zuphiinae 2. *Colasidia monteithi* sp. nov. from North Queensland, first record of the tribe Leleupidiini in Australia (Insecta: Coleoptera: Carabidae). *Memoirs of the Queensland Museum* **25**, 135–140.

Baehr M (1989) Revision of the Australian species of the genus Pogonoglossus Chaudoir (Insecta: Coleoptera: Carabidae: Helluodinae). *Invertebrate Taxonomy* **2**, 961–972. doi:10.1071/IT9880961

Baehr M (1990a) The carabid community living under the bark of Australian eucalypts. In *The Role of Ground Beetles in Ecological and Environmental Studies.* (Ed. N Stork) pp. 3–11. Intercept, Andover, UK.

Baehr M (1990b) A review of the Australian species of *Minuthodes* Andrewes, with the description of two new species (Coleoptera, Carabidae, Lebiinae). *Spixiana* **13**, 33–41.

Baehr M (1990c) Revision of the Australian ground-beetle genus *Tasmanitachoides* Erwin (Insecta: Coleoptera: Carabidae: Bembidiinae), with special regard to the tropical species. *Invertebrate Taxonomy* **4**, 867–894. doi:10.1071/IT9900867

Baehr M (1990d) Revision of the Australian species of the genus *Apotomus* Illiger (Insecta, Coleoptera, Carabidae, Apotominae). *Invertebrate Taxonomy* **3**, 619–627. doi:10.1071/IT9890619

Baehr M (1992) Revision of the Pseudomorphinae of the Australian region. 1. The previous genera *Sphallomorpha* Westwood and *Silphomorpha* Westwood. Taxonomy, phylogeny, zoogeography (Insecta, Coleoptera, Carabidae). *Spixiana* **18**, (Supplement), 1–410.

Baehr M (1994) Phylogenetic relations and biogeography of the genera of Pseudomorphinae (Coleoptera, Carabidae). In *Carabid Beetles: Ecology and Evolution*. (Ed. K Desender *et al.*) pp. 11–17. Kluwer Academic, Dordrecht.

Baehr M (1996) The ground beetle genus *Casnoidea* Castelnau. Taxonomy, phylogeny, zoogeography (Insecta, Coleoptera, Carabidae, Odacanthinae). *Invertebrate Taxonomy* **10**, 1041–1084. doi:10.1071/IT9961041

Baehr M (1997) Revision of the Pseudomorphinae of the Australian Region. 2. The genera *Pseudomorpha* Kirby, *Adelotopus* Hope, *Cainogenion* Notman, *Paussotropus* Waterhouse, and *Cryptocephalomorpha* Ritsema. Taxonomy, phylogeny and zoogeography. *Spixiana Supplement* **23**, 508 pp.

Baehr M (1999) A preliminary survey of the classification of the Psydrinae (Coleoptera: Carabidae). In *Phylogeny and Classification of Caraboidea. Proceedings of a Symposium (28 August, 1996, Florence, Italy), XX International Congress of Entomology*. (Eds GE Ball, A Casale and A. Vigna Taglianti) pp. 359–368. Museo Regionale di Scienze Naturali, Torino.

Baehr M (2003) Psydrine ground beetles (Coleoptera: Carabidae: Psydrinae) excluding Amblytelini of eastern Queensland rainforests. *Memoirs of the Queensland Museum* **49**, 65–109.

Baehr M (2004a) The Amblytelini. Taxonomy, phylogeny, biogeography (Insecta, Coleoptera, Carabidae, Psydrinae). *Coleoptera* **8**, 1–286.

Baehr M (2004b) A revision of the odacanthine ground beetles from Australia, including checklists of the species of Australia and of the Papuan subregion (Insecta: Coleoptera: Carabidae: Odacanthinae). *Memoirs of the Queensland Museum* **50**, 133–203.

Baehr M (2005) A new genus and three new species of helluonine beetles from Australia (Insecta, Coleoptera, Carabidae, Helluoninae). *Spixiana* **28**, 21–31.

Baehr M (2006a) The Australian scaritine genus *Steganomma* Macleay (Coleoptera: Carabidae: Scaritinae). *Mitteilungen der Münchener Entomologische Gesellschaft* **95**, 59–66.

Baehr M (2006b) A peculiar new genus of arboricolous lebiine ground beetles from interior Australia (Insects: Coleoptera: Carabidae: Lebiinae). *Memoirs of the Queensland Museum* **52**, 1–6.

Baehr M (2006c) Revision of the Australian species of the subgenus *Anomotarus* Chaudoir s. str. (Insecta, Coleoptera, Carabidae, Lebiinae). *Coleoptera* **9**, 11–107.

Baehr M (2007) Revision of the genus *Meonis* Castelnau (Insecta, Coleoptera, Carabidae, Psydrinae). *Annali del Museo Civico di Storia Naturale "Giacomo Doria", Genova* **99**, 563–662.

Baehr M (2008) The Australian Clivinini 1. The genera *Ancus* Putzeys, *Aspidiglossa* Putzeys, *Clivinarchus* Sloane, *Platysphyrus* Sloane, *Pseudoclivinia* Kult, *Rhysocara* Sloane, *Syleter* Andrewes, the subgenera *Paraclivina* Kult, *Semiclivina* Kult, and the *atrata-*, *biplagiata-*, *brevicornis-*, *coronata-*, *coryzoides- cribrosa-*, *denticollis-*, *grandiceps-*, *incerta-*, *lobata-*, *obliquata-*, *obsoleta-*, *orbitalis-*, *planiceps-*, *sulcaticeps-*, *tranquebaria-*, and *wurargae-* groups of the genus *Clivina* Latreille. With a note on a record of the genus *Parathlibops* Basilewsky (Scapterini) (Carabidae, Scaritinae). *Coleoptera* **12**, 1–220.

Baehr M (2009a) New and rare species of the pseudomorphine genera *Sphallomorpha* Westwood, *Adelotopus* Hope, and *Pseudomorpha* Kirby mainly from far Northern Territory, Australia (Insecta, Coleoptera, Carabidae, Pseudomorphinae). 12th supplement to the "Revision of the Pseudomorphinae of the Australian Region". *Spixiana* **32**, 231–253.

Baehr M (2009b) A new genus and two new species of the subfamily Migadopinae from Tasmania (Coleoptera, Carabidae). *Folia Heyrovskyana* **17**, 95–103.

Baehr M (2011) New species of the genera *Melisodera* Westwood, *Rhaebolestes* Sloane and *Moriodema* Castelnau from Australia (Coleoptera: Carabidae: Psydrini). *Australian Entomologist* **38**(3), 129–144.

Baehr M (2012a) Three peculiar new genera of lebiine carabid beetles from Queensland, Australia (Insecta: Coleoptera:Carabidae: Lebiini). *Memoirs of the Queensland Museum – Nature* **56**(1), 99–115.

Baehr M (2012b) The genus *Coptoglossa* Chaudoir in eastern Australia (Insecta: Coleoptera: Carabidae: Lebiini). *Memoirs of the Queensland Museum – Nature* **56**(1), 85–98.

Baehr M (2012c) A new species of the genus *Pheropsophus* (Coleoptera: Carabidae: Brachininae) from northern Australia. *Records of the Western Australian Museum* **27**, 62–67.

Baehr M (2012d) Two new Australian species of the genus *Hexagonia* Kirby, first record of the tribe Hexagoniini from Australia (Coleoptera, Carabidae). *Spixiana* **35**(1), 29–34.

Baguena Corella L (1948) *Estudio sobre los Aderidae (Coleopteros Heteromeros). Xylophilidae, Hylophilidae sive Euglenidae.* 547 pp. Instituto de Estudios Africanos, Madrid.

Baguena Corella L (1962) *Exploration du Parc National de la Garamba. Mission H. de Saeger. Fascicule 26. Aderidae (Coleoptera Heteromeroidea).* 135 pp. Institut des Parcs Nationaux du Congo et du Ruanda-Urundi, Brussels.

Bai M, Krell FT, Ren D, Yang X-K (2010) A new, well-preserved species of Glaresidae (Coleoptera: Scarabaeoidea) from the Jehol Biota of China. *Acta Geologica Sinica* **84**, 676–679. [English Edition]doi:10.1111/j.1755-6724.2010.00229.x

Bai M, Ren D, Yang X-K (2011) *Prophaenognatha*, a new Aclopinae genus from the Yixian Formation, China and its phylogenetic position based on morphological characters (Coleoptera: Scarabaeidae). *Acta Geologica Sinica* **85**, 984–993. [English Edition] doi:10.1111/j.1755-6724.2011.00532.x

Bai M, Ren D, Yang X-K (2012a) *Prosinodendron krelli* from the Yixian Formatikon, China: a mssing link among Lucanidae, Diphyllostomatidae and Passalidae (Coleoptera: Scarabaeoidea). *Cretaceous Research* **34**, 334–339. doi:10.1016/j.cretres.2011.11.017

Bai M, Ahrens D, Yang X-K, Ren D (2012b) New fossil evidence of the early diversification of scarabs: *Alloioscarabaeus cheni* (Coleoptera: Scarabaeoidea) from the Middle Jurassic of Inner Mongolia, China. *Insect Science* **19**, 159–171. doi:10.1111/j.1744-7917.2011.01460.x

Bai M, Beutel RG, Shih C-K, Ren D, Yang X-K (2012c) Septiventeridae, a new and ancestral fossil family of Scarabaeoidea (Insecta: Coleoptera) from the Late Jurassic to Early Cretaceous Yixian Formation. *Journal of Systematic Palaeontology* **11**, 359–374. doi:10.1080/14772019.2012.660995

Baker CW (1968) Larval taxonomy of the Troginae in North America with notes on biologies and life histories (Coleoptera: Scarabaeidae). *Bulletin of the United States National Museum* **279**, 1–79. doi:10.5479/si.03629236.279.1

Balfour-Browne F (1922) The life-history of the water-beetle *Pelobius tardus* Herbst. *Proceedings of the Zoological Society of London* **1922**, 79–97.

Balfour-Browne F, Balfour-Browne J (1940) An outline of the habits of the water-beetle, *Noterus capricornis* Herbst (Coleopt.). *Proceedings of the Zoological Society of London A* **15**, 105–112.

Balke M (2005) 7.6. Dytiscidae Leach, 1815. In *Handbuch der Zoologie/ Handbook of Zoology. Band/Volume IV Arthropoda: Insecta Teilband/ Part 38. Coleoptera, Beetles. Volume 1. Morphology and Systematics (Archostemata, Adephaga, Myxophaga, Polyphaga partim).* (Eds RG Beutel and RAB Leschen) pp. 90–116. W. DeGruyter, Berlin.

Balke M, Ribera I (2004) Jumping across Wallace's line: *Allodessus* Guignot and *Limbodessus* Guignot revisited (Coleoptera: Dytiscidae, Bidessini) based on molecular-phylogenetic and morphological data. *Australian Journal of Entomology* **43**, 114–128. doi:10.1111/j.1440-6055.2004.00415.x

Balke M, Watts CHS, Cooper SJB, Humphreys WF, Vogler AP (2004) A highly modified stygobiont diving beetle of the genus *Copelatus* (Coleoptera, Dytiscidae): taxonomy and cladistic analysis based on mitochondrial DNA sequences. *Systematic Entomology* **29**, 59–67. doi:10.1111/j.1365-3113.2004.00229.x

Balke M, Ribera I, Beutel R, Viloria A, Garcia M, Vogler AP (2008) Systematic placement of the recently discovered beetle family Meruidae (Coleoptera: Dytiscoidea) based on molecular data. *Zoologica Scripta* **37**(6), 647–650. doi:10.1111/j.1463-6409.2008.00345.x

Ball GE (1979) Conspectus of carabid classification: history, holomorphology, and higher taxa. In *Carabid Beetles: Their Evolution, Natural History, and Classification.* (Eds TL Erwin, GE Ball, DR Whitehead and AL Halpern) pp. 63–111. Junk, The Hague.

Ball GE (2000) 9. Trachypachidae C. G. Thomson, 1857. In *American Beetles. Volume 1. American Beetles. Archostemata, Myxophaga, Adephaga, Polyphaga: Staphyliniformia.* (Eds RH Arnett Jr and MC Thomas) pp. 144–146. CRC Press, Boca Raton, Florida.

Ball GE, Bousquet Y (2000) 6. Carabidae Latreille, 1810. In *American Beetles. Volume 1. American Beetles. Archostemata, Myxophaga, Adephaga, Polyphaga: Staphyliniformia.* (Eds RH Arnett Jr and MC Thomas) pp. 32–132. CRC Press, Boca Raton, Florida.

Ball GE, Casale A, Vigna Taglinati A (1998) Introduction. In *Phylogeny and Classification of Caraboidea. Proceedings of a Symposium (28 August, 1996, Florence, Italy), XX International Congress of Entomology.* (Eds GE Ball, A Casale and A. Vigna Taglianti) pp. 15–52. Museo Regionale di Scienze Naturali, Torino.

Ballantyne LA (1968) Revisional studies of Australian and Indomalayan Luciolini (Coleoptera: Lampyridae: Luciolinae). *Papers of the Department of Entomology. University of Queensland* **2**, 103–139.

Ballantyne LA (1987a) Further revisional studies on the firefly genus *Pteroptyx* Olivier (Coleoptera: Lampyridae: Luciolinae: Luciolini). *Transactions of the American Entomological Society* **113**, 117–170.

Ballantyne LA (1987b) Lucioline morphology, taxonomy and behaviour: a reappraisal (Coleoptera: Lampyridae). *Transactions of the American Entomological Society* **113**, 171–188.

Ballantyne LA, Lambkin C (2000) Lampyridae of Australia (Coleoptera: Lampyridae: Luciolinae: Luciolini). *Memoirs of the Queensland Museum* **46**(1), 15–93.

Ballantyne LA, Lambkin C (2009) Systematics of Indo-Pacific fireflies with a redefinition of Australasian *Atyphella* Olliff, Madagascan *Photuroluciola* Pic, and description of seven new genera from the Luciolinae (Coleoptera: Lampyridae). *Zootaxa* **1997**, 1–188.

Ballantyne LA, Mclean MR (1970) Revisional studies on the firefly genus *Pteroptyx* Olivier (Coleoptera: Lampyridae: Luciolinae: Luciolini). *Transactions of the American Entomological Society* **96**, 223–305.

Banaszkiewicz M (2006) Comparative study of female genitalia in Pedinini (*sensu* Iwan 2004) (Coleoptera: Tenebrionidae: Pedinini), with notes on the classification. *Annales Zoologici* **56**(1), 59–77.

Barbero E, Palestrini C (1993) *Chaetonyx robustus* ssp. *liguricus* Mariani, 1946: descrizione del terzo stadio larvale e note sulla morfologia preimmaginale della famiglia (Coleoptera Scarabaeoidea Orphnidae). *Bolletino della Societa Entomologica Italiano, Genova* **125**, 143–149.

Barbier R, Ferran A, Le Lannic J, Allo M-R (1992) Morphology and ultrastructure of integumentary glands of *Semiadalia undecimnotata* Schn. (Coleoptera: Coccinellidae). *International Journal of Insect Morphology & Embryology* **21**(3), 223–234. doi:10.1016/0020-7322(92)90018-I

Barker S (1986) *Stigmodera (Castiarina)* (Coleoptera: Buprestidae): taxonomy, new species and a checklist. *Transactions of the Royal Society of South Australia* **110**, 1–36.

Bartlett JS (2009a) The Cleridae of Lord Howe Island, with descriptions of two new species (Coleoptera: Cleroidea). *Records of the Australian Museum* **61**, 225–228. doi:10.3853/j.0067-1975.61.2009.1530

Bartlett JS (2009b) A replacement name for *Phlogistus blackburni* Hintz, 1908 (Coleoptera: Cleridae). *Australian Entomologist* **36**(3), 97–98.

Bartlett JS (2009c) *Scrobiger splendidus* (Newman) (Coleoptera: Cleridae) associated with Hylaeus sp. (Hymenoptera: Colletidae) in southeastern Queensland. *Australian Entomologist* **36**(2), 79–83.

Bartlett JS (2009d) Taxonomic revision of *Apteropilo* Lea, 1908 (Coleoptera: Cleridae). *Zootaxa* **2200**, 41–53.

Bartolozzi L (1996) A new species of *Hoplogonus* Parry, 1875 (Coleoptera, Lucanidae). *Mitteilungen der Schweizerische Entomologische Gesellschaft* **60**, 483–487.

Batelka J (2006) New synonymies in the family Ripiphoridae (Coleoptera) from Malaysia. *Folia Heyrovskyana, Serie A* **13**(4), 195–199.

Beal RS Jr (1985) Review of Nearctic species of *Orphilus* (Coleoptera: Dermestidae) with description of the larva of *O. subnitidus* Leconte. *Coleopterists Bulletin* **39**(3), 265–271.

Beal RS (1991) Dermestidae (Bostrichoidea) (including Thorictidae, Thylodriidae). In *Immature Insects. Volume 2.* (Ed. FW Stehr) pp. 434–439. Kendall/Hunt, Dubuque, Iowa.

Beal RS (2003) Annotated checklist of Nearctic Dermestidae with revised key to the genera. *Coleopterists Bulletin* **57**(4), 391–404. doi:10.1649/573

Beattie R (2007) The geological setting and palaeoenvironmental and palaeoecological reconstructions of the Upper Permian insect beds at Belmont, New South Wales, Australia. *African Invertebrates* **48**(1), 41–57.

Beattie RG, Avery S (2012) Palaeoecology and palaeoenvironment of the Jurassic Talbragar Fossil Fish Bed, Gulgong, New South Wales, Australia. *Alcheringa* **36**, 451–465. doi:10.1080/03115518.2012.671675

Beckemeyer RJ, Engel MS (2008) A second specimen of *Permocoleus* (Coleoptera) from the Lower Permian Wellington Formation of Noble County, Oklahoma. *Journal of the Kansas Entomological Society* **81**(1), 4–7. doi:10.2317/JKES-708.01.1

Becker EC (1991) Throscidae (Elateroidea) (including Dicronychidae, Lissomidae) – Eucnemidae (Elateroidea) (= Melasidae, including Perothopidae, Phylloceridae). In *Immature Insects. Volume 2.* (Ed. FW Stehr) pp. 410–421. Kendall/Hunt, Dubuque, Iowa.

Beeson CFC (1926) On some Indian Cleridae (Coleoptera) (Part II). Notes on the biology of the Cleridae. *Indian Forest Records – Entomology series* **12**(8), 217–231.

Beier M (1952) Bau und Funktion der Mundwerkzeuge bei den Helodiden-Larven (Col.). *Proceedings of the International Congress of Entomology* **9**(1), 135–138.

Belkaceme T (1986) Skelet und Muskulatur der Hinterhüfte von *Haliplus lineaticollis* Mrsh. (Haliplidae, Coleoptera). *Stuttgarter Beitrage zur Naturkunde. Serie A, Biologie* **393**, 1–12.

Belkaceme T (1991) Skelet und Musculatur des Kopfes und Thorax von *Noterus laevis* Sturm. Ein Beitrag zur Morphologie und Phylogenie der Noteridae (Coleoptera: Adephaga). *Stuttgarter Beitrage zur Naturkunde. Serie A, Biologie* **462**, 1–94.

Bell RT (1966) *Trachypachus* and the origin of the Hydradephaga (Coleoptera). *Coleopterists Bulletin* **20**, 107–112.

Bell RT (1967) Coxal cavities and the classification of the Adephaga (Coleoptera). *Annals of the Entomological Society of America* **60**, 101–107.

Bell RT (1991) Rhysodidae (Adephaga). *Immature Insects. Vol. 2.* (Ed. FW Stehr). pp. 304–305. Kendall/Hunt, Dubuque, Iowa.

Bell RT (1998) Where do the Rhysodini (Coleoptera) belong? In *Phylogeny and Classification of Caraboidea. Proceedings of a Symposium (28 August, 1996, Florence, Italy), XX International Congress of Entomology.* (Eds GE Ball, A Casale and A. Vigna Taglianti) pp. 261–272. Museo Regionale di Scienze Naturali, Torino.

Bell RT, Bell JR (1978) Rhysodini of the world. Part I. A new classification of the tribe, and a synopsis of *Omoglymmius* subgenus *Nitiglymmius*, new subgenus (Coleoptera: Carabidae or Rhysodidae). *Quaestiones Entomologicae* **14**, 43–88.

Bell RT, Bell JR (1979) Rhysodini of the world. Part II. Revisions of the smaller genera (Col: Carabidae or Rhysodidae). *Quaestiones Entomologicae* **15**, 377–446.

Bell RT, Bell JR (1982) Rhysodini of the world. Part III. Revision of *Omoglymmius* Ganglbauer (Col: Carabidae or Rhysodidae) and substitutions for preoccupied generic names. *Quaestiones Entomologicae* **18**, 127–259.

Bell RT, Bell JR (1991) The Rhysodini of Australia (Insecta: Coleoptera: Carabidae or Rhysodidae). *Annals of the Carnegie Museum* **60**, 179–210.

Bell KL, Yeates DK, Moritz C, Monteith GB (2004) Molecular phylogeny and biogeography of the dung beetle genus *Temnoplectron* Westwood (Scarabaeidae: Scarabaeinae) from Australia's wet tropics. *Molecular Phylogenetics and Evolution* **31**(2), 741–753. doi:10.1016/j.ympev.2003.09.010

Bellamy CL (1986) The higher classification of Australian Buprestidae, with the description of a new genus and species (Coleoptera). *Australian Journal of Zoology* **34**, 583–600. doi:10.1071/ZO9860583

Bellamy CL (1997) Phylogenetic relationships of *Xyroscelis* (Coleoptera: Buprestidae). *Invertebrate Taxonomy* **11**, 569–574. doi:10.1071/IT94026

Bellamy CL (2002) *Zoological Catalogue of Australia. Volume 29.5. Coleoptera: Buprestoidea.* 508 pp. CSIRO Publishing and Australian Biological Resources Study, Canberra, ACT.

Bellamy CL (2003) An illustrated summary of the higher classification of the superfamily Buprestoidea (Coleoptera). *Supplementum* **10**, 1–197.

Bellamy CL (2008a) A World Catalogue and Bibliography of the Jewel Beetles (Coleoptera: Buprestoidea), Volume 1: Introduction; Fossil Taxa; Schizopodidae; Buprestidae: Julodinae – Chrysochroinae: Poecilonotini. pp. 1–625. Pensoft (Series Faunistica No. 76), Sofia.

Bellamy CL (2008b) A World Catalogue and Bibliography of the Jewel Beetles (Coleoptera: Buprestoidea), Volume 2: Chrysochroinae: Sphenopterini through Buprestinae: Stigmoderini. pp. 626–1259. Pensoft (Series Faunistica No. 77), Sofia.

Bellamy CL (2008c) A World Catalogue and Bibliography of the Jewel Beetles (Coleoptera: Buprestoidea), Volume 3: Buprestinae: Pterobothrini through Agrilinae: Rhaeboscelina. pp. 1260–1931. Pensoft (Series Faunistica No. 78), Sofia.

Bellamy CL (2008d) A World Catalogue and Bibliography of the Jewel Beetles (Coleoptera: Buprestoidea), Volume 4: Agrilinae: Agrilina through Trachyini. pp. 1932–2684. Pensoft (Series Faunistica No. 79), Sofia.

Bellamy CL (2009) A World Catalogue and Bibliography of the Jewel Beetles (Coleoptera: Buprestoidea), Volume 5: Appendices, Bibliography, Indices. pp. 2685–3264. Pensoft (Series Faunistica No. 80), Sofia.

Bellamy CL, Nelson GH (2002) 41. Buprestidae Leach 1815. In *American Beetles. Volume 2. Polyphaga: Scarabaeoidea through Curculionoidea.* (Eds RH Arnett Jr, MC Thomas, PE Skelley and JH Frank) pp. 98–112. CRC Press, Boca Raton, Florida.

Bellamy CL, Volkovitsh MG (2005) 17. Buprestoidea Crowson, 1955. In *Handbuch der Zoologie/Handbook of Zoology. Band/Volume IV Arthropoda: Insecta Teilband/Part 38. Coleoptera, Beetles. Volume 1. Morphol-*

ogy and Systematics (Archostemata, Adephaga, Myxophaga, Polyphaga partim). (Eds RG Beutel and RAB Leschen) pp. 461–468. W. DeGruyter, Berlin.

Bellamy CL, Weir T (2008) The reinstatement of *Julodimorpha saundersii* Thomson 1879 (Coleoptera: Buprestidae) as a valid species. *Zootaxa* **1751**, 46–54.

Bellés X (1982) Idees sobre la classificació supragenèrica de la família Ptinidae (Coleoptera). II. *Sessió conjunta d'Entomologia. Institució Catalana d'Història Natural/Societat Catalana de Lepidopterologia* **1981**, 61–65.

Bellés X (1985) Sistematica, filogenia y biogeografia de la subfamilia Gibbiinae (Coleoptera, Ptinidae). *Treballs del Museu de Zoologi, Barcelona* **3**, 1–94.

Bellés X (1992) Sistemática, historia natural y biogeografía del género *Pitnus* Gorham, 1880 (Coleoptera, Ptinidae). *Eos* **68**, 167–192.

Bellés X (1998) A new subgenus and two new species of *Sphaericus* (Coleoptera: Ptinidae) from Western Australia. *European Journal of Entomology* **95**, 263–268.

Bellés X, Lawrence JF (1990) Notes on the genus *Neoptinus* Gahan (= *Ptinosphaerus* Belles and Lawrence) (Coleoptera: Ptinidae). *Australian Entomological Magazine* **17**, 61–63.

Benjamin RK (1973) Laboulbeniales. In *The Fungi, An Advanced Treatise. Vol. IVA. A Taxonomic Review with Keys: Ascomycetes and the Fungi Imperfecti.* (Eds GC Ainsworth, FK Sparrow and AS Sussman), pp. 223–246. Academic Press, New York.

Bernhard D, Ribera I, Komarek A, Beutel RG (2009) Phylogenetic analysis of Hydrophiloidea (Coleoptera: Polyphaga) based on molecular data and morphological characters. *Insect Systematics & Evolution* **40**, 3–41. doi:10.1163/187631209X416741

Berrioz-Ortiz A (1985) The presence of mesothoracic glands in *Epicauta segmenta* (Coleoptera: Meloidae) and other blister beetles. *Journal of the Kansas Entomological Society* **58**(1), 179–181.

Bertrand HP (1972) *Larves et Nymphes des Coléoptères Aquatiques du Globe.* 804 pp. Paillart, Paris.

Bertrand HP (1977) *Larves et Nymphes des Coléoptères Aquatiques du Globe. Errata et Addenda.* 19 pp. Paillart, Abbeville.

Bertrand HP, Watts C (1965) Les premiers états des *Sclerocyphon* Blackb. (Col. Eubriidae). *Bulletin du Museum National d'Histoire Naturelle* **37**(2), 412–435.

Besuchet C (1956a) Biologie, morphologic et systematique des *Rhipidius* (Col. Rhipiphoridae). *Mitteilungen der Schweizerische Entomologische Gesellschaft* **29**, 73–144.

Besuchet C (1956b) Larves et nymphes de Pselaphides (Coleopteres). *Revue Suisse de Zoologie* **63**, 697–705.

Besuchet C (1972) Les Coleopteres Aculagnathides. *Revue Suisse de Zoologie* **79**, 99–145.

Besuchet C (2008) Synonymies et combinaisons nouvelles, revalidations et description de taxa nouveaux de Pselaphinae (Coleoptera: Staphylinidae). *Mitteilungen der Schweizerische Entomologische Gesellschaft* **81**, 61–82.

Besuchet C, Burckhardt DH, Löbl I (1987) The "Winkler/Moczarski" eclector as an efficient extractor for fungus and litter Coleoptera. *Coleopterists Bulletin* **41**, 392–394.

Béthoux O (2009) The earliest beetle identified. *Journal of Paleontology* **83**(6), 931–937. doi:10.1666/08-158.1

Betrem G (1932) Beiträge zur Kenntnis des Tribus Horiini der Familie der Meloidae (Col.). 1. Die Systematik der Horiini. *Treubia* **14**, 85–101, 1 pl.

Betz O, Thayer MK, Newton AF (2003) Comparative morphololgy and evolutionary pathways of the mouthparts in spore-feeding Staphylinoidea (Coleoptera). *Acta Zoologica, Stockholm* **84**, 179–238. doi:10.1046/j.1463-6395.2003.00147.x

Beutel RG (1986a) Skelet und Muskulatur des Kopfes und Thorax von *Hygrobia tarda* (Herbst). Ein Beitrag zur Klarung der phylogenetischen Beziehungen der Hydradephaga (Insecta: Coleoptera). *Stuttgarter Beitrage zur Naturkunde. Serie A, Biologie* **388**, 1–54.

Beutel RG (1986b) Skelet und Muskulatur des Kopfes der Larve von *Haliplus lineatocollis* Mrsh. (Coleoptera: Haliplidae). *Stuttgarter Beitrage zur Naturkunde. Serie A, Biologie* **390**, 1–15.

Beutel RG (1989a) The head of *Spanglerogyrus albiventris* Folkerts (Coleoptera: Gyrinidae). Contribution towards clarification of the phylogeny of Gyrinidae and Adephaga. *Zoologische Jahrbucher. Abteilung fur Anatomie und Ontogenie der Tiere* **118**, 431–461.

Beutel RG (1989b) The prothorax of *Spanglerogyrus albiventris* Folkerts, 1979 (Coleoptera, Gyrinidae). Contribution towards clarification of the phylogeny of Gyrinidae. *Entomologica Basiliensia* **13**, 151–173.

Beutel RG (1990a) Metathoracic features of *Omoglymmius hamatus* and their significance for classification of Rhysodini (Coleoptera: Adephaga). *Entomologia Generalis* **15**, 185–201.

Beutel RG (1990b) Phylogenetic analysis of the family Gyrinidae (Coleoptera) based on meso- and metathoracic characters. *Quaestiones Entomologicae* **26**, 163–191.

Beutel RG (1993) Phylogenetic analysis of Adephaga (Coleoptera) based on characters of the larval head. *Systematic Entomology* **18**, 127–147. doi:10.1111/j.1365-3113.1993.tb00658.x

Beutel RG (1994) Phylogenetic analysis of Hydrophiloidea based on characters of the head of adults and larvae (Coleoptera: Staphyliniformia). *Koleopterologische Rundschau* **64**, 103–131.

Beutel RG (1995a) The Adephaga (Coleoptera): phylogeny and evolutionary history. In *Biology, Phylogeny, and Classification of Coleoptera. Papers Celebrating the 80th Birthday of Roy A. Crowson.* (Eds J Pakaluk and SA Ślipiński) pp. 173–217. Muzeum i Instytut Zoologii Polska Akademia Nauk, Warsawa.

Beutel RG (1995b) Phylogenetic analysis of Elateriformia (Coleoptera: Polyphaga) based on larval characters. *Journal of Zoological Systematics and Evolutionary Research* **33**, 145–171.

Beutel RG (1996) Study of the larva of *Nosodendron fasciculare* (Olivier 1790) (Coleoptera: Nosodendridae) with implications for the phylogeny of Bostrichiformia. *Journal of Zoological Systematics and Evolutionary Research* **34**, 121–134.

Beutel RG (1997) Über Phylogenese und Evolution der Coleoptera (Insecta), insbesondere der Adephaga. *Abhandlungen des Naturwissenschaftlichen Vereins in Hamburg NF* **31**, 1–164.

Beutel RG (1998) Trachypachidae and the phylogeny of Adephaga (Coleoptera). In *Phylogeny and Classification of Caraboidea. Proceedings of a Symposium (28 August, 1996, Florence, Italy), XX International Congress of Entomology.* (Eds GE Ball, A Casale and A. Vigna Taglianti) pp. 81–105. Museo Regionale di Scienze Naturali, Torino.

Beutel RG (1999a) Morphology and evolution of the larval head of Hydrophiloidea and Histeroidea (Coleoptera: Staphyliniformia). *Tijdschrift voor Entomologie* **142**, 9–30.

Beutel RG (1999b) Phylogenetic analysis of Myxophaga (Coleoptera) with a redescription of *Lepicerus horni* (Lepiceridae). *Zoologischer Anzeiger* **237**, 291–308.

Beutel RG (2005a) 1. Systematic position, basal branching pattern and early evolution. In *Handbuch der Zoologie/Handbook of Zoology. Band/Volume IV Arthropoda: Insecta Teilband/Part 38. Coleoptera, Beetles. Volume 1. Morphology and Systematics (Archostemata, Adephaga, Myxophaga, Polyphaga partim).* (Eds RG Beutel and RAB Leschen) pp. 1–9. W. DeGruyter, Berlin.

Beutel RG (2005b) 6. Myxophaga Crowson, 1955. In *Handbuch der Zoologie/Handbook of Zoology. Band/Volume IV Arthropoda: Insecta Teilband/Part 38. Coleoptera, Beetles. Volume 1. Morphology and Systematics (Archostemata, Adephaga, Myxophaga, Polyphaga partim).* (Eds RG Beutel and RAB Leschen) pp. 43–44, 51–52. W. DeGruyter, Berlin.

Beutel RG (2005c) 7.9. Family Rhysodidae Laporte, 1840 (= Rhysodini s. Bell). In *Handbuch der Zoologie/Handbook of Zoology. Band/Volume IV Arthropoda: Insecta Teilband/Part 38. Coleoptera, Beetles. Volume 1. Morphology and Systematics (Archostemata, Adephaga, Myxophaga, Polyphaga partim)*. (Eds RG Beutel and RAB Leschen) pp. 146–152. W. DeGruyter, Berlin.

Beutel RG, Arce-Pérez R (2005) 6.3. Sphaeriusidae Erichson, 1845 (Jäch 1999) (= Microsporidae). In *Handbuch der Zoologie/Handbook of Zoology. Band/Volume IV Arthropoda: Insecta Teilband/Part 38. Coleoptera, Beetles. Volume 1. Morphology and Systematics (Archostemata, Adephaga, Myxophaga, Polyphaga partim)*. (Eds RG Beutel and RAB Leschen) pp. 48–49. W. DeGruyter, Berlin.

Beutel RG, Arndt E (2005) 7.7. Trachypachidae C. G. Thomson, 1857. In *Handbuch der Zoologie/Handbook of Zoology. Band/Volume IV Arthropoda: Insecta Teilband/Part 38. Coleoptera, Beetles. Volume 1. Morphology and Systematics (Archostemata, Adephaga, Myxophaga, Polyphaga partim)*. (Eds RG Beutel and RAB Leschen) pp. 116–119. W. DeGruyter, Berlin.

Beutel RG, Belkaceme T (1986) Comparative studies on the metathorax of Hydradephaga and Trachypachidae (Coleoptera). *Entomologica Basiliensia* **11**, 221–229.

Beutel RG, Friedrich F (2005) Comparative study of larvae of Tenebrionoidea (Cucujiformia, Coleoptera). *European Journal of Entomology* **102**, 241–264.

Beutel RG, Gorb SN (2006) A revised interpretation of the evolution of attachment structures in Hexapoda with special emphasis on Mantophasmatodea. *Arthropod Systematics and Phylogeny* **64**(1), 3–25.

Beutel RG, Haas A (1998) Larval head morphology of *Hydroscapha natans* LeConte, 1874 (Coleoptera, Myxophaga, Hydroscaphidae) with special reference to miniaturization. *Zoomorphology* **118**, 103–116. doi:10.1007/s004350050061

Beutel RG, Haas F (2000) Phylogenetic relationships of the suborders of Coleoptera (Insecta). *Cladistics* **16**, 103–141. doi:10.1111/j.1096-0031.2000.tb00350.x

Beutel RG, Hörnschemeyer T (2002a) Description of the larva of *Rhipsideigma raffrayi* (Coleoptera: Archostemata), with phylogeneteic and functional implications. *European Journal of Entomology* **99**, 53–66.

Beutel RG, Hörnschemeyer T (2002b) Larval morphology and phylogenetic position of *Micromalthus debilis* LeConte (Coleoptera: Micromalthidae). *Systematic Entomology* **27**, 169–190. doi:10.1046/j.1365-3113.2002.00172.x

Beutel RG, Jäch MA (1995) Untersuchungen über die Proplerua der Hydraenidae (Coleoptera: Staphyliniformia). *Zeitschrift der Arbeitgemeinschaft Österreichischer Entomologen* **47**(1/2), 59–63.

Beutel RG, Komarek A (2004) Comparative study of thoracic structures of adults of Hydrophiloidea and Histeroidea with phylogenetic implications (Coleoptera, Polyphaga). *Organisms, Diversity & Evolution* **4**, 1–34. doi:10.1016/j.ode.2003.10.001

Beutel RG, Lawrence JF (2005) 4. Coleoptera, Morphology. In *Handbook of Zoology. Volume IV Arthropoda: Insecta. Part 38. Coleoptera. Volume 1: Morphology and Systematics (Archostemata, Adephaga, Myxophaga, Polyphaga partim)*. (Eds RG Beutel and RAB Leschen) pp. 23–27. W. DeGruyter, Berlin.

Beutel RG, Leschen RAB (eds) (2005a) *Handbuch der Zoologie/Handbook of Zoology. Band/Volume IV Arthropoda: Insecta Teilband/Part 38. Coleoptera, Beetles. Volume 1. Morphology and Systematics (Archostemata, Adephaga, Myxophaga, Polyphaga partim)*. W. DeGruyter, Berlin.

Beutel RG, Leschen RAB (2005b) 2. Classification. In *Handbuch der Zoologie/Handbook of Zoology. Band/Volume IV Arthropoda: Insecta Teilband/Part 38. Coleoptera, Beetles. Volume 1. Morphology and Systematics (Archostemata, Adephaga, Myxophaga, Polyphaga partim)*. (Eds RG Beutel and RAB Leschen) pp. 11–16. W. DeGruyter, Berlin.

Beutel RG, Leschen RAB (2005c) 8. Polyphaga Emery, 1886. Introduction, Phylogeny. In *Handbuch der Zoologie/Handbook of Zoology. Band/Volume IV Arthropoda: Insecta Teilband/Part 38. Coleoptera, Beetles. Volume 1. Morphology and Systematics (Archostemata, Adephaga, Myxophaga, Polyphaga partim)*. (Eds RG Beutel and RAB Leschen) pp. 153–154. W. DeGruyter, Berlin.

Beutel RG, Leschen RAB (2005d) 9. Staphyliniformia Lameere, 1900. In *Handbuch der Zoologie/Handbook of Zoology. Band/Volume IV Arthropoda: Insecta Teilband/Part 38. Coleoptera, Beetles. Volume 1. Morphology and Systematics (Archostemata, Adephaga, Myxophaga, Polyphaga partim)*. (Eds RG Beutel and RAB Leschen) pp. 155–156. W. DeGruyter, Berlin.

Beutel RG, Leschen RAB (2005e) 14. Elateriformia Crowson, 1960. In *Handbuch der Zoologie/Handbook of Zoology. Band/Volume IV Arthropoda: Insecta Teilband/Part 38. Coleoptera, Beetles. Volume 1. Morphology and Systematics (Archostemata, Adephaga, Myxophaga, Polyphaga partim)*. (Eds RG Beutel and RAB Leschen) pp. 427–429. W. DeGruyter, Berlin.

Beutel RG, Leschen RAB (2005f) 18.10. Chelonariidae Blanchard, 1845. In *Handbuch der Zoologie/Handbook of Zoology. Band/Volume IV Arthropoda: Insecta Teilband/Part 38. Coleoptera, Beetles. Volume 1. Morphology and Systematics (Archostemata, Adephaga, Myxophaga, Polyphaga partim)*. (Eds RG Beutel and RAB Leschen) pp. 543–546. W. DeGruyter, Berlin.

Beutel RG, Leschen RAB (2005g) Phylogenetic analysis of Staphyliniformia (Coleoptera) based on characters of larvae and adults. *Systematic Entomology* **30**, 510–548. doi:10.1111/j.1365-3113.2005.00293.x

Beutel RG, Molenda R (1997) Comparative morphology of selected larvae of Staphylinoidea (Coleoptera, Polyphaga) with phylogenetic implications. *Zoologischer Anzeiger* **236**, 37–67.

Beutel RG, Pollock DA (2000) Larval head morphology of *Phycosecis littoralis* (Pascoe) (Coleoptera: Phycosecidae) with phylogenetic implications. *Invertebrate Taxonomy* **14**, 825–835. doi:10.1071/IT00030

Beutel RG, Raffaini GB (2003) First record of Sphaeriusidae for Argentina (Coleoptera: Myxophaga). *Koleopterologische Rundschau* **73**, 1–6.

Beutel RG, Ribera I (2005) 7. Adephaga Schellenberg, 1806. In *Handbuch der Zoologie/Handbook of Zoology. Band/Volume IV Arthropoda: Insecta Teilband/Part 38. Coleoptera, Beetles. Volume 1. Morphology and Systematics (Archostemata, Adephaga, Myxophaga, Polyphaga partim)*. (Eds RG Beutel and RAB Leschen) pp. 53–55. W. DeGruyter, Berlin.

Beutel RG, Roughley RE (1987) On the systematic position of the genus *Notomicrus* Sharp (Hydradephaga, Coleoptera). *Canadian Journal of Zoology* **65**, 1898–1905. doi:10.1139/z87-290

Beutel RG, Roughley RE (1988) On the systematic position of the family Gyrinidae (Coleoptera: Adephaga). *Zeitschrift für Zoologische Systematique und Evolutionsforschung* **26**, 380–400. doi:10.1111/j.1439-0469.1988.tb00324.x

Beutel RG, Roughley RE (1994) Phylogenetic analysis of Gyrinidae based on characters of the larval head (Coleoptera: Adephaga). *Entomologica Scandinavica* **24**, 459–468. doi:10.1163/187631293X00217

Beutel RG, Roughley RE (2005) 7.1. Gyrinidae Latreille, 1810. In *Handbuch der Zoologie/Handbook of Zoology. Band/Volume IV Arthropoda: Insecta Teilband/Part 38. Coleoptera, Beetles. Volume 1. Morphology and Systematics (Archostemata, Adephaga, Myxophaga, Polyphaga partim)*. (Eds RG Beutel and RAB Leschen) pp. 55–64. W. DeGruyter, Berlin.

Beutel RG, Ruhnau S (1990) Phylogenetic analysis of the genera of Haliplidae (Coleoptera) based on characters of adults. *Aquatic Insects* **12**(1), 1–17. doi:10.1080/01650429009361381

Beutel RG, Ślipiński SA (2001) Comparative study of head structures of larvae of Sphindidae and Protocucujidae (Coleoptera: Cucujoidea). *European Journal of Entomology* **98**, 219–232.

Beutel RG, Maddison DR, Haas A (1999) Phylogenetic analysis of Myxophaga (Coleoptera) using larval characters. *Systematic Entomology* **24**, 171–192. doi:10.1046/j.1365-3113.1999.00075.x

Beutel RG, Anton E, Bernhard D (2001) Head structures of adults of *Spercheus* (Coleoptera: Spercheidae): their function and possible significance to staphyliniform phylogeny. *Annales Zoologici* **51**(4), 473–484.

Beutel RG, Anton E, Jäch MA (2003) On the evolution of adult head sructures and the phylogeny of Hydraenidae (Coleoptera, Staphyliniformia). *Journal of Zoological Systematics and Evolutionary Research* **41**(4), 256–275. doi:10.1046/j.1439-0469.2003.00224.x

Beutel RG, Balke M, Steiner WJ (2006) The systematic position of Meruidae (Coleoptera, Adephaga) and the phylogeny of the smaller aquatic adephagan beetle families. *Cladistics* **22**, 102–131. doi:10.1111/j.1096-0031.2006.00092.x

Beutel RG, Bocak L, Bocakova M (2007) Are Polyphaga (Coleoptera) really a basal neopteran lineage—a reply to Kazantsev. *Acta Zoologica* **88**, 153–158. doi:10.1111/j.1463-6395.2007.00266.x

Beutel RG, Ge S, Hörnschemeyer T (2008) On the head morphology of *Tetraphalerus*, the phylogeny of Archostemata and the basal branching events in Coleoptera. *Cladistics* **23**, 1–29.

Beutel RG, Leschen RAB, Lawrence JF (eds.) (2010a) *Handbuch der Zoologie/ Handbook of Zoology. Band/Volume IV Arthropoda: Insecta Teilband/ Part 38. Coleoptera, Beetles. Volume 2. Morphology and Systematics (Polyphaga partim)*. W. DeGruyter, Berlin.

Beutel RG, Ribera I, Balke M (2010b) 3. Adephaga (Addendum). Introduction and Phylogeny. In *Handbuch der Zoologie/Handbook of Zoology. Band/ Volume IV Arthropoda: Insecta Teilband/Part 38. Coleoptera, Beetles. Volume 2. Morphology and Systematics (Polyphaga partim)*. (Eds RAB Leschen, RG Beutel and JF Lawrence) p. 21. W. DeGruyter, Berlin.

Beutel RG, Balke M, Ribera I (2010c) 3.1. Aspidytidae Ribera, Beutel, Balke and Vogler, 2002. In *Handbuch der Zoologie/Handbook of Zoology. Band/ Volume IV Arthropoda: Insecta Teilband/Part 38. Coleoptera, Beetles. Volume 2. Morphology and Systematics (Polyphaga partim)*. (Eds RAB Leschen, RG Beutel and JF Lawrence) pp. 21–28. W. DeGruyter, Berlin.

Beutel RG, Balke M, Steiner WE Jr (2010d) 3.2. Meruidae Spangler and Steiner, 2005. In *Handbuch der Zoologie/Handbook of Zoology. Band/Volume IV Arthropoda: Insecta Teilband/Part 38. Coleoptera, Beetles. Volume 2. Morphology and Systematics (Polyphaga partim)*. (Eds RAB Leschen, RG Beutel and JF Lawrence) pp. 28–34. W. DeGruyter, Berlin.

Beutel RG, Friedrich F, Hörnschemeyer T, Pohl H, Hünefeld F, Beckmann F, Meier R, Misof B, Whiting MF, Vilhelmsen L (2010e) Morphological and molecular evidence converge upon a robust phylogeny of the megadiverse Holometabola. *Cladistics* **26**, 1–15.

Bianchi FA (1962) Notes on the biology of *Cissites auriculata* (Champion) (Coleoptera: Meloidae). *Proceedings of the Hawaiian Entomological Society* **18**(1), 111–119.

Bielawski R (1963) Monographie der Epilachninae (Coleoptera, Coccinellidae) der Australischen Region. *Annales Zoologici* **21**, 295–461.

Bils W (1976) Das Abdomenende weiblicher, terrestrisch lebender Adephaga (Coleoptera) und seine Bedeutung für die Phylogenie. *Zoomorphologie* **84**, 113–193. doi:10.1007/BF00999711

Bílý S (1972) Larvae of *Ptosima flavoguttata* (Illiger) (Coleoptera, Buprestidae). *Acta Entomologica Bohemoslovaka* **69**, 18–22.

Bílý S (1986) Descriptions of adult larvae of *Thrincopyge alacris* LeConte and *Aphanisticus cochinchinae seminulum* Obenberger (Coleoptera, Buprestidae). In *Entomological Papers presented to Y. Kurosawa on the Occasion of his Retirement, March 20,* 1986. (Ed. I Ueno) pp. 198–204. Coleopterist's Association of Japan, Tokyo.

Bílý S (1989) Descriptions of last instar larvae of *Polycesta porcata* and *Paratrachys hederae hederae* (Coleoptera: Buprestidae). *Acta Entomologica Bohemoslovaka* **86**, 61–66.

Bílý S (1997) *Anocisseis danieli* sp. n. from Seram and larval morphology of the genus *Anocisseis* (Coleoptera, Buprestidae: Polycestinae). *Folia Heyrovskyana* **5**(3), 115–122.

Bílý S (1999) Larvae of buprestid beetles of Central Europe. *Acta Entomologica Musei Nationalis Prage* Suppl. **9**, 1–45, 32 pls.

Bílý S (2000a) *Euleptodema sainvali* sp. n. from New Caledonia (Coleoptera: Buprestidae: Polycestinae) and larval morphology of the genus. *Folia Heyrovskyana* **8**(1), 35–46.

Bílý S (2000b) A new concept of Anthaxiini (Coleoptera: Buprestidae). *Folia Heyrovskyana* **8**(2), 109–114.

Bílý S, Volkovitsh MG (2003) Larvae of Australian Buprestidae (Coleoptera). Part 1. Genera *Austrophorella* and *Pseudotaenia*. *Acta Societatis Zoologicae Bohemoslovenicae* **67**, 99–114.

Bílý S, Volkovitsh MG (2005) Larvae of Australian Buprestidae (Coleoptera). Part 3. Genera *Maoraxia* and *Anthaxoschema* with a review of larval characters of known anthaxiine taxa. *Folia Heyrovskyana, Serie A* **13**(1–2), 7–26.

Bílý S, Fikáček M, Šipek P (2008) First record of myrmecophily in buprestid beetles: immature stages of *Habroloma myrmecophila* sp. nov. (Coleoptera: Buprestidae) associated with Oecophylla ants (Hymenoptera: Formicidae). *Insect Systematics & Evolution* **39**, 121–131. doi:10.1163/187631208788784084

Blackburn CT (1892) Further notes on Australian Coleoptera, with descriptions of new genera and species. XI. *Transactions of the Royal Society of South Australia* **15**, 20–73.

Blackwelder RE (1944–1957) Checklist of the coleopterous insects of Mexico, Central America, the West Indies, and South America. *Bulletin of the United States National Museum* **185**, xii + 1492 pp.

Blackwell M (2006) How many yeasts? *Mycology Today* **November 2006**, 14–18.

Blair KG (1919) Notes on the Pythidae with descriptions of new species. *Entomologist's Monthly Magazine* **55**, 112–124.

Blair KG (1920) Notes on the Australian coleopterous genera *Palaestra* Cast., *Tmesidera* Westw., and *Palaestrida* White (Fam. Meloidae). *Entomologist's Monthly Magazine* **56**, 28–32.

Blair KG (1925) Further notes on the Pythidae (Coleoptera). *Entomologist's Monthly Magazine* **61**, 209–219.

Blair KG (1930) *Brachypsectra,* Lec.–the solution of an entomological enigma. *Transactions of the Entomological Society of London* **78**, 45–50. doi:10.1111/j.1365-2311.1930.tb01199.x

Blair KG (1940) Some new species of Cisidae (Col.) from Australia. *Entomologist's Monthly Magazine* **76**, 131–136.

Blum P (1979) Zur Phylogenie und ökologischen Bedeutung der Elytrenreduktion und Abdomenbeweglichkeit der Staphylinidae (Coleoptera). – Vergleichend- und funktionsmorphologische Untersuchungen). *Zoologische Jahrbucher. Abteilung fur Anatomie und Ontogenie der Tiere* **102**, 533–582.

Bocák L (2002) Generic revision and phylogenetic analysis of the Metriorrhynchinae (Coleoptera: Lycidae). *European Journal of Entomology* **99**, 315–351.

Bocák L, Bocákova M (1990) Revision of the supergeneric classification of the family Lycidae (Coleoptera). *Polskie Pismo Entomologiczne* **59**, 623–676.

Bocák L, Bocákova M (2008) Phylogeny and classification of the family Lycidae (Insecta: Coleoptera). *Annales Zoologici* **58**(4), 695–720. doi:10.3161/000345408X396639

Bocák L, Bocákova M (2010a) 4.10. Omalisidae Lacordaire, 1857. In *Handbuch der Zoologie/Handbook of Zoology. Band/Volume IV Arthropoda: Insecta Teilband/Part 38. Coleoptera, Beetles. Volume 2. Morphology and Systematics (Polyphaga partim)*. (Eds RAB Leschen, RG Beutel and JF Lawrence) pp. 110–114. W. DeGruyter, Berlin.

Bocák L, Bocákova M (2010b) 4.11. Lycidae Laporte, 1836. In *Handbuch der Zoologie/Handbook of Zoology. Band/Volume IV Arthropoda: Insecta Teilband/Part 38. Coleoptera, Beetles. Volume 2. Morphology and Systematics (Polyphaga partim)*. (Eds RAB Leschen, RG Beutel and JF Lawrence) pp. 114–123. W. DeGruyter, Berlin.

Bocák L, Matsuda K (2003) Review of the immature stages of the family Lycidae (Insecta: Coleoptera). *Journal of Natural History* **37**, 1463–1507. doi:10.1080/00222930210125362

Bocák L, Bocakáva M, Hunt T, Vogler AP (2008) Multiple ancient origins of neoteny in Lycidae (Coleoptera): consequences for ecology and macroevolution. *Proceedings. Biological Sciences* **275**, 2015–2023. doi:10.1098/rspb.2008.0476

Bocák L, Branham MA, Kundrata R (2010) 4.9. Drilidae Blanchard, 1845. In *Handbuch der Zoologie/Handbook of Zoology. Band/Volume IV Arthropoda: Insecta Teilband/Part 38. Coleoptera, Beetles. Volume 2. Morphology and Systematics (Polyphaga partim).* (Eds RAB Leschen, RG Beutel and JF Lawrence) pp. 104–110. W. DeGruyter, Berlin.

Bocákova M, Bocák L, Hunt T, Vogler AP (2007) Molecular phylogenetics of Elateriformia (Coleoptera): evolution of bioluminescence and neoteny. *Cladistics* **23**, 477–496. doi:10.1111/j.1096-0031.2007.00164.x

Bocákova M, Constantin R, Bocák L (2011) Molecular phylogenetics of the melyrid lineage (Coleoptera: Cleroidea). *Cladistics* **27**, 1–13.

Bologna MA, Di Giulio A (2011) Biological and morphological adaptations in the pre-imaginal phasesw of the beetle family Meloidae. *Atti Accademia Nazionale Italiana di Entomologia* **59**, 141–152.

Bologna MA, Laurenzi M (1994) Descriptions of the triungulins of *Synhoria testacea* (Fabricius) and another undetermined African species (Coleoptera: Meloidae), with data on Horiini larvae. *African Entomology* **2**, 155–162.

Bologna MA, Pinto JD (2001) Phylogenetic studies of Meloidae (Coleoptera), with emphasis on the evolution of phoresy. *Systematic Entomology* **26**, 33–72. doi:10.1046/j.1365-3113.2001.00132.x

Bologna MA, Pinto JD (2002) The Old World genera of Meloidae (Coleoptera): a key and synopsis. *Journal of Natural History* **36**, 2013–2102. doi:10.1080/00222930110062318

Bologna MA, Olivierio M, Pitzalis M, Mariottini P (2008) Phylogeny and evolutionary history of the blister beetles (Coleoptera, Meloidae). *Molecular Phylogenetics and Evolution* **48**, 679–693. doi:10.1016/j.ympev.2008.04.019

Bologna MA, Turco F, Pinto JD (2010) 11.19. Meloidae Gyllenhal, 1810. In *Handbuch der Zoologie/Handbook of Zoology. Band/Volume IV Arthropoda: Insecta Teilband/Part 38. Coleoptera, Beetles. Volume 2. Morphology and Systematics (Polyphaga partim).* (Eds RAB Leschen, RG Beutel and JF Lawrence) pp. 681–693. W. DeGruyter, Berlin.

Booth RTG, Cox ML, Madge RB (1990) *IIE Guides to Insects of Importance to Man. 3. Coleoptera.* International Institute of Entomology, London.

Bornemissza GF (1971a) A new variant of the paracopric nesting type in the Australian dung beetle, *Onthophagus compositus. Pedobiologia* **11**, 1–10.

Bornemissza GF (1971b) Mycetophagous breeding in the Australian dung beetle, *Onthophagus dunningi. Pedobiologia* **11**, 133–142.

Bornemissza GF (1976) The Australian Dung Beetle Project 1965–1975. *Australian Meat Research Commission Review* **30**, 1–30.

Borowiec L (1987) The genera of seed-beetles (Coleoptera, Bruchidae). *Polskie Pismo Entomologiczne* **57**, 3–207.

Borowiec L (1995) Tribal classification of the cassidoid Hispinae (Coleoptera: Chrysomelidae). In *Biology, Phylogeny, and Classification of Coleoptera. Papers Celebrating the 80th Birthday of Roy A. Crowson.* (Eds J Pakaluk and SA Ślipiński) pp. 541–558. Muzeum I Instytut Zoologii Polska Akademia Nauk, Warsaw.

Borowski J, Węgrzynowicz P (2007) *World Catalogue of Bostrichidae (Coleoptera).* Wydawnictwo Mantis, Olsztyn, Poland.

Bouchard P, Steiner WE Jr (2004) First descriptions of Coelometopini pupae (Coleoptera: Tenebrionidae) from Australia, Southeast Asia and the Pacific region, with comments on phylogenetic relationships and antipredator adaptations. *Systematic Entomology* **29**, 101–114. doi:10.1111/j.1365-3113.2004.00230.x

Bouchard P, Yeates DK (2001) Phylogenetic relationships of the Australasian Coelometopini (Coleoptera: Tenebrionidae); a quantitative cladistic analysis with a review of biology. *Organisms, Diversity & Evolution* **1**, 17–43. doi:10.1078/1439-6092-00003

Bouchard P, Brooks DR, Yeates DK (2005a) Mosaic macroevolution in Australian wet tropics arthropods: community assemblage by taxon pulses. In *Tropical Rainforests: Past, Present and Future* (Eds E Bermingham, C Dick and C Moritz) pp. 425–469. University of Chicago Press, Chicago, Illinois.

Bouchard P, Lawrence JF, Davies A, Newton AF (2005b) Synoptic classification of the world Tenebrionidae (Insecta: Coleoptera) with a review of family-group names. *Annales Zoologici* **55**(4), 499–530.

Bouchard P, Bousquet Y, Davies AE, Alonso-Zarazaga MA, Lawrence JF, Lyal CHC, Newton AF, Reid CAM, Schmitt M, Ślipiński SA, Smith ABT (2011) *Family-group Names in Coleoptera (Insecta).* Pensoft, Sofia.

Boucher S (2005) Évolution et phylogénie des Coléoptères Passalidae (Scarabaeoidea). Les taxons du groupe famille. La tribu néotropicale des Proculini et son complexe *Veturius. Annales de la Société Entomologique de France (N. S.)* **41**(3–4), 239–604.

Bousquet Y (1986) Description of first-instar larva of *Metrius Contractus* Eschscholtz (Coleoptera: Carabidae) with remarks about phylogenetic relationships and ranking of the genus *Metrius* Eschscholtz. *Canadian Entomologist* **118**, 373–388. doi:10.4039/Ent118373-4

Bousquet Y (1987) Description of the larva of *Helluomorphoides praeustus bicolor* Harris with comments on the relationships of the Helluonini (Coleoptera: Carabidae). *Canadian Entomologist* **119**, 921–930. doi:10.4039/Ent119921-10

Bousquet Y (1989) A review of the North American genera of Cryptophaginae (Coleoptera: Cryptophagidae). *Coleopterists Bulletin* **43**, 1–17.

Bousquet Y (1990) *Beetles Associated with Stored Products in Canada: An Identification Guide.* Research Branch, Agriculture Canada (Publication 1837), Ottawa.

Bousquet Y (1991) Carabidae (Adephaga) (including Brachinidae, Cicindelidae, Omophronidae, Paussidae, Pseudomorphidae, Trachypachidae etc.). In *Immature Insects. Volume 2.* (Ed. FW Stehr) pp. 308–310. Kendall/Hunt, Dubuque, Iowa.

Bousquet Y (2002) 79. Monotomidae Laporte 1840. In *American Beetles. Volume 2. Polyphaga: Scarabaeoidea through Curculionoidea.* (Eds RH Arnett Jr, MC Thomas, PE Skelley and JH Frank) pp. 319–321. CRC Press, Boca Raton, Florida.

Bousquet Y (2010) 10.8. Monotomidae Laporte, 1840. In *Handbuch der Zoologie/Handbook of Zoology. Band/Volume IV Arthropoda: Insecta Teilband/Part 38. Coleoptera, Beetles. Volume 2. Morphology and Systematics (Polyphaga partim).* (Eds RAB Leschen, RG Beutel and JF Lawrence) pp. 319–324. W. DeGruyter, Berlin.

Bousquet Y, Goulet H (1984) Notation of primary setae and pores on larvae of Carabidae (Coleoptera). *Canadian Journal of Zoology* **62**(4), 573–588. doi:10.1139/z84-085

Bousquet Y, Laplante S (2000) Taxonomic review of the Canadian species of the genus *Monotoma* Herbst (Coleoptera: Monotomidae). *Proceedings of the Entomological Society of Ontario* **130**, 67–96.

Bousquet Y, Heffern DJ, Bouchard P, Nearns EH (2009) Catalogue of family-group names in Cerambycidae (Coleoptera). *Zootaxa* **2321**, 1–80.

Böving AG (1921) The larvae and pupae of the social beetles *Coccidotrophus socialis* (Schwarz and Barber) and *Eunausibius wheeleri* (Schwarz and Barber) with remarks on the taxonomy of the family Cucujidae. *Zoologica* **3**, 197–213. [pls. vii–x.]

Böving AG (1929) On the classification of beetles according to larval characters. *Bulletin of the Brooklyn Entomological Society* **24**, 55–80.

Böving AG (1936) Description of the larva of *Plectris aliena* Chapin and explanation of new terms appplied to the epipharynx and raster. *Proceedings of the Entomological Society of Washington* **38**(8), 169–185, pls. 9–10.

Böving AG (1954) Mature larvae of the beetle-family Anobiidae. *Danske Biologiske Meddelelser* **22**(2), 1–298, 50 pls.

Böving AG, Champlain AB (1920) Larvae of North American beetles of the family Cleridae. *Proceedings of the United States National Museum* **57**, 575–649, pls. 42–53.

Böving AG, Craighead FC (1931) An illustrated synopsis of the principal larval forms of the order Coleoptera. *Entomologica Americana New Series* **11**, 1–351.

Böving AG, Henriksen KL (1938) The developmental stages of the Danish Hydrophilidae. *Videnskabelige Meddelelser fra Dansk Naturhistorisk Forening I Khobenhavn* **102**, 27–162.

Bowestead S (1999) *A revision of the Corylophidae (Coleoptera) of the West Palaearctic region*. Muséum d'Histoire Naturelle (Instrumenta Biodiversitatis III), Geneva.

Bowestead S, Leschen RAB (2002) 94. Corylophidae LeConte 1852. In *American Beetles. Volume 2. Polyphaga: Scarabaeoidea through Curculionoidea.* (Eds RH Arnett Jr, MC Thomas, PE Skelley and JH Frank) pp. 390–394. CRC Press, Boca Raton, Florida.

Bowestead S, Booth RG, Ślipiński A, Lawrence JF (2001) The genus *Cleidostethus* Arrow, 1929 reappraisal and transfer from Coccinellidae to Corylophidae (Coleoptera: Cucujoidea). *Annales Zoologici* **51**(3), 319–323.

Brancucci M (1980) Morphologie comparée, evolution et systematique des Cantharidae (Insecta: Coleoptera). *Entomologica Basiliensia* **5**, 215–388.

Brancucci M (1981a) Révision du genre *Maronius* Gorham et notes sur les genres voisins (Coleoptera: Cantharidae). *Entomologica Basiliensia* **6**, 328–367.

Brancucci M (1981b) *Maranoides* n. gen., genre nouveau de la tribu des Chauliognathini (Coleoptera: Cantharidae). *Entomologica Basiliensia* **6**, 368–386.

Brancucci M (1981c) Révision du genre *Psuedolobetus* Champion (Coleoptera: Cantharidae). *Entomologische Arbeiten aus dem Museum G. Frey* **29**, 235–243.

Brancucci M (1981d) Révision du genre *Macromalthinus* Pic (Col. Cantharidae). *Entomologische Arbeiten aus dem Museum G. Frey* **29**, 259–272.

Brancucci M (1982) Révision du genre *Lobetus* Kies. (Col. Cantharidae). *Entomologica Basiliensia* **7**, 315–339.

Brancucci M (1983) La familie des Cantharidae (Coleoptera) sur l'île de Taiwan. 4. La sous-famille Chauliognathinae. *Entomologica Basiliensia* **8**, 256–309.

Brancucci M (1985) Révision du genre *Trypherus* LeConte (Coleoptera, Cantharidae). *Entomologica Basiliensia* **10**, 251–322.

Brancucci M, Monteith GB (1997) A second *Terradessus* species from Australia (Coleoptera, Dytiscidae). *Entomologica Basiliensia* **19**, 585–591.

Brandmayr P, Zetto-Brandmayr T (1979) The evolution of parental care phenomena in pterostichine ground beetles, with special reference to the genera *Abax* and *Molops* (Coleoptera, Carabidae). In *On the Evolution of Behaviour in Carabid Beetles.* (Eds. PJ Den Boer, HU Thiele and F Weber) pp. 35–49. Agricultural University (Miscellaneous Papers No. 18), Wageningen.

Branham MA (2010a) 4.8 Plastoceridae Crowson, 1972. In *Handbuch der Zoologie/Handbook of Zoology. Band/Volume IV Arthropoda: Insecta Teilband/Part 38. Coleoptera, Beetles. Volume 2. Morphology and Systematics (Polyphaga partim).* (Eds RAB Leschen, RG Beutel and JF Lawrence) pp. 103–104. W. DeGruyter, Berlin.

Branham MA (2010b) 4.15. Lampyridae Latreille, 1817. In *Handbuch der Zoologie/Handbook of Zoology. Band/Volume IV Arthropoda: Insecta Teilband/Part 38. Coleoptera, Beetles. Volume 2. Morphology and Systematics (Polyphaga partim).* (Eds RAB Leschen, RG Beutel and JF Lawrence) pp. 141–149. W. DeGruyter, Berlin.

Branham M, Wenzel JW (2001) The evolution of bioluminescence in cantharoids (Coleoptera: Elateroidea). *The Florida Entomologist* **84**, 565–586. doi:10.2307/3496389

Branham MA, Wenzel JW (2003) The origin of photic behavior and the evolution of sexual communication in fireflies (Coleoptera: Lampyridae). *Cladistics* **19**, 1–22. doi:10.1111/j.1096-0031.2003.tb00404.x

Brass P (1914) Das 10. Abdominalsegment der Käferlarven als Bewegungsorgan. Zoologische *Jahrbücher Abteilung für Systematik* **37**, 65–122, pls 4–7.

Britton EB (1957) *A revision of the Australian chafers (Coleoptera: Scarabaeidae: Melolonthinae). Vol. 1.* British Museum (Natural History), London.

Britton EB (1966) On the larva of *Sphaerius* and the systematic position of the Sphaeriidae (Coleoptera). *Australian Journal of Zoology* **14**, 1193–1198. doi:10.1071/ZO9661193

Britton EB (1970) Coleoptera (beetles). In *The Insects of Australia. A Textbook for Students and Research Workers.* (Ed. CSIRO) pp. 495–621. Melbourne University Press, Melbourne.

Britton EB (1971) A new intertidal beetle (Coleoptera: Limnichidae) from the Great Barrier Reef. *Journal of Entomology B* **40**, 83–91.

Britton EB (1973) *Hyphalus wisei* sp. n., a new intertidal beetle from New Zealand (Coleoptera: Limnichidae). *Records of the Auckland Institute and Museum* **10**, 119–122.

Britton EB (1974a) Coleoptera (beetles). In *The Insects of Australia. A Textbook for Students and Research Workers. Supplement 1974.* (Ed. CSIRO) pp. 62–89. Melbourne University Press, Melbourne.

Britton EB (1974b) *Tripectenopus occultus* sp. n., an eyeless beetle (Coleoptera: Staphylinidae) from a Western Australian cave. *Journal of the Australian Entomological Society* **13**, 85–87. doi:10.1111/j.1440-6055.1974.tb02159.x

Britton EB (1977) Three new intertidal species of Limnichidae (Coleoptera) from New Zealand. *Records of the Auckland Institute and Museum* **14**, 81–85.

Britton EB (1978) A revision of the Australian chafers (Coleoptera: Scarabaeidae: Melolonthinae) Vol. 2. Tribe Melolonthini. *Australian Journal of Zoology* **60**, (Supplement), 1–150.

Britton EB (1980) A revision of the Australian chafers (Coleoptera: Scarabaeidae: Melolonthinae). Vol. 3. Tribe Liparetrini: genus *Liparetrus. Australian Journal of Zoology* **76**, (Supplement), 1–209.

Britton EB (1981) The Australian Hygrobiidae (Coleoptera). *Journal of the Australian Entomological Society* **20**, 83–86. doi:10.1111/j.1440-6055.1981.tb01004.x

Britton EB (1986) A revision of the Australian chafers (Coleoptera: Scarabaeidae: Melolonthinae). Vol. 4. Tribe Liparetrini: genus *Colpochila. Australian Journal of Zoology* **118**, (Supplement), 1–135.

Britton EB (1987) A revision of the Australian chafers (Coleoptera: Scarabaeidae: Melolonthinae). Vol. 5. Tribes Scitalini and Comophorinini. *Invertebrate Taxonomy* **1**, 685–799. doi:10.1071/IT9870685

Britton EB (1988) Synopsis of the genera of Australian Heteronychini (Coleoptera: Scarabaeidae: Melolonthinae). *Journal of the Australian Entomological Society* **27**, 27–36. doi:10.1111/j.1440-6055.1988.tb01139.x

Britton EB (1990) A synopsis of the genera of Liparetrini (Coleoptera: Scarabaeidae: Melolonthinae). *Invertebrate Taxonomy* **4**, 159–195. doi:10.1071/IT9900159

Britton EB (1995) A synopsis of the genera of Diphucephalini (Coleoptera: Scarabaeidae: Melolonthinae) including *Watkinsia*, gen. nov. et spp. nov. *Invertebrate Taxonomy* **9**, 115–128. doi:10.1071/IT9950115

Britton EB (2000) A review of *Heteronyx* Guérin-Méneville (Coleoptera: Scarabaeidae: Melolonthinae). *Invertebrate Taxonomy* **14**, 465–589. doi:10.1071/IT98009

Britton EB, Lee B (1972) *Stethorus loxtoni* sp. n. (Coleoptera: Coccinellidae) a newly-discovered predator of the two-spotted *mite. Journal of the Australian Entomological Society* **11**, 55–60. doi:10.1111/j.1440-6055.1972.tb01604.x

Bro-Larsen E (1936) Biologische Studien über die tunnelgrabenden Käfer auf Skallingen. *Videnskabelige Meddelelser fra den Dansk Naturhistoriske Forening i Kjøbenhavn* **100**, 1–231.

Brooks JG (1965) New records of Coleoptera In Australia. *Journal of the Entomological Society of Queensland* **4**, 85. doi:10.1111/j.1440-6055.1965.tb00659.x

Brown HP (1981) Key to the world genera of Larinae (Coleoptera, Dryopoidea, Elmidae), with descriptions of new genera from Hispaniola, Colombia, Australia, and New Guinea. *The Pan-Pacific Entomologist* **57**, 76–104.

Brown HP (1987) Biology of riffle beetles. *Annual Review of Entomology* **32**, 253–273. doi:10.1146/annurev.en.32.010187.001345

Brown HP (1991) Limnichidae (Dryopoidea). In *Immature Insects. Volume 2.* (Ed. FW Stehr) pp. 401–402. Kendall/Hunt, Dubuque, Iowa.

Brown VV, Jones AJ, Lacey MJ, Moore BP (1985) The chemistry of buprestins a and b. Bitter principles of jewel beetles (Coleoptera: Buprestidae). *Australian Journal of Chemistry* **38**, 197–206. doi:10.1071/CH9850197

Browne FG (1962) *Sosylus spectabilis* Grouvelle (Coleoptera, Colydiidae), a predator and parasite of African ambrosia beetles. *Report of the West African Timber Borer Research Unit* **5**, 91–96.

Browne FG (1971) *Austroplatypus*, a new genus of the Platypodidae (Coleoptera) infesting live *Eucalyptus* trees in Australia. *Commonwealth Forestry Review* **50**, 1530

Browne FG (1972) Larvae of the principal old world genera of the Platypodinae (Coleoptera: Platypodidae). *Transactions of the Royal Entomological Society of London* **124**, 167–190. doi:10.1111/j.1365-2311.1972.tb00362.x

Browne DJ, Scholtz CH (1994) The morphology and terminology of the hindwing articulation and wing base of the Coleoptera, with specific reference to the Scarabaeoidea. *Systematic Entomology* **19**, 133–143. doi:10.1111/j.1365-3113.1994.tb00583.x

Browne DJ, Scholtz CH (1995) Phylogeny of the families of Scarabaeoidea (Coleoptera) based on characters of the hindwing articulation, hindwing base and wing venation. *Systematic Entomology* **20**, 145–173. doi:10.1111/j.1365-3113.1995.tb00089.x

Browne DJ, Scholtz CH (1996) The morphology of the hind wing articulation and wing base of the Scarabaeoidea (Coleoptera) with notes on phylogenetic trends. *Bonner Zoologische Monographien* **40**, 1–200.

Browne DJ, Scholtz CH (1998) Evolution of the scarab hindwing articulation and wing base: a contribution toward the phylogeny of the Scarabaeidae (Scarabaeoidea: Coleoptera). *Systematic Entomology* **23**, 307–326. doi:10.1046/j.1365-3113.1998.00059.x

Browne DJ, Scholtz CH (1999) A phylogeny of the families of Scarabaeoidea (Coleoptera). *Systematic Entomology* **24**, 51–84. doi:10.1046/j.1365-3113.1999.00067.x

Brues CT (1924) Triungulin larvae from the Williams Galapagos expedition. Triungulins of a meloid beetle borne by *Xylocopa*, with remarks on this type of larva in the Coleoptera and Strepsiptera. *Zoologica* **5**(11), 125–136.

Bugnion E (1910) Le *Cissites testaceus* Fabr. des Indes et de Ceylan. Métamorphoses. – Appareil génital. *Bulletin de la Société Entomologique d'Egypte* **1909**(4), 182–200, 3 pls.

Bullians MS, Leschen RAB (2004) *Notoeucinetus* new genus from New Zealand and Chile and notes on *Eucinetus stewarti* (Broun) (Coleoptera: Eucinetidae). *New Zealand Entomologist* **27**, 29–38. doi:10.1080/00779962.2004.9722121

Büning J (1979) The telotrophic nature of ovarioles of polyphage Coleoptera. *Zoomorphologie* **93**, 51–57. doi:10.1007/BF02568674

Burakowski B (1973) Immature stages and biology of *Drapetes biguttatus* (Piller) (Coleoptera, Lissomidae). *Annales Zoologici* **30**, 335–347.

Burakowski B (1975a) Descriptions of larva and pupa of *Rhysodes sulcatus* (F.) (Coleoptera, Rhysodidae) and notes on the bionomy of this species. *Annales Zoologici* **32**, 271–287.

Burakowski B (1975b) Development, distribution and habits of *Trixagus dermestoides* (L.), with notes on the Throscidae and Lissomidae (Coleoptera, Elateroidea). *Annales Zoologici* **32**, 375–405.

Burakowski B (1988) Notes on the biology of *Xylobanellus erythropterus* (Baudi a Selve) (Coleoptera, Lycidae) with description of the immature stages. *Polskie Pismo Entomologiczne* **58**, 575–585.

Burakowski B (1991): *Klucze do Oznaczania Owadów Polski. Czesc XIX. Chrzaszcze – Coleoptera. Zeszyt 35–37. Cerophytidae, Eucnemidae, Throscidae, Lissomidae.* Polskie Towarzystwo Entomologiczne, Warsaw.

Burakowski B, Ślipiński SA (1987) A new species of *Protosphindus* (Coleoptera: Sphindidae) from Chile with notes and descriptions of immature stages of related forms. *Annali del Museo Civico di Storia Naturale Giacomo Doria* **86**, 605–625.

Burakowski B, Ślipiński SA (2000) The larvae of Leiestinae with notes on the phylogeny of Endomychidae (Coleoptera: Cucujoidea). *Annales Zoologici* **50**(4), 559–573.

Burckhardt D, Löbl I (1990) Redescription of *Jacobsonium dimerum* Heller 1926, member of the rare tropical Indo-Australian genus *Sarothrias* Grouvelle 1918 (Coleoptera Jacobsoniidae). *Tropical Zoology* **3**, 237–241. doi:10.1080/03946975.1990.10539466

Burckhardt D, Löbl I (1992) A review of the Pterogeniidae (Coleoptera). *Zoological Journal of the Linnean Society* **104**, 243–291. doi:10.1111/j.1096-3642.1992.tb00924.x

Burckhardt D, Ślipiński SA (1991) A review of the Passandridae of the world (Coleoptera, Cucujoidea). III. Genera *Anisocerus, Aulonosoma, Passandrella, Passandrina, Scalidiopsis* and *Taphroscelidia. Revue Suisse de Zoologie* **98**, 453–497.

Burckhardt D, Ślipiński SA (2003) Phylogeny and taxonomy of the world Passandridae (Coleoptera). In *Systematics of Coleoptera: Papers Celebrating the Retirement of Ivan Löbl.* (Eds G Cuccodoro and RAB Leschen) pp. 753–883. Associated Publishers (Memoirs on Entomology, International, Volume 17), Gainesville, Florida.

Burckhardt D, Ślipiński A (2010) 10.19. Passandridae Blanchard, 1845/Erichson, 1845. In *Handbuch der Zoologie/Handbook of Zoology. Band/Volume IV Arthropoda: Insecta Teilband/Part 38. Coleoptera, Beetles. Volume 2. Morphology and Systematics (Polyphaga partim).* (Eds RAB Leschen, RG Beutel and JF Lawrence) pp. 363–368. W. DeGruyter, Berlin.

Burke HE (1919) Notes on a cocoon making colydiid (Coleoptera). *Proceedings of the Entomological Society of Washington* **21**, 123–124.

Burmeister EG (1976) Der Ovipositor der Hydradephaga (Coleoptera) und seine phylogenetische Bedeutung unter besonderer Berücksichtigung der Dytiscidae. *Zoomorphologie* **85**, 165–257. doi:10.1007/BF00993515

Burmeister EG (1980) Funktionsmorphologie und Evolution des Ovipositor der Adephaga (Coleoptera). *Verhandlungen des Naturwissenschaftlichen Vereins zu Hamburg. (N. F.)* **24**(1), 89–184.

Burmeister EG (1990a) On the systematic position of Amphizoidae, emphasizing features of the female genital organs (Insecta: Coleoptera: Adephaga). *Quaestiones Entomologicae* **26**, 245–272.

Burmeister EG (1990b) The female genital structures of *Spanglerogyrus albiventris* Folkerts, 1979. A contribution to the systematic position of the Gyrinidae. *Spixiana* **13**(3), 253–265.

Cai C, Short AEZ, Huang D (2012) The first skiff beetle (Coleoptera: Myxophaga: Hydroscaphidae) from early Cretaceous Jehol Biota. *Journal of Paleontology* **86**(1), 116–119.

Calder AA (1986) Classification, relationships and distribution of the Crepidomeninae (Coleoptera: Elateridae). *Australian Journal of Zoology* **122**, (Supplement), 1–220.

Calder AA (1989) The alimentary canal and nervous system of Curculionoidea (Coleoptera): gross morphology and systematic significance. *Journal of Natural History* **23**, 1205–1265. doi:10.1080/00222938900770671

Calder AA (1990a) Gross morphology of the soft parts of the male and female reproductive systems of Curculionoidea (Coleoptera). *Journal of Natural History* **24**, 453–505. doi:10.1080/00222939000770341

Calder AA (1990b) *Anthracalaus australis* Fleutiaux, a little-known Australian elaterid (Coleoptera: Elateridae), with a key to related genera. *Invertebrate Taxonomy* **6**, 437–522.

Calder AA (1992) Notes on *Parablax* schwarz and the subfamily Pityobiinae with description of *Parablax ossa* sp. n. from Tasmania (Coleoptera: Elateridae). *Journal of the Australian Entomological Society* **31**, 143–158. doi:10.1111/j.1440-6055.1992.tb00470.x

Calder AA (1996) *Click Beetles: Genera of Australian Elateridae (Coleoptera).* CSIRO Publishing (Monographs on Invertebrate Taxonomy, Vol. 2), Collingwood, Victoria.

Calder AA (1998) *Zoological Catalogue of Australia. 29.6 Coleoptera: Elateroidea.* CSIRO Publishing, Collingwood, Victoria.

Calder AA (2000) Relationships of the genera of Australian Strongylurini (Coleoptera: Cerambycidae). *Invertebrate Taxonomy* **14**, 863–870. doi:10.1071/IT00029

Calder AA, Sands DPA (1985) A new Brazilian *Cyrtobagous* Hustache (Coleoptera: Curculionidae) introduced into Australia to control salvinia. *Journal of the Australian Entomological Society* **24**, 57–64. doi:10.1111/j.1440-6055.1985.tb00185.x

Calder AA, von Hayek CMF (1992) A reappraisal and revision of the genus *Anthracalaus* Fairmaire (Coleoptera: Elateridae). *Entomologica Scandinavica* **23**, 11–62. doi:10.1163/187631292X00029

Calder AA, Lawrence JF, Trueman JWH (1993) *Austrelater*, gen. nov. (Coleoptera: Elateridae), with description of the larva and coments on elaterid relationships. *Invertebrate Taxonomy* **7**, 1349–1394. doi:10.1071/IT9931349

Callan EM (1981) Further records of *Macrosiagon* (Coleoptera: Rhipiphoridae) reared from eumenid and sphecid wasps in Australia. *Australian Entomological Magazine* **7**, 81–83.

Campau EJ (1940) The morphology of *Chauliognathus pennsylvanicus* (DeGeer) (Coleoptera: Cantharidae). *Microentomology* **5**(3), 57–85.

Carlson CD (1991) Lucanidae (Scarabaeoidea), Passalidae (Scarabaeoidea), Scarabaeidae (Scarabaeoidea) (including Acanthoceridae, Ceratocanthidae, Cetoniidae, Diphyllostomatidae, Geotrupidae, Glaphyridae, Hybosoridae, Pleocomidae, Trogidae, etc.). In *Immature Insects. Volume 2.* (Ed. FW Stehr) pp. 372–384. Kendall/Hunt, Dubuque, Iowa.

Carlson DC (2002a) 30. Ochodaeidae Mulsant and Rey 1870. In *American Beetles. Volume 2. Polyphaga: Scarabaeoidea through Curculionoidea.* (Eds RH Arnett Jr, MC Thomas, PE Skelley and JH Frank) pp. 28–31. CRC Press, Boca Raton, Florida.

Carlson CD (2002b) 33. Glaphyridae MacLeay 1819. In *American Beetles. Volume 2. Polyphaga: Scarabaeoidea through Curculionoidea.* (Eds RH Arnett Jr, MC Thomas, PE Skelley and JH Frank) pp. 37–38. CRC Press, Boca Raton, Florida.

Carlton CE, Leschen RAB (2007) Descriptions of *Soronia* complex (Coleoptera: Nitidulidae: Nitidulinae) larvae of New Zealand with comments on life history and taxonomy. *New Zealand Entomologist* **30**, 41–51. doi:10.1080/00779962.2007.9722150

Carlton CE, Leschen RAB (2008) Description of the larva of *Pselaphophus atriventris* (Staphylinidae: Pselaphinae: Pselaphini) with notes on its life history and a list of described pselaphine immature stages. *Annals of the Entomological Society of America* **101**(1), 13–19. doi:10.1603/0013-8746(2008)101[13:DOTLOP]2.0.CO;2

Carlton CE, Townsend VR, Van Zandt PA, Mopper S (2000) Description of the larva of *Loberus impressus* LeConte with notes on its natural history (Coleoptera: Languriidae: Xenoscelinae). *Annals of the Entomological Society of America* **93**, 356–361. doi:10.1603/0013-8746(2000)093[0356:DOTLOL]2.0.CO;2

Carnaby K (1986) *Jewel Beetles of Western Australia.* Privately published, Wilga WA.

Carne PB (1957a) An ecological study of the pasture scarab *Aphodius howitti* Hope. *Australian Journal of Zoology* **4**, 259–314. doi:10.1071/ZO9560259

Carne PB (1957b) A revision of the ruteline genus *Anoplognathus* Leach (Coleoptera: Scarabaeidae). *Australian Journal of Zoology* **5**, 88–143. doi:10.1071/ZO9570088

Carne PB (1957c) *A Systematic Revision of the Australian Dynastinae (Coleoptera: Scarabaeidae).* CSIRO, Melbourne.

Carne PB (1958) A review of the Australian Rutelinae (Coleoptera: Scarabaeidae). *Australian Journal of Zoology* **6**, 162–240. doi:10.1071/ZO9580162

Carne PB (1965) A revision of the genus *Elephastomus* Macleay (Coleoptera: Geotrupidae). *Journal of the Entomological society of Queensland* **4**, 3–13. doi:10.1111/j.1440-6055.1965.tb00637.x

Carne PB (ed.) (1987) *Scientific and Common Names of Insects and Allied Forms occurring in Australia.* 120 pp. CSIRO, Melbourne.

Carne PB, Taylor KL (1978) Insect pests. In *Eucalypts for Wood Production.* (Eds E Hillis and AG Brown) pp. 155–168. CSIRO, Adelaide.

Carpenter FM (1992) *Treatise on Invertebrate Paleontology. Part R. Arthropoda 4. Volume 4: Superclass Hexapoda.* Geological Society of America and University of Kansas, Boulder, Colorado and Lawrence, Kansas.

Carroll RL (1988) *Vertebrate Paleontology and Evolution.* Freeman, New York.

Carter HJ (1926) A check list of the Australian Tenebrionidae. *Australian Zoologist* **4**, 117–163, 280, 294.

Carter HJ and Zeck EH (1929) A monograph of the Australian Dryopidae. Order–Coleoptera. *Australian Zoologist* **4** (1), 49–72, pls 1–7.

Carter HJ, Zeck EH (1937) A monograph of the Australian Colydiidae. *Proceedings of the Linnean Society of New South Wales* **62**, 181–208.

Casari SA, Teixeira EP (2011) Larva of *Atractocerus brasiliensis* (Lepeletier & Audinet-Serville, 1825) (Lymexylidae: Atractocerinae). *Papéis Avulsos de Zoologia* **51**(12), 197–205.

Caterino MS (2003) New species of *Chlamydopsis* (Histeridae: Chlamydopsinae), with a review and phylogenetic analysis of all known species. *Memoirs of the Queensland Museum* **49**(1), 159–235.

Caterino MS (2006) Chlamydopsinae (Coleoptera: Histeridae) from New Caledonia. *Memoirs of the Queensland Museum* **52**(1), 27–64.

Caterino MS, Dégallier N (2007) A review of the biology and systematics of Chlamydopsinae (Coleoptera: Histeridae). *Invertebrate Systematics* **21**, 1–28. doi:10.1071/IS06017

Caterino MS, Vogler AP (2002) The phylogeny of the Histeroidea. *Cladistics* **18**, 394–415. doi:10.1111/j.1096-0031.2002.tb00158.x

Caterino MS, Shull VS, Hammond PM, Vogler AP (2002) Basal relationships of Coleoptera inferred from 18S rDNA sequences. *Zoologica Scripta* **31**, 41–49. doi:10.1046/j.0300-3256.2001.00092.x

Caterino MS, Hunt T, Vogler AP (2005) On the constitution and phylogeny of Staphyliniformia (Insecta: Coleoptera). *Molecular Phylogenetics and Evolution* **34**, 655–672. doi:10.1016/j.ympev.2004.11.012

Caveney S (1986) The phylogenetic significance of ommatidium structure in the compound eyes of polyphagan beetles. *Canadian Journal of Zoology* **64**, 1787–1819. doi:10.1139/z86-270

Chaboo CS (2007) Biology and phylogeny of the Cassidinae Gyllenhal *sensu lato* (tortoise and leaf-mining beetles) (Coleoptera: Chrysomelidae). *Bulletin of the American Museum of Natural History* **305**, 1–250. doi:10.1206/0003-0090(2007)305[1:BAPOTC]2.0.CO;2

Chaboo CS, Nguyen TC (2004) Immatures of *Hemisphaerota palmarum* (Boheman), with discussion of the caudal processes and shield architecture in the tribe Hemisphaerotini (Chrysomelidae, Cassidinae). In *New Developments in the Biology of Chrysomelidae.* (Ed. P Jolivet, JA Santiago-Blay and M. Schmitt) pp. 171–184. SPB Academic Publishing, The Hague.

Chalumeau F, Brochier B (1993) Un nouveau genre de Chaisognathinae australien. *Bulletin de la Société des Sciences Naturelles* **79**, 16–17.

Chalumeau F, Brochier B (1995) Les Chaisognathinae: genres, sous-genres et synonymies (Coleoptera: Lucanidae). *Bulletin de la Société des Sciences Naturelles* **83**, 18–24.

Chamorro-Lacayo ML, Konstantinov AS (2004) Morphology of the prothorax in the New World Cryptocephalini (Coleoptera: Chrysomelidae: Cryptocephalinae). *Zootaxa* **676**, 1–46.

Champion GC (1916a) Notes on Melandryidae (2). *Entomologist's Monthly Magazine* **52**, 1–10.

Champion GC (1916b) Notes on Melandryidae (3). *Entomologist's Monthly Magazine* **52**, 32–40, 52–59, 75–83, 99–108,144–57, pl. 2.

Champion GC (1922) The geographical distribution and synonymy of the dasytid-beetle *Acanthocnemus nigricans* Hope (= *ciliatus* Perris). *Entomologist's Monthly Magazine* **8**, 77–79.

Chandler DS (1983) Larvae of wrack Coleoptera in the families Corylophidae, Rhizophagidae, and Lathridiidae. *Psyche* **90**, 287–296. doi:10.1155/1983/36909

Chandler DS (1999) Publication dates of papers on the Anthicidae (Coleoptera) by the Marquis F. D. de Laferté-Sénectère. *Transactions of the American Entomological Society* **125**(4), 433–439.

Chandler DS (2001) *Biology, Morphology, and Systematics of the Ant-Like Litter Beetle Genera of Australia (Coleoptera: Staphylinidae: Pselaphinae).* Associated Publishers (Memoirs on Entomology, International, Volume 15), Gaineville, Florida.

Chandler DS (2002a) 117. Anthicidae Latreille 1819. In *American Beetles. Volume 2. Polyphaga: Scarabaeoidea through Curculionoidea.* (Eds RH Arnett Jr, MC Thomas, PE Skelley and JH Frank) pp. 549–558. CRC Press, Boca Raton, Florida.

Chandler DS (2002b) 118. Aderidae Winkler 1927. In *American Beetles. Volume 2. Polyphaga: Scarabaeoidea through Curculionoidea.* (Eds RH Arnett Jr, MC Thomas, PE Skelley and JH Frank) pp. 559–563. CRC Press, Boca Raton, Florida.

Chandler DS (2010) 11.26. Anthicidae Latreille, 1819. In *Handbuch der Zoologie/Handbook of Zoology. Band/Volume IV Arthropoda: Insecta Teilband/Part 38. Coleoptera, Beetles. Volume 2. Morphology and Systematics (Polyphaga partim).* (Eds RAB Leschen, RG Beutel and JF Lawrence) pp. 729–741. W. DeGruyter, Berlin.

Chandler DS, Nardi GF, Telnov D (2004) Nomenclatural notes on the Palaearctic Anthicidae (Coleoptera). *Mitteilungen des Internationalen Entomologischen Vereins e. v. Frankfurt* **29**(4), 109–193.

Chang H, Zhao Y, Ren D (2009) New fossil elaterids (Insects: Coleoptera: Polyphaga: Elateridae) from the Middle Jurassic of Inner Mongolia, China. *Progress in Natural Science* **19**, 1433–1437. doi:10.1016/j.pnsc.2009.04.006

Chang H, Kirejtshuk AG, Ren D (2011) On taxonomy and distribution of fossil Cerophytidae (Coleoptera: Elateriformia) with description of a new Mesozoic species of *Necromera* Martynov 1926. *Annales de la Société Entomologique de France (N. S.)* **47**(1–2), 33–44.

Charpentier R (1965) A monograph of the family Heteroceridae of the Ethiopian region. In *South African Animal Life. Results of the Lund University Expedition in 1950–1951. Vol. XI.* (Eds B. Hanström, P. Brinck and G. Rudebeck) pp. 215–343. Almqvist and Wiksell, Stockholm.

Charpentier R (1968) A monograph of the family Heteroceridae (Coleoptera) of the Notogean region. *Arkiv för Zoologi* **20**, 205–241.

Chatzimanolis S, Engel MS, Newton AF, Grimaldi DA (2010) New ant-like stone beetles in mid-Cretaceous amber from Myanmar (Coleoptera: Staphylinidae: Scydmaeninae). *Cretaceous Research* **31**, 77–84. doi:10.1016/j.cretres.2009.09.009

Chatzimanolis S, Grimaldi DA, Engel MS, Fraser NC (2012) *Leehermania prorova*, the earliest staphyliniform beetle, from the Late Triassic of Virginia (Coleoptera: Staphylinidae). *American Museum Novitates* **3761**, 1–28.

Chen S (1985) Phylogeny and classification of the Chrysomeloidea. *Entomography* **3**, 465–475.

Chen S and T'an CC (1973) A new family of Coleoptera from the Lower Cretaceous of Kansu. *Acta Entomologica Sinica* **16** (2), 169–178, 1 pl. [in Chinese and English]

Chiao E, McHugh JV (2000) Larval Sphindidae (Coleoptera: Cucujoidea): phylogenetic implications and new descriptions. *Invertebrate Taxonomy* **14**, 807–824. doi:10.1071/IT00026

Chisholm AH (1952) Bird-insect nesting associations in Australia. *The Ibis* **94**, 395–405. doi:10.1111/j.1474-919X.1952.tb01841.x

Chûjô M (1969) *Fauna Japonica. Erotylidae (Insecta: Coleoptera).* Academic Press of Japan, Tokyo.

Chûjô M (1987) Erotylidae from New Guinea and her adjacent islands I (Coleoptera). *Esakia* **25**, 5–36, 2 pls.

Chûjô M, Chûjô M (1988) A catalog of the Erotylidae (Insecta, Coleoptera) from the Old World (excl. the Ethiopian Region). *Esakia* **26**, 139–185.

Chûjô M, Chûjô M (1989) A catalog of the Erotylidae (Insecta, Coleoptera) from the Old World (excl. the Ethiopian Region). *Esakia* **28**, 75–96.

Chûjô M, Chûjô M (1990) A catalog of the Erotylidae (Insecta, Coleoptera) from the Old World (excl. the Ethiopian Region) III. *Esakia* **29**, 1–67.

Čiampor F Jr, Kodada J (2004) Systematic revision of the genus *Graphelmis* (Coleoptera: Elmidae) V. *Graphelmis picta* species group. *Entomological Problems* **34**(1–2), 55–102.

Clark J (1925) Forest pests. The pin-hole borer *(Atractocerus kreuslerae,* Pasc.). *Journal of the Department of Agriculture of Western Australia* **2**, 138–142.

Clark SM, Riley EG (2002a) 122. Megalopodidae Latreille 1802. In *American Beetles. Volume 2. Polyphaga: Scarabaeoidea through Curculionoidea.* (Eds RH Arnett Jr, MC Thomas, PE Skelley and JH Frank) pp. 609–612. CRC Press, Boca Raton, Florida.

Clark SM, Riley EG (2002b) 123. Orsodacnidae Thompson 1859. In *American Beetles. Volume 2. Polyphaga: Scarabaeoidea through Curculionoidea.* (Eds RH Arnett Jr, MC Thomas, PE Skelley and JH Frank) pp. 613–616. CRC Press, Boca Raton, Florida.

Clarke AC (1968) Haldane and space. In *Haldane and Modern Biology.* (Ed. KR Dronamraju) pp. 243–248. Johns Hopkins Press, Baltimore.

Cline AR (2008) Revsion of the sap beetle genus *Pocadius* Erichson, 1843 (Coleoptera: Nitidulidae: Nitidulinae). *Zootaxa* **1799**, 1–120.

Cline AR (2010) 10.27. Smicripidae Horn, 1879. In *Handbuch der Zoologie/Handbook of Zoology. Band/Volume IV Arthropoda: Insecta Teilband/Part 38. Coleoptera, Beetles. Volume 2. Morphology and Systematics (Polyphaga partim).* (Eds RAB Leschen, RG Beutel and JF Lawrence) pp. 407–411. W. DeGruyter, Berlin.

Cline AR, Shockley FW (2010) 10.6. Biphyllidae LeConte, 1861. In *Handbuch der Zoologie/Handbook of Zoology. Band/Volume IV Arthropoda: Insecta Teilband/Part 38. Coleoptera, Beetles. Volume 2. Morphology and Systematics (Polyphaga partim).* (Eds RAB Leschen, RG Beutel and JF Lawrence) pp. 306–311. W. DeGruyter, Berlin.

Cline AR, Ślipiński A (2010) 10.31. Discolomatidae Horn, 1878. In *Handbuch der Zoologie/Handbook of Zoology. Band/Volume IV Arthropoda: Insecta Teilband/Part 38. Coleoptera, Beetles. Volume 2. Morphology and Systematics (Polyphaga partim).* (Eds RAB Leschen, RG Beutel and JF Lawrence) pp. 435–442. W. DeGruyter, Berlin.

Cline AR, Goodrich MA, Leschen RAB (2010) 10.2. Byturidae Jacquelin Du Val, 1858. In *Handbuch der Zoologie/Handbook of Zoology. Band/Volume IV Arthropoda: Insecta Teilband/Part 38. Coleoptera, Beetles. Volume 2. Morphology and Systematics (Polyphaga partim).* (Eds RAB Leschen, RG Beutel and JF Lawrence) pp. 286–292. W. DeGruyter, Berlin.

Cloudsley-Thompson JL (1965) On the function of the subelytral cavity in desert Tenebrionidae (Col.). *Entomologist's Monthly Magazine* **100**, 148–151.

Cobos A (1960) Throscidae de Africa Central (Coleoptera). *Annales du Musée Royal du Congo Belge, Tervuren, Série in 8°. Sciences Zoologiques* **86**, 1–67.

Cobos A (1961) Sobre la posicion sistematica del genero *Potergus* Bonvouloir y revision de las categorias supragenericas de la familia Throscidae (Coleoptera). *Bulletin de l'Institut Royale des Sciences Naturelles de Belgique* **37** (35), 1–6.

Cobos A (1964) Materiales para el estudio de la familia Eucnemidae. Primera parte (Coleoptera). *Eos, Madrid* **40**, 289–435.

Cobos A (1967) Estudios sobre Throscidae, II (Col. Sternoxia). *Eos, Madrid* **42**, 311–351.

Coffin J (1993) "Fosses céphalique" énigmatiques découvertes chez les espèces du genre *Throscus* Latreille, 1796 (Coleoptera, Throscidae). *Bulletin de la Société Entomologique de France* **97**, 309–311.

Cogan BH, Smith KGV (1974) *Instructions for Collectors. No. 4a: Insects.* British Museum (Natural History, London.

Constantin R (1990) Descriptions des larves *d'Enicopus pyrenaeus* Fairmaire et de *Danacaea pallipes* (Panzer). Contribution à l'étude de la biologie et de la systématique larvaire des Melyridae (Coleoptera). *Nouvelle Revue d'Entomologie (N. S.)* **6**(4), 387–405.

Constantin R, Menier JJ (1990) Description des larves du genre *Gietella* et discussion sur la position systematique du genre (Coleoptera, Melyridae, Gietellinae). *Revue Française d'Entomologie (N. S.)* **12**(4), 171–182.

Cook D (1987) Sexual selection in dung beetles. I. A multivariate study of the morphological variation in two species of *Onthophagus* (Scarabaeidae: Onthophagini). *Australian Journal of Zoology* **35**, 123–132. doi:10.1071/ZO9870123

Corbiere-Tichané G (1974) Fine structure of an antennal sensory organ ('vesicule olfactive') of *Speophyes lucidulus* Delar. (cave Coleoptera of the Bathysciinae subfamily). *Tissue & Cell* **6**, 535–550. doi:10.1016/0040-8166(74)90044-5

Costa C, Fonseca CRV (1986) Larvae of Neotropical Coleoptera. XIII. Passalidae, Passalinae. *Revista Brasileira de Entomologia* **30**(1), 57–78.

Costa C, Zaragoza-Caballero S (2010) 4.13. Phengodidae LeConte, 1861. In *Handbuch der Zoologie/Handbook of Zoology. Band/Volume IV Arthropoda: Insecta Teilband/Part 38. Coleoptera, Beetles. Volume 2. Morphology and Systematics (Polyphaga partim).* (Eds RAB Leschen, RG Beutel and JF Lawrence) pp. 126–135. W. DeGruyter, Berlin.

Costa C, Casari-Chen SA, Teixeira EP (1986) Larvae of Neotropical Coleoptera XVI. Nosodendridae. *Revista Brasileira de Entomologia* **30**(2), 291–297.

Costa C, Vanin SA, Casari-Chen SA (1988) *Larvas de Coleoptera do Brasil.* Museu de Zoologia, Universidade de São Paulo, São Paulo.

Costa C, Casari-Chen SA, Vanin SA (1992) On the larvae of Tetralobini (Coleoptera, Elateridae). *Revista Brasileira de Entomologia* **36**, 879–888.

Costa C, Vanin SA, Ide S (1995) Larvae of Neotropical Coleoptera XXII. Description of adults and immatures of *Lagrioida nortoni* sp. n., and bionomics (Coleoptera, Tenebrionoidea, Anthicidae). *Iheringia, Porto Alegre. Série Zoologia* **78**, 113–126.

Costa C, Vanin SA, Ide A (1999) Systematics and bionomics of Cneoglossidae with a cladistic analysis of Byrrhoidea *sensu* Lawrence & Newton (1995) (Coleoptera, Elateriformia). *Arquivos de Zoologia* **35**, 231–300.

Costa C, Vanin SA, Lawrence JF, Ide S (2003) Systematics and cladistic analysis of Cerophytidae (Elateroidea: Coleoptera). *Systematic Entomology* **28**, 375–407. doi:10.1046/j.1365-3113.2003.00219.x

Costa C, Vanin SA, Ide S (2005) 18.8. Cneoglossidae Guérin-Méneville, 1843. In *Handbuch der Zoologie/Handbook of Zoology. Band/Volume IV Arthropoda: Insecta Teilband/Part 38. Coleoptera, Beetles. Volume 1. Morphology and Systematics (Archostemata, Adephaga, Myxophaga, Polyphaga partim).* (Eds RG Beutel and RAB Leschen) pp. 533–536. W. DeGruyter, Berlin.

Costa C, Vanin SA, Lawrence JF, Ide S, Branham MA (2006) Review of the family Brachypsectridae (Coleoptera: Elateroidea). *Annals of the Entomological Society of America* **99**(3), 409–432. doi:10.1603/0013-8746(2006)99[409:ROTFBC]2.0.CO;2

Costa C, Vanin S, Lawrence JF, Ide S, Branham MA (2010a) 4.3. Brachypsectridae Horn, 1881. In *Handbuch der Zoologie/Handbook of Zoology. Band/Volume IV Arthropoda: Insecta Teilband/Part 38. Coleoptera, Beetles. Volume 2. Morphology and Systematics (Polyphaga partim).* (Eds RAB Leschen, RG Beutel and JF Lawrence) pp. 47–54. W. DeGruyter, Berlin.

Costa C, Vanin S, Lawrence JF, Ide S (2010b) 4.4. Cerophytidae Latreille, 1834. In *Handbuch der Zoologie/Handbook of Zoology. Band/Volume IV Arthropoda: Insecta Teilband/Part 38. Coleoptera, Beetles. Volume 2. Morphology and Systematics (Polyphaga partim).* (Eds RAB Leschen, RG Beutel and JF Lawrence) pp. 54–61. W. DeGruyter, Berlin.

Costa C, Lawrence JF, Rosa SP (2010c) 4.7 Elateridae Leach, 1815. In *Handbuch der Zoologie/Handbook of Zoology. Band/Volume IV Arthropoda: Insecta Teilband/Part 38. Coleoptera, Beetles. Volume 2. Morphology and Systematics (Polyphaga partim).* (Eds RAB Leschen, RG Beutel and JF Lawrence) pp. 75–103. W. DeGruyter, Berlin.

Craighead FC (1916) The determination of the thoracic and abdominal areas of cerambycid larvae as based on a study of the muscles. *Proceedings of the Entomological Society of Washington* **18**, 129–146.

Craighead FC (1920) Biology of some Coleoptera of the families Colydiidae and Bothrideridae. *Proceedings of the Entomological Society of Washington* **22**, 1–13.

Craighead FC (1921) Larva of the North American beetle *Sandalus niger* Knoch. *Proceedings of the Entomological Society of Washington* **23**, 44–48.

Craighead FC (1923) North American cerambycid larvae. A classification and the biology of North American cerambycid larvae. *Dominion of Canada Department of Agriculture Bulletin.* New Series (Technical) **27**, 239 pp.

Crisp MD, West JG, Linder HP (1999) Biogeography of the terrestrial flora. In *Flora of Australia. Vol. 1.* (Eds. ABRS) pp. 321–367. ABRS, Canberra and CSIRO Publishing, Melbourne.

Cros A (1924) Révision des espèces africaines et orientales des genres *Horia* Fabr. et *Cissites* Latr., avec description de larves inédites. *Bulletin de la Société Royale Entomologique d'Egypte* **1924**, 24–80.

Cros A (1927) Révision des genres *Horia* Fabr. et *Cissites* Latr. (Note rectitative et complementaire). *Bulletin de la Société Royale Entomologique d'Egypte* **1927**(2–4), 103–115.

Cros A (1938) Description de la larve primaire d'une espèce indéterminée de la tribu des Horiini (Col. Meloidae). *Bulletin du Musée Royal d'Histoire Naturelle de Belgique* **14**(47), 1–13.

Crovetti A (1969) Contributi alla conoscenza dei Coleotteri Scarabeidi. I. Il genere *Pachypus* Serville (Coleoptera, Scarabaeidae, Pachypodinae). *Bolletino di Zoologia Agraria e di Bachicoltura, Serie II* **9**, 133–188.

Crowson RA (1938) The metendostemite in Coleoptera: a comparative study. *Transactions of the Royal Entomological Society of London* **87**, 397–415. doi:10.1111/j.1365-2311.1938.tb00723.x

Crowson RA (1944) Further studies on the metendostemite in Coleoptera. *Transactions of the Royal Entomological Society of London* **94**, 273–310. doi:10.1111/j.1365-2311.1944.tb01220.x

Crowson RA (1946) A revision of the genera of the chrysomelid group Sagrinae (Coleoptera). *Transactions of the Royal Entomological Society of London* **97**, 75–115. doi:10.1111/j.1365-2311.1946.tb00276.x

Crowson RA (1955) *The Natural Classification of the Families of Coleoptera.* Nathaniel Lloyd, London.

Crowson RA (1959) Studies on the Dermestoidea (Coleoptera), with special reference to the New Zealand fauna. *Transactions of the Royal Entomological Society of London* **111**, 81–94. doi:10.1111/j.1365-2311.1959.tb02277.x

Crowson RA (1960) The Phylogeny of Coleoptera. *Annual Review of Entomology* **5**, 111–134. doi:10.1146/annurev.en.05.010160.000551

Crowson RA (1961a) Considerations on the genera *Endecatomus* Mellié and *Euderia* Broun (Coleoptera: Bostrychidae), with descriptions of their larvae. *Proceedings of the Royal Entomological Society of London (B)* **30**, 113–120.

Crowson RA (1961b) On some new characters of classificatory importance in adults of Elateridae (Coleoptera). *Entomologist's Monthly Magazine* **96**, 158–161.

Crowson RA (1964a) A new genus of Australian clavicorn Coleoptera, probably of a new family. *Proceedings of the Linnean Society of New South Wales* **89**, 241–245.

Crowson RA (1964b) A review of the classification of Cleroidea (Coleoptera), with descriptions of two new genera of Peltidae and of several new larval types. *Transactions of the Royal Entomological Society of London* **116**, 275–327. doi:10.1111/j.1365-2311.1964.tb02298.x

Crowson RA (1966a) Further observations on Peltidae (Coleoptera: Cleroidea), with definitions of a new subfamily and of four new genera. *Proceedings of the Royal Entomological Society of London (B)* **35**, 119–127.

Crowson RA (1966b) Observations on the constitution and subfamilies of the family Melandryidae (Coleoptera). *Eos, Madrid* **41**, 507–513.

Crowson RA (1967) The natural classification of the families of Coleoptera. Addenda and corrigenda. *Entomologist's Monthly Magazine* **103**, 209–214.

Crowson RA (1970) Further observations on Cleroidea (Coleoptera). *Proceedings of the Royal Entomological Society of London (B)* **39**, 1–20.

Crowson RA (1971a) Observations on the superfamily Dascilloidea (Coleoptera: Polyphaga), with the inclusion of Karumiidae and Rhipiceridae. *Zoological Journal of the Linnean Society* **50**, 11–19. doi:10.1111/j.1096-3642.1971.tb00749.x

Crowson RA (1971b) Some problems in the phylogenetic classification of the Coleoptera. *Proceedings of the International Congress of Entomology* **13**(1), 235–237.

Crowson RA (1972a) On the systematic value of the alimentary canal in Cleridae. *Systematic Zoology* **21**, 339–340. doi:10.2307/2412171

Crowson RA (1972b) A review of the classification of Cantharoidea (Coleoptera), with the definition of two new families, Cneoglossidae and Omethidae. *Revista de la Universidad de Madrid* **21**(82), 35–77.

Crowson RA (1973a) Further observations on Phloeostichidae and Cavognathidae, with definitions of new genera from Australia and New Zealand. *Coleopterists Bulletin* **27**, 54–62.

Crowson RA (1973b) On a new superfamily Artematopoidea of polyphagan beetles, with the definition of two new fossil genera from the Baltic amber. *Journal of Natural History* **7**, 225–238. doi:10.1080/00222937300770181

Crowson RA (1975) The evolutionary history of Coleoptera, as documented by fossil and comparative evidence. *Atti del Congresso Nazionale Italiano di Entomologia* **10**, 47–90.

Crowson RA (1976) The systematic position and implications of *Crowsoniella*. *Bolletino del Museo Civico di Storia Naturale di Verona* **2**, 459–463.

Crowson RA (1978) Problems of phylogenetic relationships in Dryopoidea (Coleoptera). *Entomologica Germanica* **4**, 250–257.

Crowson RA (1979) Observations on Clambidae (Coleoptera), with descriptions of a new genus and species and of several larvae. *Revue Suisse de Zoologie* **86**, 611–623.

Crowson RA (1980) On amphipolar distribution patterns in some cool climate groups of Coleoptera. *Entomologia Generalis* **6**, 281–292.

Crowson RA (1981) *The Biology of the Coleoptera*. Academic Press, London.

Crowson RA (1982) On the dryopoid affinities of Buprestidae. *Coleopterists Bulletin* **36**, 22–25.

Crowson RA (1984a) The use of male terminalia in the higher classification of Coleoptera. *Entomologia Generalis* **10**, 53–58.

Crowson RA (1984b) On the systematic position of *Bruchela* Dejean (*Urodon* auctt.) (Coleoptera). *Coleopterists Bulletin* **38**, 91–93.

Crowson RA (1985) The systematic position of *Nemonyx* Redtenbacher (Coleoptera: Curculionoidea: Nemonychidae). *Entomologia Generalis* **11**, 57–58.

Crowson RA (1990) A new genus of Boganiidae (Coleoptera) from Australia, with observations on glandular openings, cycad associations and geographical distribution in the family. *Journal of the Australian Entomological Society* **29**, 91–99. doi:10.1111/j.1440-6055.1990.tb00324.x

Crowson RA (1991) The relations of Coleoptera to Cycadales. In *Advances in Coleopterology*. (Eds M Zunino, X. Bellés and M. Blas) pp. 13–28. Associacion Europea de Coleopterologia, Barcelona.

Crowson RA, de Viedma MG (1964) Observations on the relationships of the genera *Circaeus* Yablok. and *Mycterus* Clairv. with a description of the presumed larva of *Mycterus* (Col. Heteromera). *Eos, Madrid* **40**, 99–107.

Crowson RA, Sen Gupta T (1969) The systematic position of Propalticidae and of *Carinophloeus* Lefkovitch (Coleoptera, Clavicornia) with description of a new species of *Propalticus* and of its supposed larva. *Proceedings of the Royal Entomological Society of London (B)* **38**, 132–140.

Curletti G (2001) The genus *Agrilus* in Australia. *Jewel Beetles* **9**, 76 pp.

D'Hotman D, Scholtz CH (1990a) Phylogenetic significance of the structure of the external male genitalia in Scarabaeoidea. *Entomology Memoirs* **77**, 1–51.

D'Hotman D, Scholtz CH (1990b) Comparative morphology of the male genitalia of derived groups of Scarabaeoidea (Coleoptera). *Elytron* **4**, 3–39.

Dajoz R (1968) Révision des Colydiidae anophthalme de la faune paléarctique (Col.). IV. Etudes sur les genres *Anommatus* et *Langelandia*. *Annales de la Société Entomologique de France (N. S.)* **4**, 975–988.

Dajoz R (1977) *Faune de l'Europe et du Bassin Méditerranéen. 8. Coléoptères Colydiidae et Anommatidae Paléarctiques*. 275 pp. Masson, Paris.

Darlington PJ (1943) Carabidae of mountains and islands: data on the evolution of isolated faunas, and on atrophy of wings. *Ecological Monographs* **13**, 37–61. doi:10.2307/1943589

Darlington PJ (1961) Australian carabid beetles. V. Transition of wet forest faunas from New Guinea to Tasmania. *Psyche (Cambridge, Massachusetts)* **68**, 1–24.

Darlington PJ (1965) *Biogeography of the Southern End of the World. Distribution and History of Far-Southern Life and Land, with an Assessment of Continental Drift*. x + 236 pp. Harvard University Press, Cambridge, Massachusetts.

Darlington PJ (1971) The carabid beetles of New Guinea. Part IV. General considerations; analysis and history of fauna; taxonomic supplement. *Bulletin of the Museum of Comparative Zoology* **142**, 129–337.

Das GM (1937) The musculature of the mouth-parts of insect larvae. *Quarterly Journal of Microscopical Science* (N. S.) **80**, 39–80, 12 pls.

Davis JA (1986) Revision of the Australian Psephenidae (Coleoptera): systematics, phylogeny and historical biogeography. *Australian Journal of Zoology* **119**, (Supplement), 1–97.

Davis SR (2009) Morphology of Baridinae and related groups (Coleoptera, Curculionidae). *ZooKeys* **10**, 1–136. doi:10.3897/zookeys.10.47

Davis SR (2010) Delimiting baridine weevil evolution (Coleoptera: Curculionidae: Baridinae). *Zoological Journal of the Linnean Society* **161**(1), 88–156. doi:10.1111/j.1096-3642.2010.00619.x

Davis ALV, Scholtz CH, Philips TK (2002) Historical biogeography of scarabaeine dung beetles. *Journal of Biogeography* **29**, 1217–1256. doi:10.1046/j.1365-2699.2002.00776.x

De Little D (1979) A preliminary review of the genus *Paropsis* Olivier (Coleoptera: Chrysomelidae) in Tasmania. *Journal of the Australian Entomological Society* **18**, 91–107. doi:10.1111/j.1440-6055.1979.tb00819.x

De Marzo L (1979) Studi sulle larve dei coleotteri ditiscidi. X. Anatomia e funzionamento dell'apparato succhiante cibariofaringeo in alcune forme larvali delle subff. Dytiscinae, Colymbetinae, Laccophilinae e Hydroporinae. *Entomologica, Bari* **15**, 5–72.

De Marzo L (1986) Osservazioni etologiche suite larve di *Batrisodes oculatus* Aubé (Coleoptera Pselaphidae). *Frustula Entomologica (N.S.)* **7–8**, 501–506.

De Marzo L (1987) Morfologia della larva matura in alcuni Pselafidi (Coleoptera). *Entomologica, Bari* **22**, 97–135.

De Marzo L (1992) Le ghiandole odorifere in *Formicomus pedestris* (Rossi) (Coleoptera Anthicidae). *Bolletino della Società Entomologica Italiana, Genova* **124**, 12–16.

De Marzo L (2000) Larve di coleotteri in detriti vegetali di origine agricola: lineamente morfologici e presenza stagionale (Polyphaga: 20 famiglie). *Entomologica, Bari* **34**, 65–131.

De Marzo L (2009) Studio sulla diversità anatomica della spermateca nei Coleotteri. *Entomologica, Bari* **41**, 13–76.

Deane C (1930) Trichopterygidae of Australia and Tasmania. *Proceedings of the Linnean Society of New South Wales* **55**, 477–487.

Deane C (1931) Trichopterygidae of Australia and adjacent islands. *Proceedings of the Linnean Society of New South Wales* **56**, 237–242.

Delgado JA, Archangelsky M (2005) Description of the larval stages of *Gymnochthebius jensenhaarupi* and phylogenetic analysis of the relationships with other species of the subfamily Ochthebiinae (Coleoptera: Hydraenidae). *European Journal of Entomology* **102**, 231–240.

Delkeskamp K (1977) *Coleopterorum Catalogus Supplementa edita a J. A. Wilcox. Pars 165, Fasc. 1. Cantharidae*. W. Junk, Berlin.

Delkeskamp K (1978) *Coleopterorum Catalogus Supplementa edita a J. A. Wilcox. Pars 165, Fasc. 2. Cantharidae. Corrigenda et Addenda*. W. Junk, Berlin.

Dellacasa M (1988a) Contribution to a world-wide catalogue of Aegialiidae, Aphodiidae, Aulonocnemidae, Termitotrogidae (Coleoptera Scarabaeoidea). *Memorie della Società Entomologica Italiana* **66**, 1–455.

Dellacasa M (1988b) Contribution a world-wide catalogue of Aegialiidae, Aphodiidae, Aulonocnemidae, Termitotrogidae (Coleoptera Scarabaeoidea). Part II. *Memorie della Societa Entomologica Italiana* **67**, 1–229.

Dellacasa M (1989) Contribution to a world-wide catalogue of Aegialiidae, Aphodiidae, Aulonocnemidae, Termitotrogidae (Coleoptera Scarabaeoidea). Addenda et Corrigenda (first note). *Memorie della Societa Entomologica Italiana* **67**, 291–316.

Delobel A, Sembène M, Fédière G, Roguet D (2003) Identity of the groundnut and tamarind seed-beetles (Coleoptera: Bruchidae: Pachymerinae), with restoration of *Caryedon gonagra* (F.). *Annales de la Société Entomologique de France (N. S.)* **39**(3), 197–206.

Deroe C, Pasteels JM (1982) Distribution of adult defense glands in chrysomelids (Coleoptera: Chrysomelidae) and its significance in the evolution of defense mechanisms within the family. *Journal of Chemical Ecology* **8**, 67–82. doi:10.1007/BF00984006

Despins JL, Vaughan JD, Turner EC (1988) Role of the lesser mealworm, *Alphitobius diaperinus* (Panzer) (Coleoptera: Tenebrionidae), as a predator of the house fly, *Musca domestica* L. (Diptera: Muscidae) in poultry houses. *Coleopterists Bulletin* **42**, 211–216.

Dettner K (1985) Ecological and phylogenetic significance of defensive compounds from pygidial glands of Hydradephaga (Coleoptera). *Proceedings of the Academy of Natural Sciences of Philadelphia* **137**, 156–171.

Dettner K (1987) Chemosystematics and evolution of beetle chemical defenses. *Annual Review of Entomology* **32**, 17–48. doi:10.1146/annurev.en.32.010187.000313

Dettner K (2005a) 7.3. Noteridae Thomson, 1857. In *Handbuch der Zoologie/Handbook of Zoology. Band/Volume IV Arthropoda: Insecta Teilband/Part 38. Coleoptera, Beetles. Volume 1. Morphology and Systematics (Archostemata, Adephaga, Myxophaga, Polyphaga partim).* (Eds RG Beutel and RAB Leschen) pp. 72–81. W. DeGruyter, Berlin.

Dettner K (2005b) 7.4. Amphizoidae LeConte, 1853. In *Handbuch der Zoologie/Handbook of Zoology. Band/Volume IV Arthropoda: Insecta Teilband/Part 38. Coleoptera, Beetles. Volume 1. Morphology and Systematics (Archostemata, Adephaga, Myxophaga, Polyphaga partim).* (Eds RG Beutel and RAB Leschen) pp. 81–85. W. DeGruyter, Berlin.

Dettner K (2005c) 7.5. Hygrobiidae Régimbart, 1879. In *Handbuch der Zoologie/Handbook of Zoology. Band/Volume IV Arthropoda: Insecta Teilband/Part 38. Coleoptera, Beetles. Volume 1. Morphology and Systematics (Archostemata, Adephaga, Myxophaga, Polyphaga partim).* (Eds RG Beutel and RAB Leschen) pp. 85–90. W. DeGruyter, Berlin.

Dhileepan K, Taylor DBJ, Lockett C, Treviño M (2012) Cat's claw creeper leaf-mining jewel beetle *Hylaeogena jureceki* Obenberger (Coleoptera: Buprestidae), a host-specific biological control agent for *Dolichandra unguis-cati* (Bignoniaceae) in Australia. *Australian Journal of Entomology* doi:10.1111/aen.12014

Di Giulio A, Moore W (2004) The first-instar larva of the genus *Arthropterus* (Coleoptera: Carabidae: Paussinae): implications for evolution of myrmecophily and phylogenetic relationships within the subfamily. *Invertebrate Systematics* **18**, 101–115. doi:10.1071/IS03028

Di Giulio A, Moore W (2009) The first known larva of the Australian genus *Mystropomus* Chaudoir (Coleoptera: Carabidae: Paussinae). *Australian Journal of Entomology* **48**, 140–148. doi:10.1111/j.1440-6055.2009.00698.x

Di Giulio A, Vigna Taglianti A (2001) Biological observations on *Pachyteles* larvae (Coleoptera: Carabidae: Paussinae). *Tropical Zoology* **14**, 157–173. doi:10.1080/03946975.2001.10531149

Di Giulio A, Pinto JD, Bologna MA (2010) First-instar larva of *Palaestra rufipennis* (Westwood, 1841) and other Australian blister beetles (Coleoptera, Meloidae, Nemognathinae). *Australian Journal of Entomology* **49**, 332–340. doi:10.1111/j.1440-6055.2010.00765.x

Dibb JR (1938) Synopsis of Australian Passalidae (Coleoptera). *Transactions of the Entomological Society of London* **87**, 103–124. doi:10.1111/j.1365-2311.1938.tb00092.x

Dodge HR (1941) Observations on *Sandalus niger* Knoch, its egg, and first instar larva. *Annals of the Entomological Society of America* **34**, 458–466.

Dolin VG (1973) Fossil forms of Elateridae (Coleoptera) from the Lower Jurassic in Central Asia. In *Fauna and Biology of Moldavian Insecta.* (Eds MF Yaroshenko *et al.*) pp. 71–82 (in Russian). Shtüntza, Kishinev.

Dolin VG (1975a) A contribution to the systematics of Mesozoic click beetles (Coleoptera, Elateridae). *Paleontologicheski Zhurnal* **1975** (4), 51–62 (in Russian; translation in *Paleontological Journal* **9**, 474–486).

Dolin VG (1975b) Wing venation of click beetles (Coleoptera, Elateridae) and its importance for taxonomy of the family. *Зоологический журнал* **54**, 1618–1633. [in Russian]

Dolin VG (1976) Fossil click beetles (Coleoptera, Elateridae) of the subfamilies Negastriinae and Cardiophorinae of the high Jurassic period in Karatau. *Vestnik Zoologii* **1976**(2), 68–75.

Dolin VG (1978a) *Identification Table for the Larvae of Click Beetles in the Fauna of the U. S. S. R.* Urozhai, Kiev (in Russian; translation by El-Ahram Center for Scientific Translations, Washington, D. C.).

Dolin VG (1978b) Phylogeny of click beetles (Coleoptera, Elateridae). *Vestnik Zoologii* **1978**(3), 3–12. [in Russian]

Dolin VG (1980) Elateridae from the Upper Jurassic of Karatau. In *Fossil Insects of the Mesozoic.* (Eds VG Dolin, DV Panfilov, AG Ponomarenko and DN Pritykina) pp. 17–81. Naukova Dumka, Kiev.

Dolin VG (2000) The role of larval and wing venation characters in the systematics of Elateroidea (Coleoptera). In *Meetings in Memory of N. A. Cholodkovsky. Lecture at the 52nd Annual Meeting. 1 April 1999. St. Petersburg* (in Russain). Rossiiskaya Akademiya Nauk and Russoye Entomologicheskoye Obshchestvo, St. Petersburg.

Dorsey CK (1943) The musculature of the labrum, labium, and pharyngeal region of adult and immature Coleoptera. *Smithsonian Miscellaneous Collections* **103** (7), 1–42, 24 pls.

Doube BM, Macqueen A, Ridsdill Smith TJ, Weir TA (1991) Native and introduced dung beetles in Australasia. In *Dung Beetle Ecology* (Eds I Hanski and Y Cambefort) pp. 255–78. (Princeton University Press, Princeton, New Jersey.)

Doyen JT (1966) The skeletal anatomy of *Tenebrio molitor* (Coleoptera: Tenebrionidae). *Miscellaneous Publications of the Entomological Society of America* **5**, 103–150.

Doyen JT (1976) Marine beetles (Coleoptera excluding Staphylinidae). In *Marine Insects* (Ed. L Cheng) pp. 497–519. North-Holland Publishing Company, Amsterdam.

Doyen JT (1979) The larva and relationships of *Cononotus* LeConte (Coleoptera: Heteromera). *Coleopterists Bulletin* **33**, 33–39.

Doyen JT (1984) Reconstitution of the Diaperini of North America, with new species of *Adelina* and *Sitophagus* (Coleoptera: Tenebrionidae). *Proceedings of the Entomological Society of Washington* **86**, 777–789.

Doyen JT (1985) Rconstitution of the tribes Ulomini and Triboliini for North and Central America (Tenebrionidae; Coleoptera). *Proceedings of the Entomological Society of Washington* **87**, 512–524.

Doyen JT (1989) Reconstitution of Coelometopini, Tenebrionini and related tribes of America North of Colombia (Coleoptera: Tenebrionidae). *Journal of the New York Entomological Society* **97**, 277–304.

Doyen JT, Lawrence JF (1979) Relationships and higher classification of some Tenebrionidae and Zopheridae (Coleoptera). *Systematic Entomology* **4**, 333–377. doi:10.1111/j.1365-3113.1979.tb00619.x

Doyen JT, Tschinkel WR (1982) Phenetic and cladistic relationships among tenebrionid beetles (Coleoptera). *Systematic Entomology* **7**, 127–183. doi:10.1111/j.1365-3113.1982.tb00129.x

Doyen JT, Matthews EG, Lawrence JF (1990) Classification and annotated checklist of the Australian genera of Tenebrionidae (Coleoptera). *Invertebrate Taxonomy* **3**, 229–260. doi:10.1071/IT9890229

Drea JJ (1966) Studies of *Aleochara tristis* (Coleoptera: Staphylinidae), a natural enemy of the face fly. *Journal of Economic Entomology* **59**, 1368–1373.

Dressler C, Beutel RG (2010) The morphology and evolution of the adult head of Adephaga (Insecta: Coleoptera). *Arthropod Systematics & Phylogeny* **68** (2), 239–287.

Drilling K, Dettner K, Klass K-D (2010) Morphology of the pronotal compound glands in *Tritoma bipustulata* (Coloeoptera: Erotylidae). *Organisms, Diversity & Evolution* **10**, 205–214. doi:10.1007/s13127-010-0008-0

Duffy EAJ (1952) The immature stages of *Sessinia livida* (Fabricius) (Coleoptera: Oedemeridae). *Proceedings of the Hawaiian Entomological Society* **14** (3), 379–383.

Duffy EAJ (1953) *A Monograph of the Immature Stages of British and Imported Timber Beetles (Cerambycidae).* 350 pp., 8 pls. British Museum (Natural History), London.

Duffy EAJ (1963) *A Monograph of the Immature Stages of Australasian Timber Beetles (Cerambycidae).* 235 pp., 11 pls. British Museum (Natural History), London.

Duffy EAJ (1968) *A Monograph of the Immature Stages of Oriental Timber Beetles (Cerambycidae).* 434 pp., 18 pls. British Museum (Natural History), London.

Dumbacher JP, Wako A, Derrickson SR, Samuelson A, Spande TF, Daly JW (2004) Melyrid beetles (*Choresine*): a putative source for the batrachotoxin alkaloids found in poison-dart frogs and toxic passerine birds. *Proceedings of the National Academy of Sciences of the United States of America* **101** (45), 15857–15860. doi:10.1073/pnas.0407197101

Dumbleton LJ (1957) The immature stages of some New Zealand longhorn beetles (Coleoptera-Cerambycidae). *Transactions of the Royal Society of New Zealand* **84**(3), 611–628.

Duncan IJ, Briggs DEG, Archer M (1998) Three-dimensionally mineralized insects and millipedes from the Tertiary of Riversleigh, Queensland, Australia. *Palaeontology* **41**(5), 835–851.

Dunstan B (1923) *Mesozoic Insects of Queensland. Part 1–Introduction and Coleoptera.* Queensland Geological Survey (Publication No. 273), Brisbane.

Dybas HS (1976) The larval characters of featherwing and limulodid beetles and their family relationships in the Staphylinoidea (Coleoptera: Ptiliidae and Limulodidae). *Fieldiana Zoology* **70**, 29–78.

Dybas HS (1990) Insecta: Coleoptera. Ptiliidae. In *Soil Biology Guide* (Ed. D Dindal) pp. 1093–1012. John Wiley and Sons, New York.

Eberhard WG (1979) The function of horns *in Podischnus agenor* (Dynastinae) and other beetles. In *Sexual Selection and Reproductive Competition in Insects*. (Eds MS and NA Blum) pp. 231–58. Academic Press, New York.

Eberhard WG (1980) Horned beetles. *Scientific American* **242**(3), 124–131. doi:10.1038/scientificamerican0380-166

Eberhard WG (1982) Beetle horn dimorphism: making the best of a bad lot. *American Naturalist* **119**, 420–426. doi:10.1086/283920

Edwards JG (1953) Species of the genus *Syneta* of the world (Coleoptera: Chrysomeloidea). *Wasmann Journal of Biology* **11**, 23–82.

Edwards JS (1961a) On the reproduction of *Prionoplus reticularis* (Coleoptera, Cerambycidae), with general remarks on reproduction in the Cerambycidae. *The Quarterly Journal of Microscopical Science* **102**(4), 519–529.

Edwards JS (1961b) Observations on the ecology and behaviour of the Huhu beetle, *Prionoplus reticularis* White (Col. Ceramb.). *Transactions of the Royal Society of New Zealand* **88** (4), 733–741, pl. 61.

Eickwort GC (1981) Presocial insects. In *Social Insects.*Vol. II. (Ed. HR Hermann) pp. 199–280. Academic Press, New York.

Eidt DC (1959) Mode of feeding of the larva of *Ctenicera aeripennis destructor* (Brown) (Coleoptera: Elateridae). *Canadian Entomologist* **91**, 97–101. doi:10.4039/Ent9197-2

Eisner T, Van Tassell E, Carrel JE (1967) Defensive use of a "fecal shield" by a beetle larva. *The Sciences* **158**, 1471–1473. doi:10.1126/science.158.3807.1471

Eisner T, Smedley SR, Young DK, Eisner M, Roach B (1996a) Chemical basis of courtship in a beetle (*Neopyrochroa flabellata*): Cantharidin as precopulatory "enticing" agent. *Proceedings of the National Academy of Sciences* **93**, 6494–6498. doi:10.1073/pnas.93.13.6494

Eisner T, Smedley SR, Young DK, Eisner M, Roach B (1996b) Chemical basis of courtship in a beetle (*Neopyrochroa flabellata*): Cantharidin as "nuptial gift". *Proceedings of the National Academy of Sciences* **93**, 6499–6503. doi:10.1073/pnas.93.13.6499

Ekis G, Gupta AP (1971) Digestive system of Cleridae (Coleoptera). *International Journal of Insect Morphology & Embryology* **1**, 51–86. doi:10.1016/0020-7322(71)90008-0

Ekis G, Menier JJ (1980) Discovery of Chaetosomatidae in Madagascar. Systematics of the new genus *Malgassochaetus* (Col. Cleroidea). *Annales de la Société Entomologique de France* **16**, 197–208.

Elliot HJ, De Little DW (1985) *Insect Pests of Trees and Timber in Tasmania.* Forestry Commission of Tasmania, Hobart.

Elzinga RJ (1977) Observations on *Sandalus niger* Knoch (Coleoptera: Sandalidae) with a description of the triungulin larva. *Journal of the Kansas Entomological Society* **50**, 324–328.

Emden FI van (1924a) Zur Kenntnis der Sandalidae. II. u. III. *Entomologische Blätter* **20**, 86–99.

Emden FI van (1924b) Zur Biologie von *Thorictodes heydeni* Reitt. (Col. Thorictidae). *Treubia* **6**(1), 1–7.

Emden FI van (1927) Zur Kenntnis der Sandalidae VII. (Die Gattung *Chamaerhipis* Latr.). *Entomologische Blätter* **23**, 16–24.

Emden FI van (1930) Ueber die Gattung *Polymerius* R. A. Phil. *Revista Chilena de Historia Natural (Valparaiso, Chile)* **34**, 262–265.

Emden FI van (1931a) Die Sandaliden des Congo-Museums. *Revue de Zoologie et de Botanique Africaines* **20**, 183–185.

Emden FI van (1931b) Zur Kenntnis der Sandalidae XI–XVI. *Entomologische Blätter* **27**, 49–59, 107–116, 145–152, pl. 1.

Emden FI van (1932a) Die Larven der Callirhipini, eine mutmassliche *Cerophytum*-Larve und Familien-Bestimmungstabelle der Larven der Malacodermata-Sternoxia-Reihe (Coleoptera) (Zur Kenntnis der Sandalidae 18). *Bulletin et Annales de la Société Entomologique de Belgique* **72**, 199–259. [pls XI–XIII.]

Emden FI van (1932b) Zweite Revision der Gattung *Simianellus* Emd. 1924. (17. Beitrag zur Kenntnis der Sandalidae, zugleich Philippinische Sandalidae 1). *Wiener Entomologische Zeitung* **49**(4), 189–211.

Emden FI van (1933) Die philippinischen Arten der Untergattung *Callirhipis* (Coleoptera). Zur Kenntnis der Sandalidae 19; zugleich philippinische Sandalidae 2). *Philippine Journal of Science* **51**, 331–356.

Emden FI van (1935) Die Gattungsunterschiede der Hirschkäferlarven, ein Beitrag zum natürlichen System der Familie (Col. Lucan.). *Stettiner Entomologische Zeitung* **96**, 178–200.

Emden FI van (1938) On the taxonomy of Rhynchophora larvae (Coleoptera). *Transactions of the Royal Entomological Society of London* **87**, 1–37. doi:10.1111/j.1365-2311.1938.tb01800.x

Emden FI van (1942a) A key to the genera of larval Carabidae (Col.). *Transactions of the Royal Entomological Society of London* **92**, 1–99. doi:10.1111/j.1365-2311.1942.tb03318.x

Emden FI van (1942b) Larvae of British beetles – III. Keys to families. *Entomologist's Monthly Magazine* **78**, 206–226, 253–272.

Emden FI van (1946) Egg-bursters in some more families of polyphagous beetles and some general remarks on egg-bursters. *Proceedings of the Royal Entomological Society of London A* **21**, 89–97. doi:10.1111/j.1365-3032.1946.tb01081.x

Emden FI van (1952) On the taxonomy of Rhynchophora larvae: Adelognatha and Alophinae (Insecta: Coleoptera). *Proceedings of the Zoological Society of London* **122**, 651–795. doi:10.1111/j.1096-3642.1952.tb00248.x

Emden FI van (1956) The *Georyssus* larva-a hydrophilid. *Proceedings of the Royal Entomological Society of London A* **31**, 20–24. doi:10.1111/j.1365-3032.1956.tb00194.x

Emden FI van (1958) Über die Larvenmerkmale einiger deutscher Byrrhidengattungen. *Mitteilungen der Deutschen Entomologischen Gesellschaft* **17**, 39–40.

Emlen DJ, Marangelo J, Ball B, Cunningham CW (2005) Diversity in the weapons of sexual selection: Horn evolution in the beetle genus *Onthophagus* (Coleoptera: Scarabaeidae). *Evolution* **59**, 1060–1084.

Emlen DJ, Lavine LC, Ewen-Campen B (2007) On the origins and evolutionary diversification of beetle horns. *Proceedings of the National Academy of Sciences of the United States of America* **104**, (Supplement 1), 8661–8668. doi:10.1073/pnas.0701209104

Endrödi S (1985) *The Dynastinae of the World.* Junk (Series Entomologica 28), Dordrecht.

Endrödy-Younga S (1959) Systematischer Überblick über die Familie Clambidae (Col.). *Opuscula Entomologica* **24**, 81–116.

Endrödy-Younga S (1990) A revision of the Australian Clambidae (Coleoptera: Eucinetoidea). *Invertebrate Taxonomy* **4**, 247–280. doi:10.1071/IT9900247

Endrödy-Younga S (1991a) Boganiidae (Coleoptera: Cucujoidea) associated with cycads in South Africa: two new species and a new synonym. *Annals of the Transvaal Museum* **35**, 285–293.

Endrödy-Younga S (1991b) *Faune de Madagascar. 76. Insectes Coléoptères. Nosodendridae.* Muséum National d'Histoire Naturelle, Paris.

Endrödy-Younga S (1997) Microsporidae (Coleoptera: Myxophaga), a new family for the African continent. *Annals of the Transvaal Museum* **36**(23), 309–311.

Endrödy-Younga S, Crowson RA (1986) Boganiidae, a new beetle family for the African fauna (Coleoptera: Cucujoidea). *Annals of the Transvaal Museum* **34**, 253–273.

Engel MS, Grimaldi DA (2008) A jugular-horned beetle in Cretaceous amber from Myanmar (Coleoptera: Prostomidae). *Alavesia* **2**, 215–218.

Entwistle EM (1955) *The Biology and Morphology of the Fungus Beetles of the Family Cisidae and their Parasites.* Imperial College (thesis), London.

Ermisch K (1950) Die Gattungen der Mordelliden der Welt (16. Beitrag zur Kenntnis der Mordelliden). *Entomologische Blätter* **45–46**, 34–92.

Ermisch K (1962) I. Beitrag zur Kenntnis der Mordelliden Australiens (Coleopt. Heterom. Mordellidae). *Entomologische Arbeiten aus dem Museum G. Frey* **13**, 338–370.

Erwin TL (1967) Bombardier beetles (Coleoptera: Carabidae) of North America: Part II. Biology and behavior of *Brachinus pallidus* Erwin in California. *Coleopterists Bulletin* **21**, 41–55.

Erwin TL (1970) A reclassification of bombardier beetles and a taxonomic revision of the North and Middle American species (Carabidae: Brachinida). *Quaestiones Entomologicae* **6**, 4–215.

Erwin TL (1974) Studies of the subtribe Tachyina (Coleoptera: Carabidae: Bembidiini). Part II: a revision of the New World-Australian genus *Pericompsus* Leconte. *Smithsonian Contributions to Zoology* **162**, 1–96. doi:10.5479/si.00810282.162

Erwin TL (1979) A review of the natural history and evolution of ectoparasitoid relationships in carabid beetles. In *Carabid Beetles: Their Evolution, Natural History, and Classification.* (Eds TL Erwin, GE Ball, DR Whitehead and AL Halpern) pp. 479–484. Junk, The Hague.

Erwin TL (1980) Systematic and zoogeographic notes on Apotomini, an Old World faunal element new to the Western Hemisphere list, with a description of a new species from Brazil (Coleoptera: Carabidae). *Coleopterists Bulletin* **34**(1), 99

Erwin TL (1981) A synopsis of the immature stages of Pseudomorphini (Coleoptera: Carabidae) with notes on tribal affinities and behavior in relation to life with ants. *Coleopterists Bulletin* **35**, 53–68.

Erwin TL (1983) Beetles and other insects of tropical forest canopies at Manaus, Brazil, sampled by insecticidal fogging. In *Tropical Rain Forest: Ecology and Management.* (Eds SL Sutton, TC Whitmore and AC Chadwick) pp. 59–75. Blackwell Scientific, Oxford, UK.

Escalona HE (2008) Revisión del género neotropical *Serrotibia* Reitter (Coleoptera: Salpingidae: Prostominiinae). *Entomotropica* **23**(2), 105–166.

Escalona HE, Ślipiński A (2010) *Sticholotis ruficeps* Weise: new synonymies and descriptions of its mature larva (Coleoptera: Coccinellidae: Sticholotidini). *Annales Zoologici* **60**(3), 309–318. doi:10.3161/000345410X535307

Escalona HE, Ślipiński A (2011) *Australodon*, a remarkable new genus of Australian longhorned beetle (Coleoptera: Cerambycidae: Cerambycinae). *Annales Zoologici* **61**(4), 731–738. doi:10.3161/000345411X622552

Escalona HE, Ślipiński A (2012) Generic revision and phylogeny of Microweiseinae (Coleoptera: Coccinellidae). *Systematic Entomology* **37**, 125–171. doi:10.1111/j.1365-3113.2011.00601.x

Español F (1970) Notas sobre Anóbidos (Col.). XLVI. Contribución al conocimiento de la subfamilia Hedobiinae. XLVII. El género *Xenogonus* Broun. XLVIII. A propósito del *Xyletinus pauliani* Pic. *Memorias de la Real Academia de Ciencias y Artes de Barcelona* **40**(7), 457–482.

Español F (1976) Contribución al conocimiento de los *Anobium* y géneros affines de Australia y Nueva Zelanda (Col. Anobiidae, nota 71). *Miscelánea Zoologica, Barcelona* **3**(5), 141–160.

Español F (1977a) Notas sobre anóbidos. LXXIV. Sobre Anobiidae de Ghana: el género *Rhamna* Peyerimhoff. LXXV. Contribución al conocimiento de las *Caenocara* Thoms. euroasiaticas. LXXVI. Avance al estudio de la sección *Dorcatoma*. *Miscelànea Zoològica, Barcelona* **4**, 171–215.

Español F (1977b) Nuevos datos sobre la subfamilia Dorcatominae: La sección *Mirosternus* (Col. Anobiidae, nota 73). *Miscellànea Zoològica, Barcelona* **4**, 147–169.

Español F (1979) Anobiidae (Coleoptera) of the Kermadec Islands. *New Zealand Journal of Zoology* **6**, 61–67. doi:10.1080/03014223.1979.10428348

Español F (1982) Sobre *Pronus* Lea y otros generos de Anobiidae (Col.) de la fauna Australiana. Nota 95. *Miscellànea Zoològica, Barcelona* **6**, 59–70.

Evans MEG (1961a) The alimentary canal of *Atomaria ruficornis* (Marsham) (Col., Cryptophagidae). *The Entomologist* **94**, 263–273.

Evans MEG (1961b) The life history of *Atomaria ruficornis* (Marsh.) (Col., Cryptophagidae). *Transactions of the Society for British Entomology* **4**, 207–222.

Evans MEG (1961c) On the muscular and reproductive systems of *Atomaria ruficornis* (Marsham) (Coleoptera, Cryptophagidae). *Transactions of the Royal Society of Edinburgh* **64**, 297–399.

Evans MEG (1971) Morphology and phylogeny in Coleoptera – the prothorax and procoxae. *Proceedings of the International Congress of Entomology* **13**(1), 242–243.

Evans MEG (1972) The jump of the click beetle (Coleoptera, Elateridae)-a preliminary study. *Journal of Zoology* **167**, 319–336. doi:10.1111/j.1469-7998.1972.tb03115.x

Evans MEG (1973) The jump of the click beetle (Coleoptera: Elateridae)-energetics and mechanics. *Journal of Zoology* **169**, 181–194. doi:10.1111/j.1469-7998.1973.tb04553.x

Evans MEG (1974) Propleural structures in Coleoptera. *International Journal of Insect Morphology & Embryology* **3**, 67–72. doi:10.1016/S0020-7322(74)81007-X

Evans MEG (1977) Locomotion in the Coleoptera Adephaga, especially Carabidae. *Journal of Zoology* **181**, 189–226. doi:10.1111/j.1469-7998.1977.tb03237.x

Evans MEG (1982) Early evolution of the Adephaga–some locomotor speculations. *Coleopterists Bulletin* **36**, 597–607.

Evans MEG (1985) Hydradephagan comparative morphology and evolution: some locomotor features and their possible phylogenetic implications. *Proceedings. Academy of Natural Sciences of Philadelphia* **137**(1), 172–181.

Evans AV, Bellamy CL (1996) *An Inordinate Fondness for Beetles.* 208 pp. Henry Holt & Co., New York.

Evans MEG, Forsythe TG (1985) Feeding mechanisms, and their variation in form, of some adult ground-beetles (Coleoptera: Caraboidea). *Journal of Zoology* **206**, 113–143. doi:10.1111/j.1469-7998.1985.tb05640.x

Evers AMJ (1997) Die Carphuridae (Malachiidae) der Ost-Orientalis und West-Polynesiens, besonders Irian Jayas aus den Ausbeuten Alexander Riedels. Teil I. Die Gattungen *Metachoresine* n. gen., *Choresine* Pascoe, *Falsolaius* Pic und *Carphurotroglops* Pic. *Entomologische Blätter* **93**, 127–151.

Falin ZH (2002) 102. Ripiphoridae Gemminger and Harold 1870 (1853). In *American Beetles. Volume 2. Polyphaga: Scarabaeoidea through Curculionoidea.* (Eds RH Arnett Jr, MC Thomas, PE Skelley and JH Frank) pp. 431–444. CRC Press, Boca Raton, Florida.

Falin Z (2004) *Phylogenetic analysis and revision of the genera and subfamilies of the Ripiphoridae (Coleoptera).* University of Kansas (PhD Thesis), Lawrence, Kansas. University Microfilms International, Ann Arbor, Michigan.

Farrell BD (1998) "Inordinate fondness" explained: why are there so many beetles? *Science* **281**, 555–559. doi:10.1126/science.281.5376.555

Farrell BD, Sequeira AS, O'Meara BC, Normark BB, Chung JH, Jordal BH (2001) The evolution of agriculture in beetles (Curculionidae: Scolytinae and Platypodinae). *Evolution* **55**(10), 2011–2027.

Farrow RA (1974) The larva of *Laius villosus* (Coleoptera: Melyridae) feeding on the egg pods of the Australian plague locust, *Chortoicetes tenninifera* (Orthoptera: Acrididae). *Journal of the Australian Entomological Society* **13**, 185–188. doi:10.1111/j.1440-6055.1974.tb02172.x

Faustini DL, Halstead DGH (1982) Setiferous structures of male Coleoptera. *Journal of Morphology* **173**, 43–72. doi:10.1002/jmor.1051730106

Fedorenko DN (2009) *Evolution of the Beetle Hind Wing, with Special Reference to Folding (Insecta, Coleoptera).* Pensoft, Sofia and Moscow.

Ferris GF (1940) The myth of the thoracic sternites in insects. *Microentomology* **5**(3), 87–90.

Fikáček M, Barclay MVL, Perkins PD (2011) Two new species of the *Epimetopus mendeli* species group and notes on its adult and larval morphology (Coleoptera: Hydrophiloidea: Epimetopidae). *Acta Entomologica Musei Nationalis Pragae* **51**(2), 477–504.

Fiori G (1971) Contributi alla conoscenza morfologica ed etologica dei Coleotteri. IX. *Psilothrix viridicaeruleus* (Geoffr.) (Melyridae Dasytinae). *"Studi Sassaresi" Sez. III. Annali della Facolta' di Agraria dell' Università di Sassari* **19**, 1–70.

Fisher RC (1988) An inordinate fondness for beetles. *Biological Journal of the Linnean Society. Linnean Society of London* **35**, 313–319. doi:10.1111/j.1095-8312.1988.tb00473.x

Fitton MG (1975) The larvae of the British genera of Cantharidae (Coleoptera). *Journal of Entomology B* **44**(3), 243–254.

Fletcher TB (1916) One hundred notes on indian insects. *Bulletin of the Agricultural Research Institute Pusa* **59**, 1–39.

Foley IA, Ivie MA (2008) A phylogenetic analysis of the tribe Zopherini with a review of the species and generic classification (Coleoptera: Zopheridae). *Zootaxa* **1928**, 1–72.

Folkerts GW (1989) Egg guarding and its significance in the heterocerid beetle *Dampfius collaris* (Kies.). *Journal of Insect Behavior* **2**, 139–141. doi:10.1007/BF01053624

Forbes WTM (1922) The wing-venation of the Coleoptera. *Annals of the Entomological Society of America* **15**, 328–345, pls 29–35.

Forbes WTM (1926) The wing folding patterns of the Coleoptera. *Journal of the New York Entomological Society* **34**, 42–68, 91–138.

Forbes WTM (1942) The wing of Schizopini (Coleoptera: Dascillidae). *Entomological News* **53**, 101–102.

Ford EJ (1970) Some new anobiids (Coleoptera) from the Papuan subregion. *Pacific Insects* **12**, 117–132.

Ford EJ, Spilman TJ (1979) Biology and immature stages of *Dirrhagofarsus lewisi*, a species new to the United States (Coleoptera, Eucnemidae). *Coleopterists Bulletin* **33**(1), 75–83.

Forrester JA, McHugh JV (2010) 10.5. Sphindidae Jacquelin Du Val, 1861. In *Handbuch der Zoologie/Handbook of Zoology. Band/Volume IV Arthropoda: Insecta Teilband/Part 38. Coleoptera, Beetles. Volume 2. Morphology and Systematics (Polyphaga partim).* (Eds RAB Leschen, RG Beutel and JF Lawrence) pp. 300–306. W. DeGruyter, Berlin.

Forster PI, Machin PJ (1994) Cycad host plants for *Lilioceris nigripes* (Fabricius) (Coleoptera: Chrysomelidae) and *Theclinesthes onycha* (Hewitson) (Lepidoptera: Lycaenidae). *Australian Entomologist* **21**(3), 99–102.

Forster PI, Machin PJ, Mound L, Wilson GW (1994) Insects associated with reproductive structures of cycads in Queensland and northeast New South Wales, Australia. *Biotropica* **26**(2), 217–222. doi:10.2307/2388812

Forsyth DJ (1968) The structure of the defence glands in the Dytiscidae, Noteridae, Haliplidae and Gyrinidae (Coleoptera). *Transactions of the Royal Entomological Society of London* **120**, 159–181. doi:10.1111/j.1365-2311.1968.tb00352.x

Forsyth DJ (1970) The structure of the defence glands of the Cicindelidae, Amphizoidae, and Hygrobiidae (Insecta: Coleoptera). *Journal of Zoology* **160**, 51–69. doi:10.1111/j.1469-7998.1970.tb02897.x

Forsyth DJ (1972) The structure of the pygidial glands of Carabidae (Coleoptera). *Transactions of the Zoological Society of London* **32**, 249–309. doi:10.1111/j.1096-3642.1972.tb00029.x

Forsythe TG (1982) Feeding mechanisms of certain ground beetles (Coleoptera: Carabidae). *Coleopterists Bulletin* **36**, 26–73.

Foster DE (1976) North American Thaneroclerinae larvae (Coleoptera: Cleridae). *Coleopterists Bulletin* **30**, 75–80.

Foster DE, Lawrence JF (1991) Melyridae (Cleroidea) (including Dasytidae, Malachiidae, Prionoceridae, Rhadalidae). In *Immature Insects. Volume 2.* (Ed. FW Stehr) pp. 453–454. Kendall/Hunt, Dubuque, Iowa.

Fragoso SA (1985) The terminalia as a basis for the classification of Cerambycidae subfamilies. Part I, terminology and genital morphology of *Cerambyx cerdo. Revista Brasileira de Entomologia* **29**, 125–134.

Franciscolo ME (1957) Coleoptera: Mordellidae. A monograph of the South African genera and species, 1. In *South African Animal Life: Results of the Lund University Expedition in 1950–1951. Vol. 4.* (Eds B Hanström, P Brinck and G Rudebeck) pp. 207–291. Almqvist & Wiksell, Stockholm.

Franciscolo ME (1964) Nota preliminare sulla filogenia degli Scraptiidae (Coleoptera Heteromera). *Atti della Accademia Nazionale Italiana di Entomologia, Bologna* **11**, 175–181.

Franciscolo ME (1965) Coleoptera: Mordellidae. A monograph of the South African genera and species, 2. Mordellini. In *South African Animal Life. Vol. 11.* (Eds B Hanström, P Brinck and G Rudebeck) pp. 344–468. Swedish Natural Science Research Council, Stockholm.

Franciscolo ME (1967) Coleoptera: Mordellidae. A monograph of the South African genera and species, 3. Mordellistenini. In *South African Animal Life. Vol. 13.* (Eds B Hanström, P Brinck and G Rudebeck) pp. 67–203. Swedish Natural Science Research Council, Stockholm.

Franciscolo ME (1972) Su alcuni generi poco noti di Anaspidinae (47° contributo alla conoscenza degli Scraptiidae). *Memorie della Societa Entomologica Italiana* **51**, 123–155.

Francke-Grosmann H (1967) Ectosymbiosis in wood-inhabiting insects. In *Symbiosis. Volume II.* (Ed. SM Henry) pp. 141–205. Academic Press, New York.

Frank JH, Kanamasu K (1987) *Paederus,* sensu lato (Coleoptera: Staphylinidae): natural history and medical importance. *Journal of Medical Entomology Honolulu* **24**, 155–191.

Frank JH, Thomas MC (1984) Cocoon-spinning and the defensive function of the median gland in larvae of Aleocharinae (Coleoptera, Staphylinidae): a review. *Quaestiones Entomologicae* **20**, 7–23.

Frank JH, Erwin TL, Hemenway RC (2009) Economically beneficial ground beetles. The specialized predators *Pheropsophus aequinoctalis* (L.) and *Stenaptinus jessoensis* (Morawitz): their laboratory behavior and descfriptions of immature stages (Coleoptera, Carabidae, Brachininae). *ZooKeys* **14**, 1–36. doi:10.3897/zookeys.14.188

Franz J (1955) Tannenstanunlause *(Adelges piceae* Ratz.) unter einer Pilzdecke von *Cucurbitaria pithyophila* (Kze. et Schm.) de Not., nebst Beobachtungen an *Aphidoletes thompsoni* Mohn (Dipt., Itonididae) und *Rabocerus mutilatus* Beck (Col., Pythidae) als Tannenlausfeinde. *Zeitschrift für Pflanzenkrankheiten. Pflanzenpathologie und Pflanzenschutz* **62**, 49–61.

Franz H (1975) Revision der Scydmaeniden von Australien, Neuseeland und den benachbarten Inseln. *Denkschriften der Österreichischen Akademie der Wissenschaften* **118**, 1–312.

Franzmann BA (1978) Discovery of *Lema trilineata* (Olivier) in Queensland. *News Bulletin of the Entomological Society of Queensland* **6**, 119–120.

Freeman P (Ed) (1980) *Common Insect Pests of Stored Food Products: A Guide to their Identification.* 6th Edition. British Museum (Natural History) (Economic Series No. 15), London.

Freeman JC, Allemand R, Van Meer C (2003) *Odontosphindus grandis* Hampe, nouvelle espèce, nouveau genre, nouvelle sous-famille pour la faune de

France et pour l'Europe occidentale (Coleoptera, Sphindidae). *Bulletin de la Societe Entomologique de France* **108**(3), 221–232.

Freitag R (1979) Reclassification, phylogeny and zoogeography of the Australian species of *Cicindela* (Coleoptera: Cicindelidae). *Australian Journal of Zoology* **66**, (Supplement), 1–99.

Freude H (1955) Die Monommiden der Welt. I. Teil: Die Monommiden der Indo-Australischen Region. *Entomologische Arbeiten aus dem Museum Georg Frey* **6**, 1–73.

Friedrich F, Beutel RG (2006) The pterothoracic skeletomuscular system of Scirtoidea (Coleoptera: Polyphaga) and its implications for the high-level phylogeny of beetles. *Journal of Zoological Systematics* **44**(4), 290–315. doi:10.1111/j.1439-0469.2006.00369.x

Friedrich F, Beutel RG (2008a) Micro-computer tomography and a renaissance of insect morphology. In Developments in X-Ray Tomography VI (Ed. SR Stock). Proc. of SPIE Vol. 7078 70781U-1-6.

Friedrich F, Beutel RG (2008b) The thorax of *Zorotypus* (Hexapoda, Zoraptera) and a new nomenclature for the musculature of Neoptera. *Arthropod Structure & Development* **37**(1), 29–54. doi:10.1016/j.asd.2007.04.003

Friedrich F, Beutel RG (2010) Goodbye Halteria? The thoracic morphology of Endopterygota (Insecta) and its phylogenetic implications. *Cladistics* **26**, 579–612. doi:10.1111/j.1096-0031.2010.00305.x

Friedrich F, Farrell BD, Beutel RG (2009) The thoracic morphology of Archostemata and the relationships of the extant suborders of Coleoptera (Hexapoda). *Cladistics* **25**, 1–37, 209. doi:10.1111/j.1096-0031.2008.00233.x

Froggatt WW (1894) On the life-histories of Australian Coleoptera. Part II. *Proceedings of the Linnean Society of London* **9**(2), 113–125.

Froggatt WW (1923) *Forest Insects of Australia*. Government Printer, Sydney.

Froggatt WW (1927) *Forest Insects and Timber Borers*. Government Printer, Sydney.

Fukuda A (1969) Description of larva of *Aphanocephalus hemisphericus* Wollaston, with the relationship of the related genera (Coleoptera: Notiophygidae). *Kontyû* **37**, 20–26.

Funk DH, Fenstermacher H (2002) Description of the aquatic larva and pupa of *Paralichas trivittis* (Coleoptera: Ptilodactylidae), with notes on its biology. *Entomological News* **113**(5), 344–354.

Furth DG (1982) The metafemoral spring of flea beetles (Chrysomelidae: Alticinae). *Spixiana* **7**, (Supplement), 11–27.

Furth DG, Traub W, Harpaz I (1983) What makes *Blepharida* jump? A structural study of the metafemoral spring of a flea beetle. *The Journal of Experimental Zoology* **227**, 43–47. doi:10.1002/jez.1402270107

Gage JH (1920) The larvae of the Coccinellidae. *Illinois Biological Monographs* **6**, 232–294.

Gahan CJ (1906) *The Fauna of British India, including Ceylon and Burma. Coleoptera. Vol. I. (Cerambycidae)*. xviii + 329 pp. Taylor and Francis, London.

Gahan CJ (1908a) On the cervinus larvae of *Trictenotoma childreni* Gray, *Melittomma insulare* Fairm. and *Dascillus cervinus* L. *Transactions of the Entomological Society of London* **1908**, 275–282.

Gahan CJ (1908b) Notes on the coleopterous genera *Horia* and *Cissites* and a list of the described species. *Annals & Magazine of Natural History* **2**(8), 199–204.

Galian J, Lawrence JF (1993) First karyotypic data on a cupedid beetle (Coleoptera: Archostemata) showing achiasmatic meiosis. *Entomological News* **104**, 83–87.

Galian J, Moore BP (1994) Chromosome numbers and sex-determining mechanisms in 70 species of Australian Carabidae (Coleoptera). *Coleopterists Bulletin* **48**(3), 226–235.

Ganglbauer L (1903) Systematisch-koleopterologische Studien. *Münchener Koleopterologische Zeitschrift* **1**, 271–319.

Gardner JCM (1930) The early stages of *Niponius andrewesi*, Lew. (Col. Hist.). *Bulletin of Entomological Research* **21**, 15–18. doi:10.1017/S0007485300021490

Gardner JCM (1931a) Immature stages of Indian Coleoptera (8). (Cerambycidae – contd.). *Indian Forest Records* **16**(3), 162–202. [Entomology Series]

Gardner JCM (1932) Immature stages of Indian Coleoptera (11) (Platypodidae). *The Indian Forest Records*. Entomology Series. **17**(3), 1–10, 2 pls.

Gardner JCM (1933a) Immature stages of Indian Coleoptera (12) (Carabidae contd). *Indian Forest Records* **17**(8), 1–12.

Gardner JCM (1933b) Immature stages of Indian Coleoptera (13) (Bostrychidae). *Indian Forest Records* **18**(9), 1–19, pls 1–4.

Gardner JCM (1934) Immature stages of Indian Coleoptera (14) (Curculionidae). *Indian Forest Records* **20**(2), 1–48.

Gardner JCM (1935a) *Genera Insectorum de P. Wytsman. Fascicule* 98. *Coleoptera Fam. Histeridae Subfam. Niponiinae*. L. Desmet-Verteneuil, Brussels.

Gardner JCM (1935b) Immature stages of Indian Coleoptera (16) (Scarabaeoidea). *Indian Forest Records New Series* **1**(1), 1–33.

Gardner JCM (1935c) Immature stages of Indian Coleoptera (17) (Eucnemidae). *Indian Forest Records New Series* **1**, 79–94.

Gardner JCM (1935d) Immature stages of Indian Coleoptera (18) (Brenthidae). *Indian Forest Records New Series* **1**(7), 139–151.

Gardner JCM (1936) A larva of the subfamily Balginae (Col., Elateridae). *Transactions of the Royal Entomological Society of London B* **5**(1), 30–32.

Gardner JCM (1937) Immature stages of Indian Coleoptera (21) Cleridae. *Indian Forest Records New Series* **3**(2), 31–48.

Gardner JCM (1938) Immature stages of Indian Coleoptera (24, Curculionidae contd.). *Indian Forest Records New Series* **3**(12), 139–151.

Gardner JCM (1946) Larvae of Cantharoidea (Coleoptera). *Indian Journal of Entomology* **8**, 121–129.

Gardner JA (1989) Revision of the genera of the tribe Stigmoderini (Coleoptera: Buprestidae) with a discussion of phylogenetic relationships. *Invertebrate Taxonomy* **3**, 291–361. doi:10.1071/IT9890291

Ge S-Q, Beutel RG, Yang XK (2007) Thoracic morphology of adults of Derodontidae and Nosodendridae and its phylogenetic implications (Coleoptera). *Systematic Entomology* **32**(4), 635–667. doi:10.1111/j.1365-3113.2007.00392.x

Ge S-Q, Friedrich F, Beutel RG (2010) On the systematic position and taxonomic rank of the extinct myxophagan *Haplochelus* (Coleoptera). *Insect Systematics & Evolution* **41**, 329–338. doi:10.1163/187631210X537385

Geertsema H, van Dijk DE, van den Heever JA (2002) Palaeozoic insects of southern Africa: a review. *Palaeongologica Africana* **38**, 19–25.

Gerstmeier R (1990a) Short communications on systematics of Cleridae. 1. *Neoscrobiger ephippius* (Boisduval, 1835) is a *Trogodendron* Spinola, 1841. *Acta Coleopterologica* **6**(2), 80.

Gerstmeier R (1990b) Revision of the genus *Olesterus* Spinola, 1841, with description of new species from Australia (Coleoptera, Cleridae). *Mitteilungen der Münchner Entomologische Gesellschaft* **80**, 21–38.

Gerstmeier R (2002) Generic concept of clerid taxa related to *Clerus* (Coleoptera: Cleridae: Clerinae). *Entomological Problems* **32**(2), 99–111.

Ghilarov MS (1964) Order Coleoptera—Beetles. In *Determination Key to Soil-dwelling Insect Larvae*. (Ed. MS Ghilarov) pp. 49–71. "Nauka", Moscow.

Ghilarov MS, Medvedev LN (1964) Family Chrysomelidae-Leaf Beetles. In *Determination Key to Soil-dwelling Insect Larvae*. (Ed. MS Ghilarov) pp. 507–530. "Nauka", Moscow.

Giachino PM (2005) Revision of the Australian Anilina (Coleoptera, Carabidae, Bembidiini). In *Results of the Zoological Missions to Australia of the Regional Museum of Natural Sciences of Turin, Italy*. (Eds M Daccordi and PM Giachino) pp. 137–238. Museo Regionale di Scienze Naturali (Monografie 42), Turin, Italy.

Gillogly LR (1962) Coleoptera: Nitidulidae. *Insects of Micronesia* **16**, 133–188.

Gimmel ML (2009) Family Phalacridae. The shining mold beetles. In *The Coleoptera of the Seychelles Islands*. (Ed. J Gerlach) pp. 106–108. Pensoft, Sofia, Moscow.

Gimmel ML (2011) World catalogue of Propalticidae, with a replacement name for *Discogenia* Kolbe (Coleoptera: Cucujoidea). *Insecta Mundi* **0155**, 1–12.

Gimmel ML (2013) Genus-level revision of the family Phalacridae (Coleoptera: Cucujoidea). *Zootaxa* **3605**, 1–147.

Gimmel ML, Leschen RAB, Ślipiński A (2009) Review of the New Zealand endemic family Cyclaxyridae. *Acta Entomologica Musei Nationalis Pragae* **49**(2), 511–528.

Giorgi JA, Vandenberg NJ, McHugh JV, Forrester JA, Ślipiński A, Miller KB, Shapiro LR, Whiting MF (2009) The evolution of food preferences in Coccinellidae. *Journal of Biological Control* **51**, 215–231. doi:10.1016/j.biocontrol.2009.05.019

Glaister A (1985) Laboratory rearing of Australian elmid larvae (Elmidae: Coleoptera). *Australian Society of Limnology Bulletin* **10**, 51–58.

Glaister A (1999) *Guide to the Identification of Australian Elmidae Larvae (Insecta: Coleoptera).* Cooperative Research Centre for Freshwater Ecology (Identification Guide No. 21), Albury, New South Wales.

Glen R (1950) Larvae of elaterid beetles of the tribe Lepturoidini (Coleoptera: Elateridae). *Smithsonian Miscellaneous Collections* **111**(11), 1–246.

Goede A (1967) Tasmanian cave fauna: character and distribution. *Helictite* **5**, 71–86.

Gómez-Zurita J, Jolivet P, Vogler AP (2005) Molecular systematics of Eumolpinae and the relationships with Spilopyrinae (Coleoptera, Chrysomelidae). *Molecular Phylogenetics and Evolution* **34**, 584–600. doi:10.1016/j.ympev.2004.11.022

Gómez-Zurita J, Hunt T, Vogler AP (2007) Multilocus ribosomal RNA phylogeny of the leaf beetles (Chrysomelidae). *Cladistics* **23**, 1–17.

Gompel N, Barrau E (2002) Les Aderidae de la faune de France (Coleoptera). *Annales de la Société Entomologique de France (N. S.)* **38**(3), 211–238.

Good HG (1925) Wing venation of the Buprestidae. *Annals of the Entomological Society of America* **18**, 251–272, pls 8–11.

Goodrich MA (2002a) 88. Byturidae Jacquelin du Val 1858. In *American Beetles. Volume 2. Polyphaga: Scarabaeoidea through Curculionoidea.* (Eds RH Arnett Jr, MC Thomas, PE Skelley and JH Frank) pp. 354–355. CRC Press, Boca Raton, Florida.

Goodrich MA (2002b) 89. Biphyllidae LeConte 1861. In *American Beetles. Volume 2. Polyphaga: Scarabaeoidea through Curculionoidea.* (Eds RH Arnett Jr, MC Thomas, PE Skelley and JH Frank) pp. 356–357. CRC Press, Boca Raton, Florida.

Goodrich MA, Springer CA (1988) A new species of *Xerasia* (Coleoptera: Byturidae) from the Middle East, with a key to the *Xerasia* of the world. *Coleopterists Bulletin* **42**(4), 345–351.

Goodrich MA, Springer CA (1992) A Revision of the Family Biphyllidae (Coleoptera) for America, North of Mexico. *Coleopterists Bulletin* **46**, 361–377.

Goodrich MA, Springer CA (1995) A new species of *Bispinatus* Springer & Goodrich from Thailand (Coleoptera: Byturidae: Platydascillinae), with notes on the distribution of the subfamily. *Coleopterists Bulletin* **49**(2), 183–190.

Gorb SN (2004) Attachment systems of arthropods. *Arthropod Structure & Development* **33**(1), 1–2. doi:10.1016/j.asd.2003.11.002

Gordh G, Headrick D (2011) *A Dictionary of Entomology.* Second Edition. 1500 pp. CSIRO Publishing, Collingwood, Victoria.

Gordon RD (1985) The Coccinellidae (Coleoptera) of America North of Mexico. *Journal of the New York Entomological Society* **93**, 1–912.

Gordon RD, Vandenberg N (1995) Larval systematics of North American *Coccinella* L. (Coleoptera: Coccinellidae). *Entomologica Scandinavica* **26**, 67–86.

Gough N, Hamacek EL (1989) Insect inducted bud fall in cultivated hibiscus and aspects of the biology of *Macroura concolor* (Macleay) (Coleoptera: Nitidulidae). *Journal of the Australian Entomological Society* **28**, 267–277. doi:10.1111/j.1440-6055.1989.tb00897.x

Gough N, Bartreau T, Montgomery BL (1994) Distribution, hosts and pest status of the orchid beetle *Stethopachys formosa* Baly (Coleoptera: Chrysomelidae). *Australian Entomologist* **21**(4), 49–54.

Gould SJ (1993) A special fondness for beetles. *Natural History* **2**(1), 4–12.

Grandi G (1971) Comparative morphology and ethology of insects with a specialized diet, *Rhysodes germari* Ganglb. *Bolletino dell'Istituto di Entomologia della Università degli Studi di Bologna* **30**, 31–47.

Gratshev VG, Zherikhin VV (1996) A revision of the nemonychid weeevil subfamily Brethorrhininae (Insecta, Coleoptera, Nemonychidae). *Paleontological Journal* **29**(4), 112–127.

Gratshev VG, Zherikhin VV (2003) The fossil record of weevils and related beetle families (Coleoptera, Curculionoidea). *Acta Zoologica Cracoviensia* **46** Supplement (Fossil Insects. Proceedings of the 2nd Congress on Palaeoentomology, Krakow, Poland, 5–9 September, 2001), 129–138.

Gravely FH (1916) Some lignicolous beetle-larvae from India and Borneo. *Records of the Indian Museum* **12**, 137–175, pls 20–22.

Grebennikov VV (2004a) The basal phylogeny of Scarabaeoidea (Insecta: Coleoptera) inferred from larval morphology. *Invertebrate Systematics* **18**, 321–348. doi:10.1071/IS03013

Grebennikov VV (2004b) Review of larval morphology of beetles of the suborder Archostemata (Insecta: Coleoptera), including first-instar chaetotaxy. *European Journal of Entomology* **101**, 273–292.

Grebennikov VV (2005) 11. Staphylinoidea Latreille, 1802. Introduction and Phylogeny. In *Handbuch der Zoologie/Handbook of Zoology. Band/Volume IV Arthropoda: Insecta Teilband/Part 38. Coleoptera, Beetles. Volume 1. Morphology and Systematics (Archostemata, Adephaga, Myxophaga, Polyphaga partim).* (Eds RG Beutel and RAB Leschen) pp. 223–224. W. DeGruyter, Berlin.

Grebennikov VV (2008a) How small you can go: Factors limiting body miniaturization in winged insects with a review of the pantropical genus *Dicheramocephalus* and description of six new species of the smallest beetles (Pterygota: Coleoptera: Ptiliidae). *European Journal of Entomology* **105**, 313–328.

Grebennikov VV (2008b) *Tasmanitachoides* belongs to Trechini (Coleoptera: Carabidae): discovery of the larva, its phylogenetic implications and revised key to Trechitae genera. *Invertebrate Systematics* **22**, 479–488. doi:10.1071/IS07051

Grebennikov VV, Beutel RG (2002) Morphology of the minute larva of *Ptilella tenella,* with special reference to effects of miniaturisation and the systematic position of Ptiliidae (Coleoptera: Staphylinoidea). *Arthropod Structure & Development* **31**, 157–172. doi:10.1016/S1467-8039(02)00022-1

Grebennikov VV, Leschen RAB (2010) External exoskeletal cavities in Coleoptera and their possible mycangial functions. *Entomological Science* **13**, 81–98. doi:10.1111/j.1479-8298.2009.00351.x

Grebennikov VV, Newton AF (2008) Minute larvae of Leptotyphlinae (Coleoptera: Staphylinidae): description of three genera with discussion on the monophyly and phylogenetic position of the subfamily as inferred from larval morphology. *Zootaxa* **1817**, 49–58.

Grebennikov VV, Newton AF (2009) Good-bye Scydmaenidae, or why the ant-like stone beetles should become megadiverse Staphylinidae *sensu latissimo* (Coleoptera). *European Journal of Entomology* **106**, 275–301.

Grebennikov VV, Scholtz CH (2003) Larvae and pupae of Dascillidae (Coleoptera): morphological study and discussion of their relationships to Scarabaeoidea and Eulichadidae. *Insect Systematics & Evolution* **34**(1), 29–39. doi:10.1163/187631203788964908

Grebennikov VV, Scholtz CH (2004) The basal phylogeny of Scarabaeoidea (Insecta: Coleoptera) inferred from larval morphology. *Invertebrate Systematics* **18**, 321–348. doi:10.1071/IS03013

Grebennikov VV, Ballerio A, Scholtz CH (2002) Larva and pupa of *Cyphopisthes descarpentriesi* Paulian (Coleoptera: Scarabaeoidea: Ceratocanthidae) and their phylogenetic implications. *Australian Journal of Entomology* **41**, 367–374. doi:10.1046/j.1440-6055.2002.00307.x

Grebennikov VV, Ballerio A, Ocampo FC, Scholtz CH (2004) Larvae of Ceratocanthidae and Hybosoridae (Coleoptera: Scarabaeoidea): study of morphology, phylogenetic analysis and evidence of paraphyly of Hybosoridae. *Systematic Entomology* **29**, 524–543. doi:10.1111/j.0307-6970.2004.00257.x

Gressitt JL (1951) Longicorn Beetles of China. *In Longicornia – Études et Notes sur les Longicornes. Volume II.* (Ed. P. Lepesme) 667 pp., 22 pls. P. Lechevalier, Paris.

Gressitt JL (1956) Cerambycidae. *Insects of Micronesia* **17**, 61–183.

Gressitt JL (1957) Hispine beetles from the South Pacific (Coleoptera: Chrysomelidae). *Nova Guinea (N. S.)* **8**, 205–324.

Gressitt JL (1959) Longicorn beetles from New Guinea, I (Cerambycidae). *Pacific Insects* **1**, 59–171.

Gressitt JL (1963) Hispine beetles (Chrysomelidae) from New Guinea. *Pacific Insects* **5**(3), 591–714.

Gressitt JL (1984) Systematics and biogeography of the longicorn beetle tribe Tmesisternini. *Pacific Insects Monographs* **41**, 263 pp.

Gressitt JL, Rondon JA, Breuning S (1970) Cerambycid beetles of Laos (Longicornes du Laos). *Pacific Insects Monographs* **24**, vi + 651 pp.

Grove S, Bashford D, Yee M (2008) A long-term experimental study of saproxylic beetle (Coleoptera) succession in Tasmanian *Eucalyptus obliqua* logs: findings from the first five years. *Insect Ecology and Conservation* **2008**, 71–114.

Gupta AP (1971) External genitalia of Meloidae (Coleoptera). II. The genitalia and their taxonomic significance. *Misscellaneous Publications of the Entomological Society of America* **8**(1), 1–29.

Gur'yeva YeL (1969) Some trends in the evolution of click beetles (Coleoptera, Elateridae). *Entomologicheskoye Obozreniye* **48**, 263–272 (in Russian; translation in *Entomological Review*, Washington **48**, 154–159).

Gur'yeva YeL (1974) Thoracic struture of click beetles (Coleoptera, Elateridae) and the significance of the structural characters for the system of the family. *Entomologicheskiye Obozreniye* **53**, 93–113 (in Russian; translation in *Entomological Review* 53, 67–79).

Guy PL, Gibbs AJ (1985) Further studies on turnip yellow mosaic tymovirus isolates from an endemic Australian *Cardamine. Plant Pathology* **34**, 532–544. doi:10.1111/j.1365-3059.1985.tb01404.x

Gwynn-Silvey JK (1935) An investigation of the burrowing inner-beach insects of some fresh-water lakes. *Papers of the Michigan Academy of Science, Arts and Letters* **21**, 655–696.

Gwynne DT, Rentz DCF (1983) Beetles on the bottle: male buprestids mistake stubbies for females (Coleoptera). *Journal of the Australian Entomological Society* **22**, 79–80. doi:10.1111/j.1440-6055.1983.tb01846.x

Haas F, Kukalová-Peck J (2001) Dermaptera hindwing structure and folding: new evidence for familial, ordinal and superordinal relationships within Neoptera (Insecta). *European Journal of Entomology* **98**, 445–509.

Haas F, Gorb S, Blickhan R (2000) The function of resilin in beetle wings. *Proceedings of the Royal Entomological Society of London* **267**, 1375–1381. doi:10.1098/rspb.2000.1153

Habeck DH (2002a) 76. Brachypteridae Erichson 1845. In *American Beetles. Volume 2. Polyphaga: Scarabaeoidea through Curculionoidea.* (Eds RH Arnett Jr, MC Thomas, PE Skelley and JH Frank) pp. 309–310. CRC Press, Boca Raton, Florida.

Habeck DH (2002b) 77. Nitidulidae Latreille 1802. In *American Beetles. Volume 2. Polyphaga: Scarabaeoidea through Curculionoidea.* (Eds RH Arnett Jr, MC Thomas, PE Skelley and JH Frank) pp. 311–315. CRC Press, Boca Raton, Florida.

Habu A, Sadanaga K (1965) Illustrations for identification of the larvae of the Carabidae found in cultivated fields and paddyfields (III). *Bulletin of the National Institute of Agricultural Sciences Tokyo (Ser. C)* **3**(19), 81–216.

Hackman RH (1984) Chapter 30. Arthropoda: Cuticle: Biochemistry. In *Biology of the Integument. Volume 1: Invertebrates*. (Eds J Bereitter, AG Matolsky and KS Richards) pp. 583–610. Springer-Verlag, Berlin.

Hagen KS (1962) Biology and ecology of predaceous Coccinellidae. *Annual Review of Entomology* **7**, 289–326. doi:10.1146/annurev.en.07.010162.001445

Hájek J (2007) Revision of the genus *Eulichas* Jacobson, 1913 (Coleoptera: Eulichadidae) I. Introduction, morphology of adults, key to subgenera and species groups, and taxonomy of *E. funebris* species group. *Zootaxa* **1620**, 1–35.

Hájek J (2010) *Brachypsectra kadleci* sp. nov. from western Iran – the first Palaearctic member of the family Brachypsectridae (Insecta: Coleoptera: Elateriformia). *Annales Zoologici* **60**(1), 29–33. doi:10.3161/000345410X499498

Hájek J (2011) World catalogue of the family Callirhipidae (Coleoptera: Elateriformia), with nomenclatural notes. *Zootaxa* **2914**, 1–66.

Halffter G, Edmonds WD (1982) *The Nesting Behavior of Dung Beetles (Scarabaeinae): An Ecological and Evolutive Approach.* Instituto de Ecologia, Mexico, D. F.

Hall WE (1999) Generic revision of the tribe Nanosellini (Coleoptera: Ptiliidae: Ptiliinae). *Transactions of the American Entomological Society* **125**(1–2), 39–126.

Hall WE (2000a) 3. Microsporidae Crotch, 1873. In *American Beetles. Volume 1. American Beetles. Archostemata, Myxophaga, Adephaga, Polyphaga: Staphyliniformia.* (Eds RH Arnett Jr and MC Thomas) pp. 24–26. CRC Press, Boca Raton, Florida.

Hall WE (2000b) 4. Hydroscaphidae LeConte, 1874. In *American Beetles. Volume 1. American Beetles. Archostemata, Myxophaga, Adephaga, Polyphaga: Staphyliniformia.* (Eds RH Arnett Jr and MC Thomas) pp. 27–29. CRC Press, Boca Raton, Florida.

Hall WE (2000c) 17. Ptiliidae Erichson, 1845. In *American Beetles. Volume 1. American Beetles. Archostemata, Mycophaga, Adephaga, Polyphaga: Staphyliniformia.* (In *American Beetles. Volume 1. American Beetles. Archostemata, Mycophaga, Adephaga, Polyphaga: Staphyliniformia.* (Eds RH Arnett Jr and MC Thomas) pp. 233–2246. CRC Press, Boca Raton, Florida.

Hall WE (2005) 11.2. Ptiliidae Erichson, 1845. In *Handbuch der Zoologie/ Handbook of Zoology. Band/Volume IV Arthropoda: Insecta Teilband/ Part 38. Coleoptera, Beetles. Volume 1. Morphology and Systematics (Archostemata, Adephaga, Myxophaga, Polyphaga partim).* (Eds RG Beutel and RAB Leschen) pp. 251–261. W. DeGruyter, Berlin.

Hall WD, Howe RW (1953) Revised key to larvae of the Ptinidae associated with stored products in Britain. *Bulletin of Entomological Research* **31**, 331–381.

Halstead DGH (1968) Observations on the biology of *Murmidius ovalis* (Beck) (Coleoptera: Cerylonidae). *Journal of Stored Products Research* **4**, 13–21. doi:10.1016/0022-474X(68)90025-8

Halstead DGH (1973) A revision of the genus *Silvanus* Latreille (s. l.) (Coleoptera: Silvanidae). *Bulletin of the British Museum (Natural History). Entomology* **29**(2), 37–112.

Halstead DGH (1980) A revision of the genus *Oryzaephilus* Ganglbauer, including descriptions of related genera (Coleoptera: Silvanidae). *Zoological Journal of the Linnean Society* **69**, 271–374. doi:10.1111/j.1096-3642.1980.tb01126.x

Halstead DGH (1993) Keys for the identification of beetles associated with stored products – II. Laemophloeidae, Passandridae and Silvanidae. *Journal of Stored Products Research* **29**, 99–197. doi:10.1016/0022-474X(93)90030-8

Hamilton AG (1897) On the fertilization of *Eupomatia laurina*, R. BR. *Proceedings of the Linnean Society of New South Wales* **22**, 48–55, pl. 3.

Hamilton KGA (1971) The insect wing, part I. *Journal of the Kansas Entomological Society* **44**, 421–433.

Hamilton KGA (1972) The insect wing, parts II, III, IV. *Journal of the Kansas Entomological Society* **45**, 54–58. 145–162, 295–308.

Hamilton RW (2002) 128. Attelabidae Billberg 1820. In *American Beetles. Volume 2. Polyphaga: Scarabaeoidea through Curculionoidea.* (Eds RH Arnett Jr, MC Thomas, PE Skelley and JH Frank) pp. 703–710. CRC Press, Boca Raton, Florida.

Hamilton-Smith E, Adams DJH (1966) The alleged obligate ectoparasitism of *Myotyphlus jansoni* (Matthews) (Coleoptera: Staphylinidae). *Journal of the Entomological Society of Queensland* **5**, 44–45. doi:10.1111/j.1440-6055.1966.tb00678.x

Hammad SM (1953) The immature stages of *Metophthalmus serripennis* Broun (Coleoptera: Lathridiidae). *Proceedings of the Royal Entomological Society of London A* **28**, 133–138. doi:10.1111/j.1365-3032.1953.tb00641.x

Hammond PM (1979) Wing-folding mechanisms of beetles, with special reference to investigations of adephagan phylogeny (Coleoptera). In *Carabid Beetles: Their Evolution, Natural History, and Classification.* (Eds TL Erwin, GE Ball, DR Whiteheadand AL Halpern) pp. 113–180. Junk, The Hague.

Hammond PM (1985) Dimorphism of wings, wing-folding and wing-toiletry devices in the ladybird, *Rhyzobius litura* (F.) (Coleoptera: Coccinellidae), with a discussion of inter-population variation in this and other wing-dimorphic beetle species. *Biological Journal of the Linnean Society. Linnean Society of London* **24**, 15–33. doi:10.1111/j.1095-8312.1985.tb00157.x

Hammond PM, Lawrence JF (1989) Appendix: Mycophagy in insects: a summary. In *Insect-Fungus Interactions. 14th Symposium of the Royal Entomological Society of London in collaboration with the British Mycological Society.* (Eds N Wilding, NM Collins, PM Hammond and JF Webber) pp. 275–324. Academic Press, London.

Hand S, Archer M, Bickel D, Creaser P, Dettmann M, Godthelp H, Jones A, Norris B, Wicks D (2010) Chapter 5. Australian Cape York Amber. In *Biodiversity of Fossils in Amber from the Major World Deposits* (Ed. D. Penney) pp. 69–79. Siri Scientific Press, Castleton, UK.

Hangay G, Zborowski P (2010) *A Guide to the Beetles of Australia.* x + 237 pp. CSIRO Publishing, Collingwood, Vic.

Hannappel U, Paulus HF (1987) Arbeiten zu einem phylogenetischen System der Helodidae (Coleoptera) – Feinstrukturuntersuchungen an europäischen Larven. *Zoologische Beiträge (N. F.)* **31**, 77–150.

Hannappel U, Paulus HF (1992) Some undetermined Helodidae larvae from Australia and New Zealand: fine structure of mouthparts and phylogenetic position. In *Advances in Coleopterology.* (Eds M Zunino, X Bellés and M Blas) pp. 89–128. Associación Europea de Coleopterología, Barcelona.

Hannappel U, Paulus HF (1994) 29. Familie: Scirtidae. In *Die Larven der Käfer Mitteleuropas. 2. Band. Myxophaga. Polyphaga Teil 1.* (Ed. B Klausnitzer) pp. 75–87. Goecke & Evers, Krefeld.

Hansen M (1990) Australian Sphaeridiinae (Coleoptera: Hydrophilidae): a taxonomic outline with descriptions of new genera and species. *Invertebrate Taxonomy* **4**, 317–395. doi:10.1071/IT9900317

Hansen M (1991a) The Hydrophiloid Beetles. Phylogeny, Classification and a Revision of the genera (Coleoptera, Hydrophiloidea). *Biologiske Skrifter, Copenhagen* **40**, 1–367.

Hansen M (1991b) A review of the genera of the beetle family Hydraenidae (Coleoptera). *Steenstrupia (Copenhagen)* **17**(1), 1–52.

Hansen M (1995) Evolution and classification of the Hydrophiloidea a systematic review. In *Biology, Phylogeny, and Classification of Coleoptera. Papers Celebrating the 80th Birthday of Roy A. Crowson.* (Eds J Pakaluk and SA Ślipiński) pp. 321–353. Muzeum i Instytut Zoologii PAN, Warsaw.

Hansen M (1997a) Evolutionary trends in "staphyliniform" beetles (Coleoptera). *Stenstrupia* **23**, 43–86.

Hansen M (1997b) A new subfamily for a remarkable new genus and species of Hydrophilidae from New Guinea (Coleoptera: Hydrophilidae). *Annales Zoologici* **47**(1/2), 107–110.

Hansen M (1997c) Phylogeny and classification of the staphyliniform beetle families (Coleoptera). *Biologiske Skrifter, Copenhagen* **48**, 1–339.

Hansen M (1998) *World Catalogue of Insects. Volume 1. Hydraenidae (Coleoptera).* 168 pp. Apollo Books, Stenstrup.

Hansen M (1999) *World Catalogue of Insects. Volume 2: Hydrophiloidea (s. str.) (Coleoptera).* 416 pp. Apollo Books, Stenstrup.

Hansen M (2000) Observations on the immature stages of Georissidae (Coleoptera: Hydrophiloidea), with remarks on the evolution of the hydrophiloid egg cocoon. *Invertebrate Taxonomy* **14**, 907–916. doi:10.1071/IT00016

Hansen M, Richardson BA (1998) A new species of *Omicrus* Sharp (Coleoptera: Hydrophilidae) from Puerto Rico and its larva, the first known larva of Omicrini. *Systematic Entomology* **23**, 1–8. doi:10.1046/j.1365-3113.1998.00036.x

Harde KW (1966) 87. Familie: Cerambycidae, Bockkäfer. In *Die Käfer Mitteleuropas. Band 9, Cerambycidae, Chrysomelidae.* (Eds H Freude, KW Harde and GA Lohse) pp. 7–94. Goecke & Evers, Krefeld.

Harley KLS, Miller IL, Napompeth B, Thamasara S (1985) An integrated approach to the management of *Mimosa pigra* l. In Australia and Thailand. *Proceedings of the Tenth Asian-Pacific Weed Science Society Conference 1985.* 1, 209–215.

Hartley CS, McHugh JV (2010) 10.35. Latridiidae Erichson, 1842. In *Handbuch der Zoologie/Handbook of Zoology. Band/Volume IV Arthropoda: Insecta Teilband/Part 38. Coleoptera, Beetles. Volume 2. Morphology and Systematics (Polyphaga partim).* (Eds RAB Leschen, RG Beutel and JF Lawrence) pp. 481–486. W. DeGruyter, Berlin.

Hasenpusch JW, Moeseneder CH (2010) *Storeyus*, a new genus and *Storeyus pseudodipterus*, a new species of Cetoniinae (Coleoptera: Scarabaeidae) from Australia, with a redescription of *Lenosoma* Kraatz. *Australian Entomologist* **36**(4), 221–234.

Hashimoto K, Hayashi F (2012) Structure and function of the large pronotal horn of the sand-living anthicid beetle *Mecynotarsus tenuipes*. *Entomological Science* **15**(3), 274–279. doi:10.1111/j.1479-8298.2012.00517.x

Hatch MH (1925) The phylogeny and phylogenetic tendencies of Gyrinidae. *Papers of the Michigan Academy of Sciences, Arts and Letters* **5**, 429–467.

Háva J (2000) Distributional notes on some Oriental Nosodendridae (Coleoptera), with descriptions of five new species. *Acta Musei Moraviae. Scientiae Biologicae* **85**, 57–65.

Háva J (2003) *World Catalogue of the Dermestidae (Coleoptera).* Studie a zprávy Oblastního muzea Praha - východ v Brandýse nad Labem, Czech Republic.

Háva J (2004) World keys to the genera and subgenera of Dermestidae (Coleoptera), with descriptions, nomenclature and distributional records. *Acta Musei Nationalis Pragae, Series B. Natural History* **60**(3–4), 149–164.

Háva J, Löbl I (2005) A world catalogue of the family Jacobsoniidae (Coleoptera). *Studies and Reports of District Museum Prague-East. Taxonomical Series* **1**(1–2), 89–94.

Hawkeswood TJ (1982) Notes on the life history of *Aspidomorpha maculatissima* Boheman (Coleoptera: Chrysomelidae: Cassidinae) at Townsville, North Queensland. *Victorian Naturalist* **99**, 92–101.

Hawkeswood TJ (1985) The larva of *Diadoxus erythrurus* (White) (Coleoptera: Buprestidae). *Australian Entomological Magazine* **12**, 23–28.

Hawkeswood TJ (1987a) *Beetles of Australia.* Angus and Robertson, North Ryde, New South Wales.

Hawkeswood TJ (1987b) Observations on a *Dasytes* species (Melyridae: Coleoptera), a pollen-feeding beetle on daisies (Asteraceae) In northern New South Wales. *Victorian Naturalist* **104**, 76–80.

Hawkeswood TJ (1989) New host records for adults of some fungus-feeding beetles (Coleoptera) from New South Wales and Queensland, Australia. *Victorian Naturalist* **106**, 93–95.

Hawkeswood TJ (1992) Notes on the biology and host plants of the Australian leaf beetle *Lilioceris* (*Crioceris*) *nigripes* (Fabricius) (Coleoptera: Chrysomelidae). *The Entomologist* **111**(4), 210–212.

Hawkeswood TJ, Takizawa H (1997) Taxonomy, ecology andf description of the larva, pupa and adult of the Australian hispine beetle, *Eurispa vittata* Baly (Insecta, Coleoptera, Chrysomelidae). *Spixiana* **20**(3), 245–253.

Hayashi M (1961) Cerambycidae from New Caledonia (Col.), Part I. *Bulletin of the Osaka Museum of Natural History* **13**, 7–65, 3 pls.

Hayashi N (1969) On the larvae of some species of small families of Cucujoidea in Japan (Coleoptera). *Insecta Matsumurana* **7**, (Supplement), 1–9.

Hayashi N (1971) On the larvae of Mycetophagidae occurring in Japan (Coleoptera: Cucujoidea). *Kontyû* **39**(4), 361–367.

Hayashi N (1975) On the larvae of Melandryidae (Coleoptera, Cucujoidea) and some related families occurring in Japan. *Kontyû* **43**, 147–169.

Hayashi N (1978) A contribution to the knowledge of the larvae of Nitidulidae occurring in Japan (Coleoptera: Cucujoidea). *Insecta Matsumurana (N. S.)* **14**, 1–97.

Hayashi N (1986) Larvae. In The Coleoptera of Japan in Colour. Vol. 1. (Eds K Morimoto and N Hayashi) pp. 202–218, pls. 1–113. Hoikusha Publ. Co., Osaka.

Hayashi N (1994) On the larvae of three species of Passandridae, Tenebrionidae and Melandryidae occurring in Japan (Coleoptera: Cucujoidea). *Transactions of the Shikoku Entomological Society* **20**(3–4), 137–144.

Hayashi N, Takanaka H (1965) Notes on the immature stages of *Encaustes praenobilis* Lewis (Coleoptera: Erotylidae). *Mikado* **1**, 35–39.

Hayes WP (1928) The epipharynx of lamellicorn larvae (Coleop.) with a key to common genera. *Annals of the Entomological Society of America* **21**(2), 282–306.

Hayes WP, Chu HF (1946) The larvae of the genus *Nosodendron* Latr. (Coleoptera, Nosodendridae). *Annals of the Entomological Society of America* **39**, 69–79.

Hebauer F (1997) Revision der Arten der Familie Spercheidae Erichson, 1837 (Coleoptera, Hydrophiloidea). *Entomologische Blätter* **93**, 9–42.

Hemp C, Dettner K (1997) Morphology and chemistry of mesothoracic glands in anthicid beetles (Coleoptera: Anthicidae). *Entomologia Generalis* **22**(2), 97–108.

Hemp C, Dettner K (2003) Description of larvae of the genus *Formicomus* Laferté and data on the life cycles of *Omonadus floralis* (Linné) and *Notoxus monoceros* (Linné) (Coleoptera: Anthicidae). *Coleopterists Bulletin* **57**(4), 361–368. doi:10.1649/369

Hendrich L (2001) A new species of *Hygrobia* Latreille from peatlands of southwestern Australia. *Koleopterologische Rundschau* **71**, 17–25.

Hendrich L, Watts CHS (2004) Taxonomic revision of the Australian genus *Sternopriscus* Sharp, 1882 (Coleoptera: Dytiscidae: Hydroporinae). *Koleopterologische Rundschau* **74**, 75–142.

Hendrich L, Watts CHS (2009) Taxonomic revision of the Australian predaceous water beetle genus *Carabhydrus* Watts, 1978 (Col. Dytiscidae, Hydroporinae, Hydroporini). *Zootaxa* **2048**, 1–30.

Hendrich L, Hawlitschek O, Balke M (2009) The epigean Australasian species of *Neobidessodes* gen. n. diving beetles — a revision integrating morphology, cybertaxonomy, DNA taxonomy and phylogeny (Coleoptera: Dytiscidae, Bidessini). *Zootaxa* **2288**, 1–41.

Herman LH (1986) Revision of *Bedius*. Part IV. Classification of species groups, phylogeny, natural history, and catalogue (Coleoptera, Staphylinidae, Oxytelinae). *Bulletin of the American Museum of Natural History* **184**, 1–368.

Herman LH (2001) Catalog of the Staphylinidae (Insecta: Coleoptera). 1758 to the end of the second millenium. I–VII. *Bulletin of the American Museum of Natural History* **265**, i–vi, 1–4218.

Hernando C, Ribera I (2000) The first species of the intertidal genus *Hyphalus* Britton from the Indian Ocean (Coleoptera: Limnichidae: Hyphalinae). *Annales de la Société Entomologique de France (N. S.)* **36**(3), 239–243.

Hernando C, Ribera I (2003) *Babalimnichus splendens* sp. n., a new jumping shore beetle from the Tonga Islands (Polynesia) (Coleoptera: Limnichidae, Thaumastodinae). *Koleopterologische Rundschau* **73**, 269–273.

Hernando C, Ribera I (2004) *Hyphalus madli* sp. n., a new intertidal limnichid beetle from the Seychelles (Coleoptera: Limnichidae: Hyphalinae). *Koleopterologische Rundschau* **74**, 413–417.

Hernando C, Ribera I (2005) 18.5. Limnichidae Erichson, 1846. In *Handbuch der Zoologie/Handbook of Zoology. Band/Volume IV Arthropoda: Insecta Teilband/Part 38. Coleoptera, Beetles. Volume 1. Morphology and Systematics (Archostemata, Adephaga, Myxophaga, Polyphaga partim)*. (Eds RG Beutel and RAB Leschen) pp. 512–518. W. DeGruyter, Berlin.

Hernando C, Ribera I (2010) Limnichidae from New Caledonia, description of a new species, and a checklist of the taxa recorded from the Australian/Pacific region (Coleoptera). In *Water Beetles of New Caledonia (part 1)*. (Eds MA Jäch & M Balke) pp. 439–449. Zoologisch-Botanische Gesellschaft in Österreich and Wiener Coleopterologenverein (Monographs on Coleoptera. Vol. 3), Vienna.

Hickman JR (1931) Contribution to the biology of the Haliplidae (Coleoptera). *Annals of the Entomological Society of America* **24**, 129–142.

Hickman VV (1974) Notes on the biology of *Ptinus exulans* Erichson (Coleoptera: Ptinidae). *Journal of the Entomological Society of Australia (N.S. W.)* **8**, 7–14.

Hicks EA (1959) *Check-List and Bibliography on the Occurrence of Insects in Birds' Nests*. Iowa State College Press, Ames, Iowa.

Hicks EA (1962) Check-list and bibliography on the occurrence of insects in birds' nests. Supplement I. *Iowa State Journal of Science* **36**, 233–348.

Hicks EA (1971) Check-list and bibliography on the occurrence of insects in birds' nests. Supplement II. *Iowa State Journal of Science* **46**, 123–338.

Hieke F (1966) Vergleichende funktionelle Anatomie der Abdominalmuskulatur einiger männlicher Coleopteren unter besonderer Berücksichtigung des Genitoanalkomplexes. *Deutsche Entomologische Zeitschrift (N. F.)* **13**, 1–168.

Hieke F (1971) Die Evolution des 9. Abdominalsegmentes bei den Lamellicornia (Coleopteren). *Proceedings of the International Congress of Entomology* **13**(1), 252

Hiller A (1990) Notes on systematics and biology of *Trichaulax* Kraatz (Coleoptera: Scarabaeidae: Cetoniinae). *Australian Entomological Magazine* **17**, 117–130.

Hinton HE (1941) The Ptinidae of economic importance. *Bulletin of Entomological Research* **31**, 331–381. doi:10.1017/S0007485300005101

Hinton HE (1945a) *A Monograph of the Beetles Associated with Stored Products*. Vol. 1. British Museum (Natural History), London.

Hinton HE (1945b) The Histeridae associated with stored products. *Bulletin of Entomological Research* **35**, 309–340.

Hinton HE (1946) The "gin-traps" of some beetle pupae; a protective device which appears to be unknown. *Transactions of the Royal Entomological Society of London* **97**, 473–496. doi:10.1111/j.1365-2311.1946.tb00273.x

Hinton HE (1947) On the reduction of functional spiracles in the aquatic larvae of the Holometabola, with notes on the moulting process of spiracles. *Transactions of the Royal Entomological Society of London* **98**, 449–473. doi:10.1111/j.1365-2311.1947.tb01222.x

Hinton HE (1951) On a little-known protective device of some chrysomelid pupae (Coleoptera). *Proceedings of the Royal Entomological Society of London A* **26**, 67–73. doi:10.1111/j.1365-3032.1951.tb00123.x

Hinton HE (1955) On the respiratory adaptations, biology, and taxonomy of the Psephenidae, with notes on some related families (Coleoptera). *Proceedings of the Zoological Society of London* **125**, 543–568. doi:10.1111/j.1096-3642.1955.tb00617.x

Hinton HE (1963) The ventral ecdysial lines of the head of endopterygote larvae. *Transactions of the Royal Entomological Society of London* **115**, 39–61. doi:10.1111/j.1365-2311.1963.tb00814.x

Hinton HE (1965) A revision of the Australian species of *Austrolimnius* (Coleoptera: Elmidae). *Australian Journal of Zoology* **13**, 97–172. doi:10.1071/ZO9650097

Hinton HE (1966) Respiratory adaptations of the pupae of beetles of the family Psephenidae. *Philosophical Transactions of the Royal Society B* **251**, 211–245. doi:10.1098/rstb.1966.0009

Hinton HE (1967a) On the spiracles of the larvae of the suborder Myxophaga (Coleoptera). *Australian Journal of Zoology* **15**, 955–959. doi:10.1071/ZO9670955

Hinton HE (1967b) Structure and ecdysial process of the larval spiracles of the Scarabaeoidea, with special reference to those of *Lepidoderma*. *Australian Journal of Zoology* **15**, 947–953. doi:10.1071/ZO9670947

Hinton HE (1968) The subgenera of Austrolimnius (Coleoptera: Elminthidae). *Proceedings of the Royal Entomological Society of London (B)* **37**, 98–102.

Hinton HE (1969) Plastron respiration in adult beetles of the suborder Myxophaga. *Journal of Zoology* **159**, 131–137. doi:10.1111/j.1469-7998.1969.tb03074.x

Hinton HE (1981) *Biology of Insect Eggs*, 3 vols. Pergamon Press, Oxford.

Hlavac TF (1971) Differentiation of the carabid antenna cleaner. *Psyche, Cambridge* **78**, 51–66.

Hlavac TF (1973) The prothorax of Coleoptera: origin, major features of variation. *Psyche, Cambridge* **79**, 123–149. doi:10.1155/1972/31579

Hlavac TF (1975) The prothorax of Coleoptera: (except Bostrichiformia-Cucujiformia). *Bulletin of the Museum of Comparative Zoology* **147**, 137–183.

Hockey MJ (1986) *Eucalyptus maculata* Hook (Myrtaceae), a new host recorded for *Calymmaderus incisus* Lea (Coleoptera: Anobiidae). *Australian Entomological Magazine* **12**, 114

Hodek I, Honek A (1996) *Ecology of Coccinellidae*. 464 pp. Kluwer Academic Publishers (Series Entomologica 54), Dordrecht.

Hoffmann A (1958) *Faune de France. 62. Coléoptères Curculionides (Troisieme Partie)*. pp. 1208–1838. Paul Lechevalier, Paris.

Holloway BA (1960) Taxonomy and phylogeny in the Lucanidae (Insecta: Coleoptera). *Records of the Dominion Museum, Wellington* **3**, 321–365.

Holloway BA (1968) The relationships of *Syndesus* MacLeay and *Sinodendron* Schneider (Coleoptera: Lucanidae). *New Zealand Journal of Science* **11**, 264–269.

Holloway BA (1969) Further studies on generic relationships in Lucanidae (Insecta: Coleoptera) with special reference to the ocular canthus. *New Zealand Journal of Science* **12**, 958–977.

Holloway BA (1972) The systematic position of the genus *Diphyllostoma* Fall (Coleoptera Scarabaeoidea). *New Zealand Journal of Science* **15**, 31–38.

Holloway BA (1982) *Fauna of New Zealand. Number 3. Anthribidae (Insecta: Coleoptera)*. Department of Scientific and Industrial Research, Wellington, New Zealand.

Holloway BA (1984) Morphology of pronotal carinae and associated surface structures in New Zealand Anthribidae (Insecta: Coleoptera). *New Zealand Journal of Zoology* **11**, 445–450. doi:10.1080/03014223.1984.10428259

Holloway BA (1985) Comparative morphology of setiferous sex patches in male Anthribidae (Insecta: Coleoptera). *New Zealand Journal of Zoology* **12**, 159–167. doi:10.1080/03014223.1985.10428275

Holloway BA (1997) Elytral surface structures as indicators of relationships in stag beetles, with special reference to the New Zealand species (Coleoptera: Lucanidae). *New Zealand Journal of Zoology* **24**, 47–64. doi:10.1080/03014223.1997.9518105

Holloway BA (1998) A re-evaluation of the genera of New Zealand aesaline stag beetles (Coleoptera: Lucanidae). *Journal of the Royal Society of New Zealand* **28**(4), 641–656. doi:10.1080/03014223.1998.9517578

Holloway BA (2007) *Fauna of New Zealand. Number 61. Lucanidae (Insecta: Coleoptera)*. Manaaki Whenua Press, Lincoln, Canterbury, New Zealand.

Hong Y (1998) A new Early Cretaceous beetle family – Magnocoleidae fam. nov. (Insecta: Coleoptera) in Hebei Province. *Geoscience* **12**(1), 39–48.

Honomichl K (1980) Die digitiformen Sensillen auf dem Maxillarpalpus von Coleoptera. I. Vergleichend-topographische Untersuchung des kutikulären Apparates. *Zoologischer Anzeiger, Jena* **204**(1/2), 1–12.

Hörnschemeyer T (1998) New species of *Electribius* Crowson 1973 (Coleoptera: Artematopodidae) from Baltic amber. *Paläontologische Zeitschrift* **72** (3/4), 299–306.

Hörnschemeyer T (2005) 5. Archostemata Kolbe, 1908. In *Handbuch der Zoologie/Handbook of Zoology. Band/Volume IV Arthropoda: Insecta Teilband/Part 38. Coleoptera, Beetles. Volume 1. Morphology and Systematics (Archostemata, Adephaga, Myxophaga, Polyphaga partim)*. (Eds RG Beutel and RAB Leschen) pp. 29–42. W. DeGruyter, Berlin.

Hörnschemeyer T (2009) The species-level phylogeny of archostematan beetles—where do *Micromalthus debilis* and *Crowsoniella relicta* belong? *Systematic Entomology* **34**, 533–558. doi:10.1111/j.1365-3113.2009.00476.x

Hörnschemeyer T, Goebbels J, Weidemann G, Faber C, Haase A (2006) The head morphology of *Ascioplaga mimeta* (Coleoptera: Archostemata) and the phylogeny of Archostemata. *European Journal of Entomology* **103**, 409–423.

Houston KJ (1982) Immature stages of *Monolepta australis* (Jacoby) and *Candezea palmerstoni* Blackburn. *Journal of the Australian Entomological Society* **21**, 123–130. doi:10.1111/j.1440-6055.1982.tb01778.x

Houston WWK (1986) Exocrine glands in the fore legs of dung beetles in the genus *Onitis* F. (Coleoptera: Scarabaeidae). *Journal of the Australian Entomological Society* **25**, 161–169. doi:10.1111/j.1440-6055.1986.tb01096.x

Houston WWK (ed.) (1992) *Zoological Catalogue of Australia. Volume 9. Coleoptera: Scarabaeoidea*. xii + 544 pp. Australian Government Publication Service, Canberra.

Houston TF (2011) Egg gigantism in some Australian earth-borer beetles (Coleoptera: Geotrupidae: Bolboceratinae) and its apparent association with reduction or elimination of larval feeding. *Australian Journal of Entomology* **50**, 164–173. doi:10.1111/j.1440-6055.2010.00794.x

Houston TF, Bougher NL (2010) Records of hypogeous mycorrhizal fungi in the diet of some Western Australian bolboceratine beetles (Coleoptera: Geotrupidae, Bolboceratinae). *Australian Journal of Entomology* **49**(1), 49–55. doi:10.1111/j.1440-6055.2009.00720.x

Houston TF, Zhang J, Hanich BP (2010) Diet of the flightless trogid beetle *Omorgus rotundulus* (Haaf) (Coleoptera: Trogidae) in the Little Sandy Desert of Western Australia. *Australian Entomologist* **36**(4), 207–212.

Hovore FT (2002) 28. Pleocomidae LeConte 1861. In *American Beetles. Volume 2. Polyphaga: Scarabaeoidea through Curculionoidea*. (Eds RH Arnett Jr, MC Thomas, PE Skelley and JH Frank) pp. 20–22. CRC Press, Boca Raton, Florida.

Howarth FG (1983) Ecology of cave arthropods. *Annual Review of Ecology and Systematics* **28**, 365–389.

Howarth FG (1988) Environmental ecology of North Queensland caves: or why there are so many troglobites in Australia. In *Papers for the 17th Australian Speleological Federation Tropicon Conference, Lake Tinaroo, Far North Queensland, 27–31 Dec. 1988*. (Ed. L Pearson). Australian Speleological Federation, Cairns.

Howden HF (1955) Biology and taxonomy of North American beetles of the subfamily Geotrupinae with revision of the genera *Bolbocerosoma, Eucanthus, Geotrupes* and *Peltotrupes* (Scarabaeidae). *Proceedings of the United States National Museum* **104**, 151–319. doi:10.5479/si.00963801.104-3342.151

Howden HF (1979) A revision of the Australian genus *Blackburnium* Boucomont (Coleoptera: Scarabaeidae: Geotrupinae). *Australian Journal of Zoology* **72**, (Supplement), 1–88.

Howden HF (1982) Larval and adult characters of *Frickius* Germain, its relationship to the Geotrupini, and a phylogeny of some major taxa in the Scarabaeoidea (Insecta: Coleoptera). *Canadian Journal of Zoology* **60**, 2713–2724. doi:10.1139/z82-347

Howden HF (1985) A revision of the Australian beetle genera *Bolboleaus* Howden & Cooper, *Backbolbus* Howden & Cooper, and *Bolborhachium* Boucomont (Scarabaeidae: Geotrupinae). *Australian Journal of Zoology* **111**, (Supplement), 1–179.

Howden AT (1986) Notes on the biology of adult and immature Amycterinae (Coleoptera, Curculionidae). *Proceedings of the Linnean Society of New South Wales* **109**, 91–105.

Howden AT (1995) Structures related to oviposition in Curculionoidea. *Memoirs of the Entomological Society of Washington* **14**, 53–100.

Howden HF (2001) A new tribe of Hybosorinae with a description of a new species of *Callosides* Howden (Coleoptera: Scarabaeidae). *Coleopterists Bulletin* **55**, 199–204. doi:10.1649/0010-065X(2001)055[0199:ANTOHW]2.0.CO;2

Howden HF, Cooper JB (1977) The generic classification of the Bolboceratini of the Australian region, with descriptions of four new genera (Scarabaeidae: Geotrupinae). *Australian Journal of Zoology* **50**, (Supplement), 1–50.

Howden HF, Gill BD (2000) Tribes of New World Ceratocanthinae, with keys to genera and descriptions of new species (Coleoptera: Scarabaeidae). *Sociobiology* **35**, (2B), 281–329.

Howden AT, Howden HF (1981) The larva and adult biology of *Rhinosimus viridiaeneus* (Coleoptera: Salpingidae). *Canadian Entomologist* **113**, 1055–1060. doi:10.4039/Ent1131055-12

Howden HF, Howden AT (1991) Nocturnal perching of scarabaeine dung beetles (Coleoptera, Scarabaeidae) in an Australian tropical rain forest. *Biotropica* **23**, 51–57. doi:10.2307/2388687

Howden HF, Lawrence JF (1974) The new world Aesalinae, with notes on the North American lucanid subfamilies (Coleoptera, Lucanidae). *Canadian Journal of Zoology* **52**, 1505–1510. doi:10.1139/z74-193

Howden HF, Peck SB (1987) Adult habits, larval morphology, and phylogenetic placement of *Taurocerastes patagonicus* Philippi (Scarabaeidae: Geotrupinae). *Canadian Journal of Zoology* **65**, 329–332. doi:10.1139/z87-051

Howden HF, Storey RI (1992) Phylogeny of the Rhyparini and the new tribe Stereomerini, with descriptions of new genera and species (Coleoptera; Scarabaeidae; Aphodiinae). *Canadian Journal of Zoology* **70**, 1810–1823. doi:10.1139/z92-248

Howden HF, Storey RI (2000) New Stereomerini and Rhyparini from Australia, Borneo and Fiji (Coleoptera: Scarabaeidae: Aphodiinae). *Memoirs of the Queensland Museum* **46**(1), 175–182.

Howden H, Howden A, Holloway G (2007) Digging down under: Australian Bolboceratini, their habits and a list of species (Coleoptera: Scarabaeoidea: Geotrupidae). *Zootaxa* **1499**, 47–59.

Howick CD, Carr H (1971) Status of the European house borer in Australia and an example of quarantine action to prevent its establishment. *International Pest Control* **13**(6), 8–11.

Hudson GV (1934) *New Zealand Beetles and their Larvae.* Ferguson and Osborn, Wellington.

Hudson HJ (1968) The ecology of fungi in plant remains above the soil. *New Phytologist* **67**, 837–874. doi:10.1111/j.1469-8137.1968.tb06399.x

Hudson I (1975) A systematic revision of the New Zealand Oedemeridae (Coleoptera, Insecta). *Journal of the Royal Society of New Zealand* **5**, 227–274. doi:10.1080/03036758.1975.10421850

Hughes J, Longhorn SJ, Papadopoulou A, Theodorides K, de Riva A, Mejia-Chang M, Foster PG, Vogler AP (2006) Dense taxonomic EST sampling and its applications for molecular systematics of the Coleoptera (beetles). *Molecular Biology and Evolution* **23**(2), 268–278. doi:10.1093/molbev/msj041

Hünefeld F, Marvaldi A, Müller B, Lawrence JF, Beutel RG (2011) The male postabdomen of the ancestral beetle *Tetraphalerus bruchi* Heller, 1913 (Archostemata: Ommatidae) and its phylogenetic significance. *Arthropod Structure & Development* **40**(2), 146–158. doi:10.1016/j.asd.2010.12.003

Hunt T, Bergsten J, Levkanicova Z, Papadopoulou A, St. John O, Wild R, Hammond PM, Ahrens D, Balke M, Caterino MS, Gómez-Zurita J, Ribera I, Barraclough TG, Bocakova M, Bocak L, Vogler AP (2007) A comprehensive phylogeny of beetles reveals the evolutionary origins of a super-radiation. *Science* **318**, 1913–1916. doi:10.1126/science.1146954

Hunter PE, Rosario RMT (1988) Associations of Mesostigmata with other arthropods. *Annual Review of Entomology* **33**, 393–417. doi:10.1146/annurev.en.33.010188.002141

Hutchinson GE (1959) Homage to Santa Rosalia or why are there so many kinds of animals? *American Naturalist* **93**, 145–159. doi:10.1086/282070

Hyslop JA, Böving AG (1935) Larva of *Tetrigus fleutiauxi* Van Zwaluwenburg. *Proceedings of the Hawaiian Entomological Society* **9**(1), 49–61.

Iablokoff-Khnzorian SM (1974) Remarques sur les genitalia femelles des Coléoptères et leur armure. *Annales de la Société Entomologique de France (N. S.)* **10**, 467–486.

Iablokoff-Khnzorian SM (1980) Le segment genital male des Coléoptères et son importance phylogenique (Coleoptera). *Deutsche Entomologische Zeitschrift (N. F.)* **27**, 251–295. doi:10.1002/mmnd.19800270407

Ide S, Costa C, Vanin SA (2005) 18.4. Lutrochidae Kasap & Crowson 1975. In *Handbuch der Zoologie/Handbook of Zoology. Band/Volume IV Arthropoda: Insecta Teilband/Part 38. Coleoptera, Beetles. Volume 1. Morphology and Systematics (Archostemata, Adephaga, Myxophaga, Polyphaga partim).* (Eds RG Beutel and RAB Leschen) pp. 508–512. W. DeGruyter, Berlin.

International Commission on Zoological Nomenclature(2000) Opinion 1957. *Sphaerius* Waltl, 1838 (Insecta, Coleoptera): conserved; and Sphaeriidae Erichson, 1845 (Coleoptera): spelling emended to Sphaeriusidae, so removing the homonomy with Sphaeriidae Deshayes, 1854 (1820) (Mollusca, Bivalvia). *Bulletin of Zoological Nomenclature* **57**(3), 182–184.

Israelson G (1971) On the Aderidae of the Canary Islands (Coleoptera). *Eos* **46**, 109–122.

Ivie MA (1985) *Phylogenetic Studies in the Bostrichiformia (Coleoptera).* Ohio State University (PhD thesis), Columbus, Ohio.

Ivie MA (2002a) 49. Ptilodactylidae Laporte 1836. In *American Beetles. Volume 2. Polyphaga: Scarabaeoidea through Curculionoidea.* (Eds RH Arnett Jr, MC Thomas, PE Skelley and JH Frank) pp. 135–138. CRC Press, Boca Raton, Florida.

Ivie MA (2002b) 50. Chelonariidae Blanchard 1845. In *American Beetles. Volume 2. Polyphaga: Scarabaeoidea through Curculionoidea.* (Eds RH Arnett Jr, MC Thomas, PE Skelley and JH Frank) pp. 139–141. CRC Press, Boca Raton, Florida.

Ivie MA (2002c) 67. Nosodendridae Erichson 1846. In *American Beetles. Volume 2. Polyphaga: Scarabaeoidea through Curculionoidea.* (Eds RH Arnett Jr, MC Thomas, PE Skelley and JH Frank) pp. 224–227. CRC Press, Boca Raton, Florida.

Ivie MA (2002d) 69. Bostrichidae Latreille 1802. In *American Beetles. Volume 2. Polyphaga: Scarabaeoidea through Curculionoidea.* (Eds RH Arnett Jr, MC Thomas, PE Skelley and JH Frank) pp. 233–244. CRC Press, Boca Raton, Florida.

Ivie MA (2002e) 103. Colydiidae Erichson 1845. In *American Beetles. Volume 2. Polyphaga: Scarabaeoidea through Curculionoidea.* (Eds RH Arnett Jr, MC Thomas, PE Skelley and JH Frank) pp. 445–453. CRC Press, Boca Raton, Florida.

Ivie MA (2002f) 104. Monommatidae Blanchard 1845. In *American Beetles. Volume 2. Polyphaga: Scarabaeoidea through Curculionoidea.* (Eds RH Arnett Jr, MC Thomas, PE Skelley and JH Frank) pp. 454–456. CRC Press, Boca Raton, Florida.

Ivie MA (2002g) 105. Zopheridae Solier 1834. In *American Beetles. Volume 2. Polyphaga: Scarabaeoidea through Curculionoidea.* (Eds RH Arnett Jr, MC Thomas, PE Skelley and JH Frank) pp. 457–462. CRC Press, Boca Raton, Florida.

Ivie MA (2002h) Key to Families of Beetles in America North of Mexico. In *American Beetles. Volume 2. Polyphaga: Scarabaeoidea through Curculionoidea.* (Eds RH Arnett Jr, MC Thomas, PE Skelley and JH Frank) pp. 816–835. CRC Press, Boca Raton, Florida.

Ivie MA (2005) 18.11. Eulichadidae Crowson, 1973. In *Handbuch der Zoologie/Handbook of Zoology. Band/Volume IV Arthropoda: Insecta Teilband/Part 38. Coleoptera, Beetles. Volume 1. Morphology and Systematics (Archostemata, Adephaga, Myxophaga, Polyphaga partim).* (Eds RG Beutel and RAB Leschen) pp. 547–551. W. DeGruyter, Berlin.

Ivie MA (2010) Additions and corrections to Borowski and Wegrzynowicz's world catalogue of Bostrichidae (Coleoptera). *Zootaxa* **2498**, 28–46.

Iwan D (2001) A comparative study of male genitalia in Opatrinae sensu Medvedev (1968) (Coleoptera: Tenebrionidae), with notes on the tribal classification. Part I. *Annales Zoologici* **51**(3), 351–390.

Iwan D (2004) A comparative study of male genitalia in Opatrinae sensu Medvedev (1968) (Coleoptera: Tenebrionidae), with notes on the reinterpreted tribal classification. Part II. *Annales Zoologici* **54**(4), 735–765.

Iwata R, Araya K, Johnki Y (1992) The community of arthropods with spherical postures, including *Madrasostes kazumai* (Coleoptera: Ceratocanthidae), found from the abandoned part of a nest of *Coptotermes formosanus* (Isoptera. Rhinotermitidae) in Tokara-Nakanoshima Island, Japan. *Sociobiology* **20**, 233–244.

Jäch MA (1995) Hygrobiidae (Coleoptera). In *Water Beetles of China. Vol. I.* (Eds MA Jäch and L Ji) pp. 109–110. Zoologische-Botanische Gesellschaft in Österreich and Wiener Coleopterologenverein.

Jäch MA (2001) *Ochthebius* s. str. Leach in Indonesia and Australia (Coleoptera: Hydraenidae). *Koleopterologische Rundschau* **71**, 101–104.

Jäch MA, Beutel RG, Delgado JA, Diaz JA (2005) 11.1. Hydraenidae Mulsant, 1844. In *Handbuch der Zoologie/Handbook of Zoology. Band/Volume IV Arthropoda: Insecta Teilband/Part 38. Coleoptera, Beetles. Volume 1. Morphology and Systematics (Archostemata, Adephaga, Myxophaga, Polyphaga partim).* (Eds RG Beutel and RAB Leschen) pp. 224–251. W. DeGruyter, Berlin.

Jackman JA, Lu W (2002) 101. Mordellidae Latreille 1802. In *American Beetles. Volume 2. Polyphaga: Scarabaeoidea through Curculionoidea.* (Eds RH Arnett Jr, MC Thomas, PE Skelley and JH Frank) pp. 423–430. CRC Press, Boca Raton, Florida.

Jackson DJ (1956) Observations on flying and flightless water beetles. *Journal of the Linnean Society. Zoology (Jena, Germany)* **43**, 18–42.

Jacoby M, Clavareau H (1905) *Genera Insectorum. Fascicule 33. Coleoptera Phytophaga Fam. Megalopidae.* 20 pp., 2 pls. P. Wytsman, Brussels.

Jałoszyński P (2012a) Adults of European ant-like stone beetles (Coleoptera: Staphylinidae: Scydmaeninae) *Scydmaenus tarsatus* Müller & Kunze and *Scydmaenus hellwigii* (Herbst) prey on soft-bodied arthropods. *Entomological Science* **15**, 35–41. doi:10.1111/j.1479-8298.2011.00479.x

Jałoszyński P (2012b) Description of *Euroleptochromus* gen. n. (Coleoptera, Staphylinidae, Scydmaeninae) from Baltic amber, with discussion of biogeography and mouthpart evolution within Clidicini. *Systematic Entomology* **37**, 346–359. doi:10.1111/j.1365-3113.2012.00623.x

Jałoszyński P (2012c) Observations on cannibalism and feeding on dead arthropods in *Scydmaenus tarsatus* (Coleoptera: Staphylinidae: Scydmaeninae). *Genus* **23**(1), 25–31.

Jałoszyński P, Beutel RG (2012) Functional morphology and evolution of specialized mouthparts of Cephenniini (Insecta, Coleoptera, Scydmaeninae, Staphylinidae). *Arthropod Structure & Development* **41**(6), 593–607. doi:10.1016/j.asd.2012.07.002

Jałoszyński P, Kilian A (2012) Larval morphology of *Scydmaenus tarsatus* and *S. hellwigii*, with notes on feeding behavioir and a review of the bibliography on the preimaginal stages of ant-like stone beetles (Coleoptera: Staphylinidae: Scydmaeninae). *European Journal of Entomology* **109**, 587–601.

Jameson ML (2002a) 26. Glaresidae Semenov-Tian-Shanskii and Medvedev 1932. In *American Beetles. Volume 2. Polyphaga: Scarabaeoidea through Curculionoidea.* (Eds RH Arnett Jr, MC Thomas, PE Skelley and JH Frank) pp. 15–16. CRC Press, Boca Raton, Florida.

Jameson ML (2002b) 27. Trogidae MacLeay 1819. In *American Beetles. Volume 2. Polyphaga: Scarabaeoidea through Curculionoidea.* (Eds RH Arnett Jr, MC Thomas, PE Skelley and JH Frank) pp. 17–19. CRC Press, Boca Raton, Florida.

Jameson ML (2002c) 29. Geotrupidae Latreille 1802. In *American Beetles. Volume 2. Polyphaga: Scarabaeoidea through Curculionoidea.* (Eds RH Arnett Jr, MC Thomas, PE Skelley and JH Frank) pp. 23–27. CRC Press, Boca Raton, Florida.

Jameson ML (2002d) 31. Hybosoridae Erichson 1847. In *American Beetles. Volume 2. Polyphaga: Scarabaeoidea through Curculionoidea.* (Eds RH Arnett Jr, MC Thomas, PE Skelley and JH Frank) pp. 32–33. CRC Press, Boca Raton, Florida.

Jameson ML (2002e). 32. Ceratocanthidae Martinez 1968. In *American Beetles. Volume 2. Polyphaga: Scarabaeoidea through Curculionoidea.* (Eds RH Arnett Jr, MC Thomas, PE Skelley and JH Frank) pp. 34–36. CRC Press, Boca Raton, Florida.

Jameson ML, Ratcliffe BC (2002a) Superfamily Scarabaeoidea Latreille 1802. Introduction. In *American Beetles. Volume 2. Polyphaga: Scarabaeoidea through Curculionoidea.* (Eds RH Arnett Jr, MC Thomas, PE Skelley and JH Frank) pp. 1–5. CRC Press, Boca Raton, Florida.

Jameson ML, Ratcliffe BC (2002b). 24. Diphyllostomatidae Holloway 1972. In *American Beetles. Volume 2. Polyphaga: Scarabaeoidea through Curculionoidea.* (Eds RH Arnett Jr, MC Thomas, PE Skelley and JH Frank) pp. 10–11. CRC Press, Boca Raton, Florida.

Janssen W (1963) Untersuchungen zur Morphologie, Biologie und Ökologie von *Cantharis* L. und *Rhagonycha* Eschsch. (Cantharidae, Col.). *Zeitschrift fur Wissenschartliche Zoologie* **169**, 115–202.

Janssens A (1949) Contribution a l'étude des Coléoptères lamellicornes. XIII. Table synoptique et essai de classification pratique des Coléoptères Scarabaeidae. *Bulletin de l'Institut Royal des Sciences Naturelles de Belgique* **25**(15), 1–30.

Jeannel R, Paulian R (1944) Morphologic abdominale des Coléoptères et systematique de l'ordre. *Revue Francaise d'Entomologie* **11**, 65–109.

Jelinek J (1999) Contribution to taxonomy of the beetle subfamily Nitidulinae (Coleoptera: Nitidulidae). *Folia Heyrovskyana* **7**(5), 251–281.

Jelinek J, Cline AR (2010) 10.25. Kateretidae Erichson in Agassiz, 1846. In *Handbuch der Zoologie/Handbook of Zoology. Band/Volume IV Arthropoda: Insecta Teilband/Part 38. Coleoptera, Beetles. Volume 2. Morphology and Systematics (Polyphaga partim).* (Eds RAB Leschen, RG Beutel and JF Lawrence) pp. 386–390. W. DeGruyter, Berlin.

Jelinek J, Carlton CE, Cline AR, Leschen RAB (2010) 10.26. Nitidulidae Latreille, 1802. In *Handbuch der Zoologie/Handbook of Zoology. Band/Volume IV Arthropoda: Insecta Teilband/Part 38. Coleoptera, Beetles. Volume 2. Morphology and Systematics (Polyphaga partim).* (Eds RAB Leschen, RG Beutel and JF Lawrence) pp. 390–407. W. DeGruyter, Berlin.

Jell PA (2004) The fossil insects of Australia. *Memoirs of the Queensland Museum* **50**(1), 1–124.

Jendek E (2001) A comparative study of the abdomen in the family Buprestidae (Coleoptera). *Acta Musei Moraviae. Scientiae Biologicae, Brno* **86**, 1–41.

Jeng ML, Yang PS, Satô M (1998) The genus *Cyphonocerus* (Coleoptera, Lampyridae) from Taiwan and Japan, with notes on the subfamily Cyphonocerinae. *Elytra, Tokyo* **26**(2), 379–398.

Jerath ML (1960) Notes on larvae of nine genera of Aphodiinae in the United States (Coleoptera: Scarabaeidae). *Proceedings of the United States National Museum* **111**, 43–94. doi:10.5479/si.00963801.111-3425.43

Jin Z, Ślipiński A, Pang H (2013) A revision of the genus *Notodascillus* Carter (Coleoptera: Dascillidae). *Zootaxa* **3613**(3), 245–256.

John H (1959) Genera Insectorum de P. Wytsman. Fascicule 213. *Coleoptera Clavicornia Fam. Discolomidae (= Notiophygidae).* Genera Insectorum, Crainhem, Belgium.

John H (1960) Eine Übersicht über die Familie Propalticidae (Col.). *Pacific Insects* **2**, 149–170.

John A, Andreae H (1967) Revision der Gattung *Thorictus* Germar (Coleoptera, Thorictidae). III. Teil. *Entomologische Abhandlungen und Berichte aus dem Staatlichen Museum für Tierkunde in Dresden* **34**, 145–158.

Johnson C (1975) *Cortinicara,* a new genus of Corticariinae (Coleoptera: Lathridiidae). *Entomologica Scandinavica* **6**, 283–285. doi:10.1163/187631275X00127

Johnson C (1982) An introduction to the Ptiliidae (Coleoptera) of New Zealand. *New Zealand Journal of Zoology* **9**, 333–376. doi:10.1080/03014223.1982.10423865

Johnson C (1985) Revision of Ptiliidae (Coleoptera) occurring in the Mascarenes, Seychelles and neighboring islands. *Entomologica Basiliensia* **10**, 159–237.

Johnson PJ (1987) *Larval Taxonomy, Biology, and Biogeography of the Genera of North American Byrrhidae (Insecta: Coleoptera).* University of Idaho (MsC Thesis), Moscow, Idaho.

Johnson PJ (2002a) 42. Byrrhidae Latreille 1804. In *American Beetles. Volume 2. Polyphaga: Scarabaeoidea through Curculionoidea.* (Eds RH Arnett Jr, MC Thomas, PE Skelley and JH Frank) pp. 113–116. CRC Press, Boca Raton, Florida.

Johnson PJ (2002b) 55. Cerophytidae Latreille 1834. In *American Beetles. Volume 2. Polyphaga: Scarabaeoidea through Curculionoidea.* (Eds RH Arnett Jr, MC Thomas, PE Skelley and JH Frank) pp. 150–151. CRC Press, Boca Raton, Florida.

Johnson PJ (2002c) 57. Throscidae Laporte 1840. In *American Beetles. Volume 2. Polyphaga: Scarabaeoidea through Curculionoidea.* (Eds RH Arnett Jr,

MC Thomas, PE Skelley and JH Frank) pp. 158–159. CRC Press, Boca Raton, Florida.

Johnson PJ (2002d) 58. Elateridae Leach 1815. In *American Beetles. Volume 2. Polyphaga: Scarabaeoidea through Curculionoidea.* (Eds RH Arnett Jr, MC Thomas, PE Skelley and JH Frank) pp. 160–173. CRC Press, Boca Raton, Florida.

Johnson PJ (2005) 18.1. Byrrhidae Latreille, 1804. In *Handbuch der Zoologie/ Handbook of Zoology. Band/Volume IV Arthropoda: Insecta Teilband/ Part 38. Coleoptera, Beetles. Volume 1. Morphology and Systematics (Archostemata, Adephaga, Myxophaga, Polyphaga partim).* (Eds RG Beutel and RAB Leschen) pp. 469–471. W. DeGruyter, Berlin.

Jolivet P. (1957): Recherches sur l'aile des Chrysomeloidea, première partie. *Mémoires de l'Institut Royale des Sciences Naturelles de Belgique* (2) **51**, 1–180, pls 1–20.

Jolivet P (1959) Recherches sur l'aile des Chrysomeloidea, deuxième partie. *Mémoires de l'Institut Royale des Sciences Naturelles de Belgique* (2) **58**, 1–152, pls 21–40.

Jolivet P (1968) De la position systematique du genre *Microdonacia* Blackburn (Col.Chrysomelidae). *Entomologische Arbeiten aus dem Museum Georg Frey* **19**, 188–192.

Jolivet P, Verma KK (2011) Reflexions on cycloalexy among Chrysomelidae (Coleoptera). *Nouvelle Revue d'Entomologie (N. S.)* **27**(4), 311–329.

Jordal BH, Normark BB, Farrell BD (2000) Evolutionary radiation of an inbreeding haplodilpoid beetle lineage (Curculionidae, Scolytinae). *Biological Journal of the Linnean Society. Linnean Society of London* **71**, 483–499. doi:10.1111/j.1095-8312.2000.tb01270.x

Jordal BH, Beaver RA, Normark BB, Farrell BD (2002) Extraordinary sex ratios and the evolution of male neoteny in sib-mating *Ozopemon* beetles. *Biological Journal of the Linnean Society. Linnean Society of London* **75**(3), 353–360. doi:10.1111/j.1095-8312.2002.tb02076.x

Juan C and Petit Pierre E (1991) Chromosome mechanisms and sex-determining systems in Tenebrionidae (Coleoptera). In *Advances in Coleopterology.* (Eds M Zunino, X Bellés and M Blas) pp. 167–176. Associación Europea de Coleopterologia, Barcelona.

Julien MH (Ed) (1987) *Biological Control of Weeds: A World Catalogue of Agents and Their Target Weeds*, 2nd Edition. C. A. B. International, Wallingford, UK.

Käärik A (1975) Succession of microorganisms during wood decay. In *Biological Transformation of Wood by Microorganisms.* (Ed. W. Liese) pp. 39–51. Springer, Berlin.

Kaiser A, Klok CJ, Socha JC, Lee W-K, Quinian MC, Harrison JF (2007) Increase in tracheal investment with beetle size supports hypothesis of oxygen limitation on insect gigantism. *Proceedings of the National Academy of Sciences of the United States of America* **104**(32), 13198–13203. doi:10.1073/pnas.0611544104

Kamiya H (1960) A new tribe of Coccinellidae (Coleoptera). *Kontyû* **28** (1), 22–26, pl. 3.

Kamiya H (1965) Comparative morphology of larvae of the Japanese Coccinellidae, with special reference to the tribal phylogeny of the family (Coleoptera). *The Memoirs of the Faculty of Education. Fukui University Series II, Natural Science* **14**, 83–100.

Kanar P (2003) Revision of the genus *Trypeticus* Marseul (Coleoptera: Histeridae). *Zoologische Verhandelingen* **342**, 1–318.

Kapur AP (1950) The biology and external morphology of the larvae of Epilachninae (Coleoptera: Coccinellidae). *Bulletin of Entomological Research* **41**, 161–208. [pl. VI.]doi:10.1017/S0007485300027565

Kasahara S, Miyano S (1997) The larva of *Drypta fulveola* Bates (Coleoptera, Carabidae). *Elytra. Toyko* **25**(1), 147–150.

Kasap H, Crowson RA (1975) A comparative anatomical study of Elateriformia and Dascilloidea (Coleoptera). *Transactions of the Royal Entomological Society of London* **126**, 441–495. doi:10.1111/j.1365-2311.1975.tb00858.x

Kasap H, Crowson RA (1976) On systematic relations of *Oomorphus concolor* (Sturm) (Col., Chrysomeidae), with descriptions of its larva and of an aberrant cryptocephaline larva from Australia. *Journal of Natural History* **10**, 99–112. doi:10.1080/00222937600770071

Kasap H, Crowson RA (1977) A comparative study of the internal anatomy and abdominal structures of Curculionoidea. *Hacettepe Bulletin of Natural Sciences and Engineering, Ankara* **6**, 35–86.

Kasule FK (1966) The subfamilies of the larvae of Staphylinidae (Coleoptera) with keys to the larvae of the British genera of Steninae and Proteininae. *Transactions of the Royal Entomological Society of London* **118**, 261–283. doi:10.1111/j.1365-2311.1966.tb00838.x

Kaszab Z (1969) The system of the Meloidae (Coleoptera). *Memorie della Societa Entomologica Italiana* **48**, 241–248.

Kaszab Z (1978) Australische und südpazifische Tenebrioniden der Tribus Phrenapatini und Gnathidiini (Coleoptera) sowie synonymische Bemerkungen. *Annales Historico-Naturales Musei Nationalis Hungarici* **70**, 163–177.

Kaszab Z (1981) Die Gattungen und Arten der Tribus Archeocrypticini (Coleoptera: Tenebrionidae). *Folia Entomologica Hungarica* **42**, 95–115.

Kaszab Z (1982a) Die papuanisch-pazifischen Arten vder Gattung *Archeoglenes* Brown, 1893 (Coleoptera: Tenebrionidae). *Acta Zoologica Academiae Scientiarum Hungaricae* **28**, 45–55.

Kaszab Z (1982b) Revision der Australischen *Uloma*-Arten (Coleoptera: Tenebrionidae). *Acta Zoologica Academiae Scientiarum Hungaricae* **28**, 233–291.

Kaszab Z (1984) Revision der australischen Archeocrypticinen (Coleoptera, Tenebrionidae). *Annales Historico-Naturales Musei Nationalis Hungarici.* **76**, 143–163.

Kaszab Z (1987) Die Papuanisch-Australischen Arten der Gattung *Derosphaerus* Thompson, 1858 (Coleoptera: Tenebrionidae). *Acta Zoologica Hungarica* **33**, 41–85.

Kaszab Z (1988) Katalog und Bestimmungstabelle der Gattung *Promethis* Pascoe, 1869 (Coleoptera, Tenebrionidae). *Acta Zoologica Academiae Scientiarum Hungaricae* **34**, 67–170.

Katovich K (2002a) 39. Rhipiceridae Latreille 1834. In *American Beetles. Volume 2. Polyphaga: Scarabaeoidea through Curculionoidea.* (Eds RH Arnett Jr, MC Thomas, PE Skelley and JH Frank) pp. 92–94. CRC Press, Boca Raton, Florida.

Katovich K (2002b) 47. Heteroceridae MacLeay 1825. In *American Beetles. Volume 2. Polyphaga: Scarabaeoidea through Curculionoidea.* (Eds RH Arnett Jr, MC Thomas, PE Skelley and JH Frank) pp. 127–132. CRC Press, Boca Raton, Florida.

Kaufmann T, Stansly P (1979) Bionomics of *Neoheterocerus pallidus* Say (Coleoptera: Heteroceridae) in Oklahoma. *Journal of the Kansas Entomological Society* **52**(3), 565–577.

Kavanaugh DH (1986) A systematic review of amphizoid beetles (Amphizoidae: Coleoptera) and their phylogenetic relationships to other Adephaga. *Proceedings of the California Academy of Sciences* **44**, 67–109.

Kawanabe M (1994a) A new genus and species of the family Ciidae (Coleoptera) from Southwest Japan. *Japanese Journal of Entomology* **62**(1), 186–192.

Kawanabe M (1994b) Notes on the genus *Neoennearthron* (Coleoptera, Ciidae), with description of a new species from Luzon Island, the Philippines. *Japanese Journal of Entomology* **62**(4), 641–650.

Kawanabe M (2002) A revision of the genus *Octotemnus* (Coleoptera, Ciidae) from Japan. *Japanese Journal of Systematic Entomology* **8**(2), 243–267.

Kawashima I, Lawrence JF, Branham MA (2010) 4.14. Rhagophthalmidae Olivier, 1907. In *Handbuch der Zoologie/Handbook of Zoology. Band/ Volume IV Arthropoda: Insecta Teilband/Part 38. Coleoptera, Beetles. Volume 2. Morphology and Systematics (Polyphaga partim).* (Eds RAB Leschen, RG Beutel and JF Lawrence) pp. 135–140. W. DeGruyter, Berlin.

Kazantsev SV (2004a) Phylogeny of the tribe Erotini (Coleoptera, Lycidae), with descriptions of new taxa. *Zootaxa* **496**, 1–48.

Kazantsev SV (2004b) Three new lycid genera from the Old and New Worlds (Coleoptera: Lycidae). *Elytron* **17–18**, 3–10.

Kazantsev SV (2005) Morphology of Lycidae with some considerations on evolution of Coleoptera. *Elytron* **17–18**, 73–248.

Kazantsev SV (2006a) Comparative morphology of mandibular structures in lycid larvae and its phylogenetic implications (Polyphaga, Hexapoda). *Acta Zoologica* **87**, 229–238. doi:10.1111/j.1463-6395.2006.00236.x

Kazantsev SV (2006b) A new species of *Chinotaphes* from Laos and a new replacement name for Proteros Kazantsev (Coleoptera: Lycidae: Erotinae). *Zoosystematica Rossica* **14**(2), 215–218.

Kejval Z (2003) The genus *Anthephila* (Coleoptera: Anthicidae). *European Journal of Entomology* **100**, 381–392.

Kejval Z (2009) Taxonomic revision of the genus *Chileanthicus* Werner (Coleoptera: Anthicidae). *Zootaxa* **2180**, 1–82.

Kemner NA (1918) *Vergleichende Studien über das Pygopodium einiger Koleopteren-Larven*. Wretsman, Uppsala.

Kemner NA (1924) Über die Lbensweise and Entwicklung des angeblich Myrmecophilen oder Termitophilen Genus *Trochoideus* (Col. Endomych.), nach Beobachtungen über *Trochoideus termitophilus* Roepke auf Java. *Tijdschrift voor Entomologie* **67**, 180–194.

Kent DS (2010) The external morphology of *Austroplatypus incompertus* (Schedl) (Coleoptera, Curculionidae, Platypodinae). *ZooKeys* **56**, 121–140. doi:10.3897/zookeys.56.521

Kent DS, Simpson JA (1992) Eusociality in the beetle *Austroplatypus incompertus* (Coleoptera: Curculionidae). *Naturwissenschaften* **79**, 86–87. doi:10.1007/BF01131810

Kifune T (1961) Biological notes on two Japanese species of the genus *Zonitis* Fabricius (Coleoptera, Meloidae) (Studies on Meloidae–I). *Mushi* **35**, 51–61.

Kilian A (2007) Comparative morphology of larval Camiarinae (Coleoptera: Leiodidae). Part I. Genus *Paragyrtodes* Szymczakowski 1966. *Zootaxa* **1640**, 1–39.

Kilian A (2012) Phylogenetic analysis of Camiarinae (Coleoptera: Leiodidae) based on larval morphology. *Annales Zoologici* **62**(1), 11–68. doi:10.3161/000345412X633667

Kim JI, Lumaret JP (1988) Les larves du genre *Onthophagus* Latreille, 1802. I. caractères communs et variations (Coleoptera, Scarabaeidae). *Elytron* **2**, 51–61.

Kimoto S, Gressitt JL (1982) Chrysomelidae (Coleoptera) of Thailand, Cambodia, Laos and Vietnam. III. Eumolpinae. *Esakia* **18**, 1–141.

King JG, Starr JR, Lago PK (2011) Molecular data resolves relationships within Heteroceridae (Dryopoidea: Coleoptera). *Systematic Entomology* **36**, 435–445. doi:10.1111/j.1365-3113.2011.00571.x

Kingsbury PD, Beveridge JG (1977) A simple bubbler for sorting bottom fauna samples by elutriation. *Canadian Entomologist* **109**, 1265–1268. doi:10.4039/Ent1091265-9

Kingsolver JM (2002a) 68. Dermestidae Gyllenhal 1808. In *American Beetles. Volume 2. Polyphaga: Scarabaeoidea through Curculionoidea.* (Eds RH Arnett Jr, MC Thomas, PE Skelley and JH Frank) pp. 228–232. CRC Press, Boca Raton, Florida.

Kingsolver JM (2002b) 121 Bruchidae Latreille 1802. In *American Beetles. Volume 2. Polyphaga: Scarabaeoidea through Curculionoidea.* (Eds RH Arnett Jr, MC Thomas, PE Skelley and JH Frank) pp. 602–608. CRC Press, Boca Raton, Florida.

Kirejtshuk AG (1986a) Analysis of structure of genitalia for reconstruction of phylogeny and substantiation of the system of the family of sap beetles (Nitidulidae, Coleoptera). *Trudy Vsesoyuznogo Entomologicheskogo Obtchestva* **68**, 22–28. [in Russian]

Kirejtshuk AG (1986b) New genera and species of nitidulid beetles (Coleoptera, Nitidulidae) from the Australian region. I. *Entomologicheskoye Obozreniye* **65**, 559–573 (in Russian; translation in *Entomological Review,* Washington **66**(1), 147–160, 1987).

Kirejtshuk AG (1986c) On polyphyly of the Carpophilinae with description of a new subfamily, Cillaeinae (Coleoptera: Nitidulidae). *Coleopterists Bulletin* **40**, 217–221.

Kirejtshuk AG (1986d) Revision of the genus *Aethina* Er. (Coleoptera, Nitidulidae) of the fauna of the Oriental and Palaearctic regions. *Trudy Zoologicheskogo Instituta Akademia Nauk S. S. S. R.* **140**, 44–82. [in Russian]

Kirejtshuk AG (1987) New genera and species of nitidulid beetles (Coleoptera, Nitidulidae) from the Australian region. II. *Entomologicheskoye Obozreniye* **66**, 773–799 (in Russian; translation in *Entomological Review*, Washington **67**(3), 129–156, 1988).

Kirejtshuk AG (1990) New genera and species of nitidulid beetles (Coleoptera, Nitidulidae) from the Australian region. III. *Entomologicheskoye Obozreniye* **69**, 857–878 (in Russian; translation in *Entomological Review*, Washington **70**(7), 52–73, 1992).

Kirejtshuk AG (1994) Parandrexidae fam. nov., a Jurassic fossil family of beetles from the infraorder Cucujiformia (Coleoptera, Polyphaga). *Paleontologicheskii Zhurnal* **1994** (1), 57–64, pls 7–8 (in Russian; translation in *Paleontological Journal* **28**, 69–78).

Kirejtshuk AG (1996) System, evolution of mode of life and phylogeny of the order Coleoptera. II. Entomologicheskoye Obozreniye **75** (1), 39–62 (in Russian; translation in *Entomological Review, Washington* **76**(1), 1–20).

Kirejtshuk AG (1998a) *Nitidulidae (Coleoptera) of the Himalayas and northern India. Part 1: Subfamily Epuraeinae.* Koeltz Scientific Books (Theses Zoologicae. Volume 28), Koenigstein, Germany.

Kirejtshuk AG (1998b) The position of the subfamily Maynipeplinae subfam. n. (Coleoptera, Nitidulidae) in the classification and notes on the evolution and structural modifications among sap beetles. *Entomologicheskoye Obozreniye* **77**, 540–554 (in Russian, translation in *Entomological Review* **78**, 793–807).

Kirejtshuk AG (1999) *Sikhotealinia zhiltzovae* (Lafer, 1996) – recent representative of the Jurassic coleopterous fauna (Coleoptera, Archostemata, Jurodidae). *Proceedings of the Zoological Institute of the Russian Academy of Sciences* **281**, 21–26.

Kirejtshuk AG (2000) On origin and early evolution of the superfamily Cucujoidea (Coleoptera, Polyphaga). Comments on the family Helotidae. *The Kharkov Entomological Society Gazette* **8**, 8–38.

Kirejtshuk AG (2004) Four new genera of the *Soronia* complex (Coleoptera: Nitidulidae) from Australia, New Zealand, Fiji and tropical America with notes on composition of the complex and description of new species from Southern Hemisphere. *Russian Entomological Journal* **12**(3), 239–256.

Kirejtshuk AG (2006a) New species of the genus *Physoronia* from the Far East and *Kryzhanovskiella* gen. n. from Australia, with taxonomic notes on the *Pocadius* complex (Coleoptera: Nitidulidae). *Annales Historico-Naturales Musei Nationalis Hungarici* **98**, 121–132.

Kirejtshuk AG (2006b) Two new genera and three new species of the tribes Nitidulini and Cyllodini (Coleoptera: Nitidulidae) from Australia and New Zealand, with taxonomic notes. *Trudy Russkogo Entomologicheskogo Obshchestva. S. Peterburg* **77**, 166–177.

Kirejtshuk AG (2008) A current generic classification of sap beetles (Coleoptera, Nitidulidae). *Zoosystematica Rossica* **17**(1), 107–122.

Kirejtshuk AG (2009) A new genus and species of Sphaeriusidae (Coleoptera, Myxophaga) from Lower Cretaceous Burmese amber. *Denisia 26, zugleich Kataloge der oberösterreichischen Landesmuseen. Neue Serie* **86**, 99–102.

Kirejtshuk AG (2011a) A new genus of the family Nitidulidae (Coleoptera: Polyphaga) from Australia. *Zoosystematica Rossica* **19**(2), 54–61.

Kirejtshuk AG (2011b) "Paradoxal" new genus and species of the family Nitidulidae (Polyphaga: Coleoptera) from Afro-Madagascarean and Australian regions. *Zoosystematica Rossica* **20**(2), 274–298.

Kirejtshuk AG, Azar D (2008) New taxa of beetles (Insecta, Coleoptera) from Lebanese amber with evolutionary and systematic comments. *Alavesia* **2**, 15–46.

Kirejtshuk AG, Kvamme T (2001) Notes on taxonomy opf the subfamily Epuraeinae (Coleoptera: Nitidulidae). *Zoosystematica Rossica* **9**(2), 409–436.

Kirejtshuk AG, Kvamme T (2002) Revision of the subgenus *Lasiodites* Jelinek, 1999, stat. nov. of the genus *Phenolia* Erichson, 1843 from Africa and

Madagascar (Coleoptera, Nitidulidae). *Mitteilungen aus dem Museum für Naturkunde in Berlin. Zoologische Reihe* **78**(1), 3–70.

Kirejtshuk AG, Lawrence JF (1990) Revision of the Australian genus *Idaethina* Reitter (Coleoptera: Nitidulidae). *Journal of the Australian Entomological Society* **29**, 1–9. doi:10.1111/j.1440-6055.1990.tb00305.x

Kirejtshuk AG, Lawrence JF (1992a) Cychramptodini, a new tribe of Nitidulidae (Coleoptera) from Australia. *Journal of the Australian Entomological Society* **31**, 29–46. doi:10.1111/j.1440-6055.1992.tb00456.x

Kirejtshuk AG, Lawrence JF (1992b) Review of the *Thalycrodes* complex of genera (Coleoptera: Nitidulidae) endemic to the Australian region. *Journal of the Australian Entomological Society* **31**, 119–142. doi:10.1111/j.1440-6055.1992.tb00469.x

Kirejtshuk AG, Lawrence JF (1999) Notes on the *Aethina* complex (Coleoptera: Nitidulidae: Nitidulinae), with a review of *Aethina* (*Cleidorura*) subgen. nov. and *Aethina* (*Idaethina*) Gemminger et Harold. *Annales Zoologici* **49** (3), 233–254.

Kirejtshuk AG, Leschen RAB (1998) Review of the *Thalycra* complex (Coleoptera: Nitidulinae) with three new genera and notes on mycophagy. *Annales Zoologici* **48**(3/4), 253–273.

Kirejtshuk AG, Nel A (2008) New beetles of the suborder Polyphaga from the Lowermost Eocene French amber (Insecta: Coleoptera). *Annales de la Société Entomologique de France (N. S.)* **44**(4), 419–442.

Kirejtshuk AG, Nel A (2009) New genera and species of Cucujiformia (Coleoptera, Polyphaga) from lowermost Eocene French amber. *Denisia 26, zugleich Kataloge der oberösterreichischen Landesmuseen. Neue Serie* **86**, 103–118.

Kirejtshuk AG, Poinar G (2006) Haplochelidae, a new family of Cretaceous beetles (Coleoptera: Myxophaga) from Burmese amber. *Proceedings of the Entomological Society of Washington* **108**, 155–164.

Kirejtshuk AG, Ponomarenko AG (2010) A new coleopterous family Mesocinetidae fam. nov. (Coleoptera: Scirtoidea) from Late Mesozoic and notes on fossil remains from Shar-Teg (Upper Jurassic, South-Western Mongolia). *Zoosystematica Rossica* **19**(2), 301–325.

Kirejtshuk AG, James DG, Heffer R (1997) Description and biology of a new species of *Cybocephalus* Erichson (Coleoptera: Nitidulidae), a predator of Australian citrus whitefly. *Australian Journal of Entomology* **36**, 81–86. doi:10.1111/j.1440-6055.1997.tb01436.x

Kirejtshuk AG, Azar D, Tafforeau P, Boistel R, Fernandez V (2009) New beetles of Polyphaga (Coleoptera, Polyphaga) from Lower Cretaceous Lebanese amber. *Denisia 26, zugleich Kataloge der oberösterreichischen Landesmuseen. Neue Serie* **86**, 119–130.

Kirejtshuk AG, Chang H, Ren D, Kun SC (2010a) Family Lasiosynidae n. fam., new palaeoendemic Mesozoic family from the infraorder Elateriformia (Coleoptera: Polyphaga). *Annales de la Société Entomologique de France (N. S.)* **46**(1–2), 67–87.

Kirejtshuk AG, Hava J, Nel A (2010b) New genus and species of subfamily Trinodinae (Coleoptera, Polyphaga, Dermestidae) from Lowermost Eocene French amber. *Zoosystematica Rossica* **19**(1), 54–69.

Kirejtshuk AG, Ponomarenko AG, Prokin A, Chang H, Nikolajev GV, Ren D (2010c) Current knowledge of Mesozoic Coleoptera from Daohugou and Liaoning (Northeast China). *Acta Geologica Sinica* **84**(4), 783–792. [English Edition]doi:10.1111/j.1755-6724.2010.00253.x

Kirejtshuk AG, Azar D, Montreuil O (2011) First Mesozoic representative of the subfamily Liparochrinae (Coleoptera: Hybosoridae) from the Lower Cretaceous Lebanese amber. *Zosystematica Rossica* **20**(1), 62–70.

Kiselyova T (2008) Description of the larval stage of *Myrmeanthrenus frontalis* Armstrong and *Anthrenocerus stigmacrophilus* Armstrong (Coleoptera: Dermestidae), with a discussion of their phylogenetic relationships. *Coleopterists Bulletin* **62**(2), 333–343. doi:10.1649/657.1

Kiselyova T, McHugh JV (2006) A phylogenetic study of Dermestidae (Coleoptera) based on larval morphology. *Systematic Entomology* **31**, 469–507. doi:10.1111/j.1365-3113.2006.00335.x

Kissinger DG (1968) *Curculionidae Subfamily Apioninae of North and Central America. with Reviews of the World Genera of Apioninae and World Subgenera of Apion Herbst (Coleoptera). Taxonomic Publications,* South Lancaster, Massachusetts.

Kistner DH (1979) Social and evolutionary significance of social insect symbionts. In *Social Insects*. Vol. 1. (Ed. HR Hermann) pp. 339–413. Academic Press, New York.

Kistner DH (1982) The social insects' bestiary. In *Social Insects*. Vol. III. (Ed. HR Hermann) pp. 1–244. Academic Press, New York.

Kistner DH, Watson JAL (1972) A new tribe of termitophilous Aleocharinae (Coleoptera: Staphylinidae) from Australia. *Australian Journal of Zoology* **17**, (Supplement), 1–24.

Kitching RL, Allsopp PG (1987) *Prionocyphon niger* sp. n. (Coleoptera: Scirtidae) from water-filled tree holes in Australia. *Journal of the Australian Entomological Society* **26**, 73–79. doi:10.1111/j.1440-6055.1987.tb00264.x

Klausnitzer B (1983) *Beetles* (translated by S. Furness, revised by S. L. Shute). Exeter, New York.

Klausnitzer B (2006) Zur Kenntnis der larven einiger Scirtidae aus Neuseeland und Borneo mit wahrscheinlich terrestrischer lebensweise (Coleoptera). *Entomologische Nachrichten und Berichte* **50**, 141–152.

Klimaszewski J, Watt JC (1997) *Fauna of New Zealand. Number 37. Coleoptera: family group review and keys to identification.* Manaaki Whenua Press, Lincoln, New Zealand.

Klimaszewski J, Newton AF Jr, Thayer MK (1996) A review of the New Zealand rove beetles (Coleoptera: Staphylinidae). *New Zealand Journal of Zoology* **23**, 143–160. doi:10.1080/03014223.1996.9518074

Koch A (1967) Insects and their endosymbionts. In *Symbiosis. Volume II.* (Ed. M. Henry) pp. 1–106. Academic Press, New York.

Kodada J, Jäch MA (2005a) 18.2. Elmidae Curtis, 1830. In *Handbuch der Zoologie/Handbook of Zoology. Band/Volume IV Arthropoda: Insecta Teilband/Part 38. Coleoptera, Beetles. Volume 1. Morphology and Systematics (Archostemata, Adephaga, Myxophaga, Polyphaga partim)*. (Eds RG Beutel and RAB Leschen) pp. 471–496. W. DeGruyter, Berlin.

Kodada J, Jäch MA (2005b) 18.3. Dryopidae Billberg, 1820 (1817). In *Handbuch der Zoologie/Handbook of Zoology. Band/Volume IV Arthropoda: Insecta Teilband/Part 38. Coleoptera, Beetles. Volume 1. Morphology and Systematics (Archostemata, Adephaga, Myxophaga, Polyphaga partim).* (Eds RG Beutel and RAB Leschen) pp. 496–508. W. DeGruyter, Berlin.

Koegen M, Goeden RD (1970) The biology of *Lema trilineata daturaphila* (Coleoptera: Chrysomelidae), with notes on the efficiency of food utilization by larvae. *Annals of the Entomological Society of America* **63**(2), 537–546.

Koerner L, Gorb SN, Betz O (2012) Functional morphology and adhesive performance of the stick-capture apparatus of the rove beetles *Stenus* spp. (Coleoptera, Staphylinidae). *Zoology (Jena, Germany)* **115**, 117–127.

Kohlmann B (2006) History of scarabaeoid classification. *Coleopterists Society Monographs* **60**, 19–34. doi:10.1649/0010-065X(2006)60[19:HOSC]2.0.CO;2

Kolbe H (1908) Mein System der Coleopteren. *Zeitschrift für Wissenschaftlichen Insektenbiologie* **4**, 116–1, 153–162, 219–226, 246–251, 286–294, 389–400.

Kolibáč J (1987) Morphological comparison of type (or model) genera of the subfamilies of Cleridae. *Mitteilungen der Münchener Entomologischen Gesellschaft* **77**, 103–135.

Kolibáč J (1989a) Further observations on morphology of some Cleridae (Coleoptera) (I). *Acta Scientiarum Naturalium Academiae Scientiarum Bohemoslovacae, Brno (N. S.)* **23**(1), 1–50.

Kolibáč J (1989b) Further observations on morphology of some Cleridae (Coleoptera) (II). *Acta Scientiarum Naturalium Academiae Scientiarum Bohemoslovacae, Brno (N. S.)* **23**(2), 1–42.

Kolibáč J (1992) Revision of Thanerocleridae n. stat. (Coleoptera, Cleroidea). *Mitteilungen der Schweizerische Entomologische Gesellschaft* **65**, 303–340.

Kolibáč J (1997) Classification of the subfamilies of Cleridae (Coleoptera: Cleroidea). *Acta Musei Moraviae, Scientiae Biologicae, Brno* **81**, 307–361.

Kolibáč J (1998a) Classification of the subfamily Hydnocerinae Spinola, 1844 (Coleoptera: Cleridae). *Acta Musei Moraviae. Scientiae Biologicae, Brno* **83**, 127–210.

Kolibáč J (1998b) New Australian Thanerocleridae, with notes on the biogeography of the subtribe Isoclerina Kolibac (Coleoptera: Cleroidea). *Invertebrate Taxonomy* **12**, 951–975. doi:10.1071/IT96019

Kolibáč J (1998c) Notes on the classification of *Natalis* Laporte de Castelnau and Notocymatodera Schenkling, with a description of *Notocymatodera disjuncta* sp. n. (Coleoptera: Cleridae). *Acta Musei Moraviae, Scientiae Biologicae, Brno* **82**, 191–198.

Kolibáč J (1999) Comparative morphology of mandible, epipharynx and alimentary canal in larval and adult Cleroidea (Coleoptera). *Acta Musei Moraviae. Scientiae Biologicae, Brno* **84**, 11–69.

Kolibáč J (2000) Classification and phylogeny of the Buprestoidea (Insecta: Coleoptera). *Acta Musei Moraviae. Scientiae Biologicae, Brno* **85**, 113–184.

Kolibáč J (2003) A review of Australian genera of Korynetinae (Coleoptera, Cleridae). *Entomologica Basiliensia* **25**, 41–97.

Kolibáč J (2004a) An approach to the higher taxa based on transformation series: a theory and examples from Cleroidea (Insecta: Coleoptera). *Entomologica Basiliensia* **26**, 269–291.

Kolibáč J (2004b) Metaxinidae fan. nov., a new family of Cleroidea (Coleoptera). *Entomologica Basiliensia* **26**, 239–268.

Kolibáč J (2005) A review of the Trogositidae. Part 1: Morphology of the genera (Coleoptera, Cleroidea). *Entomologica Basiliensia et Collectionis Frey* **27**, 39–159.

Kolibáč J (2006) A review of the Trogositidae. Part 2: Larval morphology, phylogeny and taxonomy (Coleoptera, Cleroidea). *Entomologica Basiliensia et Collectionis Frey* **28**, 105–153.

Kolibáč J (2007) Further observations on the tribe Ancyronycini Kolibac, 2006 (Coleoptera, Trogossitidae, Peltinae). *Entomologica Basiliensia et Collections Frey* **29**, 53–76.

Kolibáč J (2008) Morphology, taxonomy and phylogeny of *Phloiophilus edwardsi* Stephens, 1830 (Coleoptera, Cleroidea). *Entomologica Basiliensia et Collections Frey* **30**, 105–133.

Kolibáč J (2009) Some nomenclatorial notes on the family Trogossitidae (Coleoptera, Cleroidea). *Entomologica Basiliensia et Collectionis Frey* **31**, 127–129.

Kolibáč J (2010) 9.6. Cleridae Latreille, 1802. In *Handbuch der Zoologie/Handbook of Zoology. Band/Volume IV Arthropoda: Insecta Teilband/Part 38. Coleoptera, Beetles. Volume 2. Morphology and Systematics (Polyphaga partim).* (Eds RAB Leschen, RG Beutel and JF Lawrence) pp. 257–261. W. DeGruyter, Berlin.

Kolibáč J, Huang D-Y (2011) *Mathesius liaoningensis* gen. et sp. nov. of Jehol Biota, a presumptive relative of the clerid or thaneroclerid branches of Cleroidea (Coleoptera). *Zootaxa* **2872**, 1–17.

Kolibáč J, Leschen RAB (2010a) 9.2. Trogossitidae Fabricius, 1801. In *Handbuch der Zoologie/Handbook of Zoology. Band/Volume IV Arthropoda: Insecta Teilband/Part 38. Coleoptera, Beetles. Volume 2. Morphology and Systematics (Polyphaga partim).* (Eds RAB Leschen, RG Beutel and JF Lawrence) pp. 241–247. W. DeGruyter, Berlin.

Kolibáč J, Leschen RAB (2010b) 9.4 Metaxinidae Kolibáč, 2004 In *Handbuch der Zoologie/Handbook of Zoology. Band/Volume IV Arthropoda: Insecta Teilband/Part 38. Coleoptera, Beetles. Volume 2. Morphology and Systematics (Polyphaga partim).* (Eds RAB Leschen, RG Beutel and JF Lawrence) pp. 250–253. W. DeGruyter, Berlin.

Kolibáč J, Leschen RAB (2010c) 9.5. Thanerocleridae Chapin, 1924. In *Handbuch der Zoologie/Handbook of Zoology. Band/Volume IV Arthropoda: Insecta Teilband/Part 38. Coleoptera, Beetles. Volume 2. Morphology and Systematics (Polyphaga partim).* (Eds RAB Leschen, RG Beutel and JF Lawrence) pp. 253–257. W. DeGruyter, Berlin.

Kolibáč J, Zaitzev AA (2010) A description of a larva of *Ancyrona diversa* Pic, 1921 and its phylogenetic implications (Coleoptera: Trogossitidae). *Zootaxa* **2451**, 53–62.

Konstantinov AS (1995) New species of *Clavicornaltica* and Orthaltica from the Australian and Oriental regions (Coleoptera: Chrysomelidae, Alticinae). *Zoosystematica Rossica* **3**, 283–288.

Konstantinov AS, Duckett CN (2005a) New species of *Clavicornaltica* Scherer (Coleoptera: Chrysomelidae) from continental Asia. *Zootaxa* **1037**, 49–64.

Konstantinov AS, Duckett CN (2005b) Correction to "A description of four new species of *Clavicornaltica* Scherer (Coleoptera: Chrysomelidae) from continental Asia". *Zootaxa* **1042**, 64

Korolev SG (1987) Morphoecological peculiarities of larvae of the beetle family Lymexylonidae (Coleoptera). In *Ecology and Morphology of Insects Inhabiting Fungal Substrates.* (Ed. FN Pravdin) pp. 101–112. Nauka, Moscow.

Korschefsky R (1944) Bestimmungtabelle der bekanntesten deutschen Dermestidenlarven. *Arbeiten über physiologische und angewandte Entomologie aus Berlin-Dahlem* **11**, 140–152, pl. 1.

Korte A, Ribera I, Beutel RG, Bernhard D (2004) Interrelationships of staphyliniform groups inferred from 18S and c28S rDNA sequences, with special emphasis on Hydrophiloidea (Coleoptera: Staphyliniformia). *Journal of Zoological Systematics and Evolutionary Research* **42**, 281–288. doi:10.1111/j.1439-0469.2004.00282.x

Kovár I (1996a) Morphology and anatomy. In *Ecology of Coccinellidae.* (Eds I Hodek and A Honek) pp. 1–18. Kluwer Academic Publishers, Dordrecht, Boston, London.

Kovár I (1996b) Phylogeny. In *Ecology of Coccinellidae.* (Eds I Hodek and A Honek) pp.19–31. Kluwer Academic Publishers, Dordrecht, Boston, London.

Kovarik PW, Caterino MS (2000) 15. Histeridae Gyllenhal, 1808. In *American Beetles. Volume 1. American Beetles. Archostemata, Myxophaga, Adephaga, Polyphaga: Staphyliniformia.* (Eds RH Arnett Jr and MC Thomas) pp. 212–227. CRC Press, Boca Raton, Florida.

Kovarik PW, Caterino MS (2005) 10.4. Histeridae Gyllenhal, 1808. In *Handbuch der Zoologie/Handbook of Zoology. Band/Volume IV Arthropoda: Insecta Teilband/Part 38. Coleoptera, Beetles. Volume 1. Morphology and Systematics (Archostemata, Adephaga, Myxophaga, Polyphaga partim).* (Eds RG Beutel and RAB Leschen) pp. 190–222. W. DeGruyter, Berlin.

Kovarik PW, Passoa S (1993) Chaetotaxy of larval histeridae (Coleoptera: Hydrophiloidea) based on a description of *Onthophilus nodatus* LeConte. *Annals of the Entomological Society of America* **86**(5), 560–576.

Krantz GW (1983) Mites as biological control agents of dung-breeding flies, with special reference to the Macrochelidae.In *Biological Control of Pests by Mites.* (Eds MA Hoy *et al.*) pp. 91–98. University of California Agricultural Experiment Station, Berkeley, California.

Kreiss EJ, Schmitz A, Schmitz H (2005) Morphology of the prothoracic discs and associated sensilla of *Acanthocnemus nigricans* (Coleoptera, Acanthocnemidae). *Arthropod Structure & Development* **34**, 419–428. doi:10.1016/j.asd.2005.06.001

Krell F-T (2006) Fossil record and evolution of Scarabaeoidea (Coleoptera: Polyphaga). *Coleopterists Society Monographs* **5**, 120–143.

Krell F-T (2007) Catalogue of fossil Scarabaeoidea (Coleoptera: Polyphaga) of the Mesozoic and Tertiary – Version 2007. *Denver Museum of Nature & Science Technical Report* **8**, 1–79.

Krikken J (1984) A new key to the suprageneric taxa in the beetle family Cetoniidae, with annotated lists of the known genera. *Zoologische Verhandlingen, Leiden* **210**, 1–75.

Kriska NL (2002) 109. Oedemeridae Latreille 1810. In *American Beetles. Volume 2. Polyphaga: Scarabaeoidea through Curculionoidea.* (Eds RH Arnett Jr, MC Thomas, PE Skelley and JH Frank) pp. 514–519. CRC Press, Boca Raton, Florida.

Kristensen NP (1991) Phylogeny of extant hexapods. In *The Insects of Australia. A Textbook for Students and Research Workers.* (2nd Editition). (Ed. Csiro) pp. 125–140. Melbourne University Press, Carlton, Victoria.

Kristensen NP (1995) Forty Years' Insect Phylogenetic Systematics. Hennig's "Krisitsche Bemerkungen…", and subsequent developments. *Zoologische Beiträge (N. F.)* **36**(1), 83–124.

Kristensen NP (1999) Phylogeny of endopterygote insects, the most successful lineage of living organisms. *European Journal of Entomology* **96**, 237–253.

Kubán V, Majer K, Kolibáč J (2000) Classification of the tribe Coraebini Bedel, 1921 (Coleoptera, Buprestidae, Agrilinae). *Acta Musei Moraviae. Scientiae Biologicae, Brno* **85**, 185–287.

Kukalová J (1969) On the systematic position of the supposed Permian beetles, Tshekardocoleidae, with a description of a new collection from Moravia. *Sbornik Geologických Ved (Paleontologie)* **11**, 139–162.

Kukalová-Peck J (1978) Origin and evolution of insect wings and their relation to metamorphosis, as documented by the fossil record. *Journal of Morphology* **156**, 53–125. doi:10.1002/jmor.1051560104

Kukalová-Peck J (1983) Origin of the insect wing and wing articulation from the arthropod leg. *Canadian Journal of Zoology* **61**(7), 1618–1669. doi:10.1139/z83-217

Kukalová-Peck J (1991) Fossil history and the evolution of hexapod structures. In *The Insects of Australia. A Textbook for Students and Research Workers.* (2nd Editition). (Ed. Csiro) pp. 141–179. Melbourne University Press, Carlton, Victoria.

Kukalová-Peck J, Beutel RG (2012) Is the Carboniferous †*Adiphlebia lacoana* really the "oldest beetle" (Coleoptera)? Critical reassessment and description of a new Permian beetle family. *European Journal of Entomology* **109**, 633–645.

Kukalová-Peck J, Lawrence JF (1993) Evolution of the hind wing in Coleoptera. *Canadian Entomologist* **125**, 181–258. doi:10.4039/Ent125181-2

Kukalová-Peck J, Lawrence JF (2004) Relationships among coleopteran suborders and major endoneopteran lineages: evidence from hind wing characters. *European Journal of Entomology* **101**, 95–144.

Kundrata R, Bocak L (2011) The phylogeny and limits of Elateridae (Insecta, Coleoptera): is there a common tendency of click beetles to soft-bodiedness and neoteny? *Zoologica Scripta* **40**, 364–378. doi:10.1111/j.1463-6409.2011.00476.x

Kurosawa Y (1985) Revisional notes on the family Lymexylonidae (Coleoptera) in eastern and southeastern Asia. *Bulletin of the National Science Museum, Tokyo* **11**, 109–119. [A]

Kuschel G (1959) Nemonychidae, Belidae y Oxycorynidae de la fauna Chilena, con algunas consideraciones biogeográficos (Coleopterfa, Curculionoidea, aporte 28). *Investigaciones Zoológicas* **5**, 229–271.

Kuschel G (1979) The genera *Monotoma* Herbst (Rhizophagidae) and *Anommatus* Wesmael (Cerylidae) in New Zealand (Coleoptera). *New Zealand Entomologist* **7**, 44–48. doi:10.1080/00779962.1979.9722328

Kuschel G (1983) Past and present of the relict family Nemonychidae (Coleoptera: Curculionoidea). *GeoJournal* **7**, 499–504.

Kuschel G (1989) The Nearctic Nemonychidae (Coleoptera: Curculionoidea). *Entomologica Scandinavica* **20**, 121–171. doi:10.1163/187631289X00276

Kuschel G (1990) *Beetles In A Suburban Environment: a New Zealand Case Study.* Department of Scientific and Industrial Research (Pant Protection. Report No. 3), Auckland, New Zealand.

Kuschel G (1992) Reappraisal of the Baltic Amber Curculionoidea described by E. Voss. *Mitteilungen aus dem Geologisch-Paläontologischen Institut der Universität Hamburg* **73**, 191–215.

Kuschel G (1993) The Palaearctic Nemonychidae (Coleoptera: Curculionoidea). *Annales de la Société Entomologique de France (N. S.)* **29**(1), 23–46.

Kuschel G (1994) Appendix. Nemonychidae of Australia, New Guinea and New Caledonia. In *Australian Weevils*. Volume I. (Ed. EC Zimmerman) pp. 563–637. CSIRO Information Services, East Melbourne, Victoria.

Kuschel G (1995) A phylogenetic classification of Curculionoidea to families and subfamilies. *Memoirs of the Entomological Society of Washington* **14**, 5–33.

Kuschel G (2003) *Fauna of New Zealand. Number 45. Nemonychidae, Belidae, Brentidae (Insects: Coleoptera: Curculionoidea).* Manaaki Whenua Press, Lincoln, Canterbury, New Zealand.

Kuschel G, Leschen RAB (2003) Appendix 1. Phylogenetic relationships of the genera of Belinae. In Kuschel, G. *Fauna of New Zealand. Number 45. Nemonychidae, Belidae, Brentidae (Insects: Coleoptera: Curculionoidea)* pp. 48–55. Manaaki Whenua Press, Lincoln, Canterbury, New Zealand.

Kuschel G, Leschen RAB (2010) Phylogeny and taxonomy of the Rhinorhynchinae (Coleoptera: Nemonychidae). *Invertebrate Systematics* **24**, 573–615.

Kuschel G, May BM (1990) Palophaginae, a new subfamily for leaf-beetles, feeding as adult and larva on araucarian pollen in Australia (Coleoptera: Megalopodidae). *Invertebrate Taxonomy* **3**, 697–719. doi:10.1071/IT9890697

Kuschel G, May BM (1996a) Discovery of Palophaginae (Coleoptera: Megalopodidae) on *Araucaria araucana* in Chile and Argentina. *New Zealand Entomologist* **19**, 1–13. doi:10.1080/00779962.1996.9722015

Kuschel G, May BM (1996b) Palophagidae, their systematic position and biology. In *Chrysomelidae Biology. Volume 3: General Studies* (Eds PHA Jolivet and ML Cox) pp. 173–185. SPB Academic Publishing, Amsterdam.

Kuschel G, May BM (1997) A new genus and species of Nemonychidae (Coleoptera) associated with *Araucaria cangustifolia* in Brazil. *New Zealand Entomologist* **20**, 15–22. doi:10.1080/00779962.1997.9722664

Kuschel G, Leschen RAB, Zimmerman EC (2000) Platypodidae under scrutiny. *Invertebrate Taxonomy* **14**, 771–805. doi:10.1071/IT00024

LaBella DM, Lloyd JE (1991) Lampyridae (Cantharoidea). In *Immature Insects. Volume 2.* (Ed. FW Stehr) pp. 427–429. Kendall/Hunt, Dubuque, Iowa.

Lacordaire JT (1868) *Histoire Naturelle des Insectes. Genera des Coléoptères.* Vol. 8. 552 pp., pls. 81–91. Libraire Encyclopédique de Roret, Paris.

Lacordaire JT (1869) *Histoire naturelle des Insectes. Genera des Coléoptères.* Vol. 9. Part 1. Pp. 1–409, pls. 92–100. Librairie Encyclopédique de Roret, Paris.

Lacordaire JT (1872): *Histoire naturelle des Insectes. Genera des Coléoptères.* Vol. 9. Part 2. Pp. 411–930, pls. 101–110. Librairie Encyclopédique de Roret, Paris.

Łączyński P, Tomaszewska W (2009) Revision of the genus *Orcus* Mulsant (Coleoptera: Coccinellidae: Chilocorini). *Annales Zoologici* **59**(4), 585–611. doi:10.3161/000345409X484955

Lafer GS (1996) Fam. Sikhotealiniidae Lafer, fam. n. In *Key to the Insects of the Russian Far East. Vol. III. Part 3.* (Ed. PA Ler) pp. 390–396. Dal'nauka, Vladivostok.

Lameere A (1903) Nouvelles notes pour la classification des Coléoptères. *Annales de la Société Entomologique de Belgique* **47**, 155–165.

Lameere A (1919) Coleoptera Longicornia Fam. Cerambycidae Subfam. Prioninae. *In* Wytsman, P. (ed.), *Genera Insectorum.* Fascicule 172. 189 pp., 8 pls. L. Desmet-Verteneuil, Bruxelles.

Langor DW (1987) Flight muscle changes in the eastern larch beetle, *Dendroctonus simplex* Leconte (Coleoptera: Scolytidae). *Coleopterists Bulletin* **41**, 351–357.

Larsén O (1966) On the morphology and function of the locomotor organs of the Gyrinidae and other Coleoptera. *Opuscula Entomologica* **30**, (Supplement), 1–242.

Larson DJ (1993) Ecology of tropical Australian Hydradephaga (Insecta: Coleoptera). Part 1. Natural history and distribution of northern Queensland species. *Proceedings of the Royal Society of Queensland* **103**, 47–63.

Larson DJ, Storey RI (1994) *Carabhydrus mubboonus,* a new species of rheophilic water beetle (Coleoptera: Dytiscidae) from Queensland, Australia. *Canadian Entomologist* **126**, 895–906. doi:10.4039/Ent126895-3

Larsson SG (1978) Baltic amber – A palaeobiological study. *Entomograph* **1**, 1–192.

Lawrence JF (1971) Revision of the North American Ciidae (Coleoptera). *Bulletin of the Museum of Comparative Zoology* **142**, 419–522.

Lawrence JF (1973) Host preference in ciid beetles (Coleoptera: Ciidae) inhabiting the fruiting bodies of Basidiomycetes in North America. *Bulletin of the Museum of Comparative Zoology. Harvard University* **145**, 163–212.

Lawrence JF (1974) The larva of *Sphindocis denticollis* Fall and a new subfamily of Ciidae (Coleoptera: Heteromera). *Breviora* **424**, 1–14.

Lawrence JF (1977a) Coleoptera associated with an *Hypoxylon* species (Ascomycetes: Xylariaceae) on oak. *Coleopterists Bulletin* **31**, 309–312.

Lawrence JF (1977b) The family Pterogeniidae, with notes on the phylogeny of the Heteromera. *Coleopterists Bulletin* **31**, 25–56.

Lawrence JF (1980) A new genus of Indo-Australian Gempylodini with notes on the constitution of the Colydiidae (Coleoptera). *Journal of the Australian Entomological Society* **19**, 293–310. doi:10.1111/j.1440-6055.1980.tb00989.x

Lawrence JF (1981) Notes on larval Lucanidae (Coleoptera). *Journal of the Australian Entomological Society* **20**, 213–219. doi:10.1111/j.1440-6055.1981.tb01035.x

Lawrence JF (1982) Coleoptera. In *Synopsis and Classification of Living Organisms*. Vol. 2 (Ed. SB Parker) pp. 482–553. McGraw-Hill, New York.

Lawrence JF (1985a) The genus *Nothoderodontus* (Coleoptera: Derodontidae) with new species from Australia, New Zealand, and Chile. In *Taxonomy, Phylogeny and Zoogeography of Beetles and Ants. A Volume dedicated to the Memory of Philip Jackson Darlington Jr (1904–1983).* (Ed. GE Ball) pp. 68–83. Junk (Series Entomologica 33), Dordrecht.

Lawrence JF (1985b) The genus *Teredolaemus* Sharp (Coleoptera: Bothrideridae) in Australia. *Journal of the Australian Entomological Society* **24**, 205–206. doi:10.1111/j.1440-6055.1985.tb00227.x

Lawrence JF (1987a) Notes on the classification of some Australian Elateriformia (Coleoptera). *Journal of the Australian Entomological Society* **26**, 360. doi:10.1111/j.1440-6055.1987.tb01983.x

Lawrence JF (1987b) Notes on the classification of some Australian Tenebrionoidea (Coleoptera). *Journal of the Australian Entomological Society* **26**, 361–362. doi:10.1111/j.1440-6055.1987.tb01984.x

Lawrence JF (1988a) Notes on the classification of some Australian Cucujoidea (Coleoptera). *Journal of the Australian Entomological Society* **27**, 53–54. doi:10.1111/j.1440-6055.1988.tb01144.x

Lawrence JF (1988b) Rhinorhipidae, a new beetle family from Australia, with comments on the phylogeny of the Elateriformia. *Invertebrate Taxonomy* **2**, 1–53. doi:10.1071/IT9880001

Lawrence JF (1989) Mycophagy in the Coleoptera: feeding strategies and morphological adaptations. In *Insect-Fungus Interactions. 14th Symposium of the Royal Entomological Society of London in collaboration with the British Mycological Society*. (Eds N Wilding, NM Collins, PM Hammond and JF Webber) Pp. 1–23. Academic Press, London.

Lawrence JF (1991a) 34. Order Coleoptera [Introduction, morphology, biology and ecology, classification, key to families and many subfamilies (worldwide)]. In *Immature Insects. Volume 2.* (Ed. FW Stehr) pp. 144–298. Kendall/Hunt, Dubuque, Iowa.

Lawrence JF (1991b) Ommatidae–Hydroscaphidae, Hygrobiidae, Eucinetidae–Clambidae, Dascillidae–Rhipiceridae, Byrrhidae–Ptilodactylidae, Heteroceridae, Cerophtyidae, Brachypsectridae–Drilidae, Telegeusidae, Omethidae–Jacobsoniidae, Bostrichidae, Ptinidae, Phloiophilidae, Chaetosomatidae, Phycosecidae–Acanthocnemidae, Protocucujidae–Endomychidae, Corylophidae–Mordellidae, Colydiidae–Monommidae, Zopheridae–Chalcodryidae, Cephaloidae, Oedemeridae–Mycteridae, Trictenotomidae, Cerambycidae, Urodontidae–Ithyceridae, Coleoptera Bibliography. In *Immature Insects. Volume 2.* (Ed. FW Stehr) pp. 298–304, 312, 364–366, 369–371, 384–394, 402–404, 409–410, 421–424, 426, 428–434, 439–441, 444–445, 447–448, 450, 452–453, 454–485, 495–509, 512–515, 518–520, 529–530, 534–537, 539, 556–561, 589–591, 617–658. Kendall/Hunt, Dubuque, Iowa.

Lawrence JF (1991c) Three new Asiatic Ciidae (Coleoptera: Tenebrionoidea) associated with commercial, dried fungi. *Coleopterists Bulletin* **45**, 286–292.

Lawrence JF (1994a) The larva of *Sirrhas variegatus* sp. n., with notes on Perimylopidae, Ulodidae (stat. n.), Zopheridae and Chalcodryidae (Coleoptera: Tenebrionoidea). *Invertebrate Taxonomy* **8**, 329–349. doi:10.1071/IT9940329

Lawrence JF (1994b) Review of the Australian Archeocrypticidae (Coleoptera), with descriptions of a new genus and four new species. *Invertebrate Taxonomy* **8**, 449–470. doi:10.1071/IT9940449

Lawrence JF (1995) Two new species of *Rhopalobrachium* Boheman (Coleoptera: Phloeostichidae: Hymaeinae) from Australia and Chile. In *Biology, Phylogeny, and Classification of Coleoptera. Papers Celebrating the 80th Birthday of Roy A. Crowson*. (Eds J Pakaluk and SA Ślipiński) pp. 433–447. Muzeum I Instytut Zoologii Polska Akademia Nauk, Warsaw.

Lawrence JF (1999) The Australian Ommatidae (Coleoptera: Archostemata): new species, larva and discussion of relationships. *Invertebrate Taxonomy* **13**, 369–390. doi:10.1071/IT99008

Lawrence JF (2001) A new genus of Valdivian Scirtidae (Coleoptera) with comments on Scirtoidea and the beetle suborders. In *Sukunahikona*. (Eds K Morimoto, K Mizuno, Y Hayashi, T Ito, N Ito, K Ando, M Tanikado and S. Shiyake) pp. 351–361. The Japan Coleopterological Society (Special Publication No. 1), Osaka.

Lawrence JF (2005a) 15.4. Scirtidae Fleming, 1821. In *Handbuch der Zoologie/Handbook of Zoology. Band/Volume IV Arthropoda: Insecta Teilband/Part 38. Coleoptera, Beetles. Volume 1. Morphology and Systematics (Archostemata, Adephaga, Myxophaga, Polyphaga partim).* (Eds RG Beutel and RAB Leschen) pp. 443–450. W. DeGruyter, Berlin.

Lawrence JF (2005b) 16.1. Dascillidae Guérin-Méneville, 1843. In *Handbuch der Zoologie/Handbook of Zoology. Band/Volume IV Arthropoda: Insecta Teilband/Part 38. Coleoptera, Beetles. Volume 1. Morphology and Systematics (Archostemata, Adephaga, Myxophaga, Polyphaga partim).* (Eds RG Beutel and RAB Leschen) pp. 451–456. W. DeGruyter, Berlin.

Lawrence JF (2005c) 16.2. Rhipiceridae Latreille, 1834. In *Handbuch der Zoologie/Handbook of Zoology. Band/Volume IV Arthropoda: Insecta Teilband/Part 38. Coleoptera, Beetles. Volume 1. Morphology and Systematics (Archostemata, Adephaga, Myxophaga, Polyphaga partim).* (Eds RG Beutel and RAB Leschen) pp. 456–460. W. DeGruyter, Berlin.

Lawrence JF (2005d) 18.9. Ptilodactylidae Laporte, 1836. In *Handbuch der Zoologie/Handbook of Zoology. Band/Volume IV Arthropoda: Insecta Teilband/Part 38. Coleoptera, Beetles. Volume 1. Morphology and Systematics (Archostemata, Adephaga, Myxophaga, Polyphaga partim).* (Eds RG Beutel and RAB Leschen) pp. 536–542. W. DeGruyter, Berlin.

Lawrence JF (2005e) 18.12. Callirhipidae Emden, 1924. In *Handbuch der Zoologie/Handbook of Zoology. Band/Volume IV Arthropoda: Insecta Teilband/Part 38. Coleoptera, Beetles. Volume 1. Morphology and Systematics (Archostemata, Adephaga, Myxophaga, Polyphaga partim).* (Eds RG Beutel and RAB Leschen) pp. 551–555. W. DeGruyter, Berlin.

Lawrence JF (2005f) *Brevipogon*, a new genus of North American Artematopodidae (Coleoptera), with a key to world genera. *Coleopterists Bulletin* **59** (2), 223–236. doi:10.1649/749

Lawrence JF (2005g) *Dasytomima*, a new genus of Australian Oedemeridae and its relationship to *Polypria* Chevrolat (Coleoptera: Tenebrionoidea). *Annales Zoologici* **55**(4), 663–676.

Lawrence JF (2010a) 4.1. Rhinorhipidae Lawrence, 1988. In *Handbuch der Zoologie/Handbook of Zoology. Band/Volume IV Arthropoda: Insecta Teilband/Part 38. Coleoptera, Beetles. Volume 2. Morphology and Systematics (Polyphaga partim).* (Eds RAB Leschen, RG Beutel and JF Lawrence) pp. 38–42. W. DeGruyter, Berlin.

Lawrence JF (2010b) 4.2. Artematopodidae Lacordaire, 1857. In *Handbuch der Zoologie/Handbook of Zoology. Band/Volume IV Arthropoda: Insecta Teilband/Part 38. Coleoptera, Beetles. Volume 2. Morphology and Sys-*

tematics (Polyphaga partim). (Eds RAB Leschen, RG Beutel and JF Lawrence) pp. 42–47. W. DeGruyter, Berlin.

Lawrence JF (2010c) 4.12. Telegeusidae Leng, 1920. In *Handbuch der Zoologie/Handbook of Zoology. Band/Volume IV Arthropoda: Insecta Teilband/Part 38. Coleoptera, Beetles. Volume 2. Morphology and Systematics (Polyphaga partim)*. (Eds RAB Leschen, RG Beutel and JF Lawrence) pp. 123–126. W. DeGruyter, Berlin.

Lawrence JF (2010d) 6.2. Endecatomidae LeConte, 1861. In *Handbuch der Zoologie/Handbook of Zoology. Band/Volume IV Arthropoda: Insecta Teilband/Part 38. Coleoptera, Beetles. Volume 2. Morphology and Systematics (Polyphaga partim)*. (Eds RAB Leschen, RG Beutel and JF Lawrence) pp. 206–209. W. DeGruyter, Berlin.

Lawrence JF (2010e) 6.3. Bostrichidae Latreille, 1802. In *Handbuch der Zoologie/Handbook of Zoology. Band/Volume IV Arthropoda: Insecta Teilband/Part 38. Coleoptera, Beetles. Volume 2. Morphology and Systematics (Polyphaga partim)*. (Eds RAB Leschen, RG Beutel and JF Lawrence) pp. 209–217. W. DeGruyter, Berlin.

Lawrence JF (2010f) 8. Lymexyloidea Fleming, 1821. In *Handbuch der Zoologie/Handbook of Zoology. Band/Volume IV Arthropoda: Insecta Teilband/Part 38. Coleoptera, Beetles. Volume 2. Morphology and Systematics (Polyphaga partim)*. (Eds RAB Leschen, RG Beutel and JF Lawrence) pp. 229–235. W. DeGruyter, Berlin.

Lawrence JF (2010g) 9.10. Mauroniscidae Majer, 1994. In *Handbuch der Zoologie/Handbook of Zoology. Band/Volume IV Arthropoda: Insecta Teilband/Part 38. Coleoptera, Beetles. Volume 2. Morphology and Systematics (Polyphaga partim)*. (Eds RAB Leschen, RG Beutel and JF Lawrence) pp. 271–273. W. DeGruyter, Berlin.

Lawrence JF (2010h) 10.21. Propalticidae Crowson, 1952. In *Handbuch der Zoologie/Handbook of Zoology. Band/Volume IV Arthropoda: Insecta Teilband/Part 38. Coleoptera, Beetles. Volume 2. Morphology and Systematics (Polyphaga partim)*. (Eds RAB Leschen, RG Beutel and JF Lawrence) pp. 374–376. W. DeGruyter, Berlin.

Lawrence JF (2010i) 11.2. Archeocrypticidae Kaszab, 1964. In *Handbuch der Zoologie/Handbook of Zoology. Band/Volume IV Arthropoda: Insecta Teilband/Part 38. Coleoptera, Beetles. Volume 2. Morphology and Systematics (Polyphaga partim)*. (Eds RAB Leschen, RG Beutel and JF Lawrence) pp. 496–501. W. DeGruyter, Berlin.

Lawrence JF (2010j) 11.3. Pterogeniidae Crowson, 1953. In *Handbuch der Zoologie/Handbook of Zoology. Band/Volume IV Arthropoda: Insecta Teilband/Part 38. Coleoptera, Beetles. Volume 2. Morphology and Systematics (Polyphaga partim)*. (Eds RAB Leschen, RG Beutel and JF Lawrence) pp. 501–504. W. DeGruyter, Berlin.

Lawrence JF, Arias ET (2009) *Valdivelater*, a new genus of Protelaterini (Elateridae: Lissominae) from the forests of central and southern Chile. *Annales Zoologici* **59**(3), 319–328. doi:10.3161/000345409X476422

Lawrence JF, Britton EB (1991) Coleoptera (beetles). In The Insects of Australia. A Textbook for Students and Research Workers. Second Edition. (Ed. CSIRO) pp. 543–683. Melbourne University Press, Carlton, Victoria.

Lawrence JF, Britton ED (1994) *Australian Beetles*. Melbourne University Press, Melbourne.

Lawrence JF, de Viedma MG (1991) Anobiidae. In *Immature Insects. Volume 2.* (Ed. FW Stehr) pp. 441–444. Kendall/Hunt, Dubuque, Iowa.

Lawrence JF, Hlavac TF (1979) Review of the Derodontidae (Coleoptera: Polyphaga) with new species from North America and Chile. *Coleopterists Bulletin* **33**, 369–414.

Lawrence JF, Leschen RAB (2003) Review of Lamingtoniidae (Coleoptera: Cucujoidea) with descriptions of two new species. In *Systematics of Coleoptera: Papers Celebrating the Retirement of Ivan Löbl.* (Eds G. Cuccodoro and RAB Leschen). pp. 905–919. Associated Publishers (Memoirs on Entomology, International, Volume 17), Gainesville, Florida.

Lawrence JF, Leschen RAB (2010a) 5.3. Jacobsoniidae Heller, 1926. In *Handbuch der Zoologie/Handbook of Zoology. Band/Volume IV Arthropoda: Insecta Teilband/Part 38. Coleoptera, Beetles. Volume 2. Morphology and Systematics (Polyphaga partim)*. (Eds RAB Leschen, RG Beutel and JF Lawrence) pp. 190–195. W. DeGruyter, Berlin.

Lawrence JF, Leschen RAB (2010b) 9.1. Phloiophilidae Kiesenwetter, 1863. In *Handbuch der Zoologie/Handbook of Zoology. Band/Volume IV Arthropoda: Insecta Teilband/Part 38. Coleoptera, Beetles. Volume 2. Morphology and Systematics (Polyphaga partim)*. (Eds RAB Leschen, RG Beutel and JF Lawrence) pp. 239–241. W. DeGruyter, Berlin.

Lawrence JF, Leschen RAB (2010c) 9.7. Acanthocnemidae Crowson, 1964. In *Handbuch der Zoologie/Handbook of Zoology. Band/Volume IV Arthropoda: Insecta Teilband/Part 38. Coleoptera, Beetles. Volume 2. Morphology and Systematics (Polyphaga partim)*. (Eds RAB Leschen, RG Beutel and JF Lawrence) pp. 262–265. W. DeGruyter, Berlin.

Lawrence JF, Leschen RAB (2010d) 9.9. Prionoceridae Lacordaire, 1857. In *Handbuch der Zoologie/Handbook of Zoology. Band/Volume IV Arthropoda: Insecta Teilband/Part 38. Coleoptera, Beetles. Volume 2. Morphology and Systematics (Polyphaga partim)*. (Eds RAB Leschen, RG Beutel and JF Lawrence) pp. 268–271. W. DeGruyter, Berlin.

Lawrence JF, Leschen RAB (2010e) 9.11. Melyridae Leach, 1815. In *Handbuch der Zoologie/Handbook of Zoology. Band/Volume IV Arthropoda: Insecta Teilband/Part 38. Coleoptera, Beetles. Volume 2. Morphology and Systematics (Polyphaga partim)*. (Eds RAB Leschen, RG Beutel and JF Lawrence) pp. 273–280. W. DeGruyter, Berlin.

Lawrence JF, Leschen RAB (2010f) 10.18. Lamingtoniidae Sen Gupta and Crowson, 1969. In *Handbuch der Zoologie/Handbook of Zoology. Band/Volume IV Arthropoda: Insecta Teilband/Part 38. Coleoptera, Beetles. Volume 2. Morphology and Systematics (Polyphaga partim)*. (Eds RAB Leschen, RG Beutel and JF Lawrence) pp. 360–363. W. DeGruyter, Berlin.

Lawrence JF, Leschen RAB (2010g) 11.1. Mycetophagidae Leach, 1815. In *Handbuch der Zoologie/Handbook of Zoology. Band/Volume IV Arthropoda: Insecta Teilband/Part 38. Coleoptera, Beetles. Volume 2. Morphology and Systematics (Polyphaga partim)*. (Eds RAB Leschen, RG Beutel and JF Lawrence) pp. 491–496. W. DeGruyter, Berlin.

Lawrence JF, Leschen RAB (2010h) 11.5. Tetratomidae Billberg, 1820. In *Handbuch der Zoologie/Handbook of Zoology. Band/Volume IV Arthropoda: Insecta Teilband/Part 38. Coleoptera, Beetles. Volume 2. Morphology and Systematics (Polyphaga partim)*. (Eds RAB Leschen, RG Beutel and JF Lawrence) pp. 514–520. W. DeGruyter, Berlin.

Lawrence JF, Leschen RAB (2010i) 11.12. Chalcodryidae Watt, 1974. In *Handbuch der Zoologie/Handbook of Zoology. Band/Volume IV Arthropoda: Insecta Teilband/Part 38. Coleoptera, Beetles. Volume 2. Morphology and Systematics (Polyphaga partim)*. (Eds RAB Leschen, RG Beutel and JF Lawrence) pp. 567–571. W. DeGruyter, Berlin.

Lawrence JF, Lopes-Andrade C (2010) 11.4. Ciidae Leach in Samouelle, 1819. In *Handbuch der Zoologie/Handbook of Zoology. Band/Volume IV Arthropoda: Insecta Teilband/Part 38. Coleoptera, Beetles. Volume 2. Morphology and Systematics (Polyphaga partim)*. (Eds RAB Leschen, RG Beutel and JF Lawrence) pp. 504–514. W. DeGruyter, Berlin.

Lawrence JF and Medvedev GS (1982) A new tribe of darkling beetles (Coleoptera: Tenebrionidae) from Australia and its systematic position. *Entomologicheskoye Obozreniye* **61**(3), 548–571 (in Russian; translation in *Entomological Review* **61**(3), 85–107).

Lawrence JF, Milner RJ (1996) Associations between arthropods and fungi. In *Fungi of Australia. Volume 1B. Introduction—Fungi in the Environment.* (Eds K Mallet and C Grgurinovic) pp. 137–201. Australian Biological Resources Study, Canberra.

Lawrence JF, Newton AF (1982) Evolution and classification of beetles. *Annual Review of Ecology and Systematics* **13**, 261–290. doi:10.1146/annurev.es.13.110182.001401

Lawrence JF, Newton AF Jr (1995) Families and subfamilies of Coleoptera (with selected genera, notes, references and data on family-group names). In *Biology, Phylogeny, and Classification of Coleoptera. Papers Celebrating the 80th Birthday of Roy A. Crowson.* (Eds J Pakaluk and SA Ślipiński) pp. 779–1006. Muzeum I Instytut Zoologii Polska Akademia Nauk, Warsaw.

Lawrence JF, Pollock D (1994) Relationships of the Australian genus *Synercticus* Newman (Coleoptera: Boridae). *Journal of the Australian Entomological Society* **33**, 35–42. doi:10.1111/j.1440-6055.1994.tb00914.x

Lawrence JF, Reichardt H (1969) The myrmecophilous Ptinidae (Coleoptera), with a key to Australian species. *Bulletin of the Museum of Comparative Zoology. Harvard University* **138**, 1–27.

Lawrence JF, Ślipiński A (2005) Three new genera of Indo-Australian Dermestidae (Coleoptera) and their phylogenetic significance. *Invertebrate Systematics* **19**, 231–261. doi:10.1071/IS04033

Lawrence JF, Ślipiński A (2010a) 6.1 Dermestidae Latreille, 1804. In *Handbuch der Zoologie/Handbook of Zoology. Band/Volume IV Arthropoda: Insecta Teilband/Part 38. Coleoptera, Beetles. Volume 2. Morphology and Systematics (Polyphaga partim).* (Eds RAB Leschen, RG Beutel and JF Lawrence) pp. 198–206. W. DeGruyter, Berlin.

Lawrence JF, Ślipiński A (2010b) 10.1. Boganiidae Sen Gupta & Crowson, 1966. In *Handbuch der Zoologie/Handbook of Zoology. Band/Volume IV Arthropoda: Insecta Teilband/Part 38. Coleoptera, Beetles. Volume 2. Morphology and Systematics (Polyphaga partim).* (Eds RAB Leschen, RG Beutel and JF Lawrence) pp. 282–286. W. DeGruyter, Berlin.

Lawrence JF, Ślipiński A (2010c) 10.13. Phloeostichidae Reitter, 1911. In *Handbuch der Zoologie/Handbook of Zoology. Band/Volume IV Arthropoda: Insecta Teilband/Part 38. Coleoptera, Beetles. Volume 2. Morphology and Systematics (Polyphaga partim).* (Eds RAB Leschen, RG Beutel and JF Lawrence) pp. 340–346. W. DeGruyter, Berlin.

Lawrence JF, Ślipiński A (2010d) 10.23. Tasmosalpingidae Lawrence and Britton, 1991. In *Handbuch der Zoologie/Handbook of Zoology. Band/Volume IV Arthropoda: Insecta Teilband/Part 38. Coleoptera, Beetles. Volume 2. Morphology and Systematics (Polyphaga partim).* (Eds RAB Leschen, RG Beutel and JF Lawrence) pp. 381–383. W. DeGruyter, Berlin.

Lawrence JF, Ślipiński A (2010e) 11.7. Mordellidae Latreille, 1802. In *Handbuch der Zoologie/Handbook of Zoology. Band/Volume IV Arthropoda: Insecta Teilband/Part 38. Coleoptera, Beetles. Volume 2. Morphology and Systematics (Polyphaga partim).* (Eds RAB Leschen, RG Beutel and JF Lawrence) pp. 533–537. W. DeGruyter, Berlin.

Lawrence JF, Ślipiński A (2010f) 11.13. Trachelostenidae Lacordaire, 1859. In *Handbuch der Zoologie/Handbook of Zoology. Band/Volume IV Arthropoda: Insecta Teilband/Part 38. Coleoptera, Beetles. Volume 2. Morphology and Systematics (Polyphaga partim).* (Eds RAB Leschen, RG Beutel and JF Lawrence) pp. 571–574. W. DeGruyter, Berlin.

Lawrence JF, Ślipiński A (2010g) 11.17. Stenotrachelidae C. G. Thomson, 1859. In *Handbuch der Zoologie/Handbook of Zoology. Band/Volume IV Arthropoda: Insecta Teilband/Part 38. Coleoptera, Beetles. Volume 2. Morphology and Systematics (Polyphaga partim).* (Eds RAB Leschen, RG Beutel and JF Lawrence) pp. 670–674. W. DeGruyter, Berlin.

Lawrence JF, Ślipiński A (2010h) 11.18. Oedemeridae Latreille, 1810. In *Handbuch der Zoologie/Handbook of Zoology. Band/Volume IV Arthropoda: Insecta Teilband/Part 38. Coleoptera, Beetles. Volume 2. Morphology and Systematics (Polyphaga partim).* (Eds RAB Leschen, RG Beutel and JF Lawrence) pp. 674–681. W. DeGruyter, Berlin.

Lawrence JF, Ślipiński A (2010i) 11.27. Aderidae Winkler 1927. In *Handbuch der Zoologie/Handbook of Zoology. Band/Volume IV Arthropoda: Insecta Teilband/Part 38. Coleoptera, Beetles. Volume 2. Morphology and Systematics (Polyphaga partim).* (Eds RAB Leschen, RG Beutel and JF Lawrence) pp. 741–746. W. DeGruyter, Berlin.

Lawrence JF, Ślipiński A (2010j) 11.28. Scraptiidae Mulsant, 1856. In *Handbuch der Zoologie/Handbook of Zoology. Band/Volume IV Arthropoda: Insecta Teilband/Part 38. Coleoptera, Beetles. Volume 2. Morphology and Systematics (Polyphaga partim).* (Eds RAB Leschen, RG Beutel and JF Lawrence) pp. 746–750. W. DeGruyter, Berlin.

Lawrence JF, Stribling JB (1992) A new genus of Ptilodactylidae (Coleoptera) from Australia, with a description of the larva. *Journal of the Australian Entomological Society* **31**, 19–27. doi:10.1111/j.1440-6055.1992.tb00454.x

Lawrence JF, Yoshitomi H (2007) *Nipponocyphon*, a new genus of Japanese Scirtidae (Coleoptera) and its phylogenetic significance. *Elytra, Tokyo* **35** (2), 507–527.

Lawrence JF, Weir TA, Pyke JE (1987) Archostemata, Myxophaga, Adephaga, Haliplidae, Hygrobiidae, Noteridae, Dytiscidae And Gyrinidae. *Zoological Catalogue of Australia. Volume 4. Coleoptera: Archostemata, Myxophaga and Adephaga.* (Ed. DW Walton) pp. 1–19, 321–66. Australian Government Publication Service, Canberra.

Lawrence JF, Kistner DH, Pasteels JM (1990) A new genus and three new species of termitophilous Aderidae (Coleoptera) from Australia, Papua New Guinea and the Philippines, with notes on their biology. *Invertebrate Taxonomy* **4**, 643–654. doi:10.1071/IT9900643

Lawrence JF, Nikitsky NB, Kirejtshuk AG (1995a) Phylogenetic position of Decliniidae (Coleoptera: Scirtoidea) and comments on the classification of Elateriformia (sensu lato). In *Biology, Phylogeny, and Classification of Coleoptera. Papers Celebrating the 80th Birthday of Roy A. Crowson.* (Eds J Pakaluk and SA Ślipiński) pp. 375–410. Muzeum I Instytut Zoologii Polska Akademia Nauk, Warsaw.

Lawrence JF, Ślipiński SA, Pakaluk J (1995b) From Latreille to Crowson: a history of the higher classification of beetles. In *Biology, Phylogeny, and Classification of Coleoptera. Papers Celebrating the 80th Birthday of Roy A. Crowson.* (Eds J Pakaluk and SA Ślipiński) pp. 87–154. Muzeum I Instytut Zoologii Polska Akademia Nauk, Warsaw.

Lawrence JF, Hastings AM, Dallwitz MJ, Paine TA, Zurcher EJ (1999a): *Beetle Larvae of the World: Descriptions, Illustrations, Identification, and Information Retrieval for Families and Subfamilies.* CD-ROM, Version 1.1 for MS-Windows. CSIRO Publishing, Melbourne.

Lawrence JF, Hastings AM, Dallwitz MJ, Paine TA, Zurcher EJ (1999b): *Beetles of the World: A Key and Information System for Families and Subfamilies.* CD-ROM, Version 1.0 for MS-Windows. Melbourne: CSIRO Publishing, Melbourne.

Lawrence JF, Muona J, Teräväinen M, Ståhls G, Vahtera V (2007) *Anischia, Perothops* and the phylogeny of Elateroidea (Coleoptera: Elateriformia). *Insect Systematics & Evolution* **38**, 205–239. doi:10.1163/187631207794761001

Lawrence JF, Archibald SB, Ślipiński A (2008) A new species of Prionoceridae (Coleoptera: Cleroidea) from the Eocene of British Columbia, Canada. *Annales Zoologici* **58**(4), 689–693. doi:10.3161/000345408X396620

Lawrence JF, Beutel RG, Leschen RAB, Ślipiński A (2010a) 1. Changes in classification and list of families and subfamilies. In *Handbuch der Zoologie/Handbook of Zoology. Band/Volume IV Arthropoda: Insecta Teilband/Part 38. Coleoptera, Beetles. Volume 2. Morphology and Systematics (Polyphaga partim).* (Eds RAB Leschen, RG Beutel and JF Lawrence) pp. 1–7. W. DeGruyter, Berlin.

Lawrence JF, Beutel RG, Leschen RAB, Ślipiński A (2010b) 2. Glossary of morphological terms. *In Handbuch der Zoologie/Handbook of Zoology. Band/Volume IV Arthropoda: Insecta Teilband/Part 38. Coleoptera, Beetles. Volume 2. Morphology and Systematics (Polyphaga partim).* (Eds RAB Leschen, RG Beutel and JF Lawrence) pp. 9–20. W. DeGruyter, Berlin.

Lawrence JF, Bocak L, Bocakova M, Beutel RG, Muona J (2010c) 4. Elateroidea. Introduction, Phylogeny. In *Handbuch der Zoologie/Handbook of Zoology. Band/Volume IV Arthropoda: Insecta Teilband/Part 38. Coleoptera, Beetles. Volume 2. Morphology and Systematics (Polyphaga partim).* (Eds RAB Leschen, RG Beutel and JF Lawrence) pp. 35–38. W. DeGruyter, Berlin.

Lawrence JF, Kawashima I, Branham MA (2010d) 4.18. Elateriformia *Incertae Sedis*. In *Handbuch der Zoologie/Handbook of Zoology. Band/Volume IV Arthropoda: Insecta Teilband/Part 38. Coleoptera, Beetles. Volume 2. Morphology and Systematics (Polyphaga partim).* (Eds RAB Leschen, RG Beutel and JF Lawrence) pp. 162–177. W. DeGruyter, Berlin.

Lawrence JF, Beutel RG, Leschen RAB (2010e) 5. Derodontiformia. Introduction, Phylogeny. In *Handbuch der Zoologie/Handbook of Zoology. Band/Volume IV Arthropoda: Insecta Teilband/Part 38. Coleoptera, Beetles.*

Volume 2. Morphology and Systematics (Polyphaga partim). (Eds RAB Leschen, RG Beutel and JF Lawrence) pp. 179–180. W. DeGruyter, Berlin.

Lawrence JF, Leschen RAB, Beutel RG (2010f) 6. Bostrichiformia Latreille 1802. In *Handbuch der Zoologie/Handbook of Zoology. Band/Volume IV Arthropoda: Insecta Teilband/Part 38. Coleoptera, Beetles. Volume 2. Morphology and Systematics (Polyphaga partim)*. (Eds RAB Leschen, RG Beutel and JF Lawrence) pp. 197–198. W. DeGruyter, Berlin.

Lawrence JF, Ślipiński A, Lee CF (2010g) 10.3. Helotidae Reitter, 1876/ Chapuis, 1876. In *Handbuch der Zoologie/Handbook of Zoology. Band/ Volume IV Arthropoda: Insecta Teilband/Part 38. Coleoptera, Beetles. Volume 2. Morphology and Systematics (Polyphaga partim)*. (Eds RAB Leschen, RG Beutel and JF Lawrence) pp. 292–295. W. DeGruyter, Berlin.

Lawrence JF, Gimmel M, Steiner WE Jr (2010h) 10.20. Phalacridae Leach, 1815. In *Handbuch der Zoologie/Handbook of Zoology. Band/Volume IV Arthropoda: Insecta Teilband/Part 38. Coleoptera, Beetles. Volume 2. Morphology and Systematics (Polyphaga partim)*. (Eds RAB Leschen, RG Beutel and JF Lawrence) pp. 368–374. W. DeGruyter, Berlin.

Lawrence JF, Pollock DA, Ślipiński A (2010i) 11. Tenebrionoidea. Introduction, Phylogeny. In *Handbuch der Zoologie/Handbook of Zoology. Band/ Volume IV Arthropoda: Insecta Teilband/Part 38. Coleoptera, Beetles. Volume 2. Morphology and Systematics (Polyphaga partim)*. (Eds RAB Leschen, RG Beutel and JF Lawrence) pp. 487–491. W. DeGruyter, Berlin.

Lawrence JF, Falin Z, Ślipiński A (2010j) 11.8. Ripiphoridae Gemminger and Harold, 1870 (Gerstaecker, 1855). In *Handbuch der Zoologie/Handbook of Zoology. Band/Volume IV Arthropoda: Insecta Teilband/Part 38. Coleoptera, Beetles. Volume 2. Morphology and Systematics (Polyphaga partim)*. (Eds RAB Leschen, RG Beutel and JF Lawrence) pp. 538–548. W. DeGruyter, Berlin.

Lawrence JF, Ślipiński A, Elgueta M (2010k) 11.11. Promecheilidae Lacordaire, 1859. In *Handbuch der Zoologie/Handbook of Zoology. Band/ Volume IV Arthropoda: Insecta Teilband/Part 38. Coleoptera, Beetles. Volume 2. Morphology and Systematics (Polyphaga partim)*. (Eds RAB Leschen, RG Beutel and JF Lawrence) pp. 563–567. W. DeGruyter, Berlin.

Lawrence JF, Ślipiński A, Pollock DA, Escalona H (2010l) 11.25. Salpingidae Leach, 1815. In *Handbuch der Zoologie/Handbook of Zoology. Band/ Volume IV Arthropoda: Insecta Teilband/Part 38. Coleoptera, Beetles. Volume 2. Morphology and Systematics (Polyphaga partim)*. (Eds RAB Leschen, RG Beutel and JF Lawrence) pp. 722–729. W. DeGruyter, Berlin.

Lawrence JF, Escalona H, Leschen RAB (2010m) 11.29. Tenebrionoidea Incertae sedis. In *Handbuch der Zoologie/Handbook of Zoology. Band/ Volume IV Arthropoda: Insecta Teilband/Part 38. Coleoptera, Beetles. Volume 2. Morphology and Systematics (Polyphaga partim)*. (Eds RAB Leschen, RG Beutel and JF Lawrence) pp. 750–760. W. DeGruyter, Berlin.

Lawrence JF, Ślipiński A, Seago AE, Thayer MK, Newton AF, Marvaldi AE (2011) Phylogeny of the Coleoptera based on morphological characters of adults and larvae. *Annales Zoologici* **61**(1), 1–217. doi:10.3161/000345411X576725

Lawrene JF, Spilman TJ (1991) Tenebrionidae (Tenebrionoidea) (including Alleculidae, Cossyphodidae, Lagriidae, Nilionidae, Rhysopaussidae, Tentyriidae). In *Immature Insects. Volume 2.* (Ed. FW Stehr) pp. 520–528. Kendall/Hunt, Dubuque, Iowa.

Lawson FA (1991) Chrysomelidae (Chrysomeloidea) (= Cassididae, Cryptocephalidae, Megalopodidae, Sagridae, etc.). In *Immature Insects. Volume 2.* (Ed. FW Stehr) pp. 568–585. Kendall/Hunt, Dubuque, Iowa.

Lea AM (1904b) Notes on Australian and Tasmanian Cryptocephalides, with descriptions of new species. *Transactions of the Entomological Society of London* **1904**, 329–461.

Lea AM (1905) On *Nepharis* and other ants' nest beetles taken by Mr. J. C. Goudie at Birchip. *Proceedings of the Royal Society of Victoria* **17**, 371–385. [pl. xxvii.]

Lea AM (1907) Catalogue of the Australian and Tasmanian Byrrhidae; with descriptions of new species. *Transactions of the Entomological Society of London* **1907**(1), 135–146.

Lea AM (1909) Revision of the Australian and Tasmanian Malacodermidae. *Transactions of the Entomological Society of London* **1909**, 45–251.

Lea AM (1910) Australian and Tasmanian Coleoptera inhabiting or resorting to the nests of ants, bees, and termites. *Proceedings of the Royal Society of Victoria* **23**, 116–230.

Lea AM (1914) Notes on some miscellaneous Coleoptera, with descriptions of new species. *Transactions of the Royal Society of South Australia* **38**, 249–344, pl. 16.

Lea AM (1915) Notes on Australian Eumolpides (Coleoptera Chrysomelidae), with descriptions of new species. *Transactions of the Royal Society of South Australia* **39**, 102–339.

Lea AM (1917a) Descriptions of *new* species of Australian Coleoptera. Part xiii. *Proceedings of the Linnean Society of New South Wales* **42**, 545–582.

Lea AM (1917b) Notes on some miscellaneous Coleoptera, with descriptions of new species.- part iii. *Transactions of the Royal Society of South Australia* **41**, 121–322.

Lea AM (1920) On Australian Coleoptera. Family Byrrhidae. Part II. *Records of the South Australian Museum* **1**, 273–290, pl. 32.

Lea AM (1922a) On Australian Anthicidae (Coleoptera). *Proceedings of the Linnean Society of New South Wales* **47**, 471–512.

Lea AM (1922b) On Australian Coleoptera. Part iv. *Records of the South Australian Museum.* **2**, 271–308.

Lea AM (1924) On Australian Anobiides (Coleoptera). *Transactions of the Royal Society of South Australia* **48**, 15–64.

Lea AM (1925) Notes on some calcareous insect puparia. *Records of the South Australian Museum* **3**, 35–36.

Lea AM (1932) The Phalacridae (Coleoptera) of Australia and New Guinea. *Records of the South Australian Museum* **4**, 433–481.

Lee JE (1990a) Description of first-instar larva of *Syneta adamsi* Baly from Japan, with notes on the systematic position of Synetinae (Coleoptera: Chrysomelidae). *Esakia* **29**, 77–81.

Lee JE (1990b) Larva of *Zeugophora annulata* (Baly) from Japan, with notes on the systematic position of Zeugophorinae (Coleoptera: Chrysomelidae). *Esakia* **29**, 73–76.

Lee JE (1991) A taxonomic study on the larvae of the subfamily Donaciinae from Japan (Coleoptera: Chrysomelidae). *Entomological Research Bulletin (Korea)* **17**, 33–46.

Lee JE (1992) Larval description of four alticine species of genera *Altica* and *Agropistes* from Japan (Chrysomelidae: Coleoptera). *Korean Journal of Entomology* **22**(4), 287–295.

Lee CY, Morimoto K (1988a) Larvae of the weevil family Attelabidae of Japan. Part 1. Subfamily Attelabinae (Insecta: Coleoptera). *Journal of the Faculty of Agriculture of Kyushu University* **32**, 215–237.

Lee CY, Morimoto K (1988b) Larvae of the weevil family Attelabidae of Japan. Part 2. Subfamily Rhynchitinae (Insecta: Coleoptera). *Journal of the Faculty of Agriculture of Kyushu University* **32**, 239–254.

Lee CF, Jäch MA, Beutel RG (2005a) 18.7. Psephenidae Lacordaire, 1854. In *Handbuch der Zoologie/Handbook of Zoology. Band/Volume IV Arthropoda: Insecta Teilband/Part 38. Coleoptera, Beetles. Volume 1. Morphology and Systematics (Archostemata, Adephaga, Myxophaga, Polyphaga partim)*. (Eds RG Beutel and RAB Leschen) pp. 521–533. W. DeGruyter, Berlin.

Lee CF, Lawrence JF, Satô M (2005b) Description of a new species of *Drupeus* Lewis (Coleoptera: Ptilodactylidae), with a discussion of the systematic position of the genus based on larval, pupal, and adult characters. *Insect Systematics & Evolution* **36**, 371–393. doi:10.1163/187631205788838546

Lee CF, Satô M, Shepard WD, Jäch M (2007) Phylogeny of Psephenidae (Coleoptera: Byrrhoidea) based on larval, pupal and adult characters. *Systematic Entomology* **32**(3), 502–538. doi:10.1111/j.1365-3113.2006.00374.x

Lefkovitch LP (1959) A revision of the European Laemophloeinae (Coleoptera: Cucujidae). *Transactions of the Royal Entomological Society of London* **111**, 95–118. doi:10.1111/j.1365-2311.1959.tb02278.x

Lefkovitch LP (1962) A revision of African Laemophloeinae (Coleoptera: Cucujidae). *Bulletin of the British Museum (Natural History). Entomology* **12**, 167–245.

Lekanicová Z (2009) *Molecular phylogeny of the superfamily Tenebrionoidea (Coleoptera: Cucujiformia)*. 126 pp. Palaký University (PhD thesis), Olomouc.

Leng CW (1920) *Catalogue of the Coleoptera of America, North of Mexico*. x + 470 pp. Cosmos Press, Cambridge, Massachusetts.

Lenko K (1967) *Chelonarium vestitum*, inquilino da formiga *Camponotus rufipes* (Coleoptera: Chelonariidae). *Studia Entomologica* **10**(1–4), 433–438.

Lepesme P (1944) Les Coléoptères des denrées alimentaires et des produits industriels entreposés. *Encyclopédie Entomologique A* **22**, 1–355.

LeSage L (1982) The immature stages of *Exema canadensis* Pierce (Coleoptera: Chrysomelidae). *Coleopterists Bulletin* **36**(2), 318–327.

LeSage L (1991) Helodidae (Eucinetoidea) (= Cyphonidae, Scirtidae). In Immature Insectgs. Volume 2 (Ed. FW Stehr) pp. 366–369. Kendall/ Hunt Publishing Co., Dubuque, Iowa.

LeSage L, Harper P (1976) Notes on the life history of the toed-winged beetle *Anchytarsus bicolor* (Melsheimer) (Coleoptera: Ptilodactylidae). *Coleopterists Bulletin* **30**, 233–238.

Leschen RAB (1996) Phylogeny and revision of the genera of Cryptophagidae (Coleoptera: Cucujoidea). *The University of Kansas Science Bulletin* **55**(15), 549–634.

Leschen RAB (1999) Systematics of Nitidulinae (Coleoptera: Nitidulinae): phylogenetic relationships, convexity and the origin of phallalophagy. *Invertebrate Taxonomy* **13**, 845–882. doi:10.1071/IT98016

Leschen RAB (2000) Beetles feeding on bugs (Coleoptera, Hemiptera): repeated shifts from mycophagous ancestors. *Invertebrate Taxonomy* **14**, 917–929. doi:10.1071/IT00025

Leschen RAB (2001) A new genus of Cryptosomatulini (Coleoptera: Cryptophaginae). *New Zealand Entomologist* **24**, 7–9. doi:10.1080/00779962.2001.9722076

Leschen RAB (2002a) 66. Derodontidae LeConte 1861. In *American Beetles. Volume 2. Polyphaga: Scarabaeoidea through Curculionoidea.* (Eds RH Arnett Jr, MC Thomas, PE Skelley and JH Frank) pp. 221–223. CRC Press, Boca Raton, Florida.

Leschen RAB (2002b) 72. Trogossitidae Latreille 1802. In *American Beetles. Volume 2. Polyphaga: Scarabaeoidea through Curculionoidea.* (Eds RH Arnett Jr, MC Thomas, PE Skelley and JH Frank) pp. 263–266. CRC Press, Boca Raton, Florida.

Leschen RAB (2003) *Fauna of New Zealand. Number 47. Erotylidae (Insecta: Coleoptera: Cucujoidea): phylogeny and review.* Manaaki Whenua Press, Lincoln, Canterbury, New Zealand.

Leschen RAB (2005a) 15.2. Eucinetidae Lacordaire, 1857. In *Handbuch der Zoologie/Handbook of Zoology. Band/Volume IV Arthropoda: Insecta Teilband/Part 38. Coleoptera, Beetles. Volume 1. Morphology and Systematics (Archostemata, Adephaga, Myxophaga, Polyphaga partim)*. (Eds RG Beutel and RAB Leschen) pp. 433–438. W. DeGruyter, Berlin.

Leschen RAB (2005b) 15.3. Clambidae Fischer, 1821. In *Handbuch der Zoologie/Handbook of Zoology. Band/Volume IV Arthropoda: Insecta Teilband/Part 38. Coleoptera, Beetles. Volume 1. Morphology and Systematics (Archostemata, Adephaga, Myxophaga, Polyphaga partim)*. (Eds RG Beutel and RAB Leschen) pp. 438–443. W. DeGruyter, Berlin.

Leschen RAB (2010a) 9. Cleroidea Latreille, 1802. Introduction and Phylogeny. In *Handbuch der Zoologie/Handbook of Zoology. Band/Volume IV Arthropoda: Insecta Teilband/Part 38. Coleoptera, Beetles. Volume 2. Morphology and Systematics (Polyphaga partim)*. (Eds RAB Leschen, RG Beutel and JF Lawrence) pp. 237–239. W. DeGruyter, Berlin.

Leschen RAB (2010b) 9.3 Chaetosomatidae Crowson, 1952. In *Handbuch der Zoologie/Handbook of Zoology. Band/Volume IV Arthropoda: Insecta Teilband/Part 38. Coleoptera, Beetles. Volume 2. Morphology and Systematics (Polyphaga partim)*. (Eds RAB Leschen, RG Beutel and JF Lawrence) pp. 247–250. W. DeGruyter, Berlin.

Leschen RAB (2010c) 10.10. Cryptophagidae Kirby, 1837. In *Handbuch der Zoologie/Handbook of Zoology. Band/Volume IV Arthropoda: Insecta Teilband/Part 38. Coleoptera, Beetles. Volume 2. Morphology and Systematics (Polyphaga partim)*. (Eds RAB Leschen, RG Beutel and JF Lawrence) pp. 327–334. W. DeGruyter, Berlin.

Leschen RAB (2011) World review of *Laricobius* (Coleoptera: Derodontidae). *Zootaxa* **2908**, 1–44.

Leschen RAB, Beutel RG (2004) Ocellar ativism in Coleoptera: plesiomorphy or apomorphy? *Journal of Zoological Systematics and Evolutionary Research* **42**, 63–69. doi:10.1046/j.0947-5745.2003.00241.x

Leschen RAB, Beutel RG (2010a) 5.1. Derodontidae LeConte, 1861. In *Handbuch der Zoologie/Handbook of Zoology. Band/Volume IV Arthropoda: Insecta Teilband/Part 38. Coleoptera, Beetles. Volume 2. Morphology and Systematics (Polyphaga partim)*. (Eds RAB Leschen, RG Beutel and JF Lawrence) pp. 180–185. W. DeGruyter, Berlin.

Leschen RAB, Beutel RG (2010b) 5.2. Nosodendridae Erichson, 1846. In *Handbuch der Zoologie/Handbook of Zoology. Band/Volume IV Arthropoda: Insecta Teilband/Part 38. Coleoptera, Beetles. Volume 2. Morphology and Systematics (Polyphaga partim)*. (Eds RAB Leschen, RG Beutel and JF Lawrence) pp. 185–190. W. DeGruyter, Berlin.

Leschen RAB, Beutel RG (2010c) 9.8. Phycosecidae Crowson, 1952. In *Handbuch der Zoologie/Handbook of Zoology. Band/Volume IV Arthropoda: Insecta Teilband/Part 38. Coleoptera, Beetles. Volume 2. Morphology and Systematics (Polyphaga partim)*. (Eds RAB Leschen, RG Beutel and JF Lawrence) pp. 265–268. W. DeGruyter, Berlin.

Leschen RAB, Buckley TR (2007) Multistate characters and diet shifts: evolution of Erotylidae (Coleoptera). *Systematic Biology* **56**(1), 97–112. doi:10.1080/10635150701211844

Leschen RAB, Early J (2004) New collection of *Brounia thoracica* (Coleoptera: Chelonariidae). *The Weta* **27**, 15–21.

Leschen RAB, Lackner T (2013) Gondwanan Gymnochilini (Trogossitidae): long-range Pacific dispersal, generic concepts, and review of New Zealand species. *Systematic Entomology* **38**, 278–304. doi:10.1111/j.1365-3113.2012.00661.x

Leschen RAB, Lawrence JF (1991) Fern sporophagy in Coleoptera from the Juan Fernandez Islands, Chile, with descriptions of two new genera in Cryptophagidae and Mycetophagidae. *Systematic Entomology* **16**, 329–352.

Leschen RAB, Lawrence JF (2010a) 10.11. Agapythidae Sen Gupta and Crowson, 1969. In *Handbuch der Zoologie/Handbook of Zoology. Band/Volume IV Arthropoda: Insecta Teilband/Part 38. Coleoptera, Beetles. Volume 2. Morphology and Systematics (Polyphaga partim)*. (Eds RAB Leschen, RG Beutel and JF Lawrence) pp. 334–336. W. DeGruyter, Berlin.

Leschen RAB, Lawrence JF (2010b) 10.12. Priasilphidae Crowson, 1973. In *Handbuch der Zoologie/Handbook of Zoology. Band/Volume IV Arthropoda: Insecta Teilband/Part 38. Coleoptera, Beetles. Volume 2. Morphology and Systematics (Polyphaga partim)*. (Eds RAB Leschen, RG Beutel and JF Lawrence) pp. 336–340. W. DeGruyter, Berlin.

Leschen RAB, Löbl I (2005) Phylogeny and classification of Scaphisomatini (Staphylinidae: Scaphidiinae) with notes on mycophagy, termitophily, and functional morphology. *Coleopterists Society Monograph* **59**, 1–63. doi:10.1649/0010-065X(2005)059[0001:PACOSS]2.0.CO;2

Leschen RAB, Marris JWM (2005) *Carpophilus (Coleoptera: Nitidulidae) of New Zealand with notes on Australian species*. Landcare Research New Zealand (Research Contract Report LC0405/153), Auckland, New Zealand.

Leschen RAB, Michaux B (2005) Biogeography and evolution of New Zealand Priasilphidae (Coleoptera: Cucujoidea). *New Zealand Entomologist* **28**, 55–64. doi:10.1080/00779962.2005.9722686

Leschen RAB, Newton AF (2003) Larval description, adult feeding, behavior, and phylogenetic placement of *Megalopinus* (Coleoptera: Staphylinidae). *Coleopterists Bulletin* **57**(4), 469–493. doi:10.1649/586

Leschen RAB, Rhode BE (2002) A new genus and species of large extinct Ulodidae (Coleoptera) from New Zealand. *New Zealand Entomologist* **25**, 57–64. doi:10.1080/00779962.2002.9722095

Leschen RAB, Skelley PE (2002a) 85. Cryptophagidae Kirby 1837. In *American Beetles. Volume 2. Polyphaga: Scarabaeoidea through Curculionoidea.* (Eds RH Arnett Jr, MC Thomas, PE Skelley and JH Frank) pp. 338–342. CRC Press, Boca Raton, Florida.

Leschen RAB, Skelley PE (2002b) 86. Languriidae Wiedeman 1823. In *American Beetles. Volume 2. Polyphaga: Scarabaeoidea through Curculionoidea.* (Eds RH Arnett Jr, MC Thomas, PE Skelley and JH Frank) pp. 343–347. CRC Press, Boca Raton, Florida.

Leschen RAB, Ślipiński A (2010a) 7. Cucujiformia Lameere, 1938. Introduction, Phylogeny. In *Handbuch der Zoologie/Handbook of Zoology. Band/Volume IV Arthropoda: Insecta Teilband/Part 38. Coleoptera, Beetles. Volume 2. Morphology and Systematics (Polyphaga partim).* (Eds RAB Leschen, RG Beutel and JF Lawrence) pp. 227–228. W. DeGruyter, Berlin.

Leschen RAB, Ślipiński A (2010b) 10. Cucujoidea Latreille, 1802. Introduction, Phylogeny. In *Handbuch der Zoologie/Handbook of Zoology. Band/Volume IV Arthropoda: Insecta Teilband/Part 38. Coleoptera, Beetles. Volume 2. Morphology and Systematics (Polyphaga partim).* (Eds RAB Leschen, RG Beutel and JF Lawrence) pp. 281–282. W. DeGruyter, Berlin.

Leschen RAB, Ślipiński A (2010c) 11.10. Ulodidae Pascoe, 1869. In *Handbuch der Zoologie/Handbook of Zoology. Band/Volume IV Arthropoda: Insecta Teilband/Part 38. Coleoptera, Beetles. Volume 2. Morphology and Systematics (Polyphaga partim).* (Eds RAB Leschen, RG Beutel and JF Lawrence) pp. 559–563. W. DeGruyter, Berlin.

Leschen RAB, Węgrzynowicz P (1998) Generic catalogue and taxonomic status of Languriidae (Cucujoidea). *Annales Zoologici* **48**(3/4), 221–243.

Leschen RAB, Lawrence JF, Ślipiński SA (2005) Classification of basal Cucujoidea (Coleoptera: Polyphaga), with a cladistic analysis, description of new genera and species, and keys to adults and larvae of cucujoid families. *Invertebrate Systematics* **19**, 17–73. doi:10.1071/IS04007

Leschen RAB, Beutel RG, Lawrence JF (eds) (2010a). *Handbuch der Zoologie/Handbook of Zoology. Band/Volume IV Arthropoda: Insecta Teilband/Part 38. Coleoptera, Beetles. Volume 2. Morphology and Systematics (Polyphaga partim).* W. DeGruyter, Berlin.

Leschen RAB, Skelley PE, McHugh JV (2010b) 10.7. Erotylidae Leach, 1815. In *Handbuch der Zoologie/Handbook of Zoology. Band/Volume IV Arthropoda: Insecta Teilband/Part 38. Coleoptera, Beetles. Volume 2. Morphology and Systematics (Polyphaga partim).* (Eds RAB Leschen, RG Beutel and JF Lawrence) pp. 311–319. W. DeGruyter, Berlin.

Leschen RAB, Gimmel M, Ślipiński A (2010c) 10.24. Cyclaxyridae Gimmel, *Leschen &* Ślipiński, 2010. In *Handbuch der Zoologie/Handbook of Zoology. Band/Volume IV Arthropoda: Insecta Teilband/Part 38. Coleoptera, Beetles. Volume 2. Morphology and Systematics (Polyphaga partim).* (Eds RAB Leschen, RG Beutel and JF Lawrence) pp. 383–386. W. DeGruyter, Berlin.

Lesne P (1897) Révision des Coléoptères de la famille des Bostrychides. 2e Mémoire. Bostrychides Hypocéphaliques, Dinoderinae. *Annales de la Société Entomologique de France* **66**, 319–350.

Lesne P (1899) Révision des Coléoptères de la famille des Bostrychides. 3e Mémoire. Bostrychinae sens. strict. – I. Les *Bostrychus*. *Annales de la Société Entomologique de France* **67**, 438–621.

Lesne P (1901) Révision des Coléoptères de la famille des Bostrychides. 4e Mémoire. Bostrychinae sens. strict. – II. Les *Xylopertha*. *Annales de la Société Entomologique de France* **69**, 473–639.

Lesne P (1906) Révision des Coléoptères de la famille des Bostrychides. 5e Mémoire. Synoxylinae. *Annales de la Société Entomologique de France* **75**, 445–561.

Lesne P (1909) Révision des Coléoptères de la famille des Bostrychides. 6e Mémoire. Dinapatinae et Apatinae. *Annales de la Société Entomologique de France* **78**, 471–574.

Lesne P (1911) Les Tristariens; leurs affinités zoologiques. Synopsis du groupe. *Bulletin du Muséum d'Histoire Naturelle* **1911**, 202–208.

Lesne P (1921) Classification des Coléoptères xylophages de la famille des Bostrychides. *Compte Rendu, Association Française pour l'Avancement des Sciences Naturelles. Congrés, Strasbourg* **1920**, 285–289.

Levey B (1985) Afreminae: a new subfamily of Anthicidae (Coleoptera) from southern Africa. *Entomologica Scandinavica* **15**, 419–422.

Li CS (1993) Review of the Australian Epilachninae (Coleoptera: Coccinellidae). *Journal of the Australian Entomological Society* **32**(3), 209–224. doi:10.1111/j.1440-6055.1993.tb00577.x

Liebherr JK (1990) A new tribal placement for the Australasian genera *Homethes* and *Aeolodermus* (Coleoptera: Carabidae: Odacanthini). *The Pan-Pacific Entomologist* **66**, 312–321.

Liebherr JK (2011) Cladistic assessment of subtribal affinities within the tribe Moriomorphini with description of *Rossjoycea glacialis*, gen. n. and sp. n. from the South Island, and revision of *Meonochilus* Liebherr and Marris from the North Island, New Zealand (Coleoptera, Carabidae). *ZooKeys* **147**, 277–335. doi:10.3897/zookeys.147.1898

Liebherr JK, Kavanaugh DH (1985) Ovoviviparity in carabid beetles of the genus *Pseudomorpha* (Insecta: Coleoptera). *Journal of Natural History* **19**, 1079–1086. doi:10.1080/00222938500770681

Liebherr JK, Will KW (1998) Inferring phylogenetic relationships within Carabidae (Insecta, Coleoptera) from characters of the female reproductive tract. In *Phylogeny and Classification of Caraboidea. Proceedings of a Symposium (30 August, 1996, Florence, Italy). XX International Congress of Entomology.* (Eds GE Ball, A Casale and AV Taglianti), pp. 107–170. Museo Regionale di Scienze Naturali, Torino.

Lin QB (1986) Early Mesozoic Fossil Insects from South China. *Palaeontologia Sinica* **170**, 1–112.

Lindroth CH (1957) The principal terms used for male and female genitalia in Coleoptera. *Opuscula Entomologica* **22**, 241–256.

Lingafelter SW, Hoebeke ER (2002) *Revision of* Anoplophora *(Coleoptera: Cerambycidae).* 236 pp. Entomological Society of Washington, Washington, D. C.

Lingafelter SW, Konstantinov AS (2000) The monophyly and relative rank of alticine and galerucine leaf beetles: a cladistic analysis using adult morphological characters (Coleoptera: Chrysomelidae). *Entomologica Scandinavica* **30**, 397–416. doi:10.1163/187631200X00525

Linnaeus C (1758) *Systema Naturae.... Tomus I Edito Decima.* Laurentius Salvius, Stockholm.

Linsley EG (1959) Ecology of Cerambycidae. *Annual Review of Entomology* **4**, 99–138. doi:10.1146/annurev.en.04.010159.000531

Linsley EG (1961) The Cerambycidae of North America. Part I. Introduction. *University of California Publications in Entomology* 18, 97 pp. 35 pls.

Linsley EG (1962a) The Cerambycidae of North America, Part II. Taxonomy and classification of the Parandrinae, Prioninae, Spondylinae, and Aseminae. *University of California Publications in Entomology* **19**, 102 pp., 1 pl.

Linsley EG (1962b) The Cerambycidae of North America, Part III. Taxonomy and classification of the subfamily Cerambycinae, tribes Opsimini through Megaderini. *University of California Publications in Entomology* **20**, 188 pp.

Linsley EG (1963) The Cerambycidae of North America, Part IV. Taxonomy and classification of the subfamily Cerambycinae, tribes Elaphidionini through Rhinotragini. *University of California Publications in Entomology* **21**, 165 pp.

Linsley EG (1964) The Cerambycidae of North America, Part V. Taxonomy and classification of the subfamily Cerambycinae, tribes Callichromini through Ancylocerini. *University of California Publications in Entomology* **22**, 197 pp., 1 pl.

Linsley EG, Chemsak JA (1984) Cerambycidae of North America. Part VII, No. 1. Taxonomy and classification of the subfamily Lamiinae, tribes Parmenini through Acanthoderini. *University of California Publications in Entomology* **102**, 1–258.

Linsley EG, Chemsak JA (1995) Cerambycidae of North America. Part VII, No. 2. Taxonomy and classification of the subfamily Lamiinae, tribes Acanthocinini through Hemilophini. *University of California Publications in Entomology* **114**, 1–292.

Linsley EG, Chemsak JA (1997) Cerambycidae of North America. Part VIII: Bibliography, index, and host plant index. *University of California Publications in Entomology* **117**, 1–534.

Linsley EG, MacSwain JW (1942) The bionomics of *Ptinus californicus*, a depredator in the nests of bees. *Bulletin of the California Academy of Sciences* **40**, 126–137.

Linsley EG, MacSwain JW, Smith RF (1952) The life history and development of *Rhipiphorus smithi* with notes on their phylogenetic significance (Coleoptera, Rhipiphoridae). *University of California Publications in Entomology* **9**, 291–314.

Liu LY, Schönitzer K (2011) Phylogenetic analysis of the family Bostrichidae auct. at suprageneric levels. *Mitteilungen des Münchener Entomologische Gesellschaft* **101**, 99–132.

Liu M, Ren D (2006) First fossil Eccoptarthridae (Coleoptera: Curculionoidea) from the Mesozoic of China. *Zootaxa* **1176**, 59–68.

Liu M, Ren D, Tan J-J (2006) New fossil weevils (Coleoptera: Curculionoidea: Nemonychidae) from the Jehol Biota of western Liaoning, China. *Annales Zoologici* **56**(4), 605–612.

Lloyd JE (2002) 62. Lampyridae Latreille 1817. In *American Beetles. Volume 2. Polyphaga: Scarabaeoidea through Curculionoidea.* (Eds RH Arnett Jr, MC Thomas, PE Skelley and JH Frank) pp. 187–196. CRC Press, Boca Raton, Florida.

Löbl I (1995) New species of terrestrial *Microsporus* from the Himalaya (Coleoptera: Microsporidae). *Entomologische Blätter* **91**(3), 128–138.

Löbl I (1997) *Catalogue of the Scaphidiinae (Coleoptera: Staphylinidae).* Museum d'Histoire Naturelle, (Instrumenta Biodiversitatis I), Genève.

Löbl I (2005) *Laenagenius apterus* gen. et sp. nov. from China (Coleoptera: Pterogeniidae). *Entomological Problems* **35**(2), 153–155.

Löbl I, Burckhardt D (1988) Revision der Gattung *Sarothrias* mit Bemerkungen zur Familie Jacobsoniidae (Coleoptera). *Stuttgarter Beitrage zur Naturkunde. Serie A, Biologie* **422**, 1–23.

Löbl I, Leschen RAB (1994) *Anogenius clypealis* (Coleoptera: Pterogeniidae), a new genus and species from Malaysia. *Entomological Problems* **25**, 53–57.

Löbl I, Leschen RAB (2003) *Fauna of New Zealand. Number 48. Scaphidiinae (Insecta: Coleoptera: Staphylinidae).* Manaaki Whenua Press, Lincoln, Canterbury, New Zealand.

Löbl I, Leschen RAB (2010) Notes on the *Toxidium* group (Coleoptera: Staphylinidae: Scaphidiinae). *Folia Heyrovskyana* **18**, 71–93.

Löbl I, Smetana A (eds) (2003) *Catalogue of Palaearctic Coleoptera. Volume 1. Archostemata – Myxophaga – Adephaga.* Apollo Books, Stenstrup.

Löbl I, Smetana A (eds) (2004) *Catalogue of Palaearctic Coleoptera. Volume 2. Hydrophiloidea – Histeroidea – Staphylinoidea.* Apollo Books, Stenstrup.

Löbl I, Smetana A (eds) (2006) *Catalogue of Palaearctic Coleoptera. Volume 3. Scarabaeoidea – Scirtoidea – Dascilloidea – Buprestoidea – Byrrhoidea.* Apollo Books, Stenstrup.

Löbl I, Smetana A (eds) (2007) *Catalogue of Palaearctic Coleoptera. Volume 4. Elateroidea – Derodontoidea – Bostrichoidea – Lymexyloidea – Cleroidea – Cucujoidea.* Apollo Books, Stenstrup. 935 pp. Apollo Books, Stenstrup.

Löbl I, Smetana A (eds) (2008). *Catalogue of Palaearctic Coleoptera. Volume 5. Tenebrionoidea.* Apollo Books, Stenstrup.

Löbl I, Smetana A (eds) (2010). *Catalogue of Palaearctic Coleoptera. Volume 6. Chrysomeloidea.* 924 pp. Apollo Books, Stenstrup.

Löbl I, Smetana A (eds) (2011). *Catalogue of Palaearctic Coleoptera. Volume 7. Curculionoidea I.* 373 pp. Apollo Books, Stenstrup.

Longhorn SJ, Pohl HW, Vogler AP (2010) Ribosomal protein genes of holometabolan insects reject the Halteria, instead revealing a close affinity of Strepsiptera with Coleoptera. *Molecular Phylogenetics and Evolution* **55**, 846–859. doi:10.1016/j.ympev.2010.03.024

Lopes-Andrade C (2008a) An essay on the tribe Xylographellini (Coleoptera: Tenebrionoidea: Ciidae). *Zootaxa* **1832**, 1–110.

Lopes-Andrade C (2008b) The first record of *Cis chinensis* Lawrence from Brazil, with the delimitation of the *Cis multidentatus* species-group (Coleoptera: Ciidae). *Zootaxa* **1755**, 35–46.

Lopes-Andrade C, Gumier-Costa F, Sperber CF (2003) Why do male *Xylographus contractus* Mellié (Coleoptera: Ciidae) present abdominal fovea? Evidence of sexual pheromone secretion. *Neotropical Entomology* **32**(2), 217–220. doi:10.1590/S1519-566X2003000200005

Lord NP, Hartley CS, Lawrence JF, McHugh JV, Whiting MF, Miller KB (2010) Phylogenetic analysis of the minute brown scavenger beetles (Coleoptera: Latridiidae), and recognition of a new beetle family Akalyptoischiidae fam. n. (Coleoptera: Cucujoidea). *Systematic Entomology* **35**, 753–763. doi:10.1111/j.1365-3113.2010.00532.x

Louw S (1993) Systematics of the Urodontidae (Coleoptera: Curculionoidea) of southern Africa. *Entomology Memoirs, Department of Agriculture. Republic of South Africa* **87**, 1–92.

Louw S (2004) Microcerini Lacordaire, 1863 (Coleoptera, Curculionoidea). In *Brentidae of the World (Coleoptera, Curculionoidea).* (Eds. A Sforzi and L Bartolozzi) pp. 905–935. Museo Regionale di Scienze Naturali (Monografie 39), Torino.

Lu W, Jackman JA, Johnson PW (1997) Male genitalia and phylogenetic relationships in North American Mordellidae (Coleoptera). *Annals of the Entomological Society of America* **90**(6), 742–767.

Lubkin SH, Engel MS (2005) *Permocoleus*, new genus, the first Permian beetle (Coleoptera) from North America. *Annals of the Entomological Society of America* **98**(1), 73–76. doi:10.1603/0013-8746(2005)098[0073:PNGTFP]2.0.CO;2

Lyal CHC (1995) The ventral structures of the weevil head (Coleoptera: Curculionoidea). *Memoirs of the Entomological Society of Washington* **14**, 35–51.

Lyal CHC, King T (1996) Elytro-tergal stridulation in weevils (Insects: Coleoptera: Curculionoidea). *Journal of Natural History* **30**, 703–773. doi:10.1080/00222939600770391

Lyal CHC, Douglas DA, Hine SJ (2006) Morphology and systematic significance of sclerolepidia in the weevils (Coleoptera: Curculionoidea). *Systematics and Biodiversity* **4**(2), 203–241. doi:10.1017/S1477200005001714

Lyubarsky GY (1998) Phylogenetics of the beetles of the family Cryptophagidae: a gradistic analysis. *Zoologicheskiye Issledovania* **1**, 1–90. [in Russian]

MacSwain JW (1956) A classification of the first instar larvae of the Meloidae (Coleoptera). *University of California Publications in Entomology* **12**, 1–182.

Maddison DR, Baker MD, Ober KA (2000) Phylogeny of the beetle suborder Adephaga and family Carabidae, as inferred using ribosomal DNA and the *wingless* gene. *Proceedings of the International Congress of Entomology* **21**(1), 930

Maddison DR, Moore W, Baker MD, Ellis TM, Ober KA, Cannone JJ, Gutell RR (2009) Monophyly of terrestrial adephagan beetles as indicated by three nuclear genes (Coleoptera: Carabidae and Trachypachidae). *Zoologica Scripta* **38**, 43–62. doi:10.1111/j.1463-6409.2008.00359.x

Magis N, Wittmer W (1974) Nouvelle répartition des genres de la sous-famille des Chauliognathinae (Coleoptera, Cantharoidea: Cantharidae). *Bulletin de la Société Royale des Sciences de Liège* **42**(1–2), 78–95.

Majer K (1987) Comparative morphology and proposed major taxonomy of the family Melyridae (Insecta, Coleoptera). *Polskie Pismo Entomologiczne* **56**, 719–859.

Majer K (1994a) Gietellidae, full family status for Melyridae: Gietellinae (Coleoptera: Cleroidea). *Entomological Problems* **25**(1), 65–72.

Majer K (1994b) Review of the classification of the Melyridae and related families (Coleoptera, Cleroidea). *Entomologica Basiliensia* **17**, 319–390.

Majer K (1995) Revision of the family Mauroniscidae (Insecta: Coleoptera: Cleroidea). *Entomologische Abhandlungen aud dem Staatliches Museum für Tierkunde in Dresden* **57**, 57–89.

Majer K (2002) Subfamilial classification of the family Malachiidae (Coleoptera, Cleroidea). *Entomologica Basiliensia* **24**, 179–244.

Malloch G, Fenton B, Goodrich MA (2001) Phylogeny of raspberry beetles and other Byturidae (Coleoptera). *Insect Molecular Biology* **10**(3), 281–291. doi:10.1046/j.1365-2583.2001.00266.x

Mamaev BM (1974) Preimaginal phase of *Syntelia histeroides* Lewis (Synteliidae) in comparison with some Histeridae (Coleoptera). *Entomologicheskoye Obozreniye* **53**(4), 866–871.

Mamaev BM (1976a) Morphological types of eucnemid larvae (Coleoptera, Eucnemidae) and their evolutionary significance. In *Evolutionary Morphology of Insect Larvae*. (Ed. BM Mamayev) pp. 136–155. Nauka, Moscow (in Russian).

Mamaev BM (1976b) Review of larvae of the family Trogossitidae (Coleoptera) in the fauna of the USSR. *Зоологический журнал* **55**, 1648–1658. [in Russian]

Mamaev BM (1977) Clerid larvae (Coleoptera, Cleridae) that prey upon forest pests and timber pests. *Entomologicheskoye Obozreniye* 54, 395–408 (in Russian; English translation in *Entomological Review* **54**, 116–124, 1978).

Mann JS, Crowson RA (1981) The systematic positions of *Orsodacne* Latr. and *Syneta* Lac. (Coleoptera Chrysomelidae), in relation to characters of larvae, internal anatomy and tarsal vestiture. *Journal of Natural History* **15**, 727–749. doi:10.1080/00222938100770531

Mann JS, Crowson RA (1983) On the internal male reproductive organs and their taxonomic significance in the leaf beetles (Coleoptera: Chrysomelidae). *Entomologia Generalis* **9**, 75–99.

Marquet J (1992) Essai sur les Euchiridae (Coleoptera: Scarabaeoidea). *Bulletin de l'Entomologie Tourangelle* **13**(1–2), 1–7.

Martin JEH (Ed) (1977) *The Insects and Arachnids of Canada. Part I. Collecting, Preparing and Preserving Insects, Mites, and Spiders.* Canadian Department of Agriculture Research Branch (Publication No. 1943), Ottawa.

Martin NA (2000) A longicorn leaf miner, *Microlamia pygmaea* (Coleoptera: Cerambycidae: Lamiinae) found in New Zealand. *New Zealand Entomologist* **23**, 86. doi:10.1080/00779962.2000.9722073

Martin SK (2010) Early Jurassic coleopterans from the Mintaja insect locality, Western Australia. *Acta Geologica Sinica* **84**(4), 925–953. doi:10.1111/j.1755-6724.2010.00276.x

Martynov AV (1926) To the knowledge of fossil insects of Jurassic shales of Turkestan. 5. On some forms of beetles (Coleoptera). *Ezhegodnik Russkogo Paleontologicheskogo Obschestva* **5**, 1–38.

Marvaldi AE (1997) Higher level phylogeny of Curculionidae (Coleoptera: Curculionoidea) based mainly on larval characters, with special reference to broad-nosed weevils. *Cladistics* **13**, 285–312. doi:10.1111/j.1096-0031.1997.tb00321.x

Marvaldi AE (2003) Key to larvae of the South American subfamilies of weevils (Coleoptera, Curculionoidea). *Revista Chilena de Historia Natural (Valparaiso, Chile)* **76**, 603–612.

Marvaldi AE (2005) Larval morphology and biology of oxycorynine weevils and the higher phylogeny of Belidae (Coleoptera, Curculionoidea). *Zoologica Scripta* **34**, 37–48. doi:10.1111/j.1463-6409.2005.00169.x

Marvaldi AE, Morrone JJ (2000) Phylogenetic systematics of weevils (Coleoptera: Curculionoidea): a reappraisal based on larval and adult morphology. *Insect Systematics & Evolution* **31**, 43–58. doi:10.1163/187631200X00309

Marvaldi AE, Sequeira AS, O'Brien CW, Farrell BD (2002) Molecular and morphological phylogenetics of weevils (Coleoptera: Curculionoidea): do niche shifts accompany diversification? *Systematic Biology* **51**(5), 761–785. doi:10.1080/10635150290102465

Marvaldi AE, Oberprieler RG, Lyal CHC, Bradbury T, Anderson RS (2006) Phylogeny of the Oxycoryninae sensu lato (Coleoptera: Belidae) and evolution of host-plant associations. *Invertebrate Systematics* **20**, 447–476. doi:10.1071/IS05059

Marvaldi AE, Duckett CN, Kjer KM, Gillespie JJ (2009) Structural alignment of 18S and 28S rDNA sequences provides insights into phylogeny of Phytophaga (Coleoptera: Curculionoidea and Chrysomeloidea). *Zoologica Scripta* **38**, 63–77. doi:10.1111/j.1463-6409.2008.00360.x

Matsuda R (1960) Morphology of the pleurosternal region of the pterothorax in insects. *Annals of the Entomological Society of America* **53**(6), 712–731.

Matsuda R (1965) Morphology and evolution of the insect head. *Memoirs of the American Entomological Institute* **4**, 1–334.

Matsuda R (1970) Morphology and evolution of the insect thorax. *Memoirs of the Entomological Society of Canada* **102**, 1–431. doi:10.4039/entm10276fv

Matsuda R (1976) *Morphology and Evolution of the Insect Abdomen, with Special Reference to Developmental Patterns and their Bearings on Systematics.* Pergamon Press, Oxford.

Matthes D (1962) Excitatoren und Paarungsverhalten Mitteleuropäischer Malachiiden (Coleopt., Malacodermata). *Zeitschrift fur Morphologie und Oekologie der Tiere* **51**, 375–546. doi:10.1007/BF00406263

Matthews EG (1972) A revision of the scarabaeine dung beetles of Australia. I. Tribe Onthophagini. *Australian Journal of Zoology* **9**, (Supplement), 1–330.

Matthews EG (1974) A revision of the scarabaeine dung beetles of Australia. II. Tribe OnthoScarabaeini. *Australian Journal of Zoology* **24**, (Supplement), 1–211.

Matthews EG (1976) A revision of the scarabaeine dung beetles of Australia. III. Tribe Coprini. *Australian Journal of Zoology* **38**, (Supplement), 1–52.

Matthews EG (1980) *A Guide to the Genera of Beetles of South Australia. Part 1. Achostemata and Adephaga.* South Australian Museum (Special Education Bulletin Series), Adelaide.

Matthews EG (1982) *A Guide to the Genera of Beetles of South Australia. Part 2. Polyphaga: Staphylinoiodea and Hydrophiloidea.* South Australian Museum (Special Education Bulletin Series), Adelaide.

Matthews EG (1984) *A Guide to the Genera of Beetles of South Australia. Part 3. Polyphaga: Eucinetoidea, Dascilloidea and Scarabaeoidea.* South Australian Museum (Special Education Bulletin Series, No. 6), Adelaide.

Matthews EG (1985) *A Guide to the Genera of Beetles of South Australia. Part 4. Polyphaga: Byrrhoidea, Buprestoidea, Dryopoidea, Elateroidea, Cantharoidea, Derodontoidea, Bostrichoidea.* South Australian Museum (Special Education Bulletin Series, No. 7), Adelaide.

Matthews EG (1986) Revision of the troglophilic genus *Brises* Pascoe, with a discussion of the Cyphaleini (Coleoptera, Tenebrionidae). *Records of the South Australian Museum* **19**, 77–90.

Matthews EG (1987) *A Guide to the Genera of Beetles of South Australia. Part 5. Polyphaga: Tenebrionoidea.* South Australian Museum (Special Education Bulletin Series, No. 8), Adelaide.

Matthews EG (1992a) *A Guide to the Genera of Beetles of South Australia. Part 6. Polyphaga: Lymexyloidea, Cleroidea and Cucujoidea.* South Australian Museum (Special Education Bulletin Series, No. 9), Adelaide.

Matthews EG (1992b) Classification, relationships and distribution of the genera of Cyphaleini (Coleoptera: Tenebrionidae). *Invertebrate Taxonomy* **6**, 437–522. doi:10.1071/IT9920437

Matthews EG (1993) Classification, relationships and distribution of the genera of Heleini (Coleoptera: Tenebrionidae). *Invertebrate Taxonomy* **7**, 1025–1095. doi:10.1071/IT9931025

Matthews EG (1997) *A Guide to the Genera of Beetles of South Australia. Part 7. Polyphaga: Chrysomeloidea: Cerambycidae.* South Australian Museum (Special Education Bulletin Series, No. 10), Adelaide.

Matthews EG (2000) Origins of Australian arid-zone tgenebrionid beetles. *Invertebrate Taxonomy* **14**, 941–951. doi:10.1071/IT00021

Matthews EG (2003) *Ulomotypus* Broun, a member of the new subfamily Palorinae, with remarks on *Aphtora* Bates and *Demtrius* Broun (Coleoptera, Tenebrionidae). *New Zealand Entomologist* **26**, 7–14. doi:10.1080/00779962.2003.9722104

Matthews EG, Bouchard P (2008) *Tenebrionid Beetles of Australia. Descriptions of Tribes, Keys to Genera, Catalogue of Species.* viii + 398 pp. Australian Biological Resources Study, Canberra, ACT.

Matthews EG, Lawrence JF (1992) A new genus and species of Heleini from Tasmania (Coleoptera: Tenebrionidae). *Journal of the Australian Entomological Society* **31**, 311–316. doi:10.1111/j.1440-6055.1992.tb00513.x

Matthews EG, Lawrence JF (2005) New taxa, new synonymy and new generic records for Australian Tenebrionidae (Coleoptera). *Annales Zoologici* **55** (4), 531–547.

Matthews EG, Monteith GB (2010) *Monteithium storeyi*, a new species of Adeliini (Coleoptera: Tenebrionidae) from the wet tropics of Queensland, Australia. *Australian Entomologist* **36**(4), 201–206.

Matthews EG, Reid CAM (2002) *A Guide to the Genera of Beetles of South Australia. Part 8. Polyphaga: Chrysomeloidea: Chrysomelidae.* South Australian Museum (Special Educational Bulletin Series No. 11), Adelaide.

Matthews EG, Stebnicka Z (1986) A review of *Dmarziella* Balthasar, with a transfer from Aphodiinae to Scarabaeinae (Coleoptera: Scarabaeidae). *Australian Journal of Zoology* **34**, 449–461. doi:10.1071/ZO9860449

Matthews EG, Lawrence JF, Bouchard P, Steiner WE Jr, Ślipiński A (2010) 11.14. Tenebrionidae Latreille, 1802. In *Handbuch der Zoologie/Handbook of Zoology. Band/Volume IV Arthropoda: Insecta Teilband/Part 38. Coleoptera, Beetles. Volume 2. Morphology and Systematics (Polyphaga partim).* (Eds RAB Leschen, RG Beutel and JF Lawrence) pp. 574–659. W. DeGruyter, Berlin.

Maulik S (1931) On the structure of larvae of hispine beetles (I). *Proceedings of the Zoological Society of London* **1931**, 1137–1162.

Maulik S (1932) On the structure of larvae of hispine beetles (II). *Proceedings of the Zoological Society of London* **1932**, 293–322.

Maulik S (1941) Biology and morphology of the Sagrinae (Chrysomelidae, Coleoptera). *Annals & Magazine of Natural History* **7**(11), 235–254.

May BM (1992) *Geochus* Broun, 1882: An enigmatic genus now known to be a leafminer (Curculionidae: Curculionini). *New Zealand Entomologist* **15**, 22–25. doi:10.1080/00779962.1992.9722623

May BM (1993) *Fauna of New Zealand. Number 28. Larvae of Curculionoidea (Insecta: Coleoptera): A Systematic Overview.* Manaaki Whenua Press, Lincoln, Canterbury, New Zealand.

May BM (1994) An introduction to the immature stages of Australian Curculionoidea. In In *Australian Weevils.* Volume II. (Ed. EC Zimmerman) pp. 365–702. CSIRO Publications, East Melbourne, Victoria.

May BM (1997) Larvae of two weevil genera (Coleoptera: Curculionidae) associated with *Crossotarsus armipennis* Lea (Platypodidae) from New South Wales. *New Zealand Entomologist* **20**, 7–13. doi:10.1080/00779962.1997.9722662

Mayor AJ (2002) 74. Melyridae Leach 1815. In *American Beetles. Volume 2. Polyphaga: Scarabaeoidea through Curculionoidea.* (Eds RH Arnett Jr, MC Thomas, PE Skelley and JH Frank) pp. 281–304. CRC Press, Boca Raton, Florida.

Mazur S (2011). *A concise catalogue of the Histeridae (Insecta: Coleoptera).* 332 pp. Warsaw University of Life Sciences, SGGW Press, Warsaw.

McCairns RF, Feitag R, Rose HA, McDonald FJD (1997) Taxonomic revision of the Australian Cicindelidae (Coleoptera), excluding species of Cicindela. *Invertebrate Taxonomy* **11**(4), 599–688. doi:10.1071/IT94011

McDermott FA (1964) The taxonomy of the Lampyridae (Coleoptera). *Transactions of the American Entomological Society* **90**, 1–72.

McDermott FA (1966) *Coleopterorum Catalogus Supplementum. Pars 9. Lampyridae (edito secunda).* W. Junk, The Hague.

McHugh JV (1993) A revision of *Eurysphindus* Leconte (Coleoptera: Cucujoidea: Sphindidae) and a review of sphindid classification and phylogeny. *Systematic Entomology* **18**, 57–92. doi:10.1111/j.1365-3113.1993.tb00654.x

McHugh JV (2002) 75. Sphindidae Jacquelin du Val 1861. In *American Beetles. Volume 2. Polyphaga: Scarabaeoidea through Curculionoidea.* (Eds RH Arnett Jr, MC Thomas, PE Skelley and JH Frank) pp. 305–308. CRC Press, Boca Raton, Florida.

McHugh JV, Kiselyova T (2003) First description for larval stages of *Eurysphindus* (Coleoptera: cucujoidea: Sphindidae). *Coleopterists Bulletin* **57**(1), 17–25. doi:10.1649/0010-065X(2003)057[0017:FDFLSO]2.0.CO;2

McHugh JV, Pakaluk J (1997) Review of the larval stages of Epipocinae (Insecta: Coleoptera: Endomychidae). *Annales Zoologici* **47**(1/2), 59–77.

McHugh JV, Wheeler QD (1991) *Notosphindus slateri,* a new genus and species of Sphindidae (Coleoptera: Cucujoidea) from Australia. *Journal of the New York Entomological Society* **99**(3), 527–537.

McHugh JV, Marshall CJ, Fawcett FL (1997) A study of adult morphology in *Megalodacne heros* (Say) (Coleoptera: Erotylidae). *Transactions of the American Entomological Society* **123**, 167–223.

McKenna DD, Farrell BD (2010) 9-genes reinforce the phylogeny of Holometabola and yield alternate views on the phylogenetic placement of Strepsiptera. *PLoS ONE* **5**, e11887. doi:10.1371/journal.pone.0011887

McKeown KC (1947) Catalogue of the Cerambycidae (Coleoptera) of Australia. *Memoirs of the Australian Museum* **10**, 1–190. doi:10.3853/j.0067-1967.10.1947.477

McKie BGL, Cranston PS (1998) Keystone coleopterans? Colonization by wood-feeding elmids of experimentally immersed woods in south-eastern Australia. *Marine and Freshwater Research* **49**, 79–88. doi:10.1071/MF97086

McQuillan PB (1985) The identification of root-feeding cockchafer larvae (Coleoptera: Scarabaeidae) found in pastures in Tasmania. *Australian Journal of Zoology* **33**, 509–546. doi:10.1071/ZO9850509

Medvedev LN (1968) Leaf-beetles of Jurassic of Karatau (Coleoptera, Chrysomeldae). In *Jurassic Insects of Karatau.* (Ed. BB Rohdendorf) pp. 155–165. Nauka, Moscow (in Russian).

Medvedev LN (1971) The ways of evolution and phylogeny of Chrysomelidae (Coleoptera). *Proceedings of the International Congress of Entomology* **13**(1), 271–272.

Medvedev GS. (1977) Taxonomic significance of the antennal sensilla of darkling beetles (Coleoptera, Tenebrionidae). *Trudy Vsesoyuznogo Entomologicheskogo Obshchestva* **58**, 61–86, 18 pls. (in Russian).

Medvedev GS and Lawrence JF (1983) Beetles of the tribe Hyocini (Coleoptera, Tenebrionide) of Australia. I. Subgenera *Nannohyocis* subgen. n. and *Neohyocis* subgen. n. (genus *Hyocis* Pasc.). *Entomologicheskoye Obozreniye* **62**(3), 569–582 (in Russian; translation in *Entomological Review* **62**(3), 117–130).

Medvedev GS and Lawrence JF (1984) Beetles of the tribe Hyocini (Coleoptera, Tenebrionidae) of Australia. II. Genera *Parahyocis* Kasz. and *Csiro* gen. n. *Entomologicheskoye Obozreniye* **63**(3), 561–581 (in Russian; translation in *Entomological Review* **63**(3), 129–148).

Medvedev GS and Lawrence JF (1986) Tenebrionidae of the tribe Hyocini (Coleoptera, Tenebrionidae) of Australia. III. Brittonina and Uptonina, new subtribes. *Entomologicheskoye Obozreniye* **65**(3), 574–591 (in Russian; translation in *Entomological Review* **65**(3), 161–179).

Medvedev GS and Lawrence JF (1987) Tenebrionid beetles of the tribe Hyocini (Coleoptera, Tenebrionidae) of Australia. IV. A key to genera and subgenera. *Entomologicheskoye Obozreniye* **66**(4), 803–816 (in Russian; translation in *Entomological Review* **67**(3), 159–171, 1988).

Medvedev LN, Zaitzev YM (1978) *The Larvae of Chrysomelid Beetles of Siberia and the Far East.* Nauka, Moscow (in Russian).

Meinwald J, Meinwald WC, Chalmers AM, Eisner T (1968) Dihidroxymatricaria acid: acetylenic acid secreted by a soldier beetle. *Science* **160**, 890–892. doi:10.1126/science.160.3830.890

Menier JJ, Ekis G (1982) Les Chaetosomatidae Malgaches. II. Description d'un genre nouveau et de quatre espèces nouvelles (Coleoptera, Cleroidea). *Annales de la Société Entomologique de France* **18**, 343–348.

Menier JJ, Jouan M (2003) Biologie et description d'*Aethina tumida* Murray, 1867, espèce invasive déprédatrice de l'Abeille mellifère (Coleoptera, Nitidulidae). *Bulletin de la Societe Entomologique de France* **108**(3), 323–328.

Méquignon A (1934) Les *Chelonarium* de l'Amérique continentale. *Annales de la Société Entomologique de France* **103**, 199–256.

Méquignon A (1935) Notes sur le genre *Pseudochelonarium*, Pic (Col. Dryopidae). *Stylops* **4**(3), 66–68.

Merkl O (1986) A review of the Australian species of the subtribe Statirina (Coleoptera, Tenebrionidae: Lagriini). *Annales Historico-Naturales Musei Nationalis Hungarici* **78**, 187–199.

Merkl O (1987) A review of the Australian species of the subtribe Lagriina (Coleoptera, Tenebrionidae: Lagriini). *Annales Historico-Naturales Musei Nationalis Hungarici* **79**, 121–166.

Merkl O (1988) Novelties of *Sivacrypticus* Kaszab, 1964 and *Enneboeus* Waterhouse, 1878 (Coleoptera, Archeocrypticidae). *Annales Historico-Naturales Musei Nationalis Hungarici* **80**, 71–78.

Merkl O, Kompantseva TV (1996) Old World *Rhipidandrus* LeConte: synonymies, faunistics, identification key and description of two new species from Australia (Coleoptera: Tenebrionidae). *Acta Zoologica Academiae Scientiarum Hungaricae* **42**, 89–109.

Mesa A, Fontanetti CS (1985) The chromosomes of a primitive species of beetle: *Ytu zeus* (Coleoptera, Myxophaga, Torridincolidae). *Proceedings. Academy of Natural Sciences of Philadelphia* **137**, 102–105.

Michener CD (1944) A comparative study of the appendages of the eighth and ninth abdominal segments of insects. *Annals of the Entomological Society of America* **37**, 336–351.

Mickoleit G (1973) Über den Ovipositor der Neuropteroidea und Coleoptera und seine phylogenetische Bedeutung (Insecta, Holometabola). *Zeitschrift für Morphologie der Tiere* **74**, 37–64. doi:10.1007/BF00291795

Mico E, Morón MA, Sipek P, Galante E (2008) Larval morphology enhances phylogenetic reconstruction in Cetoniidae (Coleoptera: Scarabaeoidea) and allows the interpretation of the evolution of larval feeding habits. *Systematic Entomology* **33**, 128–144. doi:10.1111/j.1365-3113.2007.00399.x

Miller WV (1988) Two new species of *Heterocerus* from North America, with notes on related species. *Coleopterists Bulletin* **42**, 313–320.

Miller RS (1997) Immature stages of *Plateros floralis* (Melsheimer) and discussion of phylogenetic relationships (Coleoptera: Lycidae). *Coleopterists Bulletin* **51**, 1–12.

Miller KB (2001) On the phylogeny of the Dytiscidae (Insecta: Coleoptera) with emphasis on the morphology of the female reproductive system. *Insect Systematics & Evolution* **32**, 45–89. doi:10.1163/187631201X00029

Miller RS (2002a) 59. Lycidae Laporte 1836. In *American Beetles. Volume 2. Polyphaga: Scarabaeoidea through Curculionoidea.* (Eds RH Arnett Jr, MC Thomas, PE Skelley and JH Frank) pp. 174–178. CRC Press, Boca Raton, Florida.

Miller RS (2002b) 60. Telegeusidae Leng 1920. In *American Beetles. Volume 2. Polyphaga: Scarabaeoidea through Curculionoidea.* (Eds RH Arnett Jr, MC Thomas, PE Skelley and JH Frank) pp. 179–180. CRC Press, Boca Raton, Florida.

Miller KB (2009) On the systematics of Noteridae (Coleoptera: Adephaga: Hydradephaga): phylogeny, description of a new tribe, genus and species, and survey of female genital morphology. *Systematics and Biodiversity* **7**(2), 191–214. doi:10.1017/S1477200008002946

Miller KB, Bergsten J (2012) Phylogeny and classification of whirligig beetles (Coleoptera: Gyrinidae): relaxed-clock model outperforms parsimony and time-free Bayesian analyses. *Systematic Entomology* **37**, 706–746.

Minoshima Y, Hayashi M (2011) Larval morphology of the Japanesec species of the tribes Acidocerini, Hydrobiusini and Hydrophilini (Coleoptera: Hydrophilidae). *Acta Entomologica Musei Nationalis Pragae* **51**, (Supplementum), 1–118.

Miskimen GW (1961a) A new family of beetles found in Cantharoidea. *Coleopterists Bulletin* **15**(1), 17–25.

Miskimen GW (1961b) Zoogeography of the coleopterous family Chauliognathidae. *Systematic Zoology* **10**, 140–153. doi:10.2307/2411597

Miyatake M (1957) Two genera of Mycetophagidae, *Litargus* Erichson and *Litargops* Reitter, of Japan (Coleoptera). *Transactions of the Shikoku Entomological Society* **5**(3), 33–45.

Moczek AP (1998) Horn polyphenism in the beetle *Onthophagus taurus*: larval diet quality and plasticity in parental investment determine adult body size and male horn morphology. *Behavioral Ecology* **9**(6), 636–641. doi:10.1093/beheco/9.6.636

Moczek AP, Emlen DJ (1999) Proximate determination of male horn dimorphism in the beetle *Onthophagus taurus* (Coleoptera: Scarabaeidae). *Journal of Evolutionary Biology* **12**, 27–37. doi:10.1046/j.1420-9101.1999.00004.x

Monné MA (2005) Catalogue of the Cerambycidae (Coleoptera) of the Neotropical Region. Part II. Subfamily Lamiinae. *Zootaxa* **1023**, 1–760.

Monné MA, Bezark LG (2009) *Checklist of the Cerambycidae, or longhorned wood-boring beetles, of the Western Hemisphere.* 460 pp. BioQuip Publications, Rancho Dominguez, California.

Monrós F (1947) Revision de los Megalopidae Argentinos (Col., Chrysomelidae). *Revista de la Sociedad Entomologica Argentina* **13**, 150–217.

Monrós F (1951) Consideraciones sobre "*Leasia australis*" Jac. (Coleoptera). *Revista Brasileira de Biologia* **11**, 451–456.

Monrós F (1956) Revision generica de Lamprosominae con descripcion de algunos generos y especies nuevas (Col., Chrysomelidae). *Revista Agronomica Noroeste Argentino* **2**(1), 25–77.

Monrós F (1958) Die gattung *Microdonacia* Blackburn (Col. Chrysomelidae). *Entomologische Arbeiten aus dem Museum Georg Frey* **9**, 742–749.

Monrós F (1959) Notas sobre Chrysomelidae (Coleoptera). *Acta Zoologica Lilloana* **17**, 1–24.

Monrós F (1960) Los generos de Chrysomelidae (Coleoptera). *Opera Lilloana* **3**, 1–337.

Monrós F, Monrós MM (1952) Las especies Argentinas de Cupedidae (Coleoptera). *Anales de la Sociedad Científica de Argentina* **153**, 19–41.

Monteith GB (1970) Life history of the chrysomelid, *Aproidea balyi* Pascoe. *News Bulletin of the Entomological Society of Queensland* **72**, 9–10.

Monteith GB (1991) Corrections to published information on *Johananica gemellata* (Westwood) and other Chrysomelidae (Coleoptera). *Victorian Naturalist* **21**, 147–154.

Monteith GB (2011) A giant scydmaenid from Mt. Abbott, North Queensland. *News Bulletin of the Entomological Society of Queensland* **39**(9), 116–119.

Monteith GB, Storey RI (1981) The biology of *Cephalodesmius,* a genus of dung beetles which synthesizes "dung" from plant material (Coleoptera: Scarabaeidae: Scarabaeinae). *Memoirs of the Queensland Museum* **20**, 253–277.

Monteith GB, Turco F (2010) *Mystropomus* carabids from the Mt. Etna caves at Rockhampton. *News Bulletin of the Entomological Society of Queensland* **38**(4), 66–69.

Monteith GB, Turco F (2011) Rediscovery of the enigmatic beetle family Brachypsectridae in Australia. *News Bulletin of the Entomological Society of Queensland* **39**(9), 114–116.

Moore BP (1957) The identity of *Ptinus latro* auct. (Coleoptera: Ptinidae). *Proceedings of the Royal Entomological Society of London B* **26**, 199–202.

Moore BP (1964) Present-day cave beetle fauna in Australia: a pointer to past climatic change. *Helictite* **3**, 3–9, pl. 1.

Moore BP (1967a) Hydrogen cyanide in the defensive secretions of larval Paropsini (Coleoptera: Chrysomelidae). *Journal of the Australian Entomological Society* **6**, 36–38. doi:10.1111/j.1440-6055.1967.tb02135.x

Moore BP (1967b) New Australian cave Carabidae (Coleoptera). *Proceedings of the Linnean Society of New South Wales* **91**, 179–184.

Moore BP (1972a) A revision of the Australian Trechinae (Coleoptera: Carabidae). *Australian Journal of Zoology* **18**, (Supplement), 1–61.

Moore BP (1972b) Southern cave-beetle fauna in perspective. *Proceedings of the 8th National Conference of the Australian Speleological Federation.* pp. 81–4.

Moore BP (1973) Larvae of *Zenithicola crassus* (Newman) (Coleoptera: Cleridae) feeding on termites. *Australian Entomological Magazine* **1**, 48–50.

Moore BP (1974) The larval habits of two species of *Sphallomorpha* Westwood (Coleoptera: Carabidae, Pseudomorphinae). *Journal of the Australian Entomological Society* **13**, 179–183. doi:10.1111/j.1440-6055.1974.tb02171.x

Moore BP (1977) A new cave-frequenting carabid beetle from the Snowy Mountains region of New South Wales. *Australian Entomological Magazine* **3**, 92–94.

Moore BP (1978a) A new Australian stag beetle (Coleoptera: Lucanidae) with Neotropical affinities. *Journal of the Australian Entomological Society* **17**, 99–103. doi:10.1111/j.1440-6055.1978.tb01751.x

Moore BP (1978b) A new species of the Tasmanian cave carabid genus *Idacarabus* (Coleoptera). *Australian Entomological Magazine* **5**, 23–25.

Moore BP (1979) Chemical defense in carabids and its bearing on phylogeny. In *Carabid Beetles: Their Evolution, Natural History, and Classification*. (Eds TL Erwin, GE Ball, DR Whitehead and AL Halpern) pp. 193–203. Junk, The Hague.

Moore BP (1980) *A Guide to the Beetles of South-Eastern Australia*. Fascicles 1–2. Australian Entomological Press, Greenwich, New South Wales.

Moore BP (1981) *A Guide to the Beetles of South-Eastern Australia*. Fascicle 3. Australian Entomological Press, Greenwich, New South Wales.

Moore BP (1982) *A Guide to the Beetles of South-Eastern Australia*. Fascicle 4. Australian Entomological Press, Greenwich, New South Wales.

Moore BP (1983) *A Guide to the Beetles of South-Eastern Australia*. Fascicle 5. Australian Entomological Press, Greenwich, New South Wales.

Moore BP (1984) *A Guide to the Beetles of South-Eastern Australia*. Fascicle 6. Australian Entomological Press, Greenwich, New South Wales.

Moore BP (1986) *A Guide to the Beetles of South-Eastern Australia*. Fascicle 7. Australian Entomological Press, Greenwich, New South Wales.

Moore BP (1987) *A Guide to the Beetles of South-Eastern Australia*. Fascicle 8. Australian Entomological Press, Greenwich, New South Wales.

Moore BP (1989) *A Guide to the Beetles of South-Eastern Australia*. Fascicle 9. Australian Entomological Press, Greenwich, New South Wales.

Moore BP (1990) *A Guide to the Beetles of South-Eastern Australia*. Fascicle 10. Australian Entomological Press, Greenwich, New South Wales.

Moore BP (1992) *A Guide to the Beetles of South-Eastern Australia*. Fascicle 11. Australian Entomological Press, Greenwich, New South Wales.

Moore BP (1994) A new species of *Eucarteria* Lea (Coleoptera: Lucanidae) and notes on the affinities of the genus. *Australian Entomologist* **21**(1), 1–6.

Moore BP, Brown VV (1989) Graded levels of chemical defence in mimics of lycid beetles of the genus *Metriorrhynchus* (Coleoptera). *Journal of the Australian Entomological Society* **28**, 229–233. doi:10.1111/j.1440-6055.1989.tb00889.x

Moore W, Di Giulio A (2006) Description and behaviour of *Goniotropis kuntzeni* larvae (Coleoptera: Carabidae: Paussinae: Ozaenini) and a key to genera of Paussinae larvae. *Zootaxa* **1111**, 1–19.

Moore BP, Humphreys WF (2001) Australie. In *Encyclopaedia Biospeologica*. Tome 3. (Eds C Juberthie and V. Decou) pp. 2055–2058. *Laboratoire Souterrain du CNRS,* Moulis, France.

Moore BP, Lawrence JF (1994) The extraordinary larval characters of *Carenum* Bonelli and their bearing on the phylogeny of the Scaritinae (Coleoptera: Carabidae). *Canadian Entomologist* **126**, 503–514. doi:10.4039/Ent126503-3

Moore I, Legner DF (1979) An Illustrated Guide to the Genera of Staphylinidae of America north of Mexico exclusive of the Aleocharinae (Coleoptera). 332 pp. University of Calfornia Press, Berkeley.

Moore BP, Monteith GB (2004) A second Australian species of the Gondwanan stag beetle genus *Sphaenognathus* Buquet (Coleoptera: Lucanidae). *Memoirs of the Queensland Museum* **49**(2), 693–699.

Moore BP, Weir TA, Pyke JE (1987) Rhysodidae and Carabidae. *Zoological Catalogue of Australia. Volume 4. Coleoptera: Archostemata, Myxophaga and Adephaga*. (Ed. DW Walton) pp. 20–320. Australian Government Publication Service, Canberra.

Moore W, Song X, Di Giulio A (2011) The larva of *Eustra* (Coleoptera, Paussinae, Ozaenini): a facultative associate of ants. *ZooKeys* **90**, 63–82. doi:10.3897/zookeys.90.1136

Morgan FD (1967) *Ips grandicollis* in South Australia. *Australian Forestry* **31**, 137–155.

Morgan A (1968) The occurrence of a pair of exocrine glands in the mesothorax of *Lytta nutalli* (Coleoptera: Meloidae). *Canadian Journal of Zoology* **46**, 915–917. doi:10.1139/z68-129

Morimoto K (1962) Comparative morphology and phylogeny of the superfamily Curculionoidea of Japan (comparative morphology, phylogeny and systematics of the superfamily Curculionoidea of Japan. I). *Journal of the Faculty of Agriculture of Kyushu University* **11**, 331–373.

Morimoto K (1972) A key to the genera of Oriental Anthribidae. *Government Forest Experiment Station Bulletin* **246**, 35–54, figs. 1–100.

Morimoto K (1976) Notes on the family characters of Apionidae and Brentidae (Coleoptera), with key to the related *families. Kontyû* **44**, 469–476.

Morimoto K (1978) The family Anthribidae of Japan (Coleoptera). Part. 1. *Esakia* **12**, 17–47, figs. 1–11, A–R.

Morimoto K (1979) The family Anthribidae of Japan (Coleoptera). Part. 2. *Esakia* **14**, 1–23, figs. 1–8, A–N.

Morimoto K (1980) The family Anthribidae of Japan (Coleoptera). Part. 3. *Esakia* **15**, 11–47, figs. 1–13, A–Z, a–s.

Morimoto K (1981) The family Anthribidae of Japan (Coleoptera). Part. 4. *Esakia* **17**, 1–14, figs. 1–11, A–Z, a–s.

Morimoto K, Kojima H (2003) Morphologic characters of the weevil head and phylogenetic implications (Coleoptera, Curculionoidea). *Esakia* **43**, 133–169.

Morimoto K, Kojima H (2006) Larva of *Desmidophorus crassus* and the systematic position of the Desmidophorini (Coleoptera: Curculionoidea). *Esakia* **46**, 89–100.

Morón MA (1991) Larva and pupa of *Aegidium cribratum* Bates (Coleoptera: Scarabaeidae: Orphninae). *Coleopterists Bulletin* **45**, 360–367.

Morrone JJ (1997) The impact of cladistics on weevil classification, with a new scheme of families and subfamilies (Coleoptera: Curculionoidea). *Trends in Entomology* **1**, 129–136.

Moulds MS (1990) *Australian Cicadas*. New South Wales University Press, Kensington, New South Wales.

Mroczkowski M (1968) Distribution of the Dermestidae (Coleoptera) of the world with a catalogue of all known species. *Annales Zoologici* **26**, 15–191.

Mroczkowski M, Ślipiński SA (1997) Notes on the Marioutinae (Coleoptera: Dermestidae) with a review of the described species. *Annales Zoologici* **47** (1/2), 11–16.

Muona J (1991a) The Eucnemidae of South-east Asia and the western Pacific – a biogeographical study. *Australian Systematic Botany* **4**, 165–182. doi:10.1071/SB9910165

Muona J (1991b) A revision of the Indomalesian tribe Galbitini new tribe (Coleoptera, Eucnemidae). *Entomologica Scandinavica* **39**, (Supplement), 1–67.

Muona J (1993a) Eucnemidae and Throscidae from Baltic amber (Coleoptera). *Entomologische Blätter, Krefeld* **89**, 15–45.

Muona J (1993b) Review of the phylogeny, classification and biology of Eucnemidae (Coleoptera). *Entomologica Scandinavica* **44**, (Supplement), 1–133.

Muona J (1995) The phylogeny of Elateroidea (Coleoptera), or which tree is the best? *Cladistics* **9**, 413–426.

Muona J (2002) 56. Eucnemidae Eschscholtz 1829. In *American Beetles. Volume 2. Polyphaga: Scarabaeoidea through Curculionoidea.* (Eds RH Arnett Jr, MC Thomas, PE Skelley and JH Frank) pp. 152–157. CRC Press, Boca Raton, Florida.

Muona J (2010) 4.5 Eucnemidae Eschscholtz, 1829. In *Handbuch der Zoologie/Handbook of Zoology. Band/Volume IV Arthropoda: Insecta Teilband/Part 38. Coleoptera, Beetles. Volume 2. Morphology and Systematics (Polyphaga partim)*. (Eds RAB Leschen, RG Beutel and JF Lawrence) pp. 61–69. W. DeGruyter, Berlin.

Muona J, Teräväinen M (2008) Notes on the biology and morphology of false click-beetle larvae (Coleoptera, Eucnemidae). *Coleopterists Bulletin* **62**(4), 475–479. doi:10.1649/1059.1

Muona J, Lawrence JF, Ślipiński A (2010) 4.6. Throscidae Laporte, 1840. In *Handbuch der Zoologie/Handbook of Zoology. Band/Volume IV Arthropoda: Insecta Teilband/Part 38. Coleoptera, Beetles. Volume 2. Morphology and Systematics (Polyphaga partim)*. (Eds RAB Leschen, RG Beutel and JF Lawrence) pp. 69–74. W. DeGruyter, Berlin.

Naomi SI (1985) The phylogeny and higher classification of the Staphylinidae and their allied groups (Coleoptera, Staphylinoidea). *Esakia* **23**, 1–27.

Naomi SI (1987a) Comparative morphology of the Staphylinidae and the allied groups (Coleoptera, Staphylinoidea). I. Introduction, head sutures, eyes and ocelli. *Kontyû* **55**(3), 450–458.

Naomi SI (1987b) Comparative morphology of the Staphylinidae and the allied groups (Coleoptera, Staphylinoidea). II. Cranial structure and tentorium. *Kontyû, Tokyo* **55**(4), 666–675.

Naomi SI (1988a) Comparative morphology of the Staphylinidae and the allied groups (Coleoptera, Staphylinoidea). III. Antennae, labrum and mandibles. *Kontyû, Tokyo* **56**(1), 67–77.

Naomi SI (1988b) Comparative morphology of the Staphylinidae and the allied groups (Coleoptera, Staphylinoidea). IV. Maxillae and labium. *Kontyû, Tokyo* **56**(2), 241–250.

Naomi SI (1988c) Comparative morphology of the Staphylinidae and the allied groups (Coleoptera, Staphylinoidea). V. Cervix and prothorax. *Kontyû, Tokyo* **56**(3), 506–513.

Naomi SI (1988d) Comparative morphology of the Staphylinidae and the allied groups (Coleoptera, Staphylinoidea). VI. Mesothorax and metathorax. *Kontyû, Tokyo* **56**(4), 727–738.

Naomi SI (1989a) Comparative morphology of the Staphylinidae and the allied groups (Coleoptera, Staphylinoidea). VII. Metendosternite and wings. *Japanese Journal of Entomology* **57**(1), 82–90.

Naomi SI (1989b) Comparative morphology of the Staphylinidae and the allied groups (Coleoptera, Staphylinoidea). VIII. Thoracic legs. *Japanese Journal of Entomology* **57**(2), 269–277.

Naomi SI (1989c) Comparative morphology of the Staphylinidae and the allied groups (Coleoptera, Staphylinoidea). IX. General structure, lateral plates, stigmata and 1st to 7th segments of abdomen. *Japanese Journal of Entomology* **57**(3), 517–526.

Naomi SI (1989d) Comparative morphology of the Staphylinidae and the allied groups (Coleoptera, Staphylinoidea). X. Eighth to 10th segments of abdomen. *Japanese Journal of Entomology* **57**(4), 720–733.

Naomi SI (1990) Comparative morphology of the Staphylinidae and the allied groups (Coleoptera, Staphylinoidea). XI. Abdominal glands, male genitalia and female spermatheca. *Japanese Journal of Entomology* **58**(1), 16–23.

Napp DS (1994) Phylogenetic relationships among the subfamilies of Cerambycidae (Coleoptera – Chrysomeloidea). *Revista Brasileira de Entomologia* **38**, 265–419.

Nardi G (2007) Nomenclatorial and faunistic notes on some world Aderidae. *Zootaxa* **1481**, 21–34.

Neboiss A (1960) On the family Cupedidae, Coleoptera. *Proceedings of the Royal Society of Victoria* **72**, 12–20.

Neboiss A (1967) The genera *Paracalais* gen. nov. and *Austrocalais* gen. nov. (Coleoptera: Elateridae). *Proceedings of the Royal Society of Victoria* **80**, 259–288.

Neboiss A (1968) Larva and pupa of *Cupes varians* Lea, and some observations on its biology (Coleoptera; Cupedidae). *Memoirs of the National Museum of Victoria* **28**, 17–19.

Neboiss A (1984) Reclassification of *Cupes* Fabricius (s. lat.), with descriptions of new genera and species (Cupedidae: Coleoptera). *Systematic Entomology* **9**, 443–477. doi:10.1111/j.1365-3113.1984.tb00520.x

Neboiss A (1987) A new species of *Adinolepis* Neboiss from Western Australia (Cupedidae: Coleoptera). *Records of the Western Australian Museum* **13**, 323–325.

Neboiss A (1989) New species of Archostemata (Coleoptera, Ommatidae, Cupedidae). *Revue Française d'Entomologie (N S.)* **11**, 109–115.

Nelson GH, Bellamy CL (1991) A revision and phylogenetic re-evaluation of the family Schizopodidae (Coleoptera, Buprestoidea). *Journal of Natural History* **25**, 985–1026. doi:10.1080/00222939100770651

Nelson GH, Bellamy CL (2002) 40. Schizopodidae LeConte 1861. In *American Beetles. Volume 2. Polyphaga: Scarabaeoidea through Curculionoidea.* (Eds RH Arnett Jr, MC Thomas, PE Skelley and JH Frank) pp. 95–97. CRC Press, Boca Raton, Florida.

New TR (1978) Notes on the biology of *Lemidea subaenea* (Coleoptera: Cleridae) on *Acacia* in Victoria. *Australian Entomological Magazine* **5**, 21–22.

New TR (1981) Paropsine beetle larvae as possible pollinators of *Acacia baileyana* in Victoria. *Australian Entomological Magazine* **8**, 55

Newton AF (1982) A new genus and species of Oxytelinae from Australia, with a description of its larva, systematic position, and phylogenetic relationships (Coleoptera, Staphylinidae). *American Museum Novitates* **2744**, 1–24.

Newton AF (1984) Mycophagy in Staphylinoidea (Coleoptera). In *Fungus-Insect Relationships: Perspectives In Ecology And Evolution* (Eds Q Wheeler and M Blackwell) pp. 302–53. Columbia University Press, New York.

Newton AF (1985) South temperate Staphylinoidea (Coleoptera): their potential for biogeographic analysis of austral disjunctions. In *Taxonomy, Phylogeny and Zoogeography of Beetles and Ants. A Volume dedicated to the Memory of Philip Jackson Darlington Jr (1904–1983).* (Ed. GE Ball) Pp. 180–220. Junk (Series Entomologica 33), Dordrecht.

Newton AF (1989) Review of *Dactylosternum* Wollaston Species of Australia and New Zealand (Coleoptera: Hydrophilidae). *Australian Entomological Magazine* **16**, 49–58.

Newton AF (1990) *Myrmelibia,* a new genus of myrmecophile from Australia, with a generic review of Australian Osoriinae (Coleoptera: Staphylinidae). *Invertebrate Taxonomy* **4**, 81–94. doi:10.1071/IT9900081

Newton AF (1991) Agyrtidae–Silphidae, Pselaphidae, Sphaeritidae–Histeridae. In *Immature Insects. Volume 2.* (Ed. FW Stehr) pp. 224–341, 353–355, 359–364. Kendall/Hunt, Dubuque, Iowa.

Newton AF Jr (1997) Review of Agyrtidae (Coleoptera), with a new genus and species from New Zealand. *Annales Zoologici* **47**, 111–156.

Newton AF Jr (1998) Phylogenetic problems, current classification and generic catalog of world Leiodidae (including Cholevidae). In *Phylogeny and Evolution of Subterranean and Endogean Cholevidae (= Leiodidae Cholevinae). Proceedings of a Symposium (30 August, 1996, Florence, Italy). XX International Congress of Entomology.* (Eds M Giachino and SB Peck) pp. 41–178 Museo Regionale di Scienze Naturali, Torino.

Newton AF (2000) 14. Sphaeritidae Shuckard, 1839. In *American Beetles. Volume 1. American Beetles. Archostemata, Myxophaga, Adephaga, Polyphaga: Staphyliniformia.* (Eds RH Arnett Jr and MC Thomas) pp. 209–211. CRC Press, Boca Raton, Florida.

Newton AF (2005a) 10.2. Sphaeritidae Shuckard, 1839. In *Handbuch der Zoologie/Handbook of Zoology. Band/Volume IV Arthropoda: Insecta Teilband/Part 38. Coleoptera, Beetles. Volume 1. Morphology and Systematics (Archostemata, Adephaga, Myxophaga, Polyphaga partim).* (Eds RG Beutel and RAB Leschen) pp. 183–186. W. DeGruyter, Berlin.

Newton AF (2005b) 10.3. Synteliidae Lewis, 1882. In *Handbuch der Zoologie/Handbook of Zoology. Band/Volume IV Arthropoda: Insecta Teilband/Part 38. Coleoptera, Beetles. Volume 1. Morphology and Systematics (Archostemata, Adephaga, Myxophaga, Polyphaga partim).* (Eds RG Beutel and RAB Leschen) pp. 187–190. W. DeGruyter, Berlin.

Newton AF (2005c) 11.3. Agyrtidae Thomson, 1859. In *Handbuch der Zoologie/Handbook of Zoology. Band/Volume IV Arthropoda: Insecta Teilband/Part 38. Coleoptera, Beetles. Volume 1. Morphology and Systematics (Archostemata, Adephaga, Myxophaga, Polyphaga partim).* (Eds RG Beutel and RAB Leschen) pp. 261–269. W. DeGruyter, Berlin.

Newton AF (2005d) 11.4. Leiodidae Fleming, 1821. In *Handbuch der Zoologie/Handbook of Zoology. Band/Volume IV Arthropoda: Insecta Teilband/Part 38. Coleoptera, Beetles. Volume 1. Morphology and Systematics (Archostemata, Adephaga, Myxophaga, Polyphaga partim).* (Eds RG Beutel and RAB Leschen) pp. 269–280. W. DeGruyter, Berlin.

Newton AF, Chandler DS (1989) World catalog of the genera of Pselaphidae (Coleoptera). *Fieldiana Zolology (N. S.)* **53**, 1–93.

Newton AF Jr, Franz H (1998) World catalog of the genera of Scydmaenidae (Coleoptera). *Koleopterologische Rundschau* **68**, 137–165.

Newton AF, Thayer MK (1988) A critique on Naomi's phylogeny and higher classification of Staphylinidae and allies (Coleoptera). *Entomologia Generalis* **14**, 63–72.

Newton AF, Thayer MK (1992) Current classification and family-group names in Staphyliniformia (Coleoptera). *Fieldiana Zoology* (N. S.) **67**, iv + 92 pp.

Newton AF Jr, Thayer MK (1995) Protopselaphinae, new subfamily for *Protopselaphus* new genus from Malaysia, with a phylogenetic analysis and review of the omaliine group of Staphylinidae including Pselaphidae (Coleoptera). In *Biology, Phylogeny, and Classification of Coleoptera. Papers Celebrating the 80th Birthday of Roy A. Crowson*. (Eds J Pakaluk and SA Ślipiński) pp. 219–320. Muzeum I Instytut Zoologii Polska Akademia Nauk, Warsaw.

Newton AF, Thayer MK, Ashe JS, Chandler DS (2000). 22. Staphylinidae Latreille, 1802. In *American Beetles. Volume 1. American Beetles. Archostemata, Mycophaga, Adephaga, Polyphaga: Staphyliniformia*. (Eds RH Arnett Jr and MC Thomas) pp. 272–418. CRC Press, Boca Raton, Florida.

Nichols SW (1985) *Omophron* and the origin of Hydradephaga (Insecta: Coleoptera: Adephaga). *Proceedings. Academy of Natural Sciences of Philadelphia* **137**, 182–201.

Nichols SW (compiler) (1989) *The Torre-Bueno Glossary of Entomology. Revised Edition of A Glossary of Entomology by J. R. de la Torre-Bueno, including Supplement A by George S. Tulloch*. xviii + 840 pp. The New York Entomological Society and the American Museum of Natural History, New York.

Niehuis O, Hartig G, Grath S, Pohl H, Lehmann J, Tafer H, Donath A, Krauss V, Elsenhardt C, Hertel J, Petersen M, Mayer C, Meusemnn K, Peters RS, Stadler PF, Beutel RG, Bornberg-Bauer E, McKenna DD, Misof B (2012) Genomic and morphological evidence converge to resolve the enigma of Strepsiptera. *Current Biology* **22**, 1309–1313. doi:10.1016/j.cub.2012.05.018

Nikitsky NB (1976a) Larval morphology and mode of life of Histeridae (Coleoptera) in bark-beetle passages. *Entomologicheskoye Obozreniye* **55**, 875–888 (in Russian; translation in *Entromological Review* **55**, 102–111).

Nikitsky NB (1976b) The morphology of *Sphaerites glabratus* larva and the phylogeny of Histeroidea. *Зоологический журнал* **55**(4), 531–537. [in Russian]

Nikitsky NB (1983) Species of the genus *Biphyllus* (Coleoptera, Biphyllidae) in the East Palearctic. *Зоологический журнал* **62**, 695–706. [in Russian]

Nikitsky NB (1986a) Beetles of the subfamilies Monotominae and Thioninae (Coleoptera, Rhizophagidae) from the Far East of the USSR. *Зоологический журнал* **65**, 1622–1629. [in Russian]

Nikitsky NB (1986b) The family Pilipalpidae stat. n. (Coleoptera, Heteromera), its composition and taxonomic relationships. *Зоологический журнал* **65**, 1178–1189.

Nikitsky NB (1991) News on the genus *Triphyllia* Reitter, 1898 (Coleoptera: Tetratomidae). *Elytron* **5**, 159–168.

Nikitsky NB (1993) *Mushroom Beetles (Coleoptera, Mycetophagidae) of the Fauna of Russia and Contiguous Land*. Izdatel'stvo Moskovskogo Universiteta, Moscow (in Russian).

Nikitsky NB (1996) Fam. Ischaliidae – False Fire-red Beetles. In *Keys to the Insecta of the Far Eastern U.S.S.R. in Six Volumes. Vol. 3. Coleoptera or Beetles, Part 3*. (Ed. PA Ler) pp. 427–428. Dal'nauka, Vladivostok (in Russian).

Nikitsky NB (1998) *Generic Classification of the Beetle Family Tetratomidae (Coleoptera, Tenebrionoidea) of the world, with description of new taxa*. Pensoft (Series Faunistica No. 9), Sofia and Moscow.

Nikitsky NB (2002) The larvae of the false darkling beetles (Coleoptera, Melandryidae) of Russia, with taxonomic notes. *Byulletin Moskovskogo Obshchestva Ispytatelei Prirody, Moscow* **107**(3), 8–30. [in Russian]

Nikitsky NB (2004a) The beetles of the subfamily Tetratominae Billberg, 1820 (Coleoptera, Tetratomidae) of the world fauna. *Byulleten Moskovskogo Obshestva Ispytatelei Prirody. Otdel Biologicheskii* **109**(2), 25–36. [in Russian]

Nikitsky NB (2004b) The beetles of the subfamily Piseninae Miyatake, 1960 (Coleoptera, Tetratomidae) of the world fauna. *Byulleten Moskovskogo Obshestva Ispytatelei Prirody. Otdel Biologicheskii* **109**(4), 3–7. [in Russian]

Nikitsky NB, Egorov AB (1992) Fam. Ischaliidae, stat. n. – false fire-red beetles. In *Keys to the Insecta of the Far Eastern U.S.S.R. in Six Volumes. Vol. 3. Coleoptera or Beetles, Part 2*. (Ed. PA Ler) pp. 497–498. Nauka, St. Petersburg (in Russian).

Nikitsky NB, Pollock DA (2010) 11.6. Melandryidae Leach, 1815. In *Handbuch der Zoologie/Handbook of Zoology. Band/Volume IV Arthropoda: Insecta Teilband/Part 38. Coleoptera, Beetles. Volume 2. Morphology and Systematics (Polyphaga partim)*. (Eds RAB Leschen, RG Beutel and JF Lawrence) pp. 520–533. W. DeGruyter, Berlin.

Nikitsky NB, Lawrence JF, Kirejtshuk AG, Gratshev WG (1994) A new beetle family, Decliniidae fam. n., from the Russian Far East and its taxonomic relationships (Coleoptera Polyphaga). *Russian Entomological Journal* **2** (5–6), 3–10.

Nikitsky NB, Semenov VB, Dolgin MM (1998) The beetles of the Prioksko-Terrasny Biosphere Reserve - Xylobiontes, Mycetobiontes, and Scarabaeidae (with the review of the Moscow Region fauna of these groups). Supplement 1 (with remarks on nomenclature and systematics of some Melandryidae of the world fauna). *Archives of Zoological Museum Moscow State University* **36**, (Supplement 1), 1–60. [in Russian, with English summary]

Nikolaev GV (2005) New species of the subfamily Aclopinae (Coleoptera, Scarabaeidae) from the Upper Jurassic of Kazakhstan. *Trudy Instituta Zoologii, Ministerstvo Obrazovania i Nauki Republik Zoologo-ekologicheskiye Issledovaniya* **49**, 112–114.

Nikolajev GV (1992) Taxonomic characters and generic composition of Mesozoic Scarabaeidae (Coleoptera). *Paleontologicheskii Zhurnal* **1992**, 75–88 (in Russian; English translation in *Paleontological Journal* **26**, 96–111, 1992).

Nikolajev GV (1993a) The discovery of a pectinicorn beetle (Coleoptera, Lucanidae) from the Upper Cretaceous of Kazakhstan. *Selevinia* **1993** (1), 89–92.

Nikolajev GV (1993b) The taxonomic placement in the subfamily Aphodiinae (Coleoptera, Scarabaeidae) of the new genus of Lower Cretaceous scarab beetles from Transbaykal. *Paleontological Journal* **27**, (1A), 1–7.

Nikolajev GV (1995) A new subfamily of lamellicorn beetles (Coleoptera, Scarabaeidae) from the Lower Cretaceous of Transbaikalia and its phylogenetic relationships. *Paleontologicheskii Zhurnal* **1995**(2), 147–151 (in Russian; English translation in *Paleontological Journal* **29**, 185–192, 1995).

Nikolajev GV (1996) Lamellicorn beetle species (Coleoptera, Scarabaeidae) with multistriate elytra from the Lower Cretaceous Baisa Locality, Transbaikalia. *Paleontologicheskii Zhurnal* **1996**(2), 91–99 (in Russian; translation in *Paleontological Journal* **30**(2), 217–224).

Nikolajev GV (1998) Pleurostict lamellicorn beetles (Coleoptera, Scarabaeidae) from the Lower Cretaceous of Transbaikalia. *Paleontologicheskii Zhurnal* **1998**, 77–84 (in Russian; English translation in *Paleontological Journal* **32**, 513–520, 1998).

Nikolajev GV (1999a) New data on the systematics of the scarab beetles of the subfamily Hybosorinae (Coleoptera, Scarabaeidae): the establishment of the new tribe comprising four monotypic South American genera, and description of some new taxa from the Lower Cretaceous of Transbaikalia. *Tethys Entomological Research* **1999**, 173–182. [in Russian]

Nikolajev GV (1999b) On the polyphyly of the subfamily Penichrolucaninae (Coleoptera, Lucanidae), with the erection of the new monotypic tribe Brasilucanini. *Tethys Entomological Research* **1999**, 171–172. [in Russian]

Nikolajev GV (2000a) A new subfamily of lamellicorn beetles (Coleoptera, Scarabaeoidea) from the Lower Cretaceous of Transbaikalia and Mongolia, and its position within the superfamily. *Paleontological Journal* **2000**, 63–66. [in Russian]

Nikolajev GV (2000b) New subfamily of the stag beetles (Coleoptera: Scarabaeidae: Lucanidae) from the Mesozoic of Mongolia, and its position in the system of the superfamily. *Paleontological Journal* **34**, 327–330.

Nikolajev GV (2002) The possible assignment of a new genus of lamellicorn beetles from the Lower Cretaceous of Mongolia to the subfamily Pleocominae. *Paleontological Journal* **36**, 279–282. [in Russian]

Nikolajev GV (2004) The Mesozoic stage in evolution of the subfamily Aclopinae (Coleoptera, Scarabaeidae). *Tethys Entomological Research* **10**, 33–46.

Nikolajev GV (2008) Family Geotrupidae (Coleoptera) from the Lower Cretaceous of Asia. *Tethys Entomological Research* **16**, 31–36.

Nikolajev GV, Ren D (2010) The oldest fossil Ochodaeidae (Coleoptera: Scarabaeoidea) from the Middle Jurassic of China. *Zootaxa* **2553**, 65–68.

Nikolajev GV, Wang B, Zhang H-C (2011) A new fossil genus of the family Glaphyridae (Coleoptera: Scarabaeoidea) from the Lower Cretaceous Yixian Formation. *Zootaxa* **2811**, 47–52.

Nilsson AN (2001) *World Catalogue of Insects. Volume 3: Dytiscidae Coleoptera*. 395 pp. Apollo Books, Stenstrup.

Nilsson AN, Roughley RE, Brancucci M (1989) A review of the genus- and family-group names of the family Dytiscidae Leach (Coleoptera). *Entomologica Scandinavica* **20**, 287–316. doi:10.1163/187631289X00348

Noirot C, Quennedey A (1974) Fine structure of insect epidermal glands. *Annual Review of Entomology* **19**, 61–80. doi:10.1146/annurev.en.19.010174.000425

Nutting WL, Spangler HG (1969) The hastate setae of certain dermestid larvae: an entangling defense mechanism. *Annals of the Entomological Society of America* **62**, 763–769.

Nyholm T (1972) Zur Morphologie und Funktion des Helodiden-Aedoeagus (Col.). *Entomologica Scandinavica* **3**, 81–119. doi:10.1163/187631272X00193

O'Keefe ST (2000) 20. Scydmaenidae Leach, 1815. In *American Beetles. Volume 1. American Beetles. Archostemata, Myxophaga, Adephaga, Polyphaga: Staphyliniformia.* (Eds RH Arnett Jr and MC Thomas) pp. 259–267. CRC Press, Boca Raton, Florida.

O'Keefe ST (2002a) 38. Dascillidae Guérin-Méneville 1843. In *American Beetles. Volume 2. Polyphaga: Scarabaeoidea through Curculionoidea.* (Eds RH Arnett Jr, MC Thomas, PE Skelley and JH Frank) pp. 90–91. CRC Press, Boca Raton, Florida.

O'Keefe ST (2002b) 61. Phengodidae LeConte 1861. In *American Beetles. Volume 2. Polyphaga: Scarabaeoidea through Curculionoidea.* (Eds RH Arnett Jr, MC Thomas, PE Skelley and JH Frank) pp. 181–186. CRC Press, Boca Raton, Florida.

O'Keefe ST (2005) 11.5. Scydmaenidae Leach, 1815. In *Handbuch der Zoologie/Handbook of Zoology. Band/Volume IV Arthropoda: Insecta Teilband/Part 38. Coleoptera, Beetles. Volume 1. Morphology and Systematics (Archostemata, Adephaga, Myxophaga, Polyphaga partim).* (Eds RG Beutel and RAB Leschen) pp. 280–288. W. DeGruyter, Berlin.

O'Keefe ST, Monteith GB (2000) *Clidicus abbotensis* O'Keefe, a new species of Scydmaenidae (Coleoptera: Staphylinoidea) from Australia with description of the larva. *Memoirs of the Queensland Museum* **46**(1), 211–223.

Ober KA (2002) Phylogenetic relationships of the carabid subfamily Harpalinae (Coleoptera) based on molecular sequence data. *Molecular Phylogenetics and Evolution* **24**, 228–248. doi:10.1016/S1055-7903(02)00251-8

Oberprieler RG (1993) New taxa of Ochyromerina from Africa, with comments on the subtribe and description of immature stages (Coleoptera: Curculionidae: Curculioninae: Tychiini). *Journal of African Zoology* **107**(3), 217–252.

Oberprieler RG (1995) The weevils (Coleoptera: Curculionoidea) associated with cycads. 1. Classification, relationships, and biology. In *Proceedings of the Third International Conference on Cycad Biology held in Pretoria, South Africa, 5–9 July 1993.* (Ed. P. Vorster) pp. 295–334. Cycad Society of South Africa, Stellenbosch.

Oberprieler RG (2000) The larvae of the weevil tribe Eurhynchini and the phylogeny of the Brentidae (Coleoptera: Curculionoidea). *Invertebrate Taxonomy* **14**, 755–770. doi:10.1071/IT00032

Oberprieler RG (2004a) Antliarhininae Schoenherr, 1823 (Coleoptera, Curculionoidea). In *Brentidae of the World (Coleoptera, Curculionoidea).* (Eds A Sforzi and L Bartolozzi) pp. 829–853. Regione Piemonte, Museo Regionale di Scienze Naturali, Torino.

Oberprieler RG (2004b) Eurhynchinae Lacordaire, 1863 (Coleoptera, Curculionoidea). In *Brentidae of the World (Coleoptera, Curculionoidea).* (Eds A Sforzi and L Bartolozzi) pp. 873–903. Regione Piemonte, Museo Regionale di Scienze Naturali, Torino.

Oberprieler RG (2010) A reclassification of the weevil subfamily Cyclominae (Coleoptera: Curculionidae). *Zootaxa* **2515**, 1–35.

Oberprieler RG, Marvaldi AE, Anderson RS (2007) Weevils, weevils, weevils everywhere. *Zootaxa* **1668**, 491–520.

Ocampo FC (2006a) Introduction to the scarab family Hybosoridae (Coleoptera: Scarabaeoidea). *Bulletin of the University of Nebraska State Museum* **19**, 3–6.

Ocampo FC (2006b) Phylogenetic analysis of the Anaidini based on morphological data and revision of the New World tribe Anaidini (Coleoptera: Scarabaeoidea). *Bulletin of the University of Nebraska State Museum* **19**, 13–177.

Ocampo FC, Ballerio A (2006) Catalog of the subfamilies Anaidinae, Ceratocanthinae, Hybosorinae, Liparochrinae, and Pachyplectrinae (Scarabaeoidea: Hybosoridae). *Bulletin of the University of Nebraska State Museum* **19**, 178–209.

Ocampo FC, Hawks DC (2006) Molecular phylogenetics and systematic placement of the family Hybosoridae (Coleoptera: Scarabaeoidea). *Bulletin of the University of Nebraska State Museum* **19**, 7–12.

Ocampo FC, Mondaca J (2012) Revision of the scarab subfamily Aclopinae Blanchard (Coleoptera: Scarabaeidae) in Argentina and Chile. *Zootaxa* **3409**, 1–29.

Ochs G (1949) A revision of the Australian Gyrinidae. *Records of the Australian Museum* **22**, 171–199. doi:10.3853/j.0067-1975.22.1949.599

Odnosum VK (1991) The larvae of mordellid beetles (Coleoptera, Mordellidae) of the U.S.S.R. *Entomologicheskoye Obozreniye* **71**(3), 542–556 (in Russian; translation in *Entomological Review* **71**(3), 118–132.

Ôhara M (1994) A revision of the superfamily Histeroidea of Japan (Coleoptera). *Insecta Matsumurana (N.S.)* **51**, 1–283.

Ôhara M (1999) A revision of the superfamily Histeroidea of Japan (Coleoptera). Supplementum 1. *Insecta Matsumurana (N.S.)* **55**, 75–132.

Ohaus F (1933) *Genera Insectorum de P. Wytsman. Fascicule 195. Coleoptera Lamellicornia Fam. Scarabaeidae Subfam. Euchirinae – Phaenomerinae.* 16 pp., 1 pl. L. Desmet-Verteneuil, Brussels

Oke C (1932) Notes on Australian Coleoptera, with descriptions of new species. II. *Proceedings of the Linnean Society of New South Wales* **57**, 148–172.

Olliff AS (1885) A list of Cucujidae of Australia, with notes and descriptions of new species. *Proceedings of the Linnean Society of New South Wales* **10**, 203–224.

Onore G (1994) Description of the immature stages of six species of *Sphaenognathus*, with comparative notes on phylogeny and natural history (Insecta: Coleoptera: Lucanidae). *Annals of the Carnegie Museum* **63**(1), 77–99.

Opitz W (2002) 73. Cleridae Latreille 1804. In *American Beetles. Volume 2. Polyphaga: Scarabaeoidea through Curculionoidea.* (Eds RH Arnett Jr, MC Thomas, PE Skelley and JH Frank) pp. 267–280. CRC Press, Boca Raton, Florida.

Opitz W (2010) New taxa of Epiphloeinae Kuwert (Cleridae) and Chaetosomatidae Crowson (Coleoptera: Cleroidea). *Insecta Mundi* **0123**, 1–28.

Orledge GM, Reynolds SE (2005) Fungivore host-use groups from cluster analysis: patterns of utilisation of fungal fruiting bodies by ciid beetles. *Ecological Entomology* **30**, 620–641. doi:10.1111/j.0307-6946.2005.00727.x

Orledge GM, Smith PA, Reynolds SE (2010) The non-pest Australasian fungivore *Cis bilamellatus* Wood (Coleoptera: Ciidae) in northern Europe: spread dynamics, invasion success and ecological impact. *Biological Invasions* **12**, 515–530. doi:10.1007/s10530-009-9455-y

Osborne HL, Kulhavy DL (1975) Notes on *Nosodendron californicum* Horn on slime fluxes of grand fir, *Abies grandis* (Douglas) Lindley, in northern Idaho (Coleoptera: Nosodendridae). *Coleopterists Bulletin* **29**, 71–73.

Otero JC, Gayoso A, Santamaría JM (1996) Caracteres femeninos de interés taxonómico en la familia Monotomidae Laporte, 1840 (Coleoptera). *Boletin de la Asociacion Espanola de Entomologia* **21**(1–2), 9–28.

Pace R (1976) An exceptional endogeous beetle: Crowsoniella relicta n. gen. n. sp. of Archostemata Tetraphaleridae from Central Italy. (XVI). Contribution to knowledge of endogeous beetles). *Bolletino del Museo Civico di Storia Naturale di Verona* **2**, 445–458.

Pace R (1996) *Fauna d'Italia. XXXIV. Coleoptera: Staphylinidae: Leptotyphlinae.* Edizioni Calderini, Bologna.

Pacheco FM (1964) Sistemática, filogenia y distribución de los Heterocéridos de América (Coleoptera: Heteroceridae). *Monograficias del Colegio de Post-Graduados, Escuela Nacional de Agricultura. Chapingo* **1**, 1–155.

Pakaluk J (1985) New genus and species of Corylophidae (Coleoptera) from Florida, with a description of its larva. *Annals of the Entomological Society of America* **78**, 406–409.

Pakaluk J (1991) Phylogenetic placement of *Phaenocephalus* Wollaston (Coleoptera: Phaenocephalidae & Phalacridae). *Proceedings of the Entomological Society of Washington* **93**, 317–321.

Pakaluk J, Lawrence JF (1986) *Priamima,* a new genus of Corylophidae (Coleoptera) from Australia. *Journal of the Australian Entomological Society* **25**, 81–84. doi:10.1111/j.1440-6055.1986.tb01078.x

Pal TK (1985) A revision of Indian *Psammoecus* Latreille (Coleoptera: Silvanidae). *Records of the Zoological Survey of India. Miscellaneous Publication. Occasional Paper* **71**, 1–54.

Pal TK, Lawrence JF (1986) A new genus and subfamily of mycophagous Bothrideridae (Coleoptera: Cucujoidea) from the Indo-Australian region, with notes on related families. *Journal of the Australian Entomological Society* **25**, 185–210. doi:10.1111/j.1440-6055.1986.tb01101.x

Pal TK, Sen Gupta T, Crowson RA (1985)) Revision of Uleiota (Coleoptera: Silvanidae) from India and Sri Lanka and its systematic position. *Oriental Insects* **18**, 213–233. doi:10.1080/00305316.1984.10432204

Pang H, Ślipiński A (2009) Revision of the Australian Coccinellidae (Coleoptera). Genus *Diomus* Mulsant. Part 1. *Annales Zoologici* **59**(4), 641–698. doi:10.3161/000345409X485008

Pang H, Ślipiński A (2010) Revision of the Australian Coccinellidae (Coleoptera). Genus *Diomus* Mulsant. Part 2. *Annales Zoologici* **60**(4), 493–545. doi:10.3161/000345410X550382

Parker JB, Böving AG (1914) The blister beetle *Tricrania sanguinipennis*—biology, description of different stages, and systematic relationships. *Proceedings of the United States National Museum* **64** (23), 1–49, 5 pls.

Parkin EA (1933) The larvae of some wood-boring Anobiidae (Coleoptera). *Bulletin of Entomological Research* **24**, 33–68. doi:10.1017/S000748 5300029990

Pasteels JM, Rowell-Rahier M (1989) Defensive glands and secretions as taxonomic tools in the Chrysomelidae. *Entomography* **6**, 423–432.

Paulian R (1950) Les Corylophidae d'Afrique (Coleoptera). *Mémoires de l'Institut Français d'Afrique Noire* **12**, 1–126.

Paulian R (1958) Un nouveau Coléoptère Scarabeide de Madagascar type d'une nouvelle sous-famille. *Le Naturaliste Malgache* **10**, 39–46.

Paulian R (1962) Observation sur les Saciinae d'Asie tropicale. *Reichenbachia* **1**, 73–105.

Paulian R (1976) Les Orphnidae des Madagascar (Insecta: Coleoptera: Scarabaeoidea). *Bulletin de Muséum National d'Histoire Narurelle. Zoologie* **288**, 1199–1223.

Paulian R (1977) The Australian Ceratocanthidae (Coleoptera: Scarabaeoidea). *Journal of the Australian Entomological Society* **16**, 261–265. doi:10.1111/ j.1440-6055.1977.tb00099.x

Paulian R (1978) Revision des Ceratocanthidae (Col. Scarabaeoidea). II. Les espèces Orientales et Australiennes. *Annales de la Société Entomologique de France (N. S.)* **14**, 479–514.

Paulian R (1980) Revision taxonomique des *Liparochrus* Erichson, genre Australo-Melanesien de Coléoptères Hybosoridae, Scarabaeoidea. *Annales de la Société Entomologique de France (N. S.)* **16**, 389–433.

Paulian R (1984) Les Orphnidae Améericains (Coléoptères, Scarabaeoidea). *Annales de la Société Entomologique de France (N.S.)* **20**, 65–92.

Paulian R (1988) *Biologie des Coléoptères.* xi + 719 mpp. Lechevalier, Paris.

Paulian R, Lumaret J-P (1982) Le larve des Orphnidae (Col. Scarabaeoidea). *Bulletin de la Societe Entomologique de France* **87**, 263–272.

Paulsen MJ, Mondaca EJ (2006) Revision of the South American Ceratognathini (Coleoptera: Lucanidae: Aesalinae) with the description of a new genus and a new species. *Zootaxa* **1191**, 1–19.

Paulus HF (1969) *Pseudochelonarium* (*Neochelonarium*) *kalimantanense* n. sp. aus Borneo, mit Bemerkungen zum System der Chelonariidae (Col. Dryopoidea). *Zeitschrift der Arbeitsgemeinschaft der Österreichischen Entomologen* **21**, 105–109.

Paulus HF (1970) Zur Morphologie und Biologie der Larven von *Pelochares* Mulsant & Rey (1869) und *Limnichus* Latreille (1829) (Coleoptera: Dryopoidea: Limnichidae). *Senckenbergiana Biologica* **51**, 77–87.

Paulus HF (1972) Der Stand unserer Kenntnis über die Familie Byrrhidae (Col.). *Folia Entomologica Hungarica* **25**(21), 335–348. [Series Nova]

Paulus HF (1973) Revision der Familie Byrrhidae I: Zur Systematik und Faunistik der westpaläarktischen Vertreter der Gattung *Curimopsis* Ganglbauer, 1902 (Col: Byrrhidae, Syncalyptinae). *Senckenbergiana Biologica* **54**, 353–367.

Paulus HF (1979) Eye structure and the monophyly of Arthropoda. In *Arthropod Phylogeny* (Ed. AP Gupta) pp. 299–383. Van Nostrand Reinhold, New York.

Paulus HF (2004) *Urtrea graeca* nov. gen. et nov. spec., der erste Vertreter der tropischen Atractocerinae in Europa sowie eine Beschreibung von *Homaloxylon aspoecki* nov. spec. aus Yunnan (China) (Coleoptera, Cucujiformia, Lymexylidae, Atractocerinae nov. status). *Denisia* **13**, 277–290.

Paviour-Smith K (1960a) The fruiting bodies of macrofungi as habitats for beetles of the family Ciidae (Coleoptera). *Oikos* **11**, 43–71. doi:10.2307/ 3564883

Paviour-Smith K (1960b) The invasion of Britain by *Cis bilamellatus* Fowler (Coleoptera: Ciidae). *Proceedings of the Royal Entomological Society of London* **35**, 145–155. [A]

Peacock ER (1978) *Trichelodes (= Hexanodes),* type of Trichelodini new tribe, and its affinities with the Trinodini (Coleoptera: Dermestidae). *New Zealand Journal of Zoology* **5**, 341–349. doi:10.1080/03014223.1978.10428320

Peacock ER (1987) A review of the Rhadalinae (= Aplocneminae) (Coleoptera: Melyridae). *Bulletin of the British Museum (Natural History). Entomological Series* **56**(3), 129–170.

Peacock ER (1993) *Handbooks for the Identification of British Insects. Vol. 5, Part 3. Adults and Larvae of Hide, Larder and Carpet Beetles and their relatives (Coleoptera: Dermestidae) and of Derodontid Beetles (Coleoptera: Derodontidae).* Royal Entomological Society of London, London.

Peake FGG (1953) On a bostrychid wood-borer in the Sudan. *Bulletin of Entomological Research* **44**(2), 317–325. doi:10.1017/S00074853000 23105

Pearson DL (1988) Biology of tiger beetles. *Annual Review of Entomology* **33**, 123–147. doi:10.1146/annurev.en.33.010188.001011

Peck SB (1977) A unusual sense receptor in internal antennal vesicles of *Ptomaphagus* (Coleoptera: Leiodidae). *Canadian Entomologist* **109**, 81–86. doi:10.4039/Ent10981-1

Peck SB (1990) Insecta: Coleoptera. Silphidae and the associated families Agyrtidae and Leiodidae. In *Soil Biology Guide* (Ed. D Dindal) pp. 1113–1136. John Wiley and Sons, New York.

Peck SB (2000a) 18. Agyrtidae C. G. Thomson, 1859. In *American Beetles. Volume 1. American Beetles. Archostemata, Myxophaga, Adephaga, Poly-*

phaga: Staphyliniformia. (Eds RH Arnett Jr and MC Thomas) pp. 247–249. CRC Press, Boca Raton, Florida.

Peck SB (2000b) 19. Leiodidae Fleming, 1821. In *American Beetles. Volume 1. American Beetles. Archostemata, Myxophaga, Adephaga, Polyphaga: Staphyliniformia.* (Eds RH Arnett Jr and MC Thomas) pp. 250–258. CRC Press, Boca Raton, Florida.

Peck SB (2000c) 21. Silphidae Latreille, 1807. In *American Beetles. Volume 1. American Beetles. Archostemata, Myxophaga, Adephaga, Polyphaga: Staphyliniformia.* (Eds RH Arnett Jr and MC Thomas) pp. 268–271. CRC Press, Boca Raton, Florida.

Peck SB (2001) Review of the carrion beetles of Australia and New Guinea (Coleoptera: Silphidae). *Australian Journal of Entomology* **40**, 93–101. doi:10.1046/j.1440-6055.2001.00216.x

Peck SB (2003) A review and revision of the Leiodidae of southern Africa (Coleoptera). Part I. Leiodinae: Sogdini, Estadiini, and Leiodini (in part), with notes on species from other regions. In *Systematics of Coleoptera: Papers Celebrating the Retirement of Ivan Löbl.* (Eds G Cuccodoro and RAB Leschen) pp. 103–139. Associated Publishers (Memoirs on Entomology, International Vol. 17), Gainesville, Florida.

Peck SB (2010) *Derolathrus cavernicolus* n. sp., a beetle family new for North America (Coleoptera: Jacobsoniidae). *Annals of the Entomological Society of America* **103**, 1–6. doi:10.1603/008.103.0101

Peck SB, Anderson RS (1985) Taxonomy, phylogeny and biogeography of the carrion beetles of Latin America (Coleoptera: Silphidae). *Quaestiones Entomologicae* **21**, 247–317.

Peck SB, Cook J (1992) Use of "car-nets" to sample flying micro-Coleoptera. *Canadian Entomologist* **124**, 745–749. doi:10.4039/Ent124745-5

Peck SB, Cook J (2003) Neotropical species of *Dietta* (Coleoptera: Leiodidae: Leiodinae: Estadiini). *Canadian Entomologist* **135**, 775–810. doi:10.4039/n02-111

Peck SB, Cook J (2009) Review of the Sogdini of North and Central America (Coleoptera: Leiodidae: Leiodinae) with descriptions of fourteen new species and three new genera. *Zootaxa* **2102**, 1–74.

Peck SB, Cook J (2011) Systematics, distributions and bionomics of the Catopocerini (eyeless soil fungivore beetles) of North America (Coleoptera: Leiodidae: Catopocerinae). *Zootaxa* **3077**, 1–118.

Peck SB, Davies AE (1980) Collecting small beetles with large-area "window" traps. *Coleopterists Bulletin* **34**, 237–239.

Peck SB, Kaulbars MM (1987) A synopsis of the distrtibution and bionomics of the carrion beetles (Coleoptera: Silphidae) of the conterminous United States. *Proceedings of the Entomological Society of Ontario* **118**, 47–81.

Pederzani F (1995) Keys to the identification of the genera and subgenera of adult Dysticidae (sensu lato) of the world (Coleoptera Dytiscidae). *Atti dell'Accademia Roveretana degli Agiati* **244** (1994), ser. VII, vol. IV, B, 5–83.

Penney D (Ed.) (2010) *Biodiversity of Fossils in Amber from the Major World Deposits.* 304 pp. Siri Scientific Press, Manchester, UK.

Perkins PD (1980) Aquatic beetles of the family Hydraenidae in the Western Hemisphere: classification, biogeography and inferred phylogeny (Insecta: Coleoptera). *Quaestiones Entomologicae* **16**, 3–554.

Perkins PD (1981) *Hughleechia giulianii,* a new genus and new species of intertidal beetle from Australia (Coleoptera: Hydraenidae). *The Pan-Pacific Entomologist* **57**, 295–302.

Perkins PD (1997) Life on the effective bubble: exocrine secretion delivery systems (ESDS) and the evolution and classification of beetles in the family Hydraenidae (Insecta: Coleoptera). *Annals of the Carnegie Museum* **66**(2), 89–207.

Perkins PD (2000) 16. Hydraenidae Mulsant, 1844. In *American Beetles. Volume 1. American Beetles. Archostemata, Myxophaga, Adephaga, Polyphaga: Staphyliniformia.* (Eds RH Arnett Jr and MC Thomas) pp. 228–232. CRC Press, Boca Raton, Florida.

Perkins PD (2004a) *Limnebius acupunctatus*, a new species of water beetle from Australia and Papua New Guinea (Coleoptera: Hydraenidae). *Zootaxa* **749**, 1–12.

Perkins PD (2004b) A revision of the Australian endemic water beetle genus *Gymnanthelius* Perkins (Coleoptera: Hydraenidae). *Zootaxa* **585**, 1–39.

Perkins PD (2004c) A revision of the Western Australian endemic humicolous beetle genus *Tympallopatrum* Perkins (Coleoptera: Hydraenidae). *Zootaxa* **672**, 1–16.

Perkins PD (2005) A revision of the water beetle genus *Gymnochthebius* Orchymont (Coleoptera: Hydraenidae) for Australia and Papua New Guinea. *Zootaxa* **1024**, 1–161.

Perkins PD (2006) A revision of the Australian humicolous hygropetric water beetle genus *Tympanogaster* Perkins, and comparative morphology of the Meropathina (Coleoptera: Hydraenidae). *Zootaxa* **1346**, 1–396.

Perkins PD (2007a) A revision of the Australian intertidal water beetle genus *Hughleechia* Perkins (Coleoptera: Hydraenidae). *Zootaxa* **1527**, 17–29.

Perkins PD (2007b) A revision of the Australian species of the water beetle genus *Hydraena* Kugelann (Coleoptera: Hydraenidae). *Zootaxa* **1489**, 1–207.

Perkins FO, Barta JR, Clopton RE, Pierce MA, Upton SJ (2000) Phylum Apicomplexa Levine, 1970. In *An Illustrated Guide to the Protozoa.* Second Edition, Vol. 1. (Eds JJ Lee, GF Leedale and P. Bradbury) pp. 190–369. Society of Protozoologists, Lawrence, Kansas.

Perreau M, Růžička J (2007) Systematic position of *Perkovskius* Lafer 1989 (Coleoptera: Leiodidae: Catopocerinae), with description of a second species from the Far East of Russia. *Annales de la Société Entomologique de France (N. S.)* **43**(3), 257–264.

Peters BC, Creffield JW, Eldridge RH (2002) Lyctine (Coleoptera: Bostrichidae) pests of timber in Australia: A literature review and susceptibility testing protocol. *Australian Forestry* **65**(2), 107–119.

Peterson A (1959) *Entomological Techniques: How to Work with Insects.* 9th edition. A. Peterson, Ann Arbor, Michigan.

Petitpierre E (1988) Cytogenetics, cytotaxonomy and genetics of Chrysomelidae. In. *Biology of Chrysomelidae.* (Eds P Jolivet, E Petitpierre and TH Hsiao) pp.131–59. Kluwer Academic, Dordrecht.

Petitpierre E (2011) Cytogenetics, cytotaxonomy and chromosomal evolution of Chrysomelinae revisited (Coleoptera, Chrysomelidae). *ZooKeys* **157**, 67–79. doi:10.3897/zookeys.157.1339

Petitpierre E, Segarra E, Yadav JS, Vurkki N (1988) Chromosome numbers and meioformulae of Chrysomelidae. In *Biology of Chrysomelidae.* (Eds P Jolivet, E Petitpierre and TH Hsiao) pp. 161–186. Kluwer Academic, Dordrecht.

Peyerimhoff P de (1933) Les larves des Coléoptères d'apres A. G. Böving et F. C. Craighead et les grands criteriums de l'ordre. *Annales de la Société Entomologique de France* **102**, 77–106.

Pfaffenberger GS (1991) Bruchidae (Chrysomeloidea) (= Lariidae, Mylabridae). In *Immature Insects. Volume 2.* (Ed. FW Stehr) pp. 561–568. Kendall/Hunt, Dubuque, Iowa.

Philips TK (2000) Phylogenetic analysis of the New World Ptininae (Coleoptera: Bostrichoidea). *Systematic Entomology* **25**, 235–262. doi:10.1046/j.1365-3113.2000.00102.x

Philips TK (2002) 70. Anobiidae Fleming 1821. In *American Beetles. Volume 2. Polyphaga: Scarabaeoidea through Curculionoidea.* (Eds RH Arnett Jr, MC Thomas, PE Skelley and JH Frank) pp. 245–260. CRC Press, Boca Raton, Florida.

Philips TK, Bell KL (2010) 6.4. Ptinidae Latreille, 1802. In *Handbuch der Zoologie/Handbook of Zoology. Band/Volume IV Arthropoda: Insecta Teilband/Part 38. Coleoptera, Beetles. Volume 2. Morphology and Systematics (Polyphaga partim).* (Eds RAB Leschen, RG Beutel and JF Lawrence) pp. 217–225. W. DeGruyter, Berlin.

Philips TK, Ivie MA (2002) 90. Bothrideridae Erichson 1845. In *American Beetles. Volume 2. Polyphaga: Scarabaeoidea through Curculionoidea.* (Eds RH Arnett Jr, MC Thomas, PE Skelley and JH Frank) pp. 358–362. CRC Press, Boca Raton, Florida.

Philips TK, Xie W (2000) 11. Amphizoidae LeConte, 1853. In *American Beetles. Volume 1. American Beetles. Archostemata, Myxophaga,*

Adephaga, Polyphaga: Staphyliniformia. (Eds RH Arnett Jr and MC Thomas) pp. 153–155. CRC Press, Boca Raton, Florida.

Philips TK, Young DK (2000) 2. Micromalthidae Barber, 1913. In *American Beetles. Volume 1. American Beetles. Archostemata, Myxophaga, Adephaga, Polyphaga: Staphyliniformia.* (Eds RH Arnett Jr and MC Thomas) pp. 22–23. CRC Press, Boca Raton, Florida.

Philips TK, Ivie MA, Giersch JJ (2002) 65. Jacobsoniidae Heller 1926. In *American Beetles. Volume 2. Polyphaga: Scarabaeoidea through Curculionoidea.* (Eds RH Arnett Jr, MC Thomas, PE Skelley and JH Frank) pp. 219–220. CRC Press, Boca Raton, Florida.

Philips TK, Pretorius E, Scholtz CH (2004) A phylogenetic analysis of dung beetles (Scarabaeinae: Scarabaeidae): unrolling an evolutionary history. *Invertebrate Systematics* **18**, 53–88. doi:10.1071/IS03030

Pic M (1930) Contribution a l'étude des Coléoptères malacodermes (2e article). *Annales de la Société Entomologique de France* **99**, 311–324.

Piel PO (1938) Note sur le parasitisme de *Dastarcus helophoroiodes* Fairmaire (Coléoptère Colydiide). *Notes d'Entomologie Chinoise* **4** (1937), 1–15, pls. 1–2.

Ping C (1928) Cretaceous fossil insects of China. *Palaeontologica Sinica B* **13**(1), 1–57.

Pinto ID (1987) Permian insects from Paraná Basin, south Brazil. IV Coleoptera. *Pesquisas em Geociências* **19**, 5–12.

Pinto JD, Bologna MA (2002) 111. Meloidae Gyllenhal 1810. In *American Beetles. Volume 2. Polyphaga: Scarabaeoidea through Curculionoidea.* (Eds RH Arnett Jr, MC Thomas, PE Skelley and JH Frank) pp. 522–529. CRC Press, Boca Raton, Florida.

Polilov AA (2005) Anatomy of the feather-winged beetles *Acrotrichis montandoni* and *Ptilium myrmecophilum* (Coleoptera, Ptiliidae). *Entomological Review* **85**, 467–475.

Polilov AA (2008) Anatomy of the smallest of the Coleoptera, feather-winged beetles from tribe Nanosellini (Coleoptera, Ptiliidae) and limits to insect miniaturization. *Entomological Review* **88**, 26–33. doi:10.1134/S0013873808010041

Polilov AA, Beutel RG (2009) Miniaturization effects in larvae and adults of *Mikado* sp. (Coleoptera: Ptiliidae), one of the smallest free-living insects. *Arthropod Structure & Development* **38**, 247–270. doi:10.1016/j.asd.2008.11.003

Polilov AA, Beutel RG (2010) Developmental stages of the hooded beetle *Sericoderus lateralis* (Coleoptera: Corylophidae) with comments on the phylogenetic position and effects of miniaturization. *Arthropod Structure & Development* **39**, 52–69. doi:10.1016/j.asd.2009.08.005

Pollock DA (1992) A new species of *Tydessa* Peacock (Coleoptera: Pythinae: Pilipalpinae) from Western North America. *The Pan-Pacific Entomologist* **68**, 243–247.

Pollock DA (1994) Systematic position of Pilipalpinae (Coleoptera: Tenebrionoidea) and composition of Pyrochroidae. *Canadian Entomologist* **126**, 515–532. doi:10.4039/Ent126515-3

Pollock DA (1995) Classification, reconstructed phylogeny and geographical history of genera of Pilipalpinae (Coleoptera: Tenebrionoidea: Pyrochroidae). *Invertebrate Taxonomy* **9**, 563–708. doi:10.1071/IT9950563

Pollock DA (1998) Revision of the Neotropical genus *Ischyomius* with a discussion on its systematic position (Insecta: Coleoptera: Tenebrionoidea: Pythidae). *Annals of Carnegie Museum* **67**(3), 245–264.

Pollock DA (1999) Review of the New World Hemipeplinae (Coleoptera: Mycteridae) with descriptions of ten new species. *Entomologica Scandinavica* **30**, 47–73. doi:10.1163/187631200X00200

Pollock DA (2002a) 100. Melandryidae Leach 1815. In *American Beetles. Volume 2. Polyphaga: Scarabaeoidea through Curculionoidea.* (Eds RH Arnett Jr, MC Thomas, PE Skelley and JH Frank) pp. 417–422. CRC Press, Boca Raton, Florida.

Pollock DA (2002b) 112. Mycteridae Blanchard 1845. In *American Beetles. Volume 2. Polyphaga: Scarabaeoidea through Curculionoidea.* (Eds RH Arnett Jr, MC Thomas, PE Skelley and JH Frank) pp. 530–533. CRC Press, Boca Raton, Florida.

Pollock DA (2002c) 113. Boridae C. G. Thomson 1859. In *American Beetles. Volume 2. Polyphaga: Scarabaeoidea through Curculionoidea.* (Eds RH Arnett Jr, MC Thomas, PE Skelley and JH Frank) pp. 534–536. CRC Press, Boca Raton, Florida.

Pollock DA (2002d) 114. Pythidae Solier 1834. In *American Beetles. Volume 2. Polyphaga: Scarabaeoidea through Curculionoidea.* (Eds RH Arnett Jr, MC Thomas, PE Skelley and JH Frank) pp. 537–539. CRC Press, Boca Raton, Florida.

Pollock DA (2002e) 116. Salpingidae Leach 1815. In *American Beetles. Volume 2. Polyphaga: Scarabaeoidea through Curculionoidea.* (Eds RH Arnett Jr, MC Thomas, PE Skelley and JH Frank) pp. 544–548. CRC Press, Boca Raton, Florida.

Pollock DA (2002f) 119. Scraptiidae Mulsant 1856. In *American Beetles. Volume 2. Polyphaga: Scarabaeoidea through Curculionoidea.* (Eds RH Arnett Jr, MC Thomas, PE Skelley and JH Frank) pp. 564–567. CRC Press, Boca Raton, Florida.

Pollock DA (2010a) 11.20. Mycteridae Blanchard, 1845. In *Handbuch der Zoologie/Handbook of Zoology. Band/Volume IV Arthropoda: Insecta Teilband/Part 38. Coleoptera, Beetles. Volume 2. Morphology and Systematics (Polyphaga partim).* (Eds RAB Leschen, RG Beutel and JF Lawrence) pp. 693–699. W. DeGruyter, Berlin.

Pollock DA (2010b) 11.21. Boridae C. G. Thomson, 1859. In *Handbuch der Zoologie/Handbook of Zoology. Band/Volume IV Arthropoda: Insecta Teilband/Part 38. Coleoptera, Beetles. Volume 2. Morphology and Systematics (Polyphaga partim).* (Eds RAB Leschen, RG Beutel and JF Lawrence) pp. 699–704. W. DeGruyter, Berlin.

Pollock DA (2010c) 11.23. Pythidae Solier, 1834. In *Handbuch der Zoologie/Handbook of Zoology. Band/Volume IV Arthropoda: Insecta Teilband/Part 38. Coleoptera, Beetles. Volume 2. Morphology and Systematics (Polyphaga partim).* (Eds RAB Leschen, RG Beutel and JF Lawrence) pp. 708–715. W. DeGruyter, Berlin.

Pollock DA (2012) Review of the Eustrophinae (Coleoptera, Tetratomidae) of America north of Mexico. *ZooKeys* **188**, 1–153. doi:10.3897/zookeys.188.2976

Pollock DA, Lawrence JF (1995) Review of *Anaplopus* Blackburn (Coleoptera: Pythidae), with comments on constituents and systematics of Pythidae. In *Biology, Phylogeny, and Classification of Coleoptera. Papers Celebrating the 80th Birthday of Roy A. Crowson.* (Eds J Pakaluk and SA Ślipiński) pp. 449–472. Muzeum I Instytut Zoologii Polska Akademia Nauk, Warsaw.

Pollock DA, Normark BB (2002) The life cycle of *Micromalthus debilis* LeConte (1878) (Coleoptera: Archostemata: Micromalthidae): historical review and evolutionary perspective. *Journal of Zoological Systematics and Evolutionary Research* **40**, 105–112. doi:10.1046/j.1439-0469.2002.00183.x

Pollock DA, Telnov D (2010) 11.22. Trictenotomidae Blanchard, 1845. In *Handbuch der Zoologie/Handbook of Zoology. Band/Volume IV Arthropoda: Insecta Teilband/Part 38. Coleoptera, Beetles. Volume 2. Morphology and Systematics (Polyphaga partim).* (Eds RAB Leschen, RG Beutel and JF Lawrence) pp. 704–708. W. DeGruyter, Berlin.

Pollock DA, Ide S, Costa C (2000) Review of the Neotropical genus *Physiomorphus* Pic (Coleoptera: Mycteridae: Lacconotinae), with description of the larvae of three species. *Journal of Natural History* **34**, 2209–2239. doi:10.1080/002229300750037875

Ponomarenko AG (1961) On the systematic position of *Coptoclava longipoda* Ping (Insecta, Coleoptera). *Палеонтологический журнал* **3**, 67–72.

Ponomarenko AG (1969a) Cretaceous insects from Labrador. 4. A new family of beetles (Coleoptera, Archostemata). *Psyche, Cambridge* **76**, 306–310. doi:10.1155/1969/85918

Ponomarenko AG (1969b) Historical development of the Coleoptera-Archostemata. *Trudy Paleontologicheskogo Instituta* **125**, 1–240. [in Russian]

Ponomarenko AG (1971a) The geological history and evolution of beetles. *Proceedings of the International Congress of Entomology* **13**(1), 281.

Ponomarenko AG (1971b) Systematic position of some beetles from the Solenhofen Shales of Bavaria. *Paleontologicheski Zhurnal* **1971**(1), 67–81 (in Russian; translation in *Paleontological Journal* **1971** (1), 62–75).

Ponomarenko AG (1972) On nomenclature of wing-venation in beetles (Coleoptera). *Entomologicheskoye Obozreniye* **51**, 768–775 (in Russian; translation in *Entomological Review*, Washington **51**, 454–458).

Ponomarenko AG (1973) On the division of the beetle order into suborders. In *Problems of Insect Palaeonotology. Lectures in Memoriam N. A. Cholodkowsky*. pp. 78–89. Nauka, Leningrad.

Ponomarenko AG (1977) Mesozoic Coleoptera. Descriptions of new taxa. Suborder Adephaga. The Mesozoic stage in the evolution of Adephaga. Suborder Polyphaga. Polyphaga incertae *sedis*. Infraorder Staphyliniformia. *Trudy Paleontologicheskogo Institut Akademia Nauk SSSR* **161**, 17–119 (in Russian; translation 1989. Smithsonian Institution Libraries and National Science Foundation, Washington, D. C., pp. 19–162).

Ponomarenko AG (1980) On the systematic position of beetles described by L. Deichmuller from Jurassic locality Solnhofen. In *Mesozoic Fossil Insects*. (Eds VG Dolin, DV Panfilov, AG Ponomarenko and LN Pritikina) pp. 111–119. Naukova Dumka, Kiev.

Ponomarenko AG (1985a) Coleoptera from the Jurassic of Siberia and western Mongolia. In Jurassic Insects of Siberia and Mongolia. (Ed. AP Rasnitsyn). *Trudy Paleontologicheskogo Institut Akademia Nauk SSSR* **211**, 47–87.

Ponomarenko AG (1985b) Fossil insects from the Tithonian "Solnhofener Plattenkalke" in the Museum of Natural History, Vienna. *Annalen der Naturhistorischen Museums, Wien* (Mineralogie, Petrographie, Geologie, Paläontologie, Anthropologie, Prähistorie) **87**, 135–144.

Ponomarenko AG (1986) Scarabaeida. In *Insects in the Early Cretaceous Ecosystems of the West Mongolia* (Transactions of the Joint Soviet-Mongolian Palaeontological Expedition 28). (Ed. AP Rasnitsyn) pp. 84–105. Nauka, Moscow.

Ponomarenko AG (1990) Scarabaeida. In Late Mesozoic insects of eastern Transbaikalia. (Ed. AP Rasnitsyn). *Trudy Paleontologicheskogo Institut Akademia Nauk SSSR* **239**, 39–87.

Ponomarenko AG (1993) Two new species of Mesozoic dytiscoid beetles from Asia. *Paleontological Journal* **27**, (1A), 182–191.

Ponomarenko AG (1995) The geological history of beetles. In *Biology, Phylogeny, and Classification of Coleoptera: Papers Celebrating the 80th Birthday of Roy A. Crowson*. (Eds J Pakaluk and SA Ślipiński) pp. 155–171. Muzeum i Institut Zoologii, Polska Akademia Nauk, Warsaw.

Ponomarenko AG (2000) New beetles from the Permian of European Russia. *Paleontological Journal* **34**, (Suppl. 3), S312–S316.

Ponomarenko AG (2003a) Ecological evolution of beetles (Insecta: Coleoptera). *Acta Zoologica Cracoviensia* **46**(Supplement. Fossil Insects), 319–328.

Ponomarenko AG (2003b) The first beetles (Permosynidae, Coleoptera) from the Upper Tatarian of European Russia. *Paleontologicheskii Zhurnal* **2003** (2), 65–68 (in Russian; English translation. *Paleontological Journal* **37**(2), 170–173.

Ponomarenko AG (2004) Beetles (Insecta, Coleoptera) of the Late Permian and Early Triassic. *Paleontological Journal* **38**, (Supplement 2), 185–196.

Ponomarenko AG, Martínez-Delclòs X (2000) New beetles (Insecta: Coleoptera) from the Lower Cretaceous of Spain. *Acta Geologica Hispanica* **35**, 47–52.

Ponomarenko AG, Mostovski MB (2005) New beetles (Insecta: Coleoptera) from the Late Permian of South Africa. *African Invertebrates* **46**, 253–260.

Pons J, Ribera I, Bertranpetit J, Balke M (2010) Nucleotide substitution rates for the full set of mitochondrial protein-coding genes in Coleoptera. *Molecular Phylogenetics and Evolution* **56**, 796–807. doi:10.1016/j.ympev.2010.02.007

Pooran J, Ślipiński A (2009) A revision of the genus *Scymnodes* Blackburn and *Apolinus* Pope et Lawrence (Coleoptera: Coccinellidae). *Annales Zoologici* **59**(4), 549–584. doi:10.3161/000345409X484946

Poorani J, Ślipiński A, Booth RG (2008) A revision of the genus *Synona* Pope, 1989 (Coleoptera: Coccinellidae: Coccinellini). *Annales Zoologici* **58**(3), 579–594. doi:10.3161/000345408X364427

Pope RD (1979) Wax production by coccinellid larvae (Coleoptera). *Systematic Entomology* **4**, 171–196. doi:10.1111/j.1365-3113.1979.tb00632.x

Pope RD (1981) *Rhyzobius ventralis* (Coleoptera: Coccinellidae), its constituent species, and their taxonomy and historical roles in biological control. *Bulletin of Entomological Research* **71**, 19–31. doi:10.1017/S0007485300050999

Pope RD (1989) A revision of the Australian Coccinellidae (Coleoptera). Part 1. Subfamily Coccinellinae. *Invertebrate Taxonomy* **2**, 633–735. doi:10.1071/IT9880633

Pope RD, Lawrence JF (1990) A review of *Scymnodes* Blackburn, with the description of a new Australian species and its larva (Coleoptera: Coccinellidae). *Systematic Entomology* **15**, 241–252. doi:10.1111/j.1365-3113.1990.tb00315.x

Porch N (2010) Climate space, bioclimatic envelopes and coexistence methods for the reconstruction of past climates: a method using Australian beetles and significance for Quaternary reconstruction. *Journal of Quaternary Science* **29**, 633–647.

Porch N, Elias SA (2000) Quaternary beetles: a review and issues for Australian studies. *Australian Journal of Entomology* **39**, 1–9. doi:10.1046/j.1440-6055.2000.00143.x

Porch N, Jordan GJ, Price DM, Barnes RW, Macphail MK, Pemberton M (2009) Last interglacial climatesof south-eastern Australia: plant and beetle based reconstructions from Yarra Creek, King Island, Tasmania. *Quaternary Science Reviews* **28**, 3197–3210. doi:10.1016/j.quascirev.2009.09.023

Price MB (2002) 78. Smicripidae Horn 1879. In *American Beetles. Volume 2. Polyphaga: Scarabaeoidea through Curculionoidea*. (Eds RH Arnett Jr, MC Thomas, PE Skelley and JH Frank) pp. 316–318. CRC Press, Boca Raton, Florida.

Pringle JA (1938) Obserevations on certain wood-boring Coleoptera occurring in South Africa. *Transactions of the Royal Entomological Society of London* **87**(11), 247–270, 1 pl.

Prokin A, Ren D, Fikáček M (2010) New Mesozoic water scavenger beetles from the Yixian formation in China (Coleoptera: Hydrophiloidea). *Annales Zoologici* **60**(2), 173–179. doi:10.3161/000345410X516821

Pütz A (2002) *Papuamicrochaetes* gen. nov. – eine neue Gattung der Tribus Microchaetini aus Melanesien (Coleoptera: Byrrhidae, Syncalyptinae, Microchaetini). *Entomologische Zeitschrift, Stuttgart* **112**(10), 313–318.

Rakovič M (1997) A revision of the genus *Odochilus* Harold with remarks on the tribal classification of the subfamily Aphodiinae (Coleoptera, Scarabaeidae). *Acta Entomologica Bohemoslovaka* **84**, 27–44.

Ramsdale AS (2002a) 63. Omethidae LeConte 1861. In *American Beetles. Volume 2. Polyphaga: Scarabaeoidea through Curculionoidea.* (Eds RH Arnett Jr, MC Thomas, PE Skelley and JH Frank) pp. 197–201. CRC Press, Boca Raton, Florida.

Ramsdale AS (2002b) 64. Cantharidae Imhoff 1856. In *American Beetles. Volume 2. Polyphaga: Scarabaeoidea through Curculionoidea.* (Eds RH Arnett Jr, MC Thomas, PE Skelley and JH Frank) pp. 202–218. CRC Press, Boca Raton, Florida.

Ramsdale AS (2010a) 4.16. Omethidae LeConte, 1861. In *Handbuch der Zoologie/Handbook of Zoology. Band/Volume IV Arthropoda: Insecta Teilband/Part 38. Coleoptera, Beetles. Volume 2. Morphology and Systematics (Polyphaga partim).* (Eds RAB Leschen, RG Beutel and JF Lawrence) pp. 149–153. W. DeGruyter, Berlin.

Ramsdale AS (2010b) 4.17. Cantharidae Imhoff 1856 (1815). In *Handbuch der Zoologie/Handbook of Zoology. Band/Volume IV Arthropoda: Insecta Teilband/Part 38. Coleoptera, Beetles. Volume 2. Morphology and Systematics (Polyphaga partim).* (Eds RAB Leschen, RG Beutel and JF Lawrence) pp. 153–162. W. DeGruyter, Berlin.

Ratcliffe BC (1984) A review of the Penichrolucaninae with analyses of phylogeny and biogeography, and description of a second New World species from the Amazon Basin (Coleoptera: Lucanidae). *Quaestiones Entomologicae* **20**, 60–87.

Ratcliffe BC (2002). 23. Lucanidae Latreille 1804. In *American Beetles. Volume 2. Polyphaga: Scarabaeoidea through Curculionoidea.* (Eds RH Arnett Jr, MC Thomas, PE Skelley and JH Frank) pp. 6–9. CRC Press, Boca Raton, Florida.

Ratcliffe BC, Jameson ML, Smith ABT (2002) 34. Scarabaeidae Latreille 1802. In *American Beetles. Volume 2. Polyphaga: Scarabaeoidea through Curculionoidea.* (Eds RH Arnett Jr, MC Thomas, PE Skelley and JH Frank) pp. 39–81. CRC Press, Boca Raton, Florida.

Rees BE (1941) First-instar larvae of *Buprestis rusticorum* (Kby.) and *Schizopus sallei* Horn, with notes on the classification of *Schizopus. Proceedings of the Entomological Society of Washington* **43**(9), 210–222.

Rees BE (1943) Classification of the Dermestidae (larder, hide, and carpet beetles) based on larval characters, with a key to the North American genera. *Miscellaneous Publications of the United States Department of Agriculture* **511**, 1–18.

Rees BE, Anderson DM, Bouk D, Gordon RD (1994) Larval key to genera and selected species of North American Coccinellidae (Coleoptera). *Proceedings of the Entomological Society of Washington* **96**(3), 387–412.

Regenfuss H (1975) Die Antennen-Putzeinrichtung der Adephaga (Coleoptera), parallele evolutive Vervollkommnung einer komplexen Struktur. *Journal of Zoological Systematics and Evolutionary Research* **13**, 278–299.

Reichardt H (1973) A critical study of the suborder Myxophaga, with a taxonomic revision of the Brazilian Torridincolidae and Hydroscaphidae (Coleoptera). *Arquivos de Zoologia, São Paulo* **24**, 73–162.

Reichardt H (1974) Relationships between Hydroscaphidae and Torridincolidae, based on larvae and pupae, with the description of the immature stages of *Scaphydra angra* (Coleoptera, Myxophaga). *Revista Brasileira de Entomologia* **18**, 117–122.

Reichardt H (1976) Monograph of the New World Nosodendridae and notes on the Old World forms (Coleoptera). *Papéis Avulsos de Zoologia* **29**(22), 185–220.

Reid CAM (1988) *Diachus auratus* (F.) (Coleoptera: Chrysomelidae), a recent immigrant to the south-west Pacific region, on legumes. *General and Applied Entomology* **20**, 5–8.

Reid CAM (1989) The Australian Species of the tribe Zeugophorini (Coleoptera: Chrysomelidae: Megalopodinae). *General and Applied Entomology* **21**, 39–47.

Reid CAM (1990a) *Micrantipha* Blackburn, a synonym of *Orthaltica* Crotch (Col., Chrysomelidae). *Entomologist's Monthly Magazine* **126**, 205–206.

Reid CAM (1990b) *The Systematics of the Australian Cryptocephalinae (Coleoptera: Chrysomelidae).* Australian National University (PhD thesis), Canberra.

Reid CAM (1991a) The Australian species of Chlamisini Coleoptera: Chrysomelidae). *Journal of the Australian Entomological Society* **30**, 315–323. doi:10.1111/j.1440-6055.1991.tb00445.x

Reid CAM (1991b) Immature stages of the genera *Johannica* Blackburn, *Lamprolina* Baly and *Chalcolampra* Blanchard (Coleoptera, Chrysomelidae, Chrysomelinae). *Journal of Natural History* **25**, 341–357. doi:10.1080/00222939100770251

Reid CAM (1991c) A new genus of Cryptocephalinae from Australia (Coleoptera: Chrysomeloidea). *Entomologica Scandinavica* **22**, 139–157. doi:10.1163/187631291X00020

Reid CAM (1992a) The leaf-beetle genus *Microdonacia* Blackburn (Coleoptera: Chrycomelidae: Galerucinae): revision and systematic placement. *Systematic Entomology* **17**, 359–387. doi:10.1111/j.1365-3113.1992.tb00557.x

Reid CAM (1992b) Revision of the genus *Cheiloxena* Baly (Coleoptera: Chrysomeldae: Eumolpinae). *Memoirs of the Museum of Victoria* **53**, 101–114.

Reid CAM (1993) *Donacia australasiae* Blackburn: the sole representative of the subfamily Donaciinae (Coleoptera: Chrysomelidae) in Australia and New Guinea. *Journal of the Australian Entomological Society* **32**, 103–111. doi:10.1111/j.1440-6055.1993.tb00554.x

Reid CAM (1994) Revision of the genus *Platycolaspis* Jacoby (Coleoptera: Chrysomelidae: Cryptocephalinae). *Memoirs of the Museum of Victoria* **54**, 207–220.

Reid CAM (1995) A cladistic analysis of subfamilial relationships in the Chrysomelidae sensu lato (Chrysomeloidea). In *Biology, Phylogeny, and Classification of Coleoptera. Papers Celebrating the 80th Birthday of Roy A. Crowson.* (Eds J Pakaluk and SA Ślipiński) pp. 559–631. Muzeum I Instytut Zoologii Polska Akademia Nauk, Warsaw.

Reid CAM (1999a) A new generic synonym in the Australian Lucanidae (Coleoptera). *Coleopterists Bulletin* **53**, 175–177.

Reid CAM (1999b) Revision of leaf-beetles of the genus *Cadmus* Erichson, subgenus *Lachnabothra* Saunders (Coleoptera: Chrysomelidae: Cryptocephalinae). *Invertebrate Taxonomy* **13**, 1–66. doi:10.1071/IT97026

Reid CAM (2000a) A complex of cryptic species in the genus *Coptodactyla* Burmeister (Coleoptera: Scarabaeidae: Coprini). *Memoirs of the Queensland Museum* **46**, 231–251.

Reid CAM (2000b) Spilopyrinae Chapuis: a new subfamily in the Chrysomelidae and its systematic placement (Coleoptera). *Invertebrate Taxonomy* **14**, 837–862. doi:10.1071/IT00042

Reid CAM (2001) *Galerucella placida* Baly in Australia (Coleoptera: Chrysomelidae: Galerucinae). *Australian Journal of Entomology* **40**(4), 331–334. doi:10.1046/j.1440-6055.2001.00245.x

Reid CAM (2003) Recognition of the genus *Hoplosaenidea* Laboissière in Australia, with a key to the Australian genera of Galerucini (Coleoptera: Chrysomelidae: Galerucinae). *Australian Journal of Entomology* **42**(1), 40–45. doi:10.1046/j.1440-6055.2003.00329.x

Reid CAM (2006) A taxonomic revision of the Australian Chrysomelinae, with a key to the genera (Coleoptera: Chrysomelidae). *Zootaxa* **1292**, 1–119.

Reid CAM, Beatson M (2010a) The genus *Atysa* Baly in Australia (Coleoptera: Chrysomelidae: Galerucinae). *Australian Entomologist* **36**(4), 189–200.

Reid CAM, Beatson M (2010b) Revision of the Australo-Papuan genus *Macrolema* Baly (Coleoptera: Chrysomelidae: Spilopyrinae), with description of a new genus. *Zootaxa* **2486**, 1–60.

Reid CAM, Beatson M (2010c) Revision of the Australo-Papuan genus *Spilopyra* Baly (Coleoptera: Chrysomelidae: Spilopyrinae). *Zootaxa* **2692**, 1–32.

Reid CAM, Beatson M (2010d) Erratum. Reid, C.A.M. & Beatson, M. (2010) Revision of the Australo-Papuan genus *Macrolema* Baly (Coleoptera: Chrysomelidae: Spilopyrinae), with description of a new genus. *Zootaxa* **2486**, 1–60. *Zootaxa* **2520**, 66–68.

Reid CAM, Beatson M (2011) Revision of the New Caledonian endemic genus *Bohumiljania* Monrós (Coleoptera: Chrysomelidae: Spilopyrinae). *Zootaxa* **3000**, 1–43.

Reid CAM, Beatson M (2013) A new genus and species of Bruchinae, with a key to the genera from Australia (Coleoptera: Chrysomelidae). *Zootaxa* **3599**(6), 535–548.

Reid CAM, Storey RI (2000) Revision of the dung beetle genus *Temnoplectron* Westwood (Coleoptera: Scarabaeidae: Scarabaeini). *Memoirs of the Queensland Museum* **46**, 253–297.

Reid CAM, Beatson M, Hasenpusch J (2009a) The morphology and biology of *Pterodunga mirabile* Daccordi, an unusual subsocial chrysomeline (Coleoptera: Chrysomelidae). *Journal of Natural History* **43**(7–8), 373–398. doi:10.1080/00222930802586016

Reid CAM, Jurado-Rivera JA, Beatson M (2009b) A new genus of Chrysomelinae from Australia (Coleoptera: Chrysomelidae). *Zootaxa* **2207**, 53–66.

Ren D, Tan J (2006) A new cupedid genus (Coleoptera: Archostemata: Cupedidae) from Jehol Biota of western Liaoning, China. *Annales Zoologici* **56**(3), 457–464.

Ren D, Tan J, Ge S (2006) New fossil ommatid (Coleoptera: Archostemata: Ommatidae) from Jehol Biota of western Liaoning, China. *Progress in Natural Science* **16**(6), 639–643. doi:10.1080/10020070612330046

Ren D, Shih C, Gao T, Yao Y, Zhao Y (2010) *Silent Stories—Insect Fossil Treasures from Dinosaur Era of the Northeastern China.* Science Press, Beijing.

Reyes-Castillo P, Halffter G (1983) La structure sociale chez les Passalidae (Col.). *Bulletin de la Societe Entomologique de France* **88**, 619–635.

Reyes-Castillo P, Jarman M (1983) Disturbance sounds of adult passalid beetles (Coleoptera: Passalidae): structural and functional aspects. *Annals of the Entomological Society of America* **76**, 6–22.

Ribera I, Beutel RG (2003) Aspidytidae: On the discovery of a new beetle family: detailed morphological analysis, description of a second species, and key to fossil and extant adephagan families (Coleoptera). In *Water Beetles of China. Vol. III.* (Eds MA Jäch and L Ji) pp. 53–66. Zoologisch-Botanische Gesellschaft in Österreich und Wiener Coleopterologenverein, Vienna.

Ribera I, Beutel RG, Balke M, Vogler AP (2002a) Discovery of Aspidytidae, a new family of aquatic Coleoptera. *Proceedings of the Royal Society B* **269**, 2351–2356. doi:10.1098/rspb.2002.2157

Ribera I, Hogan JE, Vogler AP (2002b) Phylogeny of Hydradephagan water beetles inferred from 18S rDNA sequences. *Molecular Phylogenetics and Evolution* **23**, 43–62. doi:10.1006/mpev.2001.1080

Richards AM (1971) An ecological study of the cavernicolous fauna of the Nullarbor Plain Southern Australia. *Journal of Zoology* **164**, 1–60. doi:10.1111/j.1469-7998.1971.tb01297.x

Richards AM (1980) Sexual selection, guarding and sexual conflict in a species of Coccinellidae (Coleoptera). *Journal of the Australian Entomological Society* **19**, 26. doi:10.1111/j.1440-6055.1980.tb00956.x

Richards AM (1981) *Rhyzobius ventralis* (Erichson) and *R. forestieri* (Mulsant) (Coleoptera: Coccinellidae), their biology and value for scale insect control. *Bulletin of Entomological Research* **71**, 33–46. doi:10.1017/S0007485300051002

Richards AM (1983) The *Epilachna vigintioctopunctata* complex (Coleoptera: Coccinellidae). *International Journal of Entomology* **25**, 11–41.

Richards AM (1985) Biology and defensive adaptations in *Rodatus major* (Coleoptera: Coccinellidae) and its prey, *Monophlebulus pilosior* (Hemiptera: Margarodidae). *Journal of Zoology* **205**, 287–295. [A] doi:10.1111/j.1469-7998.1985.tb03535.x

Richards AM, Goletsos C (1991) Feeding behaviour in Australian aphidophagous Coccinellidae. In *Behaviour and Impact of Ahidophaga.* (Eds L Polgár, RJ Chambers, A Dixon and I Hodek) pp. 227–234. SPB Publishing, The Hague.

Richmond, EA (1920) Studies on the biology of the aquatic Hydrophilidae. *Bulletin of the American Museum of Natural History* **42**, 1–94, 16 pls.

Riek EF (1955) The Australian rhipidiine parasites of cockroaches (Coleoptera: Rhipiphoridae). *Australian Journal of Zoology* **3**, 71–94. doi:10.1071/ZO9550071

Riek EF (1973) Rhipiphorid beetles of the subfamily Nephritinae (Coleoptera: Rhipiphoridae). *Journal of the Australian Entomological Society* **12**, 261–276. doi:10.1111/j.1440-6055.1973.tb01672.x

Riek EF (1974) Upper Triassic insects from the Molteno "Formation", South Africa. *Palaeontologica Africana* **17**, 19–31.

Riley EG, Clark SM, Flowers RW, Gilbert AJ (2002) 124. Chrysomelidae Latreille 1802. In *American Beetles. Volume 2. Polyphaga: Scarabaeoidea through Curculionoidea.* (Eds RH Arnett Jr, MC Thomas, PE Skelley and JH Frank) pp. 617–691. CRC Press, Boca Raton, Florida.

Riley EG, Clark SM, Seeno TN (2003) *Catalog of the leaf beetles of America north of Mexico (Coleoptera: Megalopodidae, Orsodacnidae and Chrysomelidae, excluding Bruchinae).* 290 pp. The Coleopterists Society (Special Publication No. 1), Sacramento, California.

Ritcher PO (1966) *White Grubs and Their Allies. A Study of North American Scarabaeoid Larvae.* Oregon State University Press, Corvallis, Oregon.

Roach AME (2000) Review of the Australian species of the dermestid genus *Anthrenocerus* Arrow (Coleoptera: Dermestidae). *Invertebrate Taxonomy* **14**, 175–224. doi:10.1071/IT97031

Roberts H (1977) The Platypodidae (Coleoptera) of Fiji (with descriptions of two new species). *Journal of Natural History* **11**, 555–578. doi:10.1080/00222937700770481

Roberts H (1980) Description of the developmental stages of *Sosylus* spp. (Coleoptera: Colydiidae) from New Guinea, parasites and predators of ambrosia beetles (Coleoptera: Platypodidae). *Bulletin of Entomological Research* **70**, 245–252. doi:10.1017/S0007485300007501

Robertson JA, McHugh JV, Whiting MF (2004) A molecular phylogenetic analysis of the pleasing fungus beetles (Coleoptera: Erotylidae): evolution of colour patterns, gregariousness and mycophagy. *Systematic Entomology* **29**, 173–187. doi:10.1111/j.0307-6970.2004.00242.x

Robertson JA, Whiting MF, McHugh JV (2008) Searching for natural lineages within the cerylonid series (Coleoptera: Cucujoidea). *Molecular Phylogenetics and Evolution* **46**, 193–205. doi:10.1016/j.ympev.2007.09.017

Robertson JA, Ślipiński SA, Hiatt K, Miller KB, Whiting MF, McHugh JV (2013) Molecules, morphology and minute hooded beetles: a phylogenetic study with implications for the evolution and classification of Corylophidae (Coleoptera: Cucujoidea). *Systematic Entomology* **38**, 209–232.

Rohdendorf BB (1961) Ordo Coleoptera. Beetles. In *Palaeozoic Insects of the Kuznetz Basin.* (Eds BB Rohdendorf *et al.*) pp. 393–469. Akademia Nauk S.S.S.R., Moscow.

Rohdendorf BB, Ponomarenko AG (1962) Otryad Coleoptera. Zhestkokriliye, ili Zhuki. In *Osnovy Paleontologii. Tom 9: Chlenistonogie Trakheinie i Khelitserovie.* (Ed. BB Rohdendorf) pp. 241–267. Akademia Nauk SSSR, Moscow (in Russian; translation: Order Coleoptera. Beetles. In *Fundamentals of Paleontology. Volume 9. Arthropoda, Tracheata, Chelicerata* (Ed. BB Rohdendorf) pp. 341–382 Smithsonian Institution Libraries and National Science Foundation, Washington, D.C.).

Roig-Juñent S (1998) Cladistic relationships of the tribe Broscini (Coleoptera: Carabidae). In *Phylogeny and Classification of Caraboidea. Proceedings of a Symposium (28 August, 1996, Florence, Italy), XX International Congress of Entomology.* (Eds GE Ball, A Casale and A. Vigna Taglianti) pp. 343–358. Museo Regionale di Scienze Naturali, Torino.

Roig-Juñent S (2000) The subtribes and genera of the tribe Broscini (Coleoptera: Carabidae): cladistic analysis, taxonomic treatment, and biogeographical considerations. *Bulletin of the American Museum of Natural History* **255**, 1–90. doi:10.1206/0003-0090(2000)255<0001:TSAGOT>2.0.CO;2

Rondonuwu SA, Austin AD (1988) A new species of *Uracanthus* (Coleoptera: Cerambycidae): a pest on ornamental cypresses in the Adelaide region. *Transactions of the Royal Society of South Australia* **112**, 109–117.

Roughley RE (1981) Trachypachidae and Hydradephaga (Coleoptera): a monophyletic unit? *The Pan-Pacific Entomologist* **57**, 273–285.

Roughley RE (2000a) 7. Gyrinidae Latreille, 1810. In *American Beetles. Volume 1. American Beetles. Archostemata, Myxophaga, Adephaga, Polyphaga: Staphyliniformia.* (Eds RH Arnett Jr and MC Thomas) pp. 133–1371. CRC Press, Boca Raton, Florida.

Roughley RE (2000b) 8. Haliplidae Aubé, 1836. In *American Beetles. Volume 1. American Beetles. Archostemata, Myxophaga, Adephaga, Polyphaga: Staphyliniformia.* (Eds RH Arnett Jr and MC Thomas) pp. 138–143. CRC Press, Boca Raton, Florida.

Roughley RE (2000c) 10. Noteridae C. G. Thomson, 1857. In *American Beetles. Volume 1. American Beetles. Archostemata, Myxophaga, Adephaga, Polyphaga: Staphyliniformia.* (Eds RH Arnett Jr and MC Thomas) pp. 147–152. CRC Press, Boca Raton, Florida.

Roughley RE, Larson DJ (2000) 12. Dytiscidae Leach, 1815. In *American Beetles. Volume 1. American Beetles. Archostemata, Myxophaga, Adephaga, Polyphaga: Staphyliniformia.* (Eds RH Arnett Jr and MC Thomas) pp. 156–186. CRC Press, Boca Raton, Florida.

Rozen JG Jr (1958) The external anatomy of the larva of *Nacerdes melanura* (Linnaeus) (Coleoptera: Oedemeridae). *Annals of the Entomological Society of America* **51**(3), 222–229.

Rozen JG Jr (1960) Phylogenetic-systematic study of larval Oedemeridae (Coleoptera). *Miscellaneous Publications of the Entomological Society of America* **1**, 35–68.

Rühm W, Milewski I (1996) Zur Bedeutung der larvalen Bestimmungsmerkmale beim Studium von Abbausukzessionen: Bestimmungmerkmale einiger Cisidenlarven (Cisidae, Col.). *Entomologische Mitteilungen der Zoologischen Museum der Universität Hamburg* **12**(154), 55–62.

Ruhnau S (1986) Phylogenetic relations within the Hydradephaga (Coleoptera) using larval and pupal characters. *Entomologica Basiliensia* **1**, 231–271.

Ruiz F (1924) El genero *Idiostoma* en Chile. *Revista Chilena de Historia Natural (Valparaiso, Chile)* **1924**, 99–101.

Ruta R, Yoshitomi H (2010) Revision of the genus *Exochomoscirtes* Pic (Coleoptera: Scirtidae: Scirtinae). *Zootaxa* **2598**, 1–80.

Ruta R, Thorpe S, Yoshitomi H (2011) *Stenocyphon neozealandicus*, a new species from New Zealand of a previously monotypic subfamily from Chile (Coleoptera: Scirtidae: Stenocyphoninae). *Zootaxa* **3113**, 65–68.

Rymer Roberts AW (1930) A key to the principal families of Coleoptera in the larval stage. *Bulletin of Entomological Research* **21**, 57–72. doi:10.1017/S0007485300021556

Rymer Roberts AW (1939) On the taxonomy of Erotylidae (Coleoptera), with special reference to the morphological charactrers of the larvae. *Transactions of the Royal Entomological Society of London* **88**(3), 89–117. doi:10.1111/j.1365-2311.1939.tb00251.x

Rymer Roberts AW (1958) On the taxonomy of Erotylidae (Coleoptera), with special reference to the morphological charactrers of the larvae. II. *Transactions of the Royal Entomological Society of London* **110**(8), 245–285. doi:10.1111/j.1365-2311.1958.tb00794.x

Saalas U (1936) Über das Flügelgeäder und die phylogenetische Entwicklung der Cerambyciden. *Annales Zoologici Societatis Zoologicae-Botanicae Fennicae, Vanamo* **4**(1), 1–198, 19 pls.

Sagegami-Oba R, Oba Y, Ôhira H (2007a) Phylogenetic relationships of click beetles (Coleoptera: Elateridae) inferred from 28S ribosomal DNA: Insights into the evolution of bioluminescence in Elateridae. *Molecular Phylogenetics and Evolution* **42**, 410–421. doi:10.1016/j.ympev.2006.07.017

Sagegami-Oba R, Takahashi N, Oba Y (2007b) The evolutionary process of bioluminescence and aposematism in cantharoid beetles (Coleoptera: Elateroidea) inferred by the analysis of 18S ribosomal DNA. *Gene* **400**, 104–113. doi:10.1016/j.gene.2007.06.004

Saini RS (1964) Histology and physiology of the cryptonephridial system of insects. *Transactions of the Royal Entomological Society of London* **116**, 347–392. doi:10.1111/j.1365-2311.1964.tb02302.x

Saito A (1989a) Female reproductive organs of lepturine cerambycid beetles from Japan, with reference to their taxonomic significance (Part 1). *Elytra, Tokyo* **17**(1), 65–86.

Saito A (1989b) Female reproductive organs of lepturine cerambycid beetles from Japan, with reference to their taxonomic significance (Part 2). *Elytra, Tokyo* **17**(2), 201–233.

Saito A (1990) Female reproductive organs of cerambycid beetles from Japan and the neighboring areas. I. Philini through Atimiini. *Elytra, Tokyo* **18**(2), 231–260.

Saito A (1991) Female reproductive organs of cerambycid beetles from Japan and the neighboring areas. II. Methiini through Callidiopini. *Elytra, Tokyo* **19**(2), 167–178.

Saito A (1992) Female reproductive organs of cerambycid beetles from Japan and the neighboring areas. III. Obriini through Rosaliini. *Elytra, Tokyo* **20**(1), 103–118.

Saito A (1993a) Female reproductive organs of cerambycid beetles from Japan and the neighboring areas. IV. Callichromini through Cleomenini. *Elytra, Tokyo* **21**(1), 3–25.

Saito A (1993b) Female reproductive organs of cerambycid beetles from Japan and the neighboring areas. V. General consideration. *Elytra, Tokyo* **21**(2), 199–216.

Sakai M, Satô M (1996) The coleopterous family Decliniidae (Elateriformia: Scirtoidea) new to Japan with description of its second representative. *Elytra, Tokyo* **24**, 103–111.

Sakurai K (1985) An attelabid weevil (*Euops splendida*) cultivates fungi. *Journal of Ethology* **3**(2), 151–156. doi:10.1007/BF02350306

Samuelson GA (1989) A review of the hispine tribe Apriodini (Coleoptera: Chrysomelidae). *Memoirs of the Queensland Museum* **27**, 599–604.

Samuelson GA (1996) Binding sites: elytron-to-body meshing structures of possible significance in the higher classification of Chrysomeloidea. In *Chrysomelidae Biology. Volume 1: The Classification, Phylogeny and Genetics*. (Eds PHA Jolivet and ML Cox) pp. 267–290. SPB Academic Publishing, Amsterdam.

Sanborne M (1981) Biology of *Ithycerus noveboracensis* (Forster) (Coleoptera) and weevil phylogeny. *Ecological Monographs* **4**, 1–80.

Santos-Silva A, Hovore FT (2007) Divisão do gênero *Distenia* Lepeletier & Audinet-Serville, notaas sobre a venação alar em Disteniini, homonímias, sinonímia e redescrições (Coleoptera, Cerambycidae, Disteniinae). *Papéis Avulsos de Zoologia* **47**(1), 1–29.

Santos-Silva A, Heffern D, Matsuda K (2010) Revision of Hawaiian, Australasian, Oriental, and Japanese Parandrinae (Coleoptera, Cerambycidae). *Insecta Mundi* 130, 120 pp.

Sasaji H (1967) A revision of the Formosan Coccinellidae (I) the subfamily Sticholotidinae, with an establishment of a new tribe (Coleoptera). *Etizenia* **25**, 1–28.

Sasaji H (1968a) Descriptions of the coccinellid larvae of Japan and Ryukyus (Coleoptera). *Memoirs of the Faculty of Education, Fukui University. Series II. Natural Science* **18**, 93–135.

Sasaji H (1968b) Phylogeny of the family Coccinellidae (Coleoptera). *Etizenia* **35**, 1–37.

Sasaji H (1971) *Fauna Japonica: Coccinellidae (Insecta: Coleoptera)*. Academic Press of Japan, Tokyo.

Sasaji H (1978) Notes on the Japanese Endomychidae, with an establishment of a new subfamily (Coleoptera). *Memoirs of the Faculty of Education. Fukui University. II. Natural Sciences* **28**, 1–31.

Sasaji H (1988) Contribution to the taxonomy of the superfamily Cucujoidea (Coleoptera) of Japan and her adjacent districts,IV. *Memoirs of the Faculty of Education. Fukui University. II. Natural Sciences* **38**, 13–48.

Sasaji H (1989) Discovery of the curious beetle genus *Xenoscelinus* Grouvelle (Languriidae-Cryptophilinae) from Japan. *Kanagawa-Chûhô, Yokohama* **90**, 223–228.

Sasaji H (1990) The family Mychothenidae of Japan (Coleoptera). *Esakia* **1**, (Special Issue), 65–75.

Satô M (1982) Discovery of Torridincolidae (Coleoptera) in Japan. *Annotationes Zoologicae Japonenses* **55**, 276–283.

Satô M (1994) Note on the genus *Pseudeucinetus* Heller and its new relative (Coleoptera, Limnichidae). *Special Bulletin of the Essa Entomological Society* **2**, 173–177.

Satô M, Yoshitomi H (1998) The northernmost record of *Hyphalus taekoae* M. Satô (Coleoptera, Limnichidae) from Amami-Ôshima. *Elytra, Tokyo* **26** (2), 296

Satô M, Yoshitomi H (2004) Discovery of a second representative of the genus *Horelophopsis* (Coleoptera, Hydrophilidae) from the Ryukyu Archipelago, Japan. *Elytra, Tokyo* **32**(1), 41–49.

Sawada Y, Morimoto K (1985) A revision of the genus *Euops* Schoenherr (Coleoptera: Attelabidae) from Japan, Korea and Taiwan. *Journal of the Faculty of Agriculture. Kyushu University* **30**, 175–195.

Schawaller W (1993) The genus *Prostomis* (Coleoptera: Prostomidae) in Australia and adjacent regions. *Stuttgarter Beitrage zur Naturkunde. Serie A, Biologie* **489**, 1–12.

Schawaller W (2003) New species and records of *Prostomis* Latreille, including the first fossil records from Baltic amber and a checklist of the species (Coleoptera: Prostomidae). *Stuttgarter Beitrage zur Naturkunde. Serie A, Biologie* **650**, 1–11.

Schedl KE (1962) Forstentomologische Beiträge aus dem Kongo. Rauber und Kommensalen. *Entomologische Abhandlungen und Berichte aus dem Staatlichen Museum für Tierkunde in Dresden* **28**, 37–84.

Schicha E (1974) Functional morphology of the mouth parts, and feeding habits of *Dicranolaius bellulus* (Guerin-Meneville) (Col.: Malachiidae). *Zoologischer Anzeiger* **192**, 22–26.

Schildknecht H (1970) The defensive chemistry of land and water beetles. *Angewandte Chemie International Edition* **9**, 1–9. doi:10.1002/anie.197000011

Schildknecht H, Krauss D, Connert J, Essenbreis H, Orfanides N (1975) The spreading alkaloid stenusin from the staphylinid *Stenus comma* (Coleoptera: Staphylinidae). *Angewandte Chemie International Edition* **14**, 427. doi:10.1002/anie.197504271

Schiödte JC (1862–1883) Die metamorphosi Eleutheratorum observaciones: bidrag til insecternes udviklingshistorie. Naturhistorisk Tidsskrift (3) **1**, 193–232, pls. 1–10; **3**, 131–224, pls. 1–12; **4**, 415–552, pls. 12–22; **6**, 353–378, 467–536, pls. 1–10; **8**, 165–226, 545–564, pls. 1–9, 18–20; **9**, 227–376, pls. 8–19; **10**, 369–458, pls. 12–18; **11**, 479–598, pls. 5–12; **12**, 513–598, pls. 14–18; **13**, 415–426, pl. 18.

Schmid R (1988) Morphologische Anpassungen in einen Räuber-Beute-System: Ameisenkäfer (Scydmaenidae, Staphylinoidea) und gepanzerte Milben (Acari). *Zoologische Jahrbüche (Systematik)* **115**, 207–228.

Schmidt A (1913) *Genera Insectorum. Fascicule 150. Coleoptera Lamellicornia Fam. Scarabaeidae Subfam. Aegialiinae, Chironinae, Dynamopinae, Hybosorinae, Idiostominae, Ochodaeinae, Orphninae*. 87 pp., 3 pls. V. Verteneuil and L. Desmet, Brussels

Schmitt M (1988) The Criocerinae: biology, phylogeny and evolution In *Biology of Chrysomelidae*. (Eds P Jolivet, E Petitpierre and TH Hsiao) pp. 475–495. Kluwer Academic, Dordrecht.

Schmitt M (1992) Stridulatory devices of leaf beetles (Chrysomelidae) and other Coleoptera. In *Advances on Coleopterology*. (Eds M Zunino, X Bellés and M Blas) pp. 263–280. Associación Europea de Coleopterologia, Barcelona.

Schmitt M (1994) The position of the Megalopodinae and Zeugophorinae in a phylogenetic system of the Chrysomeloidea (Insecta: Coleoptera). In *Proceedings of the Third International Symposium on the Chrysomelidae, Beujing, 1992*. (Ed. DG Furth) pp. 38–44. Backhuys Publishers, Leiden.

Schmitt M (1996) The phylogenetic system of the Chrysomelidae — History of ideas and present state of knowledge. In *Chrysomelidae Biology. Volume 1: The Classification, Phylogeny and Genetics* (Eds PHA Jolivet and ML Cox) pp. 57–96. SPB Academic Publishing, Amsterdam.

Schmitz H, Schmitz A (2002) Australian fire-beetles. *Landscope Spring* **2002**, 36–41.

Schmitz H, Schmitz A, Trenner S, Bleckmann H (2002) A new type of insect infrared organ of low thermal mass. *Naturwissenschaften* **89**, 226–229. doi:10.1007/s00114-002-0312-4

Schneider P (1975) Die Flugtypen der Käfer (Coleoptera). *Entomologica Germanica* **1**, 222–231.

Schneider P (1978) Die Flug- und Faltungstypen der Käfer (Coleoptera). *Zoologische Jahrbucher* **99**, 174–210. [Anatomie]

Scholtz CH (1986a) Phylogeny and systematics of the Trogidae (Coleoptera: Scarabaeoidea). *Systematic Entomology* **11**, 355–363. doi:10.1111/j.1365-3113.1986.tb00186.x

Scholtz CH (1986b) Revision of the genus *Trox* Fabricius (Coleoptera: Trogidae) of the Australasian Region. *Australian Journal of Zoology* **125**, (Supplement), 1–99.

Scholtz CH (1990a) Phylogenetic trends in the Scarabaeoidea (Coleoptera). *Journal of Natural History* **24**, 1027–1066. doi:10.1080/00222939000770631

Scholtz CH (1990b) Revision of the Trogidae of South America (Coleoptera: Scarabaeoidea). *Journal of Natural History* **24**(6), 1391–1456. doi:10.1080/00222939000770841

Scholtz CH (1991) Descriptions of larvae of Australian *Omorgus* Erichson, with implications for the phylogeny of the Trogidae (Insecta: Coleoptera). *Invertebrate Taxonomy* **5**, 827–835. doi:10.1071/IT9910827

Scholtz CH, Browne DJ (1996) Polyphyly in the Geotrupidae (Coleoptera: Scarabaeoidea): a case for a new family. *Journal of Natural History* **30**, 597–614. doi:10.1080/00222939600770311

Scholtz CH, Chown SL (1995) The evolution of habitat use and diet in the Scarabaeaoidea: a phylogenetic approach. In *Biology, Phylogeny, and Classification of Coleoptera. Papers Celebrating the 80th Birthday of Roy A. Crowson*. (Eds J Pakaluk and SA Ślipiński) pp. 355–374. Muzeum I Instytut Zoologii Polska Akademia Nauk, Warsaw.

Scholtz CH, Grebennikov VV (2005a) 12. Scarabaeiformia Crowson, 1960. In *Handbuch der Zoologie/Handbook of Zoology. Band/Volume IV Arthropoda: Insecta Teilband/Part 38. Coleoptera, Beetles. Volume 1. Morphology and Systematics (Archostemata, Adephaga, Myxophaga, Polyphaga partim)*. (Eds RG Beutel and RAB Leschen) pp. 345–365. W. DeGruyter, Berlin.

Scholtz CH, Grebennikov VV (2005b) 13. Scarabaeoidea Latreille, 1802. In *Handbuch der Zoologie/Handbook of Zoology. Band/Volume IV Arthropoda: Insecta Teilband/Part 38. Coleoptera, Beetles. Volume 1. Morphology and Systematics (Archostemata, Adephaga, Myxophaga, Polyphaga partim)*. (Eds RG Beutel and RAB Leschen) pp. 367–425. W. DeGruyter, Berlin.

Scholtz CH, Peck S (1990) Description of a *Polynoncus* Burmeister larva, with implications for phylogeny of the Trogidae (Coleoptera: Scarabaeoidea). *Systematic Entomology* **15**, 383–389. doi:10.1111/j.1365-3113.1990.tb00072.x

Scholtz CH, d'Hotman D, Nel A (1987a) Glaresidae, a new family of Scarabaeoidea (Coleoptera) to accommodate the genus *Glaresis* Erichson. *Systematic Entomology* **12**, 345–354. doi:10.1111/j.1365-3113.1987.tb00206.x

Scholtz CH, d'Hotman D, Nel A (1987b) Systematic position of the South American genus *Cryptogenius* Westwood (Coleoptera: Scarabaeoidea: Trogidae and Hybosoridae). *Coleopterists Bulletin* **41**, 193–199.

Scholtz CH, d'Hotman D, Evans AV, Nel A (1988) Phylogeny and systematics of the Ochodaeidae (Insecta: Coleoptera: Scarabaeoidea). *Journal of the Entomological Society of Southern Africa* **51**, 207–240.

Scholtz CH, Browne DJ, Kukalová-Peck J (1994) Glaresidae, archaeopteryx of the Scarabaeoidea (Coleoptera). *Systematic Entomology* **19**, 259–277. doi:10.1111/j.1365-3113.1994.tb00590.x

Schultz TD (1994) Predation by larval soldier beetles (Coleoptera: Cantharidae) on the eggs and larvae of *Pseudoxycheila tarsalis* (Coleoptera: Cicindelidae). *Entomological News* **105**, 14–16.

Schulze I (1975) A review of silk production and spinning activities in Arthropoda with special reference to spinning in tenebrionid larvae (Coleoptera). *Transvaal Museum Memoirs* **19**, 1–51.

Schunger I, Beutel RG, Britz R (2003) Morphology of immature stages of *Prostomis mandibularis* (Coleoptera: Tenebrionoidea: Prostomidae). *European Journal of Entomology* **100**, 357–370.

Schuster R (1966) Über den Beutefang des Ameisenkäfers *Cephennium austriacum* Reitter. *Naturwissenschaften* **53**, 113. doi:10.1007/BF00601487

Schuster JC (2002). 25. Passalidae Leach 1815. In *American Beetles. Volume 2. Polyphaga: Scarabaeoidea through Curculionoidea*. (Eds RH Arnett Jr, MC Thomas, PE Skelley and JH Frank) pp. 12–14. CRC Press, Boca Raton, Florida.

Schütz C, Dettner K (1992) Cantharidin-secretion by elytral notches of male anthicid species (Coleoptera: Anthicidae). *Zeitschrift für Naturforschung* **47C**, 290–299.

Scott H (1922). Reports of the Percy Sladen Trust Expedition to the Indian Ocean in 1905. Vol. 7, No. 4. Coleoptera: Scydmaenidae, Scaphidiidae, Phalacridae, Cucujidae (supplement), Lathridiidae, Mycetophagidae (including *Propalticus*), Bostrychidae, Lyctidae. *Transactions of the Linnean Society of London* (2nd Series. Zoology) **18**, 195–260, pls. 19–22.

Scott H (1926). Reports of the Percy Sladen Trust Expedition to the Indian Ocean in 1905. Vol. 8, No. 1. Coleoptera, Ciidae. *Transactions of the Linnean Society of London* (2nd Series. Zoology) **19**, 1–41, pl. 1.

Scott AC (1936) Haploidy and aberrant spermatogenesis in a coleopteran, *Micromalthus debilis* LeConte. *Journal of Morphology* **59**, 485–515. doi:10.1002/jmor.1050590305

Scott AC (1938) Paedogenesis in the Coleoptera. *Zeitschrift fur Morphologie und Oekologie der Tiere* **33**, 633–653. doi:10.1007/BF00407571

Scott AC (1941) Reversal of sex production in *Micromalthus. The Biological Bulletin* **81**, 420–431. doi:10.2307/1537915

Scriven GT, Reeves EL, Luck RF (1986) Beetle from Australia threatens Eucalyptus. *California Agriculture* **40**(7–8), 4–6.

Scudder GGE (1961) The comparative morphology of the insect ovipositor. *Transactions of the Royal Entomological Society of London* **113**(2), 25–40. doi:10.1111/j.1365-2311.1961.tb00800.x

Seago AE (2009) Revision of *Agyrtodes* Protevin (Coleoptera: Leiodidae). *The Coleopterists Society Monograph* **63**, 1–73. doi:10.1649/0010-065X-63.sp7.1

Seago AE, Beutel RG (2010) 11.15. Prostomidae C. G. Thomson, 1859. In *Handbuch der Zoologie/Handbook of Zoology. Band/Volume IV Arthropoda: Insecta Teilband/Part 38. Coleoptera, Beetles. Volume 2. Morphology and Systematics (Polyphaga partim).* (Eds RAB Leschen, RG Beutel and JF Lawrence) pp. 659–667. W. DeGruyter, Berlin.

Seago AE, Brady P, Vigneron JP, Schultz TD (2009) Gold bugs and beyond: a review of irridescence and structural colour mechanisms in beetles (Coleoptera). *Journal of the Royal Society, Interface* **6**, 165–184. doi:10.1098/rsif.2008.0354.focus

Seago AE, Giorgi JA, Li J, Ślipiński A (2011) Phylogeny, classification and evolution of ladybird beetles (Coleoptera: Coccinellidae) based on simultaneous analysis of molecular and morphological data. *Molecular Phylogenetics and Evolution* **60**, 137–151. doi:10.1016/j.ympev.2011.03.015

Seeger W (1971) Morphologie, Bionomie und Ethologie von Halipliden, unter besonderer Berücksichtigung Funktions-Morphologischer Gesichtspunkte (Haliplidae; Coleoptera). *Archiv fuer Hydrobiologie* **68**, 400–435.

Seeno TN, Wilcox JA (1982) Leaf beetle genera (Coleoptera: Chrysomelidae). *Entomography* **1**, 1–221.

Seevers CH (1978) A generic and tribal revision of the North American Aleocharinae (Coleoptera, Staphylinidae). *Fieldiana Zoology* **71**, i–vi, 1–289.

Seevers CH, Dybas HS (1943) A synopsis of the Limulodidae (Coleoptera): a new family proposed for myrmecophiles of the subfamilies Limulodinae (Ptiliidae) and Cephaloplectinae (Staphylinidae). *Annals of the Entomological Society of America* **36**, 546–586.

Sehnal F, Švachá P, Zrzavý J (1996) Evolution of insect metamorphosis. In *Metamorphosis. Postembryonic Reprogramming of Gene Expression in Amphibian and Insect Cells.* (Eds LI Gilbert, JR Tata and BG Atkinson) pp. 3–58. Academic Press, San Diego.

Selander RB (1957) The systematic position of the genus *Nephrites* and the phylogenetic relationships of the higher groups of Rhipiphoridae (Coleoptera). *Annals of the Entomological Society of America* **50**, 88–103.

Selander RB (1991a) Rhipiphoridae (Tenebrionoidea). In *Immature Insects. Vol. 2.* (Ed. FW Stehr) pp. 509–512. Kendall/Hunt, Dubuque, Iowa.

Selander RB (1991b) Meloidae (Tenebrionoidea). In *Immature Insects. Vol. 2.* (Ed. FW Stehr) pp. 530–534. Kendall/Hunt, Dubuque, Iowa.

Selman BJ, Lowman MD (1983) The biology and herbivory rates of *Novacastria nothofagi* Selman (Coleoptera: Chrysomelidae), a new genus and species on *Nothofagus moorei* in Australian temperate rain forests. *Australian Journal of Zoology* **31**, 179–191. doi:10.1071/ZO9830179

Sen Gupta T (1967) A new subfamily of Languriidae (Coleoptera) based on four genera, with a key to the species of *Toramus. Proceedings of the Royal Entomological Society of London B* **36**, 1676–176.

Sen Gupta T (1968a) Review of the genera of the tribe Loberini (Coleoptera: Languriidae). *Breviora* **303**, 1–27.

Sen Gupta T (1968b) Revision of the genera of Cladoxenini (= Cladoxeninae Arrow) and Thallisellini trib. nov. of the family Languriidae (Coleoptera: Clavicornia). *Journal of Natural History* **2**, 463–475. doi:10.1080/00222936800770961

Sen Gupta T (1969) On the taxonomy of Erotylidae (Insecta: Coleoptera: Clavicornia), with descriptions of two new larvae. *Proceedings of the Zoological Society of Calcutta* **22**, 97–107.

Sen Gupta T (1979) A new subfamily of Merophysiidae (Clavicornia: Coleoptera) and descriptions of two new species of *Gomya* Dajoz and its larva. *Revue Suisse de Zoologie* **86**(3), 691–698.

Sen Gupta T (1988) Review of the genera of the family Rhizophagidae (Clavicornia: Coleoptera) of the world. *Memoirs of the Zoological Survey of India* **17**(1), 1–58.

Sen Gupta T, Crowson RA (1966) A new family of cucujoid beetles, based on six Australian and one New Zealand genera. *Annals & Magazine of Natural History* **9**, 61–85. doi:10.1080/00222936608651639

Sen Gupta T, Crowson RA (1969a) Further observations on the family Boganiidae, with definition of two new families Cavognathidae and Phloeostichidae. *Journal of Natural History* **3**, 571–590. doi:10.1080/00222936900770481

Sen Gupta T, Crowson RA (1969b) On a new family of Clavicornia (Coleoptera) and a new genus of Languriidae. *Proceedings of the Royal Entomological Society of London B* **38**, 125–131.

Sen Gupta T, Crowson RA (1971) A review of classification of the family Languriidae (Coleoptera: Clavicornia) and the place of Languriidae in the natural system of Cavicornia. *Memoirs of the Zoological Survey of India* **15**(2), 1–42.

Sen Gupta T, Crowson RA (1973) A review of the classification of Cerylonidae (Coleoptera, Clavicornia). *Transactions of the Royal Entomological Society of London* **124**, 365–446. doi:10.1111/j.1365-2311.1973.tb00369.x

Sen Gupta T, Crowson RA (1979) The coleopteran family Sphindidae. *Entomologist's Monthly Magazine* **113**, 177–191.

Sen Gupta T, Pal TK (1995) Further observations on the family Rhizophagidae with descriptions of seven new genera (Coleoptera, Rhizophagidae). *Mitteilungen aus dem Zoologischen Museum in Berlin* **71**, 129–146. doi:10.1002/mmnz.19950710114

Sen Gupta T, Pal TK (1996) *Fauna of India and Adjacent Countries: Clavicornia: Coleoptera. Family Silvanidae.* Zoological Survery of India, Calcutta.

Serrano J (1981) Chromosome numbers and karyotypic evolution of Caraboidea. *Genetica* **55**, 51–60. doi:10.1007/BF00134005

Shepard WD (2002a) 43. Elmidae Curtis 1830. In *American Beetles. Volume 2. Polyphaga: Scarabaeoidea through Curculionoidea.* (Eds RH Arnett Jr, MC Thomas, PE Skelley and JH Frank) pp. 117–120. CRC Press, Boca Raton, Florida.

Shepard WD (2002b) 46. Limnichidae Erichson 1846. In *American Beetles. Volume 2. Polyphaga: Scarabaeoidea through Curculionoidea.* (Eds RH Arnett Jr, MC Thomas, PE Skelley and JH Frank) pp. 125–126. CRC Press, Boca Raton, Florida.

Shepard WD (2002c) 48. Psephenidae Lacordaire 1854. In *American Beetles. Volume 2. Polyphaga: Scarabaeoidea through Curculionoidea.* (Eds RH Arnett Jr, MC Thomas, PE Skelley and JH Frank) pp. 133–134. CRC Press, Boca Raton, Florida.

Shepard WD (2003) The biology of *Georissus californicus* LeConte (Coleoptera, Hydrophilidae, Georissinae). *Special Bulletin of the Japanese Society of Coleopterology, Tokyo* **6**, 121–128.

Shigo AL (1967) Succession of organisms in discoloration and decay of wood. In *International Review of Forestry Research* II. (Eds JA Romberger and P Mikola) pp. 237–99. New York.

Shockley FW, Tomaszewska KW, McHugh JV (2009) An annotated checklist of the handsome fungus beetles of the world (Coleoptera: Cucujoidea: Endomychidae). *Zootaxa* **1999**, 1–113.

Short AEZ, Fikáček M (2011) World catalogue of the Hydrophiloidea (Coleoptera): additions and corrections II (2006–2010). *Acta Entomologica Musei Nationalis Pragae* **51**(1), 83–122.

Short AEZ, Hebauer F (2006) World catalogue of the Hydrophiloidea (Coleoptera) – additions and corrections, I (1999–2005) (Coleoptera). *Koleopterologische Rundschau* **76**, 315–359.

Shull VL, Vogler AP, Baker MD, Maddison DR, Hammond PM (2001) Sequence alignment of 18S ribosomal RNA and the basal relationships of adephagan beetles: evidence for monophyly of aquatic families and the placement of Trachypachidae. *Systematic Biology* **50**, 945–969. doi:10.1080/106351501753462894

Sikes DS (2005) 11.6. Silphidae Latreille, 1807. In *Handbuch der Zoologie/Handbook of Zoology. Band/Volume IV Arthropoda: Insecta Teilband/Part 38. Coleoptera, Beetles. Volume 1. Morphology and Systematics (Archostemata, Adephaga, Myxophaga, Polyphaga partim)*. (Eds RG Beutel and RAB Leschen) pp. 288–296. W. DeGruyter, Berlin.

Sikes DS, Madge RB, Newton AF (2002) A catalog of the Nicrophorinae (Coleoptera: Silphidae) of the world. *Zootaxa* **65**, 1–3304.

Sivinski J (1981) The nature and possible functions of luminescence in Coleoptera larvae. *Coleopterists Bulletin* **35**, 167–179.

Skelley PE, Leschen RAB (2002) 92. Endomychidae Leach 1815. In *American Beetles. Volume 2. Polyphaga: Scarabaeoidea through Curculionoidea*. (Eds RH Arnett Jr, MC Thomas, PE Skelley and JH Frank) pp. 366–370. CRC Press, Boca Raton, Florida.

Skelley PE, Leschen RAB (2007) *Fauna of New Zealand. Number 59. Erotylinae (Insecta: Coleoptera: Erotylidae): Taxonomy and Biogeography*. Manaaki Whenua Press, Lincoln, Canterbury, New Zealand.

Skelley PE, McHugh JV (2002) 87. Erotylidae Leach 1815. In *American Beetles. Volume 2. Polyphaga: Scarabaeoidea through Curculionoidea*. (Eds RH Arnett Jr, MC Thomas, PE Skelley and JH Frank) pp. 348–353. CRC Press, Boca Raton, Florida.

Slater AE (1951) Biological problems of space flight. A report of Professor Haldane's lecture to the society on April 7,1951. *Journal of the British Interplanetary Solciety* **10**, 154–158.

Ślipiński SA (1983) A review of the Ethiopian species of *Passandra* Dalman, with notes on the constitution of the Cucujidae (Coleoptera). *Polskie Pismo Entomologiczne* **53**, 77–105.

Ślipiński SA (1984) Notes on the Lapethini with a revsion of the world *Lapethus* Casey (Coleoptera, Cerylonidae), including descriptions of related genera. *Polskie Pismo Entomologiczne* **54**, 3–104.

Ślipiński SA (1986) Description of two new species of *Sarothrias* Grouvelle (Coleoptera, Jacobsoniidae). *Revue Suisse de Zoologie* **93**, 59–62.

Ślipiński SA (1987a) A review of the Passandridae of the world (Coleoptera, Cucujoidea). I – Genus *Passandra* Dalman. *Annali del Museo Civico di Storia Naturale Giacomo Doria* **86**, 553–603.

Ślipiński SA (1987b) Studies on the Colydiidae (Coleoptera, Tenebrionidae) of New Guinea. Part II – genus *Pseudendestes* Lawrence. *Doriana* **6** (261), 5 pp.

Ślipiński SA (1988) Revision of the Australian Cerylonidae (Coleoptera: Cucujoidea). *Annales Zoologici* **42**, 1–74.

Ślipiński SA (1989) A review of the Passandridae (Coleoptera, Cucujoidea) of the world. II. Genus *Catogenus* Westwood. *Polskie Pismo Entomologiczne* **59**, 85–129.

Ślipiński SA (1990) A monograph of the world Cerylonidae (Coleoptera; Cucujoidea). Part I – Introduction and higher classification. *Annali del Museo Civico di Storia Naturale Giacomo Doria* **88**, 1–273.

Ślipiński SA (1992) Larinotinae a new subfamily of Trogossitidae (Coleoptera), with notes on the constitution of Trogossitidae and related families of Cleroidea. *Revue Suisse de Zoologie* **99**(2), 439–463.

Ślipiński SA (1998) Revision and phylogeny of Protocucujidae (Coleoptera: Cucujoidea). *Annales Zoologici* **48**(3), 275–298.

Ślipiński SA (2003) A monograph of the world Cerylonidae (Coleoptera, Cucujoidea). Part II Subfamily Loebliorylonimae. In *Systematics of Coleoptera: Papers Celebrating the Retirement of Ivan Löbl*. (Eds G Cuccodoro and RAB Leschen) pp. 921–941. Associated Publishers (Memoirs on Entomology, International, Volume 17), Gainesville, Florida.

Ślipiński SA (2004) Revision of the Australian Coccinellidae (Coleoptera). Part 2. Tribe Sticholotidini. *Annales Zoologici* **54**, 389–402.

Ślipiński A (2007) *Australian Ladybird Beetles (Coleoptera: Coccinellidae). their Biology and Classification*. Australian Biological Resources Study, Department of the Environment and Water Resources, Canberra.

Ślipiński A, Beutel RG (2010) 10.4. Protocucujidae Crowson, 1954. In *Handbuch der Zoologie/Handbook of Zoology. Band/Volume IV Arthropoda: Insecta Teilband/Part 38. Coleoptera, Beetles. Volume 2. Morphology and Systematics (Polyphaga partim)*. (Eds RAB Leschen, RG Beutel and JF Lawrence) pp. 296–300. W. DeGruyter, Berlin.

Ślipiński A, Burckhardt D (2006) Revision of the Australian Coccinellidae (Coleoptera). Part 5. Tribe Serangiini. *Annales Zoologici* **56**(1), 37–58.

Ślipiński A, Giorgi JA (2006) Revision of the Australian Coccinellidae (Coleoptera). Part 6. Tribe Chilocorini. *Annales Zoologici* **56**(2), 265–304.

Ślipiński SA, Lawrence JF (1997) Genera of Colydiinae (Coleoptera: Zopheridae) of the Australo-Pacific Region. *Annales Zoologici* **47**(3/4), 341–440.

Ślipiński SA, Lawrence JF (1999) Phylogeny and classification of Zopheridae *sensu novo* (Coleoptera: Tenebrionoidea) with a review of the genera of Zopherinae (excluding Monommatini). *Annales Zoologici* **49**(1/2), 1–53.

Ślipiński A, Lawrence JF (2010a) 10.29. Cerylonidae Billberg, 1820. In *Handbuch der Zoologie/Handbook of Zoology. Band/Volume IV Arthropoda: Insecta Teilband/Part 38. Coleoptera, Beetles. Volume 2. Morphology and Systematics (Polyphaga partim)*. (Eds RAB Leschen, RG Beutel and JF Lawrence) pp. 422–432. W. DeGruyter, Berlin.

Ślipiński A, Lawrence JF (2010b) 11.9. Zopheridae Solier, 1834. In *Handbuch der Zoologie/Handbook of Zoology. Band/Volume IV Arthropoda: Insecta Teilband/Part 38. Coleoptera, Beetles. Volume 2. Morphology and Systematics (Polyphaga partim)*. (Eds RAB Leschen, RG Beutel and JF Lawrence) pp. 548–559. W. DeGruyter, Berlin.

Ślipiński A, Lawrence JF (2010c) 11.16. Synchroidae Lacordaire, 1859. In *Handbuch der Zoologie/Handbook of Zoology. Band/Volume IV Arthropoda: Insecta Teilband/Part 38. Coleoptera, Beetles. Volume 2. Morphology and Systematics (Polyphaga partim)*. (Eds RAB Leschen, RG Beutel and JF Lawrence) pp. 667–669. W. DeGruyter, Berlin.

Ślipiński SA, Mazur S (1999) *Epuraeosoma*, a new genus of Histerinae and phylogeny of the family Histeridae (Coleoptera, Histeroidea). *Annales Zoologici* **49**(3), 209–230.

Ślipiński SA, Pakaluk J (1992) Problems in the classification of the cerylonid series of Cucujoidea (Coleoptera). In *Advances in Coleopterology*. (Eds M Zunino, X Bellés and M Blas) pp. 79–88. Associación Europea de Coleopterologia, Barcelona.

Ślipiński A, Tomaszewska W (2005) Revision of the Australian Coccinellidae (Coleoptera). Part 3. Tribe Sukunahikonini. *Australian Journal of Entomology* **44**(4), 369–384. doi:10.1111/j.1440-6055.2005.00488.x

Ślipiński A, Tomaszewska W (2010a) 10.17. Cavognathidae Sen Gupta and Crowson, 1969. In *Handbuch der Zoologie/Handbook of Zoology. Band/Volume IV Arthropoda: Insecta Teilband/Part 38. Coleoptera, Beetles. Volume 2. Morphology and Systematics (Polyphaga partim)*. (Eds RAB Leschen, RG Beutel and JF Lawrence) pp. 357–360. W. DeGruyter, Berlin.

Ślipiński A, Tomaszewska W (2010b) 20.30. Alexiidae Imhoff, 1856. In *Handbuch der Zoologie/Handbook of Zoology. Band/Volume IV Arthropoda: Insecta Teilband/Part 38. Coleoptera, Beetles. Volume 2. Morphology and Systematics (Polyphaga partim)*. (Eds RAB Leschen, RG Beutel and JF Lawrence) pp. 432–434. W. DeGruyter, Berlin.

Ślipiński A, Tomaszewska W (2010c) 10.33. Coccinellidae Latreille, 1802. In *Handbuch der Zoologie/Handbook of Zoology. Band/Volume IV Arthropoda: Insecta Teilband/Part 38. Coleoptera, Beetles. Volume 2. Morphology and Systematics (Polyphaga partim)*. (Eds RAB Leschen, RG Beutel and JF Lawrence) pp. 454–472. W. DeGruyter, Berlin.

Ślipiński A, Tomaszewska W (2010d) Revision of the family Cavognathidae (Coleoptera: Cucujoidea). *Australian Journal of Entomology* **49**, 256–267. doi:10.1111/j.1440-6055.2010.00762.x

Ślipiński SA, Pope RD, Aldridge RJW (1989) A Review of the world Bothriderini (Coleoptera, Bothrideridae). *Polskie Pismo Entomologiczne* **59**, 131–202.

Ślipiński A, Lawrence JF, Tomaszewska W (2001) The placement of *Periptyctus* Blackburn in Corylophidae (Coleoptera: Cucujoidea) with descriptions of a new genus and subfamily. *Annales Zoologici* **51**(3), 311–317.

Ślipiński A, Pang H, Pope RD (2005) Revision of the Australian Coccinellidae (Coccinellidae). Part 4. Tribe Telsimini. *Annales Zoologici* **55**(2), 243–269.

Ślipiński A, Tomaszewska W, Lawrence JF (2009) Phylogeny and classification of Corylophidae (Coleoptera: Cucujoidea) with descriptions of nee genera and larvae. *Systematic Entomology* **34**, 409–433. doi:10.1111/j.1365-3113.2009.00471.x

Ślipiński A, Tomaszewska W, Lawrence JF (2010a) 10.16. Myraboliidae Lawrence and Britton, 1991. In *Handbuch der Zoologie/Handbook of Zoology. Band/Volume IV Arthropoda: Insecta Teilband/Part 38. Coleoptera, Beetles. Volume 2. Morphology and Systematics (Polyphaga partim).* (Eds RAB Leschen, RG Beutel and JF Lawrence) pp. 354–357. W. DeGruyter, Berlin.

Ślipiński A, Lord N, Lawrence JF (2010b) 10.28. Bothrideridae Erichson, 1845. In *Handbuch der Zoologie/Handbook of Zoology. Band/Volume IV Arthropoda: Insecta Teilband/Part 38. Coleoptera, Beetles. Volume 2. Morphology and Systematics (Polyphaga partim).* (Eds RAB Leschen, RG Beutel and JF Lawrence) pp. 411–422. W. DeGruyter, Berlin.

Ślipiński A, Lawrence JF, Cline AR (2010c) 10.34. Corylophidae LeConte, 1852. In *Handbuch der Zoologie/Handbook of Zoology. Band/Volume IV Arthropoda: Insecta Teilband/Part 38. Coleoptera, Beetles. Volume 2. Morphology and Systematics (Polyphaga partim).* (Eds RAB Leschen, RG Beutel and JF Lawrence) pp. 472–481. W. DeGruyter, Berlin.

Ślipiński SA, Leschen RAB, Lawrence JF (2011) Order Coleoptera Linnaeus, 1758. In Animal biodiversity: an outline of higher-level classification and survey of taxonomic richness (ed. Z.-Q. Zhang). *Zootaxa* **3148**, 203–208.

Smith SG (1950) The cyto-taxonomy of Coleoptera. *Canadian Entomologist* **82**, 58–68. doi:10.4039/Ent8258-3

Smith JA (1981) Two Tasmanian species of *Sclerocyphon* Blackburn (Coleoptera: Psephenidae) with notes on their life history and distribution. *Journal of the Australian Entomological Society* **20**, 277–288. doi:10.1111/j.1440-6055.1981.tb01048.x

Smith ABT (2006) A review of the family-group names for the superfamily Scarabaeoidea (Coleoptera) with corrections to nomenclature and a current classification. *The Coleopterists Society Monographs* **60**, 144–204. doi:10.1649/0010-065X(2006)60[144:AROTFN]2.0.CO;2

Smith JA, Dartnall AJ (1980) Boundary layer control by water pennies (Coleoptera: Psephenidae). *Aquatic Insects* **2**(2), 65–72. doi:10.1080/01650428009361008

Smith SG, Virkki N (1978) *Animal Cytogenetics Vol. 3: Insecta 5. Coleoptera.* Gebrüder Borntraeger, Berlin.

Smith ABT, Hawks DC, Heraty JM (2006) An overview of the classification and evolution of the major scarab beetle clades (Coleoptera: Scarabaeoidea) based on preliminary molecular analyses. *Coleopterists Society Monographs* **60**, 35–46. doi:10.1649/0010-065X(2006)60[35:AOOTCA]2.0.CO;2

Smithers CN (1981) *Handbook of Insect Collecting.* A.H. and A. W. Reed, Sydney.

Snodgrass RE (1963) A contribution toward an encyclopedia of insect anatomy. *Smithsonian Miscellaneous Collections* **146**(2), 1–48.

Sokoloff A (1959) The habitat-niche of American Nosodendridae. *Coleopterists Bulletin* **13**, 97–98.

Solervicens JA (2005a) A prothoracic character that defines Korynetinae (Coleoptera: Cleridae). *New Zealand Entomologist* **28**, 49–54. doi:10.1080/00779962.2005.9722685

Solervicens JA (2005b) Descripción de estados juveniles de *Polymerius chilensis* (Laporte, 1834) (Coleoptera: Rhipiceridae). *Acta Entomologica Chilena* **29**(1), 71–75.

Solervicens JA (2007) Cladistic analysis of species of *Natalis* Laporte (1836) and related genera *Eunatalis* Schenkling (1909), *Metademius* Schenkling (1899) and *Eurymetomorphon* Pic (1950) (Coleoptera: Cleridae: Clerinae) with redescription of a restored Clerinae genus. *Zootaxa* **1398**, 1–14.

Solodovnikov A (2006) Revision and phylogenetic assessment of *Afroquedius* gen. nov. from South Africa: toward new concepts of the genus *Quedius*, subtribe Quediina and reclassification of the tribe Staphylinini (Coleoptera: Staphylinidae: Staphylininae). *Annals of the Entomological Society of America* **99**(6), 1064–1084. doi:10.1603/0013-8746(2006)99[1064:RAPAOA]2.0.CO;2

Solodovnikov A, Newton AF (2009) *Australotarsius*—a new genus of the rove beetle tribe Staphylinini from Australia (Coleoptera: Staphylinidae: Staphylininae). *Zootaxa* **2033**, 49–57.

Song H, Sheffield NC, Cameron SL, Miller KB, Whiting MF (2010) When phylogenetic assumptions are violated: base compositional heterogeneity and among-site rate variation in beetle mitochondrial phylogenomics. *Systematic Entomology* **35**(3), 429–448. doi:10.1111/j.1365-3113.2009.00517.x

Sörensson M (1997) Morphological and taxonomical novelties in the world's smallest beetles, and the first Old World record of Nanosellini (Coleoptera: Ptiliidae). *Systematic Entomology* **22**, 257–283. doi:10.1046/j.1365-3113.1997.d01-40.x

Soriano C, Delclòs X (2006) New cupedids from the lower Cretaceous of Spain and the palaeogeography of the family. *Acta Palaeontologica Polonica* **51**, 185–200.

Soriano C, Gratshev VG, Delclos X (2006a) New early Cretaceous weevils (Insecta, Coleoptera, Curculionoidea) from El Montsec, Spain. *Cretaceous Research* **27**, 555–564. doi:10.1016/j.cretres.2005.10.015

Soriano C, Kirejtshuk AG, Declos X (2006b) The Mesozoic Laurasian family Parandrexidae (Insecta: Coleoptera), new species from the Lower Cretaceous of Spain. *Comptes Rendus. Palévol* **5**, 779–784. doi:10.1016/j.crpv.2006.03.006

Soriano C, Ponomarenko AG, Delclos X (2007) Coptoclavid beetles (Coleoptera: Adephaga) from the Lower Cretaceous of Spain: a new feeding strategy in beetles. *Palaeontology* **50**(2), 525–536. doi:10.1111/j.1475-4983.2007.00642.x

Spahr U (1981a) Bibliographie der Bernstein- und Kopal-Käfer (Coleoptera). *Stuttgarter Beitrage zur Naturkunde. Serie B, Geologie und Palaontologie* **72**, 1–21.

Spahr U (1981b) Systematischer Katalog der Bernstein- und Kopal-Käfer (Coleoptera). *Stuttgarter Beitrage zur Naturkunde. Serie B, Geologie und Palaontologie* **80**, 1–107.

Spangler PJ (1980a) Chelonariid larvae, aquatic or not? *Coleopterists Bulletin* **34**(1), 105–114.

Spangler PJ (1980b) A new species of *Ytu* from Brazil (Coleoptera: Torridincolidae). *Coleopterists Bulletin* **34**, 145–158.

Spangler PJ (1983) Immature stages and biology of *Tetraglossa palpalis* Champion (Coleoptera: Ptilodactylidae). *Entomological News* **94**, 161–175.

Spangler PJ (1991) Haliplidae (Adephaga), Noteridae (Adephaga), Dytiscidae (Adephaga), Gyrinidae (Adephgaga), Hydraenidae (Staphylinoidea), Hydrophilidae (Hydrophiloidea) (including Helophoridae, Hydrochidae, Sphaeridiidae, Spercheidae), Georyssidae (Hydrophiloidea) and Chelonariidae (Dryopoidea). In *Immature Insects. Volume 2.* (Ed. FW Stehr) pp. 311–312, 314–322, 355–359, 394–395. Kendall/Hunt, Dubuque, Iowa.

Spangler PJ (1995) A review and two new species of the genus *Pseudeucinetus* Heller from Southeast Asia and a world checklist of the Thaumastodinae (Coleoptera, Limnichidae). *Special Bulletin of the Japanese Society of Coleopterology, Tokyo* **4**, 395–405.

Spangler PJ, Santiago-Fragoso S (1992) The aquatic beetler subfamily Larainae (Coleoptera: Elmidae) in Mexico, Central America, and the West Indies. *Smithsonian Contributions to Zoology* **528**, i–vi, 1–74.

Spangler PJ, Steiner WE Jr (2005) A new aquatic beetle family, Meruidae, from Venezuela. *Systematic Entomology* **30**, 339–357. doi:10.1111/j.1365-3113.2005.00288.x

Spangler PJ, Staines CL, Spangler PM, Staines SL (2001) A checklist of the Limnichidae and Lutrochidae of the world. *Insecta Mundi* **15**(3), 151–165.

Spilman TJ (1954) Generic names of the Salpingidae and their type species (Coleoptera). *Journal of the Washington Academy of Sciences* **44**, 85–94.

Spilman TJ (1959) A study of the Thaumastodinae, with one new genus and two new species (Limnichidae). *Coleopterists Bulletin* **13**, 111–122.

Spilman TJ (1967) The heteromerous intertidal beetles (Coleoptera: Salpingidae: Aegialitinae). *Pacific Insects* **9**, 1–21.

Spilman TJ (1972) A new genus and species of jumping shore beetle from Mexico (Coleoptera: Limnichidae). *The Pan-Pacific Entomologist* **48**(2), 108–115.

Springer CA, Goodrich MA (1983) A revision of the family Byturidae (Coleoptera) for North America. *Coleopterists Bulletin* **37**(2), 183–192.

Springer CA, Goodrich MA (1986) A revision of the family Byturidae (Coleoptera) in Europe. *Coleopterists Bulletin* **40**(4), 335–352.

Springer CA, Goodrich MA (1990) A revision of the family Byturidae (Coleoptera) in Asia. *Coleopterists Bulletin* **44**(4), 461–483.

Springer CA, Goodrich MA (1994) A revision of the subfamily Platydascillinae (Coleoptera: Byturidae) from Southeast Asia, with descriptions of two new genera and three new species. *Coleopterists Bulletin* **48**(1), 60–78.

Srivastava US (1959) The maxillary glands of some Coleoptera. *Proceedings of the Royal Entomological society of London (A)* **34**, 57–62. doi:10.1111/j.1365-3032.1959.tb00238.x

St. George RA (1924) Studies on the larvae of North American beetles of the subfamily Tenebrioninae with a description of the larva and pupa of *Merinus laevis* (Olivier). *Proceedings of the United States National Museum* **65**, 1–22, pls 1–3.

St. George RA (1926) Taxonomic studies of the larvae of the genera *Tenebrio* and *Neatus* Leconte (Coleoptera: Tenebrionidae). *Proceedings of the Entomological Society of Washington* **28**, 102–111.

St. George RA (1939) The larva of *Perimylops antarcticus* (Müller) and the systematic position of the family Perimylopidae (Coleoptera). *Proceedings of the Entomological Society of Washington* **41**(7), 207–214, pl. 27.

Stanger-Hall KF, Lloyd JE, Hillis DM (2007) Phylogeny of North American fireflies (Coleoptera: Lampyridae): implications for the evolution of light signals. *Molecular Phylogenetics and Evolution* **45**, 33–49. doi:10.1016/j.ympev.2007.05.013

Stebnicka ZT (1998) The Aphodiinae of New Guinea, the Bismarck Archipelago and the Solomon Islands (Coleoptera: Scarabaeoidea). *Invertebrate Taxonomy* **12**, 833–895. doi:10.1071/IT97006

Stebnicka ZT (2001) *Fauna of New Zealand. Number 42. Aphodiinae (Insecta: Coleoptera: Scarabaeidae).* Manaaki Whenua Press, Lincoln, Canterbury, New Zealand.

Stebnicka ZT (2009) *Aphodiinae of Australia (Coleoptera: Scarabaeoidea): Iconography.* 187 pp. Institute of Systematics & Evolution of Animals, Polish Academy of Sciences, Kraków, Poland.

Stebnicka ZT, Howden HF (1995) Revision of Australian genera in the tribes Aphodiini, Aegialiini and Proctophanini (Coleoptera: Scarabaeidae: Aphodiinae). *Invertebrate Taxonomy* **9**, 709–766. doi:10.1071/IT9950709

Stebnicka ZT, Howden HF (1996) Australian genera and species in the tribes Odontolochini, Psammodiini, Rhyparini, Stereomerini and part of the Eupariini (Coleoptera: Scarabaeoidea: Aphodiinae). *Invertebrate Taxonomy* **10**, 97–170. doi:10.1071/IT9960097

Stedman M (2006) Small hive beetle (SHB): *Aethina tumida* Murray (Coleoptera: Nitidulidae). *Primary Industries and Resources South Australia Fact Sheet* **03/06**, 15 pp.

Steffan AW (1967) Ectosymbiosis in aquatic insects. In *Symbiosis Volume II* (Ed. SM Henry) pp. 207–289. Academic Press, New York.

Steiner WE (1984) A review of the biology of phalacrid beetles (Coleoptera). In *Fungus-Insect Relationships: Perspectives In Ecology And Evolution* (Eds Q Wheeler and M Blackwell) pp. 424–445. Columbia University Press, New York.

Steiner WE Jr (1995) Structures, behavior and diversity of the pupae of Tenebrionidae (Coleoptera). In *Biology, Phylogeny, and Classification of Coleoptera. Papers Celebrating the 80th Birthday of Roy A. Crowson.* (Eds J Pakaluk and SA Ślipiński) pp. 503–539. Muzeum I Instytut Zoologii Polska Akademia Nauk, Warsaw.

Steiner WE Jr (2002) 84. Phalacridae Leach 1815. In *American Beetles. Volume 2. Polyphaga: Scarabaeoidea through Curculionoidea.* (Eds RH Arnett Jr, MC Thomas, PE Skelley and JH Frank) pp. 335–337. CRC Press, Boca Raton, Florida.

Steiner WE Jr, Anderson JJ (1981) Notes on the natural history of *Spanglerogyrus albiventris* Folkerts, with a new distribution record (Coleoptera: Gyrinidae). *The Pan-Pacific Entomologist* **57**(1), 124–132.

Steiner WE Jr, Singh BP (1987) Redescription of an ergot beetle, *Acylomus pugetanus* Casey, with immature stages and biology (Coleoptera: Phalacridae). *Proceedings of the Entomological Society of Washington* **87**(4), 744–758.

Steinke G (1919) Die Stigmen der Käferlarven. *Archiv für Naturgeschichte* **85**(A) 7, 1–58, 2 pls.

Stevens MM, Stanton RA, Wu H, Sampson B, Weir TA, Reid CAM, Mo J (2010) Detection of *Lema bilineata* Germar (Coleoptera: Chrysomelidae) in Australia. *General and Applied Entomology* **39**, 1–4.

Stickney FS (1923) The head-capsule of Coleoptera. *Illinois Biological Monographs* **8**(1), 1–105.

Storey RI, Howden HF (1996) Revision of *Australoxenella* Howden & Storey in Australia (Coleoptera: Scarabaeidae: Aphodiinae). *Memoirs of the Queensland Museum* **39**, 365–380.

Storey RI, Monteith GB (2000) Five new species of *Aptenocanthon* Matthews (Coleoptera: Scarabaeidae: Scarabaeinae) from tropical Australia, with notes on distribution. *Memoirs of the Queensland Museum* **46**(1), 349–358.

Stork NE (1980) A scanning electron microscope study of tarsal adhesive setae in the Coleoptera. *Zoological Journal of the Linnean Society* **68**, 173–306. doi:10.1111/j.1096-3642.1980.tb01121.x

Strand RM, Spangler PJ (1994) The natural history, distribution, and larval description of *Brychius hungerfordi* Spangler (Coleoptera: Haliplidae). *Proceedings of the Entomological Society of Washington* **96**(2), 208–213.

Straneo SL, Ball GE (1989) Synopsis of the genera and subgenera of the tribe Peleciini, and revision of the Neotropical and Oriental species (Coleoptera: Carabidae). *Insecta Mundi* **3**, 73–178.

Stribling JB (1985) World Generic Revision of Ptilodactylidae (Coleoptera: Dryopoidea). Ohio State University (PhD thesis), Columbus, Ohio.

Stribling JB (1986) Revision of *Anchytarsus* (Coleoptera: Dryopoidea) and a key to the New world genera of Ptilodactylidae. *Annals of the Entomological Society of America* **79**, 219–234.

Stride GO (1955) On the respiration of an aquatic African beetle, *Potamodytes tuberosus* Hinton. *Annals of the Entomological Society of America* **48**, 344–351.

Striganova BR (1961) Morpho-functional characters of *Prionoycphon serricornis* Müll. (Coleoptera, Helodidae) larvae in relation to their inhabitation of water. *Entomologischeskoye Obozreniye* **40**, 577–582 (in Russian; translation in *Entomological Review* **40**, 314–317).

Strohecker HF (1953) *Genera Insectorum de P. Wytsman. Fascicule 210.* Coleoptera Clavicornia Fam. Endomychidae. Genera Insectorum, Crainhem, Belgium.

Suh S, Marshall CJ, McHugh JV, Blackwell M (2003) Wood ingestion by passalid beetles in the presence of xylose-fermenting gut yeasts. *Molecular Ecology* **12**, 3137–3145. doi:10.1046/j.1365-294X.2003.01973.x

Suh SO, Blackwell M (2006) Phylogenetics of Saccharomycetales, the ascomycete yeasts. *Mycologia* **98**(6), 1006–1017. doi:10.3852/mycologia.98.6.1006

Suh SO, McHugh JV, Blackwell M (2004) Expansion of the *Candida tanzawaensis* yeast clade: 16 novel *Candida* species from basidiocarp-feeding beetles. *International Journal of Systematic and Evolutionary Microbiology* **54**(6), 2409–2429. doi:10.1099/ijs.0.63246-0

Suh SO, McHugh JV, Pollock DD, Blackwell M (2005) The beetle gut: a hyperdiverse source of novel yeasts. *Mycological Research* **109**(3), 261–265. doi:10.1017/S0953756205002388

Šulc K (1949) On the repugnatorial stink-glands in the beetles of the genus *Cantharis*. *Česká Akademie věd a Umění* **50**, 79–100.

Sumlin WD (1982) Studies on the Australian Cicindelidae new taxa from Australia (Coleoptera). *Coleopterists Bulletin* **35**, 273–280.

Süss L, Puppin O (1976) Observationi sulla morfologia e sulla biologia di *Ptilodactyla exotica* Chapin (Coleoptera, Ptilodactylidae) nelle serre della Lombardia e contributo bibliografico allo studio del gruppo. *Bolletino di Zoologia Agraria e di Bachicoltura* **13**(2), 143–165.

Suzuki K (1994) Comparative morphology of the hindwing venation of the Chrysomelidae (Coleoptera). In *Novel Aspects of the Biology of Chrysomelidae.* (Eds PH Jollivet, ML Cox and E Petitpierre) pp. 337–354. Kluwer Academic Publishers, Dordrecht.

Švácha P (1992) What are and what are not imaginal discs: reevaluation of some basic concepts (Insecta, Holometabola). *Developmental Biology* **154**, 101–117. doi:10.1016/0012-1606(92)90052-I

Švácha P (1994) Bionomics, behaviour and immature stages of *Pelecotoma fennica* (Paykull) (Coleoptera: Rhipiphoridae). *Journal of Natural History* **28**, 585–618. doi:10.1080/00222939400770271

Švácha P (1995) The larva of *Scraptia fuscula* (P.W.J. Müller) (Coleoptera: Scraptiidae): autotomy and regeneration of the caudal appendage. In *Biology, Phylogeny, and Classification of Coleoptera. Papers Celebrating the 80th Birthday of Roy A. Crowson.* (Eds J Pakaluk and SA Ślipiński) pp. 473–489. Muzeum I Instytut Zoologii Polska Akademia Nauk, Warsaw.

Švácha P (2001) 17. Überfamilie Chrysomeloidea. 114. Familie Cerambycidae. 7. Unterfamilie Lamiinae. In *Die Larven der Käfer Mitteleuropas. 6. Band. Polyphaga. Teil 5.* (Ed. B. Klausnitzer) pp. 248–298.

Švácha P, Danilevsky ML (1987) Cerambycoid larvae of Europe and Soviet Union (Coleoptera, Cerambycoidea). Part I. *Acta Universitatis Carolinae Biologica* **30**, 1–176.

Švácha P, Danilevsky ML (1988) Cerambycoid larvae of Europe and Soviet Union (Coleoptera, Cerambycoidea). Part II. *Acta Universitatis Carolinae Biologica* **31**, 121–284.

Švácha P, Danilevsky ML (1989) Cerambycoid larvae of Europe and Soviet Union (Coleoptera, Cerambycoidea). Part III. *Acta Universitatis Carolinae Biologica* **32**, 1–205.

Švácha P, Wang JJ, Chen SC (1997) Larval morphology and biology of Philus antennatus and *Heterophilus punctulatus*, and systematic position of the Philinae (Coleoptera: Cerambycidae and Vesperidae). *Annales de la Société Entomologique de France (N. S.)* **33**(3), 323–369.

Švihla V (1986) Revision of the generic classification of the Old World Oedemeridae (Coleoptera). *Sborník Národního Musea v Praze B* **41**, 141–238.

Szawaryn K, Tomaszewska W (2011) Three new species of *Myrabolia* Reitter (Coleoptera: Cucujoidea: Myraboliidae). *Annales Zoologici* **61**(2), 367–374. doi:10.3161/000345411X584834

Szpoda A (1989) Pre-imaginal stages of *Cytilus sericeus* (Forster) (Coleoptera, Byrrhidae). *Polskie Pismo Entomologiczne* **58**, 783–792.

Takami Y, Sota T (2006) Four new species of the Australian *Pamborus* Latreille (Coleoptera, Carabidae) carabid beetles. *Australian Journal of Entomology* **45**, 44–54. doi:10.1111/j.1440-6055.2006.00495.x

Takizawa H (1971) On the larvae of the genus *Chrysolina* Motschulsky in Japan (Coleoptera: Chrsomelidae). *Entomological Review of Japan* **23**(2), 102–109.

Takizawa H (1972) Descriptions of larvae of glanduliferous group of Galerucinae in Japan, with notes on subdivisions of the subfamily (Coleoptera: Chrysomelidae). *Insecta Matsumurana, Supplement* **10**, 1–14, pls 1–8.

Takizawa H (1976) Larvae of the genus *Gonioctena* Chevrolat (Coleoptera, Chrysomelidae): descriptions of Japanese species and the implications of larval characters for the phylogeny. *Kontyû* **44**(4), 444–468.

Tan J, Ren D (2006a) Jurassic and Cretaceous Cupedomorpha (Insecta: Coleoptera) faunas of China. *Progress in Natural Science* **16**, (Special Issue), 313–322.

Tan J, Ren D (2006b) New fossil Priacmini (Insecta: Coleoptera: Archostemata: Cupedidae) from the Jehol Biota of China. *Journal of Natural History* **40**(47–48), 2653–2661. doi:10.1080/00222930601121445

Tan J, Ren D (2006c) *Ovatocupes*: a new cupedid genus (Coleoptera: Archostemata: Cupedidae) from the Jehol Biota (Late Jurassic) of western Liaoning, China. *Entomological News* **117**, 223–232. doi:10.3157/0013-872X(2006)117[223:OANCGC]2.0.CO;2

Tan J, Ren D (2007) Two exceptionally well-preserved catiniids (Coleoptera: Archostemata: Catiniidae) from the Late Mesozoic of northeastern China. *Annals of the Entomological Society of America* **100**(5), 666–672. doi:10.1603/0013-8746(2007)100[666:TEWCCA]2.0.CO;2

Tan J, Ren D (2009) Mesozoic Archostematan Fauna from China (in Chinese with English summary). Science Press, Beijing.

Tan J, Ren D, Liu M (2005) New ommatids from the Late Jurassic of western Liaoning, China (Coleoptera: Archostemata). *Insect Science* **12**(3), 207–216. doi:10.1111/j.1005-295X.2005.00026.x

Tan J, Ren D, Shih C (2006a) New cupedids from the middle Jurassic of Inner Mongolia, China (Coleoptera: Archostemata). *Annales Zoologici* **56**(1), 1–6.

Tan J, Ren D, Shih C (2006b) First record of fossil *Priacma* (Coleoptera: Archostemata: Cupedidae) from the Jehol Biota of western Liaoning, China. *Zootaxa* **1326**, 55–68.

Tan J, Ren D, Shih C, Ge S (2006c) New fossil beetles of the family Ommatidae (Coleoptera: Archostemata) from the Jehol Biota of China. *Acta Geologica Sinica* **80**(4), 474–485.

Tan J, Huang D, Ren D (2007a) First record of fossil *Mesocupes* from China (Coleoptera: Archostemata: Cupedidae). *Acta Geologica Sinica* **81**(5), 688–696. doi:10.1111/j.1755-6724.2007.tb00993.x

Tan J, Ren D, Shih C (2007b) New beetles (Insecta: Coleoptera: Archostemata) from the Late Mesozoic of North China. *Annales Zoologici* **57**(2), 231–247.

Tan J, Ren D, Shih C (2007c) New ommatids of Ommatinae (Coleoptera: Archostemata: Ommatidae) from the Yixian formation of western Liaoning, China. *Progress in Natural Science* **17**(7), 803–811. doi:10.1080/10002007088537476

Tan J, Wang Y, Ren D, Yang X (2012) New fossil species of ommatids (Coleoptera: Archostemata) from the Middle Mesozoic of China illuminating the phylogeny of Ommatidae. *BMC Evolutionary Biology* **12**, 113. doi:10.1186/1471-2148-12-113

Tanner VM (1927) A preliminary study of the genitalia of female Coleoptera. *Transactions of the American Entomological Society* **53**, 5–50.

Tavares II (1979) The Laboulbeniales and their arthropod hosts. In *Insect-Fungus Symbiosis: Nutrition, Mutualism, and Commensalism.* (Ed. LR Batra), pp. 229–258. John Wiley & Sons, New York.

Telnov D (2003) Zwei neue Gatungen und eine neue Art der notogäischen Anthicidae (Coleoptera). *Mitteilungen des Internationalen Entomologischen Vereins Frankfurt am Main* **28**(1/2), 13–23.

Telnov D (2006a) The first record of Steropinae from the Oriental region, with the description of a new *Steropes* from Vietnam (Coleoptera: Anthicidae). *Journal of the Zoological Society of Wallacea* **2**, 63–66.

Telnov D (2006b) Nomenclatural notes on Anthicidae and Pyrochroidae (Coleoptera). 1. *Latvijas Entomologs* **43**, 57–77.

Telnov D (2007) A review of the genus *Lemodes* Boheman, 1858 (Coleoptera: Anthicidae: Lemodinae). *Veröffentlichungen Naturkundemuseum Erfurt* **26**, 241–258.

Telnov D (2009a) Species of *Anthicomorphus* Lewis, 1895 (Coleoptera: Anthicidae) from the Indo-Australian transition zone (Wallace), with comments on selected taxa from adjacent areas. *Vernate* **28**, 377–408.

Telnov D (2009b) Three new species of *Lemodes*, with further faunal records on Papuan and Moluccan Anthicidae (Coleoptera). *Folia Heyrovskyana, Serie A* **17**(1), 31–42.

Telnov D (2011a) New *Gilletinus* Boucomont, 1932 (Coleoptera: Geotrupidae: Bolboceratinae) from Misool, with the key to known species. In *Biodiversity, Biogeography and Nature Conservation in Wallacea and New Guinea. Vol. 1.* (Ed. D Telnov) pp. 287–290, pls. 38–39. The Entomological Society of Latvia, Riga.

Telnov D (2011b) Taxonomische Revision der Gattung *Macratria* vNewman, 1838 (Coleoptera: Anthicidae: Macratriinae) aus Wallacea, Neuguinea und den Salomonen. In *Biodiversity, Biogeography and Nature Conservation in*

Wallacea and New Guinea. Vol. 1. (Ed. D Telnov) pp. 97–2285, pls. 17–37. The Entomological Society of Latvia, Riga.

Telnov D (2011c) Tomoderinae (Coleoptera: Anthicidae) from the Indo-Australian transition zone. *Vernate* **30**, 223–248.

Thayer MK (1985) Revision, phylogeny and biogeography of the austral genus *Metacorneolabium* Steel (Coleoptera: Staphylinidae: Omaliinae). In *Taxonomy, Phylogeny and Zoogeography of Beetles and Ants. A Volume dedicated to the Memory of Philip Jackson Darlington Jr (1904–1983).* (Ed. GE Ball) pp. 113–179. Junk (Series Entomologica 33), Dordrecht.

Thayer MK (1992) Discovery of sexual wing dimorphism in Staphylinidae (Coleoptera): "*Omalium*" *flavidum*, and a discussion of wing dimorphism in insects. *Journal of the New York Entomological Society* **100**(4), 540–573.

Thayer MK (1997) *Proglypholoma aenigma* gen. et sp. nov., *Glypholoma* spp. nov. and new records, and a phylogenetic analysis of Glypholomatinae (Coleoptera: Staphylinidae). *Annales Zoologici* **47**(1/2), 157–174.

Thayer MK (2000) *Glypholoma* larvae at last: phylogenetic implications for basal Staphylinidae? (Coleoptera: Staphylinidae: Glypholomatinae). *Invertebrate Taxonomy* **14**, 741–754. doi:10.1071/IT00019

Thayer MK (2005) 11.7. Staphylinidae Latreille, 1802. In *Handbuch der Zoologie/Handbook of Zoology. Band/Volume IV Arthropoda: Insecta Teilband/Part 38. Coleoptera, Beetles. Volume 1. Morphology and Systematics (Archostemata, Adephaga, Myxophaga, Polyphaga partim).* (Eds RG Beutel and RAB Leschen) pp. 296–344. W. DeGruyter, Berlin.

Thayer MK, Lawrence JF (2002) 98. Ciidae Leach in Samouelle 1819. In *American Beetles. Volume 2. Polyphaga: Scarabaeoidea through Curculionoidea.* (Eds RH Arnett Jr, MC Thomas, PE Skelley and JH Frank) pp. 403–412. CRC Press, Boca Raton, Florida.

Thayer MK, Newton AF (1978) Revision of the South Temperate genus *Glypholoma* Jeannel, with four new species (Coleoptera: Staphylinidae: Omaliinae). *Psyche, Cambridge* **85**, 25–63. doi:10.1155/1978/29756

Thayer MK, Newton AF, Chatzimanolis S (2012) *Prosolierius*, a new mid-Cretaceous genus of Solieriinae (Coleoptera: Staphylinidae) with three new species from Burmese amber. *Cretaceous Research* **34**, 124–134. doi:10.1016/j.cretres.2011.10.010

Thomas JB (1957) The use of larval anatomy in the study of bark beetles (Coleoptera: Scolytidae). *Canadian Entomologist Supplement* **5**, 45 pp.

Thomas MC (1984) A new species of apterous *Telephanus* (Coleoptera: Silvanidae) with a discussion of phylogenetic relationships of the Silvanidae. *Coleopterists Bulletin* **38**, 43–55.

Thomas MC (1988) Generic key to the known larvae of the Cucujidae, Passandridae, and Silvanidae of America north of Mexico (Coleoptera). *Insecta Mundi* **2**, 81–89.

Thomas MC (2002a) 80. Silvanidae Kirby 1837. In *American Beetles. Volume 2. Polyphaga: Scarabaeoidea through Curculionoidea.* (Eds RH Arnett Jr, MC Thomas, PE Skelley and JH Frank) pp. 322–326. CRC Press, Boca Raton, Florida.

Thomas MC (2002b) 81. Passandridae Erichson 1845. In *American Beetles. Volume 2. Polyphaga: Scarabaeoidea through Curculionoidea.* (Eds RH Arnett Jr, MC Thomas, PE Skelley and JH Frank) pp. 327–328. CRC Press, Boca Raton, Florida.

Thomas MC (2002c) 82. Cucujidae Latreille 1802. In *American Beetles. Volume 2. Polyphaga: Scarabaeoidea through Curculionoidea.* (Eds RH Arnett Jr, MC Thomas, PE Skelley and JH Frank) pp. 329–330. CRC Press, Boca Raton, Florida.

Thomas MC (2002d) 83. Laemophloeidae Ganglbauer 1899. In *American Beetles. Volume 2. Polyphaga: Scarabaeoidea through Curculionoidea.* (Eds RH Arnett Jr, MC Thomas, PE Skelley and JH Frank) pp. 331–334. CRC Press, Boca Raton, Florida.

Thomas MC (2002e) 91. Cerylonidae Billberg 1820. In *American Beetles. Volume 2. Polyphaga: Scarabaeoidea through Curculionoidea.* (Eds RH Arnett Jr, MC Thomas, PE Skelley and JH Frank) pp. 363–365. CRC Press, Boca Raton, Florida.

Thomas MC (2004) The Brontini of the world: a generic review of the tribe (Coleoptera: Silvanidae: Brontinae). *Insecta Mundi* **17**, 1–31.

Thomas MC (2010) A new Chilean species of *Brontoliota* Thomas (Coleoptera: Silvanidae: Brontinae: Brontini). *Insecta Mundi* **0113**, 1–4.

Thomas MC, Leschen RAB (2010a) 10.14. Silvanidae Kirby, 1837. In *Handbuch der Zoologie/Handbook of Zoology. Band/Volume IV Arthropoda: Insecta Teilband/Part 38. Coleoptera, Beetles. Volume 2. Morphology and Systematics (Polyphaga partim).* (Eds RAB Leschen, RG Beutel and JF Lawrence) pp. 346–350. W. DeGruyter, Berlin.

Thomas MC, Leschen RAB (2010b) 10.15. Cucujidae Latreille, 1802. In *Handbuch der Zoologie/Handbook of Zoology. Band/Volume IV Arthropoda: Insecta Teilband/Part 38. Coleoptera, Beetles. Volume 2. Morphology and Systematics (Polyphaga partim).* (Eds RAB Leschen, RG Beutel and JF Lawrence) pp. 350–354. W. DeGruyter, Berlin.

Thomas MC, Leschen RAB (2010c) 10.22. Laemophloeidae Ganglbauer, 1899. In *Handbuch der Zoologie/Handbook of Zoology. Band/Volume IV Arthropoda: Insecta Teilband/Part 38. Coleoptera, Beetles. Volume 2. Morphology and Systematics (Polyphaga partim).* (Eds RAB Leschen, RG Beutel and JF Lawrence) pp. 376–380. W. DeGruyter, Berlin.

Thomas MC, Nearns EH (2008) A new genus of telephanine Silvanidae (Coleoptera: Cucujoidea), with a diagnosis of the tribe and key to genera. *Insecta Mundi* **0048**, 1–14.

Thomas MC, Woodruff RE (1986) Description of the larvae of two species of *Hemipeplus* Latreille (Coleoptera: Mycteridae). *Insecta Mundi* **1**(3), 121–124.

Thompson RG (1979) Larvae of North American Carabidae with a key to the tribes. In *Carabid Beetles: Their Evolution, Natural History, and Classification.* (Eds TL Erwin, GE Ball, DR Whitehead and AL Halpern) pp. 209–291. Junk, The Hague.

Thompson RT (1992) Observations on the morphology and classification of weevils (Coleoptera, Curculionoidea) with a key to major groups. *Journal of Natural History* **26**, 835–891. doi:10.1080/00222939200770511

Thompson RT, Marshall JE (1980) A taxonomic study of *Phalacrus uniformis* (Coleoptera: Phalacridae), an Australian beetle now established in New Zealand. *New Zealand Journal of Zoology* **7**, 407–416. doi:10.1080/03014223.1980.10423794

Thomson J (1864) Systema Cerambycidarum ou exposé de tous les genres compris dans la famille des Cerambycides et familles limitrophes. *Livres* **1–3**, 1–352. [H. Dessain, Liège.]

Thomson J (1865) Systema Cerambycidarum ou exposé de tous les genres compris dans la famille des Cerambycides et familles limitrophes. *Livre* **4**, 353–578. [H. Dessain, Liège.]

Thongphak D, Wang Q (2007) Taxonomic revision of the longicorn beetle genus *Uracanthus* Hope (Coleoptera: Cerambycidae: Cerambycinae) from Australia. *Zootaxa* **1569**, 1–139.

Thongphak D, Wang Q (2008) Phylogeny and distribution of the Australian longicorn beetle genus *Uracanthus* Hope (Coleoptera: Cerambycidae). *Zootaxa* **1958**, 1–16.

Tikhomirova AL (1968) Staphylinid beetles from the Jurassic of Karatau. In *Jurassic Insects of Karatau.* (Ed. BB Rhodendorf) pp. 139–154. Nauka, Moscow (in Russian).

Tikhomirova AL (1973) *Morphological and Ecological Characteristics and Phylogeny of Staphylinids.* 190 pp. Akademia Nauk SSSR, Moscow (in Russian).

Tikhomirova AL (1983) On homology of the joints of larval and imaginal legs in insects with complete metamorphosis (*Tenebrio molitor* taken as an example). *Зоологический журнал* **62**, 530–539. [in Russian]

Tillyard RJ (1924) Upper Permian Coleoptera and a new order from the Belmont Beds, New South Wales. *Proceedings of the Linnean Society of New South Wales* **49**(4), 429–435, pls 45–46.

Tillyard RJ (1926) *The Insects of Australia and New Zealand.* 560 pp. Angus and Robertson, Sydney.

Ting PC (1936) The mouth parts of the coleopterous group Rhynchophora. *Microentomology* **1**(3), 93–114.

Tomaszewska KW (2000) Morphology, phylogeny and classification of adult Endomychidae (Coleoptera: Cucujoidea). *Annales Zoologici* **50**(4), 449–558.

Tomaszewska KW (2004a) A review of the genus *Erotendomychus* Lea (Coleoptera: Endomychidae). *Annales de la Société Entomologique de France (N. S.)* **40**(3–4), 395–419.

Tomaszewska KW (2004b) Taxonomic placement of the Australian genus *Daulis* Erichson, 1842 with description of a new species (Coleoptera: Endomychidae). *Annales Zoologici* **54**(3), 553–556.

Tomaszewska KW (2005) Phylogeny and generic classification of the subfamily Lycoperdininae with a re-analysis of the family Endomychidae (Coleoptera: Cucujoidea). *Annales Zoologici* **55**, Supplement 1, 172 pp.

Tomaszewska W (2010a) 10.32. Endomychidae Leach, 1815. In *Handbuch der Zoologie/Handbook of Zoology. Band/Volume IV Arthropoda: Insecta Teilband/Part 38. Coleoptera, Beetles. Volume 2. Morphology and Systematics (Polyphaga partim)*. (Eds RAB Leschen, RG Beutel and JF Lawrence) pp. 442–454. W. DeGruyter, Berlin.

Tomaszewska W (2010b) Rhyzobius *(Coleoptera: Coccinellidae). A Revision of the World Species.* 475 pp. Museum and Institute of Zoology, Polish Academy of Sciences, Warsaw.

Tomaszewska KW, Ślipiński A (1995) A review of the family Hobartiidae (Coleoptera: Cucujoidea). *Genus* **6**(3–4), 303–325.

Tomaszewska KW, Ślipiński A (2002) A review of the genus *Pteriptyctus* Blackburn, 1895 (Coleoptera: Corylophidae). *Annales Zoologici* **52**(4), 499–532.

Tomaszewska KW, Ślipiński A (2008) Review of the genus *Myrabolia* Reitter (Coleoptera: Cucujoidea: Myraboliidae). *Australian Journal of Entomology* **47**, 32–50. doi:10.1111/j.1440-6055.2007.00625.x

Tomaszewska W, Ślipiński A (2010) 10.9. Hobartiidae Sen Gupta and Crowson, 1966. In *Handbuch der Zoologie/Handbook of Zoology. Band/Volume IV Arthropoda: Insecta Teilband/Part 38. Coleoptera, Beetles. Volume 2. Morphology and Systematics (Polyphaga partim).* (Eds RAB Leschen, RG Beutel and JF Lawrence) pp. 324–327. W. DeGruyter, Berlin.

Tomaszewska KW, Ślipiński A (2011) Revision of the genus *Rodatus* Mulsant, 1850 (Coleoptera: Coccinellidae: Coccidulini). *Annales Zoologici* **61**(4), 657–684. doi:10.3161/000345411X622516

Tôyama M, Ohbayashi N (1994) The male of *Ohananomia yagii* Tôyama, 1987 (Coleoptera, Rhipiphoridae). *Japanese Journal of Entomology* **62**(2), 252–254.

Treherne JE (1954) Osmotic regulation in the larvae of *Helodes* (Coleoptera: Helodidae). *Transactions of the Royal Entomological Society of London* **105**, 117–130. doi:10.1111/j.1365-2311.1954.tb00779.x

Trend B (2007) Small hive beetle – *Aethina tumida. Western Australia Department of Agriculture and Food Farmnote* **248**, 4 pp.

Tribe GD, Cillié JJ (1985) A device to monitor larvae of the eucalyptus tortoise beetle, *Trachymela tincticollis* (Chrysomelidae: Paropsini). *Journal of the Entomological Society of Southern Africa* **48**, 213.

Trizzino M, Jäch MA, Audisio P, Alonso R, Ribera I (2013) A molecular phylogeny of the cosmopolitan hyperdiverse genus Hydraena Kugelann (Coleoptera, Hydraenidae). *Systematic Entomology* **38**, 192–208.

Tschinkel WR, Doyen JT (1980) Comparative anatomy of the defensive glands, ovipositors and female genital tubes of tenebrionid beetles (Coleoptera). *International Journal of Insect Morphology & Embryology* **9**, 321–368. doi:10.1016/0020-7322(80)90009-4

Turco F, Ślipiński A, Lambkin CL (2012) Taxonomic revision of Australian *Pristoderus* Hope (Coleoptera, Zopheridae). *Zootaxa* **3239**, 1–34.

Turnbow RH, Thomas MC (2002) 120. Cerambycidae Leach 1815. In *American Beetles. Volume 2. Polyphaga: Scarabaeoidea through Curculionoidea.* (Eds RH Arnett Jr, MC Thomas, PE Skelley and JH Frank) pp. 568–601. CRC Press, Boca Raton, Florida.

Ueno S-I (1957) Blind aquatic beetles of Japan, with some accounts of the fauna of Japanese subterranean waters. *Archiv fuer Hydrobiologie* **53**(2), 250–296.

Ueno S-I (1996) New phreatobiontic beetles (Coleoptera, Phreatodytidae and Dytiscidae) from Japan. *Journal of the Speleological Society of Japan* **21**, 1–50, 3 pls.

Uhmann G (2007) Die Anthicidae Latreille 1819 (Coleoptera: Tenebrionoidea) von Australien ohne Formicomini. 68. Beitrag zur Kenntnis der Anthicidae. *Coleoptera* **11**, 1104

Upton MS (1997) *A Rich and Diverse Fauna. The History of the Australian National Insect Collection 1926–1991.* xx + 386 pp. CSIRO Publishing, Collingwood, Victoria.

Upton MS, Mantle BL (2010) *Methods For Collecting, Preserving, and Studying Insects.* 5th Edition. Australian Entomological Society (Miscellaneous Publication No. 3), Indooroopilly.

Vahtera V, Muona J, Lawrence JF (2009) Phylogeny of the Thylacosterninae (Coleoptera, Elateridae). *Cladistics* **25**, 147–160. doi:10.1111/j.1096-0031.2009.00239.x

Valentine BD (1960) The genera of the weevil family Anthribidae north of Mexico (Coleoptera). *Transactions of the American Entomological Society* **86**, 41–85.

Valentine BD (1999) A review of Nearctic and some related Anthribidae (Coleoptera). *Insecta Mundi* **12** (1998), 251–296.

Valentine BD (2002) 126. Anthribidae Billberg 1820. In *American Beetles. Volume 2. Polyphaga: Scarabaeoidea through Curculionoidea.* (Eds RH Arnett Jr, MC Thomas, PE Skelley and JH Frank) pp. 695–700. CRC Press, Boca Raton, Florida.

Van de Kamp T, Vagovič P, Baumbach T, Riedel A (2011) A biological screw in a beetle's leg. *Science* **333**, 52. doi:10.1126/science.1204245

Van Tassell ER (2000) 13. Hydrophilidae Latreille, 1802. In *American Beetles. Volume 1. American Beetles. Archostemata, Myxophaga, Adephaga, Polyphaga: Staphyliniformia.* (Eds RH Arnett Jr and MC Thomas) pp. 187–208. CRC Press, Boca Raton, Florida.

Van Zandt PA, Townsend VR Jr, Carlton CE, Blackwell M, Mopper S (2003) *Loberus impressus* (LeConte) (Coleoptera: Erotylidae) fungal associations and presence in the seed capsules of *Iris hexagona. Coleopterists Bulletin* **57** (3), 281–288. doi:10.1649/553

Vandenberg NJ (2002) 93. Coccinellidae Latreille 1807. In *American Beetles. Volume 2. Polyphaga: Scarabaeoidea through Curculionoidea.* (Eds RH Arnett Jr, MC Thomas, PE Skelley and JH Frank) pp. 371–389. CRC Press, Boca Raton, Florida.

Vanin SA (1976) Taxonomic revision of the South American Belidae (Coleoptera). *Arquivos de Zoologia, São Paulo* **28**(1), 1–75.

Vanin SA, Costa C, Gianuca NM (1995) Larvae of Neotropical Coleoptera XXI: Description of immatures and ecology of *Efflagitatus freudei* Pacheco, 1973 (Dryopoidea, Heteroceridae). *Iheringia. Série Zoologia* **78**, 99–112.

Vanin SA, Ide S, Casari SA, Costa C (1996) Larvae of Neotropical Coleoptera. XXIV: *Scraptia triangularis* Champion, description of immatures and redescription of adult (Tenebrionidae, Scraptiidae). *Revista Brasileira de Entomologia* **40**(1), 89–96.

Vanin SA, Costa C, Ide S, Beutel RG (2005) 18.6. Heteroceridae Macleay, 1825. In *Handbuch der Zoologie/Handbook of Zoology. Band/Volume IV Arthropoda: Insecta Teilband/Part 38. Coleoptera, Beetles. Volume 1. Morphology and Systematics (Archostemata, Adephaga, Myxophaga, Polyphaga partim).* (Eds RG Beutel and RAB Leschen) pp. 518–521. W. DeGruyter, Berlin.

Varley GC (1939) On the structure and function of the hind spiracles of the larva of the beetle *Donacia* (Coleoptera, Chrysomelidae). *Proceedings of the Royal Entomological Society of London* **14**, 115–123. [A]

Vásquez XA (1993) *Fauna Iberica. Vol. 5. Coleoptera: Oedemeridae, Pyrochroidae, Pythidae, Mycteridae.* Museo Nacional de Ciencias Naturales, Madrid.

Vásquez XA (1996) Revision of the southern African Oedemeridae (Coleoptera, Tenebrionoidea). *Mitteilungen aus dem Zoologischen Museum in Berlin* **72**, 83–147. doi:10.1002/mmnz.19960720109

Veer V, Rao KM (2000) Descriptions of adult and larval morphology of *Evorinea indica* (Arrow) (Coleoptera: Dermestidae). *Oriental Insects* **34**, 293–300. doi:10.1080/00305316.2000.10417269

Verdú JR, Galante E, Lumaret J-P (1998) Description de la larve de Bolbelasmus bocchus (Erichson) et position systématique du genre (Coleoptera: Geotrupidae: Bolboceratinae). *Annales de la Société Entomologique de France (N.S.)* **34**(3), 245–251.

Verhoeff KW (1923) Zur Entwicklung, Morphologie und Biologie der vorlarven und larven der Canthariden. *Archiv für Naturgeschichte* **83**(2), 102–140. [A]

Vidal P, Guerrero M (2007) *Los Tenebriónidos de Chile.* 478 pp. Ediciones Universidad Catolica de Chile, Santiago.

Villiers A (1978) *Cerambycidae. Faune des Coléoptères de France I.* xxvii + 607 pp. P. Lechevalier, Paris.

Vincent JFV, Wegst UGK (2004) Design and mechanical properties of insect cuticle. *Arthropod Structure & Development* **33**, 187–199. doi:10.1016/j.asd.2004.05.006

Vit S (1977) Contribution a la connaissance des Eucinetidae (Coleoptera). *Revue Suisse de Zoologie* **84**, 917–935.

Vit S (1985) Étude de la morphologie des espèces paléarctiques du genre *Eucinetus* Germar et quelques remarques sur son utilisation taxonomique (Eucinetidae). *Revue Suisse de Zoologie* **92**, 421–460.

Vives E (2012) Two new genera of myrmecomorph longicorn beetles from Australia and New Caledonia (Insecta: Coleoptera: Cerambycidae). *Memoirs of the Queensland Nature* **56**(1), 79–84.

Volkovitsh MG (2001) The comparative morphology of antennal structures in Buprestidae (Coleoptera): evolutionary trends, taxonomic and phylogenetic implications. Part 1. *Acta Musei Moraviae. Scientiae Biologicae, Brno* **86**, 43–169.

Volkovitsh MG, Hawkeswood TJ (1987) The larva of *Neocuris gracilis* Macleay (Coleoptera: Buprestidae). *Zoologischer Anzeiger* **219**(5/6), 274–282.

Volkovitsh MG, Hawkeswood TJ (1990) The larvae of *Agrilus austrlasiae* Laporte & Gory and *Ethon affine* Laporte & Gory (Insecta: Coleoptera: Buprestidae). *Spixiana* **13**(1), 43–59.

Volkovitsh MG, Hawkeswood TJ (1993) The larvae of *Anilara antiqua* Thèry and *Anilara nigrita* Kerremans (Insecta, Coleoptera, Buprestidae). *Spixiana* **16**(2), 157–167.

Volkovitsh MG, Hawkeswood TJ (1999) The larva of *Prospheres aurantiopicta* (Laporte & Gory) with comments on the larval characteristics of polycestoid taxa (Insecta, Coleoptera, Buprestidae). *Mauritiana. Altenburg* **17**, 295–314.

Volkovitsh MG, Bílý S, Hasenpush J (2003) Larvae of Australian Buprestidae (Coleoptera). Part 2. Genus *Metaxymorpha. Folia Heyrovskyana* **11**(3–4), 203–216.

Vondel BJ van (1995) Revision of the Haliplidae (Coleoptera) of the Australian region and the Moluccas. *Records of the South Australian Museum* **28**(1), 61–101.

Vondel BJ van (2004) First description of larvae of *Haliplus*-species from Australia (Coleoptera: Haliplidae). *Tijdschrift voor Entomologie* **147**, 57–61. doi:10.1163/22119434-900000140

Vondel BJ van (2005) 7.2. Haliplidae Aubé, 1836. In *Handbuch der Zoologie/Handbook of Zoology. Band/Volume IV Arthropoda: Insecta Teilband/Part 38. Coleoptera, Beetles. Volume 1. Morphology and Systematics (Archostemata, Adephaga, Myxophaga, Polyphaga partim).* (Eds RG Beutel and RAB Leschen) pp. 64–72. W. DeGruyter, Berlin.

Voss E (1932) Monographie der Rhynchiten-Tribus Rhynchitini, 2. Gattungsgruppen Rhynchitina. *Koleopterologische Rundschau* **18**, 153–189.

Voss E (1965) The zoological results of Gy. Topal's collectings in south Argentina. *Annales Historico-Naturales Musei Nationalis Hungarici* **57**, 329–332.

Vršanský P (2003) Umenocoleoidea – an amazing lineage of aberrant insects (Insecta, Blattaria). *AMBA projekty* **7**(1), 1–32.

Vrydagh JM (1958) Contribution à l'étude des Bostrychidae. 11. Les Bostrychidae de l'Australie, de la Tasmanie et la Nouvelle-Zélande. *Bulletin et Annales de la Société Royale d'Entomologie de Belgique* **94**(1–2), 35–64.

Vrydagh JM (1959a) Contribution à l'étude des Bostrychidae. 18. Additions a l'étude des Bostrychidae de l'Australie. *Bulletin et Annales de la Société Royale d'Entomologie de Belgique* **95**, 274–285.

Vrydagh JM (1959b) Contribution à l'étude des Bostrychidae. 19. Nouvelle additions et corrections à l'étude des Bostrychidae de l'Australie. *Bulletin et Annales de la Société Royale d'Entomologie de Belgique* **95**, 42–46.

Vulcano MA, Pereira FS (1975) Cupesidae (Coleoptera). *Arquivos dos Instituto Biológico São Paulo* **42**, 31–68.

Vulcano MA, Pereira FS, Martinez A (1966) Notas sobre Orphninae Neotropicos com descriçao de um genero e uma espécie novos (Coleoptera). *Papéis Avulsos do Departamento de Zoologia, Sao Paulo* **18**, 251–260.

Vulinec K (1987) Swimming in whirligig beetles (Coleoptera: Gyrinidae): a possible role of the pygidial gland secretion. *Coleopterists Bulletin* **41**, 151–153.

Wachmann E (1977) Vergleichende Analyse der feinstrukturellen Organisation offener Rhabdome in den Augen der Cucujiformia (Insecta, Coleoptera), unter besonderer Berücksichtigung der Chrysomelidae. *Zoomorphologie* **88**, 95–131. doi:10.1007/BF01880649

Wade JS, St. George RA (1923) Biology of the false wireworm *Eleodes suturalis* Say. *Journal of Agricultural Research* **36** (11), 547–566, 2 pls.

Wagner DL, Liebherr JK (1992) Flighlessness in insects. *Trends in Ecology & Evolution* **7**(7), 216–220. doi:10.1016/0169-5347(92)90047-F

Wallace MHM, Winks RG, Voestermans J (1985) The use of a beetle, *Alphitobius diaperinus* (Panzer) (Coleoptera: Tenebrionidae) for the biological control of poultry dung in high-rise layer houses. *Journal of the Australian Institute of Agricultural Sciences* **51**, 214–219.

Wallage GN, Rose HA (1982) Some aspects of the biology of *Triphyllus minor* (Lea) (Coleoptera: Mycetophagidae). *Journal of the Australian Entomological Society* **21**, 111–112. doi:10.1111/j.1440-6055.1982.tb01775.x

Waller JB, Watt JC (1979) A record of *Reesa vespulae* (Coleoptera: Dermestidae) in New Zealand. *New Zealand Entomologist* **7**, 51. doi:10.1080/00779962.1979.9722330

Walsh GB, Dibb JR (Eds) (1975) *A Coleopterist's Handbook. 2nd Edition* (Revised by J Cooter and PW Cribb). Amateur Entomologists' Society, Feltham, UK.

Wanat M (2001) *Genera of Australo-Pacific Rhadinocybinae and Myrmacicelinae with biogeography of the Apionidae (Coleoptera: Curculionoidea) and phylogeny of the Brentidae* (s. lato). 432 pp. Mantis, Oslztyn.

Wanat M (2007) Alignment and homology of male terminalia in Curculionoidea and other Coleoptera. *Invertebrate Systematics* **21**, 147–171. doi:10.1071/IS05055

Wang W (1993) On Liaoximordellidae fam. nov. (Coleoptera, Insecta) from the Jurassic of Western Liaoning Province, China. *Acta Geologica Sinica* **67**(1), 86–94.

Wang Q (1993a) A revision of *Atesta* Pascoe (Coleoptera: Cerambycidae, Phoracanthini) from Australia, with descriptions of 18 new species. *Invertebrate Taxonomy* **7**, 961–1024. doi:10.1071/IT9930961

Wang Q (1993b) *Paratesta* gen. n. (Coleoptera: Cerambycidae) from Australia. *Journal of the Australian Entomological Society* **32**, 369–377. doi:10.1111/j.1440-6055.1993.tb00606.x

Wang Q (1995a) A taxonomic revision of the Australian genus *Phoracantha* Newman (Coleoptera: Cerambycidae). *Invertebrate Taxonomy* **9**, 865–958. doi:10.1071/IT9950865

Wang Q (1995b) An Australian longicorn beetle genus *Skeletodes* Newman (Coleoptera: Cerambycidae). *Coleopterists Bulletin* **49**(2), 109–118.

Wang Q (2000) Taxonomy, phylogeny and distribution of the longicorn beetle genus *Allotisis* Pascoe (Coleoptera: Cerambycidae: Phoracanthini). *Journal of Natural History* **34**, 1684–1711. doi:10.1080/00222930050117567

Wang Q, Leschen RAB (2003) Identification and distribution of *Arhopalus* species (Coleoptera: Cerambycidae: Aseminae) in Australia and New

Zealand. *New Zealand Entomologist* **26**, 53–59. doi:10.1080/00779962.2003.9722108

Wang B, Zhang H (2011) The oldest Tenebrionoidea (Coleoptera) from the Middle Jurassic of China. *Journal of Paleontology* **85**(2), 266–270. doi:10.1666/09-088.1

Wang Q, Thornton IWB, New TR (1994) Systematics and biogeography of the Australian-New Guinean genus *Thoris* Pascoe (Coleoptera: Cerambycidae: Phoracanthini). *Invertebrate Taxonomy* **8**, 839–860. doi:10.1071/IT9940839

Wang Q, Thornton IWB, New TR (1999) A cladistic analysis of the phoracanthine genus *Phoracantha* Newman (Coleoptera: Cerambycidae: Cerambycinae), with discussion of biogeographic distribution and pest status. *Annals of the Entomological Society of America* **92**(5), 631–638.

Wang B, Ponomarenko AG and Zhang H (2009) A new coptoclavid larva (Coleoptera: Adephaga: Dytiscoidea) from the Middle Jurassic of China, and its phylogenetic implication. *Paleontologicheskii Zhurnal* **2009**(6), 49–57 (in Russian; English translation in *Paleontological Journal* **43**(6), 652–659.

Wang B, Ponomarenko AG, Zhang H (2010) Middle Jurassic Coptoclavidae (Insecta: Coleoptera: Dytiscoidea) from China: a good example of mosaic evolution. *Acta Geologica Sinica* **84**(4), 680–687. doi:10.1111/j.1755-6724.2010.00272.x

Ward RD (1979) Metathoracic wing structures as phylogenetic indicators in the Adephaga (Coleoptera). In *Carabid Beetles: Their Evolution, Natural History, and Classification.* (Eds TL Erwin, GE Ball, DR Whitehead and AL Halpern) pp. 181–191. Junk, The Hague.

Wassell JLH (1966) Coprophagous weevils (Coleoptera: Curculionidae). *Journal of the Entomological Society of Queensland* **5**, 73–74. doi:10.1111/j.1440-6055.1966.tb00688.x

Waterhouse DF (1974) The biological control of dung. *Scientific American* **230** (4), 100–109. doi:10.1038/scientificamerican0474-100

Watson JAL (1973) *Austrospirachtha mimetes, a* new termitophilous corotocine from northern Australia (Coleoptera: Staphylinidae). *Journal of the Australian Entomological Society* **12**, 307–310. doi:10.1111/j.1440-6055.1973.tb01678.x

Watson JAL, Howick CD (1975) The rediscovery of *Mastopsenius australis* Seevers (Coleoptera: Staphylinidae). *Journal of the Australian Entomological Society* **14**, 19–21. doi:10.1111/j.1440-6055.1975.tb01995.x

Watson JAL, Kistner DH (1972) The glandular anatomy and biology of the termitophilous Australian aleocharine, *Drepanoxenus* (Coleoptera: Staphylinidae). *Australian Journal of Zoology* **20**, 341–358. doi:10.1071/ZO9720341

Watson JAL, Kistner DH (1985) *Australoptochus phoreticus,* a new genus and species of Termitogastrina (Coleoptera, Staphylinidae, Aleocharinae) from nests of the Australian termite, *Nasutitermes graveolus* (Isoptera: Termitidae). *Sociobiology* **11**, 87–106.

Watt JC (1967) The families Perimylopidae and Dacoderidae (Coleoptera, Heteromera). *Proceedings of the Royal Entomological Society of London* **36**, 109–118. [B]

Watt JC (1969) Keys to genera and some species of New Zealand Lathridiidae (Coleoptera). *New Zealand Entomologist* **4**(2), 49–67. doi:10.1080/00779962.1969.9722907

Watt JC (1970) Coleoptera: Perimylopidae of South Georgia. *Pacific Insects Monographs* **23**, 243–253.

Watt JC (1974a) Chalcodryidae: a new family of heteromerous beetles (Coleoptera: Tenebrionoidea). *Journal of the Royal Society of New Zealand* **4**, 19–38. doi:10.1080/03036758.1974.10419404

Watt JC (1974b) A revised subfamily classification of Tenebrionidae (Coleoptera). *New Zealand Journal of Zoology* **1**, 381–452. doi:10.1080/03014223.1974.9517846

Watt JC (1980) *Zeonidicola* (Coleoptera: Cavognathidae)-beetles inhabiting birds' nests. *Journal of the Royal Society of New Zealand* **10**, 331–339. doi:10.1080/03036758.1980.10419454

Watt JC (1987) The family and subfamily classification and New Zealand genera of Pythidae and Scraptiidae (Coleoptera). *Systematic Entomology* **12**, 111–136. doi:10.1111/j.1365-3113.1987.tb00552.x

Watt JC (1992) *Fauna of New Zealand. Number 26. Tenebrionidae (Insecta: Coleoptera): Catalogue of Types and Keys to Taxa.* DSIR Plant Protection, Auckland.

Watt JC, Marris JWM, Klimaszewski J (2001) A new species of *Platisus* (Coleoptera: Cucujidae) from New Zealand, described from the adult and larva. *Journal of the Royal Society of New Zealand* **31**(2), 327–339. doi:10.1080/03014223.2001.9517657

Watts CHS (1963) The larvae of Australian Dytiscidae (Coleoptera). *Transactions of the Royal Society of South Australia* **87**, 23–40.

Watts CHS (1964) The larvae of Australian *Cybister* spp. Curt., *Homoeodytes* spp. Reg. and *Hyderodes shuckardi* Hope (Coleoptera: Dytiscidae). *Transactions of the Royal Society of South Australia* **88**, 145–156.

Watts CHS (1978) A revision of the Australian Dytiscidae (Coleoptera). *Australian Journal of Zoology* **57**, (Supplement), 1–166.

Watts CHS (1982) A blind terrestrial water beetle from Australia. *Memoirs of the Queensland Museum* **20**, 527–531.

Watts CHS (1987) Revision of Australian *Berosus* Leach (Coleoptera: Hydrophilidae). *Records of the South Australian Museum* **21**, 1–28.

Watts CHS (1988a) Revision of Australian Halipilidae (Coleoptera). *Records of the South Australian Museum* **22**, 21–28.

Watts CHS (1988b) Revision of Australasian *Hydrophilus* Muller, 1764 (Coleoptera: Hydrophilidae). *Records of the South Australian Museum* **22**, 117–130.

Watts CHS (1989) Revision of Australasian *Sternolophus* Solier (Coleoptera: Hydrophilidae). *Records of the South Australian Museum* **23**, 89–95.

Watts CHS (1990) Revision of Australian *Hydrobiomorpha* Blackburn (Coleoptera: Hydrophilidae). *Records of the South Australian Museum* **24**, 35–42.

Watts CHS (1999) Revision of Australian *Hydrochus* (Coleoptera: Hydrochidae). *Records of the South Australian Museum* **32**(1), 1–43.

Watts CHS (2002) The larvae of some Australian aquatic Hydrophilidae (Coleoptera: Insecta). *Records of the South Australian Museum* **35**(2), 105–138.

Watts CHS (2004) Revision of Australian *Scirtes* Illiger and *Ora* Clark (Coleoptera: Scirtidae). *Transactions of the Royal Society of South Australia* **128**(1), 131–168.

Watts CHS (2007) Revision of Australian *Pseudomicrocara* Armstrong (Coleoptera: Scirtidae). *Transactions of the Royal Society of South Australia* **131**(1), 1–82.

Watts CHS (2009) Revision of *Heterocyphon* Armstrong, *Peneveronatus* Armstrong and *Accolabass* gen. nov. (Scirtidae: Coleoptera). *Transactions of the Royal Society of South Australia* **133**(1), 108–149.

Watts CHS (2010a) Revision of the genus *Macrohelodes* (Scirtidae: Coleoptera: Insecta). *Transactions of the Royal Society of South Australia* **134**(1), 19–52.

Watts CHS (2010b) Revision of Australian *Prionocyphon* Redtenbacher (Scirtidae: Coleoptera: Insecta). *Transactions of the Royal Society of South Australia* **134**(1), 53–88.

Watts CHS (2011) Revision of Australian Scirtidae of the genera *Chameloscyphon* gen. nov., *Daploeuros* gen. nov., *Dasyscyphon* gen. nov., *Eurycyphon* gen. nov., *Macrodascillus* Carter, *Petrocyphon* gen. nov. and *Spaniosdascillus* gen. nov. (Coleoptera). *Transactions of the Royal Society of South Australia* **135**(1), 66–110.

Watts CHS, Hamon H (2010) *Pictorial Guide to the Australian Whirligig Beetles.* 27 pp. South Australian Museum, Adelaide.

Watts CHS, Humphreys WF (1999) Three new genera and five new species of Dytiscidae (Coleoptera) from underground waters in Australia. *Records of the South Australian Museum* **32**(2), 121–142.

Watts CHS, Humphreys WF (2000) Six new species of *Nirridessus* Watts and Humphreys and *Tjirtudessus* Watts and Humphreys (Coleoptera: Dytiscidae) from underground waters in Australia. *Records of the South Australian Museum* **33**(2), 127–144.

Watts CHS, Humphreys WF (2001) A new genus and six new species of Dytiscidae (Coleoptera) from underground waters in the Yilgarn palaeo-drainage system of Western Australia. *Records of the South Australian Museum* **34**(2), 99–114.

Watts CHS, Humphreys WF (2003) Twenty-five new Dytiscidae (Coleoptera) of the genera *Tjirtudessus* Watts & Hymphreys, *Nirripirti* Watts and Humphreys and *Bidessodes* Regimbart from underground waters in Australia. *Records of the South Australian Museum* **36**(2), 135–187.

Watts CHS, Humphreys WF (2004) Thirteen new Dytiscidae (Coleoptera) of the genera *Bongurrus* Larson, *Tjirtudessus* Watts & Hymphreys and *Nirripirti* Watts and Humphreys, from underground waters in Australia. *Transactions of the Royal Society of South Australia* **128**(2), 99–129.

Watts CHS, Hancock PJ, Leys R (2007) A stygobitic *Carabhydrus* Watts (Dytiscidae, Coleoptera) from the Hunter Valley in New South Wales, Australia. *Australian Journal of Entomology* **46**, 56–59. doi:10.1111/j.1440-6055.2007.00585.x

Webb GA (1994) *Parandra araucariae* Gressitt (Coleoptera: Cerambycidae: Parandrinae): a new record for Norfolk Island. *Memoirs of the Queensland Museum* **37**(1), 325–328.

Webb GA, Simpson JA (1991) Notes on some Australian fungus beetles and their hosts and parasites. *Coleopterists Bulletin* **45**, 42–44.

Węgrzynowicz P (2000) New species of the genus *Xenocryptus* Arrow from Africa (Coleoptera: Languriidae). *Zoölogische Mededeelingen, Leiden* **73**(21), 317–320.

Węgrzynowicz P (2002) Morphology, phylogeny and classification of the family Erotylidae based on adult characters (Coleoptera: Cucujoidea). *Genus* **13**(4), 435–504.

Weinreich E (1968) Über den Klebfangapparat der Imagines von *Stenus* Latr. (Coleopt., Staphylinidae) mit einem Beitrag zur Kenntnis der Jugendstadien dieser Gattung. *Zeitschrift für Morphologie der Tiere* **62**, 162–210. doi:10.1007/BF00299123

Weir TA, Monteith GB (2010) *Lepanus storeyi,* a new species of dung beetle (Coleoptera: Scarabaeidae: Scarabaeinae) from southeast Queensland. *Australian Entomologist* **36**(4), 235–241.

Werner FG (1990) Revision of the Aderidae of eastern North America. *Journal of the New York Entomological Society* **98**(2), 187–232.

Werner FG, Chandler DS (1995) *Fauna of New Zealand. Number 34. Anthicidae (Insecta: Coleoptera).* Manaaki Whenui Press, Lincoln, Canterbury, New Zealand.

Wheeler QD (1986) Revision of the genera of Lymexylidae (Coleoptera: Cucujiformia). *Bulletin of the American Museum of Natural History* **83**, 113–210.

Wheeler WC (1989) The systematics of insect ribosomal DNA. In *The Hierarchy of Life*. (Eds. B. Fernholm *et al.*) pp. 307–321. Elsevier Press, Amsterdam.

Wheeler QD, Hoebeke ER (1984) A review of mycophagy in the Eucinetoidea (Coleoptera), with notes on an association of the eucinetid beetle, *Eucinetus oviformis,* with a Coniophoraceae fungus (Basidiomycetes: Aphyllophorales). *Proceedings of the Entomological Society of Washington* **86**, 274–277.

Wheeler WC, Whiting M, Wheeler QD, Carpenter JM (2001) Phylogeny of the extant hexapod orders. *Cladistics* **17**, 113–169. doi:10.1111/j.1096-0031.2001.tb00115.x

Whelan PI, Weir TA (1987) Skin lesions caused by *Paederus australis* Guérin-Méneville (Coleoptera: Staphylinidae). *Journal of the Australian Entomological Society* **26**, 287–288. doi:10.1111/j.1440-6055.1987.tb00304.x

White RE (1974) Type-species for world genera of Anobiidae (Coleoptera). *Transactions of the American Entomological Society* **99**, 415–475.

White ME (1986) *The Greening of Gondwana*. Reed Books, French's Forest, New South Wales.

Whiting MF (1998) Phylogenetic position of the Strepsiptera: review of molecular and morphological evidence. *International Journal of Insect Morphology & Embryology* **27**, 53–60. doi:10.1016/S0020-7322(97)00032-9

Whiting MF, Carpenter JC, Wheeler QD, Wheeler WC (1997) The Strepsiptera problem: phylogeny of the holometabolous insect orders inferred from 18S and 28S ribosomal DNA sequences and morphology. *Systematic Biology* **46**, 1–68.

Wibmer GJ, O'Brien CW (1986) Annotated checklist of the weevils (Curculionidae *sensu lato)* of South America (Coleoptera: Curculionoidea). *Memoirs of the American Entomological Institute* **39**, 1–563.

Wiegmann BM, Trautwein MD, Kim J-W, Cassel BK, Bertone MA, Winterton SL, Yeates DK (2009) Single-copy nuclear genes resolve the phylogeny of the holometabolous insects. *BMC Biology* **7** (34), 16 pp.

Williams IW (1938) The comparative morphology of the mouthparts of the order Coleoptera treated from the standpoint of phylogeny. *Journal of the New York Entomological Society* **46**, 245–289.

Williams GA (1985) New larval foods plants for some Australian Buprestidae and Cerambycidae (Coleoptera). *Australian Entomological Magazine* **12**, 41–46.

Williams GA, Weir TA (1987) Four new species and new records of Australian Mastogeniinae (Coleoptera: Buprestidae). *Journal of the Australian Entomological Society* **26**, 153–159. doi:10.1111/j.1440-6055.1987.tb00277.x

Williams GA, Weir TA (1988) Further new species of Australian Mastogeniinae (Coleoptera: Buprestidae). *Journal of the Australian Entomological Society* **27**, 179–181. doi:10.1111/j.1440-6055.1988.tb01518.x

Wilson FA (1920) Notes on the Australian coleopterous genera *Palaestra* Cast., *Tmesidera* Westwood, and *Palaestrida* White (Fam. Meloidae). *Entomologist's Monthly Magazine* **6**(3), 28–32.

Wilson FA (1921) New Australian Coleoptera with notes on some previously described species, Part I. *Proceedings of the Royal Society of Victoria (N.S.)* **34**, 33–41.

Wilson EO (1971) *The Insect Societies*. Belknap Press, Cambridge, Massachusetts.

Winkler JR (1982) Subfamilies and neutral terms proposed for groups higher than subfamily in Cleridae (Coleoptera) – purpose, definitions, identification key. *Acta Universitatis Carolinae Biologica* **1980**, 517–531.

Winkler JR (1987) Berendtimiridae fam. n., new family of fossil beetles from Baltic Amber. *Mitteilungen der Münchener Entomologische Gesellschaft* **77**, 51–59.

Winkler JR (1989) *Sedlacekvia tanamica* gen.n., sp. n., an anomalous new taxon from Australia, tentatively assigned to the family Cleridae (Coleoptera). *Acta Universitatis Carolinae Biologica* **32**, 457–476.

Witt ABR, Edwards PB (2002) 2001. Aspects of the biology, distribution, and host range of *Crioceris* sp. (Col.: Chrysomelidae: Criocerinae), a potential biological control agent for *Asparagus asparagoides* in Australia. *Biological Control* **23**, 56–63. doi:10.1006/bcon.2001.0991

Wittmer W (1952) Remarks on some Australian *Laius Guer.* (Col.: Malachiidae). *Proceedings of the Linnean Society of New South Wales* **76**, 187–189.

Wittmer W (1954) Zur Kenntnis der Malachiidae (Col.) Australiens. *Annals & Magazine of Natural History* **7**, 305–310. doi:10.1080/00222935408651732

Wittmer W (1955) Ergebnisse der Neu Guinea Expedition 1938–39. Coleoptera: Cantharidae und Malachiidae (16. Beitrag zur Kenntnis der indo-malaiischen Malacodermata). *Nova Guinea* **6**, 57–86.

Wittmer W (1957) Eine neue Gattung aus der Familie Malachiidae (Col.) aus Australien. *Annals & Magazine of Natural History* **10**, 351–352. doi:10.1080/00222935708655967

Wittmer W (1959) Zwei neue *Dicranolaius* Champ. aus Australien (Col. Malachiidae). *Entomologische Arbeiten aus dem Museum G. Frey* **10**, 475–477.

Wittmer W (1964) Neue Malacodermata aus Neu Guinea. (24. Beitrag zur Kenntnis der indo-malaiischen Malacodermata, Col.). *Nova Guinea. Zoology (Jena, Germany)* **30**, 115–137.

Wittmer W (1966) Zur Kenntnis der *Chauliognathus*-Arten von New Guinea (Coleoptera: Cantharidae). 27. Beitrag zur Kenntnis der indo-malaiischen Cantharidae. *Pacific Insects* **8**, 379–388.

Wittmer W (1969) Zur Kenntnis der indo-malaiischen *Silini* unter besondered Berücksichtigung der Fauna von Neuguinea (Col.: Cantharidae). 28. Beitrag zur Kenntnis der indo-malaiischen Cantharidae. *Pacific Insects* **11**, 217–454.

Wittmer W (1971) Zur Kenntnis der Cantharidae (Col.) Neuguineas und angrenzender Gebiete. 29. Beitrag zur Kenntnis der indo-malaiischen Cantharidae. *Pacific Insects* **13**, 545–574.

Wittmer W (1972a) Synonymische und systematische Notizen sowie neue Taxa in Cantharidae (Col.). *Verhandlungen der Naturforschenden Gesellschaft in Basel* **82**(1), 105–121.

Wittmer W (1972b) Zur Kenntnis der Cantharidae (Col.) Neuguineas I. *Pacific Insects* **14**, 339–388.

Wittmer W (1979) Zur Kenntnis der Cantharidae (Col.) Neuseelands. *Entomologica Basiliensia* **4**, 275–325.

Wood SL (1952) Observations on the homologies of the copulatory apparatus in male Coleoptera. *Annals of the Entomological Society of America* **45**, 613–617.

Wood SL (1973) On the taxonomic status of Platypodidae and Scolytidae (Coleoptera). *The Great Basin Naturalist* **33**(2), 77–90.

Wood SL (1978) A reclassification of the subfamilies and tribes of Scolytidae (Coleoptera). *Annales de la Société Entomologique de France* **14**, 95–122.

Wood SL (1986) A reclassification of the genera of Scolytidae (Coleoptera). *Great Basin Naturalist Memoir* **10**, 1–126.

Wood SL (1993) Revision of the genera of Platypodidae (Coleoptera). *The Great Basin Naturalist* **53**(3), 259–281.

Woodroffe GE (1958) The mode of reproduction of *Ptinus clavipes* Panzer form *mobilis* Moore (= *P. latro* auct.) (Coleoptera: Ptinidae). *Proceedings of the Royal Entomological Society of London* **33**, 25–30. [A]

Wooldridge DP (1983) New World Limnichinae VIII. *Paralimnichus* Delève (Coleoptera: Dryopoidea: Limnichidae). *Journal of the Kansas Entomological Society* **56**(2), 232–233.

Wooldridge DP (1984) *Parathroscinus*, a new genus of beetles from Southeast Asia (Limnichidae: Cephalobyrrhinae). *Journal of the New York Entomological Society* **92**(2), 121–124.

Xie W (2000) *Classification of Amphizoidae (Coleoptera) based on Immature Stage Characters.* 326 pp. University of Manitoba, Winnipeg.

Yan EV (2009) A new genus of elateriform beetles (Coleoptera, Polyphaga) from the Middle-Late Jurassic of Karatau. *Paleontologicheskii Zhurnal* **2009**(1), 71–75 (in Russian; translation in *Paleontological Journal* **43**(1), 78–82).

Yan EV, Wang B (2010) A new genus of elateriform beetle (Coleoptera, Polyphaga) from the Jurassic of Daohugou, China. *Paleontologicheskii Zhurnal* **2010**(3), 51–56 (in Russian; translation in *Paleontological Journal* **44**(3), 297–302.

Yan EV, Wang B, Ponomarenko AG, Zhang H (2013) The most mysterious beetles: Jurassic Jurodidae (Insecta: Coleoptera) from China. *Gondwana Research.* doi:10.1016/j.gr.2013.04.002

Yeates DK, Bouchard P, Monteith GB (2002) Patterns and levels of endemism in the Australian Wet Tropics rainforest: evidence from flightless insects. *Invertebrate Systematics* **16**, 605–619. doi:10.1071/IT01032

Yeates DK, Harvey MS, Austin AD (2003) New estimates for terrestrial arthropod species-richness in Australia. *Records of the South Australian Museum. Monograph Series* **7**, 231–241.

Yoshitomi H (2005) Systematic revision of the family Scirtidae of Japan, with phylogeny, morphology and bionomics (Insecta: Coleoptera, Scirtoidea). *Journal of Systematic Entomology Monographic Series* **3**, 1–212.

Yoshitomi H (2006) A new record of *Hyphalus taekoae* (Coleoptera, Limnichdae) from Taiwan. *Elytra, Tokyo* **34**(2), 352.

Yoshitomi H, Satô M (2001) Description of the larva of *Babalimnichus masamii* M. Satô (Coleoptera: Limnichidae: Thaumastodinae). *Coleopterists Bulletin* **55**(4), 471–474. doi:10.1649/0010-065X(2001)055[0471:DOTLOB]2.0.CO;2

Young DK (1976) The systematic position of *Sphalma quadricollis* Horn (Coleoptera: Salpingidae: Pythini) as clarified by discovery of its larva. *Coleopterists Bulletin* **30**, 227–232.

Young DK (1978) A new species of *Lagriomorpha,* with observations on the systematic position of the genus (Coleoptera: Anthicidae). *Pacific Insects* **18**, 105–109.

Young OP (1983) The biology of the Silphidae (Coleoptera): a coded bibliography. *Miscellaneous Publications of the Maryland Agricultural Experiment Station* **981**, 1–47.

Young DK (1984) Cantharidin and insects: an historical review. *Great Lakes Entomologist* **17**, 187–194.

Young DK (1985a) Description of the larva of *Ischalia vancouverensis* Harrington (Coleoptera: Anthicidae: Ischaliinae), with observations on the systematic position of the genus. *Coleopterists Bulletin* **39**(3), 201–206.

Young DK (1985b) The true larva of *Lecontia discicollis* and change in the systematic position of the genus (Coleoptera: Boridae). *Great Lakes Entomologist* **18**, 97–101.

Young RM (1989) Euchirinae (Coleoptera: Scarabaeidae) of the world: distribution and taxonomy. *Coleopterists Bulletin* **43**, 205–236.

Young DK (1991) Prostomidae–Synchroidae, Boridae, Pythidae–Scraptiidae. In *Immature Insects. Vol. 2.* (Ed. FW Stehr) pp. 515–518, 535–539, 539–556. Kendall/Hunt, Dubuque, Iowa.

Young DK (2000) 1. Cupedidae Laporte, 1836. In *American Beetles. Volume 1. American Beetles. Archostemata, Myxophaga, Adephaga, Polyphaga: Staphyliniformia.* (Eds RH Arnett Jr and MC Thomas) pp. 19–21. CRC Press, Boca Raton, Florida.

Young DK (2002a) 35. Eucinetidae Lacordaire 1857. In *American Beetles. Volume 2. Polyphaga: Scarabaeoidea through Curculionoidea.* (Eds RH Arnett Jr, MC Thomas, PE Skelley and JH Frank) pp. 82–84. CRC Press, Boca Raton, Florida.

Young DK (2002b) 36. Clambidae Jacquelin du Val 1857. In *American Beetles. Volume 2. Polyphaga: Scarabaeoidea through Curculionoidea.* (Eds RH Arnett Jr, MC Thomas, PE Skelley and JH Frank) pp. 85–86. CRC Press, Boca Raton, Florida.

Young DK (2002c) 37. Scirtidae Fleming 1821. In *American Beetles. Volume 2. Polyphaga: Scarabaeoidea through Curculionoidea.* (Eds RH Arnett Jr, MC Thomas, PE Skelley and JH Frank) pp. 87–89. CRC Press, Boca Raton, Florida.

Young DK (2002d) 52. Callirhipidae Emden 1924. In *American Beetles. Volume 2. Polyphaga: Scarabaeoidea through Curculionoidea.* (Eds RH Arnett Jr, MC Thomas, PE Skelley and JH Frank) pp. 144–145. CRC Press, Boca Raton, Florida.

Young DK (2002e) 53. Artematopodidae Lacordaire 1857. In *American Beetles. Volume 2. Polyphaga: Scarabaeoidea through Curculionoidea.* (Eds RH Arnett Jr, MC Thomas, PE Skelley and JH Frank) pp. 146–147. CRC Press, Boca Raton, Florida.

Young DK (2002f) 54. Brachypsectridae Böving and Craighead 1931. In *American Beetles. Volume 2. Polyphaga: Scarabaeoidea through Curculionoidea.* (Eds RH Arnett Jr, MC Thomas, PE Skelley and JH Frank) pp. 148–149. CRC Press, Boca Raton, Florida.

Young DK (2002g) 71. Lymexylidae Fleming 1821. In *American Beetles. Volume 2. Polyphaga: Scarabaeoidea through Curculionoidea.* (Eds RH Arnett Jr, MC Thomas, PE Skelley and JH Frank) pp. 261–262 CRC Press, Boca Raton, Florida.

Young DK (2002h) 96. Mycetophagidae Leach 1815. In *American Beetles. Volume 2. Polyphaga: Scarabaeoidea through Curculionoidea.* (Eds RH Arnett Jr, MC Thomas, PE Skelley and JH Frank) pp. 399–400. CRC Press, Boca Raton, Florida.

Young DK (2002i) 97. Archeocrypticidae Kaszab 1964. In *American Beetles. Volume 2. Polyphaga: Scarabaeoidea through Curculionoidea.* (Eds RH Arnett Jr, MC Thomas, PE Skelley and JH Frank) pp. 401–402. CRC Press, Boca Raton, Florida.

Young DK (2002j) 107. Prostomidae Horn 1888. In *American Beetles. Volume 2. Polyphaga: Scarabaeoidea through Curculionoidea.* (Eds RH Arnett Jr, MC Thomas, PE Skelley and JH Frank) pp. 510–511. CRC Press, Boca Raton, Florida.

Young DK (2002k) 108. Synchroidae Horn 1888. In *American Beetles. Volume 2. Polyphaga: Scarabaeoidea through Curculionoidea.* (Eds RH Arnett Jr, MC Thomas, PE Skelley and JH Frank) pp. 512–513. CRC Press, Boca Raton, Florida.

Young DK (2002l) 110. Stenotrachelidae Thomson 1859. In *American Beetles. Volume 2. Polyphaga: Scarabaeoidea through Curculionoidea.* (Eds RH Arnett Jr, MC Thomas, PE Skelley and JH Frank) pp. 520–521. CRC Press, Boca Raton, Florida.

Young DK (2002m) 115. Pyrochroidae Latreille 1807. In *American Beetles. Volume 2. Polyphaga: Scarabaeoidea through Curculionoidea.* (Eds RH Arnett Jr, MC Thomas, PE Skelley and JH Frank) pp. 540–543. CRC Press, Boca Raton, Florida.

Young DK, Pollock D (1991) Description of the mature larva of *Pedilus flabellatus* (Coleoptera: Pyrochroidae: Pedilinae), with phylogenetic implications of the discovery. *Canadian Journal of Zoology* **69**, 2234–2238. doi:10.1139/z91-312

Young DK, Pollock DA (2002). 99. Tetratomidae Billberg 1820. In *American Beetles. Volume 2. Polyphaga: Scarabaeoidea through Curculionoidea.* (Eds RH Arnett Jr, MC Thomas, PE Skelley and JH Frank) pp. 413–416. CRC Press, Boca Raton, Florida.

Young DK, Pollock DA (2010) 11.24. Pyrochroidae Latreille, 1807. In *Handbuch der Zoologie/Handbook of Zoology. Band/Volume IV Arthropoda: Insecta Teilband/Part 38. Coleoptera, Beetles. Volume 2. Morphology and Systematics (Polyphaga partim).* (Eds RAB Leschen, RG Beutel and JF Lawrence) pp. 715–721. W. DeGruyter, Berlin.

Yu Y, Leschen RAB, Ślipiński A, Ren D, Pang H (2012) The first fossil bark-gnawing beetle from the Middle Jurassic of Inner Mongolia, China (Coleoptera: Trogossitidae). *Annales Zoologici* **62**(2), 245–252. doi:10.3161/000345412X652765

Yus-Ramos R, Kingsolver JM, Nápoles JR (2007) Sobre el estatus taxonómico actual ded los brúquidos (Coleoptera: Bruchidae) en los Chrysomeloidea. *Dugesiana* **14**(1), 1–21.

Zacharuk RY (1962) Sense organs of the head of larvae of some Elateridae (Coleoptera): their distribution, structure and innervation. *Journal of Morphology* **111**, 1–33. doi:10.1002/jmor.1051110102

Zacharuk RY, Albert PJ, Bellamy FW (1977) Ultrastructure and function of digitiform sensilla on the labial palp of a larval elaterid (Coleoptera). *Canadian Journal of Zoology* **55**, 569–578. doi:10.1139/z77-072

Zeck EH (1919) Swarming of the soldier beetle, *Telephorus pulchellus. Australian Zoologist* **1**, 205–206.

Zerche L (2004) Revision der Gattung *Aegialites* Mannerheim (Coleoptera: Salpingidae: Aegialitinae). *Stuttgarter Beitrage zur Naturkunde. Serie A, Biologie* **666**, 1–116.

Zeuner, F. E. (1958) Jurassic beetles from Grahamland, Antarctica. *Palaeontology* **1,** 407–409, pl. 72.

Zeuner FE (1961) A Triassic insect fauna from the Molteno beds of South Africa. *International Congress of Entomology* **11**(1), 304–306.

Zhang J-F (2005) The first find of chrysomelids (Insecta: Coleoptera: Chrysomeloidea) from Callovian-Oxfordian Daohugou biota of China. *Geobios* **38**, 865–871. doi:10.1016/j.geobios.2004.05.003

Zhang J-F, Rasnitsyn AP (2006) New extinct taxa of Pelecinidae *sensu lato* (Hymenoptera: Proctotrupoidea) in the Laiyang Formation, Shandong, China. *Cretaceous Research* **27**, 684–688. doi:10.1016/j.cretres.2006.03.001

Zhantiev RD (2000) Classification and phylogeny of dermestids (Coleoptera, Dermestidae). *Zoologicheski Zhurnal* **79**(3), 1–15 (in Russian; translation *Entomological Review* **80**(9), 1115–1129.

Zherikhin VV (1977) Mesozoic Coleoptera. Cerophytidae, Acanthocnemidae, Cryptophagidae, Lathridiidae, Attelabidae, Curculionidae, Remarks on the phyletic links and ecology of Baisian Rhynchophora. *Trudy Paleontologicheskogo Institut Akademia Nauk SSSR* **161**, 130–142, 176–182 (in Russian; translation 1989. Smithsonian Institution Libraries and National Science Foundation, Washington, D. C., pp. 177–191, 242–250).

Zherikhin VV (2002) Chapter 3.2. Ecological history of the terrestrial insects. In *History of Insects* (Eds AP Rasnitsyn and DLJ Quicke), pp. 331–388. Kluwer Academic Publishers, Dordrecht/Boston/London.

Zherikhin VV, Gratshev VG (1993) Obrieniidae, fam. nov., the oldest Mesozoic weevils (Coleoptera, Curculionoidea). *Paleontological Journal* **27**, (1A), 50–69.

Zherikhin VV, Gratshev VG (1995) A comparative study of the hind wing venation of the superfamily Curculionoidea, with phylogenetic implications. In *Biology, Phylogeny, and Classification of Coleoptera. Papers Celebrating the 80th Birthday of Roy A. Crowson.* (Eds J Pakaluk and SA Ślipiński) pp. 633–777. Muzeum I Instytut Zoologii Polska Akademia Nauk, Warsaw.

Zherikhin VV, Gratshev VG (1997) The Early Cretaceous weevils from Sierra del Montsec, Spain (Insecta: Coleoptera: Curculionoidea). *Cretaceous Research* **18**, 625–632. doi:10.1006/cres.1997.0084

Zilahi-Balogh GMG, Humble LM, Kok LT, Salom SM (2006) Morphology of *Laricobius nigrinus* (Coleoptera: Derodontidae), a predator of the hemlock woolly adelgid. *Canadian Entomologist* **138**, 595–601. doi:10.4039/n05-096

Zimmerman EC (1942) Rhipiceridae of Guam. *Bulletin of the Bernice P. Bishop Museum* **172**, 45–46, pl. 1.

Zimmerman EC (1991) *Australian Weevils (Coleoptera: Curculionoidea). Volume V. Colour Plates 1–304.* CSIRO Publications, East Melbourne, Victoria.

Zimmerman EC (1992) *Australian Weevils (Coleoptera: Curculionoidea). Volume VI. Colour Plates 305–632.* CSIRO Publications, East Melbourne, Victoria.

Zimmerman EC (1993) *Australian Weevils (Coleoptera: Curculionoidea).* Volume III. Nanophyidae, Rhynchophoridae, Erirhinidae, Curculionidae: Amytcterinae, Literature Cited. CSIRO Publications, East Melbourne, Victoria.

Zimmerman EC (1994a) *Australian Weevils (Coleoptera: Curculionoidea). Volume I. Orthoceri. Anthribidae to Attelabidae. The Primitive Weevils.* CSIRO Information Services, East Melbourne, Victoria.

Zimmerman EC (1994b) *Australian Weevils (Coleoptera: Curculionoidea). Volume II. Brentidae, Eurhynchidae, Apionidae and a Chapter on Immature Stages by Brenda May.* CSIRO Information Services, East Melbourne, Victoria.

Zimmermann A (1924) Die Halipliden der Welt. Bestimmungstabellen der Haliplidae mit gleichzeitig beigefügten Detailbeschreibungen der einzelnen Arten. *Entomologische Blätter* **20**, 1–16, 65–80, 129–144, 193–213.

Zocchi R (1961) Contributo alla conoscenza della morfologia larvale del coleottero Dascillidae *Dascillus cervinus* L. *Studi Sassaresi* **9**(3), 430–444.

Zwick P (1977) Australian *Hydraena* (Coleoptera: Hydraenidae). *Australian Journal of Zoology* **25**, 147–184. doi:10.1071/ZO9770147

Zwick P (1979) Contributions to the knowledge of Australian Cholevidae (Catoptidae auct.: Coleoptera). *Australian Journal of Zoology* **70**, (Supplement), 1–56.

Zwick P (2008) Das Tibiotarsalorgan, ein neues Merkmal larvaler Scirtidae (Coleoptera). *Entomologische Nachrichten und Berichte* **52**, 51–54.

Zwick P (2011) Some Scirtidae (Coleoptera) from Palawan (the Philippines), mainly from phytotelmata. *Aquatic Insects* **33**(3), 233–252. doi:10.1080/01650424.2011.614257

Zwick P (2012) Australian marsh beetles (Coleoptera: Scirtidae). 1. Additions to genus *Petrocyphon. Entomologische Blätter und Coleoptera* **108**, 159–180.

Zwick P (2013) Australian marsh beetles (Coleoptera: Scirtidae). 2. Pachycyphon, a new genus of presumably terrestrial Australian Scirtidae. *Zootaxa* **3626**(3), 326–344.

Zwolfer H, Bennett FD (1969) *Ludovix fasciatus* Gyll. (Col., Curculionidae), an entomophagous weevil. *Entomologist's Monthly Magazine* **105**, 122–123.

INDEX

Principal entries for Australian suborders series superfamilies and families are indicated by bold page numbers. Botanical and mycological names are distinguished by (BOT) or (MYC), respectively. Names of non-Coleoptera are distinguished by class or order in brackets.

www.ingramcontent.com/pod-product-compliance
Lightning Source LLC
LaVergne TN
LVHW071240210926
848486LV00007B/311
9780643097285